The SAGE Handbook of
Political Geography

The SAGE Handbook of
Political Geography

Edited by
Kevin R. Cox
Murray Low
Jennifer Robinson

SAGE Publications
Los Angeles • London • New Delhi • Singapore

First published 2008

SAGE Publications Ltd
1 Oliver's Yard
55 City Road
London EC1Y 1SP

SAGE Publications Inc.
2455 Teller Road
Thousand Oaks, California 91320

SAGE Publications India Pvt Ltd
B 1/I 1 Mohan Cooperative Industrial Area
Mathura Road
New Delhi 110 044

SAGE Publications Asia-Pacific Pte Ltd
33 Pekin Street #02-01
Far East Square
Singapore 048763

Library of Congress Control Number: 2007935210

British Library Cataloguing in Publication data

A catalogue record for this book is available from
the British Library

ISBN 978-0-7619-4327-3

Typeset by CEPHA Imaging Pvt. Ltd., Bangalore, India
Printed in Great Britain by The Cromwell Press Ltd, Trowbridge, Wiltshire.
Printed on paper from sustainable resources

Contents

Preface and Acknowledgements

In a contemporary world of multi-level governance, conflicts around cultural and other forms of identity and difference, and the threat of environmental catastrophe, the concerns of political geographers with spatial organization, variation across the globe and society/nature relationships have resonated more widely and intensely with concerns in other disciplinary contexts than perhaps ever before. This is proving an exciting opportunity for the field, expanding the range of empirical and theoretical concerns within its purview considerably. But it also makes attempts to summarize and encompass current endeavours in relation to the discipline's pasts and possible futures a hazardous enterprise. One consequence of this was the difficult choice we faced in deciding who we might ask to contribute: this was not an easy task with so much excellent work to choose from. Our degrees of freedom were of course limited by the particular slants which we thought would help achieve various sorts of balance across the book. Political geography has undergone tremendous growth in many different areas, with many original contributions; we hope colleagues who weren't involved in this particular venture will find that much of their work has received proper acknowledgement in the papers, chapters and books cited by the contributors.

The *Handbook* has been under construction for several years, with its far-flung contributors all periodically intersecting with its development in the context of their own biographies and other projects. We would like to thank them all for their contributions, patience and good humour as the project originally proposed on a few sheets of paper has emerged into the largest pile of original material any of us have assembled before. Developing the book also benefited substantially from the advice and counsel of an advisory board for which we are grateful: the members included James Anderson, Paul Claval, Ron Johnston, Anssi Paasi, Haripriya Rangan and Paul Robbins.

Robert Rojek, who originally suggested the project, was an absolute rock throughout providing a sense of calm about progress and hospitality at Sage in London. We would also like to thank Sue Searle, Jan Smith and Sheree Barboteau of the Geography Discipline at the Open University for their assistance in formatting much of the material, and the Geography Discipline there for financial support in the production of this book.

No handbook of this size can in the end aspire to being *truly* contemporary with a world that is always moving on, always in the process of construction. But we hope it can inspire reflection on where political geography has been and might be heading, and inspire others, though their reading, discussion and use of the book, to keep creatively re-thinking the subject's pasts and futures in relation to the world's challenges.

The Editors

Contributors

John Agnew is a Professor of Geography at UCLA. He is the author or co-author of such books as *Mastering Space* (1995), *Place and Politics in Modern Italy* (2002), *Geography of the World Economy* (2003) and *Hegemony* (2005).

Karen Bakker is an Associate Professor in the Department of Geography at the University of British Columbia, Vancouver, Canada. Her research interests span the fields of political ecology and political economy. Her work on water management has focused on privatization, transboundary water governance, drought vulnerability, demand management, pricing, and access to urban water supply in developing countries. She is currently completing a research project on 'after-globalization' movements and the environment. Recent book publications include *An Uncooperative Commodity: Privatizing Water in England and Wales* (Oxford University Press, 2004) and *Eau Canada: The Future of Canada's Water* (UBC Press, 2006).

Clive Barnett is Reader in Human Geography, and Associate Member of the Centre for Citizenship, Identities and Governance, in the Faculty of Social Science at the Open University, Milton Keynes, UK. He is the author of *Culture and Democracy* (Edinburgh University Press, 2003), co-editor with Murray Low of *Spaces of Democracy* (Sage 2004) and co-editor with Jennifer Robinson and Gillian Rose of *A Demanding World* (Open University, 2006).

Nick Blomley is Professor of Geography at Simon Fraser University. He has a longstanding interest in the spatiality of legal practice and representation. He is the author of *Law, Space and the Geographies of Power* (Guilford Press, 1994), co-editor (with David Delaney and Richard T. Ford) of *The Legal Geographies Reader* (Blackwell, 2001) and author of *Unsettling the City: Urban Land and the Politics of Property* (Routledge, 2004). He is currently researching the geographies of rights in relation to the regulation of street begging in Canadian cities.

Bruce Braun is Associate Professor of Geography at the University of Minnesota. His published works include *The Intemperate Rainforest: Nature, Culture and Power on Canada's West Coast* (University of Minnesota Press), two co-edited books, *Remaking Reality: Nature at the Millennium* (Routledge) and *Social Nature: Theory, Practice, Politics* (Blackwell), and numerous essays on the politics of nature. His current research examines emerging scientific and political practices of biosecurity.

Gavin Bridge is a Reader in Economic Geography in the School of Environment and Development at the University of Manchester. His published research addresses the geographical and historical dynamics of natural resource development, and contributes to theories of resource access and environmental governance. At the core of his work is an interest in the economic and cultural practices through which parts of the non-human world become enacted as 'resources' and subsequently proliferate through the economy in the form of commodities. His research on the oil, gas and metal-mining sectors has been funded by the NSF and the European Commission.

Michael Brown is an Associate Professor at the University of Washington in Seattle. His research is on urban political and cultural geography. He is the author of *Closet Space: Geographies of Metaphor from the Body to the Globe* and *RePlacing Citizenship: AIDS Activism and Radical Democracy*. He is also an editor of *Social and Cultural Geography*.

Noel Castree is a Professor in the School of Environment and Development at Manchester University. He is co-editor of *Antipode* and *Progress in Human Geography* and author, most recently, of *Nature* (Routledge, 2005). Informed by Marxian political economy, his research interests are threefold: the causes and effects of environmental change, the geographies of employment and labour struggle, and the function and politics of academic life.

Stuart Corbridge is Professor of Development Studies at the London School of Economics. He is the author or co-author of six books in the fields of development studies, geopolitics and South Asia. His most recent book, with Glyn Williams, Manoj Srivastava and René Véron, is *Seeing the State: Governance and Governmentality in India* (Cambridge University Press, 2005).

Kevin R. Cox is Distinguished University Professor of Geography at the Ohio State University. His interests include historical geographical materialism, the politics of local and regional development and South Africa. He is the author of numerous books, including Political Geography: Territory, State and Society. His latest, entitled South Africa and the History of Globalization, will appear in 2008.

Simon Dalby is Professor in the Department of Geography and Environmental Studies at Carleton University in Ottawa where he teaches courses on environment and geopolitics. He holds a PhD from Simon Fraser University and is author of *Creating the Second Cold War* (Pinter and Guilford Press, 1990) and *Environmental Security* (University of Minnesota Press, 2002). He is co-editor of *Rethinking Geopolitics* (Routledge, 1998) and *The Geopolitics Reader* (Routledge, 1998; 2nd edition, 2006). Current research interests include the debate about empire and the geopolitics of the Bush doctrine in addition to matters of environmental security and sustainability.

Elena Dell'Agnese is Associate Professor of Geography at the University of Milan-Bicocca. Her research focuses on political geography and gender issues (nation-building processes and regional iconographies in Europe and Asia; European borders and their role as identity markers; cities as symbolic landscapes and political places/spaces; the political geography of masculinity). She has been, since 2004, a member of the Steering Committee of the Commission on Political Geography of the International Geographical Union. She is the author of *Geografia politica critica* (Guerini Scientifica, 2005) and co-editor, with Enrico Squarcina, of *Europa. Vecchi confini e nuove frontiere* (Utet Libreria, 2005).

Kim England is Associate Professor of Geography at the University of Washington. Her research focuses on labour markets and women's paid/unpaid work, care work, workplaces (including the home) and the linkages between critical theories and the politics and ethics of fieldwork. She explores all these issues in the context of systems of difference, especially gender, race/ethnicity, class, sexuality, (dis)ability and national identities. She is the editor of *Who Will Mind the Baby? Geographies of Child-Care and Working Mothers* (Routledge, 1996), and the co-editor (with Kevin Ward) of *Neoliberalization: States, Networks, Peoples* (Blackwell, 2007).

David Featherstone is a lecturer in the Department of Geography, University of Liverpool. He has key theoretical interests in the geographies of resistance, subaltern political ecologies and the relations between space and politics. He has explored these theoretical concerns through research on spatially stretched forms of resistance in the eighteenth-century Atlantic world and through engagement with the geographies of counter-globalization movements. He is currently working on a book for the RGS-IBG book series called *Resistance, Space and Political Identities: The Making of Counter-Global Networks*.

Benjamin Forest is an Associate Professor of Geography at McGill University. He received his PhD in Geography from UCLA in 1997, and was an Assistant and Associate Professor at Dartmouth College from 1998 to 2006. He has published articles on identity, race and ethnicity, and political representation in *Urban Geography*, *Political Geography*, *Social and Cultural Geography*, the *Proceedings of the National Academy of Sciences* and the *Annals of the Association of American Geographers*. His current research interests include the impacts of GIS technology on political redistricting and representation.

Jouni Häkli is Professor of Regional Studies at the Department of Regional Studies, University of Tampere, Finland. His areas of research include political geography, national identities and European border regions, regionalization, and the production of geographical knowledge. Professor Häkli's publications in this field include 'In the territory of knowledge: state-centered discourses and the construction of society', *Progress in Human Geography*, 25(3), 2001; *Boundaries and Place* (co-edited with David H. Kaplan; Rowman &

Littlefield, 2002); and 'Regions, networks and fluidity in the Finnish nation-state', *National Identities*, 9, 2007 (forthcoming).

Guntram H. Herb is Associate Professor of Geography at Middlebury College, Vermont. He received a master's level degree from the University of Tübingen and a PhD from the University of Wisconsin–Madison. His publications include *Perthes World Atlas* (Klett Perthes/McGraw-Hill, 2006; editor-in-chief); *Nested Identities: Nationalism, Territory, and Scale* (Rowman & Littlefield, 1999; co-edited with David H. Kaplan); and *Under the Map of Germany: Nationalism and Propaganda, 1918–1945* (Routledge, 1997). He is currently editing a four-volume encyclopedia on nations and nationalism with David H. Kaplan for ABC-Clio Press.

Steve Herbert is Associate Professor of Geography and Law, Societies and Justice at the University of Washington. He researches social control in contemporary cities, particularly as exercised by the uniformed police. He is the author of *Policing Space: Territoriality and the Los Angeles Police Department* (University of Minnesota Press, 1997) and *Citizens, Cops, and Power: Recognizing the Limits of Community* (University of Chicago Press, 2006).

Andrew Herod is Professor of Geography, University of Georgia, USA. He has written widely on issues of globalization and labour politics. He is author of *Labor Geographies: Workers and the Landscapes of Capitalism* (Guilford Press, 2001); editor of *Organizing the Landscape: Geographical Perspectives on Labor Unionism* (University of Minnesota Press, 1998); and co-editor of *The Dirty Work of Neoliberalism: Cleaners in the Global Economy* (Basil Blackwell, 2006, with Luis Aguiar), *Geographies of Power: Placing Scale* (Basil Blackwell, 2002, with Melissa Wright) and of *An Unruly World? Globalization, Governance and Geography* (Routledge, 1998 with Gearóid Ó Tuathail and Susan Roberts).

Ron Johnston is a Professor in the School of Geographical Sciences at the University of Bristol. He has been researching in electoral studies since 1970. Recent publications include: (with D. J. Rossiter and C. J. Pattie) *The Boundary Commissions* (Manchester University Press, 1999), (with C. J. Pattie, D. F. L. Dorling and D. J. Rossiter) *From Votes to Seats* (Manchester University Press, 2001) and (with C. J. Pattie) *Putting Voters in their Place* (Oxford University Press, 2006).

Eleonore Kofman is Professor of Gender, Migration and Citizenship at Middlesex University, UK. She has published extensively on gendered migrations and policies in Europe and feminist political geography and globalization. She has co-authored *Gender and International Migration in Europe* (2000), and co-edited *Globalization: Theory and Practice* (2003) and *Mapping Women, Making Politics: Feminist Perspectives on Political Geography* (2004). She is review editor of the journal *Political Geography*.

Merje Kuus is Assistant Professor of Geography at the University of British Columbia. Her research focuses on security and militarism, nationalism and state power, and identity construction. She has also published on the concept of Europe and on discourses of regional expertise in Europe. Her recent work has appeared in *Progress in Human Geography* and *Transactions of the Institute of British Geographers*, among other venues. She is currently completing a book manuscript on the (re)construction of geopolitics and security in the context of EU and NATO enlargements in 2004.

Alan Lester is Professor of Historical Geography at the University of Sussex. Most recently, he is the author of *Imperial Networks: Creating Identities in Nineteenth Century South Africa and Britain* (Routledge, 2001) and co-editor, with David Lambert, of *Colonial Lives Across the British Empire: Imperial Careering in the Long Nineteenth Century* (Cambridge University Press, 2006). His research is on the historical geographies of connection and contestation between British colonial projects, and the shaping of colonial and metropolitan politics and cultures in the nineteenth century.

Murray Low is Lecturer in Human Geography at the London School of Economics. His work focuses on the themes of urban politics, and urban and spatial aspects of democracy. He has also written on political aspects of globalization, and concepts of power, states and political communities in geography and related disciplines. He is co-editor, with Clive Barnett, of *Spaces of Democracy* (Sage, 2004).

Becky Mansfield is an Assistant Professor of Geography at the Ohio State University. Her research is at the intersection of political geography, economic geography and nature/society relations, with a focus on political economy of natural resources. Themes of research include neo-liberalism and privatization, the role of the state, and the importance of the biophysical in political economic processes. Her articles appear in a range of journals, including *Antipode*, *Global Environmental Change* and the *Annals of the Association of American Geographers*.

Giles Mohan is a Senior Lecturer in Development Geography at the Open University. He has researched questions of local governance and participation in Africa and the dynamics of diasporic communities. He has published extensively in human geography and development studies journals and currently edits the *Review of African Political Economy*.

Richard C. M. Mole is Lecturer in the Politics of Central Europe at the School of Slavonic and East European Studies, University College London. His teaching and research focus on the relationship between national identity and political power, with particular reference to the Baltic States. He is the editor of *Discursive Constructions of Identity in European Politics* (Palgrave, 2007) and the author of the forthcoming *The Baltic States: From the Soviet Union to the European Union* (Routledge Curzon, 2008).

Joshua S. S. Muldavin, Professor of Geography and Asian Studies, Sarah Lawrence College, was recently named an SSRC/Abe Fellow for 2006–8 to continue his work analysing Japanese environmental aid to China, and was awarded an NSF grant for 2006–8 to pursue his joint research with Piers Blaikie in the Himalayas on comparative international environmental policy between China, India and Nepal. Former Chair and Director of International Development Studies at UCLA, he has conducted research in China for over 24 years, and is currently writing a book on the social and environmental impacts of China's reforms and global integration.

John O'Loughlin is Professor of Geography and Research Faculty Associate at the Institute of Behavioral Science, University of Colorado at Boulder. His research interests are in the political geography of the former-Soviet Union, including Russian and Ukrainian geopolitics and ethno-territorial nationalisms. His current major research project is an analysis of post-war outcomes in Bosnia-Herzegovina and the North Caucasus of Russia. He has also published on the diffusion of democracy, the geography of conflict, and the political geography of Nazi Germany. He is editor in-chief of *Political Geography*.

Elizabeth A. Olson is Lecturer in Human Geography at the Institute of Geography, Edinburgh University. Her research has explored the relationship between development, knowledge and organizations, with a recent emphasis on the roles of transnational and local religious institutions. She draws upon post-colonial, post-development and feminist theory for understanding the construction of new religious spaces in Peru, and for thinking through the changing partnerships and situations of faith-based development organizations.

Joe Painter is Professor of Geography and Director of the Centre for the Study of Cities and Regions at Durham University in the UK. His research interests include urban and regional governance and politics, geographies of the state and citizenship, and socio-spatial theory. He has published widely on these and related themes and is also the co-author of *Practising Human Geography* (Sage, 2004).

Sue Parnell is a Professor of Environmental and Geographical Sciences at the University of Cape Town. Her research is split, with a historical focus on the rise of racial residential segregation in South African cities and the impact of colonialism on urbanization and town planning in sub-Saharan Africa. Under a democratic South Africa, her work has shifted to more contemporary urban policy research and is primarily concerned with issues of urban reconstruction and transformation.

Charles Pattie is Professor of Geography at the University of Sheffield. He has written extensively on electoral geography, party campaigning and political participation. Among his recent publications are *Citizenship in Britain* (Cambridge University Press, 2004, co-authored with Pat Seyd and Paul Whiteley) and *Putting Voters in their Place* (Oxford University Press, 2006, co-authored with Ron Johnston).

Jan Penrose is a Senior Lecturer in Geography at the University of Edinburgh. Her work bridges cultural and political geography through considerations of how the formation and maintenance of identities intersect with power relations in a variety of contexts. In particular she is concerned with identities that have formed around the evolving categories of race, nation and indigeneity.

Clionadh A. Raleigh is a PhD Candidate in Geography at the University of Colorado, Boulder, and a researcher at the International Peace Research Institute (PRIO) at the Centre for the Study of Civil War (CSCW). Her dissertation is an exploration of the spatial and temporal patterns of civil wars in Central Africa from 1960 to 2005. This study employs disaggregated conflict data from the Armed Conflict Location and

Event Dataset (ACLED), developed by Ms Raleigh and Håvard Hegre at CSCW. Her recent publications focus on developing national and subnational explanations of onset and diffusion patterns in civil war using local-level information on war and governance patterns.

Haripriya Rangan teaches at the School of Geography and Environmental Science, Monash University, in Melbourne, Australia. Her research lies at the intersection of the fields of economic geography, development geography, urban and regional planning, and political ecology. She has worked on regional development and forestry issues in the Indian Himalayas, and on the economic geography of the medicinal plant trade in South Africa.

Paul Robbins is Associate Professor of Geography at the University of Arizona and author of the books *Political Ecology: A Critical Introduction* and *Lawn People: How Grass Weeds and Chemicals Make Us Who We Are*. His work centres on the power-laden relationships between individuals (homeowners, hunters, foresters), environmental actors (lawns, elk, mesquite trees) and the institutions that connect them. He and his students seek to explain human environmental practices and knowledges, the influence non-humans have on human behaviour and organization, and the implications these interactions hold for ecosystem health, local community and social justice.

Jennifer Robinson is Professor of Urban Geography at the Open University in the UK and has also worked at the University of KwaZulu-Natal in South Africa, and the London School of Economics. She has published widely on South African cities, especially on the spatiality and politics of apartheid and post-apartheid cities. Her most recent book, *Ordinary Cities* (Routledge, 2006), offers a post-colonial critique of urban studies.

Paul Routledge is a Reader in Human Geography in the Department of Geographical and Earth Sciences at the University of Glasgow. His research interests include critical geopolitics, geographies of resistance, political ecologies of development, and activist collaborative methodologies. He is author of *Terrains of Resistance* (Praeger, 1993), co-editor (with Jo Sharp, Chris Philo and Ronan Paddison) of *Entanglements of Power* (Routledge, 2000) and co-editor (with Gearóid Ó Tuathail and Simon Dalby) of *The Geopolitics Reader*, 2nd edition (Routledge, 2006). His current research is concerned with global justice networks and ecologically sustainable and socially just alternatives to neo-liberalism.

Yvonne Rydin is Professor of Planning, Environment and Public Policy at University College London's Bartlett School of Planning, where she specializes in environmental policy, and governance and sustainability issues at the urban level. Her recent research projects cover work on sustainable construction and planning in London and the implementation of local sustainability indicators. Other work within London has involved the study of institutional change and interest representation concerning planning for sustainability in the early days of the Greater London Authority. Her other interests include community engagement on sustainable development and the discursive aspects of environmental planning.

Arun Saldanha is Assistant Professor of Geography at the University of Minnesota. He has published on the body, music, tourism and race, and his book *Psychedelic White: Goa Trance and the Viscosity of Race* is published by the University of Minnesota Press in 2007. He is working on a second book on the historical geography of early modern Dutch travel writing.

James D. Sidaway is currently Professor of Human Geography at the University of Plymouth, UK. He has previously taught in the Geography Department and on the European Studies Programme at the National University of Singapore, and at Loughborough, Reading and Birmingham Universities in the UK. In addition to critical geopolitics and political geography, his research and teaching interests include geographies of development and the history and philosophy of geographic thought. He is associate editor of *Political Geography* and co-editor of the *Singapore Journal of Tropical Geography*.

Rachel Silvey is Associate Professor of Geography at the University of Toronto. Her researches examines the politics of migration and immigration through the lenses of feminist theory and critical development studies with a regional specialization in Indonesia. Her current work deals with transnational migration in relation to political Islam. Some of the outlets for her published work include *Gender, Place and Culture*, *Political Geography*, *International Migration Review*, *Annals of the Association of American Geographers*, *Global Networks* and *Progress in Human Geography*.

Lynn A. Staeheli is Ogilvie Professor of Human Geography at the University of Edinburgh. Her research addresses issues related to democracy, citizenship and the construction of publics. Recent empirical work has focused on gender, community activism, public space and immigration. She is co-editor, with Eleonore Kofman and Linda Peake, of *Mapping Women, Making Politics: Feminist Perspectives on Political Geography* (Routledge, 2004) and co-author, with Don Mitchell, of *The People's Property? Power, Politics, and the Public* (Routledge, 2007).

Kristian Stokke is Professor of Human Geography at the University of Oslo, specializing in movement politics and democratization in South Africa and intrastate conflict and political transformations in Sri Lanka. He is co-editor of *Politicising Democracy: The New Local Politics of Democratisation* (Palgrave Macmillan, 2004) and *Democratizing Development: The Politics of Socio-economic Rights in Sri Lanka* (Martinus Nijhoff, 2005). Recent papers include 'Building the Tamil Eelam state: emerging state institutions and forms of governance in LTTE-controlled areas in Sri Lanka' (*Third World Quarterly*, 2006) and 'Maximum working class unity? Challenges to local social movement unionism in Cape Town' (*Antipode*, 2006).

Peter J. Taylor is Professor of Geography at Loughborough University, UK and founder of the GaWC (Globalization and World Cities) Study Group and Network. His latest books are *World City Network: A Global Urban Analysis* (Routledge), *Cities in Globalization* (edited by B. Derudder, P. Saey, and F. Witlox, Routledge) and *Political Geography: World-Economy, Nation-State, Locality* (5th edition with Colin Flint, Prentice Hall). His current work focuses upon developing a geohistorical understanding of city/state relations.

Yaffa Truelove is a graduate student in the Department of Geography at the University of Colorado at Boulder. Her research focuses on issues of development, gender and political ecology in South Asia. Her current work examines the politics of water and gender in urban India.

Andrew Wood is an Associate Professor in the Department of Geography at the University of Kentucky. He is an economic, urban and political geographer whose research interests include urban and regional governance, the politics of local economic development, globalization, and issues relating to competition and collaboration between firms. His recent publications include *Governing Local and Regional Economies* with David Valler as well as articles in *Political Geography, Environment and Planning A, Area, Economic Geography, Urban Studies* and the *Journal of Economic Geography*.

Introduction
Political Geography: Traditions and Turns

Kevin R. Cox, Murray Low and
Jennifer Robinson

INTRODUCTION

Geographical perspectives seem more indispensable than ever in understanding political processes. In one of the world's longest running and intractable conflicts, Israel was invading southern Lebanon and bombarding other parts of the country to devastating effect as we sat down to write this introduction in July and August 2006. Israel, it seemed, hoped to cripple the Hezbollah militia which had been operating as a military force and quasi-state in part of a broader polity. The long-standing conflicts in this region and the efforts to resolve them, as a number of geographers have pointed out, starkly illustrate how space, place and territory are integral to the global fabric of power and conflict (see, for example, Falah, 1996; Yiftachel and Yacobi, 2002; Gregory, 2004).

Aside from the seemingly intractable politics of the region itself, there are many broader analytical issues raised by the events we were so aware of in mid-2006. These broader issues highlight the insights that geographical thinking can provide. The conflict poses several questions about the nature of states, for example. If states have often been assumed to be the primary actors on the modern world political stage, events in the Middle East expose the limits to concepts of territorialized state 'sovereignty'. Not only do warring neighbouring countries impinge on any sense that a state may have control over the territory it claims – and

indeed, occupy this territory at times and for long periods of time – but in this region state sovereignty is clearly forged only in relationship with other states, like Iran, Syria and the United States. Moreover, our understanding of the situation is severely limited by reducing it to a politics of conflict and alliance among territorial states, where 'proper' states such as the United States, Israel, Lebanon, Syria and Iran form the basis of empirical and normative judgement. Non-state political actors, such as Hezbollah, Palestinian resistance organizations, Israeli settler movements, religious movements and constituencies in the region and elsewhere, connect in important ways with the events of July and August 2006. Movements and constituencies such as these often exhibit a geography not confined within the grid laid down by the territorial state system.

Indeed, events more generally in the first decade of the twenty-first century throw some classic political geography questions into new relief. In the face of rising global affiliations and transnational organizations, for example, how do people come to feel attached to particular territories? What is a border and how should such a thing be delimited and operate in a world of wider connections and transnational flows? How far do, and should, states act on behalf of the populations and interests confined within their borders? When and in relation to what issues is a state 'sovereign'? How central is violence to state power, and can (or should) violent forms of politics be monopolized by states that mobilize these capacities far beyond their territories, and where violence is also claimed as an important weapon for non-territorial organizations?

Phenomena such as borders, territories, flows and networks can all be related to understandings of the *spatialities* of politics. Since political geographer's concerns with the spatial dimensions of politics go beyond such headline issues as war and global imperialism, to a very wide range of political processes, from the politics of bodies to uneven development, from local social movements to international trade unions, the ways in which they have come to interpret the spatialities of politics are very broad indeed. It is our hope that the contributors to this *Handbook* together amply demonstrate the diverse range of spatially inflected insights that political geography has to offer to geographers and other social scientists. The chapters reflect the field's diverse engagements with wider social and political theory as well as with an array of global political issues and controversies. They indicate many different ways in which these engagements can be taken forward.

The field of political geography has a long history, arguably longer than those of most other geographical sub-disciplines. Although there are important continuities in the themes explored over this period – like the territoriality of states, or the geographical foundations of world power – not surprisingly there have been huge changes over the one hundred years or so during which it has been established as a field of inquiry. These changes have recently occurred with increased rapidity: it is sometimes hard to maintain a sense of overall direction, or to appreciate the continuities with past traditions of scholarship as new themes emerge. This *Handbook* offers an opportunity to reach back into the traditions of the field, to re-examine conventional themes, to see how they have been re-appropriated more recently, or perhaps left behind in the quest for intellectually robust and politically engaged geographical research.

In this Introduction we take the opportunity to reflect on both these longstanding traditions and on more recent intellectual transformations or 'turns', opening up a space for charting some of the challenges facing political geographers today. Among the many pressing contemporary political issues that we feel political geographers are well placed to address, the *Handbook* will create an opportunity to consider, for example, how transnational networking of the poor and marginalized might counter the powerful capabilities of capital and dominant states more effectively; how a dispersed politics of identity and meaning could contribute to challenging such forces; how the dominant agendas of neo-liberalism and security are prosecuted by states and other agents; how power is (or is not) effectively exercised from one end of the globe to another; whether liberal democracy can and should be translated from the context of 'the West' to other, different, political and cultural contexts; and how 'natural' conditions and constraints shape politics and to what extent politics can shape them.

We use this Introduction to offer the reader a few pathways through the long and complex subdisciplinary history, its traditions and 'turns', and how these have shaped the kinds of questions geographers have addressed. For readers new to, or from outside of political geography as a scholarly community, we hope it will provide a context for the individual chapters to be better appreciated. For political geographers, we entertain the hope that our particular review of collective disciplinary traditions will illuminate some of the choices of topic in the *Handbook* as a whole. To some degree, political geography has marched in step with changes in human geography as a whole, though there have been some interesting and significant tensions with this wider disciplinary context. Developments such as the spatial-quantitative revolution (the development of new theories and agendas driven by the use of mathematical modelling and associated 'scientific' theory) in the 1960s, Marxism and other forms of social theory in the 1970s and 1980s, and the turn to cultural theory and themes in more recent years have all had their effects. A number of the chapters in this volume take issue with any simple narrativization of political geography's past, emphasizing, among other things, complications of language, country and gender. Many chapters that follow explore in some detail, and from different points of view, the history of political geography and the emergence and transformation of particular theoretical approaches or topics of interest. Our limited ambition here is to draw some historical threads together around two important developments. First, we discuss the *socialization* of political geography, by which we mean its growing engagement with social and political theory. This was a key element in its revival as a sub-discipline in the 1980s. Second, we emphasize several aspects of what could be summarized as the *diversification* of political geography. This has been particularly associated with the influence of post-structuralist and related ideas and has engendered a greater sensitivity to the complexity of the spatial organization of the world, and the multiplicity of sites and processes that political geography, traditionally rather focused on states, has to be concerned with. The emergence of 'power' as a core concept competing with, or in tension with, states as the central focus of the field is a key element of the diversification we outline. These intellectual diversifications have, not coincidentally, been accompanied and reinforced by a growing awareness of how feminist, radical and marginal voices are, and have always been, present in shaping the trajectories of political geography.

FROM 'HEARTLANDS' AND 'REGIONS' TO 'SOCIETY AND SPACE': SOCIALIZING POLITICAL GEOGRAPHY

The continuities and ruptures in the approaches and themes of political geography in the twentieth century reflect in many ways tensions felt more broadly by human geography as a discipline. From the vantage point of the early twenty-first century, efforts to isolate what counts as 'geographical' as opposed to social or political explanation have framed many key twentieth-century political geography texts and mirror the wider discipline's concern to determine its distinctive scope of analysis (see, for example, Hartshorne, 1954). Initially, political geographers were preoccupied with territorial entities, such as states and, to a lesser degree, administrative regions. A concern with relating political processes and their spatial organization to environmental conditions and physical geography more broadly was widespread and limited the attentiveness of political geographers to the significance of the emerging social sciences for their concerns. The course of Western political geography in the early twentieth century can seem, in a hindsight informed by knowledge of the interest of prominent contemporary theorists in geographical themes (such as Gramsci on politics and regional development, Louis Wirth on regions or Simmel on cities), like a strange collective evasion of social science explanation and theory (see Smith, 1989). But this risks ironing out complicated shifts of emphasis in the sub-discipline over time and also the many vital contributions that geographers have made over the years to the analysis of political themes. To dismiss too easily these early contributions may also make the ambitions of the present (where, for example, the theme of 'nature' is being imaginatively re-appropriated by many geographers in a political context) appear more radically different than is helpful. Nonetheless, the eventual embrace of descriptive and explanatory language from the other social sciences – without losing contact with the traditions of the field – marks one of the more striking transformations in political geography in the last one hundred years. This section addresses this transformation, which was catalysed by the development of a radically new, 'scientific' geography in the late 1950s, emphasizing how it has coincided with a shift away from states and their environments as political geography's key set of preoccupations.

In the first half of the twentieth century, much of the specificity of political geography consisted in its resolutely international focus. There was very little research on national, urban or regional issues of the kind that was to dominate the sub-field later in the century. Furthermore, as both Herb and Painter point out in the first section of this *Handbook*, as a study of international relations, political geography was clearly, even self-consciously, subordinated to the statecraft of particular nationstates. From American and British perspectives, the German *Geopolitik* of the 1930s was roundly condemned for its unscientific and nationalist character (Bowman, 1942), although the degree of congruence between *Geopolitik* and National Socialist strategy in Germany has been contested over the decades (see, *inter alia*, Whittlesey, 1942; Bassin, 1987). Anglo-American political geography was not much different in its commitment to national interests, although a shade less obvious, more guarded, cloaking national ambition in a veil of objectivity and universality (see Paterson, 1987). Nonetheless, Isaiah Bowman (see Smith, 2003), along with Friedrich Ratzel and Halford Mackinder, was resolutely arguing from the standpoint of a particular national interest, and within that, from a particular political position (see Bassin, 1987, on Ratzel; Toal, 1992, and Kearns, 1984, on Mackinder). Bowman, in the United States, made it plain that his 1928 revision of *The New World* was in the light of growing American trade and investment interests around the world, and the need for government officials to have knowledge of the conditions in other countries that could inform foreign policy. In the UK, Mackinder's major concern was the future of the British Empire at a time of international rivalries abroad and class conflicts at home (see, Semmel, 1960; Toal, 1992).

Although this international focus was distinctive, in common with the rest of the discipline during the early decades of the twentieth century political geographers tended to approach these themes through the rubric of people/nature relations. The influence of 'geographic conditions' on the power of states was a common analytical concern. This involved an emphasis on environmental variations and territorial configurations, such as Mackinder's concern with the distribution of landmasses and the oceans in relation to global political power. The key early writers in the field were often cautious in their attributions of causality to the environment as such, though. It was, after all, the construction of railroads that made Mackinder apprehensive of the Eurasian heartland's potential dominance of world politics. Similarly in relation to human agency, Bowman argues with respect to Britain and its imperial possessions that 'Huge coal deposits and an island base have supplied two prime geographical advantages, but they have contributed power only through English character'

(Bowman, 1928: 35). Nevertheless, as these examples suggest, the people/nature relation, so central to the political geography of the time, was rarely understood in terms of its social mediations, irrespective of the contributions of scholars in other academic disciplines and the evidence to hand. This rather conservative political geography, then, did not even engage with questions about the political nature and workings of states, even though states were seen to be the core concern of political geography insofar as they brought together a regional form and a political phenomenon. Questions of class, of social differences internal to a country, were also marginalized. The gendered nature of politics and of political geography, so important to contemporary writers, was simply incomprehensible within its frames of reference.[1]

During this early period, then, the dominant theme of Western political geography remained that of struggle between nation-states for space and natural resources, as well as, and to a lesser extent, the internal organization of what was called the state-region (see, for example, Whittlesey, 1935). To our contemporary eyes, this work was markedly a-theoretic, although valuing detailed empirical evidence. Writers might use terms like 'description' and 'explanation' but never 'theory'. There were certainly changes in the dominant framework, shifts in emphasis that could have been related to more abstract assumptions about people and nature and even society – such as Sauer's culturalist innovations (Sauer, 1925, 1952). But human geography in the middle decades of the twentieth century was a predominantly empirical pursuit, which resonated with an emphasis on careful observation and fieldwork. Even when confronted with the rise of Fascism, political geographers would reach for these useful, but limited, tools. In a book written with the collaboration of Charles Colby and Richard Hartshorne, Derwent Whittlesey concludes an interesting and in many ways still convincing discussion of the rise of Nazism in the context of German state formation and politics, and an extensive examination of both *Geopolitik* and German war strategy, with a section entitled 'What we can do'. This section (Whittlesey, 1942: 262–8) centres on a call for the building, through expeditions and associated field observations, of a greater knowledge base with which to enrich the political understanding of states and publics. Even after 1945, Hartshorne (1954) exemplifies how political geographers responded to the rise of the social sciences in the context of re-structured universities, not by acknowledging the need to take on board social and political theory and explanation as part of a new political geography, but by drawing clearer lines between geography and the other social sciences. These were fields with which he considered geographers should cooperate rather than emulate. This meant that the connections between human geography and its various sub-fields and the other social sciences remained very weak through the 1950s.

Such was Anglo-American and European political geography as it existed up to the mid-twentieth century: empirically focused; with very limited connections with the other social sciences; methodologically limited and far from reflexive about how to do research; and with a quite narrow vision when it came to understandings of space. Into this world, the spatial-quantitative revolution (henceforth SQR) broke like a clap of thunder. Starting in the late 1950s, it ushered in a period of dramatic change in the field in many different ways, and in some that were by no means intended by its protagonists at the time. To a considerable degree it bore the fruit that they were hoping for. But it was also like opening a Pandora's box, generating a new sense that radical change was possible in the discipline. Through its strong relationships with developments in some of the social sciences, and the new value placed on deductive theory, these trends shifted the discipline decisively towards engagement with social and eventually political conceptualization.

On its own terms, the SQR represented several things. First, it represented a sharp move in the direction of a more explanatory geography. This was how it defined itself in its jousting with a human geography that, it claimed, had been focused on the uniqueness of places and had spurned law-like statements.[2] Second, it established 'space' as a new core concern for human (and physical) geography. The laws sought were to be spatial laws emphasizing nearness, connection, distance, and what at the time came to be known as spatial pattern – the dispersion of settlements, segregation within cities, the localization of industry and the like. Spatial pattern, or more broadly spatial organization, was to find its immediate explanation in spatial interaction: in migration, diffusion, location, all of which were in turn seen as patterned, organized, predictable, and spatially structured by logics of distance minimization and centralization. Third (see Barnes, 2001), this analytical turn got human geographers thinking in terms of theory, situating their empirical work theoretically, critically examining concepts and reaching out to the other social sciences in search of concepts that could be spatialized. Most obviously, this created connections with a location theory that was itself rooted in neo-classical economics or could be assimilated to it: the work of Christaller, Lösch, von Thunen and Weber, to which was soon to be added that of people like Isard and Alonso. It seemed as if human geography was itself becoming a social science (Cox, 1976).

The different sub-fields developed very unevenly with respect to the SQR. Economic, urban and transportation geography were clearly to the fore, and there is no doubt that these developments posed a particularly severe challenge to political geography. In terms of the criteria of disciplinary progress current at the time among the spatial-scientific vanguard, political geography 'lagged'. It lagged no more than cultural or historical geography, but interestingly it was the sub-field that attracted the most adverse attention, perhaps because of the way in which political science was also undergoing a 'quantitative revolution' at the time, and perhaps because of the important role of Hartshorne and other political geographers in defining the terms of the discipline for the previous fifty years. To the extent that political geography did put in a showing during the 1960s, it was with respect to those political phenomena that seemed most easily measurable in quantitative terms. Political geographers turned to voting statistics to build on a theme that had only been fitfully present in the past (see Taylor and Johnston, 1979). There was also work in what became known as the geography of modernization, associated with Peter Gould and his students (Soja, 1968; Gould, 1970; Riddell, 1970), and some work on measuring the effects of political boundaries on spatial interaction' (Mackay, 1958; Minghi, 1963). There was a policy science interest in cities and regions which was highly spatial, if not quantitative, as in the work of Peter Hall (1973) and Gerald Manners (1962), and some of the work of planners like Alonso (1968).

All of these themes had to do with politics and so could be construed as contributing to the beginning of a revival in political geography. But in general this work was drained of those questions of power, conflict and struggle usually seen as central to making up 'politics'. There were good reasons for this. The first had to do with the generally positivist tenor of the SQR. Facts spoke for themselves, and the business of a new, scientific, geography consisted in generating politically neutral spatial knowledge. Livingstone (1992) claims, following comments made by David Harvey, that this scientism was related to a political threat to academia, exemplified most forcefully in the United States by McCarthyism, which conditioned a broader quantitative turn in the social sciences at that time. He also suggests that by claiming the authority of 'science', geography could try to recoup some of the prestige it had lost subsequent to the closures of departments at the prestigious Ivy League schools, particularly those at Harvard and Yale. The roots of this neutral, technocratic, impulse were also no doubt deeper than these interesting observations imply, however. The 1950s and 1960s, after all, were what has become known as the 'golden age' of capitalism. In the United States, radical politics generally retreated, with the important exception of the innovative strategies associated with the civil rights movement in southern States, although this movement would, in turn, form the basis for new waves of radicalism in the 1960s in the United States. In Britain a succession of defeats of the Labour Party in the 1950s prompted the question 'Must Labour Lose?'[3] And the 'end of ideology' was announced[4] in a social science establishment broadly committed to scientific objectivity and where US institutions had already and rapidly achieved hegemonic status.

There is another, perhaps related, reason why the SQR effected an elimination of questions of power from work in human geography. This has to do with the nature of the social theory drawn on in discussions of location. Location theory was based in neo-classical economics. But neo-classical economics is strongly a-political, its fundamental assumptions eliminating power differentials and any reason for conflict (Cox, 1995). Such were the limits of the form of social theory characteristically permissible within the rubric of the SQR. Yet, these limits were to be decisively breached at the turn of the decade, the contributions of Julian Wolpert and David Harvey being particularly important in this process. Wolpert's (1964, 1970) critique of the SQR was an internal one, focusing on the silences of location theory without calling for its wholesale rejection. His intervention in the early 1970s was particularly important, drawing attention to the role of conflict in the explanation of location.[5] Locations were contested and outcomes determined not so much by the market, as location theory suggested, as by the various forms of power – rhetorical, institutional, economic – on which the various protagonists could call.

Like Wolpert, Harvey had been an important figure in the SQR, but his later critique of it was much more far-reaching. His contribution was more holistic in its vision than Wolpert's. He showed that the SQR and the very fact of locational conflict were thoroughly incompatible with one another. Conflict is a necessary aspect of society as we know it, and struggles around location are only one of the ways in which it is expressed. Location theory, notions of equilibrium, and of exploitation as the exception rather than the rule, had, on his view, to be thrown out. Conflict was pervasive and was formative of change in society. His personal intellectual transformation from a Wolpert-like internal critique of location theory, emphasizing externalities, preference formation and the like, to a more Marxist perspective on the world is nicely documented in his book *Social Justice and the City* (1973), and that book, of course, was a landmark in what we are referring to here as the socialization of human geography. It was this socialization of human geography, furthermore, that was to make what followed in the rest of the twentieth century so different from what had dominated the field prior to the SQR. Harvey made the more far-reaching contribution at that point, but he could not have realized the changes that he would set in motion. In short, the Marxist critique of location theory resulted in a more properly political approach within geography as a whole.

For political geography specifically these developments, and especially *Social Justice and the City* (1973), issued in a raft of new themes. Marxist theories of the state enabled political geography to begin to attend to the power relations of states and their implications in processes of class struggle (Clark and Dear, 1984). Critical comparisons between liberal theories of justice, welfare economics and Marxism led to an important focus on the politics of urban location and distribution (Cox, 1973, 1979). Indeed, a strong shift towards a theoretical and research engagement with political processes at the urban and regional scales, sites of many of the political upheavals experienced by Western countries in the late 1960s and 1970s, suggests that this 'phase' in the socialization of the discipline was crucial in decisively dislodging the nation-state from its position as the primary focus of political discussion in geography. Somewhat ironically, though, and especially given the global context informing much protest in Western cities, geopolitics, notwithstanding the interventions of some key figures (see, for example, Dell'Agnese, this volume, on Yves Lacoste) was not similarly revitalized. This might have been because this was not a field in which the SQR that was being reacted against had made much of a mark (although see Harvey, 1985).

We can identify three major, interrelated changes in the practice of human geography subsequent to this shift, via neo-classical economics and then Marxism, towards a more conscious awareness of the significance of social and political theory. The first has been a fuller appreciation of the social. Through an intense interaction with social theory, particularly in its more critical versions, a variety of dimensions of society (states, firms, social reproduction, work, culture, nationalism, the law, gender, race and consumption, for example) have been incorporated as 'geographical' concerns. Gregory (1978), in particular, signalled that the socialization of human geography had to be about more than the mutual encounter of Marxism, or other pre-given bodies of theory, and spatial concerns. Through engagement with a wide field of social theory, different theorizations of social life have also been brought into a critical relationship with one another. This has involved diverse debates between, for example, Marxist geography and what emerged as humanistic geography in the 1970s, culminating in the attempted resolution of the agency/structure problem in the early 1980s and the 'cultural turn' thereafter; and between modernist and postmodernist interpretations producing new syntheses of the political economy tradition in human geography and its more cultural counterpart.[6] An important effect of these debates and their various resolutions and stalemates has been an effacing of the boundaries between the various sub-fields of human geography producing, for example, 'new' cultural geographies and 'new' economic geographies.

A second change in geography's practices, of immediate importance for political geography in particular, involved a widespread awakening to the fact that the making and re-making of human geographies involves the deployment of power. All the sub-fields of human geography have been affected by this understanding and become more 'political' in the process. This recognition owes a great deal to Marxist geography in particular, which created a receptive disciplinary audience for other approaches positioning the concept of power even more centrally as a basic prerequisite for conceptualizing social life. The post-structuralist human geography of the 1990s and early twenty-first century, in its mobilization of the work of Foucault in particular, has reinforced and extended this insight, which has in some ways reorganized political geography around the concept of power. As this is a particularly important transformation in the field, we consider it at greater length in the following section.

Third, in terms of the broadest trends in the field, the space/society debate that subsequently dominated human geography in the mid-eighties,[7] led to a longstanding emphasis on the social production of space (Gregory and Urry, 1985; see Lefebvre, 1991). This in turn has stimulated an understanding of the diverse senses in which one can talk about space. For, and following Lefebvre, produced space is a *relational* space; space is shaped by various agents and social collectivities, it reflects and is conditioned by extant norms, relations of production and discursive constructions. A wide range of new research agendas flowed from these theoretical and political shifts of the 1970s and early 1980s, which decisively shaped the political geography practised today. The following section explores these in more detail.

DIVERSIFYING POLITICAL GEOGRAPHY

The previous section detailed the transformations in human geography through the twentieth century that one might suppose have left the discipline almost unrecognizable to its early twentieth-century practitioners. However, we suggest that the legacy of earlier trends in political geography as such nonetheless remains vital: geopolitics has undergone a revival since the 1980s, states are a perennial theme, and questions of regional differentiation and uneven development have steadily preoccupied scholars. Certainly the quantitative 'turn' retains a strong legacy, not only directly in quantitatively oriented electoral geography, migration and refugee studies, or in analyses of racial, social, and economic differentiation in cities, but indirectly in the continuing interest in urban politics that, however unintentionally, it helped to inspire. But later turns in the discipline have had a further profound influence on the topics and approaches that political geography currently encompasses. A wider politics inspired the Marxist turn in geography, as well as feminist geography and the later cultural turn: anti-war and race politics in the United States, a politics of immigration and nationalist anti-colonial politics elsewhere, the enthusiasm for cultural politics in the more post-Marxist Western European intellectual context, a strong if diverse women's movement internationally and an increasingly global environmental movement (see, in this volume, Brown, Dell'Agnese, Herb, Kofman, Mansfield). More recently a politics of globalization has generated a considerable reorientation of political geography towards understanding the transnational networks and alliances as well as the complex spatialities that animate both the powerful architects of globalization and their opponents (see, in this volume, Castree et al., Dalby, Routledge).

Even longstanding traditional themes have undergone significant changes in their focus and in the terms of their analysis as a result of these theoretical and political shifts. Although during the 1970s it seemed as if the international might be lost to political geography in the wake of the quantitative revolution and a re-focusing of inquiry around urban and regional phenomena, it underwent a revival from two complementary directions. One of these drew on wider attempts to spatialize Marx, and here Wallerstein's world systems approach was introduced to the discipline by Peter Taylor (1985), whose textbooks propagating this view have for many years been a staple of teaching in the field. Second, with a strong theoretical influence from post-structuralism, geopolitics underwent a transformation, re-emerging as *critical* geopolitics (see Dalby, Dell'Agnese, this volume). In contrast to assuming positions of national self-interest, these scholars were now concerned to excavate the geopolitical discourses that shaped international relations and to consider how geographical imaginations subtended processes of intervention and engagement (Ó Tuathail, 1986; Dalby, 1991).

The post-structuralist influence on geography more generally has profoundly shaped the ways in which the sub-discipline has developed. This, as noted previously, has led to a widespread reformulation of the focus of political geography around the relations between power and space. In particular, this reformulation found fertile ground in

Foucauldian-inspired studies, both historical and contemporary, which also contributed to additional diversification of the subject matter of political geography (for example, Philo, 1992; Driver, 1997). This had already been underway with the shift from the traditional foci of states and geopolitics to concern with the regional and the urban. Now there was a detailed concern with the microspaces of power, the disciplining of bodies, the reform of individuals and the detailed arrangements of institutions, as well as with the significance of texts and discourses in constituting political and spatial phenomena (see Huxley, this volume). Along with feminism and, indeed, with the ongoing transformations of geography's Marxisms, post-structuralist perspectives have opened up an attentiveness to politics at all scales, from the body to the globe, in many different sites, from the factory floor to the home and the refugee camp, and in many different styles, from formal political movements of unions and urban movements to the diffuse forms of power represented by cultural interventions and performances of identity. The argument for seeing the nature of the relationship between power and space as core to the sub-discipline has become theoretically pervasive in political geography (see Smith, 1994; Painter, 1995). In the early 1990s, Graham Smith proposed that political geography's proper focus should be politics in the broadest sense, and he quite accurately predicted that exploring the nature of 'power in space' would bring the field into a close association with political theory and political science (see also Painter, this volume). This has led to a critical interrogation of the traditional concepts of the subfield, such as the state, power and territory, through an energetic engagement with wider social and political theory – a relationship that has seen some measure of reciprocal interest from political theorists (for example, Jessop, 1994; Rosenau, 2003).

The concept of power itself, though, has received relatively little attention in its own right within geography, although a recent intervention by John Allen (2003) brings into question the concepts of power that have held sway in political geography. Traditionally it has been the idea of power as a property, something that is possessed, which has been drawn on by geographers; or as Mackinder would have put it,

Who rules East Europe commands the Heartland:

Who rules the Heartland commands the World-Island:

Who rules the World-Island commands the World.
(1919: 194)

John Allen points out that in spatial terms this corresponds to the idea of centralized power, in which power radiates from some point where it is thought to be held. Instead he suggests that we explore other vocabularies of power; for example, power as something that emerges or is immanent in the context of particular social relations, or that has to be created, and sustained, through the work of networks and associations. Power, then, understood in more relational terms is an effect rather than an attribute or a property. As Kuus and Agnew make clear in their contribution to this book, sovereignty is something that should be regarded as socially constructed and that needs to be constantly maintained, not something that is an a priori possession of states, and certainly not a form of power that is confined to the state. There is a clear link here to contemporary ideas of governance as a joint performance of state agencies across different levels of government and non-state actors (see Rydin, this volume). Allen also proposes, though, a more nuanced and, importantly, diversified geographic understanding of power, in which spatiality is understood to be constitutive of power, and not simply an outcome of power. Thus, power is made possible, it is produced, through spatial relations of networking, proximity, reach, and more subtly the seductive powers of experience in place. (In Chapter 3, Joe Painter provides a more detailed consideration of Allen's work.)

These kinds of formulations of political geography have drawn the sub-discipline into engagements with many different aspects of human geography, as cultural politics, a politics of the body and of resistance figure alongside but also reconfigure more traditional themes. Geopolitics, for example, has to be understood to rest on the bodies of the men and women who shape and are caught up in international events (Hyndman, 2004) and the micropolitics of identification and desire shape both the routine exercise of power in institutions and the form of violent conflicts (for useful collections see Pile and Keith, 1997; Sharp et al., 1999), as well as a more diffuse politics of race and sexuality, for example (see Saldanha, Brown, this volume). Many of these currents have fed into a vital new initiative promoting 'critical' geography, a radical left geographical movement that stretches to include these diverse cultural and post-structuralist approaches to politics alongside the now established Marxist critiques.

One important consequence of the stronger engagement with a wider range of left political theory has been a growth of interest in environmental politics, and a return to questions surrounding the ways in which political conflicts and political theory might be inflected by nature, or environment. One strand of this has reached into ideas inspired by environmental politics in poorer countries, as political ecology, originally strongly influenced by Marxism (see, for example, Watts, 1983; Blaikie, 1985; Schroeder, 1993; Jewitt, 1995;

in this volume see Muldavin, Robbins). Another emerges out of a theoretically inflected engagement with the materialities of social life, drawing on the writings of Bruno Latour and Gilles Deleuze and re-imagining the ways in which the material subtends and shapes – and is indeed an integral component of – social processes (see Castree and Braun, 2001; Whatmore, 2002; Braun, Saldanha, this volume).

The more general interrogation of the society/space relation has played an important role in opening up these new prospects in political geography; in this the sub-field has shared in shifts in understanding of space that have characterized human geography as a whole. Conceptions of absolute and relative space, which dominated analyses of state-regions and early geopolitics, have been supplemented by a strong investment in more relational accounts of space, such as that proposed by John Allen (2003) (see Harvey, 2006, on these different approaches to space). One way in which one can illustrate the increasing significance of ideas of relational space is through the intersection of political geography with questions of geographic scale. An important feature of the revival of the sub-field has been the multiplication of scales at which geography seemed to involve questions of power. One initial result of this was, as Taylor remarked in 1991, the use of geographic scale as a means of organizing the subject matter of political geography: 'Nearly all books of this period used a three-scale structure of local/urban, national/state, and international/global. This organization was treated as unproblematic …' (1991: 393). Taylor's initial response to this was to draw on world systems theory and suggest how this organization could be problematized by demonstrating the complementarity and complicity of the three different scales in organizing the contemporary world and our consciousness or knowledge of it, insisting that there are interactions over space as the global scale affects the local, the national scale is structurally unable to encompass the global economy, and so on.

The major change in understandings of scale in political geography, however, was to come later. In effect, Taylor too had taken the relevant scales as given. But a decade later this was being called into question in a much more radical fashion. Rather than being given, geographic scale itself was to be understood as *produced*. The scalar division of labour of the state is transformed through conflict and struggle. The relative significance of more local and more central branches of the state could change over time – a rescaling of state power might be one response to changing political and economic circumstances (Brenner, 2004). Central branches of the state might lose functions to non-state agents, as with the privatizations of the 1980s and 1990s, or to new, supranational organizations, like the EU, or the urban scale might become far more signifi-cant as a site of accumulation and state intervention (Swyngedouw, 2004). And a relational account of power also invites one to look beyond scales – even if they are shifting and always to be understood as produced – to the diverse and complex spatiali-ties of the specific social relations being considered rather than the relatively stable and quasi-territorial formations that scales are often taken to be. These social relations are just as likely to take the form of networks, or be stretched out over space, or exhibit disconnections and dissociations suggesting that a more strongly relational approach might generate quite different geographies of politics (Low, 1997; Massey, 2005; Castree et al., Kuus and Agnew, Lester, this volume).

These interventions have all been highly con-sequential. Not the least of their implications has been a prising apart of one of the central concerns of the sub-field, namely the state. In political geog-raphy as it was practised in the earlier part of the century it had seemed as if the relationship between the state, power and territory was an internal one. Power was a question of the relationship between states and states in geopolitics; or to a lesser degree, between states and citizens in analyses of national and regional processes. Power was understood to be exercised territorially and the politics of the state referred to regulation within a particular territory and actions undertaken to defend that territory, to exclude people and threats from it, or alternatively, as far as say, trade, was concerned, to *in*clude. These tight relationships have broken down and the rela-tions between state, power and territory are now seen as more contingent. For example:

- State power is not necessarily territorial, as John Agnew (2005) has argued extensively. Territorial-ity is only one type of spatiality of state power, and the centralized power of the state, he sug-gests, is not limited to territorial blocs but can operate at a distance, and often ignores territorial boundaries. Mackinder's sea power was of this order and in a world of increasing transnational flows and the extensive reach of military and eco-nomic forces across the globe, the geography of the state is increasingly de-territorialized. Agnew also refers to what he calls 'diffused power'. This is not centred or commanded but results from influ-ence, association and interaction as, presumably, in the underpinning of state sovereignty by the mutual recognition of different states or the exten-sion of certain beliefs, perhaps in democracy or revolution, to distant places (Anderson, 2002).
- Territorial power is not necessarily exercised only by the state. This was one of the conclusions coming out of David Sack's exhaustive survey

of territoriality. Rather all manner of organizations ranging from the corporate, through households, to the ecclesiastical engage in practices designed to, in Sack's abstract terms, 'influence the content of areas'.
- Power is not confined to the state, as both Marxist geography and feminist geography indi-cated. Power is constituted and exercised in a range of different contexts without reference to the state.
- (Non-state) power is not necessarily territorially constituted, and its exercise can be in ways that are not territorial (Anderson, 1996; Agnew, 2005). Marx conceived of the power of capi-tal in the relational terms that Allen describes; surplus value was appropriated from the work-ing class. And while one could indeed argue in Foucauldian manner that the factory has territory-like properties, the flow of value can by no means be reduced to movement within those absolute spaces. Likewise the mobility of capital, that abil-ity to threaten relocation which has been so evident in an age represented as 'globalizing', is a threat of a non-territorial nature, operat-ing regardless of any territorial limits. Similar remarks apply to gender power. An English*man's* home might have been his castle, but gendered power has also had the diffuse qualities that Agnew identified as one form of the spatiality of power.

In this way, a perennial concern of geographers – the inherently geographical phenomenon of the state – has been recast within a globalized context, through strong engagement with political theory, and with a more nuanced and relational account of power and space. Thus, even as various 'turns' and theoretical fashions have inflected political geography as a field, traditional topics, like states and geopolitics, remain vital to appreciating con-temporary events but in new ways. Along with a gendered critique of the state (see England, this volume), though, must come an appreciation of the state as a diverse political formation. Within political geography, much of the analysis of the spatiality of state power and the gendered nature of the state has focused on states in relatively wealthy and Western contexts. But the state in many poorer countries, or the global South, takes a number of different forms (see Slater, 2002; Sidaway et al., 2004). Here one of the key chal-lenges for political geography is posed: many of the theoretical analyses and assumptions of the sub-discipline draw on a relatively limited range of contexts, but usually propose universal accounts of political phenomena. Increasingly, geographers writing in and about contexts beyond the wealthy or Western countries are questioning this practice and indicating ways to produce a post-colonial political geography with relevance to wider contexts and a more modest approach to generalization (Slater, 1992; Slater, 2004; Sidaway, this volume; Paasi, 2003; Robinson, 2003). One useful, if relatively small strategy that we have adopted here to support this process has been to include writers, admittedly mostly from within the Western academy, who work on politics in poorer, or developing, country contexts (see Corbridge, Mohan and Stokke, Muldavin, O'Loughlin and Raleigh, Parnell, Rangan, Robbins, Sidaway, Silvey et al., this volume) or from beyond the hegemonic Anglo-American and English-language academic contexts (Häkli, Herb, Dell'Agnese, Kuus and Agnew, this volume).

In this *Handbook*, then, we have invited authors to air their sense of the significant traditions in the sub-field, to figure ways in which political geographers have contributed to understanding their particular topic, and to suggest ways in which we might continue to do so into the near future. We have held fast to some traditional themes – including states and electoral geography – but offered new ways of interpreting these concerns, aligned with more relational accounts of space and power, and with a strong intertwining with wider social and political theory. And we have tried to reflect at least some of the new turns in political geography – in concerns with uneven development, for example, and political identities. We encouraged authors to bring into their analyses understandings of gender, race, sexuality, nature and different contexts. Across the *Handbook* there are many examples of this being successfully achieved, as in accounts of sexuality informing analyses of urban social movements (see Brown, this volume); accounts of the politics of governance strongly inflected by the experiences within environmental politics (see Rydin, this volume); and accounts of transnationalism and migration that attend to gendered processes of mobility and immobility (Silvey et al., this volume). There is much more still to be done to 'mainstream' these critical voices within political geography so that in future collections of this kind the founding fathers will no longer dominate the narrative, the Anglo-American context will not be the assumed reference point for histories of the discipline, and the core topics of the field will be strongly inflected with a sensitivity to questions of gender, sexuality and context, for example. Political geography has travelled a long way from Ratzel's expansionist state, Mackinder's imperial dominance and Hartshorne's state-region. But there are pressing and hopefully exciting agendas remaining, and we hope the chapters collected in this *Handbook* will reflect something of that vibrant future.

THE ORGANIZATION OF THIS *HANDBOOK*

From the foregoing discussion, it will be clear that political geography has come to resemble a diversified, productively unruly, social gathering. Around the room, as it were, there are at any moment noticeable clusters of substantive interest, representing growth sectors in the field: political ecology, the politics of local and regional development, critical geopolitics, a cluster of people with interests in voting studies, but with a hugely transformed approach from that prevailing in the 1960s. There also remain sharp differences in approach, style and tone of voice. Marxist understandings jostle with ones of a more post-structural provenance. The rise of gender studies in political geography has come about through political economy approaches but also approaches that have stronger relations to the cultural turn in human geography. And of course, the topics of interest to political geographers have become much more varied than they were in the first half of the last century, though the more classical subjects are still apparent and form a basis for discussion across divergent approaches and generations in the field. Political geography is a diverse world and has undergone some striking mutations: in this context, organizing a *Handbook of Political Geography* has been a challenging task.

We have been concerned to capture this diversity of approach and topic while ensuring that the *Handbook* remains open to the possible futures of political geography. Much as we might like to think that political-geographic research can be gathered and organized under some overall theoretical scheme where everything done could, as it were, be seen to peacefully co-exist, we came to the conclusion that this was not a desirable strategy. It would iron out possibly unbridgeable differences of concern, conceal the complex layering of different kinds of political geography over time and, by imposing a closed structure on the field, might foreclose the possibility of appreciating 'new' political geographies past and present.

Changes in the world of politics mean that any research field must develop an agile and ongoing critique of its own concepts, but some of these concepts have been around a long time and retain their importance. Territoriality, the state, the significance of borders, their re-invention and even hardening in the face of growing transnationalism, continue to attract the attention of political geographers many years after explorations of these concepts were first introduced into the field. The *Handbook*, then, is organized around certain topics of longstanding importance in political geography. We have used these as starting points for an exploration of connected areas of concern and for opening out the field to new directions and subject matters. One can get a sense of this from the titles of the seven sections into which we have divided the *Handbook*. After a general opening section which establishes the historiographical and theoretical scope of the field ('The Scope and Development of Political Geography'), we turn to revisit and explore new directions for two of the most longstanding topics within the field, states and nature ('States' and 'Re-Naturing Political Geography'). We then pick up on one of the most recent trends in the field, the politics of difference ('Identities and Interests in Political Organizations') before establishing the grounds for new initiatives in a field that rose to prominence since the 1960s and that has been very strongly associated with political geography ('From *La Géographie Electorale* to the Politics of Democracy'). The book concludes with sections dealing with ('Global Political Geographics' and 'The Politics of Uneven Development').

Within each section, chapter themes have been selected as vehicles for the exploration of both the old and the new. This means that titles like 'Coercion, Territoriality, Legitimacy: The Police and the Modern State' appear alongside ones like 'Theorizing the State Geographically: Sovereignty, Subjectivity, Territoriality', though the insertion of 'subjectivity' in the latter is already indicative of the sharp changes that the sub-field has experienced. Likewise, something like 'Global Environmental Politics' picks up on longstanding interests in political geography in people/nature relations and the international, but the sense of the terms 'global' and 'environmental politics' indicate that we are living in a very different world from that of Friedrich Ratzel.

The diversification of scales and spaces we drew attention to in the previous section is apparent in chapter titles like 'Polarized Cities' or 'Planning: Territories, People and Governance', as well as the whole of the section dedicated to electoral studies and democracy and Arun Saldanha's chapter on 'The Political Geography of Many Bodies'. The variety of substantive foci of the contemporary sub-field is also evident. Identity politics is a significant theme in many of the chapters (for example, Brown, Häkli, Huxley, Kofman, Lester, Penrose, Saldanha, Silvey et al., Routeldge). Many of the themes explored would, as noted above, have been quite alien to earlier generations of political geographers, but some chapters focus on older themes that have undergone a revival and reinterpretation. One of the earliest studies of gerrymandering by a geographer was by Carl Sauer (1918), for example, and Forest (this volume) picks up on this interest but also demonstrates the difference that the ensuing near 90 years of interdisciplinary work have made to the meanings that are assigned to that particular way of 'producing geographies'. Similarly, the interest in the politics of uneven development is apparent in East and Moodie's (1956) collection *The Changing World*, and even earlier

in Bowman's (1928) *The New World*. That interest is represented in the *Handbook* in the chapters by Mohan and Stokke, Rangan and Taylor, though in ways that Bowman, and later East and Moodie, might have found startling.

In our invitations to individual contributors our approach was to offer fairly detailed, but not dogmatic, instructions regarding themes that might – emphasize *might* – be highlighted. Survey has been only a part of what we were looking for, and in some instances a very minor part. More importantly, our aim has been to elicit fresh perspectives and insights, new combinations of substance and framing perhaps, or combinations that exemplified particular themes, contemporary or otherwise, with enhanced clarity. One result of this has been that the boundaries between the different sections occasionally appear somewhat blurred. Chapter headings were assigned to different sections but in the way we as editors envisaged them. In their execution by our diverse collaborators, they often overlap more than one section, and in some instances these overlaps are very strong. Yvonne Rydin's chapter on sustainable development and governance could quite easily have gone in the section on re-naturing political geography, for example. The chapter by Giles Mohan and Kristian Stokke on the politics of decentralization appears in the section on development, and their slant on the topic is that of development studies. But it *could* have gone in the section on the state, given recent arguments about state rescaling. The section on 'Global Political Geographies' contains chapters that could, without much difficulty, have been placed elsewhere. Resorting to a 'scalar' definition for this section has, at least, resulted in several other sections being less unwieldy than they might otherwise have been. Unsurprisingly, the overlaps are especially apparent with the section on the state; tallying up, it would seem that at least nine of our chapters could have been in that section rather than the one that we had in mind and for which they were, in addition, tailored. We do not regard this as a problem; rather we see it as a mark of the fascinating, and ongoing, attempt of the sub-discipline to both settle accounts with the state as an object of study and set of institutions in the world, and to define emergent areas of interest and alternative conceptualizations in relation to this most central inherited object of political-geographic tradition.

It is now just over a century since the appearance of Halford Mackinder's 'The Geographic Pivot of History' (1904). Mackinder is rightly regarded as one of the founders of modern political geography. His framework of understanding was very different from what is accepted as political geography today, as only the most cursory perusal of what follows will convince. In many ways his definitions and interpretations were at one with the times. We believe that those of our contributors are too. We also believe, however, and hope that this introduction has demonstrated, that the century or more of theoretical and political changes that have contributed to shaping political geography today give the way in which our authors address problems a perhaps more incisive, if more modest, purchase on the politics of the world that eluded our intellectual forebears. It is now up to the reader to determine just how correct we are in our assessment, and whether the diverse set of conversations around political geography's concerns can be of use in engaging with the sort of complex, intractable and destructive situations going on around us in the world as we write.

NOTES

1 In this regard there are histories of political geography that remain to be written of the contributions of women – for example, the role of women travel writers and explorers who generated and popularized geographical imaginations of empire and knowledge about different places (Domosh, 1991; Blunt, 1994), or the important contributions of those women geographers excluded from major institutions and positions and narratives of the discipline's past, but active in the production of significant geographical knowledge (see Kofman, this volume).

2 In passing, we should note that this was a little overdone. Sauerian geography might indeed have spurned morphological laws of the sort that Schaefer was interested in, but it was far from being non-explanatory. The systematic geography of the fifties, some of it quite spatial in character, had also had an explanatory emphasis. The work of people like Harris (1954) on the location of industry in the United States as a function of market potential, Smith (1955) on Weber, and Ullman (1957) on commodity flows, is exemplary.

3 There was a (1960) book with this title written by Mark Abrams, Rita Hinden and Richard Rose.

4 Most famously by Daniel Bell (1961).

5 Intriguingly, discussions of 'locational conflict' were not totally foreign to human geography. Mayer (1964) had written provocatively on a particular instance of what he called 'land use conflicts', But it did not register in terms of generating a line of related research. Wolpert's interventions did, and we would suggest that this was because he did indeed think in terms of theory, situating himself with respect to theory (as in his use of the expression 'locational conflict') and drawing out the implications of his empirical observations for the further development of theory. Again, we are drawn to the huge difference that the SQR made to the way in which human geographers thought, as Barnes has emphasized.

6 For a collection of papers on the deconstruction of some of these binaries, see Cloke and Johnston (2005).

7 Landmark contributions here were Harvey's groundbreaking work in the later sections of *The Limits to Capital* (1982), the publication of the Gregory and Urry collection in 1985 and the creation of a new journal, *Environment and Planning D: Society and Space,* in 1983.

REFERENCES

Abrams, M., Rose, R. and Hinden, R. (1960) *Must Labour Lose?* Harmondsworth, Middlesex: Penguin Books.

Agnew, J.A. (2005) 'Sovereignty regimes: territoriality and state authority in contemporary world politics', *Annals of the Association of American Geographers*, 95(2): 437–61.

Allen, J. (2003) *Last Geographies of Power.* Oxford: Blackwell.

Alonso, W. (1968) 'Urban and regional imbalances in economic development', *Economic Development and Cultural Change*, 17: 1–14.

Anderson, J. (1996) 'The shifting stage of politics: new medieval and postmodern territorialities?', *Environment and Planning D: Society and Space*, 14(2): 133–53.

Anderson, J. (2002) *Transnational Democracy.* London and New York: Routledge.

Barnes, T.J. (2001) 'Retheorizing economic geography: from the quantitative revolution to the "Cultural Turn"', *Annals of the Association of American Geographers*, 91(3): 546–65.

Bassin, M. (1987) 'Race contra space: the conflict between German *Geopolitik* and National Socialism', *Political Geography Quarterly*, 6(2): 115–34.

Bassin, M. (1987) 'Imperialism and the nation-state in Friedrich Ratzel's political geography', *Progress in Human Geography*, 11: 473–95.

Bell, D. (1961) *The End of Ideology.* New York: Collier.

Blaikie, P. (1985) *The Political Economy of Soil Erosion in Developing Countries.* Harlow, Essex: Longman.

Blunt, Alison (1994) *Travel, Gender and Imperialism: Mary Kingsley and West Africa.* New York. Guilford.

Bowman, I. (1928) *The New World.* London: George Harrap.

Bowman, I. (1942) 'Geography versus geopolitics', *Geographical Review*, 32(4): 646–58.

Brenner, Neil (2004) *New State Spaces: Urban Governance and the Rescaling of Statehood.* Oxford: Oxford University Press.

Castree, N. and Braun, B. (2001) *Social Nature.* Oxford: Blackwell.

Clark, G. and Dear, M.J. (1984) *State Apparatus: Structures and Language of Legitimacy.* London: Allen & Unwin.

Cloke, P. and Johnston, R.J. (eds) (2005) *Spaces of Geographical Thought.* London: Sage.

Cox, K.R. (1973) *Conflict, Power and Politics in the City: A Geographic Approach.* New York: McGraw Hill.

Cox, K.R. (1976) 'American geography: social science emergent', in C.M. Bonjean et al. (eds), *Social Science in America.* Austin: University of Texas Press, pp. 182–207.

Cox, K.R. (1979) *Location and Public Problems: An Introduction to Political Geography.* Chicago: Maaroufa Press.

Cox, K.R. (1995) 'Concepts of space, understanding in human geography, and spatial analysis', *Urban Geography*, 16(4): 304–26.

Dalby, S. (1991) 'Critical geopolitics: discourse, difference and dissent', *Environment and Planning D: Society and Space*, 9: 261–83.

Domosh, Mona (1991) 'Towards a feminist historiography of geography', *Transactions, Institute of British Geograpers, New Series* 16: 95–104.

Driver, F. (1997) 'Bodies in space: Foucault's account of disciplinary power', in T.J. Barnes and D. Gregory (eds), *Reading Human Geography: The Poetics and Politics of Inquiry*. London: Arnold, pp. 279–89.

East, W.G. and Moodie, A.E. (eds) (1956) *The Changing World*. London: George Harrap.

Falah, G. (1996) 'The 1948 Israeli-Palestinian war and its aftermath: the transformation and de-signification of Palestine's cultural landscape', *Annals of the Association of American Geographers*, 86(2): 256–85.

Gould, P.R. (1970) 'Tanzania 1920–1963: the spatial impress of the modernization process', *World Politics*, 22(2): 149–70.

Gregory, D. (1978) *Ideology, Science and Human Geography*. London: Hutchinson.

Gregory, D. (2004) *The Colonial Present*. Oxford: Blackwell.

Gregory, D. and Urry, J. (eds) (1985) *Social Relations and Spatial Structures*. London: Macmillan.

Hall, P. et al. (1973) *The Containment of Urban England*. London: Allen & Unwin.

Harris, C.D. (1954) 'The market as a factor in the localization of industry in the United States', *Annals of the Association of American Geographers*, 44: 315–48.

Hartshorne, R. (1954) 'Political geography', in P. James and C.F. Jones (eds), *American Geography: Inventory and Prospect*. Syracuse, NY: Syracuse University Press.

Harvey, D. (1973) *Social Justice and the City*. Baltimore: Johns Hopkins University Press.

Harvey, D. (1982) *The Limits to Capital*. Chicago: Chicago University Press.

Harvey, D. (1985) 'The geopolitics of capitalism', in D. Gregory and J. Urry (eds), *Social Relations and Spatial Structures*. London: Macmillan, chap. 7.

Harvey, D. (2006) 'Space as a keyword', in N. Castree and D. Gregory (eds), *David Harvey: A Critical Reader*. Oxford: Blackwell. chap. 14.

Hyndman, J. (2004) 'Mind the gap: bridging feminist and political geography through geopolitics', *Political Geography*, 23: 307–22.

Jessop, B. (1994) 'Post-Fordism and the state', in A. Amin (ed.), *Post-Fordism: A Reader*. Oxford: Basil Blackwell, pp. 251–79.

Jewitt, S. (1995) 'Europe's "others"? Forestry policy and practices in colonial and postcolonial India', *Environment and Planning D: Society and Space*, 13(1): 67–90.

Kearns, G. (1984) 'Closed space and political practice: Frederick Jackson Turner and Halford Mackinder', *Environment and Planning D: Society and Space*, 2(1): 23–34.

Lefebvre, H. (1991) *The Production of Space*, trans. D. Nicholson-Smith. Oxford: Basil Blackwell.

Livingstone, D.N. (1992) *The Geographical Tradition*. Oxford: Blackwell.

Low, M.M. (1997) 'Representation unbound: globalization and democracy', in K.R. Cox (ed.), *Spaces of Globalization*. New York: Guilford Press. chap. 9.

Mackay, J.R. (1958) 'The interactance hypothesis and boundaries in Canada', *Canadian Geographer*, No. 11: 99–109.

Mackinder, H.J. (1904) 'The geographic pivot of history', *Geographical Journal*, 23: 421–42.

Mackinder, H.J. (1919) *Democratic Ideals and Reality*. London: Constable.

Manners, G. (1962) 'Regional protection: a factor in economic geography', *Economic Geography*, 38: 122–29.

Massey, D. (2005) *For Space*. London: Sage.

Mayer, H. (1964) 'Politics and land use: the Indiana shoreline of Lake Michigan', *Annals of the Association of American Geographers*, 54(4): 508–23.

Minghi, J.V. (1963) 'Television Preference and Nationality in a Boundary Region', *Sociological Inquiry*, 33: 65–79.

Ó Tuathail, G. (Toal, G.) (1986) 'The language and nature of the "new geopolitics": the case of U.S.–El Salvador relations', *Political Geography Quarterly*, 5: 73–85.

Paasi, Anssi (2003) 'Region and place: regional identity in question', *Progress in Human Geography*, 27: 475–85.

Painter, J. (1995) *Politics, Geography and 'Political Geography': A Critical Perspective*. London: Arnold.

Paterson, J.H. (1987) 'German geopolitics reassessed', *Political Geography Quarterly*, 6(2): 107–114.

Philo, C. (1992) 'Foucault's geography', *Environment and Planning D: Society and Space*, 10(2): 137–61.

Pile, S. and Keith, M. (eds) (1997) *Geographies of Resistance*. London: Routledge.

Riddell, B. (1970) *The Spatial Dynamics of Modernization in Sierra Leone*. Evanston, IL: Northwestern University Press.

Robinson, J. (2003) 'Political Geography in a postcolonial context', *Political Geography*, 22(6): 647–53.

Rosenau, J. (2003) *Distant Proximities: Dynamics Beyond Globalization*. Princeton: Princeton University Press.

Sack, R.D. (1983) 'Human territoriality: a theory', *Annals of the Association of American Geographers*, 73(1): 55–74.

Sauer, C.O. (1918) 'Geography and the gerrymander', *American Political Science Review*, 12: 403–26.

Sauer, C.O. (1925) *The Morphology of Landscape*. Berkeley: University of California Press.

Sauer, C.O. (1952) *Agricultural Origins and Dispersals*. New York: American Geographical Society.

Schroeder, R. (1993) 'Shady practice: gender and the political ecology of resource stabilization in Gambian garden/orchards', *Economic Geography*, 69: 349–65.

Semmel, B. (1960) *Imperialism and Social Reform*. Garden City, NY: Doubleday.

Sharp, J.P., Philo, C., Routledge, P. and Paddison, R. (eds) (1999) *Entanglements of Power*. London: Routledge.

Sidaway, J.D., Bunnell, T., Grundy-Warr, C., Mohammad, R., Park, B. G. and Saito, A. (2004) 'Translating political geographies', *Political Geography*, 23: 1037–49.

Slater, D. (1992) 'On the borders of social theory: learning from other regions', *Environment and Planning D: Society and Space*, 10(3): 307–27.

Slater, David (2002) 'Other domains of democratic theory: space, power and the politics of democratization', *Environment & Planning D: Society and Space*, 20(3): 255–76.

Slater, David (2004) *Geopolitics and the Post-Colonial: Rethinking North-South Relations*. Oxford : Blackwell.

Smith, G.E. (1994) 'Political theory and human geography', in D. Gregory, R. Martin and G.E. Smith (eds), *Human Geography: Society, Space and Social Science*. Basingstoke: Macmillan, pp. 54–77.

Smith, N. (1989) 'Geography as museum: private history and conservative idealism in *The Nature of Geography*', in J.N. Entrikin and S.D. Brunn (eds), *Reflections on Richard Hartshorne's The Nature of Geography*. Washington, DC: Association of American Geographers.

Smith, N. (2003) *American Empire*. Los Angeles and Berkeley: University of California Press.

Smith, W. (1955) 'The location of industry', *Transactions of the Institute of British Geographers,* 21: 1–18.

Soja, E.W. (1968) *The Geography of Modernization in Kenya*. Syracuse, NY: Syracuse University Press.

Swyngedouw, E. (2004) 'Globalisation or "glocalisation": networks, territories and rescaling', *Cambridge Review of International Affairs,* 17(1): 25–48.

Taylor, P.J. (1985) *Political Geography: World-Economy, Nation-State and Locality*. London and New York: Longman.

Taylor, P.J. (1991) 'Political geography within world-systems analysis', *Review,* 14(3): 387–402.

Taylor, P.J. and Johnston, R.J. (1979) *Geography of Elections*. New York: Holmes & Meier.

Toal, G. (1992) 'Putting Mackinder in his place', *Political Geography,* 11(7): 100–18.

Ullman, E.L. (1957) *American Commodity Flow*. Seattle: University of Washington Press.

Watts, M. (1983) 'Hazards and crises: a political economy of drought and famine in Northern Nigeria', *Antipode,* 15(1): 24–34.

Whatmore, S. (2002) *Hybrid Geographies: Natures, Cultures, Spaces*. London: Sage.

Whittlesey, D. (1935) 'The impress of effective central authority upon the landscape', *Annals of the Association of American Geographers,* 25(2): 85–97.

Whittlesey, D. (1942) *German Strategy of World Conquest* (written with the collaboration of Charles C. Colby and Richard Hartshorne). New York: Farrar & Rinehart.

Wolpert, J. (1964) 'The decision process in spatial context', *Association of American Geographers,* 54: 537–58.

Wolpert, J. (1970) 'Departures from the usual environment in locational analysis', *Annals of the Association of American Geographers,* 50(2): 220–29.

Yiftachel, O. and Yacobi, H. (2002) 'Planning a bi-national capital: should Jerusalem remain united?', *Geoforum,* 33(1): 137–44.

The Scope and Development of Political Geography

Introduction

Kevin R. Cox

Our objective in this first section of the *Handbook* is to address some of the fundamental concepts of the sub-field of political geography, and the changing ways in which they have been articulated over the course of its history. We have focused on some of the more critical and recent developments, over the last couple of decades or so, which bear a close relationship to the way in which a range of different topics are analysed in the remainder of the *Handbook*. These are: the state, power, gender and positionality. In keeping with the broader brief of the *Handbook* series, to review and anticipate developments in a field, these chapters set the scene for an historical appreciation of the emergence of the core ideas of political geography, but also explore current debates that are setting the agenda for the future.

Eleonore Kofman exemplifies this as she examines the relatively contentious relationship between feminism and political geography; the theme of gender relations is one that recurs in many of the subsequent chapters (e.g., England, Brown, Silvey et al., Staeheli, Saldanha, all in this volume). In a similarly wide-ranging critique of the past trajectories of political geography, James Sidaway addresses the geography of political geography; that is, the degree to which the sub-field reflects the concerns of particular (usually Western) contexts. We have attempted to incorporate this – increasingly expressed – post-colonial critique in the *Handbook* by encouraging authors to address mainstream political geography issues from standpoints beyond the West (see among others Corbridge, Muldavin, Parnell, Rangan, Robbins, Mohan and Stokke) and from beyond the Anglo-American core (see among others Dell'Agnese, Häkli, Herb).

Both of these chapters address the situatedness of knowledge in political geography, an issue that is also evident in the other two chapters here: who, after all, can doubt the situatedness of classical geopolitics as Guntram Herb underlines? And Joe Painter's concern for the nature of power in political geography draws us to a subtle analysis of

the operation of power that can inform many of the issues that attentiveness to positionality raises. Herb and Painter each take a detailed view of central concepts of political geography as they have developed over the course of the sub-field's history. Herb focuses on the state and its relation to political geography while Painter examines the changing ways in which power has been understood, or even considered, for that matter. These themes resonate with the concerns of feminist and post-colonial critics, as reviewed by Kofman and Sidaway. Feminists have alerted us to the question of 'For whom is the state?' And Sidaway raises the question of 'For whom is political geography, power and the state understood, at all?'

Herb and Painter's treatments of their respective topics share some interesting overlaps, especially in relation to their periodizations of political geography. Herb's focus is the changing relationship between the state and political geography. From this standpoint he characterizes the history of political geography in terms of three phases. The first phase is that of political geography as handmaiden, implicitly or not, to the state and its objectives in international arenas. The assumption is of states contending with one another for dominance and the purpose of political geography is to identify the difference that resources and the advantages of geographic position make in that struggle. The works of Mackinder and German *Geopolitik* are exemplary in their different ways. There is then a second phase in which political geography becomes less concerned with issues of foreign aggrandizement and imperialism and more with the maintenance of the state as an institution. Political geography turns inwards and it is the internal structure of the state and domestic policy, that become central. This results in a focus on issues like the internal integration of the state, its administrative organization or scalar division of labor, and electoral geography. Since the 1970s, this emphasis has been overtaken by approaches that are more critical of the state and that ask the question, whose interests are served by it? This has been inspired by what in the Introduction to the *Handbook* we call the socialization of human geography, including the incorporation of work in political economy and the politics of difference.

This periodization is repeated in Painter's chapter on political geography and power, though there is much else there in addition. So for Painter, the first phase corresponds to one in which the issue of power was broached. Mackinder talked, for example, about manpower as an aspect of national strength. But the *nature* of power remained unexamined and, as he shows, largely taken for granted. Power was typically understood to mean the ability of one state to influence, or to impose its will upon, another. Corresponding to Herb's second phase where the focus switches from the international to the *intra*-national, there is a retreat from questions of power. Painter attributes this largely to the poor image that many argue political geography acquired through its association with geopolitics, particularly in its German incarnation, even though that simply represented an extreme development of forms of argumentation latent in other national geopolitical schools. He points out also how this phase was prolonged by the spatial-quantitative revolution and its embrace of neo-classical economics as its social theory – a theory that eschews power differentials through ideas of perfect competition and the elimination of tendencies to exploitation, as indeed we suggested in the Introduction.

The third and critical phase then constitutes for Painter a period in which political geographers, after a hesitant start, begin to question the nature of power. This has involved, *inter alia*, a move away from an exclusive concern with state forms of power, away from notions of power as entirely constituted by resources or as something that agents possess toward a more relational conception of power involving diverse agents, both within and outside the state. This approach has strong Foucauldian origins and signals an analysis of power in which its exercise is multiple and varied, and in which power might seem to be literally everywhere, immanent in social relations of all kinds (here Painter draws on Allen, 2003). The ties to feminist political theory are strong, bringing into view power relations associated with the personal, with identity, bodies and sexuality (see England, Kofman, Saldanha, Silvey et al., all in this volume). But it has also encouraged changes in how spatialities of state power are understood, including a move away from the older concerns with territory (Agnew, 1994) toward a concern with state/society relations (Corbridge, this volume), transnational forms of state power (Kuus and Agnew, this volume), and the multiple, often contradictory agendas of the state (Robbins, this volume).

To some extent, one might argue that the state has been 'the pivot of political geography', in more ways than the one that Herb suggests in framing his periodization. For even while showing signs of analytical and substantive 'wear and tear', insofar as they often strongly shape distinctive institutional and social contexts, states have significantly structured the development of political geography itself. Sidaway's contribution here refers to how national political geographies as schools of thought vary, and how the meanings of fundamental concepts like territory, state, nation, are always context-specific. A good instance of this is provided by the work of political geographers on the politics of local and regional development, in particular, the difficulties experienced in transferring ideas like 'growth coalition' and 'urban regime' from the United States to the United Kingdom, or for that matter, the rest of Western Europe (see Wood, this volume; also Brown on this in relation to urban social movements, this volume).

Sidaway also sees a tendency for Western political geography, particularly Anglophone political geography, to impose its categories and understandings on the rest of the world, though political geographers elsewhere may seem, for a range of different reasons, to cooperate in this (see Yeung, 2001, for a discussion of the demands of 'international' publishing in different contexts). In relation to theorizing the state, for example, very often 'failed' or 'juridical' states are discussed, implying a continuum at one end of which are the 'successful' and 'empirical' states of the West and the 'failed' and 'fictional' states of a number of developing societies (Migdal, 1988). But as Sidaway asks, for whom are these states failures, and against what yardstick are their activities to be measured? (See also Power, 2001, and Corbridge et al., 2005, on states, and Slater, 2002, for a related discussion on democracy.)

It was the feminism of the seventies, of course, that did so much to alert political geographers to questions of positionality and the situatedness of their theories and assumptions. The encounter between feminism and political geography is taken up here by Eleonore Kofman. Her contribution is a wide-ranging one and it is hard to do justice to it in a necessarily brief summary. On the one hand, she argues, the relationship between women and political geography was exclusionary and prevented many women from making a full and acknowledged contribution to the discipline. For most of the twentieth century, political geography shared with the remainder of human geography a strong gender bias in its preoccupations; there were also very few women political geographers. Kofman argues, however, that even after the feminist revolution of the seventies, political geography has only recently begun to take feminist work seriously. Parenthetically this *Handbook* marks something of the changes in the gender balance of the field. In an earlier compendium of political geography, *The Changing World*, published fifty years ago and edited by East and Moodie, only one of forty-two chapters was written by a woman. Of the 36 chapters in this *Handbook*, 13 are written by women. But clearly the task of bringing a feminist analysis to bear on all the core areas of political geography remains, as Kofman reminds us, a challenge for the future.

The contributions of feminist geographers have already set new agendas for political geography. One thinks here of the way in which feminist writers have contributed to de-centering the state from considerations of power in political geography, which both Herb and Painter indicate as a major sub-disciplinary trajectory. And it is unlikely that the whole issue of public/private space would have received the critical attention that it has or exposed the dangers of conflating the gender dualism with it without a feminist perspective. The issue of the politics of scale has been extended to include questions of social reproduction; the household and the body have been identified as scales hitherto neglected or even overlooked entirely (see Marston, 2000). In addition, new substantive areas, like the interest in care and its implications for migration, have been opened up (on these issues see England, Silvey et al., this volume).

In the context of globalization, attention to questions of geographic scale has grown considerably and moved beyond simply the scalar division of labor of the state. As Guntram Herb writes suggestively in his chapter, 'The arguably central unifying element of the current critical tradition is scale. Critical political geographers universally recognize linkages among scales, stress that scale should not be equated with pre-existing administrative units, and embrace the idea that social relations spill over state boundaries'. There is, however, an interesting history to all this and the development of political geography is worth considering from this standpoint. It revolves around the relation between the international and the national as scales of interest for political geographers.

As Herb points out, there is an initial phase which is mainly international in its focus. To the extent that the nation-state is referred to, it is in the context of international struggles. This is true, for example, of both Mackinder and Bowman, as we discussed in the Introduction to the *Handbook*. This is followed by a phase between the end of the Second World War and the beginning of the seventies in which the focus of political geography is almost exclusively national and domestic. More recently, what we have witnessed has been an analytical and substantive dismantling of the national/international dualism. The attack on state-centric understandings of power, partly of feminist provenance, as Kofman points out, and partly as a result of a more theoretically inflected discipline, has contributed to this. So too has the questioning of the idea of territorialized forms of sovereignty, so central to the accounts of state power; Herb and Painter both discuss these developments (see also Kuus and Agnew, this volume).

This softening of the national/international distinction has at least two different consequences for the way in which political geography has come to be practised, and one of these has attracted more attention than the other. Herb refers to how 'social relations spill over state boundaries'. The more general interest in globalization has surely been important here, although, as we note in the *Handbook* Introduction, the questioning of the geographical correspondence of state and society goes back to the seventies and early eighties in the work of people like Giddens, Mann and Wallerstein. One result of this for political geographers has been the interest in structures and processes like 'world cities', 'transnational urbanism', the international entanglements of 'sub-national states' (Paul, 2005; see Taylor, this volume), 'multiculturalism' and its diverse implications for, say, local politics (Mitchell, 1993; see Penrose and Mole, this volume), and, of course, the politics of 'glocalization' (Jessop, date; see Routledge, this volume).

There are clearly some political and normative concerns consequent upon these trends, as one of us (Low, 1997) has pointed out elsewhere; among other things, given the extensiveness of the geographic spillovers of state policy, what does this mean for the geography of modes of representation? Does the classic, territorially defined form of democracy suffice? Clearly this is not the case; both Barnett and Staeheli engage with these issues in this volume. So how might this be handled? And what sorts of functional substitutes might emerge?

Painter points toward an increasingly dominant focus within the sub-discipline on power as a product of networks, whose geographies might reach across scales, and stretch beyond pre-given territories, perhaps making up new forms of spatial connection and disconnection or territorialization. As Silvey et al. suggest (this volume), though, the geographies of these associations cannot be presumed and might lead, as in the case of transnational migration that they discuss, to geographies of immobility and entrapment, rather than to experiences of mobility and opportunity. The politics of transnational associations, and forms of contestation and governance that draw in a range of actors beyond the state and reach across state borders, or stretch across localities, become increasingly important (see Rydin, Routledge, Castree et al. and

Lester, this volume). They suggest that forms of governance emerge at and across all scales that bring together governments and agents of civil society, or capital, in networks of association rather than in bounded blocks of space. Distinguishing between scales becomes increasingly difficult as elements of states, for example, become significant sites for the creation of global processes, or are constituted in and through very localized arrangements (Sassen, 1996; see Corbridge, Robbins, Routledge, this volume). Consequently, questions arise over whether scale remains a pertinent framework for political geography – as Painter implies when he counterposes territorial to topological views of space, and as explored in detail by Castree et al. (this volume).

The four chapters in this section direct us towards a substantial theoretical agenda for future political geography research. Current practices will, we suggest, encounter many challenges as questions of states, power, positionality and feminism continue to be explored. These concerns will also encourage new thinking for geography at the most general level: re-thinking scale as a result of the changing politics of de/re-territorialization or because of a feminist political critique, re-imagining the spatialities of networks as we re-theorize power, or bringing new spaces such as the body or positionality into view as critical political interventions like feminism and post-colonialism disturb taken-for-granted approaches. A fully 'socialized' and theoretical political geography rightly assumes a place at the core of the wider discipline of geography.

REFERENCES

Agnew, J.A. (1994) 'The territorial trap: the geographical assumptions of international relations theory', *Review of International Political Economy*, 1: 53–80.

Alexander, T. (1996) *Unravelling Global Apartheid*. Cambridge: Polity Press.

Allen, J. (2003) *Lost Geographies of Power*. Oxford: Blackwell.

Corbridge, S., Williams, G., Srivastava, M. and Veron, R. (2005) *Seeing the State: Governance and Governmentality in India*. Cambridge: Cambridge University Press.

Low, M.M. (1997) 'Representation unbound: globalization and democracy', in K.R. Cox (ed.), *Spaces of Globalization*. New York: Guilford Press, chap. 9.

Marston, S.A. (2000) 'The social construction of scale', *Progress in Human Geography*, 24(2): 219–42.

Migdal, J. (1988) *Strong Societies and Weak States: State-Society Relations and State Capabilities in the Third World*. Princeton: Princeton University Press.

Mitchell, K. (1993) 'Multiculturalism, or the united colors of capitalism', *Antipode*, 25(4): 263–94.

Paul, D. (2005) *Rescaling International Political Economy*. New York: Routledge.

Power, M. (2001) 'Patrimonialism and petro-diamond capitalism: geo-politics and the economics of war in Angola', *Review of African Political Economy*, 90: 489–502.

Sassen, S. (1996) *Losing Control? Sovereignty in an Age of Globalisation*. New York: Columbia University Press.

Slater, D. (2002) 'Other domains of democratic theory: space, power and the politics of democratization', *Society and Space*, 20: 255–76.

Yeung, H. (2001) 'Editorial: redressing the geographical bias in social science knowledge', *Environment and Planning A*, 33(1): 1–9.

The Politics of Political Geography

Guntram H. Herb

INTRODUCTION

'La Géographie, de nouveau un savoir politique'
(Geography: once again a political knowledge).

(Lacoste, 1984)

This statement by the chief editor of *Hérodote*, intended to celebrate the politicization of French geography through the journal in the 1970s and 1980s, also, and paradoxically, captures a profound dilemma of contemporary political geography. If, as a recent academic forum showed, the political is alive and well in *all* of geography, does this not question the continued relevance and validity of having a separate sub-field of political geography (Cox and Low, 2003)? The most fruitful response to such existential questions about academic sub-disciplines is delving into the past and tracing the genesis of the subject. In what follows, I will seek to understand the meaning of political geography by analyzing the historical development and implications for present practices, in short, the politics of political geography.

The standard starting points for political histories of academic subjects are the first use of the term and the seminal first work; in the present case, the coining of 'political geography' by the French philosopher Turgot in 1750 and the publication of Friedrich Ratzel's *Political Geography* in 1897 (Agnew, 2002: 13). Yet, the majority of evolutionary approaches are limiting, to say nothing of being potentially stodgy and boring. The tendency is to present a story of progress from a benighted past to an enlightened present. In the

case of political geography, the usual story is of a heyday characterized by racism, imperialism, and war in the nineteenth and early twentieth centuries, followed by a period of stagnation and decline in the 1950s, and finally a Phoenix-like revival that started in the late 1960s and now seems to be coming to a lackluster end with the cooptation of key issues of 'politics' and 'power' by other sub-disciplines of geography. However, as David Livingstone has pointed out so aptly, the history of geography, and by extension, political geography, cannot be reduced to a single story (Livingstone, 1995). There are many stories and these stories are marked by discontinuities and contestations, in other words, 'messy contingencies', which complicate things (Livingstone, 1993: 28).

A further problem is what one should include under the rubric 'political geography': publications of scholars, the work of professional academic associations, the content of courses, textbooks and popular accounts, or the activities of practitioners in government institutions (Mamadouh, 2003: 664–5)? A promising solution to understanding the politics of political geography is to focus on its central concepts, such as power, territory, boundaries, scale, and place (Agnew et al., 2003). Yet, the difficulty remains of deciding which concepts are truly central (Mamadouh, 2004).

As an alternative to standard evolutionary and concept-based approaches, I have chosen to organize my discussion of the politics of political geography around the arguably most visible structure at the heart of the 'political': the state. This does not mean that I advocate a state-centred approach to political geography or restrict my analysis to politics with a large 'P' (Flint, 2003). Recent scholarship on the politics of identity, the role of political discourse, and changing forms of political practice have exposed such a view as short-sighted (Dalby, 1992; Kodras, 1999; Cox and Low, 2003; Pratt, 2004). Nevertheless, neither the embodied politics at the level of the individual nor the networked politics at the global scale can exclude consideration of the state. States continue to be major reference points of politics by virtue of the binding legal codes they define and enforce.

While I have singled out the state as the 'pivot' of political geography – to borrow a term from Halford Mackinder – I do so in the sense of a locus of engagement, not in the sense of the state as the exclusive locus of politics and power. Moreover, my view of the state is not restricted to the modern or territorial state that is premised on the nation-state ideal, but includes other spatially constituted structures of government and political authority, such as the early states of antiquity, the networks of medieval power or the increasingly state-like European Union. The term 'state' simply offers the most succinct way to express the institutionalized political authority and mode of social organization that is behind 'strategies of inclusion and exclusion, of territory and territoriality', and thus at the heart of political geography (Cox, 2005).

Historically, political geographers have engaged with the state in three ways: they have sought to facilitate the process of maximizing its power over space; to maintain and manage its territorial existence; and to actively resist and question its spatially manifested actions. I propose to use these three ways or traditions to achieve a deeper and more comprehensive understanding of the politics of political geography.

Political geographers that follow the first way prioritize the state as the most important actor, privilege the state or national interest, and are decidedly realist or power-oriented. They employ oppositional identities (us/them, black/white) and oppositions of power (sea vs land power) to offer representations of the world that dazzle through their simplicity. They have an activist stance and advocate change to achieve state dominance in a world characterized by competition and conflict. As a consequence, their work focuses on state and global scales for the most part, though internal divisions are recognized as important for the strength of the state. Their efforts privilege the role of the state executive.

By contrast, work in the second tradition denies political motives and professes neutrality and objectivity. The goal is to maintain a balanced and peaceful status quo or a homeostatic equilibrium in a closed system. The state is viewed as a given and its existence is not problematized. The main focus is at the scale of the state and its administrative regions. Work by political geographers in this tradition is implicated in the governance of the state and aids state administration and policy. It is inward-looking and eschews the problem of states in their relations with one another.

Political geographers in the third tradition are critical of the activity, purpose, and legitimacy of the state. They recognize multiple scales and expressions of power from the bodies of individuals to global networks. Some of them focus on class and the dominant influence of the capitalist world economy, others direct their attention to diverse groups and communities, embrace the notion of hybridity of identities, and examine the discursive power and production of knowledge. They are united in their engagement with social process, which makes them distinct from the other two traditions. Political geographers in this vein openly work toward transformation to achieve destabilization, resistance, or revolution. They are oriented toward oppositional groups and new social movements.

The advantages of organizing a history of political geography along the traditions of advocacy, governance, and critique of the state respectively are two-fold. First, they allow a consideration of political ideologies since, for the most part, these ways of engagement or traditions reflect the major political ideologies of right, center, and left. All too often political ideologies are not presented up-front in political geographic studies but brought in through the back door (Agnew, 2003: 605). Second, this approach avoids a potential silencing of alternative approaches. Histories generally focus on those perspectives that are most visible or dominant in a given time period, which gives the impression that other views are obsolete. For example, the critical view is currently the prevailing approach in the flagship journal, *Political Geography*, and a perusal of its content would not fully reflect the key role that advocacy of state power continues to play in other disciplines, in conservative think-tanks, and outside academia.

As with all forms of organizing knowledge, the focus on the three traditions I have outlined requires some caveats. The structure is necessarily arbitrary and simplistic. Within each tradition there are different expressions and one should not assume uniformity in thought or political orientation. For example, the scholars associated with the journal *Hérodote* are advocates of state power, yet fall into the Neo-Marxist camp. Likewise, a nationalist focus is not the sole prerogative of the power-oriented tradition, but can also be found among practitioners of the governance tradition. I attempt to address the plurality and hybridity that exists in the last section of the chapter, where I examine how the three traditions are reflected in maps and other forms of visualization and I identify areas of difference and cross-fertilization. As further safeguards against one-sided and facile interpretations I am including critical notes in the tradition of *Hérodote* so as to extend the discussion.

NO LIMITS? MAXIMIZING THE POWER OF THE STATE

The objective of geographic work in this tradition is to support and justify the extension of the power of the national state by outlining specific geographic features or areas that are crucial for political control. The tradition could be labeled strategic, nationalist, or power-oriented political geographies and is usually identified as 'geopolitics'.[1] It views the international system as based on competition and conflict and seeks to ensure a dominant position for the respective national state. The intellectual origins of this power-oriented and dynamic tradition are generally placed in the late nineteenth century and connected to the prevalent imperialism and its associated rivalries among states as well as to the establishment of geography as an academic discipline. However, the fundamental ideas behind it – the use of geography to project political power – can be traced back to Herodotus in the fourth century BCE and to Ibn Khaldun in the fourteenth century. Herodotus, who is considered the father of history, is also claimed as the father of geography (Gould, 1985: 11; Holt-Jensen, 1999: 11). Some scholars go so far as to label Herodotus 'an intelligence agent in the service of Athenian imperialism' (un agent de renseigne ments de l'impérialisme athénien) and stress that his work had not only a strategic function, but also an ideological one: to justify conquest (Hérodote, 1976: 59).[2] The Islamic geographer Ibn Khaldun offered similar geographic aids to statecraft and warfare. He linked the rise and fall of empires to the interaction between nomadic warrior tribes and permanently settled populations. Postulating that conquerors lose their ability to project power and maintain control over their empire after becoming settled among more docile populations, he was able to predict the collapse of the Islamic state he lived in (Holt-Jensen, 1999: 13).

The development of the tradition can be traced through three phases: (1) the formulation of fundamental concepts at the turn of the century; (2) the application of these concepts in the period 1919–1945; and (3) a rebirth and popularization after the 1980s. The context for the first period was the increased competition between European states due to rapid industrialization and anxieties about the finite nature of the world (Kearns, 1993). The uncertainty created by the ascendancy of Germany as a major challenger to the established imperial powers of Britain and France led to the development of new concepts that sought to provide guidance for political action. Of central importance was how environmental features, such as mountains, rivers, climate, and coastlines or the relative disposition of landmasses and oceans, affected the control of territories. This did not mean that the tradition employed a crude form of environmentalism, since these geographers were particularly interested in the way technology (such as railroads) or societal development (such as urbanization) affected the influence of the environment.

The key new texts were Friedrich Ratzel's *Politische Geographie* (1897) and Halford Mackinder's (1904) article 'The geographical pivot of history'.[3] Ratzel used a biological analogy and compared states to organisms formed by the interaction between a people and their territory. He posited that conflicts were inevitable since states needed to grow to survive. Germany was especially vulnerable since it was bordered by numerous states and had high population growth. German territorial expansion thus appeared as a matter of self-defense. Mackinder based his approach on an analogy with Newtonian physics and developed a 'theory of political motion' (Archer and Shelley, 1985: 17). He explained that technological advances in transport, in particular railroads, gave land power based in the unassailable citadel of Central Russia a locational advantage against Britain's sea power. It was imperative that this 'pivot' of world history not fall into the hands of a major industrial power. Russian industrialization or an alliance with rapidly developing Germany thus posed a grave danger to the future of the British Empire.

There was a clear political dimension to these works and they made the discipline indispensable for the scientific justification of territorial conquest. Mackinder was fully committed to applying geography for political ends and advocated the teaching of geography for the 'maintenance and progress of our Empire' (cited after Livingstone, 1993: 194). The projection of state power in the international arena was also accommodated to an internal vision. The nation was to be made up of organic neighborhoods, provinces, and other communities to transcend the potentially disastrous effects of class warfare (Mackinder, 1942: 186). Ratzel also had clear political motives. He sought to strengthen the German state and joined associations that propagated the acquisition of colonies (Sandner and Roessler, 1994).

The new concepts generated some lively theoretical debates regarding the place of politics in geography – Ratzel's advocacy of a separate subfield of political geography was thoroughly criticized by Vidal de la Blache in France – but neither of them was directly applied until the end of the First World War. The threats that they presented did not appear pressing at the time they published their ideas. Mackinder postulated a threat from land power at a time when Germany was challenging

Britain's naval supremacy and Russia was still lagging behind. Ratzel pointed to the potential vulnerability of Germany's territorial configuration when the primary political concern was the lack of overseas colonies. The First World War changed all that. In Britain there was concern about the vast territorial gains of Germany in the Treaty of Brest-Litovsk in 1918 – German control over the pivot now seemed a distinct possibility – and in Germany the universal outrage over the immense losses stipulated in the Treaty of Versailles generated fears that the country had received a mortal blow against its territory.

Mackinder refined his concept around the time of the peace conference in Paris and identified Eastern Europe as the key to the pivot, which he now termed 'heartland' (Mackinder, 1919). As a solution, he proposed creating a series of buffer states in Eastern Europe to prevent Germany from getting direct access to the heartland and from forming an alliance with Russia, a vision that has a striking correspondence to the newly created map of Europe in the peace treaties of 1919 (Heffernan, 2000: 38–9). The most ardent advocates of Mackinder's and Ratzel's concepts, however, were to be found in Germany. There, a network of geographers and nationalists established a school of thought that applied Mackinder and Ratzel in their analyses, developed suggestive maps, and offered their findings as 'scientific weapons' for the German cause (Herb, 1997). To identify this combination of geography and politics, they adopted the catchy term *Geopolitik*, which had been coined in 1899 by the Swedish political scientist and follower of Ratzel's ideas, Rudolf Kjellén (Holdar, 1992).

German *Geopolitik* shared many of the territorial ambitions outlined by Hitler in *Mein Kampf*, such as the unification of all Germans in one state and extension of German control into *Mitteleuropa*, though it differed significantly from National Socialist ideology in ascribing a determining influence to the environment rather than to race (Bassin, 1987). Nevertheless, the perception abroad was that the school of thought provided the blueprint for Hitler's conquests, and *Geopolitik* came to be viewed as synonymous with Nazi imperialism (Strausz-Hupé, 1942). The association had severe consequences for this particular tradition in political geography and it was essentially banished from academia after the war.[4]

At that time, geographers in Germany and elsewhere went through great pains to evade the stigma of *Geopolitik* by dissociating academic political geography from any form of political activism. They used a rhetorical maneuver and labeled works that presented geo-deterministic explanations of politics and had political motives as *geopolitics*. This deviant version was excluded from academic geography. By contrast, the term *political geography* was reserved exclusively for 'scientific' studies, which they considered 'objective' and thus ultimately 'neutral' (Troll, 1947; Hepple, 1986b). Academic work in political geography shifted wholesale into the tradition of governance.

The tradition of maximizing state power did not disappear altogether; it simply became less visible. Advocacy of projecting state power continued in military academies in different countries and the US School of Foreign Service (Hepple, 1986b; Ó Tuathail, 2000). One academic geographer was undeterred: Saul Cohen reformulated some of Mackinder's ideas and adopted them for US foreign policy recommendations during the Cold War (Cohen, 1963, 1973). Even more influential, geopolitical concepts were widely disseminated through popular media, such as *Reader's Digest* (Sharp, 2000). In South America, the tradition prospered and informed the policies of military regimes, as in Argentina, Chile, and Brazil. General Pinochet, among others, was a trained geographer (Child, 1979; Hepple, 1986a; Dodds, 1993).

In the early 1980s, the tradition of strategic political geographies once again rose to prominence in North America and Europe. In the context of nuclear parity among the superpowers, Reagan's confrontational policies during the Second Cold War, the stationing of medium-range missiles in Europe, and increased regional conflicts, numerous works appeared that re-emphasized geographic conditions as determining factors for political power (Hepple, 1986b; Ossenbrügge, 1989). The founding of the pro-NATO International Institute of Geopolitics in 1982 in Paris, which published the journal *Géopolitique*, further popularized the tradition (Hepple, 1986b). There was even a major geopolitical initiative from the Left. A group of Neo-Marxist geographers from the Université de Vincennes (Paris VIII) headed by Yves Lacoste founded the journal *Hérodote* in 1976, which demanded political action and initiative from academic geography and started using the term 'géopolitique' in its subtitle in 1983 (Claval, 2000: 245; Hepple, 2000).

An important, but hitherto neglected, intellectual context for this revival of geopolitics was the rise of the New Right in Europe which started in the 1960s. Leading proponents of this political movement, such as Alain de Benoist in France and Robert Steuckers in Belgium, returned to radical conservative ideas of the interwar period and propagated the significance of biological differences and the determining influence of the environment (Bassin, 2003: 361–62). This connection between geopolitics and the political Right should not come as a surprise since the school of German *Geopolitik* was inspired by the very same hyper-conservative interwar thinkers that the New Right rediscovered,[5] but it makes it difficult to explain the Left geopolitics of *Hérodote*. Paul Claval (2000: 255–8) has argued vehemently that Lacoste and his group are cosmopolitan and liberal, but as Mark Bassin (2003: 362–3) has shown convincingly, there are definite affinities between Lacoste's fixation on the nation and the ideas of the New Right. No matter what Lacoste's 'true' political intentions, the accolades he has received from geopoliticians of the New Right show that he – though not necessarily the editorial group of *Hérodote* – fits into this tradition of political geography (Bassin, 2003: 363).[6]

With the exception of the case of *Hérodote*, geopolitics or political geographies that advocate state power are mainly pursued outside of geography at present. It seems that the 'rhetorical space' that was opened up by Kissinger's rehabilitation of the term 'geopolitics' was filled by other disciplines and politicians. The pervasiveness of notions such as Samuel Huntington's 'clash of civilizations', Robert Kaplan's 'coming anarchy', 'rogue states', and the 'axis of evil' shows that the tradition is alive and well. While these recent concepts do not make explicit references to the determining influence of environmental conditions like the earlier examples, they base their simplistic models on regional differences that are rooted in either long-term human/environment interaction in specific realms or geographic location and territorial size. Political geographers have taken notice and, as will be discussed in the third section, are engaging with these recent concepts from a critical perspective.

IN PERFECT BALANCE? MAINTAINING THE POWER OF THE STATE

On the most fundamental level, this tradition views the state as a given. Its main objective is to maintain the status quo and to compile all facts necessary for the continued existence of a given state or the maintenance of a balanced international system. The approach is professedly neutral and objective. The state is described and dissected, but not questioned. The focus is inward. It privileges the internal structure of states and relations between state and society, rather than relations between one state and another. Its obvious usefulness for efficient state administration means that on an applied level, the tradition always has and will be influential. Its public visibility and academic role, however, have changed quite substantially over the course of its history.

As in the case of power-oriented political geographies, there are early representatives in the classical period. Chief among them is Strabo's (64 BCE–20 CE) seventeen-volume encyclopedic description of the Roman Empire (Holt-Jensen, 1999:12). Though Holt-Jensen considers Strabo's work on a par with that of Herodotus, the French geopoliticians of the journal that bears the latter's

name make a clear distinction: Herodotus was not content with mere description, he also had an ideological bent and sought to explain and justify actions (Hérodote, 1976).

Conceptual roots are also found in political arithmetic and regionalism. Political arithmetic refers to the recording, classifying, and cataloguing of information regarding states, such as William Petty's quantification of social phenomena in Ireland and England (Livingstone, 1993: 90–2). These were crucial facts that modern states needed to manage and thus became particularly important with the consolidation of national economies and the advent of popular sovereignty in the 1800s (Scott, 1998). Such statistical compendia were common in the age of Ritter and were referred to as political geography (Oberhummer, 1923: 608–9). More recently, tabular inventories have been used in power analysis approaches (Archer, 1982: 233) and still feature prominently in country studies, such as the CIA World Fact Book.[7]

Regionalism represents an alternative approach to the geographic experiment of geo-determinism (Livingstone, 1993). The key influence came from the French school, in particular Paul Vidal de la Blache. He advocated the notion of *genres de vie*, which represented the ways of life that human communities had developed over a long period in the milieu of particular places (Livingstone, 1993: 267). Vidalian regionalism was not simply a reaction to the weight ascribed to the environment in Ratzel's work, but was fostered by specific social and political contexts. When Vidal laid the foundation for his new French geography, France still had to come to terms with the defeat at the hands of the Germans in 1871 and the loss of Alsace-Lorraine. Moreover, industrialization and urbanization were encroaching on traditional French ways of life. National education, and especially geographic education, was seen as a way to unite the nation, since 'one only loves what one knows' (Livingstone, 1993: 266; Capel, 1996: 79). Vidal's work was also related to the contemporary discussion about more efficient administrative practice in France (Taylor and van der Wusten, 2004: 88). *La France de l'Est* (Vidal de la Blache, 1917) is a telling example of the confluence of policy, national education, and regional identity. In France, political geography was synonymous with regional geography and with governance of the state.[8]

Outside of France, regionalism was also influential in geography, but in places like Germany, Britain, and the United States, political geography was initially dominated by the strategic tradition, that is, geopolitics.[9] The governance tradition came to a par with its rival for the first time in the period leading up to the peace conference at Paris in 1919. As early as September of 1917, the American president instituted a commission of experts, known as the Inquiry, to study the future territorial adjustments. It included prominent geographers such as Isaiah Bowman, who was also the director of the American Geographical Society (Herb, 1997; Smith, 2003).

In light of the most sweeping redrawing of the map of Europe, academic geographers in other countries eagerly prepared work for the benefit of their nation: De Martonne for France, Marinelli for Italy, Cvijić for Serbia, Romer for Poland, Penck for Germany (Mehmel, 1995; Taylor and van der Wusten, 2004). Most became directly involved in the peace delegations of different countries and country studies abound in the academic journals of the period. Though much of this work is now easily exposed as biased and politically motivated, the general view – also perpetuated by the geographers themselves – was that they conducted 'objective scientific' studies. It is quite clear that the maps they offered as scientific evidence, such as ethnographic maps or maps of election results, bolstered their case (Herb, 1997). Maps have historically been associated with authority and are generally perceived as 'true' and 'objective' pictures of reality (Harley, 1988, 1992). The guiding premise of the new boundary delimitation was that international conflict could be avoided if all states were internally balanced, but just what constituted 'balanced' left the door open for advocacy of their own national interests. There was a clear overlap with the nationalist orientation of geographers working in the power tradition, but the main difference is that the geographers in the governance tradition assumed the mantle of neutrality.

The case of Isaiah Bowman is rather telling too. He professed to 'leaving the facts … to speak for themselves' (cited after Archer and Shelley, 1985: 18), but nevertheless was the architect of Roosevelt's empire-building, as Neil Smith (2003) has shown. Thus, in contrast to the power-oriented political geography tradition, which unabashedly celebrates power and generally acknowledged its political mission, the governance tradition is presented under the guise of being 'objective', a mere supplier of facts.

In the interwar period, the two traditions coexisted, though often in a confrontational manner. German political geographers were drawn to the strategic tradition and even became involved in the pseudo-discipline of *Geopolitik*, while French political geographers vigorously held on to their regional concepts in the rationalist tradition and sought to invalidate not only *Geopolitik*, but the entire tradition based on Ratzel (Buleon, 1992).

After the Second World War, when academic geographers tried to escape affiliation with German *Geopolitik*, the governance tradition effectively took over the sub-discipline. Instead of trying to maximize the power of states, geographers now shifted their attention to serving state and society and to aid in the development of the most efficient state apparatus. They concentrated their efforts in three areas: (1) conceptualizations of the nature and organization of states; (2) state inputs, in particular elections; (3) and state outputs in the form of planning and the location of facilities.

New concepts by Hartshorne (1950) and Gottmann (1951, 1952, 1973) drew the attention of geographers to countervailing forces that acted on states, which the former called centrifugal and centripetal forces and the latter circulation and iconography or security and opportunity.[10] The goal was to achieve a balance of these forces – an idea that is related to the French geographer Jaques Ancel's 1938 notion of borders as 'political isobars' (Parker, 2000: 960).[11]

A second set of conceptualizations applied this interchange of external and internal forces to the historical development of states. Jones' (1954) unified field theory took inspiration from Hartshorne's *raison d'être* and identified a chain of activities that started with an idea to have a state (or more accurately ideology) and culminated in the creation of a state area. Pounds and Ball's (1964) model referenced the core area concept of Whittlesey (1939: 24) and sought to show that 'most European states grew in fact by a process of accretion from germinal areas' that were environmentally favored.

Yet, despite the conceptual innovations, it seemed to some that the 'subject reverted to the status of a verbal and cartographic political arithmetic whose matrix of cells was partitioned along the boundaries between sovereign states, political dependencies, or unorganized areas' (Archer and Shelley, 1985: 16–17). Much of the work tried to prove its 'objective' stance by simply presenting 'facts' and failed to recognize that the process of collecting and classifying facts is structured by social norms and values and thus is never 'neutral' (Natter et al., 1995). Moreover the new concepts were rather vague and did not stand up to analytic scrutiny, as Burghardt's (1969) critique of the core concept revealed. According to one of the leading figures in human geography in the 1960s, Brian Berry, this lack of rigor and explanation in the, by then, dominant governance tradition had turned the sub-discipline into the oft-quoted 'moribund backwater' (Berry, 1969).

In retrospect, Berry's criticism is somewhat ironic since the spatial-analytic approach he advocated had similar shortcomings to the governance approach. Both were heavily empirical, believed in objectivity, and ultimately supported the liberal, pluralist view of the state. As David Harvey charged, spatial analysis was apolitical and hid behind the 'shield of positivism' (Harvey, 1984).

Nevertheless, the quantitative and spatial-analytical revolution that swept the field of geography in the 1960s did have positive impacts on the governance tradition in political geography.

Above all, geographers began to think in terms of theory and asked new questions that served as crucial points of departure for more radical political geographies in the 1970s. Systems theory offered a potentially sophisticated extension of Hartshorne's functionalism even if the work did not develop beyond a few isolated studies (Dikshit, 1997a: 77–9; Taylor and van der Wusten, 2004: 98–9). More importantly, the new quantitative methods allowed for more refined research on state inputs.

Large electoral data sets that were conveniently divided into existing administrative districts provided easy application for computer-based modeling. Traditional map comparisons could now be replaced by advanced statistical procedures, such as correlations and regressions (Taylor and van der Wusten, 2004: 98). Work in this vein helped refine the national electoral cleavage thesis adopted from political sociologists by revealing the continued importance of regional and place influences on voting (Archer et al., 1986; Reynolds, 1990), and offered new insights into locational conflicts, such as the placement of public facilities through analyses of electoral behavior in space (Mumphrey and Wolpert, 1972). They opened the door to new questions and several of the leading figures ended up shifting their work into the critical tradition (Cox, 1973; Archer and Reynolds, 1976; Johnston, 1979).

Studies on state outputs were far less prominent than those on conceptions of the nature and organization of states or those on state inputs. Work on outputs focused mainly on planning issues, as in the work of G. H. J Daysh, Dudley Stamp, and Peter Hall (Daysh, 1949; Stamp, 1960; Hall, 1973). These issues were picked up in earnest only by the critical tradition in its treatment of welfare geography in the 1970s, as discussed below.

On a practical or applied level, the governance tradition has always played an important role. As in the case of the Paris peace commissions, geographers working in the OSS, or the more recent Dayton peace agreement, such political geographic work is useful for state institutions, such as intelligence agencies and foreign offices (Kirby, 1994). Other examples are the area studies series published by the American University in Washington, D.C., and the CIA's World Fact Book. Increasingly, its followers employ sophisticated tools, as the application of GIS systems, such as Powerscene, in peace settlements shows (Corson and Minghi, 2000). Finally, the tradition helps instill and strengthen national identities since it provides materials for national education.[12]

In academic political geography, the governance tradition has declined in importance among Anglo-American geographers since the early 1970s, given the considerable rise of critical political geographies. Some authors have continued working on traditional themes, such as border conflicts, the administrative divisions of state territory or the evolution of state territories,[13] but others have extended the tradition into new areas in the 1980s. These authors have been inspired by behavioral and humanist concepts, such as mental maps and sense of place (Henrikson, 1980; Murphy, 1988), or have adopted sophisticated spatial-analytic methods to investigate international and civil wars, diplomatic relations, and other dimensions of state power (O'Loughlin, 1986). While the governance tradition's acceptance and implicit support of the modern state system as well as its claim to objectivity can be criticized, it offers insightful analyses that are cognizant of recent conceptual developments, such as the need to address different scales and different forms of politics.

WHAT STATE? QUESTIONING THE POWER OF THE STATE

This tradition is in many ways more complex and diverse than the other two. It is influenced by several philosophies, including Marxism, post-structuralism, anarchism, humanism, and postmodernism. These different strands are nevertheless united in a common suspicion of the true intentions of states or their governments and the belief that power emanates from a variety of groups and structures at an equal variety of geographic scales. There are no precursors from classical Greece, China, or elsewhere that fit this tradition. Early geographers were generally in the service of the ruling class and thus would have had considerable difficulty questioning the legitimacy of their patrons. On the other hand, we know from the works of feminist and other critical scholars that history silences women, indigenous people or critical voices, since history is written by the powerful; it is 'his story'.

The earliest identifiable representatives of the critical political geography tradition are, in fact, the anarchist geographers of the nineteenth century: Élisée Reclus and Peter Kropotkin. Both passionately rejected the hierarchical power structure of the state, which they considered responsible for war.[14] They advocated a decentralized anarchist society built upon a federation of small, independent cooperative communities. Their views were definitively at odds with the prevailing imperialist and nationalist attitudes of the late nineteenth century. Kropotkin railed against nationalist hatred, capitalist exploitation, and colonialism, and argued that geographic education was a road to peace (Dunbar, 1978; Kropotkin, 1996). Reclus, who was a most prolific author – he wrote well over 20,000 pages – exposed the evils of Dutch and British colonialism and paid particular attention to social inequalities and structures of exploitation (Giblin, 1987; Lacoste, 1987).[15] Though both were well respected among their peers in geography for their publications in physical and regional geography, these radical aspects of their work did not have a significant influence on political geography until the early 1970s (Blunt and Wills, 2000: 2). Following in their footsteps commanded a steep price. The authorities in Russia and France considered their anarchist views threatening; both were jailed and exiled for periods of time and forced to lead a nomadic life-style (Blunt and Wills, 2000: 4–5). It was much easier to join in the chorus of power-oriented political geographers or hide behind the neutrality of governance and be assured a prestigious position in academia.

While alternative texts undoubtedly existed elsewhere before the 1970s – an example is the work of the Marxist geographer Karl Wittfogel (1929) during the tumultuous late 1920s in Germany – they were doomed to being isolated calls in the wilderness unless they soundly resonated with the societal and intellectual contexts.[16] The late 1960s to the early 1970s, however, did indeed herald changed contexts, and ones that ultimately established a more receptive place for left politics in academia. The civil rights movement in the United States and student protests across most Western countries put social equity issues on the agenda. While social science as a whole became politicized and focused attention on local issues, such as poverty and racism, as well as global issues, such as uneven development,[17] for the critical tradition in political geography these contexts initially meant greater attention to issues below the scale of the state.

Political geographers were not only sensitized by the civil rights disturbances and other social conflicts around them, but became interested in public policy issues through work in urban geography. Rapid suburbanization and the associated need for locating new freeways, bridges, and desirable facilities, such as schools, supermarkets, and hospitals on the one hand, and noxious facilities, such as landfills and polluting industries on the other, led to locational conflicts and brought issues of social and racial equity to the forefront. Inspired by the spatial quantitative revolution to think more theoretically, but at the same time aware of the shortcomings of the dominant focus on efficiency and abstract space in neo-classical economic models, Julian Wolpert and his students and David Harvey focused on externalities and their distributional implications, such as the impact of locating a bridge in different neighborhoods in New Orleans (Mumphrey and Wolpert, 1972), urban ghettos (Harvey, 1972), and social justice in general (Harvey, 1973).[18]

Critical studies in political geography at the local scale were thus intertwined with the move in human geography toward social relevance, which led to the founding of the radical journal *Antipode* in 1969. Among other foci, research efforts revolved around residential segregation, poverty, the local state, environmental issues, urban and regional questions, and welfare geography. They were based on a broadly political economy approach and decidedly critical of the role of the state in providing equal access.[19] As discussed above, electoral geography provided an important stepping stone and supported the early focus on local issues, but engagement with the inherently uneven nature of the capitalist system and the influence of capitalism over the actions of the state quickly led to considerations of more global dimensions.[20]

However, this invigoration of the critical tradition was confined to North America and Britain. In France, governance and the geopolitics of *Hérodote* were the exclusive traditions of the field for about two decades longer, given the dominant influence of Vidalian regionalism and public interest in mapping electoral geographies (Buleon, 1992). Similarly, in Germany, political geography was a thoroughly neglected field and until the early 1990s remained mainly focused on administrative issues to avoid association with the political activism of *Geopolitik* (Tietze, 1997).[21]

While the political economy approach provided the major stimulus for a critical political geography in the 1970s and sustained a large body of work, it quickly generated critiques by humanist, post-structuralist, feminist, and postmodern geographers. While there is a clear danger in generalizing about these developments – especially because some of these critiques are premised on ideas of diversity and multivocality – three different strands of critique can be identified for heuristic purposes.

First, humanistic geographers denigrated the neglect of human agency and what they believed to be the rigid character of more structuralist interpretations. This critique originally grew out of the general frustration of humanistic geographers with the people-less nature of human geography in the 1960s whose dominant spatial-analytic models or their behavioralist variants left no room for individual creativity or action. The Marxism of the early seventies then became a new object of these concerns, which were most eloquently brought to a point by James Duncan and David Ley (1982). This led to a debate about agency and structure, which was then seemingly resolved by the mid-1980s through structurationist concepts (Thrift, 1983).

A second set of controversies were brought in through the cultural turn in geography and the stress on identity politics. Here the context was the women's movement of the 1970s, which joined up with the earlier civil rights and environmental movements to produce an academic interest in identity politics and social movements. The focus of critique was the supposed economism of political economy and its neglect of culture and other forms of social cleavage. This brought political issues into much sharper focus in what is commonly referred to as the *new* cultural geography and cultural issues into political geographic work, such as nationalism.[22]

The third source of criticism centered on the political economy's claim to a universal, scientific knowledge. Chief influences here came from post-structuralist, postmodernist, and feminist geographers. An early expression of this concern can be found in the exchange between David Sibley and Richard Walker (Sibley, 1981a, 1981b; Walker, 1981) about the role of order in centralized states and scholarly inquiry. Towards the end of the 1980s, the debate intensifies around the related notions of positionality, the inherently political character of discourse, and particularity (Dear, 1988; Soja, 1989). Feminist theorists have been highly influential in this regard (Sparke, 2004). They criticized the dominant masculinist 'view from nowhere' that privileged Western theory, drew attention to situated knowledge and practices, recovered the private as a site of politics, and stressed the crucial role of embodied politics (Staeheli and Kofman, 2004). Although feminist geographers claim that their impact on political geography has been negligible (Staeheli, 2001), many feminist ideas and concepts are closely related to those postulated in postmodernism, post-structuralism, and other social theorizations.

The three sets of debates occur against a background of significant changes in theorizations about human geography, academic climate, and real-world political geographies. The controversies were part of the general movement in human geography to reconsider the role of the social and the spatial in the discipline. The founding of the international journal *Society and Space* in 1983 and numerous sessions at the annual meetings of the AAG that had 'rethinking' as part of their titles are indicative of this development. There also was a new academic milieu due to an influx of faculty and graduate students from nontraditional backgrounds. The expansion of secondary education with the coming of age of the baby boom generation and increased affluence in Europe and North America had weakened the dominance of white males from upper- and middleclass backgrounds (Johnston 1988; Agnew, 2002: 101–2). Finally, there was a rise in social activism around feminism, race, and the environment, the end of the Cold War, and growing impacts of globalization. International boundaries were redrawn and became more pronounced in the newly independent states of the former Soviet Union and less significant in the European Union. State power was challenged by globalized production and its associated local restructuring, by international flows of capital and transnational corporations, environmental disasters, such as Chernobyl, by a commercialized global culture and media, and by regional separatism.

The impacts of these developments and associated debates on the critical tradition in political geography have been far-reaching. Political geographers have eagerly addressed new issues, such as changing forms of sovereignty, networks of power, the role of transnational corporations, telecommunications, sub-state identities, new social movements, and the politics of turf and gender.[23] More importantly, there has been a major reconceptualization of the state in contemporary critical political geography. The state is critiqued both (1) as a social construction – a notion that emerged from the three sets of debates discussed above – and (2) as no longer deserving a central role in political geography.[24]

First, the notion of social construction means that the state loses its normative edge, as a neutral body, the structure of which was supposed to reflect some national interest. The environment or space is no longer considered an objective reality. Many critical political geographers now view the world as being accessible and conveyed through descriptions, termed geographs. These are analogous to movie scripts that frame our understanding. Being an author of a script or geograph thus means commanding authority (Ó Tuathail, 1996; Dalby and Ó Tuathail, 1998). This has resulted in a devastating assessment of classical, power-oriented concepts, such as Huntington's civilizations or the simplistic land-power versus sea-power dichotomy of Mackinder – a critique that has come to be known as 'critical geopolitics'. It also has prompted the question: for whom is the state? For the capitalist class? For white, Western males?

Similarly, the idea of social construction exposes political geographic concepts, such as scale and regions, as inherently discursive and in need of being 'unpacked'. For example, Paasi (1996) and Kaplan and Häkli (2002) have elucidated our understanding of the relationship between regional identities and borders, and Howitt (1998) and Marston (2000) have exposed scale as a process and introduced the term 'scaling' to denote this dynamic character. Some works in this vein have revealed the need to consider new forms of power, such as networks and new social movements (Miller, 2000) and new ways of seeing power geographically (Allen, 2003). Others have exposed different expressions and forms of identities and introduced notions such as hybridity to go beyond simplistic dichotomies of us and them (Mitchell, 1997).

The second thing about the state is that its centrality to political geography has come into question. With new states forming and others disintegrating after the Cold War and different levels of political-territorial structure changing their relative power, such as regions in the European Union, the state has lost its sense of permanence. Globalization adds to skepticism about the central role of the state as state power seemingly diminishes relative to that of multinational corporations and the flows of international currencies around the world. The state begins to be viewed as just one expression of the political in the modern world. Class struggles, gender struggles, colonial struggles are seen to lurk behind state formation and disintegration, and investigations of the power of localities versus capital versus the state take center-stage.

The arguably central unifying element of the current critical tradition is scale. Critical political geographers universally recognize linkages among scales, stress that scale should not be equated with pre-existing administrative units, and embrace the idea that social relations spill over state boundaries. As early as 1985, Taylor's textbook, *Political Geography*, used world system theory to offer an explicit framework for integrating all three scales from local to state to global.[25] Similarly, Agnew and Corbridge (1995) and Swyngedouw (1997) tie the global to the local, and Cox (1998) has developed new concepts to break out of these existing territorial frames.[26]

Current critical political geographies are also distinguished by a renewed political activism that picks up from the revolutionary engagement of Reclus and some early Marxist geographers. These recent works can be labeled as 'anti-geopolitics'. They profess the intention to bring about change by countering the global hegemony of neo-liberal capitalism, militarism, and repressive state power. Examples include conceptualizations of resistance (Pile and Keith, 1997; Sharp et al., 2000) and strategies for executing struggle (Routledge, 2000, 2003; Featherstone, 2003). David Harvey's (2000: chap. 12) recent reflections on insurgent architects, militant particularism, and political action also belong here, even though they are presented from a more orthodox position.[27]

Currently, the critical tradition has a commanding hold on the academic field of political geography in North America and Britain and there is a similar trend in Germany, France, and some of the other European countries. It has to be credited with enhancing our understanding of the traditional center of political geography, that is, the state, and with the introduction of new concepts, such as geographs and scaling. Yet, some of its proponents are straight-jacketed in their radical views, and some hide behind obtuse language, which has limited the influence of their ideas on the rest of the field and particularly outside academia.[28]

TRANSITIONS AND VISIONS

While the different traditions are distinct in their attitude toward the state and in their respective political orientations, there are also zones of transition and continuities between them, which serve to underline the existence of a cohesive sub-discipline. There is overlap in terms of themes, methods, concepts, and individuals. The most striking case is *Hérodote*. Its geo-determinist stance and focus on strategy and power puts it squarely in the maximization-of-state-power tradition. Yet, its discussion of administrative districts in France and emphasis on the cohesion of the French nation-state appear to fit nicely in the governance tradition.[29] Finally, its innovative contributions seem to mirror the three main elements of the reinvigorated critical tradition: it has extended investigations to the local level in discussing local identity movements; it has addressed global economic processes in its treatment of development; and it has attempted a new conceptualization of scale with its notion of 'spatial ensembles' (Lacoste, 1984).[30] Moreover, *Hérodote*'s professed ideological attitude is related to the anti-geopolitics strand.

The united and at the same time pluralistic nature of the larger field of political geography can be seen in the revival of the 1980s. In each of the traditions there was renewed interest in extending studies in new directions. Political geographers who were power-oriented found fresh outlets for their studies in newly founded research institutes, those who were state-focused adopted novel techniques and concepts, and those who were skeptical of the state refined their theoretical foundations. These revitalizing efforts merged through two academic venues: the founding of a new scholarly journal, *Political Geography Quarterly*, in 1982 and the IGU Commission on the world political map in 1984.[31]

The similarities and differences between the three traditions are also reflected in visualization. Maps are of paramount importance in political geography. From the age of the Pharaohs they have been associated with central authority (Harley, 1988). Maps should be considered an explicitly political form of knowledge since they allow us to control what exists by selecting what is depicted and thus officially recognized (Latour, 1986). The most substantial cartographic contributions in political geography stem from the power-oriented tradition. These geopolitical maps are commonly associated with propaganda and are distinguished by their powerful simplicity (Herb, 1997). The maps seem to 'talk' on their own (Herb, 1989: 292) and their 'gaze from nowhere' hides their authorship (Ó Tuathail, 1996). Two illuminating examples are the maps in Langhans-Ratzeburg (1929) and Lacoste (1986) (see Figures 1.1 and 1.2). They depict large sweeping bands of contested regions

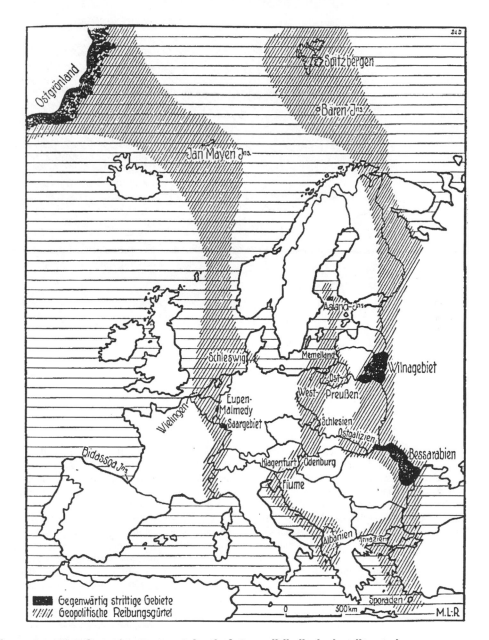

Figure 1.1   Map from the German School of *Geopolitik* displaying disputed regions and geopolitical 'zones of friction'.
Source: Langhans-Ratzeburg (1929)

or zones of tension across Europe unhampered by state borders and even the seas.[32] They illustrate the commonly held view in this tradition that political boundaries are dynamic and insignificant in light of large-scale environmental influences. The likeness of the maps shows that *Hérodote* belongs in the power-oriented tradition.

By contrast, the map of conflicting claims in Bowman (1922) timidly clings to clearly demarcated territories and exposes its state-focused character (see Figure 1.3). The precise delimitation of these areas also implies factual accuracy and thus objectivity. Other maps in the governance tradition strive for the same 'scientific' status. The maps used by the American Inquiry were authored by respected scientists (American Geographical Society, 1919), the electoral atlases that are so popular in France are based on official statistics

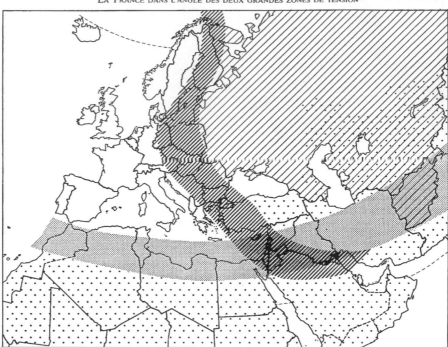

LA FRANCE DANS L'ANGLE DES DEUX GRANDES ZONES DE TENSION

Figure 1.2   Map from the journal *Hérodote* depiciting 'zones of tension'.
Source: Lacoste (1986: 27)

(Buleon 1992: 37–8), and the GIS-based visualizations for the Dayton Accord employed massive data sets and dazzled with technologically sophisticated displays (Corson and Minghi, 2000).

While the strategic and governance traditions have unique cartographic styles and plentiful examples of maps, critical political geographies mainly seem to have a unique stance toward visualizations: maps are criticized and deconstructed, but few are used to illustrate findings. Is the tradition too self-critical? How should one portray the invisible hand of the market, the multivocality of ideas, the hybridity of identities, or the palimpsests of the political landscape? The *State of ...* series, which includes the excellent atlas on women by Joni Seager (2000), is a good start, but there needs to be more intense engagement with the practice of visualization. Goodchild's (1997) and MacEachren's (1992) studies show that complex and critically informed depictions are possible. Interactive, multi-layered maps on a GIS basis would bring multiple voices to life with the click of a mouse, link points, lines, and symbols to other data sources, and allow a mixing of different genres, such as photos, film clips, interviews, poems, music or art. Even Thrift's (2000) demand to include the 'little things' of everyday life could be addressed.

Despite Agnew's (2002) well-argued claim of 'plurality' in contemporary academic political geography, a perusal of the major journal, *Political Geography*, suggests that the critical tradition is now dominant. Some major textbooks, such as Cox, Muir, and Short, even silence the other traditions by not discussing the influential roles they have played historically. As a result, they are able to present a clearly articulated and well-defined view that is unhampered by an often unsavory past (Dikshit, 1997a: 58). On the other hand, Glassner and Fahrer (2004) cover a wide array of issues in political geography but their encyclopedic breadth does not allow for sufficient depth and they end up neglecting most critical political geographies. Similarly, the textbook by Shelley et al. (1996) is impressive for its sustained engagement with electoral geography, but is slanted toward the governance tradition and only engages with the world system dimension of the critical tradition.

Taylor and Flint's (2000) text, based on world system theory, is the most systematic attempt to incorporate all traditions apart from Agnew (2002), but requires a leap of faith to believe in the global cycles of Kondratieff and Modelski. These models prescribe an astonishingly neat and structurally determined regularity for the occurrence of global conflicts and economic busts and booms, which

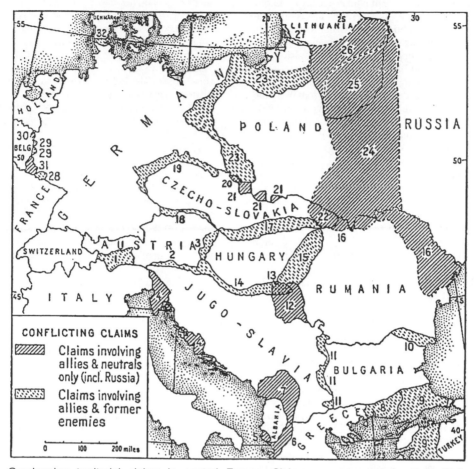

Overlapping territorial claims in central Europe. Claims are represented not in their most extreme but in their more conservative forms; in general, therefore, the ethnic line is taken as the limit of the claims of Austria and Hungary; the eastern limit of Poland's claim as shown on the map is some distance west of her boundary in 1772. The districts are numbered as follows:

1. Part of Austrian Tyrol
2. German-Slovene borderland
3. German Hungary
4. Istria and Dalmatia
5. Valona
6. Northern Epirus
7. Serbo-Albanian zone
8. Western Thrace
9. Eastern Thrace and the area claimed by
    Greece in Asia Minor
10. Southern Dobrudja
11. Western Bulgaria
12. Southern Banat
13. Northern Banat
14. Southern Hungary
15. Western Transylvania
16. Eastern Ruthenia and Bessarabia
17. Southern Slovakia
18. Southern Bohemia
19. German Bohemia
20. Czech districts in German Silesia
21. Teschen, Orawa, and Spits (named in
    order from west to east)
22. Ruthenia
23. Upper Silesia, Posen, Danzig, Marien-
    werder, and Allenstein
24. Polish-Russian border zone
25. Lithuanian-Polish-Russian border zone
26. Polish-Lithuanian border zone
27. Trans-Niemen territory
28. Saar basin
29. Malmédy, Eupen, and Moresnet
30. Southern Limburg
31. Luxemburg
32. Northern Slesvig

Figure 1.3 Map in the governance tradition showing conflicting territorial claims in Europe.
Source: Bowman (1922)

discounts the influence of human agency or unique combinations of events. These criticisms notwithstanding, the textbook scene is not stagnant, but open to new ideas as the increased publication of specialized supplemental texts shows (Storey, 2001; Allen, 2003).

To truly move toward a pluralistic sub-discipline in all regards, academic political geography should heed the call by Robinson (2003) and open the field to the diversity of insights, cases, and ideas presented in area studies. This does not only mean embracing the knowledge of the global periphery, there is even a multiplicity of views within the West that have been sorely neglected (Häkli, 2003). For example, the innovative approach of *Hérodote* is still not widely known among Anglo-American geographers.

Political geography has successfully broken out of the confines of an excessively state-centered view, and now it might be time to transcend the barriers of language and ideology. In our excitement about the unique and powerful language of GIS, we should not forget that communicating in accessible English and training in other languages is also important.[33] *Hérodote* serves as an example for ideological openness. The journal is fundamentally left, yet was not afraid to apply traditional geopolitical reasoning. Critical political geographers, in particular proponents of anti-geopolitics, might find some of the work of the power-oriented tradition quite beneficial for their cause. To echo Klaus Dodds' call for more engagement with military affairs and strategy: 'if critical geopolitics is going to be in a position to articulate alternatives to militarism then one must have some understanding of these particular organizations and cultures' (Dodds, 2001: 472). Despite overlapping interests in strategy and tactics, critical political geography and military geography still rarely engage with each other. If we want to have a truly pluralistic political geography and move forward conceptually, we cannot continue to approach work conducted from a different political viewpoint as inherently flawed.

ACKNOWLEDGMENTS

I would like to thank Jenny Robinson, Murray Dry, and Kevin Cox for their helpful comments. I am particularly indebted to Kevin for his profound insights and suggestions.

NOTES

1 Nationalist here is used to express an allegiance to and advocacy of the state. A more accurate, but also more awkward term would be 'statist'.

2 Herodotus' inquiries into the growth of the Persian empire and the causes of the war between Greeks and Persians should have been of considerable inspiration for the invasions of Asia Minor between the fifth and fourth century by Cyrus the Younger, the Spartans, and Alexander the Great. For example, Alexander the Great studied under Aristotle, who certainly knew Herodotus' *History*, and Cyrus the Younger probably was exposed to his works through the Greek mercenaries he hired for his campaign.

3 There are many examples of texts that supported such imperialist thinking in other academic disciplines, such as Frederick Jackson Turner's frontier thesis, Alfred Thayer Mahan's treatise on sea power, and Friedrich Nauman's work on Mitteleuropa, Turner (1963), Mahan (1890), Naumann (1915), Stedman Jones (1972).

4 There were attempts during the Second World War to develop geopolitical concepts in the US, such as Spykman, (1942, 1944) but they did not have a lasting influence.

5 For example, Carl Schmitt and Arthur Moeller van den Bruck.

6 The members of the editorial staff at *Hérodote* hold divergent views, as is illustrated by the critical annotations that accompany some of the articles. See, for example, Hérodote (1976).

7 http://www.cia.gov/.

8 Classical electoral geography shares elements with both of these strands. Like political arithmetic it is concerned with the recording and presentation of facts, such as the mapping of the voting, and like regionalism it seeks to identify and explain regional voting patterns. Sauer's proposed redistricting on the basis of communal regions in 1918 and André Siegfried's *Tableau Politique de la France*, which was published in 1913, illustrate this well (Shelley and Archer (1997). These works also have a common tendency to be empiricist and descriptive and to uphold the status quo (Shelley et al., 1990).

9 A classic example of British regional political geography is Fawcett's *Provinces of England* of 1919, which sought to identify more meaningful administrative districts for England. The book appeared as part of *The Making of the Future* series which was edited by Patrick Geddes and Victor Branford. It was considered such a timely piece during the heyday of the governance tradition in the post-Second World War period that it was reissued with only minor revisions by W. Gordon East and S.W. Wooldridge in 1960. (Fawcett, 1960).

10 Hartshorne termed his approach functional political geography, which in contemporary texts is usually derided as descriptive and naïve. Yet, despite or maybe because of its pedantic character, Hartshorne's approach is a terrific tool for teaching and raising awareness of the main oppositional forces in a state.

11 Gottman made exceptional contributions and his conceptualizations had great potential. He brought inter-state relations back into political geography by conceiving the state system as dynamic and also identified new forms of political processes by stressing network linkages among urban places across state borders (Agnew, 2002). However, his work was largely ignored in political geography; he was not even mentioned in Sack's (1986) seminal work on territoriality despite having written a substantive piece on the subject (Gottman, 1973).

12 For a further discussion of the relationship between geographic knowledge, education, and national identities, see Buttimer (1999).

13 See for example, Dikshit (1975), Murphy (1990), Rumley and Minghi (1991).

14 See Dunbar (1978), Breitbart (1981), Blunt and Wills (2000).

15 Reclus is also considered the founding figure of social geography. See Philo and Söderström (2004).

16 Another example is the book by Kapp (1950). It discussed the costs of pollution and natural resource depletion and thus anticipated later environmentalists' critiques of capitalism, but failed to make a significant impact during a time of unabashed economic growth.

17 A case in point is Andre Gunder Frank's core/periphery model. See Frank (1967).

18 See also the paper by Morrill (1974).

19 See, for example, the special issues of *Antipode* on access to essential public services (vol. 3, no. 1, 1971) and on the geography of American poverty in the Unites States (vol. 2, no. 2, 1970).

20 See, for example, the special issues of *Antipode* on 'underdevelopment in the Third World' (vol. 9, nos. 1 and 3, 1977).

21 The main representatives of political geography were Ulrich Ante at the University of Würzburg and K.-A. Boesler at the University of Bonn. Jürgen Ossenbrüggge at the University of Hamburg was the first to introduce critical concepts. See Ante (1981), Boesler (1983), Ossenbrügge (1983).

22 See, for example, Johnston et al. (1988), Murphy (1988).

23 This can be seen in the types of articles that appeared in *Political Geography* and *Society and Space* in the mid-1980s to 1990s.

24 The idea of social construction is connected to all three controversies. The structure/agency debate initiated by humanistic geographers pointed to people constructing social forms and relationships by drawing on existing customs, norms, and structures. The interest in identity politics gave this further momentum through questioning how people acquired particular identities. The postmodern strand emphasized the constructed nature of all

knowledge with the concept of discourse and its formative effect.

25 Electoral geography was also important for the development of this Wallerstein-inspired approach. See Archer and Taylor's 1981 study of US presidential elections, Osei-Kwame and Taylor's work on Ghana, and Taylor's dismissal of independent democratic elections outside the core. Dikshit has questioned Taylor's argument regarding elections in India. See Archer and Taylor (1981), Osei-Kwame and Taylor (1984), Dikshit (1997a, b), Taylor and Flint (2000).

26 See also Taylor (1994), Appadurai (1991).

27 Harvey arguably was the most important early political activist in geography. He started and advocated revolutionary change right after his conversion from being a theorist of spatial analysis to one of Marxist geography.

28 An illustrative case is the paper by Clarke and Doel (1995). Even the language of the abstract is daunting: 'As "political geography" searches in desperation for new (theoretical) directions to follow, this paper argues that the category of the "political" has already curved back on itself, attaining the status of the "transpolitical". This curvature is itself associated with profound shifts in the experience of history and time, of geography and space, and of the very ideas of theory, politics and events – shifts which continue to fascinate, haunt and transfix political geography in the enigmatic hereafter of the transpolitical. The paper assesses: the transpolitical figures of anomaly, ecstasy, obesity, and obscenity; the irruption of the hyperreal; the mutation of the political scene of representation into the transpolitical ob-scene of pornogeography; the fatal strategies pursued by the masses in relation to the spectre of the (trans)political; and the challenge of a transfinite universe for conjuring theoretical practice at the end(s) of political geography. Beginning with the transition from the political era the paper attempts to animate a transpolitical geography which affirms the s(ed)uction of superficial abysses and instantiates an ethics of the transpearing event'. Unfortunately, this is not a unique case, as the commentary by Patrice Nelson Limerick on Allen Pred's writing shows (Limerick, 1993).

29 See, for example, the contributions in *Hérodote* 50/51 (July–December 1988).

30 The concept of 'spatial ensembles' is premised on the idea that the world is too complex to be understood through isolating individual phenomena in an examination. It advocates investigations that look at the intersections of a phenomenon with multiple spatial sets (e.g. hydrography, geology, climate, demography, economy, etc.) and at different orders of magnitude. The graph accompanying Lacoste's 1984 article (pp. 22–3) provides an effective illustration.

31 A brief history of the Commission is accessible at: www.cas.sc.edu/geog/cpg/history.html.
32 Other examples are the European 'shatterbelts' in Cohen (1991). See also figure 13.1 in Cohen (2003).
33 Competency in foreign languages should also include awareness of the cultural context. As Sidaway et al. (2004: 1046) have pointed out: 'Languages and meanings of the political are everywhere caught up in wider cosmographies and hermeneutics'.

REFERENCES

Agnew, J.A. (2002) *Making of Political Geography*. New York: Oxford University Press.

Agnew, J.A. (2003) 'Contemporary political geography: intellectual heterodoxy and its dilemmas', *Political Geography*, 22: 603–6.

Agnew, J.A., Mitchell, K., et al. (eds) (2003) *A Companion to Political Geography*. Malden, MA: Blackwell.

Agnew, J.A. and Corbridge, S. (1995) *Mastering Space: Hegemony, Territory and International Political Economy*. London and New York: Routledge.

Allen, J. (2003) *Lost Geographies of Power*. Malden, MA: Blackwell.

American Geographical Society (1919) 'The American Geographical Society's contribution to the Peace Conference', *Geographical Review* 7, 1–10.

Ante, U. (1981) *Politische Geographie*. Braunschweig: Westermann.

Appadurai, A. (1991) 'Global ethnoscapes: notes and queries for a transnational anthropology', in R.G. Fox (ed.), *Recapturing Anthropology: Working in the Present*. Santa Fe: School of American Research Press.

Archer, J.C. (1982) 'political geography', *Progress in Human Geography*, 6: 231–41.

Archer, J.C. and Reynolds, D.R. (1976) 'Locational logrolling and citizen support of municipal bond proposals: the example of St. Louis', *Public Choice*, 27: 22–70.

Archer, J.C. and Shelley, F.M. (1985) 'Theory and methodology in political geography' in M. Pacione (ed.), *Progress in Political Geography* Dover, NH: Croom Helm, pp. 11–40.

Archer, J.C. and Taylor, P.J. (1981) *Section and Party: A Political Geography of American Presidential Elections, from Andrew Jackson to Ronald Reagan*. Chichester and New York: Research Studies Press.

Archer, J.C., Shelley, F.M., et al. (1986) *American Electoral Mosaics*. Washington, DC: Association of American Geographers.

Bassin, M. (1987) 'Race contra space: the conflict between German Geopolitik and National Socialism', *Political Geography Quarterly*, 6: 115–34.

Bassin, M. (2003) 'Between realism and the "New Right": geopolitics in Germany in the 1990s', *Transactions of the Institute of British Geographers*, 28: 350–66.

Berry, B.J.L. (1969) 'Review of Russett, B.M., International regions and the international system', *Geographical Review*, 59: 450–1.

Blunt, A. and Wills, J. (2000) *Dissident Geographies: An Introduction to Radical Ideas and Practice*. New York: Prentice Hall.

Boesler, K.-A. (1983) *Politische Geographie*. Stuttgart: B.G. Teubner.

Bowman, I. (1922) *The New World: Problems in Political Geography*. Yonkers-on-Hudson, New York: World Book Company.

Breitbart, M.M. (1981) 'Peter Kropotkin, the anarchist geographer', in D.R. Stoddart (ed.), *Geography, Ideology, and Social Concern*. New York: Barnes & Noble.

Buleon, P. (1992) 'The state of political geography in France in the 1970s and 1980s', *Progress in Human Geography*, 16: 24–40.

Burghardt, A. (1969) 'The core concept in political geography: a definition of terms', *Canadian Geographer*, 13: 349–53.

Buttimer, A., Brun, S.D. and Wardenga, U. (1999) *Text and Image: Social Construction of Regional Knowledges*. Leipzig: Institut für Länderkunde.

Capel, H. (1996) 'Institutionalization of geography and strategies of change', in J.A. Agnew, D.N. Livingstone and A. Rogers (eds), *Human Geography: An Essential Anthology*. Cambridge, MA: Blackwell, pp. 66–94.

Child, J. (1979) 'Geopolitical thinking in Latin America', *Latin American Research Review*, 14: 89–111.

Clarke, D.B. and Doel, M.A. (1995) 'Transpolitical geography', *Geoforum*, 25(4): 505–24.

Claval, P. (2000) '*Hérodote* and the French Left', in K.J. Dodds and D. Atkinson (eds), *Geopolitical Traditions: A Century of Geopolitical Thought*. New York: Routledge, 239–67.

Cohen, S.B. (1991) 'Global geopolitical change in the post-Cold War era', *Annals of the Association of American Geographers*, 81: 551–80.

Cohen, S.B. (1963) *Geography and Politics in a World Divided (1st edn)*. New York: Random House.

Cohen, S.B. (1973) *Geography and Politics in a World Divided (2nd edn)*. New York: Oxford University Press.

Cohen, S.B. (2003) *Geopolitics of the World System*. Lanham, MD: Rowman & Littlefield.

Corson, M.W. and Minghi, J.V. (2000) 'The case of Bosnia: military and political geography in MOOTW', in E.J. Palka and F.A. Galgano (eds), *The Scope of Military Geography: Across the Spectrum from Peacetime to War*. New York: McGraw-Hill, pp. 291–322.

Cox, K.R. (1973) *Conflict, Power, and Politics in the City: A Geographic View*. New York: McGraw-Hill.

Cox, K.R. (1998) 'Spaces of dependence, spaces of engagement and the politics of scale, or, looking for local politics', *Political Geography*, 17: 1–23.

Cox, K.R. (2005) 'General introduction' in K.R. Cox (ed.), *Political Geography: Critical Concepts in the Social Sciences*, Vol. 1. Abingdon: Routledge, pp. 1–31.

Cox, K.R. and Low, M.M. (2003) 'Political geography in question', *Political Geography*, 22: 599–602.

Dalby, S. (1992) 'Ecopolitical discourse: "environmental security" and political geography', *Progress in Human Geography*, 16: 503–22.

Dalby, S. and Ó Tuathail, G. (1998) *Rethinking Geopolitics*. New York: Routledge.

Daysh, G.H.J. (1949) *Studies in Regional Planning: Outline Surveys and Proposals for the Development of Certain Regions of England and Scotland.* London: G. Philip.

Dear, M.J. (1988) 'The postmodern challenge: reconstructing human geography', *Transactions of the Institute of British Geographers,* 13(3): 262–74.

Dikshit, R.D. (1975) *The Political Geography of Federalism: An Inquiry into Origins and Stability.* New York: Wiley.

Dikshit, R.D. (1997a) 'Continuity and change: a century of progress in theory and practice', in R.D. Dikshit (ed.), *Developments in Political Geography: A Century of Progress.* New Delhi: Sage, pp. 45–84.

Dikshit, R.D. (1997b) 'The world systems theory of elections and the crucial case of liberal democracy in India', in R.D. Dikshit (ed.), *Developments in Political Geography: A Century of Progress.* New Delhi: Sage, pp. 226–41.

Dodds, K.J. (1993) 'Geopolitics, cartography and the state in South America', *Political Geography,* 12: 361–81.

Dodds, K.J. (2001) 'Political geography III: critical geopolitics after ten years', *Progress in Human Geography,* 25(3): 469–84.

Dunbar, G. (1978) 'Élisée Reclus, geographer and anarchist', *Antipode,* 10: 16–21.

Duncan, J.S. and Ley, D. (1982) 'Structural Marxism and human geography: a critical assessment', *Annals of the Association of American Geographers,* 72(1): 30–59.

Fawcett, C.B. (1960) *Provinces of England: A Study of Some Geographical Aspects of Devolution.* London: Hutchinson.

Featherstone, D. (2003) 'Spatialities of transnational resistance to globalization: the maps of grievance of the Inter-Continental Caravan', *Transactions of the Institute of British Geographers,* 28: 404–21.

Flint, C. (2003) 'Dying for a "P"? Some questions facing contemporary political geography', *Political Geography,* 22: 617–20.

Frank, A.G. (1967) *Capitalism and Underdevelopment in Latin America: Historical Studies of Chile and Brazil.* New York: Monthly Review Press.

Giblin, B. (1987) 'Élisée Reclus and colonization', in P. Girot and E. Kofman (eds), *International Geopolitical Analysis.* New York: Croom Helm.

Glassner, M.I. and Fahrer, C. (2004) *Political Geography.* Hoboken, NJ: Wiley.

Goodchild, M.F. (1997) 'Geographic information systems', in S. Hanson (ed.), *Ten Geographic Ideas that Changed the World.* New Brunswick, NJ: Rutgers University Press, pp. 60–83.

Gottmann, J. (1951) 'Geography and international relations', *World Politics,* 3: 153–73.

Gottmann, J. (1952) *La politique des états et leur géographie.* Paris: Colin.

Gottmann, J. (1973) *The Significance of Territory.* Charlottesville: University Press of Virginia.

Gould, P.R. (1985) *The Geographer at Work.* New York: Routledge.

Häkli, J. (2003) 'To discipline or not to discipline, is that the question?', *Political Geography,* 22: 657–61.

Hall, P. (1973) *The Containment of Urban England.* London: Allen & Unwin.

Harley, J.B. (1988). 'Maps, knowledge, and power', in D. Cosgrove and S. Daniels (eds), *The Iconography of Landscape.* Cambridge: Cambridge University Press, pp. 277–312.

Harley, J.B. (1992) 'Deconstructing the map', in T.J. Barnes and J.S. Duncan (eds), *Writing Worlds: Discourse, Text and Metaphor in the Representation of Landscape.* New York: Routledge, pp. 231–47.

Hartshorne, R. (1950) 'The functional approach in political geography', *Annals of the Association of American Geographers,* 40: 95–130.

Harvey, D. (1972) 'Revolutionary and counter-revolutionary theory in geography and the problem of ghetto formation', *Antipode,* 4(2): 1–13.

Harvey, D. (1973) *Social Justice and the City.* Baltimore: Johns Hopkins University Press.

Harvey, D. (1984) 'On the history and present conditions of geography: an historical materialist manifesto', *Professional Geographer,* 36: 1–11.

Harvey, D. (2000) *Spaces of Hope.* Berkeley: University of California Press.

Heffernan, M. (2000) 'Fin de siècle, fin de monde? On the origins of European geopolitcs, 1890–1920, in K.J. Dodds and D. Atkinson (eds), *Geopolitical Traditions: A Century of Geopolitical Thought.* New York: Routledge, pp. 27–51.

Henrikson, A.K. (1980) 'The geographical mental maps of American foreign policy makers', *International Political Science Review,* 1: 495–530.

Hepple, L.W. (1986a) 'Geopolitics, generals and the state in Brazil', *Political Geography Quarterly,* 5(Suppl. 4): S79–S90.

Hepple, L.W. (1986b) 'The revival of geopolitics', *Political Geography Quarterly,* 5(Suppl. 4): S21–S36.

Hepple, L.W. (2000) 'Géopolitique de Gauche: Yves Lacoste, Hérodote and French radical geopolitics', in K.J. Dodds and D. Atkinson (eds), *Geopolitical Traditions: A Century of Geopolitical Thought.* New York: Routledge, pp. 268–301.

Herb, G.H. (1989) 'Persuasive cartography in Geopolitik and national socialism', *Political Geography Quarterly,* 8: 289–303.

Herb, G.H. (1997) *Under the Map of Germany.* New York: Routledge.

Hérodote (1976) 'Pourquoi Hérodote? Crise de la géographie et géographie de la crise', *Hérodote,* (1): 8–69.

Holdar, S. (1992) 'The ideal state and the power of geography: The life-work of Rudolf Kjellén', *Political Geography,* 11: 307–23.

Holt-Jensen, A. (1999) *Geography, History and Concepts: A Student's Guide.* London: Sage.

Howitt, R. (1998) 'Scale as a relation: musical metaphors of geographical scale', *Area,* 30: 49–58.

Johnston, R.J. (1978) 'Paradigms and revolution or evolution? Observations on human geography since the Second World War', *Progress in Human Geography,* 2: 189–206.

Johnston, R.J. (1979) *Political, Electoral, and Spatial Systems: An Essay in Political Geography.* New York: Oxford University Press.

Johnston, R.J., Knight, D.B. and Kofman, E. (1988) *Nationalism, Self-Determination and Political Geography.* London and New York: Croom Helm.

Jones, S.B. (1954) 'A unified field theory of political geography', *Annals of the Association of American Geographers*, 44: 111–23.

Kaplan, D. H. and Häkli, J. (2002) *Boundaries and Place: European Borderlands in Geographical Context.* Lanham, MD: Rowman & Littlefield.

Kapp, W.K. (1950) *The Social Costs of Private Enterprise.* Cambridge, MA: Harvard University Press.

Kearns, G. (1993) 'Prologue: Fin de siècle geopolitics: Mackinder, Hobson and theories of global closure', in P.J. Taylor (ed.), *Political Geography of the Twentieth Century: A Global Analysis.* London: Belhaven, pp. 9–30.

Kirby, A. (1994) 'What did you do in the war, Daddy?' in A. Godlewska and N. Smith (eds), *Geography and Empire.* Cambridge, MA: Blackwell, pp. 300–15.

Kodras, J.E. (1999) 'Geographies of power in political geography', *Political Geography*, 18: 75–9.

Kropotkin, P. (1996) 'What geography ought to be', in J.A. Agnew, D.N. Livingstone and A. Rogers (eds), *Human Geography: An Essential Anthology.* Cambridge, MA: Blackwell, pp. 139–54.

Lacoste, Y. (1984) 'Les géographes, l'action et le politique', *Hérodote*, 33–34: 3–32.

Lacoste, Y. (1986) 'Géopolitique de la France', *Hérodote*, 40: 5–31.

Lacoste, Y. (1987) 'The geographical and the geopolitical', in P. Girot and E. Kofman (eds), *International Geopolitical Analysis.* New York: Croom Helm, pp. 10–25.

Langhans-Ratzeburg, M. (1929) 'Die geopolitischen Reibungsgürtel der Erde', *Zeitschrift für Geopolitik*, 6: 158–67.

Latour, B. (1986) 'Visualization and cognition thinking with eyes and hands', *Knowledge and Society*: *Studies in the Sociology of Culture Past and Present*, 6: 1–40.

Limerick, P.N. (1993) 'Dancing with professors: the trouble with academic prose', *New York Times Book Review*, 31st October, AI823.

Livingstone, D.N. (1993) *The Geographical Tradition: Episodes in the History of a Contested Enterprise.* Oxford: Blackwell.

Livingstone, D.N. (1995) 'Geographical traditions', *Transactions of the Institute of British Geographers*, 20: 420–22.

MacEachren, A., et al. (1992) 'Visualization', in R. Abler, M. Marcus and J. Olson (eds), *Geography's Inner Worlds: Pervasive Themes in Contemporary American Geography.* New Brunswick, NJ: Rutgers University Press, pp. 99–137.

Mackinder, H.T. (1904) 'The geographical pivot of history', *Geographical Journal*, 23(4): 421–44.

Mackinder, H.T. (1919) *Democratic Ideals and Reality: A Study in the Politics of Reconstruction.* New York: Holt.

Mackinder, H.T. (1942) *Democratic Ideals and Reality: A Study in the Politics of Reconstruction.* New York: Holt, reprint of 1919 edition.

Mahan, A.T. (1890) *The Influence of Sea Power Upon History, 1660–1783.* Boston: MA, Little, Brown.

Mamadouh, V. (2003) 'Some notes on the politics of political geography', *Political Geography*, 22: 663–75.

Mamadouh, V. (2004) 'Review of Agnew, John, Katharyne Mitchell, and Gerard Toal, eds. *A Companion to Political Geography*', *Annals of the Association of American Geographers*, 94: 433–6.

Marston, S.A. (2000) 'The social construction of scale', *Progress in Human Geography*, 24: 219–42.

Mehmel, A. (1995) 'Deutsche Revisionspolitik in der Geographie nach dem Ersten Weltkrieg', *Geographische Rundschau*, 47(9): 498–505.

Miller, B.A. (2000) *Geography and Social Movements: Comparing Antinuclear Activism in the Boston Area.* Minneapolis: University of Minnesota Press.

Mitchell, K. (1997) 'Different diasporas and the hype of hybridity', *Environment and Planning D: Society and Space*, 15: 533–53.

Morrill, R.L. (1974) 'Efficiency and equity of optimum location models', *Antipode*, 6(1): 41–6.

Mumphrey, A.J. and Wolpert, J. (1972) *Equity Considerations and Concessions in the Siting of Public Facilities.* Philadelphia: University of Pennsylvania, Wharton School of Finance and Commerce.

Murphy, A.B. (1988) *The Regional Dynamics of Language Differentiation in Belgium: A Study in Cultural-Political Geography.* Chicago: University of Chicago Press.

Murphy, A.B. (1990) 'Historical justifications for territorial claims', *Annals of the Association of American Geographers*, 80: 531–48.

Natter, W., Schatzki, T.R., et al. (1995) *Objectivity and its Other.* New York: Guilford Press.

Naumann, F. (1915) *Mitteleuropa.* Berlin: G. Reimer.

Ó Tuathail, G. (1996) *Critical Geopolitics: The Politics of Writing Global Space.* Minneapolis: University of Minnesota Press.

Ó Tuathail, G. (2000) 'Spiritual geopolitics: Fr. Edmund Walsh and Jesuit anti-communism', in K.J. Dodds and D. Atkinson (eds), *Geopolitical Traditions: A Century of Geopolitical Thought.* New York: Routledge, pp. 187–210.

Oberhummer, E. (1923) 'Die politische Geographie vor Ratzel und ihre jüngste Entwicklung', in *Friedrich Ratzel, Politische Geographie.* Munich and Berlin: Oldenbourg, pp. 597–618.

O'Loughlin, J. (1986) 'Spatial models of international conflict: extending current theories on war behavior', *Annals of the Association of American Geographers*, 76: 63–80.

Osei-Kwame, P. and Taylor, P.J. (1984) 'A politics of failure: the political geography of the Ghanaian elections, 1954–1979', *Annals of the Association of American Geographers*, 74: 574–89.

Ossenbrügge, J. (1983) *Politische Geographie und Konfliktforschung. Konzepte zur Analyse der politschen und sozialen Organisation des Raumes auf der Grundlage anglo-amerikanischer Forschungsansätze.* Hamburg: Institut für Geographie und Wirtschaftsgeographie der Universität Hamburg.

Ossenbrügge, J. (1989) 'Territorial ideologies in West Germany, 1945–1985: between geopolitics and a regionalist attitude', *Political Geography Quarterly*, 8: 387–99.

Paasi, A. (1996) *Territories, Boundaries, and Consciousness: The Changing Geographies of the Finnish-Russian Border.* Chichester: John Wiley.

Parker, G. (2000) 'Ratzel, the French school and the birth of alternative geopolitics', *Political Geography*, 19: 957–69.

Philo, C. and Söderström, O. (2004) 'Social geography: looking for society in its spaces', in G. Benko and U. Strohmayer (eds), *Human Geography: A History for the 21st Century.* New York: Oxford University Press, pp. 105–38.

Pile, S. and Keith, M. (1997) *Geographies of Resistance*. London and New York: Routledge.

Pounds, N.J.G. and Ball, S.S. (1964) 'Core-areas and the development of the European states system', *Annals of the Association of American Geographers,* 54: 24–40.

Pratt, G. (2004) 'Feminist geographies: spatialising feminist politics', in P. Cloke, P. Crang and M. Goodwin (eds), *Envisioning Human Geographies*. London: Edward Arnold, pp. 128–145.

Ratzel, F. (1897) *Politische Geographie, oder, die Geographie der Staaten, des Verkehrs und des Krieges*. Munich: Oldenbourg.

Reynolds, D.R. (1990) 'Whither electoral geography? A critique', in F.M. Shelley, R.J. Johnston and P.J. Taylor (eds), *Developments in Electoral Geography,* New York: Routledge, pp. 22–35.

Robinson, J. (2003) 'Political geography in a postcolonial context', *Political Geography,* 22: 647–651.

Routledge, P. (2000) 'Our resistance will be as transnational as capital: convergence space and strategy in globalising resistance', *GeoJournal,* 52: 25–33.

Routledge, P. (2003) 'Convergence space: process geographies of grassroots globalization networks', *Transactions of the Institute of British Geographers,* 28: 333–49.

Rumley, D. and Minghi, J.V. (1991) *The Geography of Border Landscapes*. New York: Routledge.

Sack, R.D. (1986) *Human Territoriality. Its Theory and History. Cambridge Studies in Historical Geography* 7. Cambridge: Cambridge University Press.

Sandner, G. and Roessler, M. (1994) 'Geography and Empire in Germany, 1871–1945' in A. Godlewska and N. Smith (eds), *Geography and Empire*. Cambridge, MA: Blackwell, pp. 115–27.

Sassen, S. (1994) *Cities in a World Economy*. Thousand Oaks, CA: Pine Forge Press.

Scott, J.C. (1998) *Seeing Like a State: How Certain Schemes to Improve the Human Condition Have Failed*. New Haven, CT: Yale University Press.

Seager, J. (2000) *The State of Women in the World Atlas*. New York: Penguin Reference.

Sharp, J.P. (2000) *Condensing the Cold War: Reader's Digest and American Identity*. Minneapolis: University of Minnesota Press.

Sharp, J.P., Routledge, P., et al. (2000) *Entanglements of Power: Geographies of Domination/Resistance*. London and New York: Routledge.

Shelley, F.M. and Archer, J.C. (1997) 'Political sociology and political geography: a quarter century of progress in electoral geography', in R.D. Dikshit (ed.), *Developments in Political Geography: A Century of Progress*. New Delhi: Sage, pp. 205–25.

Shelley, F.M., Johnston, R.J., et al. (1990) 'Developments in electoral geography', in R.J. Johnston, F.M. Shelley and P.J. Taylor (eds), *Developments in Electoral Geography'*, New York: Routledge, pp. 1–11.

Shelley, F.M., Archer, J.C., et al. (1996) *Political Geography of the United States*. New York: Guilford Press.

Sibley, D. (1981a) 'The notion of order in spatial analysis', *Professional Geographer,* 33(1): 1–5.

Sibley, D. (1981b) 'Reply to Richard Walker', *Professional Geographer,* 33(1): 10–11.

Sidaway, J.D., et al. (2004) 'Commentary: translating political geographies', *Political Geography,* 23: 1037–49.

Smith, N. (2003) *American Empire: Roosevelt's Geographer and the Prelude to Globalization*. Berkeley: University of California Press.

Soja, E.W. (1989) *Postmodern Geographies: the Reassertion of Space in Critical Theory*. London: Verso.

Sparke, M. (2004) 'Political geography: political geographies of globalization (1) – dominance', *Progress in Human Geography,* 28: 777–94.

Spykman, N.J. (1942) *America's Strategy in World Politics*. New York: Harcourt.

Spykman, N.J. (1944) *The Geography of the Peace*. New York: Harcourt.

Staeheli, L.A. (2001) 'Of possibilities, probabilities and political geography', *Space and Polity,* 5: 177–89.

Staeheli, L.A. and Kofman, E. (2004) 'Mapping gender, making politics', in L.A. Staeheli, E. Kofman and L.J. Peake (eds), *Mapping Women, Making Politics: Feminist Perspectives on Political Geography*. New York: Routledge, pp. 1–13.

Stamp, D. (1960) *Applied Geography: How the Geographer's Survey and Analysis Can Help in Understanding the Britain of Today and in Planning for its Future*. London: Penguin.

Stedman Jones, G. (1972) 'The history of US imperialism', in R. Blackburn (ed.), *Ideology in Social Science: Readings in Critical Social Theory*. New York: Pantheon Books, pp. 207–37.

Storey, D. (2001) *Territory: The Claiming of Space*. New York: Prentice Hall.

Strausz-Hupé, R. (1942) *Geopolitics: The Struggle for Space and Power*. New York: GP. Putnam's Sons.

Swyngedouw, E. (1997) 'Neither global nor local, glocalization and the politics of scale', in K.R. Cox (ed.), *Spaces of Globalization: Reasserting the Power of the Local*. New York: Guilford Press, pp. 137–66.

Taylor, P.J. (1994) 'The state as container: territoriality in the modern world system', *Progress in Human Geography,* 18(2): 151–62.

Taylor, P.J. and Flint, C. (2000) *Political Geography: World-Economy, Nation-State, and Locality*. Harlow and New York: Prentice Hall.

Taylor, P.J. and van der Wusten, H. (2004) 'Political geography: spaces between war and peace', in G. Benko and U. Strohmayer (eds), *Human Geography: A History for the 21st Century*. New York: Oxford University Press, pp. 83–104.

Thrift, N.J. (1983) 'On the determination of social action in space and time', *Environment and Planning D: Society and Space,* 1: 23–57.

Thrift, N.J. (2000) 'It's the little things', in K.J. Dodds and D. Atkinson (eds), *Geopolitical Traditions: A Century of Geopolitical Thought*. New York: Routledge, pp. 380–87.

Tietze, W. (1997) 'Raumwirksame Staatstätigkeit. Anmerkungen zu einem Begriff, der mit den wissenschaftlichen Leistungen von Klaus-Achim Boesler eng verbunden ist. Spatially effective state activity. Notes on a concept which is closely connected with the scientific achievements of Klaus-Achim Boesler', *Colloquium Geographicum,* 23: 249–53.

Troll, C. (1947) 'Die geographische Wissenschaft in Deutschland in den Jahren 1933 bis 1945: Eine Kritik und Rechtfertigung', *Erdkunde,* 1: 3–48.

Turner, F.J. (1963) *The Significance of the Frontier in American History.* Edited, with an introduction, by Harold P. Simonson. New York: Ungar.

Vidal de la Blache, P. (1917) *La France de l'Est (Lorraine-Alsace).* Paris: A. Colin.

Walker, R.A. (1981) 'Left-wing libertarianism, an academic disorder: a response to David Sibley', *Professional Geographer,* 33(1): 5–9.

Whittlesey, D.S. (1939) *The Earth and the State: A Study of Political Geography.* New York: Holt.

Wittfogel, K.A. (1929) 'Geopolitik, geographischer Materialismus und Marxismus', *Unter dem Banner des Marxismus* 3, 5, 7: 17–51, 485–522, 698–735.

The Geography of Political Geography

James D. Sidaway

Without doubt, the German and Anglo-Saxon schools of political geography and geopolitics have been those that have most profoundly marked the profile of these [sub-]disciplines. Because of that, their influences have transcended their particular state and cultural ambits to be assumed and adapted in other contexts. However, it is possible to encounter diverse and equally interesting schools of thought and analysis. ... (Fontí and Rufí, 2001, 49–50, my translation)

Rather than speak of modernity as a monolithic structure of seamless occidental origin, we ought to turn to readings of modernity in various histories and cultures which are informed, inflected, and vectored by various traditions and which, in turn, have become syncretic, vernacular nodes for forming newer traditions and different styles. (Vigo, 2004)

INTRODUCTION

This chapter examines key political geography concepts in translation and traditions beyond the Anglo-American realm. This is a tentative move, however, confined to some political geography terms and categories as they operate in a limited selection of other languages and traditions. Moving beyond Europe and beyond written and spoken language into a range of other discourses and modes of representation, the chapter argues that there is something more at stake than supplementing the range of case studies and terminologies that characterise Anglophone political geography.

In particular, it is argued that (frequently taken-for-granted) assumptions about universality in political geography are themselves products of particular circumstances. Thus, political geography would benefit from more critical awareness of the situated basis of its claims and vantage-point.

The first major section of the chapter reconsiders the relationships between Anglophone, Francophone and German-language political geographies. Following this, the chapter addresses the complex question of translation and explores the consequences for political geography of moving beyond Euro-American texts and languages and beyond the written and spoken word to other systems of representation. With these in mind, the final part of the chapter acknowledges the potential of post-colonial critiques in 'repositioning' political geographies and sets out some critical possibilities and challenges.

THE LANGUAGES AND VANTAGE-POINTS OF POLITICAL GEOGRAPHY

Fontí and Rufí's (2001) Spanish-language textbook that was cited at the start of this chapter goes on to chart French political geography, Italian geopolitics, as well as Soviet Russian and Spanish writings on territorial logics, political geography and geopolitics. As they point out, although it has been written – in various guises – for over a century, the course of political geography (and the allied field of geopolitics) 'often considered as disciplines, sub-disciplines or scientific fields' (ibid., p. 29) has been a chequered one. Close associations with state-building and competing national and imperial projects earlier in the twentieth century provided intellectual space and patronage for political geography in a number of European countries and in the Americas. Subsequently, however, the vogue for positivism and quantitative analysis, which characterised much human geography of the 1960s, saw political geography relatively sidelined in Anglophone geography. However, in the last two decades of the twentieth century, the sub-discipline underwent a renaissance in the UK and North America. The appearance of a specialist journal in 1982 (*Political Geography Quarterly*, subsequently *Political Geography*) marked an opportunity for such a revitalised and critical political geography. The journal rapidly established a reputation within and beyond academic geography and in the 1990s it was joined by two other evidently political geography journals, *Geopolitics* and *Space and Polity*. New texts also appeared. The overall sense was of a field of research and scholarship that was relatively healthy in terms of the number of contributing academics, outlets and courses. Within this, feminist and allied scholarship has broadened the concern of political geography, in particular the connections between 'macro' issues of states and geopolitics and 'micro' politics of everyday lives, summarised by Nogué (1998: 35, my translation) as 'a much more open political geography'. Although the result is a rather unwieldy and diffuse field (Brown and Staeheli, 2003) and although feminist contributions to political geography are sometimes sidelined (according to Staeheli and Kofman, 2004, and Kofman's chapter in this volume), as many of the other chapters in this collection bear witness, boundaries were blurred as the range of theoretical inspirations, interdisciplinary crossovers and political orientations broadened.

At the same time, however, to some observers, the revitalised political geography appeared marked by a primary focus on Anglo-American cases and concerns. Writing a few years after the establishment of the journal *Political Geography*, Perry (1987: 6) claimed that 'Anglo-American political geography poses and pursues a limited and impoverished version of the discipline, largely ignoring the political concerns of four fifths of humankind'.

Kofman reiterated this in the mid-1990s, noting 'the heavily Anglocentric, let alone Eurocentric, bias of political geography writing' (1994: 437). In this political geography is not alone; the same critique has periodically been levelled at 'Anglo-American' human geography more widely (e.g. Slater, 1989; Minca, 2003; Berg, 2004). And the allied field of international relations remains configured, as it was in Hoffmann's (1977) designation of nearly thirty years ago, as 'An American Social Science' (cf. Tickner, 2003).[1]

In the case of political geography however, there is some irony in this Anglocentrism. In its century-long history as a field of study, political geography had significant roots in French and German-language debates. Political geography texts usually mention this continental genealogy. Sometimes, like the introduction to an early 1980s collection on *Developments in Political Geography* (Busteed, 1983: 7), they cite Turgot's *Géographie Politique* (1751). But the more frequent reference point is Ratzel. Both the Spanish-language textbook cited above (Fontí and Rufí, 2001) and three other recent English language texts (Agnew, 2002; Jones et al., 2004; Blacksell, 2006) attribute the terminology and foundation of political geography to Friedrich Ratzel's organic conceptions of the state (codified in his 1897 magnum opus *Politische Geographie*). Beginning with a reference to *Politische Geographie*, the chapter on political geography in a recently published *Human Geography: A History for the 21st Century,* claims that 'This long text (more than 800 pages in the extended second edition of 1903, published just before his death) effectively brings political geography to life as a sub-discipline of human geography' (Taylor and van der Wusten, 2004: 83).

Many other political geography texts open with at least a cursory citation of Ratzel. Writing in the 1970s, the Ghanaian political geographer Amano Boateng, for example, begins with Ratzel as an entrée to *A Political Geography of Africa*, focused on the location and morphology of states. Although Boateng (1978: 5) makes passing reference to non-state (for example, 'tribal') manifestations of 'politically organised areas', his text (like most others of the time) is organised around mapping states, located in the tradition attributed to Ratzel.

As Bassin (1987) has shown, the influences on Ratzel were a complex mix of Prussian-German nationalism, Hegelian philosophy and Darwinian biology. In turn, Ratzel's notions of *Lebensraum* provided ample inspiration to German nationalism. They did not stop at the German borders, however. The conservative Swedish intellectual Rudolf Kjellén reworked Ratzel's ideas, coining the term *Geopolitisk* (geopolitics) in 1899. Subsequently, in the first half of the twentieth century, work on frontiers and boundaries was a markedly Francophone and German-language affair. The use (both practical and scholarly) of this material had underlain some of the development of Anglophone political geography in the 1940s and 1950s (e.g. Jones, 1959; Kristof, 1959; Minghi, 1963), and its textbooks (e.g. Fitzgerald, 1948; East and Moodie, 1956) were replete with references to continental European political geographers, such as Ancel, Goblet, Demangeon, Obst and Maull. A few years earlier, Derwent Whittlesey's (1939) text both cited German-language material and reflected a commitment to the regional synthesis *à la* Vidal de la

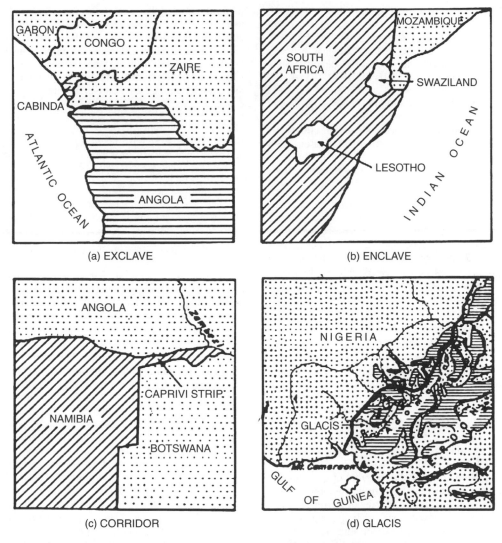

Figure 2.1   Political geography as the morphology of states: *'exclave'*, *'enclave'*, *'corridor'*, and *'glacis'* as illustrated by Boateng (1978). (The original title of the figure, which appears in the introductory chapter on 'Elements of Political Geography' and is the first of over thirty figures in Boateng's text, was 'Examples of certain special configurations'.)

Blache, the doyen of French geography in the first half of the twentieth century (Cohen, 2002). Much of this work (of the 1930s to the 1950s) focused on boundary drawing and boundary disputes. Applying this work to African states, Boateng (1978: 11, italics in the original) maps the phenomena of the *'enclave, exclave, glacis* [which refers to a relatively small portion of a state that extends across a mountain divide forming the boundary between that state and another], [and] *corridor'*, noting how they 'form part of the vocabulary of political geography'. The making and terms of this vocabulary

(illustrated by Boateng's maps, which are reproduced here as Figure 2.1) reflected a transnational conversation.

Likewise, the contested field of geopolitics has comprised a (frequently acrimonious) set of conversations and influences between Francophone, German-speaking and other (e.g. Portuguese, Romanian and Hungarian) figures and institutions (Dodds and Atkinson, 2000; Natter, 2003), and American reactions in the career and writings of the American geographer-geopolitican Isaiah Bowman (Smith, 2003), and of émigré Europeans, such as

Robert Strausz-Hupé (Crampton and Ó Tuathail, 1996), Nicholas Spykman, Hans Morgenthau, and later Henry Kissinger (Hepple, 1986) and Zbigniew Brzezinski (Sidaway, 1998). Classical geopolitics also found fruitful soil in Japan and its Korean colony in the 1930s and 1940s (Narangoa, 2004) and in Latin America in the 1960s and 1970s (Child, 1985; Kelly, 1997). Although deeply rooted in national contexts, in all cases classical geopolitics rested on the *international* circulation of geopolitical ideas, reactions, adaptations, and responses.

Stephen Kearn (1983: 228) thus describes how:

> Rivers of geopolitics coursed all over the European cultural terrain. They started in the high reaches of theoretical tomes such as Ratzel's two major works [*Anthropogeographie* (1882) and *Politische Geographie* (1887)] and cascaded through volumes of the new periodicals that were founded – *National Geographic Magazine* (1889), *Annales de Géographie* (1891), *The Geographical Journal* (1893), *Geographische Zeitschrift* (1895).

This circulation continued through the twentieth century and found many forms and expressions: classical European geopolitics, the ersatz-geopolitics of Bowman and his successors in Cold War America, and the military interest in geopolitics in 1950s Portugal (Sidaway and Power, 2005) or 1970s Latin America. Moreover, the bulk of twentieth-century political geography (especially in the case of geopolitics) was, in Taylor's (2003: 47) words, 'conspicuously conservative in orientation … not at the forefront of querying the *status quo*, rather they [political geographers] have provided spatial recipes for the powerful'.

There were exceptions, including figures like Robert Dickson and Frank Horrabin in the UK (see Hepple, 1999; Johnston, 2000) or Karl Wittfogel in Germany (see Bassin, 1996; Smith, 1987), but they were rarely influential in policy circles or, like Owen Lattimore in the US, they ran foul of the Cold War Red-baiting (see Harvey, 1983). Whilst many political geographers may have been on the political right (often linked to military cliques and nationalist parties), the consequences of their work (in terms of territorial claims and aspirations on planetary space) were often disturbing to the established international order. This conservative and right-wing legacy and lingering sense of danger was the main reason why the creation of a commission on political geography was opposed by Soviet academics when such a group was proposed at the Twenty-fifth Congress of the International Geographical Union held in Paris in August 1984. Although the Commission was established in 1984, the objections voiced in Paris to the 'fascist and Nazi connotations'

of 'political geography' (International Geographical Union, Commission on Political Geography, 2005) meant that it was termed the Commission on the World Political Map (it was only renamed as the Commission on Political Geography in August 2000). Of course, right-wing geopolitical discourses continue to flourish. For as Murphy et al. (2004: 619) point out,

> Political geographers typically invoke the term [geopolitics] with reference to the geographical assumptions and understandings that influence world politics. Outside of the academy, geopolitics often connotes a conservative or right-wing political-territorial calculus associated with the strategic designs of Henry Kissinger, Aleksander Dugin, and followers of the new *Geopolitik* in Germany.

Murphy et al. cite examples of what they term a 'neoclassical geopolitics' from a range of national contexts. Likewise, O'Loughlin (2000: 127) describes this as 'high geopolitics', 'motivated by traditional security concerns', and Ingram (2001) and Smith (1999) chart the Russian case in detail. Yet, at the same time that such 'neoclassical' geopolitics have proliferated beyond the academy, the 1980s and 1990s revival of the field in Anglophone academic geography saw political geography influenced by radical and critical currents, including Marxism, post-structuralism and feminism. Within this reinvigorated and critical frame, the histories (and current reworkings) of classical geopolitics came in for sustained and critical scrutiny (Ó Tuathail, 1996; Ó Tuathail and Dalby, 1998; Dodds and Atkinson, 2000). Ó Tuathail/Toal (2005: 65) therefore describes the emergence over the last twenty years of 'Critical geopolitics [which is] a discursive approach that has sought to rethink the meaning and analytical utility of the notion of "geopolitics". Such critical geopolitics insists that '… while there is no stable and singular notion of "geopolitics", it is a domain of foundational questions, nested plural problematics and pressing political challenges, which is now attracting considerable critical attention within geography and other social sciences' (Ó Tuathail/Toal, 2005: 70).

With a few exceptions, however, such developments have usually tended to limit themselves to interrogating English-language material (although many of the political and theoretical literatures that have inspired such moves are translated from elsewhere – Althusser, Foucault, and Gramsci, for example). Despite the presence of these theoretical inspirations from elsewhere (usually in translation), work in and from other linguistic contexts remains marginalised in a revitalised Anglophone political geography. And, in Megoran's (2005: 556) words, critical geopolitics carries a 'dominant focus on US and European case studies'.

Meanwhile, other critical schools have developed. Inspired in part by critical geopolitics (and responding also to conservative revivals of geopolitical thought in reunified Germany), German-speaking scholars have scrutinised the histories and historiographies of *Geopolitik* (see Natter, 2003, for a summary). Significant French-language schools of political geography and (critical approaches to) geopolitics have also evolved in the past three decades. In particular, work by Lacoste (1976), scholarship associated with the journal *Hérodote* (published since 1976) and writings by Claval (e.g. 1994) and Raffestin (e.g. 1995) stand out, but communication between these and Anglophone scholarship has remained fairly limited.[2] More recently an Italian journal of geopolitics has (re)appeared (www.limesonline.com), latterly also producing an associated English-language online journal (http://www.heartland.it/index.html). However, like the weekly Paris-based www.geopolitique.com, these journals seem to target broad policy and public readerships, rather than being fundamentally scholarly journals.

For a few years, especially in the 1960s and early 1970s, Jean Gottmann (Professor at the Oxford School of Geography from 1968 to 1983) was a key bridge between Francophone and Anglophone political geography, as well as an original theoretical voice at a time when the field was otherwise relatively marginal (Muscarà, 2005). Writing almost sixty years ago in *Annales de Géographie*, Gottmann (1947) had called for the geographical scrutiny of carrefours (crossroads) in a world of flows and mobility. Although better known for his account of an emerging Megalopolis, Gottmann (1961) made significant contributions to political geography over forty years, drawing on a variety of traditions. Since then, however, the depth and quantity of exchange has lessened. Hepple (2000: 269) thus notes how:

> The Anglophone geographical and geopolitical communities tend to know of the existence of the *Hérodote* school, but there has been remarkably little engagement with its ideas or referencing of *Hérodote* sources. ... This neglect appears remarkable when one starts to list some of the features of the Lacoste-*Hérodote* school: the early work by Lacoste in the Marxist tradition; the construction of *Hérodote's* analysis within the radical, post-Marxist culture of Vincennes (where Foucault and Deleuze, with other influential thinkers, worked for periods); and a direct engagement with Foucault in the first issues of *Hérodote*. ...

Hepple attributes this inattention to the linguistic limitations of many Anglophone geographers. In tandem with this, he identifies a certain linguistic imperialism on the part of the wider Anglophone academy. He also notes that the style and foci of most of the work in *Hérodote* is likely to puzzle many Anglophone political geographers. Likewise:

> ... *Hérodote* has been embedded within the Francophone world and within debates of French geography ... analysis of the development and shaping of geography and geopolitics is also set almost exclusively within the French context (with some references to the earlier history of German geography). Likewise, the contemporary debates it includes – with Lévy on Marxism, Brunet on chorèmes, and Raffestin on geopolitics – are Francophone debates. (Hepple, 2000: 270–71)

So far, there is relatively little to show for Hepple's call for dialogue. For whilst it may be informed by theories and literatures from elsewhere, Anglophone political geography (like the wider Anglophone discipline) operates predominantly in a transatlantic space of communication, conferences and journals in which the US–UK link is predominant. This is not to say that there are no differences across the Atlantic in intellectual cultures and the traditions of Anglophone geography and political geography within them (Sidaway and Johnston, 2004). Moreover, English-language journals include works published in English by authors based outside the predominantly Anglophone realms and contain some Editorial Board members and editors from elsewhere. In the past few years, the journal *Geopolitics* (which is edited from the US and Israel,[3] has published useful special issues, for example, on French (5(2), Autumn 2000) and German (7(3), Winter 2002) geopolitics. However, commentators (such as Häkli, 2003, and Mamadouh, 2003) continue to critique what they see as the restricted linguistic and geographical horizons of much Anglophone political geography, as raised earlier by Kofman (1994) and Perry (1987) and echoed in Hepple's (2000) observations. The Finnish political geographer Jouni Häkli (2003: 660), for example, claimed that, '... as a scholar writing from the North European periphery, I want to pay attention to the fact that where you write from makes a big difference'. For him, the concepts associated with political geography as developed in Anglophone literatures might carry 'the risk of an unwarranted universalism':

> A particular parochiality is thus universalized and made to pass as the best available internationally recognized scholarship. But perhaps there is a market for other parochialities. French, Mediterranean, Nordic, Iberian, South American, African, ones

that are poorly known by those who cannot read work done outside Anglo-American circles. If this is the case ... then the universalism of Anglophone geography is but an illusion caused by lack of knowledge concerning the richness of the political geographical world. (Häkli, 2003: 660).

Despite the limits that Häkli signals, it is perhaps in work on Europe (and specifically on European integration) where a more international literature has evolved. As concepts are translated across European languages and institutionalised (in European Union discourses), the field of political geography (as in other social sciences) has come to interact with and intersect with what Jensen and Richardson (2004: ix) claim is '... a new field of European spatial policy, which is embedding new ideas about relationships across space in a multi-level, transnational field of activity'. Much of this literature about the European Union project is in political science and planning, although the centrality of space and territorial questions to its concerns means that political geography has found a role (Sidaway, 2005). More widely, however, in terms of the concepts and vocabulary of political geography, Häkli's critique invites further critical reflection.

TRANSLATING POLITICAL GEOGRAPHIES

In the light of the contexts and debates sketched above, one productive way forward is to ask, what is lost (and gained) in translations? Translation is fraught with difficulties and possibilities. Reviewing these and how they are shaped by commercial, cultural, and geopolitical power, Susan Bassnett (2005: 397) describes how

Translation is a kind of journey. It is an activity that always involves motion, it is a passage from one language to another, and hence from one culture into another. There is also always a temporal dimension, for what is written in one place, in one time, is then rendered for readers in another place and another time. Translation theory today is increasingly concerned with translation as movement between different contexts, and increasingly concerned also with the consequent ideological dimension.

Attentive to translations as continuations of and alterations to texts through space and time, Sidaway et al. (2004) explore categories and concepts that have been central to much political geography (they focus on the terms for 'state', 'territory' and 'border') in Portuguese, Spanish, Malay, Korean, Japanese and Urdu. They point out that the question of what happens to even

quite fundamental political geography concepts, such as the state, sovereignty, borders and territory, in different national-linguistic contexts, begins to problematise taken-for-granted assumptions. Thus,

'Languages and meanings of the political are everywhere caught up in wider cosmographies and hermeneutics. These are usefully approached with an openness to political thought and meanings: recognizing that our categories, terms, and analyses deserve enrichment from relatively unaccustomed sources' (Sidaway et al., 2004: 10).

Likewise, Ingerflom (1993) and Kharkhordin (2001) consider the significant differences in the concept of the state between Russian and West European languages. In all cases, however, the state emerges as a complex concept rooted in residual mystical foundations of power and tied in with the national imaginaries. Critical analysis of these foundations needs to move beyond the spoken and written word to consider other representations. Political geography has long recognised that such national-sovereign representations include mappings (see Black, 1998). But it is from a historian of Thai ideas of identity/sovereignty that one of the most suggestive ways of conceptualising these has emerged. Based on his study of Thai concepts of territory and sovereignty, Winichakul (1996: 67) thus develops the concept of the sovereign 'geo-body':

Unarguably, the territory of a nation is the most concrete feature of a nation for the management of nationhood as a whole. ... For people of a nation, it is part of SELF, a collective self. It is a nation's geobody. ... Geographically speaking, the *geo-body* of a nation occupies a certain portion of the earth's surface which can be easily identified. It seems to be concrete to the eyes and having a long history as if it were natural, and independent from technology or any cultural and social construction.

Winichakul sets out the ways in which the geo-body is an effect of modern practices and technologies of mapping. In the Thai case, the earlier sense of vague frontiers and interlaced sovereign powers was transformed through contact with colonial powers who were carving out bounded territories in Southeast Asia. In response, Thai territory was mapped and demarcated and represented as the geo-body of Siam. In turn, the 'Thai-ness' of these lands was further asserted when Siam was renamed as Thailand in 1939.

However, representations of geo-bodies are not confined to conventional maps. One example is the circulation in South Asia of striking representations of *Bharat Mata* (Mother India) as the geo-body of the subcontinent. Ramaswamy (2003: 178) notes the significance of these in late colonial and post-colonial India. Thus:

> ... by the middle decades of the century, the map and the mother's body had become interchangeable, one substituting for the other. Indeed, in 1936, when the very first 'temple' to Bharat Mata was opened in Banares, it did not house an image of the goddess but a marble map of India made to scale with all its topographic features shown in great detail.

The conception of political space in these figures is thus an invocation of the supposed interface between a cosmic and a geopolitical order:

> ... these bodyscapes visually 'Hinduize' India's geo-body by resorting to the familiar image of the Hindu mother goddess such as Durga and Lakshmi on whom Bharat Mata is clearly modelled. But what is also worth noting is the inclusivistic definition of 'Hinduism' that bodyscapes appear to operate with, in that symbolic markers which we associate with other religious sensibilities such as Buddhism, Jainism and Sikhism are also incorporated into the mother's body. ... Bharat Mata's body thus becomes a microcosm of the nation's plural religious history, even while it is used to signal what can − and cannot − be included within its ample folds. ... (Ramaswamy, 2003: 180–1).

It is partly these cosmo-socio-geopolitical interfaces that lead Krishna (1996) to the concept of cartographic anxiety, describing post-colonial India's fixation with territoriality, mappings and the secure inscription of borders. Similarly, Fretag (2001: 39) points to the acts of imagining nation and community, 'where spectatorship meets creation in complex interplays between bodies, visuality, discourse, institutions and figurality', and Turner (2002: 93) suggests that '... the subaltern, however disempowered in the contexts of official institutions, public media, and government, certainly can and does speak, *but not necessarily to you*'.

In other words, there is a potential diversity of other political geographies − or ways of seeing and knowing the political. There are other ethics of power and space to acknowledge. This applies both to sites/foci and theoretical inspirations. In the latter terms, Tyner (2004: 341) calls for a 'broadening of horizons' in political geography. Tyner's focus is on the political thought of the African-American radical Malcolm X. But, writing from the US, he makes a wider point:

> Our geographies, and especially our political geographies, remain largely distant from non-European theorists and theories. Our texts on nationalism and identities, in particular, are woefully ignorant of Pan-African nationalism and other African diasporic movements. Also largely missing are the geographies of Pan-Asian, American and Chicano movements.

Sustained attention to location, translation, language, and context offer fruitful and challenging departures. Yet, as in Tyner's recognition of Malcolm X's writings, there is something more fundamental at stake here than other languages, artefacts and narratives supplementing (and sometimes disrupting) the conventional foci of much Anglophone political geography.

First, languages may sometimes be incommensurable. More specifically, translation frequently relates to and sometimes becomes part of concrete political struggles. An example is the 1840 Treaty of Waitangi, the founding document of the New Zealand state, which was signed by all parties in both English and Maori versions. The text of this Treaty/Triti (between Maori tribal representatives and the representative of the British crown) forms the basis for the settlement of New Zealand and still configures the terms of struggles over the rights and obligations it is seen to confer. However, as Lashley (2000: 4–5) notes:

> ... the English and Maori texts of the treaty differ profoundly. According to the English text, indigenous Maori ceded 'all the rights and powers of sovereignty over their respective territories' to Queen Victoria. Maori people retained 'full and undisturbed possession of their lands and estates, forests, fisheries and other properties' (Article II) in return for 'royal protection' and 'all the rights and privileges of British subjects' (Article III). According to the Maori text, Maori ceded *kawanatanga* (governance) not sovereignty and retained *rangatiratanga* (chieftainship). For Maori, this distinction is paramount. The fundamental difference between the two texts, from the Maori perspective, is that confirmation of chieftainship also confirms sovereignty, in return for a *limited* concession of power.

On similar lines, Gibson (1999) points out how the worldwide struggle for 'self-determination' by 'first [aboriginal, indigenous or native] peoples' begins to shift the meaning of self-determination, raising questions about national-state identities, structures, and the processes by which states claim authority and legitimacy. Thus, some of the political claims and languages of indigenous peoples do not simply echo the logic of territory-sovereignty of the established states (even though these states may seek to contain and police such indigenous assertions). Gibson (1999: 54) notes that

'Aboriginal peoples in Australia are also construct-ing "worlds of meaning" for self-determination based on multivalent interpretations of sovereignty and identity'.

Although alert to how they are limited, frag-mented and contained within webs of state and settler power, Gibson points to the significance of such Aboriginal territorialities as challenges to the commodifying Cartesian logic of the modern cap-italist state. Part of this is found in language and related systems of meaning and being in the world (ontology). Gibson (1999: 56, italics in the orig-inal) cites the example of Pitjantjatjara language and community, that for its speakers and members

> ... denotes a specific geography (*Pitjantjatjara* coun-try), a legal system of responsibility and action (*Pitjantjatjara* law), a language group (to speak *Pitjantjatjara*), and the identity of people '(he is, she is, they are, I am *Pitjantjatjara*)'. ... Avoiding the Cartesian divide that for over 400 years has dom-inated Western epistemology, Aboriginal asser-tions of personal identity commonly revolve around notions of the individual inseparable from the col-lective and the universal: 'The land is my spirit. ... Our land is our spirit'.

Second, certain assumptions about universality need to be treated more critically. Critical geopol-itics has been insistent that all descriptions are views from somewhere; in other words, it con-textualizes geopolitical discourse and so rejects the universalistic pretensions of earlier variants of geopolitics. But more sensitivity to the structural positionality of academic norms and suppositions is vital, including those made in critical guise. This comes more clearly into view when it is recognised that assumptions about universality in political geography themselves tend to be prod-ucts of particular circumstances. An achievement of the social sciences (and here geography plays a key role) has been to visualise how connections between diverse phenomena and places are con-figured through universal processes, narratives and logics; hence, the references in political geogra-phy (as in other social sciences), to the reach and roles of capital, modernity, and sovereignty. Yet visions of universal processes are both bound up with Western power (as Edward Said and others have pointed out) and geographically specific and situated. Commenting on this, Mitchell (2003: 167) points out how conventionally '... the diver-sity of languages in which communities articulate their political demands and identities, their visions and their revulsions, are to be translated into the universal language of political economy'.

Material imperatives of capitalism compel con-vergence in meanings (profit, price, globalisa-tion, trade enter wide translation and become commensurate in markets), yet that process is uneven and multi-faceted. It is also tied in with meanings and understandings that have a Western provenance, but are never simply reducible to a Western *telos*. As writers such as Dipesh Chakrabarty (2000) have detailed, there is no easy way around this problematic interplay of com-plexity, diversity, and abstraction. A starting point, however, is to recognise that the putatively univer-sal languages into which 'local' phenomena and terms are translated are themselves derived from Western norms and practices. Thus, although his focus is on political economy, Mitchell's (2003: 167–8) critique also holds for the ways that poli-tical geography and other social sciences conceive of norms and difference:

> The local forms of political organisation and expres-sion are understood as mere languages, meaning the cultural and 'ideational' forms for express-ing the real interests that shape their world. The language into which these expressions are trans-lated – political economy – is assumed by definition not to be an ideational form, not a cultural prac-tice, but the transparent and global terminology of economic reality.

At a more general level, Berg (2004: 555) argues that '... geographies of the United Kingdom and America are unmarked by limits – they constitute the field of geography. Geographies of other peo-ple and places become marked as Other – exotic, transgressive, extraordinary and unrepresentative'.

It is these problematics that Robinson (2003: 279) tackles in her suggestion for a political geogra-phy derived from Southeast Asia (cf. Grundy-Warr and Sidaway, 2003):

> And why not a political geography whose focus is Southeast Asia? The suggestions that I make try to escape the contemporary biases which shape the ways in which different places come into view in the western elements of the discipline: for example, Southeast Asia is seen as 'interesting' to west-ern scholars because it has tiger economies and so-called 'world cities'. We need to be constantly on the alert for such moves that reinstate a sense of 'knowledge' of other places serving 'our' purposes and concerns, whatever these might be. The orien-tation is, rather, towards learning from the complex and rich experiences and scholarship of different places.

The notion is therefore not simply that such non-Western geographies (or political geographies) should be supplements that remain as examples, footnotes or exceptions to the Anglo-American mainstream, but that the mainstream becomes more attendant to its own situatedness. Keeping these challenges and dilemmas in mind, the last section of this chapter sketches some paths towards repositioning political geographies.

REPOSITIONING POLITICAL GEOGRAPHIES

Amidst the transdisciplinary discourse of globalisation (see Dalby, this volume), political geographers have been keen to point to the contradictory dialectics of global/local/state relations (Cox, 1997; Brenner et al., 2003); the reworking (but continued roles) of borders (Newman, 2003); the dialectics of globalisation, imperialism, and resistances (Sparke, 2004); the geopolitics of diaspora (Carter, 2005); the significance and contestation of the construction and intersections of scales (Herod and Wright, 2002; Howitt, 2003) and attendant reterritorialisations (Herod et al., 1998). A parallel stress on networks – especially in work on 'world cities' and their interactions with states – has also been a productive response to tracing new political geographies (Taylor, 2000, 2005).

Considerable potential remains, however, especially in the domain of those other (hitherto relatively marginalised) political geographies sketched above. In this task I am encouraging political geographers to adopt an attitude towards systems of organising space and power that has long been advocated by anthropology (and taken up by aspects of cultural geography). That is, to adopt an attitude that, once a few core questions are asked (e.g. how does this or that group or state use some degree of control over space to maintain political cohesion), there are an open range of prospective answers and categorisations that defy hierarchical ordering. Cultural (and to some extent economic) geographers have been willing to engage with alternative organising systems coming from different parts of the world, yet Anglophone political geography has – in recent years – been less willing to engage alternatives. It is beyond the scope of this chapter (and the competence of its author) to provide anything other than some selective pointers here. In the first place, however, it is necessary to reconsider what and where have become taken for granted as the norms in literatures on the sovereign state.

Drawing on post-structuralist and post-colonial literatures, and focusing on African sovereignties, Sidaway (2003, 175) articulates:

A demand for new and unorthodox maps of flows and sovereignties, but coupled with scepticism about the ability of charts to adequately represent what these might amount to (some wariness about

the phantom objectivity of mappings); these might serve as promising points of departure towards a postcolonial political geography.

This argument is linked to a call for political geography to recognise that the diversity and character of post-colonial sovereignties (cf. Sidaway, 2000) ought not to be interpreted as a simple hierarchy with the putatively 'strong', long-established Western states at the apex and post-colonial states (especially those that are visibly fractured by insurgencies or secessionist movements) as somehow abnormal or simply lacking the features of the Western state. Instead, it is argued that the supposed weakness of some post-colonial states might be interpreted as arising not from a lack or absence of authority and connection (including the presence of the West), but rather as an excess of certain forms of them. The paper concentrates on the political geographies of Angola and Zaïre. In the former case, for example, the 'normal' experience of sovereignty has been one of dislocation, turmoil, and fractures. Yet,

The key point is not only that Angola's '(ab)normality' has been many years in the making, but it is profoundly connected with others. Indeed, it would be utterly impossible to imagine or describe without reference to Western normality, to transnational flows of oil, gems, weapons and capital. ... Angola is symptomatic of how malign combinations of imperialism, Cold War, the power of money, minerals (global demand for oil and diamonds), and violence may interact. Angola's situation is a product of these interactions. ... What has become normal in Angola emerges out of profound connections to other modern norms, to all our normalities. (Sidaway, 2003: 164)

The focus is in Africa, but these arguments also apply elsewhere. It is especially evident in those places, Afghanistan, Cambodia, and Colombia among them, that became significant domains of Cold War confrontations. In slightly different terms, Robinson (2003) – whose provocative suggestion for a political geography with a focus on Southeast Asia was introduced earlier – argues that a more nuanced sensibility towards the range of post-colonial trajectories and politics ought to introduce caution into some Western narratives about their universality and value across diverse contexts.

A second path is set out through Oren Yiftachel and As'ad Ghanem's (2004) notion of ethnocracy, where they argue (using the Israeli case, but also venturing into comparisons where similar tendencies are evident, such as Estonia and Sri Lanka) that *ethnos* and not *demos* becomes the main organising political principle. In other words, political subjects are not defined primarily according to territorial convention, but according to ethnic criteria. Elsewhere, Yiftachel (1998: 1) defines ethnocracy:

... as a regime type with several key characteristics:

- Despite several democratic features, ethnicity, not territorial citizenship, is the main logic behind resource allocation.
- State borders and political boundaries are fuzzy: there is no identifiable 'demos', mainly due to the role of ethnic diasporas inside the polity and the inferior position of ethnic minorities.
- A dominant 'charter' ethnic group appropriates the state apparatus and determines most public policies.
- Significant (though partial) civil and political rights are extended to minority members, distinguishing ethnocracies from Herrenvolk or authoritarian regimes.

This academic work has led to fierce attacks from the Israeli right wing, for clearly it poses an analytical and political challenge in the specific context of Zionism. The concepts contain more universal assertions, however, about the operation of 'democracy' and electoral geographies elsewhere. The Arab states of the Gulf spring to mind, where (as in the UAE or Kuwait for example) not only is political space for opposition and parties circumscribed by the ruling Sultanates, but citizenship (and voting) are restricted by ethnically defined (and relatively narrow) citizenship criteria that enfranchise a minority of the population. The category might be pushed further; the US South prior to the mid-1960s had ethnocratic features, as did Australia until the 1970s, as well as the more obvious case of apartheid South Africa. Close scrutiny reveals many states (in Europe and elsewhere) where *jus solis* (automatic citizenship by birthplace within a territory) is displaced by *jus sanguinis* (citizenship primarily by the parent's birthplace, nationality or 'race').

A third example of the ways in which other political geographies might be produced relates to the quite basic issue of conventional political maps of the world, especially their margins. Over recent decades the basics of (Western) cartography have certainly been questioned and problematised. The critique of the Mercator projection and the circulation of alternatives (notably the Peters projection) that portray the world from other vantage-points have developed since the 1970s, along with a broader awareness of Eurocentrism across the disciplines. In this context, Lewis and Wigen's (1997) book on *The Myth of Continents: A Critique of*

Metageography and Alastair Bonnett's (2004) *The Idea of the West* have summed up the ways that a fundamentally Eurocentric vision of the world had come to be taken for granted. Moreover, some spaces tend to remain relegated to the margins, fragmented and often forgotten. Commenting on this, in the context of the trajectory of area studies, Van Schendel (2002) notes how:

... atlases commonly have maps with the captions 'Southeast Asia' and 'South Asia'. These apparently objective visualizations present regional heartlands as well as peripheries – parts of the world that always drop off the map, disappear into the folds of two-page spreads, or end up as insets. In this way, cartographic convenience reinforces a hierarchical spatial awareness, highlighting certain areas of the globe and pushing others into the shadows.

He points to the mountainous region covering Burma, northeast India and parts of China, as sharing a certain cultural-linguistic and historical commonality. Van Schendel charts how the geopolitics of the Cold War (which fragmented the region into different spheres of influence), the ways that it included only relatively marginal areas of states (in social sciences that continued to work with states as the basic units of organisation and analysis) and the lack of an influential scholarly community committed to its study meant that the linguistic and cultural connections and affinities across the region rarely became the domain of systematic attention. Instead, other 'regions' (such as East Asia, South Asia, and Southeast Asia) emerged as supposedly natural domains of study and interaction. In turn, these regions saw intensified political, cultural and economic exchanges (and narratives) that produced the objects (Southeast Asia, for example, in the form of areas studies programmes and ASEAN after 1967) they sought to describe (Evans, 2002).

Today, however, there is a sense of critical possibility and challenge in terms of alternative meso- and macro-geographies. Each (including the conventional Western designations) carries distinctive sets of exclusions and focus points. Consider, for example, the ways that Arabic representations have long constructed the lands of the 'East Indies' as a distant archipelago of Islam, yet one deeply and closely configured by the common circulation of pilgrims and commodities (Laffan, 2004). Similarly, Halliday (2002, 15) notes how Arabic discourses describe what in English is termed Arabia as *Al-Jazira*, the peninsula. In recent years, the term has become more familiar to Anglophone audiences, since it came to be used as a name for the pan-Arab satellite TV station based in Qatar since 1996. It was also adapted in the rhetoric of

Osama bin Laden and others who now reject the legitimacy of Saudi Arabia and the political fragmentation of the peninsula (into the seven states of Saudi Arabia, Kuwait, Oman, Bahrain, Qatar and the UAE) to *jazirat Muhammad* (the peninsula of the Prophet Muhammad). The relationship of these spaces to South Asia is also charted in Ghosh's (1994) *In an Antique Land: History in the Guise of a Traveller's Tale*. Part travelogue, part history, part history, this text defies easy classification.[4] In this, Ghosh sets out how his experiences as an Indian-born anthropologist in an Egyptian village led him to examine twelfth-century documents drawn up by Jewish and Arab merchants, who travelled and traded through the Arab world and South Asia. In turn, Ghosh's narrative connects this with the contemporary flows of migrants and workers between South Asia, Arabia and the Gulf. Ghosh's account of 'transnationalism' is removed from the modern West in both time and place and therefore does not have the latter as its key reference point.[5]

Elsewhere, Halliday (2002: 214) reminds us of the significance of vantage-points in naming practices with an example from Cold War narratives: 'What the Western world calls "the Cuba missile crisis" of October 1962 is to the Russians the "Caribbean crisis" and to the Cubans, who point out none of the missiles involved was 'Cuban', it is *la crise de octubre*'.

These superficially trivial issues of nomenclature reveal more profound differences of vantage-point and scales and geometries of reference. In terms of alternatives for political geography, Van Schendel in particular points to other 'regions' cross-cutting the conventional ones, and thereby focuses on unfulfilled potential for the study of borderlands and transnational flows. In these suggestions, however, Van Schendel is not an isolated voice. The Thai scholar Thongchai Winichakul (2003: 10) (whose work on geo-bodies was introduced above), for example, notes how:

As the arbitrariness of the world of nations and the limitations of national history come under scrutiny, plenty of alternatives are emerging. Among them are a transnational history of Southeast Asia as a common trading zone … a history of areas that span or cross borders between different modern nations, especially if they shared a common history before the drawing of modern national boundaries … and subnational histories of regions and locales that once were 'autonomous' and not part of the major nations, but were 'integrated' with them later. … And with a different geography, different stories emerge.

In parallel terms, Delaney (2005: 65) points to the ways that, rather than disappearing (along the lines of corporate texts on globalisation), boundaries are proliferating and being dispersed throughout social, political, and economic spaces:

… if the US–Mexico border is to be found at various sites throughout the spaces of the US and Mexico [in multiple sites of racialized boundaries] then one might be able to consider sites of the US–Guatemalan border, the Pakistani–Canadian border, or the US–Afghani border. One might also consider territorial phenomena such as the Indian–Pakistani border as it may be differentially encountered in London, Kuwait City, or Los Angeles.

Other possibilities emerge too when political geography escapes its conventional earth-bound focus and moves into the space of flows and movement of ocean-spaces. There is a long history of work in political geography extending the study of the territorial state into the realm of competition for maritime resources and spaces, a topic that was rejuvenated in 1994 by the passage of the UN Convention on the Law of the Sea (Steinberg, 2001). But when the oceans are seen too as spaces of connection and flow (rather than merely competition and control as in classical Western geopolitics), then (continental) margins and centres are also repositioned. The Indian Ocean, for example, becomes a space of interchange that connects and configures littoral communities in Eastern Africa, the Mashriq, and South and Southeast Asia (Lewis, 1999; Lewis and Wigen, 1999; Chaturvedi, 2002). Similarly, the Atlantic, so key in Anglo-American notions of geopolitical and geocultural connection, also encloses a subaltern Black Atlantic (Gilroy, 1992), of multiple, African, American, Caribbean, and European cultural, political and commodity interchanges.

Openness to these alternatives should be productively disruptive. In such spirit, this chapter has no neat conclusions, for developing conceptual alternatives is an ongoing challenge and political geography is richest when reworked, resituated, redeployed and re-imagined.

ACKNOWLEDGEMENTS

Although begun and finished in Singapore (whilst I was based at the National University of Singapore), the bulk of this chapter was drafted in October 2004, when I was a visiting fellow at the National Centre for Research on Europe, University of Canterbury, Aotearoa/New Zealand, whose hospitality is gratefully acknowledged. Simon Dalby, Ron Johnston, Olivier Kramsch, Claudio Minca, Robina Mohammad, Eric Pawson, Phil Steinberg, Peter Taylor, and the editors provided helpful comments on earlier drafts.

The chapter was reworked following a presentation of its arguments at a meeting of the AAG Political Geography Speciality Group, held at the University of Colorado, Boulder, 3–5 April 2005. I am grateful to the hosts and audience at Boulder for their suggestions.

I dedicate this chapter to the celebration of the many border-crossings made with my daughter Jasmin Leila Sidaway (1997–2007).

NOTES

1 For details of an alternative, working with the assumptions of mainstream International Relations but from non-western vantage points, see Cicek (2004).

2 As this chapter went to press a paper by Juliet Fall (2007) was published on the circulation of ideas within Francophone political geographies. Fall provides both a nuanced account of the uneven exchanges between Anglophone and Francophone political geographies and an insight into the politics of knowledge production within the latter.

3 It is notable that, with the exception of the work of the Palestinian geographer Ghazi-Walid Falah (e.g. Falah and Newman, 1995; Falah, 2005), little political geography from the Arab world or by Arab geographers appears in English.

4 I have adapted the summary of Ghosh's book from the details provided at the post-colonial studies at Emory website (http://www.english.emory.edu/Bahri/Intro.html, accessed 24 March 2005). See too http://www.amitavghosh.com/.

5 See Roy (2002) for another readable account of the blurred connections between Islam, society and territory and Sayyid (1997) for an account of Islamism within the logics of Eurocentrism and modernity. Roy reads 'fundamentalism' as a product and agent of complex forces of globalisation and Sayyid traces the convoluted geopolitical genealogies of Islamism.

REFERENCES

Agnew, J.A. (2002) *Making Political Geography*. London: Arnold.

Bassin, M. (1996) 'Nature, geopolitics and Marxism: ecological contestations in Weimar Germany', *Transactions of the Institute of British Geographers,* NS, 21(2): 315–41.

Bassin, M. (1987) 'Imperialism and the nation-state in Friedrich Ratzel's Political Geography', *Progress in Human Geography,* 11(4): 473–95.

Bassnett, S. (2005) 'Translating terror', *Third World Quarterly,* 26: 393–403.

Berg, L.D. (2004) 'Scaling knowledge: towards a critical geography of critical geographies', *Geoforum,* 35: 553–8.

Black, J. (1998) *Maps and Politics*. Chicago: University of Chicago Press.

Blacksell, M. (2006) *Political Geography*. London and New York: Routledge.

Boateng, E.A. (1978) *A Political Geography of Africa*. Cambridge: Cambridge University Press.

Bonnett, A. (2004) *The Idea of the West: Culture, Politics and History*. Basingstoke and New York: Palgrave Macmillan.

Brenner, N., Jessop, B., Jones, M. and MacLeod, G. (eds) (2003) *State/Space: A Reader*. Malden, MA and Oxford: Blackwell.

Brown, M. and Staeheli, L.A. (2003) '"Are we there yet?" feminist political geographies', *Gender, Place and Culture: A Journal of Feminist Geography,* 10: 247–55.

Busteed, M. (1983) 'The developing nature of political geography', in M.A. Busteed (ed.), *Developments in Political Geography*. London: Academic Press, pp. 1–67.

Carter, S. (2005) 'The geopolitics of diaspora', *Area,* 37(1): 54–63.

Chakrabarty, D. (2000) *Provincializing Europe: Postcolonial Thought and Historical Difference*. Princeton: Princeton University Press.

Chaturvedi, S. (2002) 'Afro-Asian oceanic dialogue: imperatives and impediments', *Identity, Culture and Politics,* 3(2): 125–42.

Child, J.C. (1985) *Geopolitics and Conflict in South America: Quarrels among Neighbours*. New York: Praeger.

Cicek, O. (2004) 'Review of a perspective: subaltern realism', *Review of International Affairs,* 3(3): 495–501.

Claval, P. (1994) *Géopolitique et Géostratégie: la pensése politique, l'espace et le territoire au XXe siècle*. Paris: Nathan.

Cohen, S.B. (2002) 'Textbooks that moved generations. Whittlesey, D. 1939: *The Earth and the State: a Study of Political Geography*', *Progress in Human Geography,* 26(5): 679–82.

Cox, K.R. (ed.) (1997) *Spaces of Globalization*. New York: Guilford Press.

Crampton, A. and Ó Tuathail, G. (1996) 'The case of Robert Strausz-Hupé and American geopolitics', *Political Geography,* 15: 553–6.

Delaney, D. (2005) *Territory: A Short Introduction*. Malden, MA and Oxford: Blackwell.

Dodds, K.J. and Atkinson, D. (eds) (2000) *Geopolitical Traditions: A Century of Geopolitical Thought*. London and New York: Routledge.

East, W.G. and Moodie, A.E. (1956) *The Changing World: Studies in Political Geography*. London: George G. Harrap.

Evans, G. (2002) 'Between the global and local there are regions, culture areas: a review article', *Journal of Southeast Asian Studies,* 33(1): 147–62.

Falah, G. and Newman, D. (1995) 'The manifestation of threat: Israelis and Palestinians seek a "good" border', *Political Geography,* 14(8): 689–706.

Falah, G.W. (2005) 'Geopolitics of "enclavisation" and the demise of a two-state solution to the Israeli-Palestinian conflict', *Third World Quarterly,* 26(8): 1341–72.

Fall, J.J. (2007) 'Lost geographers: power games and the circulation of ideas within Francophone political geographies', *Progress in Human Geography,* 31(2): 195–216.

Fitzgerald, W. (1948) *The New Europe: An Introduction to its Political Geography*. London: Methuen.

Fontí, J.N. and Rufí, J.V. (2001) *Geopolítica, identidad y globalización*. Barcelona: Editorial Ariel.

Fretag (2001) 'Visions of the nation: theorizing the Nexus between creation, consumption, and participation in the public sphere', in D. Rachel and P. Christopher (eds), *Pleasure and the Nation: The History, Politics, and Consumption of Public Culture in India*. Oxford: Oxford University Press. pp. 35–75.

Ghosh, A. (1994) *In an Antique Land: History in the Guise of a Traveller's Tale*. New York: Vintage.

Gibson, C. (1999) 'Cartographies of the colonial/capitalist state: a geopolitics of indigenous self-determination in Australia', *Antipode*, 31(1): 45–79.

Gilroy, P. (1992) *The Black Atlantic: Modernity and Double Consciousness*. London and New York: Verso.

Gottmann, J. (1947) 'De la méthode d'analyse en géographie humaine', *Annales de Géographie*, 56: 1–12.

Gottmann, J. (1961) *Megalopolis: The Urbanized Northeastern Seaboard of the United States*. New York: The Twentieth Century Fund.

Grundy-Warr, C. and Sidaway, J.D. (2003) 'Editorial: On being critical about geopolitics in a "shatterbelt"', *Geopolitics*, 8(2): 1–6.

Häkli, J. (2003) 'To discipline or not to discipline, is that the question?' *Political Geography*, 22(6): 657–61.

Halliday, F. (2002) *Two Hours That Shook the world. September 11, 2001: Causes and Consequences*. London: Saqi Books.

Harvey, D. (1983) 'Owen Lattimore: a memoire', *Antipode*, 15: 3–11.

Hepple, L.W. (1986) 'The revival of geopolitics', *Political Geography Quarterly*, 5(4): 521–36.

Hepple, L.W. (1999) 'Socialist geography in England: J.F. Horrabin and a workers' economic and political geography', *Antipode*, 31(1): 80–109.

Hepple, L.W. (2000) '*Géopolitiques de Gauche*: Yves Lacoste, Hérodote and French radical geopolitics', in K.J. Dodds and D. Atkinson (eds), *Geopolitical Traditions: A Century of Geopolitical Thought*. London and New York: Routledge, pp. 268–301.

Herod, A., Roberts, S. and Ó Tuathail, G. (eds) (1998) *An Unruly World? Geography, Globalization and Governance*. London and New York: Routledge.

Herod, A. and Wright, M. (2002) (eds), *Geographies of Power: Placing Scale*. Malden, MA: Blackwell.

Hoffmann, S. (1977) 'An American social science: international relations', *Daedalus*, 3: 41–60.

Howitt, R. (2003) 'Scale', in J.A. Agnew, K. Mitchell and G. Toal (eds), *A Companion to Political Geography*. Malden, MA and Oxford: Blackwell, pp. 138–57.

Ingerflom, C. (1993) 'Oublier l'état pour comprendre la Russie?' *Revue des études slaves*, 66: 125–34.

Ingram, A. (2001) 'Alexander Dugin: geopolitics and neo-fascism in Europe', *Political Geography*, 20: 1029–51.

International Geographical Union, Commission on Political Geography (2005) *History of the IGU Commission on Political Geography*, http://www.cas.sc.edu/geog/cpg/History.html, accessed 29 May 2005.

Jensen, O.B. and Richardson, T. (2004) '*Making European Space: Mobility, Power and Territorial Identity*'. London and New York: Routledge.

Johnston, R. (2000) 'City-regions and a Federal Europe: Robert Dickinson and post-World War II reconstruction', *Geopolitics*, 5(3): 153–76.

Jones, S.B. (1959) 'Boundary concepts in the setting of place and time', *Annals of the Association of American Geographers*, 49(3): 241–55.

Jones, M., Jones R. and Woods, M. (2004) *An Introduction to Political Geography: Space, Place and Politics*. London and New York: Routledge.

Kelly, P. (1997) *Checkerboards and Shatterbelts: The Geopolitics of South America*. Austin: University of Texas Press.

Kearn S. (1983) '*The Culture of Time and Space*', 1880–1918. Princeton: Harvard University Press.

Kharkhordin, O. (2001) 'What is the state? The Russian concept of *Gosudarstvo* in the European context, *History and Theory*, 40: 206–40.

Krishna, S. (1996) 'Cartographic anxiety: mapping the body politic in India', in M.J. Shapiro and H. Alker (eds), *Challenging Boundaries: Global Flows, Territorial Identities*. Minneapolis: University of Minnesota Press, pp. 193–214.

Kofman, E. (1994) 'Unfinished agendas: acting upon minority voices of the past decade', *Geoforum*, 25: 429–43.

Kristof, L.K.D. (1959) 'The nature of frontiers and boundaries', *Annals of the Association of American Geographers*, 49(3): 269–82.

Lacoste, Y. (1976) *La Géographie, ça sert, d'abord, à faire la guerre*. Paris: Maspero.

Laffan, M. (2004) 'Camphor, Ka'Ba and Jâwa: shifting representations of Southeast Asia and Southeast Asians in Arabic', paper presented at the workshop on Southeast Asia and the Middle East: Islam, Movement and the Longue Durée. Asia Research Institute, National University of Singapore, 17–18 August 2004.

Lashley, M.E. (2000) 'Implementing treaty settlements via indigenous institutions: social justice and detribalization in New Zealand', *The Contemporary Pacific*, 12(1): 1–55.

Lewis, M.W. (1999) 'Dividing the ocean sea', *Geographical Review*, 89(2): 188–214.

Lewis, M.W. and Wigen, K.E. (1997) *The Myth of Continents: A Critique of Metageography*. Berkeley: University of California Press.

Lewis, M.W. and Wigen, K.E. (1999) 'A maritime response to the crisis in area studies', *Geographical Review*, 89(2): 161–8.

Mamadouh, V. (2003) 'Some notes on the politics of political geography', *Political Geography*, 22: 663–75.

Megoran, N. (2005) 'The critical geopolitics of danger in Uzbekéstan and Kyrgyzstan', *Environment and Planning D: Society and Space*, 23(4): 555–80.

Minca, C. (2003) 'Guest editorial: Critical peripheries', *Environment and Planning D: Society and Space*, 21: 160–8.

Minghi, J.V. (1963) 'Boundary studies in political geography', *Annals of the Association of American Geographers*, 53: 407–28.

Mitchell, T. (2003) 'Deterritorialization and the crisis of social science', in A. Mirsepassi, A. Basu and F. Weaver (eds), *Localizing Knowledge in a Globalizing World: Recasting the Area Studies Debate*. Syracuse, NY: Syracuse University Press, pp. 148–70.

Murphy, A., Bassin, M., Newman, D., Reuber, P. and Agnew, J.A. (2004) 'Forum: Is there a politics to geopolitics?', *Progress in Human Geography*, 28(5): 619–40.

Muscarà, L. (2005) 'Territory as a pyschomatic device: Gottmann's kinetic political geography', *Geopolitics*, 10: 26–49.

Narangoa, L. (2004) 'Japanese geopolitics and the Mongol lands, 1915–1945', *European Journal of East Asia Studies*, 3(1): 45–68.

Natter, W. (2003) 'Geopolitics in Germany, 1919–45: Karl Haushofer, and the *Zeitschrift für Geopolitik*', in J.A. Agnew, K. Mitchell and G. Toal (eds), *Companion to Political Geography*, Malden, MA and Oxford: Blackwell, pp. 187–203.

Newman, D. (2003) 'Boundaries', in J.A. Agnew, K. Mitchell and G. Toal (eds), *A Companion to Political Geography*, Malden, MA and Oxford. Blackwell, pp. 123–37.

Nogué, J. (1998) *Nacionalismo y Territorio*. Lleida: Milenio.

O'Loughlin, J. (2000) 'Geography as space and geography as place: the divide between political science and political geography continues', *Geopolitics*, 5(3): 126–37.

Ó Tuathail, G. (1996) *Critical Geopolitics: The Politics of Writing Global Space*. London: Routledge.

Ó Tuathail, G. and Toal, G. (2005) 'Geopolitics', in D. Atkinson, P. Jackson, D. Sibley and N. Washbourne (eds), *Cultural Geography: A Critical Aictionary of Key Concepts*. London and New York: I.B. Tauris.

Ó Tuathail, G. and Dalby, S. (eds) (1998) *Rethinking Geopolitics*. London and New York: Routledge.

Perry, P. (1987) 'Editorial comment. *Political Geography Quarterly*: a content (but discontented) review', *Political Geography Quarterly*, 6(1): 5–6.

Raffestin, C. (1995) *Géopolitique et Historie*. Lausanne: Payot.

Ramaswamy, S. (2003) 'Visualising India's geo-body: globes, maps, bodyscapes', in S. Ramaswamy (ed.), *Beyond Appearances? Visual Practices and Ideologies in Modern India*. New Delhi: Sage, pp. 151–89.

Robinson, J. (2003) 'Postcolonialising geography: tactics and pitfalls', *Singapore Journal of Tropical Geography*, 24(3): 273–89.

Roy, O. (2002) *Globalised Islam: The Search for a New Ummah*. London: Hurst & Co.

Sayyid, B.S. (1997) *A Fundamental Fear: Eurocentrism and the Emergence of Islamism*. London: Zed.

Sidaway, J.D. (1998) 'What is in a Gulf? From the "arc of crisis" to the Gulf war', in S. Dalby and G. Ó Tuathail (eds), *Rethinking Geopolitics*. London and New York: Routledge, pp. 224–39.

Sidaway, J.D. (2000) 'Postcolonial geographies: an exploratory essay', *Progress in Human Geography*, 24(4): 573–94.

Sidaway, J.D. (2003) 'Sovereign excesses? Portraying postcolonial sovereigntyscapes', *Political Geography*, 22(1): 157–78.

Sidaway, J.D. and Johnston (2004) 'The trans-Atlantic connection: 'Anglo-American Geography', reconsidered, *Geojournal*, 59: 585–601.

Sidaway, J.D. (2005) 'The nature of the beast. Re-charting political geographies of the European Union', *Geografiska Annaler B*, 88B(1): 1–14.

Sidaway, J.D. and Power, M. (2005) ' "The tears of Portugal": empire, identity, "race" and destiny in Portuguese geopolitical narratives', *Environment and Planning D: Society and Space*, 23(4): 527–54.

Sidaway, J.D., Bunnell, T., Grundy-Warr, C., Mohammad, R., Park, B. and Saito, A. (2004) 'Translating political geographies', *Political Geography*, 23(8): 1037–49.

Slater, D. (1989) 'Peripheral capitalism and the regional problematic', in R. Peet and N.J. Thrift (eds), *New Models in Geography*, vol. 2. London: Unwin Hyman, pp. 267–94.

Smith, G.E. (1999) 'The masks of Proteus: Russia, geopolitical shift and the new Eurasianism', *Transactions of the Institute of British Geographers*, 24: 481–94.

Smith, N. (1987) 'Essay review: Rehabilitating a renegade? The geography and politics of Karl August Wittfogel', *Dialectical Anthropology*, 12(1): 127–36.

Smith, N. (2003) *American Empire: Roosevelt's Geographer and the Prelude to Globalization*. Berkeley: University of California Press.

Sparke, M. (2004) 'Political geography: political geographies of globalization (I) – dominance', *Progress in Human Geography*, 28(6): 777–94.

Staeheli, L.A. and Kofman, E. (2004) 'Mapping gender, making politics: toward feminist political geographies', in L.A. Staeheli, E. Kofman and L.J. Peake (eds), *Mapping Women, Making Politics: Feminist Perspectives on Political Geography*. New York and London: Routledge, pp. 1–13.

Steinberg, P. (2001) *The Social Construction of the Ocean*. Cambridge: Cambridge University Press.

Taylor, P.J. (2000) 'World cities and territorial states under conditions of contemporary globalization', *Political Geography*, 19: 5–32.

Taylor, P.J. (2003) 'Radical political geographies', in J.A. Agnew; K. Mitchell and G. Toal (eds), *A Companion to Political Geography*. Malden, MA and Oxford: Blackwell, pp. 47–58.

Taylor, P.J. (2005) 'New political geographies: global civil society and global governance through world city networks', *Political Geography*, 24(6): 703–30.

Taylor, P.J. and van der Wusten, H. (2004) 'Political geography: spaces between war and peace', in G. Benko and U. Strohmayer (eds), *Human Geography: A History for the 21st Century*. London: Arnold, pp. 83–104.

Tickner, A. (2003) 'Seeing IR differently: notes from the Third World', *Millennium: Journal of International Studies*, 32(2): 295–324.

Turner, S. (2002) 'Sovereignty, or the art of being native', *Cultural Critique*, 51: 74–100.

Tyner, J. (2004) 'Territoriality, social justice and gendered revolutions in the speeches of Malcolm X', *Transactions of the Institute of British Geographers*, NS, 29: 330–43.

Van Schendel, W. (2002) 'Geographies of knowing, geographies of ignorance: jumping scale in Southeast Asia', *Environment and Planning D: Society and Space*, 20: 647–68.

Vigo, J. (2004) 'Hybrid iconographies of modernity and tradition: moroccan official portraiture of King Hassan II', *Postcolonial Text* [online], 1(1), 30 July 2004, available at http://www.pkp.ubc.ca/pocol/rst/viewarticle.php?id=17.

Whittlesey, D. (1939) *The Earth and the State: A Study of Political Geography*. New York: Henry Holt & Co.

Winichakul, T. (1996) 'Maps and the formation of the geo-body of Siam', in S. Tønnesson and H. Antølov (eds), *Asian Forms of the Nation*. Richmond: Curzon Press, pp. 67–91.

Winichakul, T. (2003) 'Writing at the interstices: Southeast Asian historians and postnational histories in Southeast Asia', in A.T. Ahmad and T.L. Ee (eds), *New Terrains in Southeast Asian History*. Athens, OH and Singapore: Singapore University Press, pp. 3–29.

Yiftachel, O. (1998) 'Democracy or ethnocracy: territory and settler politics in Israel/Palestine, *Middle East Report*', summer 1998 (http://www.merip.org/mer/mer207/yift.htm, accessed 24 March 2005).

Yiftachel, O. and Ghanem, A. (2004) 'Understanding "ethnocratic" regimes: the politics of seizing contested territories', *Political Geography*, 23(6): 647–76.

Geographies of Space and Power

Joe Painter

INTRODUCTION

Politics is about power. By extension, political geography is about the relationships between power and space and place. Yet until recently, surprisingly few political geographers paid much attention to what power means, or how it should be understood. Indeed, some even questioned whether political geographers should be concerned with the nature of power at all. In a commentary for political scientists written before the Second World War, Richard Hartshorne defined the proper scope of political geography. 'Is the political geographer', he wondered, 'concerned with the ability of a certain political territory to provide the requirements for political power, in peace or war?' He concluded that 'the geographer's emphasis is always on the area itself, with power, actual or potential, as one of its characteristics, and *not on the question of power itself*' (Hartshorne, 1935: 952, emphasis added). This strange attempt to exclude from the scope of political geography one of the core concepts of political analysis contributed to the isolation of political geography not only from the study of power, but also from the study of politics more generally.

Until recently, if political geographers have discussed power, they have for the most part treated it as transparent or self-evident – something that might be explained but that did not need to be defined or theorized. Peter Taylor (2003) has pointed out that for much of the twentieth century, mainstream political geography was weakly developed, because of a lack of attention to questions of power. In 1964, for example, the editor of a collection of readings surveying the field of political geography had to turn for his chapter on 'Approaches to the Study of Political Power' to the work of Franz Neumann, a political theorist (and member of the Frankfurt School of critical social theory), instead of including a discussion of the subject by a geographer (Jackson, 1964; Neumann, 1950). Saul Cohen remarked that 'much of what is termed "political geography" lacks political substance. It is, in fact, cultural geography organized according to political units' (1963: v). However, the founders of political geography did not shy away from questions of power, and it is to their writings that we turn first.

ASSERTING POWER: EARLY POLITICAL GEOGRAPHY

Ideas about power were at the heart of the early development of political geography by Friedrich Ratzel (1844–1904), Halford Mackinder (1861–1947), and Isaiah Bowman (1878–1950), often regarded as the 'founding fathers' of German, British, and American political geography, respectively, (Taylor, 1993: 50). In different ways, all three understood political power to be closely related to the occupation and control of territory and geographic space. Ratzel's *Politische Geographie* (Ratzel, 1897) was effectively an extended treatise on what he understood to be the intimate ties between people, earth and state. According to Ratzel the state should be seen as an organism whose potential life and growth is closely related to the qualities of the earth it occupies. Ratzel's ideas formed the basis for the development of German geopolitics (*Geopolitik*) under the direction of Karl Haushofer between the First and Second World Wars. Although Haushofer's political links to the Hitler government during the Third Reich are the subject of debate (Bassin, 1987; Heske, 1987), he put forward a highly nationalistic vision of the relations between geography and international politics that helped to legitimize the Nazi regime.

Whatever their formal political affiliations, the promoters of the nationalistic perspectives associated with *Geopolitik* adopted an uncompromisingly 'realist' approach to the study of international relations. As a theory of international relations, *realism* holds that individual states represent the highest level of political authority and that therefore international politics is made up only of interactions between sovereign states that are the principal agents in international relations. States are assumed to act rationally to further their own interests (and particularly their security) in a world in which other states' true intentions can never be known, but may be hostile. In the realist view of the world, the ability of a state to promote its own interests depends on its power relative to other states, which is related in turn to the resources under its control.

Realism also characterized Halford Mackinder's political geography. Mackinder (1904) argued that control over territory was a crucial determinant of political power. With the growth of more efficient land-based transport, he suggested, continental states (and particularly those occupying the Eurasian heartland of human history) would catch up with and surpass maritime powers, such as Britain. Some have taken his arguments as predicting the rise in power of the Soviet Union, others have suggested that his emphasis on the strategic importance of rail travel has turned out to be misplaced. What is important for the present argument is that, like Ratzel, Mackinder stressed the close relationship between the occupation of territory and political strength. He was also concerned with the relationship between power and the spatial organization of the world – its division into what he called the 'pivot area' of continental Eurasia, the 'inner crescent' around its fringes and the 'outer crescent' consisting of the Americas and the lands of the southern hemisphere. This way of thinking about the global spatial arrangement of political power, with its immanent possibility of confrontation between 'sea-powers' and 'land-powers', was later taken up and adapted by Nicholas Spykman in his discussion of 'heartland and rimland' (Spykman, 1944).

In a later work, written just after the end of the First World War, Mackinder predicted (correctly as it turned out) the re-emergence of international tensions that were to lead in due course to the Second (Mackinder, 1942: 1). He went on:

> The great wars of history – we have had a world war about every hundred years for the last four centuries – are the outcome, direct or indirect, of the unequal growth of nations, and that unequal

growth is not wholly due to the greater genius and energy of some nations as compared with others; in large measure it is the result of the uneven distribution of fertility and strategical opportunity upon the face of the globe. (Mackinder, 1942: 2)

Mackinder called his book *Democratic Ideals and Reality* and it was intended as a warning to those idealists who sought to prevent further wars through the establishment of a democratic international order embodied in a League of Nations. For Mackinder, idealism had to be tempered with realism since geographical realities ('uneven distribution') would inevitably threaten democratic ideals unless deliberate steps were taken to counter them. Indeed, many 'idealist' political leaders were strongly committed in practice to the promotion of their particular national interests, by force if necessary. As Noam Chomsky points out, for example, the idealism of US President Woodrow Wilson, a champion of the League of Nations after the end of the First World War, went hand-in-glove with a hard-nosed and frequently undemocratic and violent promotion of American interests overseas (Chomsky, 2003: 46–8).

Mackinder's arguments anticipated the principal division in twentieth-century international relations theory between realism and idealism. Idealism acknowledges the possibility of an international order based on universal principles and the rule of law, whereas the realist view of the world is characterized by inter-state anarchy. Idealism provided a guiding framework for the work of Isaiah Bowman, the third of political geography's 'founding fathers'. Bowman was a strong supporter of the League of Nations, formed after the First World War, and espoused international cooperation. He argued that armed conflict is neither inevitable nor a desirable mechanism for settling disputes, given that the application of science to warfare had 'made it possible to conduct war to the point of self-destruction' (Bowman, 1928: 4–5). Although his idealism sounded high-minded, Bowman was hardly disinterested. Rather, he was an active proponent of American influence overseas. For example, writing about the Philippines he averred that:

> There can be little question that the Philippines should not be left to themselves either to frame a foreign policy independent of American interests and possibly inimical to them, or to become the prey of an unscrupulous power. [... T]he duty of the United States is clear. It was through us that the Islands were freed from Spanish influence; it is to us that the thought of the world turns for a safe and fair solution of the Philippine problem, no matter if we be charged with the crimes of

imperialism and selfish business purpose. (Bowman, 1928: 722–3)

The affinities with early twenty-first-century American foreign policy in Iraq are clear: American power is still exercised (unsurprisingly) in pursuit of American interests, but is dressed up as benevolent paternalism. Bowman's understanding of the relation between space and power, then, saw the development of national territories as inextricably interlinked. The need to project American power over great distances conjured a global vision of world politics, but one in which national exceptionalism retained its force (see also Smith, 2003).

Power and its relationship with territory was central to the early development of political geography as a distinctive sub-discipline, focused as it was on questions of rivalry between states. Peter Taylor has called this the 'power-political heritage of geopolitics' (Taylor, 1993: 52–64). However, while control of territory was taken to be an important *source* of power, the *nature* of power remained unexamined and largely taken for granted. Power was typically understood by these early political geographers to mean the ability of one state to influence, or to impose its will upon, another, but even this simple definition was rarely made explicit.

LEGITIMATING POWER: POLITICAL GEOGRAPHY IN THE MID-TWENTIETH CENTURY

In the 1940s, many Anglophone political geographers were shocked by what they saw as the alliance between German *Geopolitik* and National Socialism and the corruption of geographical science for ultra-nationalist political purposes. So political geography retreated from an explicit engagement with questions of state power vis-à-vis other states, largely abandoning the field to the neighbouring discipline of international relations. Partly as a result, political geography fell into the doldrums in the 1950s and 1960s (Taylor, 1994: 448–9), producing limited amounts of research of even more limited interest. The state remained the primary focus of political-geographic study, but the main concern became the functional integration of the territories of individual states (Hartshorne, 1950). Hartshorne argued that 'the fundamental purpose of any state […] is to bring all the varied territorial parts, the diverse regions of the state-area, into a single organized unit' (1950: 104) and the objective of political geography was to analyse the barriers to this integration and how they may be overcome. Hartshorne eschewed any explicit discussion of the nature of political power or its relationship with space and territory. Nevertheless, he implies that

state power is not only at the heart of his analysis, but is effectively endorsed as legitimate and necessary. State power is the mechanism through which territory is brought under unified political control:

> In all cases, [the state] attempts to *establish complete and exclusive control over* internal political relations – in simplest terms, the creation and maintenance of law and order. Local political institutions *must conform* with the concepts and institutions of the central, overall, political organization. [… M]ost importantly, because we live in a world in which the continued existence of every state-unit is subject to the threat of destruction by other states, every state *must strive* to secure the supreme loyalty of the people in all its regions, in competition with any local or provincial loyalties, and in definite opposition to any outside state-unit. (Hartshorne, 1950: 104–5, emphasis added)

This kind of legitimation of potentially quite unjust and anti-democratic actions on the part of state authorities justifies Peter Taylor's suggestion that mainstream political geography was a deeply conservative discourse (Taylor, 2003: 48). Be that as it may, Hartshorne's account implied that power is related to the state's ability to extend its authority across all of its territory, and this is one of the main ways in which the relationship between power and space has been understood in social theory (Allen, 2003).

One exception to the general neglect of the issue of the *nature* of power in the writings of mid-twentieth-century political geographers is found in the work of Stephen B. Jones, Professor of Geography at Yale University. Writing in 1954, Jones was unusual in setting out an explicit, if brief, *definition* of power:

> Power is here defined as 'that which permits participation in the making of decisions' […]. This is perhaps not truly a definition; it tells not what power is, but what makes it possible. It has the virtues of including constructive uses and of saying that power is not solely material or possessed only by those who have a lot of it. Power, like radiant energy, can move in many directions at once. (Jones, 1954: 422)

Jones links together two different understandings of power: power as a stock of resources and power as an aspect of political practice (or, as Jones puts it, 'national strategy'):

> An estimate of national power has two aspects which are related, in a figurative way, like the two rays of a triangulation. […] One ray or beam is the conventional inventory of the elements of factors

of power. It gives the power resources of a nation, using 'resource' in a broad sense. The other ray is here called 'national strategy'. (Jones, 1954: 421)

For John Allen (2003), the idea that power can flow is another of the principal ways that social theorists have linked power and space.

The concept of 'national power' can be traced to the work of the leading realist international relations scholar Hans Morgenthau. In his classic work *Politics Among Nations*, Morgenthau argued that a variety of elements contribute to the power of one nation relative to others, including 'geography', natural resources, industrial capacity, military preparedness, population, national character, national morale, the quality of a nation's diplomacy, and quality of government (Morgenthau, 1949: 80–108). Jones adopted Morgenthau's approach, proposing the idea of a 'power inventory' (1954: 424), an enumeration of the resources of a state. The notion that a state's power is strongly influenced by, or is even the same thing as, its stock of resources, has been the most common understanding of power in most political-geographic writing, reflecting both the realist underpinnings of mainstream political geography as well as its lack of theoretical sophistication.

In *Geography and Politics in a World Divided*, Saul Cohen (1963) identifies 'the power analysis approach' as only one of four possible approaches to the study of political geography. As Cohen acknowledges, the term 'power analysis' in this context comes from Hartshorne (1954), although Hartshorne did not elaborate its implications in any detail. Like Jones, power analysis for Cohen entails the construction of a comparative 'power inventory' for different states or other political areas. He suggests that the inventory should include:

1 *The physical environment* (landforms, climate, soils, vegetation, waterbodies, etc.);
2 *Movement* (the directional flow of the transportation and communication of goods, men, [*sic*] and ideas);
3 *Raw materials, semi-finished and finished goods* (employed and potential, in both time and space terms);
4 *Population* (in its various characteristics, particularly qualitative and ideological);
5 *The body politic* (its various administrative forms, ideals, and goals in their areal expression, such as county, state, national, and international bloc frameworks.) (Cohen, 1963: 8)

In addition, Cohen suggests that 'space' should be a sixth distinctive category: 'in this sense the location, shape, and boundaries of political entities are analyzed, as well as the impact of space upon the internal character and external relations of such political entities' (1963: 9). Although Cohen is careful to note the limitations of the approach, he suggests that data from these various categories can be tabulated and converted into an index that would provide insights into power differences between states or other political entities. A similar method is advocated in Harm de Blij's US college text *Systematic Political Geography* (de Blij, 1973: 55–79), and Richard Muir's *Modern Political Geography* (Muir, 1981).

To those familiar with recent geographical writings informed by social and political theory and continental philosophy, these inventory-based approaches are painfully unsophisticated. However, their basic assumptions were not very different from those underlying many more recent discussions of space and power, even if those assumptions were not made explicit. At least five of these assumptions are worth highlighting. First, power can be quantified so that the different levels of power of different states can be enumerated in a table and compared. Second, power can be possessed and held in reserve awaiting deployment. Thus 'potential raw materials', such as mineral reserves, add to the power of the state, even though they may never be exploited. Third, power is equated with resources (physical, material, human and organizational). Fourth, different resources can be related to each other as equivalents (by converting them to a common index scale), suggesting that there is a general 'essence' of power. Fifth, space affects power in various ways, so that the spatial characteristics of a state can increase or reduce its political power.

The concern with quantification evident in Cohen's version of political geography was shared by wider changes in geography in the 1950s and 1960s. The reframing of geography as 'spatial science' or 'spatial analysis' was not uncontested, but, for a period at least, quantitative methods, formal spatial models and a broadly positivist epistemology became dominant in Anglo-American geography. For political geography these developments had two main consequences. First, the locational models associated with spatial science were underpinned by the theories of neo-classical economics. Grounded in methodological individualism and the assumption that market competition leads to equilibrium conditions and Pareto optimality, the neo-classical approach leaves little room for considerations of political power. Actions of governments and other political actors are reduced to 'distortions' or, in more sophisticated accounts, efforts to correct for market failure. Moreover, one of the paradoxes of spatial analysis was its attenuated understanding of space. As Kevin Cox has argued, notions of territory and place, which are intimately linked to power, had no place in the conceptual framework of spatial analysis (Cox, 1995).

Second, the new orthodoxy's emphasis on quantitative methods produced a narrow vision of political geography which limited it to the study of quantifiable political phenomena such as election results. While the methodology of electoral geography has undoubtedly gained much from the development of statistical techniques, the difficulty of applying quantitative approaches to other kinds of political processes resulted in the sidelining of political geography throughout the 1960s.

From its nineteenth-century beginnings, then, political geography was concerned with power, but for much of the twentieth century little attention was paid to the nature of that power. Power appeared as a largely unexamined concept in most political-geographic writing. Insofar as it was theorized at all, it was typically equated with state power understood loosely as national strength. During the heyday of spatial analysis, questions of power were neglected or taken for granted. Since the 1970s, though, there has been a transformation in the theories and practices of political geography – and questions of power and its relationship with space have been at their heart.

QUESTIONING POWER: CRITICAL POLITICAL GEOGRAPHY

As the dominance of spatial-analytic and quantitative geography was challenged by a variety of anti-positivist alternatives, the 1980s and 1990s saw a major revival of political geography. Research expanded, new textbooks appeared, new courses were developed, and a specialist journal, *Political Geography Quarterly* (now called simply *Political Geography*), was launched. Political geographers started to engage with social and political theory and developed a diverse and still growing range of perspectives and approaches to the subject. These developments paralleled similar changes in other areas of human geography, but the parlous state of political geography in the preceding decades meant that its resurgence was more striking. A central feature of this resurgence has been the adoption of a variety of critical approaches drawn from Marxism and post-Marxism, world-systems theory, feminism, post-structuralism, post-colonialism, and political ecology. Each of these has brought into political geography a particular view of power and, in some cases, a distinctive approach to thinking about the relationship between power and space. Yet when it comes to grasping the underlying nature of power there are some striking similarities with earlier ideas: that power can be possessed, for example, or that it derives from resources, or that it can flow across space.

The world-systems approach

One of the most influential works of this period was Peter Taylor's *Political Geography: World-Economy, Nation-State and Locality*, first published in 1985 and now in its fifth edition (Taylor, 1985; Flint and Taylor, 2006). Taylor's book was one of the first by a political geographer to consider the nature of power in any detail, with a substantial section of the opening chapter devoted to a discussion of 'power and politics in the world-economy' (Taylor, 1993: 23–48). Taylor argues that politics is often 'equated with activities surrounding states', but that these 'do not constitute the sum of political activities' (1993: 24). He continues: 'If we equate politics with [the] use of power then we soon appreciate that political processes do not begin and end with states: all social institutions have their politics' (Taylor, 1993: 24).

Taylor then proposes a simple model of power involving two protagonists ('A and B') in a conflict, whose power ranking can be inferred from the conflict's outcome. (A similar model is used by Ronan Paddison (1983: 3).) 'When we inquire why A was able to defeat B', suggests Taylor, 'we would expect to find that in some sense A possessed more resources than B' (1993: 25). Here power is seen as a product of the possession of resources. However, because politics cannot usually be reduced to a series of simple one-on-one conflicts, Taylor then proposes a development of the model in which the protagonists seek to widen the scope of the dispute in order to change the balance of power. He suggests that this might involve an appeal to an external authority. Although Taylor does not discuss the distinction, as we shall see later this mention of authority invokes a somewhat different meaning of power than the notions of coercion and domination underlying his initial A versus B model.

Taylor then goes on to explain that there is evidence to suggest that the hierarchy of power is not an adequate predictor of the outcome of political conflicts and that 'the nature of power is much subtler than we have so far supposed' (1993: 33). Part of the problem is that '"power" is one of those concepts which cannot be directly measured' (1993: 33) and attempts to find proxy measures along the lines of the power inventory or index are unsatisfactory. What 'we need', says Taylor, is 'a completely fresh approach to studying state power' (1993: 34). He goes on:

This can be achieved through the world-systems approach by employing the concepts of overt and covert power relations. Whereas the former is what we normally understand by power as reflected in conflict, we shall argue that covert power, the ability to forward particular interests without resort to coercion or threat, is much more pervasive and important. (Taylor, 1993: 34)

Overt power may be 'actual' or 'latent', says Taylor. Actual overt power involves the use of force or coercion to decide the outcome of a conflict, the most obvious example being warfare. Latent overt power involves the threat of force. Taylor illustrates the point with reference to research identifying '215 incidents between 1945 and 1976, when armed forces were used politically to further US interests but without appreciable violence' (1993: 35).

There are also two types of covert power, 'non-decision-making' and 'structural position' (Taylor, 1993: 36–8). Non-decision-making refers to the idea that some issues and conflicts never become the object of a formal political decision, so that the status quo is preserved despite the existence of interests that might challenge it. Some political actors are able to exert more influence than others over the political agenda and thereby ensure that conflicts that might challenge them are avoided entirely. Although non-decision-making can be important in particular circumstances, Taylor argues that 'the most important form of power relation is structural' (1993: 37). In this case, certain political actors have more power because of their structural position in the world-system. Because the world-economy is structured unfairly in ways that result in the economic exploitation of peripheral and semi-peripheral countries by core countries, countries in the core do not have to engage in the overt use of force, or even the manipulative agenda-setting involved in 'non-decision-making', in order to see their interests prevail. According to world-systems theorists like Taylor, the system is already structured in such a way as to favour the interests of some countries over others. Core countries are therefore said to enjoy power over others as a result of their structural position.

For Taylor, the three-tier spatial structure of the world-economy described by world-systems theory helps to explain why some states adopt liberal political regimes and others are more authoritarian. States in the core benefit substantially from covert power. 'Generally speaking', says Taylor, 'there is a tendency for core states to be relatively liberal in their characteristics since their power is based primarily on their economic prowess' (1993: 38). In other words, core states can afford to allow their citizens civil and political rights. Semi-peripheral states, by contrast, tend to be authoritarian 'strong states' because, while they are economically dynamic, they have to use state power in attempts to 'restructure the [world-]system in their favour', although this strategy is usually unsuccessful (1993: 38). Finally, peripheral states have been

subject to colonialism throughout most of their history and remain highly economically dependent. Despite substantial shows of state power (and especially its military trappings) and frequent resort to internal repression, these states are essentially weak (1993: 39).

Taylor's discussion of power represented a major advance in the literature of political geography. It moved the debate beyond simplistic 'power inventory' approaches and emphasized that there are different forms of power (overt and covert, military and economic, and so on). It also laid stress on the relationship between spatial organization and the operation of different forms of power (1993: 40–8). Although Taylor acknowledges the importance of differences in resource endowment among conflicting actors, he does not reduce power to resources, but recognizes other sources of power such as 'structural position' and 'non-decision-making'. He also cites examples of conflicts whose outcomes seem to run counter to what might have been expected from initial assessments of power differentials, such as the defeat of the United States in Vietnam (1993: 39–40). Although Taylor does not express it in these terms, such counter-intuitive examples suggest that power cannot be understood as something that actors possess; rather it is an effect of their practices (Allen, 2003: 2). Moreover state governments and armed forces are not the only actors involved in determining the outcomes of international conflicts. As Taylor points out, 'internal' political processes, such as popular dissent, are also influential (1993: 39). We also need to recognize that states are not the only *international* actors: multinational corporations, intergovernmental bodies (such as the World Trade Organization, the IMF, the World Bank and the United Nations), NGOs, transnational political movements, world religions, and transnational criminal networks are all important. Recognizing that such diverse entities exercise power internationally may not accord with the realist principles that have historically underpinned most political geography, but it is surely essential if we are to make any sense of the contemporary world (Strange, 1996; Kaldor, 1999).

Marxist political economy and the state

Another major current in critical political geography has been derived from historical materialism and Marxist political economy. In the 1970s and 1980s several geographers, including Ron Johnston, Michael Dear, Gordon Clark, and John Short, drew on Marxist concepts in their writings on the geography of state power (Dear and Clark, 1978; Clark and Dear, 1981, 1984; Johnston, 1982, 1989; Short, 1982). Short, for example, emphasized the role of uneven capitalist development in driving processes of inter-state rivalry. His account of the functions of the state under capitalism stressed the contributions of Marxist state theorists such as Ralph Miliband (1973) and Nicos Poulantzas (1973, 1978) as well as those of Jürgen Habermas (1976) and James O'Connor (1973), who are not Marxists in a strict sense, but whose work forms part of a broader historical materialist tradition. Johnston argued that the weakness of much previous political geography stemmed from its failure to develop a 'sensible treatment of the central element in its work – the state' (Johnston, 1982: ix). He used similar sources to those cited by Short to explain the existence of the state and many of its basic activities, and also drew on Clark and Dear (1981). Clark and Dear elaborated their 1981 arguments in their book *State Apparatus: Structures and Languages of Legitimacy* (1984). Here they argued that the state must be understood in relation to the capitalist mode of production, but its workings cannot be reduced to an immanent 'logic of capital'. Rather the state is institutionally separate from the economic sphere and able to operate with considerable autonomy (1984: 33–5). Clark and Dear argue that this provides for a 'state-centred' rather than a 'society-centred' theory of the state.

None of these authors develops an explicit theory of power in any detail, although Clark and Dear do consider the specific question of the legal powers of the local state (1984: 138–44). Questions of power are, though, central to Marxism, which sees power inequalities as endemic in the capitalist mode of production. For one strand of Marxist political analysis, the key question is how processes of capital accumulation are related to the forms and functions of the state. This question has no simple answer, as Clark and Dear explain, but broadly speaking there are two kinds of approaches to it. First, there are those that suggest that political power derives directly from the social relations of capital accumulation, and that the state exercises power in a fairly straightforward way to favour capital accumulation and the class interests that benefit most from it. The second set of approaches emphasizes that the state is both a medium for and an outcome of social struggles, including struggles between classes, and that its institutional capacities and actions have complex dynamics of their own. This approach sees capitalism as simultaneously a political and an economic system – a complex, differentiated totality in which political power cannot be reduced to economic power or vice versa.

A rather different approach is taken by David Harvey, unarguably the leading geographical interpreter of Marx. For Harvey, the workings of the capitalist mode of production not only lead to uneven economic development, but also produce tendencies towards a more general territorial coherence at the scale of the urban region (Harvey, 1985a). Capitalist production (in fact any production) operates under spatial constraints. Because production requires the bringing together of labour, materials and technology in particular places, there is a limit, at least in the short term, to the spatial flexibility of capitalism. Therefore, Harvey says, there are processes at work that define regional spaces within which production and consumption, supply and demand (for commodities and labour power), production and realization, class struggle and accumulation, culture and life-style, hang together as some kind of structured coherence within a totality of productive forces and social relations (1985a: 146).

Moreover, the resulting 'territorial coherence becomes even more marked when formally represented by the state' (1985a: 146). The tendency to structured coherence opens up a space within which, for a time at least, a 'relatively autonomous' urban politics can arise (Harvey, 1985b: 125–64). This relative autonomy (from the logic of capital accumulation) allows the formation of 'regional class alliances' that bring together otherwise opposed political interests. Thus for Harvey, the power of the local state is not immediately determined by the logic of capital accumulation, but arises from the necessarily territorial character of the conditions that make accumulation possible.

Within political geography, Harvey's arguments have been taken up by a number of writers on the local state (e.g. Duncan et al., 1988; Painter, 1997). In recent years, however, the geographical literature on urban and regional governance has tended to draw more heavily on the 'strategic-relational' state theory developed in the work of the political theorist Bob Jessop. Jessop offers the following summary of the approach:

[The state] can be defined as an ensemble of socially embedded, socially regularized, and strategically selective institutions, organizations, social forces, and activities organized around (or at least actively involved in) making collectively binding decisions for an imagined political community. A relational approach implies that the exercise of state power (or, better, state powers in the plural) involves a form-determined condensation of the changing balance of forces. In other words, state power reflects the prevailing balance of forces as this is institutionally mediated through the state apparatus with its structurally-inscribed strategic selectivity. This refers to the ways in which the state qua social ensemble has a specific, differential impact on the ability of various political forces to pursue particular interests and strategies in specific spatio-temporal contexts through their access to and/or control over given state capacities – capacities that always depend for their effectiveness on links to forces and powers that exist and operate beyond the state's formal boundaries. (Jessop, 2003: 145)

Thus the strategic-relational approach rejects the idea that power is 'possessed' or 'exercised' by the state:

It follows that to talk of state managers, let alone of the state itself, exercising power is at best to perpetrate a convenient fiction that masks a far more complex set of social relations that extend far beyond the state apparatus and its distinctive capacities. Thus we should always seek to trace the circulation of power through wider and more complex sets of social relations both within and beyond the state. (Jessop, 2003: 45)

Jessop's approach has been influential in political geography partly because of its emphasis on the importance of spatial (and especially scalar) relations and their transformation over time. For example, he suggests that 'the national scale has lost the taken-for-granted primacy it held in post-war Atlantic Fordist regimes' (Jessop, 2002: 179) but that no other scale has emerged as the primary focus of organization. Instead there are now 'continuing struggles over which spatial scale should become primary and how scales should be articulated' (Jessop, 2002: 179). This 'rescaling of the state' is reflected in the rise of international trading blocs, cross-border regions, and cities (Jessop, 2002: 177–87). These arguments have been widely applied in political geography and urban and regional studies and have also been subject to critical debate (e.g. Jones, 1997, 2001; Jones and MacLeod, 1999; MacLeod, 1999; MacLeod and Goodwin, 1999; MacLeod and Jones, 2001; Brenner and Theodore, 2002; Cox, 2002a; Peck, 2002; Brenner et al., 2003; Brenner, 2004). For the most part, though, these applications and debates have focused on the more complex geographies of power arising from scalar restructuring and have not explicitly addressed the nature of power or how power itself should be theorized (Allen, 2003: 34–6).

Jessop's work and associated research in geography have been strongly influenced by regulation theory, which is a particular variant of Marxist political economy. The validity of regulation theory is disputed by Marxist scholars and political geographers. Kevin Cox grounds his approach to political geography in an analysis of the political economy of capitalism (Cox, 2002b) but is explicit in his rejection of regulationist approaches (Cox, 2002a). Although the exercise of power is central to Cox's concerns, he offers only a brief discussion of power as such, but does emphasize that 'power comes in different forms' (Cox, 2002b: 7).

He briefly identifies the power of money and the power of norms as important influences before turning to the relations between power and territory. Cox argues that 'territorial strategies are always exercises in power' and suggests that territoriality is quintessentially associated with state power. Even territorial strategies undertaken by private actors (such as private schools or gated communities) in the end depend on the state for their legality (Cox, 2002b: 8). The state's territorial sovereignty means that it can adopt a number of strategies for the regulation of space – for Cox, space and power come together in the shape of territory (Cox, 2002b: 367). We will return to the issue of territory later, after considering two further strands in critical political geography: feminism and critical geopolitics.

Feminist political geography

Most of our attention so far has been focused on power related to the state – either power relations between states or the power exercised by states over their territories. This is in accordance with the state-centrism that has characterized political geography for much of its history. World-systems theory has provided one kind of challenge to state-centrism, with its emphasis on the principal role of the capitalist world-economy. A rather different challenge arises from the growing influence of feminist approaches to political geography. Feminist writers have argued strongly that our definition of politics (and hence of political geography) should not be limited to formal political institutions such as the state, but should extend to include all sorts of social relations in which power is exercised. Feminism is centrally concerned with the nature and effects of gender relations and their potential transformation. One implication of a focus on gendered power relations is that interpersonal interactions come to be seen as just as much part of 'politics' as the actions of governments and conflicts between states. The feminist slogan 'the personal is political' captures this argument concisely. At the same time, feminism does not address only interpersonal power relations, but is concerned with the gendering of all kinds of social processes and institutions operating at a wide variety of geographical scales. Among many other topics, feminist geographers have examined the politics of nationalism, militarism, electoral representation, local governance, privatization, welfare reform, social movements, indigenous rights, citizenship and migration (Staeheli et al., 2004). Much of this work has contributed to a productive blurring of sub-disciplinary boundaries, so that political geographers are now taking seriously topics that were previously seen as the domain of urban, social, cultural or economic geography, despite their evidently political character.

Notwithstanding these developments, and the fact that more than two decades have passed since feminism began to influence geography, an explicitly feminist political geography is by no means firmly established, as Jennifer Hyndman notes: 'Geographers who find themselves at the crossroads of feminist and political geography have lamented the paucity of scholarship that links the two [and] despite on-going work to advance a thoroughly feminist geography, the intersections between these two sub-disciplines are relative[ly] few' (Hyndman, 2004: 307–8).

Despite these limitations, it is already clear that feminist work in political geography has considerable implications for our understanding of the relations between power and space. For example, two of the major debates in political geography from the 1990s have been about the politics of scale and the geographies of citizenship, and feminist theory has played a major role in each. In the case of scale, feminism draws attention to a range of other spatial scales, besides the conventional global/national/local triad. In particular, feminists emphasize the scales of the body and the household as important sites of gendered and generational power relations. Sallie Marston, for example, identifies the household as a neglected arena in debates on scale, arguing that an account of gendered practices at the household scale is essential to understanding the sphere of consumption and social reproduction (Marston, 2000). Second, feminism has had considerable influence on studies of the geographies of citizenship (see, e.g. Staeheli and Cope, 1994; Kofman, 1995; McEwan, 2000; Secor, 2003; Staeheli and Clarke, 2003). Work in this area has drawn on feminist political theory to suggest that dominant definitions of citizenship are strongly gendered. One consequence of this is that in the West the public sphere and the public spaces within which citizenship is enacted have been encoded as rational, masculine and political while the private realm and the private space of the home has been encoded as emotional, feminine and outside politics. Feminist political geographers have contested these binary divisions and assumed equivalences and have sought to disrupt the rigid separation of the public and the private that they assume.

While feminist geographers in no sense deny the importance of the state or of inter-state relations in shaping political geographies (Hyndman, 2004; Staeheli et al., 2004), their work has opened up political geography to a wider range of understandings of power and a more plural sense of spatiality than that of state territoriality alone, while also unsettling conventional accounts of the geographies of the state (Desbiens et al., 2004).

Critical geopolitics

The close connections between knowledge, space and power that are the focus of much work in feminist political geography also lie at the heart of the study of 'critical geopolitics'. Critical geopolitics emerged during the 1990s as a way of rethinking the concept of geopolitics to move it beyond traditional realist theories and the polarized East–West political confrontation of the Cold War (Ó Tuathail and Dalby, 1991, 1998b; Dodds and Sidaway, 1994; Ó Tuathail, 1996, 2000; Sparke, 2000). Gearóid Ó Tuathail and Simon Dalby, two of the pioneers of the approach, have identified five arguments associated with critical geopolitics. First, geopolitics does not refer only to realist power-politics, but is 'a broader cultural phenomenon' encompassing the 'cultural mythologies of the state'. Second, critical geopolitics does not take for granted the boundaries of states and the separation between the state's inside and outside. States do not exist as distinct actors prior to the development of relations between them; rather states are 'perpetually constituted by their performances in relation to an outside against which they define themselves'. Third, geopolitics is not limited to practices of statecraft conducted by political leaders, but involves 'a plural ensemble of representational practices that are diffused throughout societies'. These can be divided into 'formal geopolitics', involving theories and strategies developed in think-tanks, strategic institutes and academia, the 'practical geopolitics' conducted by state institutions, and 'popular geopolitics', which refers to everyday cultural constructions in the mass media, films and so on. Fourth, the study of geopolitics 'can never be politically neutral'. Critical geopolitics insists on the 'situated, contextual and embodied nature of all forms of geopolitical reasoning'. Finally, critical geopolitics seeks to understand geopolitical practices in their broader socio-technical context. Geopolitics is understood as part of a wider process of government and control and is therefore concerned with 'the historical expansion of states, techniques of governmentality and histories of technology and territoriality' (Ó Tuathail and Dalby, 1998a: 2–7).

These arguments provide the basis for a radically different conception of power and of power's relationship to space from the one that dominated traditional studies of geopolitics. Three differences stand out in particular. First, power does not inhere (only) in material resources, but is also exercised culturally, and in particular through practices of representation. Critical geopolitics draws attention to the power of discourse and particularly of narrative to shape understandings of political events and of relations between places, and thus to influence those events and relations. Studies have focused on all three categories of geopolitics described by Ó Tuathail and Dalby, covering the formal texts of strategists and academics, the practical discourses of political leaders and diplomats and the every-day narratives of the mass news and entertainment media. Joanne Sharp, for example, has written of the crisis in American culture and national identity generated by the loss of its geopolitical Other, the Soviet Union. She argues that the resulting sense of disarray was evident in the narratives of many pop-ular films in the 1990s. But movies do not simply express a feeling of chaos and disorder, they also respond to it, and they do so in particular, argues Sharp, though a discourse of re-masculinization, thereby contributing to the construction of new forms of national identity (Sharp, 1998).

The second difference is that critical geopolitics adopts a more dispersed and plural view of power than traditional geopolitics. Power is not located exclusively in the state, but is diffused through society. This view owes a lot to the work of Michel Foucault, and many writers on critical geopolitics have indeed been much influenced by Foucault's ideas. Foucault urges us to 'keep it clearly in mind that unless we are looking at it from a great height and from a very great distance, power is not some-thing that is divided between those who have it and hold it exclusively, and those who do not have it and are subject to it' (Foucault, 2004c: 29). He continues:

> Power must, I think, be analyzed as something that circulates, or rather as something that functions only when it is part of a chain. It is never localized here or there, it is never in the hands of some, and it is never appropriated in the way that wealth or a commodity can be appropriated. Power functions. Power is exercised through networks, and individu-als do not simply circulate in those networks; they are in a position to both submit to and exercise this power. They are never the inert or consenting tar-gets of power; they are always its relays. In other words, power passes through individuals. It is not applied to them. (Foucault, 2004c: 29)

This 'capillary' view of power is a radically different conception of what power entails from the 'power inventory' approach of traditional geopolitics. It has been taken up very widely in geography, though it has been far more influen-tial in social, cultural, and historical geography than in political geography. This neglect stems in part from the legacy of earlier 'power poli-tics' approaches and the popularity of Marxism. It may also reflect an assumption that political geography is predominantly concerned with state power and that therefore Foucault's ideas about the dispersal of power through populations are not directly relevant. This assumption seems mistaken

on two counts. First, political geography has in recent years begun to move away from its state-centric view of the world, and second, Foucault's approaches do address questions of statehood, sovereignty and government. How and to what extent Foucault addressed these questions has only become fully apparent with the recent publication of his lecture courses (Foucault, 2004a, 2004b, 2004c), although his concept of 'governmentality', drawing attention to the constitution of objects of governance and to the practices and technologies of government, has been widely applied in political geography since the 1990s (for example, Murdoch and Ward, 1997; MacKinnon, 2000; Moon and Brown, 2000; Painter, 2002; Raco, 2003).

The third difference is that critical geopolitics seeks to disrupt the spatial orderings of conven-tional geopolitics. Conventional geopolitics took the territorial sovereignty of the state for granted, and assumed that it was possible in theory and in practice to draw a clear distinction between domestic and foreign policy. Critical geopolitics contests these assumptions. In his commentary on the bombing of the US Federal government office in Oklahoma in 1995, Matthew Sparke notes that 'the spatial dualisms of "outside" and "inside" are a blindingly sharp force of division in mod-ern Western thought, ultimately governing even moral considerations of good and bad' (Sparke, 1998: 198). Sparke points out that the bombing, which was perpetrated by an American citizen and avowed patriot, Timothy McVeigh, overturned the easy associations of 'inside equals good' and 'out-side equals bad'. Moreover McVeigh was a former soldier who had fought in Iraq against the bad outside. This kind of interpenetration of spaces formerly thought to be clearly distinct is a recurrent theme in critical geopolitical writing. It has also become a leitmotiv of geographical theory more generally and central to recent attempts to spatial-ize theories of power. It is to these spatializations of power that we now turn.

SPATIALIZING POWER

Until now, the dominant spatial concept in politi-cal geography to date has been 'territory' (Cox, 2002b). This accords with dominant understand-ings of the spatiality of the nation-state, which has been the prime focus of political geography. In recent years it has become clear to many politi-cal geographers that this fixation with the state and its territoriality has been profoundly limiting, and critiques of state-centrism have begun to emerge (Taylor, 1996). This is not to deny the importance of either the state or territory, but it is to insist that neither can be assumed to be natural, universal or pre-given, and that both must be placed in question and seen not as the timeless foundations of politics but rather as two of politics' outcomes.

John Agnew (1994, 1999) has coined the term 'territorial trap' to challenge the 'state-centred account of the spatiality of power' (Agnew, 2003: 51). The territorial trap is underpinned by three assumptions. These are that 'modern state sovereignty requires clearly bounded territorial spaces', that 'there is a fundamental opposition between domestic and foreign affairs in the modern world', and that 'the territorial state acts as the geographical container of modern society' (Agnew, 2003: 53). Agnew argues that this state-centric view of the world is sustained by particular understandings of power. There is a tendency to understand state power in terms of coercion ('power over') exercised within discrete blocks of space. This view, Agnew suggests, both neglects infrastructural power (Mann, 1984), through which states make collective provision for their populations, but also 'completely ignores the extent to which power is inherent in all human agency' (Agnew, 2003: 55) and thus that power can also be 'power to' and 'power from below'. The territorial trap also takes a limited view of the exercise of power between states, with coercion being the dominant understanding of power in the international as well as the domestic arena. Drawing on the Gramscian concept of hegemony, Agnew argues that this view is mistaken because 'power can involve gaining assent, defining expectations, and co-opting others as much as or more than simply coercing them' (2003: 57). Finally, Agnew argues that the territorial understanding of power is challenged by the state's own powers in relation to property rights. State-centred accounts, he says, are 'silent as to the role that states have played in the growth of certain basic social practices of capitalism – defining and protecting private property rights – that have inexorably led beyond state boundaries in pursuit of wealth' (2003: 59). Agnew is at pains to point out that none of this means that the state is necessarily withering away. However, neither the state nor its territorial spatiality can be simply taken for granted in the context of processes such as the growth of international migration and the liberalization of financial markets, which are generating 'new spatialities of power' (Agnew, 2003: 61–5).

Agnew's work is both important and relatively unusual in geography in drawing attention to the *variety* of forms that power can take and the different spatialities with which they are associated. Surprisingly few political geographers have tried to set out a systematic account of different types of power. Ronan Paddison's *The Fragmented State* (1983), which is subtitled 'the political geography of power', discusses the distribution of power spatially and among hierarchical divisions within the state, but does not dwell on the nature of power itself at any length. Peter Slowe's *Geography and*

Political Power (1990) is organized around five sources of power ('might', 'right', 'nationhood', 'legality' and 'legitimacy'). This seems promising as it explicitly recognizes that power comes in many forms, but Slowe provides little discussion of the conceptual distinctions between these different types.

In fact it was not until John Allen's *Lost Geographies of Power* (2003) that a sustained attempt was made to examine both the diversity of modes of power and their relationship to space and place. In *Lost Geographies*, Allen provides a multi-dimensional account of power that recognizes that different kinds of power have different geographies. He suggests that despite widespread acceptance of the close connection between space and power, existing scholarship has 'lost the sense in which geography makes a difference to the exercise of power' (Allen, 2003: 1). Allen's account includes critiques of most of the assumptions about power that underlie virtually all the scholarship I have discussed in this chapter. Whether it is the view that power is a matter of resources that we can hold, the Foucauldian concept of power as dispersed and all around us, or the suggestion that power flows through networks, Allen argues that all the widely used understandings of power fail to capture fully how space and place make a difference to power's role in our lives.

Allen starts by insisting that power is not a thing or a substance, it does not travel or flow and it cannot be stored or saved up or possessed. Instead, power 'is a relational effect of social interaction' (2003: 2) and arises only as it is practised or exercised: it does not exist prior to and separate from its use. Noting that many recent studies have emphasized the importance of space to the exercise of power, Allen sets out three 'spatial vocabularies of power' (2003: 6–9, 13–91). These vocabularies represent the three main ways in which the relationship between power and space has been understood in contemporary social theory – and Allen's analysis challenges each of them. First, power may be said to radiate out from a centre. Such a view of power treats it as a something that can be held or possessed and is located in a central institution, such as the apparatus of the state. From this perspective, power can be 'decentralized' or 'delegated' through the action of the central body; it also spreads out across space. As may be seen from the previous sections in this chapter, this vocabulary of power is the one that has dominated political geography throughout the twentieth century, and is, in fact, the main everyday, common-sense view of power in the West.

Secondly, and, Allen argues, less intuitively, power can be seen as a *medium*, that is, as the means to achieve outcomes. In this view, power is generated through networks of social interaction involving the mobilization of collective resources.

The spatial vocabulary of power involved here is that of flows of power (or more accurately resources) over, or through, social networks that are stretched across space. Such a conception of power is evident, to take a particularly prominent example, in the work of Manuel Castells and his idea of the networked society (Castells, 1996). The difficulty with this conception, argues Allen, is that the spatial distanciation involved in networked social relations seems to make remarkably little difference to the exercise of power. Networks are viewed as unproblematically transmitting power from here to there, without really affecting its nature (2003: 61–3).

The third vocabulary of power understands it as immanent; that is, power is not something external to human subjects that affects them by imposing from outside, but 'is implicated in all that we are and all that we inhabit' (2003: 65). In this case the key question is not 'Who has the power?', but 'How does power work?' The 'how' of power is exactly how Foucault describes his central concern (Foucault, 2004c: 24), and for Allen the writings of Foucault and associated writers such as Gilles Deleuze provide the clearest expression of the idea of power as immanent, in which there is no enduring capacity to power that may or may not be realized, only the routine deployment of techniques – spatial, organizational, classificatory, representational, ethical or otherwise, depending upon the forms of power involved – that seek to mould the conduct of specific groups or individuals and, above all, limit their possible range of actions. (Allen, 2003: 67)

As we saw above, this view of power is increasingly influential within political geography. For Allen, though, the idea that power is immanently everywhere, leads us to lose sight of the spatiality of power. This may seem a surprising claim, given the widespread perception that geography and space are at the very heart of Foucault's project (e.g. Philo, 1992). However, Allen argues that, if power is immanent, it is essentially unmediated, 'it works on and through everyone and every individual, but without spatial reference' (Allen, 2003: 89). It seems that if power is immanently everywhere, it is nowhere in particular.

In order to recover the geographies of power that he claims we have lost through the spread of the these three spatial vocabularies (centred power, networked power and immanent power), Allen emphasizes the diverse range of modalities that power can take, each of which relates rather differently to space and place (or to 'proximity and reach' as he puts it). He discusses at least eight different modalities of power: authority, domination, inducement, coercion, manipulation, seduction, negotiation and persuasion, each of which involves a different kind of social relation. Power as authority, for example, works through relations of recognition. Authority is obeyed insofar as it is recognized as legitimate, and laughter in the face of authority, Allen suggests, is the quickest way to destabilize it. Because of its dependence on recognition, authority is most effective in close proximity, particularly in face-to-face encounters. Anyone who has worked in a large organization will know that the authority of a manager is much more effective when manager and employee are in the same room. When the manager is absent, his or her authority dissipates and employees frequently take advantage of this to subvert managerial power. A different example is that of seduction. Seduction, Allen says, is a 'gentle form of power', which works by operating suggestively on our existing preferences and desires. A particularly clear example of seductive power is the advertising industry. Nothing in an advertisement coerces us into purchasing a product, rather advertisements draw us in by stimulating our curiosity. Unlike authority, seduction works well at a distance. We do not need to be physically in the presence of a copywriter or a salesperson for advertising to have its effect:

> The fact that seduction works on curiosity, seeking to take advantage of attitudes and values that are already present, leaving open the possibility of rejection or indifference, is what gives it its considerable reach, yet at the same time curbs its intensity. *These are not accidental features of the way in which seduction works; they are qualities that distinguish seduction from other modes of power and mark it out as a distinctive way of exercising power.* (Allen, 2003: 103, emphasis added)

The second sentence in this quotation is italicized because it captures Allen's key argument about the geographies of power. Each of the different modalities of power is bound up with a different relationship to space and it is these geographies that help to make the different forms of power what they are. For Allen, then, geography is constitutive of power, but in a different way in each case. Coercion, involving the use of physical force, requires co-presence, or technological systems capable of overcoming distance. Manipulation involves a geography of concealment, because it relies on misrepresentation, and so on.

One of the striking things about Allen's discussion of different kinds of power is that it reveals the limited nature of the conceptions of power underlying most work in political geography, and the lack of explicit attention paid to the variety of power. As we have seen, individual political geographers have typically discussed at most one or two forms of power, insofar as they have discussed the nature of power at all. Most commonly in political geography power has been defined as domination or coercion (these are often considered to be the same thing). Less often power has been considered in terms of negotiation, manipulation or inducement, though again without clear distinctions being made between these. On the other hand, while explicit discussions of power in political geography have not covered the full range of modalities of power identified by Allen, in practice much political-geographic research has shown how these various forms of power operate geographically.

Specific practices can involve more than one modality of power. Allen stresses that there is no rigid separation between different forms, and these can co-exist in place. For example, the ongoing US-led war in Iraq involved power in numerous guises – not just coercion, but also manipulation, inducement, persuasion, domination and so on. One of the great strengths of Allen's approach is that it allows subtle distinctions to be made between the ways in which power is exercised, without ignoring 'stronger' forms of power such as domination and coercion.

POWER, SPACE AND POLITICAL GEOGRAPHY

What are the implications for political geography of thinking power in a fully spatialized way, and giving full weight to its multiple modalities? World events at the beginning of the twenty-first century have undoubtedly prompted renewed interest in the exercise of political power. The conflicts in Palestine, Iraq and Afghanistan, the so-called 'war on terror', flows of migrants and refugees, lack of progress on slowing climate change, national liberation struggles, the emergence of new forms of territorial governance, such as those associated with the growth of the European Union, and the relentless push by economically dominant states – especially the US – to open new markets for investment, production and consumption (called 'free' trade by its proponents, though the freedoms involved are rarely equally distributed); all these developments focus attention on practices of power, and they all, of course, have distinctive geographies. The challenge for political geography, then, is not only to write those geographies, but also to consider how space and place are implicated in the 'hows' and 'whys' of power in ways that do not just reproduce familiar assumptions about power being a resource that is held or transmitted – or the newer assumption that power is immanently everywhere.

One approach that has begun to inform recent work in political geography adopts a topological understanding of space. Topology allows us to understand spatial relations not in terms of fixed distances over a flat surface, but in terms of simultaneous or real-time connections in which the distant is drawn near and the near is made distant (Allen, 2003: 191–3). In a topological view of the world, then, Washington is much closer to Baghdad than it is to Tripoli, even though Tripoli appears nearer to Washington on a conventional map. Technologies of war, surveillance and communication have all functioned to render Baghdad close at hand. Political leaders from the occupying countries pop up unannounced in Baghdad with unnerving regularity. Officials thousands of kilometres away in the Pentagon can direct bombing raids or covert operations as if they were battlefield commanders. And the 'green zone' in Baghdad, housing the political and military administrations of the United States, operates as a kind of Washington-on-Tigris. At the same time, such topological 'bendings' of conventional spaces do not eliminate geography and nor do they guarantee the effective exercise of power in all its modes. Notwithstanding the technological capacity of the US, it has not (so far) been able to impose its will even on the whole of Baghdad, let alone Iraq. In Allen's terms, it lacks authority – and it may be argued it has also failed at persuasion and inducement.

There is an emerging divide within political geography between perspectives that emphasis 'hard' forms of state territorial power and those that focus on power as a more diffuse cultural phenomenon. What I have been able to show in this chapter, I hope, is that, given a nuanced account of power and its relations to space, this divide is more apparent than real. Territory is not a source of power, but one of power's many possible effects. The production of territory may involve all the different modalities of power, which in turn owe their effectiveness in different ways to their spatialities, many of which may be non-territorial or topological in form. From this perspective, topology is not a departure from territory; on the contrary, territory is just another topological twist.

ACKNOWLEDGEMENTS

I am grateful to Bryant Allen, Kevin Cox, Kathie Gibson and Jennifer Litau for their comments on this chapter. Responsibility for its remaining inadequacies rests with me.

REFERENCES

Agnew, J.A. (1994) 'The territorial trap: the geographical assumptions of international relations theory', *Review of International Political Economy*, 1: 53–80.

Agnew, J.A. (1999) 'Mapping political power beyond state boundaries: territory, identity, and movement in world politics', *Millennium: Journal of International Studies*, 28: 499–521.

Agnew, J.A. (2003) *Geopolitics: Revisioning World Politics* (2nd edn). London: Routledge.

Allen, J. (2003) *Lost Geographies of Power*. Oxford: Blackwell.

Bassin, M. (1987) 'Race contra space: the conflict between German *Geopolitik* and National Socialism', *Political Geography Quarterly*, 6: 115–34.

Bowman, I. (1928) *The New World: Problems in Political Geography*. Yonkers-on-Hudson, NY and Chicago: World Book Company.

Brenner, N. (2004) *New State Spaces: Urban Governance and the Rescaling of Statehood*. Oxford: Oxford University Press.

Brenner, N. and Theodore, N. (eds) (2002) *Spaces of Neoliberalism: Urban Restructuring in North America and Western Europe*. Oxford: Blackwell.

Brenner, N., Jessop, B., Jones, M. and MacLeod, G. (2003) 'Introduction', in N. Brenner, B. Jessop, M. Jones, G. MacLeod (eds), *State/Space: A Reader*, Oxford: Blackwell, pp. 1–26.

Castells, M. (1996) *The Information Age: Economy, Society and Culture*, Vol 1: *The Rise of the Network Society*. Oxford: Blackwell.

Chomsky, N. (2003) *Hegemony or Survival: America's Quest for Global Dominance*. New York: Metropolitan Books.

Clark, G.L. and Dear, M.J. (1981) 'The state in capitalism and the capitalist state', in M.J. Dear and A.J. Scott (eds), *Urbanization and Urban Planning in Capitalist Societies*. London: Methuen, pp. 45–62.

Clark, G.L. and Dear, M.J. (1984) *State Apparatus: Structures and Languages of Legitimacy*. Winchester, MA: Allen & Unwin.

Cohen, S.B. (1963) *Geography and Politics in a World Divided*. New York: Random House.

Cox, K.R. (1995) 'Concepts of space, understanding in human geography, and spatial analysis', *Urban Geography*, 16: 304–26.

Cox, K.R. (2002a) '"Globalization", the "regulation approach", and the politics of scale', in A. Herod and M.W. Wright (eds), *Geographies of Power: Placing Scale*. Oxford: Blackwell, pp. 85–114.

Cox, K.R. (2002b) *Political Geography: Territory, State, and Society*. Oxford: Blackwell.

de Blij, H.J. (1973) *Systematic Political Geography* (2nd edn). New York: John Wiley.

Dear, M.J. and Clark, G.L. (1978) 'The state and geographic process: a critical review', *Environment and Planning A*, 10: 173–83.

Desbiens, C., Mountz, A. and Walton-Roberts, M. (eds) (2004) *Reconceptualizing the state* (special issue of *Political Geography*, vol. 23, no. 3). Amsterdam: Elsevier.

Dodds, K.-J. and Sidaway, J.D. (1994) 'Locating critical geopolitics', *Environment and Planning D: Society and Space*, 12: 515–24.

Duncan, S., Goodwin, M. and Halford, S. (1988) 'Policy variations in local states: uneven development and local social relations', *International Journal of Urban and Regional Research*, 12: 107–27.

Flint, C. and Taylor, P.J. (2006) '*Political Geography: World-economy, Nation-state and Locality*. Harlow: Prentice Hall.

Foucault, M. (2004a) *Naissance de la biopolitique. Cours au Collège de France* (1978–1979). Paris: Seuil/Gallimard.

Foucault, M. (2004b) *Sécurité, territoire, population. Cours au Collège de France* (1977–1978). Paris: Seuil/Gallimard.

Foucault, M. (2004c) *Society Must be Defended. Lectures at the Collège de France, 1975–76*. London: Penguin Books.

Habermas, J. (1976) *Legitimation Crisis*. London: Heinemann.

Hartshorne, R. (1935) 'Recent developments in political geography, II', *American Political Science Review*, 29: 943–66.

Hartshorne, R. (1950) 'The functional approach in political geography', *Annals of the Association of American Geographers*, 40: 95–130.

Hartshorne, R. (1954) 'Political geography', in P.E. James and C.F. Jones (eds), *American Geography: Inventory and Prospect*. Syracuse, NY: Syracuse University Press for the Association of American Geographers, pp. 167–225.

Harvey, D. (1985a) 'The geopolitics of capitalism', in D. Gregory and J. Urry (eds), *Social Relations and Spatial Structures*. London: Macmillan, pp. 128–63.

Harvey, D. (1985b) *The Urbanization of Capital*. Oxford: Blackwell.

Heske, H. (1987) 'Karl Haushofer: his role in German geopolitics and Nazi politics', *Political Geography Quarterly*, 6: 135–44.

Hyndman, J. (2004) 'Mind the gap: bridging feminist and political geography through geopolitics', *Political Geography*, 23: 307–22.

Jackson, W.A.D. (ed.) (1964) *Politics and Geographic Relationships: Readings on the Nature of Political Geography*. Englewood Cliffs, NJ: Prentice Hall.

Jessop, B. (2002) *The Future of the Capitalist State*. Cambridge: Polity.

Jessop, B. (2003) 'Putting hegemony in its place', *Journal of Critical Realism*, 2: 138–48.

Johnston, R.J. (1982) *Geography and the State: An Essay in Political Geography*. Basingstoke: Macmillan.

Johnston, R.J. (1989) 'The state, political geography, and geography', in R. Peet and N.J. Thrift (eds), *New Models in geography: The Political-Economy Perspective*, Vol. 1. London: Unwin Hyman, pp. 292–309.

Jones, M. (2001) 'The rise of the regional state in economic governance: "partnerships for prosperity" or new scales of state power?' *Environment and Planning A*, 33: 1185–211.

Jones, M.R. (1997) 'Spatial selectivity of the state? The regulationist enigma and local struggles over economic governance', *Environment and Planning A*, 29: 831–64.

Jones, M.R. and MacLeod, G. (1999) 'Towards a regional renaissance? Reconfiguring and rescaling England's economic governance', *Transactions of the Institute of British Geographers*, 24: 295–313.

Jones, S.B. (1954) 'The power inventory and national strategy', *World Politics*, 6: 421–52.

Kaldor, M. (1999) *New and Old Wars: Organized Violence in a Global Era.* Cambridge: Polity.

Kofman, E. (1995) 'Citizenship for some, but not for others: spaces of citizenship in contemporary Europe', *Political Geography*, 14: 121–37.

Mackinder, H.J. (1904) 'The geographical pivot of history', *Geographical Journal*, 23: 421–44.

Mackinder, H.J. (1942) *Democratic Ideals and Reality*, (reissued edn). New York: Henry Holt & Co.

MacKinnon, D. (2000) 'Managerialism, governmentality and the state: a neo-Foucauldian approach to local economic governance', *Political Geography*, 19: 293–314.

MacLeod, G. (1999) 'Space, scale and state strategy: rethinking urban and regional governance', *Progress in Human Geography*, 23: 503–27.

MacLeod, G. and Goodwin, M. (1999) 'Reconstructing an urban and regional political economy: on the state, politics, scale, and explanation', *Political Geography*, 18: 697–730.

MacLeod, G. and Jones, M. (2001) 'Renewing the geography of regions', *Environment and Planning D: Society and Space*, 19: 669–95.

Mann, M. (1984) 'The autonomous power of the state: its origins, mechanisms and results; *Archives Européennes de Sociologie*, 25: 185–213.

Marston, S.A. (2000) 'The social construction of scale', *Progress in Human Geography*, 24: 219–42.

McEwan, C. (2000) 'Engendering citizenship: gendered spaces of democracy in South Africa', *Political Geography*, 19: 627–51.

Miliband, R. (1973) *The State in Capitalist Society.* New York: Basic Books.

Moon, G. and Brown, T. (2000) 'Governmentality and the spatialized discourse of policy: the consolidation of the post-1989 NHS reforms', *Transactions of the Institute of British Geographers*, 25: 65–76.

Morgenthau, H.J. (1949) *Politics Among Nations: The Struggle for Power and Peace.* New York: Knopf.

Muir, R. (1981) *Modern Political Geography*, (2nd edn). Basingstoke: Macmillan.

Murdoch, J. and Ward, N. (1997) 'Governmentality and territoriality: the statistical manufacture of Britain's "national farm"', *Political Geography*, 16: 307–24.

Neumann, F. (1950) 'Approaches to the study of political power', *Political Science Quarterly*, 65: 161–80.

Ó Tuathail, G. (1996) *Critical Geopolitics: The Politics of Writing Global Space.* London: Routledge.

Ó Tuathail, G. (2000) 'The postmodern geopolitical condition: states, statecraft, and security at the millennium', *Annals of the Association of American Geographers*, 90: 166–78.

Ó Tuathail, G. and Dalby, S. (1991) 'Critical geopolitics: discourse, difference, and dissent', *Environment and Planning D: Society and Space*, 9: 261–83.

Ó Tuathail, G. and Dalby, S. (1998a) 'Introduction: rethinking geopolitics', in G. Ó Tuathail and S. Dalby (eds), *Rethinking Geopolitics*, London: Routledge, pp. 1–15.

Ó Tuathail, G. and Dalby, S. (eds) (1998b) *Rethinking Geopolitics.* London: Routledge.

O'Connor, J. (1973) *The Fiscal Crisis of the State.* New York: St Martin's Press.

Paddison, R. (1983) *The Fragmented State: The Political Geography of Power.* Oxford: Basil Blackwell.

Painter, J. (1997) 'Local politics, anti-essentialism and economic geography', in R. Lee and J. Wills (eds), *Geographies of Economies.* London: Arnold, pp. 98–107.

Painter, J. (2002) 'Governmentality and regional economic strategies', in J. Hillier and E. Rooksby (eds), *Habitus: A Sense of Place.* Aldershot: Ashgate pp. 115–39.

Peck, J.A. (2002) 'Political economics of scale: fast policy, interscalar relations, and neoliberal workfare', *Economic Geography*, 78: 331–60.

Philo, C. (1992) 'Foucault's geography', *Environment and Planning D: Society and Space*, 10: 137–61.

Poulantzas, N. (1973) *Political Power and Social Classes.* London: New Left Books.

Poulantzas, N. (1978) *State, Power, Socialism.* London: New Left Books.

Raco, M. (2003) 'Governmentality, subject-building, and the discourses and practices of devolution in the UK', *Transactions of the Institute of British Geographers*, 28: 75–95.

Ratzel, F. (1897) *Politische Geographie.* Munich: Oldenbourg.

Secor, A.J. (2003) 'Citizenship in the city: identity, community, and rights among women migrants to Istanbul', *Urban Geography*, 24: 147–68.

Sharp, J.P. (1998) 'Reel geographies of the new world order: patriotism, masculinity, and geopolitics in post-Cold War American movies', in G. Ó Tuathail and S. Dalby (eds), *Rethinking Geopolitics.* London: Routledge, pp. 152–69.

Short, J.R. (1982) *An Introduction to Political Geography.* London: Routledge & Kegan Paul.

Slowe, P.M. (1990) *Geography and Political Power.* London: Routledge.

Smith, N. (2003) *American Empire: Roosevelt's Geographer and the Prelude to Globalization.* Berkeley: University of California Press.

Sparke, M. (1998) 'Outsides inside patriotism: the Oklahoma bombing and the displacement of heartland geopolitics', in G. Ó Tuathail and S. Dalby (eds), *Rethinking Geopolitics*, London: Routledge, pp. 198–223.

Sparke, M. (2000) 'Graphing the geo in geo-political: critical geopolitics and the re-visioning of responsibility', *Political Geography*, 19: 373–80.

Spykman, N.J. (1944) *The Geography of the Peace.* New York: Harcourt, Brace.

Staeheli, L.A. and Cope, M. (1994) 'Empowering women's citizenship', *Political Geography*, 13: 443–60.

Staeheli, L.A. and Clarke, S.E. (2003) 'The new politics of citizenship: structuring participation by household, work, and identity', *Urban Geography*, 24: 103–26.

Staeheli, L.A., Kofman, E. and Peake, L. (2004) *Mapping Women, Making Politics: Feminist Perspectives on Political Geography*, London: Routledge.

Strange, S. (1996) *The Retreat of the State: The Diffusion of Power in the World Economy.* Cambridge: Cambridge University Press.

Taylor, P.J. (1985) *Political Geography: World-Economy, Nation-State and Locality.* Harlow: Longman.

Taylor, P.J. (1993) *Political Geography: World-Economy, Nation-State and locality,* (3rd edn). Harlow: Longman.

Taylor, P.J. (1994) 'Political geography', in R.J. Johnston, D. Gregory and D.M. Smith (eds), *The Dictionary of Human Geography* (3rd edn). Oxford: Blackwell, pp. 447–51.

Taylor, P.J. (1996) 'Embedded statism and the social sciences: opening up to new spaces', *Environment and Planning A,* 28: 1917–28.

Taylor, P.J. (2003) 'Radical political geographies', in J.A. Agnew, K. Mitchell and G. Toal (eds), *A Companion to Political Geography.* Oxford: Blackwell, pp. 47–58.

Feminist Transformations of Political Geography

Eleonore Kofman

INTRODUCTION

A vibrant and burgeoning contribution by feminist scholars to political geography (Brown and
Staeheli, 2003; Staeheli et al., 2004;[1] Sharp,
2003a, 2003b), two successive presentations at the
Political Geography lecture of the Association of
American Geographers in 2003 (Marston, 2004)
and 2004 (Smith, 2005b) by geographers strongly
influenced by feminism, and the inclusion of a
section on state/nation in a recent *Companion to
Feminist Geography* (Nelson and Saeger, 2005),
all these developments would seem to augur well
for closer intellectual interaction between feminist and political geographies. Among other things
they seem to have begun to lay the basis of a
closer engagement between the two fields. Yet this
is in contrast to a more pessimistic evaluation of
their interaction that emerged from a review of
feminist and political geography journals and a
survey of feminist scholars. Staeheli (2001) found
that feminist political geography continued to be
marginal in both feminist geography and political
geography taken separately, and that the differences in the ways in which the two sub-disciplines
are perceived and practised are daunting. Furthermore, this situation prevailed after over a decade
during which feminist geographers had made substantial, but largely unacknowledged, contributions
in arguing for the significance of gender relations
in shaping political activity and in rethinking the
notion and boundaries of the political.

As Peter Taylor (2000) had commented, 'the
sub-discipline has still to meet the challenge of
feminist geography whose concerns for power in

place and space from a gender perspective have
only appeared intermittently in contemporary political geography'. Whilst this is a step forward from
its total absence in the research agenda set out in
the first issue of the journal *Political Geography
Quarterly* in 1982, one might want to ask why
feminist analyses continue to be so marginal compared to their influence in other areas of geography
(Kofman, 1994; Staeheli, 2001). In response to
Stanley Waterman's (1998) analysis of the contents
of the journal, and the degree to which it reflects
the sub-discipline, Janet Kodras (1999: 387) notes
that 'it is not typically the place to find theoretical leaps in feminism, anti-racism and sexuality
and other facets of political identity'. More trenchantly, Michael Dear' commented that there is a
lack of theoretical interrogation and 'one would
be hard-pressed to know from the pages of the
journal, exactly what political geographers think
about feminist theory, postcolonialism ...'. This
failure to engage with feminist theorizations is not
confined to journals such as *Political Geography,
Space and Polity* and *Geopolitics*. Key texts at best
briefly mention feminism without asking what it
actually has contributed (Agnew, 2002) or treat it
as a social movement but not a mode of analysis
(Painter, 1995). In other instances, gender is located
in a specific site such as the household[2] (Taylor
and Flint, 2000), a formulation that reproduces the
private/public divide, and what a feminist political
geography has striven to disrupt. In some cases,
gender relations are taken into account before
retreating to business as usual, as has happened
in much critical geopolitics (O' Tuathail, 1996a,
1996b).

Furthermore, since the 1980s feminists have focused on developments that are only now being addressed more comprehensively in political geography. These include the meaning of the political, a term few political geographers have examined critically (Kofman, 1994: 430; Agnew, 2002: 20). This is beginning to change with the recent questioning of what the political (Dikeç, 2005) means, how space is central to the negotiation of rights, privileges, and obligations (Isin, 2002; Drummond and Peake 2005) and the various discussions of how the political is used by political geographers (Agnew, 2003, Cox and Low, 2003). But prior to this, it was feminists who had been at the forefront in challenging the concept of the political and its constitution in different sites, spaces and scales, and especially the demarcation between the public and private that underpins so much political theory. Likewise it is as a result of feminist influence that recent concern with scale has emphasised smaller units such as the body, household and localities and the role of consumption and reproduction (Staeheli, 1994; Marston, 2000; Hyndman, 2001; England, 2003) rather than a preoccupation with production and global and scalar hierarchies (Taylor, 1982; Brenner, 2001).

Feminists have also been concerned with epistemological issues (Staeheli and Kofman, 2004) and, in particular, the role of situated knowledges and standpoints (Haraway, 1991; Harding, 2001). This is in contrast to disembodied views from nowhere usually presented as universal visions in much of political geography. An interest in research methods (Sharp, 2004) and in how political practices are enacted has also characterized feminist political geography. However, as a result of the slowness with which these feminist insights have been absorbed into political geography, many feminist geographers have looked elsewhere for an understanding into the relationship between space and politics. Indeed, in the survey conducted by Lynn Staeheli (2001), a number of the respondents considered that feminist theory and feminist political writings (Martin, 2004), such as those of Judith Butler (1990), Butler and Scott (1992), Nancy Fraser (1989), Carol Pateman (1988) and Iris Marion Young (1990), were a more useful way into political geography than the writings of mainstream political geographers. Hence Hyndman's (2004: 308) comment that the intersections between feminist and political geographies are relatively few.

The discussion about the relationship between feminism and political geography is relatively recent, but this is at least an advance on the exclusion of women from the creation of political geography as knowledge and practice until the 1970s and the marginalization of feminist perspectives until the 1990s. As a subject dominated by its founding fathers, women nevertheless played a part in the production and presentation of knowledge, for example in their role as editors of geographical journals. Though usually not specifically identified as political geographers, some of them wrote on political topics. A clearer interest in the political domain by feminist geographers can be traced to the emergence of the new radical geography of the 1970s, yet here too their contribution has not been considered an integral element of it (Taylor, 2003). Hence in this chapter I start by casting a backward look at the earlier exclusion of women and outlining some of the initial feminist critiques of political geography. Second, I turn to the emergence and development of gendered agendas in the 1990s and explore some of the reasons for the continuing masculinity of political geography. I proffer some answers to the continuing reluctance to acknowledge the relevance of feminist perspectives and their potential for offering new ways of approaching political geographies. Finally, I suggest some areas where feminist and political geographers have much to say to each other, though they often do not recognize it.

FOUNDING FATHERS AND EXCLUSION FROM THE POLITICAL

The emergence of political geography as a discipline came under the aegis of those we know as the founding 'fathers' and was hegemonically masculine (Connell, 1987: 183). This is in the sense that it was based on a 'strong and dominant masculinity constructed in relation to subordinated masculinities and women'. The creation of geographical knowledge was conducted without women, who were officially excluded from organized and funded explorations and from the new institutional bodies (Domosh, 1991), especially from metropolitan geographical societies such as the Royal Geographical Society (until 1913) and the Association of American Geographers. They therefore had to set up their own organisations, such as the Washington-based Society of Women Geographers in 1925, to enable them to undertake fieldwork (Rossler, 1996).

Though thinly represented at the pinnacle as intellectuals of statecraft, women were clearly involved in the production and reproduction of political geography as a set of power-imbued social practices. Feminist scholars of international politics have demonstrated the role of the support cast and those who have enabled the key processes behind the scenes to occur – the migrant workers, the nurses, the military administrators and the sex workers (Enloe, 1989). So too do we have to consider the handmaidens[3] whose bodies are usually consigned to the zones of non-belonging (Sylvester, 1998) but who ensure the smooth running of the corridors of power. This alternative analysis leaves us, however, with a polarized scenario in the production of political geography, omitting the middle echelons of professionals and administrators in which women have been prominent for some time. As with other professional hierarchies, women often ran the bureaucratic apparatuses as its administrators. In the US, a number of them worked in government agencies that are normally associated with statecraft, security, and foreign policy, such as the CIA, the Office of Strategic Services (OSS), and the State Department. For example, Sophia Saucerman was for years at the US Department of State and Betty Didcoct Burrill was Head of the Latin American section at the CIA (Monk, 2003).[4] Lois Olson started during the Second World War at the OSS in London, Washington, and Paris and, when the CIA was founded in 1947, served as its Chief Geographical Editor until her retirement in 1962. They, and many others, were involved in the compilation of politically pertinent facts, and in the production and presentation of knowledge about places and peoples in the world (Thrift, 2000).

Some women did act as editors of journals and therefore made decisions about how geographical knowledge was presented and reproduced (Monk, 2003). Gladys Wrigley, for example, was the editor of the *Geographical Review* (from 1915 to 1949) in the US and worked closely with Isaiah Bowman, the Director of the American Geographical Society, and probably the most prominent of all American political geographers (Smith, 2003). In the UK, too, women were admitted to non-metropolitan geographical societies, and were prominent in the Scottish geographical world. Marion Newbigin became assistant editor of the *Scottish Geographical Magazine* from 1902 and then editor until her death in 1934. She had an enormous influence on the journal and also wrote numerous articles and books (Maddrell, 1997) and was followed by Harriet Wanklyn, who wrote a widely acknowledged biography of Friedrich Ratzel in 1961 and was a specialist at Cambridge University on Eastern Europe (1941, 1954). We need therefore to broaden our understanding of knowledge creation beyond the great male thinkers to the institutional sites and networks in which that knowledge has been and continues to be produced.[5]

It took a number of decades before the history of geography, geopolitics, and imperialism would be brought under critical scrutiny in the 1970s (Hudson, 1977). At the same time, the role of women and patriarchal relations in geography began to receive attention. In the 1970s, *Antipode*, the journal of radical geography, had published a wide range of articles on topics that today would be recognised as political geography: the state, anarchist movements in Spain, nineteenth-century

anarchist geographers – Kropotkin and Reclus – American socialism and women. Even before women's groups in geography were established in the UK and North America, a special issue of *Antipode* in 1974 addressed the issue of women (Kofman, 2005). Hayford (1974) focused on the shift of political power under capitalism from the household to the state and the consequent loss of power for women. Helms (1974) wrote on the marginalization of older women. In general, much early feminist writing, especially in the UK, stemmed from socialist feminists who were active in social and urban geography study groups, which gave them support in the setting up of a Women and Geography Study Group in 1982.

In contrast, political geography, which also established its institutional credentials in 1982, through the formation of the Political Geography Study Group in the UK, the launch of *Political Geography Quarterly* (jointly edited in the UK and US), and the setting up of the World Political Map Working Party (International Geographical Union), was very masculine in its outlook and composition. *Political Geography Quarterly*, later to be known simply as *Political Geography*, drew up a research agenda in which the only woman cited was Cynthia Cockburn (1977). This was for her book on the *Local State*. Intellectually the blindness to gendered dimensions of political issues meant that for a small number of feminist geographers, their feminist and political research interests within the discipline were kept separate.[6] Some of the reasons for the lack of engagement at this time were probably the emphasis on traditional topics and approaches, the perceived masculinity of the sub-discipline and the lack of interest in what politics meant. There were, nonetheless, some critical voices for whom the amalgam of the old and the new in political geography consisted of 'capricious eclecticism' and presented a 'profoundly traditional and anti-theoretical view of the social process' (Dear, 1982).

The failure to address sexist bias in the research agenda drew a sharp comment from Drake and Horton (1983). They pointed out how knowledge was a social creation that has produced, in the case of political geography, a male interpretation of the organization of political space and one that often projects the assumption that women are 'passive and do not possess political identity'. For them, 'political geography would be far richer as a discipline if it addressed the issue of male bias in all its aspects and ramifications'. The one area in which there was some engagement between feminist and political perspectives was in urban political geography. Here, feminist inquiries into the political asked what would happen to our understanding of urban politics if it was assumed that women were as interested in and as competent to exercise political power as men. It was argued that gender relations in urban political geography were invisible because of our initial conceptualization of terms such as collective consumption (Peake, 1986) and that we had to challenge the notion that politics is exclusively undertaken as a public activity within the state arena. Political discourse could be understood as the search for new identities, and identity is shaped by what we can do and how others act towards us. Hence masculinity and femininity are constituted through the meanings we ascribe to these identities and how these meanings are forged through struggles.

EMERGENCE AND DEVELOPMENT OF GENDERED AGENDAS

The first comprehensive attempt to gender the agenda took a few years to emerge. In 1990, Kofman and Peake sought to expose the vanishing acts and tricks of the trade (Thiele, 1986) common in political theory and political geography. In doing this, they turned to developments in feminist political science (political participation, elections) and political theory – work that, among other things, challenged the dichotomy of the public and private. This distinction, it was argued, had formed the basis of the social and political contract underpinning political life, on the one hand, denying rationality to women, and thus the right to participate in the public sphere, and on the other, relegating their bodies to the private sphere (Pateman, 1988). One of the key objectives of this exploratory agenda was to demonstrate how gender relations and issues could be incorporated into mainstream political geography – the local and national state, urban politics, and service provision, and international politics and conflicts – as well as opening up new areas, such as gay involvement in urban land markets (Knopp, 1990), the participation of Australian aboriginal women (Gayle, 1990), and the intersection of race, class, and gender in Brazilian women's lives (Alves Calio, 1990). The starting point was to interrogate the meaning of the political so as to uncouple it from particular institutions and places. In this way it could be contrasted with mainstream political geography, which continued to avoid the study of what constitutes the political and lacked normative theories, a critique that was also made by some minority voices (Logan, 1978; Dear, 1982, 1999; Howell, 1994).

In the following decade, feminists did not generally engage with topics on the mainstream agenda. Electoral geography had traditionally been a major theme, but there continue to be few feminist studies (but see Webster, 2000; Secor, 2004; Cupple and Larios, 2005). Much more surprisingly, interest in the state, the prime topic of

the 1980s, has ebbed and flowed (Mountz, 2004). Feminists had for long challenged the gender-neutral character of the state (Kofman and Peake, 1990; Chouinard, 2004: 229), but it was not until the 1980s that radical feminists, such as Catherine MacKinnon (1989), argued that the state was systemically male in its interests and formally through its judicial procedures. Feminist geographers drew primarily from Marxist and socialist feminist perspectives, conceptualising the state within the wider dynamics of a capitalist society and as a terrain of struggle over class and gender inequalities (Chouinard and Fincher, 1987).

The most sophisticated and wide-ranging analysis of the state as a gendered institution and constellation of practices was developed by Connell (1990), who emphasised the state as process rather than thing and the necessity of an institutional analysis that makes regulation possible. The state, as the central organizing power of gender relations, has a specific gender regime, which can be defined 'as the historically produced state of play in gender relations within an institution which can be analyzed by taking a structural inventory'. The three key components involved a gender division of labour, a structure of power and a structure of cathexis or the gender patterning of emotional attachments. Though the interests that interact with the state are not fixed, the state is involved in the overall patterning of the gender order.

However, at more or less the same time, post-structuralist and postmodernist critiques were arguing against theorising the state (Kofman, 1993; Chouinard, 2004), contending that we should rather be concerned with mechanisms of power and the diverse and local sites of women's oppression. The state was not monolithic and homogeneous but was fragmented, contradictory, and inconsistent. Pringle and Watson (1992), for example, viewed the state as disconnected and erratic, a site where interests are constantly changing, constructed historically, and manifested in the meanings arrived at through discursive practices and strategies. Despite these critiques, some scholars felt that the state should not be rejected altogether because it highlighted 'particular linkages, connections, and intensifications' and it does appear to act through individual institutions and as overall entity (Cooper, 1993: 258–9). It remains the most organized institution in society (Knuttila and Kubik, 2000) and is a site of co-ordination and the playing out of diverse and often opposing strategies, with which feminists interact in different ways and with varying outcomes (Chappell, 2000). Similarly, Jessop (2001) has recently applied his strategic relational analysis of the state to the 'manner in which the state transforms, maintains, and reproduces modes of domination (or institutionally and discursively materialised, asymmetrically structured power relations) between men and women'. The state itself is seen as an ensemble of power centres that offers unequal chances to different forces within and outside the state to act for different political purposes. There is no single form of masculinity and femininity, whilst gender regimes intersect with class, nation, ethnicity, and 'race'.

Yet despite these more considered reflections on gender regimes, sexuality and governance, feminist geographers did not pursue theorizations of the state. To some extent this was partly due to the shift of attention to the differential effects of state regulation, changing welfare regimes and the disciplining of subjects (Cope, 1997). Thus the absence of new thinking on state theorization more generally led Janet Kodras (1999) to conclude at the end of the 1990s that there was a dearth of cutting-edge theoretical treatments of the state.

In the past few years, interest has been revived among political and feminist geographers in the state (Flint, 2003), its multiple axes of differentiation (Chouinard, 2004) and 'the role of patriarchy, difference and identity within state processes and structures, including how government institutions and practices are produced and contested in concrete ways' (Desbiens et al., 2004). Many of the papers published in the special issue of *Political Geography* entitled 'Reconceptualising the state from the margins of political geography', focused on the power of the state. This included ethnographic studies of the mundane practices of bureaucracies (Mountz, 2004), which reflects a concern for the statization of everyday life or the prosaic state (Painter, 2005), and the interpenetration of different spheres of the state, civil society, and the family. In particular, the state reveals its masculine practices in its relationship to the reproductive arena (Connell, 1995), which is devalued compared to productive activities.

Nonetheless, despite the waning of direct interest in the state in the 1990s, engagement with it did not entirely disappear; it was present more obliquely through studies of gendered citizenship and inequalities of access to economic and social resources and political participation (Walby, 1994). Until more recently, citizenship has been identified with membership of a national community and its attendant rights and obligations. There is a long-standing critique of the constitution of the Western model of citizenship. This is based on the presumed independence of the individual embodying male norms and attributes (Pateman, 1989). This would include the worker acting as the breadwinner for his dependants and in possession of his body, as well as the soldier prepared to sacrifice himself for his country (Yuval-Davis, 1997). Though gender divisions in relation to the different dimensions of citizenship are now less pronounced, as women have entered formal employment, the military and the public sphere, these differences have

certainly not disappeared. And within the broader literature on inequalities and citizenship, feminist geographers have tended to focus on empowering women and spaces of citizenship at the local level (Staeheli and Cope, 1994; Fincher and Panelli, 2001) with some studies of the national construction of citizenship (Marston, 1990) and the variable geometry of citizenship in Europe (Kofman, 1995). Although most studies have focused on Western states with strong welfare provision, a few studies have looked at citizenship practices in Third World states such as McEwan (2000) on South Africa, Nelson (2004) on Mexico and Peake and Trotz (1999) on Guyana.

An underlying theme of feminist studies of citizenship is the distinction and relationship between public and private spaces and how different activities are associated with distinct types of space (Staeheli, 1996; Fincher, 2004). In the past decade, thinking about the relationship between the private and the public has moved from conceiving it as a continuum to conceptualising it as the mutual constitution of a multiplicity of spaces whose use depends on context (Fincher, 2004). Staeheli (1996) has argued that the spaces used in politics are constructed according to ideas of both publicity and privacy. Moreover, both terms have multiple meanings. One of the common meanings of the public, derived from Habermas (1989), is a space accessible to all, where citizens can discuss their common affairs. In reality, different interpretations and ways of using public spaces circulate to include what Nancy Fraser (1990) has termed subaltern counter-publics in which members of marginalized social groups can articulate interests and strategies. 'Private', on the other hand, usually connotes a space, often associated with the home and the domestic, in which particular, rather than general, interests are addressed, the irrational and emotional expressed, and from which outsiders can be excluded.

Drawing on Staeheli, Anderson and Jacobs (1999) provided a concrete instance of the permeability of different types of spaces by showing how women activists in the 1970s constituted a movement to fight against the demolition by the state and big developers of areas of inner Sydney. The women residents occupied a counter or alternative public sphere rather than being situated in the privacy of their homes. Into this public arena, they brought feelings and emotions associated with women and domesticity, such as care, which also may be expressed in public and by public agencies (Fincher, 2004: 53). The strategic planning of actions in the public sphere may also of course take place in the proverbial kitchen (Staeheli, 1996). Hence neither public nor private spaces can be reduced to specified types of actions and emotional responses. Furthermore, the significance of each spatial term varies across social locations and identities, demanding the recognition of the limits of universalizing geographies (Pratt, 2004). So whilst the private may be viewed suspiciously by white middle-class women as depoliticizing their actions, for many black women privacy may represent a positive claim (Williams, 1991).

Discussions about the multiple meanings and spaces of public and private are contributing to a broader debate about the relationship of the different spheres of the state, market, and civil society, which cannot be as clearly demarcated as a lot of research seems to have assumed (Fincher, 2004; Palmer, 2003). These sentiments strongly echo those of Susan Smith (2005a: 1–2) in her counsel to consider the 'persistent divide between states which manage politics, markets which perform the economy, and caring communities whose work is anchored in the spaces of the home'. In an era when the principle of competitive individualism that is associated with the market is being increasingly transferred to the operation of the state as part of welfare reforms and state restructuring, we may well want to transfer notions of care, applied primarily to personal relations, to the impersonal world of social policy as well as the market.

Staeheli and Brown (2003) have also argued that critiques of neo-liberal welfare reform do not question the liberal political subject – atomized, rational and the bearer of rights in a pre-political zone, or the kind of person who exemplifies bourgeois masculinity (Hooper, 1999). They contend that a feminist ethics of care highlights the complex web of social relationships that bind people together in space and time and construct political subjects. Smith, Staeheli, and Brown all draw upon the extension to non-familial spheres of an ethics and politics of care that has been advocated by feminist political theorists (Tronto, 1993; Sevenhuijsen, 1998, 2000) and by social policy scholars (Daly and Lewis, 2000).[7] Care of course looms large in the everyday lives of women and its geographies are generating complex care-giving relationships (Conradson, 2003).

Fisher and Tronto (1990: 40) define care as 'species activities that include everything that we do to maintain, continue and repair our "world" so that we can live in it as well as possible. The world includes our bodies, ourselves and our environment'. Care as a concrete activity and moral orientation challenges the idea of the independent citizen who denies the support received from the caring work of others, or treats dependants as inferior. Care work is carried out not only in the confines of the home and through intimate relations but also through intermediate institutions of civil society and the state. However, developing care based on mutual obligations would demand a transformation of gender relations and policies that would bring about greater equality between men and women (McDowell, 2004).

Yet at present, care of the self and others has become progressively commodified and globalized, resulting in complex global chains of care that can be defined as 'a series of personal links between people across the globe based on the paid or unpaid work of caring' (Hochschild, 2000). Within a profoundly unequal global economy, shortages in welfare sectors and unmet demand in wealthy countries in Europe, North America, and parts of Asia draw in large numbers of female migrants to undertake primarily lesser skilled work, not just in the household but also in state, private and non-profit-making institutions (Yeates, 2004). This represents the globalisation of social reproduction. There is not of course a homogeneous incorporation of Third World women but rather a complicated and dynamic hierarchy shaped by social class, ethnicity, and 'race' (Andall, 2003). And though the Third World migrant provides devalued forms of work in the First World, she also transfers emotional labour, a topic that has received increasing attention in geography (Davidson and Milligan, 2004).

International migration has become a new area for political geography that has been heavily invested by feminist geographers (Hyndman, 2000; Nagel, 2002; Kofman, 2004; Raghuram, 2004), including the role of the state in the management of female mobility (Mountz, 2004; Silvey, 2004; Walton-Roberts, 2004). State policies contribute to the creation of stratified rights that differentiate female migrants according to skills, race, and position of the country within the global economic and geopolitical order, as well as regulating domestic labour (England, 2003; Huang and Yeoh, 2003). Gender, race, age and other social divisions intersect, rather than pile up in a series of disparate identities (Anthias and Yuval-Davis, 1992; Yuval-Davis et al., 2005).

In contrast to international migration, the field of critical geopolitics has not generated a substantial body of feminist work within it as might have been expected (Dowler and Sharp 2001). Critical geopolitics has sought to 'understand and question the relationship between, on the one hand, power, and on the other, discourses about the spatiality of international politics, particularly those developed by intellectuals of statecraft', a group which encompasses the 'community of bureaucrats, leaders, foreign policy experts and advisors who comment upon, influence and conduct the activities of statecraft' (O' Tuathail and Agnew, 1992: 193). The first essay on a gendered critical geopolitics emanated from one of its leading practitioners, Simon Dalby (1994). He saw such a geopolitics as exploring other forms of political communities, and which recognised that insecurity could arise from forms of violence other than those implied by territorial sovereignty, such as those associated with patriarchy. One of the key issues was therefore to understand the various forms of insecurities in the new world order and the different constraints facing women in different places.

However, by the end of the 1990s, the promise of this initial agenda had not resulted in a fruitful dialogue between feminist work and critical geopolitics but had stopped short at the point of critique. The fascination of critical geopolitics with texts was paralleled by a lack of attention to everyday events (Thrift, 2000). Its strength in deconstructing classifications and categories was not matched by reconstruction or alternative scenarios for changes. It is a story of the big men who script the world in a way that displaces others who are concerned with its production. Thus, rather than challenging the masculinist tradition of geopolitics, the effect of critical geopolitics is to perpetuate it (Sharp, 2000: 363). As noted previously, there is scant consideration of the array of people, clerical and professional, many of them female, who compile and synthesize data to reinforce political and spatial imaginaries. Only those at the apex of knowledge production enter into the hallowed domain of geopolitical imaginations. There are also few voices of resistance included in dominant geopolitical practices. Where they are, as in the inclusion of Maggie O'Kane's reportage on Bosnia (O' Tuathail, 1996b), it seems to re-inscribe the classic binary of the rational and calculating male theoretician and his universal gaze, contrasted with the emotional female in contact with the messiness of people's lives at the local scale.

Hyndman (2001, 2004), however, has recently suggested ways of advancing critical geopolitics in conjunction with feminist theory to provide a more embodied vision and an agenda based on human rather than state security. The term 'human security' was, in fact, introduced by the UN in 1992 to draw attention to insecurity within states and to disaggregate the notion. For Hyndman, this move would involve shifting scales of analysis away from the preoccupation with the nation-state whilst transposing approaches normally reserved for lower scales, such as the private/public divide, to a transnational scale. Other feminist scholars too have pursued a more embodied and human dimension for gendered geopolitics. Smith (2001: 231) seeks to refigure geopolitics in the context of post-Communist transition and to repopulate geopolitical landscapes 'from deterritorialized spatialities of globalisation to the embodied geography of gendered subjects'. For Secor (2001), too, geopolitical knowledge is not only produced at the global scale but encompasses every level from the neighbourhood to global considerations of nationalism and modernity, or what she calls the counter-geography of those who are uncounted.

Some critical geopolitics has focused on social movements, turning attention away from elites and states to those who challenge state-centred notions of power and the colonization of the political by the state (Routledge, 1996: 509). Here resistance strategies encompass a multiplicity of possibilities and movements in which gender plays a key role. Many studies have examined the conditions of how and why women get involved and participate (Radcliffe and Westwood, 1993; Fairhurst et al., 2004). Nevertheless, and though participating massively in grassroots movements, their roles have often been circumscribed, as in the US civil rights movements or the South African national liberation struggle.

An understanding of embodied geopolitics, and participation in social and political movements would benefit from ethnographic studies. These could also help to counteract the depoliticizing effects of institutional abstractions, such as the state. Use of the latter, for example, tends to marginalize the agency prevalent in the workings of state bureaucracies (Mountz, 2004). On the contrary, feminists have preferred methods that start from women's lives and their social location (Secor, 2004), hence opting for qualitative methods. Feminist methodology has stressed the voices of the subjects of its research as a starting point of research rather than relying exclusively on the view of the expert (Moss, 2002), though the relationship of the researcher and the researched may certainly be problematic (Kobayashi, 2005). It should not be assumed that all subjects are powerless; indeed, many of those being studied may belong to the elite (Commode and Hughes, 1999). Yet we should bear in mind that political geography has not, unlike other sub-disciplines, reflected on the methodologies it has used or considered them of importance (Sharp, 2004). As Sallie Marston (2003: 635) notes, if political geographers paid more attention to matters of agency, the everyday and the micro level of social life, our students (and ourselves) would have to be as adept at ethnography as they are at geopolitical analysis.

TOWARDS CLOSER ENGAGEMENT BETWEEN FEMINIST AND POLITICAL GEOGRAPHERS

Though critical geopolitics did not initially bring about a closer engagement between feminist and political geography scholars, a number of developments, some of them external to the discipline, have pushed gender analyses more to the fore. In this section I outline some of the areas in which a number of theoretical concerns and political developments are beginning to bring about a closer encounter between feminist and political geographies. These encompass empire-building and imperialism and the interplay of cultural and economic dimensions in the display of hegemonic power; the gendered basis of empire and nation-building, including the significance of masculinities; the gendered analysis of the security state and the intersection between geopolitics and domestic politics in the 'war against terror'; and lastly, interest in the body in the literature on scale and as a site of resistance and oppression.

As empire and imperialism have come back into intellectual fashion at the beginning of the twenty-first century, so has America's supremacy been portrayed as imperialism (Harvey, 2003; Arrighi, 2005). In the 1990s, American power was discussed more in terms of 'hegemony' (Arrighi, 2005: 23). Hegemonic powers are not just hegemonic through coercion but through their ability to convince others of the desirability of a political system, forms of consumption, and socio-cultural practices. An understanding of the quotidian image of modernity represented by the prevailing hegemonic power has been approached separately from Taylor's (1999) world systems approach and Domosh's (2005) feminist analysis. For Taylor, hegemons have been in the forefront of the creation of modern practices in the household. The Dutch invented the modern house and through the private/public distinction developed the notion of domesticity celebrated in Dutch paintings of the seventeenth century. Here women were surrounded by homely objects, whilst carrying out household tasks within a contained space. The shift to British hegemony from the late eighteenth century brought about increasing comfort as well as the separation of work and home in the Victorian era. For the working class, and for the man who was increasingly the breadwinner, the home was supposed to be his haven, while for the woman it remained a place of work. As Mona Domosh has emphasised, however, and as became increasingly apparent in the late nineteenth century, American hegemony introduced a consumer modernity within a suburban location based on the dynamic and productivity of its industry. American products, such as McCormick harvesters and Singer sewing machines (Domosh, 2005), represented the nation commercially through images of the patriarchal white family, with the male in his proper productive role and the female in her domestic reproductive sphere. These products were part of an American national identity and civilization that underpinned and legitimated American imperialism at the end of the nineteenth and beginning of the twentieth century. Thus, both political and feminist geographers have deepened our understanding of the cultural and social dimensions of hegemonic power, especially the historical underpinning of American hegemony.

Feminist studies in the 1990s also drew attention to the involvement of women and men in empire (McClintock, 1995; Stasiulis and Yuval-Davis, 1995) and nation-building (Mayer, 2000) and their gendered representations. At the same time throughout the 1990s, the significance of masculinities generated growing interest (Connell, 1995). With the current militarization of society, especially that of the United States in its renewed imperial expansion and punitive expeditions, these two concerns have combined in the attention currently being given to the play of cultures of masculinities (Enloe, 1993; Kimmel, 1996). Hannah (2005), for example, highlights the historical origins of American manhood in the dream of the self-made man and the frontier myth in the making of the nation. As in the Western film, the hero confronts the enemy in a showdown that follows a ritual of direct confrontation, escalation and climax. However, in the case of the 'terrorist' this scenario has been short-circuited so that manhood and virility cannot be realized. The inability to avenge in 'normal' manly ways leads to revenge based on a 'savage war' of 'infinite justice' in which punishment is inflicted disproportionately (Ratner and Ray, 2004) and upon those who happen to find themselves in the wrong place at the wrong time. Punishment is meted out on the sequestered body, while a strategy of containment is deployed towards other groups depicted as threats to state sovereignty and national identity, such as asylum-seekers (Hyndman, 2005).

Young (2003) argues that while feminists have generally emphasised the dominating and predatory male subject, there is another side of masculinity that has come to the fore. This is that of the male in his role of head and protector of the household, which is needed to 'make a home, a haven', whether it be against immigrants or terrorists. In the post-September 11 (2001) era and the reinforcement of the security state, the world is presented in Manichean terms as in the way in which George Bush stated categorically (20 September 2001) that in the 'war against terrorism', 'if you are not with us, you are against us' (cited in Hyndman, 2005: 569). In this, the leader creates himself as a virtuous male who protects his citizens against aliens – whether they be fearsome outsiders or alien insiders – in exchange for obedience. This, in turn, justifies the abrogation of certain rights and the creation of a state of exception (Agamben, 2000). At the same time, certain outsiders, such as Afghani women, became convenient pawns in the 'war against terror' (Young, 2003; Hyndman, 2005), hence putting the 'woman question' onto the world stage. The knight in shining armour was to charge, or rather fly, to their defence and liberate them from oppression by their menfolk.

The model of the male head of family is also commonly present in many of those nationalist and far-right discourses predicated upon the traditional family of male breadwinner and housewife. The (national) male is portrayed as the heroic defender of the nation and protector of women against external violation (Sharp, 2000) or from foreign men within the nation (Kofman, 1997). Women, on the other hand, as the cultural and biological reproducers of the nation, play a symbolic but subordinate role (Yuval-Davis and Anthias, 1989). Accordingly, many nations are portrayed as female. In the Irish case, the nation is the passive feminine, violated, avenged, or inseminated, but with little space for her own needs or desires (Johnson, 1995). But as Kearns (2004: 443) highlights, these symbols are frequently ambivalent, whilst real women participated in revolutionary struggles and fought against their exclusion from public spaces and citizenship rights (Ryan and Ward, 2004). Especially in times of political transition, conflict and war, women may indeed assume more public roles. At the same time, in situations of violent conflict, women's bodies may be raped (in order) to undermine the virility of men and their ability to protect the reproducers of the nation and its honour (Mayer, 2004).

Whilst feminists have been at the forefront of putting bodies on the map (Longhurst, 2005) and recognising that it is 'central to an understanding of gender relations at every spatial scale' (McDowell, 1999), it is surprising that it is only recently that a corporeal geography has come to the fore in political geography (Mountz, 2003, 2004; Hyndman, 2004). As with critical geopolitics, there is at the same time a growing critique of the tendency in much of the work on the body to abstract the subject from 'personal, lived history, as well as from its historical and geographical embeddedness' (Nelson, 1999). Bodies are after all both the sites of oppression and resistance. They are used to assert power over, and to control, weaken, and demean the enemy. The war on terror has, as noted above, led to the incarceration of bodies and their surveillance. This has primarily been directed towards visible Muslim male bodies. The oppressed too seek to deploy their bodies as resistance, as in the most extreme case of suicide bombers, a tactic recently adopted by Palestinian and Chechen women or with asylum-seekers who partially destroy and mutilate their bodies to draw attention to their cause. These are instances of the bodily, global politics and the geopolitical interpenetrating each other (Hannah, 2005).

CONCLUSION

After almost two decades of feminist work in political geography, there still seems to exist two parallel political geographies. This is not to suggest two homogeneous schools; and indeed some (Brown and Staeheli, 2003) emphasize the disparateness of feminist political geography. What it does indicate is a lack of sustained engagement and very different views of what feminist perspectives might contribute. For feminists have not only

demonstrated the significance of gender relations in the constitution of political geography and challenged taken-for-granted concepts, but have opened up new ways of approaching topics, as I have highlighted in this chapter. In many instances, they have sought to transcend dichotomies and dualisms and suggest ways of connecting disparate spheres and processes. Mainstream political geography has generally not acknowledged the extensive insights feminist political geographers bring to the field as a whole and simply, at best, tacks on women or gender differences. Discussing the meaning of the political, considerations of normative aspects of political concepts and an awareness that political geography needs to be more theoretical may, though not necessarily, bring feminist and political geographers closer together.

There are, nevertheless, some indications that the previous lack of engagement is slowly breaking down. Some political geographers, such as Michael Brown and Matthew Sparke (2004), are now collaborating with feminist geographers and appreciating the transformatory potential of feminist contributions. At the same time, a number of feminist geographers are publishing in mainstream political geography journals and situating themselves within the field. They are seeking to bring the margins into the centre, as the editors of the special issue of *Political Geography* on 'Reconceptualising the State' affirm. It is important that we once again seriously engage with the state without treating it as a homogeneous entity confronting fixed and pre-constituted interest groups, whilst recognising that one of its key functions is to co-ordinate disparate activities and engender a sense of belonging and togetherness, whether it be fictional or real. In the face of globalizing processes the state is 'managing, shaping, regulating and supporting complex and often contradictory circuits of capital and people' (Mitchell et al., 2003). Economic and political insecurity and uncertainty are also propelling the state to provide protection and greater certainty for its citizens. We have seen this in its reinforcement of border controls and security measures, the reassertion of national identity, and the globalisation of social reproduction. All of these aspects have a gender dimension in the interaction of state, society, and space to which feminist perspectives have much to contribute.

Closer engagement too may accelerate with the recently renewed visibility of social geography and an interest in social justice and ethics. The dominance of cultural geography and the impact of the cultural turn (Sayer, 1998) unfortunately led to the relegation of the social, which either became the junior partner of cultural geography or disappeared altogether.[8] As the social returns, some of the boundaries within geography are also becoming more permeable, thus facilitating conversations between a range of perspectives, including those derived from feminist standpoints. Care, the politics of reproduction and welfare; modernities and hegemonic powers; politics of gendered borders and migrations; femininities, masculinities and national identities; the meaning of (in)security and racialization of populations and citizenship, are just some of the areas in which hopefully this dialogue will continue and will help to break down the silences between feminist and political geographies.

NOTES

1 The book was awarded the Julian Minghi prize in 2005 for outstanding research by the Association of American Geographers Political Geography Speciality Group.
2 At the same time others have belittled the attempt by feminists to incorporate the household within a wider series of scales (see Brenner, 2001; Marston and Smith, 2001).
3 A term borrowed from Margaret Attwood's fictional handmaids from Gilead in her novel The Handmaid's Tale, published in 1985.
4 Thanks to the research of Jan Monk for the information on women geographers in the US and to Avril Maddrell for the UK without whom this section could not have been written.
5 Though Wrigley and Newbigin were not primarily political geographers, they did write articles on the military campaigns against Germany's African colonies (Wrigley, 1918) and the Balkans (Newbigin, 1915a, 1915b). In particular, Newbigin (1917) noted the lack of attention paid to women in analyses of forced migration and naturalisation in her paper on 'Race and nationality' and recognised the political nature of women's actions, raising the issue of the private/public dichotomy (Maddrell, 1997: 39).
6 This certainly applied to me. I was one of the founding members of the Women and Geography Study Group and a co-author of the first text on Gender and Geography (Women and Geography Study Group, 1984). Whilst I participated in political geography events in the 1980s and wrote on regionalism and nationalism, I only brought the two areas together through an article on women in the French Revolution, followed by the special issue of Political Geography Quarterly edited with Linda Peake in 1990.
7 This perspective also complements a renewed interest in social justice, ethics (Harvey, 1996; Smith, 2000) and more recently the moral economy (Sayer, 2004).
8 During the 1990s social geography disappeared from reports in Progress in Human Geography until it was reinstated in 1999.

REFERENCES

Agamben, G. (2000) Beyond human rights, Means Without End. Notes on politics, trans. V. Bienitti and C. Casarino. Minneapolis: University of Minnesota Press, pp. 14–25.
Agnew, J.A. (2002) Making Political Geography. London: Arnold.
Agnew, J.A. (2003) 'Contemporary political geography: intellectual heterodoxy and its dilemmas', Political Geography, 22: 603–606.
Alves Calio, S. (1990) 'The Brazilian economic crisis and its impact on women', Political Geography Quarterly, 9(4): 415–24.
Andall, J. (ed.) (2003) 'Hierarchy and interdependence: the emergence of a service caste in Europe', in J. Andall (ed.), Gender and Ethnicity in Contemporary Europe. Oxford: Berg, pp. 39–60.
Anderson, K. and Jacobs, J. (1999) 'Geographies of publicity and privacy: residential activism in Sydney in the 1970s', Environment and Planning A, 31: 1017–30.
Anthias, F. and Yuval-Davis, N. (1992) Racialized Boundaries. London: Routledge.
Arrighi, G. (2005) 'Hegemony unravelling – 1', New Left Review, 32: 23–80.
Brenner, N. (2001) 'The limits to scale? Methodological reflections on scalar structuration', Progress in Human Geography, 25(4): 591–614.
Brown, M. and Staeheli, L.A. (2003) 'Are we there yet? Feminist political geographies', Gender, Place and Culture, 10(3): 247–55.
Butler, J. (1990) Gender Trouble, Feminism and the Subversion of Identity. London: Routledge.
Butler, J. and Scott, J. (eds) (1992) Feminists Theorize the Political. London: Routledge.
Chappell, L. (2000) 'Interacting with the state', International Journal of Feminist Politics, 2(2): 244–76.
Chouinard, V. and Fincher, R. (1987) 'State formation in capitalism: a conjunctural approach to analysis', Antipode, 19: 329–53.
Chouinard (2004) 'Making feminist sense of the state and citizenship', in L.A. Staeheli and E. Kofman (eds), Mapping Women, Making Politics. Feminist Perspectives on Political Geography. New York: Routledge, pp. 227–343.
Cockburn, C. (1977) The Local State. Management of Cities and People. London: Pluto Press.
Commode, L. and Hughes, A. (1999) 'The economic geographer as a situated researcher of elites', Geoforum, 30: 299–300.
Connell, R. (1987) Gender and Power: Society, the Person and Sexual Politics. Cambridge: Polity Press.
Connell, R.W. (1990) 'The state, gender and sexual politics: theory and appraisal', Theory and Society, 19(5): 507–44.
Connell, R.W. (1995) Masculinities. Cambridge: Polity Press.
Conradson, D. (2003) 'Geographies of care: spaces, practices, experiences', Social and Cultural Geography, 4(4): 451–4.
Cooper, D. (1993) 'An engaged state: sexuality, governance, and the potential for change', Journal of Law and Society, 20(3): 257–75.
Cope, M. (1997) 'Responsibility, regulation and retrenchment: the end of welfare?', in L.A. Staeheli, J. Kodras and

C. Flint (eds), *State Devolution in America: Implications for a Diverse Society*. Thousand Oaks, CA: Sage.

Cox, K.R. and Low, M.M. (2003) 'Political geography in question', *Political Geography*, 22: 599–602.

Cupple, J. and Larios, I. (2005) 'Gender, elections, terrorism: the geopolitical enframing of the 2001 Nicaraguan elections', *Political Geography*, 24(3): 317–39.

Dalby, S. (1994) 'Gender and critical geopolitics: reading security discourse in the new world disorder', *Environment and Planning D: Society and Space*, 12: 595–612.

Daly, M. and Lewis, J. (2000) 'The concept of social care and the analysis of contemporary welfare states', *British Journal of Sociology*, 51(2): 281–98.

Davidson, J. and Milligan, C. (2004) 'Embodying emotion sensing place: introducing emotional geographies', *Social and Cultural Geography*, 5(4): 523–32.

Dear, M.J. (1982) 'Research agendas in political geography – a minority', *Political Geography Quarterly*, 1(2): 179–80.

Dear, M.J. (1999) 'Telecommunications, gangster nations and the crisis of representative democracy', *Political Geography*, 18(1): 81–3.

Desbiens, C., Mountz, A. and Walton-Roberts, M. (2004) 'Introduction: reconceputalizing the state from the margins of political geography', *Political Geography*, 23(3): 241–3.

Dikeç, M. (2005) 'Space, politics and the political', *Environment and Planning D: Scoiety and Space*, 23: 171–88.

Domosh, M. (1991) 'For a feminist historiography of geography', '*Transactions of the Institute of British Geographers*, 16: 95–104.

Domosh, M. (2005) 'Gender, race and nationalism: American identity and economic imperialism at the turn of the twentieth century', in L. Nelson and J. Saeger (eds), *A Companion to Feminist Geography*. Oxford: Blackwell, pp. 534–49.

Dowler, L. and Sharp, J.P. (2001) 'A feminist geopolitics?', *Space and Polity*, 5(3): 165–76.

Drake, C. and Horton, J. (1983) 'Comment on editorial essay: sexist bias in political geography', *Political Geography Quarterly*, 2: 329–37.

Drummond, L. and Peake, L. (2005) 'Introduction to Engin Isin's *Being Political: Genealogies of Citizenship*', *Political Geography*, 24(3): 341–3.

England, K. (2003) 'Towards a feminist political geography', *Political Geography*, 22: 611–16.

Enloe, C. (1989) *Beaches, Bananas and Bases: Making Feminist Sense of International Politics*. London: Pandora.

Enloe, C. (1993) *The Morning After: Sexual Politics at the End of the Cold War*. Berkeley: University of California Press.

Fairhurst, J., Ramutsindelka, M. and Jumilla, B. (2004) 'Social movements, protest and resistance', in L.A. Staeheli, E. Kofman and L. Peake (eds), *Mapping Women, Making Politics: Feminist Perspectives on Political Geography*. New York: Routledge, pp. 199–208.

Fincher, R. and Panelli, R. (2001) 'Making space, Women's urban and rural activism and the Australian state', *Gender, Place and Culture*, 8: 129–48.

Fincher, R. (2004) 'From dualisms to multiplicities: gendered political practices', in L.A. Staeheli, E. Kofman and L. Peake (eds), *Mapping Women, Making Politics: Feminist*

Perspectives on Political Geography. New York: Routledge, pp. 49–70.

Fisher, B. and Tronto, J. (1990) 'Toward a feminist theory of caring', in E. Abel and M. Nelson (eds), *Circles of Care, Work and Identity in Women's Lives*. Albany: State University of New York Press.

Flint, C. (2003) 'Political geography II: terrorism, modernity, governance and governmentality', *Progress in Human Geography*, 27(1): 97–106.

Fraser, N. (1989) *Unruly Practices: Power, Discourse and Gender in Contemporary Social Theory*. Minneapolis: Minnesota University Press.

Fraser, N. (1990) 'Rethinking the public sphere: a contribution to the critique of actually existing democracy', *Social Text*, 25/26: 56–80.

Gayle, F. (1990) 'The participation of Australian Aboriginal women in a changing political environment', *Political Geography Quarterly*, 9(4): 381–95.

Habermas, J. (1989) *The Structural Transformation of the Public Sphere*, trans. T. Burger. Cambridge, MA: MIT Press.

Hannah, M. (2005) 'Virility and violation in the US "war on terrorism"', in L. Nelson and J. Saeger (eds), *A Companion to Feminist Geography*. Oxford: Blackwell, pp. 550–64.

Haraway, D. (1991) 'Situated knowledges: the science question in feminism and the privilege of partial knowledge', in *Simians, Cyborgs and Women*. London: Routledge.

Harding, S. (2001) 'A response to Walby's "Against epistemological chasms: a standard misreading"', *Signs*, 26: 511–25.

Harvey, D. (1996) *Justice, Nature and the Geography of Difference*. Oxford: Blackwell.

Harvey, D. (2003) *The New Imperialism*. Oxford: Oxford University Press.

Hayford, A. (1974) 'The geography of women: an historical introduction', *Antipode*, 6(2): 1–19.

Helms, J. (1974) 'Old women in America: the need for social justice', *Antipode*, 6(2): 26–32.

Hochschild, A. (2000) 'Global care chains and emotional surplus value', in W. Hutton and A. Giddens (eds), *On the Edge: Living with Global Capitalism*. London: Jonathan Cape.

Hooper, C. (1999) 'Disembodiment, embodiment and the construction of hegemonic masculinity', in G. Youngs (ed.), *Political Economy, Power and the Body*. Basingstoke: Macmillan.

Howell, P. (1994) 'The aspiration towards universality in political theory and political geography', *Geoforum*, 25(4): 429–43.

Huang, S. and Yeoh, B. (2003) 'The difference gender makes: state policy and contract migrant workers in Singapore', *Asian and Pacific Migration Journal*, 11(1): 13–46.

Hudson, B. (1977) 'The new geography and the new imperialism: 1870–1918', *Antipode*, 9: 12–19.

Hyndman, J. (2000) *Managing Displacement. Refugees and the Politics of Humanitarianism*. Minnneapolis: University of Minesota Press.

Hyndman, J. (2001) 'Towards a feminist geopolitics', *Canadian Geographer*, 45: 210–22.

Hyndman, J. (2004) 'Mind the gap: bridging feminist and political geography through geopolitics', *Political Geography*, 23: 307–22.

Hyndman, J. (2005) ' Feminist geopolitics and September 11', in L. Nelson and J. Saeger (eds), *A Companion to Feminist Geography*. Oxford: Blackwell, pp. 565–77.

Isin, E. (2002) *Being Political: Genealogies of Citizenship*. Minneapolis: Minnesota University Press.

Jessop, B. (2001) 'The gender selectivities of the state', Department of Sociology, Lancaster University, http://www.comp.lanc.ac.uk/sociology/papers/jessop-gender-selectivities.pdf.

Johnson, N. (1995) 'Caste in stone: monuments, geography and nationalism', *Environment and Planning D: Society and Space*, 13: 51–65.

Kearns, G. (2004) 'Mother Ireland and the revolutionary sisters', *Cultural Geographies*, 11: 459–83.

Kimmel, M. (1996) *Manhood in America: A Cultural History*. New York: Free Press.

Knopp, L. (1990) 'Some theoretical implications of gay involvement in an urban land market', *Political Geography Quarterly*, 9(4): 337–52.

Knutilla, M. and Kubik, W. (2000) *State Theories: Classical, Global and Feminist Perspectives*. London: Zed.

Kobayashi, A. (2005) 'An anti-racist feminism in geography: an agenda for social action', in L. Nelson and J. Saeger (eds), *A Companion to Feminist Geography*. Oxford: Blackwell, pp. 32–41.

Kodras, J. (1999) 'Geographies of power in political geography', *Political Geography*, 18: 75–9.

Kofman, E. (1993) 'Vers une théorisation féministe de l'état: complexité, contradictions, confusions', in A. Gautier and J. Heinen (eds), *Le sexe des politiques sociales*. Paris: Côté Femmes, pp. 25–36.

Kofman, E. (1994) 'Unfinished agendas: acting upon minority voices of the past decade', *Geoforum*, 25(4): 429–43.

Kofman, E. (1995) 'Citizenship for some but not for others: spaces of citizenship in contemporary Europe', *Political Geography*, 14(2): 121–38.

Kofman, E. (1997) 'When society was simple: the far and new right on gender and ethnic divisions in France', in N. Charles and H. Hintjens (eds), *Gender, Ethnicity and Political Ideologies*. London: Routledge, pp. 91–106.

Kofman, E. (2004) 'Gendered global migrations: diversity and stratification', *International Feminist Journal of Politics*, 6(4): 643–65.

Kofman, E. (2005) 'Feminist political geographies', in L. Nelson and J. Saeger (eds), *A Companion to Feminist Geography*. Oxford: Blackwell, pp. 519–33.

Kofman, E. and Peake, L. (1990) 'Into the 1990s: a gendered agenda for political geography', *Political Geography Quarterly*, 9(4): 313–36.

Logan, W. (1978) *Post-Convergence Political Geography: Death or Transfiguration?* Melbourne: Monash Publications in Geography 18.

Longhurst, R. (2005) 'Situating bodies', in L. Nelson and J. Saeger (eds), *A Companion to Feminist Geography*. Oxford: Blackwell, pp. 337–49.

MacKinnon, C. (1989) *Towards a Feminist Theory of the State*. Cambridge, MA: Harvard University Press.

Maddrell, A. (1997) 'Scientific discourse and the geographical work of Marion Newbigin', *Scottish Geographical Magazine*, 113: 33–41.

Mank, J. (2003) 'Women's worlds at the American Geographical Society', *Geographical Review*, 93(2): 237–57.

Marston, S.A. (1990) 'Who are "the people"? Gender, citizenship and the making of the American nation', *Environment and Planning D: Society and Space*, 8: 229–58.

Marston, S.A. (2000) 'The social construction of scale', *Progress in Human Geography*, 24(2): 19–42.

Marston, S.A. (2003) 'Political geography in question', *Political Geography*, 22: 633–6.

Marston, S.A. (2004) 'Space, culture, state: uneven developments in political geography', *Political Geography*, 23(1): 1–16.

Marston, S.A. and Smith, N. (2001) 'States, scales and households: limits to scale thinking', *Progress in Human Geography*, 25(4): 615–19.

Martin, P. (2004) 'Contextualizing feminist political theory', in L.A. Staeheli, E. Kofman and L. Peake (eds), *Mapping Women, Making Politics: Feminist Perspectives on Political Geography*. New York: Routledge, pp. 15–30.

Mayer, T. (ed.) (2000) *Gender Ironies of Nationalism: Sexing the Nation*. London: Routledge.

Mayer, T. (2004) 'Nation, gender and boundaries: feminist political geography and the study of nationalism', in L.A. Staeheli, E. Kofman and L. Peake (eds), *Mapping Women, Making Politics: Feminist Perspectives on Political Geography*. New York: Routledge, pp. 153–68.

McClintock, A. (1995) *Imperial Leather: Race, Gender and Sexuality in the Colonial Context*. London: Routledge.

McDowell, L. (1999) *Gender, Identity and Place: Understanding Feminist Geographies*. Cambridge: Polity Press.

McDowell, L. (2004) 'Work, workfare, work/life balance and an ethic of care', *Progress in Human Geography*, 28(2): 145–63.

McEwan, C. (2000) 'Engendering citizenship: gendered spaces of democracy in South Africa', *Political Geography*, 19: 627–51.

Mitchell, K., Marston, S.A. and Katz, C. (2003) 'Life's work: an introduction, review and critique', *Antipode*, 35: 415–42.

Moss, P. (ed) (2002) *Feminist Geography in Practice: Research and Methods*. Oxford: Blackwell.

Mountz, A. (2003) 'Human smuggling: the transnational imaginary, and everyday geographies of the nation-state', *Antipode*, 35(3): 622–44.

Mountz, A. (2004) 'Embodying the nation-state: Canada's response to human smuggling', *Political Geography*, 23(3): 323–45.

Nagel, C. (2002) 'Geopolitics by another name: immigration and the politics of assimilation', *Political Geography*, 21: 971–87.

Nelson, L. (1999) 'Bodies (and spaces) do matter: the limits of performativity', *Gender, Place and Culture*, 6(4): 321–53.

Nelson, L. (2004) 'Transnational topographies of gender and citizenship. Purhépechan Mexican women claiming political subjectivities', *Gender, Place and Culture*, 11(2): 163–87.

Nelson, L. and Saeger, J. (eds) (2005) *A Companion to Feminist Geography*. Malden MA: Blackwell.

Newbigin, M. (1915a) *Geographical Aspects of Balkan Problems*. London: Constable.

Newbigin, M. (1915b) 'The Balkan peninsula: its peoples and its problems', *Scottish Geographical Magazine*, 32: 57–69.

Newbegin, M. (1917) 'Race and nationality', *Geographical Journal*, 50: 313–35.

O' Tuathail and Agnew, (1992) 'Geopolitics and discourse: practical geopolitical reasoning in American foreign policy', *Political Geography*, 11: 19–204.

O' Tuathail, G. (1996a) *Critical Geopolitics*. London: Routledge.

O' Tuathail, G. (1996b) 'An anti-geopolitical eye: Maggie O'Kane in Bosnia 1992–3', *Gender, Place and Culture*, 3: 171–85.

Painter, J. (1995) *Politics, Geography and 'Political Geography'*. London: Arnold.

Painter, J. (2005) 'Prosaic states', paper presented at AAG Conference, Denver, 5–9 April.

Pateman, C. (1988) *The Sexual Contract*. Cambridge: Polity Press.

Pateman, C. (1989) *The Disorder of Women: Democracy, Feminism and Political Theory*.

Peake, L. (1986) 'A conceptual enquiry into urban politics and gender', in K. Hoggart and E. Kofman (eds), *Politics, Geography and Social Stratification*. London: Croom Helm, pp. 62–85.

Peake and Trotz (1999) *Gender, Ethnicity and Place: Women and Identities in Guyana*. London: Routledge.

Pratt, G. (2004) 'Feminist geographies: spatialising feminist politics', in P. Cloke, P. Crang and M. Goodwin (eds), *Envisioning Human Geographies*. London: Arnold, pp. 128–45.

Pringle and Watson (1992) 'Women's interests and the post-structuralist state', in M. Barrett and A. Phillips (eds), *Destabilizing Theory. Contemporary Feminist Debates*. Cambridge: Polity.

Radcliffe, S. and Westwood, S. (1993) *Viva! Women and Popular Protest in Latin America*. London: Routledge.

Raghuram, P. (2004) 'Crossing borders: gender and migration', in L.A. Staeheli and E. Kofman (eds), *Mapping Women, Making Politics. Feminist Perspectives on Political Geography*. New York: Routledge, pp. 185–98.

Ratner, M. and Ray, E. (2004) *Guantanamo: What the World Should Know*. White River Junction, VT: Chelsea Green.

Rossler, M. (1996) 'From the Ladies program to the feminist session', in M.C. Robic, A.M. Briend and M. Rossler (eds), *Géographes Face au Monde*. Paris: L'Harmattan, pp. 259–67.

Routledge, P. (1996) 'Critical geopolitics and terrains of resistance', *Political Geography*, 15(6/7): 509–31.

Ryan, L. and Ward, M. (eds) (2004) *Irish Women and Nationalism: Soldiers, New Women and Wicked Hags*. Dublin: Irish Academic Press.

Sayer, A. (1998) 'Critical and uncritical cultural turns', Department of Sociology, Lancaster University.

Sayer, A. (2004) 'Moral economy' Department of Sociology, Lancaster University.

Secor, A.J. (2001) 'Towards a feminist counter-geopolitics: gender, space and Islamist politics in Istanbul', *Space and Polity*, 5(4): 191–211.

Secor, A.J. (2004) 'Feminizing electoral geography', in L.A. Staeheli, E. Kofman and L. Peake (eds), *Mapping Women, Making Politics: Feminist Perspectives on Political Geography*. New York: Routledge, pp. 261–72.

Sevenhuijsen, S. (1998) *Citizenship and the Ethics of Care*. London: Routledge.

Sevenhuijsen, S. (2000) 'Caring in the third way: the relation between obligation, responsibility and care in Third Way discourse', *Critical Social Policy*, 62: 5–37.

Sharp, J.P. (2000) 'Re-masculinisng geo-politics? Comments on Gearoid O'Tuathail's critical geopolitics', *Political Geography*, 19: 361–4.

Sharp, J.P. (2003a) 'Gender in a political and patriarchal world', in M. Domosh, S. Pile and N.J. Thrift (eds), *Handbook of Cultural Geography*. London: Sage, pp. 473–84.

Sharp, J.P. (2003b) 'Feminist and postcolonial engagements', in J.A. Agnew, K. Mitchell and G. Toal (eds), *A Companion to Political Geography*. Malden, MA and Oxford: Blackwell.

Sharp, J.P. (2004) 'Doing feminist political geographies', in L.A. Staeheli, E. Kofman and L. Peake (eds), *Mapping Women, Making Politics: Feminist Perspectives on Political Geography*. New York: Routledge, pp. 87–98.

Silvey, R. (2004) 'Transational domestication: state power and Indonesian migrant women in Saudi Arabia', *Political Geography*, 23(3): 245–64.

Smith, D. (2000) *Moral Geographies: Ethics in a World of Difference*. Edinburgh: Edinburgh University Press.

Smith, D. (2001) 'Refiguring the geopolitical landscape: nation, transition and gendered subjects in post-Cold War Germany', *Space and Polity*, 5(4): 213–35.

Smith, N. (2003) *American Empire: Roosevelt's Geographer and the Prelude to Globalization*. Berkeley: University of California Press.

Smith, S. (2005a) 'States, markets and an ethic of care', *Political Geography*, 24(1): 1–20.

Smith, S. (2005b) 'Care-full markets? A reply to John Christman and Veronica Crossa', *Political Geography*, 24(1): 35–8.

Sparke, M. (2004) 'Political geography: political geographies of globalisation (1) – dominance', *Progress in Human Geography*, 28(6): 777–94.

Staeheli, L.A. (1994) 'Empowering political struggle: spaces and scales of resistance', *Political Geography*, 13(5): 387–91.

Staeheli, L.A. and Cope, M. (1994) 'Empowering women's citizenship', *Political Geography*, 13: 433–60.

Staeheli, L.A. (1996) 'Publicity, privacy and women's political action', *Environment and Planning D: Society and Space*, 14: 601–19.

Staeheli, L.A. (2001) 'Of possibilities, probabilities and political geography', *Space and Polity*, 5(3): 177–89.

Staeheli, L.A. and Brown, M. (2003) 'Where has welfare gone? Introductory remarks on the geographies of care and welfare', *Environment and Planning A*, 35: 771–7.

Staeheli, L.A., Kofman, E. and Peake, L. (eds) (2004) *Mapping Women, Making Politics: Feminist Perspectives on Political Geography*. New York: Routledge.

Stasiulis, D. and Yuval-Davis, N. (eds) (1995) *Unsettling Settler Societies: Articulations of Gender, Race, Ethnicity and Class*. London: Sage.

Sylvester, C. (1998) '"Handmaids" tales of Washington power: the abject and the real Kennedy White House', *Body and Society*, 4(3): 39–66.

Taylor, P.J. (1982) 'Materialist framework for political geography', *Transactions IBG*, 7: 15–34.

Taylor, P.J. (1999) 'Places, spaces and Macy's: place-space tensions in the political geography of modernities', *Progress in Human Geography*, 23: 7–26.

Taylor, P.J. (2000) 'Political geography', in R.J. Johnston, D. Gregory, G. Pratt and M. Watts (eds), *Dictionary of Human Geography* (4th edn). Oxford: Blackwell, pp. 594–7.

Taylor, P.J. (2003) 'Radical political geographies', in J.A. Agnew, K. Mitchell and G. Toal (eds), *A Companion to Political Geography*. Oxford: Blackwell, pp. 47–58.

Taylor, P.J. and Flint, C. (2000) *Political Geography* (4th edn). Harlow: Pearson Education.

Thiele, B. (1986) 'Vanishing acts in social and political thought: tricks of the trade', in C. Pateman and E. Gross (eds), *Feminist Challenges*. Sydney: George Allen & Unwin, pp. 30–43.

Thrift, N.J. (2000) 'It's the little things', in K.J. Dodds and D. Atkinson (eds), *Geopolitical Traditions: A Century of Geopolitical Thought*. London: Routledge, pp. 380–7.

Tronto, J. (1993) *Moral Boundaries: A Political Argument for an Ethic of Care*. New York: Routledge.

Walton-Roberts, M. (2004) 'Rescaling citizenship: gendering Canadian immigration policy', *Political Geography*, 23(3): 265–81.

Walby (1994) 'Is citizenship gendered?' *Sociology*, 28(2): 379–95.

Wanklyn, H. (1941) *The Eastern Marchlands of Europe*. London: G. Philip & Son.

Wanklyn, H. (1954) *Czechoslovakia*. New York: Praeger.

Wanklyn, H. (1961) *Friedrich Ratzel: A Biographical Memoir and Bibliography*. Cambridge: Cambridge University Press.

Waterman, S. (1998) 'Political geography as a mirror of political geography', *Political Geography*, 17: 373–88.

Webster, G. (2000) 'Women, politics, elections and citizenship', *Journal of Geography*, 99(1): 1–10.

Williams, P. (1991) *The Alchemy of Race and Rights*. Cambridge, MA: Harvard University Press.

Women and Geography Study Group (1984) *Geography and Gender: an introduction to feminist geography*. London: Hutchinson.

Wrigley, G. (1910) 'The military campaigns against Germany's African colonies', *Geographical Review*, 5(1): 44–65.

Yeates, N. (2004) 'Global care chains: critical reflections and lines of enquiry', *International Feminist Journal of Politics*, 6(3): 369–91.

Young, I.M. (1990) *Justice and the Politics of Difference*. Princeton: Princeton University Press.

Young, I.M. (2003) 'The logic of masculinist protection: reflections on the current security state', *Signs*, 29(1): 1–25.

Yuval-Davis, N. and Anthias, F. (eds) (1989) *Woman-Nation-State,* London: Routledge.

Yuval-Davis, N. (1997) *Gender and Nation,* London: Sage.

Yuval-Davis, N., Anthias, F. and Kofman, E. (2005) 'Secure borders and safe haven and the gendered politics of belonging: beyond social cohesion', *Ethnic and Racial Studies*, 28(3): 513–15.

States

Introduction

Kevin R. Cox

To have a section of a *Handbook of Political Geography* devoted to the state, and to give it priority of place, should need little in the way of explanation. The state has been a central concept in political geography. The characteristic concerns of political geography make little sense outside of it. As Herb pointed out in Section 1, the state has been a major point of reference for the sub-field, albeit in changing ways. It has been and continues to be a pervasive presence, despite the recent awareness that there is much more to the core concepts of power and space in political geography than what is, somehow, mediated by the state. But, as Painter showed in his chapter in the last section, many questions remain in understanding the relationship between power, the state and space, suggesting its continuing analytical importance. And as the authors in this section (and other places in the *Handbook*) make clear, many contemporary political issues continue to entrain the state, ensuring that the project of excavating the spatiality of state power is as vital as ever.

As we saw in the first section of this *Handbook*, how political geographers have related to the state has changed very considerably over the last century. Not least, analyses of the state registered the changes ushered in the early seventies by what we termed in our Introduction to this *Handbook* 'the socialization of human geography'. The politics and internal dynamics of the state along with its apparently fixed territorial form were taken for granted by political geographers for decades; looking back, it is remarkable that political geographers could hold off a concern for the contested relationship between state and society for as long as they did. By contrast, one of the hallmarks of the way the state is now approached in political geography – and reflected by the authors in this section – is through careful engagement with social and political theory, including situating state power in relation to wider social processes. The authors here take sight of the state in different ways, but together they highlight the achievements and remaining challenges for a geographical account

of state power: a nuanced account of the relationship between sovereignty and space, attentiveness to the diversity of state forms, the social processes that make up and flow from state power (its gendered form), and the importance of space within many state projects, especially those of policing, law, and governing space itself.

The theoretical points of departure for these authors and for political geography in general have been numerous and can be thought of along various dimensions, each disclosing a characteristic approach to the state. There are more totalizing versions originating in human geography's earlier interest in Marxism. According to this, the state is an essential moment of the accumulation process. The critique of Marxism led to conceptions of a more pluralistic sort: an interest in the role of a diversity of social movements, challenges to the gendered, raced, and colonial (or imperial) character of the state. More recently, there has been a major expansion of interest in Foucault's governmentality, and therefore the roles of discourse, expert knowledge, normalization, the construction of identities, induction into technologies of self-regulation, as mediating the relation between state and society. In part it is this that has produced some skepticism toward the centrality of the state in political geography, since these practices of governmentality stretch across a range of different institutions and agents; they are far from unique to the state. It has also shifted the axis of interest away from the material or economic conditions of state interest toward a range of different foundations for state actions, including the discursive constitution of state power – from analytical interest in structures to the processes of social construction.

Thus, what took shape through the individual chapters in this section is something that looks very different from what might have transpired if this *Handbook* had been written twenty years ago. One can imagine then, for example, chapters on the local state, public choice approaches to the state's territorial organization, and perhaps something on normative approaches to devising service areas: a language that would have been more economistic in tone, therefore, dealing variously with externalities, class, even utility functions. Instead, threading through the chapters that follow are concerns for processes of governmentality, subjectification, construction and deconstruction, the relation between the state and power more generally, and the difference that makes to how specifically *state* power gets translated over space.

Political geographers have certainly stretched more general accounts of state power considerably in their attentiveness to the spatiality of the state. They have asked, what difference it makes to appreciate the state/society relation as deeply embedded in space, place, nature, wider social connections, and to be sure, different conceptions of space? Traditionally, political geographers assumed a tight association between states and their bounded territories; the state-area formed an analytical focus for geographers in the first half of the twentieth century. But for political geographers today, that association is deeply in question. As Kuus and Agnew demonstrate, territorial sovereignty is an achievement, dependent on many different processes and power relations both within and beyond a state's borders. Their chapter highlights the traffic between political geographers concerned with the state and wider geographical thinking, as understandings of space respond to empirical shifts in the spatialities of state power (especially in a context of globalization), and in turn inform how states are interpreted. Together, the authors in this section direct us toward a number of ways in which states bring the spatiality of political power into view.

Perhaps most striking is the sheer multiplicity of states. The partitioning of the world into the territorial jurisdictions of so many states is sometimes contrasted with claims about social processes that are more global in character. This may be achieved through attention to the world market, as in Wallerstein's world systems, Giddens' time–space distanciation, which refers to both economic and social processes, or Michael Mann's differentiated, 'messy' networks. This is a contrast that can be overdone. We can certainly point to social relations that are more globalized, if only in the minimal sense of stretching across state boundaries. But equally we can point to social relations that remain more specific to the territorial spaces of states, or localized at some sub-state level. The social relations that cross state boundaries include those among states themselves; in forming 'the inter-state system', states are crucial actors in the production of 'global' politics.

In noting the multiplicity of states, we should also be cognizant of the diversity of forms assumed by states. In territorial terms at least, the range is immense: unitary as opposed to federal states, for sure, but also colonial states, Mamdani's (1996) bifurcated states dependent upon varied forms of

rule, war-torn and weakened states with limited reach, and curiosities like the EU, particularly since its recent constitutional changes enhanced its powers vis-à-vis its constituent states. Broad contrasts like these, however, hardly do justice to the sheer variety. Canada and the United States, for example, are both federal, but beyond that they are very different. The same sort of diversity is evident in the forms of political power that characterize states – from centralized party control, to the fragmentation of power across religious authority and elected representatives, to clientelist power bases dependent upon disbursal of benefits or personal enrichment. And as Corbridge explores, there are various projects through which states try to construct their own purpose and unity: the welfare state, the developmental state, the neo-liberal state.

The underlying variety of forms suggests that there should be quite severe limits to generalizing across states with respect to the political geographies of state projects. Following Sidaway's contribution in Section 1, bringing this diversity of state forms to bear on broader theoretical insights is an enormous project for political geographers of the future, one that Corbridge's wide-ranging and careful analysis here suggests will be exciting, but whose difficulties we should not underestimate.

For example, even in relation to the regions most studied by political geographers, building generalizations is not straightforward. Much has been made recently of the idea of glocalization, the thesis that states are hollowing out through simultaneous processes of devolution and yielding formal powers to supranational organizations. In turn this has been linked to the emergence of neo-liberalism as a particular form of state project. But as a generalization applying to the advanced industrial societies, it needs considerable qualification. The American case seems quite different. Quite apart from the fact that NAFTA is in no way the EU, the sort of devolution that is being noted in France and the United Kingdom, and other EU members, has been a fact of life in the US for a very long time indeed, and the same goes for the sorts of practices celebrated by the glocalization thesis, like territorial competition. And as for the degree of devolution that has occurred in the Western Europe, it is of a very meager sort when compared with what has long been the case in the US.[1]

This does not mean that different state forms may not fulfill similar functions. The East Asian tigers are all developmental states, but the state structures mobilized for that purpose are all quite different (Henderson, 1999). The point is that any classification of states, any attempt to postulate similar patterns or tendencies across them, should be regarded with caution. States are formed in distinct and different geohistorical circumstances. This produces unique state structures, modes of organization, of representation, and so on, and to draw on Jessop's term, different strategic selectivities with respect to who or what is more or less likely to prevail, even when at more abstract levels there are strong similarities, for example, among neo-liberal or colonial states.

A second aspect of the relationship between states and space is the idea of territory, or territoriality. States are obviously defined territorially. They have boundaries, they exclude, they include, and they manage various movements, and so try to influence what goes on within their boundaries. But within that space also, state structures and practices have clear territorial referents. States are usually organized territorially, as in the distinction between central and local branches. Systems of representation can be conceived in territorial terms: legislators represent territorially defined constituencies or districts at one end of the spectrum and the whole country in systems of pure proportional representation. Likewise, the various interventions into civil society often assume quite stark territorial forms through various areally specific policies or regulatory practices – urban policy, agricultural policy, national parks, enterprise zones, and so on. The chapters here by Nick Blomley and Steve Herbert, in relation to the law and policing, especially consider these 'spatializations' of state power. And Margo Huxley's chapter explores how the centrality of space to many different state projects saw the emergence of a specific governmental rationality in planning.

However, if there is an emergent direction in geographic theorization of the state, the authors in this section suggest it is to be found in an emphasis on networked forms of power. The contrast is with more territorialized conceptions of state power, where power proceeds homogeneously from some center until it reaches the boundaries of the state, imposing itself like a wave that is indifferent to any obstacles in its way, to any structuring forces that might guide it. This is a distinction that Kuus and Agnew prioritize in their chapter, but the emphasis on networked forms of power is very clear in the others. Kuus and Agnew focus on the inter-state system. As they argue, and by way of example: 'Today relatively powerful states can supervise security threats in distant places and financial transactions in "offshore" centers (even as they have little direct regulatory control) by rewarding and/or punishing actions that they judge in relation to maintaining or enhancing the spatial efficacy of their authority' (p. 101). It is clear, however, that similar logics apply within territorial boundaries. People are governed, enrolled into state projects, through strategies that inevitably have to assume a channeled and therefore networked form: the mobilization of organizations into a set of relations through which the state can cooperate with representatives of civil society in governance (as in the case of sustainable development discussed by Yvonne Rydin in this volume); relays of instructions through a web of officials, terminating in the contact of an inspector or one of Stuart Corbridge's 'street-level bureaucrats' with the governed; little wonder, therefore, that, as James Scott (1998) has pointed out, states intrinsically require people to have distinct names and identifiable addresses in order that they can be incorporated into a web of relations involving the state.

The idea that power might be conveyed through some sort of network raises questions, however. As John Allen (2003) has pointed out, there is no reason to believe that the translation of instructions, the definition of abstract state objectives, and their ultimate realization in the action intended can proceed unproblematically; that there will not be re-interpretations, intended or otherwise, and that the concrete circumstances of the action might not result in some deviation from the state's original intent. The possibilities here are very clear. As Huxley points out, it is in these terms that we can make sense of the development of institutions to allow for public participation in the planning process, to accommodate the (im)possibility of state interventions in space effectively realizing hoped-for social outcomes. Likewise, applying the abstract language of the law in the diverse and complex contexts of practical policing is a constant challenge for the police. Some of this concreteness can be accommodated by developing strategies to respond to the spatially differentiated nature of social life. Nevertheless, but contradictions remain, as Herbert makes clear in his chapter.

Beyond a common concern with the geography of state power, these chapters also share some broader theoretical points of reference. The first is the multiplicity of power centers outside of any particular state that shape state actions. As we discussed in the Introduction to the *Handbook*, the state has been somewhat decentered in the work of political geographers. Rather the view now is of a dispersal of power among corporations, households, and organizations and, to be sure, other states. The reference to the inter-state system signals that this dispersal is literal as well as metaphoric; that, for example, states have to contend with competing centers of power in the localities and regions as well as in the form of other states. But even the idea of the inter-state system underestimates the dispersion of power, since it suggests that the state is a unified subject, with clear projects in mind, and behind which all its various branches and agencies unite, and that is far from the case. There are hierarchies within the state but they are constantly subject to challenge, as agencies build alliances with organizations in civil society in order to rework structures and practices.

The second point of reference is the idea of governmentality, mentioned earlier. It is not, however, just states and their agencies that engage in the 'conduct of conducts'; rather it is corporations, political parties, institutions, families, organizations of all sorts, as well. These too seek to regulate others, through discourse, the institutionalization of norms, the mobilization of 'expert knowledge', and the construction of identities. People are subjectified in various ways and this serves to underpin a particular social status quo. Welfare states have always drawn a line between the deserving and undeserving poor. As England shows in her chapter here, in the American case this has been expressed through the distinction between those eligible for state insurance and those who only qualify for tax-supported public assistance. The land use planning system also works to reproduce the social stratification system; the hierarchical character of land use zoning in the US is extraordinarily evocative in this regard: single-family housing intended for owner occupation can be built in areas zoned for apartments but not the other way around. This means that discourse of various sorts can profitably be examined from the standpoint of what has come to be known as 'positionality' – the idea, that is, as a representation of the world, it also expresses the interests and identities of who is doing the representing.

Third, there is the question of discourse itself, the significance of representation, of meaning, and how a set of meanings gets to be understood as authoritative, as a regime of truth, so as to structure action. A crucial task for the political geographer is then to examine the meanings around which discourses are constructed in terms of the hidden assumptions behind them; in other words, to refuse to take their taken-for-grantedness for granted. A recurrent theme in the chapters here, therefore, is a critical examination of the taken-for-granted binaries around which discourse is often structured. Critiquing the distinction between public and private space is a key underlying concern in England's chapter on the welfare state, as it has been in a good deal of feminist writing on the state. She is particularly keen to show how it has been lined up with other dualisms aside from gender and including production/reproduction and, of course, the political and the non-political; how therefore it has been mobilized in assumptions about the scalar aspects of state activity in order to subordinate women. The notion that care work is private and therefore not a legitimate area for state provision is particularly troubling at a time when the poorest of the poor among single female parents, unable to afford childcare, struggle to respond to state injunctions to be self-supporting.[2]

Binaries are always highly selective interpretations of the world that order it in a way that frequently advances interests and identities, again in a highly selective manner. In this regard, James Scott's widely cited book *Seeing Like a State*

(1998) receives echoes in several of the chapters to follow. His distinction between expert knowledge and lay knowledge and the way it has been mobilized in state-sponsored modernist projects is explicitly taken up in Corbridge's discussion of the developmental state. It is more implicit in Huxley's discussion of planning. Among other things, she focuses on the society/space dualism and why, therefore, spatial plans often simply move problems around rather than address their social causes. This separation has also been instrumental in justifying the technocratic interventions of planners.

Indeed, the idea of space as a thing-in-itself is a key one for Blomley in his discussion of the law. As he points out, 'space itself, imagined as an objective surface, was a product of law'. The law homogenized land into pieces of private property, to which universally recognized rights and obligations attached. It conditioned the formation of a land market and allowed one to talk about its spatial properties. Accordingly, the legal and the spatial are not separate, thing-like, but aspects of each other. Spatial orderings, like those incorporated into Huxley's land use plans, are always legal orderings: property maps to be negotiated in eminent domain proceedings, or land use zonings that are inevitably internalized in dealings in the land market.

And then, of course, there is the whole vexed question of the state/society distinction itself. States are socio-historical in character. Lucy Mair (1964) long ago talked about peoples without governments, but what she was really talking about were peoples without states. So how indeed does power in society get separated off and given a distinct institutional form as a branch of the division of labor? Political geographers have often taken the sovereign state with its territorial boundaries for granted. In their chapter here, Kuus and Agnew show why we should not and why we should understand state sovereignty as socially constructed.[3] But as Corbridge and Herbert point out in their chapters, and in their different ways, maintaining the state/society distinction is something that has to be worked at, and occasionally has to be softened if the state is to achieve its objectives, at least in part.

As a concluding reflection on these chapters, one might remark that the contrast with how the state might have been treated in political geography twenty years ago is a stark one. Instead of an emphasis on the material, there is now one on the discursive. The objective nature of the world fades into the background as ideas about social construction and subjectification take center stage. 'Structural' is a word that rarely appears any more, unless preceded by the prefix 'post-'. States are 'raced', 'gendered' (though seldom 'classed'), but the idea of 'social base' is rarely encountered.

Networks are in and territory is looked at quite a bit more critically. So although the question 'What next?' has some unfortunate associations of opportunism, it must inevitably be raised; just what might work in political geography or the state look like when it is reviewed in a new *Handbook*, say ten years hence? Or what *should* it look like?[4]

One potential line of inquiry could involve bringing the old into a relationship with the new. This is a cliché, of course, until one turns to concrete possibilities. The critique of binaries is nothing new, for example. They are central to Marx's understanding of the separations induced by capitalist society and the development of the division of labor.[5] Likewise, as these chapters indicate, the territorial is not going to go away. More than this, as Harvey has shown, the tension between fixity and mobility is a central contradiction of capitalist space, and one that is expressed at all manner of geographic scales, both inter-state and intra-state. One could argue, then, that territory and territoriality have material bases in persistent capitalist dynamics as, therefore, does the territorial character of the state. This suggests that our enthusiasm for the discursive as the essential moment in state formation might be usefully tempered by stronger attention to the material. There is still a lot of work to do and scope for developments that will hopefully be just as exciting as those exemplified in the chapters in this section of the *Handbook*.

NOTES

1 This sort of overextension of pattern has been picked up in another context by Alex Callinicos (2004). His point is with reference to the failure of Jessop's *The Future of the Capitalist State* to say much about the international system and how this may limit the sorts of generalizations he arrives at. As he continues: 'Much of his discussion of rescaling and metagovernance is concerned with the European Union. But it may well be that these processes are particularly pronounced in the EU because of its determinate place in the state

system – as a region where the leading capitalist states have chosen partially to share national sovereignty and in a geopolitical context where the United States has sought with great determination to retain its primacy in Eurasia since the end of the Cold War ...' (p. 429)

2 The public/private dualism also appears in Herbert's chapter on crime and the police but from a rather different standpoint – that of state legitimacy. As he notes: 'Excessive police intrusiveness ... threatens their legitimacy as much as the excessive use of force, because it symbolizes an overreaching of state authority into sanctified private territory', though who finds it illegitimate can clearly, though not necessarily, depend on gender.

3 For an excellent case study of just how sovereignty got constructed by the colonial authorities in South Africa, see Crais (2006).

4 For a Marxist treatment of the political/economic distinction, see Ellen Meiksins Wood (1981).

5 As, for example, in *The German Ideology* (1978).

REFERENCES

Allen, J. (2003) *Lost Geographies of Power*. Oxford: Blackwell.

Callinicos, A. (2004) 'Marxism and the international', *British Journal of Politics and International Relations*, 6: 426–33.

Crais, C. (2006) 'Custom and the politics of sovereignty in South Africa', *Journal of Social History*, 39(3): 721–40.

Henderson, J. (1999) 'Uneven crises: institutional foundations of East Asian economic turmoil', *Economy and Society*, 28(3): 327–68.

Jessop, B. (2002) *The Future of the Capitalist State*. Cambridge: Polity Press.

Mair, L. (1964) *Primitive Government*. Baltimore: Penguin Books.

Mamdani, M. (1996) *Citizen and Subject*. Princeton: Princeton University Press.

Marx, K. and Engels, F. (1978) *The German Ideology*. New York: International Publishers.

Scott, J. (1998) *Seeing Like a State*. New Haven: Yale University Press.

Wood, E. M. (1981) 'The separation of the economic and the political', *New Left Review*, No. 127: 66–95.

Theorizing the State Geographically: Sovereignty, Subjectivity, Territoriality

Merje Kuus and John Agnew

INTRODUCTION

The spatiality of state formation and administration is a central theme in political geography. Within this scholarship, the present chapter focuses on the concept of sovereignty as the presumably constitutive characteristic of the state and the apparent foundational principle of the state system. It is in relation to sovereignty that other crucial concepts – like power, authority, community, and obligation – have acquired their present meaning (Walker, 1993: 164). The chapter takes the pivotal importance of sovereignty to contemporary understandings of the state – and politics more generally – as its starting point. It foregrounds the conceptual premises of the traditional debates on sovereignty, and it highlights key strands in the recent rethinking of these premises. It bypasses as inherently problematic theoretically the familiar questions about what is the role of the state in world politics or whether the state is withering away. It instead examines the unspoken assumptions that make it possible to discuss the state in terms of its capacity to act and control territory. In so doing, it disentangles sovereignty, subjectivity, and territoriality, and identifies avenues for a more flexible conceptualization of agency and spatiality in the state system. This, we hope, offers insights not only into the domain of state authority, but into the spatial operations of power more broadly.

The presentation proceeds in three steps. The first section 'State sovereignty as social construct'

highlights the central, if problematic, role of the principle of sovereignty in the fields of international relations and political geography. The second section 'Sovereignty and subjectivity' focuses on two interlinked assumptions within that scholarship: that states are subjects that express their interest and identity through foreign policy, and, addressed in more detail also in the third section, that state power is exercised territorially over blocks of space. A widely cited definition of sovereignty as 'a normative conception that links authority, territory (population, society), and recognition … in a particular place' (Biersteker and Weber, 1996: 3) well illustrates the conceptual bundle of subjectivity and territory that underpins much understanding of the meaning of sovereignty. The two constituent assumptions of that bundle reinforce each other: the conception of the state as an autonomous subject strengthens the territorial conception of power and, conversely, the conception of the sovereign state as a territorial unit bolsters the assumption that the state is the singular subject of international politics. We elaborate on why these assumptions are problematic and how they have been challenged in recent years. The second section draws particularly from the interface of international relations and cultural theory to expose the ontological emptiness of the state. We argue that a state is not an autonomous subject pursuing its interests and expressing its identity through action. Rather, the state as a subject is defined through policies operating under its name. Discourses of state sovereignty, security, and identity thus constitute the category – the territorial sovereign state – that they supposedly describe.

The third section 'Sovereignty and territoriality' concentrates on the spatiality of power. More specifically, we argue that political authority is not necessarily predicated on and defined by strict and fixed territorial boundaries, and we propose conceptual tools for theorizing the spatiality of power in terms other than the territorialized sovereign state. Drawing evidence from the actual exercise of state authority, or what we term effective sovereignty, we propose the concept of sovereignty regimes as a more flexible tool for understanding the spatiality of power and authority.

Together, the conclusions of the second and third sections underscore the view that state sovereignty is best understood as an effect or outcome of the territorial and non-territorial practices of the state. They also elaborate on how further research might go about investigating these practices without anchoring the analysis to the sovereign state as a given category. The concluding section foregrounds the implications of our analysis for political geography. We emphasize the need to study states not as autonomous subjects but as processes of subject-making, and we urge more attention to the spatiality of power beyond that of territoriality.

STATE SOVEREIGNTY AS SOCIAL CONSTRUCT

The Westphalian ideal of sovereignty that envisions the world as a patchwork of states each exercising ultimate control over its territory has always been just that: an ideal (Biersteker, 2002: 162). It has never captured the actual spatiality of power. Yet mainstream theories of international relations (IR), whether realist, neo-realist, or idealist, take the territorial state as their starting point (Biersteker and Weber, 1996). While in principle acknowledging that states have never practiced absolute authority over their territory, these theories nonetheless approach state sovereignty as the necessary organizing principle of politics (Biersteker, 2002: 167). As a consequence, voluminous IR research investigates the types of states and sovereignties and the 'true' extent of sovereignty today, but there is much less critical reflection on the construction, operation, and political effects of the category of state sovereignty itself.

Realist claims about the overwhelming necessity for, and primacy of, the territorial state in contemporary political life are less influential in political geography than in IR. However, despite the varied and sophisticated work on the state in political geography, relatively few studies focus explicitly on the principle of sovereignty (for exceptions, see

Agnew and Corbridge, 1995; Murphy, 1996, 1999). True, sovereignty in the sense of the actual exercise of state power is touched upon in several strands of analysis. For example, there is work on the historical formation of the Westphalian state system (Painter, 1995; Taylor, 1996a). Substantial research investigates the exercise of state sovereignty in different societal spheres such as the economy, immigration, human rights, and the environment (O' Tuathail et al., 1998; Dalby, 2002; Kofman, 2002; Mountz, 2004). Much has been written on the identity narratives, especially nationalism, that provide the state with legitimacy and thereby reinforce the notion of sovereign statehood (Sharp, 2000; Dodds, 2002; Agnew, 2004; Marston, 2003). Political geography is attuned to the operation of state sovereignty in different societal contexts such as the global South vs the global North and to the fact that states are far from equal or identical in the power they exercise within and outside their territorial boundaries (Glassman, 1999; Sidaway, 2002). The rejuvenated, more theoretical, border studies offer fresh insights into the external bounding of states (Newman, 2001). A number of inquiries touch explicitly or implicitly on emerging forms of territorialities and spatialities, and urge us to think of the spatiality of power beyond state boundaries (Anderson, 1996; Agnew, 1999; O' Tuathail, 2000; Dalby, 2002). There is now also a body of work analyzing the state explicitly from a feminist perspective (Staeheli et al., 2004). These lines of scholarship show how state sovereignty intersects with, conditions, and is constituted by a number of other power relationships such as colonialism, class, and gender relations. They demonstrate that states are highly unequal in terms of their exercise of power, and they show that state power is exercised very differently in different spheres of social life. Last, but not least, they expose the statist tendencies in much social science research, including geography (*Environment and Planning A*, 1996; Häkli, 2001).

Political geography thus offers ample evidence of how the discourse of state sovereignty enables and conditions the spatial exercise of power. Yet that discourse itself tends to remain invisible within the sub-discipline. Most political geography, apart from the work drawing from world systems theory, still uses the individual nation-state or local state, 'the state' in the singular, as the primary unit of analysis (Chouinard, 2004; Gilmartin and Kofman, 2004). What Alex Callinicos (2004: 428) notes for Marxist state theory applies more generally: 'it brackets the state system.' State institutions and state power are generally studied in terms of their internal organization rather than their external definition and legitimation (see Jessop, 2002; Brenner et al., 2003; *Political Geography*, 2004). In other words, it is the state as a bureaucracy or apparatus of rule rather than states as territorial polities endowed with popular sovereignty that has been the focus for much of the most innovative theoretical work in the field over the past thirty years (e.g. Clark and Dear, 1984; MacLeod and Goodwin, 1999; Jessop, 2002). The tendency in this literature has been to see a state in functional terms as a bureaucracy performing certain tasks for the accumulation of capital or the reproduction of certain social relations rather than in terms of how a state is constituted or takes on meaning as an effective agent. In other words, authors probe the internal workings of the state as a bounded unit and an actor, but not the assumptions and modes of analysis that make it possible to conceive of the state in this way.

In relation to state territoriality, although there are detailed analyses of sovereignty discourses in relation to particular states, these analyses usually do not tackle the category of sovereignty more broadly (see Kemp and Ben-Elizer, 2000; Kuus, 2002a). Given that sovereignty presumes and justifies an alignment between territory, identity, and political community, it ought to receive more attention as one of the core concerns of political geography. Quite regardless of whether the sovereign state is an adequate conceptual tool for understanding contemporary world politics, it still underpins much analysis and understanding of power and political spatiality (Giddens, 1987). In the international sphere, the regulation of issues such as immigration or the environment is based on international law, which, in turn, is premised on the principle of state sovereignty (Murphy, 1999; Stacy, 2003). In the domestic sphere, narratives of identity, homeland, and borders likewise rest on claims about 'the people', defined territorially by the borders of the sovereign state. Effective critique of the state-centered assumptions in policy-making and academic research therefore requires that we unravel the discourse of state sovereignty as the conceptual encasing that enframes and enables the narratives of borders, identity, and society that are well studied by political geographers.

To date, most of such unraveling has been done in IR and among a small number of political geographers. Labeled variously as constructivism, postmodernism, or post-structuralism, this critical research approaches state sovereignty not as a universal and foundational principle of politics, but as a historically specific construct, whose effects vary across space (for a discussion of the differences between the above theoretical positions, see Campbell, 1998: chap. 8). This general argument about the social construction of the state system is indeed quite well rehearsed by now in the critical strands of IR theory (cf. Walker, 1993; Biersteker and Weber, 1996). We thus turn directly to the two aspects of that construction that are especially relevant to political geography and deserve more critical reflection within the sub-discipline.

Drawing from the broadly post-structuralist strands in political geography and IR, we approach state sovereignty both as the dominant normative framework of the operation of power (how power is supposed or assumed to work) and also as the actual practice of state power (how power is exercised on a daily basis). We argue that the mainstream discourse of sovereignty as a territorial expression of a pre-existing state interest is wholly inadequate for understanding the daily practice of state power. At the same time, and quite regardless of its empirical inadequacy, this discourse still reifies and reproduces the state system. Even though some recent research in political geography and IR has moved beyond such state-centrism, further work still needs to be done to effectively undermine statist assumptions about state sovereignty, subjectivity, and territoriality.

SOVEREIGNTY AND SUBJECTIVITY

Contemporary discourses of sovereignty rely on specific assumptions about subjectivity and agency in the international arena. These assumptions, in turn, reinforce mainstream discourses of state sovereignty as a natural and necessary framework of power. Thus, the notion that the state exercises sovereign control over its territory and advances its sovereign interests through foreign policy rests on an assumption that the state is the subject of international politics. It conceives the state as an entity presiding over its society and making decisions in the interests thereof (Ashley, 1988a: 227). The habitual statements about the interests of a state, such as 'America' or 'Russia', as the causes of their foreign policy illustrate the conception of the state as a pre-existing subject analogous to an autonomous individual. This conception, with its roots in the political tradition flowing from Machiavelli to Hobbes and Locke, is indeed a crucial pillar of the state system.

The view of the state as the subject of international politics is based on the modernist conception of the autonomous self as 'the heroic figure of reasoning man who is himself the origin of language, the maker of history, and the source of meaning in the world' (Ashley, 1988b, quoted in Edkins and Pin-Fat, 1999: 3). It presumes that state sovereignty reflects the interests of a pre-existing subject. It sets an autonomous subject apart from the objective world, at once conceiving agency as prior to action and separating agents from their action (Campbell, 1993: 82). While rendering state action as contingent and problematic, it conceives its subjectivity as natural and given. It thereby frames the study of the state in terms of its 'real' identity and interest – its subjectivity – because state action is assumed to flow from its subjectivity. It encourages questions about the effectiveness of the state as an actor, but simultaneously suppresses questions about how is it possible to conceive the state as an autonomous actor. It thereby makes possible the dichotomies of inside/outside and Self/Other that are central to contemporary conceptions of politics (Walker, 1993: 8).

This is not a static view of the state. Individual states as well as the concept of the state are certainly acknowledged to have histories, and these histories are widely discussed and debated. Struggle among states is indeed a key premise of political realism (Ashley, 1987: 413). However, sovereignty is studied primarily as an attribute of the state, and this framework assumes that the state exists prior to conceptions of sovereignty. Put differently, it presupposes the state as a pre-existing unit to which the various socially constructed conceptions of sovereignty (or interest or identity) refer (Weber, 1998: 82–3). It is also not a monolithic view of the state, as it readily acknowledges that state interests are always contested. However, it presumes that the competing views of that interest flow from pre-given individual or collective subjects – be it nations, individuals, or specific social groups such as elites. As an example, consider Biersteker and Weber's (1996: 3) claim that 'the ideal of state sovereignty is a product of the actions of powerful agents and the resistances to those actions by those located at the margins of power'. Although this claim acknowledges that state sovereignty never reflects a static or monolithic identity of interest, it nonetheless presumes pre-given subjects, particularly that of the state.

In political geography, this organic conception of the state was most prevalent in the early twentieth century, when social phenomena were increasingly approached in the naturalized terms of the physical and life sciences. This was an era of intense interest in the nature of state formation and inter-state competition (Livingstone, 1993: 200; Agnew, 2003: 58–9). States were seen as organisms competing for survival. The state was connected to the nation as a mythic pre-political community, and the principle of natural selection was used to advance organic conceptions of nation-statehood, racial competition, degradation, and domination. While social Darwinism has been disgraced since, conventional conceptions of the state, including international law, continue to approach the state as a pre-given autonomous subject in international politics (Murphy, 1999). Indeed, while figures like Ratzel explicitly studied the state, much of later political geography simply took the nation-state for granted (Livingstone, 1993; Taylor, 1996b; Agnew, 2003). It embedded a kind of banal statism, not an explicit glorification of the state, but a more insidious common-sense equation of state and society (*Environment and Planning A*, 1996). While in principle recognizing that the state is not a unitary actor, and while compellingly showing the internal fragmentation and contestation of the state, political geography rarely challenged the conception of the state as the subject of international politics.

The strongest critique of the subject-based conception of the state comes from the broadly 'post-structuralist' strand of IR. This scholarship posits that a subject, such as the state, is constituted within and through practices operating in its name. The sovereign state then does not pre-exist its actions, which include all acts of the state apparatus in the domestic and the international arenas. Rather, it is through such actions that the alignment between territoriality and identity is effected, so that it becomes possible to speak of a particular state as a unit or a subject with a definitive character. In terms of sovereignty, foreign policy is especially important because it is through foreign policy that the state is made into a coherent subject on the international arena. Inter-state interaction may undermine the sovereignty of specific states, but it strengthens the notion that such interaction is based on states as already existing subjects. Thus, one of the most significant global impacts of the United Nations has been extending states' sovereignty (Giddens, 1987: 283). As foreign policy is conducted in the name of 'the people', it is also central to the domestic consolidation of the nation (Campbell, 1998).

The state therefore has no ontological status apart from the practices that constitute its reality. In other words, a state is not a thing-in-itself but is constituted out of the representations and practices that are associated with it. It is not the source but the effect of power (Marston, 2003: 4). Sovereignty, then, does not express an already existing subjectivity and agency of the state. Rather, the category of the sovereign state is constructed and reconstructed through practices operating in its name. It follows, then, that states are necessarily always in the process of being represented and hence constituted as such. For a state to end its practices of representation would be to expose its lack of pre-discursive foundations (Campbell, 1993: 24; Weber, 1998; Aretxaga, 2003). This conception explicitly counters the conceptualization of the state and the nation as identical and autonomous actors – even if socially constructed and changing through time. Its shifts the focus from the state as a self-evident subject to the practices of making the state as a subject. It urges us to ask not just how state interests are expressed through state practices, but also how these practices construct the very interests in whose name they are undertaken.

To say that states have no separate ontological existence is not to say that they have no materiality. State institutions and state power surely exist materially and have material effects, and they are not transformed simply as different actors shift their identity claims. What is affirmed by the above argument is not so much that objects cannot exist externally to thought or that they are not produced through material forces, although some writers do imply as much (e.g. Wendt, 1999). What is challenged, rather, is that objects can be represented as outside any discursive formation (Campbell, 1993: 9). Thus, the materiality of state power is a part and parcel of, not prior to, the discourse of sovereignty, as our understanding of material reality is formed within this discourse. Likewise, what is challenged is not that states' foreign policy is shaped by material interests, but that these interests necessarily exist prior to foreign policy. The above argument rejects the notion that the material effects of states are derived from pre-given identities and interests that one should uncover in order to understand state action.

To give specific examples of *how* the notion of state as an autonomous subject is produced, we next touch on two aspects of the process of subjectification or subject-making: (1) the relationship between sovereignty and nationalism, and (2) the constitutive role of security threats in producing the state. Nationalism is important here because sovereignty claims cannot be neatly confined to considerations of statehood alone. The view that sovereign states are the subjects of international politics relies on the assumption that the legitimacy of the state is derived from the people. The doctrines of popular sovereignty posit that states are the masters of territory and peoples are the masters of states (Yack, 2001). It is this presumed internal homogeneity that allows the state to act in the name of 'the people'. However, there is a crucial ambiguity as to how 'the people' should be defined and demarcated. On the one hand, the doctrines of popular sovereignty conceive 'the people' as a territorial community, defined by the state. On the other hand, these doctrines also evoke an image of the people as a pre-political community that establishes state institutions and has the final say on their legitimacy 'The people' is thus imagined both as existing in an already constituted state and also as pre-existing the state. The state is consequently framed not only as a political but also as an organic cultural community. Together, the state and the nation constitute 'the people's two bodies' (Yack, 2001: 519) that together make possible the conception of the state as a subject. Nationalism thus both legitimizes the state and rests on the alliance of putative nation with state-organized territory (Dallmyr and Rosales, 2001; Agnew, 2004: 225). Even when nationalism undermines a particular state, it strengthens the hegemony of territorial states in general as the presumed containers of politics. The implication for political geography is to remain keenly aware of, and wary of, how state and nation are conflated in both policy-making and academic research. This is not to imply that either nation or state is a pre-existing ontological

category, but to urge a clearer analytical distinction between them.

Security threats or, more precisely, the discourse of such threats, constitute another key mechanism in producing the state. Whereas the conflation of the state and the nation inscribes the 'inside' of the state as a subject, evoking national security threats designates the anarchical 'outside' against which the state is defined. State legitimacy requires moral boundaries of identity just as much as it requires political boundaries of states, and these identity boundaries are produced and maintained through invoking threats from the outside. Moreover, framing an entity – such as the state – as under threat casts it as natural and unproblematic, and legitimizes emergency measures deemed necessary to deal with the alleged threat. Security thereby functions as a powerful depoliticizing or mobilizing tool. The whole inter-state system is based on the fusing of identity and security so that each state supposedly protects its territorially defined national identity. Insecurity is therefore not external to the state, but is an integral part of the process of establishing the state's identity. It is not a threat to but a precondition for the sovereign state (Campbell, 1998; Weldes et al., 1999). In the context of globalization, as the state's ability to control territory declines in response to an array of material forces, an outside threat becomes even more important for the reproduction and consolidation for the putative national community. Issues of identity and cultural survival are consequently moved to the forefront of security debates.

Of course, security and threat are also not monolithic, static, or uncontested categories. One could argue that broader or alternative definitions of security, which frame that category in terms of individual quality of life rather than state survival, do not reify the state. Instead, this argument posits, the broader definitions of security fragment the 'inside' and the 'outside' and problematize the distinction between them. They thereby open up security debates to non-state actors such as individuals or international and non-governmental organizations. Environmental protection could serve as an example of such a challenge to the traditional statist conceptions of security. However, a closer empirical investigation of the effects of the broadened security agenda yields a gloomier picture. When a societal issue is framed as a matter of security, it is also framed as a matter in which the state should intervene, and urgently. The broader definitions of security thus end up not redefining security but only applying it to more spheres. These definitions abound with 'not only', 'also', and 'more than' clauses. In terms of subjectivity, they contemplate increasing cooperation between groups of citizens – states – but do not imagine the possibility of another subject of international politics. Security can thereby encompass ever larger parts of the social and political agenda, including domestic issues that have hitherto been defined as matters of 'normal boring' politics (see Wæver, 1995; Dalby, 2002, for in-depth discussions of these arguments). Framing a domestic issue in terms of national security brings the anarchical 'outside' into the 'inside' of the state and expands governmental power over this issue (Kuus, 2002b). Broadening the security agenda to non-military spheres therefore does not necessarily undermine, but could instead reinforce, the state system, as it construes ever wider areas of social life in terms of state survival and state intervention. The 'war on terror' well illustrates how issues such as ethnic minorities or education are framed as matters of security and hence of direct state action.

Security discourses produce the sovereign state also in the sense that framing an issue in terms of security casts that issue in territorial terms. These discourses do not simply reify the idealized 'inside', but also demarcate the 'inside' and the 'outside' as territorial categories. They thereby reproduce the state as an inherently territorial subject. Although here too one could argue that the broadening of the security agenda decouples security from the territorial state, such a shift has not actually happened. Once an issue, such as immigration or the environment, is framed in terms of security, it is shifted to the realm of territorial defense and border controls (Bigo, 2002; Dalby, 2002). The 'war on terror' is again a good example of how casting complex problems in terms of national security prioritizes the territorial nation-state over other scales such as the local or the global.

The productive and mobilizing functions of outside threats underscore the need to take security claims seriously as objects of analysis. Given that threat – be it from other states, 'globalization' or 'terrorists' – gives the state its most powerful justification, political geography needs to pay closer attention to the constitutive role of security in producing the categories of state, identity, and community. It needs to investigate not only how states respond to various threats but also how security discourses produce and bolster a particular geopolitical norm – the state system.

SOVEREIGNTY AND TERRITORIALITY

Linked to the view that the state is an autonomous subject of international politics is the assumption that political authority is invariably exercised territorially. The success of state sovereignty as an organizing principle of politics has much to do with its territorial underpinnings (Murphy, 1996: 119). Many of the activities and externalities of modern life have taken territorial forms. Examples would include the delivery of many public goods

(education, police, etc.) and the provision of regulations governing property rights and contracts. Consequently, the merging of state power with a clearly bounded territory as a basic theoretical assumption is still prevalent in both international relations and political geography. This section examines why we need to rethink the spatiality of power in international affairs, how we might go about this task, and what are the implications of such a rethinking to political geography. In particular, we develop the concept of sovereignty regimes as spatial systems of rule to analyze the spatiality of state power beyond the habitual territorial assumptions.

Modern political theory tends to understand geography entirely as territorial: the world is divided up into contiguous spatial units with the territorial state as the basic building block from which other territorial units (such as alliances, spheres of influence, and empires) derive or develop (Agnew, 1994). This is the reason why much of the speculation about 'the decline of the state' or 'sovereignty at bay' is posed as the 'end of geography'. Yet, the historical record suggests that there is no necessity for polities to be organized territorially. As Spruyt (1994: 34) claims:

> If politics is about rule, the modern state is verily unique, for it claims sovereignty and territoriality. It is sovereign in that it claims final authority and recognizes no higher source of jurisdiction. It is territorial in that rule is defined as exclusive authority over a fixed territorial space. The criterion for determining where claims to sovereign jurisdiction begin or end is thus a purely geographic one. Mutually recognized borders delimit spheres of jurisdiction.

Territoriality, the use of territory for political, social, and economic ends, is in fact a strategy (Sack, 1986) that has developed more in some historical contexts and places than in others. Thus, the territorial state as it is known to contemporary political theory developed initially in early modern Europe with the retreat of non-territorial dynastic systems of rule and the transfer of sovereignty from the personhood of monarchs to discrete national populations (Davies, 2003; Reynolds, 2003). That modern state sovereignty did not occur overnight or completely following the Peace of Westphalia in 1648 is now well established (e.g., Osiander, 2001; Teschke, 2003). Territorialization of political authority was further enhanced by the development of mercantilist economies and, later, by an industrial capitalism that emphasized capturing powerful contiguous positive externalities from exponential distance-decay declines in transportation costs and from the clustering of external economies (material mixes, social relations, labor pools, etc.) within national-state boundaries (Kratochwil, 1986; Teschke, 2002).

Absent such conditions, sovereignty – in the sense of the hegemony of socially constructed practices of political authority – could be exercised non-territorially or in scattered pockets connected by flows across space-spanning networks. From this viewpoint, sovereignty can be practiced in networks across space with distributed nodes in places that are either hierarchically arranged or reticular (without a central or directing node). In the former case authority is centralized whereas in the latter it is essentially shared across the network. All forms of polity – from hunter-gatherer tribes through nomadic kinship structures to city-states, territorial states, spheres of influence, alliances, trade pacts, seaborne empires – therefore occupy some sort of space (Agnew and Corbridge, 1995; Smith, 2003). What is clear, however, if not widely recognized within contemporary debates about state sovereignty, is that political authority is not necessarily predicated on and defined by strict and fixed territorial boundaries.

The terms 'territory' and 'space' need to be very carefully distinguished from one another (Durand et al., 1992). Territory refers to the units of a partitioned space, not to spatial organization in its entirety. Simply because territory might be superseded or supplemented in the organization of political authority does not mean that spatial organization is altogether eclipsed – that the 'end of geography' is nigh. 'States' can continue to serve as the main locus of political authority even as their power, for example, is deployed by networked flows penetrating other nominally sovereign states rather than by direct territorial control. Today relatively powerful states can supervise security threats in distant places and financial transactions in 'offshore' centers (even as they have little direct regulatory control) by rewarding and/or punishing actions that they judge in relation to maintaining or enhancing the spatial efficacy of their authority (e.g. Palen, 2002). Of course, in such situations state sovereignty is not absolute but divisible for all of the states involved because it is specific to different issue areas (money, security, trade, etc.) and involves the assent and cooperation of a wide range of non-state actors such as banks, international organizations, and investors.

The main problem is that territoriality is only one type of spatiality or way in which space is constituted socially and mobilized politically (Durand et al., 1992; Offner and Pumain, 1996; Agnew, 1999; Allen, 2003). Territoriality always has two features: blocks of rigidly bordered space and domination or control as the modality of power upon which the bordering relies (Sack, 1986). This may well be legitimate power, that is, exercised with authority (either bureaucratic or charismatic), but it ultimately rests on demarcation through domination. Yet both space and power have other possible modalities (Mann, 1993; Allen, 2003). *Centralized* power, involving command and obedience, can also operate over long distances as well as over territorial blocs, for example through the deployment of military assets, but this may have less possibility of sustained, and thus legitimate, impact on the people with whom it comes into contact (Delbruck, 2003). This is a networked form of domination. It is based on control over flows through space-spanning networks, not control over blocks of space. In contrast to centralized power, *diffused* power refers to power that is not centered or directly commanded but that results from patterns of social association and interaction in groups and movements or through market exchange (credit-rating agencies such as Moody's are one example; see Sinclair, 2000). Diffused power can be territorialized and authoritative, but only insofar as the networks it defines are territorially constrained by centralized power. Otherwise, networks are limited spatially only by the purposes for which they are formed. In this way, power is generated through association and affiliation (from the bottom up) rather than through command or domination (from the top down) (e.g. Arendt, 1958). When not sustained through collective action, however, as with so-called centralized power, the power networks thus created will disintegrate. Today, both centralized and diffused powers are arguably less territorialized by state boundaries than at any time since the late nineteenth century (Hardt and Negri, 2000; Stacy, 2003).

Sovereignty as the *legitimate* exercise of power (or authority), therefore, is necessarily about diffused as well as centralized power. The precise combination of power mechanisms (coercion, association, affiliation) involved in the exercise of authority can be the same for all issue areas (trade, security, currency, etc.) or differ across them and operate solely within a single state territory or more widely. But without at least a modicum of active collaboration on both sides of borders, state sovereignty can be neither sustained nor undermined. As John Allen (2003: 159) remarks, 'domination is not everywhere'. Indeed, even demarcation through borders, the essence of state territoriality, relies to a considerable degree on the extent to which networks of association and affiliation parallel the boundaries of domination. Consider, for example, the capacity of the International Monetary Fund to persuade and cajole states into adopting specific economic policies as well as to threaten to cut off loans. Even the seemingly most Westphalian of states, then, are riddled with power networks whose extension beyond territorial boundaries can render claims to absolute sovereignty moot but whose continuing presence inside the boundaries is critical to their credibility.

Political authority qua sovereignty requires (1) a governmental apparatus to serve as a final seat of authority and (2) an accepted definition of functional and geographical scope (territorial and non-territorial) beyond which its commands go unheeded and unenforced. Yet, lacking in discussing the range of representations and practices associated with sovereignty has been precisely a means of identifying the effects of co-variation in the effectiveness of state authority, on one hand, and its relative reliance on territoriality, on the other (e.g. Hurd, 1999). A useful approach to doing so comes from writing on the historical sociology of power. Specifically, the terms *despotic* and *infrastructural* power have been used by the sociologist Michael Mann (1984) to identify the two different ways in which a governmental apparatus acquires and uses centralized power. In other words, these terms identify, respectively, the two different functions that states perform and that underpin their claims to sovereignty: the struggle for power among elites and interest groups in one state in relation to elites and interest groups elsewhere, and the provision of public goods that are usually provided publicly (by states). In Mann's (1984: 188) words:

Let us clearly distinguish these two types of state power. The first sense [despotic power] denotes power by the state elite itself over civil society. The second [infrastructural power] denotes the power of the state to penetrate and centrally co-ordinate the activities of civil society through its own infrastructure.

Historically, infrastructural power has risen in importance since the eighteenth century. This is because elites have been forced through political struggles to become more responsive to their populations and, as a result, rising pressure groups have demanded more infrastructural goods. In turn, this gave a boost to the territorialization of sovereignty. Until recently, the technologies for providing public goods have had a built-in territorial bias, not least relating to the capture of positive externalities. Increasingly, however, infrastructural power can be deployed across networks that, though located in discrete places, are not necessarily territorial in the externality fields that they produce. Thus, currencies, systems of measure, trading networks, educational provision, and welfare services need not be associated with exclusive membership in a conventional nation-state (e.g. Rosenau, 2003). New deployments of infrastructural power can both de-territorialize existing states and re-territorialize membership around cities and hinterlands, regions, and continental-level political entities such as the European Union (Cox, 1998; Scott, 1998). International organizations, both private and state-run,

likewise have developed the capacity to deliver a wide range of public goods associated with infrastructural power. There is a simultaneous scaling-up and scaling-down of the relevant geographical fields of infrastructural power depending on the political economies of scale of different regulatory, productive, and redistributive public goods (Ilgen, 2003). Consequently,

... the more economies of scale of dominant goods and assets diverge from the structural scale of the national state – and the more those divergences feed back into each other in complex ways – then the more the authority, legitimacy, policymaking capacity, and policy-implementing effectiveness of the [territorial] state will be eroded and undermined both within and without. (Cerny, 1995: 621)

At the same time, despotic power (in Mann's sense) need not always involve a singular focus on fixed state territories if elites and pressure groups adjust their identities and interests to other territorial levels (such as city-regions, localities, and empires) or shift loyalties to non-territorial entities such as international organizations, corporations, social movements, or religious groupings (e.g. Gill, 1994; Kobrin, 1997; Cutler et al., 1999; Hall and Biersteker, 2002; Cutler, 2003). This could involve the enhancing of territorial hierarchy in pursuit of, for example, an imperium, or the attenuation of territorial sovereignty in the form of the diffusion of authority across a multinodal financial network involving banks, other states, and debt-rating agencies. There is no necessary association, therefore, between despotic power and central state authority. Both despotic-governmental and diffuse social power can work together to challenge central state authority. Consequently, either an up-scaling or a fragmentation of sovereignty can result as elites and social groups pursue their goals in ways that potentially territorially expand or undercut the authority of the central governmental apparatus, respectively. Governance cannot be reduced simply to the actions of governments nominally invested with state sovereignty within rigidly defined territorial boundaries.

But the recasting of the territorial basis to sovereignty and the challenge to central state authority through de-territorialization at the state level and re-territorialization at local and supranational scales of infrastructural and despotic power are uneven around the world (Albert et al., 2001). And, as noted previously, such trends are not invariably equivalent to the erosion of state sovereignty *tout court*. What is needed, therefore, is a typology of the main ways in which sovereignty is currently exercised to take account of (1) its social construction; (2) its association with hierarchical subordination (the fact that states are far from equal in the powers they can deploy and many are thus differentially subject to the actions of other states and non-state actors); and (3) its redeployment in territorial and non-territorial forms (Agnew, 2005). The two basic dimensions to the typology are defined by the relative strength of central state authority (state despotic power) on one axis and its relative consolidation in state territoriality (state infrastructural power) on the other. Regarded as social constructions, these define both the extent of state autonomy and the degree to which it is territorial in practice. Intersecting continua rather than discrete categories, four extreme cases can be identified nevertheless as ideal types for purposes of theoretical discussion and empirical analysis. These are relational in character, referring to how sovereignty is exercised effectively over time and space, rather than discrete territorial categories into which existing states can be neatly slotted. We refer to these four ideal types as sovereignty regimes, recognizing that any actual real-world situation might not exactly conform totally to a particular regime but involve several (Table 5.1).

Of the four exemplary categories, the *classic* example is the one closest to the story frequently told about absolute state sovereignty (the combination of strong central state authority and consolidated territoriality), although even here there can be complications (for example, on Hong Kong and Taiwan for China). The sense is one of despotic and infrastructural power still largely deployed within a bounded state territory (even if increasingly dependent on foreign direct investment and overseas markets for its exports) and a high degree of effective central state political authority. Contemporary China is a good test case, however, for how long absolute sovereignty can survive pressures for divisibility when increasingly open to the rest of the world.

The second case resembles most a story that emphasizes hierarchy in world politics but with networked reach over space as much as direct territorial control. This *imperialist* regime is in all respects the exact opposite of the classic case. Not only is central state authority seriously in question to those living in places where it prevails because of external dependence and manipulation as well as governmental corruption and chronic

mismanagement; state territoriality is also subject to separatist threats, local insurgencies, and poor infrastructural integration. Infrastructural power is weak (the state does little or nothing for its citizens) and despotic power is often effectively in outside hands (local elites are the clients of external authorities). This is an imperialist regime, if also reliant on the assent and cooperation of local elites, because the practice of sovereignty is tied ineluctably to the dependent political-economic status that many states endure in the regions where it prevails, such as that between the US and the states of Central America and that between France and many states in West Africa.

The other two cases are less familiar in relation to both conventional and critical perspectives on state sovereignty. The third regime is the *integrative*, represented today most obviously by the European Union. In this case, sovereignty has complexities relating to the coexistence between different levels or tiers of government and the distinctive functional areas that are represented differentially across the different levels, from EU-wide to the national-state and subnational-regional. But the territorial character of some of its infrastructural power is difficult to deny (consider the Common Agricultural Policy, for example), even if central state authority for both the entire EU and the member states is weaker than when each of the states was an independent entity. Quite clearly, many of the founding states of the Westphalian system have thrown in their lot with one another to create a larger and, as yet, politically unclassifiable entity that challenges existing state sovereignty in functionally complex and oftentimes non-territorial ways.

Finally, the fourth regime is the *globalist*. Perhaps the best current example of this is the effective sovereignty currently exercised by the United States within and beyond its territorial boundaries. Certainly, Britain in the nineteenth century also followed a version of this regime. But in both cases attempts have been made to recruit other states' elites into the regime, by co-optation and assent as much as by coercion. Indeed, globalization can be seen as the process (along with necessary technological and economic changes) of enrolling states in the globalist sovereignty regime marked by floating currency exchange rates, increased openness to foreign trade, and diminution of independent policy initiatives. From this viewpoint, the directing globalist state relies on hegemony, in the sense of a mix of coercion and active consent, to bring others into line with its objectives. The need to accept compromise in order to enrol others distinguishes this regime fundamentally from the coercive dependency of the imperialist regime.

Although US central state authority is relatively strong (notwithstanding the problems of its republican constitutionalism in coping with its global role), its centrality to world politics catches it

Table 5.1   Sovereignty regimes

		State territoriality	
		Consolidated	Open
Central state authority	Stronger	Classic	Globalist
	Weaker	Integrative	Imperialist

between two conflicting spatial impulses: one that presses towards a scattered imperium (as in contemporary Iraq) and one that pushes towards keeping the US as an open economy. The basis of its hegemony is welcoming of immigrants and foreign investment and goods and encouraging of these tendencies elsewhere. But, at the same time, the US is increasingly subject to fiscal over-extension as it endeavors to intervene globally yet also serve the demands of its population for pensions and healthcare benefits. It is this regime that is most closely identified with globalization and with the need to substitute the de-territorialized legitimacy of a wide range of actors (both public and private) for the classic link between territorialized popular legitimacy and absolute state sovereignty. The threat that helps maintain demand for at least a modicum of classic sovereignty is that the globalist regime will so undermine central state authority at home and abroad that borders must be, at least ritually, protected from the 'worst excesses' of globalization lest the regime is transformed into an imperialist one.

CONCLUSIONS AND IMPLICATIONS

This chapter has foregrounded the central importance of the discourse of state sovereignty to prevailing conceptions of political power and authority. We have argued that a naturalized conception of the state as an autonomous territorial subject still underpins many analyses of power, and we critiqued this conception on two counts. First, we countered the assumption that the state can be viewed as the singular subject of international politics. The sovereign state, we posited, is not the basis for, but the effect of, discourses of sovereignty, security, and identity. State identity and interest do not precede foreign policy, but are forged through foreign policy. This approach moves away from the usual focus on the international relations of states as the interaction between pre-given subjects, and instead emphasizes processes of subject-making in international politics. It dislodges discussions of the state from often under-theorized notions of identity and interest, and links these discussions to the increasing interest in processes of subjectification in both political geography and IR. We secondly showed that state power in the sense of effective sovereignty is not necessarily territorial. Sovereignty regimes, we suggest, offer a more differentiated view of how authority is actually exercised as they help us to understand the *extent* of state autonomy and the *degree* to which it is territorial in practice. This approach takes us beyond the question of states versus globalization as it is posed in much recent discussion in political geography, to the question of the different ways in which

political authority has been reconfigured and is now operating globally (Ferguson and Gupta, 2002).

For political geography, then, the key questions to address are not about the 'real' meaning or extent of state sovereignty in some general or universal sense, but, more specifically, about how state power is discursively and practically produced and spatially operationalized in both territorial and non-territorial forms. This line of investigation would advance political geography further toward studying the state as a process in its own right rather than as a pre-existing entity solely defining the geography of other social and political processes. It would enable political geography to demystify more effectively the practices of state power in the international sphere. It would also erode the unjustifiable intellectual division of labor whereby political geography tends to concern itself primarily with the internal workings of the state and leaves the state system to international relations.

REFERENCES

Agnew, J.A. (1994) 'The territorial trap: the geographical assumptions of international relations theory', *Review of International Political Economy*, 1: 53–80.

Agnew, J.A. (1999) 'Mapping political power beyond state boundaries: territory, identity, and movement in world politics', *Millennium: Journal of International Studies*, 28(3): 499–521.

Agnew, J.A. (2003) *Making Political Geography*. London: Arnold.

Agnew, J.A. (2004) 'Nationalism', in J.S. Duncan, N. Johnson and R. Schein (eds), *A Companion to Cultural Geography*. London and New York: Blackwell, pp. 223–37.

Agnew, J.A. (2005) 'Sovereignty regimes: territoriality and state authority in contemporary world politics', *Annals of the Association of American Geographers*, 95(2): 437–61.

Agnew, J.A. and Corbridge, S. (1995) *Mastering Space: Hegemony, Territory and International Political Economy*. New York: Routledge.

Albert, M. et al. (eds) (2001) *Identities, Borders, Orders: Rethinking International Relations Theory*. Minneapolis: University of Minnesota Press.

Allen, J. (2003) *Lost Geographies of Power*. Oxford: Blackwell.

Anderson, J. (1996) 'The shifting stage of politics: new medieval and postmodern territorialities?', *Environment and Planning D: Society and Space*, 14: 133–53.

Arendt, H. (1958) *The Human Condition*. Chicago: University of Chicago Press.

Aretxaga, B. (2003) 'Maddening states', *Annual Review of Anthropology*, 32: 393–410.

Ashley, R.K. (1987) 'The geopolitics of geopolitical space: toward a critical social theory of international politics', *Alternatives*, 12: 403–34.

Ashley, R.K. (1988a) 'Untying the sovereign state: a double reading of the anarchy problematique', *Millennium: Journal of International Studies*, 17(2): 227–62.

Ashley, R.K. (1988b) 'Living on border lines: man, poststructuralism, and war', in J. Der Derian and M. Shapiro (eds), *International/Intertextual Relations*. New York: Lexington Books, pp. 263–75.

Biersteker, T.J. (2002) 'State, sovereignty and territory', in W. Carlsnaes, T. Risse and B. Simmons (eds), *Handbook of International Relations*. London: Sage, pp. 157–76.

Biersteker, T.J., and Weber, C. (eds) (1996) *State Sovereignty as Social Construct*. Cambridge: Cambridge University Press.

Bigo, D. (2002) 'When two become one: internal and external securitizations in Europe', in M. Kelstrup and M. Williams (eds), *International Relations Theory and the Politics of European Integration: Power, Security and Community*. London and New York: Routledge, pp. 171–204.

Brenner, N., Jessop, B., Jones, M. and Macleod, G. (eds) (2003) *State/Space: A Reader*. Malden, MA: Blackwell.

Callinicos, A. (2004) 'Marxism and the international', *British Journal of Politics and International Relations*, 6: 426–33.

Campbell, D. (1993) *Politics Without Principle: Sovereignty, Ethics, and the Narratives of the Gulf War*. Boulder, CO: Lynne Rienner.

Campbell, D. (1998) *Writing Security: United States Foreign Policy and the Politics of Identity* (2nd ed.), Minneapolis: University of Minnesota Press.

Cerny, P.G. (1995) 'Globalization and the changing logic of collective action', *International Organization*, 49: 595–625.

Chouinard, V. (2004) 'Making feminist sense of the state and citizenship', in L.A. Staeheli, E. Kofman and L. Peake (eds), *Mapping Women, Making Politics: Feminist Perspectives on Political Geography*. New York: Routledge, pp. 227–43.

Clark, G.L. and Dear, M.J. (1984) *State Apparatus: Structures and Language of Legitimacy*. London: Allen & Unwin.

Cox, K.R. (ed.) (1998) *Spaces of Globalization*. New York: Guilford Press.

Cutler, A.C. (2003) *Private Power and Global Authority: Transnational Merchant Law in the Global Political Economy*. New York: Cambridge University Press.

Cutler, A.C. et al. (1999) 'Private authority and international affairs', in A.C. Cutler et al. (eds), *Private Authority and International Affairs*. Albany, NY: SUNY Press.

Dalby, S. (2002) *Environmental Security*. Minneapolis: University of Minnesota Press.

Dallmyr, F. and Rosales, J.M. (eds) (2001) *Beyond Nationalism? Sovereignty and Citizenship*. Lanham, MD: Lexington Books.

Davies, R. (2003) 'The medieval state: the tyranny of a concept', *Journal of Historical Sociology*, 16: 280–300.

Delbruck, J. (2003) 'Exercising public authority beyond the state: transnational democracy and/or alternative legitimation strategies?', *Indiana Journal of Global Legal Studies*, 10: 29–44.

Dodds, K. J. (2002) *Pink Ice: Britain and the South Atlantic Empire*. London and New York: I.B. Tauris.

Durand, M.-F. et al. (1992) *Le monde, espaces et systèmes*. Paris: Dalloz.

Edkins, J. and Pin-Fat, V. (1999) 'The subject of the political', in J. Edkins, N. Persram and V. Pin-Fat (eds), *Sovereignty and Subjectivity*. Boulder, CO: Lynne Rienner, pp. 1–18.

Environment and Planning A (1996) 'On the nation-state, the global, and social science' (Discussion forum), 28: 1917–84.

Ferguson, J. and Gupta, A. (2002) 'Spatializing states: toward an ethnography of neoliberal governmentality', *American Ethnologist*, 29: 981–1002.

Giddens, A. (1987) *The Nation-State and Violence*. Berkeley and Los Angeles: University of California Press.

Gilmartin, M. and Kofman, E. (2004) 'Critically feminist geopolitics', in L.A. Staeheli, E. Kofman and L.J. Peake (eds), *Mapping Women, Making Politics: Feminist Perspectives on Political Geography*. London: Routledge, pp. 113–26.

Gill, S. (1994) 'Structural change and the global political economy: globalizing elites and the emerging world order', in Y. Sakamoto (ed.), *Global Transformation: Challenges to the State System*. Tokyo: United Nations University.

Glassman, J. (1999) 'State power beyond the "territorial trap": the internationalization of the state', *Political Geography*, 18: 669–96.

Häkli, J. (2001) 'In the territory of knowledge: state centered discourses and the construction of society', *Progress in Human Geography*, 25(3): 403–22.

Hall, R.B. and Biersteker, T.J. (eds) (2002) *The Emergence of Private Authority in Global Governance*. New York: Cambridge University Press.

Hardt, M. and Negri, A. (2000) *Empire*. Cambridge, MA: Harvard University Press.

Hurd, I. (1999) 'Legitimacy and authority in international politics', *International Organization*, 53: 379–408.

Ilgen, T.L. (ed.) (2003) *Reconfigured Sovereignty: Multi-layered Governance in the Global Age*. Aldershot: Ashgate.

Jessop, B. (2002) *The Future of the Capitalist State*. London: Polity Press.

Kemp, A. and Ben-Elizer, U. (2000) 'Dramatizing sovereignty: the construction of territorial dispute in the Israeli–Egyptian border at Taba', *Political Geography*, 19: 315–44.

Kobrin, S.J. (1997) 'The architecture of globalization: state sovereignty in a networked global economy', in J.H. Dunning (ed.), *Governments, Globalization, and International Business*. Oxford: Oxford University Press.

Kofman, E. (2002) 'Contemporary European migrations, civic stratification and citizenship', *Political Geography*, 21: 1035–54.

Kratochwil, F. (1986) 'Of systems, boundaries, and territoriality: an inquiry into the formation of the state system', *World Politics*, 34: 27–52.

Kuus, M. (2002a) 'Sovereignty for security? The discourse of sovereignty in Estonia', *Political Geography*, 21(3): 393–412.

Kuus, M. (2002b) 'Toward co-operative security? International integration and the construction of security in Estonia', *Millennium: Journal of International Studies*, 31(2): 297–317.

Livingstone, D.N. (1993) *The Geographical Tradition: Episodes in the History of a Contested Enterprise*. Oxford and Cambridge, MA: Blackwell.

MacLeod, G. and Goodwin, M. (1999) 'Reconstructing an urban and regional political economy: on the state, politics, scale, and explanation', *Political Geography*, 18: 697–730.

Mann, M. (1984) 'The autonomous power of the state: its origins, mechanisms and results', *European Journal of Sociology*, 25: 185–213.

Mann, M. (1993) *The Sources of Social Power*, Vol. II: *The Rise of Classes and Nation-States, 1760–1914*. Cambridge: Cambridge University Press.

Marston, S.A. (2003) 'Space, culture, state: uneven developments in political geography', *Political Geography*, 23: 1–16.

Mountz, A. (2004) 'Embodying the nation-state: Canada's response to human smuggling', *Political Geography*, 23(3): 323–45.

Murphy, A.B. (1996) 'The sovereign state system as political-territorial ideal: historical and contemporary considerations', in T. Biersteker and C. Weber (eds), *State Sovereignty as Social Construct*. Cambridge: Cambridge University Press, pp. 81–120.

Murphy, A.B. (1999) 'International law and the sovereign state system: challenges to the status quo', in G.J. Demko and W.B. Wood (eds), *Reordering the World: Geopolitical Perspectives on the Twenty-First Century* (2nd ed.). Boulder, CO: Westview Press, pp. 227–45.

Newman, D. (2001) 'Boundaries, borders and barriers: changing geographic perspectives on territorial lines', in M. Albert, D. Jacobson and Y. Lapid (eds), *Identities, Borders, Orders: Rethinking International Relations Theory*. Minneapolis: University of Minnesota Press, pp. 137–51.

Offner, J.-M. and Pumain, D. (eds) (1996) *Territoires et réseaux. Significations croisées*. Tour d'Aigues: L'Aube.

Osiander, A. (2001) 'Sovereignty, international relations, and the Westphalian myth', *International Organization*, 55: 251–87.

O' Tuathail, G. (2000) 'De-territorialised threats and global dangers: geopolitics and risk society', *Geopolitics*, 3(3): 17–29.

O' Tuathail, G., Herod, A. and Roberts, S. (eds) (1998) *Unruly World? Globalisation, Governance and Geography*. London and New York: Routledge.

Painter, J. (1995) *Politics, Geography and Political Geography*. London: Arnold.

Palen, R. (2002) 'Tax havens and the commercialization of state sovereignty', *International Organization*, 56: 151–76.

Political Geography (2004) 'Reconceptualizing the state' (special issue), 23(3): 241–366.

Reynolds, S. (2003) 'There were states in medieval Europe: a response to Rees Davies', *Journal of Historical Sociology*, 16: 550–5.

Rosenau, J.N. (2003) *Distant Proximities: Dynamics Beyond Globalization*. Princeton: Princeton University Press.

Sack, R.D. (1986) *Human Territoriality: Its Theory and History*. Cambridge: Cambridge University Press.

Scott, A.J. (1998) *Regions and the World Economy*. Oxford: Oxford University Press.

Sharp, J.P. (2000) *Condensing the Cold War: Reader's Digest and American Identity*. Minneapolis: University of Minnesota Press.

Sidaway, J.D. (2002) *Imagined Regional Communities: Integration and Sovereignty in the Global South*. London: Routledge.

Sinclair, T.J. (2000) 'Reinventing authority: embedded knowledge networks and the new global finance', *Environment and Planning C: Government and Policy*, 18: 487–502.

Smith, A.T. (2003) *The Political Landscape: Constellations of Authority in Early Complex Polities*. Berkeley: University of California Press.

Spruyt, H. (1994) *The Sovereign State and its Competitors: An Analysis of Systems Change*. Princeton: Princeton University Press.

Stacy, H. (2003) 'Relational sovereignty', *Stanford Law Review*, 55: 2029–59.

Staeheli, L.A., Kofman, E. and Peake, L.J. (eds) (2004) *Mapping Women, Making Politics: Feminist Perspectives on Political geography*. London: Routledge.

Taylor, P.J. (1996a) *The Way the Modern World Works: World Hegemony to World Impasse*. Chichester and New York: John Wiley.

Taylor, P.J. (1996b) 'Embedded statism and the social sciences: opening up to new spaces', *Environment and Planning A*, 28: 1917–28.

Teschke, B. (2002) 'Theorizing the Westphalian system of states: international relations from absolutism to capitalism', *European Journal of International Relations*, 8: 5–48.

Teschke, B. (2003) *The Myth of 1648: Class, Geopolitics and the Making of Modern International Relations*. London: Verso.

Wæver, O. (1995) 'Securitization and desecuritization', in R.D. Lipschutz (ed.), *On Security*. New York: Columbia University Press, pp. 46–86.

Walker, R.B.J. (1993) *Inside/Outside: International Relations as Political Theory*. Cambridge and New York: Cambridge University Press.

Weber, C. (1998). 'Performative states', *Millennium: Journal of International Studies*, 27(1): 77–95.

Weldes, J., Laffey, M., Gusterson, H. and Duvall, R. (1999) 'Introduction: constructing insecurity', in J. Weldes, M. Laffey, H. Gusterson and R. Duvall (eds), *Cultures of Insecurity: States, Communities and the Production of Danger*. Minneapolis: University of Minnesota Press, pp. 1–33.

Wendt, A. (1999) *Social Theory of International Politics*. Cambridge: Cambridge University Press.

Yack, B. (2001) 'Popular sovereignty and nationalism', *Political Theory*, 29(4): 517–36.

State and Society

Stuart Corbridge

INTRODUCTION

This chapter reviews a small part of an enormous and growing literature on the topic of state and society. The focus is not on theories of the state in a conventional sense. Marxist, Foucauldian, pluralist and neo-liberal models of the state are all mentioned here, but emphasis is placed firmly on recent accounts of state/society interactions and on the constitution of what Partha Chatterjee has called 'political society'. The bias is to state/society relations in poorer countries, or the global South.

The chapter begins with the classic and enduring work of Max Weber. It considers ideal-typical models of how the state might solve the collective action problems that necessarily arise in society. It also reviews a body of work that has concerned itself with planning and developmental states. The darker side of unrestrained state actions is then explored with reference to the work of Steven Lukes, James Scott, Amartya Sen, and others. The second and larger part of the chapter reviews an emerging body of work on what might be called the anthropology of the state, or what Joel Migdal calls the 'state in society' tradition. To impose order on an unwieldy literature, this section is divided into sub-sections that deal with: (a) territoriality, boundaries, distance, and state failure; (b) rent-seeking, translation, and the actions of street-level bureaucrats; (c) democratization, decentralization, and good governance; and (d) civil and political society. A short conclusion considers avenues for future research.

SEEING LIKE A STATE

Max Weber famously argued that 'A state is a human community that (successfully) claims the monopoly of the legitimate use of physical force within a given territory' (Weber, 1958: 82). This definition reaches back to Hobbes' account of *Leviathan*. It confers legitimacy on a human institution that imposes order on a society that otherwise tends to anarchy or a brutal state of nature. Weber believed, however, that what made the modern state distinctive was its sense of service to the nation or the public. The emerging states of northern Europe were meant to respond efficiently to the demands of what Foucault (1997) would later call 'populations': to the needs of urbanites for clean water and sanitation, for example, or of children for state-provided education.

These states were to be elevated above society. Whether they were linked to a more or less restricted range of democratic practices and citizenship rights was not the main issue; the qualities and roles of public officials were to be essentially the same. Officials were to be recruited through competitive public examinations. They were then expected to rule without fear or favour, which is to say they were to behave impersonally and as bureaucrats. Family, community, or ethnic backgrounds were not meant to affect the activities of state personnel. Government servants would be properly rewarded. They would also inhabit defined public buildings or spaces (and perhaps uniforms), and would answer to their equally well-trained and gentlemanly superiors.

Weber knew that this model of the bureaucratic society was an ideal-typical model that was not met in full even in northern Europe. The idea that public officials should treat strangers on the same basis as close family members, which is the idea of generalized morality (Platteau, 1994), was a goal for states to aspire to rather than a description of actually existing affairs. It was also a goal that helped to defuse demands for the greater transfer of powers to ordinary men and women, or citizens. Weber, like Keynes and the British Fabians, not to mention Wilsonians in the United States, sometimes expressed the view that power needed to be exerted on behalf of ordinary people (for Keynes, less educated people: Skidelsky, 1992) by members of a governing caste that could think and act dispassionately. The state could then act for the general good over the long run.

This model of state/society separation proved to be an especially compelling ideology, or state idea, in the period following 1945. Marx might have looked forward to a classless and stateless society, but the Second International appealed to a vanguard state to effect the transition from capitalism to a dictatorship of the proletariat (Balibar, 1977). A much weaker version of the same ideology was advanced by many social democratic societies in Western Europe, including in the United Kingdom where the first governments of Harold Wilson (1964–70) looked to planning and a scientific elite to fuel the 'white heat of technological revolution' in British industry (Clarke, 1997).

These Promethean visions transferred easily to many ex-colonial countries, nearly all of which were charged with the twin tasks of nation-building and economic development. One of the conventional axioms of the development economics that emerged in the 1950s was that poorer countries had to generate sufficient savings to pay for a 'big push' in the direction of industrialization (Rosenstein-Radan, 1943; Nurkse, 1953; Gerschenkron, 1962). These savings mainly had to be generated from the populations of the poorer or developing countries. Foreign aid would play its part, but citizens of the 'Third World' had to be persuaded to discount the present in favour of the future. This is where the developmental state came in. A strong state was required to command or cajole poorer people to accept a regime of deferred gratification.

This ideology was most obviously on display in a socialist country like China, but it also informed the Five-Year Plans that began in democratic India from 1951. The state promised to act wisely in return. In India, key decisions were taken away from elected representatives, including Members of Parliament, and entrusted to the Planning Commission. This body was dominated in the 1950s by professional economists, and was headed by the Prime Minister, Jawaharlal Nehru (Zachariah, 2004). Armed with the new figures of state (data on GNP per capita, trend rates of the economy, foreign exchange reserves, monetary multipliers, input/output models, and so on), the Planning Commission presented Indian society with a road map to its future. The Second and Third Five-Year Plans, especially the (1956–66), under the leadership of physicist turned econometrician P.C. Mahalanobis), were the embodiment of an Olympian vision wherein an elevated state – a state that commanded the heights of the economy, to use the language of the time – was meant to look past present society toward its future.

The problem was that it did not. Lloyd Rudolph and Susanne Hoeber Rudolph (1987) have argued that the state in India was in 'command' of society up to the time of Nehru's death in 1964. Since that time, with the exception of the years when democratic rule was suspended (the Emergency, 1975–77), the state has been forced to respond to 'demand groups', including farmers, unionized workers and students (Rudolph and Rudolph, 1987). Its policies have become more sectional and focused on the short term as a result. This analysis chimes with work carried out by Pranab Bardhan (1984) and Atul Kohli (1990), among others.

In East Asia, however, it has been equally widely remarked that developmental states have been in place in some countries – certainly Taiwan and South Korea, and possibly China – since the 1950s. Bruce Johnston and Peter Kilby argued in the 1970s that the foundations of economic success in South Korea and Taiwan were laid by the radical land-to-the-tiller land reforms that were enacted after the Second World War (Johnston and Kilby, 1975). These reforms eliminated parasitic landlordism. They also led to rapid improvements in on-farm productivity, and ensured a mass rural market for goods like radios, bicycles and two-wheel tractors, all of which could be made by domestic industry. Johnston and Kilby maintained that the reforms were made possible by the high and unusual degree of separation from society (from domestic interest groups) that had been achieved by the governing states in both countries. Agrarian reform in South Korea was pushed through by the Americans. In Taiwan it was sponsored by the government of Chiang Kai-shek, itself incoming from mainland China.

Later work by Alice Amsden (1989) and Robert Wade (1990) confirmed this analysis. Amsden and Wade further argued that the governments of South Korea and Taiwan had overseen high and largely continuous rates of growth of exports and GNP per capita by getting relative prices wrong. That is to say, they took issue with the World Bank's account of the East Asian miracle. This account sought to portray the East Asian newly industrialized countries (NICs) as export platforms, or as countries that had prospered by opening themselves up to the disciplines of the world market (see World Bank, 1993). Amsden and Wade also rejected the argument, later to be restated by Paul Krugman (1994), that South Korea and Taiwan had achieved one-off gains from employing labour more effectively (long hours, low pay), rather than by raising the productivity of labour on a sustained basis. They suggested that the NICs had prospered by 'governing the market', to use Wade's felicitous phrase.

Developmental states in South Korea and Taiwan had successfully used their powers of separation from sectional interests to specify a social welfare function that could be fulfilled by an activist industrial policy.

This policy evolved over time and was based around careful sequencing. In the 1950s it involved autarkic trade policies for both countries, a high measure of import-substitution industrialization, and undervalued currencies (Akyuz et al., 1998). Industrial units, mainly in the private sector, were assured of government support for specified periods of time. They were also left in no doubt that protective measures would be withdrawn at an agreed date. In other words, a strong state insisted that industrial units should form rational expectations about future state actions. The state was not for turning, nor was it for sale (at least not in theory: see Khan and Jomo, 2000, for more considered discussion of so-called crony capitalism in East Asia).

Ha-Joon Chang (2002) has argued that the G8 powers are now trying to 'kick away the ladder' used successfully by South Korea and Taiwan, as well as by Germany, France, the US and Japan in the nineteenth and early twentieth centuries. In a provocative and often witty essay on the 'good governance' agenda, he argues that none of these countries developed (that is, industrialized) as democracies, or without corruption and healthy measures of government support. Chang returns to many of the arguments made by Friedrich List in the mid-nineteenth century (List, 1966). He also revisits some of the key ideas advanced by development economists in the 1950s and 1960s. Countries must industrialize to develop; they need to generate savings to industrialize; this is difficult to effect when countries are in a low-level equilibrium trap or where isolated agents maximize their short-run self-interest; a strong state is required to see beyond a 'society of individuals'; a developmental state makes this sighting and imposes order, deferred gratification, and direction to secure long-run economic growth; democratization may or may not follow development.

Other scholars disagree with some or all parts of this argument. Amartya Sen maintains that:

> We cannot really take the high economic growth of China or South Korea as a definitive proof that authoritarianism does better in promoting economic growth – any more than we can draw the opposite conclusion on the basis of the fact that the fastest-growing African country (and one of the fastest growing in the world), viz., Botswana, has been an oasis of democracy on that troubled continent. Much depends on the precise circumstances. (Sen, 2000: 149–50)

He further insists, in contradistinction to institutionalists like Wade and Chang, that:

> There is nothing whatsoever to indicate that any of these policies ['openness to competition, the use of international markets, a high level of literacy and school education, successful land reforms and public provision of incentives for investing, exporting, and industrialization'] is inconsistent with greater democracy and actually had to be sustained by the elements of authoritarianism that happened to be present in South Korea or Singapore or China. (ibid.: 150)

This is a comparatively weak argument, however, as I have shown elsewhere (Corbridge, 2002). Not only does Sen's list of policies miss out some initiatives where authoritarian (or developmental) regimes might have a comparative advantage (for example, in terms of discounting the immediate demands of social elites), but there is evidence to suggest that successful land reforms have not generally been carried out in 'democracies', where they have usually been blocked by landowning groups (Herring, 1983). A stronger argument, which Sen also makes at times, is that we cannot assume that most authoritarian states will be benign or developmental. Most authoritarian regimes are not imposed from outside, as was largely the case in post-war Taiwan and South Korea. They emerge from powerful domestic groups and maintain links with those communities. In addition, as Steven Lukes (1985) showed in his account of *Marxism and Morality*, and as was discussed by novelists including Arthur Koestler (1976) and George Orwell (1980), the juxtaposition of centralized power and Promethean ideology is often very dangerous. A philosophy in which the ends justify the means was partly responsible for some of the worst tragedies of the twentieth century (see also Peffer, 1990).

James Scott has generalized this argument in his account of *Seeing Like a State* (Scott, 1998). Scott does not discuss the case of successful developmental states. They are likely to be exceptions to a more general rule of failure. Everyday life, Scott suggests, is based around complexity and is aided by diffuse flows of information, a point on which neo-classical economists will agree. When states brush this complexity aside, they encourage the production of 'transforming visions' that result in dangerous 'simplifications'. Compulsory villagization in Tanzania, scientific forestry, single-crop agricultural systems, and Le Corbusier's urban experiments in Brasilia and Chandigarh are all cited as examples of 'schemes that failed to improve the human condition' (the subtitle of Scott's book). Each of these ventures was at odds with the ways that ordinary people wanted and needed to run their lives – on the basis, in part, of diversity and the security that comes from not putting all one's eggs in one basket. Far worse, in terms of preventable deaths, were the grotesquely misnamed Great Leap Forward in China (1958–61, which led to 30 million famine deaths: Sen, 1989) and the forcible collectivization of parts of the Soviet countryside.

In Scott's view, moreover, these tragedies were enforced by high modernist regimes that sought precisely to uncouple the state from society, as many political ideologies were then suggesting (or demanding) should be the case. What he calls 'full-fledged disasters' occurred when four elements came together: first, an administrative ordering of nature and society, or a set of transformative state simplifications; second, a high modernist ideology that evinced a muscle-bound self-confidence in scientific and technical progress; third, an authoritarian state that was willing and able to bring high modernist designs into being; and fourth, a prostrate civil society that lacks the capacity to resist these plans.

It is important to note that Scott is not opposed to 'development', or even state-directed development. He also accepts that all states have to engage in the 'simplifications' that produce citizens as taxpayers (through various codes) or as engineers who need scientific training. Scott's work is important, nonetheless, because it begins to point to a political geography of extreme violence in the twentieth century: a geography that is able to locate some of that century's worst tragedies against the four-point grid that is provided in *Seeing Like a State*. It also points in the direction of a theory of 'sightings', in this case of present and future societies by the state. Scott argues that the high modernist state is dangerous because it arrogates to itself the powers of sight, including those of hindsight and foresight. Catastrophes follow in the wake of this 'monocularism'. What Scott does not properly discuss, however, is the possibility that the connections between states and societies in many poorer countries today are typified less by an excess of state vision than by its relative absence. Nor, in this book, does he discuss the question of how poorer people see the state, although this theme is developed in two earlier works, *Weapons of the Weak* (1985) and *Domination and the Arts of Resistance* (1990).

SEEING THE STATE

The traditions of political science and sociology that are most indebted to Max Weber are inclined to present the state as a unitary actor with far-reaching, if hopefully well-disciplined, powers. This is true, moreover, and perhaps by definition, of many state-centred social theories, including realist theories in international relations (see Agnew, 1998, for a review).

In contrast, Western Marxism has insisted on the priority of class, both ontologically and epistemologically. Some of its proponents have been keen to portray the state as an executive agency of the bourgeoisie (for key debates, see Miliband, 1969; Poulantzas, 1978; Therborn, 1978). The state is required to legitimize capitalism as it enters periodic crises of profitability and adjustment. It also has to ensure that property rights and the rule of law are directed to the long-run interests of capital. But if Marxism has resisted the idea that states can be fully autonomous of class interests, it has often assumed, nonetheless, that the state can be made to do the bidding of various ruling elites. The capacity of the state, in other words, has not been in doubt; indeed, very often, its repressive capabilities, in terms of the disciplining of labour or the promotion of accumulation by dispossession, have been highlighted and properly critiqued (Thompson, 1968; Breman, 1985; Perelman, 2000). Foucauldians, too, for all that they conceive of 'the state' in more dispersed terms – in terms, indeed, of a set of apparatuses, practices, and technologies of rule that emerge from governmental obligations to populations – focus strongly on the state's far-reaching powers of surveillance and social reach (McNay, 1994). In human geography, especially, the emphasis that some followers of Foucault have placed on the Panopticon has encouraged a view of the state as an object, or set of relations, that tends to fill social and geographical space (see Crampton and Elden, 2005).

Pluralists, for their part, as well as many public choice theorists, have been reluctant to grant the state such powers. Robert Dahl (1961, 1971, 1985) prefers to think of state units, from the local to the national, as polyarchies, in which power is dispersed among contending social groups. Politics is then about negotiation, and the politician serves as a bridge between the state and the society. Mancur Olson (1982), meanwhile, from the perspective of public choice theory, places emphasis on the actions of individual state agents and on an associated range of principal/agent problems. Principals need unclogged information flows and enforcement mechanisms to ensure that agents do their jobs.

These perspectives have the effect of decentring the state, both positively and normatively. To some degree they influence the bodies of work we consider below. This is particularly so in regard to a prospectus for decentralization and good governance. Before we turn to these issues, however, it will be helpful to say something about the territoriality of the state, and what might be called the cultures of translation that exist within 'it'. Recent work on these issues has challenged some prevailing assumptions about the power of the state, on the one hand, and about the languages that might link principals to agents on the other.

Territoriality, Boundaries, Distance and State Failure

The so-called war on terrorism has prompted renewed interest in the failure of some states to govern the territories that fall under their jurisdiction (Bobbitt, 2002; Harvey, 2003). The work of Benedict Anderson has provided a starting point for some of these discussions. In addition to the classic machineries of state that are required to 'make' nations – well-policed boundaries, a functioning bureaucracy, a common currency, and so on – Anderson insists that nations must be imagined as 'communities' that link strangers to one another. Nations hold together to the extent that their members come to believe that they share 'a deep horizontal comradeship. Ultimately it is this fraternity that makes it possible over the past two centuries for so many millions of people, not so much to kill, as be willing to die for such limited imaginings' (Anderson, 1983: 7; see also Mann, 1984).

Anderson suggests that citizens were linked to each other in early-modern Europe by the spread of print capitalism. Newspapers and journals were read, and were discussed in coffee houses or at the workplace. More recently, the development of the railway system helped turn peasants into Frenchmen, as Eugene Weber put it in 1976, and a continuing annihilation of space by time is evident in countries that have well-developed road networks and functioning telecommunications systems (including access to the Web). People are said to interact closely even when they are at a distance. They consume events like an international cricket match as part of a collective libidinal experience (see Rajagopal, 2001). A series of public rituals are put into play that establishes a sense of shared history and a common identity. Significantly, many of these rituals are aimed at men, as the cricket match perhaps suggests (not to mention the idea of turning peasants into Frenchmen). The production of national anthems and flags is just the most obvious moment in this process of political production.

Anderson's work has been linked to a body of work on social memory. The idea that memories can be held collectively was developed in the inter-war years by Maurice Halbwachs and Frederic Bartlett, as Steven Legg has shown in an important essay. This work was later extended by Pierre Nora (1989). Working with a large group of fellow French scholars, Nora and his team published seven volumes of essays between 1984 and 1992 that focused 'on French memories of the republic,

the nation, and France itself' (Legg, 2005: 4). These memories were embodied in particular sites, places, or realms of memory (Nora's *lieux de mémoires*). 'These sites could be "physical", such as commemorated locations or statues, "symbolic", including ceremonies and pilgrimages, or "functional", for instance associations or dictionaries' (ibid.).

As Legg remarks, Nora has placed particular emphasis on the ways in which these sites of memory – war memorials, for example – establish a continuing and in this case nostalgic sense of 'Frenchness' in a country that has lost its peasant culture and 'the quintessential repository of collective memory' (Nora, 1989: 7) that supposedly went with it. Nora positions the state's attempts at promoting sites of memory against an imagined past in which memories of Frenchness – *milieux de mémoire* – were passed seamlessly down the generations.

Other students of social memory have been less keen to hitch their work to a project that celebrates a distant, less plural, and often frankly masculine sense of nationhood. Recent work has sought to establish the contested nature of memories, and thus of social identities. Nuala Johnson, for example, in her study of *Ireland, the Great War and the Geography of Remembrance*, reminds us that the nationalist community fought strongly against British memorials to the Irish war dead in Dublin. She also draws attention to the creation of 'anti-monuments' in symbolically important localities, and thus to the fact that 'Social memory is never a simple empty space awaiting manipulation by the powerful. Rather it is a messy space where competing and at times conflicting memories are accumulated, accreted, refined and sometimes challenged' (Johnson, 2003: 170).

Debates over the location and substance of public memorials continue to attract scholarly attention in a range of countries, including at Ground Zero in New York City. Derek Gregory reminds us that 'Ground Zero' was first named for Hiroshima in a very different geopolitical conjuncture (Gregory, 2004; see also Abu El-Haj, 2001; Simpson and Corbridge, 2006). There is also considerable academic work on the role of boundary markers, and on the production of people as citizens, refugees, immigrants and asylum-seekers. Not surprisingly, a lot of this work has focused on the US–Mexico border and on entries to Fortress Europe (for overviews, see Durand and Massey, 1992; Davis, 2001; Spencer, 2003; Cornelius et al., 2004). This work links to a body of work on transnational citizenship, and on the multiple links that people increasingly are forced to negotiate to states and societies in the plural (see Gilroy, 1995; Ong, 1998; Brettell, 2000). Important work is also emerging on the developmental significance of international migration. Attention is shifting away from remittances to what is now being called 'brain circulation' and the investment decisions of diasporic communities (see Borjas, 1994; Mountford, 1997; Meyer and Brown, 1999; House of Commons, IDC, 2004). Still other commentaries consider the shaming inequities of a regime of 'globalization' that promotes the free flows of goods and services but not of people (see Rodrik, 1999; Walmsley and Winters, 2002; Hayter, 2004).

For many people in the global South, however, the prospect of joining what Hardt and Negri (2000) call the international 'migrant hordes' (or the 'multitude', a political collectivity that is romanticized here to the point of absurdity) is a long way off. Migration is central to the livelihood strategies of people throughout Latin America, Africa, and Asia, but it is most often conducted within the borders of a given region. If borders are crossed, this is because they are porous and not well marked. This is the case even between parts of West Bengal, India, and Bangladesh, despite the presence in India of a well-staffed Border Security Force (Véron et al., 2003). In many parts of sub-Saharan Africa the construction of borders, and thus in one sense of territorial states, is less well established. Metsola (2004) and Zeller (2007) are carrying out work in classic anthropological fashion to examine the ways in which the production, policing and use of border crossing points between Namibia and Zambia is helping (or not) to manufacture a sense of the nation as an imagined community and as a functioning territorial state.

This work contributes to a rich debate on the construction of 'states' and 'societies' in post-colonial Africa. Recent studies have probed the so-called criminalization of the state in West Africa (Bayart, 1993; Bayart et al., 1999; and, more critically, Chabol and Daloz, 1999), and the ways that different ethnic groups that were invented or set in stone by the colonial powers act as gatekeepers to government resources (Cooper, 2002). Mahmood Mamdani (1996) has argued that the apartheid state in South Africa was the logical culmination of forms of indirect rule in Africa that were pioneered by the British, and that amplified divisions along the lines of 'race' and geography (urban/rural). The construction of a sense of citizenship in these circumstances has proved difficult. Achille Mbembe (2001), meanwhile, has challenged accounts of the African 'tragedy' that fail to engage with the 'political economies of brutality' that emerged under colonial rule. His important study of the post-colony resonates with further work on violence and youth militancy in some African countries (Richards, 1996), and on the ways in which politicized Islam is becoming inflected in local struggles over resources (Roy, 1994; Watts, 1996). A further corpus of work has examined the ways in which state formation in Africa has been affected by Cold War geopolitics (French, 2004), or has been

damaged by continuing problems of debt and trade discrimination (Arrighi, 2002).

Perhaps the most ambitious work on 'states and power in Africa', however, has been conducted by the American political scientist, Jeffrey Herbst (2000). Herbst argues that sub-Saharan Africa faces a future of deepening 'state failure' for two reasons. First, because of geography and demography. Many African societies, he argues, exist at low population densities and are stretched over enormous distances. The political centre finds it difficult to maintain control over the periphery of 'the country', particularly where geographical conditions conspire against the building of efficient transportation systems. Second, because of the post-1945 international settlement, as embodied in the idea of the United Nations. Herbst maintains that political elites are able to access foreign aid only if they present themselves as the rulers of a country. The international system, that is to say, is not set up to recognize the sorts of regional kingdoms or tribal territorial units that Herbst believes would have provided a better foundation for security and state-building in post-colonial Africa. States and societies are at odds with one another.

Rent-Seeking, Cultures of Translation and Street-Level Bureaucrats

Herbst's work has been criticized for paying too little attention to the colonial legacy in sub-Saharan Africa, and for paying too much attention to brute environmental conditions. Similar criticisms have been leveled at Jeffrey Sachs' work on 'tropicality'. Sachs maintains that economic and political development in non-coastal Africa have been impeded by extreme climatic conditions, poor soils, and costly access to the international market-place (Sachs, 2000; Sachs et al., 2001; see also Watts, 2003).

More so than Sachs, however, Herbst is able to direct attention to the ways in which some states in Africa have developed in the service of recognizably urban constituents, a point made in the 1970s by Michael Lipton (1977) and again in the 1980s by Robert Bates (1981, 1989). Bates argued that the crisis of food production in sub-Saharan Africa was centrally bound up with this 'urban bias'. Governments used Marketing Boards and rigged pricing systems to suck food cheaply from the countryside to the city. They did so because they feared urban unrest and possible coup attempts. In time, however, this caused equally rational economic agents in the countryside to either grow or sell less food. This induced the very food riots that ruling elites feared. It also created a moment of opportunity for external agencies, including the IMF and the World Bank. During these emergency situations, Bates argued, the Bretton Woods institutions could link concessional loans to adjustment programmes that would renegotiate (that is, shrink) the role of the state in African societies.

Bates' work can be linked to a much broader body of work on the ostensibly 'over-developed' nature of the state in the ex-colonial world (and not just there, of course). This corpus of work emerged out of a critique of mainstream development economics. Albert Hirschman, for example, argued in 1970 that the inefficiency of the Nigerian railway system was caused by a 'combination of exit and voice [that] was particularly noxious' (1970: 45).

> [E]xit did not have its usual attention-focusing effect because the loss of revenue was not a matter of gravity for management [which could dip into the public treasury in times of deficit], while voice was not aroused and therefore the potentially most vocal customers were the first ones to abandon the railroads for the trucks. (ibid.)

Fellow Nigerians were then saddled with an inefficient and over-subsidized public railway system, and an arena of exchanges between officialdom and ordinary citizens that encouraged 'an oppression of the weak by the incompetent and an exploitation of the poor by the lazy' (ibid.: 59).

Less generous critics than Hirschman would generalize this critique of the developmental state. In a landmark article published in the *American Economic Review*, Anne Krueger, later a Vice President of the World Bank at the time of the second Reagan administration, wrote of the political economy of the rent-seeking society. Krueger argued that the systems of protection (tariffs and quotas) and industrial support (licences and permits) that had been set up by many post-colonial states in the 1950s and 1960s had encouraged the growth of unproductive economic behaviour. Prospective entrepreneurs and traders were required to spend time and money chasing the permits that would allow them to do business. The process corrupted many of the public officials who distributed these permits, as they began to charge fees for what strictly were unnecessary services. The scale of this fee-collection business later made economic reforms difficult. In India, for example, where Krueger (1974: 295) estimated that 'the total value of rents' (mainly from import licences) was equivalent to '7.3% of national income' in 1964, civil servants were reluctant to see the License and Permit Raj dismantled.

Other studies have concluded that the level of rent-seeking in countries like India and Turkey has been less than Krueger suggests (Waterbury, 1993; though see Osborne, 2001). Mushtaq Khan and K.S. Jomo (2000) have further argued that it is the quality, not the quantity, of rent-seeking (or corruption) that matters. In their view, rent-seeking in the East Asian NICs was just as rampant as in South Asia, but fees were collected in support of an industrial policy that made good economic sense. There is some evidence to suggest, as well, that reforming governments can achieve many of their goals, even in a country like India. Rob Jenkins (1999) has argued that effective political leaders make deals behind the backs of the electorate, using side-payments to weaken (or buy off) political opposition. In his view, the process of economic liberalization in India since 1991 has in large part been devolved to State governments, the most effective of which (such as Andhra Pradesh) have accessed funds from the donor community to soften the costs of their 'adjustment' policies.

This may well be so. Jenkins would certainly recognize, however, that states do not always get their own way. (The architect of the reforms in Andhra Pradesh, Chandrababu Naidu, was removed from power in 2004 by voters disenchanted with his bias in favour of the urban middle classes.) There is now a significant body of work which suggests that states should not be thought of as unitary actors, even where the borders of the state have been secured. Joel Migdal, most notably, has called for a 'state-in-society approach that differs from Weberian political science'. The state must be seen 'as a field of power marked by the use and threat of violence and shaped by (1) the image of a coherent, controlling organization in a territory, which is a representation of the people bounded by that territory, and (2) the actual practices of the multiple parts' (Migdal, 2001: 15–16; see also Abrams, 1988). These 'actual practices', Migdal contends, are often far removed from official mythologies of the state. Few state agencies achieve the separation of the public from the private that Weber called for. This might happen to some degree in the 'commanding heights' of the state, where policies are drawn up, but it is less evident as policies are moved down through the 'agency's central offices' to the 'dispersed field offices' and finally to the 'trenches'. Particularly in the trenches, men and women who are tax collectors, police officers, teachers, and healthcare workers are required to balance the pressures placed upon them by supervisors (the principals) with those brought to bear by powerful figures in political society (ibid.: 118).

There is now an impressive body of work on these lower-level public officials – the men and women that Lipsky (1980) called 'street-level bureaucrats'. Vasan (2002) offers an interesting ethnography of the life of a forest guard in central India. Villagers often keep the forest guard at a distance. They fear that he might convict them of a forest offence. For his part, the guard learns not to enforce his power too much, or arbitrarily. The guard often comes from the same community as some of the villagers, and he might take food or accommodation from them. The division between public and private space becomes blurred. The guard also knows that he might be attacked in the forest if he fails to establish some degree of rapport with local people.

Other work has begun to explore what Fuller and Benei (2001) call the 'anthropology of the state' in terms of the movement of files (Tarlo, 2003), practices of queueing and queue-jumping (Corbridge, 2004), the organization and iconography of public buildings (Humphrey, 2002), and the ways that public officials themselves 'see' the state (in terms, perhaps, of local understandings of 'corruption', or even of dealing with superiors: Gupta, 1995; Robbins, 2000; Jeffrey, 2001; Corbridge et al., 2005). This work disrupts the idea that it is the state that mainly constructs 'subject positions', in the sense that some Foucauldians would understand that term. Most states have an agenda, explicit or otherwise, for the production of modern citizens – men and women who will govern themselves – but this agenda has to be put in play by people who might not share the ambitions, or perhaps even the vocabularies, of those who first enact public policy.

The possibility of 'slippage' has been a consistent refrain in the work of Sudipta Kaviraj (1984, 1991) and Partha Chatterjee (1993, 1997) in India. Kaviraj maintains that the developmental ambitions of the central Indian state, particularly as they were set out, in English, by Nehru and other members of elite society, were never shared by men and women schooled in the values of vernacular society. He suggests that the world-views of elite and subaltern Indians co-exist like oil and water, and that the proposals of the former are sometimes translated out of sight by the latter. The result, Kaviraj suggests, is that an apparently strong state has feet made of vernacular clay. Seen from this perspective, the disappointments of India's experiments with planned development were caused not only by '[a] reliance on the bureaucracy to translate the plans of the intellectuals to fulfil a cultic role as bearers of Reason's flame to the society' (Khilnani, 1997: 88), but also, and rather more so, by the unwillingness of ordinary men and women in India to see anything special, and far less anything sacred, in the new temples of industrial modernity that the government set before them as markers of their own futures.

What is required, Kaviraj suggests, with support from colleagues as diverse as Barbara Harriss-White (2003) and Ashis Nandy (2003), is a much closer understanding of the social construction of the state practices to which Migdal refers, and which Foucauldians are more inclined to call human technologies of rule (Dean, 1999; Rose, 1999). These practices and technologies deploy a hybrid understanding of state-in-society relationships. Especially in states where principals are weak, and where accountability mechanisms are not generalized, it is unrealistic to assume that lower-level state agents will behave as the rulebook says they should.

Democracy, Decentralization, and Good Governance

The apparent failure of some states, and the overbearing nature of others, has prompted a considerable and diverse literature on new pathways to good – or at any rate better – governance. These pathways propose new relationships between the state and society, or between governmental agencies and the populations/citizens/subjects/clients they are meant to serve. In many of the world's richest countries the debate has been led for the past thirty years by a loose collection of Austrian and constitutional economists, and monetarists. For convenience, perhaps, these scholars and activists have sometimes been labelled 'neo-liberals'. Margaret Thatcher, famously, was said to be in awe of the monetarist ideas propounded by her closest intellectual confidante, Keith Joseph (Hirst, 1989; Young, 1990). She also commended the work of Friedrich von Hayek and Milton Friedman to members of her Cabinet. In some degree, the ideas of these scholars stood behind the programmes of privatization, council house sales and published monetary targets that we associate with her governments.

Many of these ideas are being continued today. Hayek believed that unrestricted markets should be valued less for their efficiency effects than as a defence against state power. Markets promote liberty and the freedom of the individual (Hayek, 1960). State interference in the economy would lead inevitably to serfdom, a point he made in 1944 in opposition to Keynesian policies of demand management (Hayek, 1944). Milton Friedman, too, became famous in the 1970s for his aphorism, 'Inflation is always and everywhere a monetary phenomenon' – by which he meant that governments alone must bear responsibility for putting too much money in circulation. The central failing of governments in the post-1945 era was a willingness to fund social programmes and public-sector pay awards with borrowed or simply printed money (Friedman, 1968). The proper job of government was to set consistent and responsible monetary targets and to abide by them. Governments should also take steps to ensure that all citizens will bear at least some of the cost of major social programmes, such as refuse collection, education or health (to the extent that these are provided in the public sector). Margaret Thatcher's justification for the council charge (poll tax) in the late 1980s was precisely that it engineered a new, and in her view more responsible, framework for state/society interactions at the local government level (see also Buchanan 1967). Under the previous system of rateable (i.e. property) value assessments, the cost of providing local services fell very largely on homeowners and businesses. Some of the major beneficiaries of local government spending had no incentive to vote for fiscal restraint. A graduated council charge that would fall on everyone was meant to change this.

Opposition to the council charge/poll tax did much to bring down Mrs Thatcher. But the idea that governments should be creating 'opportunity' – rather than 'welfare' – societies has become a consistent refrain in many richer countries, and has been the subject of considerable research. As Jamie Peck (2001) explains, the rhetoric continues to centre on the empowerment of individuals, and on the idea that individuals can make better use of their money than can governments. It is also linked to an ideology of choice: that parents should be free to choose a school for their child, for instance, perhaps with the help of an education voucher; or that working men and women should be free to choose a pension scheme or medical plan that will best meet their future needs (see Smith and Easterlow, 2004, and consider proposals emanating from the government of George W. Bush for Health Savings Accounts, Individual Retirement Accounts, and Personal Reemployment Accounts). In some respects, the Web has become an ally of this 'new public administration'. Individuals are constituted as reasoning men and women, who will make informed choices on the basis of cost comparisons that are now available online. The state is constituted in turn as something of a dinosaur. It is a relict of a previous age, in urgent need of being slimmed down and reinvented as an agency of enablement and the last resort (see Mead, 1986; Kettl, 1993).

The adoption of neo-liberal policies has in most countries been linked to a significant increase in inequality in income and wealth. This is most evident in the United States, where it has been accompanied by the rise of gated communities, privatized security systems and a frankly carceral society (particularly for African–Americans: Brown, 1999; Wacant, 2002). A policy of gambling on the rich was taken to extremes with the tax cuts of 2002, about one-third of the value of which went to the richest 1 per cent of taxpayers. President George W. Bush presided over a tax policy reform that shifted the burden of taxation to working Americans and away from those with dividend incomes (Friedman, 2004). In theory, this was meant to encourage Americans to save, particularly at a time of growing budget and trade deficits. In practice, however, the benefits went almost exclusively to wealthy Americans; middle-income Americans (those on $30,000–70,000 a year) found it difficult to save anything at all after paying for housing, health-care, insurance, college tuition fees and the daily

necessities of life. Cuts in welfare were particularly acute for single or lone mothers and their children (Edin and Lein, 1997).

In poorer countries, the new public administration has been concerned less with privatization programmes, at least since the problems in the ex-Soviet Union became apparent, than with a programme of government reform that is intended to bring the state closer to the people. Agendas for good governance have taken shape in the wake of the crisis of food production in sub-Saharan Africa, and following the end of the Cold War. They embrace a number of elements, of which five stand out.

First, there is a call for democratization, particularly but not exclusively in erstwhile authoritarian states. Democracy is understood to be a continuous, not a discrete variable (Prezworski et al., 2000). In sub-Saharan Africa, especially, it has been suggested that urban bias can be cured in part by multi-party elections that would enfranchise voters in rural areas (the mass of the population in most cases). Amartya Sen has also argued that famines do not occur in democracies (Sen, 2000). In countries with elected assemblies, an independent judiciary and a free and activist press, there is little chance that stories of incipient famine will remain unfiled, or that crisis situations will be ignored (for a partial critique, see Sainath, 1996.)

Second, there is the linked suggestion that democratic government should be brought closer to the people. This is the agenda of decentralization, broadly conceived. As James Manor (1999) points out, however, it is important to distinguish between deconcentration, decentralization and devolution, all of which may or may not proceed together. Deconcentration involves the abolition, or more usually contracting out, of services provided directly by government agencies. Immunization programmes can be handed over to non-governmental organizations, refuse collection services can be provided by private firms or cooperatives. The decentralization of government properly involves the relocation of governmental authority structures. This might take the form of regional assemblies in the UK, or the system of Panchayati Raj that has been put in place in India following the 92nd and 93rd Constitutional Amendment Acts. In the latter case it also involves a measure of affirmative action for women. Devolution then refers to the delegation of powers and funds to these new bodies. Considerable research is now being devoted to the economic and political effects of 'decentralization' programmes. (Useful reviews can be found in Rodríguez-Pose and Busire, 2004, and Crook and Manor, 1998.)

Third, there is an injunction in favour of participation or participatory development. Parents are encouraged to be active in school oversight committees in the US and India, or in the management of the hospital trusts set up recently in the UK's National Health Service. In poorer countries, an ideology of participatory development has helped to promote development projects that draw on participatory poverty assessments (Chambers, 1994). These are conducted by or for the intended beneficiaries of an aid project or government scheme. In India in the 1990s the largest targeted poverty-alleviation programme was the Employment Assurance Scheme (EAS). This scheme offered up to 100 days of paid employment to up to two members of households that were defined as Below the Poverty Line (BPL). Since most of the payments were in the form of wages for labouring men and women, it was assumed that the rate of leakage of funds would be low. Better-off families would not be seeking paid work. More innovative, however, was the requirement of the EAS that members of poor labouring households should demand work from the local council or Block Development Office. Government officials were then required (or meant) to respond to the 'demands from below' by authorizing public funds for labour-intensive building projects. Significantly, again, the type of project to be undertaken was meant to arrive at an official's desk after it was first discussed in a village open meeting. Development was defined as a process of participation as much as a set of tangible outcomes. (For discussion of these and similar schemes, see Mohan and Stokke, 2000; Corbridge and Kumar, 2002; Platteau and Abrahams, 2002; Srivastava et al., 2002.)

Fourth, there is the question of accountability. There is broad agreement that public servants must be accountable for their decisions, and for their use of public funds. In India, the Mazdoor Kisan Shakti Sanghatan (MKSS) in Rajasthan has pioneered the use of village-based 'social hearings', or Jan Sunwais, to review the activities of public and elected officials who might be thought to be embezzling funds, or favouring one village unfairly over another (Jenkins and Goetz, 1999). The MKSS has also invested in photocopiers in an effort to create a form of paper memory, in this case of how local government says it intends to spend funds on the EAS or some other scheme. In the second half of the 1990s the organization scaled up its demands for information circulation and accountability to the State level. Partly in response to pressures from the MKSS, the Rajasthan State Legislature passed a Right to Information Law on 1 May 2000. Perhaps more significantly, the State Assembly moved in the same month to pass an Amendment to the Panchayati Raj Act that created the legal entity of the Ward Sabha and vested it with powers of social audit. The MKSS has also joined forces with the National Campaign for the People's Right to Information (NCPRI) in New Delhi.

Lastly, there have been initiatives to improve the pay and conditions, training and career trajectories, and general professionalism, of public officials. The World Bank has invested heavily in South Asia in surveys of public attitudes to government servants. It has also commissioned surveys of the ways that lower-level public servants react to the various pressures placed upon them. Elsewhere, Merilee Grindle and Judith Tendler have produced pioneering studies of what might be called 'good governance' success stories. Grindle (2000, 2004) has focused on the contexts for 'audacious reforms' in Latin America, including in the education sector. Tendler (1997), for her part, has demonstrated that the recruitment on merit of highly motivated female nurse-supervisors in Ceará State, Brazil, had remarkably positive effects on local rates of vaccination and infant mortality. (For broad critical commentary, see Leftwich, 1993; World Bank, 1994; Jenkins, 2002.)

Civil Society and Political Society

The new public administration is mainly concerned with the reform of government. This has the aim of making public servants more responsive to their clients. It is not unusual, however, to find prescriptions for greater decentralization or accountability linked to paeans to the virtues of civil society (Escobar, 1995). Civil society has been defined as that 'arena where numerous social movements (including community associations, women's groups, religious bodies, and intellectual currents) and civic organizations (of lawyers, journalists, trade unions, entrepreneurs and so on) strive to constitute themselves into an ensemble of arrangements to express themselves and advance their interests ... Civil society ... functions as the citizen's curb on the power of the state' (Haynes, 1997: 16, drawing on Stepan, 1988).

Civil society can also be defined in more expansive terms. In the 1990s, Robert Putnam offered accounts of why Americans bowl alone (Putnam, 1995), and why democracy seems to work better, and more effectively for the promotion of economic growth, in north Italy as compared to south Italy (Putnam et al., 1993). Putnam focused on social capital as his independent variable, just as the World Bank would later hold up social capital as the 'missing link' in studies of differential development (see Woolcock, 1998). Putnam argued that social capital resided in communities. It was not, as both Robert Coleman (1988) and Pierre Bourdieu (1986) had argued previously, a private resource of households. He further claimed that the formation of dense networks of horizontal ties between family members (bonding social capital) and between people from different community backgrounds (bridging social capital) helped to produce patterns of trust that underpinned high levels of economic performance. People who socialized together did business together, at least in northern Italy. The formation of social capital also helped to protect communities against civic breakdown and anomie, a finding that seemingly was confirmed by Asutosh Varshney (2001) in his study of ethnic conflict and associational life in modern India. Varshney found that Hindu–Muslim riots were most likely to break out in cities with low levels of bridging social capital across the religious divide.

Putnam's work has been severely criticized by a number of scholars, including, most notably, Sidney Tarrow (1996). It has been suggested that social capital is difficult to define and even more difficult to measure. It is also unclear why it should be counted as the independent rather than as the dependent variable. Varshney's work has also been criticized for playing down the role of public policy in guarding against 'ethnic riots' (see Brass, 2003). The State of Bihar used to have a bad reputation for communal violence. Since 1990, however, the governments of Laloo Prasad Yadav have protected Muslims in the State. Needing the Muslim vote, Mr Yadav made it clear to the police in Bihar that they were to do everything necessary to deter religious violence. In other parts of the country, as in Gujarat at the time of the anti-Muslim pogrom in 2002, the police have been deeply implicated in the organization of ethnic conflicts (see also Hansen, 2002.)

John Harriss (2001) has generalized this argument. In an engaging polemic, Harriss argues that the World Bank has engaged Putnam's work on social capital precisely because it takes politics out of 'development'. In this argument he follows the earlier work of James Ferguson (1990) on the depoliticization of development issues in Lesotho. This was effected by the international aid agencies and by what Ferguson calls the 'discourse of development'. Both Harriss and Ferguson accept, of course, that civil society organizations can bring political pressures to bear on government agencies, but they have also called for much greater attention to be paid to politics in the old-fashioned sense: as a series of conflicts around the production and distribution of resources.

Partha Chatterjee has also voiced this call to politics in his recent essays on the politics of the governed. Chatterjee maintains that civility in countries like India and Brazil is a luxury enjoyed only by the ruling elites. The concept, indeed, has applicability mainly in the West, and not in countries where governments have been required to invent civil societies. Ordinary people have to get their hands dirty in the rough-and-tumble worlds of political society, where they are required to engage with a wide range of political brokers (from criminals and fixers to party leaders) in order to get services from government. They often do so

by stretching or breaking the law, and by joining together in groups that take shape with regard to some governmental function – as slum-dwellers, for example, or as members of the Below Poverty Line population.

Chatterjee warns against judging these actions against the standards expected of people in more civil societies. 'It is morally illegitimate to uphold the universalist ideals of nationalism [and citizenship] without simultaneously demanding that the [group-based] politics of governmentality be recognized as an equally legitimate part of the real time-space of the modern political life of the nation' (Chatterjee, 2004: 25). In opposition, then, to Benedict Anderson's suggestion that 'the unrelieved nastiness of ethnic politics' (ibid.: 6, summarizing Anderson, 1998) is a product of the bound serialities that produce governmental populations (as caste groups with rights to employment, say, or as BPLs), Chatterjee maintains that most poorer people do not live in the empty and homogeneous space-time of capitalism/modernity/the nation that Anderson celebrates. Instead, 'the real space of modern life consists of heterotopia' (ibid.: 7), and this is where we find the grimy and everyday worlds of politics, or the politics of the governed. To ignore this, or to write against it only in terms that condemn, is itself a political act, says Chatterjee, and one that is dismissive of 'popular politics in most of the world' (the subtitle of his 2004 book).

CONCLUSION

It is to be hoped that more studies will take a lead from Chatterjee, and will probe the workings not just of civil societies but also of political societies. Roberto DaMatta notes that in Brazil a popular saying declares, 'To our enemies the law; to our friends everything' (DaMatta, 1991: 168). This resonates with the thrust of Chatterjee's remarks, and calls attention to the dangers that can face an individual who does not enjoy the protection of membership in a social group. More work is required on the use or threat of violence in gaining access to the state (itself, for some, an originary source of violence; see also Taussig, 1997). The question of how women see the state is also under-researched. Many of these studies will work as anthropologies or geographies of governmental practices, and some will challenge the strict civil/political divide that Chatterjee is proposing (see also Slater, 2002).

Elsewhere, we can hope for more work on the transition to workfare regimes/opportunity societies in richer countries (and not just in these countries). This shift has enormous implications for the structuring of state/society relations, and for relations between the sexes and different generations. Careful assessments of the construction of political coalitions for this agenda are required, along with robustly empirical evaluations of their likely distributional effects. In the United States, the Republican Right supports these and other reforms as part of their agenda to 'shrink the beast'. A fresh generation of social scientists will likely join with Karl Polanyi (2001), John Gray (2000), and Susan Buck-Morss (2000) in seeking to expose the dangerous contradictions of this free-market utopia.

REFERENCES

Abrams, P. (1988) 'Notes on the difficulty of studying the state', *Journal of Historical Sociology*, 1: 58–89.

Abu El-Haj, N. (2001) *Facts on the Ground: Archaeological Practice and Territorial Self-Fashioning in Israeli Society*. Chicago: University of Chicago Press.

Agnew, J.A. (1998) *Geopolitics: Revisioning World Politics*. London: Routledge.

Akyuz, Y., Chang, H.J. and Kozul-Wright, R. (1998) 'New perspectives on East Asian development', *Journal of Development Studies*, 34(6): 4–36.

Amsden, A. (1989) *Asia's Next Giant: South Korea and Late-Industrialization*. Oxford: Oxford University Press.

Anderson, B. (1983) *Imagined Communities: Reflections on the Origin and Spread of Nationalism*. London: Verso.

Anderson, B. (1998) *The Spectre of Comparisons: Nationalism, Southeast Asia and the World*. London: Verso.

Arrighi, G. (2002) 'The African crisis', *New Left Review*, NS15: 5–38.

Balibar, E. (1977) *On the Dictatorship of the Proletariat*. London: New Left Books.

Bardhan, P. (1984) *The Political Economy of Development in India*. Oxford: Clarendon.

Bates, R. (1981) *Markets and States in Tropical Africa*. Berkeley: University of California Press.

Bates, R. (1989) *Beyond the Miracle of the Market: The Political Economy of Agrarian Change in Kenya*. Cambridge: Cambridge University Press.

Bayart, J.F. (1993) *The State in Africa: The Politics of the Belly*. Harlow: Longman.

Bayart, J.F., Ellis, S. and Hibou, B. (1999) *The Criminalization of the State in Africa*. Oxford: James Currey.

Bobbitt, P. (2002) *The Shield of Achilles: War, Peace and the Course of History*. London: Allen Lane.

Borjas, G. (1994) 'The economics of immigration', *Journal of Economic Literature*, 32. 1667–1717.

Bourdieu, P. (1986) 'The forms of capital', in J. Richardson (ed.), *Handbook of Theory and Research for the Sociology of Education*. New York: Greenwood Press.

Brass, P. (2003) *The Politics of Hindu–Muslim Violence in Contemporary India*. Seattle: University of Washington Press.

Breman, J. (1985) *Of Peasants, Migrants and Paupers: Rural Labour Circulation and Capitalist Production in West India*. New Delhi: Oxford University Press.

Brettell, C. (2000) 'Theorizing migration in anthropology: the social construction of networks, identities, communities, and globalscapes', in C. Brettell and J. Hollifield (eds), *Migration Theory: Talking Across Disciplines*. London: Routledge, pp. 97–136.

Brown, M. (1999) *Race, Money and the American Welfare State*. Ithaca, NY: Cornell University Press.

Buchanan, J. (1967) *Public Finance in the Democratic Process: Fiscal Institutions and Individual Choice*, Durham, NC: University of North Carolina Press.

Buck-Morss, S. (2000) *Dreamworld and Catastrophe: The Passing of Mass Utopia in East and West*. Cambridge, MA: MIT Press.

Chabol, P. and Daloz, J.P. (1999) *Africa Works: Disorder as Political Instrument*. Oxford: James Currey.

Chambers, R. (1994) 'The origins and practice of participatory rural appraisal', *World Development*, 22: 953–69.

Chang, H.J. (2002) *Kicking Away the Ladder: Development Strategy in Historical Perspective*, New York: Anthem.

Chatterjee, P. (1993) *The Nation and Its Fragments: Colonial and Postcolonial Histories*. Princeton: Princeton University Press.

Chatterjee, P. (1997) *A Possible World: Essays in Political Criticism*. New Delhi: Oxford University Press.

Chatterjee, P. (2004) *The Politics of the Governed: Reflections on Popular Politics in Most of the World*. New York: Columbia University Press.

Clarke, P. (1997) *Hope and Glory: Britain, 1900–1990*. London: Allen Lane.

Coleman, J. (1988) 'Social capital and the creation of human capital', *American Journal of Sociology*, 94(1): S95–120.

Cooper, F. (2002) *Africa Since 1940: The Past of the Present*. Cambridge: Cambridge University Press.

Corbridge, S. (2002) 'Development as freedom: the spaces of Amartya Sen', *Progress in Development Studies*, 2: 183–217.

Corbridge, S. (2004) 'Waiting in line, or the moral and material geographies of queue-jumping', in R. Lee and D.M. Smith (eds), *Geographies and Moralities*. Oxford: Blackwell, pp. 183–98.

Corbridge, S. and Kumar, S. (2002) 'Community, corruption, landscape: tales from the tree trade', *Political Geography*, 21: 765–88.

Corbridge, S., Williams, G., Srivastava, M. and Véron, R. (2005) *Seeing the State: Governance and Governmentality in India*. Cambridge: Cambridge University Press.

Cornelius, W., Tsuda, T., Martin, P. and Hollifield, J. (eds) (2004) *Controlling Immigration: A Global Perspective*, (2nd edn). Stanford: Stanford University Press.

Crampton, J. and Elden, S. (eds) (2005) *Foucault and Geography*. London: Routledge.

Crook, R. and Manor, J. (1998) *Democracy and Decentralization in South Asia and West Africa: Participation, Accountability and Performance*. Cambridge: Cambridge University Press.

Dahl, R.A. (1961) *Who Governs? Democracy and Power in an American City*. New Haven: Yale University Press.

Dahl, R.A. (1971) *Polyarchy*. New Haven: Yale University Press.

Dahl, R.A. (1985) *A Preface to an Economic Theory of Democracy*. Cambridge: Polity.

DaMatta, R. (1991 [1978]) *Carnival, Rogues and Heroes: An Interpretation of the Brazilian Dilemma*. South Bend, IN: University of Indiana, Press.

Davis, M. (2001) *Magical Urbanism*. London: Verso.

Dean, M. (1999) *Governmentality: Power and Rule in Modern Society*. London: Sage.

Durand, J. and Massey, D. (1992) 'Mexican migration to the United States: a critical review', *Journal of Economic History*, 38: 901–17.

Edin, K. and Lein, L. (1997) *Making Ends Meet: How Single Mothers Survive Welfare and Low-Wage Work*. New York: Russell Sage Foundation.

Escobar, A. (1995) *Encountering Development: The Making and Unmaking of the Third World*. Princeton: Princeton University Press.

Ferguson, J. (1990) *The Anti-Politics Machine: 'Development', Depoliticization and Bureaucratic Power in Lesotho*. Cambridge: Cambridge University Press.

Foucault, M. (1997) *The Essential Works 1954–1984*, Vol. 1: *Ethics, Subjectivity and Truth*, (ed.) P. Rabinow. Berkeley: University of California Press.

French, H.F. (2004) *A Continent for the Taking: The Tragedy and Hope of Africa*. New York: Knopf.

Friedman, B. (2004) 'Bush and Kerry: a big divide', *New York Review of Books*, October 21: 27–9.

Friedman, M. (1968) 'The role of monetary policy', *American Economic Review*, 58: 1–17.

Fuller, C. and Benei, V. (eds) (2001) *The Everyday State and Society in Modern India*. London: Hurst.

Gerschenkron, A. (1962) *Economic Backwardness in Historical Perspective*. Cambridge, MA: Harvard University Press.

Gilroy, P. (1995) *The Black Atlantic: Modernity and Double Consciousness*. Cambridge, MA: Harvard University Press.

Gray, J. (2000) *False Dawn: The Delusions of Global Capitalism*. London: Granta.

Gregory, D. (2004) *The Colonial Present: Afghanistan, Iraq, Palestine*. Oxford: Blackwell.

Grindle, M. (2000) *Audacious Reforms: Institutional Invention and Democracy in Latin America*. Baltimore: Johns Hopkins University Press.

Grindle, M. (2004) *Despite the Odds: The Contentious Politics of Education Reform*. Princeton: Princeton University Press.

Gupta, A. (1995) 'Blurred boundaries: the discourse of corruption, the culture of politics and the imagined state', *American Ethnologist*, 22: 375–402.

Hansen, T.B. (2002) *Wages of Violence: Naming and Identity in Postcolonial Bombay*. Princeton: Princeton University Press.

Hardt, M. and Negri, A. (2000) *Empire*. Cambridge, MA: Harvard University Press.

Harriss, J. (2001) *Depoliticizing Development: The World Bank and Social Capital*. New Delhi: Leftword.

Harriss-White, B. (2003) *India Working: Essays on Economy and Society*. Cambridge: Cambridge University Press.

Harvey, D. (2003) *The New Imperialism*. Oxford: Oxford University Press.

Hayek, F. von (1944) *The Road to Serfdom*. London. Routledge & Kegan Paul.

Hayek, F. von (1960) *The Constitution of Liberty*. London: Routledge & Kegan Paul.

Haynes, J. (1997) *Democracy and Civil Society in the Third World*. Cambridge: Polity.

Hayter, T. (2004) *Open Borders: The Case Against Immigration Controls* (2nd edn). London: Pluto.

Herbst, J. (2000) *States and Power in Africa: Comparative Lessons in Authority and Control*. Princeton: Princeton University Press.

Herring, R. (1983) *Land to the Tiller: The Political Economy of Agrarian Reform in South Asia*. New Haven: Yale University Press.

Hirschman, A. (1970) *Exit, Voice and Loyalty*. Cambridge, MA: Harvard University Press.

Hirst, P. (1989) *After Thatcher*. London: Collins.

House of Commons, International Development Committee (2004) *Migration and Development: How to Make Migration Work for Poverty Reduction*. London: The Stationery Office.

Humphrey, C. (2002) *The Unmaking of Soviet Life*. Ithaca, NY: Cornell University Press.

Jeffrey, C. (2001) '"A fist is stronger than five fingers": caste and dominance in rural north India', *Transactions of the Institute of British Geographers*, 25(2): 1–30.

Jenkins, R. (1999) *Democratic Politics and Economic Reform in India*. Cambridge: Cambridge University Press.

Jenkins, R. (2002) 'The emergence of the governance agenda: sovereignty, neo-liberal bias and the politics of international development', in V. Desai and R. Potter (eds) *The Companion to Development Studies*. London: Edward Arnold, pp. 485–9.

Jenkins, R. and Goetz, A.M. (1999) 'Accounts and accountability: theoretical implications of the right to information movement in India', *Third World Quarterly*, 20: 589–608.

Johnson, N. (2003) *Ireland, the Great War and the Geography of Remembrance*. Cambridge: Cambridge University Press.

Johnston, B. and Kilby, P. (1975) *Agriculture and Structural Transformation*. Oxford: Oxford University Press.

Kaviraj, S. (1984) 'On the crisis of political institutions in India', *Contributions to Indian Sociology*, 18: 223–43.

Kaviraj, S. (1991) 'On state, society and discourse in India', in J. Manor, (ed.) *Rethinking Third World Politics*. Harlow: Longman, pp. 72–99.

Kettl, D. (1993) *Sharing Power: Public Governance and Private Markets*. Washington DC: The Brookings Institution.

Khan, M. and Jomo, K.S. (eds) (2000) *Rents, Rent-Seeking and Economic Development: Theory and Evidence in Asia*. Cambridge: Cambridge University Press.

Khilnani, S. (1997) *The Idea of India*. London: Hamish Hamilton.

Koestler, A. (1976 [1944]) *Darkness at Noon*. Harmondsworth: Penguin.

Kohli, A. (1990) *Democracy and Discontent: India's Growing Crisis of Governability*. Cambridge: Cambridge University Press.

Krueger, A. (1974) 'The political economy of the rent-seeking society', *American Economic Review*, 64: 291–303.

Krugman, P. (1994) 'The myth of Asia' miracle', *Foreign Affairs*, (Nov/Dec): 62–78.

Kumar, S. and Corbridge, S. (2002) 'Programmed to fail? Development projects and the politics of participation', *Journal of Development Studies*, 39(2): 73–103.

Leftwich, A. (1993) 'Governance, democracy and development in the Third World', *Third World Quarterly*, 14: 605–24.

Legg, S. (2005) 'Contesting and surviving memory: space, nation and nostalgia in Les Lieux de Mémoire', *Environment and Planning D* 23(4): 481–504

Lipsky, M. (1980) *Street-Level Bureaucracy: Dilemmas of the Individual in Public Services.* New York: Russell Sage Foundation.

Lipton, M. (1977) *Why Poor People Stay Poor: A Study of Urban Bias in World Development.* London: Temple Smith.

List, F. (1966 [1885]) *The National System of Political Economy.* New York: Augustus Kelly.

Lukes, S. (1985) *Marxism and Morality.* Oxford: Oxford University Press.

Mamdani, M. (1996) *Citizen and Subject: Contemporary Africa and the Legacy of Late-Colonialism.* Princeton: Princeton University Press.

Mann, M. (1984) 'The autonomous power of the state', *European Journal of Sociology,* 25: 185–213.

Manor, J. (1999) *The Political Economy of Democratic Decentralization.* Washington DC: The World Bank.

Mbembe, A. (2001) *On the Postcolony.* Berkeley: University of California Press.

McNay, L. (1994) *Foucault: A Critical Introduction.* Cambridge: Polity.

Mead, L. (1986) *Beyond Entitlement: The Social Obligations of Citizenship.* New York: The Free Press.

Metsola, L. (2004) 'Reintegration of Namibian ex-combatants and former fighters: limits of governmentality', mimeo, University of Helsinki.

Meyer, J.B. and Brown, M. (1999) *Scientific Diasporas: A New Approach to the Brain Drain,* Geneva. UNESCO, Management of Social Transformation, Discussion Paper No. 41.

Migdal, J. (2001) *State in Society: Studying How States and Societies Transform and Constitute One Another.* Cambridge: Cambridge University Press.

Miliband, R. (1969) *The State in Capitalist Society.* New York: Basic Books.

Mohan, G. and Stokke, K. (2000) 'Participatory development and empowerment: the dangers of localism', *Third World Quarterly,* 21: 247–68.

Mountford, A. (1997) 'Can a brain drain be good for growth in the source economy?', *Journal of Development Economics,* 53: 287–303.

Nandy, A. (2003) *The Romance of the State and the Fate of Dissent in the Tropics.* Delhi: Oxford University Press.

Nora, P. (1989) 'Between memory and history', *Representations,* 26: 7–25.

Nurkse, R. (1953) *Problems of Capital Formation in Underdeveloped Countries.* Oxford: Blackwell.

Olson, M. (1982) *The Rise and Decline of Nations.* New Haven: Yale University Press.

Ong, C. (1998) *Flexible Citizenship: The Cultural Origins of Transnationality.* Durham, NC: Duke University Press.

Orwell, G. (1980) *Animal Farm.* Harmondsworth: Penguin.

Osborne, E. (2001) 'Culture, development and government: reservations in India', *Economic Development and Cultural Change,* 49: 671–88.

Peck, J. (2001) *Workfare States.* New York: Guilford Press.

Peffer, R. (1990) *Marxism, Morality and Social Justice.* Princeton: Princeton University Press.

Perelman, M. (2000) *The Invention of Capitalism: Classical Political Economy and the Secret History of Primitive Accumulation.* Durham, NC: Duke University Press.

Platteau, J.P. (1994) 'Behind the market stage where real societies exist', Parts I and II, *Journal of Development Studies,* 30: 533–77, 753–817.

Platteau, J.P. and Abrahams, A. (2002) 'Participatory development in the presence of endogenous community imperfections', *Journal of Development Studies,* 39: 104–36.

Polanyi, K. (2001 [1944]) *The Great Transformation: The Political and Economic Origins of Our Times.* Boston: Beacon Press.

Poulantzas, N. (1978) *State, Power, Socialism.* London: New Left Books.

Prezworski, A., Alvarez, M., Cheibub, J. and Limongi, F. (2000) *Democracy and Development.* Cambridge. Cambridge University Press.

Putnam, R. (1995) 'Bowling alone: America's declining social capital', *The American Prospect,* 13: 35–42.

Putnam, R. with Leonardi, R. and Nanneti, R. (1993) *Making Democracy Work: Civic Traditions in Modern Italy.* Princeton: Princeton University Press.

Rajgopal, A. (2001) *Politics After Television: Hindu Nationalism and the Reshaping of the Public in India.* Cambridge: Cambridge University Press.

Richards, P. (1996) *Fighting for the Rain Forest; War, Youth and Resources in Sierra Leone.* London: Heinemann.

Robbins, P. (2000) 'The rotten institution: corruption in natural resource management', *Political Geography,* 19: 423–43.

Rodríguez-Pose, A. and Busire, A. (2004) 'The economic (in)efficiency of devolution', *Environment and Planning A* 31: 1907–20.

Rodrik, D. (1999) *The New Global Economy and Developing Countries: Making Openness Work.* Washington, DC: Overseas Development Council.

Rose, N. (1999) *Powers of Freedom: Reframing Political Thought.* Cambridge: Cambridge University Press.

Rosenstein-Radan, P. (1943) 'Problems of industrialization of eastern and south-eastern Europe', *Economic Journal,* 53: 202–11.

Roy, O. (1994) *The Failure of Political Islam.* Cambridge, MA: Harvard University Press.

Rudolph, L. and Rudolph, S.H. (1987) *In Pursuit of Lakshmi: The Political Economy of the Indian State.* Chicago: University of Chicago Press.

Sachs, J. (2000) 'The geography of economic development', United States Naval War College, Newport, RI, Jerome E. Levy, Occasional Paper No. 1

Sachs, J., Gallup, J. and Mellinger, A. (2001) 'The geography of poverty', *Scientific American,* March: 70–5.

Sainath, P. (1996) *Everybody Loves a Good Drought.* New Delhi: Penguin.

Scott, J.C. (1985) *Weapons of the Weak: Everyday Forms of Peasant Resistance.* New Haven: Yale University Press.

Scott, J.C. (1990) *Domination and the Arts of Resistance: Hidden Transcripts.* New Haven: Yale University Press.

Scott, J.C. (1998) *Seeing Like a State: How Certain Schemes to Improve the Human Condition Failed.* New Haven: Yale University Press.

Sen, A.K. (1989) 'Food and freedom', *World Development.* 17: 769–81.

Sen, A.K. (2000) *Development as Freedom.* New York: Praeger.

Simpson, E. and Corbridge, S. (2006) 'The geography of things that may become memories'. *Annals of the Association of American Geographers,* 96(3): 566–85.

Skidelsky, R. (1992) *John Maynard Keynes: The Economist as Saviour, 1920–1937.* London: Macmillan.

Slater, D. (2002) 'Other domains of democratic theory: space, power and the politics of democratization', *Environment and Planning D: Society and Space,* 20: 255–76.

Smith, S. and Easterlow, D. (2004) 'The problem with welfare', in R. Lee and D.M. Smith (eds) *Geographies and Moralities.* Oxford: Blackwell, pp. 100–19.

Spencer, S. (ed.) (2003) *The Politics of Migration.* Oxford: Blackwell.

Srivastava, M., Corbridge, S., Véron, R. and Williams, G. (2002) 'Making sense of the local state: rent-seeking, vernacular society and the Employment Assurance Scheme in eastern India', *Contemporary South Asia,* 11(3): 267–89 .

Stepan, A. (1988) *Rethinking Military Politics: Brazil and the Southern Cone.* Princeton: Princeton University Press.

Tarlo, E. (2003) *Unsettling Memories: Narratives of the Emergency in Delhi.* London: Hurst.

Tarrow, S. (1996) 'Making social science work across time and space: a critical reflection on Robert Putnam's "Making Democracy Work"', *American Political Science Review,* 90: 389–97.

Taussig, M. (1997) *The Magic of the State.* London: Routledge.

Tendler, J. (1997) *Good Government in the Tropics.* Baltimore: Johns Hopkins University Press.

Therborn, G. (1978) *What Does the Ruling Class Do When It Rules?* London: New Left Books.

Thompson, E.P. (1968) *The Making of the English Working Class.* Harmondsworth: Penguin.

Varshney, A. (2001) *Ethnic Conflict and Civic Life: Hindus and Muslims in India.* New Haven: Yale University Press.

Vasan, S. (2002) 'Ethnography of the forest guard: contrasting discourses, conflicting roles and policy implementation', *Economic and Political Weekly,* 5 October.

Véron, R., Corbridge, S., Williams, G. and Srivastava, R. (2003) 'The everyday state and political society in eastern India: structuring access to the Employment Assurance Scheme', *Journal of Development Studies,* 39(5): 1–28.

Wacant, L. (2002) 'From slavery to incarceration: rethinking the "race" question in the US', *New Left Review,* NS13: 41–60.

Wade, R. (1990) *Governing the Market: Economic Theory and the Role of Government in East Asian Industrialization.* Princeton: Princeton University Press.

Walmsley, T. and Winters, A. (2002) 'Relaxing the restrictions on the temporary movement of natural persons: a simulation analysis' Centre for Economic Policy Research, Discussion Paper No. 3719.

Waterbury, J. (1993) *Exposed to Innumerable Delusions: Public Enterprise and State Power in Egypt, India, Mexico and Turkey.* Cambridge: Cambridge University Press.

Watts, M. (1996) 'Islamic modernities?', *Public Culture,* 19: 251–90.

Watts, M. (2003) 'Development and governmentality', *Singapore Journal of Tropical Geography,* 24(1): 6–34.

Weber, E. (1976) *Peasants into Frenchmen: The Modernization of Rural France 1870–1914.* Stanford: Stanford University Press.

Weber, M. (1958) *From Max Weber: Essays in Sociology,* ed. and trans. by H.H. Gerth and C.W. Mills. London: Routledge & Kegan Paul.

Woolcock, S. (1998) 'Social capital and economic development: toward a theoretical synthesis and policy framework', *Theory and Society,* 27: 151–208.

World Bank (1993) *The East Asian Miracle: Economic Growth and Public Policy.* Washington, DC: The World Bank.

World Bank (1994) *Governance: The World Bank's Experience.* Washington, DC: The World Bank.

Young, II. (1990) *One of Us: A Biography of Margaret Thatcher.* London: Pan.

Zachariah, B. (2004) *Nehru.* London: Routledge.

Zeller, W. (2007) 'Chiefs, policing and vigilantes: "cleaning up" the Caprivi borderland of Namibia', in L. Buur and H.M. Kyed (eds), *A New Dawn for Traditional Authorities? State Recognition and Democratization in Sub-Saharan Africa.* Basingstoke: Palgrave (to be published in November).

Planning, Space and Government

Margo Huxley

INTRODUCTION

Any delineation of space or territory and prob-lematization of its characteristics implies the need for some form of administration (Hebbert, 1998: 100–2). This means that land use or spatial planning systems are pivotal elements in the government of territories and citizens. Conversely, almost all forms of state policy can be seen to have spa-tial implications of some sort. But the policies and practices, variously known as spatial, urban, regional or city planning, land use planning or development control, are some of the most explic-itly spatial forms of state attempts to manage social and economic relations.

Spatial planning and land use control systems aim to allocate land uses to spaces, to order boundaries and connections between them, and to manage the political consequences of such allo-cations and delineations in endeavours to influ-ence interactions between individuals, populations, environments and spaces within their territories. Conflicts focused on the built or natural envi-ronment, on processes of spatial development, or spatial inequalities in distributions of resources or decision-making power, involve challenges to the aims, forms, procedures or outcomes of planning systems at specific times and places. Studies of spa-tial or land use planning are therefore central to the concerns of political geography.

This chapter aims to provide an overview of some key implications for planning that can be drawn from different theoretical perspectives and the significance of such conceptualizations for thinking about the relations between states, territories and citizens. The structuring axes of this overview are: first, a discussion of the aims of planning to produce economic efficiency, social harmony and/or healthy and aesthetic environments; and second, an outline of different theoretical analyses of planning's capacity to produce effects.

While, on the one hand, some positions sup-port calls for planning to operate with properly nuanced concerns for space, there are equally stringent critiques of what is seen to be plan-ning's naïve environmental determinism. But, as I suggest in the final section of the chapter, a third axis of analysis can be derived from the concept of 'governmentality' (Foucault, 1991a). That is, rather than either uncritically accepting the spatial assumptions underpinning planning, or dismissing planning's conceptions of space and environment as inadequate or misguided, a differ-ent understanding might be gained from examining its theories and practices as rationalities of gov-ernment. From this perspective, assumptions about spatial causality in the management of popula-tions and individuals are seen as rationalities that give shape to particular projects and practices of government.

I argue, in particular, that this conception takes planning's spatial or environmental determinisms seriously as a 'mentality of government' and enables an analysis of the 'operative rationales' (Osborne and N. Rose, 1998a: 3) on which planning policies, regulations and practices rest, and which in turn contribute to governmental aspirations for the 'making up' of governable subjects and spaces. In tracing spatial rationalities and the projects of government they inform, studies of planning as a form of governmentality make contributions to conceptualizations of the relations between states, territories and citizens.

So to begin the discussion, the next section examines various aspirations and aims of planning that can be identified in (mainly British and US) post-war planning literature.

PLANNING AND
SPATIAL/ENVIRONMENTAL CAUSALITY

Spatial planning and development control can be
seen as territorially specific and historically conditioned ways in which, since the beginning of the
twentieth century, states have sought to organize
divisions of space towards particular ends.[1] Spatial
planning usually refers to regional or metropolitan
policies that attempt to bring about certain distributions and synergies of resources, communications
and populations. Land use regulations or development control concern local-level allocations of
activities to locations (or of locations to activities),
often in conjunction with classificatory systems
such as zoning, that seek to group 'compatible'
uses and activities within defined spatial boundaries (e.g. residential or industrial zones), together
with systems of permissions and monitoring to
ensure compliance. In addition, such regulations
often seek to impose aesthetic conditions on the
design and layout of developments. Since around
the 1970s, planning in most 'western' polities has
also involved, to a greater or lesser degree, some
form of 'public participation' and dispute adjudi-
cation procedures in relation to decisions affecting
the built and natural environment and the allocation
of spatial benefits and disadvantages.

In his coverage of (mainly British) planning
theory since 1945, Taylor (1998) shows that for
the first part of the twentieth century and into
the 1970s, planning operated with a number of
largely unstated assumptions about the causal efficacy of spatial order and environmental quality
in producing social and/or economic outcomes. It
was believed that good sanitation and clean water,
houses designed for light and air, rational street layout and traffic management, access to green space
and the countryside, and good design and aesthetic
surroundings would not only eradicate the physical
form of slums and their threats to health, but also
reform the kinds of behaviours and (anti-) social
interactions that flourished there. The attribution
of social and economic effects to urban and environmental reforms was further reinforced by the
work of the Chicago School of sociology and the
explicit theorization of an urban ecological system that underlay and influenced individual and
social interactions in the city (Park et al., 1925; see
Saunders, 1981; Kilmartin et al., 1985).

These underlying assumptions about the causal
effects of planned and managed spaces and environments in bringing about the well-being of
society were a common thread in town and country, city and regional planning as it emerged
as a state function in the early twentieth century (Fishman, 1982; Boyer, 1983; Taylor, 1998:
17–18). In the visionary plans of Le Corbusier or

Ebenezer Howard; in Keeble's (1952) mundane
*Principles and Practice of Town and Country Plan-
ning* or Chapin's (1965) *Urban Land Use Planning*;
in quotidian regulation of the built environment,
zoning controls, granting of permissions, or in
metropolitan and regional policies, spatial planning and design play a central part in producing
spaces and places that aim to enhance the social
and economic relations of society.

However, in the 1960s and 1970s, this early
twentieth-century focus on spatial design and
detailed regulation was overlain by academic interest in both procedural decision-making theories
(e.g. Faludi, 1973) and cybernetic systems theories (e.g. McLoughlin, 1969; Chadwick, 1971; see
Taylor, 1998: chap. 4). Procedural planning concentrated on normative models of rational decision-
making that could be carried out in any situation or
location. Systems theories also claimed a universalized applicability, seeking to maintain optimum
conditions of dynamic equilibrium between all
parts of the urban system. The task of the plan-
ner was to monitor and manage a holistic feedback
system that paid little attention to the particularities of ordering and designing specific spaces and
places.

Despite systems theory's dismissal of physical
planning and design, an assumption of environmental causality was secreted in its attempts to
manage spatial systems on the model of bodily
and ecological bio-systems (McLoughlin, 1969:
chap. 1). Town planning, operating as a form of
ecological/spatial systems analysis and cybernetic
control, could bring about the proper functioning of environmental/social/economic interactions
(Taylor, 1998: 62).

Intertwined in these developments in planning
thought, there were grassroots reactions against
physical and technocratic images of the ideal city.
Although not always attributable to the planning
system per se (for instance, in many jurisdictions,
'slum clearance' was the responsibility of state-
based housing administrations), visions of the modernist city of clean, efficient, high-rise towers had
influenced projects of urban renewal that played a
part in the destruction of inner city environments.
Residents' protests forcefully drew attention to the
'dark side' of attempts to create better environments – attempts that were seen as the products
of the drawing board and technical calculations.
Planning's claims to deploy neutral expertise for
the benefit of a generalized 'public interest' resulted
in conditions that often failed to improve the lives
of those most affected (see Taylor, 1998: chap. 3).
Jacobs' (1961) *The Death and Life of Great American Cities*, Goodman's (1971) *After the Planners*, and similar studies across 'western' cities,[2]
highlighted the social value of supposedly 'slum'
areas in the city and the negative consequences of

attempts to create sanitized, orderly environments in their place.

Protests and publications critical of the lack of concern for the people caught up in urban redevelopment schemes contributed to demands for greater public participation in the planning system. These demands ranged from calls for direct democracy at the local level, to the formulation of administrative procedures for enabling 'the public' to comment on plans and make formal objections to their effects, and to theoretical reformulations of planning's aims and purposes that stressed the importance of participatory practices in achieving the social, political and environmental goals of planning.

So, for instance, Sherry Arnstein's (1969) influential 'ladder of citizen participation' proposed various levels of public involvement, from 'non-participation' through 'tokenism' to 'degrees of citizen power' that ultimately implied a form of direct, participatory local democracy.[3] Government policy, on the other hand, tended to translate such concerns into proposals for managing public participation in relation to specific plans or developments. In England, the Skeffington Report (1969) advocated active public involvement in planning decisions, but held that the final responsibility must rest with the trained professionals.

Like procedural approaches to rational comprehensive planning, this emphasis on processes of decision-making was at one remove from the explicit focus on ordering or designing space, place and environment that continued as the rationale for planning policies and regulations. In the 1980s and 1990s, concerns with participatory counters to technical and bureaucratic dominance gave rise to what became known as 'communicative' or 'collaborative' planning theories (e.g. Forester, 1989; Innes, 1995). Drawing on Habermasian ideas of truths expressed in speech acts as the prerequisites for any human communication, communicative planning theories position planners as enablers of 'undistorted' communication between groups in conflict. Nevertheless, in these approaches, there is an underlying implication – seldom fully spelled out – that participatory or communicative planning processes will contribute to producing healthy and aesthetic places and environments that are both the result of, and will help to foster, democratic, inclusive social relations.

In her book, *Collaborative Planning*, Healey (1997) connects participatory communicative processes and environmental and place-based decision-making to state and institutional contexts. She outlines ways in which (British) planning practice and regulation might be reformed to allow community groups and individuals to determine the futures of their neighbourhoods and localities. In parallel, Graham and Healey (1999) suggest that in order to bring about positive outcomes,

planners need to have theoretical understandings of space-time as relational and dynamic. They 'criticize the legacy of object-oriented, Euclidean concepts of planning theory and practice and their reliance on "containered" views of space and time' (p. 623).[4] They propose instead that planning practice will be more effective if it is grounded in theoretical understandings of relational space-time and 'multiplex' places, 'new' urban and regional socio-economics, and theories of social agency and institutions. From this perspective, spatial management can 'get it right', or at least do it a bit better, if planners have more adequate theoretical conceptions of space and better practical understanding of processes of place-making.

This view of planning, therefore, reaffirms the central idea of the positive effects of spatial organization on groups and individuals. But, like procedural or communicative theories, it also locates some of the principal causes of planning's failures in the minds and capacities of individual planners. That is, there seem to be few inherent economic, political, social or discursive constraints on the achievement of planning goals that cannot be overcome by better planning education and better planning practice.

While it is surely important for planners to be aware of developments in theories of society and space, it seems that spatial and land use planning, by and large, continues to take for granted that spatial order, environments and milieux have significant effects on social relations. The corollary of this assumption is that planners, and the institutional systems and regulations they work with, can influence these causal links to bring about desired outcomes. Hence, understandably, most planning theories argue for the positive role of planning in improving environments and the lives of citizens. In summary, then, planning theory and practice are posited upon (largely unexamined) suppositions that space and environment have causal effects on individuals and citizens, populations and territories; and that spatial planning systems, and planners as individuals, are (potentially) capable of bringing about significant positive social, economic and environmental outcomes.

However, assumptions about the effects of spatial arrangements and environmental qualities in producing social outcomes, and the possibilities available for planning to achieve its positive social and economic aims through the regulation, or participatory creation, of places and milieux, have been subjected to critical examination from a number of perspectives. The next section sets out some of these theoretical frameworks that touch on (either explicitly or implicitly) the possibility or impossibility of spatial and land use planning contributing to the realization of social and economic improvements.

THE (IM)POSSIBILITIES OF SPATIAL PLANNING

The problems and possibilities of planning, as detailed local regulation or as regional and national policy, as state spatial management or as participatory empowerment, have long been subject to dissection and debate from various perspectives. It is not my intention to discuss these approaches in great detail. Rather, I want to give an overview of the questions they raise about spatial and land use planning. Does space make a difference? Do planning theory and practice rest on forms of misguided environmental determinism that should be discarded altogether? Or are they in need of more sophisticated theories of space, everyday place-making or power to be more effective? Are the various practices, policies and theories of planning capable of producing, or do they in fact produce, better or worse environments for different groups, or make no difference at all?

These critical approaches can be grouped under the following broad headings in relation to their implications for planning theory and practice: (1) perspectives that, while critical of the institutions and practices of planning, suggest possibilities for planning to make a difference for the better; and (2) perspectives suggesting that planning is severely limited in its ability to effect change – either it has no effects other than to support the status quo or if it has any effects, they are largely negative, exacerbating existing inequalities.

THE POSSIBILITIES OF PLANNING

The frameworks of analysis considered under this heading include those that posit the possibility for planning to make a difference, however limited. In general, such perspectives examine the role of the state and bureaucracy as pivotal points of struggle around issues of economic and spatial redistribution and the recognition of social inequalities. They focus on the conditions in which change can be brought about, and therefore these approaches have proved much more amenable to incorporation into planning theories, policies and practices.

Planning, the State and Urban Reform

Perspectives examining the role of the state – for instance, urban managerialist or urban regime theories – indicate that while the scope of manoeuvre for planning may be limited, it is not fully determined, and that policies of spatial redistribution or development can make a difference. Spatial distributions are important aspects of social relations with potential positive or negative effects on individual life chances. Thus, the social democratic welfare state

and/or networks of local governance are seen as potential vehicles for social improvement, and planning's attention to space holds out possibilities for minimizing distributional inequalities.

In the debates of the 1970s and 1980s over the significance of 'the urban' in the study of capitalist formations, neo-Weberian and some neo-Marxian approaches to the study of the state and bureaucracy pointed to the inevitably political nature of planning (and in the UK, of local government housing provision) in the allocation of scarce resources in the built environment. This was so, whether as the outcome of class struggles (Marxian-derived analyses) or as the effects of social power and status groups outside the sphere of production (Weberian-inspired approaches).[5] These examinations highlighted the inequitable outcomes of supposedly neutral administrative decision-making and technical plan-making, despite the claims of the welfare state to reduce poverty and provide a decent standard of living for all. That is, if the state in general, or urban planning in particular, was ostensibly working to ameliorate the inequalities of capitalism, it was signally failing to do so.[6]

Much of this failure could be traced to the practices and institutional cultures of state and private organizations concerned with the spatial allocation of resources. The inequity of these allocations in turn sparked protests and movements related to consumption and status issues that, from Weberian perspectives, could have effects unrelated to the sphere of production (see Saunders, 1981).

In addition, Pahl (1975: 249) argued that inequalities of spatial distribution were inevitable, regardless of the economic or political organization of society, and were present in capitalist and socialist formations alike. Nevertheless, *how* such resources are distributed is 'largely a function of the actions of those individuals who occupy strategic allocative locations in the social system' (Saunders, 1981: 117). So, if 'urban managers' and 'gatekeepers' can bring about outcomes that are not necessarily in the immediate interests of the capitalist system, they must also be able to exert positive reformist or ameliorative influences.

Pahl stresses that urban managers and planners alone cannot solve the problems of the modern big city, regardless of the economic system it is embedded in. Nevertheless, 'the role of state bureaucrats and technical experts must be central to an understanding of urban outcomes and regional development' (Pahl, 1977: 56–7). In this way, as Kilmartin et al. (1985: 54) argue, Weberian approaches seem to hold out hope for the role of state-employed practitioners

> The welfare state is not fundamentally flawed, it simply needs to be improved in order to produce desired policy results. The solution implied by the

urban managerial analysis lies, therefore, in the production of more able managers and gate keepers and more effective decision making procedures. (Kilmartin et al., 1985: 54)

If planning practitioners have better understandings of economic and social contexts of practice, better appreciations of the institutional and political operations of power, and better-researched analyses of the spatial implications of resource distributions, 'planning as such can be both more efficient and more sensitive to local needs' (ibid).

In sum, from Weberian perspectives, space is a physical limit that operates as a substrate to the political and economic conditions within which spatial planning is able to make constrained, but nevertheless worthwhile, differences. Urban managerialist perspectives did not resolve questions of the limits to urban planning in a capitalist society, nor of the limits to spatial equality, but did open up possibilities for asking how the power of the state, institutional processes and individual practices might be put to work towards the improvement of spatial resource distributions.

The questions of the possibilities of planning raised in debates about 'the urban' and the relations between general social/economic processes and local conditions (see Saunders, 1981: chap. 4), also find expression in urban regime approaches to local governance and economic development. Urban regime theory links the legacy of US community power debates about the government of the city to a greater awareness of the economic and institutional contexts of urban governance. In urban regime theory, the term 'governance' entails more than just formal structures and powers of government and political responses to pluralist group pressures. In order to get things done, alliances of interest groups and government institutions need to be brought together to produce a capacity to govern.[7]

Despite the rational choice model of behavioural motivations that underpins the original regime formulations (Painter, 1997: 132–4), urban regime theory has been regarded as capable of providing a framework for empirical enhancement of both pluralist political theories of government and political economic theories of capitalism (such as regulation theory), by including detailed studies of how coalitions of local business, community and political interests are put together and maintained as stable regimes of 'governance' (see essays in Lauria, 1997).

Even though originally developed to analyse the conditions of urban governance in the US, urban regime theory has also been applied to the operations of UK local government under the economic and political conditions of the 1980s and 1990s. It was during this time that central government

policies of contracting out local services to private firms, and promoting 'governance' through public/private/voluntary-sector partnerships, had begun to take hold, and made urban regime theory seem particularly applicable (Taylor, 1998: 141; but see note 6 on the uptake of theories in different contexts).

In urban regime approaches, planning is still understood to have the ability to produce positive localized environments through urban design, and zoning or development control are still necessary regulatory tasks. But these traditional functions must take place in the context of the bargaining and negotiation among local coalitions of business and voluntary organizations that are held to be necessary for reviving urban economies (Fainstein, 1995; Healey et al., 1995; Brindley et al., 1996).

These frameworks – from urban managerialism to urban regime theory – see space and the quality of environments as factors in the local and contingent working through of economic and social processes. Spatial planning has some capacity for influencing distributional effects within localities and creating spaces of social well-being, as well as enhancing local attractiveness to economic investment. The relations between spaces, environments and citizens are thus not determining, and planning can mediate them in particular ways in particular places to bring about meaningful improvements (see, e.g., Brindley et al., 1989).

Diversity and Identity Politics

The body of work around identity politics also indicates that changes to spatial organization and distribution of resources can lead to positive outcomes for marginalized groups in society. Studies of identity politics – movements claiming the rights and recognition of groups formed around particular cultural identities – have noted the involvement of planning in perpetuating spatial inequalities in relation to race, gender, sexuality, bodily capacities and post-colonial identities. But, it is suggested, with greater awareness of the complexity of multicultural identities and their relationships to places, planning can be reformed in order to foster democratic, multicultural, inclusive communicative decision-making fora.

So, for instance, feminist accounts of the everyday lives of women in the city have been particularly alert to the spatial inequalities reinforced by land use planning.[8] Spatial divisions are seen to be congealed and exacerbated by design practices and planning regulations, especially those of US-style zoning. The regulation of the built environment, it is argued, from house design to zoning, produces spatial arrangements that impose mobility configurations and time-budget restrictions on women, making the work of household management – which is still very much women's responsibility – even more onerous. However, in contrast to the theorization of systemic patriarchy (discussed below), these analyses have been undertaken in the spirit of liberal equal opportunities, and suggest possibilities for the elimination of both spatial inequalities and professional and institutional gender bias. Studies specifically concentrating on planning have similarly suggested that gender-sensitive practice, better education and further (action-based, participatory) research will have positive effects in bringing about improved spatial outcomes.

Subsequently, the concept of 'identity politics' broadened the field of study to examine the exclusion, discrimination and disadvantages experienced by people of colour, of different sexualities and bodily capacities, and by different cultural groups. The identification of the inequalities experienced by groups on the basis of identifications as black, female, gay or disabled challenged both liberal political constructions of the universal subject of democracy (and thus also questioned Habermas's notions of transparent 'communicative action'), and collective politics based on class identification and interests. Such challenges raised doubts about the possibilities of achieving equality through redistribution based on a supposed uniformity of needs.[9]

Identity politics thus implies the need for inclusive decision-making procedures that enable a multiplicity of voices to be heard and their claims for equity to be attended to (Tully, 1995). But such groupings and differences are in danger of being endlessly proliferated to the point of parody: multiplying inventories of groups with supposed commonalities of interests, 'colour, sexuality, ethnicity, class and able-bodiedness, invariably close with an embarrassed "etc." at the end of the list' (Butler, 1990: 143). There have also been stringent critiques of, on the one hand, the danger of formless relativism and the undecidability of claims based on subjective positions (e.g. Harvey, 1996: chap. 12), and on the other, the homogenizing assumptions that categories of sex, colour, bodily capacity or class represent the sum total of identity (see discussion of performativity below). Nevertheless, however such conceptual difficulties may be debated, identity politics have unsettled assumptions about the ways in which cities, regions and territories and their citizens can be managed, and have also prompted rethinking the relationships between place and identity (Keith and Pile, 1993; Thrift and Pile, 1995; Fincher and Jacobs, 1998).

The planning literature has most readily taken up identity politics in the form of quests for the multicultural city that fosters an 'openness to unassimilated otherness' (Young, 1990: 227) – a 'being together of strangers' (p. 237) in public spaces that 'embody the kind of free access and freedom to speak which are necessary conditions of democratic citizenship … a spatialized ideal of political life, a return to the politics of the *polis*' (Lechte, 1995: 119).[10] The aspiration for planning to act as midwife to the birth of a multicultural cosmopolis (Sandercock, 2003: 212) has been explored at length by Sandercock (1998, 2003). Here, planning is seen to have the potential to mediate between the differences, inequalities and conflicts of spatial identity politics through situated, embodied practices and imagining audacious, creative 'songlines' for alternative futures (Sandercock, 2003: chap. 9).[11] In Sandercock's depictions, planning is not necessarily only a state-based activity, nor do its practices rely on narrowly spatialized conceptions of the good city (see Sandercock, 1998: appendix). Rather, planning can work on behalf of dominated and marginalized groups to transform the relations between states and citizens and the management of territories. 'The most promising experiments in insurgent planning have involved mobilized communities forging coalitions to work for broad objectives of economic, environmental, and social and cultural justice, and in the process resisting, engaging with, and participating in "the state"' (Sandercock, 1998: 218).

The foregoing discussion has examined frameworks in which planning is seen to have capacities to influence relations between spaces and citizens, and consequently has the potential to make contributions to projects of social change. But a different set of social and geographical perspectives implies that planning may be mistaken in these attempts to produce good citizens in good cities, and examples of these positions will be examined next.

THE LIMITS TO PLANNING

The perspectives discussed under this heading regard spatial and land use planning as having little influence on the essential causes of economic and/or social relations. Of these positions, the discussion focuses on Marxian urban theories of the 1970s and 1980s, and feminist theories of patriarchy; neo-liberal economic theories; and post-structural conceptions of performative identities.

Planning Constrained

The approaches being considered here, questioning whether it is possible for planning to achieve its aims, include what might be called 'structural' or 'essentialist' perspectives. In these, planning is an expression of, and support for, inherent social and economic divisions and inequalities of, for instance, capitalist or patriarchal society. That is, meaningful social change can only be brought about by transformation of the fundamental nature of economic and/or social relations. Planning's attempts at spatial redistribution or environmental improvement are thus, at best, only short-term localized reforms that merely relocate problems; and at worst, planning serves to mystify and perpetuate fundamental relations of power and exploitation.

Among the reappraisals of planning that emerged in the 1970s was a strand of critique based in explorations of the significance of space, land and 'the urban' in neo-Marxian theory. Castells (1977), for instance, saw urban planning as part of the structural logic of capitalism (e.g. Castells, 1977: 322). As an element of the state apparatus, planning works to avert crises in social reproduction, and at the same time attempts to create and maintain spatial and physical conditions for continued capital accumulation. How this is brought about is particular to each case, but despite the 'relative autonomy' of the planning system (p. 323), in essence planning is always constrained by the limits set by capital accumulation and dominant class positions.[12]

Hence, the planning and management of space cannot, in and of itself, effect permanent improvements in people's lives, since it cannot address the fundamental contradictions of capitalism. Since space is produced out of the economic and social relations of a given society, planning's attention to 'spatial fixes' for economic inequality and social conflict fails to address their basic causes and is a further deflection from the aims of social transformation. Planning's attempts to manipulate spaces and environments are thus mistaken as to the causes of urban problems and are hence likely to be superficially remedial at best, and fundamentally an ideological smokescreen for structurally generated inequalities.[13]

This position was forcefully expressed by Harvey (1985) in his essay 'On planning the ideology of planning' in which he analysed class relations in the built environment and showed how planning signally fails to engage with these structuring processes in its quest for environmentally induced social harmony:

> The history of capitalist societies these last two hundred years suggests … that certain problems are endemic, problems that simply will not go away, no matter how hard we try. Consequently, we find that the shifting world view of the planner exhibits an accumulation of technical understandings combined with a mere swaying from side to side in ideological stance from which the planner appears to learn little or nothing. (Harvey, 1985: 178)

In a sense, such blanket condemnations imply that identifying planning with dominant ideologies

exhausts all there is to say about its functions. But the development of regulation theory enables more detailed attention to be paid to planning as an element in the achievement of a 'mode of regulation' that facilitates the persistence of capitalism despite its inherent contradictions (Boyer, 1990; Jessop, 1990):

> According to regulation theory, continued accumulation depends on a series of social, cultural, and political supports that are only contingently co-present. Moreover, there is not just one possible pattern of regulation but, at least in principle, many alternative contingent combinations of 'noneconomic' factors that might operate to support accumulation, with varying degrees of effectiveness. (Goodwin and Painter, 1997: 15)

Spatial and land use planning can thus be seen as part of a mode of regulation that brings together the institutions, practices and norms required to produce economic stability and therefore requires examination to understand how this stability is brought about. This framework for the study of broad economic and social tendencies and local public and private, informal and institutional practices gives rise to a number of studies of spatial management and urban planning.[14] Analysis of planning in 'post-Fordist' regimes suggests that, in the face of shifts in global economic relations, planning has lost legitimacy to plan comprehensively (Prior, 2005). As a result, it is now mainly concerned with piecemeal bargaining over specific projects, albeit with different capacities and effects in different national and local contexts (Leo, 1997). Ultimately, planning's attempts to produce environmental and spatial improvements are seen as serving to stabilize capitalist 'regimes of accumulation'.

A similar prognosis for the inability of planning to bring about significant change is implied by radical feminist analyses of dominant patriarchal power and its reflection in the built environment. In these approaches, valuations attributed to sex/gender differences are systematic sources of social and economic inequality.

> Patriarchy is a structural mode of organization which places men and women in different social positions in the social order. Rather than consisting of visible acts, patriarchy is a latent system which organizes, makes possible, and gives support to, individual acts of sexism. It provides the context, support, and meaning for these empirical acts. (Gross (Grosz), 1987: 131)

In versions of radical feminist theories positing male power as a fundamental attribute of social relations, spatial inequalities are seen as essentially gendered. Planning – both as a function of the masculinist state and as the product of gendered professional training – inevitably serves to perpetuate the patriarchal nature of spatial arrangements. The divisions of space between the city as the locus of the masculine world of (paid) work and the suburbs as sites of feminine (unpaid) domestic labour reinforce the power of men over women, as do the design of houses and the layout of suburban housing developments.[15]

According to this view, planning's aspirations to create 'community' and 'efficiency' through spatial regulation are incapable of producing environments suitable for the lived needs of women. Both 'community' – with its emphasis on domestic harmony – and 'efficiency' – with its support for production-oriented transport and investment – trap women in subordinate roles and under-resourced spaces. Only a revolution in gender relations can bring about real change.

Here again, in the absence of wider social transformation, planning's concerns with the reform of the city and improvement of spaces and milieux of social and economic activities are held to be misplaced. Planning is always ultimately constrained to operate in accordance with dominant power relations, and thus is largely discounted as having a positive part to play in transforming the spatial relations between states and citizens.

The Excesses of Planning

Neo-liberal objections to planning see government interference in land markets and property rights distributions as misguided. In this case, however, it is because of the way in which the constraints on individual choice implied by such interference result in less than optimal outcomes. From this perspective, planning represents the excesses of government, and the solution to its failures is to greatly reduce the scope of planning's regulatory remit.

Neo-liberal opposition to planning includes its supposed distortions of housing, property and land markets, and hence its production of inefficiencies and extensions of undemocratic bureaucratic (mis)management. However, planning's failures are attributed not to the constraints of fundamental social or economic structures, but to the misguided expansion of the social democratic welfare state into civil society and the economy. Without state and bureaucratic interference, economic interactions would operate to maximize individual choice and well-being by enabling economically active individuals to respond to the price signals of markets – whether economic markets, political markets (e.g. public choice theories of politics) or environmental markets (e.g. carbon emissions trading).

These rationales for attempting to minimize government in the 1980s and 1990s in many 'western' countries claimed affinity with (versions of) classical liberal political economy, such as Adam Smith's, and their later manifestations in the works of Hayek and Friedman, and in the philosophy of Robert Nozick. These ideas have been variously interpreted (e.g. see Pennington, 2002, on the misinterpretation of Hayek), and have given rise to different assessments of the links between economic and political theories of liberal provenance and the practices of governments (most notably the Thatcher government in the UK, and the Reagan government in the US). But a common strand to all such theories and policies is the need to reorient the role of government in relation to the economy and the individual towards less regulation of economic activity and more emphasis on individual self-help (by regulation, if necessary).

Accordingly, and in line with the removal of restraints on the economy (accompanied by active fostering of business enterprise), it is argued that planning must be pared back to the most minimal amelioration of spatial externality effects; all other urban and environmental problems can be dealt with through economic and political market preferences. In this market-based view, space may make a difference to the way market signals operate, but the answer to less than optimal outcomes is not more comprehensive planning, but less.[16]

As Pennington (2002: 194) puts it:

> To suggest that *because* social and ecological systems are complexly related entities they *must* be managed on a similarly holistic basis is a *non sequitur*. The normative conclusion that deliberative democratic control is required does not follow from the premise that social and ecological systems are holistically related. On the contrary, from a Hayekian perspective, it is precisely because these systems are complexly related wholes that conscious social control is a problematic concept.

In short, planning's attempts to reform spaces and places, and regulate the production of the built environment, are doomed to failure, not only because they interfere with the operations of the market, but also because no one bureaucrat or organization can possibly have enough knowledge to plan for all eventualities and forestall all unintended consequences. Individual actions in the market, however imperfect, will bring about better outcomes than any government policy.

These neo-liberal critiques, then, share with structural perspectives on planning the dismissal of planning's search for spatial and environmental solutions in social and economic management. For a variety of reasons, neo-liberal approaches suggest that while spatial and environmental aspects of the relations between government, markets and citizens may be real, their effects on the conduct of social and economic relations are relatively insignificant.[17] A focus on spatial and environmental regulation therefore is likely to do more harm than good.

Thus, in assessing the aspirations of planning, these otherwise disparate perspectives can be seen to broadly discount attempts to regulate the built and natural environment. This is because regulation produces either negative effects and/or perpetuates the inequalities of the status quo, or at best produces only short-term ameliorations. Hence, and seemingly paradoxically, neo-liberal critiques of planning come to very similar conclusions to those of radical theories of structured social inequalities: on balance, a focus on spatial organization and land use control is ineffectual or even detrimental.

In contrast, post-structural approaches might seem to open out all sorts of potentials for alternative planning practices, but to date, such possibilities have not been explicitly explored. Instead, the unsettling of stable and predictable/environment identity relations once again casts doubt on planning's ability to achieve its spatial-social aims.

Fostering Fixity

Post-structuralist unpackings of the categories of class, race and gender and their reproduction over time and in place, have directed analysis to the multiple ways in which subjects may be positioned or may take up and perform identities, which through repetition, become taken as essential characteristics of persons (Butler, 1990, 1993). If identities are mobile and polymorphous, spaces too are conceived as permeable and dynamic and the relations between places and identities are multiple and overlapping. The result is that they cannot be mapped directly onto, nor read off from, each other (as for instance, the Chicago School wished to do: Jacobs and Fincher, 1998: 4–10). As Gillian Rose (1999: 248) suggests:

Space then is not an anterior actant to be filled or spanned or constructed, and to claim it is runs the risk of making a contingent spatial articulation of relationality foundational. Instead, space is practised, a matrix of play, dynamic and iterative, its forms and shapes produced through the citational performance of self-other relations.

But citational enactments also include regulatory repetitions of boundaries and limits that reinforce the power invested in the production of specific spaces (G. Rose, 1999: 248–9; see also Pratt, 1998).

Relating these conceptions directly to a history of planning, Hooper (1998) has examined the

regulatory 'fantasies of male desires' for order and control reiterated since planning's inception.[18] In a similar fashion, theories relating spatial segregation to the working out of structures of the psyche, and to processes of 'abjection' that project despised aspects of the self onto the other (e.g. Sibley, 1995), associate planning's desires for spatial and social order with dominant exclusionary impulses (see also Sennett, 1970).

Thus, the idea that both subjectivities and spaces are potentially unstable and mutable and their boundaries incipiently porous and osmotic suggests that, as a regulatory practice of the state, planning's attempts to bring about durable environments and uniform orderings of spaces are constantly being threatened with slippage and disintegration and challenged by alterity and resistance. Planning is caught up in the discursive materiality of struggles over attempts to fix identities in place and limit open-ended possibilities for variation and difference to emerge. Nevertheless, as Watson and Gibson (1995a: 259) suggest, the idea of a 'postmodern' planning is not entirely impossible:

So bound up in the belief in determinate outcomes, planning (and, indeed, the politics of which it is born) almost seems impossible from a postmodern point of perspective. Yet if we switch to the idea of planning as action without determination, what do we lose? If indeterminacy is at the core of the social, is planning any less possible?

However, it could be asked whether, without aspirations to produce certain kinds of spaces with particular social effects, would planning still be 'planning'? The question remains as to how planning – whether as state-based regulation or participatory place-making – could dream of a 'space and a subject which we cannot yet imagine' (G. Rose, 1995: 354; see also Pinder, 2005).

While in other aspects very different, the three standpoints discussed here – structural, neo-liberal and post-structural (for want of better terms) – imply that planning: (a) operates with mistaken or discursively limited assumptions about the role of space in social and/or economic relations; and (b) is unlikely to achieve the positive outcomes that it claims for itself because of its interrelations (either dominant or ineffectual) with wider discursive, cultural, social or economic processes.

In examining the implications of different frameworks for planning's causal spatial assumptions, two main points emerge. Either planning is a waste of time, since space is produced out of economic processes and/or systematic relationships that are beyond planning control; or, in order to achieve democratic, open-ended, equitable cities and places, planning needs to reassess or abandon its notions of spatial order and regulation.

In the final section of this chapter, then, I want to turn to Foucauldian perspectives that offer a different approach to the question of spaces and subjects, and in particular, ones that can be put to work in examining *how* taken-for-granted suppositions about spatial causalities have operated in planning discourses. That is, rather than adjudicating on the adequacy or otherwise of planning's assumptions of spatial and environmental causality, planning's spatial aspirations can be examined as a form of governmental rationality. A Foucauldian-derived approach might take these assumed causalities or spatial rationalities as a starting point of analysis of how planning thought and practice aim to create particular environments and foster certain kinds of subjects.

GOVERNMENTALITY, RATIONALITIES, SPACES

Governmentality

In contrast to conceptions of the state found in some of the theories discussed above, a governmentality approach unsettles the assumption of massive state power – the *'monstre froid* that confronts us' (Foucault, 1991a: 103) – such as found in Lefebvre (1991) or de Certeau (1984), for instance. Rather than a state that 'neutralizes whatever resists it by castration or crushing' (Lefebvre, 1991: 23), from a governmentality point of view, the state is a heterogeneous, contingent and unstable, conflicting and converging, assemblage of institutions, regulations, practices and techniques of rule, and the exercise of power is neither automatic nor uniform nor carried out in the furtherance of particular or general 'interests' (e.g. Foucault, 1982; N. Rose and Miller, 1992; Dean, 1999: chap. 1). The power of government is already present in practices, organizations and institutions beyond the state, and is thus more dispersed, malleable and susceptible to localized resistances than monocentric, unidirectional understandings of 'power' or 'the state' would indicate.

But more especially, for Foucault (see 1981), not only is 'government' a dispersed form of power that neither originates from, nor is confined to, the state, but governmental aims and practices are not directed at the repression of pre-existing free individuals. Rather, through education, fostering and incitement, as much as through training, control and regulation, liberal governmentality aims at the production of differentiated subjects capable of exercising certain forms of freedom.

In this sense, government is 'irreducibly utopian' (Dean, 1999: 33):

> Every theory or programme of government presupposes an end of some kind – a type of person, community, organization, society, or even world which is to be achieved. ... Even at its apparently most bureaucratic and managerial, or its most market-inspired, government is a fundamentally utopian activity. It presupposes a better world, society, way of doing things or way of living. (p. 33)

The term 'governmentality', then, draws attention to two aspects of governmental power: government as dispersed attempts to act on the action of others – the conduct of conducts (Gordon, 1991: 2); and government as 'mentality' – historically specific aspirations and rationalities about aims and methods for conducting conducts. Broadly, Foucault (1986a, 1988, 1991a, 2000) identifies three main forms of governmentality that have contingently, and often contradictorily, converged around various aims and practices of the state: discipline and police power; classical liberal; and advanced liberal.

Historically, in contrast to the police power and discipline of the administrative monarchies that sought to regulate and control all aspects of life, classical liberal government was concerned with the bio-social qualities of a population, its health and reproduction, as well as the bodily comportments of individuals – what Foucault (1981: 138–43) sees as an intersection of 'bio-power' and 'anatomo-power' – the management of an 'individual-population-environment nexus' (Burchell, 1991: 142). Through the creation of appropriate behaviours in individuals and the management of populations, liberal government has sought to foster conditions under which quasi-natural and semi-autonomous economic and social processes will be maximized and made self-reinforcing. These conditions are best approximated if subjects come to see themselves as exercising individual autonomy.[19]

Different forms of liberal government can be seen in historically specific transformations, such as the appearance in the early twentieth century of welfare or progressivist 'government from a social point of view' (N. Rose, 1999: chap. 3). In this, different programmes of reform of the Victorian industrial city contingently and haphazardly gave rise to projects for the management of a generalized social sphere. Similarly, N. Rose (1999: chap. 4) suggests that a regime of advanced liberal government emerged in the last quarter of the twentieth century, in which the rationalities of government have been realigned to foster the self-management of 'enterprising selves'.

Nevertheless, 'discipline' and 'police power' continue to be present in attempts to 'make up' different subjects in regimes of liberal rule. Indeed, Foucault's (1979) study of the prison is also a study of how the 'modern soul' comes to be constituted as amenable to government through the micro-practices and knowledges, and especially forms of spatial disposition that find their origins in institutions of confinement and training (see Dean, 1999, 2002; Elden, 2001).

Rationalities of Government

The importance of seeing government as 'mentality' lies in the ways the truths of the aims and objects of government are constructed – not as programmes, the outcomes of which can be measured and tested against 'reality', but as discursively produced and circulated rationalities (Foucault, 1974: xv–xxiv; 1984, 1991a, 1991b, 1996a). As Dean (1994a: 158) notes, 'governmentality' defines

'a thought-space across the domains of ethics, government and politics, of the government of self, others, and the state, of practices of government and practices of the self, of self-formation and political subjectification, that weaves them together without a reduction of one to the other'.

Such rationalities answer to no explicit, determinative or singular principles, but are assemblages of 'heterogeneous discursive and non-discursive practices, and regimes of truth and conduct, which possess an overall coherence' (Dean, 1994b: 223, note 5; see also Dreyfus and Rabinow, 1982). Rationalities of government, however, cannot guarantee outcomes, and do not emanate from a single centre, such as the state. Governmental rationalities and technologies are 'multiform instrumentations' made up of 'bits and pieces', employing 'disparate sets of tools and methods' (Foucault, 1979: 26), connecting 'the regular application of some form of relatively systematized knowledge to the pragmatic problems of the exercise of authority ...' (Dean, 1996: 59; N. Rose, 1999: 51–5).

Rationalities connect material and discursive elements into forms of *dispositif* (variously translated as 'disposition', 'arrangement', 'grid of intelligibility', 'mechanism', 'apparatus': see Elden, 2001: 110) – an historically specific 'heterogeneous ensemble consisting of discourses, institutions, architectural forms, regulatory decisions, laws, administrative measures, scientific statements, philosophical, moral and philanthropic propositions – ... the said as much as the unsaid' (Foucault, 1980d: 196; 1986a: 255–6).[20] A *dispositif* is thus simultaneously discursive, material and spatial.

Spatial Rationalities

So, like his attention to mentalities and rationalities of government, Foucault's references to space (e.g. 1976, 1979, 1980a, 1980b, 1980c, 1986a, 1986c) include not only the construction of buildings and the disciplinary effects of the layout of rooms, for instance, but also the ways spaces figure in 'thought' about the conduct of conducts (Elden, 2001: chap. 5).

Hence, for example, a possible history of spaces indicated by Foucault (1980a: 149) – 'which would at the same time be the history of *powers* (both these terms in the plural) – from the great strategies of geo-politics to the little tactics of the habitat' – suggests that how spaces are conceived in 'strategies' and 'tactics' is as important as the 'reality' of spaces themselves.[21] In this way (and contrary to Smith and Katz, 1993), the question of space in Foucault is far more than merely metaphorical; it is raised, 'not as an ontological issue, but as a political and analytic one' (Rabinow, 1982: 269). In this framework, then, a spatial rationality of government is any attempt to direct human conduct that has at its base, 'thought' about relations between individuals, populations, spaces and environments, what is wrong and how it can be improved (Dean's (1999: 33) 'utopian element' of government; see, N. Rose, 1999: 24–8).[22]

The multiple causal connections between space, environments and subjects that express the truths embedded in spatial rationalities can be crystallized as a 'governing statement' (Dean, 1992: 220–1) about 'what is' and 'what ought to be', which gives rise to multiple solutions. Such a governing statement encapsulating spatial and environmental rationality might be constructed as a problematizing question that asks: 'What kinds of spaces, milieux and environments contribute to physical, social and moral problems of individuals and populations?' And as a consequence: 'What kinds of spaces, milieux and environments need to be created in order to solve these problems?'

Subjects come to be governed in relation to particular kinds of problematic spaces (rural areas, 'the countryside', 'the bush', regions, neighbourhoods, villages, suburbs, cities, slums, streets, buildings, styles of houses) and undesirable environments (ugly, disorderly, dark, diseased, degenerate, miasmic, tropical, desert, neglected). These spaces and environments are causally linked with types of subjects and behaviours, and attempts to institute reforms (slum clearance, sanitation, public housing, estate plans and layouts, rural settlements and colonies, home ownership, street plantings, parks, allotment gardens) involve the creation of *dispositifs* – clusters of materials, practices, buildings, institutions, regulations, and disciplines and knowledges – that aspire to the 'fabrication of virtue' (Evans, 1982; Osborne and N. Rose, 1999). These discursive material assemblages aim to produce desired qualities in populations and perpetuate self-reinforcing cycles of appropriate behaviours in individuals.[23] And finally, the aspiration of these rationalities is the creation of orderly, transparent, productive and enriching environments that will foster the good citizen in the good city or nation – disparate aims that inform often conflicting interpretations of what these 'eudaemonic' configurations of happiness might be (see Osborne and N. Rose, 1999).

In this way, an analytics of spatial government has the potential to open up different routes for examining the relations between states and citizens, and especially, to offer alternative approaches to the study of the politics and powers of spatial planning. Planning can be seen as a spatial technology of liberal government implicated in projects of bio-power and subjectification, and the historical accretion of its 'operative rationales', the problems it identifies and the technologies and practices to solve them can be traced. Examination of the historical contingency of the rationalities at work in practices like town planning can prompt an assessment of what one can 'still accept of this system of rationality. What part, on the contrary, deserves to be set aside, transformed, abandoned?' (Foucault, 1996b: 424).

This is a limited aim set within an anti-humanist philosophical framework that does not hold out promises of general societal transformation. It therefore can seem pessimistic and unhelpful to those who work for change on a wider scale. Nevertheless, it does confront the thorny problem of whether the effects of thought and action can be completely understood in retrospect, let alone known in advance. Analysing rationalities of government fosters caution in reaching for prescriptive solutions and encourages active alertness to a diversity of possibilities. This is not a recipe for 'paralysis' or 'anaesthesia', but an incitement to contribute to 'a long work of comings and goings, of exchanges, reflections, trials, different analyses' (Foucault, 1991b: 82–86); because if nothing is guaranteed, then everything is potentially otherwise – 'everything is dangerous [and] we always have something to do' (Foucault, 1986b: 343).

CONCLUSION

A constant thread runs through the changing justifications and aims of spatial or land use planning: the aspiration to create spaces and places that have positive effects on persons and populations. However, these assumptions about the beneficial outcomes of planning policies and practices have been subjected to scrutiny from various theoretical perspectives, some of the main strands of which have been discussed in this chapter.

Approaches suggesting that planning possesses capacities for making a difference to unequal resource distributions or environmental qualities attribute its failings to institutional or individual attributes, practices and world-views; for instance, as bureaucrats, technocrats, as white, as male, as heterosexual. The spaces that result from these institutional and individual powers and practices are seen to work in the interests of dominant groups by homogenizing difference and suppressing the multiplicities of social and cultural diversity. But with better technical training; or with better theoretical knowledge of economy, society and space; or with better skills at mediation, negotiation and communication, or so it is argued, planners can begin to bring about positive outcomes for environments and citizens.

However, other theories suggest that the spaces that planning produces, or contributes to perpetuating, inevitably reflect and support dominant economic or social power relations. Theories of capitalism and patriarchy have been touched on here, but planning is also seen to be implicated in the perpetuation of other dominations, such as racism, heteronormativity, 'ethnocracy', colonialism, ableism. This is because, it is argued, the state and its institutions are embedded in, or reproduce, those structures. Planning's concerns to produce ordered, healthy, inspiriting or empowering environments only serve to mask structural inequalities; and its ambitions for social improvement are doomed to failure, if it does not, at the same time, address the structural causes of exploitation or marginalization. Or, in neo-liberal analyses, planning's aims will never be achieved because such 'interventions' inhibit social and economic processes that would otherwise produce more beneficial outcomes than planning is capable of.

Post-structural challenges to notions of the undifferentiated subject and the 'there-ness' of space reinvigorate examinations of the complexity of relations between identities and places, but in ways that disturb the more simplistic spatial and environmental aspects of planning's ambitions. Dislodging assumptions about the nature of the subject – as agent or as bearer of structures – and about the transparency of projects of reform would seem to undermine the foundations on which planning (as we know it) rests.

The Foucauldian notion of governmentality as the 'conduct of conducts' and as a 'mentality' of government, enables an examination of the spatial rationalities underpinning attempts to manage an individual/population/environment complex. Spatial rationalities underpin the 'operative rationales' that link spaces and environments to attempts to fabricate virtue and make up subjects. However, outcomes are not necessarily those predicted and projects of government are continually and congenitally failing, producing further attempts at solutions to problems arising from the practices of government itself.

Spatial and land use planning can be seen to deploy truths and rationalities of spatial causality that have effects 'in the real' in terms of divisions of space that are expected to have certain outcomes, and the qualities of environments that aspire to foster subjects who will exercise liberal freedoms in appropriate ways. Examining the spatial rationalities that inhabit planning practice and theory gives some purchase on understanding how they operate and how they persist in the face of theoretical and political critique and with, at best, intermittent success.

This has been a very partial overview of how relations between states, territories, citizens and planning have been conceptualized and might be re-conceptualized. I want to conclude by suggesting that in analysing struggles – whether as localized resistances to planning projects; or as claiming rights in the name of marginalized identities; or as the uncovering of spatial inequalities; or as performances of difference; or as the 'motor-force of all analysis … it is not always easy to tell who are the good guys and who are the bad guys; most of the forces are too complex to diagnose in a zero-sum logic of power' (Osborne and N. Rose, 1998b: 51).

It is a correspondingly complex task to analyse the entangled strands of happenstance that may have enabled any given configuration of spatial regulation, or any particular struggle around the natural or built environment or local or national space. But theoretical practices and engaged research in political geography would seem to form an arena in which such a task can be fruitfully undertaken.

NOTES

1 The terms 'town and country', 'urban and regional', 'city', 'spatial' planning and permutations thereof, serve to distinguish between state-based spatial regulation and broader conceptions of 'planning' as a generic human activity, or as generalized policy planning by the state, or as forms of community self-organization. The terms 'land use planning' or 'development control' refer to the detailed allocation of land for specified purposes and the administration of development permissions or refusals. Since this chapter is about space and government, 'planning' is used as shorthand for diverse forms of spatial and land use regulation.

'Environmental' planning is also associated with these practices, but has a slightly different trajectory that is not followed here. In this chapter, 'environment' mainly refers to the 'surroundings' or 'milieux' of individuals, groups or populations – the physical, social and aesthetic qualities of spaces and places that planning hopes either to eradicate (negative qualities) or to create (positive qualities).

2 For example, Davies' (1972) study of planning in Newcastle-upon-Tyne, *The Evangelistic Bureaucrat*, or Roddewig's (1978) account of the Sydney *Green Bans*.

3 See also, for example, John Friedmann's influential *Retracking America* (1973). For an alternative view of public participation at the time, see Davidoff's 1965 paper, 'Advocacy and pluralism in planning', which sees planners' roles as being similar to legal advocacy on behalf of those unable to represent themselves.

4 And, one might add here, that a similar criticism from this point of view could be made of some design-based concepts and policies currently proposed as solutions to urban problems and lack of 'community' (e.g. Duany, 1997, cited in Harvey, 2000: 169–73; Rogers Report, 1999).

5 This literature is voluminous, and can only be indicated here. Contributions to debates about the role of the state and planning in capitalist society include, for example, Dear and Scott (1981), Harvey (1973), McDougall (1979), Pahl (1975), Paris (1982), Scott and Roweis (1977), Simmie (1974), Stilwell (1980); see also Kilmartin et al. (1985), Saunders (1981).

6 Urban managerialist perspectives were most predominant in literature seeking to understand the British post-war welfare state, and had less purchase in the US, where community power debates flourished. This may be because of differences in the structures and remits of urban government, and the extent and methods of social provision between the two polities. But Australian researchers drew on Weberian and managerialist, as well as Marxist, positions in their studies of state allocations in a federal system having very different institutional and economic dimensions from the British (e.g. Kilmartin et al., 1985; McLoughlin and Huxley, 1986); so differences between the objects of study may not provide a complete account of different theories.

7 See, for example Fainstein (1994, 1995), Judge et al. (1995), Painter (1997: 128–34), Stoker and Mossberger (1994), Stone (1989, 1993).

8 The English-speaking literature on women, cities, urban policies and planning proliferated from the late 1970s onwards, and it is only possible to give a hint of its extent here: for instance, on socio-spatial inequalities in the city, see Ferrier (1983), Hayden (1984), Little et al. (1988), Stimpson et al. (1981), Wekerle et al. (1980), Women and Geography Study Group (1984); on strategies for

counteracting planning's gender bias, see Hayden (1981), Saegert (1985), Sandercock and Forsyth (1992).

9 This debate is well illustrated in the exchanges sparked off by Fraser's (1995, 1997) distinction between 'recognition' and 'redistribution' and 'affirmative' and 'transformational' justice. Examples of studies examining the politics of identity include: Gibson-Graham (1997) on re-thinking class politics; Gleeson (1998) on justice and the disabling city; Knopp (1998) on social justice and sexuality; Yiftachel (2000), and Yiftachel and Yacobi (2003) on 'ethnocracy'; Watson (1999) on city politics.

10 See also Watson and Gibson (1995b) on postmodern cities; Watson and McGillivray (1994) on multiculturalism; and the essays in Douglass and Friedmann *Cities for Citizens* (1998).

11 Sadly, this hopeful imagining of the future of cities and the capacity for planning to assist in bringing about open-ended processes for the 'being together of strangers' seems even further from possible enactments in the wake of the destruction of the Twin Towers in New York, the London bombings of 2005 and the rise of 'Islamophobia' on the part of 'western/northern' governments.

12 For overviews, see Gottdiener (1985), Kilmartin et al. (1985), Saunders (1981), Taylor (1998: chap. 6).

13 For instance, see Badcock (1984), Dear and Scott (1981), Fainstein and Fainstein (1983), Fogelsong (1986), Gottdiener (1985), Kirk (1980), Lefebvre (1991), Sandercock (1977), Scott and Roweis (1977), Tabb and Sawers (1978).

14 On the role of urban planning, urban management and urban regimes in the production of stability, see Brindley et al. (1989), Fainstein (1995, esp. chap. 5), Lauria and Whelan (1995), Prior (2005), and the essays in Lauria (1997).

15 See, for example, Breugal and Kay (1975), England (1993), Greed (1994), Matrix (1984), Roberts (1991), Ritzdorf (1986), Saegert (1980), Wajcman (1991, chap. 5).

16 See Pennington (2000, 2002) and Sorensen and Day (1981), for versions of these arguments; and Taylor (1998) and Thornley (1991), for overviews of debates about markets vs government in relation to planning.

17 This is a formulation adapted from Saunders' (1981: 99–109) discussion of the significance of space in classical social theory.

18 From a different angle, Huxley 2002 draws on Butler's ideas of performativity and the regulative repetition of gender to examine planning as a technology of gendered governmentality.

19 Once again, there is a large literature developing this notion of the formation of subjects of freedom. A brief selection might include Burchell (1991), Dean (1999), Gordon (1991), Hindess (1996), and Rose (1999).

20 A detailed and convincing study of such *dispositifs* is Rabinow's (1989) *French Modern: Norms and Forms of the Social Environment*.

21 See also the discussion of changing conceptions of space in 'Space, Knowledge, Power' (Foucault, 1986a).

22 My interpretation of governmental and spatial rationalities here is slightly different from some other uses of the term, such as Murdoch's (2000; Murdoch and Abram, 2002). Murdoch deploys the notion of governmental rationalities in connection with versions of actor network theory and social constructionism to show how space and time are constituted as 'entities' within policy discourses and how these interact with 'spatial formations' (Murdoch, 2000: 504).

23 For example, see Hannah (2000) and Matless (1998).

REFERENCES

Allmendinger, P. and Tewdwr-Jones, M. (eds) (2002) *Planning Futures: New Directions for Planning Theory*. London and New York: Routledge.

Arnstein, S. (1969) 'A ladder of citizen participation', *Journal of the American Institute of Planners*, 35: 216–24.

Badcock, B. (1984) *Unfairly Structured Cities*. Oxford: Blackwell.

Boyer, M.C. (1983) *Dreaming the Rational City: The Myth of American City Planning*. Cambridge, MA: MIT Press.

Boyer, R. (1990) *The Regulation School: A Critical Introduction*. New York: Columbia University Press.

Breugal, I. and Kay, A. (1975) 'Women and planning', *Architectural Design*, August: 499–500.

Brindley, G., Rydin, Y. and Stoker, G. (1996) *Remaking Planning: The Politics of Urban Change* (2nd edn). London: Routledge.

Burchell, G. (1991) 'Peculiar interests: civil society and governing "The System of Natural Liberty"', in Burchell et al. (eds), pp.119–50.

Burchell, G., Gordon, C. and Miller, P. (eds) (1991) *The Foucault Effect: Studies in Governmentality*. London: Harvester/Wheatsheaf.

Butler, J. (1990) *Gender Trouble: Feminism and the Subversion of Identity*. London and New York: Routledge.

Butler, J. (1993) *Bodies That Matter: On the Discursive Limits of 'Sex'*. London and New York: Routledge.

Castells, M. (1977) *The Urban Question: A Marxist Approach*. London: Edward Arnold.

Chadwick, G. (1971) *A Systems View of Planning*. Oxford: Pergamon Press.

Chapin, F. (1965) *Urban Land Use Planning*. Urbana: University of Illinois Press.

Davidoff, P. (1965) 'Advocacy and pluralism in planning', *Journal of the American Institute of Planners*, 31: 331–8.

Davies, J.G. (1972) *The Evangelistic Bureaucrat: A Study of a Planning Exercise in Newcastle-upon-Tyne*. London: Tavistock Press.

Dean, M. (1992) 'A genealogy of the government of poverty', *Economy and Society*, 21(3): 215–51.

Dean, M. (1994a) '"A social structure of many souls": moral regulation, government, and self-formation', *Canadian Journal of Sociology*, 19(2): 145–68.

Dean, M. (1994b) *Critical and Effective Histories: Foucault's Methods and Historical Sociology*. London: Routledge.

Dean, M. (1996) 'Putting the technological into government', *History of the Human Sciences*, 9(3): 47–68.

Dean, M. (1999) *Governmentality: Power and Rule in Modern Society*. London: Sage.

Dean, M. (2002) 'Liberal government and authoritarianism', *Economy and Society*, 31(1): 37–61.

Dear, M.J. and Scott, A.J. (eds) (1981) *Urbanization and Urban Planning in Capitalist Society*. London and New York: Methuen.

de Certeau, M. (1984) *The Practice of Everyday Life*. Berkeley: University of California Press.

Douglass, M. and Friedmann, J. (eds) (1998) *Cities for Citizens: Planning and the Rise of Civil Society in a Global Age*. Chichester: John Wiley.

Dreyfus, H. and Rabinow, P. (1982) *Michel Foucault: Beyond Structuralism and Hermeneutics*. Brighton: Harvester.

Elden, S. (2001) *Mapping the Present: Heidegger, Foucault and the Project of a Spatial History*. London: Continuum.

England, K. (1993) 'Suburban pink collar ghettoes: the spatial entrapment of women', *Annals of the Association of American Geographers*, 83: 225–42.

Evans, R. (1982) *The Fabrication of Virtue: English Prison Architecture 1750–1840*. Cambridge: Cambridge University Press.

Fainstein, N. and Fainstein, S. (eds) (1983) *Restructuring the City*. New York: Longman.

Fainstein, S. (1994) *The City Builders: Property, Politics and Planning in London and New York*. Cambridge, MA and Oxford: Blackwell.

Fainstein, S. (1995) 'Politics, economics and planning: why urban regimes matter', *Planning Theory*, 14: 34–43.

Faludi, A. (ed.) (1973) *A Reader in Planning Theory*. Oxford: Pergamon Press.

Ferrier, M. (1983) 'Sexism in Australian cities: barriers to employment opportunities', *Women's Studies International Forum*, 6(1): 73–84.

Fincher, R. and Jacobs, J.M. (eds) (1998) *Cities of Difference*. New York: Guilford Press.

Fishman, R. (1982) *Urban Utopias in the Twentieth Century: Ebenezer Howard, Frank Lloyd Wright, Le Corbusier*. Cambridge, MA: MIT Press.

Fogelsong, R. (1986) *Planning the Capitalist City: The Colonial Era to the 1920s*. Princeton, NJ: Princeton University Press.

Forester, J. (1989) *Planning in the Face of Power*. Berkeley: University of California Press.

Foucault, M. (1974) *The Order of Things: An Archaeology of the Human Sciences*. London: Routledge.

Foucault, M. (1976) *The Birth of the Clinic: An Archaeology of Medical Perception*. London: Tavistock.

Foucault, M. (1979) *Discipline and Punish: The Birth of the Prison*. New York: Vintage Books.

Foucault, M. (1980a) 'Questions on Geography', in Gordon (ed.), pp. 63–76.

Foucault, M. (1980b) 'The eye of power', in Gordon (ed.), pp. 146–165.

Foucault, M. (1980c) 'The politics of health in the eighteenth century', in Gordon (ed.), pp. 166–82.

Foucault, M. (1980d) 'The confessions of the flesh', in Gordon (ed.), pp. 194–228.

Foucault, M. (1981) *The History of Sexuality*, Vol. I: *An Introduction*. London: Penguin.

Foucault, M. (1982) 'The subject and power', Afterword in H. Dreyfus and P. Rabinow (eds), *Michel Foucault: Beyond Structuralism and Hermeneutics*. London: Harvester/Wheatsheaf, pp. 208–26.

Foucault, M. (1984) 'The order of discourse', in M. Shapiro (ed.), *Language and Politics*. Oxford: Blackwell, pp.108–38.

Foucault, M. (1986a) 'Space, knowledge and power', in Rabinow (ed.), pp. 239–56.

Foucault, M. (1986b) 'On the genealogy of ethics: an overview of work in progress', in Rabinow (ed.), pp. 340–72.

Foucault, M. (1986c) 'Of other spaces', *Diacritics*, 16: 22–7.

Foucault, M. (1988) 'Politics and reason', in L. Kritzman (ed.), pp. 57–85.

Foucault, M. (1991a) 'Governmentality', in Burchell et al. (eds), pp. 87–104.

Foucault, M. (1991b) 'Questions of method', in Burchell et al. (eds), pp. 73–86.

Foucault, M. (1996a) 'Clarifications on the question of power', in Lotringer (ed.), pp. 255–263.

Foucault, M. (1996b) 'What calls for punishment', in Lotringer (ed.), pp. 423–31.

Foucault, M. (2000) 'Security, territory, population', in P. Rabinow (ed.), *Michel Foucault: Ethics, Subjectivity and Truth. Essential Works of Foucault 1954–1984*, Vol. 1, London: Penguin, pp. 67–71.

Fraser, N. (1995) 'From redistribution to recognition? Dilemmas of justice in a post-socialist age', *New Left Review*, 212: 68–93.

Fraser, N. (1997) *Justice Interruptus: Critical Reflections on the 'Postsocialist' Condition*. London and New York: Routledge.

Friedmann, J. (1973) *Retracking America*. New York: Doubleday Anchor.

Gibson-Graham, J.K. (1997) 'Re-placing class in economic geographies: possibilities for a new class politics', in R. Lee and J. Wills (eds), *Geographies of Economies*. London: Arnold.

Gleeson, B. (1998) 'Justice and the disabling city', in Fincher and Jacobs (eds), pp. 89–119.

Goodman, R. (1971) *After the Planners*. New York: Simon & Schuster.

Goodwin, M. and Painter, J. (1997) 'Concrete research, urban regimes, and regulation theory', in Lauria (ed.), pp. 13–29.

Gordon, C. (1991) 'Governmental rationality: an introduction', in Burchell et al. (eds), pp. 1–52.

Gottdiener, M. (1985) *The Social Production of Urban Space*. Austin: University of Texas Press.

Graham, S. and Healey, P. (1999) 'Relational concepts of space and place: issues for planning theory and practice', *European Planning Studies*, 7(5): 623–46.

Greed, C. (1994) *Women and Planning: Creating Gendered Places*. London: Routledge.

Gross (Grosz), E. (1987) 'Discourses of definition: philosophy', in Women's Studies Course Team, *Feminist Knowledge as Critique and Construct*. Geelong: Deakin University Press, pp. 129–52.

Hannah, M. (2000) *Governmentality and the Mastery of Territory in Nineteenth-Century America*. Cambridge: Cambridge University Press.

Harvey, D. (1973) *Social Justice and the City*. London: Edward Arnold.

Harvey, D. (1985) 'On planning the ideology of planning', in *The Urbanization of Capital*. Oxford: Blackwell, pp. 165–84.

Harvey, D. (1996) *Justice, Nature and the Geography of Difference*. Oxford: Blackwell.

Harvey, D. (2000) *Spaces of Hope*. Edinburgh: Edinburgh University Press.

Hayden, D. (1981) 'What would a non-sexist city be like? Speculations on housing, urban design and human work', in Stimpson et al. (eds), pp. 167–84.

Hayden, D. (1984) *Redesigning the American Dream: The Future of Housing, Work and Family Life*. London and New York: WW. Norton.

Healey, P. (1997) *Collaborative Planning: Shaping Spaces in Fragmented Societies*. Basingstoke: Macmillan.

Healey, P., Cameron, S., Davoudi, S., Graham, S. and Madani-Pour, A. (eds) (1995) *Managing Cities: The New Urban Context*. Chichester: John Wiley.

Hebbert, M. (1998) *London: More by Fortune Than Design*. Chichester: John Wiley.

Hindess, B. (1996) *Discourses of Power: From Hobbes to Foucault*. Oxford: Blackwell.

Hooper, B. (1998) 'The poem of male desires: female bodies, modernity and "Paris, the capital of the nineteenth century"', in L. Sandercock (ed.), *Making the Invisible Visible: A Multicultural Planning History*. Berkeley: University of California Press, pp. 227–54.

Huxley, M. (2002) 'Governmentality, gender and planning: a Foucauldian perspective', in Allmendinger and Tewdwr-Jones (eds), pp. 136–54.

Innes, J. (1995) 'Planning theory's emerging paradigm: communicative action and interactive practice', *Journal of Planning Education and Research*, 14(3): 189–90.

Jacobs, J. (1961) *The Death and Life of Great American Cities*. New York: Vintage Books.

Jacobs, J.M. and Fincher, R. (1998) 'Introduction', in Fincher and Jacobs (eds), pp. 1–25.

Jessop, B. (1990) 'Regulation theories in retrospect and prospect', *Economy and Society*, 19(2): 153–216.

Judge, B., Stoker, G. and Wolman, H. (eds) (1995) *Theories of Urban Politics*. London: Sage.

Keeble, L. (1952) *Principles and Practice of Town and Country Planning*. London: Estates Gazette.

Keith, M. and Pile, S. (eds) (1993) *Place and the Politics of Identity*. London and New York: Routledge.

Kilmartin, L., Thorns, D. and Burke, T. (1985) *Social Theory and the Australian City*. Sydney: Allen & Unwin.

Kirk, G. (1980) *Urban Planning in a Capitalist Society*. London: Croom Helm.

Knopp, L. (1998) 'Sexuality and urban space: gay male identity politics in the United States, the United Kingdom, and Australia', in Fincher and Jacobs (eds), pp. 149–77.

Lauria, M. (ed.) (1997) *Reconstructing Urban Regime Theory: Regulating Urban Politics in a Global Economy*. Thousand Oaks, CA: Sage.

Lauria, M. and Whelan, R. (1995) 'Planning theory and political economy: the need for reintegration', *Planning Theory*, 14: 8–33.

Lechte, J. (1995) '(Not) belonging in postmodern space', in Watson and Gibson (eds), pp. 99–112.

Lefebvre, H. (1991) *The Production of Space*. Oxford: Blackwell.

Leo, C. (1997) 'City politics in an era of globalization', in Lauria (ed.), pp. 77–98.

Little, J., Peake, L. and Richardson, P. (eds) (1988) *Women in Cities: Gender and the Urban Environment*. Basingstoke: Macmillan.

Matless, D. (1998) *Landscape and Englishness*. London: Reaktion Books.

Matrix (1984) *Making Space: Women and the Man-Made Environment*. London: Pluto Press.

McDougall, G. (1979) 'The state, capital and land: the history of town planning revisited', *International Journal of Urban and Regional Research*, 3(3): 361–80.

McLoughlin, J.B. (1969) *Urban and Regional Planning: A Systems Approach*. London: Faber & Faber.

McLoughlin, J.B. and Huxley, M. (eds) (1986) *Urban Planning in Australia: Critical Readings*. Melbourne: Longman Cheshire.

Murdoch, J. (2000) 'Space against time: competing rationalities in planning for housing', *Transactions of the Institute of British Geographers*, 25: 503–19.

Murdoch, J. and Abram, S. (2002) *Rationalities of Planning: Development Versus Environment in Planning for Housing*. Aldershot: Ashgate.

Osborne, T. and Rose, N. (1998a) 'Governing cities', in E. Isin, T. Osborne and N. Rose (eds), *Governing Cities: Liberalism, Neoliberalism, Advanced Liberalism*, Working Paper No. 19, Urban Studies Programme, York University, Toronto, pp. 1–32.

Osborne, T. and Rose, N. (1998b) 'Spatial thinking without maps: a response to Engin Isin', in E. Isin, T. Osborne and N. Rose (eds), *Governing Cities: Liberalism, Neoliberalism, Advanced Liberalism*, Working Paper No. 19, Urban Studies Programme, York University, Toronto, pp. 47–52.

Osborne, T. and Rose, N. (1999) 'Governing cities: notes on the spatialisation of virtue', *Environment and Planning D: Society and Space*, 17: 737–60.

Pahl, R. (1975) *Whose City?* Harmondsworth: Penguin.

Pahl, R. (1977) 'Managers, technical experts and the state: forms of mediation, manipulation and dominance in urban and regional development', in M. Harloe (ed.), *Captive Cities: Studies in the Political Economy of Cities and Regions*. London: John Wiley, pp. 49–60.

Painter, J. (1997) 'Regulation, regime and practice in urban politics', in Lauria (ed.), pp. 122–44.

Paris, C. (ed.) (1982) *Critical Readings in Planning Theory*. Oxford: Pergamon.

Park, R. and Burgess, E. with McKenzie, R. and Wirth, L. (1925) *The City: Suggestions for Investigations of Human Behavior in the Urban Environment*. Chicago: University of Chicago Press (reprinted Midway, 1984).

Pennington, M. (2000) *Planning and the Political Market: Public Choice and the Politics of Government Failure*. London: Athlone Press.

Pennington, M. (2002) 'A Hayekian liberal critique of collaborative planning', in Allmendinger and Tewdwr-Jones (eds), pp.187–205.

Pinder, D. (2005) *Visions of the City: Utopianism, Power and Politics in Twentieth-Century Urbanism*. Edinburgh: Edinburgh University Press.

Pratt, G. (1998) 'Grids of difference: place and identity formation', in Fincher and Jacobs (eds), pp. 26–48.

Prior, A. (2005) 'UK planning reform: a regulationist interpretation', *Planning Theory and Practice*, 6(4): 465–84.

Rabinow, P. (1982) 'Ordonnance, discipline, regulation: some reflections on urbanism', *Humanities in Society*, 5(3/4): 267–78.

Rabinow, P. (1989) *French Modern: Norms and Forms of the Social Environment*. Chicago: University of Chicago Press.

Ritzdorf, M. (1986) 'Women and land use zoning', *Urban Resources*, 3(2): 23–7.

Roberts, M. (1991) *Living in a Man-Made World: Gender Assumptions in Modern Housing Design*. London: Routledge.

Roddewig, R. (1978) *Green Bans: The Birth of Australian Environmental Politics*. Sydney: Hale & Iremonger.

Rogers Report (1999) *Towards an Urban Renaissance* (Final Report of the Urban Task Force). London: Department of Environment, Transport and the Regions.

Rose, G. (1995) 'Making space for the female subject of feminism', in S. Pile and N.J. Thrift (eds), *Mapping the Subject: Geographies of Cultural Transformation*. London and New York: Routledge.

Rose, G. (1999) 'Performing space', in D. Massey, J. Allen and P. Sarre (eds), *Human Geography Today*. Cambridge: Polity Press, pp. 247–59.

Rose, N. (1999) *Powers of Freedom: Reframing Political Thought*. Cambridge: Cambridge University Press.

Rose, N. and Miller, P. (1992) 'Political power beyond the state: problematics of government', *British Journal of Sociology*, 43(2): 172–205.

Saegert, S. (1980) 'Masculine cities and feminine suburbs: polarised ideas, contradictory realities', *Signs*, 5 (special supplement): S96–S111.

Saegert, S. (1985) 'The androgynous city: from critique to practice', *Sociological Focus*, 18: 161–76.

Sandercock, L. (1977) *Cities for Sale: Property, Politics and Urban Planning in Australia*. Melbourne: Melbourne University Press.

Sandercock, L. (1998) *Towards Cosmopolis: Planning for Multicultural Cities*. Chichester: John Wiley.

Sandercock, L. (2003) *Cosmopolis II: Mongrel Cities in the 21st Century*. London: Continuum.

Sandercock, L. and Forsyth, A. (1992) 'A gender agenda: new directions for planning theory', *Journal of the American Planning Association*, 58(1): 49–59.

Saunders, P. (1981) *Social Theory and the Urban Question*. London: Hutchinson.

Scott, A.J. and Roweis, S. (1977) 'Urban planning in theory and practice: a reappraisal', *Environment and Planning A*, 10: 1097–1119.

Sennett, R. (1970) *The Uses of Disorder: Personal Identity and City Life*. London: Faber & Faber.

Sibley, D. (1995) *Geographies of Exclusion: Society and Difference in the West*. London: Routledge.

Simmie, J. (1974) *Citizens in Conflict: The Sociology of Town Planning*. London: Hutchinson.

Skeffington Report (1969) *People and Planning* (Report of the Committee on Public Participation on Planning). London: HMSO.

Smith, N. and Katz, C. (1993) 'Grounding metaphor: towards a spatialized politics', in Keith and Pile (eds), pp. 67–83.

Sorensen, A. and Day, P. (1981) 'Libertarian planning', *Town Planning Review*, 52(4): 390–402.

Stilwell, F. (1980) *Economic Crises, Cities and Regions*. Sydney: Pergamon Press Australia.

Stimpson, C., Dixler, E., Nelson, M. and Yatrakis, K. (eds) (1981) *Women and the American City*. Chicago: University of Chicago Press.

Stoker, G. and Mossberger, K. (1994) 'Urban regime theory in comparative perspective', *Environment and Planning C: Government and Policy*, 12: 195–212.

Stone, C. (1989) *Regime Politics: Governing Atlanta 1946–1988*. Lawrence: University of Kansas Press.

Stone, C. (1993) 'Urban regimes and the capacity to govern: a political economy approach', *Journal of Urban Affairs*, 13(3): 289–97.

Tabb, W. and Sawers, L. (eds) (1978) *Marxism and the Metropolis: New Perspectives in Urban Political Economy*. New York: Oxford University Press.

Taylor, N. (1998) *Urban Planning Theory Since 1945*. London: Sage.

Thornley, A. (1991) *Urban Planning Under Thatcherism: The Challenge of the Market*. London: Routledge.

Thrift, N.J. and Pile, S. (eds) (1995) *Geographies of Resistance*. London and New York: Routledge.

Tully, J. (1995) *Strange Multiplicity: Constitutionalism in an Age of Diversity*. Cambridge: Cambridge University Press.

Wajcman, J. (1991) *Feminism Confronts Technology*. Cambridge: Polity Press.

Watson, S. (1999) 'City politics', in S. Pile, C. Brook and G. Mooney (eds), *Unruly Cities? Order/Disorder*. London and New York: Routledge/The Open University, pp. 201–46.

Watson, S. and Gibson, K. (1995a) 'Postmodern politics and planning: a postscript', in Watson and Gibson (eds), pp. 254–64.

Watson, S. and Gibson, K. (eds) (1995b) *Postmodern Cities and Spaces*. Oxford: Blackwell.

Watson, S. and McGillivray, A. (1994) 'Stirring up the city: housing and planning in a multicultural society', in K. Gibson and S. Watson (eds), *Metropolis Now: Planning and the Urban in Contemporary Australia*. Sydney: Pluto Press, pp. 203–16.

Wekerle, G., Peterson, R. and Morley, D. (eds) (1980) *New Space for Women*. Boulder, Co: Westview Press.

Women and Geography Study Group (1984) *Geography and Gender: An Introduction to Feminist Geography*. London: Hutchinson.

Yiftachel, O. (2000) 'Social control, urban planning and ethno-class relations: Mizrahi Jews in Israel's "development towns"', *International Journal of Urban and Regional Research,* 24(2): 418–38.

Yiftachel, O. and Yacobi, H. (2003) 'Urban ethnocracy: ethnicization and the production of space in an Israeli "mixed city"', *Environment and Planning D: Society and Space,* 21: 673–93.

Young, I.M. (1990) *Justice and the Politics of Difference.* Princeton, NJ: Princeton University Press.

8

Welfare Provision, Welfare Reform, Welfare Mothers

Kim England

INTRODUCTION

Today a hope of many years' standing is in large part fulfilled. The civilization of the past hundred years, with its startling industrial changes, has tended more and more to make life insecure. ... We can never insure one hundred percent of the population against one hundred percent of the hazards and vicissitudes of life, but we have tried to frame a law which will give some measure of protection to the average citizen and to his family against the loss of a job and against poverty-ridden old age. (President Franklin D. Roosevelt, 14 August 1935)

What we are trying to do today is to overcome the flaws of the welfare system for the people who are trapped in it. ... From now on our nation's answer to this great social challenge will no longer be a never-ending cycle of welfare: it will be the dignity, the power, and the ethic of work. Today we are taking an historic chance to make welfare what it was meant to be: a second chance, not a way of life. ... Today, we are ending welfare as we know it. (President William J. Clinton, 22 August 1996)

Sixty years separate these remarks by Franklin Roosevelt and Bill Clinton, made as they signed into law the Acts that are often described as marking the beginning and end of 'welfare as we kn[e]w it' in the US: Roosevelt was signing the 1935 Social Security Act, while Clinton was signing the 1996 Personal Responsibility and Work Opportunity Reconciliation Act (PRWORA). Roosevelt and Clinton's comments are suggestive of the ways that the role of the state,

the contours of citizenship, and the relationship between the collective and the individual have shifted over this sixty-year period. Roosevelt's words reflect the formation of a particular relationship between the state and 'the average citizen and his family' – one where state intervention could help 'protect' against 'the hazards and vicissitudes of life'. A very different sort of relationship informed Clinton's speech: welfare was to be restructured to be 'what it was meant to be: a second chance, not a way of life'. By then 'welfare' had come to have a very narrow meaning, focused primarily on Aid to Families with Dependent Children (AFDC), and sometimes a few other public assistance programmes like Food Stamps.[1] The reconfiguration of responsibilities between the state and civil society involved not only the neo-liberal restructuring of the welfare state, but also a turn away from even a weak sense of collective responsibility and towards neo-liberal values of individualism in which people should take more 'personal responsibility' and be more 'self-sufficient' to use the language contained in the PRWORA. In this chapter I investigate feminist political geographies associated with the relationship between states, markets and families by examining the rise and demise of the legislative programme put in place for 'needy dependent children' – Aid to Dependent Children, which later became Aid to Families with Dependent Children when coverage was extended to two-parent families in 1962. A(F)DC became the primary form of cash assistance for poor families in the US until it was abolished with the PRWORA in 1996.

Traditionally, political geographers have explored questions related to the formal institutional arrangements associated with geopolitics, state sovereignty, state regulation, state formation and electoral politics – what several commentators call the big 'P' politics associated with 'the state' (Kofman and Peake, 1990; Flint, 2003; Fincher, 2004). Influences from cultural and feminist geographies have brought a range of small 'p' politics to the table of political geography – the everyday, identity politics, and contestations over meanings and representations. And with the broader embrace of what counts as 'political' in recent years there has been increasing traffic between feminist and political geographies (see, for example, Brown and Staeheli, 2003; Staeheli et al., 2004; Kofman, 2005). In this chapter my attention is on both the big 'P' politics of 'the Welfare State' and the small 'p' politics associated with meanings and representations of the bodies and households of 'welfare mothers'. More specifically, I argue that to understand the formal institutions of social provision in the US requires a consideration of the relations and practices of the state as well as those in sites other than the state, in this instance the body and the household. I do this by reflecting on two pivotal themes that have captured the attention of feminist and political geographers (and those who identify as both) – the public and the private, and the politics of scale – and use these concepts to frame a discussion of gender, paid and unpaid work and citizenship in the context of welfare provision and welfare reform.

PUBLIC/PRIVATE AND THE POLITICS OF SCALE

Feminist scholars have long argued that a major discursive boundary and potent dichotomy is the public/private divide (see Fincher, 2004).The argument is that the boundary between public and private works to mark those issues allocated to the 'private' as not appropriate for inclusion on the 'political agenda'. Feminist scholars have long problematized the dichotomy between the 'private' sphere of reproduction, consumption, home, family and domesticity, and the 'public' sphere of politics, production and waged work. Taken for granted and deeply naturalized, the public/private dichotomy suggests that the spheres are mutually exclusive, fixed, stable and bounded. Instead, feminists point out, they are interdependent, culturally constructed, fluid and relational (see, for example: Fraser, 1989; Pateman, 1989), and it is more analytically satisfying to think of public and private in terms of 'multiplicities, not binaries' (Fincher, 2004: 50).

The ongoing feminist project of redefining the 'political' has dovetailed with the spirited debates among political geographers about theorizing scale. In fact I consider many of the feminist critiques of political geography to be profoundly scalar because they trouble the assumptions around what scales count as 'political' (England, 2003). Scale is currently conceptualized by critical geographers as actively produced and fluid rather than pre-given and static. This social constructivist understanding of scale (which is closely related to ideas about the production of space) emphasizes the ways that social, economic and political processes are constantly produced and (re)produced through spatial practices and discourses. Erik Swyngedouw (1997) notes that this re-conceptualization of scale means that the 'theoretical and political never resides in a particular scale, but rather in the processes through which particular scales become (re)constituted' (1997: 140); and that scale, moreover, is 'not socially or politically neutral, but embodies and expresses power relationships' (1997: 141). Recent thinking on scale has also moved the focus away from a view of scale as a hierarchical division of bounded territories (global, national, local, etc.) where particular activities, practices and processes are allocated to territories at particular scales. Instead, as Neil Brenner (2001: 605–6) argues, scales 'evolve relationally within tangled hierarchies and dispersed inter-scalar networks'. Scale, then, is increasingly theorized as relational, such that practices and discourses at one scale are implicated in, and overlap with, those at other scales.

Until recently, little attention has been paid by political geographers, and human geographers more generally, to the scales of households and bodies (but see, for example, Marston, 2000; McDowell, 2001; England, 2003; Laurie et al., 2003; Mountz, 2004). However, the cultural turn, feminist theories and post-structructuralism are producing new understandings of power, ones that clearly show that power is also expressed, (re)constructed and maintained at these finer scales. These scales are deeply political and demand analysis by political geographers. Moreover, and as Linda McDowell (2001) points out in her intervention on scale, gender and firms, there is a habit of associating certain issues with certain scales, which restricts analysis to that scale, meaning that interesting questions and alternative interconnections get overlooked. She argues that unless connections are made between 'so-called private or apparently small-scale processes' (2001: 244) – such as gender, sexuality and embodiment – and a range of economic (and in the current context I would add political) processes across a multitude of scales, explanations of change are left wanting.

Addressing the household scale is pivotal to Sallie Marston's (2000, 2004) important critical reading of the construction of scale literature. She argues that this literature places a heavy emphasis on the state, labour and capital, or some combination of them (i.e. the public sphere). For her, social reproduction, especially when viewed from the household scale, is deeply interconnected with capitalist production in scale construction, yet the 'questions now driving the scholarship on scale tend to focus on capitalist production while, at best, only tacitly acknowledging and, at worst, outrightly ignoring social reproduction and consumption' (Marston, 2000: 219). Thinking through how scales are relational, Marston builds a case for the household as a scale; she sees 'the home as a socially produced scale … that is thoroughly implicated in wider social, political and economic processes' (Marston, 2000: 232). So Marston brings the theme of finer scales and their constitution by power relations together with the idea of inter-scalar relations. More generally, an increasing number of scholars from across the social sciences and humanities are calling for renewed attention to social reproduction, especially unpaid care work (e.g. Folbre, 2001; Bakker and Gill, 2003; Mitchell et al., 2003; England, 2007). Exploring these sorts of linkages would be fruitful and, as Elenore Kofman (2005: 528) suggests, 'Opening up the scales of social reproduction and consumption represents a nexus of exchange between feminist and political geographies'.

WELFARE STATES, SOCIAL REPRODUCTION AND CITIZENSHIP

The earliest work on theorizing and analysing the welfare state focused on class as the social relation of choice – with analysis usually prioritizing the relationship between the state, the market and inequalities deriving from income (as a proxy for class). Scant, if any, attention was paid to issues of gender, race or other systems of difference, processes that also shape labour markets and a range of other cultural and social inequalities too. Feminists, however, have exposed the welfare state as a set of gendered (as well as raced, heterosexed and able-ist) institutions with spatialized social practices that differently situate and impact women compared with men. More recently, feminists have offered broader critiques of state power, offering post-structural musings that draw on Foucault's concept of power and governmentality, that suggest that rather than being something 'out there' acting upon society and individuals, the state is mutually constituted with a range of social relations and materialized through socio-spatial practices in a variety of scales, sites and spaces (Larner, 2000; Brodie, 2002; Kingfisher, 2002; Chouinard, 2004; Mitchell et al., 2003).

As a state form, the welfare state modifies social or market forces in order to meet basic social needs, and minimizes the risks of unemployment, ill-health and old age through the provision of some level of income security for individuals and households. What counts as a 'social need', which social needs are met, and how they are met is highly variable across welfare states (Esping-Anderson, 1990; O'Conner et al., 1999; Kingfisher, 2002; Jenson, 2004). For the majority of people, the market meets most of their needs, meaning they rely on earned income, either directly (as paid workers) or through their parents, partners or adult children. Families and households are also a source of 'welfare' in terms of providing a non-marketized range of social reproduction activities, including relational caregiving, cooking, cleaning and laundry. However, some welfare states choose to minimize, to a greater or lesser extent, the degree to which 'the market' determines access to all goods and services by providing, for example, some sort of health insurance for all; and some provide subsidized services for families, such as early childhood education and child care, and ensure that parents have adequate income to care for their children as in family allowance benefits. Still others encourage parents to purchase what they need through the market (Jenson, 2004).

The specific form of the welfare state varies across different places and different times, but it generally involves social programmes created around collective responsibility for social reproduction. This in turn means intervening into various private spaces to so alter social and market forces as to mitigate social risk. In what one might call post-welfare states, which also vary over time and space, social policy is increasingly marked by a neo-liberal rhetoric of free(r) markets, fiscal austerity, privatization and marketization. Subsequently the emphasis in social policy has shifted to encourage sources other than the state to meet social needs. Accordingly, more public functions and services are transferred to markets, to the volunteer sector, to community groups and, of course, to families and households. I am interpreting welfare provision, therefore, as embedded in a multiplicity of publics and privates. Moreover, in addition to feminist interpretations of the relations among economic production, social reproduction and consumption, there are also the liberal theory distinctions between the state as public and the market economy as private, as well as the state/individual, state/civil society, and society (collective)/individual distinctions.[2]

The rise of the welfare state was also associated, at least in its 'generic' state form, with the emergence of ideas of social citizenship and social rights – 'the whole range from the right to a modicum of economic welfare and security to the right to share to the full in the social heritage and to live the life of a civilized being according to the standards prevailing in society' as T.H. Marshall (1949/1992) famously defined them.

Welfarism then is organized around social citizenship rights that allow citizens to make claims on the state for particular benefits and services. Feminist scholars, on the other hand, challenge the meanings and boundaries of social citizenship, especially in terms of the uncritical and disembodied claims of impartiality, universalism and the common good (e.g. Fraser, 1989; Pateman, 1989; Lister 2003). As Joan Tronto (2001: 66) notes, 'models of citizenship define the boundaries between public and private life and determine which activities, attitudes and possessions, and so forth are to be considered worthy in any given state'. By insisting on the embodiment of the subjects and spaces of citizenship, feminist scholars have opened up a range of new themes for analysis, including the private versus public, inclusion versus exclusion, and care work versus paid work.

My argument is that a complete analysis of social welfare provision requires consideration of a variety of publics and privates at a range of scales, as well as highlighting the linkages among (and boundaries between) them. As O'Conner et al. (1999: 2) point out, '[the] restructuring of state social provision, transformations of labour markets, changing families and households, and political changes all influence each other'. Thus, it is important to take account of state interventions not only into the (labour) market but also into the family, household and market/family relationships. Care work, reproduction and paid work are not hermetically sealed off one from the other, but are profoundly interconnected. Welfare provision depends on a mix of state practices, markets, communities and households, which themselves are socially constructed by social relations of difference. Gender relations are co-constitutive of welfare states and social citizenship, as well as the institutions of social provision more broadly conceived. From a feminist perspective, then, at least three interwoven discourses shape social policy: the gendering of 'work', the gendering of citizenship, and the gendered responses of government officials and policy-makers. In the remainder of the chapter I explore these discourses against the backdrop of the politics of the public and private and scale politics.

BEGINNING AND ENDING 'WELFARE AS WE KNOW IT' IN THE UNITED STATES

A key idea in the fast-growing literature on welfare regimes and comparative social policy is that the specific form of the welfare state varies across time and space. In his classic *Three Worlds of Welfare Capitalism*, Gösta Esping-Anderson (1990) categorized countries into different welfare regimes. He defines the US, along with Australia, Canada

and the UK, as liberal welfare regimes with modest, residual welfare rights and means-tested benefits. This is liberal in the Lockean sense in that they champion the rights of individuals and have welfare systems shaped by the logic of 'the market'. Even within this group of countries, what developed in the US became what Michael Katz (1986) calls a 'semi-welfare state', based on especially modest social welfare rights and stingy means-tested benefits, including Aid to Dependent Children, rather than the fuller social citizenship rights and economic redistributive welfare provision that was more common in other national contexts. Significantly, Nancy Fraser and Linda Gordon (1992) argue that while there is a rich discourse on civil citizenship in the US – civil rights, individual liberties, freedom of speech – there is almost a total absence of the term 'social citizenship' (also see Handler, 2004). The reason is that social citizenship is associated with 'citizenship entitlements': universal eligibility for benefits that are not means-tested, meaning that anybody, regardless of their status, may receive them (for example, health care and family allowance). Instead, in the US context, 'entitlement' specifically means 'legal entitlement' as in benefits or services guaranteed by law to those who meet established eligibility criteria.[3] As Nancy Fraser and Linda Gordon (1992: 45–6) go on to argue:

> People who enjoy 'social citizenship' get 'social rights', not 'handouts'. They receive aid while maintaining their status as full members of society entitled to 'equal respect'. ... The expression 'social citizenship' is almost never heard in public debate in the United States today. Receipt of 'welfare' is usually considered grounds of disrespect, a threat to, rather than a realization of citizenship. ... The connotations of citizenship are so positive, powerful and proud, while those of 'welfare' are so negative, weak, and degraded, that 'social citizenship' here sounds almost oxymoronic.

In fact, the term 'welfare state' has also not been in common usage in the US. Americans, as Fraser and Gordon suggest, understand 'welfare' to refer to means-tested social assistance, or what used to be called 'relief' or 'dole'. Rather, a distinction is made between 'welfare' and 'social security'[4] for retired workers and their 'dependants' which can be received while remaining respectable 'full citizens'.

The particularities of the development of this 'semi-welfare state' in the US have implications for the ways in which welfare provision has been refashioned in recent years. In what follows I explore the welfare provisions outlined in the Social Security Act of 1935, and the wave of welfare reform associated with the 1996 Personal

Responsibility and Work Opportunity Reconciliation Act, paying particular attention to the policies and discourses associated with mothers, especially 'welfare mothers'.

'MEASURES OF PROTECTION', 'THE HAZARDS AND VICISSITUDES OF LIFE' AND AID TO DEPENDENT CHILDREN

Readers of this volume will be familiar with the idea that the US welfare state was built around a class compromise. In brief, the argument is that a limited social contract was struck between capital and the working class after years of labour struggle around the conditions of paid work. Based around the Fordist–Keynesian reorganization of capitalism, this social contract was a spatio-institutional fix to preserve social stability and enhance economic productivity. It was legitimized through a welfare state making some provisions for social security, minimum wages and unemployment benefits – F.D. Roosevelt's 'measures of protection' against 'the hazards and vicissitudes of life' (although other pieces of legislation were also important, such as the 1938 Fair Labor Standards Act). In the language of the politics of scale, the class conflict and subsequent compromise were negotiated, and regulations and conditions were set primarily at the national scale.

By taking seriously the call to take account of social reproduction, what becomes apparent is that another sort of contract was also necessary for the Keynesian–Fordist fix to work: the gender contract, struck at the scale of the 'private' household; or more precisely, a very particular type of gender contract formed around a normativized gendered division of paid and unpaid work, with an economically dependent mother-caregiver and a breadwinner father (Pateman, 1989; Fraser and Gordon, 1992; Tronto, 2001; Mink 2002). This particular sort of gender contract is *assumed* in Roosevelt's promise that the Social Security Act would 'give some measure of protection to the average citizen and to *his* family' (*my emphasis*).

Gender contracts, lived out at the scales of the body and household, reverberate through the scales of the community, city and nation, and the US is no exception. For instance, the consumption part of the national-scale 'class compromise' was played out at the household level through privatized consumption based on a 'family wage' and the promotion of single-family housing, through federally insured FHA mortgages, for instance. In terms of scale, then, the household was just as significant as the nation (and others besides) in the (re)production of the Keynesian–Fordist system. However, most of the attention of the architects

of the Social Security Act focused on class-based, paid-work-related policies, which is to say men's inclusion in the public realm. Policies that would be more likely to impact women and children received less attention. Feminist scholars point out that policy-makers and politicians presumed a particular model of citizenship based on paid work; at the centre of the US version of the Keynesian Welfare State, more so than in other national contexts, was the male 'worker-citizen' making wage-related contributions to the system. But as Joan Tronto remarks 'what this model fails to see ... is that workers require care to remain capable of [paid] work; this *care work is an invisible prerequisite for worker-citizens*' (2001: 67, *emphasis added*). It may well be that the gender consequences of the policies were largely unconscious – part of the taken-for-granted understandings of the time. But by codifying already existing normative expectations about gender and sexuality, women were positioned as less than 'full citizens', as 'unemployable' (i.e. not expected to be in paid work) economically dependent caregivers in the home (Pateman, 1989; Fraser and Gordon, 1992; Gordon, 1994; Abramovitz, 1996; Kessler-Harris, 1999; Mink 2002).

A familiar theme among feminist scholars of the American welfare state is that the Social Security Act created 'two tracks' – social insurance and public assistance – each with different goals, target groups and funding mechanisms (Nelson, 1990; Fraser and Gordon, 1992; Gordon, 1994). Assumptions about the body, the household and 'work' were deeply implicated in both tracks – assumptions that were themselves deeply gendered, classed and raced. As Joan Tronto notes, both tracks 'can be understood as replacing either the earnings, or actual person, of the breadwinner male. Thus just as housewives were dependent on husbands, so too recipients of [public assistance] are perceived as dependent on state allocations and, in this way, not fully citizens' (Tronto, 2001: 68).

The social insurance programmes were designed as 'contributory' programmes and included insurance for both unemployment and old age. Reflecting Fraser and Gordon's (1992) description of 'social citizenship', these programmes offered benefits as legal entitlements, without the stigma and surveillance associated with the public assistance programmes. These were the sorts of programmes that the working class and unions had fought hard to get. Based on payroll contributions (paid by the employer and employee into a general trust fund from which benefits were drawn), the benefits were firmly tied to an individual's employment history in the formal paid labour market. The specificities of the social insurance programmes were the stuff of heated debate in both the Committee on Economic Security (set up by Roosevelt in the summer of 1934 to study economic insecurity and make recommendations for legislation) and Congress once they received the draft legislation in early 1935 (see Kessler-Harris, 1999, for a feminist interpretation; and Quadagno, 1994, and Neubeck and Cazenave, 2001, for a race-centred analysis of these debates). Ultimately the typical beneficiaries of these programmes were white, male industrial workers, especially in the role of 'breadwinners' earning a 'family wage'. The Committee on Economic Security (henceforth CES) and subsequently Congress fashioned legislation in ways that upheld, and were intended to uphold, the 'dignity' of (male) workers by allowing them to maintain their sense of 'self-sufficiency' and the 'capacity to provide for his family'.[5] The rhetoric was steeped in assumptions about a particular form of the gender contract, and its implications for the household and unpaid care work. Married women, especially mothers, were seen not as autonomous individuals, but rather as auxiliaries to their husbands. A sort of 'trickle-down' theory was applied to the home: if the state helped men as 'breadwinners', they would then be able to support their wives and children.

The second track was public assistance. In formal terms this included Aid to Dependent Children and Old Age Assistance; the latter was intended for those not covered by Old Age Insurance, if they met certain criteria. Public assistance was marked as 'non-contributory' and (unlike social insurance) was funded through general tax revenues and designed as means-tested assistance benefits. The public assistance programmes took the form of federal grants-in-aid; funding came via a federal/state matching ratio to states that administered them in line with federal requirements. Here I focus on ADC, which like social insurance was shaped by a very particular (white, middle-class) feminine ideal that revolved around the socially sanctioned unpaid work of full-time child-rearing and home-making that meant certain mothers of young children were 'unemployable'.

Earlier I suggested that relative to the social insurance track, policies aimed at women and children received scant attention. One reason for that was that ADC was based on, and extended the framework of, Mothers' Aid,[6] which was already state law in all but Georgia and South Carolina (Skocpol, 1992; Gordon, 1994). The architects of the Social Security Act took these state-level programmes and incorporated them virtually as they were into the federal legislation. In terms of the politics of scale, then, Mothers' Aid was rescaled to the national level.[7] Like Mothers' Aid, ADC was shaped by particular (raced, classed and hetero-sexed) discourses of domesticity and 'motherhood as sacrosanct'. ADC was intended to allow lone mothers with young children to care full-time for their children (Gordon, 2001; Mink, 2002). Not all lone mothers, however. As J. Douglas Brown (who served on the Committee on Economic Security

and later became the first Executive Director of the Federal Board overseeing ADC) remarked, 'The ADC example we always thought about was the poor lady in West Virginia whose husband was killed in a mining accident, and the problem of how she could feed those kids' (quoted in Neubeck and Cazenave, 2001: 50). ADC then ended up being primarily for the children of white widows, who were single mothers through 'no fault of their own'.

I want to argue that while ADC was basically Mothers' Aid rescaled to the national level, in practice its administration remained at the state level, which had a significant impact on poor mothers' lives. Here I find the work of the political scientist Suzanne Mettler (1998) useful. Although not couched in the language of scale as recently theorized by geographers, Mettler offers a helpful framework for thinking about gender, the New Deal and scale. She interprets the Social Security Act as setting up a system of expanded polity via national citizenship for white male workers, whereas women and people of colour were incorporated into this national polity in a far more fragmented way. The nationally administered social insurance programmes incorporated male workers, especially white industrial workers, 'into policies to be administered in a centralized, unitary manner through standardized, routinized procedures [to be] governed as rights-bearing individuals, members of a liberal regime' (1998: xi).[8] To be sure, and as I will elaborate below, the social insurance programmes usually did not explicitly exclude white women and people of colour; if they met the appropriate eligibility requirements they could be covered, but the point was that few did. White women and people of colour were, Mettler argues, 'more likely to remain state citizens, subject to policies whose development was hindered by the dynamics of federalism and which were administered with discretion and variability' (Mettler, 1998: xii). The state-administered public assistance programmes tended to:

'incorporate women as social citizens according to nonliberal criteria that regarded them in relational, role oriented, or difference-based terms, rather than as abstract individuals. ... State governments also administered programs in a manner distinct from the national government. They added layers of eligibility requirements to those stated in federal law and permitted local officials to implement rules with ample discretion. Beneficiaries came to be treated as dependent persons who required supervision and protection rather than as bearers of rights'. (Mettler, 1998: 24)

Mettler's distinction between national and state citizenship is useful in understanding the ways in which the geographies of the US 'semi-welfare state' unfolded in the 1930s. Initially Roosevelt, in line with his desire to broaden the polity and offer income security, pushed for national policies, including the centralization of the public assistance programmes. However, the final legislation involved uneven geographies reflecting several compromises reached in the face of opposition from the business community and from within his own (Democratic) party. Indeed, political and social geographies were the basis of many of the debates and subsequent compromises over the various programmes. And the deployment of scale politics by various members of Congress shaped the debate and eventual content of the Social Security Act. Finally, despite claims of gender-neutrality, the policies had very clear gendered and raced consequences. In particular, conservative Southerners, who were disproportionately represented on important committees addressing the legislative proposals for social security, such as the Ways and Means Committee, wanted to maintain their regional competitive advantage relative to the industrial North – cheap African-American agricultural workers, especially men, but women too. The fear was that extending and federalizing Social Security programs would limit Southern agricultural landowners' access to workers by offering alternatives to the low agricultural wages and the sorts of paternalistic in-kind 'benefits' Southern landlords offered their tenants and workers instead of insurance. In particular, there were concerns that programmes intended for particular family members would be 'improperly' used to support the entire household, making poorly paid agricultural work even less attractive.

One way these fears got filtered into the legislation was through the exclusion of various occupations from social insurance coverage, notably occupations that were over-represented in the Southern states. The CES excluded agricultural workers and domestic workers, thus denying access to three-fifths of African-American women and men, and Congress later excluded small businesses, most government workers, and part-time and temporary workers. Taken *in toto*, these exclusions eliminated three-quarters of all women in paid work, and an incredible 87 per cent of all African-American women workers (Kessler-Harris, 1999; Neubeck and Cazenave, 2001).[9] Occupations are gendered and raced, so a boundary of eligibility drawn around certain 'core' occupations had gendered and raced implications; even if the origin of the pressures for these exclusions was regional, they drew upon conceptions of gender and race relations that were more widely held.

Members of Congress also deployed profoundly scalar tactics – mounting an attack on the Federal Government's efforts to 'jump scale' and take greater responsibility for 'welfare' via national programmes and national standard guidelines for the states. They argued that the draft Social Security Act violated the 'States' Rights' as set out in the Constitution, which assigned responsibility for public welfare to the states and local governments (Skocpol, 1992; Quadagno, 1994; Neubeck and Cazenave, 2001; Handler, 2004). For example, Southerners blocked an early version of the Social Security Act until it was amended to allow the states to determine eligibility and set benefit for ADC (and Old Age Assistance), and as the initial anti-race discrimination provision had also been removed from the Act, state discretion meant local race discrimination could remain part of official practice.

Once the legislation passed, the individual states, in Mettler's parlance, could administer public assistance programmes according to 'nonliberal criteria' to 'state citizens'. In other words, they had a great deal of discretion when it came to exactly which bodies and which households within the state would receive ADC. So even though African-American mothers and children, especially in the South, were far more likely to be impoverished and live in single-parent families, a Social Security Board Report indicated that between 1937 and 1940, only 14–17 per cent of ADC recipients were African-American (Neubeck and Cazenave, 2001). African-American mothers, local officers often determined, had plenty of employment opportunities as seasonal agricultural workers or domestic workers and so were 'employable mothers' in ways white mothers were not (Quadagno, 1994). In western US, similar arguments were made about Latinas and Native Americans (Quadagno, 1994; Kessler-Harris, 1999; Gordon, 2001; Mink, 2002).

The states' laws were replete with exclusionary rules based on assumptions about the homes of welfare mothers. Several states had 'suitable home' requirements for recipients of ADC, rules that meant local officials could choose to cut off ADC to families if there was evidence of a 'man in the house' or a 'substitute father'. The assumption here was that any man associated with the household must (or should) be providing for the children, or just as likely, that the man might be being supported, or at least subsidized, by ADC, rather than making himself available as cheap agricultural labour.[10] In some states, women in non-marital relationships were automatically deemed to be keeping an 'unsuitable home' and denied aid. To ensure the 'good moral character' of recipients, local officials could make surprise home visits, often early in the morning or late at night, to look for evidence of 'inappropriate' behaviour. There is ample evidence that local officials were more diligent in enforcing these rules with households headed by women of colour (Quadagno, 1994; Neubeck and Cazenave, 2001; Mink, 2002). The compromises reached in order to pass the Social Security legislation, undergirded by federalism, produced a more uneven geography of public assistance than would have been the case with a more centralized 'national' system and may well have reproduced existing gender and racial relations to a degree that might not otherwise have occurred.

'NEVER-ENDING CYCLE OF WELFARE', 'THE ETHIC OF WORK' AND TEMPORARY AID TO NEEDY FAMILIES

Sixty years after the passage of the Social Security Act, the political culture of the US has changed.[11] Now everyone, whether or not they are an African-American welfare mother, is encouraged to look to sources other than the state to meet their needs. As Janine Brodie (2002) remarks:

> the consensus about the role of the state, the nature of citizenship, and popular understandings of the appropriate relationship among the public and the private and the collective and individual have been incrementally and progressively recast into a model of governance which would have been inconceivable a half century ago. (Brodie, 2002: 90)

The neo-liberalized post-welfare state is not about collective responsibility and ameliorating social risk, but about individual responsibility – as signalled, for instance, by the title Personal Responsibility and Work Opportunity Reconciliation Act. Nikolas Rose (1996) uses the term 'responsibilization' to describe how shifts in social policy have encouraged a culture of surveillance and self-regulation of the 'individual'. The effect is that social problems are redefined as failures of the individual, which in turn become personal problems requiring individual, 'private' solutions, not collective, public ones. The goal of public assistance for poor families now is no longer about the contemporary version of the West Virginia miner's wife and how the state should address the question of 'how she could feed those kids'; rather it is about reducing 'welfare dependence', work activation, and promoting self-sufficiency. The goal is quantitative, as in reducing benefit expenditures and case loads, rather than qualitative, that is, reducing social insecurity.

The rhetoric around the 'problem' of 'welfare dependency' has been an important part of a shifting consensus about the relationship between the state, citizenship, and collective versus individual responsibilities. It has developed against the background of the longstanding (though changing) discourse about privileging paid employment as respectable and dignified. Paid employment is no longer only respectable and dignified for men; increasing numbers of women, who in Roosevelt's time would have been regarded as 'unemployable' (i.e. white mothers of young children), are entering paid employment. In 1940, 10 per cent of children had mothers in paid employment; in 2000, 70 per cent did. And as more women have entered paid employment, so the socially sanctioned and rigid gender division of labour (the gender contract) at the household scale has lost some of its rigidity. Certainly the institutionally enshrined model of motherhood, that is, full-time at home, that informed the provisions of the Social Security Act, made much less sense by the late twentieth century. This relates in part to the decline of the 'family wage' subsequent to the unravelling of the Fordist class compromise, and to women's increased educational attainment and better career prospects, which it should be noted have particularly benefited white, middle-class women.

For many groups of women this is a good thing and the outcome of years of struggle for equality. Of course, women are still far from equal participants in the labour market relative to men, as indicated by the persistence of the gender-wage gap and occupational segregation. However, various legislative changes have made women eligible for the social insurance programmes[12] and they are recognized as individual wage workers in ways they were not in the original 1935 legislation. They can even be classified as rights-bearing 'national citizens', to evoke Suzanne Mettler's (1998) description. However, in the changing context of welfare provision I want to emphasize that these gains are not universally enjoyed by all women. Moreover, as Gwendolyn Mink (2002) argues, the recent gains for some groups of women (more education, better child care options, less job discrimination) have even been appropriated and used against poor mothers because, relative to other mothers, they can then be marked as being personally 'irresponsible' for not finding paid employment and for having children they cannot 'afford'.

Bill Clinton's comments when signing the 1996 Personal Responsibility and Work Opportunity Reconciliation Act (PRWORA) captured widespread sentiments on the Right and even the Left about the 'great social challenge' of 'a never-ending cycle of welfare' that discouraged labour market participation and self-sufficiency. The provisions in the PRWORA offered 'the dignity, the power, and the ethic of work' as 'our nation's answer to this great social challenge' and in turn this would 'make welfare what it was meant to be: a second chance, not a way of life'. The PRWORA abolished AFDC, and replaced it with Temporary Aid to Needy Families (TANF). This is not, however, a legal entitlement. Rather it involves far tighter eligibility restrictions, lifetime limits for receipt of assistance, and requires immediate work preparation leading to a job within two years. As of 2001, lone parents must work 30 hours a week, and in two-parent families, one parent must work at least 35 hours a week.[13]

Like the Social Security Act before it, public assistance policy in the post-welfare state is supposed to be gender-neutral. Perhaps like the architects of the Social Security Act, those developing contemporary policies believe their policies to be gender-neutral, but, then as now, there are clear gendered (and raced) consequences. Susan Moller Okin (1989) troubled the use of gender-neutral language ('he or she', 'persons', 'individuals', for instance) in both academic and public policy discourse, finding that often it is merely what she called 'false gender-neutrality'. She remarked that 'gender-neutral terms frequently obscure the fact that so much of the real experience of "persons", so long as we live in gender-structured societies, *does* in fact depend on what sex they are' (1989: 11, *emphasis in original*). Certainly, the language in PRWORA is assiduously gender-neutral, referring for example to "a pregnant individual"; and the tables in the Department of Health and Social Services Annual Reports to Congress do not break single families down by gender (Boyer, 2003). But despite this seeming gender-neutrality, single mothers remain the primary recipients of welfare and thus the primary targets of welfare reform.

Public assistance is now about coaxing seemingly work-shy 'individuals' into paid work and to take 'personal responsibility' for themselves and their family. Yet, as was the case for the breadwinner–worker model of citizenship of the 1930s, and to paraphrase Joan Tronto's (2001) remarks that I quoted earlier, TANF recipients, like other paid workers, require care to remain capable of fulfilling their employment obligations. And by definition TANF recipients are parents, and just like other parents, their children require care in order that their mothers can take on paid work. In fact, because in most cases TANF families are headed by lone mothers, this is particularly acute as there is not another parent to share care responsibilities. Ignoring care work, which dismisses social reproduction and privileges paid work (production even), means that 'welfare mothers' can be despised as 'free riders' because they are not in paid employment and therefore are failing to meet their (neo-liberalized) obligation to contribute to society. Gwendolyn Mink (2002) finds a very clear racist overtone to the debate, and argues that racist stereotypes of laziness and immorality have been used to devalue the care work of African-American and Latina mothers: 'broad support for disciplinary welfare reform is rooted in the view that mother's poverty flows from moral failing … their "unwillingness to work", their failure to marry (or stay married), their irresponsible sexuality and childbearing' (2002: 4).

Clinton argued that the PRWORA would 'overcome the flaws of the welfare system for the people who are trapped in it'. What these flaws supposedly were and 'what [welfare] was meant to be', requires a brief consideration of events from around the mid-life of 'welfare as we kn[e]w it'. Having been denied ADC either by eligibility requirements or straightforward discrimination, by the 1950s increasing numbers of African-American women, especially in the North, challenged their exclusion from, and began enrolling in, ADC (known as AFDC after 1962), as did other groups of women of colour. During the 1960s and 1970s, welfare rights social movements pushed for major changes in the rules governing welfare eligibility. In a series of landmark US Supreme Court decisions in the late 1960s and early 1970s, AFDC was finally established as a statutory entitlement. As a result, states' decisions to deny or reduce assistance became subject to due process of law protections and benefits could no longer be arbitrarily removed or reduced, as in the 'man in the house' rules (Piven and Cloward, 1979; Quadagno, 1994; Gilbert, 2001; Neubeck and Cazenave, 2001).[14] Although the states continued to determine the actual cash amount of AFDC, and for sure that varied a great deal, there was at least less spatial variation in the administrative procedures and eligibility requirements. And AFDC came to be governed by nearly national standards of administration. Increasing proportions of African-American women and Latinas, with their higher poverty rates compared to white mothers, joined the AFDC rolls, although, contrary to popular belief, it was not until after the passage of the PRWORA that recipients were slightly more likely to be African-American than white.[15] Accordingly, AFDC arguably began to approximate a national programme, blurring the boundary between national and state citizenship as defined by Mettler (1998).

Ending welfare as we knew it began well before 1996. For instance, work incentives for some AFDC recipients date back to the late 1960s. But it was the federal Family Support Act of 1988, signed by Republican Ronald Reagan towards the end of his second term, that first required AFDC parents to take on paid work, or participate in expanded federally funded job-training programmes designed to reduce 'welfare dependency'. By then, conservative critics were mounting a multiscalar attack on AFDC. The scales of the body and the household saturated conservative critics' moralizing claims that, for instance, AFDC incentivized recipients to have children 'out of wedlock', discouraged parental reconciliation, and fuelled the 'collapse of fatherhood'. By the 1980s the bodies of welfare mothers were represented not only as an enormous drain on the public purse and as morally reprehensible, but also as a threat to society at large, linked to rising crime rates, juvenile delinquency and sundry other social problems (Gordon, 2001; Mink 2002). In particular, the discursive (and material) body of the inner-city, African-American, teenage welfare mother fanned the flames of anti-welfare discourses, even though teen mothers, whether African-American or not, were a statistically small group of recipients – about 6 per cent.

The structure of AFDC was also faulted for being too centralized and rigid (surely a coded way of attacking the national-level gains made by the welfare rights movement), and as an example of the ills of 'big government' and (nation) state intervention (Neubeck and Cazenave, 2001; Peck, 2001, 2002, Handler, 2004). In the context of the broad historical sweep of the US's 'semi-welfare state', the proposed solution was a familiar one: allow the states greater flexibility in designing their own programmes for needy families. Adapting Mettler's framework to the contemporary era, poor women and their children were to be firmly returned to being state citizens. Reminiscent of the strategies deployed during the 1935 Congress debates over the Social Security Act, the debates congealed around how locally designed welfare was more sensitive to local conditions in ways that 'big government' was not. The 1988 Family Support Act provided for this by enabling individual states to apply for 'legislative waivers' of Federal Rules (decided by the US Department of Health and Human Services) that allowed for greater state and, in some instances, municipal flexibility in the provision of AFDC. The Republican George H.W. Bush administration (1989–93) actually granted relatively few waivers, but the Clinton administration granted so many that by the mid-1990s most of the states were 'experimenting' with welfare reform, often introducing some sort of workfare programme.

The provisions in the PRWORA extended 'welfare experimentation' to all the states by legislatively devolving (or rescaling) the administrative responsibilities from the federal to state and local levels of government. Now there are fifty TANF programmes in the US and this rescaling has produced an uneven geography of how states are instituting welfare reform and practising various parameters of TANF – time limits, benefit levels, sanctions and so on. This uneven neo-liberal terrain has been an important topic of investigation for feminist and political geographers (see, for example, Cope, 1997; Kodras, 2001; Peck 2001; Wolch and Dinh, 2001; Boyer, 2003; Haylett, 2003), with some invoking the narrative of rescaling (Peck, 2002). As Jamie Peck (2002: 356) notes:

> in contrast to the ordered and regularized topography of the welfare state, the landscape of workfare is a restless one. Uneven geographical development is being established as an intentional, rather than merely incidental feature of the delivery of

workfare programs ... workfare makes a virtue of geographic differentiation, sub-national competition and personally tailored circumstance-specific interventions.

Put differently, the rescaling of 'welfare' and the elimination of (hard fought-for) welfare entitlement amounts to a return to state and local discretion in the administration of welfare programmes and Mettler's notion of 'state citizenship' for women and people of colour. Jennifer Wolch and Sissi Dinh (2001: 482) describe this rescaling of welfare as 'harkening back to the Elizabethan Poor Laws [as] states and even many localities are now charged with responsibility for controlling their "own" poor'. Certainly the return of authority to the states (and municipalities) has decentralized responsibility for impoverished women and their children and revives a political geography not seen since 1935.

CONCLUSION

The shifting contours of welfare provision and welfare reform in the US have had profound effects on the relationship between states, markets and households, the parameters of citizenship, and the responsibilities of the collective and the individual. In this chapter I have explored how welfare provision and welfare reform are multiscalar, encompassing interplays of various processes and relations that unfold at numerous intersecting, but different, scales. I have emphasized the interlinkages between paid work and unpaid work, workplaces and home places, and citizens and various levels of government. Like many other feminist scholars I argue that all these relationships are deeply gendered, and are also shaped by other hierarchies of social difference. The boundaries between public and private, between care work and paid work, and deeply gendered understandings of appropriate behaviour have informed social policy, which in turn produce outcomes that are equally deeply gendered. The explanation I offer blends together the big 'P' political geographies of the state with the small 'p' political geography of meanings and representations in a consideration of the material and discursive spaces of welfare provision and reform.

My chapter also provides an instance of a trend that Eleonore Kofman (2005: 528) identifies as feminist political geographers 'beginning to return to considerations of gendering the state'. This return is, however, one that takes account of the important transformations that feminists have already brought to political geography by redefining what counts as 'political' and what counts as 'geography'. The implications for political geography of that redefining and the return to feminist theorizing of the state have been and will continue to be profound. For instance, later in her appraisal of the contributions of feminist political geographers, Kofman notes that 'political geographers and feminist geographers recognize the significance of the home and household for wider economic, social and political processes' (2005: 528). While my chapter is intended to bolster that claim, I also want to take the argument a step further. By analysing the state from the vantage point of the body and the household, certain aspects of wider processes, and alternative meanings and explanations, are revealed that might otherwise have been all too easily overlooked from the more typical vantage points adopted in political geographies of the state – the global, national, regional and urban scales. By opening up the range of spaces, scales and sites of analysis, feminist interventions make possible a whole host of new research questions for political geographers, while further sophisticating understandings of politics, power and space.

NOTES

1 This is in contrast with the European and even Canadian contexts, where, even with the recent Americanized 'fast policy transfers', the meaning of 'welfare' is more broadly understood as personal, material and collective well-being and social inclusion.

2 See Fraser (1989) for a discussion of the multiple meanings, sometimes overlapping, sometimes contradictory, of public and private.

3 The closest the US has to universal social benefits is public education. Both Roosevelt and Clinton floated the idea of a universal health care system; it did not get very far.

4 The latter refers to a programme introduced in 1935 as Old Age Insurance and since expanded to Old Age, Survivors' and Disability Insurance and Medicare.

5 This is the language used either in the legislation or the speeches, writings and memoirs of the architects of the Social Security Act (see http://www.ssa.gov/history/).

6 Also known as 'Mothers' Pensions' and 'Widows' Pensions'.

7 To demonstrate the utility of her claims about social reproduction, scale construction and the household, one of Sallie Marston's (2004) examples is the Progressive Era white, middle-class women's activism to introduce Mothers' Aid at the local and state levels. She notes that by 'linking the private with the public, [Progressive Era] social and policy advocates connected the health of the polity to the quality of motherhood, and demanded that government provide economic assistance to poor women and children' (2004: 183). By doing so, these women were able 'to configure a scalar fix premised on maternalism and domesticity' (Marston, 2004: 184).

8 Mettler's argument needs some clarification as the only major programme in the original 1935 Act that was entirely federal in funding and administration was Old Age Insurance. Unemployment Insurance was initially a state programme, but by mid-century had developed into what she describes as 'a hybrid program … [with] nearly national standards of administration' (1998: 25). She also includes the Fair Labor Standards Act of 1938 under 'national citizenship' policies.

9 Subsequent amendments removed many of these restrictions. For example, in 1939, 55 per cent of the civilian labour force was covered by Old Age, Survivors' and Disability Insurance (OASDI) programmes, by the 1970s, 90 per cent were, and in 2002, 96 per cent were (2004 Green Book, Committee on Ways and Means, 2004 House of Representatives, Table 1–7).

10 This was the same sort of argument that was also made about social insurance.

11 Although, as I will argue, certain traces of the political imperatives that shaped the 1935 Act remain.

12 For example, the 1978 Pregnancy Discrimination Act forbids denying Unemployment Insurance to pregnant women.

13 The George W. Bush administration is currently proposing to increase work requirements to 40 hours a week for both single and two-parent families.

14 The politics of AFDC were also interwoven with legal rights gained via the Great Society/War on Poverty programmes and Civil Rights movements. Quadagno (1994) argues that the War on Poverty was an effort to eliminate the racial barriers put in place by the New Deal programmes, and an attempt to integrate African-Americans into the national political economy.

15 Of TANF families in 2002, 38 per cent were African-American, 32 per cent were white and 25 per cent were Latino/a (Sixth Report to Congress, 2004, Office of Family Assistance, US Administration for Children and Families, http://www.acf.hhs.gov/programs/ofa/annualreport6/chapter10/chap10.htm#figb).

REFERENCES

Abramovitz, M. (1996) *Regulating the Lives of Women: Social Welfare Policy from Colonial Times to the Present*. Boston: South End Press.

Bakker, I. and Gill, S. (2003) *Power, Production and Social Reproduction: Human In/security in the Global Political Economy*. New York: Palgrave Macmillan.

Boyer, K. (2003) 'At work, at home? New geographies of work and care-giving under Welfare Reform in the US', *Space and Polity*, 7(1): 75–86.

Brenner, N. (2001) 'The limits to scale? Methodological reflections on scalar structuration', *Progress in Human Geography*, 25(4): 591–614.

Brodie, J. (2002) 'The great undoing: state formation, gender politics, and social policy in Canada', in C. Kingfisher (ed.), *Western Welfare in Decline: Globalization and Women's Poverty*. Philadelphia: University of Philadelphia Press.

Brown, M. and Staeheli, L.A. (2003) 'Are we there yet? Feminist political geographies', *Gender, Place and Culture*, 10(3): 247–55.

Chouinard, V. (2004) 'Making feminist sense of the state and citizenship', in L.A. Staeheli, E. Kofman and L.J. Peake (eds), *Mapping Women, Making Politics: Feminist Perspectives on Political Geography*. New York: Routledge.

Cope, M. (1997) 'Responsibility, regulation and retrenchment: the end of welfare?', in L.A. Staeheli, J. Kodras and C. Flint (eds), *State Devolution in America: Implications for a Diverse Society*. Thousand Oaks, CA: Sage.

England, K. (2003) 'Towards a feminist political geography?', *Political Geography*, 22(6): 611–6.

England, K. (2007) 'Spaces of paid care work', unpublished document available from the author.

Esping-Anderson, G. (1990) *The Three Worlds of Welfare Capitalism*. Princeton, NJ: Princeton University Press.

Fincher, R. (2004) 'From dualisms to multiplicities: gendered political practices', in L.A. Staeheli, E. Kofman and L.J. Peake (eds), *Mapping Women, Making Politics: Feminist Perspectives on Political Geography*. New York: Routledge.

Flint, C. (2003) 'Dying for a "P"? Some questions facing contemporary political geography', *Political Geography*, 22(6): 617–20.

Folbre, N. (2001) *The Invisible Heart: Economics and Family Values*. New York: The New Press.

Fraser, N. (1989) *Unruly Practices: Power, Discourse, and Gender in Contemporary Social Theory*. Minneapolis: University of Minnesota Press.

Fraser, N. and Gordon, L. (1992) 'Contract versus charity: why is there no social citizenship in the United States?', *Socialist Review*, 22(3): 45–67.

Gilbert, M. (2001) 'From the "walk for adequate welfare" to the "march for our lives": welfare rights organizing in the 1960s and 1990s', *Urban Geography*, 22(5): 440–56.

Gordon, L. (1994) *Pitied But Not Entitled: Single Mothers and the History of Welfare, 1890–1935*. Toronto: Maxwell Macmillan Canada.

Gordon, L. (2001) 'Who deserves help? Who must provide', *Annals of the American Academy of Political and Social Science*, 577: 12–25.

Handler, J. (2004) *Social Citizenship in Workfare in the United States and Western Europe: The Paradox of Inclusion*. Cambridge: Cambridge University Press.

Haylett, C. (2003) 'Remaking labour imaginaries: social reproduction and the internationalising project of welfare reform', *Political Geography*, 22(7): 765–88.

Jenson, J. (2004) *Canada's New Social Risks: Directions for a New Social Architecture*. Ottawa: Canadian Policy Research Networks Inc.

Katz, M.B. (1986) *In the Shadow of the Poorhouse: A Social History of Welfare in America*. New York: Basic Books.

Kessler-Harris, A. (1999) 'In the nation's image: the gendered limits of social citizenship in the Depression era', *Journal of American History*, 86(3): 1251–79.

Kingfisher, C. (2002) *Western Welfare in Decline: Globalization and Women's Poverty*. Philadelphia: University of Philadelphia Press.

Kodras, J.E. (2001) 'State, capital, and civil society in the battle over welfare reform: the contribution of geographic perspectives', *Urban Geography*, 22(5): 499–502.

Kofman, E. (2005) 'Feminist political geographies', in L. Nelson and J. Seager (eds), *A Companion to Feminist Geography*. Malden, MA: Blackwell.

Kofman, E. and Peake, L. (1990) 'Into the 1990s: a gendered agenda for political geography', *Political Geography Quarterly*, 9(4): 313–36.

Larner, W. (2000) 'Post-welfare state governance: towards a code of social and family responsibility', *Social Politics*, 7(2): 244–65.

Laurie, N., Andolina, R. and Radcliff, S. (2003) 'Indigenous professionalization: transnational social reproduction in the Andes', *Antipode*, 35(3): 463–91.

Lister, R. (2003) *Citizenship: Feminist Perspectives* (2nd edn). New York: New York University Press.

Marshall, T.H. and Bottomore, T. (1949/1992) *Citizenship and Social Class*. London: Pluto Press.

Marston, S.A. (2000) 'The social construction of scale', *Progress in Human Geography*, 24(2): 219–42.

Marston, S.A. (2004) 'A long way from home: domesticating the social production of scale', in E. Sheppard and R.B. McMaster (eds), *Scale and Geographic Inquiry: Nature, Society and Method*. Malden, MA: Blackwell.

McDowell, L. (2001) 'Linking scales: or how research about gender and organizations raises new issues for economic geography', *Journal of Economic Geography*, 1: 227–50.

Mettler, S. (1998) *Dividing Citizens: Gender and Federalism in New Deal Public Policy*. Ithaca, NY and London: Cornell University Press.

Mitchell, K., Marston, S.A. and Katz, C. (2004) Special issue on '"Life's work": social reproduction and the state', *Antipode*, 35(3).

Mink, G. (2002). *Welfare's End* (2nd edn). Ithaca, NY: Cornell University Press.

Mountz, A. (2004) 'Embodying the nation-state: Canada's response to human smuggling', *Political Geography*, 23: 323–45.

Nelson, B.J. (1990). 'The origins of the two-channel welfare state: workers' compensation and mothers' aid', in L. Gordon (ed.), *Women, the State, and Welfare*. Madison: University of Wisconsin Press.

Neubeck, K.J. and Cazenave, N.A. (2001) *Welfare Racism: Playing the Race Card Against America's Poor*. New York and London: Routledge.

O'Conner, J., Orloff, A.S. and Shaver, S. (1999) *States, Markets, Families: Gender, Liberalism and Social Policy in Australia, Canada, Great Britain and the United States*. Cambridge: Cambridge University Press.

Okin, S.M. (1989) *Justice, Gender, and the Family*. New York: Basic Books.

Pateman, C. (1989) 'Feminist critiques of the public/private dichotomy', in The Disorder of Women: Democracy, Feminism, and Political Theory. Cambridge: Polity Press.

Peck, J. (2001) Workfare States. New York and London: Guilford Press.

Peck, J. (2002) 'Political economics of scale: fast policy, interscalar relations, and neoliberal workfare', Economic Geography, 78(3): 331–60.

Piven, F.F. and Cloward, R.A. (1979) Poor People's Movements: Why They Succeed, How They Fail. New York: Vintage Books.

Quadagno, J. (1994) The Color of Welfare: How Racism Undermined the War on Poverty. New York: Oxford University Press.

Rose, N. (1996) 'Governing "advanced" liberal democracies', in A. Barry, T. Osborne and N. Rose (eds), Foucault and Political Reason: Liberalism, Neo-Liberalism and Rationalities of Government. Chicago: University of Chicago Press.

Skocpol, T. (1992) Protecting Soldiers and Mothers: The Political Origins of Social Policy in the United State. Cambridge, MA: Harvard University Press.

Staeheli, L.A., Kofman E. and Peake L.J. (eds) (2004) Mapping Women, Making Politics: Feminist Perspectives on Political Geography. New York: Routledge.

Swynegedouw, E. (1997) 'Neither global nor local: "glocalization" and the politics of scale', in K.R. Cox (ed.), Spaces of Globalization: Reasserting the Power of the Local. New York and London: Guilford Press.

Tronto, J. (2001) 'Who cares? Public and private caring and the rethinking of citizenship', in N.J. Hirschmann and U. Liebert (eds), Women and Welfare: Theory and Practice in the United States and Europe. New Brunswick, NJ: Rutgers University Press.

Wolch, J. and Dinh, S. (2001) 'The new poor laws: welfare reform and the localization of help', Urban Geography, 22(5): 482–89.

9

Making Space for Law

Nicolas Blomley

INTRODUCTION

If law were a state, it would be America. Like the United States, law can lay claim to a clear self-identity, historical longevity and a powerful centralized authority. It has rigorously policed academic boundaries, with admittance only given to recognized delegates from other intellectual places, that enter on Law's terms. Geography, however, is more like the United Kingdom: while reasonably sure of itself, it laments its lost glory days, and needs to reassure itself of its standing in the academic world.

Where then is 'law and geography'? It has not attained statehood. Its boundaries are vague and overlapping, and its population is diverse and unconsolidated. Some of its residents, in fact, have not even been told that they are citizens. There is no capital city or centralized administration. Its internal affairs seem to be of little interest to powerful global players – it doesn't get much coverage on CNN. Perhaps then it is best to think of law and geography as a border zone, like the untidy space between the US and Mexico. Yet such liminal and often extra-jurisdictional zones, while fuzzy, are nevertheless, like law and geography itself, creative and productive sites.

'Law and geography' is both everywhere and nowhere. That is, a growing number of scholars in a range of fields – geography, anthropology, cultural studies, philosophy, literature and law – have explored the intersection of law and space. Yet at the same time, 'legal geography' remains a fluid, mobile and non-institutionalized field. There are no journals, specialty groups or (with a few exceptions) conferences or courses. Universities have yet, to my knowledge, sought to hire legal geographers. Yet, as an absent presence, law

and geography has created an interesting and productive space.

This, however, makes my job – identifying established and emergent themes in law and geography, as they relate to political geography – somewhat harder. I begin by providing a brief review of political geography's engagement (or disengagement) with law, noting law's obvious relevance to core concerns of the sub-discipline. By way of example, I look briefly at the ways in which an analysis of legal rights can be spatialized, before considering some contemporary writing on colonialism, a traditional concern of political geography, noting in both cases the productive conjunctions of law and geography. I draw from this to consider some challenges for political geographers interested in law.

LAW AND POLITICAL GEOGRAPHY

Although it would seem of central concern to political geography, the sub-discipline has had an ambivalent and spotty relation with law. There are some important exceptions: Kropotkin (1903/1946), for example, placed the law and its political geographies at the centre of his history of the European state. Unusually, he thought of law not simply as an outcome, or technical manifestation of state power, but as integral to the very formation of the state. He also offered an astute analysis of the geographies of legal liberalism, particularly the antipathy towards intermediate agents such as the city. In this, he departed from subsequent scholarship. As I have noted elsewhere (Blomley, 1994a), analyses tended to divide into those, such as Ellen Semple (1918), who sought to ground law in place, and others, like Derwent Whittlesey (1935), who focused on what he termed the 'impress of effective central authority upon the landscape'. Whether law is explained by reference to space, or space is seen as produced by law, the tendency was to impose separations on law, space and society (Blomley, 1994a; Platt, 2004).[1]

From the 1980s onward, when the 'spatial turn' in social theory made such a separation untenable, these categories came under attack. Drawing broadly, though not exclusively, from post-structuralist concerns with interpretation and the production of social meaning, critical legal geographers insisted that:

(a) law and space be seen both as socially produced and socially productive;

(b) space, or more accurately, spatiality, be seen as in turn productive of law, as well as produced, in part, by law;

(c) this process be acknowledged as deeply implicated with power relations, given law's importance to resistance and domination; and

(d) as a consequence, law be acknowledged as of crucial importance to geography.[2]

This interrogation of geography's basic terms was mirrored by similar re-examinations within law: critical legal studies, socio-legal studies and critical race theory, for example, insisted on the politics and sociality and, occasionally, the spatiality of law (Blomley and Bakan, 1992).

The ensuing two decades have seen a growing body of research in law and geography, although fewer contributions have been made with a primary focus on the theorization of the field. Several special issues of journals have been produced, including *Urban Geography* (1990, 11 (5–6)), *Stanford Law Review* (1996, 48), *Journal of Historical Geography* (2000, 28) and *Society and Space* (2004, 22, 4). Book-length overviews include *Law, Space and the Geographies of Power* (Blomley, 1994a), *Law and Geography* (Holder and Harrison, 2003), *The Legal Geographies Reader* (Blomley et al., 2002) and *The Place of Law* (Sarat et al., 2003). Sustained case-studies include Cooper (1998), Whatmore (2002), Olwig (2002), Delaney (1998, 2003), Stychin (1998), Maurer (2000), Darian-Smith (1999), Blomley (2004a) and Mitchell (2003).

This scholarship is remarkably diverse and lively, engaging with topics such as nature, landscape, state practice, nationalism and boundaries. Scholars draw from a range of theoretical sources, including queer theory, urban political economy, actor-network theory and cultural studies. That said, it is still relatively unusual for political geographers to write expressly on the law. A quick review of recent articles in *Political Geography*, for example, finds only a few papers that appear directly concerned with law and space. Recent textbooks on political geography, while making frequent reference to particular forms of law (zoning, sovereignty and so on), give the law in general short shrift. Although the relevance of law to political geography might seem obvious, this does not seem to be realized in the literature. Why then should political geographers, in particular, be concerned with law? Some quick responses might include the following:

1 The state, as an abstraction, institution and bundle of ideologies, is inseparable from the law. If we are interested in issues of state formation, state practice, ideology and legitimation, we necessarily bump up against the law. Many of the classical theorists of the state, such as Hobbes, Mill, Marx or Locke, were also, necessarily, legal thinkers (Ryan, 1984).

2 Law, both formal and informal, is thus a critical manifestation of state power. Yet law has its own particular logic.[3] When state initiatives become legalized, in other words, they are caught up in and transformed by legal institutions and ideologies including liberalism, the world of rights (of which more later), and the workings of the legal apparatus. Political concepts such as freedom or citizenship are indelibly legal concepts.[4] The specificities of law, as a site of power, need to be acknowledged.

3 More particularly, spatial arrangements of interest to political geographers are, in part, legal arrangements, such as sovereignty (Borrows, 1999), territory (Delaney, 1998), political districting (Forest, 2001, 2004), jurisdiction (Ford, 1999), urban rule and local autonomy (Clark, 1985), land development (Blomley, 2004c) and globalization (Twining, 2000).

4 Of course, law, just like politics, does not reduce to the state. Political conflicts and struggles, as well as forms of government at a distance, also implicate law, whether directly or through more informal, yet no less significant means. As Twining (2000: 1) notes, law is omnipresent.[5]

5 Law can usefully be understood not only as a set of operative controls, but also as a repertoire of cultural and political meanings through which citizens can negotiate and interact with one another. Legal discourses offer a 'relatively plastic medium for refiguring the terms of past settlements over legitimate expectations and for expressing aspirations for new forms of entitlement' (McCann, 1994). These discourses are important, shaping

the ways in which people conceive of the political realm, and their place within it.

Let me offer a few examples of areas of critical scholarship, of interest to political geography, in which an analysis of the law/space relation begins to prove its usefulness. I draw in part from the work of non-geographers: this reflects my desire to profile work that may be less familiar to geographers, as well as my sense that some of the best legal geographies are, ironically enough, being written by non-geographers. I begin with rights, before moving on to analyses of colonial dispossession.

THE POLITICAL GEOGRAPHIES OF RIGHTS

Rights are central to law, statecraft and social life, and have become a standard feature of the legal apparatus of the modern state. Rights-claims carry special force: in ascribing rights to certain relationships, we 'shift them out of the realm of the merely desirable and into the domain of the morally essential' (Jones, 1994: 4). Claiming a right, in this sense, can mobilize the state. More discursively, they signal an acknowledgment by the collective that certain issues are worthy of special protection.

Within liberal theory, rights tend to be thought of (a) as the entitlements of autonomous individuals, where personhood is a priori, and (b) as outside the realm of politics, serving as a trump against the state realm. Yet critical scholars (Sarat and Kearns, 1997b) argue, conversely, that rights (a) help constitute us as political subjects and (b) are deeply implicated in the political realm. Yet rights can appear forbiddingly abstract. Not only are they embedded within law, which claims a degree of detachment from society, but they also appear, metaphorically, to be 'above' law. Dworkin (1977) talks of rights as 'trumps' that have priority over other rules and principles. To that extent, it becomes harder to think through the geographical and empirical dimensions of rights. Perhaps for this reason, much scholarship on rights tends to be abstract. The problem is that legal academics 'prefer to pitch their tents in the shadow of the Supreme Court rather than on Main Street ...' (Engel, 1995: 124; Selznick, 2003). As a result, 'we know relatively little about why and when a person adopts rights talk' (Merry, 2003: 346). Yet, as Wendy Brown points out, an informed discussion of rights requires an 'analysis of the historical conditions, social powers, and political discourses with which they converge or which they interdict' (Brown, 1997: 88).

For rights seem to shape everyday moral and political thinking: 'Claiming or asserting rights has become the common mode by which people seek to promote an interest or advance a cause' (Jones, 1994: 3). Rights talk is 'one way people frame and assess their worlds' (Milner, 1992: 322). So, for example, recent debates in Canada about urban poverty and the state regulation of begging are, in part, conducted in the language of rights, whether in the courtroom or the op-ed article. In so doing, representations of space and political subjectivity (the street, public space, the interactions between panhandler and 'the public') are deployed. Bartholomew and Hunt's (1990) suggestion that we move away from 'abstract speculation about the nature of rights-in-general' (p. 52) toward a more nuanced discussion of 'the strategic value of particular rights struggles' (p. 4) alerts us to the need for a more contextual treatment of the politics of rights. I draw below from an emergent literature on rights to make a series of points concerning rights. I then conclude by noting their crucial spatiality.

Rather than thinking of rights as the entitlements of autonomous individuals, we can think of them as constitutive of subjectivity, calling certain forms of personhood into being. Rights-talk entails certain characterizations of political subjectivity: to characterize a panhandler as engaged in expressive conduct, for example, is to imagine her as a citizen rather than a nuisance.

Yet rather than being pre-political, rights-talk can reproduce certain visions of the social and political world, casting political subjectivity in particular and limited ways. This, notes Gerry Pratt (2004), is the paradox of rights. They work rhetorically through their universalistic claims, yet are intrinsically particularistic. For these reasons, some argue, we must be skeptical of the political potential of rights. Joel Bakan (1997), for example, has noted the ways in which Canadian constitutional jurisprudence reproduces a restrictively liberal world-view, premised on atomism and anti-statism. Nedelsky (1990) characterizes property rights, with their emphasis on the self as 'separative' and detached from and suspicious of the polity, as impoverished and partial. Brown (2004) points to the ways rights are entangled with contemporary forms of domination, such as militarism and capitalism. She also worries (Brown, 1997), drawing from Marx's classic account, that rights may be Faustian, promising a certain limited 'freedom' while installing a privatized and depoliticized world-view. For as Marx (1975), noted, the emancipation promised by rights is 'devious' to the extent that freedom is granted to abstract rather than concrete subjects, thus denying the material constituents of personhood, constrained by the workings of civil society. Partly for these reasons, some argue for a post-rights politics, such as Duncan Kennedy's call for 'transgressive performance' (Roithmayr, 2001).

Yet others insist on the progressive potential of rights, noting that they are not limited to formal state forums, such as the courts (Blomley, 1994b). Rights can also entail a 'wide continuum of value claims ranging from the most justiciable to the most aspirational' (Weston, 1992: 17). Rights in this sense are pluralistic, spoken in multiple vocabularies. Because of their significance, rights are constantly fought over, being invoked by 'outsiders' who seek to extend, rework or entirely remake rights in the service of particular struggles (Williams, 1991). Rights are 'protean and irresolute signifiers' (Brown, 1997: 86) whose varying and often expansionary meanings can be put to work in diverse political sites (Laclau and Mouffe, 1985; McClure, 1997; Goldberg-Hiller and Miner, 2003). Rights can provide a powerful resource for certain interests, allowing them to re-imagine, for example, women's experience of sexual harassment as that of rights-bearers, facing injustices (Marshall, 2003). In so doing, political subjectivity can be redrawn and redefined. Cornell (1997) defends abortion rights for their value in sustaining women's 'individuated selves'; those who would limit abortion rights oppose the conditions of women's existence as individuals in the world.

Rights are thus clearly important in constituting as well as resolving political struggles. A contest over rights is, in part, a contest over political subjectivity. To that end they can become a battleground; they 'seldom end political contests; instead they themselves become objects of struggle' (Sarat and Kearns, 1997a: 12). 'Examining [rights'] constitutive effects', then, 'means enquiring into the way rights call into being, and enable, particular forms and expressions of personhood, as well as the way they disable others' (Sarat and Kearns, 1997a: 3). However, such a project must acknowledge that the meanings of rights 'are not determined in the abstract, but rather in practice' (McCann, 1994: 283; Mitchell, 2003). In particular, rights, subjectivity and politics are all inseparable from geography. Space shapes the ways in which rights are construed, contested and put to work. Liberal rights help produce, and operate within, sharply demarcated spaces. Michael Walzer (1984: 315) defines liberalism as 'a certain way of drawing the map of the social and political world'. Confronting the feudal polity, liberals practiced the 'art of separation. They drew lines, marked off different realms, and created the sociopolitical map with which we are still familiar' (Dimock, 1998; Nedelsky, 1990). The way this 'sociopolitical map' is produced can affect the ways in which subjectivity is produced, as well as shaping the political possibilities associated with rights.

So, for example, as Gerry Pratt (2004) points out in the case of Filipino domestic workers within Canada, space can be used to close rights down as well as open them up. The spatial bounding of the body, the public/private divide and the mapping of the nation can all serve to delimit or deny rights, she argues. The working conditions of Filipino workers in Canada, for example, are not always subject to the same constitutional scrutiny as that given to those of Canadian citizens. Yet sending countries are also often ineffectual in defending the citizenship rights of domestic workers abroad. And yet, Pratt argues, geographic scale, the spatial contingencies of rights, and the 'empty space' associated with liberal universalism have all been put to productive work by Filipino activists, seeking to push rights-based claims. So, for example, Filipino activists in Canada have refused the legal identity imposed upon them, and are working to remake the definitional terms by which they enter Canada. As rights-claims are 'so often about access, protection, defence against incursion, it is a congenial discourse for a group of workers whose claims to a place and privacy are so fragile', she notes (p. 115). In a similar vein, others note that the peripheral and exclusionary locations to which marginalized groups have been assigned can be used as a space from which to contest and challenge the ordering of rights (Chouinard, 2001; Peake and Ray, 2001). The contradictory ways in which the space of rights is constructed can also be used to press rights-claims. As Klodawsky (2001) demonstrates, the mismatch between international human rights codes, and the diminished realities of domestic social and economic rights protections, has been used by Canadian anti-poverty activists to powerful effect. While 'the geographies written into liberalism make it very difficult to make rights claims in some spaces' (Pratt, 2004: 115), prevailing geographies of rights can be turned on their head.

One site in which the relation between rights and the production of space and subjectivity takes on an added dimension is public space. Rights and public space are historically predicated on particular and related forms of exclusion: only certain subjects were deemed appropriate rights-bearers in relation, in part, to the degree to which they were imagined as appropriate public citizens. Public space, Don Mitchell (2003) argues, is in this sense produced through political struggle, as outsider groups, such as women, the working class and ethnic minorities, have fought their way into the public realm. Such a struggle necessarily entails rights. Again, rights can be used to deny public space to certain populations or can be used as a tool to pry public space, and thus, citizenship, open. Rights, in this sense, are 'a moment in the production of space – especially material, physical space' (Mitchell, 2003: 28). Both rights and space, then, are co-produced: social action 'always operates simultaneously to influence the production of law and the production of space' (Mitchell, 2003: 29). Mitchell demonstrates this co-production, and traces its grounding and effects upon prevailing social hierarchies within the urban United States through a number of controversies, including the struggles over 'People's Park' in Berkeley, US Supreme Court decisions concerning protests in public space, and the field of 'anti-homeless' laws (Mitchell, 2003). From a somewhat different tack, Sallie Marston (2004) reflects upon US court decisions concerning the use of rights-based arguments by gay, lesbian and transgendered Irish-Americans, attempting to participate in St Patrick's Day parades in South Boston. Marston argues for the critical role of state interpretive practice in producing particular political subjectivities but also, crucially, points to the particular ways in which the courts produce spaces within which these subjectivities are to be realized.[6]

It is because of the political valence of rights, and their spatiality, that scholars have become interested in tracing the geography of rights in more than an empirical or descriptive sense. Lefebvre's call for a 'right to the city', as well as an earlier tradition within urban analysis (Merrifield and Swyngedouw, 1996), have been invoked by those who seek to articulate a normative vision of space and social justice (Harvey, 2000). Staeheli and Dowler (2002) argue that the city and its spaces have seen the expression of new claims that seek to forge a new politics of inclusion, citizenship and new ways of being. Drawing upon Lefebvre, Isin (2000: 14) argues that the 'right to the city' entails 'the right to claim presence in the city, to wrest the use of the city from privileged new masters and democratize its space'. However, much of this work, while interesting and even inspirational, remains under-theorized, and fails to engage the broader critical literature on rights, their politics and their geographies. Several questions pose themselves: Given their provenance and political effects, should rights be included in any program that seeks to advance social justice? If so, how should rights be configured? Does the 'right to the city' offer a useful political basis? Are rights best conceived in terms of their judicial effects or outside the courtroom? If we accept that rights have an expansionary logic, where can we look, within liberalism, for a basis for such an expansion? What conception of social justice can inform such a rights-claim? And what difference to all these questions does space make?

POST-COLONIAL STATE, SPACE AND LAW

If rights are a more recent concern to political geographers, then colonialism is surely a more traditional interest. Here again, though, colonialism reveals itself as a thoroughly legal and spatial affair. Thus, colonialism entails a remaking not only of governance but also of ownership. For Said, the relation is a fundamental one: 'At some very basic level, imperialism means thinking about, settling on, and controlling land that you do not possess, that is distant, that it is lived on and often involves untold misery for others' (Said, 1993: 7). Law plays a crucial role in this transfer. The irony is not lost on at least one observer: 'law, regarded by the West as its most respected and cherished instrument of civilization, was also the West's most vital and effective instrument of empire' (Williams, 1990: 6). While different legal arrangements were deployed, 'in all cases, the end result was the same: the legal form of colonization … effected the translation of newly acquired territories into exploitable property' (Patton, 2000: 27).

Law, it has been noted, serves as a powerful instrument of colonial dispossession (Comaroff, 2001). Dispossession often occurred (and continues to occur) under the sign of law, through military force, legal fraud, state expropriation, forced extinguishment, treaty abrogation and the non-enforcement of protective legislation, for example (Deloria and Lytle, 1983; Berger, 1991; Daes, 1999). Judicial practice also plays a crucial role. Thus, in deciding whether indigenous claims to land constitute a proprietary relation, common-law courts have tended to adopt a categorization that compares them with the assumed substance of English concepts of property – with predictable results (Patton, 2000). Forman and Kedar (2004, 22, 6, 809–830) carefully document the ways in which Israeli law was used to render the displacement of Palestinians, forced from their lands after the 1948 war, permanent and legitimate. A series of legal categories, such as 'abandonment', 'absentee', and 'Israeli lands', were put to work through a series of enactments that, over time, hardened state control over Arab land. Legal categorization is a form of spatialization, turning on the creation of clear boundaries, and the marking of inside and outside. Here, however, that which is inside must be placed outside. Thus Arab property-owners within the state of Israel become, in one telling legal phrase, 'Present Absentees', suspended in a paradoxical state of exception. However, the law not only facilitates dispossession but also conceals, legitimizes and depoliticizes that process, attributing 'to the new land arrangements an aura of necessity and naturalness that protects the new status quo and prevents further redistribution' (Kedar, 2001: 928).

Law, in this sense, emerges from these studies not simply as the neutral medium through which colonialism was (and is) realized, but rather as a means through which colonialism has itself been produced (Comaroff, 2001; Harris, 2002). Law, put simply, makes a difference to the ways colonialism unfolds. But if this process entails law, it also requires the remaking of space. And again, legalized space is not simply the surface upon which colonialism unfolds, but is rather produced by and productive of colonialism. So, for example, Russell Hogg (2002) characterizes the creation of courthouses in the interior of New South Wales as a symbolic statement that served to announce the 'majesty, supremacy and neutrality of a new legal order, presupposing the spaces they occupied to be empty wilderness and thereby erasing (or seeking to erase) existing landscapes, cultures and legal and spatial orders' (p. 38).

Spatial representations – notably cartography – play a critical role in colonialism. Surveyors have been termed 'the point men of British imperialism' (Edney, 1993: 62) for good reason, given their role in the 'imposition of a new economic and spatial order on 'new territory', either 'erasing pre-capitalist indigenous settlement or confining it to particular areas' (Kain and Baigent, 1992: 329). The particular, and often partial, technologies of record-keeping and land titling have been identified as significant in not only securing the territory of some, but in dispossessing others (Luna, 1998).

But if the geographies of law played a significant role in colonialism, they may have done so in complex and often contradictory ways. A discerning paper on the English colonization of North America in the seventeenth century, for example, characterizes proprietorial charters as a particular form of 'legal cartography' that sought 'to elaborate the precise statements of relationships between places and people, existing or desired, that were crucial to success' (Tomlins, 2001: 321) in the creation of colonial plantations. The charters sought to mobilize legal discourses so as 'to imprint England on America' (p. 331) (or, more bluntly, 'to hack territory out of space', p. 332). The charters, moreover, represented colonial space in ways that justified its appropriation, with recourse to evangelical claims, notions of productive activity, and renderings of native people (as savage, prenatural and so on). However, Tomlins carefully unpacks the Englishness of the day – especially in relation to law – and identifies a remarkable polyphony. Legal culture within England was diverse and pluralist. When settlement occurred, these localized variations in law and custom, he argues, were reproduced in North America, prompting tensions between singular forms of colonial authority, operating under charters, and local legal cultures. This, in turn, provided a basis for resistance to colonial rule itself. 'Law, in short, colonized early America. But it did so in multiple and sometimes conflicting ways' (p. 364).

The centrality of a particular geographic model of ownership can produce one-sided colonial interactions. Stuart Banner (1999) discusses the cultural specificities of property's geographies in his exploration of the colonization of New Zealand. A powerful 'geographic paradigm was embedded in settler consciousness', he argues, whereby ownership rights were understood to be allocated on a geographic basis (p. 832). 'Land was divided into spaces, each piece was assigned to an owner, and

the owner was ordinarily understood to command all the resources within that geographic area' (p. 810). The Maori, however, allocated property rights on a functional rather than a geographical basis. Thus, a person was not imagined to own 'a zone of space', but would rather have the right to use a particular resource in a particular way. The right to trap birds in a certain tree, for example, did not necessarily entail other rights exercised within the same space. There were apparent cultural similarities however: the Maori cultivated and fenced land, for example. Yet these similarities, which encouraged settlers to imagine a common vocabulary of ownership, if operating at different levels of civilization, were cross-cut by differences. The effect, argues Banner, was mistranslation and a cultural confusion that only dissipated when the colonial authorities were able to force the Maori to reconceptualize land as composed of geographic spaces rather than use rights: 'The colonization of land, the physical substance, could not have proceeded without the simultaneous colonization of property, the mental structure for organizing rights to land' (p. 847).

For such legal geographies continue to colonize. Ashinabek legal scholar John Borrows (1997) describes the collision between land-use planning and indigenous geographies in contemporary Ontario (cf. Harris, 2000). Aboriginal spaces, he argues, have been mapped out by a Euro-Canadian 'conceptual grid … which divides, parcels, registers and bounds peoples and places in a way that is often inconsistent with Indigenous participation and environmental integrity' (p. 430). 'The law', he argues, 'has put a culturally exclusive vision of geography in its service' (p. 431). Sherene Razack (2000) argues that contemporary forms of violence, directed against native women, continue to draw from a whole series of colonial dispossessions and displacements. Not only are such failures unjust, they also efface alternative, and potentially valuable, legal imaginaries and orderings. Dispossession as a legal practice is in this sense incomplete, but dependent on iteration and hegemonic production (Blomley, 2004c). Thus, Bhandar (2004, 22, 6, 831–845) points to the ways in which dominant legal conceptions of aboriginal title in Canada continue to rely upon powerful stories of settlement and sovereignty, even at the same moment as they call upon a self-congratulatory language of reconciliation, the effect of which is to imagine native claims as a 'burden' upon state title. However, law itself – like legal rights – can be used as a cultural resource by the dispossessed. So, for example, Kosek (2004) describes the ways in which Hispanic land activists in New Mexico drew from legal texts, notably the 1848 Treaty of Guadalupe Hidalgo between the United States and Mexico, legal artifacts, such as eviction and impoundment notices, and legal practice, such as a community-based trial, to enact their claim to land grant territory appropriated by the US federal government.

The complex intersections of law, space and, as Banner characterizes them, 'mental structures' are the focus, in part, of Timothy Mitchell's (2002) account of the nineteenth-century colonization of Egypt. In so doing, he makes some important points concerning both colonialism and the geographies of law more generally. But rather than simply showing the ways in which, for example, mapping and legal practices were instruments of colonial rule, Mitchell's interest is with the ways in which law and geography helped produce, while drawing from, a crucial categorical division of the world between the ideality of representation and an object world of 'reality'. Law in general, and property in particular, derive their power from the degree to which they appear to stand as a universal abstraction, set apart from the messy realities of local particularities.

Thus, colonial administrators sought to introduce 'modern' principles of private property to Egypt on the assumption that, as universals, they were the opposite of the arbitrary and coercive rules said to be operative in indigenous society. Mitchell shows, however, that the creation of the property regime that replaced pre-colonial forms of land-holding was itself highly particularistic as well as deeply coercive. Land seizures, the violent suppression of rural uprisings and the creation of private villages effected a transfer of land ownership that was as violent as it was arbitrary: the private village, for example, produced 'territories of arbitrary power within the larger space of legal reason and abstraction' (p. 71). The power of property, in this regard, lay in the degree to which these processes were reorganized so that some processes began to appear particular, and others general:

> ... [S]ome appear fixed, singular, anchored to a specific place and moment, like objects, while others appeared mobile, general, present everywhere at once, universal, unquestionably true at every place, and therefore abstract. One set of actions, people and sites were fixed in position as 'land' and 'peasants', made into objects to supervise and control. At the same time a series of removals, rearrangements, delaying maneuvers, simplifications, and silences established other sites and other actions as what seem the opposite of this: nonlocal, outside actuality, and therefore universal. This was to create the effect of a fundamental difference: land versus law, the particular versus the general, the physical versus the abstract, thing versus idea, force versus order. (Mitchell, 2002: 59).

So space itself, imagined as an objective surface, was a product of law. Colonialism sought to replace the tensions and interactions of rural life with a

world resolved into two spheres: 'the inert materiality of land on one side, legal codes and property rights on the other. Thing versus idea, reality versus abstraction, space versus its meaning' (p. 78).[7]

Yet Mitchell refuses to take these divisions as 'social constructions', as such a characterization still supposes the distinction he seeks to unravel, and leaves aside 'the remainder or excess that the work of social construction works upon – the real, the natural, the nonhuman' (p. 2). Put another way, to characterize the remaking of property relations in Egypt as an attempt by the state to increase its powers over rural society presumes a state/society distinction that is itself 'an uncertain outcome of the historical process' (p. 74). Rather than existing prior to them, state and society, in this sense, emerged from these practices. This emergence, he suggests, is often a messy one. Mitchell enjoins us to understand how this crucial difference between, for example, land and law is made:

> ... [W]e have to reopen the connections between what was separated, follow the links from one action to another, and see how one set of elements in this relationship was subordinated, removed, reserved, or silenced. This will bring to light what is buried when we write theories of 'property', 'law', and 'the state', when we begin with metaphysical abstractions rather than asking what methods of politics and expertise divide the world into metaphysics on the one hand and mere physics on the other. (Mitchell, 2002: 59).

RESEARCH CHALLENGES

Mitchell's account forces some important theoretical questions to the fore in the analysis of law and space. For example, his injunction to avoid metaphysical distinctions alerts us to the dangers of carving off analytical categories, such as law. Rather than accepting the universality and abstraction of law, he asks: How is legal 'universality' produced? How did modern law acquire its power and authority? How, moreover, is 'space' produced as an abstraction? The danger is that, as analysts, we reproduce these politically loaded distinctions in our work. In thinking through carefully the ways in which 'law' and 'space' intersect, it is tempting to do so in ways that treat either as autonomous and separable. There is a tendency to rely, implicitly, upon a tripartite split, where 'law' affects 'space', both of which have a relation to something called 'society'. But on closer examination, the relation may be a closer one: '"law" and "geography" do not name discrete factors that shape some third pre-legal, aspatial entity called society. Rather the legal and the spatial are, in significant ways, aspects of each other ...' (Delaney et al., 2001: xviii).

One way to acknowledge this is to think through the importance of law and space to order. If we accept that the world is in part produced through various orderings and categorizations, it becomes clear that law offers one crucial vocabulary through which this order is produced (Cohen and Hutchinson, 1990). Thus, law allows us to distinguish 'citizen' from 'alien', 'employee' from 'employer', 'husband' from 'wife' and so on. Similarly, space offers us a powerful ordering framework (Sack, 1986). But, of course, legal orderings are simultaneously spatial orderings. Thus, the 'owner' is to 'land' as the 'citizen' is to 'state territory' and so on, such that the two are inseparable. A prison is neither a legal category nor a space, but both simultaneously. As a result, it is perhaps useful to run law and space together, and think of 'splices' (Blomley, 2003). A city, for example, is a splice: it is both a particular set of socio-spatial arrangements (with neighbourhoods, scalar associations, landscapes and so on) and a legal creation (produced through particular sets of police powers, devolved legal capacities, property arrangements etc.).[8] War is also a splice: it is underpinned by and partly regulated by legal notions such as sovereignty, yet is also clearly spatialized, reliant on spatial practices, categories and arrangements. Guantanamo Bay, for example, is a particularly vivid example of such spatialized legalities.[9]

The 'splice' seems useful: as a noun it alerts us to the particular and apparently stable arrangements produced through law and space. In part, perhaps, because of the apparent stability and objectivity of law and space, such arrangements acquire an air of fixity and naturalness. Property boundaries, for example, can appear natural and pre-political. However, as a verb, splice also points to the necessary mobilizations, enrollments and everyday 'doings' that such arrangements require. The boundaries of property depend upon all the continued production and success of all sorts of complicated and often unpredictable networks, relationships, hegemonic productions, quotidian practices and so on. Once recognized as achievements, rather than facts, it begins to be possible to think of the ways splicings may be respliced, whether intentionally or accidentally. This recognition may militate against the tendency, noted by Allen (2003), to think of power as something 'stored' or held in reserve. Citing Arendt, he argues that power is always 'of the moment, something that has to be continuously reproduced over time' (p. 111).

In thinking of the splice it is tempting to focus on the role of discourse, that is, on law as a set of meanings. This reflects the tendency of much legal geography. For example, Gordon Clark's book, *Judges and the Cities* (1985), noted the important role of the judiciary in producing interpretive determinacy. While obviously important, treating law just as 'words' was criticized within law by Robert Cover (1986), who noted some years ago that law needed also to be thought of as set of practices, often entailing corporeal violence. Within geography, Vera Chouinard (1994: 420) was also critical of the interpretive turn within legal-geographic inquiry, arguing that 'texts are not enough' for the critical geographic study of law, but must be supplemented by careful attention to the material grounding of those 'texts' in lived relations of power, oppression and resistance. Timothy Mitchell (2002) also cautions against an emphasis on ideality at the expense of an examination of how human intentions and the world of things commingle in intricate and often unpredictable technopolitical amalgams. Here, some intriguing and important work has been done. David Delaney has done much to think through the ways in which legal spaces are spliced together with and through both things and representations. His notion of the 'nomosphere' begins to point toward these possibilities (Delaney, 2004). An attention to the material geographies of law has also encouraged some scholars to grapple with the ways in which law is a site for hybridity, a commingling of the 'natural' and the 'social'. Thus, for example, Sarah Whatmore (2002) characterizes law – property, in particular – as a zone in which this distinction is made and remade in unsettling ways. Such distinctions, when imposed upon contemporary struggles over colonialism, or property rights in genetic resources, can have deeply political effects, she argues, in constituting the subjects and spaces of political community (M'Gonigle, 2002; Delaney, 2003).

Chouinard's criticism of legal-geographic research was fuelled by a sense that it had lost (if it ever had) its ethical and political edge. Certainly, some early scholarship was explicitly 'critical' in orientation, it being argued that orthodox formulations of law and geography were not only wrong-headed, but also politically dangerous in their reliance upon legal closure and a restrictive form of liberalism (cf. Blomley, 1994b). Such clarion calls echoed the bracing language of critical legal studies as well as some neo-Marxist analyses of law. However, this language has become more muted of late, in line, perhaps, with a more generalized political drift.[10] Certainly, the denaturalization of legal categories continues, as does the identification of forms of resistance (though on legal resistance, see Brown's (1996) worries that resistance has become a new disciplinary hegemony). However, the scholarship on rights as well as some post-colonial writing urges new forms of political engagement. I find the work of scholars such as Don Mitchell particularly refreshing here, particularly his careful attention to the use of law in attempts at the 'cleansing' of urban public space. His recent book, *The Right to the City* (2003), offers one form of political engagement. No doubt there are others, whether reformist or more radical in orientation. Such academic work is valuable, for lawyers, politicians and others are constantly splicing law and geography together in the world, often with punitive, discriminatory or oppressive effect. Zero-tolerance policing and related notions of 'broken windows' and public order, for example, are a clear case, reliant upon particular understandings of law, space and order (cf. Blomley, 2004b). To find ways, within the academy and beyond, to join forces with those who seek to un-splice or re-splice such oppressive legal geographies seems a critical task.

Legal geographers need also to more carefully recognize the particularities of power (Allen, 2003). Power, Allen argues, is never power in general, but always power of a particular kind, expressed as domination, authority, coercion, seduction, and so on. Each has its own relational logic and spatialized specificity. To say that 'much of social space represents a materialization of power, and much of law consists in highly significant and specialized descriptions and prescriptions of power' (Delaney et al., 2001: xix) begins, rather than concludes, the analysis. What sort of power is law? Or, more likely, what sorts of power? Viewed more closely, law can entail both forms of power over others as well as associational power (power with others). If we are to understand a legal world that includes execution, ideology, enrollment, myth, map-making, narrative, courtroom layout, faith, and the construction of difference, for example, we will need a flexible and nuanced analysis of power. Legal violence, for example, can be experienced as a direct form of domination (war, execution or imprisonment), as an internalized form of self-government, or as an ideological leitmotif in legal myth-making (Blomley, 2003). Law itself is differentiated: property law, for example, works differently from refugee law. And in all cases, different modalities of space/power are at work. How does 'splicing' differ when it is done by a judge, a police officer, a property surveyor or a tenant? What complications emerge?

Here the work of Ben Forest is exemplary: he asks how judges make up legal space. For example, in a paper on US electoral districting decisions (2001, 2004), he argues that the lack of a clear judicial consensus concerning how best to draw electoral districts reflects an underlying contradiction between two opposing visions of the relation between space and political identity. Thus courts have struck down 'irregular' districts, he argues, because of a belief in an individualistic political identity. Yet they have also evaluated redistricting plans on the assumption that spatial propinquity is a vital basis for the formation of political interests. Forest demonstrates that spatial organization is not a dispassionate or simply technical issue. Rather, the courts see space as having the capacity to modify political behavior.

Further, the different ways in which the courts construct space (as geometric, or organic, for example) is not a disinterested exercise, but a consequential one.

Allen's claim (1999: 205, my emphasis) that space be recognized as 'an *integral*, rather than an additional part' of any analysis of power is welcome. But when we begin disaggregating law's power, and recognize its diversity, we must also ask more carefully how the diversities of space – organized into networks, landscapes, places, scales, flows and topologies – affect the reach and effects of law. What difference, put bluntly, does space make to law? We need, as Allen puts it, 'to be a little more curious about power's spatial constitution' (2003: 4). For example, one tendency has been to think of legal liberalism as intolerant of spatial difference, preferring an isotropic national surface in which legal interpretation and practice are, in theory at least, uniform and consistent (Pue, 1990). In this sense, law appears intrinsically aspatial. However, more careful reflection reveals that law can tolerate, indeed occasionally embrace, spatial difference. Indeed, difference can become an instrument of rule (Agamben, 1998).

Another way of trying to get at the difference that space makes is to explore the spatialized meanings of law in the everyday world. On the principle that 'a rule without a materialization is just a formless formalism' (Delaney et al., 2001: xix), the spatial manifestations of law – the 'No Trespassing' sign, or the barbed wire fence, for example – are said to be as important to law's reproduction and enactment as the pronouncements of a judge, or the writings of John Locke. Such spatializations partly work through the communication of legal meanings. Robert Sack argues that along with its classificatory advantages, human territoriality (including, by extension, legal territorialization) is important as a communication to others. The boundary, he argues, is a remarkably succinct and efficient statement, serving as perhaps the only symbolic form 'that combines a statement about direction in space and a statement about possession or exclusion' (1986: 21). Critical legal geographers similarly argue that:

> Boundaries *mean*. They signify, they differentiate, they unify the insides of the spaces that they mark. … And the form that this meaning often takes – the meaning that social actors confer on lines and spaces – is *legal* meaning. How they mean is through the authoritative inscription of legal categories. … The trespasser and the undocumented alien, no less than the owner and the citizen, are figures who are located within circuits of legally defined power by reference to physical location vis-à-vis bounded spaces. (Delaney et al., 2001: xviii, original emphasis)

Yet Allen's (2003: 142) criticism of some treatments of power – notably governmentality – for confusing intent with actual effect is relevant to research in law and geography. It is crucial that more empirical research be undertaken to explore the ways that, or the degree to which, splices produce their intended effects. My own research on residential property boundaries – both between neighbours and between public and private spheres – suggests that everyday practices and understandings are much more complicated and fluid. If boundaries mean, that meaning is open to conversation, ambiguity and nuance (Blomley, 2004a, 2004b). Ewick and Silbey (1998) similarly suggest that people may use and experience law in very different ways. Law may be experienced, they argue, as remote, as arbitrary, or as a manipulable game.

Legal geography offers important and valuable insights. Thinking the legal – such as property, or crime – in terms of space, or the spatial – such as the boundary, or place – in terms of law promises new and enriched insight. Understandings of both space and law are changed, and new questions posed. This is valuable in an analytical sense: if we are interested in how political geographies are put together or sustained over time, for example, recognizing the role of law is important. However, for better or worse, legal geographies also have political and moral effects. As political geographers, we are interested both in the analytical and ethical evaluation of space/power. Given its deep implication in both, it is imperative that we take law seriously.

NOTES

1 See the analysis of Timothy Mitchell (2002), discussed below. He points to the ways in which the state's production of legal and spatial arrangements themselves contribute to the creation of an epistemic divide between categories such as 'law', 'space' and 'society'. Law and geography, in other words, may help create 'law' and 'space'.

2 I leave unexplored the question of how 'society' is itself theoretically conceived. In part, this reflects the diversity of theoretical analyses drawn upon by legal geographers. However, it must also be acknowledged that more careful analytical work needs to be done here. I am indebted to Kevin Cox for raising this question.

3 For example, as E.P. Thompson (1975) noted in *Whigs and Hunters*, the ideologies of law, although often a form of legitimation, can force state officials to act in ways that complicate notions of straightforward class rule. Shamir (2001: 135) similarly insists that law is not a 'mere arm of the state', but rather has its own cultural logic.

4 To the extent that some observers note (and worry at) the legalization of political discourse. Rights in general are thus conceived as judicially accepted constitutional rights, at the expense of other, often more progressive possibilities (Bakan, 1997). I return to this point below.

5 'One way of introducing students to the study of law is to ask them to read every word in a newspaper and to mark all the passages that they think deal with law or are "law-related". … Even those who adopt a narrow conception of law find it on every page. … On the arts page pornography, copyright, defamation, and other issues relating to freedom of expression arise along with licensing, charities law, taxation, and the ubiquitous contract. The advertisements are permeated with legal words and phrases. And so on through the paper' (Twining, 2000: 1).

6 Another area of rights analysis, productively informed by a spatial analysis, concerns the politics of mobility, including in relation to disability. See, for example, Chouinard (2001) and Cresswell (2006).

7 Mitchell (2002) also describes the process of cadastral mapping in Egypt as 'a method for staging the world as though it were divided into two, and as means of overlooking this staging, and taking the division for granted' (pp. 82–3). We need to be cautious, I think, in emphasizing the singularity of such 'stagings' . Elsewhere, I have pointed to the particular confluence of law and cartography in the production of modern notions of property in early modern England (Blomley, 1994a).

8 On the city as a legal form, see Frug (1980).

9 Obviously, law and space do not exhaust the meanings and logics of war or cities. My intent here is, rather, to point to the important ways in which law and space meld in the production of such phenomena.

10 Although (see Hogg, 2002), 'To begin to explore the spatiality of law is one path to subverting its imperial claims to objectivity, generality and sovereignty and to recognizing the subsistence of other legal orders and other legal possibilities' (p. 39).

REFERENCES

Agamben, G. (1998) *Homo Sacer: Sovereign Power and Bare Life*. Stanford: Stanford University Press.

Allen, J. (1999) 'Spatial assemblages of power: from domination to empowerment', in D. Massey, J. Allen and P. Sarre (eds), *Human Geography Today*. Cambridge: Polity Press.

Allen, J. (2003) *Lost Geographies Of Power*. Malden, MA: Blackwell.

Bakan, J. (1997) *Just Words: Constitutional Rights and Social Wrongs.* Toronto: University of Toronto Press.

Banner, S. (1999) 'Two properties, one land: law and space in nineteenth century New Zealand', *Law and Social Inquiry,* 24(4): 807–52.

Bartholomew, A. and Hunt, A. (1990) 'What's wrong with rights?', *Law and Inequality,* 30(3): 1–58.

Berger, T.R. (1991) *A Long and Terrible Shadow.* Vancouver: Douglas & McIntyre.

Bhandar, B. (2004) 'Anxious reconciliation(s): unsettling foundations and spatializing history', *Environment and Planning A: Society and Space,* 22(6): 831–45.

Blomley, N. (1994a) *Law, Space and the Geographies of Power,* New York: Guilford Press.

Blomley, N. (1994b) 'Mobility, empowerment and the rights revolution', *Political Geography,* 13(5): 407–22.

Blomley, N. (2003) 'From "what?" to "so what?": law and geography in retrospect', in J. Holder and C. Harrison (eds), *Law and Geography.* Oxford: Oxford University Press, pp. 17–33.

Blomley, N. (2004a) 'The boundaries of property: lessons from Beatrix Potter', *Canadian Geographer,* 48(2): 91–100.

Blomley, N. (2004b) 'Un-real estate: proprietary space and public gardening', *Antipode,* 36(4): 614–41.

Blomley, N. (2004c) *Unsettling the City: Urban Land and the Politics of Property.* New York: Routledge.

Blomley, N. and Bakan, J. (1992) 'Spacing out: towards a critical geography of law', *Osgoode Hall Law Journal,* 30(3): 661–90.

Blomley, N., Delaney, D. and Ford, R.T. (eds) (2002) *The Legal Geographies Reader: Law, Power and Space.* Oxford: Blackwell.

Borrows, J. (1997) 'Living between water and rocks: First Nations, environmental planning, and democracy', *University of Toronto Law Journal,* 47: 417–68.

Borrows, J. (1999) 'Sovereignty's alchemy: an analysis of Delgamuukw v. British Columbia', *Osgoode Hall Law Journal,* 37(3): 537.

Brown, M.F. (1996) 'On resisting resistance', *American Anthropologist,* 98(4): 729–35.

Brown, W. (1997) 'Rights and identity in late modernity: revisiting the "Jewish Question"', in A. Sarat and T.R. Kearns (eds), *Identities, Politics and Rights.* Ann Arbor: University of Michigan Press, pp. 85–130.

Brown, W. (2004) '"The most we can hope for. ...": human rights and the politics of fatalism', *South Atlantic Quarterly,* 103(2): 451–63.

Chouinard, V. (1994) 'Geography, law and legal struggles: which ways ahead?' *Progress in Human Geography,* 11(5): 415–40.

Chouinard, V. (2001) 'Legal peripheries: struggles over disabled Canadians' place in law, society and space', *Canadian Geographer,* 45(1): 187–93.

Clark, G.L. (1985) *Judges and the Cities: Interpreting Local Autonomy,* Chicago: Chicago University Press.

Cohen, D. and Hutchinson, A.C. (1990) 'Of persons and property: the politics of legal taxonomy', *Dalhousie Law Journal,* 13(1): 20–54.

Comaroff, J.L. (2001) 'Colonialism, culture and the law: a foreword', *Law and Social Inquiry,* 26(2): 305–14.

Cooper, D. (1998) *Governing out of Order: Space, Law and the Politics of Belonging.* London: Rivers Oram Press.

Cornell, D. (1997) 'Bodily integrity and the right to abortion', in A. Sarat and T.R. Kearns (eds), *Identities, Politics and Rights.* Ann Arbor: University of Michigan Press, pp. 21–83.

Cover, R. (1986) 'Violence and the word', *Yale Law Journal,* 95: 601–29.

Cresswell, T. (2006) 'The right to mobility: the production of mobility in the courtroom'. *Antipode,* 38(4): 735–54.

Daes, E.-I. E. (1999) 'Human rights of indigenous peoples: United Nations Commission on Human Rights', from http://www.unhchr.ch/Huridocda/Huridoca.nsf/TestFrame/154d71ebbbdc126a802567c4003502bf?Opendocument.

Darian-Smith, E. (1999) *Bridging Divides: The Channel Tunnel and English Legal Identity in the New Europe.* Berkeley: University of California Press.

Delaney, D. (1998) *Race, Place and the Law, 1836–1948.* Austin: University of Texas Press.

Delaney, D. (2003) *Law and Nature.* Cambridge: Cambridge University Press.

Delaney, D. (2004) 'Tracing displacements: or evictions in the nomosphere', *Environment and Planning A: Society and Space,* 22(6): 847–60.

Delaney, D., Blomley, N. and Ford, R.T. (2001) 'Where is law?', in N. Blomley, D. Delaney and R.T. Ford (eds), *The Legal Geographies Reader: Law, Power, and Space.* Oxford: Blackwell, pp. xiii–xxii.

Deloria, V. and Lytle, C.M. (1983) *American Indians, American Justice.* Austin: University of Texas Press.

Dimock, W.C. (1998) 'Rethinking space, rethinking rights: literature, law and science', *Yale Journal of Law and Humanities,* 10(2): 487–504.

Dworkin, R.M. (1977) *Taking Rights Seriously.* Cambridge, MA: Harvard University Press.

Edney, M. (1993) 'The patronage of science and the creation of imperial space: the British mapping of India, 1799–1843', *Cartographica,* 30: 61–7.

Engel, D. (1995) 'Law in the domains of everyday life: the construction of community and difference', in A. Sarat and T.R. Kearns (eds), *Law in Everyday Life.* Ann Arbor: University of Michigan Press, pp. 123–70.

Ewick, P. and Silbey, S.S. (1998) *The Commonplace of Law: Stories from Everyday Life.* Chicago: University of Chicago Press.

Ford, R.T. (1999) 'Law's territory (a history of jurisdiction)'. *Michigan Law Review,* 97(4): 843–930.

Forest, B. (2001) 'Mapping democracy: racial identity and the quandary of political representation', *Annals of the Association of American Geographers.* 91(1): 143–66.

Forest, B. (2004) 'The legal (de)construction of geography: race and political community in Supreme Court redistricting decisions', *Social and Cultural Geography,* 5(1): 55–73.

Forman, G. and Kedar, A. (2004) 'From Arab land to "Israel lands": the legal dispossession of the Palestinians displaced by Israel in the wake of 1948', *Environment and Planning A: Society and Space,* 22(6): 809–30.

Frug, G. (1980) 'The city as a legal concept', *Harvard Law Review*, 93(6): 1059–1154.

Goldberg-Hiller, J. and Miner, N. (2003) 'Rights as excess: understanding the politics of special rights', *Law and Social Inquiry*, 28(4): 1075–1120.

Harris, C.D. (2002) *Making Native Space: Colonialism, Resistance, and Reserves in British Columbia*. Vancouver: University of British Columbia Press.

Harris, D.C. (2000) 'Territoriality, Aboriginal rights, and the Heiltsuk spawn-on-kelp fishery', *University of British Columbia Law Review*, 34: 195–238.

Harvey, D. (2000) *Spaces of Hope*, Edinburgh: University of Edinburgh Press.

Hogg, R. (2002) 'Law's other spaces', *Law/Text/Culture*, 6: 29.

Holder, J. and Harrison, C. (eds) (2003) *Law and Geography*. Oxford: Oxford University Press.

Isin, E.F. (2000) 'Introduction: democracy, citizenship and the city', in E.F. Isin (ed.), *Democracy, Citizenship and the Global City*. Cambridge, MA: Blackwell, pp. 1–21.

Jones, P. (1994) *Rights*. Basingstoke: Macmillan.

Kain, R.J.P. and Baigent, E. (1992) *The Cadastral Map in the Service of the State: A History of Property Mapping*. Chicago: University of Chicago Press.

Kedar, A. (2001) 'The legal transformation of ethnic geography: Israeli law and the Palestinian landholder, 1948–1967', *New York University Journal of International Law and Politics*, 33(4): 923–1000.

Klodawsky, F. (2001) 'Recognizing social and economic rights in neo-liberal times: some geographic reflections', *Canadian Geographer*, 45(1): 167–72.

Kosek, J. (2004) 'Deep roots and long shadows: the cultural politics of memory and longing in northern New Mexico', *Environment and Planning A: Society and Space*, 22: 329–54.

Kropotkin, P. (1903/1946) *The State; Its Historic Role*. London: Freedom Press.

Laclau, E. and Mouffe, C. (1985) *Hegemony and Socialist Strategy: Toward a Radical Democratic Politics*. London: Verso.

Luna, G.T. (1998) 'En el nombre de Dios Todo-Poderso: the Treaty of Guadalupe Hidalgo and Narrativos Legales', *Southwestern Journal of Law and Trade in the Americas*, 5: 45–75.

M'Gonigle, M. (2002) 'Between globalism and territoriality: the emergence of an international constitution and the challenge of ecological legitimacy', *Canadian Journal of Law and Jurisprudence*, 25(2): 159–74.

Marshall, A.-M. (2003) 'Injustice frames, legality, and the everyday construction of sexual harassment', *Law and Social Inquiry*, 28(3): 659–90.

Marston, S.A. (2004) 'Space, culture, state: uneven developments in political geography', *Political Geography*, 23: 1–16.

Marx, K. (1844[1975]) 'On the Jewish question', in L. Colletti (ed.), *Early Writings*. Harmondsworth: Penguin.

Maurer, B. (2000) *Recharting the Caribbean: Land, Law, and Citizenship in the British Virgin Islands*. Ann Arbor: University of Michigan Press.

McCann, M.W. (1994) *Rights at Work: Pay Equity Reform and the Politics of Legal Mobilization*. Chicago: University of Chicago Press.

McClure, K.M. (1997) 'Taking liberties in Foucault's triangle: sovereignty, discipline, governmentality and the subject of rights', in A. Sarat and T.R. Kearns (eds), *Identities, Politics and Rights*. Ann Arbor: University of Michigan Press, pp. 149–92.

Merrifield, A. and Swyngedouw, E. (eds) (1996) *The Urbanization of Injustice*. London: Lawrence & Wishart.

Merry, S.E. (2003) 'Rights talk and the experience of law: implementing women's human rights to protection from violence', *Human Rights Quarterly*, 25(2): 343–81.

Milner, N. (1992) 'The intrigues of rights, resistance and accommodation', *Law and Social Inquiry*, 17: 313–33.

Mitchell, D. (2003) *The Right to the City: Social Justice and the Fight for Public Space*. New York: Guilford Press.

Mitchell, T. (2002) *Rule of Experts: Egypt, Techno-politics, and Modernity*. Berkeley: University of California Press.

Nedelsky, J. (1990) 'Law, boundaries and the bounded self', *Representations*, 30: 162–89.

Olwig, K.R. (2002) *Landscape, Nature and the Body Politic: From Britain's Renaissance to America's New World*. Madison: University of Wisconsin Press.

Patton, P. (2000) 'The translation of indigenous land into property: the mere analogy of English jurisprudence', *Parallax*, 6(1): 25–38.

Peake, L. and Ray, B. (2001) 'Racializing the Canadian landscape: whiteness, uneven geographies and social justice', *Canadian Geographer*, 45(1): 180–7.

Platt, R. (2004) *Land Use and Society: Geography, Law, and Public Policy*. Washington: Island Press.

Pratt, G. (2004) *Working Feminism*. Philadelphia: Temple University Press.

Pue, W. (1990) 'Wrestling with law: (geographical) specificity vs. (legal) abstraction', *Urban Geography*, 11(6): 566–85.

Razack, S. (2000) 'Gendered racial violence and spatialized justice: the murder of Pamela George', *Canadian Journal of Law and Society*, 15(2): 91–130.

Roithmayr, D. (2001) 'Left (over) rights', *Law/Text/Culture*, 5: 407.

Ryan, A. (1984) *Property and Political Theory*, Oxford: Blackwell.

Sack, R.D. (1986) *Human Territoriality: Its Theory and History*. Cambridge: Cambridge University Press.

Said, E. (1993) *Culture and Imperialism*. New York: Alfred A. Knopf.

Sarat, A. and Kearns, T.R. (1997a) 'Editorial introduction', in A. Sarat and T.R. Kearns (eds), *Identities, Politics and Rights*. Ann Arbor: University of Michigan Press, pp. 1–20.

Sarat, A. and Kearns, T.R. (eds) (1997b) *Identities, Politics, and Rights*. Ann Arbor: University of Michigan Press.

Sarat, A., Douglas, L. and Umphrey, M.M. (eds) (2003) *The Place of Law*, The Amherst Series in Law, Jurisprudence and Social Theory. Ann Arbor: University of Michigan Press.

Selznick, P. (2003) 'Law and society revisited', *Journal of Law and Society*, 30(2): 177–86.

Semple, E.C. (1918) 'The influences of geographic environment on law, state and society', in A. Kocourek and J.H. Wigmore (eds), *Formative Influences of Legal Development*. Boston: Little, Brown, pp. 215–33.

Shamir, R. (2001) 'Suspended in space: Bedouins under the law of Israel', in N. Blomley, D. Delaney and R.T. Ford (eds), *The Legal Geographies Reader*. Oxford: Blackwell, pp. 134–42.

Staeheli, L.A. and Dowler, L. (2002) 'Introduction', *GeoJournal*, 58: 73–5.

Stychin, C.F. (1998) *A Nation by Rights: National Cultures, Sexual Identity Politics and the Discourse of Rights*. Philadelphia: Temple University Press.

Thompson, E.P. (1975) *Whigs and Hunters: The Origins of the Black Act*. London: Allen Lane.

Tomlins, C. (2001) 'The legal cartography of colonization, the legal polyphony of settlement: English intrusions on the American mainland in the seventeenth century', *Law and Social Inquiry*, 26: 315–72.

Twining, W. (2000) *Globalization and Legal Theory*. London: Butterworths.

Walzer, M. (1984) 'Liberalism and the art of separation', *Political Theory*, 12(3): 315–30.

Weston, B.H. (1992) 'Human rights', in R.P. Claude and B.H. Weston (eds), *Human Rights in the World Community*, Issues and Action 14. Philadelphia: University of Pennsylvania Press.

Whatmore, S. (2002) *Hybrid Cultures: Natures, Cultures, Spaces*. London: Sage.

Whittlesey, D. (1935) 'The impress of central authority upon the landscape', *Annals of the Association of American Geographers*, 25: 85–97.

Williams, P.J. (1991) *The Alchemy of Race and Rights*. Cambridge, MA: Harvard University Press.

Williams, R.A.J. (1990) *The American Indian in Western Legal Thought*. Oxford: Oxford University Press.

Coercion, Territoriality, Legitimacy: The Police and the Modern State

Steve Herbert

INTRODUCTION

Two different places, two similar stories.

The first place: the English Midlands in the mid-1980s. Protests over closures of coal-mining operations led to a protracted labour struggle. One key site of the struggle was Nottinghamshire, where leaders of the National Union of Mineworkers sought to wage a massive strike. Determined to prevail, the British government unleashed a police response of unprecedented size and scope. Thousands of officers mobilized a military-style operation, including 'a centralized command structure, para-military crowd dispersal tactics, and a massive system of roadblocks' (Blomley, 1994: 156). These roadblocks were set up primarily to prevent coal workers from effecting a 'flying picket' strategy, whereby they would move from coalfield to coalfield to buttress protests. The police response was extensive and expensive: 11,312 arrests were made, £192.3 million was spent. In the words of Blomley (1994: 169), 'Britain had never seen policing like this'. Not surprisingly, distress about the tactics emerged from many quarters.

The second place: Los Angeles in March 1992. The streets of the city, particularly those in the neighbourhoods of South Central, exploded in the most extensive civil unrest in American history. More than fifty people died, buildings and other infrastructure sustained millions of dollars of damages. The spark that ignited the unrest was a jury verdict that acquitted four Los Angeles Police

Department officers charged with excessive use of force in the apprehension of Rodney King. The case was brought to public attention by an amateur videotape of the arrest of King for speeding and for resisting capture. King, an African-American, was beaten and kicked nearly seventy times in less than two minutes. The videotape seemed to provide conclusive proof of a pattern of excessive police force long alleged by Los Angeles blacks. When the jury disagreed, violence and destruction ensued.

Two stories of two different places, but each testifies to the unquestioned material and symbolic significance of the police in contemporary society. When the police assert their authority boldly and violently – by marching into the presumably idyllic English countryside, by swatting and kicking a prone African-American – they raise hackles. The coercive capacity of the state is on full display, and the effect is disquieting, even revolting.[1] The legitimacy of the state and its authority hangs in the balance. And whatever restorations of that legitimacy ultimately ensue, they lie susceptible to challenge when the next crisis of order emerges.

This tenuous reality derives from the paradoxical nature of coercive force, and its uneasy relationship to state legitimacy. Because the state exalts itself as a guarantor of public order, it can legitimately use force to restore that order. Yet legitimate authority, by most definitions (see Barker, 1990; Hurd, 1999), is characterized by a *lack* of coercion. Citizens *acquiesce* with a state whose dictums they perceive as legitimate. Coercion is thereby unnecessary.

So, whenever a state exercises coercion, it implicitly recognizes its lack of legitimacy with at least some member(s) of the populace. The challenge for the state, and its police, is to use coercive force under such circumstances that it will not be questioned by large segments of society. This is not an easy challenge. Further, it is a challenge that implicates a number of critical questions about the spatial reach and form of the state's power.

In short, these stories from the English Midlands and from Los Angeles are dramatic reminders of the centrality of the police to the operations and imaginaries of the modern state. Given the significance of the police to the modern state, they deserve the attention of political geographers. Most crucially, as noted, the police stand as the key repository of the state's coercive power, the actual and symbolic backbone of state authority (Bittner, 1974; Klockars, 1985). This coercive power is employed most commonly in pursuit of the reduction of crime and the maintenance of order. Crime and order are ongoing preoccupations of the state, and stand central in electoral politics (Beckett, 1997; Beckett and Sasson, 2004). A robust and seemingly effective police force is highly desired by any government seeking legitimacy and electoral support. Yet the police's coercive power is precisely why they perpetually threaten the state's legitimacy. If this awesome power is deployed in questionable ways, as in the cases described above, public disquiet is likely. As the police intrude into matters of daily life, they raise provocative and continuing questions about the line between public authority and private life. As 'street-level bureaucrats' (Lipsky, 1980; Brown, 1981), the police operate in a regulatory limbo, caught between the formal dictums of the law and the informal dynamics of the situations to which they are summoned. The sensitive use of the resultant discretionary authority can win the support of the population, but abuse of that authority can generate alienation. Police officers thus possess both discretion *and* coercive authority, and thereby raise questions, more than any other agency, about the legitimacy of the state's power. In short, anyone interested in the basal nature of state power and legitimacy learns much by examining the police.

My principal contention in this chapter is that these questions of state power and legitimacy are helpfully magnified under a *geographic* lens (see also Fyfe, 1991). Such a lens, for example, draws attention to the contexts of police action. Police behaviour is not spatially uniform, but is shaped by the geographic milieu where it transpires. Such spatial variance is understandable and defensible; places vary. Yet widely variant police behaviour threatens the ideal of equality, and thus threatens police legitimacy. Context also matters in terms of regulation. Officers' actions are shaped by dictums that vary from station to station, from department to department, from region to region, from country to country. The geographic backdrop of police action thus contributes significantly to the structure of that action, to the mechanisms by which officer's actions are, or are not, regulated. Further, the police themselves actively shape their geographies through the exercise of territorial authority. The power of the police to influence action is intrinsically connected to their capacity to regulate space (Herbert, 1997). An emphasis on territoriality also lays bare the question of the spatial reach of the police. How far into the private lives and spaces of citizens should the police be allowed to probe? How extensive should be the nets generated by surveillance mechanisms and data collection? As these and related questions illustrate, attention to the geographies of policing are essential for a comprehensive understanding of this most critical state institution.

I develop each of these points in more detail in what follows. In the first section, I discuss the police as a unique agency. I focus, in particular, on two critical components of the police – their coercive power and their intrusiveness – and explain how each produces subsidiary effects of enormous consequence. In the second section, I explore the knotted question of the regulation of the police. There are three critical issues here: the extent of police discretion and the attendant spatial variation of officer behaviour; the scale division of police labour and regulation; and the tumultuous politics of crime control, which involve the police in longstanding debates about the proper reach of state authority. Each of these issues is consequential to efforts to control and legitimate the police; each is also intrinsically geographical. In the third section, I take up the issue of legitimation in more detail, through an exploration of the most recent paradigmatic model for structuring the police, community policing. At first blush a seemingly benign and embraceable attempt to restore trust between the police and the citizenry, community policing raises in another guise the key questions of regulation that have historically accompanied the reality of police authority. The fourth section outlines some other issues of current and future interest to political geographers of policing and social control. The fifth section is a conclusion.

THE UNIQUE AND INTRUSIVE POWER OF THE POLICE

Egon Bittner (1974) isolated long ago the central defining characteristic of the police: that they possess the capacity to exercise legitimate coercive force. Here Bittner echoes Weber's famous definition of the state, which also emphasizes legitimate coercion (Weber, 1964). The police are the state's prime agency for the use of coercion, and this deeply influences their symbolic place in society. As van Maanen (1980: 298) puts it, 'The officer at the street level symbolizes the presence of the Leviathan in the everyday lives of the citizenry'. The police, in Manning's words (1977: 4), project the 'awesome power of the state' and personify the abstract authority the state asserts for itself. No other institution in society quite replicates the police in terms of the reach and symbolism of political authority.

Although the police appear to quite rarely exercise force, their capacity to do so possesses enormous implications. One implication, as discussed, is the tension between legitimacy and coercion faced by the state. This tension is largely isolated in the police, which perhaps explains why it is *the* state agency most prone to touching off legitimation crises. This is a tension experienced not just by the state as a whole, or by the police as an organizational component of the state, but by individual officers. All of the ethical issues that attend to using force against another person fall on the shoulders of police officers, who live with the daily reality that they might possibly kill. How officers deal with this reality is central to their overall orientation to the work (Muir, 1977).

Certainly, critical aspects of the police's subculture[2] are tied to their potential use of force. Officers recognize that they will be called by the public to enter situations where violence might be necessary to achieve some resolution. A potent impulse within the subculture emphasizes this danger and the masculinist prowess to embrace it (Skolnick, 1966; Reiner, 1978; Holdaway, 1983; Hunt, 1984; Fielding, 1994; Herbert, 2001a) a valued officer is one who can calm the chaos of confrontational situations and emerge victorious. The inherent dangerousness of the work explains, further, the subculture's strong focus upon safety, which structures many of the territorial tactics officers employ (Herbert, 1997). This sense of imminent danger is also connected to the common valorization by officers of their authority. Officers expect to reorder potentially threatening situations through assertions of their hegemony. Not surprisingly, this often takes territorial forms. Officers create order by asserting their capacity to dictate action through controlling space (Holdaway, 1983; Herbert, 1997). Indeed, for many officers, this becomes a deeply personalized issue. As Rubinstein (1973: 166) puts it, 'For the patrolman, the street is everything: if he loses that, he has surrendered his reason for being what he is'.

The police's capacity to exercise coercive force also explains much about the tasks they assume. Notably, it means that the police are allocated the responsibility of apprehending criminal wrongdoers, who sometimes resist capture. The connection between the police and the stanching of criminal activity, in turn, implicates them in the fervid politics of crime control (about which more below). It also helps explain why the military has historically been used as an example for police departments to follow; officers are at 'war' with criminals, and structure their operations to help win that war (Lea and Young, 1984).

The coercive capacity of the police is tied, as well, to their second key characteristic – their intrusiveness. The police, Bittner (1974) persuasively argued, are summoned to a wide array of situations because they can, at the limit, force some resolution. This fact, coupled with the onset of the emergency dispatch system, means that the police are brought into a broad range of situations, and expected to restore them to order. Further, the police's desire to reduce crime leads them to emphasize the importance of random patrol. This allegedly accomplishes two purposes: it enables the police to respond more rapidly to calls for service, and it works as a visible deterrent to potential criminals. Although the police largely fail in these two goals (Kelling, 1974; Sherman, 1974; Moore, 1990), they do make themselves available and visible like no other agency of the state. They thus enter into people's lives – and spaces – with the full force and costumed majesty of the state. They intrude, both physically and symbolically, in a way not replicated by other state actors (Loader, 1997), and typically act to assert territorial hegemony over the spaces they enter (Werthman and Piliavin, 1967; Rubinstein, 1973; Holdaway, 1983; Herbert, 1997). In so doing, they literally stand at the threshold of one of the more significant spatial divides in liberal societies – between public and private space. Excessive police intrusiveness thus threatens their legitimacy as much as the excessive use of force, because it symbolizes an overreaching of state authority into sanctified private territory.

It is difficult to overstate the historic significance of the creation of the police as a state agency able to coerce and intrude in this fashion. Clearly, the police are a critical component of what Giddens (1985) refers to as the 'internal pacification' apparatus of the modern state, the overt attempt to extend the spatial reach of state power. This territorial hegemony is a base characteristic of the modern state, connected to the importance of state boundaries and the rise of various mechanisms to assert the state as the hegemonic centre of power. As Mann (1988) notes, no other social institution is spatially bounded in the same way, and thus no other institution can make the same claim to oversight of such activities as census-taking, taxation and conscription. The modern state is unique in its institutionalization and boundedness, and differs remarkably from the less centralized and poorly circumscribed political structures that preceded it.[3] The modern state thus makes an explicit connection between its power and its territorial hegemony, and thereby presumes to create order throughout civil society. State power is expected to quell internal disturbances when they arise.[4]

Such disturbances, according to many historians (Richardson,1974; Schneider, 1980; Monkkonen, 1981), help explain the genesis of the police. Officers were charged with the task of preserving boundaries between the thickly mixed spaces of the industrial city. Congestion, immigration and the onset of industrial capitalism generated increased insecurity for the wealth-accruing bourgeoisie. This insecurity, in turn, generated support for an ostensibly neutral coercive force to reduce disorder, and to ensure the protection of property and the emerging capitalist market (Cohen, 1979; Spitzer, 1979). It was a landmark achievement, characterized well by Silver (1967: 13): 'The growth of the police represented the penetration and continual presence of central political authority throughout daily life; the police extended through the periphery both as agent of legitimate coercion and as a personification of the values of the centre'. This assertion of state power, and its insertion deeper into civil society, was not received unquestioningly by the population. As Silver (1967: 8) explains, 'All of these characteristics struck contemporary observers as remarkable'. Indeed, early officers in the United States were often reluctant to wear uniforms, for fear of attracting abuse from the populace (Fogelson, 1977). The potentially unstable relation between the police's authority and the state's legitimacy was recognized at the outset.

In short, the police's coercive power and intrusiveness establish them as a unique agency, and are critical to their legitimacy. Yet they also potentially threaten their legitimacy, if the use of force and the extent of intrusiveness exceed certain bounds. Importantly, these are deeply spatial issues. Where is force exercised? Is there spatial variance in its exercise? What conditions its use? Where do the police roam? What lines do they cross? At what level are they controlled? How are police activities folded into wider political discourses about space and its control? Through their exercise of territorial authority, whose interests do the police serve? Each of these questions attends to the difficult challenge of maintaining police legitimacy, and each is explored in the next section.

THE GEOGRAPHIES OF POLICE POWER AND REGULATION

The police are thus unusual for the geographic extent of their reach and their capacity to compel citizen compliance. Because they enter such a wide variety of spaces, and encounter such a wide variety of social situations, they are difficult to regulate. On the one hand, the police are a legal creation, and are ostensibly regulated by the legal code. Such regulation should help ensure that the police do not abuse the awesome power they possess. Yet, on the other hand, the law is incapable of defining precisely the circumstances under which its mandates should be exercised. The police must thus improvise solutions to unique situations. The law may help guide such improvisations, but cannot determine them. Further, the police, working at the extension of the state's reach, are difficult to oversee.

The police, as a consequence, stand in a middle position between law and everyday socio-spatial life. Formal dictates possess some significance, but yet cannot regulate all that officers do. Officers possess much discretion to choose from a range of possible courses of action. These decisions have geographic predicates and geographic consequences. They pose significant dilemmas for the legitimacy of the state. I probe these dilemmas in this section. First, I discuss the extent of police discretion, its geographic situatedness, and the role of police culture in determining its use. Second, I review dilemmas concerning the regulation of the police and their discretion, focusing particularly on the nettlesome question of *where* to locate the authority to oversee officers. Third, I tie these questions of oversight to classic debates over the proper state response to criminality. Each of these discussions makes plain the importance and symbolic significance of the police to wider state challenges to ensure legitimacy.

Discretion, Geography, Subculture

The law cannot completely regulate the police (Banton, 1964; Skolnick, 1966; Bittner, 1967). The complicated nature of everyday socio-spatial life is not reflected well in the abstract language of law (Pue, 1990; Blomley, 1994). The police are thus confronted with a messy reality that their formal dictates do not anticipate well. In addition, the capillary power the police assert means that they can evade these formal dictates with some ease. In the insightful words of Skolnick (1966: 14), 'Police work constitutes the most secluded part of an already secluded system and therefore offers the greatest opportunity for arbitrary behaviour. As invokers of the criminal law, the police frequently act in practice as its chief interpreter. Thus, they are necessarily called upon to test the limits of their legal authority'.

Skolnick thus accurately captures the extent and significance of police discretion. But he overstates his case when he describes discretion as 'arbitrary'. Even if formal rules cannot determine their actions, officers behave in patterned ways. Two factors appear particularly critical in so shaping what officers do: the geographical context of police action, and the potency of the police subculture.

Like other social actors (Lofland, 1973), the police pay close attention to the spatial situatedness of the situations to which they are summoned. Several factors are potentially significant to a police officer: past experience with a locale, particularly violence; visible cues of 'disorder', such as graffiti and abandoned cars or buildings; the presence of loiterers, most notably teenagers, panhandlers, prostitutes, and those engaged in the consumption or sale of drugs and alcohol; the extent of owner-occupied housing; the reputation of a space as a host site for criminality; the racial and class composition of the population (Bittner, 1967; Sacks, 1972; Rubinstein, 1973; Sherman, 1974; van Maanen, 1974; Magahan, 1984; Reuss-Ianni, 1984; Punch, 1985; Smith, 1986; Alpert and Dunham, 1988; Fyfe, 1992; Keith, 1993). Such assessments of space inform the way officers 'read' the situation before them, and the strategies they employ to handle those situations. Police discretion, in other words, is fundamentally geographical; it varies spatially based upon the presumed characteristics of place.

Of course, officers' perceptions of a given space are shaped by more than just the material conditions that exist there. These perceptions are moulded by the culture that officers construct. Several observers of the police note the importance of this culture for shaping officers' understandings of themselves and their work (Skolnick, 1966; Black, 1980; Reuss-Ianni, 1984; Kappeler et al., 1994; Chan, 1997). These more informally created structures perform the work of shaping officer behaviour that formal regulations cannot. Considerations of possible criminality, of possible danger, of the moral taint of society's less advantaged all shape the representations of space the police employ, as well as the territorial tactics they deploy (Magahan, 1984; Herbert, 1997). These tactics become increasingly intensified in neighbourhoods where police legitimacy is at its lowest. As places fall in their moral assessment by officers, they increase in the extent of police attention and suspiciousness they receive. Police use these assessments of place to justify increasingly intrusive and brusque tactics (Bittner, 1967; Keith, 1991; Skolnick and Fyfe, 1993), which causes police legitimacy to deteriorate further (Werthman and Piliavin, 1967). This helps explain why assessments of the police vary so significantly from place to place, most notably between places dominated by white residents and those dominated by blacks (Bayley and Mendelsohn, 1968; Jacob, 1971; Hagan and Albonetti, 1982; Kinsey et al., 1986; Weitzer, 1999).

Indeed, it is the police's collective efforts in constructing their 'symbolic assailants' (Skolnick, 1966) that gives many commentators greatest pause. The police always act against some people's interests (Manning, 1977), and sometimes use force against fellow citizens. Such coercive action is easier to justify if it is employed against those who are defined as deviant, as beyond the pale of normal society. In Los Angeles, for example, officers frequently invoke the discourse of evil to explain the existence of the 'bad guys' they are forced to confront (Herbert, 1996). By so castigating their alleged enemies, the police implicitly justify any uses of violence against them (Keith, 1991); lawlessness presumably understands no other constraint. This taint of evil is often place-based; certain areas are 'dirty' and thus their inhabitants deserve the greatest suspiciousness and surveillance (Herbert, 1996). The police thus perpetuate an ongoing cycle: suspiciousness breeds greater intrusiveness, which breeds greater resident resentment and resistance, which breeds greater police suspiciousness, and on and on.[5]

In sum, the reality of police discretion persists as a nettlesome dilemma. Its use and its geographic variance are understandable, and arguably unavoidable (Bittner, 1967). Yet this very geographic variance violates hegemonic understandings in liberal societies of the importance of equal treatment under the law. This variance also reveals the potency of the informal strictures of the police culture, whose power makes obvious the inability of formal regulations to control police action completely. Yet regulators continue to try to exert dominion over officers. It is to a review of such efforts, and their geographic variance, that I now turn.

The Nature and Scale of Police Oversight

The police occupy a position between the law and the community, shuttling between the formal, invariant proscriptions of the former and the informal, highly variant realities of the other (Banton, 1964; Reiss, 1971; Brown, 1981; Grimshaw and Jefferson, 1987). They thus face the challenge of trying to strike a sensible middle position, as neither too distant nor too close. A more formal, aloof approach helps guarantee greater uniformity and fairness, yet potentially alienates citizens who desire a more personal connection. However, a police force that is too chummy may become too partisan, too susceptible to misuse by certain segments of the population.

This tension is illustrated well by the effort, beginning in the early 1900s, to 'professionalize' police forces in the United States. Most urban police departments at the time were home to significant amounts of corruption and excess brutality. Police forces were loosely regulated, thus officers were able to abuse their authority with little fear of sanction. Advocates of professionalization sought to rectify this by creating more uniform and strictly enforced bureaucratic mechanisms to hold officers accountable. Officers were to follow abstract rules of procedure, and not be influenced by the dynamics on the streets they patrolled. This move to professionalize took much time, but ultimately worked to reduce the corruption that was previously endemic (Fogelson, 1977; Walker, 1977). Yet this more aloof police force arguably increased the degree of alienation between officers and the communities they patrolled, and was cited as a factor in the 'police riots' that erupted in many US cities in the 1960s (Moore, 1990).

There is no magical solution to this dilemma. Indeed, what is notable about efforts to regulate the police is their variance. Rules that seek to shape officers' actions vary widely, at a variety of scales. Foster (1989), for instance, showed how two London police stations, just miles apart and facing similar urban realities, displayed two distinct styles of policing, one significantly more confrontational than the other. Such variance reflects the power of local precinct commanders, who can strongly determine enforcement priorities and practices. Thus, within a given police department, officers can define problems and possible solutions quite differently. Part of this is attributable to an officer's position within the bureaucracy. This determines the focus of an officer's concern and, often, the range of available tactics (Reuss-Ianni, 1984; Herbert, 1997).

Even greater variance exists from department to department. Not surprisingly, the size of the department is a critical variable in determining the width of the latitude given officers and the closeness of their connection to the community. Brown (1981), for instance, analyzed three Southern California police departments of varying size, and found that in the largest, Los Angeles, officers operated with the least fear of sanction. In a department that large, officers knew that supervision was a more challenging task, and acted with a corresponding degree of freedom. Similarly, in her comparison between an urban and a rural police force, Cain (1971) found a greater degree of familiarity with the community in the latter, and a greater degree of internal cohesion among officers.

The national context matters as well. Nations differ notably in the extent to which they seek to centralize authority over the police (Bayley, 1985). In some cases, the nation-state seeks to govern the police from the centre and to enforce a standard that is ostensibly uniform. In other cases, control of the police is devolved to the local level. This enables local units to craft enforcement strategies that are sensitive to the idiosyncracies of place. It also prevents the concentration of police power, and thus reduces the possibility of centralized authoritarianism. In his comparison of policing in the United States and Japan, Bayley (1976) argued that attitudes toward the state largely determined the regulatory structures of policing. In the United States, suspicions of an overly strong state motivated the push toward local power, and created an ever-present vigilance against an overly intrusive police presence. By contrast, in Japan, the police were much more deeply enmeshed in everyday life. As state actors, they were not seen as distinct forces about whom one must be perpetually fearful. Rather, the public viewed the police as expressions of a shared culture, and did not question their ubiquitous presence.

Even if it remains difficult to regulate police officers, whose status as street-level bureaucrats means that they are always improvising solutions to the myriad problems they encounter, still efforts to do so proceed apace. Clearly, the police's coercive force mandates that the state impose some set of schemes to regulate its use; otherwise, the spectre of a police state emerges. Yet the difficulty of this challenge helps explain the variety of means to address it. This variety of patterns of regulation is every bit as geographically conditioned as are the conditions officers confront; cultures of control vary with place.

These cultures of control are also deeply conditioned by the political potency of the issues of crime and disorder. The latitude granted police officers is very much connected to the politics of crime control, as the next section makes plain.

Crime, Politics and Policing

The police's coercive role means that they possess the responsibility to capture those suspected of criminality. The police are thus the 'front line' in the effort to reduce crime, the state agency responsible for initiating the criminal process through apprehension and arrest. Over time, this emphasis on crime reduction has emerged as the primary mission of the police. Justifications to the public for police resources, for instance, are largely couched in terms of the alleged possibility that they could help lower the crime rate (Fogelson, 1977; Walker, 1977).

This emphasis on crime fighting places the police at the centre of the increasingly potent politics of crime. Indeed, crime now possesses a political importance that it lacked a generation ago. Crime is now often constructed as a 'moral panic' (Cohen, 1972). Street muggings in England (Hall, 1978) and the emergence of crack cocaine in the United States (Chiricos, 1998) are examples of crime problems that drew much political attention. As Beckett (1997) shows, this political attention amplifies the crime problem, and legitimates a 'get tough' approach that increases the flow of resources toward the criminal justice system.

Such an approach helps legitimate the construction of the police as analogous to the military. As such, the police are understood as engaged in 'wars' against crime. This reinforces an aggressive cant to policing, as well as the infusion of high-tech infrastructure that is sometimes literally acquired from the armed forces. In the United States, for instance, the contemporary period witnessed the massive growth of Special Weapons and Tactics (SWAT) teams, even in police departments of minuscule size (Kraska and Kappeler, 1997). These teams are often equipped with sophisticated and muscular apparatuses, such as tanks, battering rams and helicopters.

Yet this 'get tough' approach raises provocative issues with implications for the spatial behaviour and reach of the police. One way of capturing these issues is with the distinction that Packer (1968) draws between two models for criminal justice: the due process model and the crime control model. The former draws its philosophical orientation from liberalism, and its stress on the protection provided to individual citizens by rights. In the due process model, the burden of proof for sustaining a criminal conviction falls to the state. Further, the accused citizen is provided any number of means to ensure a proper defence against the state, primarily through procedures mobilized at criminal trials. In the due process approach, considerable suspicion is cast toward the state, which must justify its action at every stage of the criminal process, including the initial apprehension by the police. If the state violates any of the procedures that regulates it, its ability to secure a conviction is compromised. In this fashion, abuses of discretionary power are presumably minimized.

In the crime control model, by contrast, the state is trusted rather more, and thus given wide latitude to exercise discretion. The police, for example, are presumed to act only with founded suspicion, and should be allowed the freedom to intrude as they best see fit. The crime control model emphasizes the swiftness and surety of punishment, to ensure the full deterrent impact of the criminal sanction. A robust police presence in society is thereby valued, because it stands as the best guarantor of an ordered society.

One's view of the possible spatial reach of the police differs with one's allegiance to one or the other of these two models. The due process model seeks to keep the state at arm's length, in part by constructing and reinforcing the distinction between public and private. The police are expected to respect this boundary (Stinchcombe, 1963; Fyfe, 1992), to engage in only as much surveillance as absolutely necessary, and to minimize the exercise of discretion. The police should be more *reactive*; they should wait until the public summons them, and then follow proper procedure in choosing how to act. The crime control model endorses a more intrusive, surveillant and uninhibited police force, one that can nip criminal activity in the bud. This force is more *proactive*, more vigilant in assessing the possibilities for emergent criminality and acting boldly to stanch it.

Again, the metaphor of distance emerges in evaluating strategies for regulating the police: how 'close' should the police be? Yet this is not merely a metaphoric discussion, but one that affects the material geographies of the state's presence. A police force that is encouraged to be proactive, to fight wars against crime with militaristic aggressiveness, to survey its subject population with a focused eye, represents a territorially invasive state. And this invasiveness is not spatially uniform, but concentrated in neighbourhoods of disadvantage, which are also disproportionately peopled by minorities (Baldwin and Kinsey, 1982). This may explain, for example, why African-Americans frequently express greater antipathy toward the police than do whites (Weitzer, 1999) and claim, in particular, that the police engage in 'profiling' of them for possible criminal activity (Harris, 1997; Kennedy, 1997).

To be sure, such profiling does go on, at least in terms of isolating the 'hot spots' around which crime often congregates (Sherman et al., 1989). Increasingly, police departments in the United States and elsewhere use sophisticated geographic information system software to map the type and prevalence of crime. Once gathered, this data is often used to spur 'problem-solving' police operations (Goldstein, 1990; Sparrow et al., 1990) targeted at locations that host repeated criminality, such as the sales of drugs. The use of GIS was touted most loudly in New York City, where police precinct commanders are now held accountable if crime statistics show an unbroken spatial pattern (Silverman, 1999). Police attention is now unusually well focused in its geographic orientation, and this focus is loudly legitimated as a critical component in the reduction of crime. Yet this attention is concentrated in neighbourhoods of disadvantage, whose residents are often resentful of the intrusiveness that results.

So, the politics of police legitimacy emerge again with great power. The symbolically significant issue of crime, and the police's role in fighting it, means that a charged politics emerges around the everyday spatial tactics of officers. Because due process and crime control considerations both possess merit in liberal societies with contested electoral politics, this debate will always lack an easy resolution. The geopolitics of the police thus promise to continue to affect state efforts to achieve legitimacy, efforts that will likely never attain any point of longstanding equilibrium.

THE DILEMMA OF LEGITIMACY: THE CASE OF COMMUNITY POLICING

That these intractable issues of governance will likely persist is evident when one closely examines the now hegemonic model for orienting police departments: community policing. This reform movement developed as a reaction against the apparent failures of the professional model, most notably its tendency to alienate the citizenry. Any police success in isolating and apprehending criminal suspects relies heavily on active assistance from the public (Sherman, 1974; Moore, 1990). Community policing was developed to repair the damage to police/citizen relations caused by the professional movement. Although it varies widely from place to place, community policing emphasizes overt efforts by the police to improve relations with citizens. These include patrolling by means other than a car – by foot, bicycle, horse – to decrease the distance between cop and resident; creating police units with an overt emphasis on community outreach; opening police substations in neighbourhoods, so that community members have ready access to an officer; and supporting the creation of neighbourhood watches and other forms of anti-crime citizen activity. The resultant improved police/community relations, ideally, will work to both reduce crime and increase the degree of citizen efficacy. Community members can emerge as significant political actors in their own right, working closely with each other and with the police to improve the quality of life of their neighbourhoods.

These goals seem largely unobjectionable. Crime and disorder are legitimate concerns for many urban residents, cooperative relations between state and society are presumably better than the alternative. Why not a more active citizenry, a more responsive state?

In practice, however, community policing merely replicates the inherent tensions that attend to the creation and legitimation of a coercive police force. For starters, the police largely resist the notion that the public should possess meaningful oversight of their activities (Sadd and Grinc, 1994; Lyons, 1999; Reed, 1999). Instead, they expect deference to their professional judgement, and preserve the right to exercise their discretionary authority as they see fit. This persistence of the legacy of professionalism, and its exaltation of public subservience, exemplifies yet again the challenge of exerting oversight of the police's actions at the peripheries of state power.

The preferred role for the public, in the police's eyes, is to act as their 'eyes and ears' (Saunders, 1999). The police understand that their uniformed presence puts wrongdoers on their best behaviour. They thus argue that they need citizens to report instances of criminality that officers will likely never see. Yet this role for the public simply makes them extensions of the police's net of surveillance, not active citizens with the capacity to regulate state action. In this way, community policing works to extend the spatial reach of the state's intrusiveness

far more than it succeeds in reviving localized political activity.

Further, to the extent that citizen groups are able to influence police action, this capacity is not uniform. Instead, those groups that are well organized, who understand the operations of local government, and who share the police's basic orientation toward crime are more likely to compel the police to accede to their requests (Lyons, 1999; Reed, 1999). By contrast, poorer communities are less well poised to engage the police productively (Grinc, 1994), and largely fail to broaden the discourse about crime as they would like – to include discussions of employment and social service provision (Miller, 2001). This reality suggests, yet again, that the police's discretionary authority is not geographically invariant, but conditioned by the socio-economic realities of given neighbourhoods.

Finally, it is striking that the era of community policing also witnessed two other notable developments: the hegemony of the theory of 'broken windows', and the rise of militarized units. Each of these developments suggests that the 'soft' approach of community policing is something less than hegemonic.

The popularity of the broken windows logic is impressive. The theory, first articulated by James Wilson and George Kelling (1982; see also Kelling and Coles, 1996), suggests that urban neighbourhoods must fix visible signs of 'disorder', lest they invite increased criminal activity. A broken window symbolizes a neighbourhood unable to exert informal social control, and sends a signal of opportunity to a would-be offender. Increased criminality, in turns, fuels a further weakening of social control, and a spiral of decay ensues. Yet it is not deteriorating buildings that concern Wilson and Kelling, but rather people. And not just any people, but 'disreputable or obstreperous or unpredictable people: panhandlers, drunks, addicts, rowdy teenagers, prostitutes, loiterers, the mentally disturbed'. When neighbourhood residents allows these to persist on their streets, they allegedly initiate the cycle of decay that results in increased crime.

Here, the police enter the picture. Although the essay is ostensibly about the essential need for informal social control, the authors end up endorsing a robust role for the formal apparatus represented by the police. Wilson and Kelling urge a reorientation of the police away from a reactive model that focuses largely on apprehending felony offenders after the fact, and toward a proactive model that emphasizes the need to make arrests for misdemeanour offences such as loitering and panhandling. These arrests remove the 'broken windows' and thus prevent the possibility of further criminality. Though roundly, and rightly, critiqued from a range of academic quarters (Sampson and Raudenbush, 1999; Harcourt, 2001;

Mitchell, 2001; Taylor, 2001; Herbert and Brown, 2006), the broken windows logic has been used to legitimate police crackdowns on street-level activity. The most notable instance of this is New York City, whose 'zero tolerance' approach was linked to reductions in crime to great political effect (Bratton, 1998; Silverman, 1999).[6] That such a robust role for the police has been endorsed in an era of community policing – and, in fact, is often conflated with community policing (Herbert, 2001b) – is remarkable. The police emerge, again, as the indispensable mechanism for reducing crime, an agency whose discretionary authority should extend to a close monitoring of often victimless street-level activity.

This robust role is celebrated even more grandly by the increased number and use of SWAT teams and their military-style equipment. Although the popularity of these units is perhaps a paradoxical reality in an era of community policing (Kraska, 2001), it illustrates the continued potency of the crime control rhetoric and the symbolic power of the war imagery. Even if community policing does not necessarily represent a support of due process rights, it does tout the ability of the citizenry to influence the state's priorities and practices. Yet the war metaphor works to divide the population into two groups: predators and prey. In mobilizing to oppose the former, the police implicitly construct the latter as passive potential victims, not active citizens. As the well-armed, expert professional crime fighters, the police need not bother with excessive citizen input, unless it helps them to isolate criminal threats. Tough crime control action trumps democratic process.

In short, community policing's promise to reorient police practices and reinvigorate police/citizen interactions lies unrealized. The story of this lack of realization illustrates all of the persistent problems with constructing and regulating a legitimate agency of coercive force: the extent and geographic variation of police discretionary authority; the spatial extent of the police's surveillance and enforcement capacity; the balance between robust crime control and the need for due process and other checks on state action. That these problems persist testifies to the intractable nature of the dilemmas that a coercive agency poses to the modern state.

FUTURE ISSUES IN THE GEOGRAPHY OF POLICING

There is no reason to believe these dilemmas will yield to clear resolution. Rather, they will persist and continue to roil political dynamics around crime, order, and the legitimacy of coercive force. The particular form of these dynamics will likely be influenced by two emerging, and related, trends: neo-liberalism and privatization.

The term neo-liberalism is used to describe an ascendant regime of economic and political relations. The chief characteristic of neo-liberalism is the lionization of the economic market. Indeed, the market is meant to be society's ur-metaphor. As Brenner and Theodore (2002: 3) put it, neo-liberalist policies seek to 'extend market discipline, competition, and commodification throughout all sectors of society'. Such policies pursue several shifts in state policy, including: reducing state control of the economy, to enable a less fettered market; weakening trade union protections; reducing state provisions for social welfare, either through their elimination or by turning them over to private operations; and reinforcing the principles of free trade (Peck and Tickell, 1994; Gough, 2002; Jessop, 2002). The neo-liberal state, in other words, withdraws from many of the obligations it embraced in the Fordist period, most notably those that protect workers from excessive exploitation and that provide assistance to the economically disenfranchised (Lipietz, 1992); deregulation and privatization replace welfare provision as central prongs of state policy (Larner, 2000; Peck and Tickell, 2002).

This celebration of the parochial and private is not inconsequential for policing. Indeed, one can read community policing as another instance of a neo-liberal attempt at state offloading, an effort to make urban neighbourhoods assume greater responsibility for their own security (Garland, 1996; O'Malley and Palmer, 1996). The neo-liberal era has also witnessed a massive growth in the private policing apparatus (Shearing, 1990; Johnston, 1992), as property owners and privatized suburban developments seek a level of protection unavailable from public police agencies (Davis, 1990; McKenzie, 1994; Blakely and Snyder, 1997; Low, 2003). Like other preserves of the state, social control and coercive force appear to fall prey to forces of devolution.

Yet it would be a mistake to overstate these trends, or to view them as solutions to the tensions surrounding the legitimate exercise of state coercion. The significance of crime control for the contemporary state's legitimacy means that the formal social control apparatus is unlikely to fade (Herbert, 1999). Indeed, the United States increased its rate of incarceration by 400 per cent – it now leads the world – during the ascent of neo-liberalism. The private security industry may be large and growing, but its power is minuscule compared to the awesome state capacity to arrest, convict, incarcerate and execute. And, as we saw above in our discussion of community policing, state agents of coercion remain reluctant to surrender their authority to communities and the citizens who compose them.

CONCLUSION

So, the dilemmas that attend to the state's exercise of legitimate coercive authority will persist, and will deeply implicate the politics of the spatial deployment of the police's power. Even though they are essential to the modern state, and are deeply embedded in the public imagination, the police nevertheless can all too quickly liquify the state's legitimacy through a seeming abuse of their authority, through an overextension of their reach. Police authority always walks a very fine line between formal regulation and everyday reality, between the majesty and seemingly uniform sanctity of the law and the art of street-level improvisation. Its exercise must necessarily vary across space, but excessive variation stands open to potential challenge for violating the promise of equality contained within legitimations of the law. Police activity also sits uncomfortably between different ideals for the crime control apparatus. Endorsements of robust, even militaristic policing to curb crime suggest that officers should exercise their discretion freely and intrude deeply. Yet this prospect raises for others the spectre of an authoritarian, unfettered state, and generates calls for greater protection of due process rights and other means by which the citizenry can check police power. Whether, how, and how far the police can intrude into everyday socio-spatial life will remain as irresolvable questions for the modern state.

These questions are extremely relevant to geographers of state power; they throw into open relief the central tensions of the modern state project. Our capacity to probe and illustrate the spatial dynamics of the modern state – its abilities to survey, penetrate, control, apprehend, punish – is best enabled by taking the police and their geographies seriously.

NOTES

1 The connection in Los Angeles between civil unrest and a police encounter with a citizen is not unusual. Each of a series of urban uprisings in US cities in the 1960s was sparked by a police/citizen interaction (United States National Advisory Commission on Civil Disorders, 1968).

2 To discuss a police subculture is not necessarily to imply that it is a coherent, unified entity. To the contrary, recent studies of police culture make clear that is fragmented, and varied in the extent to which it captures individual officers (see Herbert, 1998; Paoline, 2004). That said, students of the police consistently catalogue certain common elements within the informal world that officers construct. I argue here that many of these are tied directly to the police's coercive capacity.

3 The ideal of a centralized state emerged in Western Europe, and is most well developed in the context of Western capitalist countries. Yet the notion that a state should be able to pacify its population is frequently used to assess governments in other locations. As of this writing, the daily chaos in Iraq is invoked as the central indicator of an incompetent state.

4 This is not to suggest that the state assumed responsibility for all forms of policing. Various private institutions for policing persist, as do various forms of informal social control. Indeed, in this era of neoliberalism, the state often seeks to offload responsibility for policing to these private and parochial mechanisms, although without ever surrendering significant hegemony over coercive force and its mobilization against crime and disorder.

5 The perception that certain places are 'dirty' and thus deserving of police attention is one that is held not just by the police but by large segments of the population (Beckett, 1997). This pervasive sense that criminality is tied to certain places thus helps the police obviate the potential tension between targeted policing and the ideal of equal treatment under the law.

6 The connection between stanching disorder and the reduction of crime does not hold up under empirical scrutiny (Sampson and Raudenbush, 1999; Harcourt, 2001; Taylor, 2001). Other explanations for New York City's reduction in crime are explored by Harcourt (2001) and Greene (1999).

REFERENCES

Alpert, G. and Dunham, R. (1988) *Policing Multi-ethnic Neighborhoods.* New York: Greenwood Press.

Baldwin, R. and Kinsey, R. (1982) *Police Powers and Politics.* London: Quartet Books.

Banton, M. (1964) *The Policeman in the Community.* London: Tavistock.

Barker, R. (1990) *Political Legitimacy and the State.* Oxford: Clarendon Press.

Bayley, D. (1976) *Forces of Order: Police Behavior in Japan and the United States.* Berkeley: University of California Press.

Bayley, D. (1985) *Patterns of Policing: A Comparative International Analysis.* New Brunswick, NJ: Rutgers University Press.

Bayley, D. and Mendelsohn, H. (1968) *Minorities and the Police.* New York: Free Press.

Beckett, K. (1997) *Making Crime Pay: Law and Order in Contemporary American Politics.* New York: Oxford University Press.

Beckett, K. and Sasson, T. (2004) *The Politics of Injustice: Crime and Punishment in America.* Thousand Oaks, CA: Sage.

Bittner, E. (1967) 'The police on skid row: a study of peace-keeping', *American Sociological Review,* 32(4): 699–715.

Bittner, E. (1974) *The Functions of Police in Modern Society.* New York: Jason Aronson.

Black, D. (1980) *The Manners and Customs of the Police*. New York: Academic Press.

Blakely, E. and Snyder, M. (1997) *Fortress America: Gated Communities in the United States*. Washington, DC: Brookings Institution Press.

Blomley, N. (1994) *Law, Space and the Geographies of Power*. New York: Guilford Press.

Bratton, W. (1998) *Turnaround: How America's Top Cop Reversed the Crime Epidemic*. New York: Random House.

Brenner, N. and Theodore, N. (2002) 'Cities and the geographies of "actually existing neoliberalism"', in N. Brenner and N. Theodore (eds), *Spaces of Neoliberalism: Urban Restructuring in North America and Western Europe*. Oxford: Blackwell, pp. 2–32.

Brown, M. (1981) *Working the Street: Police Discretion and the Dilemmas of Reform*. New York: Russell Sage Foundation.

Cain, M. (1971) 'On the beat: interactions and relations in rural and urban police forces', in S.B. Cohen (ed.), *Images of Deviance*. Harmondsworth: Penguin, pp. 62–98.

Chan, J. (1997) *Changing police culture: Policing in a Multi-Cultural Society*. Cambridge: Cambridge University Press.

Chiricos, T. (1998) 'The media, moral panics and the politics of crime control', in G. Cole and M. Gertz (eds), *The Criminal Justice System: Politics and Policies*. Belmont, CA: Wadsworth, pp. 58–75.

Cohen, P. (1979) 'Policing the working-class city', in R. Fine (ed.), *Capitalism and the Rule of Law*. London: Hutchinson, pp. 118–36.

Cohen, S.B. (1972) *Folk Devils and Moral Panics*. London: MacGibbon & Kee.

Davis, M. (1990) *City of Quartz: Excavating the Future in Los Angeles*. London: Verso.

Fielding, N. (1994) 'Cop canteen culture', in T. Newburn and E. Stanko (eds), *Just Boys Doing Business? Men, Masculinities and Crime*. London and New York: Routledge, pp. 163–84.

Fogelson, R. (1977) *Big-City Police*. Cambridge, MA: Harvard University Press.

Foster, J. (1989) 'Two stations: an ethnography study of policing in the inner city', in D. Downes (ed.), *Crime and the City*. London: Macmillan, pp. 128–53.

Fyfe, N. (1991) 'The police, space and society: the geography of policing', *Progress in Human Geography*, 15(3): 249–67.

Fyfe, N. (1992) 'Space, time and policing: toward a contextual understanding of police work', *Environment and Planning D: Society and Space*, 10(3): 469–81.

Garland, D. (1996) 'The limits of the sovereign state: strategies of crime control in contemporary society', *British Journal of Criminology*, 26(3): 445–71.

Giddens, A. (1985) *The Nation-State and Violence*. Berkeley and Los Angeles: University of California Press.

Goldstein, H. (1990) *Problem-Solving Policing*. New York: McGraw-Hill.

Gough, J. (2002) 'Neoliberalism and socialization in the contemporary city: opposites, complements and instabilities', in N. Brenner and N. Theodore (eds), *Spaces of Neoliberalism: Urban Restructuring in North America and Western Europe*. Oxford: Blackwell, pp. 58–79.

Greene, J. (1999) 'Zero tolerance: a case study of police policies and practices in New York City', *Crime and Delinquency*, 45(2): 171–87.

Grimshaw, R. and Jefferson, T. (1987) *Interpreting Policework: Policy and Practice in Forms of Beat Policing*. London: Allen & Unwin.

Grinc, R. (1994) 'Angels in marble: problems in stimulating community involvement in community policing', *Crime and Delinquency*, 40(3): 437–68.

Hagan, J. and Albonetti, C. (1982) 'Race, class and the perception of criminal injustice in America', *American Journal of Sociology*, 88(3): 329–55.

Hall, S. (1978) *Policing the Crisis: Mugging, the State, and Law and Order*. New York: Holmes & Meier.

Harcourt, B. (2001) *The Illusion of Order: The False Promise of Broken Windows Policing*. Cambridge, MA: Harvard University Press.

Harris, D. (1997) 'Driving while black and all other traffic offenses: the Supreme Court and pretextual traffic stops', *Journal of Criminal Law and Criminology*, 87(3): 544–78.

Herbert, S. (1996) 'Morality in law enforcement: chasing "bad guys" with the Los Angeles Police Department', *Law and Society Review*, 30(4): 799–818.

Herbert, S. (1997) *Policing Space: Territoriality and the Los Angeles Police Department*. Minneapolis: University of Minnesota Press.

Herbert, S. (1998) 'Police subculture reconsidered', *Criminology*, 36(2): 343–70.

Herbert, S. (1999) 'The end of the territorially sovereign state? The case of crime control in the United States', *Political Geography*, 18(2): 149–72.

Herbert, S. (2001a) 'Hard charger or station queen? Policing and the masculinist state', *Gender, Place and Culture*, 8(1): 55–71.

Herbert, S. (2001b) 'Policing the contemporary city: fixing broken windows or shoring up neo-liberalism?', *Theoretical Criminology*, 5(3): 445–66.

Herbert, S. and Brown, E. (2006) 'Conceptions of space and crime in the punitive neo-liberal city', *Antipode*, 38(4): 755–77.

Holdway, S. (1983) *Inside the British Police: A Force at Work*. Oxford: Basil Blackwell.

Hunt, J. (1984) 'The development of rapport through the negotiation of gender in fieldwork with the police', *Human Organization*, 41(2): 283–96.

Hurd, I. (1999) 'Legitimacy and authority in international politics', *International Organization*, 53(3): 379–408.

Jacob, H. (1971) 'Black and white perceptions of justice in the city', *Law and Society Review*, 6(1): 69–90.

Jessop, B. (2002) 'Liberalism, neoliberalism and urban governance: a state-theoretical perspective', in N. Brenner and N. Theodore (eds), *Spaces of Neoliberalism: Urban Restructuring in North America and Western Europe*. Oxford: Blackwell.

Johnston, L. (1992) *The Rebirth of Private Policing*. New York: Routledge.

Kappeler, V., Sluder, R. and Alpert, G. (1994) *Forces of Deviance: Understanding the Dark Side of Policing*. Prospect Heights, IL: Waveland Press.

Keith, M. (1991) 'Policing a perplexed society: no-go areas and the mystification of police–black conflict', in E. Cashmore and E. McLaughlin (eds), *Out of Order? Policing Black People*. London: Routledge, pp. 189–214.

Keith, M. (1993) *Race, Riots and Policing: Lore and Disorder in a Multi-Racist Society*. London: University College London Press.

Kelling, G. (1974) *The Kansas City Preventative Patrol Experiment*. Washington, DC: Police Foundation.

Kelling, G. and Coles, C. (1996) *Fixing Broken Windows: Restoring Order and Reducing Crime in Our Cities*. New York: Free Press.

Kennedy, R. (1997) *Race, Crime and the Law*. New York: Pantheon Books.

Kinsey, R., Lea, J. and Young, J. (1986) *Losing the Fight Against Crime*. Oxford: Basil Blackwell.

Klockars, C. (1985) *The Idea of Police*. Beverly Hills, CA: Sage.

Kraska, P. (2001) *Militarizing the American Criminal Justice System: The Changing Roles of the Armed Forces and the Police*. Boston: Northeastern University Press.

Kraska, P. and Kappeler, V. (1997) 'Militarizing American police: the rise and normalization of para-military units', *Social Problems*, 44(1): 1–18.

Larner, W. (2000). 'Neo-liberalism: policy, ideology, governmentality', *Studies in Political Economy*, 63(1): 5–25.

Lea, J. and Young, J. (1984) *What's to Be Done About Law and Order?* Hammondsworth: Penguin.

Lipietz, A. (1992) *Towards a New Economic Order*. Cambridge: Polity Press.

Lipsky, M. (1980) *Street-Level Bureaucracy: Dilemmas of the Individual in Public Services*. New York: Russell Sage Foundation.

Loader, I. (1997) 'Policing and the social: questions of symbolic power', *British Journal of Sociology*, 48(1): 1–18.

Lofland, L. (1973) *A World of Strangers: Order and Action in urban Public Space*. New York: Basic Books.

Low, S. (2003) *Behind the Gates : Life, Security, and the Pursuit of Happiness in Fortress America*. New York: Routledge.

Lyons, W. (1999) *The Politics of Community Policing: Rearranging the Power to Punish*. Ann Arbor: University of Michigan Press.

Magahan, P. (1984) *Police Images of a City*. New York: Peter Lang.

Mann, M. (1988) *States, War and Capitalism*. Oxford: Basil Blackwell.

Manning, P. (1977) *Police Work*. Cambridge, MA: MIT Press.

McKenzie, E. (1994) *Privatopia: Homeowner Associations and the Rise of Residential Private Government*. New Haven: Yale University Press.

Miller, L. (2001) *The Politics of Community Crime Prevention: Implementing Operation Weed and Seed in Seattle*. Burlington, VT: Ashgate.

Mitchell, D. (2001) 'Postmodern geographical praxis? The postmodern impulse and the war against homeless people in the "post-justice" city', in C. Minc (ed.), *Postmodern Geography: Theory and Praxis*. Oxford: Blackwell.

Monkkonen, E. (1981) *Police in Urban America, 1860–1920*. Cambridge: Cambridge University Press.

Moore, M. (1990) 'Problem-solving and community policing', in M. Tonry and N. Morris (eds), *Modern Policing*. Chicago: University of Chicago Press, pp. 99–158.

Muir, J. (1977) *Police: Streetcorner Politicians*. Chicago: University of Chicago Press.

O'Malley, P. and Palmer, D. (1996) 'Post-Keynesian policing', *Economy and Society*, 25(2): 137–55.

Packer, H. (1968) *The Limits of the Criminal Sanction*. Stanford: Stanford University Press.

Paoline, E. (2004) 'Shedding light on police culture: an examination of officers' occupational attitudes', *Police Quarterly*, 7(2): 205–36.

Peck, J. and Tickell, A. (1994) 'Searching for the new institutional fix: the after-Fordist crisis and the global-local disorder', in A. Amin (ed.), *Post-Fordism: A Reader*. Oxford: Blackwell.

Peck, J. and Tickell, A. (2002) 'Neoliberalizing space', *Antipode*, 34(3): 380–404.

Pue, W. (1990) 'Wrestling with law: (geographical) specificity vs. (legal) abstraction', *Urban Geography*, 11(4): 566–85.

Punch, M. (1985) *Conduct Unbecoming: The Social Construction of Police Deviance and Control*. London: Tavistock.

Reed, W. (1999) *The Politics of Community Policing: The Case of Seattle*. New York: Garland Publishing.

Reiner, R. (1978) *The Blue-Coated Worker*. Cambridge: Cambridge University Press.

Reiss, A. (1971) *The Police and the Public*. New Haven: Yale University Press.

Reuss-Ianni, E. (1984) *Two Cultures of Policing*. New Brunswick, NJ: Transaction Books.

Richardson, J. (1974) *Urban Police in the United States*. Port Washington, NY: Kennkat.

Rubinstein, J. (1973) *City Police*. New York: Farrar, Straus & Giroux.

Sacks, H. (1972) 'Notes on police assessment of moral character', in D. Sudnow (ed.), *Studies in Social Interaction*. New York: Free Press, pp. 280–93.

Sadd, S. and Grinc, R. (1994) 'Innovative neighborhood oriented policing: an evaluation of community policing programs in eight cities', in D. Rosenbaum (ed.), *The Challenge of Community Policing: Testing the Promises*. Thousand Oaks, CA: Sage, pp. 57–82.

Sampson, R. and Raudenbush, S. (1999) 'Systematic social observation of public spaces: a new look at disorder in urban neighborhoods', *American Journal of Sociology*, 105(4): 603–51.

Saunders, R. (1999) 'The space community policing makes and the body that makes it', *Professional Geographer*, 51(2): 135–46.

Schneider, J. (1980) *Detroit and the Problem of Order, 1830–1880*. Lincoln: University of Nebraska Press.

Shearing, C. (1990) 'The relation between public and private policing', in M. Tonry and N. Morris (eds), *Modern Policing*. Chicago: University of Chicago Press, pp. 399–434.

Sherman, L. (1974) 'The sociology and the social reform of the American police: 1950–73', *Journal of Police Science and Administration*, 2(3): 255–62.

Sherman, L., Gartin, P. and Buerger, M. (1989) 'Hot spots of predatory crime: routine activities and the criminology of place', *Criminology*, 27(1): 27–55.

Silver, A. (1967) 'The demand for order in civil society', in D. Bordua (ed.), *The Police: Six Sociological Essays*. New York: Wiley.

Silverman, E. (1999) *NYPD Battles Crime: Innovative Strategies in Policing*. Boston: Northeastern University Press.

Skolnick, J. (1966) *Justice Without Trial*. New York: Wiley.

Skolnick, J. and Fyfe, J. (1993) *Beyond the Law: Police and the Excessive use of Force*. New York: Free Press.

Smith, D. (1986) 'The neighborhood context of police behavior', in A. Reiss and M. Tonry (eds), *Communities and Crime*. Chicago: University of Chicago Press, pp. 313–41.

Sparrow, M., Moore, M. and Kennedy, D. (1990) *Beyond 911: A New Era for Policing*. New York: Basic Books.

Spitzer, S. (1979) 'Rationalization of crime control in capitalist society', *Contemporary Crises*, 3(2): 187–206.

Stinchcombe, A. (1963) 'Institutions of privacy in the determination of police administrative practice', *American Journal of Sociology*, 64(2): 150–60.

Taylor, R. (2001) *Breaking Away from Broken Windows: Baltimore Neighborhoods and the Nationwide Fight Against Crime, Grime, Fear, and Decline*. Boulder, CO: Westview Press.

United States National Advisory Commission on Civil Disorders (1968) *Report of the National Advisory Commission on Civil Disorders*. Washington, DC: US Government Printing Office.

van Maanen, J. (1974) 'Working the street: a developmental view of police behavior', in H. Jacob (ed.), *The Potential for Reform of Criminal Justice*. Beverly Hills, CA: Sage.

van Maanen, J. (1980) 'Street justice', in R. Lundman (ed.), *Police Behavior: A Sociological Perspective*. Oxford: Oxford University Press, pp. 296–311.

Walker, S. (1977) *A Critical History of Police Reform*. Lexington, KY: Lexington Books.

Weber, M. (1964) *The Theory of Social and Economic Organization*. New York: Pantheon.

Weitzer, R. (1999) "'Citizens" perceptions of police misconduct: race and neighborhood context', *Justice Quarterly*, 16(4): 819–46.

Werthman, C. and Piliavin, I. (1967) 'Gang members and the police', in D. Bordua (ed.), *The Police: Six Sociological Essays*. New York: Wiley.

Wilson, J. and Kelling, G. (1982) 'Broken windows', *Atlantic Monthly,* March: 29–38.

Re-Naturing Political Geography

Introduction

Jennifer Robinson

The environment is, in many ways, one of the core traditional themes of political geography. And yet, as this section title suggests, and as most of the authors in this section propose, there is a pressing need for political geography to re-engage with themes of nature, environment and non-humans. For some of the earliest writers in this field, like Halford Mackinder (1904) or Friedrich Ratzel (1963), the environment held the promise of explaining the world of politics, including the dominance of certain states in international affairs, or the imperatives of colonization (see Bassin, 2003, and Braun, this volume, for reviews). The metaphorical use of environmental or biological terminology to explain social processes, like the expansion of states, spatial differentiation of cities or the cohesion of society, persisted for some time in the discipline. However, criticisms of environmental determinism meant that the environment became a more marginal explanatory presence within geographical accounts of politics, which instead came to focus on the social production of nature (for example, David Harvey's (1974) important critique). Environmental determinism has subsequently shadowed the formulation of political geography's agendas largely by its determined absence. Bassin echoes a commonly held view that what he calls 'the argument from nature' is inevitably deployed towards politically programmatic ends, making any suggestions that the environment determines social life broadly politically motivated. As a result, and especially within a Marxist rubric, geographers have agonized about the theoretical relationship between nature and society (Smith, 1996; Castree, 2000), seeking a way to appreciate the material nature of social life without falling prey to environmental determinism. In Braun's opinion, though, this has not appreciably displaced what he sees as the now dominant focus on the social production of nature to the detriment of understanding how it is that human relationships are in fact dependent upon complex alliances with the more-than-human (this volume).

Two recent theoretical developments beyond the discipline have posed serious challenges to this state of affairs. A recent resurgence of rather blunt applications of environmental analysis to politics (Diamond, 1999; Landes, 1999; Huntington, 1996) sets out a direct challenge to political geographers committed to a non-determinist approach to the environment. Suggesting that the relative powers of contemporary states have their origins in environmental differentiation and geographical location, these authors make uncomfortable reading within the political geography tradition, at least in its more recent incarnations. And from the sociology of science comes a line of analysis eager to demonstrate the agency of non-human things, from technology to animals and including the constitutive role of the material in all social activity (Latour, 1993; Whatmore, 2002). Perhaps echoes from an earlier environmental determinism are evident here too, although, as with Marxist geographers, the advocates of actor-network theory work very hard to generate a robust and lively approach that avoids reductionism and a *priori* political or ideological investments. Philosophically, there has emerged a thorough questioning of the intellectual and political habit of seeing (human) society and nature as separate, pre-given entities, and this invites a substantial rethinking of the terms of debate about human/nature relations. Bruce Braun's review in this section of the *Handbook* makes a strong case for the potential of such approaches to bring considerations of the non-human into political geography while avoiding these pitfalls. This enables a robust new agenda for an engagement between research on the politics of nature, environment and technology and the sub-discipline of political geography. The authors in this section all take up this challenge – albeit in a range of different ways – as does Saldanha, elsewhere in this volume.

The authors in this section reflect on the possible 're-naturing' of political geography here by exploring some important and longstanding, if less than central, preoccupations of political geography. Certainly, the politics of water use, the regulation of the seas, conflict over resource use or the politics of famine have remained perennial themes within the field. These have been of particular importance to geographers working in regions where food or water scarcity, for example, offered a significant opportunity for regional conflict (for example, Watts, 1983 Brown and Purcell, 2005). More recently, there has been a flourishing of research on environmental politics within geography, although not often closely tied to the sub-field of political geography – and often in areas of the discipline with stronger ties to cognate fields than to geography per se. At stake here has been the rise of environmental issues on international policy agendas, and the significant increase in environmental movements, both within particular countries and as transnational political networks.

As a consequence of these political and intellectual trends, political geographers have increasingly found that they have much to learn from the fields of political ecology and science studies (in their respective areas, Blaikie and Brookfield (1987) and Latour (1993) have become classic texts for geographers); and political geographers have in turn generated insights for both these fields (for example, Geof paper; Murdoch, 1997). The judgement of the authors of this section (see also Rydin, this volume) is that there is much to be gained from a stronger interaction between researchers within geography as a whole working on environmental politics, and political geography. Accordingly, these trends are increasingly apparent within political geography, including research in poorer country contexts. In consequence, a wider range of environmental political concerns and approaches to environmental politics are now being discussed. The chapters in this section reflect these trends.

Bruce Braun offers an overview of how geographers have contributed to theorizing relations between society and nature: from early determinists, through cultural ecologists eager to demonstrate how cultures were deeply imbricated with local environmental systems, and perhaps notably in Marxist political economy. Some major geographical theorists have developed nuanced and careful Marxist arguments to suggest that people and nature are not separately produced. Instead, people are inherently a part of nature, both subject and object in the production of nature, just as nature itself can be understood as inherently social. Although adopting a strongly relational approach to nature, the continuing Marxist emphasis on human labour makes their accounts, in Braun's view, somewhat anthropocentric as well as economically reductionist. More importantly for him, they continue to assume the existence of the categories, society and nature, even as they want to demonstrate that these categories are socially produced, mutually constructed and mutable. He turns finally to a range of new approaches to nature, strongly inspired by writers within science studies, such

as Bruno Latour, Anne-Marie Mol and John Law. Their approaches have influenced a new generation of work on nature in geography that disavows any idea that people and nature can be assumed to be ontologically separable. Instead, they seek to track the emergence of agent-ful entities through associations and networks. The constitution of an entity or an action takes place, then, through a myriad of practices drawing on a diverse range of elements, not limited to either 'nature' or 'society'. He questions the implications of this ontological shift, profound as it is, for a politics of nature, and concludes that these are as yet unclear. But certainly, this important initiative, within geography and beyond, holds up much scope for re-imagining the nature of political relations.

Another important initiative for political geography has come from an engagement with political ecology. Strongly influenced by Marxist analysis as well, political ecologists have been most represented in poorer country contexts, where the intertwining of the politics of environmental regulation and poverty has inspired attention to the complexity of political and ecological relations. Important themes for political geographers here have been understanding the state, and appreciating the differential reach of various social and ecological phenomena, that is, the question of scale. Paul Robbins recounts some of the important ways in which work in political ecology has challenged certain mainstream accounts of state power. In a chapter rich in evidence drawn from a range of different contexts, Robbins develops a sophisticated account of the state as involved in strategies to simplify landscapes (trying to model and fix inherently unstable ecologies), to (re-)territorialize activities in the face of divergent processes of de-territorialization, and to generate crisis narratives that support ambitions for intervention. Corbridge (this volume) describes similar strategies to enhance legibility and facilitate state actions in his discussion of James Scott's influential *Seeing Like a State*. However, as with Corbridge's account of the state as fundamentally shaped through its involvement with society, Robbins explores how the state's interface with environmental politics also generates states that are: (1) co-operative (with agricultural producers, traders, international actors); (2) porous in the sense of responding to local struggles and global changes; (3) that must negotiate divergent understandings of nature with a range of different agents both within and beyond the state; and that must (4) involve themselves in mutual learning both with local people and with the non-human environment if they are to successfully involve themselves in managing very diverse ecosystems.

The state, then, from the view point of political ecology, has to be understood as in a tension between being a relatively unified actor across a clearly defined territory with strong narratives and means for intervention, and a porous, fractured agent that has to actively engage with local environments and people, as well as maintain effective presence in global contexts if it is to be effective. Political ecology, then, Robbins argues, has much to contribute to political geographers concerned with theorizing the state. The intellectual learning, though, can equally run from political geography to political ecology, as Robbins suggests along with Joshua Muldavin and Becky Mansfield in this section. The question of scale is an important issue here, as writers in this field have to attend to local difference as well as to the ambitions of states and the often transnational social relations that globalization entails. In contrast with easy assumptions that political ecologists have been prone to make concerning scales as hierarchically arranged, relatively fixed and pre-given (the state, for example, being seen as a single, situated agent operating at a specific scale), writers in this section draw on geography's account of scales as historically produced and existing in relation to one another (see Swyngedouw, 1997). Increasingly, though, geographers are suggesting that focusing on scale, at least in the conventional areal sense, is at odds with the complex networks of association and influence that structure the field of environmental politics. A relational view of scale, then, could be seen to draw analysts towards an even more subtle account of the diverse spatialities of social and ecological relations, as Robbins concludes here (see also Castree et al., this volume).

Environmental politics poses other significant theoretical and political challenges to political geography, including the need to appreciate and account for the significance of actors beyond the state or inter-state organizations (as Bakker and Bridge, and also Mansfield, discuss here); and to attend to the compatibility between different theoretical approaches and the wider politics of neo-liberalism (which Muldavin considers at some length, along with Bakker and Bridge). Bakker and Bridge outline how resource management emerged historically as a dominant approach to resource use, and vested the control and regulation of resources in the state. However, over the last few decades, the range of non-state actors involved in resource regulation has proliferated, including markets, firms, indigenous communities and NGOs. The state's role has also been reconfigured to attend to wider environmental management and to a more varied range of epistemologies. For these authors, resource regulation brings the politics of governance to the fore and demands a more complex approach to resource regulation, concerned with how resources are enacted through various economic and institutional processes as well as through discursive shifts. The place of metabolism – the mutual transformation of society-natures – is also important. The state, then, can no longer be prioritized in these processes and instead analysts and activists must look to the multiple sites and institutional forms as well as the unruly materialities that mediate resource regulation.

For Bakker and Bridge, an important concern is the normative valuation placed on the role of non-state actors in resource regulation, especially within the context of neo-liberal emphases on privatization and the regulative role of the market. The importance of appreciating the intersections between neo-liberalism and environmental politics cascades through environmental research in different contexts – as Becky Mansfield indicates for global environmental politics, and Joshua Muldavin for understanding environmental aspects of the transition from socialism in China. Mansfield tracks debates about the politics of balancing development and the environment in an international context, through both formal and informal arenas. In the most recent rounds of international policy formulation on the environment, complex North–South divisions on this issue (where the South demands more development, even as the North is broadly refusing to take responsibility for the impacts of its level of development on the global environment) have intersected with the broadly neo-liberal context of international policy-making. Here, assumptions concerning the continuing importance of economic growth support Southern concerns for ongoing development and have shaded the nature and scope of international agreements. Close links can be found here with the arguments of ecological modernization. This is a currently fashionable approach to environmental problems which, as outlined by Joshua Muldavin, suggests that economic growth will generate the technological capacities and inventions to cope with its own environmental consequences. In both cases, the overwhelming neo-liberal policy environment operates so as to favour continuing economic growth and to diminish attention to the environmental consequences of this growth. Both Mansfield and Muldavin provide examples of active resistance to these approaches from affected communities and from transnational activist and advocacy networks.

Certainly, the chapters collected here indicate a creative interface between research on the environment and the themes of political geography. As these two fields become more closely intertwined, we anticipate that unruly materialities, ecological change and emergent environmental politics will instigate new concerns and more subtle approaches within political geography. And we very much hope that these conversations, once begun, will flourish, and enable political geographers to play a stronger role in informing political analyses of the people/nature interface.

REFERENCES

Bassin, M. (2003) 'Politics from nature', in J.A. Agnew et al. (eds), *A Companion to Political Geography*. Malden, MA: Blackwell, pp. 13–29.

Blaikie, P. and Brookfield, H. (1987) *Land Degradation and Society*. London: Methuen.

Castree, N. (2000) 'The production of nature', in E. Sheppard and T.J. Barnes (eds), *A Companion to Economic Geography*. Oxford: Blackwell, pp. 275–89.

Diamond, J. (1999) *Guns, Germs and Steel: The Fates of Human Societies*. New York: W.W. Norton.

Harvey, D. (1974) 'Population, resources and the ideology of science', *Economic Geography*, 50: 256–77.

Landes, D. (1999) *The Wealth and Poverty of Nations*. New York: W.W. Norton.

Latour, B. (1993) *We Have Never Been Modern*. Cambridge, MA: Harvard University Press.

Mackinder, H. (1951 [1904]) *The Geographical Pivot of History*. London: The Royal Geographical Society.

Murdoch, J. (1997) 'Towards a geography of heterogenous associations', *Progress in Human Geography*, 21(3): 321–37.

Ratzel, F. (1963) 'The laws of the spatial growth of states', in R. Kasperson and J.V. Minghi (eds), *The Structure of Political Geography*. London: University of London Press, pp. 17–28.

Smith, N. (1996) 'The production of nature', in G. Robertson, M. Mash, L. Tickner, J. Bird, B. Curtis and T. Putnam (eds), *Future Natural: Nature, Science, Culture*. London: Routledge, pp. 35–54.

Swyngedouw, E. (1997) 'Neither local nor global', in K.R. Cox (ed.), *Spaces of Globalization*. New York: Guilford Press, pp. 137–66.

Whatmore, S. (2002) *Hybrid Geographies: Natures, Cultures, Spaces*. London: Sage.

Theorizing the Nature–Society Divide

Bruce Braun

INTRODUCTION

In March 2003, doctors at a Toronto hospital attended to a patient who had a high fever and was having difficulty breathing. Within days she had died, as had her son. Many others had died or were seriously ill in cities as distant as Hong Kong, Hanoi and Singapore. The cause of their deaths, the SARS-corona virus, is now considered by medical researchers to have been only the most recent of a lengthy series of zoonotic diseases passed from animals to humans (and vice versa). In the case of SARS, the virus is believed to be endemic in horseshoe bats in China, and to have spread from this 'reservoir' to the live animal markets of Guangdong Province, and from there to Hong Kong and through air travel to far-flung places like Toronto. These diseases have caused immense concern, helped along by dire prediction of an inevitable 'next pandemic' (Osterholm, 2005). Among the many profound questions that this event raised – such as the relation between air travel and the rapid spread of infectious diseases; globalization and its relation to (bio-)security; neoliberalism and the restructuring of public health systems; new technologies for the surveillance of humans and non-humans alike – the most wide-reaching may have to do with the challenge SARS posed for how we conceptualize the 'social' life of cities. For if residents of cities like Hong Kong and Toronto learned one thing from the SARS crisis, it was that human bodies were far from discrete, autonomous or closed off from the world, but instead were composed, and at times placed at great risk, through the continuous exchange of matter and information with other entities and organisms.

While the city had at least since Aristotle been considered the site where humans decided their affairs, it was no longer clear that this 'polis' consisted merely of 'humans amongst themselves'.

The significance of SARS, then, had in part to do with how it troubled distinctions between nature and society. Were cities cultural spaces or ecological spaces? Within the polyrhythmic spaces of the city, were bodies the outcome of physical processes or cultural practices? Was SARS a product of nature (spreading 'naturally' from animals to humans), or an effect of culture, economy and politics (resulting from the intensification of food production, the crowded and unregulated spaces of live food markets, and the restructuring of public health)? Could the world be so clearly divided into nature and society, subject and object, human and non-human?

Viewed in light of SARS, such divisions seem increasingly absurd, for how can the world possibly be reduced to either? Yet, for much of the past three centuries, we – or at least those of us in the West – have been asked to engage in precisely such 'labors of division', separating the world into two separate ontological domains, nature here and society there, even as we continuously confront – and produce – entities that cannot possibly be assigned to either. Indeed, that these antinomies remain a pervasive part of Western thought has less to do with a 'reality' out there than with the daily reiteration of these divisions as part of 'good sense'. We find it, for instance, in the separation of intellectual labor into disciplines such as sociology (society), anthropology (culture) and literature (signs/texts) in the social sciences and humanities, and biology (organic life), geology (inorganic nature) and physics (force) in the physical sciences. We find it reproduced in nature documentaries, written into social studies textbooks and manifest in the spatial practices of zoos and national parks. Yet, despite its persistent hold, this dualism has in recent years begun to creak and groan under the accumulating weight of entities and places that resist categorization within these age-old divides. With the proliferation of GMOs, growing concerns over global warming, and the emergence of new food-borne diseases, it has become ever more difficult to imagine a world that consists of discrete and static essences, or that is divisible into two separate ontological domains. Increasingly, we are asking how it is that we came to conceive of the world in these dualist terms to begin with.

But why discuss the nature–society dualism in a handbook of political geography? In what ways is political geography caught up with this dualism? And in what ways might this dualism – and attempts to overcome it – be understood as political? As we will see at the end of this chapter, dissolving the division between society and nature may force us to confront an initially disconcerting reality: that far from naming an external and immutable realm prior to politics, nature – understood as those hybrid assemblages within which 'life' is produced, sustained or endangered – is something that is continuously composed by a multiplicity of actors (human and non-human) and thus suffused with politics and power from the outset. Whether we see this as a disturbing or joyful condition may ultimately have a great deal to do with how wedded we are to the nature–society dualism to begin with.

We might conclude, then, that dissolving the nature–society dualism invariably leads us to the politics of nature. This would be a grave mistake, or, at the very least, a conclusion too hastily drawn. For, as I will show in this chapter, some attempts to dissolve the nature–society dualism have led to the complete effacement of politics. We must therefore proceed cautiously and by way of a different question: In the history of geography, how has the nature–society dualism, and attempts to overcome it, been related to, and implicated in, the somewhat tortured career of the 'political' in geographical theory and practice? Or, conversely, in what ways has the 'political' in political geography turned on how the nature–society dualism has been dissolved? It is the history of this dualism in geographical thought, then, that we must carefully examine.

ENVIRONMENTAL DETERMINISM IN EARLY GEOGRAPHY, OR, HOW TO COLLAPSE SOCIETY INTO NATURE

From geography's inception as a discipline in the late nineteenth century, the relationship between 'nature' and 'society' has been one of its central concerns. Indeed, David Livingstone (1992) has gone so far as to describe the attempt to combine the human and natural world within the same analytical frame as 'the geographical experiment'. This concern set it apart from other disciplines, which tended to take either 'society' or 'nature' as their objects of inquiry. However, while the holism of early geographers made the discipline unique, how they understood it would eventually become an embarrassing stain on the discipline's reputation.

To understand this statement we need to know something of the early history of the discipline. As geography emerged in Europe and the United States, one of its chief concerns was the description of geographical difference and, as important, the explanation of these differences. In part this impulse came from European exploration and the colonial projects of various European powers, which forced Europeans to confront a differentiated world, and to come to terms with the moral and political problem of European domination. Why did other peoples have different beliefs, customs and technologies? On what basis could European control of foreign lands be justified? One of the most popular answers to these questions took the form of environmental determinism, which in general proceeded in two directions: either it looked to the environment for the 'underlying conditions' that purportedly explained social and political phenomena, or it explained social and political forms through analogy with nature (and in particular biology).

One of the reasons for the popularity of these approaches was that they appeared to go beyond mere description to provide explanations, and thus held out the promise that geography could be a predictive science. This effort at explanation is perhaps best illustrated by the maps and 'climographs' produced by Ellsworth Huntington and Griffith Taylor in the 1920s, which claimed to show correlations between climate and 'civilization' (or, in Taylor's case, race). As David Livingstone (1992) explains, these maps were based on induction from what were at best dubious data. Huntington asked colleagues to rank countries in terms of such things as 'level of civilization' – an arbitrary classification – and correlated the results with a chart of what purported to show 'climatic energy' – an equally arbitrary and selective compilation of climatic conditions. Not surprisingly, Huntington produced maps that showed the regions around the North Sea to have both the most advantageous climate and the highest level of society. Hence, he would conclude, 'climate influences health and energy, and these in turn influence civilization' (quoted in Livingstone, p. 225).[1] Notably, the power of their arguments was in part an effect of the cartographic form in which they presented their conclusions. As a form of visual representation, maps and graphs had certain advantages. Trading on the privileged relation between vision and truth, they could appear more 'objective' than written or verbal descriptions. Likewise, once detached from the many mediations that produced them, maps appeared to 'speak for themselves', even if the methods used were contentious. Indeed, it is noteworthy that environmental determinism flourished alongside an explosive growth in visual technologies – and spectacles such as world fairs – that presented European observers with the world as an ordered (and often hierarchical) whole that could be grasped as a totality, part of the 'modern metaphysics' explored with such brilliance by Timothy Mitchell (1988; see also Heidegger, 1977).

For my purposes it is not enough to show Huntington's maps and Taylor's climographs to be flawed, or to trace the effects of realist epistemologies. What is equally significant is that these forms of knowledge were far from innocent. It is not difficult, for instance, to see how Huntington's maps labored in the service of Empire, providing justification for European and American geopolitical and economic dominance. After all, had nature not determined that Europeans should be more advanced? So, while Huntington and Taylor are rarely discussed as political geographers – they did not explicitly study the state, territory or geopolitics – they provided ways of thinking about the earth and its people that were political through and through, providing part of the intellectual apparatus of imperialism (see Said, 1993). By collapsing society into nature, social conditions and formations, such as imperialism, could be seen to have natural causes, not political ones. Hence, environmental determinism could be deeply political even as it banished politics from its explanations.

Huntington and Taylor were not political geographers, but this does not mean that the emerging field of political geography was immune to determinist thought. Quite the opposite: many of the key figures in early political geography traded in a similar holism, and with similar effects. In 1887, Halford Mackinder had argued that 'no rational geography can exist which is not built upon and subsequent to physical geography' (1996 [1887]: 157). While Mackinder sought to explain explicitly political geographical phenomena – the state, their powers, and their territorial form and extent – he argued that merely describing or locating these without explaining them was insufficiently scientific, merely 'a topography … with the "reasons why" eliminated' (p. 158). The 'reasons why' were often physical. Hence, he claimed, 'everywhere political questions will depend on the results of the physical inquiry', although he was careful to note that the 'relative importance' of physical features 'varied from age to age' according to the state of knowledge and technology (p. 170).

Arguably, Mackinder's most famous application of his method is found in his 1904 paper 'The geographical pivot of history'. In it Mackinder reiterated that he was interested in finding 'geographical causation in universal history' (1951 [1904]: 30) and sought to describe 'those physical features of the world which I believe to have been most coercive of human action' with the aim to 'exhibit human history as part of the life of the world organism' (p. 31). Again, he was quick to qualify: it was man, and not nature, that initiated historical change, although 'nature in large measure controls' (p. 31). In his attempts to both explain and predict the future course of global geopolitics, therefore, Mackinder looked to physical geography and geographical location to establish 'the natural seats of power', from which he ultimately determined that whoever controlled the large landmasses at the middle of Eurasia would likely control the world.[2]

It merits comment that Mackinder imagined nation-states to be in continuous competition, echoing a Darwinian sentiment common at the time. Indeed, the displacement of natural theology by evolutionary theory – Darwinian and Lamarckian – in the late nineteenth century provided conceptual resources for a considerable range of determinist arguments. Natural theology had imagined a transcendental cause for the differentiated earth. Evolutionary theory in contrast ushered in a naturalism that taught that earthly variations had natural or material causes. Moreover, evolution provided mechanisms to explain why particular variations existed and a teleology that imagined a progressive history where the present existed as the pinnacle of evolutionary processes. In Darwinian theory those organisms that existed did so because they had adapted better to environmental change (and had more success in leaving progeny). In a world of continual population growth, and constant struggle over scarce resources – ideas Darwin had borrowed from Malthus – only the fittest survived, while others were destined to disappear.

Evolutionary ideas have only a faint echo in Mackinder. But they were present much more explicitly in the geopolitical writings of the German geographer Friedrich Ratzel. His *Anthropogeography* was heavily influenced by notions of adaptation and natural selection and, like Mackinder, he argued that the course of history was determined by the physical features of the earth.[3] Likewise, in his *Politische Geographie* he adapted evolutionary arguments to the state, which he treated as organisms that struggled for land – 'living space' – thereby naturalizing political forms, and, by doing so, finding a basis in nature for territorial growth (a convenient justification for antagonism over territory between nation-states, and for colonialist expansion). As Mark Bassin (2003) explains, in the latter text Ratzel argued through direct analogy to the plant and animal world. While Mackinder focused more on the ways physical geography affected such things as the relative power of states, for Ratzel the state was like an organism, and thus was subject to the same natural laws. His notion of *Lebensraum* was thus in many ways an evolutionary and biogeographical concept applied to human communities, derived from the assumption that living organisms required specific amounts of territory, over which they competed (Bassin, 2003). In Ratzel's day this became a justification for Germany joining the 'scramble for territory' in Africa. Later, it would inform the territorial ambitions of Germany's Third Reich.

Bassin (2003) has rightly described such approaches as deriving 'politics from nature'. In the United States, geographers such as Ellen Semple developed similar arguments, although her writings were arguably more ambivalent.[4] Others were still more cautious. Possibilists, such as Vidal de la Blache, suggested that humans could choose from a range of responses to the physical environment, while Lucien Febvre vigorously contested the determinism of writers like Ratzel, at times replacing the latter's environmental determinism with a social determinism and thereby, as is sanctioned by the underlying nature–society dualism, reversing the direction of causality.

Although today environmental determinism is widely discredited within geography, it has not disappeared entirely. Ironically, the desire to 'integrate' society and environment, and the search for general historical laws, has given environmental determinism new life outside geography, in hugely popular works such as Jared Diamond's *Guns, Germs and Steel* (1999) and David Landes' *The Wealth and Poverty of Nations* (1999). In the Eurocentric tradition of earlier determinists, Diamond lodges explanation for Europe's dominance in its physical geography. Indeed, the determinist language is surprisingly explicit: while he notes many 'proximate' causes for why certain regions gained geopolitical dominance (such as technology and writing), these are ultimately seen to be the secondary effects of what he terms 'ultimate' causes. In the final analysis, Diamond reduces the global dominance of Europe to two: the East–West orientation of Eurasia, which in Diamond's view allowed for a greater diversity of animals and plants that could be domesticated, and Europe's fragmented coast, which he claims resulted in both chronic disunity and centres of innovation, and thus enabled Europe to advance while China, with its regular coastline, languished under despotic rule. Around these conditions, Diamond writes, 'turned the fortunes of history' (p. 191).[5]

As with earlier determinists, this kind of search for 'ultimate factors' and general principles of causation relies heavily on correlations and inductive reasoning, and far less on careful explication of the links between specific physical properties of the earth and the communities and practices that were articulated with them. As Sluyter (2003) explains, this becomes tantamount to the argument that 'Eurasia's unique environment has caused the G-8's dominance, the proof being Eurasia's unique environment and the G-8's dominance' (p. 815), and goes on to condemn Diamond for a 'quick and dirty integration of the natural and social sciences' (p. 817). Others have noted that Diamond's attempt to find 'general principles of causation' leaves him unable to explain why there might be disparate social and political forms in similar environments (Merrett, 2003), each of which must somehow be explained away. Robbins (2004: 19) notes that this was a strategy already found in Huntington's work, for whom 'harsh climates were simultaneously used to explain the ingenuity of some groups and the cultural limits of others'. Diamond could be criticized on the same grounds: the East–West orientation of parts of North America is not so different than Eurasia, and if a fragmented coast is the 'ultimate' cause for innovation and political power, then it is not clear why indigenous people of the Pacific Northwest do not rule the world!

Much more can be said about the attempts by environmental determinists to 'relate' society and nature. For our purposes it is worth underlining that environmental determinism was thoroughly political, and that it was so in a number of ways. First, although it cloaked itself in scientific objectivity – 'just the facts' – it provided a political reading of the nature–society relationship that banished politics from its explanation of social geographical phenomena. This reading provided a basis for forms of racism and justifications for colonial projects, although it could at times also give credence to arguments opposing each. Second, and as important, it provided key terms and concepts for the nascent field of geopolitics. Indeed, Mackinder and Ratzel were both primarily concerned with understanding the political geography of the earth – its division into nations, states and empires – and turned to the environment for explanation; geography dictated the relative strengths and weaknesses of nation-states, and explained why some nation-states dominated others. It is perhaps not too great an exaggeration to say that political geography – and in particular, geopolitics – was weaned on environmental determinism, a remarkable erasure of politics from political geography, and a 'stain' that the field has since labored hard to remove. But, as we will see later, in some respects the cure has been as bad as the original disease, for arguably it was in response to the scandal of environmental determinism that much of political geography – and human geography more generally – banished talk of the non-human altogether in the last decades of the twentieth century, thereby deepening the nature–society divide that the 'geographical experiment' had sought to overcome.

TOWARD A UNIFIED THEORY OF CULTURE AND NATURE: CULTURAL ECOLOGY AND THE 'POLITICAL'

Given its reduction of political phenomena to environmental phenomena, we should not be surprised that scholars in other disciplines – and many geographers too – rejected environmental determinism. But this does not mean that attempts to overcome the nature–culture dualism came to an end. The years after World War II, and especially the 1960s and 1970s, saw renewed attempts to relate culture and nature, the most important of which became known as 'cultural ecology' or 'ecological anthropology'.

In part, cultural ecology emerged in reaction to the hardening division between the social and physical sciences within the academy, as well as developments within the social sciences where 'culture' and 'society' were increasingly understood in terms of their internal dynamics, entirely separate from their surrounding physical environment. Thus, Marvin Harris (1974) argued that anthropology had too readily subscribed to the position that 'culture begot culture' (see also Steward, 1955). For Harris the 'culturalism' of anthropology meant that researchers were unable to adequately understand the material conditions under which particular cultural or social forms emerged; removing the environment from the analysis, he and others argued, led to an impoverished understanding of cultural practices (see also Vayda and Rappaport, 1968). Some, such as Clifford Geertz (1963), suggested that the environmental determinists' goal of drawing nature and society into a single analytical frame had indeed been laudable, but that they had pitched their analyses at too grand a scale, and too great a level of abstraction; to speak of a relationship between the 'polar regions' and 'Eskimos', for instance, told us nothing about the way that specific practices of arctic peoples articulated with specific non-human elements of the landscape. A more exhaustive, fine-grained analysis of communities and their environments was necessary. As Geertz put it rather optimistically, the emerging field of cultural ecology could achieve an 'exact specification' of the 'relation between selected human activities, biological transactions, and physical processes' (p. 2). From this perspective, what was needed was not to reject attempts to relate nature and society, but to change the scale at which the relation was examined, and the methods used to examine it.

For my purposes, the efforts by cultural ecologists to 'overcome' the nature–culture dualism are important so far as they relate to the fortunes of the 'political' within geography. Before I turn to this, we need to know more about cultural ecology as a field.[6] One of the hallmarks of cultural

ecology was fieldwork. Indeed, determining the 'exact specifications' of the connections between culture and environment required immense effort, and it was not uncommon for researchers to spend many years carefully collecting data on agricultural practices, ecological conditions, tools, food preparation, rituals and social structure, so as to map, measure and diagram the 'complex, systemic interrelationships' (Butzer, 1989) between people and ecologies in particular places. For cultural ecologists like Geertz (1963) and Julian Steward (1955), this required strict methodological principles: one began by isolating the cultural practices most closely tied to the environment – cultivation, for instance – then documented its consequences on soils, vegetation and animal life, noted feedback loops, and ultimately came to understand how the environment shaped the cultural practices of specific groups. Certain cultural practices – from cultivation to food preparation to tool-making – could therefore be understood as 'adaptations' to environmental conditions.

For Steward, comparative studies between cultures promised to bring to light universal laws that would explain how human interaction with the environment – necessitated by the work of survival (subsistence) – led to particular social and institutional forms. Like the environmental determinists earlier, it could therefore establish cultural ecology as a predictive science. As Robbins (2004) notes, this meant considerable emphasis on establishing metrics by which to measure environmental practices – energy flows, calories and nutrients, for instance – and opened the door to the study of human populations as if they were the same as plant or animal populations. Many cultural ecologists – including Steward and Geertz – were careful not to extend their analysis to all cultural phenomena. Steward, for instance, left room for a certain amount of 'latitude' in how cultures adapted to environment, and allowed that some cultural traits and practices might be the result of diffusion, although he thought that this had been 'greatly overestimated' (1955: 42). Others were less cautious. Most famously, Roy Rappaport (1967) extended his analysis to include even religious beliefs and rituals, imagining that these existed because they maintained ecological, cultural and political balance. The conclusions he drew about pig-killing rituals in New Guinea are worth quoting at length:

> The Tsembaga ritual cycle has been regarded as a complex homeostatic mechanism, operating to maintain the values of a number of variables within 'goal ranges' (ranges of values that permit the perpetuation of a system, as constituted, through indefinite periods of time). It has been argued that the regulatory functions of ritual among the

Tsembaga and Maring helps to maintain an unde-
graded environment, limits fighting to frequen-
cies that do not endanger the existence of the
regional population, adjusts man–land ratios, facilitates trade, distributes local surpluses of pig in
the form of pork throughout the regional population, and assures people of high-quality protein
when they most need it. ... The Tsembaga, designated a 'local population', have been regarded
as a population in the animal ecologist's sense: a
unit composed of an aggregate or organisms having in common certain distinctive means whereby
they maintain a set of tropic relations with other
living and nonliving components of the biotic community with which they exist together. (Rappaport,
1967: 224)

This passage is striking for a number of reasons.
It assumes that Tsembaga 'culture' was homeostatic or balanced, varying only within a set of 'goal
ranges'. To this it adds a functional reading of cultural practices (they exist to maintain this balance),
and, finally, it treats human populations in the 'animal ecologist's' sense, as a population that acts in
the aggregate to maintain a set of tropic relations
with other populations.

Works like these revealed the strong influence of
systems theory, cybernetics and ecosystem ecology. This led many cultural ecologists to understand cultures and their environments as 'integrated
wholes', with complex feedback loops that bound
the two together. It also opened the door to functionalist and teleological readings of cultural practices,
whose sole purpose was to retain equilibrium in
the system as a whole. The problem, as various
critics noted, was that this falsely imputed the ecological effects of cultural practices as their cause,
a tendency reinforced by the historicism of much
cultural ecology (see Zimmerer, 1996). While ecologists influenced by Sauer's landscape studies or
Steward's more cautious approach often avoided
such functionalist traps, many did not. The result
was that even as the field promised to overcome
the society–nature divide, it did so by reducing
cultural practices to nature: people did things –
they plowed fields, told stories, performed rituals –
but as 'adaptive' responses that ultimately could
be reduced to environmental causes. To be sure,
cultural ecologists focused on the specific practices that articulated humans and non-humans,
avoiding the sweeping generalizations of Ratzel
or Huntington, but as Robbins (2004: 28–9) suc-
cinctly concludes, within much cultural ecology,
'humans would be seen as part of a larger system, controlled and propelled by universal forces,
energy, nutrient flows, calories and the material
struggle for existence'.

To these concerns, critics added others. Cultural ecologists, they argued, invariably studied
'traditional' or 'archaic' societies, presented culture as monolithic, static and bounded, and failed
to examine processes occurring at larger spatial
scales (see Duncan, 1980; Cosgrove and Jackson,
1987; Gupta and Ferguson, 1997). Most important for my purpose is that in their enthusiasm
for systems theory and cybernetics as answers to
the nature–society riddle, they frequently left no
place for history and politics. Indeed, it would
not be an exaggeration to say that cultural ecology, along with systems theory more generally,
at once reflected – and furthered – the marginalization of political geography in the period, not
only relegating the study of political forms and
institutions (state, nation, territory) to the periphery, but contributing to the erasure of 'politics'
and 'power' from cultural and social explanation.
Although early proponents like Geertz had sought
to distance cultural ecology from environmental
determinism, the field's solution to the nature–
culture dualism came perilously close to following
in its tracks.

TOWARD A POLITICAL THEORY OF NATURE: HISTORICAL MATERIALISM, DIALECTICS AND THE PRODUCTION OF NATURE

By the late 1970s, other approaches to the study
of society–environment relations began to displace cultural ecology as scholars increasingly ran
up against its limits.[7] Many researchers rejected
the notion of bounded and balanced cultures
and sought to contextualize people's environmental practices within wider social and political-
economic forces, an approach that would eventually come to be known as 'political ecology'. I am
less interested here in the obvious point that political ecologists restored 'politics' to understandings
of society and environment (we will later have
occasion to ask to what extent this was actually
true), than in interrogating the ways that the nature–
culture divide came to be rethought by political
ecologists, and the consequences of this for how
we think about politics and power.

A place to begin is with Robbins' (2004) observation that Julian Steward had posited the work
of subsistence as the starting point from which
to analyze how cultures were constituted in relation to their environments. For Steward, work
was the locus of adaptation, and thus culture
was ultimately what was at stake, as communities
responded to environmental conditions. Steward's
focus, however, opened the door to a different
reading of the relation between work, culture and
environment that reversed the arrow of causality: work could be seen as transformative of the

environment, as communities responded to changing political and economic conditions. Two of the earliest practitioners of political ecology defined the emerging field in precisely these terms: as the political economy of the environment, understood in terms of the environmental practices of land managers and the forces shaping them (Blaikie and Brookfield, 1987).

With this in mind, it should come as no surprise that many political ecologists turned to historical materialism for an understanding of the forces shaping social and environmental practices, including some of Steward's protégés (Sidney Mintz, Eric Wolf). This mirrored the growing appeal of Marxism in geography and the social sciences during the 1970s and 1980s, but it was also reinforced by economic changes occurring in the mainly rural, 'Third World' sites that political ecologists studied, where non-market economies and forms of 'traditional' tenure (such as common property) were being replaced by private property or state ownership and the mediation of money.

An important component of the turn to historical materialism is found in the answers it offered to the nature–society riddle and what this meant for the place of the 'political' within geography. The outlines of a historical materialist understanding of nature and society have been widely rehearsed by geographers elsewhere and can be summarized quickly (see Smith, 1984; Castree, 1995). One of the hallmarks of historical materialism is its relational ontology: within its terms the world is understood as a dynamic totality in which entities (plants, animals, humans, buildings and texts) are effects of the relations that constitute them (Harvey, 1996; Castree, 2002). It follows, then, that 'Society' and 'Nature' are not fixed entities but historical outcomes, produced through their interaction.

The implications of this are at once various and consequential. To begin, such an ontology situates humanity within nature, as both subject and object. In his *Economic and Philosophical Manuscripts*, Marx (1975[1844]: 328) put it the following way:

Nature is man's inorganic body. ... Man lives from nature, i.e. nature is his body, and he must maintain a continuing dialogue with it if he is not to die. To say that man's physical and mental life is linked to nature simply means that nature is linked to itself.

Marx restated the theme in *Capital*, Vol. 1, this time emphasizing labor as that which 'mediated' nature and society:

Labor is, in the first place, a process in which both man and Nature participate, and in which man of his own accord starts, regulates, and controls the material reactions between himself and Nature. He opposes himself to Nature as one of her own forces, setting in motion arms and legs, head and hands, the natural forces of his body, in order to appropriate Nature's productions in a form adapted to his own wants. By thus acting on the external world and changing it, he at the same time changes his own nature. (Marx, 1967 [1887]: 173).

Nature and society exist as a unity; the relation between them is internal (see Smith, 1984; Harvey, 1996). Hence, humanity is not only constituted *within* nature, but through its productive labor it is constitutive *of* nature, including humanity's own 'inner' nature. Labor is thus conceived as one of nature's own forces (albeit a force that objectifies and transforms non-human nature or, as Neil Smith suggests of capitalism, externalizes it).

For Smith, human labor was simply the actualization of a potential that existed in nature, not something foreign to nature. Hence, Smith argued, nature itself is something 'produced' through the actions of society. Smith readily admitted that this sounded quixotic, but argued that:

What jars us so much about this idea ... is that it defies the conventional, sacrosanct separation of nature and society, and it does so with such abandon and without shame. We are used to conceiving of nature as external to society, pristine and pre-human, or else a grand universal in which human beings are small and simple cogs. (Smith, 1984: xiv)

Smith's 'production of nature' thesis would be developed and debated by a generation of radical geographers and sociologists throughout the 1980s and 1990s. While it was not without its critics (more on this later), it brought with it a number of distinct advantages.

One of its most important contributions was to pull the rug out from under arguments that found in 'external' nature immutable laws governing social life. This is significant, since within Western political philosophy, as Bruno Latour (2004: 28) has noted, 'not a single line has been written ... in which the terms "nature", "natural order", "natural law", "natural right", "inflexible causality" or "imprescriptible laws", have not been followed, a few lines, paragraphs, or pages later, by an affirmation concerning the way to reform public life'. By presenting nature and society as an internal relation, Castree (2000: 277) notes, nature can no longer be invoked 'as a source of authority to legitimate ... economic, social and environmental arrangements'.

In this regard, David Harvey's (1974) critique of Malthus remains exemplary. As is well known, Malthus had argued that scarcity was a fact of nature, since in his view natural resources were finite and populations grew exponentially. In his critique, Harvey noted that Malthus had arrived at his conclusion by adhering to an empiricist position – the 'self-evidence' of scarcity – while at the same time assuming two a priori postulates: that food is necessary but limited (finite resources), and that the passion between the sexes was constant. On this basis, the empiricist Malthus was able to explain what he and others 'saw' in the world, scarcity, to be the outcome of natural laws rather than social relations, and thus inevitable. No action by individuals or the state could change this. Against Malthus, Harvey argued that scarcity was specific to a particular mode of production, and was thus social, not natural. This was true for two reasons. First, resources were not finite because nature could itself be transformed by labor and technology. And second, surplus population did not have a natural cause ('passion of the sexes'), but was an effect of capitalist production, which in its substitution of fixed capital for variable capital, continuously produced an industrial reserve army of poor and unemployed. Hence, as Harvey reiterated in *Justice, Nature and the Geography of Difference*: 'To say that scarcity resides in nature and that natural limits exist is to ignore how scarcity is socially produced and how "limits" are a social relation within nature (including human society) rather than some externally imposed necessity' (1996: 147).

A second contribution of Smith's 'production of nature' thesis was found in its challenge to dominant strains of environmentalism, especially 'preservationism' in North America. As writers like William Cronon (1995) and Richard White (1995) have noted, the nature–society divide has long provided support for the view that nature is truly 'itself' only in the absence of humans. Hence, to 'save' nature – the objective for many Western environmentalists at the time – it followed that one must remove humans. This fetishization of wilderness has at times resulted in immense violence to people and communities whose livelihoods are most closely intertwined with apparently 'natural' landscapes, and whose labor has in many cases constituted the nature that environmentalists have sought to save (see Neumann, 1998; Braun, 2002). But, if nature is 'social', the conceptual scaffolding for preservation becomes shaky, for it is not clear what nature is without humans.

Cronon and White push this further, suggesting that the idea that nature is a separate realm leads us to overlook the ways in which nature is continuously produced in the banal spaces and practices of everyday life. For, if society is 'here' and nature is 'over there', it becomes difficult to fully comprehend how social practices – in cities and their suburbs, for instance – are simultaneously ecological practices. Moreover, by fetishizing wilderness,

attention and resources are diverted from the environments in which we live, a very real problem for many poor and minority communities, for whom environmentalism is more frequently about toxicity and health than about big trees and mega-fauna (Di Chiro, 1995). Indeed, in the United States the 'externalization' of nature – and the constitution of environmentalism around wilderness – has arguably resulted in an environmental movement centered on the concerns of white, middle-class citizens who have the means to physically distance themselves from environmental risks, and to pursue recreation in 'wild' nature. In important respects Smith's argument, along with the concerted efforts of anti-racist and labor activists, has borne fruit, as environmental NGOs and environmental scholars have increasingly brought the city into view as a vital, and highly politicized, socio-ecological space (see Harvey, 1996; Gandy, 2002; Keil and Desfor, 2004; Swyngedouw, 2004).

Finally, the production of nature thesis has enabled scholars and activists to understand specific environments as the contingent outcomes of wider social, economic and political forces. This has increased attention to how particular socionatures are produced in advanced capitalism, and with what consequences for humans and non-humans alike. Equally as important, it has led scholars and activists to recognize that the future of nature remains open, to be determined by history and politics, rather than given in advance. Stated differently, Smith's 'production of nature' thesis demands that we come to terms with the reality that to live, humans must 'leave marks on the world' (Cronon, 1995) and take responsibility for producing natures that are socially just and ecologically sustainable. As Smith (1996: 50) puts it, ecopolitics must address 'how, by what social means and through what social institutions, is the production of nature to be organized? How are we to create democratic means for producing nature?'

These strengths notwithstanding, Marxist approaches to nature and society have been subjected to considerable criticism. One of the most common complaints targets the reductionism and economism of many Marxist accounts. The production of nature, critics contend, is not solely determined by the imperatives of capitalism, since the 'economy' is itself embedded in numerous non-economic practices (see Gibson-Graham, 1996; Haraway, 1997; Braun, 2002). When analysis is pitched in terms of grand abstractions such as 'capital' or the 'state', critics charge, the micro-practices that constitute socio-nature fall from view. The result is that, despite adding the adjective 'political', political ecology has often had precious little politics!

An equally common complaint is that the production of nature thesis is deeply anthropocentric since it places humanity – and work – at the heart of its analysis. This can be understood in two ways. First, it is often anthropocentric in its ethical–political orientation, since it privileges the needs and desires of humans (some use the phrase 'productivist' for this). But it is frequently anthropocentric in an analytical sense too, for it tends to put human action at the center of its accounts of nature's production. The problem here is that by focusing on human labor the production of nature thesis risks 'overcoming' the nature–society divide merely by inverting the solution of the environmental determinists. Where the latter imagined society to be determined by the environment, historical materialists often imagine the environment to be determined by society. In their accounts it is only humans, or capital, that acts; everything else is simply subject to the work of humanity or, more abstractly, the dictates of capital. In the words of David Harvey (1996), 'Capital circulation … has made the environment what it is'. This comes perilously close to reasserting the subject/object dichotomy of the Enlightenment, where human ingenuity is played out on an earth that is imagined as static and inert. Certainly not all Marxists fall into this trap. Richard Lewontin (1982), for instance, understands non-human nature to have its own history which interacts 'dialectically' with humans, and Marxist geographers like George Henderson (1999) and Gavin Bridge (2000) have emphasized what Castree (2003) calls the 'intransigence' of nature, although these accounts still begin with the acts of humans and bring non-humans into the picture only to the extent to which they 'resist' the intentions of humans. Political ecology, then, may have even less ecology than politics.

From this follows a third criticism – that dialectics is much less successful at overcoming the nature–society divide than it appears to be. While its metaphors – metabolism, mediation, interaction – insist on a relational approach to nature and society, in important respects they still presume the prior existence of the two categories even as they seek to multiply the connections between them. In many ways this remains too crude an analytical device that at best renders the divide more permeable and at worst deepens the original error. It is, in a sense, still too committed to the initial categories: in the words of Sarah Whatmore (1999: 25), 'far from challenging this a priori categorization of the things of the world, dialectics can be seen to raise its binary logic to the level of a contradiction and engine of history'. For critics like Bruno Latour (2004), its categories are still too abstract and too grand. For as he puts it, there is no 'Nature in general' any more than there is a 'Society' that exists as a single unified totality; there are only specific networks composed of various human and non-human actants, which are of greater or shorter length, more or less dense, and 'hold together' for longer or shorter periods of time.

OVERCOMING THE LABOR OF DIVISION? AMODERN ONTOLOGIES AND THE MICRO-POLITICS OF NATURE

Clearly the 'geographical experiment' has proved challenging. Most attempts to overcome the nature–society dualism have either reinforced the distinction or collapsed one pole of the binary into the other. The difficulty may arise in part from how the problem has been posed. Many geographers have sought to understand how nature and society 'relate', or how to 'bridge' physical and human geography. The point of departure therefore presupposes the categories it seeks to overcome.

With this in mind, we can perhaps appreciate the significance of recent attempts to get beyond the nature–society divide by refusing the categories altogether. Variously referred to as amodern ontologies, new materialisms, or actor-network theories, these approaches posit a world that cannot be divided into 'nature' and 'society', 'non-human' and 'human', since the nature of being is such that all entities are hybrids, composed through multiple connections with other entities and through continuous exchanges of information and matter. From this perspective the world has never taken the form of Enlightenment antinomies, despite our enduring belief that it has (Latour, 1993). As we will see, these approaches have also brought renewed attention to the relation between politics and nature, since 'being' is now seen as fluid, and entities – from bodies to ecosystems – are now viewed as the precarious outcomes of multiple forces. The future of these hybrid assemblages is thus conceived in terms of an 'ontological politics', a curious phrase that disrupts the association of ontology with the immutable and eternal, and the associated assumption that ontological investigations are, as Smith (2001) suggests, the enemy of revolution.

What does it mean to conceive the world in terms of an amodern ontology?[8] We can identify several key propositions. To begin, it places emphasis on becoming rather than being; that is, it rejects the idea that there is a fixed 'ground' or 'order' to the cosmos, and instead understands being as dynamic and historical. This amplifies a strain of Western philosophy that reaches back to Lucretius and the Epicureans and includes diverse figures such as Baruch Spinoza, Henri Bergson, Gilles Deleuze and Michel Serres. Amodern ontologies also place a premium on connectivity: objects are understood as composites rather than 'things-in-themselves', constituted as part of what Michel Callon and John Law (1995) call a 'hybrid *collectif*'.[9] This is true of the physical composition of things, and, as will see later, their capacities to 'act'. Human bodies, for instance, are composed through their relationships with other bodies; without swapping properties with non-humans – through breathing and eating, or through interactions with viruses and bacteria or articulation with machines – human bodies would not be what they are. Nobody – and no body – is a monad. The same is true for any entity or organism. Notably, from this perspective the moral panic today over the 'mixing' of humans and non-humans in biotechnology is misplaced (cf. Fukuyama, 2002). This is not because there are no good reasons to be concerned about how bodies are being composed in techno science, but because by definition the body cannot pre-exist its construction (figuratively and materially; see Haraway, 1997). What critics like Fukuyama (2002) miss in their worry that we are becoming 'post-human' is that from the very outset the human is post-human, since it is composed of more than itself, and thus always 'becoming other' than what it is. Worse, in the name of defending a human essence, they dodge what is arguably the more important ethical– political task: vigilant attention to the making of bodies. In other words, from the perspective of an amodern ontology, the pressing question is not 'How do we preserve the body (or nature) from its (technological) outside?' but 'What is to become of the body?' and 'How shall we compose bodies and with what consequences?' Politics and ethics come to be about composition, not essence.

These are important ethical–political matters to which I will return. Before I do, let me identify some additional characteristics of amodern or non-foundational ontologies. First, they displace the classic subject/object dichotomy of the Enlightenment. Entities are simultaneously subjects and objects, or, to borrow Michel Serres' phrase, they are 'quasi-subjects, quasi-objects' (Serres and Latour, 1995). This is because agency – the capacity to cause affect – is not conceived as a property inherent to objects, but an emergent effect, achieved through connection with other things (Callon and Law, 1995; Whatmore, 1999). Further, because agency is distributed across the networks in which entities are constituted – albeit unevenly – it neither belongs solely to humans, nor is defined solely by consciousness or intentionality. Indeed, from within the terms of amodern ontology it makes little sense to imagine the world in terms of discrete subjects that exist separate from and opposed to a world of inert objects. Like everything else, humans are acted upon as much as they act. Latour offers the word 'actant' to capture this relational understanding of agency, and to challenge our habit of conceiving of agency only in terms historically applied to humans (will, decision, desire).

A second common characteristic of these so-called new materialisms is a commitment to a philosophy of immanence or, in Deleuze's (1988) terms, a 'practical philosophy'. This simply refers to the idea that philosophy must avoid taking recourse to any supplemental dimension, or transcendental cause – God, History, Capital, Spirit, Nature – to explain what exists. Being is an effect of the material practices that constitute it. A philosophy of immanence, Michael Hardt (1993: xiii) explains, 'refuses any deep or hidden foundation of being' since being 'is fully expressed in the world' and is 'cause of itself'. Or, as Deleuze (1988: 122) puts it, there is only one 'common plane of immanence on which all bodies, all minds, and all individuals are situated'. Thinking in this way can be unsettling, since it refuses the age-old categories Nature and Society. Moreover, it provides no firm ground for being and allows for no hidden mechanisms that lie 'beneath' or 'above' the world and determine its course. Ultimately, it brings us to a critical insight: that any ontology that thinks being as becoming grounds being in politics (see Casarino, 2002: xviii).

Given their long engagement with the 'geographical experiment', it should not surprise us that environmental geographers have been quick to recognize the promise of such ontological investigations. Sarah Whatmore (2002), for instance, has drawn directly on this tradition to explore what she calls 'topologies of wildlife', by which she means the diverse practices, forces and connections that constitute and differentiate 'wild' animals and endow them with the capacity to act. As she explains, leopards in the Roman Coliseums of the second century CE, or elephants today in Botswana's wildlife refuges or the Paignton Zoo in England, are each unique outcomes of particular spatio-temporal networks that at once constitute and mix together people, animals, machines and texts. These 'wildlife' networks are topological – they fold and refold space and time and undermine all efforts to situate plants, animals and humans on a fixed Cartesian grid (see also Serres and Latour, 1995). The implications for geographical research are enormous. For, as Whatmore (2002: 3) explains, if one accepts these amodern ontologies, one no longer begins with the division of things into classes, or with the categories of nature and society, but with a recognition of the intimate, sensible, hectic bonds through which people and plants, devices and creatures, documents and elements take and hold their shape in relation to each other in the fabrications of everyday life.

Thus, to take one example, any accounting of 'becoming elephant' must begin with the myriad practices and relations through which any specific elephant is composed – from the biologist's stud books and the conservation practices of environmental NGOs, to the condition of specific habitats and the forms of communication between individual animals. Today, many of these networks are increasingly global, and increasingly dense. They are suffused with power and politics. Whether the topic is the health of fish in the Mekong River, GMO food, or even apparently 'social' issues like public education in the United States (which today is inconceivable without enrolling the force of atoms, the flow of rivers, fossil fuels, or wind to generate electricity), each is composed of 'integrated networks' in which not all the actors are human.

Viewed in this light, categories such as Nature and Society are not found in the world of matter; they exist as effects of the 'labors of division' by which we separate the world into distinct classes of being and assign them to one pole (elephants and SARS to nature) or to the other (cities and classrooms to society). Further, amodern ontologies challenge the anthropocentrism of much social theory, including Marxism. Humans are no longer understood as the only actors in these hybrid collectives; our 'social' worlds are always already 'more-than-human'.[10] Ironically, this brings us back to the concerns of environmental determinists and cultural ecologists, who insisted that the non-human be taken into account before any explanation of human affairs be considered adequate. In contrast to those approaches, however, today's new materialisms do not collapse the social into a transcendent nature, they displace the question of nature and society by abandoning the categories entirely.

What is gained by this conceptual turn? One of the clear advantages of amodern ontologies is that they allow us to think about entities and processes that dualist thought had rendered invisible. As Latour (1993) notes, the 'modern Constitution' that divided the world into nature and society had sanctioned an immense ignorance: it led us to imagine a world consisting of two distinct ontological domains, all the while leaving us blind to the hybrid networks in which people, things and politics were continuously composed. Because critical explanation began from the poles, 'the middle was simultaneously maintained and abolished, recognized and denied, specified and silenced' (quoted in Whatmore, 1999: 24). This was far from an innocent oversight, since it consistently produced 'blind spots' in political thought and action (see Braun and Disch, 2002). Amodern thought, in contrast, attends to the 'missing masses' and the 'excluded middle'; it brings out of hiding all that which could not be conceived within dualist conceptions of nature and society and gives them standing within political discourse.

It is not difficult to see why Latour (2004) has recently argued that political ecology – by which he means green political movements – must have nothing to do with nature. But this goes far beyond being philosophically consistent; it has everything to do with how we imagine political life. For Latour, abandoning nature requires that we tackle head-on the task of 'cosmopolitics' (see also Stengers, 1996–7). Politics, for Latour, is about composition. There are similarities here with Smith's call for a 'political theory' of nature: the form of the world

is not given in advance or dictated by a transcendent nature, it must be made. For Latour, though, the world is not composed by human actions alone: non-human entities – like viruses, prions and stem cells – continuously clamor to be taken into account and given standing within these collectives (or excluded from them). Cosmopolitics, then, is about the composition of a 'common world', about deciding what to take into account, how to order it, and what to exclude from it, in the ongoing process of constituting the collectives within which 'life' is at once enabled and shaped. For Latour, this process never ends; it is continuously renewed as new realities or previously excluded voices demand to be heard. The problem with appeals to transcendent nature, then, is that they short-circuit the politics of composition.

Like other approaches to resolving the society–nature divide, the turn to amodern ontologies has its critics. Some worry that this philosophical tradition is too assured about its ontological claims, and evades responsibility for the initial act of positing that founds any ontology (Derrida, 1994). Others worry that amodern approaches leave us unable to speak of 'macro' processes or structures, since the emphasis is placed resolutely at the level of specific practices (Castree, 2002). Still others have suggested that these approaches 'flatten' the world, and leave us unable to understand qualitative differences between, say, human and non-human agency, since all are treated alike. These latter criticisms, I would suggest, may be less damning than they appear. The first because amodern ontologies do not foreclose the possibility of large-scale phenomena, they only refuse to understand them as standing apart from the micro-practices that constitute them (see Mitchell, 2002; Read, 2003). Hence, in Latour's words, capitalism may be pervasive, but it remains 'local at all points' in the sense that even its 'global' forms consist of specific situated practices. Likewise, advocates of amodern ontologies do not argue that all actants are the same, only that the capacity of an actant to produce effects is not a property of the actant 'itself' but the networks that produce them as more or less powerful. Humans may have more power than lions, but this is an effect of the retinue of objects that humans can enrol. Strip humans of these objects and the tables are decidedly turned.

Finally, many critics have worried over the lack of normative foundations, and the lack of guidance for what an amodern politics might look like. To be sure, approaches such as actor-network theory give us tools to interrogate networks and their consequences, and to carefully consider our connections to other organisms and things. As Whatmore (1999: 30) notes, it calls us to think carefully about the 'everyday business of living in the world'. But this still fails to answer the questions of ends, and it leaves the question of means equally unanswered.

How is this 'good common world' to be composed, by whom, and toward what ends? What institutional forms are necessary? And what does it mean, as Latour suggests, to 'allow the collective to proceed according to due process to the exploration of the common world' (2004: 109). How are excluded voices heard? And what is 'due process' in a world where power is so unevenly distributed? Without a thorough analysis of the possibilities for the sort of deliberative politics that Latour envisions, his political philosophy remains hopelessly idealist and avoids the hard questions of interrogating the present historical moment for both its limits and its potential (see Wainwright, 2005). So, while Latour refuses to separate politics from nature, geographers and political theorists may find his understanding of politics far too naïve.

CONCLUSION

It has been common in geography over the past two decades to decry the nature–culture dualism and its manifestation as an internal divide between human and physical geography. It is argued that this divide must be 'overcome' in order to achieve a more holistic and integrated understanding of the world. Less often noted is that geographers have called for this – and announced its achievement – many times in the past. There are good reasons to be wary of such announcements. As I have shown, efforts to overcome the divide have often merely subsumed one pole of the binary under the other. Environmental determinists imagined nature as a transcendent realm that determined in advance the course of human history. Cultural ecologists doubted the grand abstractions of the environmental determinists but ultimately subsumed culture under nature, as a subtle and slow adaptation to environmental conditions. Historical materialists were justifiably sceptical of the determinism of both (whether or not the determinism was explicit or implicit), and rightly revealed that such arguments labored on behalf of existing social and economic hierarchies, since it assumed that they had natural rather than political causes.

Yet, in their desire to bring politics back to center stage, they were often guilty of merely reversing the environmental determinist's arrows of causation, substituting a social determinism for the environmental determinism of early geographers. At best they posited nature and society as a unity, where the two realms were in a continuous dialectical 'interaction'. More recent scholars have turned to actor-network theory and its amodern ontologies to imagine a different solution, one that rejects attempts to 'relate' Nature and Society and dispenses with the categories altogether. Whether the amodern ontologies of scholars like Latour, Stengers and Whatmore manage to displace the nature–culture dualism from its central place in Western thought and culture – and in Anglo-American geography – remains to be seen. Certainly in the United States, if not elsewhere, the insistent materialism of these writers is just as insistently opposed by resurgent appeals to God, Morality and Nature – often together. It is not at all clear that the dualism is on its last legs.

These debates are not merely ontological and epistemological debates over the nature of 'reality' and how we know it. They are intricately entangled with politics and how we conceive of political life, for the nature–society dualism has always been part of Western political thought, and how it has been resolved has had immense consequences. For this reason my account has been an interested account, since it holds some approaches more promising than others. Over the course of this chapter I traced a movement from a 'politics from nature' to a 'politics of nature'. This movement is as significant as my account is interested. The former – politics from nature – imagines nature as a transcendent realm that determines social and political forms. While it promises to overcome the nature–society dualism of the Enlightenment, it does so at the cost of politics. Ultimately, nature stands as legislator and judge, a sovereign power whose eternal and universal laws determine human life, and against which nothing can be done. Even apparently 'human' constructions – like nations, states and territory – can be subsumed under this transcendental naturalism. The latter – politics of nature – imagines the 'cosmos' as a hybrid assemblage of heterogeneous entities and relations that cannot be divided into Nature and Society and that has no pre-given form. Where the former presupposes Nature and Society as oppositions, and solves their relation by privileging the first term over the second, the former dispenses with the categories altogether and instead imagines politics to be about the future arrangement of this cosmos – a politics of composition or, more properly, cosmopolitics. Arguably this puts the 'politics' back in political geography, for the organization of the cosmos – the shape of its collectivities – can no longer be explained through appeal to a dimension prior to, or above, the plane of composition within which common worlds are constituted. This suggests that the present is always a moment of great danger and immense hope, for the future shape of these common worlds – and what strategies, institutions and practices are necessary to create them – is what remains to be determined.

NOTES

1 Huntington at times qualified such claims, leaving room for culture and politics, especially when

dealing with apparent anomalies. See Huntington (1915: 279).

2 In 1919 Mackinder summarized his conclusions in a now oft-quoted verse:

> Who rules East Europe commands the heartland.
> Who rules the Heartland commands the World Island.
> Who rules the World Island commands the world.

3 There is some debate over whether Ratzel was influenced more by Darwin or Lamarchk (see Livingstone, 1992).

4 Semple's *Influence of Geographic Environment* (1911) is arguably her most determinist text, where the arrow of causation runs regularly, and directly, from nature to politics: 'Back of Massachusetts' passionate abolition movement it sees the granite soil and boulder-strewn fields of New England; back of the South's long flight for the maintenance of slavery, it sees the rich plantations of tide water Virginia and the teeming fertility of the Mississippi bottom lands' (p. 257).

5 Diamond's environmental determinism is a methodological presupposition: 'The book's subject matter is history, but the approach is that of science – in particular, that of historical sciences such as evolutionary biology and geology' (1999: 26).

6 Cultural ecology was not a unified field. For the geographer Carl Sauer the point was to interpret the *morphology of landscapes* in order to show how they bore the imprint of culture. Many ecological anthropologists, on the other hand, were more interested in explaining the existence of particular cultural forms, and wondered about how these forms revealed the influence of the environment, rather than the reverse. This is too stark a division – cultural geographers and ecological anthropologists borrowed freely from each other and often moved between positions – but it suggests that cultural ecology was at once multiple and internally divided.

7 For surveys see Watts (2001) and Robbins (2004).

8 For more, see Whatmore (1999, 2002), Latour (2004) and Deleuze and Guattari (1987).

9 Callon and Law use the term *collectif* rather than 'collectivity' in order to avoid the notion that a collectivity consists of pre-existing entities that come together to form something larger. A *collectif* refers to the relations and exchanges that constitute entities and imbue them with agency. Latour (2004) uses the phrases 'collectivity' and 'common world' interchangeably with *collectif*.

10 The same is true for how we understand knowledge, and the relation between science and politics. Knowledge – like all other entities – comes to be seen also as the outcome of material

practices, and hence defined by, and constituted within, actor-networks. Within these networks, non-human actants (machines, organisms) are not passive or inert. Crucially, this approach to knowledge sidesteps the endless debates between 'constructivism' and 'realism'. Our ideas are not 'all in our heads' because knowledge is an effect of practices: we know the world not by separating ourselves from it, but through our connections to it. At the same time, knowledge does not simply 'mirror' the world, because it can be known only through processes of 'translation'. Thus, science and politics, or truth and power, are inseparable.

REFERENCES

Bassin, M. (2003) 'Politics from nature', in J.A. Agnew et al. (eds), *A Companion to Political Geography*. Malden, MA: Blackwell, pp. 13–29.

Blaikie, P. and Brookfield, H. (1987) *Land Degradation and Society*. London: Methuen.

Braun, B. (2002) *The Intemperate Rainforest: Nature, Culture and Power on Canada's West Coast*. Minneapolis: University of Minnesota Press.

Braun, B. and Disch, L. (2002) 'Radical democracy's "modern Constitution"', *Environment and Planning D: Society and Space*, 20(5): 505–11.

Bridge, G. (2000) 'The social regulation of resource access and environmental impact: production, nature and contradiction in the US copper industry', *Geoforum*, 31: 237–56.

Butzer, K. (1989) 'Cultural ecology', in C.J. Willmott and G. Gaile (eds), *Geography in America*. Washington, DC: Association of American Geographers and National Geographic Society.

Callon, M. and Law, J. (1995) 'Agency and the hybrid collectif', *South Atlantic Quarterly*, 94(2): 481–507.

Casarino, C. (2002) *Modernity at Sea: Marx, Melville, Conrad in Crisis*. Minneapolis: University of Minnesota Press.

Castree, N. (1995) 'The nature of produced nature: materiality and knowledge construction in Marxism', *Antipode*, 27(1): 12–48.

Castree, N. (2000) 'The production of nature', in E. Sheppard and T.J. Barnes (eds), *A Companion to Economic Geography*. Oxford: Blackwell, pp. 275–89.

Castree, N. (2002) 'False antithesis? Marxism, nature and actor-networks', *Antipode*, 34(1): 111–46.

Castree, N. (2003) 'Commodifying what nature?', *Progress in Human Geography*, 27(3): 273–97.

Cosgrove, D. and Jackson, P. (1987) 'New directions in cultural geography', *Area*, 19: 95–101.

Cronon, W. (1995) 'The trouble with wilderness; or, getting back from wrong nature', in W. Cronon (ed.), *Uncommon Ground: Toward Reinventing Nature*. New York: W.W. Norton, pp. 69–90.

Deleuze, G. (1988) *Spinoza: Practical Philosophy*, trans. R. Hurley. San Francisco: City Lights Books.

Deleuze, G. and Guattari, F. (1987) *A Thousand Plateaus: Capitalism and Schizophrenia*. Minneapolis: University of Minnesota Press.

Derrida, J. (1994) *Specters of Marx: The State of Debt, the Work of Mourning, and the New International*, trans. Peggy Kamuf. New York: Routledge.

Diamond, J. (1999) *Guns, Germs and Steel: The Fates of Human Societies*. New York: W.W. Norton.

Di Chiro, G. (1995) 'Nature as community: the convergence of environment and social justice', in W. Cronon (ed.), *Uncommon Ground: Toward Reinventing Nature*. New York: W.W. Norton, pp. 298–320.

Duncan, J.S. (1980) 'The superorganic in American cultural geography', *Annals of the American Association of Geographers*, 70: 181–98.

Fukuyama, F. (2002) *Our Posthuman Futures: Consequences of the Biotechnology Revolution*. New York: Farrar, Straus & Giroux.

Gandy, M. (2002) *Concrete and Clay: Reworking Nature in New York City*. Cambridge, MA: MIT Press.

Geertz, C. (1963) *Agricultural Involution: The Process of Ecological Change in Indonesia*. Berkeley: University of California Press.

Gibson-Graham, J.-K. (1996) *The End of Capitalism (As We Knew It): A Feminist Critique of Political Economy*. Cambridge, MA: Blackwell.

Gupta, A. and Ferguson, J. (eds) (1997) *Culture, Power and Place: Explorations in Critical Anthropology*. Durham, NC: Duke University Press.

Haraway, D. (1997) *Modest Witness@Second Millenium.Female Man© Meets OncoMouse™*. New York: Routledge.

Hardt, M. (1993) *Gilles Deleuze: An Apprenticeship in Philosophy*. Minneapolis: University of Minnesota Press.

Harris, M. (1974) *Cows, Pigs, Wars and Witches: The Riddles of Culture*. New York: Random House.

Harvey, D. (1974) 'Population, resources and the ideology of science', *Economic Geography*, 50: 256–77.

Harvey, D. (1996) *Justice, Nature and the Geography of Difference*. Oxford: Blackwell.

Heidegger, M. (1977) *The Question Concerning Technology and Other Essays*, trans. William Lovitt. New York: Harper & Row.

Henderson, G. (1999) *California and the Fictions of Capital*. Oxford: Oxford University Press.

Huntington, E. (1915) *Civilization and Climate*. New Haven: Yale University Press.

Keil, R. and Desfor, G. (2004) *Nature and the City: Making Environmental Policy in Toronto and Los Angeles*. Tucson: University of Arizona Press.

Landes, D. (1999) *The Wealth and Poverty of Nations*. New York: W.W. Norton.

Latour, B. (1993) *We Have Never Been Modern*. Cambridge, MA: Harvard University Press.

Latour, B. (2004) *Politics of Nature: How to Bring the Sciences into Democracy*. Cambridge, MA: Harvard University Press.

Lewontin, R. (1982) 'Organism and environment', in H. Plotkin (ed.), *Learning, Development and Culture*. Chichester: John Wiley.

Livingstone, D.N. (1992) *The Geographical Tradition*. Oxford: Blackwell.

Mackinder, H. (1951 [1904]) *The Geographical Pivot of History*. London: The Royal Geographical Society.

Mackinder, H. (1996 [1887]) 'On the scope and methods of geography', in J.A. Agnew et al. (eds), *Human Geography: An Essential Anthology*. Cambridge, MA: Blackwell, pp. 155–72.

Marx, K. (1967 [1887]) *Capital*, Vol. 1. London: Lawrence & Wishart.

Marx, K. (1975[1844]) 'Early economic and philosophical manuscripts', in L. Colletti (ed.), *Karl Marx: Early Writings*, ed. Harmondsworth: Pelican.

Merrett, C. (2003) 'Debating destiny: nihilism or hope' in *Guns, Germs and Steel*?' *Antipode*, 35(4): 801–7.

Mitchell, T. (1988) *Colonizing Egypt*. Cambridge: Cambridge University Press.

Mitchell, T. (2002) *Rule of Experts: Egypt, Technopolitics, Modernity*. Berkeley: University of California Press.

Neumann, R.P. (1998) *Imposing Wilderness: Struggles over Livelihood and Nature Preservation in Africa*. Berkeley: University of California Press.

Osterholm, M. (2005) 'Preparing for the next pandemic', *New England Journal of Medicine*, 352(18): 1839–42.

Rappaport, R. (1967) *Pigs for the Ancestors: Ritual in the Ecology of a New Guinea People*. New Haven: Yale University Press.

Read, J. (2003) *The Micro-Politics of Capital: Marx and the Pre-History of the Present*. Albany: State University of New York Press.

Robbins, P. (2004) *Political Ecology*. Oxford: Blackwell.

Said, E. (1993) *Culture and Imperialism*. New York: Knopf.

Semple, E. (1911) *Influences of Geographic Environment*. New York: Russell & Russell.

Serres, M. and Latour, B. (1995) *Conversations on Science, Culture and Time*, trans. R. Lapidus. Ann Arbor: University of Michigan Press.

Sluyter, A. (2003) 'Neo-environmental determinism, intellectual damage control and nature/society science', *Antipode*, 35(4): 813–17.

Smith, N. (1984) *Uneven Development: Nature, Capital and the Production of Space*. Oxford: Blackwell.

Smith, N. (1996) 'The production of nature' in G. Robertson, M. Mash, L. Tickner, J. Bird, B. Curtis and T. Putnam (eds), *Future Natural: Nature, Science, Culture*. London: Routledge, pp. 35–54.

Smith, N. (2001) 'New geographies, old ontologies: optimism of the intellect', *Radical Philosophy*, 108: 21–30.

Stengers, I. (1996–7) *Cosmopolitiques*, 7 vols. Paris: La Découverte.

Steward, J. (1955) *Theory of Culture Change: The Methodology of Multilinear Evolution*. Urbana: University of Illinois Press.

Swyngedouw, E. (2004) *Social Power and the Urbanization of Water*. Oxford: Oxford University Press.

Vayda, A. and Rappaport, R. (1968) 'Ecology, cultural and non-cultural', in J. Clifton (ed.), *Introduction to Cultural Anthropology: Essays in the Scope and Methods of the Science of Man*. Boston: Houghton Mifflin, pp. 477–97.

Wainwright, J. (2005) 'Politics of nature: a review of three recent works by Bruno Latour', *Capitalism Nature Socialism*, 16(1): 115–27.

Watts, M. (2001) 'Political ecology', in E. Sheppard and T.J. Barnes (eds), *The Companion of Economic Geography*. Oxford: Blackwell, pp. 257–74.

Whatmore, S. (1999) 'Human geographies: rethinking the "human" in human geography', in D. Massey, J. Allen and P. Sarre (eds), *Human Geography Today*. Cambridge: Polity Press, pp. 22–39.

Whatmore, S. (2002) *Hybrid Geographies: Natures, Cultures, Spaces*. London: Sage.

White, R. (1995) 'Are you an environmentalist or do you work for a living?', in W. Cronon (ed.), *Uncommon Ground: Toward Reinventing Nature*. New York: W.W. Norton, pp. 171–85.

Zimmerer, K. (1996) 'Ecology as cornerstone and chimera in human geography', in C. Earle et al. (eds), *Concepts in Human Geography*. Lanham, MD: Rowman & Littlefield, pp. 161–88.

The State in Political Ecology: A Postcard to Political Geography from the Field

Paul Robbins

INTRODUCTION

Between the years 1975 and 2000, India became the second-largest borrower from the World Bank Group[1] in the forestry sector, amassing debts of $830 million to fund programs in the areas of 'technical assistance', 'community forestry', and 'social forestry'. These programs are administered largely through state forestry agencies under the umbrella direction of the Government of India's Ministry of Environment and Forests, since, like most underdeveloped nations, roughly 90 percent of the country's 64 million hectares of forest are under state control. The collective goals of this state effort – to halt forest degradation, enhance and extend forest cover, and achieve sustainable use of forests, especially by the poor – have been largely successful according to internal Bank oversight (Kumar et al., 2000).

But the effects of these state efforts have been highly uneven throughout the country. While reforestation (recovering lost forest) and afforestation (planting forests where they have never been before) are ongoing, some of the emerging forests are effectively 'green deserts' in terms of either biodiversity or useful species for the poor, owing to the predominance of single-species monoculture. While in some states public participation has meant the increased involvement of marginal communities, in most others corruption among local officials has made forestry an expanding form of state enterprise to extract the already thin surpluses of rural producers. Many forest projects have provided new employment opportunities and resources for rural communities; others have actually appropriated lands necessary for the survival of rural livelihoods. The outcomes of state/nature interaction seem multiple and contradictory.

The explanation for all of this complexity is of course that the state itself is not monolithic, coherent, or temporally and spatially stable, a well-known fact often touted in the field of political geography. As described by Sidaway (2003), the state as a territorial sovereign power is always contested and unevenly developed. This tells us quite a bit about the chaotic outcomes of Indian forestry. Some local states are captured more fully by powerful timber interests, for example, while others have more powerful representations of minorities in both local and central institutions.

A second reason for this complexity, however, and one frequently overlooked, is that the state as an environmental actor is arguably inherently more internally fractured than it is as a provider of welfare services, paver of roads, or even maker of wars. Because forest/people relationships in India are spatially varied and tied not only to differences in production but also differences in soils, water, and floral/faunal interactions, the nature/state relationship must play itself out unevenly; many schisms and complications are unique to governing non-human nature.

This second reality, coupled with the overall complexity of the state more generally, has led to careful investigation of environmental governance in the field of political ecology, a research tradition here defined as 'empirical, research-based explorations to explain linkages in the condition and change of social/environmental systems, with explicit consideration of relations of power' (Robbins, 2004: 12). Given the inherent complexity in the relationship of sovereign governance and non-human nature, the resulting portrait of the state is unsurprisingly contradictory, with the apparatus of government sometimes appearing strong, other times weak, in some cases appearing menacing, in others sympathetic. Even so, the empirical studies provided in political ecology, usually local, often practical, and always sensitive to power, may provide several general lessons.

On the other hand, ongoing theoretical efforts within political geography – exploring the construction and effects of territorialization, revealing the structure and character of the state, and understanding the scaling of political power and the strategic use of scale in political action – provide tools that are almost universally overlooked in political ecology. These conceptual debates provide huge potential contributions explaining the problems and pitfalls of state management of nature, so evident in the case of India outlined above, and so open the door to mutual learning on the part of political geography and political ecology.

This chapter reviews the characterizations of the state in political ecology research, describing three distinct, dominant, and somewhat contradictory depictions. First, research shows that the state tends toward ecological simplification, even while non-human responses to state efforts almost always result in increased ecological complexity. Second, political ecology has demonstrated that the state interacts with diverse human and non-human players to produce networks that mediate between multinational capital, local producers, and soil, water, and plant ecologies. Third, work shows that the state is an institutionalized knower, with its own fragmented epistemology, which simultaneously erases and produces environmental knowledge.

Reviewing these three approaches to the state reveals areas where political geographic theory can provide lessons for investigation in political ecology. At the same time, however, the review also shows areas where the empirical insights of political ecology can shed light on debates on the complex and poorly understood nature of state power, capacity, and territoriality in political geography. In particular, debates over the nature of scale sit at the heart of these questions, and the serious engagement of each field with the other is prerequisite to future progress in understanding socio-environmental change.

THE PLANTATION AND MONOCULTURE STATE: PRODUCING SIMPLIFIED LANDSCAPES

In his treatise *Seeing Like a State: How Certain Schemes to Improve the Human Condition Have Failed*, James Scott articulates a dominant characterization of the state in political ecology. He suggests that it is a central propensity of the state to tend towards simplification and legibility (Scott, 1998). In this sense, he suggests that:

Certain forms of knowledge and control require a narrowing of vision. The great advantage of such tunnel vision is that it brings into sharp focus certain limited aspects of an otherwise far more complex and unwieldy reality. This very simplification, in turn, makes the phenomenon at the center of the field of vision more legible and hence more susceptible to careful measurement and calculation. (p. 12)

This observation Scott extends to all kinds of governance, from the taking of a population census to the creation of new modernist cities. But no clearer imperative to simplification exists than the complexity of the natural world. To map, categorize, and take a census of a discreet non-human nature, an effort that is prerequisite to commanding, controlling, and parceling land, water, plants and animals, is actually something of an absurdity. When counting nature, in all of its infinite complexity, what do you count? When mapping nature, where should bounds be drawn? Under high modernism, in order to govern nature, the state must simplify it, by creating codes, records, maps, and categories that are legible.

This argument can be further specified. It is possible to argue that the metrics used in this mapping and categorizing of nature, further create incentives for the behaviors of state agents. The resulting patterns layered into the natural world come to closely resemble the categories, measures, and organization of the state agents assigned to assess it. In this way, landscapes are 'reverse engineered' to fit the simplified categories used in their description (after Winner, 1977; Veregin, 1995).

Legible Forests

This is perhaps nowhere more clearly seen than in Scott's premier example, one seen repeatedly in colonial and post-colonial environmental governance: state forestry. Forests and trees themselves are infinitely complex systems, with a vast range of ecosystem services (soil binding, carbon sequestration, runoff control, faunal and avian habitat, etc.) as well as human uses (shade, fruits, medicinal bark, construction materials, etc.). But in developing modern state forestry, the state historically came to require statistical abstractions to stand in for actual trees. Typically, with the fiscal demands of revenue capture in mind, the state chose volumes of lumber or firewood (board-feet, cubic meters, etc.) for the representation of the forest and its production.

German forestry science, which came to dominate the environmental logic of states throughout Europe and North America United States, is most outstanding and precedent-setting in this regard. Applying the scientific standards of 'sustained yield' (a measure of optimal levels of extraction before the reproduction of the system is surpassed) to forests in the late eighteenth century, European foresters came to measure forest productivity, and direct forest usage, through the statistical abstraction of 'yield'. But more than this, in seeking to optimize 'yield', they further changed forest structure, literally altering the shape, regimentation, and species composition of complex systems, reducing them to ordered row crops of even-age monoculture (Scott, 1998).

Such abstractions in forestry, even when they are not capitalized or fiscal in nature, tend to drive a trend to reverse engineer the landscape to suit state metrics. By measuring success in forestry, for example, in terms of 'extent of land cover under dense canopy', there is an incentive for foresters to plant and protect tree species with wide canopies, tending to extend aerial coverage rather than ecosystem or soil health on the ground. The result is a forest that actually comes to resemble the categories of measurement, rather than vice versa (Robbins, 2001a).

Simplified Fields

Much the same can be said of agriculture itself, especially in the last half of the twentieth century. The Green Revolution is a shorthand term for the multinational effort to achieve global food security by transferring technology from wealthy donor nations and international research institutions (with an alphabet soup of acronymic names from CIAT to WARDA) through states to local farmers. The aims and goals of these projects were unquestionably about bettering the social good, as even the harshest critics of the Revolution concede. It was also largely a state-driven effort, despite the influence of multinational capital in the form of agrochemical companies and farm machinery firms (Shiva, 1991).

As is by now well known, the Green Revolution succeeded insofar as agricultural yields, for a time, climbed at unprecedented rates, especially in wheat and rice production in Asia. Overall, however, these efforts had a number of pernicious effects; land

consolidation and the decline of smallholding farmers were most evident. But more disturbingly, in many areas crop pest incidence climbed, water and soil quality deteriorated, and nutrition often declined (Lal et al., 2004).

These failures largely grew from the ecological simplifications required to make the Green Revolution function as a technical effort. To increase yields of wheat, for example, competing species needed to be expunged, the size and height of crop plants needed to be truncated, and the genetic variability of crops needed to be reduced. All of these simplifications, however, undermined its more complex ecological realities; the competing 'weed' species removed were often beneficial plants, decreased crop size destroyed the fodder-base for livestock, and the reduced genetic variability of crops increased vulnerability to disease and pests (Shiva, 1993).

These simplifications in part grew from the naïveté of agricultural science and the overconfidence of technicians in the few factors that needed to be manipulated in order to bring about agricultural change. To a great degree, however, the extension of agriculture in this form and the prolonged ignorance of local and traditional cultivation techniques and biodiversity were a product of the state's necessary tendency to equate agriculture with crop yield, and measure successes or failures precisely by this metric. In taking a census of agriculture in terms of yield and percentage of high-yielding varieties planted, the governments of India, the Philippines, Egypt, and Mexico created an incentive for state agents to see the land in a new way, and to make their simplifications real on the ground.

Mutual Lessons: Understanding Territorialization and Nature's Agency

Of course, not all state approaches to nature are crude simplifications. The increased attention to biodiversity preservation, as a prominent example, imposes a mandate on states to deal with complex and nonlinear systems, inviting more complex and subtle institutions, with more attention to simultaneously ecological and social goals (Adams et al., 2004). The rise of 'adaptive management' and other conservation goals, moreover, stresses state approaches to nature that are flexible and multipartite (Walters, 2002). There is new-found interest in the conservation of complex, human-influenced diversity, including crop diversity (Zimmerer and Young, 1998).

Even where the state has intervened in complex and subtle approaches to conservation, however, its efforts have tended toward problematic and undesired ends (Zimmerer et al., 2004). Specifically, this is because the modern trend in conservation is one of territorial partition, a practice that commonly has precisely the reverse effect of that intended. The forested plains of Tsavo national park in Kenya, for example, were enclosed precisely to protect diversity, including especially East African elephants and the open savanna forests that support their kind. Between 1948, when the park was enclosed, and the mid-1970s, the animals died in great numbers, the ground cover declined, and the noble goals of the state and its supporting international environmental donors were foiled. What had gone wrong? In part, the problem was one of imagining human influences out of the park. The historical land covers of the region were in part maintained by complex interactions with humans and human activity (e.g. burning, hunting, harvesting, etc.), which were banned as part of establishing the park (Botkin, 1990).

This partitioning of those things 'natural' from those things 'social' is a simplification typical of state management and of the modern imagination more generally (Latour, 1993). It has foiled attempts at conservation throughout Africa (Neumann, 1998), Asia (Brower and Dennis, 1998), and even in the United States (Jacoby, 2001). Certainly it may point to a tendency within the state as described by Scott and heralded in political ecology. A veritable cottage industry of research in political ecology supports this thesis, with case after case demonstrating the simplification of conservation to be the result of imagining polygons in the landscape. The resulting obsession with boundaries for conservation territories is inherently anti-ecological, insofar as natural processes poorly follow discreet lines on the map and real ecological processes follow complex and non-connected topologies. But such thinking continues to prevail in the spatial imaginary of the conservation state (Zimmerer, 2000).

Yet, there is something more basic here. In selecting the specific enclosure of space that doomed the elephants of Tsavo, the state's hands were tied in advance. Given the international and intra-national boundaries across which animals migrate, what sort of polygon was possible that met the needs of both complex sovereignties and elephants? Is simplification in ecological management inherent in the state itself, as suggested by Scott, or is it a product of a more fundamental political geographic problem of territorialization?

Certainly at the subnational level, the multiple units of management, each with differing and contrasting geographies of authority, produce overlapping mandates, studded with gaps, some of which follow ecological form and function and others that cannot. The demand for the political imposition of 'buffer zones' around conservation areas is as much a product as a cause of such overlapping patterns of authority and ecology (Shafer, 1999).

Such problems are equally emblematic at the transnational scale. As political geographer Peter Taylor points out, coherent state territories as 'containers' of state power and sovereignty are a problem, 'since states are not ecosystems' (Taylor, 1994: 161). As a result, efforts to impose coherent conservation corridors – enclosed paths following ecosystem functions, especially faunal migration – have politically failed throughout the world, causing managers to settle for more spatially fragmented constellations of enclosure areas, more closely resembling a 'braided network' than a corridor (Zimmerer et al., 2004: 525)

These findings in political ecology underline an emerging theoretical truism of political geography: territories are not given, or fixed, but are instead themselves the product of spatial political work; territory is a 'political strategy for social [and ecological] control' (Johnston, 2001: 690). In this sense, there is no fixed state scale of territory that the state has imposed against a more 'natural' ecological one because of simplistic territorial assumptions. Instead the conservation territories of the state are themselves a product of political action and struggle, along with interpretation and reinterpretation of ecological space. In the reverse sense, however, the fact that boundaries are dynamic, multiple, and transitory does not negate the fact that there are very real territorial mismatches between differing boundaries (human and non-human). As Newman and Paasi (1998) have argued for geography more generally, political ecologists need to move beyond simply critiquing the ephemerality of boundaries, and redouble efforts at empirically documenting their specific character and transformation as a window into political and ecological change. These lessons from political geography are crucial to political ecology.

At the same time, however, the specific ecological patterns and processes of natural systems are not inert in these political struggles over territory. Where foresters and state extension agents in India worked to separate jungle from agriculture, they set into motion processes of species invasion that have altered the conservation authority of forestry departments (Robbins, 2001a). Where human use of fire is quashed in South Africa to preserve native ground cover, undesired invasive species has actually proliferated, changing the political conditions of that struggle (Le Maitre et al., 2004). State efforts to domesticate the Egyptian desert while attempting to control insects, resulted contradictorily in the explosion of malarial mosquitoes, which created opportunities for intervention by new and different cadres of colonial experts (Mitchell, 2002). These myriad unintended consequences invariably force a change of direction on state agents, no matter how stubborn, altering their strategies and their relationships with other human and non-human entities.

Whether these feedbacks and unintended consequences will lead to a reconfiguration of the territorial imperative in environmental governance, or in a disavowal of simplification, remains entirely unclear. A further lesson, however, is clear: nature talks back. This does not mean that nature is simply a singular stubborn force that the state tries to overcome, although the discourse of man versus nature is important to state power. Rather, the inherent tendency toward simplification and territorialization causes the state to mix its efforts in a particular way with non-human nature, which responds and in turn reforms the state. As Timothy Mitchell observes, nature speaks, and in so doing helps to remake the simplifying state in new forms: 'Although technical development portrayed the world as passive, as nature to be overcome or material resources to be developed, the relations of science and development came into being only by working with such forces' (Mitchell, 2002: 51).

Whether this lesson might influence the way political geographers think about the emergence of territories, their ideological and legal institutionalization, and their disintegration in friction with other forces, is unclear. Research must first take the non-human seriously in its encounter with state power.

THE NETWORK STATE: MEDIATING SOIL, PRODUCERS, AND CAPITAL

Traditionally, political ecological accounts follow a 'chain of explanation'. To explain any situation or event (e.g. soil erosion in a specific Nigerian district), research proceeds by explaining the local actor who precipitated it, the social and economic forces acting on them, and finally the political forces that set those conditions. As originally outlined by Piers Blaikie and Harold Brookfield in the watershed book, *Land Degradation and Society*, this chain:

> ... starts with the land managers and their direct relations with the land (crop rotations, fuelwood use, stocking densities, capital investment and so on). The next link concerns their relations with each other, other land users, and groups in the wider society who effect them in any way, which in turn determines land management. The state and the world economy constitute the last links in the chain. (Blaikie and Brookfield, 1987: 27)

Such an account of the state portrays it as an autonomous, distant player, on a par with global capital, with the causal power to situate lower-level actors (communities and land managers) and direct their practical range of choice. Such an account is clearly in line with the image of the state portrayed previously, a modernist menace with autonomous logic, but one constantly thwarted by the agency of non-human nature.

Actual research in political ecology, however, has consistently undermined this picture of the state, and raised questions about the explanatory practice of 'chains' of explanation. Specifically, the state has shown itself to simultaneously work at multiple levels, producing political economic conditions that better resemble networks than linear chains. Moreover, the state has demonstrated itself to be highly porous to both the influences of capital and local producer communities, but also natural objects like trees, fields, and cattle. These form, and are in turn reformed, by state players and institutions. This image of a 'network state' has come to dominate political ecological explanation, with implication for thinking about governance more generally.

The Parasitic State: Building Extractive Ecologies

This is especially evident in development efforts geared toward paying national debts through reconfigurations of export economies. In Côte d'Ivoire, as a prominent example, fiscal crises in the 1970s and 1980s resulted from climbing debt payments and resulting austerity measures imposed by the International Monetary Fund. To finance development loans and recover traction in the face of spiraling debt, the state began to establish 'parastatals', where foreign investors entered joint ventures with the Ivorian state to expand export earnings. Generally, these new formalized relationships with international capital sought to expand promising but previously undercapitalized sectors.

One such sector, livestock, was particularly attractive, since it appeared to be easy to develop at minimal cost. Taking advantage both of its ability to subsidize investment and its monopoly on the establishment of property rights, Côte d'Ivoire established the parastatal SODEPRA (Société pour le Développement de la Production Animale) to foster the creation of a meat-processing economy, while simultaneously distributing grazing rights to Fulani migrant herders, to lure them in from adjacent countries, and provide raw materials for this growing export economy. The porousness of the state to foreign investors resulted in a novel network of flows of surplus and institutionalized connections between state and capital (Bassett, 1988).

The effort was not without unforeseen consequences, however. The SODEPRA plan initiated a massive migration of large cattle herds from Mali and Burkina Faso, establishing extensive Fulani grazing in densely populated agricultural areas of ethnically distinct Senufo farmers. These dense herds resulted in extensive crop damage, when cattle entered and grazed in Senufo fields. The flow of value, out of fields, through herds, outward toward meat exports, and into the pockets both of the state and of foreign investors, represents a novel ecological/economic structure, one that put the state at the center of connections between multilateral lenders, international investors, herders, cattle, and crop fields.

In the end, this structure could not of course be maintained. Ongoing crop damage by herds led to retaliations by Senufo farmers against Fulani herders, encroachment, conflict, and ultimately violence. These often brutal upheavals were commonly described at the time as ancient ethnic conflict exacerbated by population growth. The parasitic role of the ecological network state, connecting simultaneously players at multiple levels ('global', 'regional', and 'local'), and regulating the flow of value, went largely unremarked, except in political ecological analysis.

The Cooperative State: Achieving Commodity Autonomy

To say that the state is porous to the influence of multinational capital is by no means a novel claim in political geography, of course. As most critical realist and world systems theory accounts stress, one of the central functions of the state is to provide for the conditions for the accumulation of capital (Taylor and Flint, 2000).

Even so, the study of resource flows in commodity economies demonstrates considerably more complex networks, which at times produce contradictory outcomes, especially when local producers and producer institutions form key nodes in state relationships. Consider the case of contract farming for bananas in the Caribbean. Contract farming is a rapidly emerging form of agricultural organization where local producers, usually smallholding peasants, directly contract to buyer firms rather than selling products on an uncertain open market or being themselves directly vertically integrated as workers on company-owned commodity production land.

For the most part, critical ecological scholars have viewed such arrangements with reasonable suspicion. Most importantly for the present discussion, contract arrangements appear to be a case where the anemic state, especially in underdeveloped contexts, retreats from its position between international capital and local labor. This in turn allows powerful firms to directly set the terms of trade with disaggregated farmers, a monopsony relationship where farmers receive poorer returns, live with higher risks, and tend to externalize costs

into the environment, leading to erosion, pesticide overuse, and other environmental problems (Pred and Watts, 1992). Generally, political economic analysis of such relationships has reasonably suggested that conditions under contract farming will erode; peasant labor either becomes more commodified or deskilled; food production falls away in favor of cash crop production; and careless and indiscriminant use of toxic inputs becomes the norm (Grossman, 1998).

Yet, research in political ecology suggests something rather different. As Grossman (1998) has demonstrated on the Caribbean island of St Vincent, the formation of quasi-state institutions sometimes produces exactly the opposite result. Specifically, the St Vincent Banana Growers' Association (SVBGA), a statutory corporation formed in the 1950s, continues to intervene in the production and exchange process in numerous ways. The SVBGA provides extension, subsidized credit, and pesticide controls, and markets the island's banana crop as a parastatal monopsony. While by no means an ideal or utopian arrangement – the association is flawed by ineptitude, mismanagement, and other problems typical of the development state – its complex, multi-nodal connection to producers and producer communities at a range of moments in the production process results in a socio-ecological outcome that defies critical expectations. St Vincent banana growers experience sustained prices, some maintenance of food production, and low environmental impact overall.

The central reasons for this success, according to Grossman, have a great deal to do with the ways in which the SVBGA connects to producers across a range of local ecological conditions and ad hoc and experimental solutions that producers apply to solving problems in the allocation of labor, time, and chemicals. This outcome is distinct from that more typically seen in industrial contexts, moreover, specifically because of the linkage between environmental factors, producers, and parastatal interventions. In adapting to opportunities and barriers represented by SVBGA structures, the 'environmental rootedness of the farming experience encourages growers to develop more flexible labor patterns than industrial workers' (Grossman, 1998: 212). These adaptations, in turn, influence how the parastatal functions, and what new opportunities or barriers it presents.

While the case of St Vincent may be remarkable and unusual in several respects, it does represent a more general principle. Rather than being a single entity, with a discreet position, either in a 'world system' or a 'chain of explanation', the state appears as a network of connections, whose specific form is itself the product of the flows and interactions between civil society, ecosystem process, and worldwide markets.

The Porous State: Reforming Amidst Livelihood Struggles

Similarly, and perhaps more dramatically, political ecological research has shown that the form both of the state and its policies shows surprising pliability in the face of farmer collective action. The results of such porosity, where agrarian policy may go through constant reinvention as producer communities struggle internally while continuing to adapt to changing ecological and market conditions, flies in the face of traditional accounts. These more familiar theories hold that states tend to act with largely unchallenged brazen power (see the 'Plantation State' above) throughout the developing world. This is especially true of rural areas, since poorer smallholding agrarian producers are numerous and scattered and so are theorized to be politically inert (Bates, 1981), except perhaps where resisting on a small scale (Scott, 1985).

And yet the state appears far more porous to local political ecological conditions, at least in its governance of primary production, agrarian systems, and environmental goods and services more generally. Again taking the case of West Africa, local producers have on many occasions demonstrated a remarkable ability to actively organize and so influence governance of resources. More to the point, states typically characterized as 'heavy-handed' in the past, show a flexibility and vulnerability to interest 'from below'.

Côte d'Ivoire is one such case. As demonstrated above, the parastatal systems typical in West African development and economic restructuring have shown a tendency toward facilitating the extraction of natural capital from local producers. Bassett (2001) demonstrates, however, that state cotton policy in this region, even during the pre-colonial and colonial eras, was heavily influenced by local ecological and economic conditions. Specifically, colonial and post-colonial states and firms, like those of African empires before them, had difficulty in controlling and capturing local cotton production. Indigenous handicrafts and strong local markets represented some of the barriers. More than this, however, local-level producers directly resisted state programs and parastatal monopsonies because their priority was to save *labor* in managing cotton production sustainably, whereas state priorities were towards the saving of *land*. This led to divergent pressures on the state in research and development for agrarian technology and institutions. Though the state pressed for specific varieties of cotton and specific cropping methods that would facilitate them, producers organized against them, not only at the farmstead level (where they diversified production and abandoned cotton) but also through producer organizations that directly challenged and overturned state policies.

As a result, Côte d'Ivoire has undergone what has been repeatedly described as a 'cotton revolution' characterized by a growth in intensity, production, and market share on global cotton markets, all of which are unprecedented regionally. Despite their typical characterization as a product of Western technology and state intervention, such revolutions are increasingly appearing to be the result of local producer adaptations and the demands these producers make on both states and markets. Indeed, underdeveloped states like Côte d'Ivoire and Brazil have begun to press the interests of these producer groups in the international arena, specifically calling into question state subsidies for agricultural production in the United States and other developed nations. They often do so by coming to the table with state export production practices devised by and geared around peasant production systems and locally defined conceptions of sustainability. This outcome is as unpredicted by triumphalist neo-liberal theorists as it is by critical world systems theorists. The porous environmental state can embody the voices of local producers (made strong by collective action) and convey those voices to a surprisingly pliable global community. The mediating state is therefore coming to play an altogether new role in the negations of global economy and regional ecology.

Mutual Lessons: State Capacity in Expansion and Contraction

These case studies in political ecology suggest somewhat contradictory interpretations. The picture of the state in political ecology is confused by the appearance of a state that, while it is at once a significant entity, is simultaneously a dwindling power in the face of debt and global flows of capital, as in the case of Côte d'Ivoire's quasi-corporate livestock management regime, SODEPRA. There, demands for revenue continue to push the state to parasitic practice, increasingly hamstringing the enforcement practices of nations for environmental protection and amelioration and driving unsustainable and conflict-ridden extraction.

So too, however, while the capacities of the state seem to be governed from within, they are simultaneously directed by localized nature/society collectives, like St Vincent's banana growers' association. These parastatals function to minimize pesticide use and to sustain producer livelihoods, expanding the role of the nation in commodity trade.

And while the state's influence as an environmental agent often seems constrained, parochial, and confined to territorial regimes, its reach seems to have found new lengths, as where Côte d'Ivoire's cotton institutions extend the grower's power and authority to represent and negotiate in global trade agreements. Despite more general predictions of the demise of the state in an increasingly global-ized world, dominated by multilateral agencies and multinational capital, it seems as though the state remains a mediator between capital and nature, regulating the metabolism between farmers and fields, developers and soil, cities and water. Is the environmental state expanding or contracting?

Debate in political geography is enlightening here again, and theory in the field echoes that of current political ecological research. For political geographers, a split exists between those who argue that the state is in decline, especially in the face of market-driven globalization (Bhagwati, 1988), ver-sus those who insist that in fact the state remains a key facilitator and essential mediator of market integration (Wade, 1996). Certainly, the conflicts and extractive regimes of Côte d'Ivoire's livestock industry reflect the former characterization, while the successes of St Vincent's banana industry seem to show the latter.

As political geographer Glassman (1999) has observed, the appearance of contradiction is under-lain by a common mistaken assumption on both sides of the debate in the conception of the territo-rial nature of states – 'namely that states and their powers can for all practical purposes be thought of as contained within the bounded territories over which they have formal sovereignty' (p. 670). This 'territorial trap', he argues, disallows a clear view of the *internationalization* of the state, a process where the state actually expands its power as it ori-ents itself toward supporting international investors and transnational elites.

Viewed in this way, the state's reach as an eco-logical actor certainly seems to have expanded, reinforcing Glassman's interpretation of the state amidst globalization. This is ironically all the more true in an era of apparently autonomous envi-ronmental actors, including environmental NGOs (e.g. the World Wildlife Fund) and lenders with an environmental agenda (e.g. the World Bank Group). This is because such agents can only function through the action of states, which bro-ker and enforce global practices, whether through implementing international debt-for-nature swaps (Thapa, 1998) or global carbon markets (Bonnie et al., 2002). The 'global commodity economy' and the 'global environment', abstract conceptual arenas of political struggle, provide novel oppor-tunities for the extension of state reach into new areas of governance and action, though largely in the service of narrow interests.

And yet this is not the whole story either. The case of Côte d'Ivoire's cotton growers, along with a growing body of accounts of livelihood movements in political ecology, suggests that the extensive and internationalized state may also represent the institutionalized power of local growers and other primary producers. More than this, it suggests the way that local communities leverage their influ-ence in the international sphere by networking with and within the state. As producer livelihood strate-gies set the terms of state commodity development schemes, the state is simultaneously transformed and extended. Rather than simply being the tool of multinational elites, therefore, the state becomes the site of struggle over the internationalization of many different interests and forms of power. In this sense, the case materials of political ecology, with its focus on the agency and power of grassroots livelihood struggles in producing and reproducing the state, may yet provide challenges to theorization in political geography.

THE KNOWING STATE: PRODUCING AND ERASING ECOLOGICAL KNOWLEDGE

These complex pictures of the state in politi-cal ecology (menacing territorial conservation machine and porous livelihood ecology network) are joined by a further image, that of the state as a way of knowing, as a body of knowledge makers, and as a location for knowledge contests. The 'epistemological state' is of course in no way a concept unique to political ecology. The state is, in every regard and field of operations, an avid producer of narratives. Consider state narratives about poverty and unemployment, for example, which are as essential to governance as actual policy, restrictions, or enforcement.

It is possible to argue that the state is especially driven to produce, erase, and struggle over envi-ronmental narratives. This is because the state in its various forms, even including the neo-liberal min-imal state, continues to retain a near monopoly on responsibility (if not power) over what is broadly accepted as non-human nature. Even in the United States, for example, where privatization has been extended to a huge range of objects and services, states or the federal government retains sovereignty over wildlife/biodiversity, ecosystem conditions, and water/air quality.

The Crisis State: Producing Environmental Narratives

Since the state is held responsible for such matters, one might at first be tempted to think that ecologi-cal problems might be minimized in state discourse. Certainly where state actions in one sphere, military development for example, have led to ecological externalities and serious risks, enforcement has been historically lax and the narratives of risk have been minimal. Before the Environmental Protec-tion Agency was created in the wake of a number of regulatory failures, for example, a sense of crisis in the environment was rarely fostered at the federal level (Colten and Skinner, 1996).

Nevertheless, the realities of state obligations and imperatives for governing both humans and non-humans, tend to engender action on the part of those doing the governing. More subtly, sovereign state agents, including scientific experts in a range of environmental fields, are charged with action for the reproduction of the authority of their professional distinction (Mitchell, 2002). Such actions require legitimacy and funding. To justify budgets and, where appropriate and possible, to find support for spending from stingy international aid communities, there needs to be an environmental problem, or crisis, preferably one of some magnitude.

Since it commonly tends to focus on the vindication of the practices of local people, political ecological research has, as a response, frequently directed its inquiry to exposing and debunking the 'crisis' character of environmental stories narrated by the state and other bodies of authority, and limiting the efficacy of such narratives as well as their problematic effects. Examples are too many to document, but include exaggerated ecological 'crises' as far-ranging as fire in Madagascar (Kull, 2000), overfishing in New England (St Martin, 2001), and desertification throughout the Sahel (Reynolds and Stafford Smith, 2002).

Perhaps the most indicative of such stories are those focused on the deforestation of West Africa, a singularly powerful set of narratives that have together justified funding and intervention throughout colonial/post-colonial rule. In their definitive history of such narratives, *Reframing Deforestation*, James Fairhead and Melissa Leach examine the histories of deforestation claims in six nations in West Africa, exploring the statistical evidence of the problem, in terms of their origins, their usage in justifying action, and their power in framing the debate over environmental quality and governance (Fairhead and Leach, 1998).

Their results are revealing. In every case, the baselines for measuring environmental change (how much forest there supposedly used to be) are grounded in questionable observations, mostly by colonial officials charged with 'fixing' forest problems in their own time. More contemporary methods are also called into scrutiny, including the standard practice of comparing remotely sensed data from satellites with earlier air photographs, which is plagued by problems of seasonality (when seasonally was the photo taken?) and inter-annual variability (under what conditions and how much rainfall?). Most trenchantly, the work shows the way the categories used to identify land covers (forest, savanna, etc.) tend to be variably applied and interpreted, a root cause of divergent interpretations of change (Robbins, 2001a). They conclude, therefore, that deforestation has been 'massively' exaggerated, and that the convergence of crisis narratives with state interests and expert power, and the coherence of such stories with other popular narratives (overpopulation, as a prominent example), explains their persistence through the contemporary period. Similar empirical findings in political ecological research again and again demonstrate the imperative of the state to produce and defend environmental narratives.

The Expert State: Erasing Environmental Knowledge

But the state is not merely a system that produces and defends stories about environmental crises. Since such narratives flow not into a vacuum, but instead into a noisy field of claims made by differing interests and players, the state must also act as a rigorous *eraser* of knowledge. Some accounts about the environment must vanish as others achieve hegemony. This 'crossing out' of competing environmental knowledge is as much an epistemological habit as it is an actively pursued state task, but it applies as much to long-forgotten narratives (e.g. the world is flat) as it does to those later vindicated (e.g. DDT is bad for waterfowl).

Drawing heavily on certain parts of the field of science studies for inspiration, most political ecological inquiries proceed from the assumption that scientific practice, no matter how rigorous, is inevitably inflected by the social, cultural, and political conditions under which it is conducted (Hess, 1997). Typically proceeding from an epistemology of critical realism,[2] however, most political ecology also assumes that some extra-human systems and rules generally exist and persist, and can be known, albeit with some degree of uncertainty.

Even so, political ecology has been aggressive in pursuing the processes through which alternative accounts of ecological process have been actively displaced by state accounts. This work has shown how tendencies in state thinking, usually predicated on the imperatives of legibility and simplification described above, have come to dominate public discourse and policy-making, and even to permeate local thought, by denigrating, dismissing, or actively attempting to refute other accounts. Such work has shown the state's epistemological energies exercised in attacking or dismissing claims as different as those of US women assessing toxic risks (Seager, 1996) and those of farmers in the Himalayas assessing land degradation (Forsyth, 1996). Epistemological work not only renders such knowledges and claims invisible, it can actually cause them to be forgotten over time, making them disappear altogether.

One of the clearest examples of such erasure is the area of ethnoveterinary knowledge.

Generally, state veterinary services and extension are designed to dispel 'crude' local practices and extend the most contemporary and scientific information to less knowledgeable local producers. Efforts to displace local knowledge are especially ironic given the fact that rural producers (not state veterinarians) originally domesticated all forms of livestock, breeding them from wild stock and observing their behaviors and habits for the last 10,000 years. Such knowledge and practice is important in herding communities from Nigeria (Alawa et al., 2002) and Saudi Arabia (Abbas et al., 2002) to Afghanistan (Davis et al., 1995).

Among countless such local systems, that of the Raika herders of Rajasthan in India is typical in its vastness, coherence, and practical application, As Ellen Geerlings' study of the ovine ethnoveterinary knowledge of the Raika has shown, pastoralists possess a system of disease classification that includes most of the major ailments of formal veterinary science (including bottleneck, enterotoxemia, foot and mouth disease, sheep pox, hematuria, and other diseases). They have a range of ready treatments, including protein treatments for strength, and plant and oil treatments using locally available species (Geerlings, 2001). Nevertheless, state veterinarians are adamant in their active rejection of these knowledges and practices, and only vigorous efforts on the part of advocacy non-governmental organizations has begun to reconcile the expert and local communities to make space for non-state ways of knowing and doing (Kohler-Rollefson and Rathore, 1998).

Of course, zeal in political ecology to recover the dismissed knowledges of marginalized communities runs its own risks. The porousness and interrelationship between 'state/expert' knowledge and 'local/indigenous' knowledge makes both such categories analytical functions rather than real entities (Agrawal, 1995; Robbins, 2000). By valorizing such knowledges and assuming them to have discreet ontological status, moreover, a romantic political ecology of knowledge hazards the possibility of reproducing the colonial fictions of 'noble savages' that actually reinstates the unequal power relations between hegemonic actors and less powerful people (Braun, 2002; Briggs and Sharp, 2004). Nevertheless, political ecology has helped to expose the crucial task of epistemological erasure as a component of environmental governance.

The Negotiating State: Contesting Environmental Truth

In part as a result of its continued investigation of expert knowledges, and in part as a response to the oversimplification of any state/local knowledge binary, other work in political ecology has begun to explore the way in which the state is neither an actor nor a knower, per se, but instead a process or location in epistemological struggle.

This is in part because, as noted previously, the state exists in a range of locations, all differentially open to the kinds of political and economic conditions outlined above for the production of knowledge. Within the state, differing agents (e.g. ecological modelers versus field ecologists), agencies (e.g. Fish and Wildlife Service versus Animal Plant Health Inspection Service), as well as differing disciplines (e.g. biologists versus limnologists) all come to different conclusions about the nature of nature, as each have differing cultures and differing contexts and pressures acting upon them.

So, too, the state is porous to differing external forces and actors, sometimes simplistically described as 'interest groups' or 'stakeholders'. These actors participate in the processes in state laboratories, courtrooms, and statehouses, which together act to constitute the 'truth' and 'reliability' of differing claims. As such, the process of environmental knowledge production and erasure never occurs without a fight (or 'contestation' in the more abstract language of the field). This is all the more true as decision-making has become increasingly decentralized and 'participatory', with a number of new actors actively invited by the state to participate in the process of constituting facts and reviewing information. These decentralized approaches encourage diverse interests to come to narrative compromises and agreements over 'what is the case' about nature, referred to as 'discourse coalitions' by theorist Maarten Hajer (1997). A wealth of political ecological case studies traces such processes, revealing emerging knowledge coalitions from land uses in America's Appalachian region (Nesbitt and Weiner, 2001) to regulation of biotechnology in the European Union (MacMillan, 2003).

Maddock's (2004) analysis of water quality regulation in the United States is representative in this regard. Assessing the effects of participatory approaches to managing non-point source water quality regulation, she queries the relationship between agreement and disagreement on fundamental questions about the sources of water pollution and about the soundness of environmental science to direct policy in the first place. The results surprisingly show that some state actors and some interest groups (farming and environmental groups) come to agree on the causes of pollution, reaching common environmental narratives, while disagreeing on the degree of trust in formal scientific techniques for regulating the outcomes of these effects. These interests, moreover, collectively come to altogether different conclusions from government researchers themselves, as well as industrial interests. This complex constellation of narratives leads to new political alliances, and changes in which facts are taken for granted as 'true'.

Mutual Lessons: Governance and the Regulation of Environmental Knowledge

Here again, political ecological characterizations provide somewhat contradictory pictures of the state. First there remain specific capacities and motivations that are unique to the state as an institution; it appears to retain a monopoly of environmental knowledge enforcement and a dominant position in institutionalizing practices relative to non-human nature; though a 'leaky' vessel (Taylor, 1994), the state is a container that encloses exclusive forms of epistemological power. On the other hand, the state appears as a somewhat dispersed and ephemeral actor, its forms of knowing contested and negotiated at diverse locations.

Theory in political geography, again somewhat overlooked in political ecology, explicitly addresses this apparent contradiction and provides an interpretive road map for scholars in the field. Here, work has pointed to a tension between interpretations of the state as an institution, one that exercises power on behalf of class elites (following Poulantzas, 1973), and the state as a 'unified symbol', a dispersed and ephemeral ideology that produces subjection (following Abrams, 1988).

This ambiguity can be seen throughout political ecology. On the one hand, research in the field affords to state actors unique power, indeed monolithic and predatory authority, to make certain ways of knowing about nature normal and hegemonic. Certainly as an actor with unified and enforceable goals, the state can and does act with often ferocious and coordinated epistemological force, as in the modernization of the resource frontiers of China (Jiang, 2004) or Brazil (Hecht and Cockburn, 1989), where certain ways of viewing nature were paved over diverse local epistemologies.

Nevertheless, the exigencies of actually managing the environment, where animals, producers, soils, and other actors interact on a daily basis, produces constantly changing conditions and disperses the knowledge power of state agents. As a result, the knowledges of state agents become the product of countless local power struggles as much as their agents. Political ecology is littered with numerous accounts of the epistemological state occurring not as a monolithic agent, but instead as *immanent*, a crystallized form of power emerging from production conditions at a number of locations, through, for example, the interaction of local extension agents, conservation officers, and producer associations (Sivaramakrishnan, 1998; Robbins, 2000). The state here appears, as Foucault observes, only as an extension of already existing relationships, and therefore it 'can only operate on the basis of other, already existing power relations. The State is superstructural in relation to a whole series of power networks that invest the body, sexuality, the family, kinship, knowledge, technology, and so forth' (Foucault, 1980: 122).

This debate is far from resolved in political geography, but its rudiments help to locate the ambivalence of political ecological pictures of state knowledge and power. Whether as the producer of claims, the eraser of knowledges, or the location of new settlements of environmental epistemology, the state maintains a crucial role in political ecology. Or perhaps put better, governance of the environment is necessarily a process of producing, adjudicating, and eliminating truths about nature, which in turn produce new state forms and actors.

What these revelations in political ecology might in turn suggest for the theory of political geography is somewhat unclear. Certainly, work in the field might pay increased empirical attention to the way specific state institutions think, and why. Quite often, as some political ecology has shown, the kinds of knowledges that are either extolled or preserved by state actors are not those of simple instrumental value for powerful interests (i.e. capital). Rather, logics within state bureaucracies (e.g. forestry, energy, agriculture) and across state relationships to specific non-humans (e.g. trees, nuclear waste, Bt or Bacillus thuringiensis corn), direct the flow of truth effects – 'the types of discourse that it accepts and makes function as true' (following Foucault, 1980: 131). This somewhat inductive approach might help to untangle the 'sovereign' and 'immanent' qualities of the state, as well as underline the very material conditions under which it is forced to operate, a problem common to all political geographies, not simply those pertaining to 'nature'.

BEYOND STACKED BOXES: RETHINKING OR ABANDONING SCALE IN POLITICAL ECOLOGY

These multiple images of the state in political ecology sit somewhat uncomfortably alongside one another, but do reflect conversations parallel to those in with the field of political geography. The simultaneous image of the state in both fields is increasingly one of (1) a set of changing territorial strategies, (2) a collection of political capacities of increasingly international reach, and (3) an epistemological system that is, at least in part, an effect of power at multiple scales. Reconciling these images remains a challenge for both fields. How can the state be both a product and driver of territorialization? How can it be both expanding and contracting? How can it be both sovereign and immanent?

Nor do these images reconcile well with the picture of the state available in political ecological theory, such as it is. This theoretical view of the state, to the degree it has been articulated at all, has largely been conceived of as a single, situated, and coherent agent, located at a specific scale. Following the instructions of Blaikie and Brookfield so many years ago (see above), the state has become a theoretical location in a scaled hierarchical chain of explanation, a box stacked somewhere between multinational capital and local producers. The research outlined here, on the other hand, suggests a state whose position is far more complex, diverse, amorphous, and *ascalar* than that suggested by stacked boxes implied in this theoretical configuration.

One possible way forward has been proposed by Brown and Purcell (2005), who suggest that the field might better incorporate political geography's recent approaches to the problem of scale. Reminding political ecologists that scale is produced, rather than given, they suggest that the field might escape its 'local' trap, which tends to romanticize, fetishize, and otherwise locate most progressive action at the local scale. While this characterization of political ecology is somewhat contentious and may not perfectly reflect the diversity of the field (as this review suggests), the central point, that scale is far more slippery than often portrayed, clearly has application. Following the work of Eric Swyngedouw, therefore, 'scalar configurations' can be seen as 'the outcome of sociospatial processes that regulate and organize power relations' (Swyngedouw, 2004: 132). In considering the state as a kind of institutionalized but malleable scalar effect, the nature of state power becomes clearer and the multiple empirical images of the state begin to meld into one. Beginning from this understanding, political ecology might therefore proceed as a kind of study of scalar politics, exploring how various political boxes get stacked the way they do in scalar hierarchy through historical and economic processes.

But if the research reviewed here has shown anything, it is that the very notion of scalar hierarchy, whether politically constituted or not, may be a dead end. A more radical approach might reject the notion of scalar hierarchy altogether. Marston et al. (2004: 3) argue 'that attempts to refine or augment the hierarchical model cannot escape a series of problems immanent to that model; instead, we argue, geography – and by implication all of social science – should reject hierarchical versions of scale'.

They argue instead for an altogether different ontology, 'one that so flattens scale as to render the concept unnecessary'. Such a way of thinking about the players in political ecology (e.g. states, plants, households, animals, markets, pollutants, etc.) as vast flat networks, pushing and pulling

countless localities in different directions, constituting and reconstituting themselves through territorializing, controlling, extracting, and narrating environmental change, might square far better with political ecology than even the best reformed hierarchical model.

It is beyond the scope of this chapter and the abilities of its author to adjudicate sophisticated arguments such as these in political geography regarding scale, territory, or anything else. It is clear, however, that the picture of the state emerging in political ecology has, perhaps inadvertently, produced an extremely large and rich body of empirical data relevant to the discussion. Future progress in understanding political and environmental change will require that political geographical theorists acquaint themselves with the work of political ecologists (and not merely characterizations or summaries of that work), while political ecologists begin to seriously engage the tools and insights of political geography, which remains a blank space on the maps of most practitioners in the field (Robbins, 2003). For now, however, there is plenty of evidence that, unbeknownst to both bodies of scholarship, the discussions and debates of theory in political geography resonate in the case research of political ecologists, a tribute to the epistemological symmetry of contemporary critical research.

NOTES

1 The World Bank Group includes the International Bank for Reconstruction and Development, the International Development Association, the International Finance Corporation, and the Multilateral Investment Guarantee Association (IBRD, IDA, IFC, and MIGA).
2 This is not an unproblematic epistemological position for political ecology, nor is it one taken by all practicing political ecologists. For further discussion, see Demeritt (1994,1998), Robbins (2004).

REFERENCES

Abbas, B., Al-Qarawi, A.A. et al. (2002) 'The ethnoveterinary knowledge and practice of traditional healers in Qassim Region, Saudi Arabia', *Journal of Arid Environments*, 50(3): 367–79.

Abrams, P. (1988) 'Notes on the difficulty of studying the state', *Journal of Historical Sociology*, 1: 58–89.

Adams, W.H., Aveling, R. et al. (2004) 'Biodiversity conservation and the eradication of poverty', *Science*, 306: 1146–9.

Agrawal, A. (1995) 'Dismantling the divide between indigenous and scientific knowledge', *Development and Change*, 26(3): 413–39.

Alawa, J.P., Jokthan, G.E. et al. (2002) 'Ethnoveterinary medical practice for ruminants in the subhumid zone of northern Nigeria', *Preventitive Veterinary Medicine*, 54(1): 79–90.

Bassett, T.J. (1988) 'The political ecology of peasant-herder conflicts in the northern Ivory Coast', *Annals of the Association of American Geographers*, 78(3): 453–72.

Bassett, T.J. (2001) *The Peasant Cotton Revolution in West Africa: Côte d'Ivoire, 1880–1995*. Cambridge: Cambridge University Press.

Bates, R. (1981) *Markets and States in Tropical Africa: The Political Basis of Agricultural Policies*. Berkeley: University of California Press.

Bhagwati, J. (1988) 'Export-promoting trade strategy: issues and evidence', *World Bank Research Observer*, 3(1): 27–57.

Blaikie, P. and Brookfield, H. (1987) *Land Degradation and Society*. London and New York: Methuen.

Bonnie, R., Carey, M. et al. (2002) 'Protecting terrestrial ecosystems and the climate through a global carbon market', *Philosophical Transactions of the Royal Society of London Series A: Mathematical, Physical, and Engineering Sciences*, 360(1797): 1853–73.

Botkin, D.B. (1990) *Discordant Harmonies: A New Ecology for the Twenty-First Century*. New York: Oxford University Press.

Braun, B. (2002) *The Intemperate Rainforest: Nature, Culture, and Power on Canada's West Coast*. Minneapolis: University of Minnesota Press.

Briggs, J. and Sharp, J.P. (2004) 'Indigenous knowledges and development: a postcolonial caution', *Third World Quarterly*, 25(4): 661–76.

Brower, B. and Dennis, A. (1998) 'Grazing the forest, shaping the landscape? Continuing the debate about forest dynamics in Sagarmantha National Park', in K.S. Zimmerer and K.R. Young (eds), *Nature's Geography: New Lessons for Conservation in Developing Countries*. Madison: University of Wisconsin Press, pp. 184–208.

Brown, J.C. and Purcell, M. (2005) 'There's nothing inherent about scale: political ecology, the local trap, and the politics of development in the Brazilian Amazon', *Geoforum*, 36(5): 607–24.

Colten, C.E. and Skinner, P.N. (1996) *The Road to Love Canal: Managing Industrial Waste before the EPA*. Austin: University of Texas Press.

Davis, D.K., Quraishi, K. et al. (1995) 'Ethnoveterinary medicine in Afghanistan: an overview of indigenous animal health care among Pashtun Koochi nomads', *Journal of Arid Environments*, 31(4): 483–500.

Demeritt, D. (1994) 'Ecology, objectivity, and critique in writings on nature and human societies', *Journal of Historical Geography*, 20: 22–37.

Demeritt, D. (1998) 'Science, social constructivism and nature', In B. Braun and N. Castree (eds), *Remaking Reality: Nature at the Millennium*. New York: Routledge, pp. 173–93.

Fairhead, J. and Leach, M. (1998) *Reframing Deforestation: Global Analysis and Local Realities: Studies in West Africa*. New York: Routledge.

Forsyth, T. (1996) 'Science, myth, and knowledge: testing Himalayan environmental degradation in Thailand', *Geoforum*, 27(3): 375–92.

Foucault, M. (1980) 'Truth and power', in C. Gordon. (ed.), *Power/Knowledge: Selected Interviews and Other Writings 1972–1977*. New York: Pantheon, pp. 109–33.

Geerlings, E. (2001) 'Sheep husbandry and ethnoveterinary knowledge of Raika sheep pastoralists in Rajasthan, India', MSc thesis, Dept of Environmental Sciences, Wageningen University.

Glassman, J. (1999) 'State power beyond the "territorial trap": the internationalization of the state', *Political Geography*, 18: 669–96.

Grossman, L. (1998) *The Political Ecology of Bananas: Contract Farming, Peasants, and Agrarian Change in the Eastern Caribbean*. Chapel Hill: University of North Carolina Press.

Hajer, M.A. (1997) *The Politics of Environmental Discourse: Ecological Modernization and the Policy Process*. Oxford: Oxford University Press.

Hecht, S. and Cockburn, A. (1989) *The Fate of the Forest: Developers, Destroyers and Defenders of the Amazon*. London: Verso.

Hess, D.J. (1997) *Science Studies: An Advanced Introduction*. New York: New York University Press.

Jacoby, K. (2001) *Crimes Against Nature: Squatters, Poachers, Thieves, and the Hidden History of American Conservation*. Berkeley: University of California Press.

Jiang, H. (2004) 'Cooperation, land use, and the environment in Uxin Ju: the changing landscape of a Mongolian–Chinese borderland in China', *Annals of the Association of American Geographers*, 94(1): 117–39.

Johnston, R. (2001) 'Out of the "moribund backwater": territory and territoriality in political geography', *Political Geography*, 20(6): 677–93.

Kohler-Rollefson, I. and Rathore, H.S. (1998) *NGO Strategies for Livestock Development in Western Rajasthan (India): An Overview and Analysis*. Ober-Ramstadt, Germany: League for Pastoral Peoples.

Kull, C.A. (2000) 'Deforestation, erosion, and fire: degradation myths in the environmental history of madagascar', *Environment and History*, 6(4): 423–50.

Kumar, N., Saxena, N. et al. (2000) *India: Alleviating Poverty through Forest Development*. Washington, DC: World Bank.

Lal, R., Hobbs, P., Uphoff, N. and Hansen, D. (eds) (2004) *Sustainable Agriculture and the Rice-Wheat System*. New York: Marcel Dekker.

Latour, B. (1993) *We Have Never Been Modern*. Cambridge, MA: Harvard University Press.

Le Maitre, D.C., Richardson, D.M. et al. (2004) 'Alien plant invasions in South Africa: driving forces and the human dimension', *South African Journal of Science*, 100: 103–13.

MacMillan, T. (2003) 'Tales of power in biotechnology regulation: the EU ban on BST', *Geoforum*, 34(2): 187–201.

Maddock, T.A. (2004) 'Fragmenting regimes: how water quality regulation is changing political-economic landscapes', *Geoforum*, 35(2): 217–30.

Marston, S.A., Jones, J.P. et al. (2004) 'Human geography without scale', presented at annual meeting of the Royal Geographical Society/Institute of British Geographers. Glasgow, Scotland.

Mitchell, T. (2002) *Rule of Experts: Egypt, Techno-Politics, Modernity*. Berkeley: University of California Press.

Nesbitt, J.T. and Weiner, D. (2001) 'Conflicting environmental imaginaries and the politics of nature in Central Appalachia', *Geoforum*, 32(3): 333–49.

Neumann, R.P. (1998) *Imposing Wilderness: Struggles over Livelihood and Nature Preservation in Africa*. Berkeley: University of California Press.

Newman, D. and Paasi, A. (1998) 'Fences and neighbours in the postmodern world: boundary narratives in political geography', *Progress in Human Geography*, 22(2): 186–207.

Poulantzas, N. (1973) *Political Power and Social Classes*. London: New Left Books.

Pred, A. and Watts, M.J. (1992) *Reworking Modernity: Capitalisms and Symbolic Discontent*. New Brunswick, NJ: Rutgers University Press.

Reynolds, J.F. and Stafford Smith, M. (eds) (2002) *Global Desertification: Do Humans Cause Deserts?* New York: John Wiley.

Robbins, P. (2000) 'The practical politics of knowing: state environmental knowledge and local political economy', *Economic Geography*, 76(2): 126–44.

Robbins, P. (2001a) 'Fixed categories in a portable landscape: the causes and consequences of land cover classification', *Environment and Planning A*, 33(1): 161–79.

Robbins, P. (2003) 'Political ecology in political geography', *Political Geography*, 22: 641–5.

Robbins, P. (2004) *Political Ecology: A Critical Introduction*. New York: Blackwell.

St Martin, K. (2001) 'Making space for community resource management in fisheries', *Annals of the Association of American Geographers*, 91(1): 122–42.

Scott, J. (1998) *Seeing Like a State: How Certain Schemes to Improve the Human Condition Have Failed*. New Haven and London: Yale University Press.

Scott, J.C. (1985) *Weapons of the Weak: Everyday Forms of Peasant Resistance*. New Haven: Yale University Press.

Seager, J. (1996) '"Hysterical housewives" and other mad women: grassroots environmental organizing in the United States', in D. Rocheleau, B. Thomas-Slayter, and E. Wangari (eds), *Feminist Political Ecology: Global Issues and Local Experiences*. New York: Routledge, pp. 271–83.

Shafer, C.L. (1999) 'US national park buffer zones: historical, scientific, social, and legal aspects', *Environmental Management*, 23(1): 49–73.

Shiva, V. (1991) *The Violence of the Green Revolution*. Penang, Malaysia: Third World Network.

Shiva, V. (1993) *Monocultures of the Mind*. Penang, Malaysia: Third World Network.

Sidaway, J.D. (2003) 'Sovereign excesses? Portraying postcolonial sovereigntyscapes', *Political Geography*, 22(2): 157–78.

Sivaramakrishnan, K. (1998) 'Modern forestry: trees and development spaces in south-west Bengal, India', in L. Rival (ed.), *The Social Life of Trees: Anthropological Perspectives on Tree Symbolism*. New York: Oxford International, pp. 273–98.

Swyngedouw, E. (2004) 'Scaled geographies: nature, place and the politics of scale', in E. Sheppard and R. McMaster (eds), *Scale and Geographic Inquiry*. Oxford: Blackwell, pp. 129–53.

Taylor, P.J. (1994) 'The state as container: territoriality in the modern world-system', *Progresses in Human Geography*, 18: 151–62.

Taylor, P.J. and Flint, C. (2000) *Political Geography: World-Economy, Nation-State and Locality*. New York: Prentice Hall.

Thapa, B. (1998) 'Debt-for-nature swaps: an overview', *International Journal of Sustainable Development and World Ecology*, 5(6): 249–62.

Veregin, H. (1995) 'Computer innovation and adoption in geography: a critique of conventional technological models', in J. Pickles (ed.), *Ground Truth: The Social Implications of Geographic Information Systems*, New York: Guilford Press, pp. 88–112.

Wade, R. (1996) 'Globalization and its limits: reports of the death of the national economy are greatly exaggerated', in S. Berger and R. Dore (eds), *National Diversity and Global Capitalism*. Ithaca, NY: Cornell University Press, pp. 60–88.

Walters, C. (2002) *Adaptive Management of Renewable Resources*. Caldwell (New Jersey): Blackburn Press.

Winner, L. (1977) *Autonomous Technology: Technics-Out-of-Control as a Theme in Political Thought*. Cambridge, MA: MIT Press.

Zimmerer, K., Galt, R.E. et al. (2004) 'Globalization and multispatial trends in the coverage of protected-area conservation (1980–2000)', *Ambio*, 33(8): 520–9.

Zimmerer, K.S. (2000) 'The reworking of conservation geographies: nonequilibrium landscapes and nature-society hybrids', *Annals of the Association of American Geographers*, 90(2): 356–69.

Zimmerer, K.S. and Young, K.R. (1998) *Nature's Geography: New Lessons for Conservation in Developing Countries*. Madison: University of Wisconsin Press.

Regulating Resource Use

Karen Bakker and Gavin Bridge

INTRODUCTION

Our objectives in this chapter are to resuscitate the concept of resource regulation and to explore how such a concept might contribute to a revived political geography of resources. We begin from the premise that resources are inherently political. By this, we mean that resources are an epistemologically specific outcome of competing claims over access to, control over, and definitions of nature. 'Resources' bespeak the operation of geographical networks of knowledge and control, through which a radically heterogeneous world of nature is ordered, fractured and delivered up to the economy. Resource extraction, therefore, occurs as a process of social negotiation of access to, and control over, an already politicized landscape of uneven development (Roberts and Emel, 1992) and, as such, the production of nature as resources inevitably entails an exercise of power.

As we explore in greater detail in the second section, our starting point is somewhat different from that of much geographical literature on resources, which circumscribes resource regulation as resource management, enacted primarily by an administrative state. Although the political dimensions of resource production and consumption have long been discussed within the geographical literature, many studies imply that natural resources 'become political' only at particular moments: specifically, at times of crisis, or conflict over resource allocation and control. From the perspective of these conventional approaches, the extensive systems of resource provisioning that underpin contemporary economic life are assumed to be quiescently apolitical for much of the time, so that resources only enter the realm of the political at moments of crisis. We argue in this chapter that this treatment of the political dimension of resources

is overly restrictive, and that effacing the politics of resource definition effectively naturalizes and thereby (selectively) depoliticizes resources.

If we begin from the claim that resources are always already political, the apparent quiescence of resources demands our attention; the periodic failure of resources to erupt as the object of overt political struggle is something that demands explanation. In the third section, we begin to rework the concept of regulation to describe the processes by which the inherent tensions over resource definition, resource access and resource use are managed and contained. While regulation often has a constrained meaning – as sets of rules, enforced by the state – we deploy the term more broadly to encompass the ways in which institutions (in the sociological sense of rules, norms, customs) are enacted. Moreover, we argue that such an approach to regulation is increasingly justified by three interrelated transformations in resource sectors around the world in the past two decades: a transformation in scientific and lay knowledges of nature (from 'resources' to 'environment' as the targets of science, resource management and politics); a political economic transition (from states to markets); and an institutional transition (from administrative to market-based regulatory instruments).

In the fourth section, we argue that our conceptualization of regulation shares much in common with concepts of environmental governance and eco-governmentality (although we explain how these terms are not of a piece) in their focus on the relationship between multi-scalar 'communities of rule' and resources (Watts, 2004). We identify what these terms have to offer an account of regulation and, drawing both from these perspectives and from work on the production of nature, we develop a reading of regulation as the processes by which resources are performed (against alternative claims) and through which the metabolism of dynamic resource landscapes (which can resist their enrolling as resources) is negotiated. The fifth section illustrates this approach in a brief case study. In the concluding section of the chapter we reflect on this more expansive approach to regulation. We sketch the outlines of this approach as the basis for a revived resource geography, discuss its implications for what 'doing resource geography' entails and, returning to our opening theme, contrast this reading of resource regulation with contemporary work on 'resource politics'.

RESOURCE REGULATION AS 'RESOURCE MANAGEMENT'

For much of the twentieth century, 'resource regulation' has been equated with the role of the administrative state in managing natural resources such as soil, water, forests, wildlife and minerals. The state's extensive involvement with the management of resources is expressed today in the familiar panoply of acronym-inscribed agencies (the USFS, USGS and USFWS in the US, DEFRA and the EA in the UK, and DAFFA in Australia, for example). Yet the framing of regulation in terms of a science of resource management – and the conjoining of this science with the administrative authorities and capacities of the state – are historically distinctive moments associated with the growth and 'modernization' of national economies. The outlines of contemporary, state-centric modes of resource regulation first emerged at the end of the nineteenth century in response to the massive appropriation of natural resources associated with industrialization and colonial trade. The embrace of resource *management* as a legitimate state function constitutes a significant moment in the evolution of the administrative state (Hays, 1959; Worster, 1985; Lipietz, 1987). The objective of this section is to outline these historical circumstances, and to show that their legacy is a habit of mind that defines resource regulation narrowly as the management of resources by the state.

From Resource Acquisition to Resource Management

States have long held an interest in natural resource issues: consider, for example, the Spanish Crown's sponsorship of Pizarro, Cortés and Alvarado, the charter from King Charles II granting monopoly rights to the Hudson Bay Company, or the US Public Land Survey initiated in 1785. Prior to the late nineteenth century, however, the state's primary interest in resources lay in securing and expanding the supply of raw materials to the domestic economy through territorial means. The state assisted primitive accumulation and extended monopoly control by defining and defending private and public property rights (the conclusion of land treaties and concessions, for example), subsidizing the costs of resource exploration by privateers, and deploying legal, political and sometimes military means to deny access by competing nations to critical or strategic resources.[1] Foster (1999: 71), for example, captures the simultaneous interests of the state in territorial expansion/acquisition and resource allocation (into private hands): he cites Veblen's (1923) commentary on the role of the US government during this phase in 'converting all public wealth to private gain on a plan of legalized seizure' and links this to the precipitous decline in fur-bearing mammals, the depletion of Southern soils through cotton farming, timber looting in the Great Lakes and Pacific Northwest, and 'rip-and-run' mineral extraction. Thus, regulation in this 'pre-managerial' phase was an extension of the territorial function of the state, and centred primarily on acquiring, defending and allocating access to land and resources.

In the early twentieth century a significant shift in the state's role in resource regulation occurred as resource management began to supplement the more traditional role of securing and expanding territorial control over resources.[2] This effectively supplemented a longstanding interest in the political control of resources over space with a concern for the stewardship and sustained use of natural resources over time. The emergence of the Conservation Movement in the US at the end of the nineteenth century exemplifies this growing interest in the temporal management of resource stocks and their orderly flow as commodity inputs to the economy (Hays, 1959). Initially centred on scientific forestry and multi-purpose river development (through the work of Gifford Pinchot, and the incorporation of this work into federal agencies during the administration of Theodore Roosevelt), the Conservation Movement sought to transform 'a decentralized, nontechnical, loosely organized society, where waste and inefficiency ran rampant, into a highly organized, technical and centrally planned and directed social organization which could meet a complex world with efficiency and purpose' (Hays, 1959: 265).

The state's expanding interest in regulating the flow of resources into the economy, and in subsidizing the production of environmental goods that until then had been taken as 'free' inputs (such as the fertility of soil, seed stocks or game species), are the institutional markers of a structural shift in the organization of the economy taking place in the early twentieth century. As Shiva (1992: 207) has argued in the context of developing countries, the exploitation of nature appears to have taken place in two phases:

> ... in the first phase, when nature's wealth was considered abundant and freely available, 'resources' were exploited rapaciously. They were not husbanded. In the second phase, once exploitation had created degradation and scarcity, the 'management' of 'natural resources' became important in order to maintain continued supplies of raw material for commerce and industry.

This transition can be characterized as one from an extensive regime of accumulation, where economic growth is achieved through increasing the throughput of resources, to an intensive regime, in which the opportunities for growth centre on improvements in labour and resource productivity. Prudham (2004: 637), for example, documents the evolution of an intensive regime of managed resource inputs (centred on the management of seed stock, selective breeding and tree farming) in the forest industry of the Pacific Northwest. He describes how this regime emerged from the exhaustion of a system of production centred on the felling of old growth timber and how, by the 1950s, timber production had moved from 'mining the accumulated natural capital of the region's old growth forests to the industrial cultivation of forest trees, including systematic intensification and rationalization of their growth'.

Management as the Social Production of Nature

Although resource appropriation continued to expand prodigiously, the objectives of resource management during the intensive regime came to centre on the 'efficiency' of material conversion.[3] Resource *management* – as Williams (1976: 190) carefully observes – incorporates the related yet distinct meanings of trainer or director (*managgiare*) and careful housekeeper (*ménager*). As an 'applied science of possibilities' (Hays, 1959), resource management gave rise to analytical techniques, administrative methods, and decision-rules for extending active control (via systems of monitoring and feedback) over the rate of factor inputs in an effort to maximize a given objective; resource management, then, was the means for transforming nature into an 'organic machine' (White, 2000). A world in which resources can be considered the object of management reflects a subtle yet significant change in the conceptualization of resource scarcity. Rather than scarcity being a wholly external condition (a limit to which the only effective responses are geographical expansion or economic retrogression),[4] scarcity arises

from the intersection of social organization and physical resources. Scarcity and abundance, therefore, are socially produced conditions rather than externally imposed limits.

A managerial approach to resources, therefore, relaxes the absolute view of physical scarcity by recognizing how society can produce – and historically has produced – supply shortages (although resources themselves remain understood in resolutely physical terms). The pioneering works on resource management – consider Zimmerman's (1931) dictum that 'resources are not: they become' – represent an initial step towards denaturalizing resource scarcity, albeit in a rather 'weak' way (cf. Harvey, 1974 and Yapa, 1993, for 'strong' approaches), followed by stronger assertions of the role of humans in construction or 'co-producing' natural environments (McKibben, 1998; Ross, 1994). The emergence of resource management as a science therefore reflects a transition from an 'empty world' (an expanding, abundant frontier) to a 'full world', in which the possibilities for expansion are limited and the focus shifts towards utilizing existing resource supplies productively into the future (Daly, 1992). It is no coincidence that this transition began at a time of the first 'global closure' at the end of the nineteenth century (Taylor, 1993: 1), a period in which 'expanding empires [began to face] one another in a geographical zero-sum game', and famously marked in the US context by the Bureau of Census's announcement of the closing of the frontier in 1890.

Resource Management as a State Function

That resource management science developed in the late nineteenth century in response to rampant resource consumption does not explain why these capacities became so thoroughly vested in the state. The emergence of state-based regulation in the early twentieth century (and its subsequent temporal and spatial expansion) constitutes a critical moment in the Fordist production of nature, a point at which the state intervened to manage resource stocks and flows in order to shore up the supply (and quality) of 'ecological' goods that, until that point, had been treated as free inputs. State intervention was not a necessary outcome of the degradation of resources, however; as the quality and extent of formerly free goods (such as rangeland or old growth timber) became increasingly scarce, the capitalization of these inputs and their production by capital (in the form of feedlots or timber plantations, for example) became a possibility. The institutional form of state-based regulation, therefore, was determined by the balance of political and economic interests concerned and the ease/costs of producing formerly free inputs (see Hays, 1959). Migration of resource decisions up the administrative hierarchy was facilitated in many instances by the epistemology of rational, modernist science, which – with its commitment to a rigid separation of fact and value (and of science from politics) – effectively reserved managerial decisions for a centralized cadre of experts independent of the political process. In addition, state-based regulation found broad political support as a means of limiting the damaging effects of inter-capitalist competition (by, for example, controlling well-spacing, pooling and production for oil during the 1930s). State resource planning also drew support as an efficient means of co-ordinating private resource development efforts over space and time in order to maximize productivity and avoid wasteful overinvestment in infrastructure (for example, irrigation, oil and gas pipelines, electricity transmission networks and river-basin planning).

The model of functional control, technical expertise and centralized resource decisions exemplified by scientific forestry and water-basin planning in the US was widely exported to the developing world. Indeed, the technical capacities for mobilizing and harnessing natural resources were a key component in the programmes of structural change and technological upgrading (in short, 'modernization') of the developmental state that emerged in many newly decolonized countries in Asia and Africa in the second half of the twentieth century (Corbridge, 1993; Scott, 1998). Cold War politics also exacerbated the extent to which resource regulation has been understood in state-centric terms, adding a specific polarity to longstanding concerns about international competition for strategically significant resources (Krasner, 1978). In the US, Cold War concerns over access to sufficient supplies of strategic resources such as tungsten, cobalt and uranium supported an increasing role for the government in regulating the flow of minerals into and within the national economy. Recognition of the extent to which the US had relied on uranium from the Belgian Congo for the Manhattan Project, for example, drove a government-initiated uranium exploration boom in the US intermountain west during the 1950s to create sufficient domestic supply (Mogren, 2002). A lasting legacy of these state-centric approaches is the rich statistical archives of the state resource agencies (such as the US Bureau of Mines, or the Bureau of Reclamation) charged with collecting, inventorying and analysing resource data at the national scale.

In summary, the popular association of resource regulation with resource management actions of the state is grounded in the historical expansion of the state's authority and capacity for resource (and later, environmental) management in the twentieth century. The centring of resource questions on the state has been reinforced by both economic (national modes of economic regulation such as Fordism) and political developments (national efforts at modernization and development, national defence) that have implicitly defined resource regulation as planned action by the state to shape the allocation, use and conservation of resources across space and over time. One finds this definition reflected in work by political geographers on resources: resources are understood primarily through the lens of the nation-state, as either commercially valuable economic inputs or as objects and means of political power.

Young (1981), for example, outlines three distinctive roles for the state in the management of natural resources: the state as a sovereign power devolving authority over resource management to private individuals or other entities; the state as a resource operator, such as the nationalization of resource companies in many countries in the postwar period; and the state as a regulator, intervening in resource use decisions to increase allocative efficiency. Rees's (1991) substantial contribution to the understanding of resource allocation similarly treats regulation as the formal policy mechanisms adopted by the state. Her work is significant for its explicit recognition of the political economy of resource production, the way that resource corporations often have greater influence than governments in determining distributive patterns, and her critical perspective on the effectiveness of public policy to effect significant changes in the resource and welfare allocation. Along with work by Peluso (1993), Blowers (1998) and Lake and Disch (1992), Rees exemplifies how some of the most insightful critiques of contemporary resource management richly problematize idealized notions of administration and planning, yet still equate regulation with the actions (or inactions) of the administrative state.

RETHINKING RESOURCE REGULATION

In the previous section we showed how the popular understanding of resource regulation as a state-centred, functional approach to the management of resource stocks and flows over time and space underpinned by scientific principles is a legacy of the historic expansion of the state into the management of natural resources during the twentieth century. In this section we argue that the equation of resource regulation with state-based institutions of resource management is increasingly anachronistic and unsatisfactory. We outline three reasons for seeking a more expansive conceptualization of regulation and, in the process, lay the foundations for a resuscitation of resource regulation in the final three sections.

First, a primary objection to state-centric approaches to regulation is that a proliferation of non-state actors (including non-governmental organizations, supranational institutions and transnational corporations) now challenges the authority and legitimacy of the state on resource questions. To focus solely on the administrative state as the primary actor in regulation, therefore, is an unwarranted blinder to the contemporary politics and practices by which resource decisions are made. The reasons behind the emergence of a plurality of resource actors are complex, but schematically it is possible to identify how authority has been both wrested (through legal and political challenge) and ceded (via the state's transfer of resource allocation decisions to the market) from and by the state. In the former category are the challenges to the state posed by indigenous resource and conservation politics: from the Ecuadorian Oriente (oil) and northern Wisconsin (copper), to Weipa (bauxite) or Kakadu (uranium) in Australia, struggles by indigenous peoples over the right to control decisions, conditions and rents associated with resource extraction or the designation of conservation areas have undermined the popular legitimacy and legal authority of the state to manage 'national' resources. The expansion of community resource management and other collaborative approaches has also worked away at the edges of state control, demonstrating the possibility and efficacy of resource management regimes that lie outside the stark dichotomy of state versus market suggested by Hardin's influential 'The tragedy of the commons' (Hardin, 1968; Berkes, 1989; Ostrom, 1990).

In tandem with the proliferation of non-state actors is an increasing prevalence of market-based mechanisms for resource management. This 'neoliberalization of nature' has occurred through both a rapid expansion of longstanding, non-state forms of resource management (such as the growth of land trusts; see Raymond and Fairfax, 2003) and via action by the state assigning to the market allocative decisions previously held by the state. Geographers have paid increasing attention to the marketization of a striking array of nonhuman natures (fisheries, water, minerals, wetlands and genes) and to the consequences of increasing involvement of private-sector actors, replacement of public policy with market mechanisms, the uptake of environmental valuation methodologies, and the commercialization and privatization of resource management institutions (Blowers and Leroy, 1994; Bakker, 2000, 2001; Walker et al., 2000; Gibbs and Jonas, 2001; Bridge and Jonas, 2002; Castree, 2003; Johnston, 2003; McAfee, 2003; Bridge, 2004; Maddock, 2004; Mansfield, 2004a, 2004b; McCarthy, 2004; McCarthy and Prudham, 2004; Prudham, 2004; Robertson, 2004; Smith, 2004).

Second, a narrow focus on resource management overlooks the expansion of the state into environmental science and environmental management in the last thirty years. The key distinction here is between the 'conservation state' that emerged at the end of the nineteenth century (and which is described in the second section) and a secondary, late-Fordist expansion of the administrative state into the maintenance of environmental conditions and services (such as the nutrient cycling, pollution absorption and flood control capacities of wetlands). National governments have adopted administrative forms of regulation aimed at maintaining environmental functions and processes considered desirable for humans since the late 1960s in industrial countries (and the 1980s in many developing economies). The modality of regulation integrates environmental scientific knowledge based on ecology and systems-thinking, and is based on intervention in ecosystems under conditions of limited predictability and incomplete knowledge (see, for example, the literature on adaptive management: Botkin, 1992, and Gunderson and Holling, 2002).

Third, the instrumentalism and commodity-logic that dominates resource management (and which historically has characterized resource geography) is now frequently challenged from positions that reflect a broader range of epistemologies than that of Western science. This began in a modest way with the incorporation of social knowledge and non-economic factors into floodplain and river-basin planning (White, 1945; Emel and Peet, 1989), but has subsequently expanded to embrace the tangled epistemologies animating the wilderness movement (see Nash, 1967), manifold indigenous knowledges and the alternative subject positions of women, minorities and future generations (see Emel, 1990; Howitt, 2001 and Plumwood, 2002, for different critiques of the instrumentalism of resource management). Plural epistemologies are increasingly mobilized in struggles over resource development and, in some cases, have begun to intrude 'into the policies and practices of many international agencies, transnational resource companies and inter-governmental and non-government bodies' (Howitt, 2001: xiii). At their most potent, the effect of these different ways of knowing is to radically blur the definitions and practices through which conventional 'resources' are produced.

In sum, these developments (1) pry open the equation of 'politics' with the state, and free a political geography of resource regulation from the prevailing state-centric perspective by emphasizing the multiple institutions and sites of regulation; (2) problematize the question of precisely 'what' is being regulated by 'historically falsifying' (Beck 1992: 981) the notion of nature (resources/environment) as a discrete entity situated external to society (explicitly 'produced' environmental phenomena make this point with particular force, such as the bioaccumulation of pesticides or heavy metals in fish); and (3) expose the particularity (and therefore the politics) of the epistemologies that define resources qua resources. In the next section we argue that, taken together, these developments open up a more expansive interpretation of 'resource regulation': one that acknowledges the multiple sites and scales of regulation, and that understands regulation not as a way of ensuring inputs 'from' nature or controlling pollution flows 'to' nature, but as the negotiation of a dynamic landscape that is neither fully natural nor fully social.

GOVERNMENT, GOVERNANCE, GOVERNMENTALITY

The concept of *environmental governance* provides a useful entry point for examining the changing modalities of regulation identified in the previous section. The term is widely used in policy and academic debates to refer to the problems of achieving organization and co-ordination in a society composed of heterogeneous interests and capacities. Thus, governance is often referred to as the 'art of steering societies and organizations' (Pierre, 2000). Although governance can have a variety of specific meanings, it is most frequently used in ways that explicitly problematize state-centric conceptualizations of regulation and power; for example, it is often deployed to describe a shift in the scale and modality of regulation away from *government* (in which the administrative state is dominant) towards *governance* (which foregrounds the relationships among variously scaled non-state actors and between these actors and a 'rump' state) (Jessop, 1995).[5] Studies emphasize the radical reshaping of government services that has introduced private-sector norms and market principles into the civil service (sometimes referred to as 'New Public Management'), the contracting-out of state administrative functions to non-governmental actors (sometimes referred to as the 'hollowing-out of the state'), and the introduction of market-based performance measures and incentive structures and/or the creation of quasi- or simulated markets. Governance, then, captures how formal state authority has increasingly been supplemented or supplanted by a reliance on informal authority, particularly in forms of negotiated patterns of public/private/community co-operation. Roles previously allocated to governments are now (controversially) categorized as more generic social activities that may be carried out by political institutions, but which can also be carried out – and perhaps more appropriately so – by other actors (Pierre, 2000).

De-centring Regulation: Governance in the Mineral Sector

The experience of the oil, gas and mining sectors suggests that the concept of 'governance' captures something significant about the way resource production is regulated by a proliferation of non-state actors and an intensification of their interrelationships with one another and the state. In most jurisdictions, mineral resources are public resources to which the state grants access and development rights on specific terms; the state, therefore, retains a significant formal role in most mineral resource regimes around the world. In many countries, however, the reform of neo-liberal mineral investment codes over the last ten to fifteen years has removed the state's monopoly on mineral development, while preserving public ownership of the sub-surface. These reforms have made it possible for non-state actors (i.e. individuals and private corporations, including foreign corporations) to participate in mineral resource development, and in some cases have also introduced market-based elements (such as auctions) into the process for allocating exploration and development rights. One dimension of the shift to governance, then, is the increase in non-state actors and market mechanisms in the allocation and development of natural resources and, with the decline of the state as a mineral producer, the increasing penetration of market logics into the timing of mineral investment and production decisions.

There is a second and equally significant dimension to governance, however, that concerns the way decisions over mineral production are increasingly negotiated (i.e. contested) outcomes involving a range of actors (stakeholders) rather than the action of administrative fiat. Until recently, the granting of mineral rights by the state (together with the state powers that these legal rights afford, including the power to expropriate property and remove people by force) has typically been enough to ensure mineral access; in other words, state sanction (legal legitimacy) has been a sufficient condition for mineral development. Over the last decade or so, however, the access guarantee that legal legitimacy provided has been progressively eroded; state sanction may be a necessary condition of mineral development, but it is no longer sufficient. The political economy of mineral access is now one in which a range of new actors and institutions loom large: non-governmental organizations intervene in decisions traditionally reserved for corporate management, often by harnessing the informational, financial or product networks in which multinational resource firms are embedded (Pratt, 2001);[6] multinational mining firms conclude agreements with organizations such as the IUCN (on no-go areas for mining[7], Wilson, 2003) that stand outside any state-based framework; and codes of conduct or environmental commitments developed by international organizations (such as the International Council on Mining and Metals, the Equator Principles, CERES, or those emerging from the World Bank-sponsored Extractive Industries Review) assume increasing significance vis-à-vis state environmental regulations for determining where and how mineral extraction takes place. In short, it is through the workings of these institutions that resource production is now 'regulated'. The multiple dimensions to this fragmentation of the state's authority to regulate resource production are captured (albeit obscurely) by the notion of a 'new reality of mineral development' (Clark and Clark, 1999), while its power geometries are reflected (however opaquely) in the extractive industries' new credo of 'winning a social licence to operate'.

From Governance to Governmentality

Its capacity for capturing something significant about the contemporary reworking of resource institutions notwithstanding, we approach 'governance' with both caution and a call for greater precision in its usage. Work on environmental governance focuses primarily on the institutional superstructure of regulation, as opposed to the underlying political economy of resource production. The environmental governance literature thus often identifies changing patterns, forms and scales of regulation without exploring the political-economic conditions that may drive these shifts. Changing forms of governance in the mining sector, for example, cannot be fully understood without addressing the economic and spatial expansion of production during the late twentieth century. This expansion has required increasing inputs of raw materials, a demand met in part through a 'scaling up' of resource acquisition as states and corporations have increasingly reached to the 'ends of the earth' to meet raw material needs. Moreover, the term 'environmental governance' is often employed in a normative sense, naïvely celebrating the rise of non-state actors in 'civic' or 'public' environmentalism without questioning the reasons for – or the implications of – the 'retreat of the state'. There are, for example, reasons to be cautious about what the 'fluidization' and 'NGO-ization' of politics – trends so characteristic of environmental governance – imply for democratic participation and/or the possibilities for radical politics (see Faust and Nagar, 2003).

A more fundamental difficulty with the concept of environmental governance is that it fails to adequately address underlying epistemologies of environmental knowledge. While discussions of governance may acknowledge different valuations of nature (for example, as the basis for a conflict over resource use), most work on environmental governance does not pursue the full implications of this perspective for understanding how governance – the social co-ordination of the inherently political nature of resources – is achieved. Deploying a concept of *governmentality*, as distinct from governance, may be helpful here. As developed initially by Foucault, governmentality refers to a particularly modern (and Western) form of rationality, which emerged in Europe in the sixteenth and seventeenth centuries in the context of population growth, urbanization and the decline of feudalism. The term denotes a transition away from conceptions of power based on sovereignty – the hierarchical exercise of power by a ruler preoccupied with the maintenance of control over a defined territory – to governmentality, a more diffuse form of power through which an increasingly administrative, bureaucratic state came to manage its population and resources by employing a new set of *savoirs* or rationalities (such as statistics), which enabled an unprecedented degree of control and surveillance over individual lives. The twinned concerns of this new form of rationality were population and resources. Foucault refers to the simultaneous construction of an interdependent population/resources complex as the 'imbrication of men and things', the management and control of which becomes the central concern of political economy, and which lies at the core of governmentality as a form of emergent 'biopower'.

The concept of biopower was applied by Foucault largely to questions concerning human populations, leaving aside questions of the non-human, or the environment (Rutherford, 1999). Yet subsequent researchers (Braun, 1997; Luke, 1999; Drayton, 2000; Mitchell, 2000) have applied concepts of governmentality to resources and environment (Foucault's 'things'), positioning governmentality as a rationality of management through which the domination of people is achieved via the control over biological/ecological processes. The focus of this work, therefore, is on the dense interweavings of knowledge and power that produce assemblages of plants and animals as political units ('forests' or 'wetlands', for example) or the knowledge systems and calculations through which non-human populations are physically, legally and economically defined and their control and regulation achieved (via the science of tree growth, animal husbandry or wildlife counts, for example). Peluso and Vandergeest (2001), for example, document how the practices and rationalities of forest management in Southeast Asia – mapping, zoning, and the enacting of land and forest laws – provided the colonial state with a way to control populations by criminalizing previously common practices.

Enacting Resources

We argue that governance and governmentality can be mobilized as complementary concepts in redefining resource regulation. An analytic of governmentality focuses on the 'conduct of conduct' – the rationalities of rule through which we govern ourselves, one another, and 'things'. Governmentality, in other words, leads to an understanding of regulation as the rationalities and practices (laws, norms, rules and customs) through which specific natures are enacted and performed. Governance, on the other hand, places greater emphasis on the organizational aspects of regulation: the actors and loci of decision-making, the competing political and economic agendas embedded in decision-making processes, and the highly variable resource and property management regimes in and through which the political struggles over resources take place. Governance and governmentality, in this framing, are complementary concepts underpinned by the understanding that resources are not only physically produced, but socially enacted. A resource, in other words, is achieved through processes of work, where 'work' refers not only to material transformations and the historical 'dead-labour' embodied in the commodity, but also ideological transformations and the continual discursive boundary work necessary to shore up the category of 'resource', to preserve the privileged status of particular resource definitions in the face of alternative claims (Bridge, 2001).

This notion of 'enactment' underpins our approach to resource regulation as the negotiation of the metabolism of a dynamic resource landscape. The usage of the term 'metabolism' is drawn from political ecology: 'metabolism' is understood not as the act of digestion or consumption, but rather as the practice of mutual transformation of socio-natures (Swyngedouw, 1999). This revival of the term 'metabolism' draws on insights from political ecology: that our production of nature involves not only the transformation of nature, but also adaptation to biophysical conditions, in which the resources can resist human intentions and defy the properties attributed to them. In other words, socio-economic change and environmental change are mutually constitutive, and regulation is the act of mediating this relationship. Regulation, then, describes the processes by which resources are enacted (against alternative claims) and through which the metabolism of dynamic resource landscapes are negotiated. These processes are simultaneously material and discursive and extend to the enactment of institutional frameworks that embody the rules that define knowledge and legitimize authority (Bakker, 2004). In the next section, we apply this analytical framework to the case of water supply management in England and Wales.

RE-REGULATING WATER IN ENGLAND AND WALES

Throughout much of the twentieth century, water supply was mobilized as a strategic resource for societies undergoing modernization, industrialization, urbanization and agricultural intensification. Britain, like many OECD countries, adopted a 'state hydraulic' regulatory regime of water management characterized by: planning for growth and supply-led solutions; a focus on social equity and universal provision; command-and-control regulation; a discursive representation of nature as a 'resource'; and state ownership and/or strict regulation of water resources development based on a desire to provide sufficient quantities of water, where and when needed, such that economic growth could proceed unconstrained (Goubert, 1986; Hassan, 1998; Coutard, 1999; del Moral and Sauri, 2000; Bakker, 2004). Given high capital costs and long infrastructure lifetimes, public financing was critical for the development of water supply. The state played a key role as a facilitator of growth and promoter of technological progress across utility and resource sectors (Wescoat, 1987; Chant, 1989; Graham and Marvin, 2001).

Yet by the late twentieth century, water use and investment patterns had begun to change due to de-industrialization, increasing technical efficiency in an era of heightened concern over resource scarcity, and changing patterns of domestic water use (Gleick, 2000). Overall demand for water stagnated and even dropped in the UK (Bakker, 2004). Greater awareness emerged about the (still hotly debated) effects and (often unquantified) costs of hydraulic development, particularly large dams, extirpation of species, displacement of communities, flooding of cultural sites, contamination of water sources, disruption of ecological processes, and environmental degradation (Gleick, 2000; WCD, 2000; Graf, 2001; Ortolano and Cushing, 2002; Biswas, 2004). With threats to human health from waterborne diseases such as typhoid having been brought under control, concern began to focus on non-point sources of pollution and other contaminants, a concern heightened by growing realization that the post-war economic boom had obscured systematic deterioration of water supply infrastructure in many countries (Kinnersley, 1988; Melosi, 2000).

In England and Wales this deterioration was particularly acute, with significant effects on environmental water quality. Throughout most of the twentieth century, water planners had focused on developing new water sources such as reservoirs, pursuing a supply-led strategy to anticipate increasing water demands stemming from economic and population growth. Under-investment in infrastructure (to minimize public-sector borrowing for macro-economic reasons, and to maintain low water bills for political reasons) and sustained industrial water pollution contributed to the continued decline of river and tap water quality in Britain for decades (Pearse, 1982; Kinnersley, 1988, 1994; Summerton, 1998). The much-lauded integration of water supply and regulatory functions in basin-wide Regional Water Authorities according to the principle of Integrated River Basin Management had the unforeseen side-effect of discouraging enforcement of water quality regulation (particularly sewage works), further aggravating environmental degradation. In short, universal access to water supply networks had been achieved, but the costs were partially externalized onto the environment.

The state hydraulic regulatory regime remained stable in Britain throughout much of the mid-twentieth century. By the 1980s, however, political and social consensus around this regulatory regime began to break down. The enormous cost of replacing ageing infrastructure and addressing the backlog in maintenance became apparent. Media and public attention focused increasingly on the decline in water quality. The European Union announced it was suing the British government for failure to comply with its water quality legislation. These events occurred in the context of an acute public-sector fiscal crisis and dramatic shift in political direction with the election of Margaret Thatcher's Conservative government in 1979. Strict monetary controls meant that central government was no longer willing to subsidize the costs of water supply, and was unable to finance environmental clean-up on the scale required. The state hydraulic regulatory regime thus faced a multidimensional crisis, ecological, cultural, ideological and socio-economic.

Accordingly, the government initiated commercialization of the water supply sector in the early 1980s, transforming the water industry 'from a public service to a business organization' (Penning-Rowsell and Parker, 1983: 170). Labour levels and investment were reduced, tight financial controls were introduced, price increases were mandated (with bills rising above inflation), and increasing emphasis was placed on economic, as distinct from technical, performance indicators. Commercialization and subsequent privatization 'thrust [water companies] into a more commercially orientated world, wherein the organization was under pressure first to show, and then to continually expand, a return on capital employed' (O'Connell-Davidson, 1993: 191). By the late 1980s the water utilities, along with other nationalized industries, were best characterized as private monopolies, publicly regulated, operating on modified market principles (Parker and Sewell, 1988; Hay, 1996).

Privatization, through flotation of the public water authorities on the London Stock Exchange in 1989, consolidated this transformation (Bakker, 2004), introducing market-simulating regulatory mechanisms such as cost–benefit analysis into both economic and environmental regulation. Little over a decade after privatization, labour levels have been dramatically reduced, collective bargaining mechanisms dismantled, and out-sourcing 'non-core' functions has significantly changed labour relations and practices in the industry (O'Connell-Davidson, 1993). Investment levels have increased, with companies spending £31 billion from 1990 to 2000; investment over the period from 1991 to 1996 was twice the levels prior to 1989 (Kinnersley, 1998), much of it focused on sewage works and environmental quality improvements. In pricing, economic equity is prioritized over social equity (Bakker, 2001). In economic regulation, efficiency is prioritized, although the increase in efficiency of water supply management is disputed.

Water supply system management practices have evolved significantly: rather than engineering-driven approaches prioritizing redundancy and interconnection in the storage and distribution networks (and hence security of supply), economics-driven approaches prioritizing economically efficient management of the network and demand management (and hence on cost minimization for given output) are increasingly central to water resource management policies (Guy and Marvin, 1996a, 1996b; Mitchell, 1999). This shift stems in part from growing concerns about the impacts of climate change on water resource security, particularly in southern England (Arnell et al., 1994; DOE, 1996; Marsh, 1996), and an increasingly dominant discursive depiction of water as a scarce resource (noteworthy in such a 'wet' country) – which has recently been enshrined in UK legislation with the designation of official 'Areas of Water Scarcity'.

This new regulatory regime can be characterized as a variant of 'market environmentalism' (Bakker, 2004) or 'green neo-liberalism' (Goldman, 1998). 'Market environmentalism' is a mode of resource regulation which promises both economic and environmental ends via market means (Anderson and Leal, 2001). As a variant of ecological modernization, market environmentalism offers hope for a virtuous fusion of economic growth, efficiency, and environmental conservation (Hajer, 1995; Mol, 1996; Christoff, 1996; Hawkins et al., 1999). Through establishing private property rights, employing markets as allocation mechanisms and incorporating environmental externalities through pricing, proponents of market environmentalism assert that environmental goods, rescripted as commodities, will be more efficiently allocated – thereby responding to resource scarcity through addressing concerns over environmental degradation and inefficiency. Markets, in other words, will be deployed as the solution, rather than being blamed as the cause of environmental problems.

An important feature of market environmentalism is the prioritization of environmental concerns and the development of new techniques of environmental management. Environmental issues have been formally integrated into water resources planning, an environmental regulator has been created, and the water industry has to some degree reinvented itself as an 'environmental services' industry. Much greater emphasis is placed on aesthetics, amenity value of landscape, and value of 'natural landscapes' – instrumentalized through changes to pricing of water abstraction, and valorized through environmental economic valuation techniques, now widely applied in the industry. Water quality and environmental expenditure are key drivers of capital expenditure programmes in the industry. Partly as a result, chemical and biological river water quality has improved, although some levels are still viewed by the government as unsatisfactory (DEFRA, 1999). Drinking water quality has also improved significantly (DWI, 2003). Much of this improvement is driven by increasingly comprehensive European Union water quality legislation governing beaches and bathing waters, drinking water quality, and environmental quality of both surface and groundwater (Walker, 1983; Buller, 1996; Kallis and Butler, 2001; Kaika, 2003). Water companies in England and Wales are to a much greater extent guided and constrained by environmental regulations than they were three decades ago. So too are managers, whose performance-based incentive schemes now routinely incorporate environmental performance criteria (Hopkinson et al., 2000), backed up by the threat of prosecution or 'naming and shaming' by the environmental regulator.

The increasing dominance of environmental concerns in Britain over the past two decades is characteristic of a shift in the relative influence of different stakeholders under market environmentalism – with labour unions sidelined, and consumers' interests to be balanced with, or trumped by, environmental concerns. Environmental externalities are addressed within the water policy framework, and backed up in most instances by legal obligations. In contrast, social externalities are now, to a greater degree than in the past, excluded from the water policy framework (Bakker, 2001, 2004). These shifting power geometries are most clearly observed in the formal structure of regulation: whereas the environmental regulator is a separate well-funded entity, the regulatory body responsible for consumers has, until recently, operated under the aegis of the economic regulator, with a highly constrained role (Page and Bakker, 2005). A significant proportion of the increases in domestic water users' bills post-privatization has been due to environmental expenditure. Environmental and drinking water quality have improved; according to the

environmental regulator of the industry, river water quality in Britain is at its highest level since the Industrial Revolution (DEFRA, 2001; EA, 2001; DWI, 2003). Decision-making on capital investment in the industry balances the interests of consumers 'willingness to pay' against environmental protection and rehabilitation requirements – a cost–benefit exercise that minimizes the participation of labour and attempts to exclude questions of ability to pay, in distinct contrast to pre-privatization (Bakker, 2001). Market environmentalism has thus produced clear gains for the environment in some cases, but at the apparent cost of consumers: hence the frequent disagreements between environmental groups and consumers' groups over water policy, particularly given the highly controversial impacts of water debt and 'water poverty' on public health (Drakeford, 1997).

The shift from state hydraulic to market environmentalist regulatory regime has involved several interlinked transitions: political-economic (a shift from state to private-sector ownership); institutional (the supplanting of command-and-control with market-based regulatory instruments); scientific (such as new techniques of environmental valuation, concepts of water scarcity, and approaches to meteorological variability); and discursive (the environment is rescripted as legitimate user, citizens rescripted as customers). These transitions have had important distributive implications. Whereas the social costs of water production were previously externalized from the sphere of the politicized citizen to the environment, the environmental costs of water production are now externalized from the sphere of capitalized environment to consumers. This shift in power geometries has led to heightened conflict among environmental and consumer interests (Bakker, 2004). This conflict is mediated through a complex regulatory 'game' between formal regulators and organized NGOs (Lise and Bakker, 2005) in which customers have few avenues for participation and little influence (Page and Bakker, 2005).

The case of British water supply is often cited as an example of the erosion of the state's legitimacy and scope for resource management, and of the displacement of the state by private-sector actors in resource management. This shift from 'government' to 'governance' has been accompanied by a change in the nature of regulation – with the increasing use of market-based regulatory instruments in both economic and environmental regulation. As this case study has demonstrated, the shift to market environmentalism has entailed re-regulation rather than de-regulation. Bringing a concept of governmentality together with an analysis of shifting patterns of governance underscores the multiple dimensions of re-regulation: simultaneously material, discursive and social. Whereas a state-centric focus on regulation might interpret the case of British water supply as a 'retreat of the state' or 'de-regulation', our approach characterizes market environmentalism as a new mode of resource regulation. This mode of resource regulation has multiple dimensions: rescripting the identities of water users and the environment; reallocating decision-making power and entitlements between users; and altering the material ways in which water is produced, consumed and disposed of in Britain. This approach focuses attention not on the relative weakness or strength of the state, but rather on the changing power geometries of regulation – and how our collective commitment to socio-environmental justice has been transformed as a result.

RENEWING RESOURCE GEOGRAPHY

The most recent round of imperial adventures in the Persian Gulf have thrust 'resource politics' into the popular imagination (Kaplan, 2001; Klare, 2002, 2004). In parallel – although over a slightly longer time horizon – a discourse of environmental security has gathered steam that now positions resources such as water, timber and even soil as the basis for regional conflicts (Homer-Dixon, 1991, 1999). This resurgence of the 'politics of resources' is problematic. With a few notable exceptions (for a critique of the environmental security discourse, see Peluso and Watts, 2001), it tends to treat resources as a contingent mode of transmission for (national) rivalries, the current vehicle through which (national) political and economic tensions play out. At some point (although in the most apocalyptic of the 'resource war' imaginaries, this point may constitute more of a final period than a transitional comma), it is often sincerely hoped (and occasionally feared) that these tensions will be displaced from natural resources, and the hotly political will revert to the simply economic: books like *Resource Wars* will be quietly withdrawn from the shelves, the reliably quiescent terrain of oil fields, water systems and working forests will have been restored. One does not have to entertain bellicose flights of fancy to indulge this only contingently political view of resources: technocratic approaches tend to regard resources in the same way, averring that there may be moments of crisis (price rises for critical resources, for example, at which point resources 'get political'), but that technological change and/or market adjustment will ensure that resources slip back across the threshold to the realm of normal provisioning.

We have argued in this chapter for a rather different understanding of the political character of resources. We focused on regulation as a central problematic in the political economy of resources because of the way that regulation (in all but the crudest accounts) emphasizes the socio-political conditioning of resource production and consumption, and the institutional mechanisms by which resources are commanded over time and space. We have sought to pry regulation free from its traditional association with the formal rules and procedures administered by the state for managing resources, arguing on both empirical and theoretical grounds why this restriction was no longer tenable (if it ever was). We have emphasized the erosion of the state's legitimacy and scope for action on resource and environmental issues in the face of a broad range of non-state actors, and the ceding by the state to the market of some of its authority and influence over the location, timing and rate of resource production.

This has allowed us to conceptualize regulation in terms broadly complementary to those of a 'regulationist' approach to resources, with its recognition of how social and political relations (codified as institutions or cultural norms) provide a coherence to the technological and organizational aspects of production (Drummond and Marsden, 1995; Gibbs, 1996; Gandy, 1997; Cocklin and Blunden, 1998; Bridge, 2000; Bakker, 2004). Significantly, however, we have moved beyond the strong institutionalism of the regulationist account to argue that regulation is a multi-sited and irreducibly social practice through which resources are enacted or performed. In developing this account we engaged with two recent bodies of work in political ecology: on (1) governmentality, and (2) the metabolism of the production of nature. These open up a rather different conception of regulation, as the processes by which resources are performed (against alternative claims) and through which the metabolism of dynamic resource landscapes are negotiated (landscapes that can resist their enrolling into econo-natural networks). Redefining regulation in this way displaces the analyst's gaze. Contemporary processes of economic deregulation – a reduction or qualitative shift in the formal role of the administrative state in the management of resources – become framed as 're-regulation': an effort to craft new institutional arrangements to manage inherent contradictions associated with the development of a resource. Rather than an analytical focus on the 'frontier' between state and market and a reliance on the public/private binary, attention is brought to bear on the ways in which the state and private actors strategically re-position their allegiances and commitments.

The implications of this approach to regulation for resource geographies are three-fold. First, and most obviously, the analysis of regulation outlined above adopts a critical rather than instrumental position towards resource development. Critical here denotes a commitment to probing the foundations of claims over resources (including claims to *know* resources – whether based on the epistemologies of science, manual work, or other sensory experience), to demonstrating the fluid boundaries (and boundary-making exercises) between social and environmental costs, and to illustrating the social and institutional practices through which resources are not only allocated but also defined – the discursive, social, material practices through which resources qua resources come to be constituted. Second, this approach does not privilege one particular location or scale for the practice of regulation. It is initially agnostic about the role of the state in regulation, and seeks to identify the multiple sites and institutional forms through which regulation is achieved. And third, this approach proceeds from the position that resource production is an achievement (in the broadest sense), something won in the face of considerable political tensions and by working with potentially 'unruly' materialities that can fail to conform to the economic requirements of a resource/commodity. It understands a resource not as a natural permanence, but as a temporary stabilization at the nexus of a set of political, economic and technical relations that is always potentially subject to dissolution and challenge. Regulation, then, is a dual-faceted concept: it describes both the process of negotiating or working with the metabolism of dynamic landscapes, and the rationalities and calculations by which this is achieved.

Finally, we suggest that the explicit conceptualization of regulation in these terms is distinctive, but note that many of its underlying elements are not new. The instrumentalism of conventional resource geography has been punctuated from time to time by sharp, illuminating critiques: the early work of Ciriacy-Wantrup, (1969) and Innis (1942), for example, or the more recent interventions by Roberts and Emel (1992), Howitt (2001), Peluso (1993) and Watts (2004). Our aim in this chapter has been to sketch the outlines of a revived resource geography that captures something of the spirit of this heterodox tradition, and which positions 'resources' as a central problematic in the regulation of political economy.

NOTES

1 Intellectual work on resources during this period is characterized by what Zimmerman (1931) terms an 'encyclopaedic' method (see, for example, von Humboldt's reflections on the resources of New Spain), a commitment to cataloguing, delineating, inventorying and mapping physical resource occurrences, a process of gathering and ordering resource data that was frequently conducted in conjunction with cadastral surveys and land titling (Kain and Baigent, 1992).

2 Titanic struggles for control would continue (in the context of oil, see Yergin, 1991; for a discussion of aluminium, see Barham et al., 1992); the point here is that a distinctive, additional function regarding resource provisioning emerged for the state.

3 As Frederick W. Taylor noted in the introduction to his *Principles of Scientific Management* (1911), the pursuit of efficiency in resource use championed by Pinchot and Roosevelt provided a model for the application of optimization techniques in industrial production more generally.

4 Consider Malthus's writings on population growth in Europe or Jevons' concerns about the degeneracy of British coal reserves.

5 For an articulation of this transition in the different contexts of urban regime theory and international relations, see Jessop (1995) and Rosenau (1992) respectively.

6 In March 2004 Tiffany and Co., the high-end US jewellery maker, took out a full-page, open letter to the Chief of the US Forest Service in the *Washington Post*, calling for reconsideration of its approval in 2003 of a silver mine under the Cabinet Mountain Wilderness of north-west Montana.

REFERENCES

Anderson, T. and Leal, D. (2001) *Free-Market Environmentalism*. New York: Palgrave.

Arnell, N.W., Jenkins, A. and George, D.G. (1994) *The Implications of Climate Change for the National Rivers Authority*. NRA Research and Development Report, 12. Bristol, NRA.

Bakker, K. (2000) 'Privatising water, producing scarcity: the Yorkshire drought of 1995', *Economic Geography*, 76(1): 4–27.

Bakker, K. (2001) 'Paying for water: water charging and equity in England and Wales', *Transactions of the Institute of British Geographers*, 26(2): 143–64.

Bakker, K. (2004) *An Uncooperative Commodity: Privatizing Water in England and Wales*. Oxford: Oxford University Press.

Barham, B., Bunker, S. and O'Hearn, S. (1992) *States, Firms and Raw Materials: The World Economy and Ecology of Aluminium*. Madison: University of Wisconsin Press.

Beck, U. (1992) *Risk Society: Towards a New Modernity*. London: Sage.

Berkes, F. (ed.) (1989) *Common Property Resources: Ecology and Community-Based Sustainable Development*. London: Belhaven Press.

Biswas, A. (2004) 'Dams: cornucopia or disaster?', *International Journal of Water Resources Development*, 20(1): 3–15.

Blowers, A. (1998) 'Power, participation and partnership: the limits of co-operative environmental management', in P. Glasbergen (ed.), *Environmental Governance: Public–Private Agreements as a Policy Strategy*. Dordrecht: Kluwer Academic Publications.

Blowers, A. and Leroy, P. (1994) 'Power, politics and environmental inequality: a theoretical and empirical analysis of the process of "peripheralisation"', *Environmental Politics*, 3(2): 197–228.

Botkin, D. (1992) *Discordant Harmonies: A New Ecology for the Twenty-First Century*. New York: Oxford University Press.

Braun, B. (1997) 'Buried epistemologies: the politics of nature in (post)colonial British Columbia', *Annals of the Association of American Geographers*, 87(1): 3–31.

Bridge, G. (2000) 'The social regulation of resource access and environmental impact: production, nature and contradiction in the US copper industry', *Geoforum*, 31: 237–56.

Bridge, G. (2001) 'Resource triumphalism: post-industrial narratives of primary commodity production', *Environment and Planning A*, 33: 2149–73.

Bridge, G. (2004) 'Editorial: Gas, and how to get it', *Geoforum*, 35: 395–7.

Bridge, G. and Jonas, A.E.G. (2002) 'Governing nature: the re-regulation of resource access, production and consumption', *Environment and Planning A*, 34: 759–66.

Buller, H. (1996) 'Privatization and Europeanization: The changing context of water supply in Britain and France', *Journal of Environmental Planning and Management*, 39(4): 461–82.

Castree, N. (2003) 'Environmental issues: relational ontologies and hybrid politics', *Progress in Human Geography*, 27(2): 203–11.

Chant, C. (ed.) (1989) *Science, Technology and Everyday Life 1870–1950*. London: Routledge.

Christoff, P. (1996) 'Ecological modernisation, ecological modernities', *Environmental Politics*, 5(3): 476–500.

Ciriacy-Wantrup, S.V. (1969) Natural Resources and Economic Growth: the role of institutions and policies, *American Journal of Agricultural Economics*, 51(5): 1314–24.

Clark, A.L. and Clark, J.C. (1999) 'The new reality of mineral development: social and cultural issues in Asia and Pacific nations', *Resources Policy*, 25(3): 189–96.

Cocklin, C. and Blunden, G. (1998) 'Sustainability, water resources and regulation', *Geoforum*, 29(1): 51–69.

Corbridge, S. (1993) 'Colonialism, post-colonialism and the political geography of the Third World', in P.J. Taylor (ed.), *Political Geography of the Twentieth Century: A Global Analysis*. London: Belhaven Press, pp.171–206.

Coutard, O. (1999) *The Governance of Large Technical Systems*. London: Routledge.

Daly, H.E. (1992) 'From empty-world economics to full-world economics: recognizing an historical turning point in economic development', in R. Goodland et al. (eds), *Population, Technology, and Lifestyle: The Transition to Sustainability*. Washington, DC: Island Press, pp. 23–37.

DEFRA (1999) *Raising the Quality*. London: Department of Environment, Food and Rural Affairs.

DEFRA (2001) 'UK maintains record-breaking performance for river quality', (press release). London: Department for Environment, Food and Rural Affairs.

del Moral, L. and Sauri, D. (2000) 'Recent developments in Spanish water policy: alternative and conflicts at the end of the hydraulic age', *Geoforum*, 32(3): 351–63.

DOE (1996) *Review of the Potential Effects of Climate Change in the United Kingdom*. London: HMSO.

Drakeford, M. (1997) 'The poverty of privatization: poorest customers of the privatized gas, water and electricity industries', *Critical Social Policy*, 17: 115–32.

Drayton, R. (2000) *Nature's Government*. New Haven: Yale University Press.

Drummond, I. and Marsden, T.K. (1995) 'Regulating sustainable development', *Global Environmental Change: Human and Policy Dimensions*, 5(1): 51–65.

DWI (2003) *Drinking Water 2002: A Report by the Chief Inspector*. London: Drinking Water Inspectorate.

EA (2001) 'Decade of clean-up brings best-ever river and estuary quality results (press release). Bristol: Environment Agency.

Emel, J. (1990) 'Resource instrumentalism, privatization and commodification', *Urban Geography*, 11(6): 527–47.

Emel, J. and Peet, R. (1989) 'Resource management and natural hazards', in R. Peet and N.J. Thrift (eds), *New Models in Geography*, Vol 1. London: Unwin Hyman.

Faust, D. and Nagar, R. (2003) 'Third World NGOs and US academics: dilemmas and politics of collaboration', *Ethics, Place and Environment*, 6(1): 73–8.

Foster, J.B. (1999) *The Vulnerable Planet: A Short Economic History of the Environment*. New York: Monthly Review Press.

Gandy, M. (1997) 'The making of a regulatory crisis: restructuring New York City's water supply', *Transactions of the Institute of British Geographers*, 22(3): 338–50.

Gibbs, D. (1996) 'Integrating sustainable development and economic restructuring: a role for regulation theory', *Geoforum*, 27(1): 1–10.

Gibbs, D. and Jonas, A.E.G. (2001) 'Rescaling and regional governance: the English Regional Development Agencies and the environment', *Environment and Planning C*, 19: 269–88.

Gleick, P. (2000) 'The changing water paradigm: a look at twenty-first century water resources development', *Water International*, 25(1): 127–38.

Goldman, M. (1998) *Privatizing Nature: Political Struggles for the Global Commons*. London: Pluto Press.

Goubert, J.P. (1986) *The Conquest of Water*. London: Polity.

Graf, W. (2001) 'Damage control: restoring the physical integrity of America's rivers', *Annals of the Association of American Geographers*, 91(1): 1–27.

Graham, S. and Marvin, S. (2001) *Splintering Urbanism*. London: Routledge.

Gunderson, L. and Holling, C. (2002) *Panarchy: Understanding Transformations in Human and Natural Systems*. Washington, DC: Earth Island Press.

Guy, S. and Marvin, S. (1996a). 'Managing water stress: the logic of demand side infrastructure planning', *Journal of Environmental Planning and Management*, 39(1): 123–8.

Guy, S. and Marvin, S. (1996b). 'Transforming urban infrastructure provision: the emerging logic of demand side management', *Policy Studies*, 17(2): 137–47.

Hajer, M. (1995) *The Politics of Environmental Discourse: Ecological Modernization and the Policy Process*. Oxford and New York: Clarendon Press.

Hardin, G. (1968) 'The tragedy of the commons', *Science*, 162(3859): 1243–8.

Harvey, D. (1974) 'Population, resources, and the ideology of science', *Economic Geography*, 50(3): 256–77.

Hassan, J. (1998) *A History of Water in Modern England and Wales*. Manchester: Manchester University Press.

Hawkins, P., Lovins, A. and Hunter, L. (1999) *Natural Capital-ism: Creating the Next Industrial Revolution*. Boston: Little, Brown.

Hay, C. (1996) *Re-Stating Social and Political Change*. Milton Keynes: Open University Press.

Hays, S. (1959) *Conservation and the Gospel of Efficiency: The Progressive Conservation Movement, 1890–1920*. New York: Atheneum.

Homer-Dixon, T. (1991) 'On the threshold: environmental change as causes of acute conflict', *International Security*, 16(2): 76–116.

Homer-Dixon, T. (1999) *Environment, Scarcity and Violence*. Princeton. Princeton University Press.

Hopkinson, P., James, P. and Sammut, A. (2000) 'Environ-mental performance evaluation in the water industry of England and Wales', *Journal of Environmental Planning and Management*, 43(6): 873–95.

Howitt, R. (2001) *Rethinking Resource Management: Jus-tice, Sustainability and Indigenous Peoples*. London and New York: Routledge.

Innis, H. (1942) *The Cod Fisheries: the history of an international economy*. New Haven, Conn: Yale University Press.

Jessop, B. (1995) 'The regulation approach, governance and post-Fordism: alternative perspectives on economic and political change?', *Economy and Society*, 24(3): 307–33.

Johnston, B.R. (2003) 'The political ecology of water: an introduction', *Capitalism Nature Socialism*, 14(3): 73–90.

Kaika, M. (2003) 'The Water Framework Directive: A New Direc-tive for a Changing Social, Political and Economic European Framework', *European Planning Studies*, 11(3): 299–316.

Kain, R. and Baigent, E. (1992) *The Cadastral Map in the Ser-vice of the State: A History of Property Mapping*. Chicago: University of Chicago Press.

Kallis, G. and Butler, D. (2001) 'The EU Water Framework Directive: measures and implications', *Water Policy*, 3: 125–42.

Kaplan, R. (2001) *The Coming Anarchy: Shattering the Dreams of the Post Cold War*. New York: Vintage Books.

Kinnersley, D. (1998) Privatized water services in England and Wales: a mixed verdict after nearly a decade, *Water Policy*, 1(1): 67–71.

Kinnersley, D. (1988) *Troubled Water: Rivers, Politics and Pollution*. London: Hilary Shipman.

Kinnersley, D. (1994) *Coming Clean: The Politics of Water and the Environment*. London: Penguin.

Klare, M. (2002) *Resource Wars: The New Landscape of Global Conflict*. New York: Owl Books.

Klare, M. (2004) *Blood and Oil: The Dangers and Consequences of America's Growing Dependency on Imported Petroleum*. New York: Metropolitan Books.

Krasner, S. (1978) *Defending the National Interest: Raw Mate-rials Investments and US Foreign Policy*. Princeton: Princeton University Press.

Lake, R. and Disch, L. (1992) 'Structural constraints and pluralist contradictions in hazardous waste regulation', *Environment and Planning A*, 24: 663–81.

Lipietz, A. (1987) *Mirages and Miracles: The Crisis in Global Fordism*. London: Verso.

Lise, W. and Bakker, K. (2005) 'Economic regulation of the water supply industry in the UK: A Game Theoretic

Consideration of the Implications for Managing Drought Risk', *International Journal of Water*, 3(1): 18–37.

Luke, T.W. (1999) 'Environmentality as green governmentality', in E. Darier (ed.), *Discourses of the Environment*. Oxford: Blackwell, pp. 121–51.

Maddock, T.A. (2004) 'Fragmenting regimes: how water qual-ity regulation is changing political-economic landscapes', *Geoforum*, 35(2): 217–30.

Mansfield, B. (2004a) 'Rules of privatization: contradictions in neoliberal regulation of north Pacific fisheries', *Annals of the Association of American Geographers*, 94(3): 565–84.

Mansfield, B. (2004b) 'Neoliberalism in the oceans: "ratio-nalization", property rights, and the commons question', *Geoforum*, 35(3): 313–27.

Marsh, T.J. (1996) 'The 1995 UK drought – a signal of climatic instability?', *Proc. Instn Civ. Engrs Wat., Marit. & Energy*, 118: 189–95.

McAfee, K. (2003) 'Neoliberalism on the molecular scale: eco-nomic and genetic reductionism in biotechnology battles', *Geoforum*, 34(2): 203–19.

McCarthy, J. (2004) 'Privatizing conditions of production: trade agreements as neoliberal environmental governance', *Geoforum*, 35: 327–41.

McCarthy, J. and Prudham, S. (2004) 'Neoliberal nature and the nature of neoliberalism', *Geoforum*, 35: 275–83.

McKibben, B. (1998) *The End of Nature*. New York: Random House.

Melosi, M. (2000) *The Sanitary City: Urban Infrastructure in America from Colonial Times to the Present*. Baltimore and London: Johns Hopkins University Press.

Mitchell, G. (1999) 'Demand forecasting as a tool for sustain-able water resource management', *International Journal of Sustainable Development and World Ecology*, 6(1): 231–41.

Mitchell, T. (2000) *Rule of Experts: Egypt, Techno-Politics, Modernity*. Berkeley and Los Angeles: University of California Press.

Mogren, E. (2002) *Warm Sands: Uranium Mill Tailings Policy in the Atomic West*. Albuquerque: University of New Mexico Press.

Mol, A. (1996) *The Refinement of Production: Ecological Modernisation Theory and the Chemical Industry*. Utrecht: Van Arkel.

Nash, R.F. (1967) *Wilderness and the American Mind*. New Haven: Yale University Press.

O'Connell-Davidson, J. (1993) *Privatization and Employment Relations: The Case of the Water Industry*. London: Mansell.

Ortolano, L. and Cushing, K. (2002) 'Grand Coulee Dam 70 years later: what can we learn?', *International Journal of Water Resources Development*, 18(3): 373–90.

Ostrom, E. (1990) *Governing the Commons: The Evolution of Institutions for Collective Action*. Cambridge: Cambridge University Press.

Page, B, and Bakker, K. (2005) 'Water governance and water users in a privatized water industry: Participation in policy-making and in water services provision – a case study of England and Wales', *International Journal of Water*, 3(1): 38–60.

Parker, D.J. and Sewell, W.R.D. (1988) 'Evolving water insti-tutions in Britain: An assessment of two decades of experience', *Natural Resources Journal*, 28(4): 751–8.

Pearse, F. (1982) *Watershed: The Water Crisis in Britain*. London: Junction Books.

Peluso, N. (1993) 'Coercing conservation? The politics of state resource control', *Global Environmental Change*, June: 199–217.

Peluso, N. and Vandergeest, P. (2001) 'Genealogies of the political forest and customary rights in Indonesia, Malaysia, and Thailand', *Journal of Asian Studies*, 60(3): 761–812.

Peluso, N. and Watts, M. (2001) *Violent Environments*. Ithaca, NY: Cornell University Press.

Penning-Rowsell, E.C. and Parker, D.J. (1983). 'The changing economic and political character of water planning in Britain', *Progress in Resource Management and Environmental Planning*, 4: 169–99.

Pierre, J. (ed.) (2000) *Debating Governance: Authority, Steering, and Democracy*. Oxford: Oxford University Press.

Plumwood, V. (2002) *Environmental Culture: The Ecological Crisis of Reason*. London and New York: Routledge.

Pratt, D.J. (2001) 'Corporations, communities, and conservation: the Mountain Institute and Antamina Mining Company', *California Management Review*, 43(3): 38–43.

Prudham, W.S. (2004) 'Taming trees: capital, science and nature in Pacific Slope tree improvement', *Annals of the Association of American Geographers*, 93(3): 636–56.

Raymond, L. and Fairfax, S. (2003) 'The "shift to privatization" in land conservation: a cautionary essay', *Natural Resources Journal*, 42(3): 599–640

Rees, J. (1991) *Natural Resources: Allocation, Economics and Policy*. London: Routledge.

Roberts, R. and Emel, J. (1992) 'Uneven development and the tragedy of the commons: competing images for nature-society analysis', *Economic Geography*, 68: 249–71.

Robertson, M. (2004) 'The neoliberalization of ecosystem services: wetland mitigation banking and problems in environmental governance', *Geoforum*, 35(3): 361–75.

Rosenau, J. (1992) 'Governance, order, and change in world politics', in J. Rosenau and E. Czempiel (eds), *Governance Without Government: Order and Change in World Politics*. Cambridge: Cambridge University Press.

Ross, A. (1994) *The Chicago Gangster Theory of Life: Nature's Debt to Society*. London: Verso.

Rutherford, P. (1999) 'The entry of life into history', in E. Darier (ed.), *Discourses of the Environment*. Oxford: Blackwell, pp. 37–62.

Scott, J. (1998) *Seeing Like a State: How Certain Schemes to Improve the Human Condition Have Failed*. New Haven: Yale University Press.

Shiva, V. (1992) 'Resources', in W. Sachs (ed.), *The Development Dictionary: A Guide to Knowledge and Power*. London: Zed Books, pp. 206–18.

Smith, L. (2004) 'The murky waters of the second wave of neoliberalism: corporatization as a service delivery model in Cape Town', *Geoforum*, 35(3): 375–93.

Summerton, N. (1998) 'The British way in water', *Water Policy*, 1(1): pp. 45–65.

Swyngedouw, E. (1999) 'Modernity and hybridity: nature, regeneracionismo, and the production of the Spanish waterscape, 1890–1930', *Annals of the Association of American Geographers*, 89(3): 443–65.

Taylor, F.W. (1911) *Principles of Scientific Management*. New York and London: Harper & Brothers.

Taylor, P.-J. (1993) 'A century of political geography' in P.J. Taylor (ed.), *Political Geography of the Twentieth Century: A Global Analysis*. London: Belhaven Press, pp. 1–7.

Veblen, T. (1923) *Absentee Ownership and Business Enterprise in Recent Times: The Case of America*. New York: B.W. Huebsch.

Walker, D.L. (1983) 'The effect of European Community directives on water authorities in England and Wales', *Aqua*, 4: 145–7.

Walker, L., Cocklin, C. and Le Heron, R. (2000) 'Regulating for environmental improvement in the New Zealand forestry sector', *Geoforum*, 31: 281–97.

Watts, M. (2004) 'Antinomies of community: some thoughts on geography, resources and empire', *Transactions of the Institute of British Geographers*, 29(2): 195–216.

WCD (2000) *Dams and Development*. New York: World Commission on Dams.

Wescoat, J. (1987) 'The "practical range of choice" in water resources geography', *Progress in Human Geography*, 11: 41–59.

White, G.F. (1945) *Human Adjustment to Floods: A Geographical Approach to the Flood Problem in the United States*. Chicago: Department of Geography, University of Chicago.

White, R. (2000) *The Organic Machine: The Remaking of the Columbia River*. New York: Hill & Wang.

Williams, R. (1976) *Keywords: A Vocabulary of Culture and Society*. Oxford: Oxford University Press.

Wilson, R. (2003) 'The extractive industries and protected areas'. Chairman's Speech, World Parks Congress, Durban, South Africa, 16 September, http://www.icmm.com/news/161WPC-RPWpresentation160903.pdf

Worster, D. (1985) *Rivers of Empire: Water, Aridity and the Growth of the American West*. New York and Oxford: Oxford University Press.

Yapa, L. (1993) 'What are improved seeds? An epistemology of the Green Revolution', *Economic Geography*, 69(3): 254–73.

Yergin, D. (1991) *The Prize: The Epic Quest for Oil, Money and Power*. New York: Simon & Schuster.

Young, O. (1981) *Natural Resources and the State: political economy of resource management*. Berkeley: University of California Press.

Zimmerman, E.W. (1931) *World Resources and Industries: A Functional Appraisal of the Availability of Agricultural and Industrial Resources*. New York and London: Harper & Brothers.

Global Environmental Politics

Becky Mansfield

INTRODUCTION

Global environmental politics has a very long history. It stretches back at least through the period of European colonialism, when colonizers appropriated land, created large mines, turned functioning ecosystems into agricultural plantations, and transported plants and animals around the world (Crosby, 1972; Mintz, 1985; Juma, 1989). Some of this was done explicitly in the name of 'conservation', for example as people were removed from land to create parks in eastern Africa or forbidden to use forests in Indonesia (Peluso, 1992; Neumann, 1998). In another sense, however, global environmental politics is a much more recent phenomenon, stretching back less than forty years, rooted in modern environmentalism with its emphasis on the 'global environment' as an object of concern. This idea arose in public consciousness, particularly in the US, as the result of several factors, including fears about 'global' (i.e. Third World) population growth, concern about the effects of industrialization, and images of Earth from space (McCormick, 1989; Cosgrove, 1994). In this chapter I focus on the latter form of global environmental politics, following developments from the 1960s onward, but show that the longer history of global relations, particularly between the North and South, informs current debates.

Governance is a central theme in global environmental politics today (Dalby, 2002a). In addition to general calls for global responses to address global environmental problems such as climate change, there is also proliferation of new actors such as non-governmental organizations, debate about the relationship between trade and environment, and new environmental regimes that encompass both specific international laws and inter-governmental organizations. Transnational institutions, such as the United Nations and the World Trade Organization, have become increasingly involved broadly in environmental debates and more narrowly in environmental management. There are also new challenges related to environmental security and the ecological politics of empire (Klare, 2001; Dalby, 2002b; *Global Environmental Politics*, 2004). Commentators often discuss this combination of trends in terms of challenges to the traditional nation-state framework and the rise of 'global governance' as an alternative (e.g. *Global Environmental Politics*, 2003; Lifton, 2003). However, it is important to be cautious about making statements regarding overarching change. Political geographical approaches to global processes greatly enhance our understanding of the dynamics of environmental governance and politics more broadly.

Geographers have contributed not only to our knowledge of 'trans-state organization' (Roberts, 2002) but have been among those who have most strongly challenged the notion that globalization entails the end of the nation-state (e.g. Jessop, 2002; Dicken, 2003, chap. 5). Alternative notions of scale are particularly important in this regard: moving from a static notion of scales as ontologically given objects that impact each other to a relational notion in which scales are fluid, contested, and simultaneously material and discursive significantly impacts how we understand and analyze the 'global' (Herod and Wright, 2002). Not a thing or unified movement, the global is uneven and multiply articulated, in that its existence is always already a relation with multiple other scales. The global is not only about linkages that connect the world into a single place, but is simultaneously about differentiation and disconnection among people and places (Mansfield, 2003). The 'global' of 'global environmental politics', then, does not indicate a particular arena for political struggle that dominates regional, national, or local arenas, but is rather about how these all are produced and come together (or not) in environmental conflicts.

Given this expanded definition of the global, understanding issues regarding environmental governance requires addressing the larger context within which struggles over governance are conducted and examining issues that have animated global debates about the environment. My approach is to focus on the relationship between environment, economic development, and equity, and in particular to address neoliberalism and the environment through the dynamic of North–South relations. This approach combines political ecological perspectives, which have long focused on issues of environment and development in the global South, with more recent political economic perspectives on neoliberalism broadly conceived (*Antipode*, 2002; *Geoforum*, 2004; Peet and Watts, 2004; Robbins, 2004). I examine these issues within both a 'formal' environmental politics that occurs within the confines of multilateral negotiations and an 'informal' environmental politics of activism and social movements.

Formal politics is the environmental politics of UN conferences and reports and which, in recent years, has explicitly extended into the environmental politics of free trade. Key issues are about who is actually responsible for environmental degradation, what are the most appropriate measures for achieving environmental goals, and who should pay for them (with cash or lost development opportunities). Within this politics, the North is often presented as protector of the environment and the South as the protector of the poor, and economic growth is offered as the primary solution to both economic and environmental problems. This is 'politics of neoliberalism'.

Informal environmental politics generally occurs outside of official settings and is carried out by grassroots groups. Within this activist politics, a crucially important theme has been about the negative impacts of both conservation *and* development on both people *and* the environment. The activist discourse exposes the North for degrading the environment and the South for promoting policies that exacerbate problems for the world's poor. In this view, much economic development is bad for the environment and leads to greater inequities between rich and poor. This is 'politics against neoliberalism'. As will become clear, the distinction between formal and informal does not map onto North and South. Further, using shorthand such as formal/informal (or North/South) risks presenting viewpoints as though they are monolithic; the point, however, is to use this shorthand as a lens that brings into focus complex issues that comprise global politics of the environment.

In making this contrast between formal and informal, my argument is that both types of environmental politics involve and raise key issues not only about environmental protection but also about equity, global power relations, and the relationship between environment and development. Although the issues each raises are different, and in some ways contradictory, in both formal and informal politics people challenge dominant frameworks, whether those are frameworks created by Northern governments and corporations (often criticized for putting their own power and profits first) or frameworks created by Northern environmentalists (often criticized for putting environment first and ignoring the needs of people, especially the poor). Rather, in focusing on global environmental *politics*, it is precisely the contestation of these dominant frameworks that is at work. Although both are important, the formal lens is dominant and issues raised from within this debate are not exhaustive. The informal, activist lens provides important perspectives about the larger framework and addresses issues that are not generally up for discussion within formal politics. Thus, both formal and informal politics show that the global is uneven and contested, while explicitly addressing informal politics highlights that the global is produced through both linkage and differentiation. The point of this chapter, then, is not to discuss formal politics simply to point out its flaws and dismiss it, but to take seriously issues raised within this framework while also examining other factors that are raised by people within activist social movements.

FORMAL POLITICS OF NEOLIBERALISM

As environmental awareness rose in the 1960s, people increasingly began to push not only for country-by-country environmental laws, but also for international solutions to environmental problems. One notable outcome is a dramatic rise in international environmental regimes since the 1970s. Certainly some environmental regimes existed prior to this time, such as the International Whaling Convention concluded in 1946, but the majority date to the post-1960s era, and there are now conventions addressing a range of specific issues from acid rain and the ozone hole to trade in ivory from elephants and management of fish stocks that straddle international borders. There is an interesting history not only to each particular regime but also to the institutional context of these regimes as a whole, including the role of different actors and the effectiveness of regimes (Young, 1989, 1994, 1999; Porter et al., 2000).

At the same time that governments were negotiating these international agreements to solve specific environmental problems, they were also engaged in a broader discussion about environmental problems and their solutions. This discussion, conducted particularly in a series of UN-sponsored conferences and reports that extend from 1972 to the present, was and continues to be quite contentious, as representatives from different countries disagree profoundly on what counts as environmental issues, underlying causes of environmental problems, solutions to these problems, and who should pay. Despite these disagreements, one result of over thirty years of international discussion is that there seems to be a fairly widespread consensus among government officials, as well as many representatives from business and NGOs, that economic development is the key to solving both environmental and economic problems. While disagreement definitely still exists on specific problems, the neoliberal solution of expanding markets and using market-based mechanisms is now the dominant model for change. The history of these conferences and reports has been recounted elsewhere (e.g. Soroos, 1999; Adams, 2001; Middleton and O'Keefe, 2003); I draw on these works to highlight the rise of neoliberalism and changing ideas about links among environment, economy, and equity.

The first of these conferences was the UN Conference on the Human Environment, held in Stockholm in 1972. This meeting was the first international gathering to focus on global environmental problems, and was originally designed to address the concerns of environmentalists (largely from the North) about negative effects of both industrialization and population growth, including pollution and scarce resources. Countries of the South, however, saw their main problem not as too much industrialization but as too little. To them, the problem was poverty and global inequity: the vast disparities in wealth between North and South. Further, by casting conservation as global – and especially by raising the specter of population growth – environmentalists seemed to be evading responsibility for existing problems; while the North consumed most of the world's resources and produced most of the world's pollution, they presented the problem as equally shared by all. Southern countries feared they would be forced to forgo industrialization in the name of environmental protection, paying for problems they did not create. Global environmentalism seemed like an attempt to keep Southern countries in poverty and take away sovereign control over their land and resources. Because Southern countries forcefully raised these issues, a central theme of the meeting was that environment and development are not opposed: environmental protection need not hinder development, and development need not harm the environment. Thus, the main success of the Stockholm conference is that it changed the emerging global agenda from being one strictly of 'environmental conservation',

to being one of 'environment and development'. The meeting seems to be a real success for countries of the South, yet it is also important to note that one outcome of this meeting was the impression, still strong today, that the North was concerned about environmental issues while the South was not.

The link between environment and development was subsequently institutionalized in 'sustainable development', a term that originally emerged in the late 1970s, but that was popularized and brought firmly onto the international agenda in the 1987 UN-commissioned report *Our Common Future* (also known as the 'Brundtland Report'). The report defined sustainable development as 'development that meets the needs of the present without compromising the ability of future generations to meet their own needs'. This fairly unobjectionable definition – still in use today – masks the politics of the report, which are in its treatment of the relationship between environment and development. Rather than trying to bring development into a conservation framework, which was the strategy in Stockholm, the Brundtland Report treated environment and development as inseparable. Not only is it possible to have development without environmental degradation, but development is a necessary precursor to environmental sustainability. The basis for this argument was that poverty is the main cause of environmental degradation, because poverty forces people to engage in harmful activities to survive. Policy-makers continued to focus on population as a key problem, but instead of arguing that 'over-population' causes both poverty and environmental degradation, they argued that poverty causes over-population (as people have children in order to support themselves), which leads to further environmental degradation. Given this formulation of the problem, development becomes the obvious solution: it is only through economic growth, including international trade, that a country can hope to close the gap with industrialized countries of the North, thereby reducing poverty and alleviating pressure on resources.

This, again, seems like a major achievement for the South. Poverty and global inequities were recognized as major problems, and developing countries were no longer being asked to sacrifice development for environmental protection. Yet it is also important to note that this new paradigm further distracted attention from the role of the North in causing problems. Policy-makers no longer gave attention to the negative effects of industrial activity – which was instead seen as the engine of economic growth – and paid little attention to the fact that, on average, people in the North cause far more environmental damage than people in the South. Instead of using inequities in consumption to argue that the North should take greater responsibility for environmental problems, the gap between rich and poor was used to argue for economic development. Further, even while identifying global inequities and arguing that they must be addressed, little attention was given to the role of the North in creating those inequities in the first place, through its colonial and imperialist activities. Therefore, while acceding to the South's desire to focus on development, *Our Common Future* also treated 'global' environmental problems as problems of the South: it is the poor – not the rich – who degrade the environment. Also, how economic development could be done in ways that were not environmentally harmful remained unclear. As a new paradigm, 'sustainable development' seemed to offer something for everyone, but it only did so by avoiding some of the politically contentious issues about responsibility for environmental problems and their solutions.

Implementing sustainable development was the main topic of the UN Conference on Environment and Development (the Earth Summit), held in Rio de Janeiro in 1992. The goal was to produce action plans on a wide range of environmental problems, including binding international regimes on several key issues (climate change, biodiversity, and deforestation). To do so required confronting head-on many of the political issues that had been avoided in the Brundtland Report, and as a result the Earth Summit involved a 'mutual bludgeoning' between countries and, particularly, between the North and South (Adams, 2001: 83). Once again, divisions centered on what counted as important issues (the North focused on climate change and deforestation while the South focused on poverty), and on responsibility for problems. In terms of general principles, the South again made important gains, including acknowledgment of national sovereignty over resources, the idea of 'common but differentiated responsibilities', and the notion of international responsibility for conservation.

In terms of actual means of implementation, the main outcomes of the meeting included two conventions (on Biological Diversity and Climate Change), one non-convention (the Forest Principles, which was supposed to be a convention, but delegates could not agree on binding principles), and Agenda 21, which is a massive document outlining what are supposed to be actual measures for achieving sustainable development. On the one hand, Agenda 21 and the conventions are major accomplishments, in that they are the first, formal global agenda for achieving sustainable development, and they represent a politically fraught compromise among various interests. On the other hand, a close look at the documents themselves reveals important shortcomings, of which I will mention just two.

First, the language of the documents (including the binding conventions, but especially Agenda 21) is not only bland, but much of it is lacking specificity; most of the actions proposed are several steps removed from concrete actions that might achieve results. Second, the few actual conservation activities that are endorsed in many ways reassert global control over resources. For example, documents on biological diversity recommend 'in-situ' (i.e. protected areas) and 'ex-situ' (i.e. seed banks/zoos) conservation, increased use of genetic resources for biotechnological development, and 'plantation forestry' (i.e. tree farms, often using introduced or even genetically modified species). Because much concern about 'global' biodiversity centers on areas in the South – particularly tropical forests – it becomes clear that these seemingly benign measures in fact echo colonial control over resources. Protected areas and seed banks (including botanical gardens) were part of colonial strategies not only to control territory but to profit from biological resources of the colonies, and local people have long contested such areas (Juma, 1989; Neumann, 1998). Use of 'global' genetic resources by biotech corporations (mainly from the North) for private profit again reproduces colonial power relations regarding control over and benefit from genetic resources, and the contemporary version has been labeled by opponents from the South as 'biopiracy' (Shiva, 1997; Adger et al., 2001). Plantations are used to grow trees for industrial use (i.e. paper pulp), are generally capital-intensive, and often replace existing forest – and, because they are a form of modern agriculture, are *not* diverse biologically (Marchak, 1995). Thus, much of the language of the Rio documents seems to reflect interests of the South, yet many of the actual conservation measures reflect the concerns of environmentalists of the North and work to benefit business interests, also largely from the North.

There is much more that could be said about the Earth Summit (especially on NGOs and efforts to increase involvement of women, indigenous peoples, and other marginalized groups), but the final point here is that it was at this meeting that policy-makers began to make explicit links between sustainable development and neoliberalism. Within the Rio documents, policy-makers cited capital accumulation (i.e. profits) as a tool for achieving sustainable development (e.g. biodiversity will be conserved if it is made valuable by making it available for biotechnological development). Further, the Rio documents began to explicitly tie sustainable development to a free trade agenda. For example, the Forest Principles state that 'unilateral measures, incompatible with international obligations or agreements, to restrict and/or ban international trade in timber or other forest products should be removed or avoided, in order to attain long-term sustainable forest management' (paragraph 14). Here, not only is free trade treated as an important goal, but it is offered as part of the solution for protecting forests. Radical critiques of the Earth Summit have focused precisely on these economic themes and have argued that sustainable development promotes business as usual, including the enclosure of the 'global' commons for commercial interests (Sachs, 1993b; *The Ecologist*, 1993). Vandana Shiva has argued that focusing on ' "global" environmental problems has in fact narrowed the agenda' and 'transforms the environmental crisis from being a reason for change into a reason for strengthening the status quo' (Shiva, 1993: 149, 151). (For more on radical, activist critiques, see the following section.)

Ten years later, the World Summit on Sustainable Development (WSSD), held in Johannesburg, South Africa, in 2002, further entrenched the idea that sustainable development should be linked to neoliberal free trade (Barber, 2003; Pallemaerts, 2003; Wapner, 2003). The intervening decade in many ways saw free trade and 'globalization' eclipse sustainable development as the global hot topic, and the WSSD reflects this in several ways. First, unlike past decades in which each new conference and report pushed the environmental agenda in new directions, the purpose of the WSSD was simply to assess progress implementing Agenda 21, and to create an action plan for further implementation. The WSSD did have some successes, such as new targets on key problems (e.g. on providing sanitary drinking water to the world's people), and, for the first time, mentioning ethics and corporate responsibility. However, these successes were tempered by an overall weakness of approach. As one commentary put it, subtle wording changes shifted the Plan of Implementation from 'a promising document outlining commitments and obligations to one filled with voluntary options and choices, and may actually have watered down principles affirmed in the Rio declaration' (La Vina et al., 2003: 64).

Second, the approach institutionalized in WSSD documents shows that sustainable development is increasingly being subordinated to neoliberalism – or, rather, proponents promote neoliberalism as synonymous with sustainable development. Following on from Rio, policy-makers emphasize free trade as a means to achieve sustainable development. Not only are commitments to trade sprinkled throughout the WSSD documents, but there is a section of the Plan of Implementation devoted explicitly to 'sustainable development in a globalizing world' (United Nations, 2002: 37–9). While this section starts by saying 'globalization offers opportunities and challenges for sustainable development', the discussion goes on to treat the challenges not as ones that result from 'globalization' but those to which globalization can be the solution. The plan promotes free trade and investment, and in particular encourages developing countries to increase their level of participation in free trade. Indeed, there is an explicit commitment to 'implement the outcomes of the Doha Ministerial

Conference by the members of the World Trade Organization' – in other words, we can only achieve sustainable development if we implement WTO agreements. From this perspective, free trade does not present any potential challenges to the sustainable development agenda, but is simply a means to achieve this end.

Another way the WSSD shifted the sustainable development agenda more firmly in the direction of neoliberalism was through 'voluntary partnerships', in which governments work with the private sector (primarily businesses, but also NGOs) to achieve particular goals (Pallemaerts, 2003). The formal recognition and endorsement of such partnerships is often cited as a success of the WSSD, because such partnerships can move beyond the gridlock that occurs when governmental negotiations stagnate (La Vina et al., 2003). Such partnerships are neoliberal in several senses. First, they represent a mistrust of government and work to 'downsize' government by shifting state activities to non-state actors. Second, by bringing in private business they require a basic trust that goals of the private sector are congruent with larger societal goals. Third, by emphasizing the private sector – both business and NGOs – they decrease public accountability and get governments 'off the hook'. Also, to the extent they are financed with public funds, partnerships can divert funds from existing programs. Thus, partnerships, in connection with the emphasis on encouraging free trade and implementing the WTO agreement, are emblematic of a private, market-based approach to environmental protection.

It seems, then, that the WSSD represents the triumph of neoliberalism as a framework for sustainable development – what Steven Bernstein (2001) calls the 'compromise of liberal environmentalism'. This shift within global environmental politics toward neoliberalism is consistent with a general trend toward neoliberal environmentalism within countries of the North, especially the US (*Geoforum*, 2004). The WSSD also represents the triumph of 'development' over 'environment'. Paul Wapner (2003) argues that at the WSSD, the North and South to some extent swapped positions. Governments of the South increasingly expressed concern about environmental issues (while not abandoning concerns about development), and the North largely abandoned its 'environmentalist' cloak and argued explicitly for economic development, particularly in the form of economic globalization-cum-free-trade (this was particularly true of the US). Whereas the Stockholm conference thirty years earlier was mainly about conservation with a developmental angle, the Johannesburg conference was mainly about development, with an environmental angle.

This does not mean that the North and South agree now on all issues regarding environment and development. Many governments of the South continue to highlight the North's contribution to environmental problems, insist that the South be allowed to 'develop', and argue that traditional environmentalism is itself a form of intrusion and neo-colonialism. However, within this formal politics, the emphasis on development by the South has also served the purposes of the North (Sachs, 1993a). The North continues to abdicate responsibility for environmental problems, even as the South has tried to bring issues of unequal consumption onto the agenda (this is a large part of what animates the politics of climate change). The South has had much less success in raising such issues of responsibility than it has in arguing for economic growth, and the reason seems to be that unimpeded economic activity is also good for Northern governments and businesses, whereas blaming the North for environmental problems is not. That the environmental agenda has shifted over time to take into account the economic needs of Southern countries represents a real win on the part of the South, yet it seems to have come at the cost of not blaming the North for any environmental or economic problems, with the result that today sustainable development is subordinate to the 'free' market. This is politics within neo-liberalism; few in formal politics really question the economic frame, instead raising questions about how to develop, who is responsible, who pays, and so forth. That sustainable development has become neoliberal should be no surprise, as it was this possibility that made it so attractive in the first place.

INFORMAL POLITICS AGAINST NEOLIBERALISM

Formal environmental politics raises a host of important questions about environment, economy, and equity, yet these do not exhaust global environmental politics. Much political activity happens outside of governmental negotiation, carried out by various non-state actors, including non-governmental organizations (NGOs) and citizen groups. These groups often try to influence what governments do, but they also target other groups, including inter-governmental organizations, corporations, and individual people.

The rise of NGOs, in particular, has received a lot of academic attention, and many commentators have argued that there is a new 'global civil society', comprising not just environmental groups, but also groups active on issues such as human and women's rights (Fisher, 1993; Princen and Finger, 1994; Newell, 2000; Tamiotti and Finger, 2001; Warkentin, 2001; Wapner, 2002). Many NGOs argue for a more substantial role for civil society in formal politics, particularly at the global

level. Much attention has been given to the role of NGOs in forums such as the 1992 Earth Summit and the 2002 WSSD, and there is now a significant literature on the direct effectiveness of NGOs in these settings (Arts, 1998, 2001; Betsill and Corell, 2001; Humphreys, 2004). Many international NGOs also work directly on conservation projects. For example, well-known groups such as WWF and Conservation International not only advocate protected areas in environmentally sensitive areas, but actively work to establish them around the world (especially in the South). As would be expected given debates that drive formal environmental politics, these activities are politically charged, as some see them as a form of neo-colonialism – an 'ecologically updated version of the White Man's Burden' – that places environmental demands above the needs of people and takes control of land and resources away from local people and governments (Guha and Martinez-Alier, 1997: 104). Thus, while NGOs are outside formal multilateral proceedings, they do not necessarily challenge dominant frameworks.

Indeed, institutionalized NGOs, which try to engage formal politics, can be co-opted and are 'not necessarily a democratizing force within global governance', while an alternative rests in the grassroots movements that resist the dominant framework, and even call for dismantling existing governance systems (Williams and Ford, 1998: 276). Many citizens' groups work outside formal politics and openly challenge environmental and developmental frameworks. There are many thousands of such groups around the world, working on a very wide range of issues and holding diverse perspectives, yet they are often linked in 'transnational advocacy networks'; in these networks, activists 'try not only to influence policy outcomes, but to transform the terms and nature of the debate' (Keck and Sikkink, 1998: 2). Here, I focus especially on groups that broadly can be grouped into the 'grassroots globalization' movement (or 'globalization from below') (Evans, 2000; Gill, 2000; Graeber, 2002).

Movement activists have largely engaged in the politics of protest, with the immediate goal of stopping particular meetings and projects and a long-term goal of raising public awareness and undermining the neoliberal project. Because such change requires alternative visions, grassroots globalization activists have also created events, the most prominent of which is the World Social Forum, at which myriad activists gather to 'coordinate actions and articulate shared visions for global change' (Smith, 2004: 413; see also Fisher and Ponniah, 2003, for voices from the WSF). Although this is not a single movement, as it comprises groups working on issues ranging from working conditions to reproductive rights, from land control to environmental regulation, activists increasingly recognize certain common-alities among their goals; primary among these is an opposition to neoliberalism, with its empha-sis on profits and the private sector above all else (Routledge, 2003). As they recognize com-monalities, heterogeneous activist groups at times come together into larger, transnational coalitions. A prominent example is Peoples' Global Action, which is composed of groups from all continents (except Antarctica) and has been a major orga-nizer of some of the most visible protests against corporate globalization (Williams and Ford, 1998; Routledge, 2003).

In connecting what seem to be disparate move-ments, activists both explicitly and implicitly criti-cize the idea of sustainable development, especially as it is promulgated within formal governmental politics. As with formal politics, activists show that environment and development are linked, and pro-mote the idea that what happens to the poor of the world is directly related to what happens to the environment – yet the argument is in many ways the inverse of that within formal politics. Rather than arguing that poverty causes environmental degra-dation and so economic development is the answer, many activists argue that economic development (in its dominant form) increases both poverty and environmental degradation. As authors from the Indian Centre for Science and Environment put it:

the Western economic and technological model is highly material and energy-intensive. It metabolizes huge quantities of natural resources, leaving a trail of toxins and highly degraded, transformed ecosys-tems in its wake. It is this very model that today's poor cousins, the developing nations, are follow-ing for economic and social growth, leading to an extraordinary cocktail of poverty and inequality side by side with growing economies, pollution and large-scale ecological destruction. (Agarwal et al., 1999: 1)

Activists argue that offering neoliberal economic growth as the solution to all problems is simply a justification for the status quo: the solution is the very actions that created problems in the first place.

Much global activism has come under the anti-free-trade rubric, and while the WTO has been the most famous target, activists have also gathered to protest a variety of other pro-free-trade meetings, such as the Global Economic Forum and negoti-ations for the Free Trade Area of the Americas (Weber, 2001; Klein, 2002). Activists raise a variety of specific concerns about free trade and the WTO, from poor working conditions and low wages to the possible effect of WTO rules on biodiversity (Brecher et al., 2002; Friends of the Earth, 2002). One theme is concern about the effect on national sovereignty of free trade rules, as codified in the WTO and other, regional free trade agreements. The free trade agenda is an overarching frame-work to which all trade-related actions (including national laws) must conform; a WTO dispute res-olution panel can rule individual laws to be illegal under WTO rules. While some governments from the South are concerned that WTO agreements could – in the name of global harmonization – force labor or environmental standards on them, many activists are more concerned that WTO rules could undermine existing (or future) protective regula-tions. Both labor and environmental activists are concerned about this possibility, yet one of the most prominent cases in which this happened was with a US environmental law that required all shrimp imported into the US to be caught with methods that do not harm sea turtles (see more below). Rul-ings such as this seem to place free trade above all other concerns and give the WTO unprecedented power to essentially dictate many national laws. People from the North and South have different per-spectives on specific cases such as the shrimp/turtle issue, but many activists worldwide share a general concern about the power of the WTO and impli-cations for national sovereignty. Concern about sovereignty derives less from a profound trust in the state (many activists simultaneously criticize their own governments' actions) and more from a sense that the needs of corporations should not dominate global decision-making.

As these concerns indicate, within and alongside the anti-free-trade movement is an anti-corporate movement. In their response to the rising power of corporations within both the free trade and sustainable development frameworks, many envi-ronmental activists actively challenge the idea that major corporations are generally benign. One way environmental activists do this is by showing that corporate globalization – free trade – has actu-ally been bad for the environment. For example, activists point out that when US and European com-panies build plants in developing countries – or contract their work to companies in these coun-tries – they are able to avoid more stringent environ-mental laws (e.g. regarding water or air pollution) in their home countries (this is true even if lax laws are not the primary reason these companies moved in the first place) (*Global Environmental Politics*, 2002). Additionally, long-distance trade has its own environmental costs, particularly as large quanti-ties of fossil fuels are used to transport inputs and finished goods around the world.

Another way activists challenge corporations is to question their own representations of them-selves as environmentally friendly. Many of the world's largest corporations use advertisements to convince consumers that they are 'green', and many of these same corporations – including those in chemicals, oil, automobiles, and agriculture – have joined together in the World Business

Council for Sustainable Development (WBCSD). The stated mission of the WBCSD is 'to provide business leadership as a catalyst for change toward sustainable development, and to promote the role of eco-efficiency, innovation and corporate social responsibility' (World Business Council for Sustainable Development, 2004). However, activists have shown that, rhetoric aside, the environmentalism of many of these companies is a thin veneer on an otherwise environmentally damaging record; in other words, it is 'greenwashing' (Bruno and Karliner, 2002; see also Athanasiou, 1996). To take but one example, the oil company BP talks about being 'innovative, progressive, performance driven and green' (BP, 2004) because it is involved in renewable energy – yet it spent more on developing its new eco-friendly logo than on renewable energy itself, which is only a tiny fraction of the billions of dollars the company continues to spend on oil and gas exploration and production (Bruno and Karliner, 2002: 82–5).

In addition to explicitly environmental activism, there is also a wide range of activism that is about socio-economic and livelihood issues, but that has clear environmental dimensions to it. Surveying 'anti-corporate movements', Amory Starr (2000) describes 'environmental' movements, but also discusses many other movements that have environmental dimensions: movements for land reform, against genetically modified organisms, for true sustainable development (e.g. permaculture), and for indigenous sovereignty. Livelihood struggles in the South, in particular, have 'ecological content, with the poor trying to retain under their control the natural resources threatened by state takeover or by the advance of the generalized market system' (Guha and Martinez-Alier, 1997: xxi). One prominent example involves protest against the construction of dams along the Narmada River in India, which is supported by the Indian government, private companies, and, at one time, the World Bank (which pulled out under intense pressure from activist groups) (Roy, 2001). These dams have and will continue to displace millions of people from land and other resources and destroy both ecosystems and historical cultural sites. Other prominent examples include the movement of the Ogoni people against Shell Oil and the Nigerian state (Watts, 2004), the movement against privatization of water in Cochabamba, Bolivia (Barlow and Clarke, 2002), the movement for compensation by those affected by the Union Carbide disaster in Bhopal, India (Fortun, 2001), and the movement for landless rights in Brazil (Wright and Wolford, 2003).

In all of these movements, poor people are demanding the right to livelihoods against activities that encroach on their resources and impair their ability to support themselves; all of these movements also have profound environment implications, as they involve land use, pollution, and/or habitat destruction. Environmental alterations such as these generally affect the poor more than the rich, as the poor lose access to clean water, vegetated hillsides that protect them from landslides, forest products, and so on – all while the profits from such projects accrue to other people. While many of these movements seem quite local, they are a part of global environmental politics because they challenge global models of development, target transnational organizations, and join together in transnational coalitions – even using international pressure as a way to influence their own governments (Glassman, 2001; Hochstetler, 2002). For example, Southern activists have long challenged the World Bank, with its emphasis on large-scale development projects such as dams, as being driven by (Northern) economic interests and for being coercive (Fox and Brown, 1998). Anti-World Bank activism predates the anti-free-trade movement by many years and helped lay the groundwork for the present grassroots globalization movement.

Livelihood struggles are often aimed as much at local and national governments as they are at governments and NGOs from the North; activists do not necessarily see their own governments as allies in their struggles (Agarwal et al., 1999). This is because it is often governments themselves (working with international agencies) that promote actions that dispossess people of their land, access to resources and, ultimately, their livelihoods – and they often do so in the name of sustainable development. In other words, from the perspective of the poor and marginalized, it is governments and corporations of the North and South that are the problem; that is, the problem is not just the North, but is the neoliberal model that puts economic growth first and only. As the activist journal *The Ecologist* (1993, vi–vii) put it, 'the top-down, technocratic policies that have increasingly come to characterize the "greening" of development are depressingly similar to those that have characterized the development process' from the beginning; 'sustainable development ... would appear to cloak an agenda that is just as destructive, just as undermining of peoples' rights and livelihoods as the development agenda of old'. From this activist perspective, then, the real issues are about access to – and dispossession from – the commons: 'what matters most is ... rights to equitable sharing of the earth's ecological commons' (Agarwal et al., 1999: 2). In challenging the general idea of neoliberalism being good for environment and people, activists also undermine the notion that the North is pro-environment and the South is pro-poor. Not only do activists show that governments of the South embrace ideas and actions that are in fact detrimental to the poor, but their activism broadens what counts as environmental concern. As Ramachandra Guha (2000) argues, it is socio-environmental issues that are at the heart of environmentalism in the South; 'environmentalism of the poor' is fundamentally about social justice and livelihoods.

These movements are exciting because activists address concrete local concerns while at the same time building explicit interconnections among various movements of both the South and North. This should certainly not be taken to mean that all such movements are 'progressive' or that there are no divisions within the grassroots globalization movement. In particular, even as activists increasingly recognize connections among issues of concern in the North and South, there are important differences of both perspective and power that cannot be glossed over (Mertes, 2002). Issues such as language, technology, and access to media tend to privilege Northern activists even within transnational networks (Routledge, 2003). And Southern issues – particularly those of livelihood – are still not foremost for many Northern activists, which calls into question a cohesive 'globalization from below' (Glassman, 2001).

For example, in contrast to Guha's work cited above, in which he places livelihood issues at the center, in their book *Globalization from Below*, activists Jeremy Brecher et al. (2002) do not directly address issues of livelihood and dispossession; they allude to them in their discussions of debt and debt relief, but nowhere do they make them explicit or put them at the center. Northern activists, it seems, are being educated about global issues by being in this transnational movement; this is especially true of Northern environmentalists, many of whom tend to see livelihood concerns as threats to the environment, rather than as environmental concerns. In these ways, the transnational, grassroots globalization movement still struggles with North–South issues, even while subverting the dominant imaginary of formal politics, in which governments from the North and South clash over particular measures while agreeing on the larger framework. In other words, activists broaden the discussion by raising a host of socio-environmental issues that are never considered – and are actively suppressed – within the neoliberal framework promoted by governments of both the North and South.

DISCUSSION

While it is clear that informal, activist politics *against* neoliberalism and neoliberal models of sustainable development has a critical edge that is missing from formal, governmental politics *of* neoliberalism, my aim is not to romanticize activist politics, nor to argue that formal politics should be dismissed as ideological. Discussions within both are important, and show the impossibility of a completely hegemonic position of any kind; there is indeed a politics of the global environment. It is essential to understand 'critical' positions within both debates if we are ever to address problems of environment, economy, and equity. Certainly if one wants to understand global politics of the environment, one must understand both debates, even if at times they are contradictory. But in a larger sense, both formal and informal global politics of the environment importantly show that issues of environment, economy, and equity cannot be divorced; they are inherently intertwined. Not only are environmental and equity issues influenced by economic decisions, but both also influence economic outcomes. This highlights the significance of 'the global environment' not only for those who care directly about the environment/nature, but for those who aim to understand capitalism and uneven development: environmental questions are at the center of the politics of capitalism today.

A brief example – the shrimp/turtle case mentioned earlier – can illustrate the utility of looking through both formal and informal lenses to understand the linked politics of the environment and capitalism. The US has domestic laws requiring shrimp fishers to use 'turtle excluder devices' so that sea turtles – some of which are endangered – are not caught in shrimp nets. The US, under pressure from environmentalists, extended these laws to the international arena by requiring all countries that wished to export shrimp to the US to be able to document that fishing practices are not harmful. Affected countries in Asia – Malaysia, India, Pakistan, and Thailand – argued that these laws were illegal under WTO rules because they were not strictly environmental laws, but acted as trade barriers. The WTO agreed, and forced the US to change its laws and the ways they applied them. Several years later the WTO did uphold modified versions of these laws (a fact that is often not mentioned by anti-WTO activists) (see DeSombre and Barkin, 2002, for discussion). That these laws could be undermined in the name of free trade was a major blow to environmentalists, given that such laws were considered major accomplishments of the US environmental movement. These cases have since been used by US environmentalists and various (mainly Northern) anti-free-trade activists as examples of the ways that the free trade agenda is inherently anti-environmental, especially as free trade rules conflict with multilateral environmental agreements (Eckersley, 2004). This suggests that political action should be oriented toward supporting the nation-state from this assault on its sovereignty. Here we have debates about governance.

But how does this issue look when viewed from within the larger context of both formal and informal global environmental politics? *Formal global environmental politics* brings into focus the long history of unfair trade, in which Southern countries (and colonies before that) were to be markets for products of the US and European countries, but not the other way around. Viewed from this perspective, these US environmental laws regulating imports do seem protectionist and unfair. The technology required to meet the requirements of the US law may be prohibitively expensive (and may even come from the US), which would make it impossible for industries to compete in the US market. Thus, in the name of environmental protection, the US continues to exclude products from Southern countries. This means that Southern countries are right to argue that they are forced to pay (in lost development opportunities) for environmental protection and economic development in the US – while the environmental effect of US consumption patterns is completely ignored.

Informal global environmental politics, however, raises additional issues. The type of development being promoted by Southern governments – that is, export-oriented, industrial fishing – is exactly the type of activity that leads to displacing people from their resources. Viewed from this perspective, US environmental laws are not the only problem, rather it is an entire development model that is not only bad for sea turtles, but is also bad for local people, who face enclosure of the commons and new rounds of dispossession. Further, once environmental issues are recognized to include not only species protection but also control over resources, the claim that the US is pro-environmental becomes even less tenable. (Similar points could be made about Northern agricultural subsidies, controversy over which brought down the Cancun round of WTO negotiations in 2003.) These activist perspectives are largely absent from discussion about trade and environment; even when commentators do not accept that trade is inherently good for the environment and are critical of existing free trade agreements and the way they have been implemented, they still do not talk about linked socio-environmental dimensions of livelihood issues (Deere and Esty, 2002; Gallagher and Werksman, 2002; Sampson and Chambers, 2002).

Thus, formal/governmental and informal/activist politics both cast light on these complex issues. Whereas environmentalists have often treated environmental issues separately from – or dominant to – socio-economic ones, both formal and informal politics have challenged this view by linking socio-economic and environmental issues in a variety of ways. In other words, it is not wrong for Southern governments to call Northern governments on their double standards. On the other hand, this is not the entire story, as the policies promoted by Southern governments might be equally bad for marginalized people and environments, as is a major argument of many activists. Thus, activists raise important issues that

challenge the neoliberal assumptions that underlie sustainable development. As I have shown, the idea of sustainable development, while initially showing some promise and expressing some major wins for countries of the South, has largely embraced and been defined by a neoliberal approach to the environment, in which the 'free market' is the solution to all problems, both economic and environmental. This is where the official, institutional debate stands today, as different players argue for different approaches to sustainable development, free trade, and so on. This is also where the activist perspective takes off, challenging the whole notion of sustainable development in its neoliberal guise. This is the complexity of global environmental politics today. With its emphasis on relationships between places and scales (rather than on the erasure of place in the face of globalization), political geography, broadly defined, has much to offer for understanding these complex debates. This is a field with much room for growth, and as political geographers continue to engage these issues they will yield new insights not only about environmental conflict, but also global power relations, inequality, and uneven development.

REFERENCES

Adams, W.M. (2001) *Green Development: Environment and Sustainability in the Third World* (2nd edn). London: Routledge.

Adger, W.N., Benjaminsen, T.A., Brown, K. and Svarstan, H. (2001) 'Advancing a political ecology of global environmental discourses', *Development and Change*, 32: 681–715.

Agarwal, A., Narain, S. and Sharma, A. (eds) (1999) *Green Politics*. New Delhi: Centre for Science and Environment.

Antipode (2002) Special Issue on 'From the "new localism" to the spaces of neoliberalism', *Antipode*, 34(3): 341–624.

Arts, B. (1998) *The Political Influence of Global NGOs: Case Studies on the Climate and Biodiversity Conventions*. Utrecht: International Books.

Arts, B. (2001) 'The impact of environmental NGOs on international conventions', in B. Arts, M. Noortman and B. Reinalda (eds), *Non-State Actors in International Relations*. Ashgate: Aldershot, pp. 195–210.

Athanasiou, T. (1996) *Divided Planet: The Ecology of Rich and Poor*. Boston: Little, Brown.

Barber, J. (2003) 'Production, consumption and the World Summit on Sustainable Development', *Environment, Development, and Sustainability*, 5: 63–93.

Barlow, M. and Clarke, T. (2002) *Blue Gold: The Fight to Stop the Corporate Theft of the World's Water*. New York: New York Press.

Bernstein, S. (2001) *The Compromise of Liberal Environmentalism*. New York: Columbia University Press.

Betsill, M. and Corell, E. (2001) 'NGO influence on international environmental negotiations: a framework for analysis', *Global Environmental Politics*, 1(4): 65–85.

BP (2004) *BP Global 2004* [cited September 2004]. Available from http://www.bp.com/home.do.

Brecher, J., Costello, T. and Smith, B. (2002) *Globalization from Below: The Power of Solidarity* (2nd edn). Cambridge, MA: South End Press.

Bruno, K. and Karliner, J. (2002) *earthsummit.biz: The Corporate Takeover of Sustainable Development.* Oakland, CA: Food First Books.

Cosgrove, D. (1994) 'Contested global visions: *one-world, whole-earth*, and the Apollo space photographs', *Annals of the Association of American Geographers,* 84(2): 270–94.

Crosby, A.W.Jr. (1972) *The Columbian Exchange: Biological and Cultural Consequences of 1492.* Westport, CT: Greenwood Press.

Dalby, S. (2002a) 'Environmental governance', in R.J. Johnston, P.J. Taylor and M.J. Watts (eds), *Geographies of Global Change: Remapping the World.* Malden, MA: Blackwell, pp. 427–39.

Dalby, S. (2002b) *Environmental Security.* Minneapolis: University of Minnesota Press.

Deere, C.L. and Esty, D.C. (eds) (2002) *Greening the Americas: NAFTA's Lessons for Hemispheric Trade.* Cambridge, MA: MIT Press.

DeSombre, E.R. and Barkin, J.S. (2002) 'Turtles and trade: the WTO's acceptance of environmental trade restrictions', *Global Environmental Politics,* 2(1): 12–8.

Dicken, P. (2003) *Global Shift: Reshaping the Global Economic Map in the 21st Century* (4th edn). New York: Guilford Press.

Eckersley, R. (2004) 'The big chill: the WTO and multilateral environmental agreements', *Global Environmental Politics,* 4(2): 24–50.

Evans, P. (2000) 'Fighting marginalization with transnational networks: counter-hegemonic globalization', *Contemporary Sociology,* 29(1): 230–41.

Fisher, J. (1993) *The Road from Rio: Sustainable Development and the Nongovernmental Movement in the Third World.* Westport, CT: Praeger.

Fisher, W.F. and Ponniah, T. (eds) (2003) *Another World Is Possible: Popular Alternatives to Globalization at the World Social Forum.* New York: Zed Books.

Fortun, K. (2001) *Advocacy after Bhopal: Environmentalism, Disaster, New Global Orders.* Chicago: University of Chicago Press.

Fox, J.A. and Brown, L.D. (eds) (1998) *The Struggle for Accountability: The World Bank, NGOs, and Grassroots Movements.* Cambridge, MA: MIT Press.

Friends of the Earth (2002) *Implications of WTO Negotiations for Biodiversity.* Amsterdam: Friends of the Earth International.

Gallagher, K.P. and Werksman, J. (eds) (2002) *The Earthscan Reader on International Trade and Sustainable Development.* London: Earthscan.

Geoforum (2004) Special Issue on 'Neoliberal Nature and the Nature of Neoliberalism', *Geoforum,* 35: 275–393.

Gill, S. (2000) 'Toward a postmodern prince? The battle in Seattle as a moment in the new politics of globalization', *Millennium: Journal of International Studies,* 29(1): 131–40.

Glassman, J. (2001) 'From Seattle (and Ubon) to Bangkok: the scales of resistance to corporate globalization', *Environment and Planning D: Society and Space,* 19: 513–33.

Global Environmental Politics (2002) Special Issue on 'Pollution Havens', *Global Environmental Politics,* 2(2): 1–36.

Global Environmental Politics (2003) Special Issue on 'Conceptualizing Global Environmental Governance', *Global Environmental Politics,* 3(2): 1–134.

Global Environmental Politics (2004) Special Issue on 'Ecological Politics, Violence, and the Theme of Empire', *Global Environmental Politics,* 4(2): 1–23.

Graeber, D. (2002) 'The new anarchists', *New Left Review,* 13 (Jan.–Feb.): 61–73.

Guha, R. (2000) *Environmentalism: A Global History.* New York: Longman.

Guha, R. and Martinez-Alier, J. (1997) *Varieties of Environmentalism: Essays North and South.* London: Earthscan.

Herod, A. and Wright, M.W. (eds) (2002) *Geographies of Power: Placing Scale.* Malden, MA: Blackwell.

Hochstetler, K. (2002) 'After the boomerang: environmental movements and politics in the La Plata River basin', *Global Environmental Politics,* 2(4): 35–57.

Humphreys, D. (2004) 'Redefining the issues: NGO influence on international forest negotiations', *Global Environmental Politics,* 4(2): 51–74.

Jessop, B. (2002) *The Future of the Capitalist State.* Cambridge: Polity Press.

Juma, C. (1989) *The Gene Hunters: Biotechnology and the Scramble for Seeds.* Princeton: Princeton University Press.

Keck, M.E. and Sikkink, K. (1998) *Activists Beyond Borders: Advocacy Networks in International Politics.* Ithaca, NY: Cornell University Press.

Klare, M.T. (2001) *Resource Wars: The New Landscape of Global Conflict.* New York: Henry Holt.

Klein, N. (2002) *Fences and Windows: Dispatches from the Front Lines of the Globalization Debate.* New York: Picador USA.

La Vina, A.G.M., Hoff, G. and DeRose, A.M. (2003) 'The outcomes of Johannesburg: assessing the World Summit on Sustainable Development', *SAIS Review,* 23(1): 53–70.

Lifton, K. (2003) 'Planetary politics', in J.A. Agnew, K. Mitchell and G. Toal (eds), *A Companion to Political Geography.* Malden, MA: Blackwell, pp. 470–82.

Mansfield, B. (2003) 'Spatializing globalization: a "geography of quality" in the seafood industry', *Economic Geography,* 79(1): 1–16.

Marchak, M.P. (1995) *Logging the Globe.* Montreal and Kingston: McGill-Queens University Press.

McCormick, J. (1989) *Reclaiming Paradise: The Global Environmental Movement.* Bloomington: Indiana University Press.

Mertes, T. (2002) 'Grass-roots globalism', *New Left Review,* 17 (Sept.–Oct.): 101–10.

Middleton, N. and O'Keefe, P. (2003) *Rio Plus Ten: Politics, Poverty, and Environment.* London: Pluto Press.

Mintz, S.W. (1985) *Sweetness and Power: The Place of Sugar in Modern History.* New York: Viking.

Neumann, R.P. (1998) *Imposing Wilderness: Struggles over Livelihood and Nature Preservation in Africa.* Berkeley: University of California Press.

Newell, P. (2000) *Climate Change: Non-State Actors and the Global Politics of the Greenhouse.* Cambridge: Cambridge University Press.

Pallemaerts, M. (2003) 'Is multilateralism the future? Sustainable development or globalisation as "a comprehensive vision of the future of humanity"', *Environment, Development, and Sustainability,* 5: 275–95.

Peet, R. and Watts, M. (eds) (2004) *Liberation Ecologies: Environment, Development, Social Movements* (2nd edn). London: Routledge.

Peluso, N.L. (1992) *Rich Forests, Poor People: Resource Control and Resistance in Java.* Berkeley: University of California Press.

Porter, G., Welsh Brown, J. and Chasek, P.S. (2000) *Global Environmental Politics* (3rd edn). Boulder, CO: Westview Press.

Princen, T. and Finger, M. (1994) *Environmental NGOs in World Politics.* London: Routledge.

Robbins, P. (2004) *Political Ecology: A Critical Introduction.* Malden, MA: Blackwell.

Roberts, S.M. (2002) 'Global regulation and trans-state organization', in R.J. Johnston, P.J. Taylor and M.J. Watts (eds), *Geographies of Global Change: Remapping the World.* Malden, MA: Blackwell, pp. 143–57.

Routledge, P. (2003) 'Convergence space: process geographies of grassroots globalization networks', *Transactions of the American Fisheries Society,* 28: 333–49.

Roy, A. (2001) *Power Politics.* Cambridge, MA: South End Press.

Sachs, W. (1993a) 'Global ecology and the shadow of "development"', in W. Sachs (ed.), *Global Ecology: A New Arena of Political Conflict.* London: Zed Books, pp. 3–21.

Sachs, W. (ed.) (1993b) *Global Ecology: A New Arena of Political Conflict.* London: Zed Books.

Sampson, G.P. and Chambers, B. (eds) (2002) *Trade, Environment, and the Millennium* (2nd edn). Tokyo: United Nations University Press.

Shiva, V. (1993) 'The greening of global reach', in W. Sachs (ed.), *Global Ecology: A New Arena of Political Conflict.* London: Zed Books, pp. 149–56.

Shiva, V. (1997) *Biopiracy: The Plunder of Nature and Knowledge.* Boston: South End Press.

Smith, J. (2004) 'The World Social Forum and the challenges of global democracy', *Global Networks,* 4(4): 413–21.

Soroos, M.S. (1999) 'Global institutions and the environment: an evolutionary perspective', in N. Vig and R. Axelrod (eds), *The Global Environment: Institutions, Law, and Policy.* Washington, DC: Congressional Quarterly Press, pp. 27–51.

Starr, A. (2000) *Naming the Enemy: Anti-Corporate Movements Confront Globalization.* London: Zed Books.

Tamiotti, L. and Finger, M. (2001) 'Environmental organizations: changing roles and functions in global politics', *Global Environmental Politics,* 1(1): 56–76.

The Ecologist (1993) *Whose Common Future? Reclaiming the Commons.* Philadelphia: New Society.

United Nations (2002) *Report of the World Summit on Sustainable Development,* Johannesburg, South Africa, 26 August–4 September 2002. New York: United Nations.

Wapner, P. (2002) 'Horizontal politics: transnational environmental activism and global cultural change', *Global Environmental Politics,* 2(2): 38–62.

Wapner, P. (2003) 'World Summit on Sustainable Development: toward a post-Jo'burg environmentalism', *Global Environmental Politics,* 3(1): 1–10.

Warkentin, C. (2001) *Reshaping World Politics: NGOs, the Internet, and Global Civil Society.* Lanham, MD: Rowman & Littlefield.

Watts, M. (2004) 'Violent environments: petroleum conflict and the political ecology of rule in the Niger Delta, Nigeria', in R. Peet and M. Watts (eds), *Liberation Ecologies: Environment, Development, Social Movements* (2nd edn). London: Routledge, pp. 273–98.

Weber, M. (2001) 'Competing political visions: WTO governance and green politics', *Global Environmental Politics,* 1(3): 92–113.

Williams, M. and Ford, L. (1998) 'The World Trade Organization, social movements, and global environmental management', *Environmental Politics,* 8(1): 268–89.

World Business Council for Sustainable Development (2004) *About the WBCSD 2004* [cited 7 October 2004]. Available from http://www.wbcsd.ch.

Wright, A. and Wolford, W. (2003) *To Inherit the Earth: The Landless Movement and the Struggle for a New Brazil.* Oakland, CA: Food First Books.

Young, O. (1989) *International Cooperation: Building Regimes for Natural Resources and the Environment.* Ithaca, NY: Cornell University Press.

Young, O. (1994) *International Governance: Protecting the Environment in a Stateless Society.* Ithaca, NY: Cornell University Press.

Young, O. (ed.) (1999) *The Effectiveness of International Environmental Regimes: Causal Connections and Behavioral Mechanisms.* Cambridge, MA: MIT Press.

The Politics of Transition: Critical Political Ecology, Classical Economics, and Ecological Modernization Theory in China

Joshua S. S. Muldavin

INTRODUCTION

This chapter considers societies experiencing a major transition in political and economic regime, and the mutual implications for nature and politics. Re-naturing political geography may allow us to reconsider theories of large-scale social transformation. In theorizing 'socialist' transition, this essay compares historical materialist approaches with others that emphasize the broader impacts of industrialism on resource use and the environment. A particular contrast is drawn with ecological modernization approaches, which have strong European and Scandinavian roots and emphasize the slow gains for the environment resulting from technological innovation in mature economies. The reason for this focus here is the institutional dominance of ecological modernization theory in contemporary development discourse on the environment, and its current application to countries undergoing socialist transition. Case study material is drawn primarily from the author's work in China.

What are the environmental impacts of socialist transition? How might we theorize this transition in ways that help us more clearly understand these environmental changes? How do environmental issues inform our understandings of socialist transition overall, or even challenge its most fundamental assumptions? These are just a few of the questions raised by the transition of former planned socialist societies to market-oriented societies. While this chapter will not attempt to answer all of them because of space and scope concerns, they provide a context for conversation between political geography and political ecology.

THEORIZING THE SHIFT FROM PLAN TO MARKET

As Watts asserts, we can productively locate the socialist transition debate 'against the backdrop of the particularities of socialist legacies and models of transition, the distinctive role of the party state, and the classical agrarian question debates …' (1998: 156). Kautsky posed the agrarian question as 'whether and how is capital seizing hold of agriculture, revolutionizing it, making old forms of production and property untenable and creating the necessity for new ones' (1988 [1899]: 12). But key to Kautsky's approach was his interest not only in the scale of production, but also rural social structure and most importantly the *politics* of transition. Brought to contemporary socialist transition we might rephrase his questions as follows: How has socialist agriculture in its myriad forms been subjected to privatization, decollectivization and the introduction of new forms of property rights? How has capital seized hold of socialist forms of production and property while creating new ones? In this way the agrarian question is, as Watts notes, 'framed by the particular structures of socialist economy – and the diverse forms of socialist agriculture that evolved in the post revolutionary period: agrarian reform from below versus collectivization from above, or state farms versus Maoist collectives' (1998: 154). While socialist transition may not be comparable to the transition from feudalism to capitalism, this historical parallel offers potentially fruitful avenues for our further inquiry into its environmental consequences, particularly in the context of changing patterns of production and resource use (Muldavin, 1992).[1]

A key point of differentiation with the historical parallel is the contemporary role of the party state in engineering socialist transitions to market economies. As Watts and others (Walder, 1994; Oi, 1999; Wang, 2003) have argued, the 'success' of the Chinese case and the 'failure' of the Russian case turn less on the relative merits of gradualism versus 'shock therapy', 'as on the necessity and capacity of the party-state to shape the transition through the decentralization of property rights, the promotion of competition and the construction of "socialist markets"' (Watts, 1998: 156). This is a political challenge to the state and relies on the ongoing legitimacy of state action to achieve societal changes often at odds with the needs of the majority. Thus the ways in which the legitimacy was historically obtained (in China and Vietnam, for example, through extremely popular 'land to the tiller' rural transformations) sets the boundaries for acceptable forms of struggle in the contemporary period. In China's transition experience, the historical role of the party in liberating peasants from landlordism and extreme forms of oppression prior to the revolution allowed the central government and party state to maintain immense legitimacy despite implementing policies that directly harmed the long-term interests of China's vast rural peasant majority.

In Vietnam, a principally agrarian country, liberalization of the economy happened quickly. The agricultural production crisis drove the reform process, leading to a complete collapse of the cooperative structure, and providing what Watts refers to as the 'Thermidor of the Vietnamese return to capitalism' (Watts, 1998: 151). Contrary to Kornai's claim (Kornai, 1992) that fundamental reform of former socialist economies is not possible without the destruction of the Communist Party, Vietnam

(like China) confirms the neoliberal paradox that the transition requires a strong (preferably one-party-dominated) state to negotiate the difficult but necessary process of property rights reform, market liberalization, and insertion into the global economy with its heightened competitive influence on the national economy (Watts, 1998: 151).

The concept of 'transition' itself is problematic. Many analysts' share the normative assumption that the process is needed, and that it is linear in form (Lin, 1990; Sachs and Woo, 1993). This assumption obliterates the immense range of historical contexts, as well as the nonlinear, highly political character of the lived experiences of change. That is, there are numerous meanings and contradictions of transition. As one example, primitive accumulation is not supposed to happen but often does on a vast scale. The resulting political economies rarely fit into the expected categories but rather each reflects their specific national and regional histories. Who the transition is for, who will have the power in its implementation, what will be the means by which transition is carried out, and toward what ultimate goals, are all highly contentious and political questions that point to important struggles, both ideational and material. As Wang (2003) convincingly argues, China's transition is a myth promoted by the party state and other actors (economic, academic, political) to serve specific ends including state legitimacy, rapid private accumulation, and a highly unequal set of social outcomes. Rather than engage in a debate as to whether a transition is actually occurring, and the concurrent assumption that it embodies a move toward a preferred situation, I will continue to make use of the term while attempting to integrate into this essay a nonlinear vision of the changes taking place.

There is much variation in socialist transition experience across the world. In the former Soviet Union and Eastern Europe, the legacy of widespread severe industrial pollution, produced under top-down state-led modernization strategies of the previous decades, has loomed large in the complex re-emergence of environmental politics and regulation. Beyond the classic trade-off between economic development and environmental protection, substantial challenges face the nascent environmental NGO movements (Turncock, 2001). Furthermore, the immense variation of history, geography, culture, and economy make overgeneralization immediately suspect. Still, as Staddon and Turncock (2001: 233) state, 'post-communist entrepreneurs and their western partners have been quick to realize that there are large profits to be had, not just from the pillage of the region's natural resource endowment, but also from the translocation of pollution costs to a region desperate for foreign exchange and lacking strong environmental controls'. Examples include the shift of pollution-intensive industries to Eastern Europe, toxic waste dumping, and so forth. Hilary French (1999) refers to this environmental exploitation as a form of 'ecological colonialism'. Furthermore, rapid marketization and privatization, often imposed and led by the dominant development institutions of the West (IMF, World Bank) through 'shock therapy', have created distinct environmental challenges and ongoing problems (Pickles and Pavlinek, 2000).[2]

Vietnam and China adopted decollectivization strategies with the following attributes: a relatively egalitarian distribution of land and some form of private property rights, a hybrid form of state and market regulation, substantial local state intervention in the reform process, and a shift in responsibility from the state to individual peasant families to support rural surplus labor (Watts, 1998: 180). But in Vietnam, as land and markets were fully privatized, the cooperative structure also dissolved. Unlike the two-tier property system of private plots existing simultaneous with varying but continuing forms of village and township collectives in China's rapid rural industrialization process (Oi and Walder, 1999; Gilley, 2001), full privatization in Vietnam eliminated the positive potential provided by continued forms of collective organization. As Watts points out, the fact that the transitions in Vietnam and China are happening in different ways and with different consequences 'alerts us to the complex and differentiated ways in which politics and economics – rather than some undifferentiated capitalism – are being decomposed, reconstituted and refigured in the post-socialist order' (1998: 182).[3]

In China the state-led transition focused first on rural decollectivization (with much later urban reforms), and without complete privatization of natural resources and the commons, resulting in a substantially different experience than other cases of socialist transition, though environmentally problematic as detailed below. Very rapid and relatively unregulated growth led to intensification of old environmental problems of the former socialist state (Muldavin, 1992, 1998b), and the creation of new forms of resource destruction, pollution, and environmental degradation structurally enshrined in the newly evolved institutions of China's 'market socialist economy'. China's contemporary paradox – growth built on decay – is the focus of my inquiry into the 'transition' process. Building a detailed historical analysis across multiple scales of the environmental and social consequences of the reform era in rural China, I argue that a historical critical political ecology approach provides a unique optic to understand the impact of policy changes on rural livelihoods, and the important and underrated role that rural hinterland areas play in

simultaneously challenging globalization, national development objectives, and regional transformation, while being used as fundamental resources for these very processes. Though it is not the focus of my discussion here, I argue that contemporary China's transition is best explored through an analysis of China's integration into the global economy and the impact of this both in China and in the world (Muldavin, 2000a, 2005b, 2006). The recent rise in social unrest associated with the much-discussed environmental crisis in China is forcing the state to re-evaluate its current development path as it struggles to maintain legitimacy with the country's rural majority (Muldavin, 2005b, 2006).

THREE BROAD APPROACHES: KEY ARGUMENTS AND ASSUMPTIONS

In this chapter I shall lay out three generalized approaches of inquiry on transitional economies and the question of environmental disruption, which I term classical, ecological modernization, and Marxian. This is not to say that other approaches are unimportant, but more to provide clear definitions of dominant strands of engagement that have strong local and national followings within the communities and regions where they are applied and contested, as well as in the Northern countries from which they have primarily drawn inspiration and institutional backing. While certainly not exhaustive, the following typology of key positions on transitional economies and environmental outcomes is a starting point for further detailed discussion.

The Classical Approach

The first broad approach to environmental impacts of socialist transition is classical. Within the classical literature, two variants demand space. The dominant Smithian variant associated closely with the rise and subsequent hegemony of neoliberalism looks first and foremost to a *laissez-faire* model in which efficient markets resolve all environmental problems. In orthodox neoliberal theory, the proper pathway in the transition from socialism to capitalism rests on a set of key assumptions to best achieve human well-being: that private ownership, free prices, and free trade are necessary preconditions for increasing productivity. Thus, the four key processes to focus on are property rights reform, privatization, price and fiscal reform, and setting an appropriate pace of 'liberalization', that is, 'big bang' or gradualist (Watts, 1998: 155; Harvey, 2005: 2). Therefore, in terms of the environment, the goal is not to hinder industrialization, but to improve market efficiency so that it will guide resource use in ultimately the most environmentally sound manner. Promoters of this neoliberal model assert that proper pricing through well-functioning markets will bring rational use of resources as opposed to the inefficiencies of the previous planning model. Efficiency gains will be broadly shared through trickle-down and ripple effects, assuming proper utilization rates of resources. Externalities, such as environmental degradation, may require the creation of new markets or some limited state regulation, but it is the primacy of private property rights in a competitive capitalist economy that unleashes entrepreneurial freedom and skills that will best ensure proper long-term resource use. Freedom and democracy will flow from free markets and free trade and create appropriate civil society institutions and market-driven mechanisms to help fill any voids left by state withdrawal from regulation of resource use.

The Ricardian variant takes a further step by asserting that the ability to produce with negative environmental consequences provides a comparative advantage during 'transition' in terms of global competition. To hinder this comparative advantage with unnecessary regulation limits the proper functioning of markets in allocating and utilizing resources at maximum efficiency on a global basis between nation-states. Thus, for example, China or Russia may choose to undertake environmentally unsound production practices as a comparative advantage, and the costs and benefits of this will be appropriately distributed through subsequent economic growth.

Somewhat paradoxically, the classical approach uses formalist economics to assert an apolitical, even anti-historical, approach, while depending upon state economic policy and power to enforce its method and practice in achieving transition despite the political contradictions that develop. As Wang clearly argues, the attainment of hegemonic status for classical neoliberalism in China was part and parcel of the state's use of 'economic liberalization to overcome its crisis of legitimacy' (Wang, 2003: 44). Furthermore, its discursive and ideological dominance in China allows the state and other actors to use 'transition' and 'development' to 'patch up internal contradictions' (ibid.). In terms of problematizing the environment, this has legitimized the market as the primary answer, and the framing of all negative environmental outcomes as the result of continued adherence to other failed alternatives.

The Ecological Moderization Approach

Within the second broad approach to environmental impacts of socialist transition, ecological modernization, there are two strong variants or currents that I will briefly discuss here: Keynesian and Cornucopian. In the Keynesian variant it is asserted that a strong regulatory and interventionist state can limit the negative environmental externalities of the transition to a capitalist economy. This position is put forward at various times by a wide range of actors, including among others the World Bank (2002), the Chinese state, and policymakers around the globe. The state, they argue, can promote industrial modernization with positive environmental outcomes through investment in cooperative research, tax policies promoting environmental technologies, and strict regulation of corporate malfeasance and misbehavior – for example, the development of powerful national environmental protection agencies.

Within the Cornucopian variant of ecological modernization, the primary assertion is that the technological advances that accompany the transition to an advanced industrialized country will resolve the environmental impacts of rapid industrialization. This increasingly popular analysis depends on technology to solve all environmental problems, hence the use of the term Cornucopian.[4] More specifically, the new environmental problems created by industrialization will push the emergent civil society and an increasingly institutionalized environmental protection hierarchy to enable (through tax policies, for example) or force (through legal and regulatory means) industry to create new technologies to resolve the environmental problems it creates. Thus, positive development and environmental outcomes can result from rapid industrialization in a win-win scenario in which the application of advanced environmental technology and introduction of cleaner technologies helps ameliorate the environmental impacts of rapid development. The rise of independent civil society and NGOs accompanying the transition to a market economy will provide key pressure in this process, as well as providing actors who demand accountability of both the state and private sector.

Promoters of this ecological modernization model within social democracies in northern Europe argue that it is applicable throughout the world. They argue that its success in northern Europe (particularly in Scandinavia) provides a guide to its implementation in Southern nations, with great potential in China and other countries in transition from socialism to capitalism (Mol and Spaargaren, 2000). For example, in a recent collection in Ho (2006: 37) and Mol (2006: 37), ecological modernization theorists argue for the application of the theory in China. In their scenario, China's rapid modernization with technological advancement will create environmental problems but will also create the technological advances, innovations, and needed capital investment to resolve them, along with the institutional ability to regulate the environment. The reversal and clean-up of any environmental damage and the subsequent decrease in harmful environmental trends will be achieved through rapid adoption of energy-efficient technologies, investment in advanced pollution control equipment, and so forth. The win-win will also be achieved through the rapid expansion of the environmental technology market, which provides new markets for First World environmental technology exports, while simultaneously helping build up new environmental technology manufacturing capacity in transitional economies, such as China. Everyone benefits from the combination of modernization, technological advancement, and resolution of environmental problems. A key assumption concerning China is the simultaneous development of a well-informed public and democratic expression, that is, an active civil society to work in tandem with a free press in enforcing accountability of the state and private sector, thus ensuring the positive outcome of rapid development in a relatively benign environmental manner, overseen by a technocratic elite.[5]

It is not just a matter of chance that the ecological modernization vision has taken hold in the dominant development institutions during the era of neoliberal hegemony. Built on a belief in markets and growth as means to rationally appraise development success or failure, there is little conflict with classical neoliberal assumptions in this regard. The faith in science and technology, properly applied by a technocratic elite of environmental managers, to overcome the natural limits to growth through the creation of new resources, also carries a strong suspicion of broader democratic participation in decision-making around environmental issues. This parallels the neoliberal economic discourse that paternalistically speaks of the need for objective and efficient management that can carry through the painful structural adjustment process of communities that might resist this necessary medicine if given the choice through participation and democratic decision-making. As discussed above, this is part of the neoliberal paradox in which strong states are a prerequisite for successful implementation of economic liberalization.

As Blaikie (1999: 136) points out, ecological modernization is now the dominant narrative for analysis of environmental issues at the global level and by international institutions. He cites Spaargaren and Mol's 'rather optimistic view' of a technology-dependent ecological revolution (1992). Blaikie's careful assessment of ecological modernization theory concludes with a negative critique for the following reasons. First, he argues that it has 'little to say about power relations which permeate the socioeconomic process which help to shape environmental processes and outcomes, as well as their representations and framings' (2000: 138). Second, he argues that it ignores the South, talking exclusively in 'terms of existing and feasible technologies of the industrial North'. He points

to a series of publications that challenge its transferability to the South, but nonetheless concludes that it remains 'the dominant environmental discourse in spite of all the criticisms from postmodern, poststructural, and populist perspectives'. Given this domination, he points to its continuity with what he terms ecological modernization's 'colonial style', despite its participation in critiques of this very approach.

Redclift (1997) argues that the focus on internalizing externalities ignores their distributive causes and consequences and fails to acknowledge the essential confirmation of existing power relationships via the pathway of continued industrialization as the answer to environmental crisis. Redclift also argues that neoliberal economic policies have increased externalities, highlighting the basic contradictions between economic growth and conservation that ecological modernization fails to acknowledge. He draws attention to a fundamental paradox missing in ecological modernization's economic valuation of the environment: distribution and equity, both inter- and intragenerational (1997: 338–9).

The Marxian Approach

The third broad approach to environmental impacts of socialist transition is principally, though not exclusively, Marxian. It has a number of different strands, epistemological and ideological as well as 'disciplinary', including critical political ecology, radical sociology (Buttel, 2000; Szelenyi, 1998), and poststructuralism (Escobar, 1995). I will focus on one variant: critical political ecology.

What distinguishes critical political ecology approaches to the environment? First, critical political ecology is a multiscalar analysis, methodologically integrating local, regional, and global actors, institutions, and structures. Second, it is a historical analysis concerned with processes of change over time, and particularly during systemic and often dramatic shifts in political economy such as from precolonial to colonial, or colonial to postcolonial. Third, it emphasizes political economy analysis, such as in Blaikie's early definition of political ecology as ecology plus political economy (1985). The evolving roles of the state and the market in influencing environmental outcomes are central. Fourth, most political ecology is ethnographic, but unlike cultural anthropology and cultural geography that earlier provided local ethnographies and village studies, political ecologists do ethnography at multiple scales, from local to international, and inclusive of actors and institutions far beyond peasant households. Thus, Blaikie's insistence 'that policy makers, government officials and scientists be scrutinized as closely as Third World peasants' opened the door to questions of power, knowledge, and ideology and their relation to claims about environmental degradation, 'foreshadowing the incorporation of feminist, postcolonialist and postmodernist challenges to European Enlightenment notions of rationality and universal truths' in contemporary environmental debates (Neumann, 2005: 31).

Fifth, critical political ecology productively grapples with poststructuralism and discourse analysis, and hence the material analyses that dominated earlier work have been replaced to a large extent by analyses that critically interrogate the relationship between power and knowledge constructions, particularly in science. Many critical political ecologists questioned the construction of dominant scientific knowledge long before the poststructural turn in social theory (Carr 1977, Watts, 1983; Blaikie, 1985), but the recent analyses have both expanded and clarified this particular analytic (Peet and Watts, 1996; Forsyth, 2003; Blaikie and Muldavin, 2004b). The essentially political and ideological struggles over objectivity and rationality, carefully deconstructed earlier by Harvey (1974), have now been recast integrating the 'contingent and dynamic nature of environmental change … bound up with social and cultural processes' (Scoones, 1999: 493). These insights of critical social theory have paralleled and reinforced the reassessment of equilibrium models in ecology. Thus, the replacement of long-held static views of nature as a system in balance with nonequilibrium approaches has been mirrored by a rejection of static notions of knowledge construction struggles over environmental narratives, and thus the growing emphasis upon discourse analysis (Stott and Sullivan 2000; Forsyth, 2003). A persistent critique of political ecology is its limited integration of sound ecological science. This is partially answered by the current productive exchange between ecologists and political ecologists (Neumann, 2005).

Critical political ecology is a form of radical critical realism. Its research agenda focuses on uncovering the links between broad changes in the political economy and environmental degradation. Its activist agenda focuses on social change to counter perceived environmental degradation and social injustice. Still marginal in institutional terms, being limited primarily to university settings, it argues that a fundamental change must occur in social relations of production, as well as control over the world's resources at multiple scales, in order to resolve environmental problems. Its promoters assert that a multi-scaled structural analysis is needed to understand completely the environmental outcomes of various 'transitions'. Critical political ecologists generally assert that the process of market penetration and capital penetration of the Third World leads to privatization of previous collective goods with subsequent overexploitation and negative environmental impacts through rapid resource degradation. That is, the penetration of global market forces to the local level, accompanied by the commodification of natural resources, leads to intensified resource destruction and environmental degradation.

In the case of former state socialist transitional economies, environmental impacts reveal structural problems of the transition. These will not disappear with transition to a capitalist economy, as often asserted by classical and ecological modernization theorists, but will be further amplified (Muldavin, 1992, 1997; Bryant and Bailey 1997) The withdrawal of the state and regulatory structures during transition, combined with the destruction of collective and community institutions of resource management, leads to a free-for-all of environmental destruction in the subsequent institutional vacuum. The result of these structural environmental problems is declining sustainability, rising risk and vulnerability of the majority, and their increasing resistance to state policies and practices. The benefits of this process are garnered by a new elite while the problems are spread upon the vast impoverished majority, particularly the most vulnerable among them with the least access to market opportunities, either because of limited household labor or other intangibles such as requisite social networks.

As a critical and generally negative assessment of capitalist modernization, critical political ecology is highly suspect of purely technological or managerial solutions to what are fundamentally viewed as complex social processes – including historically unequal access to resources based on class, caste, age, gender, and race (Rocheleau et al., 1996). Thus a critical political ecology view of transitional economies' environments focuses our attention on the driving structural forces of global integration, transferring price signals and capital demands from distant unaccountable places in First World capitals to local environments, and utilizing a geographic fix to escape old, heavily regulated (environmentally, socially, and occupationally) economies, workers, and communities in the First World (Western Europe, Japan, and the US). These same economies have benefited (as ecological modernization theorists theorize and hope is transferable to the Third World) from environmental technology and strong institutional and regulatory frameworks. Yet, in the current context of globalized production complexes that span regions and reduce the importance of the nation-state and its regulatory potential, the result is globalized production processes driving a new race to the bottom worldwide (Gibson-Graham, 2006).

These complex interconnections challenge earlier visions of First and Third Worlds, and require a more sophisticated understanding of regional political and economic formations that are transnational in practice, if not in national politics. It is here that political geography has much to offer political ecologists as they struggle to integrate these complexities in their analyses of environmental change. Political ecologists have principally focused on the questions of market penetration into formerly peasant societies, and the political economic structures and agents affecting nature and resource use, and have under-theorized socialist transition in general. But a number of scholars (Muldavin, 1992, 1997, 2000a; Watts ,1995, 1998; Bryant and Bailey, 1997) have attempted to better understand the ways in which political ecology might be applied in this transitional context. Political geographers' focus on the political institutions and actors at play, rather than nature per se, can reinvigorate the discussion of politics in critical political ecology.

A key challenge for many political ecologists is whether theoretically and methodologically they can persuasively answer the following dominant assertion of the classical and ecological modernization approaches: the environment is best served when societies are guided in their resource use practices by market mechanisms. It is not only in the ecological realm that critical political ecologists find some of their strongest counter-arguments, but particularly in the political realm where resistance and outright rebellion to state modernization strategies encourages a careful re-evaluation of the role of politics, formal and informal, local and national, as well as the rapidly changing geopolitical context in which globalization strategies are both challenged and reconstituted on a daily basis by a wide range of actors. Here too political geography may contribute to political ecologists' attempts to integrate these political moments, actions, and structures into an overall analysis of environmental change (see Peet and Watts, 1996).

A CRITICAL POLITICAL ECOLOGY OF CHINA

China's four thousand years of resource use has led in many places to severe environmental degradation and complete environmental transformation. Each subsequent form of political and economic organization inherited this legacy. The socialist planned economy following 1949 inherited environments in many regions that were often already highly degraded, challenging planners with basic questions of how to enable sustained production in such contexts. Mass campaigns had the potential to both improve and destroy the environment. Shapiro (2001) strongly argues that the cult of Mao led to severe environmental destruction in the 1950s and

1960s in China. But simultaneous to this were many land improvement strategies that drew upon centuries of knowledge and specific understandings of local environments (Muldavin, 1997, 2000a; Ho, 2001). While Shapiro's rather lopsided account of the environment under Mao has many strong components, it fails to acknowledge the positive potential of mass mobilization for environmental preservation, soil erosion control, tree planting, water conservation, and so forth. Certainly many of these activities failed to reach their goals, but others did achieve measurable environmental improvement and thus cannot be completely disavowed. Thus, the methodological question of how to separate out what are long-term historical aspects of environmental degradation, from those associated with the more recent histories of socialist planning models, and the current transition to market-oriented economies, is key and not as simple to answer as one would hope.

In China, recent state and World Bank interest in 'third wave' environmental regulation, named because it follows earlier attempts at first, command and control, and second, market-based approaches, combines ecological modernization theory's optimistic expectations of market solutions to environmental problems via technological innovation, with the rise of civil society as the low-cost alternative to a formal regulatory process (Wang et al., 2002). A key problem identified by ecological modernization theorists is the limited role of civil society in China, and thus, predictably, the answer is its expansion (Mol, 2006).[6] The three waves of environmental regulation in China represent in essence the former state socialist approach, the classical neoliberal approach championed over the last two and a half decades, and the new ecological modernization approach that asserts a new solution to the lived environmental destruction of contemporary China. This provides a useful backdrop for the alternative analysis I will present here of the environmental outcomes of China's socialist transition.

Driving 'Mechanisms' of Transition

Beginning in 1978, the driving mechanisms of China's socialist transition experience included decollectivization, the introduction of the household responsibility and contracting system (HRS), privatization and enclosure of the commons, state withdrawal from the rural economy and entitlement provision, and ultimately global integration of China.[7] Market penetration occurred in a highly uneven and unusual fashion following the introduction of reforms. Initially limited to rural areas, the state focus in the first five years was to increase grain production by 'unleashing' individual incentives, though still heavily tied to state-controlled pricing systems. The state significantly increased grain procurement prices and improved provision of fertilizer, allowing peasants to more than double their use during this time. The resulting rapid rise in grain production further legitimated Deng's call for complete decollectivization and transformation of rural as well as other parts of society, including urban areas and state industries. While these changes were slow to take hold, not gaining real traction until the early 1990s, the successful increase in rural production was key to the ultimate introduction of the contracting system, as well as privatization and deregulation, in the urban industrial core (Muldavin, 1983, 1986, 1998a; Selden, 1998).

By the early 1980s the state had introduced free farmers' markets in the rural areas as a way for peasants to dispose of excess production beyond state quotas. This two-tier system provided peasants with new avenues to increase household income, and there was rapid diversification of the household economy into sideline production of animals and vegetable and fruit crops, as well as small industries for those lucky enough to gain access to the former collective assets. While production increased, the result was rapid economic and social stratification in a rural landscape that had remained relatively egalitarian for the previous thirty years. Even small machines, such as a grain grinder, provided a huge advantage to a family, allowing them to join the ranks of the 10,000-yuan newly rich households, widely celebrated in the Chinese official press. In one of Deng's most memorable aphorisms, 'to get rich is glorious!', this process of societal stratification was given the state's vigorous stamp of approval (Blecher, 1997).

The HRS played out in diverse ways depending on the history of the particular commune, its leadership, and available resources and assets to contract out to households. The large collective fields were divided into narrow strips of various qualities, with every family receiving a plot of each kind of land, visually transforming the countryside overnight from a mechanized large-scale farm economy to small noodle strips of land worked by hand. Where once there had been large continuous fields of corn, now there was a patchwork of crops. Where before you would have seen specialized teams of peasants plowing fields with large tractors, now each household returned to draft animals, owned or hired, to plow their tiny plots. Where before crops were rotated across the landscape through state-led 'scientific' management, now peasant households individually decided how best to manage the land to achieve the quota production demanded by the state, and produce surpluses to sell to the state or in the local free markets (Muldavin, 1992).

The technological devolution in agricultural production practices was mirrored by a devolution in gender relations (Jacka, 1997). As collective institutions declined, patriarchy was re-strengthened, and peasant households chose to send boys to school, while sending girls to the fields to take up the new intensified labor demands. The result was a rapid devaluation of women and women's labor, and rising intra-household conflict as young women saw their own situation worsen relative to that of their mothers. As sideline and village industries expanded, it was principally young girls that were sent to work in these new small factories, often facing horrendous occupational hazards and working conditions (Muldavin, 1992, 2000a).

The state also shifted control over other resources from the collective to the individual. Collective commons, such as grasslands and forests, were enclosed, often fenced, and contracted out to peasant households. Collective regulation of these former commons was further weakened, simultaneously with a generalized increase in risk, as social institutions connected with the collective, focused on long-term production and security, ceased to function now that the resources formerly used to support them had all been contracted out to households. Peasants responded by intensifying grazing, logging, and other extractive practices in a risky deregulated context where success was judged only on increased output and fulfilling contractual obligations. Time horizons in production shortened significantly throughout this period, leading to rapid degradation as discussed below (Cannon and Jenkins, 1990; Muldavin, 1992, 1997).

Impacts on Environment and Nature

The impact of socialist transition on the environment and nature in China can be assessed through any number of examples. I have argued extensively that, with decollectivization in the first decade following the reforms, the mining of communal capital was a key component of rapid environmental degradation and undermining of rural sustainability (Muldavin, 1986, 1992, 1997). I define communal capital as the resources preserved, built up, and collectively managed during the commune period with the expectation of long-term benefits accruing to all commune members. As the state shifted decision-making and control from the collective to the household and individual, and social welfare and entitlements disappeared, peasants operating in this riskier environment moved to rapidly utilize the former assets of the commune, including natural assets such as forests, grasslands, and agricultural lands as described above. This enabled a rapid rise in production that paradoxically further legitimated the reform process, though at a great cost.

Hence the term 'mining communal capital' invokes the unsustainability of the process, as well as the source of the mined 'assets'. In addition, the state mandated distribution of many village enterprises to a few lucky households in this first phase of reform. With little environmental or occupational regulation to hinder production practices, these industrial enterprises became subcontractors for the most toxic components of local, national, and even international industries. Damaging the health of workers and delivering large quantities of untreated pollutants onto the land, into the water and air of the rural villages, they simultaneously became the much touted engine of growth in China's rural economy, while destroying the people and resource base upon which they depend.

In China's vast grasslands, decollectivization led to rapid increases in animal numbers and intensification of grazing, with severe negative results including rapid desertification and sodic alkalinization of huge areas of formerly productive land (Muldavin, 1986, 1992, 1997). Forests contracted out to households were quickly felled, leading to severe soil erosion, impacting nearby arable lands and waterways, and destroying diverse habitats that had formerly supplied many important products to local villages. Cropland practices were altered and intensified, eliminating green manuring, rotation, and fallow periods, and quadrupling the use of fertilizer as well as other chemicals. After an initial rapid rise in yields, the resulting decline in soil fertility and structure as organic matter declined led to stagnating yields, as well as diminishing returns as the costs of formerly subsidized inputs rose. In the risky context of household production, these declining returns only pushed peasant farmers to intensify further as they struggled to increase crop outputs, often shifting to double cropping systems and the adoption of other unsustainable practices (Muldavin, 1997).

As this array of environmentally destructive practices intensified, the infrastructure built up during the collective period began to fall apart. The weakened collectives could no longer enforce collective labor quotas in the emergent market context, and thus irrigation infrastructure, conservation plantings, and other maintenance of critical infrastructure such as levees, was neglected if not abandoned altogether. State investment in agriculture, including infrastructure, simultaneously fell from 13 percent to 5 percent of the state budget in the first ten years of reform. As a result, levees failed, irrigation canals no longer functioned to deliver water to fields, erosion from denuded hillsides and grasslands covered arable lands with poor-quality soils, and overuse of pesticides and fertilizers led to new pest outbreaks and 'burned' soils, all corresponding with a rise in 'natural' disasters such as floods, droughts, dust storms, plagues, and crop failures (Muldavin, 1992, 1997, 2000a).

State Delegitimation, Resistance, and Political Crisis: Environmental Justice and Rising Protests

State legitimation has been a key issue throughout China's reforms and socialist transition. Whereas the initial successful rise in grain production (and rural productivity overall) legitimated state calls for complete decollectivization and application of the reforms to the urban industrial cores, subsequent stagnation in production, rapid inflation, growing socioeconomic inequality, and widespread corruption in the reform process during the later half of the 1980s, undermined state legitimacy and was at the heart of the Tian'anmen uprising in 1989. In fact, contrary to popular perception, the rural roots of Tian'anmen were substantial, and were key both to the movement's impetus as well as its ultimate failure as students were unable to forge links with China's peasants and workers in meaningful ways (Muldavin, 1999).

Beginning in this period, and intensifying through the 1990s and to the present, peasants have increased direct and indirect acts of resistance to the transition process and its outcomes. Indirectly, peasants have hidden surpluses, lied to state authorities about actual natural conditions, and otherwise attempted to limit state claims on their produced surpluses. Claims for relief from 'natural' disasters were both a product of the environmental degradation and infrastructure decline discussed above, and also were a weapon for peasants to clearly test and push the boundaries of state legitimacy (Scott, 1985; Muldavin, 1997). Following 'disasters', peasants demanded debts to be written off, new free credit, lower quota requirements, increased aid, and cheaper subsidized inputs. These ongoing indirect acts emanate from the problems discussed above, as well as excessive taxation (fees, fines, levies, etc.) imposed upon peasants by local governments no longer able to call upon the central government for financial support. The result has been a widening gulf between local political leadership and peasant producers, with the central state and the 'party of the peasants' increasingly viewed as the only remaining legitimate voice to settle disputes.

Simultaneously there has been a rise in direct forms of resistance and outright militant acts of rebellion. By the late 1980s, inflation had reduced peasant incomes in many regions to the point where they were losing money by planting crops, yet they were contractually obliged to continue producing to fulfill state quotas. Resulting sporadic demonstrations, along with the widespread indirect forms of resistance, were a wake-up call for the state. In the 1990s I witnessed numerous demonstrations against state reform policies, with violent responses by local state authorities (Muldavin, 1997, 2000a). These rural challenges came at an inopportune time for the state, as its shift to urban reforms was creating a fiscal crisis and limiting its ability to respond to peasant demands with new investments. The increasing primacy of the urban industrial coastal regions, led by the highly successful special economic zones, shifted the balance of power and thus surplus distribution from rural demands to urban concerns. Ongoing difficulties in reforming large-scale state-run enterprises, rapid increases in urban unemployment, and the creation of a new urban middle class, combined to force the state to further withdraw from rural development issues at a time of growing rural crisis. The state's response was to promote more rapid growth (averaging near 10 percent a year for the last twenty years) in the hope that trickle-down effects would help alleviate growing rural complaints (Muldavin, 2000a, 2005b). Paradoxically, this growth-based strategy has only served to amplify environmental destruction and socioeconomic inequality, thus leading to rising incidents of unrest (Muldavin, 1997, 2006). In 2004 there were 74,000 officially recorded incidents. In 2005 this grew to 87,000, widely understood as a direct result of widening gaps between rich and poor, between urban and rural areas, and between the rapidly growing industrial east and the stagnating agricultural hinterlands (Muldavin, 2006). In this context, the environment has become part of the political terrain in which Chinese peasants began to negotiate their relations with the state (Muldavin, 1996, 2000a). In essence, the state's legitimation crisis is now increasingly entwined with environmental decline, not only in rural areas, but in urban regions as well. I will provide here just a few recent examples.

There are now thousands of documented cases of peasant protests over pollution from factories causing severe air, land, and water pollution. Crop losses, health impacts, and loss of access to potable water have become part of the daily news in China (Muldavin, 2005b). The dam-building frenzy in China has displaced millions, and new plans for dozens of large dams in the southwest along the Salween, Mekong, and Yangtze rivers have led to large-scale protests and confrontations with local and central authorities. This has also led to the rise of an energized environmental NGO movement in China, though severely hampered by state limitations on the development of an independent civil society (Muldavin, 2005a).

Toxic spills are so common that it is only the largest and most destructive ones that are reported. In November 2005, 100 tons of benzene spilled into the Songhua River in northeast China, not only threatening urban water supplies in the large city of Harbin, but also passing through hundreds of rural villages with no alternative source of water. This is a direct result of the highly unregulated industrialization in China, particularly acute in the rural hinterlands where peasants labor in some of the world's dirtiest and most dangerous conditions. Many of these rural factories are subcontractors to Chinese and international companies, tying these environmental injustices to the rising consumption of Chinese-made goods globally (Muldavin, 2005b).

Other examples of environmental protests concern the state's logging ban and slope-land conversion project following record flooding in the Yangtze River valley in 1998. Peasants are increasingly criminalized for accessing traditional forest reserves. State programs requiring conversion of slope fields to grassland and forest has been very unpopular, as promised compensation for lost income and livelihoods has failed to materialize, leaving households without adequate food and fodder (Muldavin, 2005a). Rural mining is another major source of environmental pollution, as well as occupational hazards. Rural China's thousands of small coal mines have the worst safety record in the world, killing 5,986 peasant workers in 3,341 accidents in 2005 (State Administration of Worker Safety, 2006). In addition, mine effluent is improperly dumped into valleys and open areas creating toxic mountains that pollute ground water and spread onto fields, decreasing productivity. Driven by China's ever-expanding energy requirements as factory floor to the world, the combined impacts are a widely reported major source of rural conflict.

Farmers are also demonstrating against land losses, as real estate development, factory building, waste dumps, and large infrastructure projects such as roads, airports, and power plants have too often been carried out through land seizures without proper compensation or any process for dealing with peasant complaints. The land question in China is a key political problem for the state, and fundamental in the current challenge to its legitimacy. In December 2005 in China's Guangdong Province, after villagers' peaceful protests over land seized by local authorities for a power plant went unheard for years, local militia killed more than a dozen peasants during a tense standoff. I have argued that

[w]hile avoiding full land privatization and, until recently, massive landlessness of the rural majority, the state still allows unregulated rural land development for new industries and infrastructure. Land seized from peasants reduces their minimal subsistence base, leaving them with what is called 'two-mouth' lands insufficient to feed most families, thus forcing members of many households to join China's 200 million migrants in search of work across the country. In many areas where I have carried out research, some households have lost even these small subsistence lands, swelling the ranks of China's landless peasants, who number 70 million according to official estimates. (Muldavin, 2006).

Given the loss of collective welfare entitlements, peasants are forced to desperate measures to try to reverse their rising vulnerability.

As China's transition has created serious problems, and more obvious contradictions of the process have emerged, China's emergent New Left has taken up the banner of environmental and social justice. Ranging from liberal nationalists to neo-populists to more radical internationalists, these varied intellectual currents provide new voice to what is already a well-developed resistance (Muldavin, 2005a; Pocha, 2005; Wang, 2005).[8] Connected with the expanding organs of civil society, these activist intellectuals increasingly point to environmental justice narratives as the basis of legitimacy and power for various protests. Whether such a movement can coherently challenge state policies, channeling grassroots energy into an effective national debate, is still unclear. The historic divide between urban elites and rural peasants continues, and limits many collaborations to patronizing acts of charity. Still, there is no question that this is a very exciting time in modern China's political evolution, and as such the ties between environment and politics provide critical political ecologists and political geographers with much common ground for joint analysis – the subject of the next section.

WHERE TO FROM HERE? THEORY, METHOD, AND PRACTICE FOR RESEARCH, ANALYSIS, AND ADVOCACY

I initially laid out classical, ecological modernization, and critical political ecology approaches to assessing the environmental impacts of socialist transition. Given the prior dominance of the classical neoliberal approach, and the current hegemony of new ecological modernization, the alternative approach of critical political ecology is primarily offered as a counterpoint to these dominant narratives and guiding ideologies for assessing the process of transition, as well as determining the appropriate path to follow in the future. Despite its social democratic roots in the Scandinavian context, the new ecological modernization theory provides a new seductive twist on the dominant neoliberal narrative by setting up the problem as not enough rather than too much unfettered industrialization, thus simultaneously justifying the status quo of continued unsustainable growth and adherence to modernization strategies, with the equally questionable assertion that the environmental problems inherent in such strategies will force societies to confront the disasters at hand and evolve, with some limited government intervention and support, principally technological means to overcome these problems and repair the damage already done. Shared belief in free markets, rapid industrialization, and the adequacy of technological advance to resolve all environmental challenges, provides the non-conflictive context that explains the rapid adoption of the new ecological modernization by the hegemonic development institutions principally guiding the socialist transitions in question. Neoliberalism's declining star has necessitated its subtle replacement by the new ecological modernization in a form that does not challenge the fundamental assumptions and practices of the past three decades. Its successful integration at the World Bank, among other key institutions, is a clear indication not only of institutional flexibility that has allowed for the adoption (if not cooptation) of each environmental approach from sustainable development to livelihood analysis, but also of its inherent lack of critical analytical rigor.

This argument suggests that the transition from socialism to capitalism requires analysis at multiple scales, and I have argued in this chapter that a nuanced structural analysis of changes in daily practices provides an alternative approach to classical economics and ecological modernization theories for understanding this historical transformation. Important avenues for this research are assessments of the institutional means by which these transitions are carried out, in particular the powerful development organizations and international financial institutions (World Bank, IMF, UN, etc.), and the policy process through which their approaches are translated into national and then local actions. A key and under-researched aspect of socialist transition, then, is the role of international financial institutions (IFIs) and bilateral aid organizations. Surveying international development aid contributes to a clearer understanding of the various roles of aid institutions in the 'transitional' process, particularly in reference to environmental concerns. Both environment and development aid policy provide windows into the socialist transition process – its actualization and its limitations. Development aid provides massive transfers of capital and technology that underpin and enable the transition process (Muldavin, 1993, 1995, 2000b). Thus a major focus for future research should be an empirical assessment of these institutions' patterns and practices, and ethnographic investigations of the local impacts of implemented polices.

While much of the literature on socialist transition has emphasized a more urban-biased analysis around industrial transformation, the oft-overlooked aspect in transitional economies of agrarian change and the environment provides unique insights. Approaching agrarian change through an analysis of daily practices requires detailed and ethnographic accounts of people's changing perceptions and decision-making processes. To assess such subtle alterations in decision-making, recent research focuses on the issue of land use, particularly though not exclusively in agriculture. In rural China, for example, the key issue is how to identify the ways in which the constellation of forces affecting land use often move rural producers toward less sustainable practices. The long-term environmental impacts of these changing land use practices – from declining crop rotation to overuse of fertilizers to lack of maintenance of productive infrastructure – can be termed 'mining communal capital'. This rather unorthodox position asserts that decollectivization was paradoxically achieving increased production through the creation of conditions that would undermine long-term rural productivity (Muldavin 1997).

Yet this compelling dialectic can only partially explain a fundamental weakness to this transition process – the decline in the legitimacy of reform as that reform proceeds. The essentially *political* character of transition requires identifying aspects of agrarian change outside of land use in the social sphere. To this end, much new work identifies the basis of resistance to reform and transition (Unger and Barme, 1991; Chan et al., 1992; Solinger, 1993; Chan, 2001; Bernstein et al., 2003). This resistance is highly differentiated in its expression – from indirect acts of daily resistance to open and organized rebellion and struggle. Such power struggles in China have at their root the changing vulnerabilities of those actually living this particular transition. Declines in social welfare and access to entitlements, combined with rapid socioeconomic and regional differentiation, have brought rising rural tensions and declining state legitimacy. Added to the long-term undermining of land productivity, as earlier posited, and the state finds itself at a difficult moment in the transition process.

As opposed to the rosier views of reform that dominated analyses in the 1980s and 1990s, I attempt here to explain the heterogeneity of China's transition experience as a necessary and structural component of its history and geography, contextualizing the environmental and social aspects (Muldavin, 1992, 1997). This overall analysis questions the prevailing market triumphalism of the 1990s, arguing instead for a more tempered view of China's transitional process that incorporates many of the long-term challenges that some analysts downplay or ignore. To better understand such challenges, regional work in new areas and at different scales is needed, creating the basis for a new line of argument concerning 'vulnerability analysis' – a fertile theoretical arena within political ecology and development studies (Muldavin, 2000a; Wisner et al., 2004) that may find a ready audience among political geographers.

In terms of environmental policy in countries undergoing socialist transition, there are a number of crucial, under-researched, and ill-understood aspects in the environmental policy process that can be fruitful focal points for future policy-relevant research on transitional economies. In addition to an analysis of environmental policy interfaces (between international actors, such as multilateral and bilateral agencies and national governments, and between the latter and local institutions and, more broadly, civil society) as discussed above, assessing policy impacts through an analysis that begins with policy texts and moves to impacts on the ground is key. And further, analyzing the gaps between policy rhetoric and reality, and bottom-up influences on policy and implementation, is needed to broaden the scope of research from conventional concerns of formal processes of the state, to civil society, social movements, and local politics. And finally, assessing the geopolitical contexts of environmental policy will allow us to address issues of the militarization of the environment and the political control of people and resources in the name of conservation and environmental protection (see Blaikie and Muldavin, 2004a, 2004b, 2006).

CONCLUSION: THE NEW POLITICS OF CHINA'S TRANSITION

In China the authoritarian state claims the mantle of workers' rights promotion, and asserts that state unions, laws, and regulations will protect workers. The reality more closely approaches hypercapitalist labor relations in new factories and special economic zones, as well as in the far-flung suburban subcontractors and rural township and village enterprises in the hinterlands. The resulting worker uprisings, resistance, and calls for independent labor organizing and representation are repeatedly contained or crushed by state action, local and national. Paradoxically, the party state's rhetoric as a communist organ, while often appearing capitalist in action, undermines independent political action to some degree. On the other hand, state delegitimation is leading to ever-larger numbers of workers and peasants participating in uprisings and actions (direct challenges and resistance), as well as refusals to participate in processes counter to their own interests, and even acts of sabotage (indirect challenges and everyday forms of resistance). Thus the state finds itself now in a complex 'firefighting' position as hundreds of demonstrations against perceived injustices – social, economic, political, and environmental – erupt daily as small brush fires around the country challenging the political status quo. This is stretching to the breaking point the state's ability to efficiently organize and effectively respond with existing institutional structures and assets.

Here then lies a key point for active research, analysis, and advocacy on the part of political geographers. The clear ties of the social and environmental injustice and 'violence' to political decisions and subsequent policies provides opportunities to re-theorize the changing nature of post-socialist state/society relations around pressing current issues. Furthermore, the specific yet complex spatial patterns of challenges to the state create a mosaic of political geographies that at once reflect localized historical realities, while pointing to emergent patterns that reveal tenuous lines of cooperation across regions and at multiple scales. The potential political crisis this poses for the state is reflected in its heightened rhetorical campaign denouncing anything but passive, generally legalistic complaints, and attempting to isolate emergent leadership and destroy it before these brush fires join into a true conflagration. China's near 200 million migrant peasant workers, second-class undocumented 'citizens' in the urban growth cores, are both central to China's dynamism while challenging its very continuation through increasingly vocal demands for greater rights – including entitlements, occupational protection, and escape from the class-based environmental inequality of China's transition process. Thus the state spasmodically cracks down on this politically underrepresented group, using them as a scapegoat for larger social ills in a haunting parallel to the position of undocumented migrants in the US and Europe.

As this description augurs, the socioenvironmental impacts of socialist transition in China provide potential insights into the issues raised at the beginning of this chapter; namely, the rapidly changing political landscape of state/society/nature relations as China decollectivizes, privatizes, deregulates, and commodifies nature, natural resources, and land-based assets of all kinds. The speed, scale, and scope of this process in China overwhelms most analyses, and challenges us to find new analytical windows to shed light on this shifting terrain, and provide signposts of possible future trends. For critical political ecologists and political geographers, such a challenge is tied to the clear connections between environmental destruction and social decline, both in countries undergoing transition and in the rest of the world, and the pressing need for political alternatives supported by a rigorous research agenda. Critical political ecology – multi-scaled, focused on the nature/society relationship, and inclusive of a wide range of theoretical frameworks and methodological tools – offers potential insights built on fundamental concerns for environmental and social justice. Political geography can provide much-needed rigor to political ecologists' conceptualization of state/society relations, as well as the politics of non-state actors, and their spatial and scalar expressions, informing and being informed by critical political ecology's primary modalities of inquiry.

Specifically, at a global and regional scale, political geographers' insights on geopolitics can help strengthen this under-theorized aspect of political ecology (Nevins, 2005). At a national and subnational scale, political geographers' analysis of citizenship, political rights, political participation, and the role of civil society in the shifting state/societal relationship can foster a clearer institutional understanding of these dynamic actors and the power relations within which they are embedded (Varsanyi, 2006). And finally, at the local, even household and individual level, political geographers can provide insights on the particular ways politics are infused with the questions of gender, race, and class, and how these and other cross-cutting thematics influence political moments and movements and the evolving relations to 'nature' and the environment.

In summary, in former state socialist societies now undergoing various forms of 'transition', the challenge to political ecologists, as well as political geographers, is how to further develop this evolving multi-scaled and interdisciplinary approach to understanding the transition from socialism to capitalism, and in particular its environmental aspects. The hybrid and often derivative forms of political economy that have emerged, further challenge the dominant and relatively simple analyses prevalent in mainstream economics and political science. Contextualized within the mosaic of forces generally described as globalization, political ecologists and political geographers can provide particular insights into how to conceptualize political processes increasingly beyond the nation-state, and yet still rigorously tied to the spatial realities and reconfigurations of political, economic, and environmental boundaries.

ACKNOWLEDGEMENTS

I would like to thank the editors of the *Handbook*, and in particular Jennifer Robinson, for providing me the opportunity and encouragement to further refine my ideas on the subject of the environmental impacts of socialist transition. I would like to thank Piers Blaikie for his insightful critical reading of an earlier draft of the manuscript. I would also like to thank my research assistant, Julie Klinger, for her excellent work on citations, style, and her attention to crucial details that helped clarify my prose and intent. And I would like to thank Monica Varsanyi for engaging me in overarching discussions about the piece. Of course, all omissions, errors, and other shortcomings of the final manuscript are solely the responsibility of the author.

NOTES

1 For classic texts to inform further inquiry along the lines of this proposed parallel, see, among others, Goodman and Redclift (1981) on the transition of peasants to workers; the Brenner Debate (1985) for the historical transition debate; and Little (1989) for application of the classical peasant debates to the Chinese context (Scott [1976] and Popkin's [1979] moral versus rational peasant debate, the Lenin and Chayanov debate, and so forth).

2 In Cuba, a very different process occurred as the withdrawal of subsidized oil and oil products by the former USSR in 1989 forced Cuba toward an ecologically improved agriculture based on organic principles, expansive urban gardens, and a restructuring of the rural economy (Deere, 1998).

3 The Vietnamese state's courting of 'foreign direct investment' is part of a competitive process with China that may make Vietnam the new low-cost toxic industrial platform of choice for 'transnational corporations' with operations in Asia (Perlez, 2006).

4 Pepper's (1984:39) original division of technocentric approaches to the environment between Cornucopians and Environmental Managers, I would argue, has now converged in a new form of Cornucopian belief integrated into the environmental management strategy approach – what Blaikie calls the New Ecological Modernization as opposed to the old variety associated with colonialism (Blaikie, 1999: 137).

5 The nation-state-based vision of the classical and ecological modernization approaches misses fundamental dynamics both above and below this scale of assessment, for example the globalized production systems that do not rely on any single national agenda or context so as to spread risk and increase flexibility and power in determining the form of production whatever its overall environmental impact. The ecological modernization approach, in particular, misses the fact that China's growth success is partially built on the state's willingness to accept a role for China as a subcontractor in the global economy for some of the world's dirtiest industries, with subsequent environmental destruction, and that in any turn to cleaner technologies, some of this competitive advantage is lost.

6 In the Chinese case the 'third wave' proponents are experimenting with formalizing a system of shaming corporations for poor environmental performance as a means to improve environmental outcomes. This mimics to some degree the application of ecological modernization theory in Japan (Barrett, 2005).

7 For much more detailed accounts of the summaries provided below, including results based on long-term village-level research, see Muldavin (1992, 1997, 2000a).

8 For a detailed discussion of the evolving intellectual currents within the party and intelligentsia, see Yan Sun's early book, *The Chinese Reassessment of Socialism: 1976–1992* (1995), and for an assessment of China's New Left, see Chaohua Wang's diverse collection entitled *One China, Many Paths* (2005), and also Wang Hui's cogent analysis in a set of essays entitled *China's New Order* (2003).

REFERENCES

Barrett, B. (ed.) (2005) *Ecological Modernization and Japan.* London: Routledge.

Bernstein, T.P., Lü, X. and Kirby, W. (2003) *Taxation Without Representation in Rural China.* Cambridge: Cambridge University Press.

Blaikie, P. (1999) 'A review of political ecology', *Zeitschrift fur Wirtschaftsgeographie,* 43(3–4): 131–47.

Blaikie, P. and Muldavin, J. (2004a) 'Upstream downstream, China, India: The politics of environment in the Himalayan region', *Annals of the Association of American Geographers,* 94(3): 520–48.

Blaikie, P. and Muldavin, J. (2004b) 'Policy as warrant: environment and development in the Himalayan region', East-West Center Working Papers, Environmental Change, Vulnerability, and Governance Series, 59. East-West Center, University of Hawaii.

Blaikie, P. and Muldavin, J. (2006) 'Creating a usefull dossier for policy work', co-authored with Piers Blaike, *Policy working paper,* International Center for Intergrated Mountain Research and Development (ICIMOD), Katmandu, Nepal.

Blaikie, P. (1985) The Political Economy of Soil Erosion in Developing Countries. Longman Development Series, No. 1, Longman, London, pp. 188. Reprinted by Pearson Education in 2000.

Blecher, M. (1997) *China Against the Tides.* London: Pinter.

Brenner, R. Aston, T.H. and Philpin, C.H.E. (eds) (1985) *The Brenner Debate: Agrarian Class Structure and Economic Development in Preindustrial Europe.* Past and Present Publications Series. Cambridge: Cambridge University Press.

Bryant, R.L. and Bailey, S. (1997) *Third World Political Ecology.* London: Routledge.

Buttel, F.H. (2000) 'Ecological modernization as social theory,' *Geoforum,* 31: 57–65.

Cannon, T. and Jenkins, A. (eds) (1990) *The Geography of Contemporary China: The Impact of Deng Xiaoping's Decade.* London: Routledge.

Carr, C.J. (1977) *Pastoralism in Crisis: The Dasenetch and Their Ethiopian Lands.* Chicago: University of Chicago Press.

Chan, A. (2001) *China's Workers under Assault: Exploitation and Abuse in a Globalizing Ecconomy.* Armonk, NY: M.E. Sharpe.

Chan, A., Madsen, R. and Unger, J. (1992) *Chen Village under Mao and Deng.* Berkeley: University of California Press.

Deere, C. (1998) 'Cuba: the reluctant reformer', in I. Szelenyi (ed.), *Privatizing the Land: Rural Political Economy in Post-Communist Societies.* London: Routledge.

Escobar, A. (1995) *Encountering Development.* Princeton: Princeton University Press.

Forsyth, T. (2003) *Critical Political Ecology*. London: Routledge.

French, H.F. (1999) *Green Revolutions: Environmental Reconstruction in Eastern Europe and the Soviet Union*. Paper 99. Washington, DC: Worldwatch Institute.

Gibson-Graham, J.K. (2006) *A Postcapitalist Politics*. Minneapolis: University of Minnesota Press.

Gilley, B. (2001) *Model Rebels: The Rise and Fall of China's Richest Villages*. Berkeley: University of California Press.

Goodman, D. and Redclift, M. (1981) *From Peasant to Proletarian: Capitalist Development and Agrarian Transitions*. Oxford: Basil Blackwell.

Harvey, D. (1974) 'Population, resources, and the ideology of science', *Economic Geography*, 50(3): 256–77.

Harvey, D. (2005) *A Brief History of Neoliberalism*. Oxford: Oxford University Press.

Ho, P. (2001) 'Greening without conflict? Environmentalism, Green NGOs and civil society in China', *Development and Change*, 32: 893–921.

Ho, P. (ed.) (2006) Special Issue of *Development and Change*, 37(1): 1–271.

Jacka, T. (1997) *Women's Work in Rural China: Change and Continuity in an Era of Reform*. Cambridge: Cambridge University Press.

Kautsky, K. (1988 [1899]) *Agrarfrage: The Agrarian Question*, trans. Peter Burgess. London and Winchester, MA: Zwan Publications.

Kornai, J. (1992). *The Socialist System: The Political Economy of Communism*. Princeton: Princeton University Press.

Lin, J.Y. (1990) 'Institutional reforms in Chinese agriculture: retrospect and prospect', in J.A. Dorn and Wang Xi (eds), *Economic Reform in China: Problems and Prospects*. Chicago: University of Chicago Press.

Little, D. (1989) *Understanding Peasant China*. New Haven: Yale University Press.

Mol, A.P.J. (2006) 'Environment and modernity in transitional China: frontiers of ecological modernization', *Development and Change*, 37(1): 29–56.

Mol, A.P.J. and Spaargaren, G. (2000) 'Ecological modernization theory in debate: a review', in A.J.P. Mol and D.A. Sonnenfeld (eds), *Ecological Modernization Around the World: Perspectives and Critical Debates*. London: Frank Cass, pp. 17–49.

Muldavin, J. (1983) *15-Year Plan for the Agricultural and Agro-Industrial Development of Zhaozhou County*. Beijing: Ministry of Agriculture.

Muldavin, J. (1986) 'Mining the Chinese earth', Master's thesis, Department of Geography, University of California at Berkeley.

Muldavin, J. (1992) 'China's decade of rural reforms: the impact of agrarian change on sustainable development', PhD dissertation, Department of Geography, University of California at Berkeley.

Muldavin, J. (1993) 'A survey of international aid to China', paper presented at the Annual Meetings of the Association of American Geographers, 6–10 April.

Muldavin, J. (1995) Collected notes from meetings with development aid officials, Beijing, China.

Muldavin, J. (1996) 'The political ecology of agrarian reform in China: the case of Heilongjiang Province', in R. Peet and M. Watts (eds), *Liberation Ecologies: Environment, Development, social Movements*. London: Routledge, pp. 227–59.

Muldavin, J. (1997) 'Policy reform and agrarian dynamics in Heilongjiang Province', *Annals of the Association of American Geographers*, 87(4): 579–613.

Muldavin, J. (1998a) 'Agrarian change in contemporary rural China', in I. Szelenyi (ed.), *Privatizing the Land: Rural Political Economy in Post-Communist Societies*. London: Routledge.

Muldavin, J. (1998b) 'The limits of market triumphalism in rural China', *Geoforum*, 28: 289–312.

Muldavin, J. (1999) 'The world should help to avert turmoil in China', *International Herald Tribune*, 1 June.

Muldavin, J. (2000a) 'The paradoxes of environmental policy and resource management in reform era China', *Economic Geography*, 76(3): 244–71.

Muldavin, J. (2000b) 'The geography of Japanese development aid to China, 1978–1998', *Environment and Planning A*, 32: 925–46.

Muldavin, J. (2005a) Field notes, Yunnan Province.

Muldavin, J. (2005b) 'Beyond the Harbin chemical spill', *International Herald Tribune*, 30 Nov.

Muldavin, J. (2006) 'In rural China, a time bomb is ticking', *International Herald Tribune*, 1 Jan.

Neumann, R.P. (2005) *Making Political Ecology*. London: Hodder Arnold.

Nevins, J. (2005) *A Not-So-Distant Horror: Mass Violence in East Timor*. Ithaca, NY: Cornell University Press.

Oi, J. (1999) *Rural China Takes Off*. Berkeley: University of California Press.

Oi, J. and Walder, A. (eds) (1999) *Property Rights and Economic Reform in China*. Stanford: Stanford University Press.

Peet, R. and Watts, M. (1996) *Liberation Ecologies*. London: Routledge.

Pepper, D. (1984) *The Roots of Modern Environmentalism*. London & New York: Routledge.

Perlez, J. (2006) 'US competes with China for Vietnam's allegiance', *New York Times*, Late Edition-Final, A:3, 19 June.

Pickles, J. and Pavlinek, P. (2000) *Environmental Transitions: Transformation and Ecological Defence in Central and Eastern Europe*. London: Routledge.

Pocha, J.S. (2005) 'Letters from Beijing: China's New Left', *The Nation*, 9 May.

Popkin, S. (1979) *The Rational Peasant: The Political Economy of Rural Society in Vietnam*. Berkeley: University of California Press.

Rocheleau, D., Thomas-Slayter, B. and Wangari, E. (eds) (1996) *Feminist Political Ecology: Global Issues and Local Experiences*. New York: Routledge.

Redclift and Woodgate, G. (1997) The International Handbook of Environmental Sociology, London: Edward Elgar, Publishing ltd.

Sachs, J. and Woo, M. (1993) 'Structural factors in the economic reforms of China, Eastern Europe and the Former Soviet Union', paper presented to the Economic Policy Panel, Brussels, Belgium, 22–23 October.

Scoones, I. (1999) 'New ecology and the social sciences: what prospects for fruitful engagement?', *Annual Review of Anthropology*, 23: 479–507.

Scott , J. (1976) *The moral economy of the peasant: rebellion and subsisitence in Southeast Asia.* London: Yale University Press.

Scott, J.C. (1985) *Weapons of the Weak: Everyday Forms of Peasant Resistance.* New Haven: Yale University Press.

Selden, M. (1998) 'After collectivization: continuity and change in rural China', in I. Szelenyi (ed.), *Privatizing the Land: Rural Political Economy in Post-Communist Societies.* London: Routledge.

Shapiro, J. (2001) *Mao's War Against Nature: Politics and Environment in Contemporary China.* Cambridge: Cambridge University Press.

Solinger, D. (1993) *China's Transition from Socialism: Statist Legacies and Market Reforms.* Armonk, NY: M.E. Sharpe.

Spaargaren, G. and Mol, A. (1992) 'Sociology, environment, and modernity: ecological modernization as a theory of social change', *Society and Natural Resources,* 5: 323–44.

Staddon, C. and Turncock, D. (2001) 'Environmental geographies of post-socialist transition', in D. Turncock (ed.), *East Central Europe and the Former Soviet Union: Environment and Society.* London: Arnold.

State Administration of Worker Safety (SAWS) (2006) 'Coal mining and accident deaths in China', *China Labour Bulletin,* No. 60. Hong Kong.

Stott, P. and Sullivan, S. (eds) (2000). *Political Ecology: Science, Myth and Power.* London: Arnold.

Sun, Yan (1995) *The Chinese Reassessment of Socialism: 1976–1992.* Princeton: Princeton University Press.

Szelenyi, I. (ed.) (1998) *Privatizing the Land: Rural Political Economy in Post-Communist Societies.* London: Routledge.

Turncock, D. (ed.) (2001) *East Central Europe and the Former Soviet Union: Environment and Society.* London: Arnold.

Unger, J. and Barme, G. (eds) (1991) *The Pro-Democracy Protests in China: Reports from the Provinces.* New York: East Gate Books.

Varsanyi, M. (2006) 'Interrogating urban citizenship vis-à-vis undocumented migration', *Citizenship Studies,* 10(2): 229–49.

Walder, A. (1994) 'Corporate organization and local property rights in China', in V. Milor (ed.), *Changing Political Economies.* Boulder, CO: Westview Press.

Wang, Chaohua (2005) *One China, Many Paths.* London: Verso Books.

Wang, Hui (2003) *China's New Order.* Cambridge, MA: Harvard University Press.

Wang, H., Bi, J., Wheeler, D., Wang, J., Cao, D., Lu, G. and Wang, Y. (2002) *Environmental Performance Rating and Disclosure: China's Green-Watch Program.* Policy Research Working Paper 2889. Washington, DC: The World Bank.

Watts (1983) *Silent violence: food, famine, & peasantry in northern Nigeria/Michael Watts.* Berkeley: University of California Press.

Watts, M. (1995) *Agrarian Thermidor: Rural Dynamics and the Agrarian Question in Vinh Phu Province, Vietnam.* Working Paper 2. Institute of International Studies, University of California at Berkeley.

Watts, M. (1998) 'Agrarian Thermidor: state, decollectivization, and the peasant question in Vietnam', in I. Szelenyi (ed.), *Privatizing the Land: Rural Political Economy in Post-Communist Societies.* London: Routledge.

Wisner, B., Davis, I., Cannon, T. and Blaikie, P. (2004) *At Risk.* London: Routledge (1st edn. 1994).

World Bank (2002) *World Report: Development and the Environment.* Oxford: Oxford University Press.

Yan Sun's (1995) *The chinese Reassessment of Socialism: 1976–1992.* Princeton, NJ: Princeton Unversity Press.

Identities and Interests in Political Organizations

Introduction

Jennifer Robinson

Bringing together a concern with identities and interests and their implications for politics signposts an important set of intersections for political geography. One is theoretical – how is the relationship between identities and interests to be understood analytically? Another is disciplinary – how is the domain of political geography to be defined after the substantial turn to a critical cultural geography? Cutting across both the theoretical debates and the sub-disciplinary interfaces, though, is a robust field of geographical inquiry. This is an area in which geographers, and geographical thinking, can quite properly claim to have had a strong and wide influence: writers from many different backgrounds have drawn on geographical thinking in their explorations of the politics of identity (e.g. Mohanty, 1987; Gilroy, 1993; Friedman, 1998; Ong, 1999). Although this is a very recent theme in the context of the long history of political geography, over the last two decades the topic has come to occupy an important place within the field (e.g. Jackson, 1987; Keith and Pile, 1993; Painter, 1995; Fincher and Jacobs, 1998; McDowell, 1999; Sharp, 2000).

However, the theoretical analysis of the politics of identity has been contentious. For many Marxist theorists and political activists, attending to identity has been seen as obscuring the ongoing challenges of inequality, exploitation and what is often considered to be a more transformative politics of class (Fraser, 2000). In geography the classic contribution here came from David Harvey (1993), who was concerned to assert the shared class politics of co-workers who might otherwise have diverse identifications across race, nationality, ethnicity, gender and so on. The left within geography is still struggling to come to terms with the theoretical and political differences signposted by a politics of identity, along with a range of related theoretical and political issues (see Amin and Thrift, 2005; Smith, 2005).

The authors in this section take different perspectives on these issues, but all are mindful of the serious analytical challenges for political geographers who wish to address the politics of oppression, inequality, exclusion and marginalization. Iris Marion Young's proposal to consider the several '"faces" of oppression' – for her exploitation, marginalization, powerlessness, cultural imperialism, and violence – might be helpful here (1997: 151). Reminding us that economic exploitation and hence class mobilization is itself always cultural, she stresses the importance of identifying the opportunities for political engagement and coalition building in the face of the astonishing ascendancy of contemporary forms of exploitation, violence and dominance on the part of capitalism, states and non-state actors. In this state we hope these debates (aired most directly in Castree et al. in this section) will yield creative political outcomes rather than internecine academic conflict.

In developing the analysis of identities, political geography has seen an immensely fertile engagement with other sub-fields of the discipline, especially social and cultural geography. As we noted in the introduction to this *Handbook*, through the 1980s and 1990s social and cultural geography saw a substantial resurgence, as issues such as gender, sexuality, race, bodies and visual culture came to the fore in theoretical and political agendas in Anglophone geography (e.g. Duncan, 1996; Jacobs, 1996; Nash, 1996; Brown, 1997). The vitality of this crossover field of research, between cultural and political concerns – for example, in topics such as the politics of race, post-colonial critiques, feminist politics, the complex relations between place and identity politics – speaks to our suggestion in this *Handbook* that any attempt to draw firm boundaries around what should count as political geography might actively undermine its own intellectual ambitions (Cox and Low, 2003). Rather, proliferating these points of intersection between social, cultural and political geography, and opening up the traditional themes of the field to these new kinds of politics, is, we suggest, the appropriate way forward.

Consequently, though, the politics of identity and formation of political interests represents a very large, dynamic and rich field of inquiry. In order to make some specific inroads into such a large body of literature in the space available here, the chapters in this section focus our attention on the role that identities play in political organizations, and in that way tie the question of political identity more closely to that of interests. This is not to suggest that political geography's contribution ends with formal political organizations. As Arun Saldanha's chapter on 'The political geography of many bodies' clearly indicates, important dimensions of politics and power are constituted and played out beyond any identifiable social organization in the multiplicity of interactions that shape social life more generally. But for our purposes, focusing on questions of political organization brings some longstanding concerns of political geography – states, social movements, local/global interactions – into conversation with wider trends in geographical research that foreground questions of identity.

Together the chapters in this section demonstrate that geographers are well placed to make a distinctive contribution to understanding some prominent political trends, including: the apparent demise of the nation-state as the association between nations and states fragments in favour of political pluralism; the complexities of local organizing in response to global issues and the challenges of global networking involving diverse local movements; the continuing significance of place in motivating political organization; and a growing awareness of the politics of the body as a site for mobilization. We should like to have seen a contribution that addressed the intersection between class politics and the politics of identity more closely (e.g. Harvey, 1993; Silvey, 2003; Wright, 2003), although Castree, Featherstone and Herod explore class politics in relation to place-based identities. Also, the identity politics associated with migration or diasporic networks deserved a stronger airing here (see Silvey et al., this volume).

We should also note that the impact of these developments on research in political geography has been quite uneven. With notable exceptions, such as the work of Anna Secor (2001), studies in voting geography remain attached to quite traditional concepts of social interest. Likewise, as Wood notes in his chapter, the very large area of study that falls under the heading of the politics of local and regional development – a major growth area in political geography – has very little to say on these questions. There are, therefore, a whole series of potentially useful applications still to be made, though this should be predicated on a careful understanding of not only what might be gained, but also what might be lost; not either identities or interests, therefore, but an approach that interrogates the relations between them, and not least, how interests can

often fruitfully be viewed as identities; class being an interesting case in point.

For geographers, what is at stake in research on political identities and the articulation of political interests is the spatiality of the formation and contestation of political identities. In early work in this field, geographers focused on the ways in which cultural and political identities are forged in relation to particular contexts, and thus vary considerably over time and from place to place (Massey, 1984; Agnew, 1987; Jackson, 1991). A related strand engaged with landscape studies and historical geography to track the formation of locally specific, if highly contested, political identities (e.g. Johnson, 2002; Duncan, 2004). More recent research is eager to trace the diasporic connections that shape political identities stretching across localities, and transnationally (Yeoh and Willis, 1999; Mohan, 2006), or to explore the spatiality of identity formation in relation to the more intimate geographies of the body and performativity (McDowell and Court, 1994; Longhurst, 2001).

Another strand, particularly important in research on place-based identities, as well as on gender and on democracy, has been the spatiality of the theoretical analysis of political identities. Are identities to be thought of as strongly bounded, perhaps as places had once been characterized, or are they open, contestable, always subject to change and thus a source of radical political possibility (see Massey, 1995; Natter and Jones, 1997)? Politically, these approaches address the possibilities for moving beyond framing place-based identities in opposition to external processes, or defensively against outsiders. Many of the central concerns of the politics of identity and the formation of political interests can be usefully articulated through the spatial analysis offered by political geographers.

So while the substantive focus of the chapters in this section ranges from the very local (neighbourhoods and towns) to movements that aim to span the globe in their interests, and from the formal realm of the politics of the state to more informal mobilizations around issues including sexuality and workers' rights, all of them address the theme of the spatiality of political identities. Theorizing political identities requires a subtle and complex account of space, not only to appreciate how it is that identities are forged in relation to particular places or contexts, but also to be able to contribute to a growing sensitivity to the dynamic and often unstable nature of identities themselves, including ways in which they are fractured and uncertain, or perhaps rigid and unmoving but certainly contestable. A spatial account of political identities is essential to appreciate how subjects – political or otherwise – are generated in complex relations to others, to local places, and, as the chapters here make clear, to distant realities.

Perhaps the most obvious geographies at work in shaping identities are those where people's immediate environments and local contexts configure ways in which people understand themselves and offer opportunities for formulating political projects and organizations (see e.g., Escobar, 2001; Castree, 2004). The chapters here offer numerous examples of local or neighbourhood-based activities, such as the role of different venues and locally generated organizations in the gay, lesbian, bisexual, trans-sexual and 'queer' social movements in Seattle reviewed by Michael Brown. Localized political issues certainly motivate organizations to formulate opposition around place-based identities, as Routledge and Castree et al. explore here. Places also enter into the formulation of identities in more instrumental ways as part of projects of governance, as Margo Huxley explores in Chapter 7. Assessing that subjects might be shaped through the environments in which they live, various state projects – here she explores the case of planning – make interventions into designing environments so as to 'make up subjects' in ways consistent with governmental ambitions. There is much evidence, then, that locality and context matter in a very immediate way in shaping political identities. However, places also enter into the formation of political identities at a distance in a variety of ways.

Imaginative associations with place have been identified as crucial to sustaining some forms of political identification, perhaps most consistently in creating the possibility of the nation. As Benedict Anderson (1983) most famously points out, the nation is an imagined community, not one that can ever be part of an individual's directly lived experience in its entirety, but that generally individuals imagine they belong to through the means of various media, initially print, but also radio and television. Significantly, both radio and television have been strongly associated with strategic nation- and state-building projects. Chatterjee (1993) importantly countered, though, that in different contexts both the means and the form of the nation might be quite different, as he cites the importance of spiritual experience in supporting the development of a more personal form of Indian nationalism. Geographers have pointed out that these imaginative projects of nation-building have also been facilitated by interventions in the physical landscape (Knight, 1982). Piers Gruffudd (1995) demonstrates in the case of Wales how the development of a Welsh national identity was encouraged by debates over the construction of a cross-country road network, supporting an imaginative sense of an integrated and distinctive region.

Penrose and Mole point out in their chapter here, however, that the primacy of the nation in the politics of states is declining, and the political rationale of states is increasingly associated with pluralism and a more cosmopolitan political identity rather than with nationalism. In this, changes in the media through which identities are sustained at a distance have been important. Beyond the national media – newspapers, radio and television – there are many new opportunities to establish, share and broadcast identities that are therefore dispersed and transnational, facilitated by the Internet and other mobile digital technologies (Silverstone, 2005). These media are enabling new kinds of political identities and organizations to emerge, stretching across localities in various ways. As Paul Routledge discusses in his chapter here, transnational forms of grassroots organization have been emerging to counter powerful global actors and neo-liberalizing initiatives whose impact is felt and opposed in many different contexts. But building transnational political solidarity, even with the technology available to enable global networks, poses significant political challenges. He suggests that one of these is negotiating political identities across multiple locations, such as when quite different approaches to gender politics in localities drawn to work together on common political concerns threaten to destabilize the association.

The difficulties of articulating collective visions across diverse localities raises some profound analytical questions about how to theorize the relationships between the local and 'global' or spatially extensive social relations. While this emerges directly out of attempts to understand political identities in an era of globalization, it is also a much more general concern for geographers: how to think about the spatiality of political identities and political projects across localities. Noel Castree, David Featherstone and Andrew Herod offer here a review of some different approaches taken by geographers to this core issue in contemporary politics. As they make clear, key to negotiating the aporias between local concerns and translocal organizing, or between particular interests and broader ambitions for progressive change, is a relational view of space. Places are not separate and bounded, although they are clearly distinctive and often strongly identifiable. But usually places are constituted in relation to places and people elsewhere. The links and networks that stretch out from and into different contexts are as important in shaping the political identities that emerge there as are particularly local dynamics. Reviewing these analyses through the work of Kevin Cox, David Harvey and Doreen Massey, Castree et al. offer a very useful statement about how geographical insights can be important in analysing some of the central political issues of the moment. As Alan Lester (this volume) reminds us the circulation of political practices has long shaped the experience of and responses to imperial ventures in different contexts; the translocal geographies of political identity have a long reach historically (Lambert, 2005), and Lester indicates that there is much to learn from accounts of earlier periods in reflecting that on the forms of international power and popular protest in the present (see also Featherstone, 2005).

These geographies of political organization start to bring into view some aspects of the spatial dynamics of the politics of identities that arguably present agendas for future research in this area. For example, David Harvey's search for militant particularisms, political organizations that both address local issues and progress broader radical ambitions, speaks not only to aspects of translocal organizing, but also to a world of imagined political identification. Local struggles could perhaps be thought of as part of a wider movement for change, even if there are no particular organizational links trying them together. This draws us towards a whole arena of international politics, where broad public support for political interventions – humanitarian, for example, or developmental – can exert pressure on governments and international bodies to act, or refrain from taking action (Dodds, 1999; Barnett et al., 2005). Drawn to an often emotional identification with distant strangers (Corbridge, 1993) through the media or publicity efforts of campaigning NGOs, the ties across localities that enable political change can be affective as much as organizational.

This example signals the need for much more research on the dynamics, or spatiality, of subjectivity from a political perspective. If subjects should be understood, as most research in this field indicates, as fractured, uncertain achievements, and profoundly porous in relation to the external environment, then the consequences for political identities are indeed important to consider. An earlier round of political theorizing, influenced by Althusser and Lacan through the writings of Ernesto Laclau (1994), suggested that the dynamic relation between political identities and their constitutive outside supported the possibility for their disruption and renegotiation, hopefully in the direction of radical transformation (Massey, 1995; Natter and Jones, 1997). The rise of post-humanist approaches in geography has brought to the fore the challenge of a stronger appreciation of the decentred nature of the subject – and indeed, calls into question where and what subjects are, and therefore where one might locate a politics associated with identity formation. Braun (this volume) and Saldanha (this volume) both point to a politics that engages more fully with the non-human constituents of political identities, thereby stretching our understandings of the dynamics of political issues, the nature of political interests and the opportunities for intervention.

In addition to this new strand of inquiry, a more longstanding psychoanalytic approach to subjectivity still holds out an invitation to political geography to attend to a wider range of political contestations in the realms of cultural phenomena, collective imaginaries, and everyday experiences and encounters (see, e.g., Pile and Nast, 1997; Callard, 1998; Wilton, 1998). Not only do these initiatives signpost an important set of substantive topics for political geographers to engage with, they also contribute to our understanding of the spatiality of political identities, as forged in personal histories as well as wider cultural contexts and physical settings, and as deeply riven by anxieties, contradictions and antagonisms. Theories of identity draw heavily on a geographically inflected language in framing their analyses – identities are, for example, variously bounded or open, fluid or fixed, divided and networked; indeed, since Freud's highly spatialized account of the psyche and its operations (see Pile, 1996), geographical metaphors have been deployed to map this complex and barely accessible zone of human experience. If we understand political identities to play a crucial part in much of political life, then there is certainly scope for a stronger investigation of the dynamic spaces of subjectivity that frame political processes as diverse as geopolitics and development, the politics of immigration and racialization, gendered power relations and urban regeneration. This section of the *Handbook* charts a small portion of this dynamic field of inquiry, and we hope that the inevitable gaps in its coverage of such a vast field hold the promise of future research agendas.

REFERENCES

Agnew, J.A. (1987) *Place and Politics: The Geographical Mediation of State and Society*. London: Allen & Unwin.

Amin, A. and Thrift, N.J. (2005) 'What's left? Just the future'. *Antipode*, 37: 220–38.

Anderson, B. (1983) *Imagined Communities: Reflections on the Origin and Spread of Nationalism*. London: Verso.

Barnett, C., Robinson, J. and Rose, G. (eds) (2005) *A Demanding World*. Milton Keynes: The Open University.

Brown, M. (1997) 'The cultural saliency of radical democracy: moments from the AIDS Quilt', *Ecumene* 4(1): 27–45.

Callard, F.J. (1998) 'The body in theory', *Environment and Planning D: Society and Space*, 16(4): 387–400.

Castree, N. (2004) 'Differential geographies: place, indigenous rights and "local" resources' *Political Geography*, 22(9): 1012–29.

Chatterjee, P. (1993) *The Nation and its Fragments: Colonial and Postcolonial Histories*. Princeton: Princeton University Press.

Corbridge, S. (1993) 'Marxisms, modernities, and moralities: development praxis and the claims of distant strangers', *Society and Space*, 11(4): 449–72.

Cox, K.R. and Low, M.M. (2003) 'Political geography in question', *Political Geography*, 22(6): 599–602.

Dodds, K.J. (1999) *Geopolitics in a Changing World*. London: Longman.

Duncan, J.S. (2004) *The City as Text: The Politics of Landscape Interpretation in the Kandyan Kingdom*. Cambridge: Cambridge University Press.

Duncan, N. (ed.) (1996) *Body/Space*. London: Routledge.

Escobar, A. (2001) 'Culture sits in places: reflections on globalism and subaltern strategies of localization', *Political Geography*, 20: 139–74.

Featherstone, D. (2005) 'Atlantic networks, antagonisms and the formation of subaltern political identities', *Social and Cultural Geography*, 6(3): 387–404.

Fincher, R. and Jacobs, J.M. (1998) *Cities of Difference*. London: Guilford Press.

Fraser, N. (2000) 'Rethinking recognition', *New Left Review*, 3: 107–20.

Friedman, S. (1998) *Mappings: Feminism and the Cultural Geographies of Encounter*. Princeton: Princeton University Press.

Gilroy, P. (1993) *Black Atlantic*. London: Verso.

Gruffudd, P. (1995) 'Remaking Wales: nation-building and the geographical imagination, 1925–50', *Political Geography*, 14: 219–39.

Harvey, D. (1993) 'Class relations, social justice and the politics of difference', in M. Keith and S. Pile (eds), *Place and the Politics of Identity*. London: Routledge, pp. 41–66.

Jackson, P. (1987) *Race and Racism: Essays in Social Geography*. London: Routledge.

Jackson, P. (1991) 'The cultural politics of masculinity: towards a social geography', *Transactions of the Institute of British Geographers*, NS, 16(2): 199–213.

Jacobs, J. (1996) *Edge of Empire: Postcolonialism and the City*. London: Routledge.

Johnson, N.C. (2002) 'Mapping monuments: the shaping of public space and cultural identities', *Visual Communication*, 1: 293–8.

Keith, M. and Pile, S. (1993) *Place and the Politics of Identity*. London: Routledge.

Knight, D.B. (1982) 'Identity and territory: geographical perspectives on nationalism and regionalism', *Annals of the Association of American Geographers*, 72: 514–31.

Laclau, E. (1994) *New Reflections on the Revolution of Our Time*. London: Verso.

Lambert, D. (2005) *White Creole Culture, Politics and Identity During the Age of Abolition*. Cambridge: Cambridge University Press.

Longhurst, R. (2001) *Bodies: Exploring Fluid Boundaries*. London: Routledge.

Massey, D. (1984) *Spatial Divisions of Labour*. London: Methuen.

Massey, D. (1995) 'Thinking radical democracy spatially', *Society and Space*, 13(3): 283–8.

McDowell, L. (1999) *Gender, Identity and Place*. Cambridge: Polity Press.

McDowell, L. and Court, G. (1994) 'Performing work: bodily representations in merchant banks', *Society and Space*, 12(6): 727–51.

Mohan, G. (2006) 'Embedded cosmopolitanism and the politics of obligation: the Ghanaian diaspora and development', *Environment and Planning A*, 38(5): 867–83.

Mohanty, C. (1987) 'Cartographies of struggle: Third World women and the politics of Feminism', in C. Mohanty et al., (eds), *Third World Women and the Politics of Feminism*. Bloomington: Indiana University Press.

Nash, C. (1996) 'Reclaiming vision: looking at landscape and the body', *Gender, Place and Culture*, 3(2): 149–69.

Natter, W. and Jones, J.P. (1997) 'Identity, space, and other uncertainties', in G. Benko, and U. Strohmayer (eds), *Space and Social Theory*. Oxford: Blackwell, pp. 141–61.

Ong, A. (1999) *Flexible Citizenship: The Cultural Logics of Transnationality*. Durham, NC and London: Duke University Press.

Painter, J. (1995) *Politics, Geography and 'Political Geography': A Critical Perspective*. London: Arnold.

Pile, S. (1996) *The Body and the City: Psychoanalysis, Space and Subjectivity*. London: Routledge.

Pile, S. and Nast, H. (eds) (1997) *Places Through the Body*. London: Routledge.

Secor, A.J. (2001) 'Ideologies in crisis: political cleavages and electoral politics in Turkey in the 1990s', *Political Geography*, 20(5): 539–60.

Sharp, J.P. (2000) *Condensing the Cold War: Reader's Digest and American Identity*. Minneapolis: University of Minnesota Press.

Silverstone, R. (2005) 'Media and communication in a globalised world', in C. Barnett, J. Robinson and G. Rose (eds), *A Demanding World*. Milton Keynes: The Open University, pp. 55–102.

Silvey, R. (2003) 'Spaces of protest: gendered migration, social networks, and labor activism in West Java, Indonesia', *Political Geography*, 22(2): 129–55.

Smith, N. (2005) 'What's left? Neo-critical geography, or, the flat pluralist world of business class', *Antipode*, 37: 887–99.

Wilton, R. (1998) 'Disability, identity and exclusion: community opposition as boundary maintenance', *Geoforum*, 29: 173–85.

Wright, M. (2003) 'Factory daughters and Chinese modernity: a case from Dongguan', *Geoforum*, 34(3): 291–301.

Yeoh, B. and Willis, K. (1999) '"Heart" and "wing", nation and diaspora: gendered discourses in Singapore's regionalisation process', *Gender, Place and Culture*, 6(4): 355–72.

Young, I.M. (1997) 'Unruly categories: a critique of Nancy Fraser's dual systems theory', *New Left Review*, 222: 147–60.

16

Nation-States and National Identity

Jan Penrose and Richard C.M. Mole

INTRODUCTION

Given that the discipline of geography is primarily concerned with territory and the diverse forms of power associated with it, it is surprising that geographers have had so little influence on theories related to the most dominant geopolitical unit the world has ever known – the nation-state. Geographers have had surprisingly little impact on understandings of either its core concept of nation or its progenitive ideology of nationalism (cf. Knight, 1982; Mikesell, 1983: 257; Johnson, 1995: 53). With some notable exceptions (e.g. Williams and Smith, 1983; Anderson, 1986; MacLaughlin, 1986; Johnston et al., 1988; Agnew, 1994; Johnson, 1995, 2002; Penrose, 1995, 2002; Smith, 1996; Marden, 1997; Withers, 2001), the tendency has been to rely on nations, states and nation-states as units of analysis – as a *context* for, rather than a *subject* of, intellectual inquiry. These efforts have produced some outstanding work; yet, it is because geographers have so much to offer that it is important for them to intensify their engagement with ongoing interdisciplinary debates about the conceptualization of nations, the relationship between nations and nation-states, and the personal and collective identities that both kinds of entities inspire.

This chapter offers a platform for extending geographical contributions to these debates by providing a synthesis of ideas that are fundamental to the study of nations, nationalism and national identity. Accordingly, we begin by providing an overview of theories about the formation of nations and the ideology of nationalism. The nationalist belief that the boundaries of nations and states should coincide is then shown to be instrumental in the emergence of nation-states and their rise to prominence in the global geopolitical order. This discussion makes it clear that there are two main ways of pursuing the convergence of nation and state boundaries, both of which involve active human engagement in nation-state building activities. After outlining these activities, we show how they have helped to shape the construction, experience and performance of national identities.

Consideration of the form and function of national identities marks a shift in the focus of the chapter towards an argument about the contemporary relevance of nationalism and nation-states. More specifically, we suggest that the capacity for nations to legitimize states is undermined in a world where pluralism is, increasingly, the defining characteristic of most states. We build this argument by showing how the role of hegemonic groups in the formulation of national identity reveals internal divisions within nations. These divisions are manifested in the unequal positions that individuals and groups occupy within both the nation and the nation-state and in terms of their access to material and symbolic resources. This evidence that nations are not uniform introduces some of the fundamental problems that have emanated from attempts to use this concept as the basis for allocating legitimate political power in the form of nation-states. Finally, we identify and evaluate solutions that have been advanced to deal with these problems by examining European Union and post-communist responses to recent pressures for both increased international integration and the reassertion of national distinctiveness. In showing how the concept of nation and the ideology of nationalism restrict ability to adapt to changing political realities, the importance of geographical input about the flexibility of spatial political constructs (in both time and place) and their mutual constitution with a wide range of social and cultural groups becomes clear.

THEORIES OF NATION FORMATION

While the study of nations has generated hundreds of books and articles, there is still no established consensus on the definition, origins or future of this concept. All theories of nation formation rely on different definitions of the nation, the main axes of debate being whether the nation is essential or constructed, ancient or modern, political or cultural. These debates have spawned three main bodies of nationalism theory, commonly understood as (1) primordialist/perennialist, (2) ethno-symbolist and (3) modernist/instrumentalist (Özkirimli, 2000, 2005; Day and Thompson, 2004; Lawrence, 2005; Hearn, 2006).

When most people first come across nations, the explanation that is usually given for their existence is the primordial argument, as this is the approach taken by many nationalists themselves. Given that nationalists use the existence of *their* nation as the basis for claims to an independent state, it is not surprising that many of them view nations as natural phenomena that have existed for centuries, if not millennia. While perennialists do not share the primordialist view that nations are natural or essential categories – considering them, instead, to be social and historical phenomena – they do share the belief in the continuous or, at least, continually recurring existence of nations throughout history.

A more widely accepted conceptualization of nations is the ethno-symbolist approach commonly associated with the work of Anthony Smith. In general, ethno-symbolists share the perennialist view that nations are social and historical phenomena (rather than 'given'), but they reject stark 'continuism' by acknowledging the transformative impact that the modern era has had on 'the complex social and ethnic formations of earlier epochs' (Smith, 1995: 59–60). Nevertheless, ethno-symbolists argue that nations and national identities have strong roots in pre-modern *ethnies* (ethnic communities) and that they cannot be understood without reference to a living legacy of symbols, myths, memories and so on that defined the core ethnic group before modernity.

Finally, the modernist or instrumentalist conception of the nation is best explained through the ideas of its most famous exponent, Ernest Gellner (1964, 1983). Gellner understands nationalism as the product of modern industrial society. He argues that state education produced a standardized form of language, history and culture to create the idea that all inhabitants of a particular territory were part of a single community. This construction was important for two reasons: first, because it created loyal members of society, whose ability to function as such would not be hampered by attachments to sub-groups within or beyond state boundaries; and second, because it created culturally standardized, interchangeable populations who were capable of achieving high productivity in industrialized societies.

Each of these three theories conceptualizes the nation as a fundamentally cultural entity. In contrast, proponents of these perspectives disagree, sometimes vehemently, about whether the *significance* of nations is cultural or political, or both. For primordialists, the view that nations are 'given' connotes an unalienable capacity (and right) for nations to rule themselves. For them, and for most perennialists, it is the cultural unit of the nation that both predates and justifies a state and not vice versa. Ethno-symbolists share the view that culture has value in its own right, but they would also argue that the politicization of culture has granted nations much of their significance and power. Proponents of the modernist or instrumentalist schools view nations as modern entities that were conceptualized and constructed to achieve particular socio-economic and political ends. From this perspective, the state predates the nation and the function of the nation is to improve the cohesiveness of the state and the efficiency of its economy. While culture is seen as important in defining nations, the significance of nations themselves is confined to their functional, political, usefulness.

Scholars agree that nations are important because they are seen to constitute a unique cultural identity. The culture of a nation and the national identity that it fosters combine to produce the mythical qualities necessary to inspire a sense of belonging and this, in turn, is essential to the fostering of loyalty and support. The mobilization of nations involves the ideology of nationalism and it is this political doctrine that is largely responsible for the formation of nation-states. Not surprisingly, different views about the origins, qualities and significance of nations are paralleled by different views on nationalism. It is to this issue that we now turn.

NATIONALIST IDEOLOGY AND NATIONALISM: FROM NATION TO NATION-STATE

Different understandings of the concept of nation complicate the study of this phenomenon, but the situation becomes even more confused by the tendency to use the term 'nationalism' to refer to both a political ideology and a type of political movement. In an attempt to overcome this terminological laxity, we will use 'nationalist ideology' to refer to the core conviction that the boundaries of a nation (however defined) should coincide with those of a state (following Weber, 1947). In contrast, 'nationalism' will be used to refer to attempts to implement nationalist ideology in practice; nationalism is a political movement. In this section, we would like to illustrate how nationalist ideology was mobilized, through nationalism, to produce nation-states. The key point here is that it is possible to identify two trajectories of nation-state formation but that these should not be confused with two different types of nation. Once the two trajectories have been outlined, we will highlight the key mechanisms that have been deployed, with remarkable consistency, to merge the cultural unit of the nation with the political unit of the state to form the new and quintessentially modern political entity called the nation-state.

Given its incredible pervasiveness, it can be surprising to realise just how recent nationalist ideology is – a little over two hundred years old. In pre-modernity, political legitimacy was not derived from popular consent or shared culture but from divine right; whether the ruler and the ruled shared a common culture, language or ethnicity was immaterial. In the Middle Ages the development of national consciousness was hampered by the feudal structure of society and by the power and aspirations of the Church (Anderson, 1996). The clergy exercised complete control over education and the written word, the exclusive language of which was Latin. The immense prestige that this language enjoyed prevented vernaculars from gaining general acceptance and being standardized in written form, thus hampering the development of national tongues. It was not until the Reformation in the sixteenth century that the standardization of vernaculars began to engage with nascent national consciousness by gradually increasing feelings of community among people who shared a language (Mann, 1993: 217; cf. Billig, 1995: 29–36). Nevertheless, Church and monarchy continued to hold sway until the Enlightenment, when new philosophical and social conditions enabled the concept of nation to become, in time, widely accepted as the legitimate source of political power.

Nation-State Formation Take One: State + Nation = Nation-State

The first trajectory of nation-state formation emerged during the Enlightenment and involved the construction of a state prior to the formation of a nation within its boundaries. This process began with the ideas of political philosophers such as Locke, Rousseau and Mill, which came to have almost unprecedented transformative power over the societies in which they were developed. These ideas included the concept of the general will, popular sovereignty and a revaluation of democracy that included development of the notion of majority rule and the concept of representative government based on individual self-determination (i.e. allowing people to decide collectively who should represent them). It is noteworthy that Rousseau made no explicit reference to the concept of nation as a legitimating principle, although he did imply that the 'social groups from which a general will can most effectively emerge will be genuine cultural communities and not casual dynastic accumulations of mutually unsympathetic people' (Quinton, 1994: 332; see Rousseau, 1947, [1762] book II, chap. X: 41). For Rousseau, the idea of a 'general will' – as the moral personality of the state – was necessary before the idea of a nation could have any reality (Cobban, 1964: 108). As this suggests, Rousseau consistently privileged the political entity of the state, and the political principles that defined his new conception of a state, over the cultural composition or characteristics of its inhabitants (cf. Penrose, 2002: 287–9).

These priorities are reflected in one of the first attempts to apply Rousseau's ideas: the French Revolution of 1789. The French Revolution was nationalist in that its proponents wrested political legitimacy from the King and placed it in the hands of *la nation*. Importantly, this 'nation' was understood by the revolutionaries to mean *all* people who lived within the territory of the French Republic – regardless of former rank or title or place of birth (Hampson, 1991; Kristeva, 1991; Gildea, 2002). In this context, the nation was seen as a collectivity of free individuals with equal rights based on citizenship, and nationalism was synonymous with liberalism, democracy and popular sovereignty based on the principle of consent.[1]

Initially, then, the French Revolution did not promote a nation in any cultural sense, but rather a new form of political unit that was defined by citizenship and legitimized by principles of popular sovereignty and self-determination. However, in post-revolutionary France there were good reasons – ideological, psychological and functional – for promoting cultural cohesiveness within the borders of France. Ideologically, the need for homogeneity was based on the legitimizing power of nationalist ideology: for a state to be legitimate, it had to (be seen to) represent a single nation. As this suggests, if other nations existed within the borders of a state they could, in theory at least, claim a right to a state of their own. In a France that was characterized by numerous alternative nations (e.g. Bretons, Normans, Basques, Alsatians and so on) this was a real concern and the creation of a common culture based on the dominant French model became imperative to the survival of the state. Psychologically, then, the French state had to find ways of erasing or at least overriding existing loyalties to other nations (within or along its borders) if it was to retain political legitimacy. The rational appeal of citizenship and self-government remained evident, but post-revolutionary experience clearly demonstrated that political doctrines were incapable of generating the same depth of loyalty as that associated with non-rational allegiance to nations (Connor, 1994; Fine, 1999). Finally, there were very sound functional motivations for pursuing some measure of cultural uniformity within the state's borders. Simply put, it was much easier to govern a homogeneous community with a single identity than a disparate collection of heterogeneous collectivities.

For all of the reasons just outlined, post-revolutionary France began to be constructed itself as a nation-state. Crucially, however, the focus on creating cultural uniformity within the boundaries of the state only began to occur *after* the modern French state had been established. The French Revolution was not about nationalism. It was about republicanism, and the country only began to promote nationalist ideology when its leaders realized that a homogenous and unified nation was essential to the attainment of political goals, including the legitimacy of the state itself (Weber, 1977). Somewhat ironically, attempts to construct a distinctive French nation drew on the experiences of those who had pursued the second trajectory of nation-state formation. It is to this path of nation-state formation that we now turn.

Nation-State Formation Take Two: Nation + State = Nation-State

In large part, the second trajectory of nation-state formation was born of resistance to some of the key ideas advanced by Enlightenment thinkers. Romanticists such as Fichte (1922 [1806]), Hamman (1967) and Herder (1968 [1784]) emphasized the primordial elements of nationhood, arguing that the world was divided naturally into communities that were inscribed in space and defined by culture, ethnicity, tradition and history rather than politics and citizenship. They reified the concept of *Volk* (sometimes using it interchangeably with nation), generating the idea that collectivities were entitled to power and resources on the basis of shared culture.

As this suggests, the second trajectory of nation-state formation began with the cultural unit of the nation and sought to ensure that it was able to develop according to its own internal logic and values. In this case, the purpose of a state was to protect the nation – as a fundamental unit of humanity – and in doing so, the state also served as a manifestation of the nation's right to self-determination. For nation-states that were formed by following this trajectory, the cultural unit of a nation both preceded and was prioritized over the political unit of a state. Thus, territories that were inhabited by groups sharing a common language and/or culture could merge to form a single nation-state (e.g. Germany and Italy). Alternatively, larger (often imperial) territories that were home to numerous cultural groups were divided into a number of smaller polities, each representing or seeking to represent a single nation (e.g. Estonia, Bulgaria and Slovenia).

Clearly, this second trajectory's ideal of privileging culture over politics is the converse of the first trajectory of nation-state formation, which began with a state based on new political doctrines and then sought to create a nation within its boundaries. Yet, despite their antithetical priorities and processes, both trajectories were nationalist (and profoundly territorial) in that they sought to make the boundaries of the nation and the state coincide. It was out of the fusion of the idea that government should be by and for 'the people' and the idea that 'the people' should be defined by cultural communities, that nationalist ideology emerged with the goal of encouraging the formation of nation-states. Simply put, nationalism involved the politicization of culture (the nation) and the cultural codification of the state. As these new ideas and the ideologies that they supported gained prominence, state behaviour could no longer be legitimated in religious or dynastic terms but only by the nation, a cultural community. By the early twentieth century, national self-determination had become a universal principle, recognizing only one type of polity – the nation-state – whose borders were no longer determined by 'the courses of rivers, the direction of mountains, or the chances of war, but according to races or rather [ethno]nations' (Cobban, 1970: 109).

As geographers have pointed out, these dominant theoretical explanations of the formation of nation-states do a good job of outlining general processes and experiences, but they often do so at the expense of spatial and contextual sensitivities that are the hallmark of geographical analysis. For example, Agnew and Corbridge (1995: 80) explore the limitations that come from relying on 'ideal types' that are 'fixed representations of territorial or structural space … irrespective of historical context'. This is most obvious in the often neglected fact that the model of the nation-state that most scholars accepted as the norm until the mid-twentieth century was confined almost exclusively to the industrialized world (Claval, 2001: 35–6). Geographers have also argued that understandings of the formation of nation-states can be enhanced by viewing then as entities *in process* – as units that are produced and reproduced through a whole host of uneven power relations that extend from the political and socio-cultural to the economic and environmental (cf. Katz, 2003). Not surprisingly, then, it is in the realm of nation-state building that geographical perspectives have made some of their most important contributions to the understanding of nations and nationalism.

FROM NATION-STATE FORMATION TO NATION-STATE BUILDING

Just as most commentators agree that nation-states can be formed in the two ways outlined above (e.g. Connor, 1980; Smith, 1991; Ignatieff, 1994), they also agree that the fit between political and cultural boundaries was seldom, if ever, perfect and that loyalties to the new unit of the nation-state had to be developed (cf. Connor, 1972: 319). For the most part, this process of unifying a group of people within a state, as defined (in principle, at least) by those same people, has been termed nation-building. We would argue that this is often a misnomer for at least two reasons. First, as Connor (1972) has also noted, the cultural pluralism that characterizes most states means that the promotion of one nation has frequently occurred at the expense of another. Thus, within the process of nation-state formation, one nation's building often involves another's destruction. Second, the process of building a nation-state involves both of this unit's constitutive entities, namely, the nation and the state. The key quality of this new political unit is the joining together of an explicitly cultural entity with an explicitly political one, to form something brand-new. As indicated above, it is almost inevitable that, in the process of combining these two entities, the nation will become increasingly politicized and the state will become culturally encoded. Accordingly, it seems appropriate to refer to the processes of constructing, unifying and solidifying the nation-state as 'nation-state building'.

In general terms, nation-state building always involves at least some of five main processes that have been well rehearsed in the literature but which warrant a brief summary here.[2] Given that the modern state is characterized by a set of institutions and a regularized staff to administer them (Weber, 1947:143), it is not surprising that one of the key functions of nation-state building is to establish these institutions as well as the bureaucracy capable of running them. By the same count, the nation is a community defined in general terms by shared culture and meanings and this means that nation-state building is also geared to unifying the population of the new geopolitical entity. Both of these processes, creating institutions and inspiring loyalty to them, are apparent in each of five key processes commonly associated with nation-state building.

The first of these processes involves establishing the overtly political structures of the nation-state, namely, institutions of government and systems of representation. In most cases, the ideal is to develop a centralized form of government for the simple reason that this offers greatest power, and security of power, for those who command it. The next step is usually to establish national (culturally distinctive) and state-wide (territorially universal) political parties that have the advantage of forcing political issues and perspectives to be conceptualized in ways that reify the cohesiveness of the nation-state. Thus, even though people may disagree about the identification of problems and/or their solutions, the acts of engaging in debate about key societal issues and of supporting one party over others implicitly reinforce a sense of belonging and loyalty to a single geopolitical unit.

The second main mechanism of nation-state building involves establishing a monopoly over the legitimate use of force (a key characteristic of a modern state) by creating 'national' military and police forces. These institutions reinforce the power of the state but they also help to unify the nation by bringing together diverse segments of the population in ways that help to break down prejudices and nurture an overriding allegiance to their common nation-state, which they share a duty to defend. Similarly, the establishment of a national system of education – the third main process – contributes to this project by promoting shared experiences and encouraging individual identification with the new polity. Through a standardized curriculum it is possible to emphasise collective, national interpretations of both historical and current events and, perhaps even more importantly, to ensure that this shared knowledge is communicated in a common language.[3]

Standardized language is also fundamental to the fourth and fifth mechanisms of nation-state building because it establishes the means of developing and communicating shared meanings and, in the process, it can both arouse and convey ideas of a common identity (cf. Johnson, 2002: 132). As scholars like Anderson (1991) and Billig (1995) have so convincingly demonstrated, national media are capable of moulding their audience into an imagined community and thereby encouraging feelings of affinity among its members. As such, national media work to reify the existence of a given nation-state; they constitute mechanisms for promulgating particular understandings of what the nation-state is (or ought to be); and they are powerful means of inspiring personal loyalty to the polity (cf. Robins, 1995).

Finally, the building of an effective nation-state is aided by symbols, shared meanings and memories that are identified, and/or created, to confirm the existence of the nation-state and to invite personal allegiance to it (as well as performance of it). These symbols can be both material and symbolic representations of the nation-state and are thus very powerful in their own right. However, they also have the capacity to highlight connections between all mechanisms of nation-state building, producing synergies that enhance the power of all constituent elements. For example, a national anthem is a symbol of the nation-state that can inspire and mobilize emotive responses to the country it represents, making the nation state a source of community, personal identity and belonging. When the anthem is sung in school, at a Remembrance Day service, or as part of an Olympic medal ceremony, its symbolic and emotional power enhances the capacity for these events to confirm the existence of the nation-state and to invite – sometimes even demand – personal allegiance and loyalty to it.

The discipline of geography has made several important contributions to the understanding of nation-state formation, three of which we would like to accentuate here. First, geographers have demonstrated the importance of space and place to the construction of nation-states. For Johnson (2002: 141), the territorial dimension of nationalism cannot be underestimated, not least because 'the occupation of, and control over, space and the delineation of boundaries has been the source of many regional, national, and international conflicts'. Such conflicts have influenced the rise and fall of particular nations and nation-states as well as the relations within and between them. For others, the importance of space and place to nation-state building is evidenced in the capacity for geographical perspectives to challenge the dominant, but largely unquestioned, view that territorial borders are fixed and physically defined.

Instead, geographers like Paasi (1995) and Smith (1993b) have shown that, by drawing selectively on various dimensions of boundaries – historical, natural, political, cultural, economic, psychological, sensual and so on – it is possible for very different places to be constructed and for diverse ideological agendas to be advanced. For others still, the usefulness of quintessentially geographical concepts is apparent in their demonstrations of how and why specific places and territories are constructed (e.g. Tuan, 1974; Johnston, 1991; Rose, 1995), and in their explorations of how particular visions of the nation produce, and are produced by, the iconography of landscape (e.g. Cosgrove and Daniels, 1988; Lowenthal, 1994). In all of these ways, conceptions of space and place and territory help to explain the construction of nation-states and to reveal just how contested these construction processes can be.

The importance of conscious construction processes is also apparent in the second main geographical contribution to understandings of nation-state building. Here, we are thinking of arguments about the centrality of immigration policy to nation-building: in Smith's words, '[t]oday, immigration controls, at least as much as territorial extent, are an indicator of where the boundaries of a nation-state lie' (1993a: 50–1). In a world where more and more people are on the move, the importance of immigration policies in shaping nation-states – their composition in terms of age, gender, sexuality, 'race', health and so on – cannot be overstated and geographers are well placed to spearhead work in this area. Finally, geographers have attempted to show that dominant theories of nations, nation-states and nationalism have themselves been influential in the formation of these things (Livingston, 1992; Penrose, 2002). For example, the efforts of French geographers to document and explain the rise of France becomes part of the evidence that such an entity exists (Hooson, 1994: 4). Similarly, Agnew (2003) shows how very specific Euro-American experiences gave rise to a hegemonic geopolitical discourse that was then projected on the rest of the world. Both geography and geographical knowledge are important elements of nation-state building.

In summary, nation-states are constructed as empirical manifestations of nationalist ideology in practice; state borders are established and defended and within these borders constant efforts are made to encourage and maintain the cohesiveness of the population. As this suggests, nation-state building involves both the construction of an object of loyalty – the nation-state – and the inspiration of loyalty and belonging as qualities in their own right. These qualities are often referred to as 'national identity' and it is to this subject that we now turn.

FORMATION AND FUNCTION OF NATIONAL IDENTITY

Identity, like nation and nationalism, is a term that seems self-explanatory and unproblematic until people really stop and think about it. At its simplest, identity is who we are. More accurately, if more complexly, it is how we understand and construct who we are (Katz, 2003: 249). Identity is the way in which we more or less self-consciously locate ourselves in our social world and this process of location relies heavily on social roles and categories (Preston, 1997). However, knowing who we *are* also implies knowledge of who we *are not* and this makes it clear that all identities are relational (Massey, 2004: 5). All identities – both individual and collective – are thus defined with reference to both Self and Other, to a 'me' and a 'you' or an 'us' and a 'them'. The designation of who we are does not relies on a whole range of categories that serve as means of making sense of the world and communicating it to others. According to social psychologists, human beings have an instinctive need to categorize humanity into distinct social groups and to ascribe each with a unique identity.[4] They view this process of categorization as necessary because the social world has very few explicit lines of division and it helps render 'our experience of the world subjectively meaningful' (Hogg et al., 1995: 261).

Importantly, even though the *process* of category formation may be instinctive, the categories that are produced are not 'givens' (Penrose, 1995). Instead, they are a reflection of the perceptions, priorities and aspirations of those people who have the power to both construct categories and promote them as 'natural' or superior. The same is true of representations of those people who are associated with specific categories. These representations are not 'givens' but constructions that reflect the power relations that exist within any specific social, geographical and historical context. In both cases, power relations reflect a process of hegemony whereby those in positions of power have the capacity to convince subordinate others to accept the dominant group's moral, political and cultural values as the 'natural' order (Jackson, 1989: 52–3, after Gramsci).

Although attempts at persuasion are always met with some resistance, some categories of identity can become entrenched through hegemony. In part, this is because even though the content and/or significance of categories can be contested, the very act of doing so only reifies the category as a legitimate division of the world. Thus, even though the meaning of identities is not fixed, the dominant categories of identity – things like 'race', gender, class, religion and nation – have proven very difficult to challenge, let alone dislodge. Such categories of identity can also acquire a relative fixity because of individual tendencies to take on specific identities for themselves. People do this largely because it is a fundamental mechanism for generating a sense of belonging and for maximizing self-esteem. This process entails identification with an in-group (often defined by hegemonic categories or a combination of categories) and with this in-group's dominant group norms and its differentiation from the out-group. The importance of group membership for self-definition means that human beings internalize their own group categorization; as the individual becomes part of the group, the group becomes part of the individual. It is this internalization of identity that makes the categories that support them so important, and consequently, so powerful. Those categories that inspire the greatest internalisation, that become personal and perceived as key to the survival of the self, are those that assume the greatest significance in structuring divisions of people and space as well as the power relations and the structures of power that mediate them.

National identity is one such category. It constitutes one of, if not *the*, most important identities in the modern world. Simply put, it is the identity that is born of the category 'nation' and supported by personal identification with a specific nation. As this suggests, national identity can be understood in two complementary ways. First, it reflects the constitutive elements of nationhood (language, collective memory etc.), and this permits 'snapshots' of a nation's identity that suggest which cultural symbols and conventions are most salient at any particular time. In this sense, national identity is the identity of any specific nation; it is what the nation is. Second, national identity is also a psychological condition whereby 'a mass of people have made the same identification with national symbols – have internalized the symbols of the nation – so that they may act as one psychological group when there is a threat to, or the possibility of enhancement of, these symbols of national identity' (Bloom, 1993: 52). In this sense, national identity is personal; it says something about who individuals think that they are.

The successful construction of a nation fuses both elements of national identity; it inspires personal identification with the constitutive elements of the nation such that its members believe that they *are* the nation. Once this belief has been inculcated, it is likely that people who identify with the nation will defend it at all costs, for to do so is to defend themselves. This imperative is especially strong where the nation is imagined as an extension of family or kin networks because defence of the nation becomes synonymous with defence of family – past, present and future (Penrose, 2002). As this suggests, the strength of national identity stems from the tendency for members of a given nation to imagine that all of its other members view and experience their shared nation – and its associated identity – in very similar if not identical ways. These processes of convergence are aided by the mechanisms of nation-state building described above.

National identity provides individuals with objective and subjective dimensions to their sense of self, of who they are in the world. This is valuable in its own right, but the significance of national identity is magnified through other associations and functions. The fact that nation-states are the only legitimate geopolitical unit in the current world order means that national identity has become a key means of regulating access to resources. Thus, on a global scale, association with a specific national identity is key to inclusion within the space and resources defined by that nation, and it also defines all of those other spaces and resources from which one is excluded. Where national identity is expressed geographically, in natural resources, landscapes, architecture, monuments and so on, it connotes rights to a share in the material and symbolic resources that define the nation (Johnson, 1995; Penrose, 2002). The important thing here is that all of the symbols and institutions that are developed as expressions of national identity, and that serve as contexts for its performance and representation, feed back into the construction of a specific nation and the identity that it fosters. For example, playing the bagpipes while wearing a tartan kilt can reinforce dominant constructions of Scottishness and justify continued identification with this nation. Alternatively, playing an electric guitar while sporting a Mohawk haircut and Doc Martin boots, along with the same tartan kilt, can advance a less orthodox construction of Scottishness that is equally capable of inspiring personal identification with, and loyalty to, something that is universally recognized as Scotland.

The continuous and overlapping processes of constructing a nation and stimulating identification with it emphasize unity and shared experience, but this glosses over the ways in which different interests, often defined by other categories of identity (like gender, age, religion and class), can position people very differently within the nation and the state. For example, the promotion of national unity and a singular national identity works well for those who have the power to direct this process and, until very recently, this hegemonic group was constituted almost exclusively by men, particularly those who possessed wealth and/or particular social status. As long as the nation was constructed in their own image, their positions of dominance within it were secured. In contrast, the very same nation became a context in which women's marginality could be reinforced, and by identifying with this nation, women became complicit in their own marginalisation (cf. Yuval-Davis and Anthias, 1989; Yuval-Davis, 1997; Hall, 1999).

Despite these obvious disadvantages, women remain relatively privileged as long as they, like most of the population of a nation-state, are accepted as legitimate members of the nation. Everyone who holds membership in a nation is granted a share of the nation's resources and the freedom to participate actively in processes that reify the nation and that encourage internalization of its symbols. As indicated above, however, being unequally positioned within the nation (both spatially and socially) produces different experiences of these processes and of just what the national identity entails. This unfairness is compounded by contemporary processes of migration. Legal immigrants can be granted formal citizenship in a nation-state but still be denied membership in the associated nation (cf. Parekh, 1999: 71), and even more serious problems of inequality emerge

for refugees and asylum-seekers who are regularly deprived of both sources of identity within their host society.

The fact that one can hold citizenship of a state without sharing in its national identity, underscores the importance of distinguishing the two concepts. It also underscores the fact that the vast majority of nation-states do not comprise a single, unified nation, even though they continue to rely on the ideology of nationalism to legitimize their existence. This fundamental inconsistency raises serious questions about the robustness of the category 'nation', the contemporary relevance of the ideology of nationalism, and the viability of nation-states as the cornerstone of the global geopolitical order.

CHALLENGES TO NATIONS AND NATIONALIST IDEOLOGY

When the ideology of nationalism was first formulated and applied, the world was a very different place than it is today. For a start, significant parts of the globe remained uncharted and largely unknown to Western Europeans, but the evidence that did exist (and that continued to accumulate) revealed extraordinary diversity in human appearance, culture and social organization that seemed to support the view that humanity was divided into distinctive groups. This is a classic example of Alexander Pope's adage that 'a little learning is a dangerous thing', and part of the power and endurance of the concept of nation is that it continues to be legitimized by its apparently accurate reflection of reality. Because individuals could *see* that people and territories were different from one another, they seldom stopped to think about how these differences could be translated into finite categories (Penrose, 1994: 163). This meant that it was possible to avoid confronting the fact that concrete boundaries between people and environments cannot be identified in practice. It also meant that the tendency for one territory to be claimed by multiple groups could be overlooked, along with the tendency for people to hold fluid and/or multiple (sometimes even contradictory) identities.

As the world became more thoroughly known and as people began to move around it with increasing speed and ease, these comforting illusions about the integrity (and hence, usefulness) of the category of nation were harder to maintain. Indeed, and as discussed above, attempts to apply nationalist ideology and to define political units according to coinciding boundaries of nations and states were proving impossible without significant human intervention and 'national modification'. Then, as now, there were only two real options. The first involves imposing homogeneity on the population of a given state, and this can be achieved through policies that promote assimilation to the hegemonic norm or, even more objectionably, through policies of genocide and/or ethnic cleansing. In both cases, the nationalist ideals of a state serving all of its people and of the people legitimizing the state are woefully abandoned in favour of an illusion of homogeneity. If this is to be the cost of living according to nationalist ideology, then very few people are likely to be willing to pay the price.

The only other option for applying nationalist ideology is to change what is meant by the category of nation. It could be argued that this strategy was first implemented when French revolutionaries used the term 'nation', as defined by citizenship, to identify 'the people' that the new government and state would serve (Connor, 1978). Crucially, however, this political conceptualization of the nation was not effective in inspiring a loyalty that overrode existing allegiances to more culturally defined nations that existed within the boundaries of the new state (Weber, 1977). It was only when the dominant French nation, culturally defined, was transmitted to (or imposed upon) the rest of the population through active practices of nation-state building, that allegiance to the nation-state called France began to inspire loyalties that came close to pre-existing affinities with older nations associated with various regions of the country. Over time, the politicisation of the French nation and its dissemination across civil society and into institutions of state have produced a national identity that is capable of unifying what is almost universally recognized as the nation-state called France. The persistence of nationalist movements within the boundaries of this state is, however, a salutary reminder that the loyalties of citizenship seldom override the loyalties of culture if people are forced to choose between the two dimensions of identity.

The key point here is that nationalist ideology has been successful to date because personal identification with a culturally constructed nation has inspired loyalty to the associated nation-state. Despite its inadequacies, the ideology of nationalism has convinced most people that they are members of a nation – a cultural community – that is either represented and protected by a state, or worthy of acquiring such a state. It is extremely doubtful that nations defined by membership in a civil society can be as persuasive – or inspire the same depth of loyalty – as those defined by shared culture, let alone those based on assumptions of shared descent. In part, this is because the lack of distinctiveness between states (increasingly defined as liberal and democratic) would reduce their capacity to inspire personal identification with, and loyalty to, any given state. Moreover, if this were the case, then there would be good reasons for redrawing the boundaries of states so that they become the most efficient administrative units for

delivering democracy alongside individual rights and freedoms. The fact that this is not going to happen is, in itself, powerful testament to the poor logic of using exclusively civic constructions of nations to legitimize states.

The reason that this is not going to happen, aside from the reluctance of the powerful to empower the disinherited, and of the wealthy to share with the poor, is that culture has huge, inherent value to most people. Although more and more states are defining themselves as multicultural, few people – especially the immigrants that add this new 'complexion' to states – believe that this has altered the power and prominence of hegemonic nations within states (Ignatieff, 1994). Multicultural definitions of societies do reflect a welcome acknowledgement of diversity and, often, the desire to develop tolerance (if not acceptance) of difference. This may even result in some tempering of the hegemonic nation's self-definition (e.g. curry as the British 'national dish'), but this is not the same thing as relinquishing the nation as the basis of self-identification and political legitimacy or abandoning claims to privileges, power and resources that are based upon membership in the dominant nation.

As all of this suggests, the contemporary world is trapped by a reliance on the concept of nation to legitimize its division into states, at the same time as this concept is becoming increasingly indefensible as a reflection of these states. In other words, there is a profound contradiction between the growing cultural pluralism of virtually all nation-states and the ongoing ideological investment of power in the idea of a single, hegemonic nation along with the use of this 'ideal' nation to justify statehood. This unacknowledged contradiction masks the fact that the celebration of cultural diversity continues to be paralleled by an often profound (and sometimes violent) fear of cultural difference. These inconsistencies between nationalist ideology, political rhetoric and practice represent a time bomb that is in urgent need of defusing.

So far, two general strategies have been advanced for overcoming this disjuncture between the ideology that structures the global geopolitical order and the realities of growing cultural diversity within nation-states. The first strategy involves the promotion of nation-states as primarily civic entities, adopting policies of multiculturalism, and exploring responses to internal demands for minority group rights. All of these developments are cause for hope. Eventually, human beings may even learn that our similarities are much, much greater than our differences and come to live the belief that 'they are us'. At the same time though, the mobilizing power of (exclusive) cultural conceptions of the nation remains a force to be reckoned with. This is evident in the rhetorical power of Bush's 'war on terror', which is calculated to inspire fear of difference and which has already led to increasingly restrictive immigration legislation in both North America and Western Europe. The UK has recently introduced citizenship tests and linguistic requirements for new immigrants – both of which reinforce the values of the hegemonic culture of the nation-state and have the potential to undermine the concomitant push for greater inclusion.

These responses reflect a potential to slide into a second, less encouraging strategy for dealing with the contradictions between nationalist ideology's reliance on a single, homogeneous nation and the spread of cultural pluralism. This is the strategy of reasserting culturally distinct nations at the expense of internal diversity and minority rights. Throughout the 1990s the government of almost every Central and East European state set itself the task of 'returning to Europe', seeking EU membership as confirmation of its European heritage. The push for EU membership was driven as much (if not more) by issues of identity and geographical and historical self-perception as it was by a rational consideration of economic and political benefits. Yet, the Europe to which the post-communist states sought to return and the Europe created by the Treaty of Rome were entirely different entities. To the new political elites, the return to Europe was the return to the Europe of the pre-communist, inter-war period. This era is highly important in the historical memory of most of the East European nations, because the inter-war republics provided the political space within which the national identity of the people was first disseminated among, and internalized by, the population at large. While the states of Western Europe have gone to great lengths to promote identities that are at least ostensibly inclusive, to develop European citizenship, and to establish strict minority rights regimes, the newly sovereign states of the former Soviet bloc have sought to rebuild their nation-states on the basis of the ethnic cultures that had for decades been repressed by political elites and supplanted by communism as the core state identity.

Following the collapse of state socialism in the late 1980s and early 1990s, the discrediting of communist ideology left the nations of Eastern Europe with an identity crisis. To understand why – despite embracing democracy – so many post-communist nations turned to ethnicity to fill their identity vacuum, it is important to grasp the complex interplay of forces underpinning nationhood in democratic societies. George Schöpflin (2000: 35) argues that democratic nationhood comprises three essential elements that exist in a mutually interdependent relationship: civil society, ethnicity and the state. While not exactly a zero-sum game, ethnicity will, he suggests, play a greater role in the composition of national identity when the state and civil society are weak. After the collapse of communism, the absence of firmly established political, legal and military institutions meant that the states of Eastern Europe were unable to provide, by civic means, the cohesion necessary to make society function. Political elites instead sought to generate cohesion, and at the same time legitimate their claims to power, by appealing to ethnicity and 'historical rights'. And in the absence of civil society, they were left unchallenged to do so.

Thus, the harmonization process that the states of Central and Eastern Europe underwent to bring their political, economic and social structures into line with those of the European Union conflicted with many of the nation-state building measures that they were implementing at the same time. The resurgence of ethnicity and culture as primary collective resources is consistent with Max Weber's (1947) idea of monopolistic closure. He argues that when resources are scarce, titular nationalities use ethnicity to press for privileged access to economic and especially political rights. The 'natural' preference for maximizing gains for members of the in-group can result in the maximization of difference between 'us' and 'them'. Yet, ongoing developments in Europe offer post-communist states an opportunity to revive civil society and refine their states such that these forces become more effective in balancing the power of ethnicity and culture. Most obviously, the prospect of EU membership has encouraged new member-states to adhere to emerging European norms with regard to minority rights, language and citizenship. This strategy of tolerance, recognition and accommodation is clearly the most dominant moral force in the current world order and it has the capacity to refine nationalist ideology from the inside. Of course, it is possible that the states of Central and Eastern Europe (and Western Europe for that matter) may not have undergone a genuine shift in their attitudes towards minorities at all and that they have simply proclaimed support for liberal values as a rhetorical device for securing (or retaining) EU membership and its benefits. Similarly, responses to terrorism, including restrictive immigration and notions of citizenship that are heavily inflected with culture, have the potential to provoke a swing back towards the more exclusive and isolationist strategy of coping with the inadequacies of nationalist ideology in the contemporary world.

Ultimately, it remains to be seen how the world will deal with the limitations of the political ideology that currently structures and legitimizes the geopolitical order. What seems much clearer is that there is an urgent need to expose and address these limitations. As we have attempted to show, the concept of nation is unsound, the pursuit of nationalist ideology is inherently divisive, and both nations and nationalism(s) severely constrain options for adapting political power to the changing composition of states, let alone the changing demands that are being placed upon them. Geographers are well placed to help the world move beyond this deadlock because the key elements – nations and nationalism, nation-states and national identity – are all quintessentially geographical phenomena. They are spatial constructs, both grounded and imagined, that continue to draw selectively on particular understandings of boundaries and power in order to lay exclusive claim to the loyalties born of cultural affinity. Geographers have the capacity to show how the promotion of different kinds of boundaries can produce different kinds of places, and they can do so over a whole range of geographical scales and contexts. Moreover, different kinds of places can, in turn, be incorporated into the promotion of different kinds of identities and ideological agendas. In the process, geographers can help to develop new and liberating forms of geopolitical organization that work with, rather than against, evolving geopolitical realities.

NOTES

1 For some scholars, this use of the term 'nation' has been used to justify the idea that there are two types of nation, that the French Revolution gave rise to a 'civic' or political nation that could be contrasted with the original understanding of nations as 'ethnic' or cultural units. However, recent writing has questioned the logic and usefulness of the concept of a 'civic' nation (e.g. Seymour et al., 1996; Xenos, 1996; Yack, 1996; Schulman, 2002).

2 For overviews of these processes, see Hayes (1945 [1931]) and Penrose (1997); for a related discussion with specific reference to political geography, see also MacLaughlin (1986) and Johnson (2002).

3 As Grano (1981) suggests, geography and history were especially well suited to transmitting the new secular religion of nationalism, and this position has been corroborated by the work of numerous historical geographers (e.g. Livingstone, 1992; Godlewska and Smith, 1994; Hooson, 1994; Withers, 2001).

4 Two of the better-known psychological theories of identity are social identity theory and self-categorization theory (see, for example, Tajfel, 1981; Tajfel and Turner, 1986 and Turner, 1987).

REFERENCES

Agnew, J.A. (1994) 'The territorial trap: the geographical assumptions of international relations theory', *Review of International Political Economy*, 1: 53–80.

Agnew, J.A. (2003) *Geopolitics: Revisioning World Politics*, (2nd edn). London and New York: Routledge.

Agnew, J.A. and Corbridge, S. (1995) *Mastering Space: Hegemony, Territory and International Political Economy*. London and New York: Routledge.

Anderson, B. (1991) *Imagined Communities: Reflections on the Origin and Spread of Nationalism*. London: Verso.

Anderson, J. (1986) 'Nationalism and geography', in J. Anderson (ed.), *The Rise of the Modern State*. Brighton: Harvester Press, pp. 115–42.

Anderson, J. (1996) 'The shifting stage of politics: new medieval and postmodern territorialities', *Environment and Planning D*, 14(2): 133–53.

Billig, M. (1995) *Banal Nationalism*. London: Sage.

Bloom, W. (1993) *Personal Identity, National Identity and International Relations*. Cambridge: Cambridge University Press.

Claval, P. (2001) 'Identity and politics in a globalising world', in G. Dijkink and H. Knippernberg (eds), *The Territorial Factor. Political Geography in a Globalising World*. Amsterdam: University of Amsterdam Press, pp. 31–50.

Cobban, A. (1964) *Rousseau and the Modern State*, (rev. edn), London: Allen & Unwin.

Cobban, A. (1970) *The Nation-State and National Self-Determination*. New York: Thomas Crowall.

Connor, W. (1972) 'Nation-building or nation-destroying?', *World Politics*, 24: 319–55.

Connor, W. (1978) 'A nation is a nation, is a state, is an ethnic group, is a ...', *Ethnic and Racial Studies*, 1: 377–400.

Connor, W. (1980) 'Nationalism and political illegitimacy', *Canadian Review of Studies in Nationalism*, II (Fall): 201–28.

Connor, W. (1994) *Ethnonationalism: The Quest for Understanding*. Princeton: Princeton University Press.

Cosgrove, D. and Daniels, S. (eds) (1988) *The Iconography of Landscape: Essays on the Symbolic Representation, Design and Use of Past Environments*. Cambridge: Cambridge University Press.

Day, G. and Thompson, A. (2004) *Theorizing Nationalism*. London: Palgrave Macmillan.

Fichte, J.G. (1922 [1806]) *Addresses to the German Nation*. Chicago: Open Court.

Fine, R. (1999) 'Benign nationalism? The limits of the civic ideal', in E. Mortimer (ed.), *People, Nation and State: The Meaning of Ethnicity and Nationalism*. London: I.B. Tauris, pp. 149–61.

Gellner, E. (1964) *Thought and Change*. London: Weidenfeld & Nicholson.

Gellner, E. (1983) *Nations and Nationalism*. Oxford: Basil Blackwell.

Gildea, R. (2002) 'Province and nation', in M. Crook (ed.), *Revolutionary France*. Oxford: Oxford University Press, pp. 151–77.

Godlewska, A. and Smith, N. (eds) (1994) *Geography and Empire*. Oxford: Blackwell.

Grano, O. (1981) 'External influences and internal change in the development of Geography', in D.R. Stoddart (ed.), *Geography, Ideology and Social Concern*. Oxford: Blackwell, pp. 17–36.

Hall, C. (1999) 'Gender, nations and nationalism', in E. Mortimer (ed.), *People, Nation and State: The Meaning of Ethnicity and Nationalism*. London: I.B. Tauris, pp. 45–65S.

Hamann, J.G. (1967) *Schriften zur Sprache*. Frankfurt am Main: Suhrkamp.

Hampson, N. (1991) 'The idea of the nation in Revolutionary France', in A. Forest and P. Jones (eds), *Reshaping France: Town, Country and Region During the French Revolution*. Manchester: Manchester University Press, pp. 13–25.

Hayes, C.J.H. (1945 [1931]) *Historical Evolution of Nationalism*. New York: Macmillan.

Hearn, J. (2006) *Rethinking Nationalism: A Critical Introduction*. London: Palgrave Macmillan.

Herder, J.G. von (1968 [1784]) *Reflections on the Philosophy and History of Mankind*, abridged and with an introduction by Frank E. Manuel. London and Chicago: University of Chicago Press.

Hogg, M. et al. (1995) 'A tale of two theories: a critical comparison of identity theory with social identity theory', *Social Psychology Quarterly*, 58(4): 255–69.

Hooson, D. (ed.) (1994) *Geography and National Identity*. Oxford: Blackwell.

Ignatieff, M. (1994) *Blood and Belonging: Journeys into the New Nationalism*. London: Vintage.

Jackson, P. (1989) *Maps of Meaning*. London: Unwin Hyman.

Johnson, N. (1995) 'Cast in stone: monuments, geography, and nationalism', *Environment and Planning D: Society and Space*, 13: 51–65.

Johnson, N.C. (2002) 'The renaissance of nationalism', in R.J. Johnston, P.J. Taylor and M.J. Watts (eds), *Geographies of Global Change* (2nd edn). Oxford: Blackwell, pp. 130–42.

Johnston, R.J. (1991). *A Question of Place: Explaining the Practice of Geography*. Oxford: Blackwell.

Johnston, R.J, Knight, D.B. and Kofman, E. (eds) (1988) *Nationalism, Self-Determination and Political Geography*. London: Routledge.

Katz, C. (2003) 'Social formations: thinking about society, identity power and resistance', in S.L. Holloway, S.P. Rice and G. Valentine (eds), *Key Concepts in Geography*. London: Sage, pp. 249–65.

Knight, D.B. (1982) 'Identity and territory: geographical perspectives on nationalism and regionalism', *Annals of the Association of American Geographers*, 62: 514–31.

Kristeva, J. (1991) *Strangers to Ourselves*. New York: Columbia University Press.

Lawrence, P. (2005) *Nationalism: History and Theory*. Harlow: Pearson Longman.

Livingstone, D.N. (1992) *The Geographical Tradition*. Oxford: Blackwell.

Lowenthal, D. (1994) 'European and English landscapes as national symbols', in D. Hooson (ed.), *Geography and National Identity*. Oxford: Blackwell, pp. 15–38.

MacLaughlin, J. (1986) 'The political geography of nation-building and nationalism in social sciences: structural vs. dialectical accounts', *Political Geography Quarterly*, 5: 299–329.

Mann, M. (1993) *The Sources of Social Power*, Vol. II: *The Rise of Classes and Nation-States, 1760–1914*. Cambridge: Cambridge University Press.

Marden, P. (1997) 'Geographies of dissent: globalization, identity and the nation', *Political Geography*, 16(1): 37–64.

Massey, D. (2004) 'Geographies of responsibility', *Geografiska Annaler*, 86(B): 5–18.

Mikesell, M.W. (1983) 'The myth of the nation state', *Journal of Geography*, 82(6): 257–60.

Özkirimli, U. (2000) *Theories of Nationalism*. London: Macmillan.

Özkirimli, U. (2005) *Contemporary Debates on Nationalism: A Critical Engagement*. London: Palgrave Macmillan.

Parekh, B. (1999) 'Defining national identity in a multicultural society', in E. Mortimer (ed.), *People, Nation and State: The Meaning of Ethnicity and Nationalism*. London: I.B. Tauris, pp. 66–74.

Paasi, A. (1995) 'Constructing territories, boundaries and regional identities', in T. Forsberg (ed.), *Contested Territory. Border Disputes at the Edge of the Former Soviet Empire*. Aldershot: Edward Elgar, pp. 42–61.

Penrose, J. (1994) ' "Mons pays ce n'est pas un pays" full stop. The concept of nation as a challenge to the nationalist aspirations of the Parti Québécois', *Political Geography*, 13(2): 195–203.

Penrose, J. (1995) 'Essential constructions? The "cultural bases" of nationalist movements', *Nations and Nationalism*, 1(3): 391–417.

Penrose, J. (1997) 'Construction, de(con)struction and reconstruction: the impact of globalization and fragmentation on the Canadian nation-state', *International Journal of Canadian Studies/Revue internationale d'études canadiennes*, 16 (Fall/Autumn): 15–49.

Penrose, J. (2002) 'Nations, states and homelands: territory and territoriality in nationalist thought', *Nations and Nationalism*, 8(3): 277–97.

Preston, P.W. (1997) *Political/Cultural Identity. Citizens and Nations in a Global Era*. London: Sage.

Quinton, A. (1994) 'Political philosophy', in A. Kenny (ed.), *The Oxford Illustrated History of Western Philosophy*. Oxford: Oxford University Press, pp. 275–362.

Robins, K. (1995) 'New spaces of global media', in R.J. Johnston, P.J. Taylor and M.J. Watts (eds), *Geographies of Global Change*. Oxford: Blackwell, pp. 248–62.

Rose, G. (1995) 'Place and identity: a sense of place', in D. Massey and P. Jess (eds), *A Place in the World? Places, Culture and Globalization*. Oxford: Oxford University Press/Open University Press, pp. 87–118.

Rousseau, J.-J. (1947 [1762]) *The Social Contract and Discourses*, translated and with an introduction by G.D.H. Cole. London: J.M. Dent.

Schöpflin, G. (2000) 'Civil society, ethnicity and the state' in *Nations, Identity, Power*. London: Hurst, pp. 35–50.

Schulman, S. (2002) 'Challenging the civic/ethnic and west/east dichotomies in the study of nationalism', *Comparative Political Studies*, 35(5): 554–85.

Seymour, M., Couture, J. and Nielson, K. (1996) 'Questioning the ethnic/civic dichotomy', in M. Seymour, J. Couture and K. Nielson (eds), *Rethinking Nationalism. Canadian Journal of Philosophy*, Suppl. Vol. 22. Calgary: University of Calgary Press, pp. 1–61.

Smith, A.D. (1991) *National Identity*. London: Penguin.

Smith, A.D. (1995) *Nation and Nationalism in a Global Era*. Cambridge: Polity Press.

Smith, G.E. (1996) 'The nationalities question in the Soviet Union', in G.E. Smith (ed.), *The Nationalities Question in the Post-Soviet States*. London: Longman, pp. 2–22.

Smith, S.J. (1993a) 'Immigration and nation-building in Canada and the United Kingdom', in P. Jackson and J. Penrose (eds), *Constructions of Race, Place and Nation*. London: UCL Press, pp. 50–77.

Smith, S.J. (1993b) 'Bounding the borders: claiming space and making place in rural Scotland', *Transactions of the Institute of British Geographers*, 18: 291–308.

Tajfel, H. (1981) *Human Groups and Social Categories*. Cambridge: Cambridge University Press.

Tajfel, H. and Turner, J.C. (1986) 'The social identity theory of intergroup behaviour', in S. Worchel and W.G. Austin (eds), *Psychology of Intergroup Relations*. Chicago: Nelson Hall, pp. 7–24.

Tuan, Y-F (1974) *Topophilia: A Study of Environmental Perception, Attitudes and Values*. Englewood Cliffs, NJ: Prentice Hall.

Turner, J.C. (1987) *Rediscovering the Social Group: A Self-Categorization Theory*. Oxford: Basil Blackwell.

Weber, E. (1977) *Peasants into Frenchmen: The Modernization of Rural France 1870–1914*. London: Chatto & Windus.

Weber, M. (1947) *The Theory of Social and Economic Organization*, trans. A.R. Henderson and Talcott Parsons. London and Edinburgh: William Hodge & Co.

Williams, C.H. and Smith, A.D. (1983) 'The national construction of social space', *Progress in Human Geography*, 7: 502–18.

Withers, C.W.J. (2001) *Geography, Science and National Identity: Scotland since 1520*. Cambridge: Cambridge University Press.

Xenos, N. (1996) 'Civic nationalism: oxymoron?', *Critical Review*, 10(2): 213–31.

Yack, B. (1996) 'The myth of the civic nation', *Critical Review*, 10(2): 193–211.

Yuval-Davis, N. (1997) *Gender and Nation*. London: Sage.

Yuval-Davis, N. and Anthias, F. (1989) *Woman-Nation-State*. London: Macmillan.

17

Working Political Geography Through Social Movement Theory: The Case of Gay and Lesbian Seattle

Michael Brown

WHERE AND WHAT ARE POLITICS?

'Politics today', writes Magnusson (1992: 69), 'is everywhere and nowhere, and this is very confusing for all of us'. Such a broad statement may seem trite, but it insists from the start that any understanding of politics must be sensitive to multiple geographies, and that certainly has not always been the case in the social sciences (Saunders, 1985; Agnew, 1989). Social movements (usually defined as organized groups of social actors that pursue collective political goals in society) have been one of the most compelling challenges to aspatial social science. And they are a big reason why Magnusson (1996) prods scholars away from an exclusively state-centric view of politics. Conceptualizing or operationalizing states as the natural and putative containers of politics obviously has its advantages and resonance with 'the real world' (Taylor, 1994), but it also has its limitations and blind spots (Taylor, 1995).

These blind spots lie in both empirical and theoretical lines of sight. Empirically, it is patently obvious that politics do not only take place in formal, public political venues like the legislature, city hall, or the voting booth. Thus the rise of distinct forms of social movements in the latter part of the twentieth century in Western democracies was a real-world challenge to the theoretical certainties of state theory. Social movements of various stripes challenged what gets defined as 'politics' or 'political'. But the blind spots also have broader intellectual ramifications. For if we do unhinge politics from the state, then just *where* do we – or should we – look for them? How do we map 'the political' if we jettison the state as our exclusive point of reference? Deeper still is the question, what is the proper, logical relationship between the political and the geographic? It is on this question that this chapter pivots.

While I agree with Magnusson that politics are indeed everywhere, and this ubiquity is indeed confusing, as a political geographer I decidedly reject the corollary that they are 'nowhere'.[1] Politics surely take place – quite literally – in particular times and spaces, even if the issues at stake seem timeless and ubiquitous, like the search for social justice, equality, or interest-group advancement.[2] Politics can also transgress particular places or particular spatial scales in complicated and confusing ways. That does not, however, mean that they are nowhere. It does mean that a careful attention to the relationships between politics and space must be central to social movement research.

The rise of an intellectual focus on social movements over the past thirty years or so has made the geographic exploration of politics challenging and exciting both empirically and theoretically. Yet this voluminous and international literature is a clutch of confusing theoretical perspectives, empirical case studies, terms and definitions (for helpful overviews see Touraine, 1985; Fainstein and Hirst, 1995; Miller, 2000). In order to bring some clarity to all this confusion, I will focus my discussion in this chapter on unpacking the 'everywhere' in Magnusson's above characterization of social movements. With that point of focus I want to convey a sense of how political geographers have extended and challenged social movement research in interesting and productive ways; how we sensitize non-geographers (and each other) to the spatialities of social movements. But as a geographer I think theoretical concepts are best understood not in some rarefied platonic form but through immediate deployment and use, so I will take these insights and illustrate them empirically through the case study of the rise of the gay and lesbian movement in Seattle, Washington, over the past fifty years.[3]

Now while it is certainly true that the gay movement has been an empirical mainstay of social movement research, its focus has typically been in often-studied cities like San Francisco and New York (Castells, 1983; Duberman, 1993). And it is true that the iconic hallmark of the gay rights movement has been the 1969 Stonewall Inn riots in New York. The social movement globally is often defined around that flashpoint. But for theory to be intellectually useful it must travel and shed insight into different places, often with surprising insights. For example, it has been empirically demonstrated that Seattle had a strong and vibrant social movement that had its beginnings in the 1950s – well before Stonewall 'started' the movement. The Seattle movement's relationship with state power and authority was extremely different from those of other cities (Atkins, 2003). And while Seattle is typically best known in political geography for its role in anti-globalization protests against the World Trade Organization (e.g. McFarlane and Hay, 2003), it also has been home to an impressively diverse and vibrant set of activisms around gay, lesbian, bisexual, transgendered – and more recently 'queer' – (hereafter GLBTQ) issues. In other words, I want to work social movement theory through the Seattle case study because while it illustrates key dimensions of that theory, it also provides interesting challenges to the geographical bias that focuses on places like New York or San Francisco (see also Knopp and Brown, 2003).[4] By demonstrating how theory travels (Said, 1983), I want scholars interested in social movements to appreciate more deeply the utility of bringing a geographical imagination to bear on political movements that have had both theoretical and empirical significance over recent history.

WHAT WAS SO NEW, OR SO URBAN, ABOUT SOCIAL MOVEMENTS?

Early work on this topic often spoke of '*new*' social movements (Offe, 1985; Pinch, 1985; Touraine, 1985, 1992) or '*urban* social movements' (e.g. Cox, 1984, 1986, 1988; Lowe, 1986; Fincher, 1987; Harris, 1987; Fincher and McQuillen, 1989). These terms' theoretical saliency is a bit fuzzy and often underappreciated, though, so I think it is useful to begin by explicating them.

New

The adjective 'new' bewilders many (especially students) because (as defined above) social movements were obviously a part of liberal democratic society before the tumultuous sixties. The American Suffragettes or the early German homophile movement, for example, were earlier forms of gender and sexual politics that were undeniably movements. So what was so new about the movements of the sixties? There seem to be three ways in which these social movements are treated as new by scholars interested in politics. The first is that they are not based in conventional locations of politics – party or class politics. So they were new because existing ways of thinking about the usual framings of politics were not helping scholars understand them. Hence, 'new' social movements are put in tacit contrast to 'old' ones (Katznelson, 1981). 'Old' social movements refer to working-class-based political organizations directed at capital owners or the state (Offe, 1985). Based unambiguously in politics centered on class struggles, these might appear on the shop floor, or in the form of political parties in the public sphere of politics. In this respect we might recognize trades or public-service unions, employee associations, or workplace affiliations, as well as political parties.

Seattle's local politics throughout the twentieth century could certainly be sketched out along these lines (e.g. Banfield, 1965). With a local economy based on natural resource extraction (fishing, forestry) and international air, rail, and ocean transport (the Port of Seattle) but also heavy-duty manufacturing (Boeing aircraft), a politics of class conflict is obvious (Berner, 1991). These forms of political associations were clearly based at the juncture of what Marxists would call 'class for itself' with 'class in itself'. Individuals organized on the basis of a mutually recognized and shared set of interests stemming more or less directly from the exploitation structurally ensured by the capitalist production process. Indeed, a great deal of theorizing in geography and elsewhere went toward connecting social movements' politics back to either class politics, or structural elements of late twentieth century capitalism (e.g. Cox, 1978; Lowe, 1986; Fincher, 1987; Scott, 1990).

Clearly this 'old versus new' dichotomy is intellectually and geographically problematic, not least because to say that social movements are not 'based' in class politics does not mean they are disconnected from them. It means they cannot solely be explained by referring back to the articulation or obfuscation of class-based interests. There are also strong links between middle-class identities and interests and new social movements, which also sometimes involve cross-class alliances of one kind or another. But social-movement theorists insist that if we trace causality between class movements and other social movements, the arrows must go both ways and in perhaps other directions too (Miller, 2000). So in Seattle's gay movement, for example, we might note that the Dorian Society, an early, pre-Stonewall 'homophile' organization formed in 1967, was composed largely of upper-class professional white men and their politics largely reflected their class (and gender and race) privileges. Members of the Dorian Society affected a middle-or upper-class air of respectability (often in the form of conservative dress, speech, and manners) that was more than just image; it was a political tactic designed to downplay difference from, and highlight accommodation to, 'normal' American culture. Class and respectability were tightly intertwined for the Dorians. Thus, many critics in Seattle traced the Dorian Society's form of politics

to their privileged class location (Atkins, 2003). They were criticized for being centrist rather than radical, accommodationist rather than conflictual, and assimilationist rather than oppositional. The Dorian members wanted to put a respectable face on homosexuality and their efforts produced some real gains – recall that in the mid-1960s sodomy was still illegal, and so anyone suspected of homosexuality was a potential criminal. Homosexuality was also considered an illness, and so people were not just legally but medically stigmatized as well.

By the 1970s, however, many other activists – especially those from working-class backgrounds – rejected the Dorian Society for more radical formations. The Gay Liberation Front (GLF) was one such organization formed in 1970 that was inflected with more working-class activists and allies. The GLF was a direct-action, rights-claiming organization that was highly visible and confrontational in its politics. In the words of one member, 'We looked down on the Dorian Society as a bunch of closet cases, which were afraid to push'. Later he would add, 'Pushy jerks is what we were. I was so much better because I was wearing a big Gay Power T-shirt, and they wouldn't be caught dead in anything but a three-piece suit' (quoted in Atkins, 2003: 125; NWLGHMP, 2000). The GLF tried to link anti-gay oppression with capitalist oppression in Seattle. And so in this 1977 photo of a gay rights march (Figure 17.1), we can see not just the joining but a clear problematizing of the 'old versus new'

Figure 17.1 The politics of sexuality and the working class connected. During this Gay Pride march in 1977, protesters hold a banner that reads 'Gay, Proud, and Anti-Capitalist'.

dichotomy territorially on the streets of Seattle. The marchers were both anti-capitalist *and* pro-gay.

A second sense of a social movement's supposed 'newness' stems from changes in the state itself. The rise of the welfare state during the postwar era meant that there were more interest groups self-directed toward getting or securing collective-consumption goods from the state, in part because there was more stuff available from the state to get. People organized into movements to acquire these resources in ways that, again, did not appear bound by traditional class or party lines. This distributional politics took many forms, from improving public services such as schools or public health, to 'securing the blessing of liberty', such as civil rights struggles where simply getting the allegedly universal right to franchise (simple voting) was at stake. As the postwar welfare state burgeoned in advanced democracies, this sort of politics of collective consumption for many theorists defined what local politics was about (Dunleavy, 1980; Pinch, 1985), and so it fit well within extant theoretical frameworks of urban politics. During the Fordist era of mass production and accommodation between big capital and big labor, local politics was often about the allocation of scarce urban resources (Pahl, 1970) through municipal service delivery systems. And thus a great deal of urban political geography traced how the consumption-based groups – which were often manifested territorially in neighborhood-based coalitions – struggled for increased or equitable service delivery in the form of schools, transportation, and health and safety (Katznelson, 1981; Castells, 1983; Cox, 1986; Harris, 1987).

For Seattle's gay movement in the 1970s, struggles over collective consumption took several different forms, though in ways that also reveal the limits of the reductionist collective-consumption view of urban politics. For instance, the movement was not as neighborhood-based as we would expect from the literature. A longstanding point of conflict between the City of Seattle and the gay community was over the itinerant harassment of gay bar owners and patrons. Thus much of the social movement's efforts – especially by the 1970s – were directed at reforms across various municipal government departments: police, public health, licensing, and (eventually) human rights, to name but a few. In 1973, a group of activists from the Lesbian Resource Center in Seattle testified at the City's Judiciary and Personnel Committee ostensibly about gender discrimination in hiring throughout the city. They pressed the need not only for an ordinance that would prevent the Seattle employers from discriminating against women in employment, but that would also protect gays and lesbians from such discrimination. Through a great deal of political intrigue, the City Council passed the ordinance and the Mayor signed it

that year. Two years later, activists would press City Hall for protection against housing discrimination in Seattle. Again, after much political wrangling between the gay community and City Hall (as well as within it), the measure passed in 1975. Securing these collective-consumption goods (housing and employment protection) produced a backlash by a nascent conservative movement that put 'Initiative 13' (a ballot measure to repeal those progressive measures) on the 1978 city ballot. The movement coalesced into three major organizations: Seattle Committee Against 13 (SCAT), Woman Against Thirteen, and CRFE (Citizens to Retain Fair Employment) which, despite clashes, managed to convince electors to defeat the anti-gay ballot initiative. While the Initiative ultimately failed, the intense politics around it became synonymous with the gay and lesbian social movement in the city, thus exemplifying the 'newness' of social movements in both senses described so far.

A third dimension of social movements' alleged newness was their magnitude and their confluence in time and space for most advanced Western democracies. The origins of new social movement studies may be found in the turbulent decade of the 1960s, with its foment spilling over into the 1970s (Gitlin, 1987; see also Adams, 1973). In advanced Western democracies, despite an expanding Fordist economy with its social compromise between the state, big capital, and big labor, all was not right. The student movements, civil rights struggles, women's and gay liberation, an emerging disabled-rights movement, environmentalism, indigenous and post-colonial movements (to name but a few) all took civil society (and the state) by storm during the 1960s and 1970s.[5] These movements went far beyond just wanting things from government. They challenged the political, economic, and cultural status quo in direct and conflictual ways that took many urban theorists by surprise. They were about challenging and changing the meanings of who 'we the people' are (see Marston, 1990). They reconfigured the terrain of what was normal, possible, fair, and equal in Western societies.[6] Beyond the goods they brought to activists and those who shared their identities, they changed culture. They altered and weakened hegemonic ideologies about what in society was good or true or beautiful. Even for those who disagreed with them, such organizations changed the political terrain regarding what was possible or reasonable in political debate and effects. This efficacy can be as subtle as it is overwhelming.

A pre-Stonewall example from Seattle is telling. At a time when the term 'homosexual' universally signified deviant, disgusting, mentally ill, and dangerous, Figure 17.2 challenged the city to rethink the category. It is the 1967 cover of a local lifestyle magazine in the city. It shows Peter Wichern, a young member of the Dorian Society. By agreeing

SEATTLE

THE PACIFIC NORTHWEST MAGAZINE   60 CENTS · NOVEMBER 196

This is Peter Wichern.
He is a local businessman.
He is a homosexual.
(For his story, see page 35.)

ALSO: NEW BLIGHT ON / WHERE TO GO FOR / AN ABOMINABLE
          OUR SKYLINE   / ITALIAN FOOD   / SNOWMAN, *HERE*?

Figure 17.2   Peter Wichern on the cover of 'Seattle' magazine, 1967.

to be featured on the cover of a citywide periodical, Wichern effectively came out of the closet at a time when most queer folks were deeply in the closet and leading double lives, and when the culture held negative stereotypes of gays. The goal was visibility and affirmation, but toward the end of making his fellow Seattleites realize that 'homosexuals' were indeed everywhere, throughout the city. They were not necessarily reduced to stereotypes or immediately identified that way. Never before had Seattleites been shown a respectable, ordinary likeness of 'a homosexual'. The goal with such a courageous move was to change culture: to increase the tolerance and respect for gays in the city.

Urban

If the heuristic virtues of calling some social movements 'new' are now clearer, what does their description as 'urban' add? In part, I think this term was used because that was where such movements were the most visible and undeniable. Still, some scholars sought a deeper meaning. It was Castells (1977, 1983) who identified these movements as especially 'urban'. He argued that these social movements were often expressed as territorially based groupings in specific city neighborhoods. Such movements were recasting 'the urban' away from its economistic meaning in capitalist society as 'exchange value' toward meanings that resonated with 'use value'. This use value in urban life was composed, in part, of a purposeful search for meaningful cultural identity for groups like gays and lesbians. Here there was a will to resist dominant inscriptions of meaning for a more autonomous and organic formulation of a shared and subcultural identity that somehow would be productive of a new nature of urbanism itself. These movements agonistically struggled to transform the meanings of social identities and relationships cross the entire culture, and to scholars of the time their flashpoint seemed to be located in the city.

The relationship between urbanity and social movements, however, should not be overdrawn. In retrospect, while the size, density, and heterogeneity that typify the urban may constitute a sufficient context of social movements, geographers have long since recognized that it does not constitute a necessary one (e.g. Brown, 2005). Indeed, in geography there has been ongoing attention toward rural social movements (Routledge, 1993; Cresswell, 1996; Reed, 2000; Fincher and Panelli, 2001; McCarthy, 2002). And so we should consider the limits to the urban geography of Seattle's gay movement. While its activism may have been formed and expressed within the city limits (see below), those boundaries never defined it exhaustively. For example, there were lesbian collectives and farms on Vashon Island and Lopez Island in Puget Sound. Gay Community Social Services had a 'back to the land' project at Elwah on the Olympic Peninsula. For nearly twenty years there has been a GLBTQ campground in rural Snohomish County, northeast of Seattle, which has provided a respite from urban life, but also a place of activism and solidarity for Northwest queers. In the late 1950s and early 1960s a gay men's social club called 'Jamma Phi' held drag balls and summer picnics in Maple Valley, a very rural town in southeastern King County, Washington. More recently, Knopp and Brown (2003) have noted several sites of queer activism across the rural Northwest. Still, just because social movements are not congruent with 'the urban' does not mean that they don't have important theoretical relations with space and place. And thus it is to those relations that I turn next.

POLITICAL GEOGRAPHY AND SOCIAL MOVEMENTS

Over the past twenty-five years, political geography has witnessed an explosion of interest, enthusiasm, and critique. A helpful chronicle is offered by Agnew (2002), who notes the empiricist, atheoretical tenor of much of political geography through the 1960s and 1970s. This pattern was reflected in early research on social movements in political geography, which tried to describe them or traced their spatial lineaments (Adams, 1973). Later, Marxism (as I noted above) became a leading means of theorizing these movements, but ultimately it proved only partially useful (see Miller, 2000, for a helpful review). As political geographers became more interested in working through increasingly diverse and sophisticated intellectual frameworks to understand their empirical interests, so too did they realize that the theories they were using could stand a good injection of 'the geographical imagination'. In other words, political geographers began to export their own theoretical insights toward political and social theory (Rasmussen and Brown, 2004). At least two insights about the geography of social movements are especially useful in terms of thinking through social movements generally and the Seattle GLBTQ movement specifically.

Place as a Constitutive Political Force

It is by now a disciplinary truism that 'all social relations are inexorably spatial, and vice versa'. For social-movement scholarship, however, geographers had demanded that these links be explicated. And by far the most careful, deliberate, and convincing example has been Byron Miller's (2000) excellent research on the anti-nuclear movement in metropolitan Boston during the 1980s. Miller deftly compares and contrasts four Boston, Massachusetts, area organizations, each situated in very different parts of the metropolitan area (Cambridge, Waltham, and Lexington). He convincingly documents how the particular forms of anti-nuclear protest made sense in places proximate to each other, but were meaningfully different from each other. While all of the locales indisputably benefit from the high-tech, defense-related industries commonplace in Greater Boston, each place had different class demographics and different employment conditions around the defense industry that affected the form the anti-nuclear movements took. The movements gained traction in these places because they made sense there, but also faced important impediments (either in the form of structure or agency of opponents) that were no less ontologically grounded in space and time. He shows how differences in both the local states and civil societies of each of these communities created opportunities and constraints for anti-nuclear activism. In his own summary:

> We can clearly see place-specific conditions shaping peace campaigns in Cambridge, Lexington, and Waltham. The class composition, political opportunity structures, education levels, bases of solidarity, economic histories, and histories of activism vary considerably among these three municipalities.

Freeze organizations in each municipality developed place-specific strategies for mobilizing against actions of the central state. Cambridge Freeze activists, responding to already widespread sympathies for peace in Cambridge and the city's class divisions, emphasized the lobbying of congressmen and senators and de-emphasized cross-class, citywide sociospatial recruitment and alliance building. Lexington Freeze activists, responding to the relative social homogeneity of their town and the strong political networks established through the town-meeting system, built a strong organization reaching all neighborhoods of Lexington.... Waltham activists approached a situation that seemingly defied successful peace mobilization – a working-class rather than a middle-class composition; a long history of industrial decline, making residents especially sensitive to threats to the employment base; an unfavorable political opportunity structure; little history of peace activism; and a poorly developed activist network – and developed their own place-specific strategy. (Miller, 2000: 167)

Sensitive to the structuration of place and politics, and refusing to divorce space from social processes, Miller has shown how place is a constitutive political force in social movements. Place brings people and issues together, of course, but it is the particular way that universals combine in specific locations that must be carefully traced.

Places, then, can be thought of as unique geographical bundles of social and environmental relationships (Cresswell, 2004), and social movements have productive and reciprocal relationships with the places they inhabit. So in Seattle (Figure 17.3) we can note that there have been three neighborhoods that helped constitute a GLBTQ social movement. The earliest gay bars and venues were located in Pioneer Square, just south of the central business district.[7] During the mid-twentieth century, gay and lesbian identities could only largely be expressed privately in private space, or in concealed ways furtively in public space. Thus, bars, taverns, and bathhouses provided precious, hard-to-see spaces in which shared identity could be recognized and sexual desire could be manifest.[8] They became the most public spaces in which gays and lesbians could meet one another and recognize that they were not alone.

For the most part, these bars and other venues were not randomly dispersed in the city but concentrated in Pioneer Square. Known as a working-class, red-light district of gambling, saloons, and prostitution since the Gold Rush era of the late nineteenth century, Pioneer Square's status as 'Skid Row' ironically helped homosexuals to connect and interact with one another socially, sexually, and eventually politically. Placed in the midst of such social vices, it was hardly surprising that

people with same-sex desires coalesced there for social contact, since they were already ostracized from mainstream culture. Its heyday spanned a remarkable pre-Stonewall period from 1933 to 1958 (Atkins, 2003), though several gay and lesbian bars opened (and closed) up through the 1970s. While there were few residences, it became a place where both women and men who pursued same-sex desires met in saloons, bars, taverns, and later cabarets. The earliest activist groups and community centers were located there. The earliest gay parades were there.

Throughout the 1970s and into the mid-1980s, by contrast, lesbian activism was constituted in and through the University District (U-District), the Northeast Seattle neighborhood adjacent to the University of Washington. A rising feminist consciousness in the student movement, the founding of a Department of Women's Studies, and the creation of a Women's Center on campus all helped constitute the U-District as a central place for lesbian politics in the city. Frustrated with the masculinism in the Gay Liberation Front, some lesbian activists formed the Gay Women's Alliance in 1970. Originally the political movement centered in the U-District YWCA, and it quickly grew into the Lesbian Resource Center (LRC) in 1971. Other venues included a women's bookstore and coffee house. The women's newsletter *Pandora* was readily available in these venues, and eventually in 1976 it begot the lesbian paper *Out and About*. For these women, the U-District allowed a women-friendly, feminist space where lesbians could organize and express themselves. The LRC provided a space where lesbians and women questioning or exploring their sexuality could freely meet with others like them and exchange their thoughts, feelings, and mutual support. The 'CR' (consciousness-raising) groups worked because of the safe environment lesbians had created in their U-District spaces. Furthermore (and like the other areas), while never exclusively gay, the U-District became a geographic destination, where they could arrive and be supported and accepted.

Finally, Capitol Hill became the 'gay ghetto' or 'gay Mecca' by the 1980s. It is now synonymous with queer culture and politics in Seattle. It is a dense retail and residential area that rapidly gentrified throughout the 1980s. Its retail strips allowed for gay-friendly businesses like bars, restaurants, and bathhouses to open. Its mixture of high-density rental and condominium housing with old Victorian housing stock allowed single and coupled gays and lesbians to locate there, as well as renters and owners. Indeed it presently has the highest proportion of same-sex couples in the city (Gates, 2003).[9] It also now has a number of both lesbian and gay venues. For most people in Seattle, Capitol Hill is now a spatial metonym for the GLBTQ community in the city (see Brown, 2000). Each of these

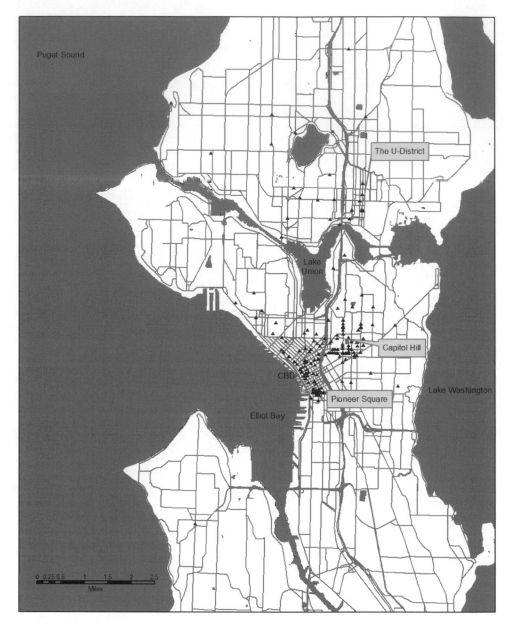

Figure 17.3   Gay and lesbian neighborhoods in Seattle, over the twentieth century. Triangles represent sites of significance for the GLBTQ community.

neighborhoods (Pioneer Square, the U-District, and Capitol Hill), not only situated social movements, they fueled them by drawing people to live, work, and play there. They were beacons for sexual dissidents to move toward, and oral histories are replete with the symbolic, political drawing power these areas had for people looking to leave home in search of identity and acceptance (Pettis, 2002).

Without such territoriality, the social movement in Seattle would not have obtained such prominence.

Place can also be used as a means of resistance. Indeed, a panoply of geographers have demonstrated the myriad ways this is possible (see, for instance, Routledge, 1993; Pile and Keith, 1997; Sharp et al., 2000). One of the most important parts of the geography of the U-District was simply that

Figure 17.4   Place as a means of resistance. Gay Pride march, downtown Seattle, 1977.

it was not Pioneer Square (or Capitol Hill). Lesbians felt they needed a space away from what
Nast (2002) now calls the 'queer patriarchy' that
grew synonymous with gay men's territoriality
in Pioneer Square. For most of its pre-Stonewall
history, Pioneer Square was a place where gays
and lesbians commingled, sometimes in the same
bars. The women-centered spaces in the U-District,
however, promised something different. Many U-
District lesbians also objected to the lesbians who
socialized in Pioneer Square, with their own patriarchal stereotypes of 'butch' and 'femme'. They
were seen as an older depoliticized generation that
this new generation rejected. We can also see elements of this rejection spatially when we note that
there were elements of lesbian separatism in the
community. A bookstore downtown, and collective
farms in the hinterland, for example, barred men
(gay or otherwise) from entering. Here territoriality is directed inward at building solidarity, but also
outward as a means of resistance.

Perhaps the most common form of this oppositional political geography in Seattle manifested in
the form of Pride Parades and protest marches (see
Figure 17.4). Occupying central spaces of the city,
these acts of resistance claimed heteronormative
space and thereby resisted homophobia and closeting. Thus if the already constituted elements of
place (such as the character of the housing stock
or of the commercial activities in an area) provide bases for social-movement politics, so too
can struggles over what these elements are or

should become. Tim Cresswell's (2004) fascinating work on what might be called 'anachorism'
(literally being out of place) illustrates this point
beautifully. The most obvious example of such
anachorism is visible, undeniable protest, where
social movements disrupt and usurp public space
by taking to the streets to protest an existing injustice and inescapably challenge those around them
to not just be aware of it, but to help address it.
Protest marches have a longstanding history in
queer Seattle's history. Early gay rights marches
took place from Pioneer Square down 1st Avenue,
a major downtown thoroughfare. This kind of physical occupation challenges the spatial structures of
heteronormativity. It exemplifies territorial claims
of existence, recognition, and voice. Think, for
example, of the popular chant: 'We're here! We're
queer! Get used to it!' These marchers are not just
protesting, they are demanding that heterosexual
(and even closeted homosexual) Seattleites change
their views, and thereby change the culture. They
resisted the closeting by which heteronormativity
erases and demands the concealment and erasure
of queer folks. What better way to resist – to resist
closeting – than to territorially claim space?

New Locations and Scales of Politics

While debates over the ontology and epistemology
of social movements were raging in urban political
studies, political geographers were focusing great

attention on the state – and specifically the nature of the local state. Dear (1986) went so far as to insist that it be the fundamental unit of inquiry. Social movements, of course, were already challenging that anchor point theoretically and empirically. They were about things beyond the state, and they often took place in locations far afield from city hall or even the community center. They suggested, as this chapter's opening quote witnesses, that politics went far beyond the single point of the state, no matter how variegated and complex it was. And it was scholars of identity-based movements like feminism (Brownhill and Halford, 1990) and gay rights who showed us that concentrating solely on formal political spaces around the state left us with a narrow and incomplete understanding of the full range of political geographies across civil society and the home and family (Brown, 1997). My discussion of Seattle already has demonstrated this point. A great deal of activism occurred in places we do not necessarily think of as political (like bars or lifestyle magazine covers). Still, it is worth elaborating on this point further.

If we look 'beyond' the state, we typically find spaces of *civil society* (areas of interaction with strangers, participation in organizations, transacting economic relations, etc.). Civil society

is typically spatialized in the form of public spaces in the city, or places of business or social clubs. In Seattle, spaces in civil society and the home and family were extremely important for the GLBTQ social movement. Early on, bars provided much needed social interaction, but they also provided sources of community and strength upon which politics flourished. They were places of affirmation and hope that infused the politics of the early movement. They inculcated a sense of community and belonging that formed a 'we', so necessary to social movement formation. They became sources of political action as well, when we consider efforts to combat police harassment and an ongoing racketeering scheme that police had running with the gay bars for decades.[10] If community organizations like the Lesbian Resource Center and the Gay Community Center are the obvious spatializations of social movements in civil society, we should not forget that more recreational and sexualized spaces were vital too (Figure 17.5).

But other kinds of spaces in civil society were important too. It is true that much of lesbian and gay activism took place in and around (or was directed against) the state. But that was never the whole story. An especially iconic space of civil society has been Seattle Center, former site of the 1964

Figure 17.5   Interior of Golden Horseshoe Bar, Pioneer Square. Because homophobia and heteronormativity lead to intense feelings of isolation and alienation, place was integral in bringing people together to resist such closeting. Bars were especially important places for GLBTQ people to build community and act on their stigmatized desire. In Seattle, they also became points of focus for the gay movement itself, because of police harassment.

Figure 17.6   Protesters outside of Seattle television studio, protesting Lloyd Cooney's anti-gay editorials, 1977. Spaces of civil society allow private issues to become publicized and thereby provide effective means of resistance to those who would closet GLBTQ people.

World's Fair and home to the city's iconic Space Needle. There, in 1978, the Imperial Court (a drag organization) held its coronation ball at the Center's exhibition hall, which was attended by Mayor Charles Royer and his wife. In the early 1990s, members of the AIDS Coalition to Unleash Power protested at the Needle, in order to advocate for a needle-exchange program to prevent the spread of HIV/AIDS among intravenous drug users. Safer-sex campaigns there also featured illustrations with the phallic Space Needle sheathed in a giant condom. Yet, such activism in civil society also takes place in rather ordinary sites too. Figure 17.6, for example, shows a demonstration against a local CBS television affiliate and its general manager, Lloyd Cooney, an ardent Mormon convert who editorialized on the air repeatedly against gay people and gay rights. Used to 'going out' and reporting on the news elsewhere in Seattle, suddenly the television station's studio itself became the news because of its treatment of angry gay and lesbian protestors.

Social movements also inhabit the *home and family* in a wide variety of ways. This geography has been especially confusing for social-movement theorists, since it complicates the assumption that politics is properly placed in public spaces and settings. Yet because social movements are about cultural meaning and self-identity, intimate and private spaces of the home cannot be ignored

(Brownhill and Halford, 1990; Staeheli, 1996; Brown, 1997). Again the Seattle case provides some fascinating examples of the social movement's political geography here.

While there is a raging debate over gay marriage presently across the United States, in Seattle the issue appeared as early as 1971 when two members of the Seattle Gay Alliance, John Singer and Paul Barwick, fronted up at City Hall to get a marriage license, in order to demonstrate the extent of heteronormativity in our culture. When the office refused to issue the license, the pair sued in court. In 1974 the Washington State Court of Appeals ruled that the denial was neither discriminatory nor unconstitutional. Nevertheless, social movements were clearly politically active in this sphere. The home also constituted politics by germinating them. Old-timers, for instance, remember the political significance of lesbian collective houses like 'Dyke Corner' or 'Hespera House' near the U-District or 'The Red Hen', 'Pud Street', and 'The Sorority House' up on Capitol Hill in the seventies and eighties. These residences were places thoroughly infused with politics. Marches and protests were organized there. Debates flourished over political possibilities around kitchen tables and in bedrooms. Posters were painted on living-room floors. Most of all, new ways of performing family and kinship were lived. New forms of loving and caring relationships were literally put into place. These domestic spaces brought people and ideas together and organizing for political action followed.

Control over the home and family themselves was surely a political geography in Seattle in which the GLBTQ movement figured. Lesbians who divorced their husbands, for instance, were (and in many places still are) automatically denied custody of their children. Such was the context in 1972 for Sandra Schuster and Madeleine Isaacson, a Seattle lesbian couple. Their ex-husbands both filed for custody of their children in 1974 on the grounds that not only their lesbianism, but also – ironically – their shared Christian fundamentalist faith, would harm the children. In his ruling, King County Superior Court Judge James Noe ruled that the women could each keep custody of their respective children. Astonishingly, however, he actually ordered the women to break up their home and live separately. It is not surprising that a Gay Mothers' Union formed at the LRC in 1972 and a Lesbian Mothers' National Defense Fund (see Figure 17.7) began in Seattle two years after Noe's ruling. The women ingeniously retaliated in 1978 by each renting apartments across the hall from one another in the same building. Eventually, the Washington State Supreme Court decided not to break up the family.

Political geographers have also prodded social-movement scholars to consider new *scales* of political action too – both in terms of the scale at which politics occurs, and the scale that scholars use to frame their analysis (Delaney and Leitner, 1997; Kurtz, 2003). For while Tip O'Neil's aphorism, 'All politics is local', is certainly true, local politics can often affect politics elsewhere. Local politics can be quite national or even global. A handful of early GLF and Lesbian activists in the city, for instance, were involved in progressive and radical politics in Cuba and elsewhere. Seattle queer politics produced several firsts and initiatives that have diffused across the nation and even internationally. The Lesbian Mothers' National Defense Fund was the first organization in the country to assist lesbians in custody battles with their ex-husbands. In the late 1960s, the Seattle Counseling Service for Sexual Minorities was a pre-Stonewall mental-health clinic for the GLBTQ community. 'Stonewall' was a Seattle organization founded in 1971 as a gay residential treatment center for substance abuse. It was the first of its kind in the nation. Lambert House on Capitol Hill opened in 1991 as the first drop-in center for GLBTQ youth, and it remains open today. My point here is that the Seattle movement produced these firsts, which undoubtedly have diffused insights to other places elsewhere. And thus an important, but often hard to trace, geography of social movement lies in these spatial diffusions and connections.

LESBIAN MOTHERS'

NATIONAL DEFENSE FUND

''Raising our Children is a Right, Not a Heterosexual Privilege''

Figure 17.7 Flyer for the Lesbian Mothers' National Defense Fund. Social movements cross-cut public/private divides in complicated ways. While we typically do not think about the private sphere of the home and family as politicized space, the state has been quite active in structuring 'proper' and 'normal' forms of kinship and families. Seattle lesbians have been active since the 1970s in fighting to defend lesbian mothers from custody battles that would remove their children from them by state power.

The micro scale is also quite important to consider, geographers have recently argued, and the Seattle case also illustrates this point. What about less considered scales like the body, for instance (Longhurst, 2000)? In Seattle, Atkins (2003) argues that its importance to the gay movement cannot be underemphasized. Dressing against one's gender was illegal in Seattle until the 1960s, and it was certainly cause for harassment and violence for both gays and lesbians in the city. Yet Atkins (2003; see also Newton, 1972) argues that the appearance of drag (both male and female) in the 1950s Seattle gay bars was a means of political resistance toward heteronormativity and even homophobia. It deliberately flouted conventional heteronormative comportment and dress (Figure 17.8). Oddly, dance was no less a political geography of the body throughout the gay movement. In the late 1950s and early 1960s, same-sex couples began to dance

Figure 17.8   Paris DeLair, a popular female impersonator in Seattle's gay bars, poses with two uniformed servicemen in 1955 at the Garden of Allah. Female and male impersonators, and drag artists, performed resistance to heteronormativity with their bodies and dress. They remind us that the body is also a scale of political geography.

together in Seattle bars, first at the Madison and the Casino.[11] Dancing at the Madison (a lesbian bar) was especially salient since it was above ground, while the Casino was in a basement. Police had to be paid off to let women dance together above ground. Later, the Golden Horseshoe (Figure 17.5 above) became the first regular-hours bar where men could dance openly with one another. While such moves (literally) might seem trivial, Atkins (2003: 77) notes that:

… dance often replaces an otherwise forbidden verbal expression. That which cannot yet be spoken

publicly can instead be danced, and so, to dance together in any sort of public setting was both to enjoy a self-acceptance and to risk giving more offense to outsiders who happened to learn about it.

Even when formal political organizations manifested in the gay movement during the 1970s, Atkins argues, the physical movement of dance continued to be important. He interprets the rise of disco as exemplifying the movement, freedom, publicity, and energy of one's sexuality. Seattle discos were also noteworthy for their ability to draw

**Figure 17.9   Advertisement for a University of Washington Gay Students'
Alliance dance from 1974. For a movement focused on issues of desire, sex,
love, and intimacy, dance provided an important means by which 'the body'
could combat homophobia and challenge heteronormativity. This poster
captures those resistances well.**

straight folks in to dance – but clearly on queer
folks' turf (Figure 17.9).

CHALLENGES AND OPPORTUNITIES

In the previous section, I have attempted to demonstrate some of the insights political geographers
have offered other political scholars, with specific reference to the GLBTQ social movement in
Seattle. I would like to leave Seattle behind now
and reflect on some of the current challenges that
face thinking about political geography and social

movements generally. These challenges are both
intellectual and practical, and there are no simple
solutions to them, which makes confronting them
all the more urgent. My hope is to build on what
has come before in political geography and social
movement theory, and push our understandings
further.

The Movement Versus Movements

Perhaps the grandest and most vexing challenge
is postmodernism's influence. If we define it
broadly as 'incredulity to metanarrative', in other

words a position of skepticism, nonplussedness or suspicion toward universal truths that hold across all times and spaces, how are we to understand and respond to social movements if they are each so idiographic and identity-based? This issue is particularly problematic because social movements are based in particular identities or positionalities. Such situatedness raises the political dilemma of how different agendas can be integrated and harmonized when they are often, at best, going in different directions and at worse in irreconcilable tension. Was the Seattle narrative I offered above about a single movement, or a series of cross-cutting, often irreconcilable ones? What are the intellectual and political costs to answering either way to that question? The narrative was partial, but which parts were left out and what are the costs? Scholars and activists have tried to textually solve these postmodern intractable dilemmas with the new appellation 'queer', meant as sort of a catch-all for sexual dissidents, but even its usage can be quite exclusionary and fracturing (see Sothern, 2004).

On what grounds can these very different identities and agendas come together? Political theorists have noted that claims to abstract universal principles are often part of the problem, since the ways those principles inevitably get grounded in real places and times often limit their progressive potential (Laclau and Mouffe, 1985; Rasmussen and Brown, 2005). Recognizing these postmodern implications for political action, however, can also be an opportunity: a means to pry open closures and gaps in already existing movements or situations. It is an opportunity to further democratize politics and movements by articulating new or ignored struggles. Here, recent feminist and queer political theory has raised the specter of identitarian politics. Both Halberstram (2005) and Wendy Brown (2001) mark the limitations of a liberal identity politics by drawing attention to the problems of rights-based claims for the cultural valorization of heretofore disvalued identities. In Brown's (2001) words:

> contemporary identity based institutions, born of social critique, invariably become conservative as they are forced to essentialize the identity and naturalize the boundaries of what they once grasped as a contingent effect of historically specific social powers. (2001: 35)

Later she defines the problem more specifically:

> identitarian political projects are very real effects of late modern modalities of power, but as effects, they do not fully express its character and so do not adequately articulate their own condition; they are symptoms of a certain fragmentation of suffering lived as identity rather than as a general injustice or

domination – but suffering cannot be resolved at the identitarian level. (2001: 39)

Queer geographers have only just begun to address the difficult consequences of such identitarian consequences of social movements (see, for example, Sothern, 2004, 2007).

Grassroots, Sequoias, and Astroturf

A second challenge has to do with the organization and morphology of social movements themselves. The success of a social movement does not come without its own costs or the creation of new political issues and dynamics, and so a great challenge may be the growth – both organizationally and geographically – of movements themselves. In other words, movements that start as small, itinerant grassroots entities sometimes grow into enormous, highly structured bureaucracies that bear little resemblance to their humble roots. Many activists grow disillusioned with the larger, more impersonal environments that highly successful grassroots organizations often become. Organizations that started out as 'grassroots' organizations have grown over the years into large, highly bureaucratized, well-financed, towering 'sequoias'.[12] There is often a danger in their size and morphology working against the very political values and principles upon which they were founded.[13]

The term 'Astroturf' refers to coalitions and political machinations that appear to be coming from grassroots, but are really highly organized, well-funded, and deliberately obfuscating about their position. It is 'fake grassroots' activism. Cultural conservatives in the United States have been masters at this tactic. For example, highly organized and well-orchestrated letters to the editor in a local paper may make it seem like the community is broadly opposed to gay rights, when really it is a small cadre of dedicated activists with a word processor who have as many pseudonyms as they do postage stamps. It pivots on the success that social and cultural movements have had, and the goodness they presumably embody. Astroturf tactics are designed to make very small movements seem as if they were in the majority, or to make groups larger and more influential and ubiquitous than they really are. They can affect popular opinion or pressure media to frame an issue a particular way. Most dangerous, however, is that their form of politics conceals the magnitude of their own power and influence. At a time when geographers are calling for careful research into new spaces and scales of politics, at a time when cyberspace, telespace, and mediascapes increasingly become the geography of the public sphere, Astroturf must be carefully tread.

Bringing the State Back In

If there was a turn away from state-centric political geographies for those who studied social movements, new directions in state theory may be forging some new and interesting switchbacks. As noted earlier, social movements very often act as pressure points on various apparatuses of the state, but their members often inhabit the state, and vice versa. Part of the success of the gay movement in Seattle involved it not just opposing the state, but having GLBTQ folks and their allies at key nodes within the state itself. In this way, Duggan's (1995) admonition that we 'queer the state' remains an interesting and ongoing line of research for queer scholars (for some excellent examples, see Cooper, 1995, 1998, 2004). Furthermore, despite the retrenchment of the welfare state, social movements around queer issues remain directed at or through the state. Witness issues like census categories, marriage, taxation, immigration, and sexual health. Each of these (and more) suggests that 'old' locations like the state are not to be ignored by social-movement scholars. Moreover, theoretical advances such as governmentality (Brown, 2000) and actor-network theory (Knopp, 2004) might provide helpful theoretical frameworks in studies that bring state theory and social-movement theory together in political geography. Empirically, it appears (at least in the United States) that issues such as gay marriage and civil unions, adoption and family law, immigration, international aid, public health, sex education, *inter alia*, will certainly continue to require further inquiry into state/social movement relations in political geography.

CONCLUSION

My purpose in this chapter has been twofold. First, I have tried to explicate the concept of social movements. I did this by tracing out the lines of consensus and debate that historically evolved around thinkers and theorists. Second, I wanted to bring a geographical imagination to bear on the topic. I did that by spatializing the GLBTQ movement historically in Seattle throughout the twentieth century. I structured that discussion by foregrounding the important ways that political geographers have insightfully contributed to thinking about the relations between politics and space/place in relation to social movements.

When I teach about social movements in my political geography course, students often seem nonplussed. Most of them have grown up in a world where such forms of politics have been the rule, rather than the exception. Indeed, it is harder for them in 'middle-class' Seattle, and the US (where everyone is deemed middle-class), to understand or appreciate old class-based social movements, even in spite of the WTO protests here. Most of them accept social movements, like the one from gay Seattle narrated above, as quotidian to politics in their lives. Their attitude is sort of, 'What's the big deal?' As I hope I have shown in this chapter, the big deal is that the relative importance of social movements has only recently been recognized in political geography. Indeed, taking social movements seriously has required a profound rethinking of the sub-discipline. Furthermore, even if we handily accept such a rethinking that politics is indeed everywhere, the challenge becomes understanding the nature of politics and the relevance of the relationship between the political and the spatial. We must always ask, in what sense are these movements political? How does their spatiality enable or constrain their movements, their aims, and their consequences? And we must look beyond both the immediately local scale and that of the state for cause and effect, even for the seemingly most locally attuned social movement.

One of the costs of framing my account of social movements around a specific movement (GLBTQ politics) in a particular time and place (postwar Seattle, Washington, USA) is that it curtails imagining other movements elsewhere. And so readers who are interested in other popular social movements (say, ones in the global South), or in issues not directly related to sexuality or gender (say, around food security, or 'race' issues), may be left wondering if this discussion has any bearing on the political geographies that concern you. Remember, my point in focusing on the GLBTQ movement in Seattle was not to suggest that it was the best example of a social movement, but rather to show the importance of developing a geographic imagination around that topic. Magnusson is right: politics *are* indeed everywhere. Yet they are never nowhere (*sic*), and that's precisely why political geographers have so much to say about them.

ACKNOWLEDGMENTS

I want to thank Dr. Pat Freeman, Larry Knopp, Ruth Pettis, Jennifer Robinson, and Matt Sothern for their help in clarifying my ideas and reading previous drafts of this chapter. Thanks also go to the Northwest Lesbian and Gay History Museum Project, and to Gary Atkins, for the empirical material in this chapter. Any mistakes or omissions are mine alone. I also want to thank the Project for permission to reproduce the images that really bring a geographic imagination to bear on social movements.

NOTES

1 To be fair, I suspect Magnusson would agree, given the serious and careful attention his work (e.g. Magnusson, 1996) gives to the geographies of politics and the political. I believe he is trying to signal that politics are not in any *single* or *particular* location per se, not that they are placeless, or timeless.

2 Consider, for example, Cox's (2002) assertion that territoriality (the exercise of power or control through space) is the fundamental point of focus for political geography. Social movements often 'take place' as a means to press their aims and goals, as I demonstrate later in this chapter.

3 Data for this chapter come from three sources. First is Atkins' (2003) historiography of gay and lesbian Seattle. While there have been other pieces of scholarship and popular history on the topic (gay bar study; Simpson and Paulson, 1996), Atkins' is the most comprehensive. Second is my own participatory research and cartography with the Northwest Lesbian and Gay History Museum Project (Brown, 2004). Third is a collection of oral histories of GLBTQ folks from Seattle and the Northwest United States (Pettis, 2002).

4 This theme of a Stonewall/New York-centric reductionist metanarrative for American lesbian and gay history comes out well in an edited volume of oral histories from GLBTQ Seattleites . See Pettis (2002) for their own voices on this bias.

5 These movements were obviously based in an oppositional progressive politics of the left (largely as a reaction to the right-wing tenor of the postwar era and a failure of the status quo politics of the era to have social inequalities addressed and have society live up to its ideals). These new political factions and fissures were by no means confined to the left. Neoconservatives, family values, Christian fundamentalist, white supremacist, and libertarian movements also took hold of the map and began redrawing lines of political geography (see Flint, 2004). Typically in geography, however, scholarly attention has been paid rather more to progressive social movements.

6 Truly, it is because they have been so successful in many respects that students nowadays have such difficulty appreciating their theoretical import. That you-reading this chapter can even be considered normal (never mind perhaps 'good'), that African-Americans now have the right to vote, serious consideration of non-human life as having integrity and worth, debates about the destruction of our own species through nuclear warfare, women receiving equal pay for work, and the fact that 'gay marriage' is on the agenda of national and local politics in the US are all testaments to what is so significant about new social movements.

7 Yesler Way is a central east–west artery that defines Pioneer Square. It is where the metaphor 'Skid Row' comes from, since during pioneer days, skids used to be rolled down its steep incline toward Elliot Bay, where the sawmills and ships were. See Morgan (1982).

8 Of course, whether bars are 'public' spaces of interaction with strangers, or 'private' spaces of business property, depends on how they get defined. Geographers have explored these hybrid public/private geographies. For an excellent example, see Staeheli (1996).

9 By highlighting these three areas I do not mean to suggest that they are the only areas that queer Seattleites inhabit. For broader views, see Brown (2004) and Gates (2003).

10 The police payoff system was a fascinating and complex story that is too intricate to detail here, but is certainly demonstrates Seattle queer politics' very specific and unique dimensions. Readers should consults Atkins (2003, esp. chaps 4 and 12).

11 The Casino was jocularly known as 'Madame Peabody's School of Dance'.

12 Sequoias are, of course, extremely tall cypress trees native to California that typically reach over 300 feet (90 meters) in height.

13 This was certainly apparent during my ethnographic research on AIDS activism in Vancouver, Canada, in the early 1990s. The rapid growth of the disease and various responses to it were changing the form of politics in Vancouver considerably before activists' eyes. It was extremely common for early activists to leave their organizations in frustration because they felt they had become too impersonal, too politically conservative, and less gay and compassionate. See Brown (1997) for an account.

REFERENCES

Adams, J. (1973) 'The geography of riots and civil disorders in the 1960's', in M. Albaum (ed.), *Geography and Contemporary Issues*. New York: Wiley, pp. 542–64.

Agnew, J.A. (1989) 'The devaluation of place in social science', in J.A. Agnew and J.S. Duncan (eds), *The Power of Place*. London: Unwin Hyman, pp. 9–29.

Agnew, J.A. (2002) *Making Political Geography*. London: Arnold.

Atkins, G. (2003) *Gay Seattle: Tales of Exile and Belonging*. Seattle: University of Washington Press.

Banfield, E. (1965) *Big City Politics*. New York: Random House.

Berner, C. (1991) *Seattle in the 20th Century*. Seattle: Charles Press.

Brown, M. (1997) *Replacing Citizenship: AIDS Activism and Radical Democracy*. New York: Guilford Press.

Brown, M. (2000) *Closet Space: Geographies of Metaphor from the Body to the Globe*. London: Routledge Press.

Brown, M. (2004) *Claiming Space: An Historical Geography of Lesbian and Gay Seattle*, map available through the Northwest Lesbian and Gay History Museum Project. PMB 797, 1122 E. Pike Street, Seattle, WA 98122, USA.

Brown, M. (2005) 'Letter to Engin', *Political Geography*, 24: 361–4.

Brown, W. (2001) *Politics Out of History*. Princeton: Princeton University Press.

Brownhill, S. and Halford, S. (1990) 'Understanding women's involvement in local politics: how useful is the formal/informal dichotomy?', *Political Geography Quarterly*, 9: 396–414.

Castells, M. (1977) *The Urban Question*. Cambridge, MA: MIT Press.

Castells, M. (1983) *The City and the Grassroots*. Berkeley: University of California Press.

Cooper, D. (1995) *Power in Struggle*. New York: New York University Press.

Cooper, D. (1998) *Governing Out of Order*. New York: New York University Press.

Cooper, D. (2004) *Challenging Diversity*. Cambridge: Cambridge University Press.

Cox, K.R. (1978) 'Local interests and urban political processes in market societies', in K.R. Cox (ed.), *Urbanization and Conflict in Market Societies*. Chicago: Maaroufa, pp. 94–111.

Cox, K.R. (1984) 'Neighborhood conflict and urban social movements, questions of historicity, class, and social change', *Urban Geography*, 5: 343–55.

Cox, K.R. (1986) 'Urban social movements and neighborhood conflicts: questions of space', *Urban Geography*, 7: 536–46.

Cox, K.R. (1988) 'Urban social movements and neighborhood conflicts: mobilization and structuration', *Urban Geography*, 8: 416–28.

Cox, K.R. (2002) *Political Geography*. Oxford: Blackwell.

Cresswell, T. (1996) *In Place/Out of Place*. Minneapolis: University of Minnesota Press.

Cresswell, T. (2004) *Place: A Short Introduction*. Oxford: Blackwell.

Dear, M.J. (1986) 'Theory and object in political geography', *Political Geography Quarterly*, 5: 295–7.

Delaney, D. and Leitner, H. (1997) 'The political construction of scale', *Political Geography*, 16: 93–7.

Duberman, M. (1993) *Stonewall*. New York: Dutton Press.

Duggan, L. (1995) 'Queering the state', in L. Duggan and N. Hunter (eds), *Sex Wars*. London: Routledge, pp. 179–93.

Dunleavy, P.J. (1980) *Urban Political Analysis: The Politics of Collective Consumption*. London: Macmillan.

Fainstein, S. and Hirst, C. (1995) 'Urban social movements', in D. Judge, G. Stoker and H. Wolman (eds), *Theories of Urban Politics*. London: Sage, pp. 181–204.

Fincher, R. (1987) 'Defining and explaining urban social movements', *Urban Geography*, 8: 152–60.

Fincher, R. and McQuillen, J. (1989) 'Progress report: women in urban social movements', *Urban Geography*, 10: 604–13.

Fincher, R. and Panelli, R. (2001) 'Making space: women's urban and rural activism and the Australian state', *Gender, Place, and Culture*, 8: 177–89.

Flint, C. (ed.) (2004) *Geographies of Hate*. London: Routledge.

Gates, G. (2003) *The Gay and Lesbian Atlas*. Washington, DC: The Urban Institute.

Gitlin, T. (1987) *The Sixties: Years of Hope, Days of Rage*. New York: Bantam.

Halberstram, J. (2005) *In a Queer Time and Place*. New York: New York University Press.

Harris, R. (1987) 'A social movement in urban politics', *International Journal of Urban and Regional Research*, 11: 363–77.

Katznelson, I. (1981) *City Trenches: Urban Politics and the Patterning of Class in the United States*. Chicago: University of Chicago Press.

Knopp, L. (2004) 'Ontologies of place, placelessness, and movement: queer quests for identity and their impacts on contemporary geographic thought', *Gender, Place, and Culture*, 11. 121 34.

Knopp, L. and Brown, M. (2003) 'Queer diffusions', *Society and Space*, 21: 409–24.

Kurtz, H. (2003) 'Scale frames and counter-scale frames', *Political Geography*, 22: 887–916.

Laclau, E. and Mouffe, C. (1985) *Hegemony and Socialist Strategy*. London: Verso.

Longhurst, R. (2000) *Bodies: Exploring Fluid Boundaries*. London: Routledge.

Lowe, S. (1986) *Urban Social Movements: The City after Castells*. Basingstoke: Macmillan.

Magnusson, W. (1992) 'Decentering the state or looking for politics', in W. Carroll (ed.), *Organizing Dissent: Social Movements in Theory and Practice*. Toronto: Garamond.

Magnusson, W. (1996) *The Search for Political Space*. Toronto: University of Toronto Press.

Marston, S.A. (1990) 'Who are "the people"? Gender, citizenship, and the making of the American nation', *Society and Space*, 8: 449–58.

McCarthy, J. (2002) 'First world political ecology: lessons from the wise-use movement', *Environment and Planning A*, 34: 1281–1302.

McFarlane, T. and Hay, I. (2003) 'The Battle for Seattle: protest and popular geopolitics in *The Australian newspaper*', *Political Geography*, 22: 211–32.

Miller, B. (2000) *Geography and Social Movements: Comparing Antinuclear Activism in the Boston Area*. Minneapolis: University of Minnesota Press.

Morgan, M. (1982) *Skid Road: An Informal Portrait of Seattle*. Seattle: University of Washington Press.

Nast, H. (2002) 'Queer patriarchies, queer racisms international', *Antipode*, 34: 874–909.

Newton, E. (1972) *Mother Camp*. Englewood Cliffs, NJ: Prentice Hall.

Northwest Lesbian and Gay History Museum Project (NWLGHMP) (2000), Interview with Paul Barwick by Ruth Pettis, January 2000.

Offe, C. (1985) 'New social movements: challenging the boundaries of institutional politics', *Social Research*, 52: 817–68.

Pahl, R. (1970) *Whose City?* Harmondsworth: Penguin.

Pettis, R. (2002) *Mosaic No. 1: Life Stories: From Isolation to Community: Oral History from the Northwest Lesbian and Gay History Museum Project*. Seattle: NWLGHMP, PMB 797 1122 E, USA. Pike Street, Seattle, WA 98122.

Pile, S. and Keith, M. (eds) (1997) *Geographies of Resistance*. London: Routledge.

Pinch, S. (1985) *Cities and Services: The Geography of Collective Consumption*. London: Routledge & Kegan Paul.

Rasmussen, C. and Brown, M. (2004) 'Amidst political theory and geography: radical democratic citizenship', in E. Isin and B. Turner (eds), *Handbook of Citizenship Studies*. London: Sage.

Rasmussen, C. and Brown, M. (2005) 'The body politic as spatial metaphor', *Citizenship Studies*, 9: 469–84.

Reed, M. (2000) 'Taking stands: a feminist perspective on "other" women's activism in forestry communities of Northern Vancouver Island', *Gender, Place, and Culture*, 8: 363–87.

Routledge, P. (1993) *Terrains of Resistance: Nonviolent Social Movements and the Contestation of Place in India*. Westport, CT: Praeger.

Said, E. (1983) *The Word, the Text, and the Critic*. Cambridge, MA: Harvard University Press.

Saunders, P. (1985) 'Space, the city, and urban sociology', in D. Gregory and J. Urry (eds), *Social Relations and Spatial Structures*. London: Macmillan, pp. 67–89.

Scott, A.J. (1990) *Ideology and New Social Movements*. London: Unwin Hyman.

Sharp, J.P., Routledge, P., Philo, C. and Pattison, R. (eds) (2000) *Entanglements of Power: Geographies of Domination and Resistance*. London: Routledge.

Simpson, R. and Paulson, D. (1996) *An Evening at The Garden of Allah*. New York: Columbia University Press.

Sothern, M. (2004) 'Unqueer patriarchies, or, "what do we think when we fuck?"', *Antipode*, 36: 183–91.

Sothern, M. (2007) 'You could truly be yourself if you just weren't you: sexuality, disabled body space and the politics of (neo)liberalism', *Society and Space*, 25: 144–59.

Staeheli, L.A. (1996) 'Publicity, privacy, and women's political action', *Society and Space*, 14: 601–19.

Taylor, P.J. (1994) 'The state as container: territoriality in the modern world system', *Progress in Human Geography*, 18: 151–62.

Taylor, P.J. (1995) 'Beyond containers: internationality, interstateness, interterritoriality', *Progress in Human Geography*, 19: 1–15.

Touraine, A. (1985) 'An introduction to the study of social movements', *Social Research*, 52: 749–88.

Touraine, A. (1992) 'Beyond social movements', *Theory, Culture, and Society*, 9: 125–45.

18

Contrapuntal Geographies: The Politics of Organizing Across Sociospatial Difference

Noel Castree, David Featherstone and Andrew Herod

Difference is as much about geography … as it is about 'race', class, gender, ethnicity and the like.

David Harvey (1998: 727)

As geographers are prone to emphasize, differing groups of people in differing sociospatial locations around the world have distinct, place-related interests and identities, so that notions of global [solidarity] … will at best display a vision and at worst betray an illusion.

Jim Glassman (2003: 514)

INTRODUCTION

This chapter is written against the background of two closely interlinked developments. The first is the increase in the number and type (or at least visibility) of transborder political movements this last decade or so, particularly during the years of what David Slater (2003: 84) calls 'the post-Seattle conjuncture'. The second is a sharp increase in geographical writing on these multifarious attempts to bridge sociospatial difference in order to challenge neo-liberal versions of 'globalization'. To oversimplify matters, we can say that this literature relates to two groups of space-spanning social actors: those associated with the labour movement

(broadly conceived) and those who are part of the New Left social (and environmental) movements that emerged in the 1980s and 1990s.

A central aspect of many projects seeking to understand alternatives to neo-liberal globalization is an exploration of how acts of international solidarity have dealt with the ineradicable fact of *geographical difference*. 'Difference', of course, has been 'on the agenda' of the social sciences and humanities for well over a decade, but often in highly ageographical or only metaphorically spatial ways (Smith and Katz, 1993). Thus, although many geographers have argued that difference is *necessarily* spatial and that this fact affects the content and form of politics, it is far from self-evident what exactly 'geographical difference' is and there is more than one way to understand how, why and with what effects it may be 'crossed'. As if this didn't complicate matters enough, there is no disciplinary consensus on how best to conceptualize any such 'crossing of difference'. Broadly speaking, two alternatives have presented themselves. On the one side, a great many human geographers have found the language of *geographical scale* to be a key resource for studying the politics of sociospatial difference in the New Labour Internationalism (NLI) and Transnational Social Movements (TSMs). Here, 'scale politics' is comprehended through such (now) familiar ideas as 'up-scaling', 'down-scaling' and, more recently, 'scale-bending'. On the other side, this scalar language has recently been challenged by those enamoured with the idea of *sociospatial networks*. Interestingly (and perhaps significantly), whilst it would certainly be overly simplistic to suggest that much of the NLI research has used a scalar framework whilst the newer work on TSMs has favoured network thinking, there is nonetheless more than a grain of truth to this claim. Heuristically, then, we are dealing here with what have been called 'topographical' and 'topological' ways of thinking, respectively (Marston et al., 2006; see also Amin, 2002, 2004).

In seeking to understand how various social actors have sought to engage with geographical difference, we outline how these two ways of spatial thinking have shaped contemporary political geography. Our discussion proceeds as follows. In the next section we review some of the key theoretical contributions on geographical difference that have underpinned the more empirical research discussed later in the chapter. Since much of that research has, as we shall see, accented the importance of place, the theoretical contributions discussed relate to this way of characterizing geographical difference (rather than to, say, 'region' or 'nation', even though these two terms are sometimes used by human geographers interchangeably with the term 'place'). Specifically, we examine the ideas of Kevin Cox, David Harvey and Doreen Massey on why place and locality matter a great deal to the content and form of political projects. The differences between Cox and Harvey, on the one side, and Massey, on the other, are as instructive as are the similarities. Thus, Cox and Harvey have been read as offering a view of inter-place politics that suggests geographical scale is the most productive way to grasp how, why and with what results spatial difference may be crossed. Massey's ideas, by contrast, have been appropriated by those researchers favouring the less 'territorial' language of networks. Yet these differences aside, all three authors insist on something that has permeated the research on the NLI and TSMs; namely, that we adopt a *relational* perspective on place difference and thus on geographical politics. For several reasons, as we shall see, this is thought to be a preferable way of conceiving of geographical difference compared to others currently abroad in academe or the wider society. Following from this, in the third and fourth sections we review theoretically informed but empirically grounded studies of the NLI and TSMs. The third section summarizes the increasingly complex disciplinary debates on geographical scale and links these to empirical studies of labour activism across space. The fourth section, which discusses recent geographical research into TSMs, explores the differences (by no means obvious) between network and scalar ways of analysing the politics of inter-place solidarity. Are these differences significant and, if so, why?

Does network analysis betoken a new way of thinking about spatial politics? If so, is Colin Flint (2003: 628) – who has insisted that current geographical research on transborder solidarity 'offers knowledge that is essential to political actors, especially the marginalized' – right to think that labour and TSM activists can usefully learn from academic studies of their geographical organizing?

What emerges from our review, as readers will discover, is that whilst there is no consensus on the raft of issues we consider, research into the politics of translocal organizing by unions and TSMs nonetheless has a fuzzy coherence. It is a coherence captured in our chapter title, 'contrapuntal geographies'. As the musically inclined among our readers will know, counterpoint is the art of harmonizing different melodies. In the present context, we use the term to refer to efforts to conjoin local place-based actors with distant others in order to serve common interests yet without sacrificing local commitments and aspirations – that is to say, to describe how social actors seek to maintain their local uniqueness within any broader effort to develop anti-neo-liberal commonalities across global space.

GEOGRAPHICAL DIFFERENCE, GEOGRAPHICAL DISTANCE AND POLITICS

Jean Gottman (1951) memorably observed that there would be no need for a discipline called 'geography' if the world were a smooth sphere. As we noted above, a resurgent interest in place and locality is one of the principal ways that human geographers have registered their continuing preoccupation with geographical difference – a preoccupation briefly thwarted during geography's 'nomothetic turn' in the 1950s and 1960s. In relation to politics, John Agnew's (1987) formative book *Place and Politics* was an early explication of why place matters to the formation and enactment of political projects – be they state-led, workerled or a product of civil society groups. Agnew identified three dimensions of place, each of which specifies a particular locality's relative difference from others near and far. These were place as *location* (a distinct point on the Earth's surface), as *locale* (a physical arena for everyday life) and as a *locus of identity* (a focus for personal and collective loyalty, affect and commitment).

Some years after *Place and Politics* was published, a flurry of contributions appeared on why place-based actors who are formally outside the state apparatus often need to reach out across space in their efforts to defend or enhance local interests. Much of this early work did not thematize place in terms of 'difference' (except implicitly), but it did initiate what is now a well-developed understanding of how scale (including the 'local' scale, which we tend to synonymize with 'place') is both a resource and constraint in place-based struggles (see Jonas, 1994; Miller, 1994; Staeheli, 1994). As such (and as we shall see in the next section), it took forward the suggestive ideas of Neil Smith (1984) and Peter Taylor (1981, 1982) on the 'politics of scale'. A decade on, scalar understandings of geographical politics are being challenged by network forms of analysis (presaged in Murdoch and Marsden's [1995] essay on 'actor-spaces'; see also Routledge, 1993) in which place difference and the contrapuntal politics of interlocal organizing is a core preoccupation. However, as we mentioned in the introduction, what both scalar and network analyses of geographical politics have in common is a commitment to understanding place in relational terms.

Though by no means the only relational theorists of place in human geography, Cox, Harvey and Massey have been uncommonly influential in this area. Whilst we are not suggesting that all the researchers whose work is discussed in the next two sections have formally appropriated Cox's, Harvey's or Massey's thinking, this thinking has been difficult to ignore for three reasons. First, it offers prima facie reasons why, in Sophie Watson's (2004: 209) words, geographical difference 'is here to stay'. Second, it deals with the two main dimensions of place covered in Agnew's tripartite definition; that is, the 'objective' dimension (places as distinct locations and locales) and the 'subjective' dimension (places as the sites and subjects of identity, belonging and affect). Finally, Cox, Harvey and Massey all give a strong steer in their writing to which *kinds* of geographical difference – at the level of both thought and practice – are 'good' and which are 'bad'. Given that 'critical geographers' have undertaken the lion's share of empirical research into translocal solidarity in geography, the highly normative cast of Cox's, Harvey's and Massey's argument has helped to orientate the kinds of judgements these geographers make about the geographical politics analysed in their case studies.

Cox: Spaces of Dependence and Engagement

In his work on urban politics, Kevin Cox (1998) has made an instructive distinction between 'spaces of dependence' and 'spaces of engagement'. Building on earlier work into growth coalitions and cross-class alliances in Western cities (e.g. Cox and Mair, 1988), he has offered a plenary account of why locally based actors (e.g. particular workforces or small firms) *necessarily* have a stake in the nature and future of their locality. Taking these actors'

perspective, he argues that whilst their political activity is fundamentally *about* place, it routinely extends *beyond* place into a geographically wider sphere of influence and action.

The theoretical basis of Cox's argument is Harvey's germinal account of the geography of capital accumulation (Harvey, 1982, chap. 13; 1985a). For Cox (1998: 2), the spatio-temporal tension between fixity and motion that is endemic to capitalism explains why many locally based actors focus their political energies on what he calls their 'space of dependence' (those more-or-less localized social relations [and institutions] upon which we depend for the realization of essential interests and for which there are no substitutes elsewhere [and which] define place-specific conditions for our material well-being and our sense of significance). In Agnew's terms, these spaces are not simply the locations in which certain actors happen to be but, rather, the locales upon which they depend for their daily survival. Yet Cox also argues that locally based actors must necessarily have an eye to their 'space of engagement' ('the space in which the politics of securing a space of dependence unfolds [and which] may be at a more global scale than the space of dependence' [ibid.]). For Cox, such a dependence on particular locales feeds into place identity (the third of Agnew's place-dimensions) not simply because people are socialized in specific places but also because the material character of those places decisively shapes people's economic and other interests. As Cox (ibid.: 5) puts it, '[Certain] agents have [specifically] local interests. These are interests in appropriating/realizing profits, rents, wages [or] taxes in particular places. Changing economic geographies at more global scales [can] threaten the realization of these local interests'. Hence, locally dependent actors clearly have a stake in controlling, or at the least influencing, other actors whose actions can decisively affect the fate of their locality. Though some of the latter will be found in the same location as they live, many will be national trade unions, transnational firms, (supra)national state bodies or international civil society organizations. Harnessing these non-local actors to local advantage can help place-dependent groups address the problem of 'ensuring that [economic] value ... continues to flow through their [locality]' (ibid.: 5). Clearly, then, whether they act alone or in alliance, these place-dependent groups will have very different spaces of engagement depending on the place in question. Likewise, the precise forms of 'political action' across space will vary too, depending on the non-local actors who are being influenced or enrolled.

Although we have simplified what, in truth, is a rather dense argument, Cox's normative stance on the politics of geographical difference is subtle. Certainly, it respects the tension between fixity and motion that underpins the nature of capitalism. Thus, on the one side, he is critical of understandings that reify or hypostatize place, for such understandings (which Merrifield [1993] calls 'Cartesian') presume both that places have 'internal' characteristics that are readily distinguishable from 'external' ones, and that places are things (infrastructures and peoples) that are ontologically distinct from the relations and processes that link them with other places. Hence, Cox argues that it is no surprise that Cartesian conceptions of place die hard, since it is precisely the necessary physical fixity of portions of capitalism that create the *appearance* of diverse locations and locales set against a more abstract realm of 'space' or 'the global' (see also Cox, 2005).

On the other side, though, Cox is neither dismissive of political projects that aim to engender certain forms of local difference, nor does he envisage the ultimate eclipse of geographical differentiation by political or any other means. Thus, his 1997 edited book *Spaces of Globalization* (subtitled 'Reasserting the power of the local') was a critique of the 'globaloney' that equates heightened inter-place connectivity with placelessness and homogeneity. This critique was normative as much as cognitive, arising not simply from the abundant evidence that place difference is here to stay (because capitalism thrives on the creation of geographical differentiation, not its elimination), but also from Cox's conviction that relatively disadvantaged place-based actors (e.g. the unemployed or already straitened workforces) have much to gain by creating a space of dependence that works in their favour in the medium and long term. The difficulty in this, however, is that the rules of the capitalist game often mean that these actors' space of engagement is harnessed in ways that encourage 'beggar-thy-neighbour' tactics, a kind of geographical practice in the face of difference that is always about 'winners' and 'losers'. For his part, though, Cox would prefer forms of inter-place alliance that do not play locally dependent groups against one another but which instead, in Massey's (1993) terms, would encourage a 'progressive sense of place' that would lead locally marginal or vulnerable groups to engage *cooperatively* with distant others – that is, promote an engagement that is undertaken in recognition of the fact that creating 'desirable' sociospatial *differences* for all depends upon agreeing to *common* goals and *joined-up* strategies.

David Harvey: Militant Particularism and Global Ambition

We have already mentioned in passing Harvey's early-to-mid-1980s thinking about the inconstant geography of capitalism. Detailing his argument in *The Limits to Capital, The Urbanization of Capital* and other publications, Harvey maintains that geographical difference is actively wrought out of the interconnectivity in a capitalist universe, making it internal (though not *reducible*) to this mode of production. From a 'capital-logic' perspective, then, uneven geographical development is not only necessary but functional since, among other things, it fragments working-class opposition to exploitation, creative destruction, unemployment and the like. From the perspective of 'people on the ground', though, Harvey's 1980s works recognized that oppressed groups have material and emotional investments in place that cannot be dismissed as mere ideological constructs (see the essays in *Consciousness and the Urban Experience* [1985b], for instance). Consequently, the conceptual challenge, as he saw it, was to take seriously the place-based nature of everyday life (i.e. the general need of all of us to have an adequate space of dependence) whilst reckoning with the increasingly distanciated ties that condition the fate of any given locality. Whereas some authors (e.g. Taylor, 1981) have considered the 'local' as the scale at which people construct their identities, Harvey has noted on several occasions that 'There are no ... grounds for considering face-to-face relations more pure [and] authentic ... than relations mediated across time and distance' (Harvey, 1993: 15).

This theme of a place-based but space-spanning politics among working-class and other oppressed groups became central to Harvey's 1990s writings. Many of Harvey's publications since his trenchant critique *The Condition of Postmodernity* (1989b) can be seen as a Marxian attempt to explain how and why 'good' forms of geographical difference might flourish – particularly the serially reprinted chapters 'Militant particularism and global ambition' and 'Class relations, social justice and the political geography of difference' (found together in *Justice, Nature and the Geography of Difference* [Harvey, 1996]). In the former, Harvey reflects on his disagreements with Teresa Hayter, co-editor of *The Factory and the City* (Harvey and Hayter, 1993):

> The view that what is right and good from the standpoint of militant shop stewards in Cowley, is right and good for the city [of Oxford], and by extension for society at large is far too simplistic. Other levels and kinds of abstraction have to be deployed if socialism is to break out of its local bonds and become a viable alternative to capitalism. ... *But* there is something equally problematic about imposing a politics guided by abstractions upon people who have given of their lives and labour for many years in a particular way in a particular place. (ibid.: 23, emphasis added)

Through a discussion of Raymond Williams's novels, Harvey tries to lay bare the problems and dilemmas confronting a 'both/and' politics that remains rooted in place yet can reach out positively across a wider landscape. Two loom large. The first is that because grassroots politics – workerist or otherwise – is inevitably conducted *from* place and is frequently *about* place, the question arises of how solidarity is to be constructed with others whose commitments, aspirations and goals are forged in quite different local contexts. For Harvey, the only way to bridge difference is to emphasize commonality (e.g. common interests, common enemies). But commonality does not speak for itself: it must be made visible and represented to those it aims to unite. In so doing, however, the risk becomes that commonality gets defined in such abstract terms (e.g. workers in place A and place B recognize that they share a 'working-class' identity) that the visceral, lived experiences that motivate political action become weakened or subsumed. As Miller (2004: 227) puts it:

Harvey and Williams call attention to the declining resonance of movement messages as they become further removed from the identities and everyday concerns of place-specific collectivities. This represents the central challenge that transnational [organizers] ... face. [They] ... must find ways to frame and reframe broad messages so they will resonate with a diverse and fluid array of collectivities in a wide range of place-specific and not so place-specific circumstances.

This links to a second problem, which is that the sorts of place-based interests and identities that often drive oppositional movements are forged under a social order (capitalist, racist, patriarchal) that is fundamentally oppressive. The challenge, then, is to forge translocal political action that can topple this order, rather than simply ameliorate its operations here and there. Aside from the logistical difficulties of this undertaking, Harvey also rightly notes how threatening it is to the very groups who have most to gain from making it happen, for it entails eradicating, in significant measure, the very sociogeographical differences that form the starting point for oppositional action in the first place. This is why, Harvey (op. cit.: 40) avers, both worker and TSM politics often acquire 'a conservative edge'.

Harvey's widely read essay 'Class relations, social justice and the political geography of difference' connects these issues of interlocal organizing with those of how social difference can be productively negotiated by oppositional movements. His starting point is to discuss a 1991 fire in a chicken-processing plant in a small North Carolina

town (Hamlet). With the exit doors locked, 25 of the 200 employees died, of whom 18 were women and 12 African-Americans. Although he does not wish to collapse differences of gender and 'race' (or ethnicity) into class difference, Harvey does note that the failure to link all three in the Hamlet case – the response of the US left generally focused upon issues facing female and minority workers in the South, rather than upon issues facing all workers, regardless of gender or 'race' – constituted a failure of political imagination and organizing. Like so many other towns in the US 'broiler belt', Hamlet had no tradition of trade unionism and its workers received minimal salaries and few benefits in a context where few other local employment opportunities presented themselves. As Harvey (ibid.: 338) argued, 'The commonality that cuts across race and gender lines in this instance is quite obviously that of class and it is not hard to see the immediate implication that a simple, traditional form of class politics could have protected the interests of women and minorities as well as those of white males'. Such a politics would, ideally, have been both local (e.g. by formally organizing Hamlet's wage workers regardless of job, gender, age or ethnicity) and translocal (e.g. established national unions putting pressure on federal and state regulators to discipline Imperial Foods, owner of the Hamlet plant). But, apropos Harvey's discussion of Williams, such organizing is by no means easy when so many different issues (of gender, 'race', etc.) have to be meaningfully expressed in a language (class) that has the capacity to be what Laclau and Mouffe (1985) called a 'hegemonic articulation'. Thus, Harvey's (op. cit.: 359) (re)definition of class in terms of 'positionality in relation to processes of capital accumulation' (rather than in terms of *identity*) ultimately ducks the important issues. Whilst it usefully reminds us that social differences irreducible to capital are nonetheless 'internalized' within the global labour force, it leaves unanswered all the important questions about precisely *how* class and identity politics, union and TSM politics, can be articulated both locally and translocally.

Even so, Harvey's thinking in this area has usefully identified some of the central issues for any politics that aims to organize across sociogeographical difference. True, some commentators (e.g. Braun, 1998) believe that Harvey ultimately wishes to *subordinate* non-class forms of social difference to class, whilst others (e.g. Routledge, 2003: 354) maintain that he wishes to *transcend* geographical difference. Even if these claims are valid, they do not detract from the fact that Harvey has highlighted some of the key tensions arising from the relational constitution of interests and identities that are lived locally and often expressed, in the first instance, in local terms. What, in his more recent book *Spaces of Hope*, he calls 'the work of synthesis' (Harvey, 2000: 72) is the difficult intellectual and practical work of conjoining sociogeographic differences even as, in the process, these differences are unlikely to survive intact. For Harvey, then, these differences *must* be conjoined because capitalism thrives on their apparent dissociation. But, at the same time, when oppositional groups undertake acts of translocal solidarity it is not in order to vanquish difference altogether. The fundamental question for Harvey is, therefore, this: what *sorts* of inter-place organizing can confront the totalizing powers of global capitalism whilst producing new *forms* of sociogeographical difference that are 'progressive' (in Massey's sense of being outward-looking) *and* not simply a new means for existing power relations to be instantiated?

Doreen Massey: The Politics of Place Unbound

Notwithstanding *Justice, Nature and the Geography of Difference*, many of Doreen Massey's influential writings on place, space and politics are more abstract than either Cox's or Harvey's. This partly explains their appeal. By writing in very philosophical but relatively jargon-free terms, Massey has managed to shape research agendas across critical human geography as a whole (and, indeed, beyond it). Another reason why her work has enjoyed a wide appeal is that it is heterodox: unlike Cox and Harvey, her Marxism has given way to post- and non-Marxist modes of thought (for instance, the work of Ernesto Laclau and Chantal Mouffe has been an important influence on her).

In a recent reflection on over a decade of her writing on place, Massey has summarized her agenda as follows: 'Thinking place relationally was designed to intervene in a charged political arena. The aim initially was to combat localist or nationalist claims to place based on eternal, essential and … exclusive characteristics of belonging: to retain, while reformulating, an appreciation of the specific and the distinctive while refusing the parochial' (Massey, 2004: 6). Like Cox and Harvey, then, Massey has long taken issue with Cartesian conceptions of place (what she calls 'the billiard ball world of an essentialist Newtonianism' [Massey, 1999a: 38]) and the politics they can mandate. However, she advocates a more overdetermined understanding of how geographical difference is produced out of geographical interconnectivity across space. As she put it in *Power-Geometries and the Politics of Space-Time*:

> … 'places' may be imagined as particular articulations of … social relations, including local relations 'within' the place and those many connections that stretch way beyond it. … This is a notion of place where specificity (local uniqueness, a sense of place) derives not from some mythical internal roots nor from a history of relative isolation – now to be disrupted by globalization – but precisely from the absolute particularity of the mixture of influences found there. (ibid.: 22)

In an indirect swipe at what she argues is the sort of 'muscular' Marxism favoured by Cox and Harvey, she has contrasted this conception of place to one underpinned by 'a closed holism … where there is no space for anything new' (ibid.: 38). But her particular conception of geographical difference is ranged against other targets too, not just the antinomies of Cartesian (non-relational) and holistic (relational but closed) understandings. It is *also* forged in opposition to a powerful geographical imaginary expressed in hegemonic uses of the term 'globalization': what she calls an 'aspatial imaginary' in which geographical difference is temporalized such that 'underdeveloped' places in the 'Third World' are seen to be 'behind' the West yet potentially able to become 'like us'. Thus, as the champions of neo-liberal globalization would have it, those localities and nations who have yet to benefit from free trade soon will, for their temporalized differences will eventually be erased as globalization leads to a convergence around Western levels of development.

Massey's three-pronged critique of certain geographical imaginations is thus undertaken in the name of 'a different kind of difference' (Massey, 1999b: 281): what she rather tendentiously calls 'real difference' (op. cit.: 21), by which she means the *contingent*, *singular* and *unpredictable* geographical differences that continually emerge out of the many connections (some yet to be made, some never to be made) that wire different places to different other places. What has all this got to do with politics? A great deal. In the first place, it is by no means easy for local actors to agree upon the sorts of difference (both 'objective' and 'subjective') that should be promoted into the future. The fact of geographical propinquity, Massey has rightly argued, does not produce any ready consensus about what local interests and identities are or ought to be. After all, many twenty-first-century places are intensely differentiated internally (in social and land-use terms) and so, often, there is local struggle over what sorts of place-projects are to be favoured in any given locality (Massey, 2004: 7).

Second, like Cox and Harvey, Massey insists that a 'politics of propinquity' can only be undertaken in relation to a 'politics of connectivity' (the terms are those of Amin [2004]). This follows from the ontological fact that contemporary places are more than they contain. This politics is also contested, since certain local actors will favour, say a competitive inter-place politics whilst others might prefer a more cooperative politics. Moreover, because the precise network of translocal connections that constitute a place varies from locality to locality, the content and form of such a connective politics will vary too. This means that grassroots attempts to challenge the wider 'power-geometries' that constitute place both objectively and subjectively will be immensely varied and complex. For Massey, this is more than a case of identifying the broad 'commonalities that cross-cut difference'. Because the precise nature of these commonalities will, *themselves*, vary depending on the cluster of places involved, the politics of organizing across difference becomes exceedingly complicated. Hence, as she puts it, 'one implication of the ... inequality inherent within capitalist globalization is that the local relation to the global will also vary, and in consequence so too will the coordinates of any potential local politics of challenging that globalization. Moreover, "challenging globalization" might precisely in consequence mean challenging, rather than defending, certain ... places' (op. cit.: 13).

In sum, Massey advocates a move from a 'politics *of* place' to a 'politics *in* place'. As Amin (2002, 2004) explains, the latter is a 'non-territorial politics' that involves the negotiation of difference *both* within and between places whereas the former is 'a politics of place in which relations within localities are cast as good and felt, separate from bad and[/or] remote external happenings' (Amin, 2002: 388). This links to the above-mentioned problematization of just how 'common' inter-place goals, enemies or aspirations can be when oppositional groups are concerned. Miller (2004: 235) summarizes this well: 'different bases of spatial interaction produce different kinds of collective [aims] ... which, in turn, present different mobilization challenges'.

As we have already noted, the fact that Massey's writings on place are often highly philosophical helps to explain their wide influence: many different human geographers have been able to appropriate her ideas for their specific needs. But their abstract nature is also their weakness. As with Cox and Harvey's writings, Massey's seem to us to avoid most of the crucial questions one might reasonably want to answer about the politics of oppositional groups organizing across difference. For instance, exactly *how* can place-based groups forge connections with distant others? *How and with what specific effects* are place differences transformed through acts of interlocal solidarity? What criteria should be used to *evaluate* these transformations? Are 'progressive' and 'regressive' place projects distinguishable *in practice* or can even the

most 'extrovert' place politics lead to unintended consequences that compromise the interests of distant others? We could go on, but the point is clear.

To summarize, then, Kevin Cox, David Harvey and Doreen Massey have developed among the best-known conceptions of geographical politics – ones in which the thematic of place difference looms large, where interlocal organizing is deemed to be essential, and where the problems of engaging constructively across space are fully acknowledged. It is no surprise that critical geographers casting around for frameworks to approach recent experiments in translocal organizing have, in part, drawn upon one or more of these three theorists' work. Whereas Cox and Harvey have commended a scalar framework as the appropriate way to understand inter-place connections (see Cox, 1998; Harvey, 1996: 352–4; 2000: 75–7), Massey has favoured network metaphors and has largely avoided the 'topographical' language of scale – although all three have largely left it to others to flesh out empirically how these connections might be made.

The 'divide' between scalar ('topographical') and network ('topological') approaches to 'crossing difference', however, is not always easy to divine – this despite some commentators (e.g. Amin, 2002, 2004) emphasizing its signal importance for analysis and practice. Such difficulty is evident in the numerous empirically grounded studies of both the NLI and TSMs that use the terms 'scale' and 'networks' interchangeably (e.g. Routledge's [2003] essay on global 'anti-capitalist' movements; see also Murdoch and Marsden, 1995). This suggests that notions of scale and network are not simply 'reflections' of the varied translocal movements whose actions are being analysed. Rather, they are also notions constructed actively within the academic community: ones designed to inform the 'real world' realities they describe (Flint's point) *and* to dictate the terms of debate within a community of critical geographers subject to the mundane imperatives of promotion, tenure and professional recognition. The differences between scalar and network imaginaries of translocal connectivity are thus part of, without being reducible to, academic struggles for citations and influence in relation to one of critical geography's 'hot topics': the topic of transnational resistance to transnationally organized relations of power. For some commentators (notably Marston and Jones, 2005), the differences are irreducible and upon them hang important issues of real-world analysis and political possibility. We are not so sure. Our tack here is simply to present both sides of the argument, since a properly comparative analysis and/or synthesis has barely been attempted by any of the authors whose work we discuss in the pages to come.

THE POLITICS OF GEOGRAPHICAL SCALE: THE NEW LABOUR INTERNATIONALISM AND BEYOND

The so-called 'new labour internationalism' is new in relative but not absolute terms. Beginning over a century ago, wage workers in several Western countries have made concerted efforts to organize their interests and actions across national borders. Despite some notable successes, however, the Marxian dictum 'Workers of all lands unite!' tended to be honoured in the breach throughout most of the twentieth century. Relatively well-off Western wage workers mostly accented their national interests and were organized through strong national trade unions. By contrast, many workers in the developing world remained non-unionized. Meanwhile, the Cold War meant that 'second world' workers were separated from their capitalist equivalents on ideological grounds. Given this history, then, the NLI is new in two ways. First, quantitatively speaking, the last fifteen years or so have seen a fairly sharp increase in the number of translocal worker campaigns. Second, many of these campaigns have breached the 'divides' between the developed world, the former communist countries and the developing world.

A good deal of the geographical literature on the NLI has applied and developed theoretical ideas about geographical scale and scale politics. This is what makes it distinctive. However, whilst the NLI has received much attention from non-geographers such as Peter Waterman, Ronaldo Munck and Richard Hyman, such authors have rarely thematized geographical difference explicitly and have tended towards a thin conception of spatial scale. In contrast, a number of geographers have expressly linked scale to questions of the production and transcendence of geographical difference. Most notably, Neil Smith (1992) has argued that 'the local' is one of *several* scales at which geographical difference is produced, such that 'crossing difference' is a multiscalar activity and not simply a 'jump' from the local to an undifferentiated space of the global. Despite Smith's call for a multiscalar sensibility to issues of crossing difference, much of the geographical research into the NLI has nevertheless had a strong local sensibility, for the reasons that Cox explains so well: that all production and reproduction *necessarily* occurs in place so that worker politics, whatever else they are about, *must* be about the local in some measure, even as they may also be about the extra-local. The key issue, then, becomes, what sorts of links at what geographical scales can further local interests without compromising those of other constituencies enrolled in these linkages? In relation to the NLI, we can therefore pose the questions: how have contemporary wage-workers organized themselves across space, why and with what outcomes?; and what is 'political' about the NLI? To answer these questions we need first to summarize the theoretical debates on geographical scale and scale politics. Since these debates have been reviewed elsewhere (see Herod and Wright, 2002; Howitt, 2003), we offer a highly stenographic account here.

Theorizing Geographical Scale

As we have intimated above, issues of geographical scale and the politics thereof began to impact geographical theorizing in the 1980s. Two of the key discussions of scale came from Smith and Taylor, though these writers had quite different approaches to the topic. For his part, Taylor (1981, 1982) outlined a vision in which particular scales of social life took on specific roles under capitalism; he viewed the global scale as that at which capitalism operated *in toto*, whereas the national and local scales were those at which, respectively, ideology (primarily that of nationalism) operated and everyday life was lived. Smith (1984), on the other hand, focused not so much upon the functions of different scales within the operation of capitalism but upon how scales were actively produced under capitalism. Hence, he viewed the global scale as the product of the universalization of the wage-labour form, whereas the national scale was solidified by the need for nationally organized capitalists to regulate 'unfair' competition among themselves and to defend their interests against those of other nationally organized capitalists, the regional scale reflected the territorial division of labour (initially based upon the natural distribution of the Earth's resources but later reworked by processes of capitalist accumulation), and the urban scale was an expression of the spatial extent of the daily labour market and of commuting patterns. Writing at the height of the 1980s debate concerning the 'new regional geography', Smith's take on scales reflected not only his Marxist analysis of why the capitalist landscape seemed to be pockmarked at a number of different spatial resolutions (the urban, the regional, the national, the global) by the processes of spatially uneven development, but also a reaction to the neo-Kantian Hartshornian 'traditional' regional geography in which space was conceptualized as something that could simply be carved up into larger or smaller bits to suit the whims of the individual researcher, a carving-up that paid no attention to the 'necessary' spatial relationships between particular places (e.g. see Smith's [1989] critique of Hartshorne).

If such early theoretical forays into the politics of geographical scale were marked by a certain functionalism and capital-centrism, the development of what has come to be called 'labour geography' in the early 1990s was tied up, in part, with a growing concern to theorize the politically *contested* nature of the production of the geographical scales of social life. Place, this literature argued, matters to workers immensely – even migrants are place-based in order to secure a living. However, acting *in* place is often insufficient for workers to realize their material and emotional interests. Rather, they invariably must cross space to do so. For example, unions may seek to develop national contracts with spatially uniform wage rates to prevent capital from playing workers in different communities against each other, whereas capital may prefer decentralized wage bargaining to take advantage of the landscape's economic and institutional and legal differentiation to 'whipsaw' workers. Equally, lest the implication be drawn that particular scales of organization are theorized as being somehow inherently more beneficial to capital whereas others are to labour, it is important to recognize that some firms may, for similar reasons, prefer to negotiate at a national scale whilst some workers prefer to do so locally: high-wage firms may seek a national agreement so that they are not outbid by their low-wage competitors, whilst workers may favour negotiating locally to take advantage of their position in the labour market.

Central in all this, though, is an implicit understanding that the construction of scale is a way to negotiate spatial difference, such that at times the construction of scale revolves around efforts to minimize difference between places (as when unions successfully negotiate nationally uniform wage rates) and at others it involves actually stressing difference (as when groups of workers construct scales of organization that emphasize variations in local conditions and/or seek to take advantage of them). Furthermore, such an engagement with the spatiality of difference can have significant impacts upon how unions' organizational structures themselves develop. Hence, Southall (1988) notes that in Britain the relative geographical mobility of different groups of workers contributed to the emergence of particular types of union consciousnesses and political structure in different places and industries. Thus, in the mining industry, the particularities of geology meant that miners needed locally specific skills to work successfully in different pits (coal seams in the very same region can vary immensely, let alone between regions), a reality that both discouraged their mobility and encouraged the development of localist consciousness based on the pit community.

The New Labour Internationalism and Beyond

Significant research effort has been expended in trying to understand how workers and their organizations overcome spatial barriers that divide them so that they might present a unified face to their employers. Growing concerns with 'neo-liberal globalization' in the 1990s seemed to make efforts to develop solidarity links (especially international ones) take on greater urgency, both within the labour movement itself and within that part of the academy that was labour-sympathetic. Thus, the more trade unionists called for greater international solidarity to confront the power of corporations, the more critical human geographers seemed to be interested in the geographical conundrums that such efforts might both face and spawn (e.g. Herod, 1995, 1997; Castree, 2000; Sadler, 2000; Wills, 2002a). Much of this work focused upon either the use of formal trade union organizations such as the International Confederation of Free Trade Unions (ICFTU), various Global Union Federations (GUFs) and European Works Councils – all of which, significantly, have particular and formalized spatialities to them – or upon less formal worksite-to-worksite connections that often develop in response to specific needs (such as the threatened closure of particular factories).

There are myriad examples within the critical geographical literature examining labour's political responses to the new spatiality of global capitalism, but, at the risk of charges of self-promotion, we will use the work of one of us to explore here briefly a number of issues relating to workers' construction of new geographic scales of organization. One of the earliest pieces to consider an international solidarity campaign confronting a globally organized corporation was Herod's (1995) investigation of a dispute involving aluminium workers in Ravenswood, West Virginia. In October 1990, some 1,800 members of United Steelworkers of America Local 5668 were locked out of their plant in a dispute over health and safety and pension issues. The plant had recently been purchased by an international consortium of investors, following which workplace injuries at the plant had soared, with five workers killed within eighteen months of the new managers taking over. When the union decided to make health and safety a central issue in upcoming contract negotiations, the company locked the unionists out and brought in non-union 'replacement' workers. Local 5668 officials soon discovered that the plant's new owners were tied to international commodities trader and fugitive from US authorities Marc Rich, who was living in Zug, Switzerland. Upon learning this, they worked with representatives of their national union, of the AFL-CIO, and of a number of international trade union organizations – most notably two GUFs, the International Metalworkers' Federation (the 'other' IMF!) and the International Confederation of Chemical, Energy, Mine, and General Workers' Unions (ICEM) – to develop links with unions and workers abroad, unions and workers who then brought pressure to bear upon the constituent elements of Rich's multi-billion-dollar global empire. As a result of such transnational strategies, after some twenty months the Ravenswood workers were successful in forcing Rich to permit the locked-out workers to return to work.

Most observers, we think, would view the Ravenswood campaign as a textbook example of how workers might transcend geographical difference and distance to challenge capitalist social relations by organizing internationally for the purposes of protecting workers in a particular place, in this case a small town in rural Appalachia. Such instances of workers engaging in international solidarity, however, raise important issues for efforts to construct (and to provide a narrative about) what political geographers might take to be a spatially and scalar-sensitive, anti-capitalist political project. Specifically, there has been a tendency in orthodox Marxist accounts to assume that instances of workers constructing scales of organization that stretch beyond the local is automatically 'progressive' politically, for it is taken to represent a great proletarian conjunction that has overcome the spatial barriers and allegiances to place that so often divide workers.

However, as Johns (1998) notes, such an assumption fails to appreciate the subtleties of geographical interconnectivity and difference. Thus, some international solidarity campaigns are clearly waged to challenge directly capitalist social relations, as when workers in one place (such as Switzerland) act to help those in other places (Ravenswood) without any expectation that they themselves will gain any immediate benefits – a practice that Johns calls 'transformatory' solidarity. However, it is important to recognize that what may, on the surface, appear to be altruistic reasons for instigating a solidarity campaign (the desire to help workers overseas increase their wages) can sometimes be more about defending particular privileged places in the global economy than challenging capitalist social relations: by increasing the wages of workers overseas, particularly those in developing countries, those workers' communities become less attractive to capital that may otherwise choose to flee the global North. Paradoxically, this kind of trans-spatial solidarity action – what Johns calls 'accommodationist solidarity' – actually divides workers *in* space under the guise of uniting them *across* space, a phenomenon that illustrates the challenges of realizing the delicate interplay of 'progressive' forms of worker localism with the kinds of wider inter-place solidarities sketched by Cox, Harvey and Massey.

In discussing workers' political praxis it is important to understand not just the reasons for which workers seek to cross difference but also how discourses of scale are implicated in different models of organizing. Thus the model used in the type of international solidarity organizing outlined above – whether that be for transformatory or accommodationist purposes – is typically one built upon a nested hierarchy of discrete levels, where pleas for help are passed up the organizational ladder domestically and then back down it overseas. In the US and Britain this has been a fairly common approach to organizing, especially because in the twentieth century unions often engaged full-time professional organizers who worked out of the union's national headquarters and served as the conduits between local branch unions and the appropriate international labour institutions; the Ravenswood local, for instance, worked up the chain of its union's organization, with its international union in Pittsburgh and its national federation (the AFL-CIO) in Washington, DC, engaging with international union organizations such as the ICEM and the ICFTU, who themselves made contacts with local unions in various countries around the world who agreed to help the Ravenswood unionists. Implicitly, then, this model is one in which labour organizers have viewed their scalar organization in terms both hierarchical and areal, focusing upon what Bruno Latour (1996: 370) calls 'notions of levels, layers, territories, [and] spheres', such that the practice of developing geographically extended solidarity is seen as a process of linking one discrete level (the local) to others (the regional, the national, the global) and back again through a quite hierarchically structured set of organizational linkages. Although the disintegration of highly vertically integrated firms and the growth of 'networked firms' have encouraged some unions recently to explore 'network-based bargaining' (see Wial, 1994; Herod, 2006), it is fair to say that during the past half century at least international solidarity actions on the part of unionized workers have typically incorporated fairly hierarchical spatial imaginaries.

In concluding this section, we would like to bring the issue back to that of place and organizing within the context of geographical difference. How we understand place (in its objective and subjective dimensions) will have implications for how we comprehend and portray scale: are scalar relations about the connections between points in a network or are they about the ways in which different areal units are considered to be linked together in some spatial *gestalt*? Moreover, without an appreciation of the language of scale, what does it even mean to talk of 'place'? Where, in fact, does one place end and another begin? In areal approaches to scale this is relatively easy to determine, but in a fibrous, thread-like, wiry, stringy, ropy and capillary Latourian (or Masseyan) world it becomes much more difficult: Are places simply the intersections of a network or are they something different? Where, exactly, does a place end if it is viewed as an intersection within a network (given that, mathematically speaking, points of intersection would be infinitely small)? If the world is stringy, is place merely one end of this piece of string and, if so, what is at the other end? Equally, what does it mean to talk about 'organizing across space' and 'organizing across difference' in light of the discussion above? What does workers' construction of new scales of organization so as to overcome the distances between them mean for issues of difference? For sure, an areal view of scale clearly incorporates within it a sense that constructing new scales of organization is a process either of eradicating, 'suspending' or of amplifying differences between places; developing a national wage rate is clearly about ensuring uniformity of remuneration across the national economic space, no matter where workers labour, whilst engaging in local bargaining is about taking advantage of the differentiation of the landscape and, perhaps, even intensifying it. But what about the issue of networked approaches to scale? Do these approaches to scale necessarily imply the same kinds of eradication, suspension and/or generation of difference? Or do they imply something different? It is to questions such as these that we now turn, even as we acknowledge that human geographers have yet to provide satisfactory answers to them.

TRANSNATIONAL SOCIAL MOVEMENTS: THE POLITICS OF NETWORKED SOLIDARITY

In this section we examine the research of those inclined to conceive of TSMs in network terms. The geographical analysis of TSMs has mushroomed in the wake of the now-famous Zapatistas struggle in southern Mexico and the anti-capitalist protests in Seattle, Prague, Genoa and elsewhere. Unlike Hardt and Negri's (2000) celebrated but abstract account of 'the multitude', this research accents the uneven topology of inter-place organizing. In part, this is because of the disciplinary influence of Massey's 'fibrous' conception of place, going back over a decade. TSMs have diverse constituencies and agendas, notwithstanding their coming together in what Routledge (2003) calls 'convergence spaces'. They include indigenous peoples' organizations, advocacy groups for women and children, anti-nuclear bodies, human rights groups and environmental organizations, among others (see Cohen and Rai, 2000). Despite the prevalence of network metaphors of organization and political activity among TSMs, geographers have approached the study of these movements from a diversity of theoretical perspectives and concerns. Thus, Merrifield (2002) has persuasively drawn on the writings of Lefebvre on everyday life

to engage with the alternative forms of urbanism and protest deployed through the Seattle protests, whilst Glassman (2003: 513) has explored how activists struggling against corporate globalization (e.g. through opposition to river damming in Thailand) have 'jumped scales' and utilized various international connections to improve 'their visibility and strengthen their prospects for success'. Glassman is one of a number of writers who have adopted scalar thinking and conceptual metaphors to make sense of non-trade-union resistance to neo-liberal globalization (see also Smith, 2000; Wainwright et al., 2000; Wills, 2002b; Ashman, 2004; Mamadouh et al., 2004; Miller, 2004; Novelli, 2004). As already noted, other geographers have suggested that such scalar analyses of political activity have serious drawbacks (Amin, 2002; Gibson-Graham, 2002; Massey, 2005).

Scales and Networks: An Ideal-Typical Comparison

Before discussing network analyses of TSMs in any detail, it is useful to delineate four key differences between these accounts of geographically stretched soldarities and scalar approaches. The complicating factor here is that various authors use the terms 'scale' and 'network' interchangeably to refer to quite different things. However, rather than playing the role of disciplinary enforcers by suggesting that some uses of these terms are appropriate whilst others are not, here we explore how the *differences* between the two approaches to geographical politics have been presented by advocates of network approaches (such as Ash Amin), for to do so throws light on how different authors conceive of the practice and implications of crossing difference.

First, writers like Amin see relations between places and networks as co-constituted rather than privileging boundaries between scales, or counterposing the local and global. Second, according to their advocates, networked theories develop a more generative role for place-based political activity. In these accounts, such activity becomes reconfigured as always already the product of different trajectories and involved in negotiating spatially stretched relations of power. In this regard, place-based movements can become the means through which transnational articulations are made and crafted rather than bounded militant particularisms (see Featherstone, 2005). Third, it has been argued that networked accounts see solidarities as constituted by dynamic trajectories and identities, rather than (relatively) fixed interests. This can account for how new political possibilities and subjectivities are produced through the coming together of different political trajectories. Finally, some authors maintain that network approaches challenge dominant accounts of what counts as political subjects. In particular, they challenge human-centred accounts of the political that see the social as the dominant actor (Murdoch, 1998; Hinchliffe, 2000; Braun and Disch, 2002; Featherstone, 2004). This reconfigures the activity of solidarities as being about bringing together articulations of humans and non-humans and intervening in the constitution of networks within and among places.

Scalar and network approaches, however, are not just about different ways of making sense of political movements. Rather, they also have markedly different conceptions of politics, both cognitively and normatively (Featherstone, 2003). Hence, whilst scalar approaches often follow Harvey's (1996: 400) argument that class is the key node around which solidarities/differences can be articulated, network approaches tend to trace how sociogeographical differences in a more-than-capitalist world are both made and co-related. Most specifically, they have arisen out of political engagements that have been intensely suspicious of hierarchies and, as such, are wary of Marxian arguments that class should be prioritized as the social division that is 'first among equals' when it comes to political identity and struggle. Instead, they see networked forms of political activity as ways of generating non-hierarchical forms of organization. Consequently, networks such as the Internet have been celebrated as a key metaphor for the kinds of non-hierarchical forms of mobilization that are central to the political practices and identities of many activists involved in counter-globalization movements (see Graeber, 2002; Klein, 2002; Hardt and Negri, 2004: 288). The political inclination of these accounts and movements is generally hostile to some of the implicit forms of vanguardism seen to be present in the analyses of political economic geographers like Harvey and Smith. Finally, network accounts, through *following* political activity rather than confining it into particular scales, can suggest more generous accounts of the relations between place-based politics and political imaginaries, according to authors like Amin. Rather than place-based political action being neatly separate from 'abstract' ideas, these approaches suggest much more dynamic and generative relations between place-based activity and networks. This opens up possibilities for different assessments of what constitutes political agency or effectiveness.

Place and the Making of Transnational Networks

In a recent paper on the geographies of transnational social movements, Byron Miller (2004: 224) has argued that TSMs represent a 'fundamental shift away from place-based forms of political organizing and towards transnational mobilization networks'. Although such a counterposing of 'place-based activity' and the formation of transnational mobilizations is fairly common in the literature, from a network point of view this is problematic, for one of the key insights of conceptualizing TSMs in network terms has been a focus on how places and networks are *co-constituted*. Hence, the growing significance of transnational forms of organizing does not inevitably mean an erosion of the importance of place, nor does it imply (as some have suggested) that resistance now inhabits a smooth and undifferentiated space of opposition to neo-liberalism (Hardt and Negri, 2000, 2004).

Indeed, recent work in geography on the formation of transnational networks of mobilization has sought to highlight the mutual constitution of place-based political activity and transnational networks in three main ways. First, it has emphazised how transnational mobilization is constituted through particular physical (local) sites (Routledge, 2003; Amin, 2004; Bunnell and Nah, 2004; Townsend et al., 2004; Featherstone, 2005 and this volume). Second, it has demonstrated how place-based political identities can be actively reshaped and reworked through involvement in spatially stretched networks of activity and can exert pressure on the formation of transnational solidarities through such networks (Oza, 2001; Slater, 2002; Featherstone, 2003; Massey, 2005). Finally, whether acting alone or together, many individual TSMs put issues of place and locality at the core of their agenda. This is particularly the case with indigenous peoples' movements, where translocal action is often a means for enacting resolutely local struggles over land and natural resources (Castree, 2004; see also Escobar, 2001; Perreault, 2003; Oslender, 2004: 958). At the limit (as we suggested in the previous section), network approaches, then, raise the vexing question of what, precisely, is 'place' (point? location? locale?).

These arguments about the co-constitution of place and transnational networks can be illustrated briefly through the activity of the European Campaign for Nuclear Disarmament (ENDS), which managed to construct spaces of organizing that cut across the entrenched geopolitical divisions and differences of the Cold War. Following the networks formed through ENDS's political activity illustrates how network approaches can foreground the relationships between place and transnational forms of mobilization. Formed in response to the late 1970s/early 1980s resurgence of Cold War hostilities, the ENDS campaign brought together peace activists from both the Eastern bloc and Western and Northern Europe to campaign for a Europe free of nuclear weapon deployment. This was a particularly urgent task, as Europe was the main proposed 'theatre of war' in the event of any nuclear exchange. Significantly, ENDS was an attempt to organize across the crude spatial division of Europe into 'East' and 'West'. Drawing heavily on the intellectual and political projects of the New Left, ENDS, then, attempted to develop political imaginaries that reunited Europe through common struggles for peace and democracy (though the European-focused nature of the campaign had its critics; see Davis, 1982).

For the purposes of this discussion, the following elements of ENDS's activity and spaces of politics are particularly salient and demonstrate how political networks and place-based political activity can be mutually constitutive. First, ENDS represented an explicit attempt to construct transnational networks and articulations of opposition to the Cold War to bring together peace activists from both sides of the Iron Curtain. As the ENDS Appeal of 1980 implored:

> We appeal to our friends in Europe, of every faith and persuasion, to consider urgently the ways in which we can work together for these common objectives. We envisage a European-wide campaign, in which every kind of exchange takes place; in which representatives of different nations and opinions confer and co-ordinate their activities; and in which less formal exchanges, between universities, churches, women's organizations, trade unions, youth organizations, professional groups and individuals, take place with the object of promoting a common object: to free all of Europe from nuclear weapons. (ENDS, 1980: 225)

Through bringing together different activists from both Eastern and Western Europe, ENDS sought to contest the spatial imaginaries of Cold War geopolitics and to develop a critique of the polarizing ideological work upon which the Cold War depended.

Second, the activities of ENDS produced an 'actually existing alternative' to that which it criticized by producing spaces for different political trajectories to come together. This allowed dissidents from Eastern Europe, such as Adam Michnik, Rudolf Bahro and Roy Medvedev, to exchange and debate ideas with figures from the peace movement and the New Left in Western and Northern Europe (see Bahro, 1982; Medvedev and Medvedev, 1982; for a more critical Eastern European perspective, see Racek, 1982). Through these practices, ENDS's networks brought together political activists from distinctive places and political traditions, including: the independent Marxism of Rudolf Bahro, which was formed through bitter critique of 'actually existing

socialism' (Bahro, 1977); the experiments with radical feminist organizing for peace that culminated in the peace camp at Greenham Common (see Cook and Kirk, 1983; Cresswell, 1996); and the distinctively 'English' New Left politics of E. P. Thompson (see Thompson, 1980, 1982). Through exchanges, debates, conflicts and arguments between activists from different places, ENDS produced a transnational political movement, not so much through activity that might be defined as 'jumping scales' (which would imply that such activists had captured, say, the national state in various countries) as by bringing together dissidents and peace activists in different countries through common networks and dialogues.

Here we can see a further way in which place matters in the construction of solidarities. Whereas distinct, place-related interests and identities are often seen as barriers to be overcome through political organizing, places can also be significant as the sites where different political trajectories are brought together for, as Latour reminds us, networks remain local at all points (Latour, 1993: 117–20). The activity of ENDS, then, was a networked *achievement* or a fabricated *articulation* of sites. These articulations were produced through spaces such as offices and homes in England, often under security service surveillance, and the flats of dissidents and peace activists in Prague and Budapest, such as those visited by Dorothy and E. P. Thompson (Palmer, 1994: 134–5). Equally, the annual ENDS conventions were events that brought together in specific places activists from all over Europe, though this was commonly a fractious, contested and difficult process (Kaldor, 2003: 63). This phenomenon emphasizes that 'what is involved' in constituting the relationships between the 'local' and 'global' 'is an immensely difficult, always grounded, and "local" […] negotiation' (Massey, 2005: 182; see also Gibson-Graham, 2002) that calls into question the stark counterposition of local particularist struggles and global universalist ones present in Harvey's (1996) accounts of the relations between local and global struggles.

The fact that these alliances were productive of new political spaces, identities and possibilities shows how place-based identities do not remain static and unchanged through involvement in transnational political networks but can be challenged, unsettled and reconfigured through such activity. The networks produced through articulations of these different place-based struggles and political cultures were also significant, as ENDS allowed dissidents in Eastern Europe to make connections with each other, albeit mediated through Western European cities such as London and Stockholm (Kaldor, 2003). As a result, the networks that ENDS produced and generated popularized across

Europe discourses and repertoires of activity associated with what has now become termed 'global civil society'.

In this section, then, we have sought to show how geographers and others have approached TSMs using a variety of analytical perspectives. Although it is an open question as to whether there is a material networked spatiality to TSMs that demands that network approaches should necessarily be used for analyses of such movements, advocates of network approaches have certainly shown that such approaches have a number of strengths that are worth considering seriously including the fact that they are alive to the productive and multiple character of political activity and resonate with the anti-hierarchical forms of organization that characterize many TSMs, whilst they are suggestive of the co-production of place-based political identities and transnational networks. Whether such approaches can generate the forms of political prescription and strategic direction demanded by some is, however, another question.

CONCLUSION

In our review of geographic research into the NLI and TSMs we have considered two ways of analysing the politics of organizing difference, all the while accenting the important of place as both a condition and an outcome of translocal solidarity. As we have seen, the research explored in this chapter has been conducted in response to the recent florescence of 'borderless' organizing by workers and social movements worldwide. A major theme of this research is how and with what consequences 'local difference' is an outcome and condition of wider geographical connectivities. But this research is not simply a reflection of the 'real world' actualities of which it seeks to make sense. Rather, the language used to describe such campaigns also has the power to help shape them materially, in much the same way as neo-liberal discourses of globalization as an unstoppable force encourage the adoption of policies that dismantle barriers to global capital flows (what we might call the 'self-fulfilling prophecy' phenomenon). However, as Miller (2004: 224) has argued – rightly, in our view – 'the work of building an emancipatory world order has just begun'. For all their media visibility, protests in places like Seattle are no substitute for the patient, complex work of challenging transnational power relations without effacing the geographical differences that such challenges attempt to unite in joint undertakings. In this light, we might reasonably ask what insights the research reviewed in this chapter yields for activists seeking to make transborder resistance truly effective. Our answer,

regrettably, is 'very few at this stage' (*contra* Flint's judgement). There are two reasons why.

First, though many of the studies reviewed in this chapter yield concrete insights into how, why and with what effects borderless solidarity occurs, it seems to us that a key task remains: to parse these insights across the literature *as a whole* so that some clear lessons can be drawn about aims, strategies and outcomes (Marston and Jones, 2005, remains almost a lone effort in this regard). This involves more than identifying the differences and commonalities between scalar and network approaches. More specifically, it involves synthesizing the literature so that we can answer the following sorts of highly specific questions: What is the most *effective* basis for otherwise place-based constituencies to unite? Is it around common *identities*, common *enemies*, common *goals*, all three or something else? What sorts of *resources* are required to make transnational organizing more than an unstable, fleeting phenomenon? What sorts of *tactics* lead (or do not lead) to the goals of those involved in the NLI and TSMs being realized? What are the unintended consequences of the NLI and TSMs? We could go on, but the point is clear enough. *Individual* studies of the NLI and TSMs answer these and other questions, but the *wider* lessons to be drawn remain unclear.

Second, Andrew Kirby (2002) has noted that the 'comfortable heuristics around which much academic discourse ... revolve[s] ... [do] not represent [the] ways that most people outside the academy think'. We agree. We earlier made fleeting reference to struggles for recognition within the community of critical geographers as one factor explaining why the rival discourses of scale and networks have risen to prominence in recent years. If Colin Flint's vision is to be realized, the researchers whose work has been discussed here (our own included) will have to make greater efforts to speak a language that activists can understand. This will, among other things, involve rendering arcane disputes about scale politics and networked solidarity into useful insights for actors seeking to make common cause across the relative and absolute distances that separate them. Despite new-found excitement about the idea of 'public geographies', it is fair to say that much of the research reviewed here remains encased in books and journals where only the cognoscenti can access it.

REFERENCES

Agnew, J.A. (1987) *Place and Politics*. London: Allen & Unwin.

Amin, A. (2002) 'Spatialities of globalization', *Environment and Planning A*, 34(3): 385–99.

Amin, A. (2004) 'Regions unbound: towards a new politics of place', *Geografiska Annaler B*, 86(1): 33–44.

Ashman, S. (2004) 'Resistance to neoliberal globaliza-tion: a case of militant particularism?', *Politics*, 24(2): 143–53.

Bahro, R. (1977) *The Alternative in Eastern Europe.* London: Verso.

Bahro, R. (1982) 'A new approach for the peace movement in Germany', in E.P. Thompson et al., *Exterminism and Cold War.* London: Verso, pp. 87–116.

Braun, B. (1998) 'The politics of possibility without the pos-sibility of politics?', *Annals of the Association of American Geographers*, 88(4): 712–18.

Braun, B. and Disch, L. (2002) 'Radical democracy's "mod-ern constitution"', *Environment and Planning D: Society and Space*, 20: 505–11.

Bunnell, T. and Nah, A. (2004) 'Counter-global cases for places: contesting displacement in globalizing Kuala Lumpur Metropolitan Area', *Urban Studies*, 41(12): 2447–67.

Castree, N. (2000) 'Geographic scale and grass-roots interna-tionalism', *Economic Geography*, 76(2): 272–92.

Castree, N. (2004) 'Differential geographies', *Political Geogra-phy*, 22(9): 1012–29.

Cohen, R. and Rai, S. (eds) (2000) *Global Social Movements.* London: Athlone Press.

Cook, A. and Kirk, G. (1983) *Greenham Women Every-where: Dreams, Ideas and Actions from the Women's Peace Movement.* London: Pluto Press.

Cox, K.R. (1998) 'Spaces of dependence, spaces of engage-ment and the politics of scale', *Political Geography*, 17(1): 1–23.

Cox, K.R. (2005) 'Local:global', in P. Cloke and R. Johnston (eds), *Spaces of Geographical Thought.* London: Sage, pp. 175–98.

Cox, K.R. and Mair, A. (1988) 'Locality and community in the politics of local economic development'. *Annals of the Association of American Geographers*, 78: 307–25.

Cresswell, T. (1996) *In Place/Out of Place.* Minneapolis: Minnesota University Press.

Davis, M. (1982) 'Nuclear imperialism and extended deter-rence', in E.P. Thompson et al., *Exterminism and Cold War.* London: Verso, pp. 35–64.

European Campaign for Nuclear Disarmament (1980) 'Appeal for European Nuclear Disarmament', in E.P. Thompson and D. Smith (eds), *Protest and Survive.* Harmondsworth: Penguin, pp. 223–226.

Escobar, A. (2001) 'Culture sits in places', *Political Geography*, 20(2): 139–74.

Featherstone, D.J. (2003) 'Spatialities of transnational resis-tance to neo-liberal globalisation: the maps of grievance of the inter-continental caravan', *Transactions of the Institute of British Geographers,* 28(4): 404–21.

Featherstone, D.J. (2004) 'Spatial relations and the materialities of political conflict: the construction of entangled political identities in the London and Newcastle Port Strikes of 1768', *Geoforum*, Special Issue on Material Geographies, 35(6): 701–11.

Featherstone, D.J. (2005) 'Towards the relational construction of militant particularisms: or why the geographies of past struggles matter for resistance to neo-liberal globalization', *Antipode*, 37(2): 250–71.

Flint, C. (2003) 'Political geography: context and agency in a multiscalar framework', *Progress in Human Geography*, 27(5): 637–48.

Gibson-Graham, J.K. (2002) 'Beyond global vs. local: eco-nomic politics outside the binary frame', in A. Herod and M.W. Wright (eds), *Geographies of Power: Placing Scale.* Oxford: Blackwell, pp. 25–60.

Glassman, J. (2003) 'From Seattle (and Ubon) to Bangkok: scales of resistance to corporate globalization', *Society and Space*, 19(4): 513–33.

Gottmann, J. (1951) 'Geography and international relations', *World Politics*, 3(1): 153–73.

Graeber, D. (2002) 'The new anarchists', *New Left Review*, 13: 61–73.

Hardt, M. and Negri, A. (2000) *Empire.* Cambridge, MA: Harvard University Press.

Hardt, M. and Negri, A. (2004) *Multitude: War and Democracy in the Age of Empire.* New York: Penguin Press.

Harvey, D. (1982) *The Limits to Capital.* Oxford: Blackwell.

Harvey, D. (1985a) *The Urbanization of Capital.* Oxford: Blackwell.

Harvey, D. (1985b) *Consciousness and the Urban Experience.* Oxford: Blackwell.

Harvey, D. (1989a) 'From urban managerialism to urban entrepreneurialism', *Geografiska Annaler B*, 71(1): 3–17.

Harvey, D. (1989b) *The Condition of Postmodernity.* Oxford: Blackwell.

Harvey, D. (1993) 'From space to place and back again', in J. Bird et al. (eds), *Mapping the Futures.* London: Routledge, pp. 3–29.

Harvey, D. (1996) *Justice, Nature and the Geography of Difference.* Oxford: Blackwell.

Harvey, D. (1998) 'The Humboldt connection', *Annals of the Association of American Geographers*, 88(4): 723–30.

Harvey, D. (2000) *Spaces of Hope.* Edinburgh: Edinburgh University Press.

Harvey, D. and Hayter, T. (eds) (1993) *The Factory and the City.* London: Mansell.

Herod, A. (1995) 'The practice of international labor solidar-ity and the geography of the global economy', *Economic Geography*, 71(4): 341–63.

Herod, A. (1997) 'Labor as an agent of globalization and as a global agent', in K.R. Cox (ed.), *Spaces of Globalization: Reasserting the Power of the Local.* New York and London: Guilford Press, pp. 167–200.

Herod, A. (2006) 'Geographies of labour organizing in the new economy: examples from the United States', in P. W. Daniels, J. V. Beaverstock, M. J. Bradshaw and A. Leyshon (eds), *Geographies of the New Economy.* Routledge: London.

Herod, A. and Wright, M. (2002) 'Placing scale', in A. Herod and M. Wright (eds), *Geographies of Power.* Oxford: Blackwell, pp. 1–14.

Hinchliffe, S. (2000) 'Entangled humans specifying powers and their spatialities', in J.P. Sharp, C. Philo, R. Paddison and P. Routledge (eds), *Entanglements of Power: Geographies of Domination/Resistance.* London: Routledge, pp. 219–37.

Howitt, R. (2003) 'Scale', in J.A. Agnew et al. (eds), *A Compan-ion to Political Geography.* Oxford: Blackwell, pp. 138–57.

Johns, R. (1998) 'Bridging the gap between class and space: US worker solidarity with Guatemala', *Economic Geography*, 74(3): 252–71.

Jonas, A.E.G. (1994) 'Editorial', *Society and Space*, 12(3): 257–64.

Kaldor, M. (2003) *Global Civil Society: An Answer to War*. Cambridge: Polity Press.

Kirby, A. (2002) 'Popular culture, academic discourse, and the incongruities of scale', in A. Herod and M. Wright (eds), *Geographies of Power*. Oxford: Blackwell, pp. 171–91.

Klein, N. (2002) 'Farewell to "the end of history": organization and vision in anticorporate movements', in L. Panitch and C. Leys (eds), *Socialist Register 2002: A World of Contradictions*. London: Merlin Press, pp. 1–14.

Laclau, E. and Mouffe, C. (1985) *Hegemony and Socialist Strategy*. London: Verso.

Latour, B. (1993) *We Have Never Been Modern*, trans. C. Porter. Cambridge, MA: Harvard University Press.

Latour, B. (1996) 'On actor-network theory: a few clarifications', *Soziale Welt*, 47: 369–81.

Mamadouh, V., Kramsch, O. and van der Velde, M. (2004) 'Articulating local and global scales', *Tijdschrift voor Economische en Sociale Geografie*, 95(5): 455–66.

Marston, S.A., Jones, J.P. III and Woodward, K. (2006) 'Human geography without scale', *Transactions of the Institute of British Geographers*, 30(4): 416–32.

Massey, D. (1993) 'Power-geometry and a progressive sense of place', in J. Bird et al. (eds), *Mapping the Futures*. London: Routledge, pp. 59–69.

Massey, D. (1999a) *Power-Geometries and the Politics of Space-Time*. Department of Geography, University of Heidelberg.

Massey, D. (1999b) 'Politics and space-time', in D. Massey et al. (eds), *Human Geography Today*. Cambridge: Polity Press, pp. 279–94.

Massey, D. (2004) 'Geographies of responsibility', *Geografiska Annaler B*, 86(1): 5–18.

Massey, D. (2005) *For Space*. London: Sage.

Medvedev, R. and Medvedev, Z. (1982) 'The USSR and the arms race', in E.P. Thompson et al., *Exterminism and Cold War*. London: Verso, pp. 153–74.

Merrifield, A. (1993) 'The struggle over place: redeveloping American Can in southeast Baltimore', *Transactions of the Institute of British Geographers*, 18(1): 102–21.

Merrifield, A. (2002) 'Seattle, Quebec, Genoa: après le déluge ... Henri Lefebvre?', *Environment and Planning D: Society and Space*, 20: 127–34.

Miller, B. (1994) 'Political empowerment, local-central state relations, and geographically shifting political opportunity structures', *Political Geography*, 13(5): 393–406.

Miller, B. (2004) 'Spaces of mobilization', in C. Barnett and M.M. Low (eds), *Spaces of Democracy*. London: Sage, pp. 223–46.

Murdoch, J. (1998) 'The spaces of actor-network theory', *Geoforum*, 29(4): 357–74.

Murdoch, J. and Marsden, T. (1995) 'The spatialization of politics', *Transactions of the Institute of British Geographers*, 20(3): 368–80.

Novelli, M. (2004) 'Globalizations, social movement unionism and new internationalisms: the role of strategic learning in the transformation of the Municipal Workers Union of EMCALI1', *Globalization, Societies and Education*, 2(2): 161–90.

Oslender, U. (2004) 'Fleshing out the geographies of social movements', *Political Geography*, 23(7): 957–85.

Oza, R. (2001) 'Showcasing India: gender, geography and globalization'. *Signs: Journal of Women in Culture*, 26(4): 1067–95.

Palmer, B.D. (1994) *E.P. Thompson: Objections and Oppositions*. London: Verso.

Perreault, T. (2003) 'Changing places: transnational networks, ethnic politics, and community development in the Ecuadorian Amazon', *Political Geography*, 22(1): 63–88.

Racek, V. (1982) 'Letter to E.P. Thompson', in E.P Thompson, *Zero Option*. London: Merlin Press, pp. 81–5.

Routledge, P. (1993) *Terrains of Resistance: Nonviolent Social Movements and the Contestation of Place in India*. Westport, CT: Praeger.

Routledge, P. (2003) 'Convergence space: process geographies of grassroots globalization networks', *Transactions of the Institute of British Geographers*, 28(3): 333–49.

Sadler, D. (2000) 'Organizing European labour: governance, production, trade unions and the question of scale', *Transactions of the Institute of British Geographers*, 25(2): 135–52.

Slater, D. (2002) 'Other domains of democratic theory: space, power, and the politics of democratization', *Environment and Planning D: Society and Space*, 20: 255–76.

Slater, D. (2003) 'Geopolitical themes and postmodern thought', in J.A. Agnew et al. (eds), *A Companion to Political Geography*. Oxford: Blackwell, pp. 75–92.

Smith, N. (1984) *Uneven Development*. Oxford: Blackwell.

Smith, N. (1989) 'Geography as museum', in N. Entrikin and S. Brunn (eds), *Reflections on 'The Nature of Geography'*. Washington, DC: AAG Publications, pp. 89–120.

Smith, N. (1992) 'Geography, difference and the politics of scale', in J. Doherty, E. Graham and M. Malek (eds), *Postmodernism and the Social Sciences*. London: Macmillan, pp. 57–79.

Smith, N. (2000) 'Global Seattle', *Environment and Planning D: Society and Space*, 18: 1–3.

Smith, N. and Katz, C. (1993) 'Grounding metaphor', in M. Keith and S. Pile (eds), *Place and the Politics of Identity*. London: Routledge, pp. 67–83.

Staeheli, L.A. (1994) 'Empowering political struggle', *Political Geography*, 13(5): 387–91.

Taylor, P.J. (1981) 'Geographical scales within the world-economy approach', *Review*, 5(1): 3–11.

Taylor, P.J. (1982) 'A materialist framework for political geography', *Transactions of the Institute of British Geographers*, NS, 7(1): 15–34.

Thompson, E.P. (1980) *Writing by Candlelight*. London: Merlin Press.

Thompson, E.P. (1982) *Zero Option*. London: Merlin Press.

Townsend, J.G., Porter, G. and Mawdsley, E. (2004) 'Creating spaces of resistance: development NGOs and their clients in Ghana, India and Mexico', *Antipode*, 36(4): 871–89.

Wainwright, J., Prudham, S. and Glassman, J. (2000) 'The battles in Seattle: micro-geographies of resistance and the challenge of building alternative futures', *Environment and Planning D: Society and Space*, 18(1): 5–13.

Watson, S. (2004) 'Cultures of democracy', in C. Barnett and M.M. Low (eds), *Spaces of Democracy*. London: Sage, pp. 207–22.

Wial, H. (1994) 'New bargaining structures for new forms of business organization', in S. Friedman, R.W. Hurd, R.A. Oswald and R.L. Seeber (eds), *Restoring the Promise of American Labor Law*. Ithaca, NY: ILR Press, pp. 303–13.

Wills, J. (2002a) 'Bargaining for the space to organize in the global economy: a review of the Accor-IUF trade union rights agreement', *Review of International Political Economy*, 9(4): 675–700.

Wills, J. (2002b) 'Political economy III: neo-liberal chickens, Seattle and geography', *Progress in Human Geography*, 26(1): 90–100.

19

The Political Geography of Many Bodies

Arun Saldanha

THE BODY POLITIC

In Stanley Kubrick's classic satirical movie on the Cold War, *Dr Strangelove* (1964), a rogue General Ripper authorizes nuclear annihilation on the basis of his paranoid conviction that the Russians were penetrating, polluting and exhausting the 'precious bodily fluids' of the American population. General Ripper's view on the pervasiveness of communist infiltration is interesting in that it understands geopolitics as at least partly unconscious, inescapably embodied in entire populations. For Ripper, geopolitics is more akin to sexual intercourse than a 'discourse'. In fact, the general *feels* the presence of the enemy. The fatigue of his body is telling him that dramatic military action is now called for, and only pre-emptive attack will save the bodily fluids of his fellow Americans.

Though General Ripper's theories are idiosyncratic, I want to suggest that his intense paranoia alerts him to the fact of embodiment often forgotten in geopolitical analysis. Much of the humour of *Dr Strangelove* is established through the personalization of the political geographies of the Cold War into comical bodies such as General Ripper's, the President's and Dr Strangelove's. Kubrick's film presents the political geography leading to nuclear annihilation not only as discursive, technological and economic, but also as a collectively embodied process of affects, prejudices, anticipations and negotiations. Geopolitical embodiment becomes for General Ripper something to be dealt with in its own terms. Thinking political geography through the body (or rather, through many

bodies) in this way is not meant to *complement* a political-economic or discursive approach; it completely changes the way we think geography and politics. This chapter suggests that there are theoretical sources for a political geography of many bodies, though it will reach conclusions very different from General Ripper's.

Despite the absence of the body in political geography, there is a long tradition, almost coterminous with Western political philosophy, of thinking the city and the state as analogous to the human body. As Richard Sennett (1994) has argued, the *polis* has never simply been the space of politics; this politics was from the beginning imagined in bodily terms. Plato and Aristotle thought that, just like the body, the state should be protected from illness and injury. Just like the body, the state has a particular spatial constellation, consisting of different parts with different functions: a king, judges, soldiers, priests, etc. And just like the body, the state has to be governed by reason and care. What was by the seventeenth century called 'the body politic' was the kingdom or state itself; it then came to stand for the geographically defined population in its political capacity. Thomas Hobbes famously regarded the (man-made, not God-given) body politic as the inevitable and desirable outcome of humans living together without succumbing to their instinct to kill each other: 'This union so made, is that which men call now-a-days a BODY POLITIC or civil society; and the Greeks call it polis, that is to say, a city, which may be defined to be a multitude of men, united as one person by a common power, for their common peace, defense, and benefit' (1650: XIX, 8).

It was Spinoza, however, who developed a political philosophy in which the body was treated not as metaphor, but as the material substratum of political society. A wave of chiefly Australian feminist political philosophy, including Genevieve Lloyd, Moira Gatens, Elizabeth Grosz and Rosalyn Diprose, has over the last decade or so ingeniously drawn from Spinoza's materialist legacy. They do so to both argue for the political-philosophical importance of bodily difference, and to expose the deep masculinist bias in the Western body politic. In other words, if 'the body' in the body politic is taken seriously, these feminists say, the very concepts of statehood, government, citizenship, nation, geopolitics and democracy have to be reworked. Though the valorization of embodiment has had a profound impact across academia (Williams and Bendelow, 1998), its central insights have yet to be brought to bear on political geography. This chapter seeks to tease out the spatiality implicit in Spinozism and suggests a political geography that is rigorous about including bodies.

Four political-geographical dimensions of human bodies will be addressed: their sensuousness, their variation, their locatedness and what is going to be called their *viscosity*. This embodied perspective is meant to depart from an individualist and mentalist way of understanding citizenship, political mobilization and consumerism. While sensory perception and bodily location have been extensively conceptualized by humanistic geographers and feminists, what happens when many bodies come together *as bodies* – viscosity – has escaped attention. Yet to appreciate the corporeality of politics, it seems difficult to ignore the problem of viscosity. The basic argument is that the body politic (governments, social movements, corporations) consists of the largely unconscious sticking together of bodies according to their differences. Viscosity refers to this dynamic emergence, at scales from the street to the planet, of collectivities of people based on attributes like sex, skin colour, nationality, economic power or fear. That these attributes should be seen as corporeal, not simply psychological, goes a long way to explaining what is meant by viscosity. Thus an attempt is made to bridge humanistic and feminist concerns with embodiment on the one hand, and Spinozist definitions of the nation-state on the other. Following the suggestion of popular science writer Philip Ball (2004) of a 'physics of society', this chapter argues for an analytical framework that does not simply append, but starts out from human embodiment. Along the way, some key philosophers and fields of inquiry pertaining to the body are indicated. The purpose is to both demonstrate the many ways that embodiment has already been conceptualised, and suggest some fruitful congruities under the umbrella of political geography.

SENSUOUS BODIES

Embodiment has until recently been neglected by geographers more or less in line, it would seem, with the foundational Western philosophical undermining of the somatic and sensual. Humanistic geographers writing in the 1970s and 1980s deserve credit, therefore, for elucidating the relationship between the embodied self and space and place. In *Space and Place: The Perspective of Experience*, Yi-Fu Tuan (1977), for example, showed how not just vision, but hearing, touch, smell, taste and motion together place humans in their physical and social environments. As David Seamon (1979) acknowledged, the phenomenology of Maurice Merleau-Ponty had already attacked the dualistic doctrines of mind versus matter and space versus things inaugurated by the likes of Descartes, Newton and Hobbes. These dualistic doctrines, epitomized in positivist science, strive for ostensibly objective and disinterested knowledge at the expense of the immediate relationships the body develops with its surroundings. 'The perception of space is not a particular class of "states of consciousness" or acts', wrote Merleau-Ponty. 'Its modalities are always an expression of the total life of the subject, the energy with which he tends towards a future through his body and his world' (1945: 283).

With Merleau-Ponty, the richness of how one *inhabits* space – through the inner ear, muscles, erotic desire, digestion, dreams, intuition, fatigue, etc. – is affirmed as primary in relation to all other knowledge. But who is this 'one'? Implicitly, Merleau-Ponty and the humanistic geography he inspired were approaching embodiment from a male, able-bodied and European point of view. In political theory, Iris Marion Young (1990) famously offers a feminist re-reading of Merleau-Ponty's ideas on comportment and spatiality by considering how girls tend to throw things differently than boys. Young shows that throwing like a girl emerges through the learned coordination between brain, musculature and, crucially, social and geohistorical location. Physiology and patriarchy thus form an inseparable site for political intervention.

Strongly influenced by Merleau-Ponty, Alphonso Lingis's rich work shows that before we can even speak of a conscious subject (citizen, politician, analyst), there is already a plethora of pre-personal biophysical processes, which may or may not coagulate into rational decision-making (e.g. Lingis, 1994). It is sometimes forgotten just how much Merleau-Ponty's phenomenology was at least partly inspired by findings in physiology and neuropsychology. The scientific study of the human brain is increasingly open to affective and emergent qualities, as popularly described by Antonio Damasio (2003) and

Francisco Varela (1992). Judging from Andy Clark (2001), Merleau-Ponty would have been enthusiastic about the phenomenological lessons of robotics and artificial life. Clark argues that computer intelligence, like human intelligence, adapts and evolves only insofar as it moves around its environment *as* body. The phenomenological insistence that the *lived* aspect exceeds the contemplated and geometric aspects of space defined the humanistic-geographical critique of the positivist striving for detachment in spatial science. But quantitative and experimental science do not necessarily entail a reductionist or dehumanized understanding of human embodiment.

In geography, the hegemony of disembodied vision has been elaborately discussed by geographers such as Paul Rodaway (1994), Gillian Rose (2003) and Susan Smith (1997). As the titles of the collection *Places Through the Body* (Nast and Pile, 1998) or Edward Casey's *Getting Back Into Place* (1993) attest, the inevitably geographical nature of embodiment is beginning to be addressed (see also Pile, 1994; Butler and Parr, 1999; Teather, 1999; Longhurst, 2000; Aitken, 2001). What in human geography and everyday language is called a 'place', such as a basketball court, parliament or mall, emerges and changes only because human bodies move around in it in habitual ways. Places are also woven together through moving bodies, thereby simultaneously delimiting possible future movements. This was understood long ago by environmental psychologists and urban sociologists like Kevin Lynch (1960), as well as the time-geography of Torsten Hägerstrand (Pred, 1977). However, these theorists stopped short of addressing the power relations involved in everyday life. This feat was attempted by Henri Lefebvre (1974), whose Marxist philosophy of body and urban space influenced a whole generation of theorists.

The visual hegemony of cartography has tended to go hand in hand with masculinism, colonialism, state surveillance and racism. The turn towards multisensory corporeality, away from vision and detachment, has enabled a re-politicization of human geography itself. Like Merleau-Ponty and Lefebvre, critical geography now argues for an array of bodily senses in the understanding of space. In a Merleau-Pontian (and, more fundamentally, Spinozist) take on politics, the very existence of political entities depends on bodily sensation and exchange (Diprose, 2002). For political geography this means more than simply a shift of empirical focus from maps and rhetoric to the concrete and fleshy. A serious appreciation of sensuousness gives a very different twist to the concept of the 'body politic'. The patrolling of state boundaries, international 'security', suicide bombing, the saluting of a tank, patriotic sentiment, diplomatic ceremony, biometrics, anti-globalization riots, a close-up of the president in *Newsweek*: in each of these it is obvious that political geography is materialized in particular bodies, only partly consciously.

What difference does it make to call these phenomena embodied? Space allows for only an indication of the most important reasons for bringing the senses into political geography, before moving on to the other aspects of embodiment. First, an insistence on sensory perception highlights certain crucial components of political geography that are more often than not obliterated from purview for ideological reasons. The nation-state is often embodied through physical violence and pain (and their painful recollections). The importance of rape for ethnic cleansing and warfare has until recently escaped attention (Mayer, 2004); every year, terror and torture are becoming more important to the geopolitical order. Second, the cultural reproduction of geopolitical relations and entities (through parades or border patrols) can be more concretely studied if what bodies do and feel enters the discussion. Third, the affective qualities of leaders and rhetoric, circulating within constituencies, can be meticulously analysed as legible corporeal signs (think of Hitler's body language). Fourth, more and more, geopolitics itself depends on the biometrical management of bodies through what Michel Foucault called biopolitics (Foucault, 2003; Sparke, 2006). What many are arguing in the wake of Foucault is that the intrusion of the state into the biological being of people (refugees, convicts, citizens) is accompanied by a whole new range of emotions and embodied skills. If it seeks to understand the concrete nature of these political changes, political geography has to take into account their targets and vehicles: bodies of flesh, blood, senses, genes and memories.

DIFFERENT BODIES

Embodiment is fundamentally about difference, and we need a section to elucidate this point. Iris Marion Young's feminist elaboration of Merleau-Ponty has already been mentioned. One body is not another: is this not feminism's central concern? Feminist politics starts from the *difference* between women and men. All of Western thought, including political geography, has not only overwhelmingly been conjured by men, among men and for men, but men have also always talked on behalf of everyone else: women, children, slaves, foreigners, the old and sick. The philosophy of Luce Irigaray has exposed the implicit privileging in European philosophy of the average male body and its particular morphology. The specific characteristics of femininity and masculinity have been consistently ignored in the centuries-old conceptualization of 'man' (political man, economic man, etc.). Irigaray's ethics and politics follow from what she sees as the irreducibility of sexual difference (Irigaray, 1984). Though she concurs with Merleau-Ponty's celebration of corporeality, especially touch, Irigaray insists that men and women are fundamentally different, and therefore experience the world in fundamentally different ways. What is required for both man and woman is not an erasure of difference, as demanded by egalitarian feminism, but an opening-up towards one another that acknowledges and fosters variation and transformation. It might seem that Irigaray espouses essentialism. But while her writing defends woman, it does not present her as eternal, universal or inherently oppositional. Like Derrida, Irigaray thinks of (sexual) difference as relational, provisional and open. Irigaray *re-invents* difference by suggesting fresh and often ironic understandings of bodily materialities. Biology is then not destiny, but the playground for ethical and political experimentation.

What Elizabeth Grosz (1994) called 'corporeal feminism' derives much force from Irigaray. In geography, Irigaray inspired Gillian Rose (1993) to call for a radical thinking and inhabiting of 'paradoxical space'. Subsequent feminist geographers have not only shown that geography is a profusely disembodied and masculinist tradition, but have also, in Irigarayan and Derridean fashion, recuperated from that tradition the concepts they need to further a feminist project. Paradoxical geography, in Rose's words, 'is a geography structured by the dynamic tension between such poles [the centre and the margin, the inside and the outside], and it is also a multidimensional geography structured by the simultaneous contradictory diversity of social relations' (1993: 155). More than Irigaray herself, feminist geography has been open to the complexity of intersecting dimensions of oppression and resistance (cf. Deutscher's critique, 2002). A white working-class woman is different from a black middle-class woman. Bodies are therefore different not just sexually, but racially too (Saldanha, 2007). Bodies furthermore differ in terms of age, fitness, sexual desires, height and weight, mental health, hereditary diseases and cultural attractiveness. As developmental biology indicates (Leroi, 2003), there is just no sense in speaking of a phenotypic standard.

The sensual and differential nature of bodies needs to be seen as a question of *many* bodies. How to conceive of collective embodiment – a question addressed at scales from the street to the geopolitical – has not been satisfactorily explored. Simply put, we hear of 'the body', but seldom about bod*ies*. This de facto methodological individualism is foreign to the life sciences. In evolutionary biology, what Ernst Mayr (1963) called

'population thinking' is thoughtfully opposed to the essentialism of eighteenth- and nineteenth-century 'typological thinking'. In population biology and biogeography, species and subspecies groups do not have essences but are dynamic outcomes of many interacting individual organisms. Evolution itself is understood as the net effects of many organisms coping with their environment. This is not radically different for humans.

So both biologically and culturally, embodiment is *already* differentiated. With Merleau-Ponty and the anthropology of Edward T. Hall (1966) and Pierre Bourdieu (1977), we can assert that every human body accumulates tastes, attitudes, habits and gestures. A human body becomes 'multiple', in Annemarie Mol's terminology (2002), as it goes to certain places and engages with things and other bodies. Bourdieu's critique of phenomenology entails that the differential between bodies is thought not just in individual terms (the difference between a woman and a man, the difference between two neighbours, the difference between you and me) but in populational terms. In a way, population thinking is central to the social sciences too, although, as in biology, essentialism always lurks around the corner. What Bourdieu's statistical studies (1979) get at is masses of bodies that, unbeknown to each other, feel, think and do things that gradually distinguish them from other collectivities. Bourdieu and his collaborators showed not that there are confluences of individual *minds*, but that sensuousness is an emphatically demographic process, especially if it concerns consumption. The geography of embodiment should therefore not just tell of how 'a' body is enfolded in a place or flow, or how two bodies exchange affects. It should explain how networks of bodies make up larger geodemographic distinctions through the sharing and excluding of affects and practices. Mathematics and GIS can be used to ascertain regularity and predictability. They can also be used, more tentatively, to sense the tendencies and shifting of masses of bodies. What does voting behaviour say about collective fear? How many protesters can move through this street without getting clogged up? How many bodies were likely to be exposed to this populist propaganda? The social sciences have long studied the distinctions between populations. The challenge is now to think those populations not as mere 'minds', decisions or sets of data, but thinking and feeling organisms continually on the move, mixing with each other in certain physical settings.

Political geographers have not yet considered the Irigarayan and biological explorations of difference; in fact, political geography seems on the whole averse to thinking the political at the level of the fleshy and the sensual. There are precise historical reasons for this, of course, not in the least to do with the masculine and Newtonian biases of geography explored elsewhere in this book. The feminist critique of geography has now made a productive exchange possible between political geography and explorations of embodiment elsewhere in academia.

LOCATED BODIES

Bodies differ not just anatomically, neurologically, medically and sociologically. As geographers know only too well, bodies are always *somewhere*. What a body is politically – what a body *can do* – depends on its location in distributions of resources and media imageries, and even, as the recent Indian Ocean tsunami and New Orleans flood showed, seismological and meteorological processes. 'Location', as it is used here, is then more than longitude and latitude. What matters is one's 'topological' position in all the uneven constellations that human and physical geographers have mapped. Feminists have taken up location to describe how uneven development and sociospatial boundaries delimit choices, and more so for women than men, more so for the poor and the black than for the white and wealthy. Adrienne Rich called on feminists already in the 1970s to *think through their body*. As she suggests in these much-quoted sentences from 'Notes towards a politics of location' (1984: 215):

> To write 'my body' plunges me into lived experience, particularity: I see scars, disfigurements, discolorations, damages, losses, as well as what pleases me. Bones well nourished from the placenta; the teeth of a middle-class person seen by the dentist twice a year from childhood. White skin, marked and scarred by three pregnancies, an elected sterilization, progressive arthritis, four joint operations, calcium deposits, no rapes, no abortions, long hours at a typewriter – my own, not in a typing pool – and so forth.

In contrast to some psychoanalytical feminists who followed Rich, such as Jane Gallop (1988), embodiment was for Rich a question of location as much as phenotype and psyche. Black, immigrant, diasporic and Third World feminists have likewise emphasized the tremendous importance of bodily specificity to the politics of location, so that the exploration of location multiplied feminism itself. Recent historical work on colonialism like Ann Laura Stoler's (2002) shows that corporeal differences have always been interlaced with the historical geography of empire (see also Ballantyne and Burton, 2005). As with history and feminism, political geography will also become richer once the interplay of bodily specificity and location is studied.

Economic geographers have of course always known that location matters. Doreen Massey has been arguing for decades that space itself is constructed differentially (see Massey, 2005). Globalization affects and is affected by different groups in glaringly different ways, according to what Massey has called power-geometries. Feminist geographers of global processes (Pratt, 2004; Writers and Nagar, 2006) offer a rich conceptualization of the manners in which national boundaries, postcoloniality, poverty, migration, multicultural policy and other political-geographical phenomena are entangled with encounters and struggles between differing bodies 'on the ground' – including the body of the researcher.

And yet what is sometimes forgotten is that it is *because* bodies vary sexually, racially and economically, that they get differently located in geographies of power and possibility. The sensuous, voluminous, phenotypic materiality of bodies should be explored as expressions of global capitalism and colonialism. It is not simply a question of bodies being 'inscribed' by geography (Pratt, 1998), but geography *materially* constituting bodies and their capacities. This is evident for the geography of food, but can also be imagined in the geography of education and religion. When trained as an electrician, one literally incorporates and transports certain skills; when facing Mecca, one literally aligns one's body with millions of others. It is equally crucial to understand, conversely, that location does not follow deterministically from phenotypic particularity. How a body is intercepted by global flows depends on what bodily features matter in everyday interaction and in the reproduction of social relations and institutions. For example, sexual and racial segregation show that it is the materiality of bodies itself that makes particular spatialities endure, though the segregation follows contingently from the interplay of urban planning, architecture, surveillance cameras, urinals, etc. The bottom line about location is that bodies are not blank surfaces. Their genitals, hair and health need to be moulded and charged by power-geometries; they need to be pushed and pulled to certain places. For political and economic geographers, the conclusion is that there is no spatial inequality without the investment in the particularities of bodies.

It is worth repeating that bodies are not only located in institutional and cultural geographies. Global warming, the spread of AIDS and bird flu, drought, so-called natural disasters and, ultimately, the possibility of collision with meteorites or asteroids, demand a more-than-human conception of 'location' (Clark, 2005). The politics of location thus feeds into epidemiological, ecological, geological and cosmological considerations. Environmental history offers compelling examples of how human life – migration, the formation of cities and states, cultural conflict – is intrinsically interwoven with ecosystems at large (McNeill, 1977; Crosby, 1986). It is unfortunate that there is hardly anyone writing on the biogeography of globalization in the popular realm except Jared Diamond (see Diamond, 1997). Though he is careful enough to avoid racism, most political geographers would read Diamond as an environmental reductionist. A more nuanced approach to human association is forming in the wake of complexity theory; another popular science author critically addressing globalization is Fritjof Capra (2002). Historical and political geographers have yet to fully acknowledge the force of microbes, animals and rocks in human culture and the body politic. The problem at the root of all the evils of early twentieth-century geography – racism, positivism, regionalism (see Peet, 1985) – was an essentialist, Newtonian conception of nature. Population biology, environmental history and complexity theory have meanwhile shown there are non-essentialist ways of studying the biophysical – and new ways of bringing it back into human geography.

The conceptual messiness of 'nature' leads Bruno Latour (2004) to suggest that political ecology abandon the very concept. Latour only exacerbates the problem. Keeping in line with Spinoza, and following Genevieve Lloyd (1994), what is required instead is a more rigorous and critical conception of nature, which at once encompasses and problematizes the human. Arguing for the biological is meant to challenge political geography into reconsidering what it has become used to excluding. Political geography has understandably been traumatized by the derivations of Nazi ideology and policy from the biological determinism of one of the founders of the sub-discipline, Friedrich Ratzel (1897). In critical geopolitics, any mention of human biology would be subjected to immediate deconstruction. But is it biology that is the problem, or an underlying assumption that all biology is equally deterministic? Writing at the height of European imperial and nationalistic strife, the kind of biogeography Ratzel adhered to was indeed evolutionist and simplistic, informed by nineteenth-century taxonomy and Romanticism. In contrast, the perspective on the biological (and the biological in capitalism and geopolitics) coming from Spinoza, Merleau-Ponty and much complexity theory is decidedly non-essentialist, relational, impure, critical and contestable. This complex biology allows no identification of place and people, for example. It only emphasizes that political geography cannot – and, given the problems we are faced with, should not – separate the human from the rest of nature. Globalization and democracy would then be contingent outcomes of biophysical processes long predating even the appearance of *Homo sapiens*. The politics of location would have to investigate inequalities held in place by

vastly more than only discourse, public opinion and capitalism.

VISCOUS BODIES

We have so far discussed three facts about embodiment: their sensuality (through phenomenology), their differences (through feminism and ecology) and their locatedness (through, again, feminism, as well as economic geography). It is not that political geographers have not noticed the turn towards embodiment. There are recent contributions signalling a change, such as Alison Mountz's analysis of Canadian refugee policy (2003). Mountz reveals the critical benefits of grasping the importance of embodiment in political geography: refugee policy and policing are aimed at bodies with particular phenotypes, moving in particular ways along particular routes, embodying particular hopes, memories and a lack of citizenship. She shows that thinking through bodies moves political geography forward not simply by including another scale of analysis. A conceptually and empirically rich understanding of networked corporealities changes the ways we conceive of such processes as the state, international relations and globalization.

Sensation and location are fundamental aspects of human embodiment. But as I noted earlier with Ernst Mayr, they do not apply to isolated bodies. Bodies feel together, they form populations. This brings us again to Spinoza. For Spinoza, the human body naturally tends towards interacting with others. This is more profound than saying that humans are social animals. According to Gilles Deleuze's reading of Spinozist metaphysics, the seventeenth-century philosopher constructs a much more general 'physics of bodies'. When Spinoza talks of 'a body', he means anything, human or not, that is more or less delineated from its background. For Spinoza, therefore, there is but 'one Nature for all bodies, one Nature for all individuals, a Nature that is itself an individual varying in an infinite number of ways' (Deleuze, 1981: 122). 'The important thing is to understand life, each living individuality, not as a form, or a development of a form, but as a complex relation between differential velocities, between deceleration and acceleration of particles. A composition of speeds and slownesses on a plane of immanence' (ibid.: 123). In Spinozism, all bodies move, and all bodies are moved by other bodies. Spinoza understood clearly the consequence of this: bodies can form dynamic, though distinct, *larger* bodies. When talking about humans, we might call these larger bodies *collectivities*. Indeed, how we might analyse and change these collectivities is the question driving what are probably the best-known titles on globalization from a Spinozist perspective: *Empire* (2000) and

Multitude (2004) by Michael Hardt and Antonio Negri. Notwithstanding their important contribution to the current recasting of Spinoza, I want to focus more than Hardt and Negri on the fleshy and concrete nature of collectivities. For this, as said, I will move from biology to physics.

Deleuze pinpoints Spinoza's great innovation in philosophy: 'what the body can do no one has hitherto determined, that is to say, experience has taught no one hitherto what the body, without being determined by the mind, can do and what it cannot do from the laws of nature alone, in so far as nature is considered merely as corporeal' (1677: 101; see Deleuze, 1968: 217–34). And, to be consistently Spinozist, we should also ask what *many bodies* are capable of. Whether human or not, many bodies together are very rarely chaotic: they are capable of 'communicating' movement to each other, becoming *sticky* relative to one another, although they keep moving and remain a plurality. This process of becoming-sticky is what I want to call *viscosity*. The concept has much affinity with the better-known term in science and analytic philosophy, now quickly becoming popularized (e.g. Johnson, 2001; Capra, 2002), of *emergence*, the spontaneous appearance of coherent group motion irreducible to what the many particles do individually. Mark Bonta and John Protevi (2004) have recently published a geographer's guide to the ontological politics of Deleuze and Félix Guattari (1980) using emergence as a cornerstone concept. Though this chapter does not aim to be exegetical, it overlaps with this call for a new and radical physicalism. The reason why I want to introduce the hydrodynamical figure of viscosity is, basically, that I think it is more geographical and versatile than Deleuzo-Guattarian terminology.

Viscosity is inevitably about large numbers of particles, and one can appreciate the many-ness of viscosity only in physical terms. A classic example sometimes used in explanations of complexity theory is the unison of flying birds. The bird bodies stick together as a temporary system, yet remain *many*, flying in order to avoid their neighbours. 'Viscosity' wants to emphasize this mobile spatiality of togetherness, of coordinated flowing and relative stability. Unlike gases, viscous liquids do not move randomly; unlike solids, they are not static. Now, viscosity is more than a metaphor; it does not only apply to fluids or birds. Human flows become viscous in crowds, in large airports and in traffic jams. Crowds are easily appreciated as being sticky, because they are about concentration in one place. But human bodies stick together in more durable ways: packs, cities, nation-states, social classes and racial formations are also examples of human viscosity. If thinking the viscosity of bodies on scales larger than the city is difficult, this only proves how unaccustomed we are to thinking bodies *as bodies*: as shifting and linked masses

subject to gravity and expulsion. Like emergence, viscosity is a quintessentially spatial process, but it need not depend on either geographical proximity or actual deceleration. It is about flows and networks, and how they intersect, before it is about places. Remember that the flock of birds can be both moving erratically in real space, and remain viscous in systemic terms.

More and more, human viscosities are studied by physicists. They are discovering that human flows might not be ontologically different from other matter, just more complex. Sadly unaware of Spinoza or Deleuze, Philip Ball's *Critical Mass: How One Thing Leads to Another* (2004) usefully summarizes recent advances in what he hopes will be a 'physics of society'. For instance, discussing the Helbing–Treiber model in 'traffic theory', Ball (2004: 213) writes:

> They dispensed with individual cars and treated the traffic as a smooth liquid. But it was a most peculiar fluid. In the traditional theory of fluid motion, called hydrodynamics, each little 'parcel' of fluid affects those around it through viscous drag: it exerts a frictional force which slows down the movement of the surrounding fluid. In Helbing and Treiber's model the interaction between parcels of 'traffic fluid' are more complex [...]. [D]rivers are assumed to speed up and slow down as they react to what is happening ahead, aiming to reach a particular velocity and to avoid collisions. It is a fluid with a mind of its own; indeed, with a multitude of minds.

Strange as it may seem, Ball's book is in effect a physicist's answer to Hobbesian political philosophy. While rejecting Hobbes's bleak assumption of human selfishness, Ball revives the early modern enthusiasm of approaching human aggregation physically, like grains of sand or droplets of water. Of course, it is today's powerful computation technology that enables the probing into the mathematical depths of collective behaviour. It is far from certain to what extent computer modelling of complex behaviour hints at physical *laws* shared by humans and everything else. What is more certain is that much of contemporary physics, like biology, is less mechanistic and positivist than the social sciences make of it. On a par with Deleuze and Irigaray, science after complexity theory grapples with the *creativity* of matter. Some new conceptualizations arising from the recent physical sciences could inspire debate in critical political geography.

If the sticking together of crowds in public space is relatively easy to appreciate, a brief elaboration on the viscosity of racial formations will allow for imagining what the concept can do for political geographies at multiple and interacting scales (see Saldanha, 2007). Population geography demonstrates that migration (for example, during the European colonial era) follows spatial *patterns*. Indigenous populations remain in a certain region for many centuries, others emigrate in large numbers (the Irish or the Polish to the United States, as well as millions of refugees), go on holiday, or maintain networks of diasporic kinship (Jews, present-day South Asians). Within countries, within cities, these populations furthermore coalesce in certain landscapes, neighbourhoods, churches or camps. All these degrees of movement and stasis is what the concept of viscosity tries to designate. However, the concept demands in addition that we imagine the phenotypic, sensory and cultural specificity of these masses of bodies: their skin colour, diet, health, reproduction rate, religious affects. And there are bodies not covered in population and urban geography, even more difficult to map: business travellers, for example, who are usually white, male, heterosexual and able-bodied. Through their high mobility, businessmen keep corporate capitalism and patriarchal power-geometries in place. In the framework of viscosity, the *formation* of racial difference becomes a literal term, an extremely complex configuration of bodies moving on a range of scales with a range of speeds, with shapes that can only be approximated in maps.

In *Global Complexities* (2003), John Urry has urged for a turn towards complexity theory in the social sciences. He writes that flows of humans are inexorably linked to flows of capital, electricity, food, and other matter and energy. Unlike much of the literature on globalization, however, the concept of viscosity insists that the flows of humans can and do slow down, capturing bodies like spider webs. In Urry's words:

> The 'particles' of people, information, objects, money, images, risks and networks move within and across diverse regions forming heterogeneous, uneven, unpredictable and often unplanned waves. Such waves demonstrate no clear point of departure, deterritorialized movement, at certain speeds and at different levels of viscosity with no necessary end state or purpose. (2003: 60)

It is difficult to appreciate the viscosity of populations if one treats people as so many minds, subjectivities or discursive effects. The collective behaviour of *many bodies* – the workings of sensuousness and interaction in aggregate, in and to certain places and not others – would escape analysis. But when Spinoza argues that a body's (a collectivity's) capacities to affect and be affected are central to what it is, it follows that the relative holding together and forcefulness of a viscosity is what matters, not just the actual velocities of the particles composing it. An engagement with complexity theory will certainly contribute to a

non-human sociology of planetary fluidity. However, as can be intuited from this quote, Urry chooses to over-emphasize chance and amorphousness at the expense of friction and slowness. Unlike Capra (2002), on whom he draws heavily, Urry does not elaborate on the systemic viscosity of power-geometries. Viscosity regulates speeds and slownesses of bodies so that a system (such as patriarchy, or the European Union) stays 'in place'. One could say the political geography of many bodies is geography *all the way*. Rather than propose a radical break with existing approaches to geopolitical process, what the concept of viscosity does is open up space for a more rigorously materialist understanding of global interconnectedness.

DEMOCRATIC BODIES

The founding drive of critical political geography is an astute sense of the spatiality of politics and injustice. The argument in this chapter has been that political geography would need to consider the physics of bodies, or viscosity, if it wants to take seriously the sensuality and differentiality of populations. But is *recognizing* viscosity all political geography can do? Would that not testify to political geography's own 'viscosity'? In this closing section, I want to briefly develop some ways through which the political geography of many bodies can change the viscid realities it studies. How does political geography become political?

First, it needs to be noted that not all viscosity is bad. Crowd behaviour can be destructive and ecstatic at the same time. And who is to say that sticking to one place or a way of life is less desirable than going elsewhere and becoming something else? 'Viscosity' strives to be a neutral ontological concept, like 'emergence' and 'force'. This does not preclude taking feelings and ethical imperatives into account. If an actual or topological viscosity of bodies prevents those bodies *or other bodies* from realizing their hopes and talents, the overall situation could be improved by introducing more fluidity. Returning to Spinoza, if nature in essence tends towards the skilful dodging of constraints, then democracy is for humans the 'natural' political system in which freedom of thought and movement can be cultivated. 'The "art" of wise government is to structure the field of human interaction in ways which minimize fear and suspicion whilst maximizing the trust and confidence necessary for the flourishing of mutual aid and civic friendship' (Gatens and Lloyd, 1999: 118). With the regimes of predatory capitalism and racist paranoia in place today, it is obvious that we are a long way from democracy. Wise leadership will need to harness far more fluidity and intermixing to counter the general tendency towards solidification. This sounds vague, and it is. Past predictions of present possibilities were probably also vague. I think that critical geography is pragmatic, however, and its research agenda can be open enough to face the relevant issues.

As Gatens and Lloyd argue, like Irigaray, the democratic body politic should work to accommodate more than one kind of body – the white, healthy, heterosexual male. 'Indeed, the very notion of harmony assumes variation, not sameness' (1999: 128). In contrast to what a lot of human geography and cultural studies say, it is the materiality, not the discourse, of politics and the law that should be understood and experimented with. 'Politics is not simply a struggle over *ideas* and formal status. It also involves the struggle to embody and embed the desires, the needs and imaginings of those whom democratic political structures in the present fail to adequately represent' (ibid.: 130). If political geography is going to tackle the age-old problem of embodiment, so central to thinking the political (as well as the spatial) at all, it will have to welcome not just phenomenological and feminist concerns, but draw from biological and physical discourses to re-imagine the parameters it takes for granted. The nation-state, the 'body politic', geopolitics, economic policy, administration, unionism, environmental activism and multiculturalism start to feel very differently when the variation of actual human physique comes at the centre of analysis. This variation, I have proposed, needs to be envisaged in terms of populations and viscosity.

The turn towards 'the body' in social and political theory is intrinsically both ontological and interventionist. This is exactly how Spinoza's influence has been vital (Gatens, 1996). It is ontological, because it forces us to approach subjectivities as things in the world. As Etienne Balibar writes:

> social relationships too must be imagined as both ideological relationships (in souls) and physical relationships (in bodies) that are exactly correlated with each other and that express the same desire for self-preservation on the part of the individual, whether that desire is compatible or not with the desires of other individuals and complexes of individuals (such as the nation or the State). (1985: 106–7)

And the ontological turn towards bodies is political. How to create a difference? How to make more people happy?

> The order of causes is therefore an order of composition and decomposition of relations, which infinitely affects all of nature. But as conscious beings, we never apprehend anything but the *effects* of these compositions and decompositions: we experience *joy* when a body encounters ours

and enters into composition with it, and *sadness* when, on the contrary, a body or an idea threaten our own coherence. (Deleuze, 1981: 19)

The reduction of the human affects to sadness, joy and an increase in power seems utilitarian and egotistical. There is no denying, however, that pleasure and suffering are integral to social relationships. There is also no denying that fruitful companionship and coordination means growth in an individual's capacity to act. Damasio (2003) contends that, understood with current neurobiological research, Spinoza seems to have had a good intuition about the primacy of feeling in sociability and personal development. Seen in the light of his theory of human existence as *inevitably* social and latently democratic, Spinoza suggests an ethics of encounter through which the participating bodies and ideas flourish, instead of trying to exploit each other. Feeling is what makes us human and what makes us both anxious and stick together. This mutual reinforcement/challenge is exactly what politics is about. Today's global geographies of exploitation and ignorance do produce joy for a few, but they create disproportionately large amounts of sadness. This sadness compels many to nihilism, and some to violence. On the brink of abandoning the Cartesian split between body and mind, the discipline of political geography, with its ongoing theorisations of space, privilege and contestation, would seem at the right conjuncture to start addressing these sticky geopolitical complexes of joy and sadness.

ACKNOWLEDGEMENTS

I am deeply grateful for the extensive criticisms provided by Murray Low and Jennifer Robinson on a previous draft of this chapter.

REFERENCES

Aitken, S.C. (2001) *Geographies of Young People: The Morally Contested Spaces of Identity*. London: Routledge.

Balibar, E. (1985) *Spinoza and Politics*, trans. Peter Snowdon. London: Verso, 1998.

Ball, P. (2004) *Critical Mass: How One Thing Leads to Another. Being An Enquiry into the Interplay of Chance and Necessity in the Way That Human Culture, Customs, Institutions and Conflict Arise*. London: Arrow Books, 2005.

Ballantyne, T. and Burton, A. (eds) (2005) *Bodies in Contact: Rethinking Colonial Encounters in World History*. Durham, NC: Duke University Press.

Bonta, M. and Protevi, J. (2004) *Deleuze and Geophilosophy: A Guide and Glossary*. Edinburgh: Edinburgh University Press.

Bourdieu, P. (1977) *Outline of a Theory of Practice*, trans. Richard Nice. Cambridge: Cambridge University Press.

Bourdieu, P. (1979) *Distinction: The Social Critique of the Judgment of Taste*, trans. Richard Nice. London: Routledge & Kegan Paul, 1984.

Butler, R. and Parr, H. (eds) (1999) *Mind and Body Spaces: Geographies of Illness, Impairment and Disability*. London: Routledge.

Capra, F. (2002) *The Hidden Connections: A Science for Sustainable Living*. New York: Anchor Books, 2004.

Casey, E. (1993) *Getting Back Into Place: Toward a New Understanding of the Place-World*. Bloomington: Indiana University Press.

Clark, A. (2001) *Being There: Putting Brain, Body and World Together Again*. Cambridge, MA: MIT Press.

Clark, N. (2005) 'Ex-orbitant globality', *Theory, Culture and Society*, 22(5): 165–85.

Crosby, A.W.Jr. (1986) *Ecological Imperialism: The Biological Expansion of Europe, 900–1900*. Cambridge: Cambridge University Press.

Damasio, A. (2003) *Looking for Spinoza: Joy, Sorrow and the Feeling Brain*. New York: Harvest.

Deleuze, G. (1968) *Expressionism in Philosophy: Spinoza*, trans. Martin Joughin. New York: Zone Books, 1990.

Deleuze, G. (1981) *Spinoza: Practical Philosophy*, trans. Robert Hurley. San Francisco: City Light Books, 1988.

Deleuze, G. and Guattari, F. (1980) *A Thousand Plateaus: Capitalism and Schizophrenia*, trans. Brian Massumi. Minneapolis: University of Minnesota Press, 1987.

Deutscher, P. (2002) *A Politics of Impossible Difference: The Later Work of Luce Irigaray*. Ithaca, NY: Cornell University Press.

Diamond, J. (1997) *Guns, Germs and Steel: The Fates of Human Societies*. New York: W.W. Norton.

Diprose, R. (2002) *Corporeal Generosity: On Giving with Nietzsche, Merleau-Ponty and Levinas*. Albany: State University of New York Press.

Foucault, M. (2003) *Society Must Be Defended: Lectures at the Collège de France, 1975–1976*, trans. David Macey. London: Picador.

Gallop, J. (1988) *Thinking Through the Body*. New York: Columbia University Press.

Gatens, M. (1996) *Imaginary Bodies: Ethics, Power and Corporeality*. London: Routledge.

Gatens, M. and Lloyd, G. (1999) *Collective Imaginations: Spinoza, Past and Present*. London: Routledge.

Grosz, E. (1994) *Volatile Bodies: Toward a Corporeal Feminism*. Bloomington: Indiana University Press.

Hall, E.T. (1966) *The Hidden Dimension: Man's Use of Space in Public and Private*. London: The Bodley Head.

Hardt, M. and Negri, A. (2000) *Empire*. Cambridge, MA: Harvard University Press.

Hardt, M. and Negri, A. (2004) *Multitude: War and Democracy in the Age of Empire*. New York, Penguin.

Hobbes, T. (1650) *The Elements of Law Natural and Politic: I Human Nature, II De Corpore Politico*. Oxford: Oxford University Press, 1994.

Irigaray, L. (1984) *An Ethics of Sexual Difference*, trans. Carolyn Burke and Gillian C. Gill. Ithaca, NY: Cornell University Press, 1993.

Johnson, S. (2001) *Emergence: The Connected Lives of Ants, Brains, Cities and Software*. London: Penguin.

Latour, B. (2004) *Politics of Nature: How to Bring the Sciences into Democracy*. Cambridge, MA: Harvard University Press.

Lefebvre, H. (1974) *The Production of Space*, trans. Donald Nicholson-Smith. Oxford: Blackwell, 1991.

Leroi, A.M. (2003) *Mutants: On the Form, Varieties and Errors of the Human Body*. London: HarperCollins.

Lingis, A. (1994) *Foreign Bodies*. New York: Routledge.

Lloyd, G. (1994) *Part of Nature: Self-Knowledge in Spinoza's Ethics*. Ithaca, NY: Cornell University Press.

Longhurst, R. (2000) *Bodies: Exploring Fluid Boundaries*. London: Routledge.

Lynch, K. (1960) *The Image of the City*. Cambridge, MA: MIT Press.

Massey, D. (2005) *For Space*. London: Sage.

Mayer, T. (2004) 'Embodied nationalisms', in L.A. Staeheli, E. Kofman and L.J. Peake (eds), *Mapping Women, Making Politics: Feminist Perspectives on Political Geography*. New York: Routledge.

Mayr, E. (1963) *Animal Species and Evolution*. Cambridge, MA: Harvard University Press.

McNeill, W. (1977) *Plagues and People*. New York: Doubleday.

Merleau-Ponty, M. (1945) *Phenomenology of Perception*, trans. Colin Smith. London: Routledge & Kegan Paul, 1962.

Mol, A. (2002) *The Multiple Body: Ontology in Medical Practice*. Durham, NC: Duke University Press.

Mountz, A. (2003) 'Embodying the nation-state: Canada's response to human smuggling', *Political Geography*, 23(3): 323–45.

Nast, H.J. and Pile, S. (eds) (1998) *Places Through the Body*. London: Routledge.

Peet, R. (1985) 'The social origins of environmental determinism', *Annals of the Association of American Geographers*, 75(3): 309–33.

Pile, S. (1994) *The Body and the City: Psychoanalysis, Space and Subjectivity*. London: Routledge.

Pratt, G. (1998) 'Inscribing domestic work on Filipina bodies', in S. Pile and H.J. Nast (eds), *Places Through the Body*. London: Routledge.

Pratt, G. (2004) *Working Feminism*. Philadelphia: Temple University Press.

Pred, A. (1977) 'The choreography of existence: comments on Hägerstrand's time-geography and its usefulness', *Economic Geography*, 53(2): 207–21.

Ratzel, F. (1897) *Politische Geographie*. Munich: Oldenbourg.

Rich, A. (1984) 'Notes towards a politics of location', in *Blood, Bread and Poetry: Selected Prose 1979–1985*. New York: W.W. Norton.

Rodaway, P. (1994) *Sensuous Geographies: Body, Sense and Place*. London: Routledge.

Rose, G. (1993) *Feminism and Geography: The Limits of Geographical Knowledge*. Cambridge: Polity.

Rose, G. (2003) 'On the need to ask how, exactly, is geography "visual"?', *Antipode*, 35(2): 212–21.

Saldanha, A. (2007) *Psychedelic White: Goa Trance and the Viscosity of Race*. Minneapolis: University of Minnesota Press.

Sangtin Writers and Nagar, R. (2006) *Playing with Fire: Feminist Thought and Activism Through Seven Lives in India.* Minneapolis: University of Minnesota Press.

Seamon, D. (1979) *A Geography of the Lifeworld: Movement, Rest and Encounter.* New York: St Martin's Press.

Sennett, R. (1994) *Flesh and Stone: The Body and the City in Western Civilization.* New York: W.W. Norton.

Smith, S.J. (1997) 'Beyond geography's visible worlds: a cultural politics of music', *Progress in Human Geography*, 21(4): 502–29.

Sparke, M.B. (2006) 'A neoliberal nexus: economy, security and the biopolitics of citizenship on the border', *Political Geography*, 25(2): 151–80.

Spinoza, B. (1677) *Ethics*, trans. W.H. White and A.H. Stirling. Wase Herts: Wordsworth, 2001.

Stoler, A.L. (2002) *Carnal Knowledge and Imperial Power: Race and the Intimate in Colonial Rule.* Berkeley: University of California Press.

Teather, E.K. (ed.) (1999) *Embodied Geographies: Spaces, Bodies and Rites of Passage.* London: Routledge.

Tuan, Y.-F. (1977) *Space and Place: The Perspective of Experience.* Minneapolis: University of Minnesota Press.

Urry, J. (2003) *Global Complexities.* London: Polity.

Varela, F.J., Evan T.T. and Eleanor R. (1992) *The Embodied Mind: Cognitive Science and Human Experience.* Cambridge, MA: MIT Press

Williams, S.J. and Bendelow, G. (1998) *The Lived Body: Sociological Themes, Embodied Issues.* London: Routledge.

Young, I.M. (1990) 'Throwing like a girl: a phenomenology of feminine body comportment, motility, and spatiality', in *Throwing Like a Girl and Other Essays in Feminist Philosophy and Social Theory.* Bloomington: Indiana University Press.

Transnational Political Movements

Paul Routledge

INTRODUCTION

[W]e will make a collective network of all our partic-ular struggles and resistances. An intercontinental network of resistance against neoliberalism, an intercontinental network of resistance for human-ity. This intercontinental network of resistance, recognizing differences and acknowledging similar-ities, will search to find itself with other resistances around the world. This intercontinental network of resistance is not an organizing structure; it doesn't have a central head or decision maker; it has no cen-tral command or hierarchies. We are the network, all of us who resist. (Subcommandante Marcos, quoted in *We Are Everywhere*, 2001: 13)

On 1 January 1994, media vectors around the world carried the dramatic news that ski-masked guerrillas had captured the town of San Cristóbal de las Casas (in the Mexican state of Chia-pas) and declared war on the Mexican state. As the drama unfolded, it became apparent that the EZLN (Ejército Zapatista de Liberacion Nacional) or Zapatistas, as they became known, differed from the recent guerrilla movements such as the FMLN (Frente Farabundo Martí para la Libera-cion Nacional) movement in El Salvador, and the FSLN (Frente Sandanista de Liberacion Nacional) in Nicaragua. Unlike them the EZLN did not see itself as the vanguard directing a struggle to seize state power. Rather, they demanded the democratic revitalization of Mexican civil and political society, and autonomy for, and recognition of, indigenous culture.[1] Their struggle ignited the imagination of many on the Left, prematurely buried by the 'end

of history' (Fukuyama, 1992). As a result, in 1996, when the Zapatistas organized an international encounter in Chiapas, it was attended by activists, intellectuals and journalists from around the world. At the encounter, Subcommandante Marcos – the articulate, humorous and poetic spokesperson of the Zapatistas – issued the above call for a network of transnational resistance against neoliberalism.

At least two important transnational initiatives were to emerge from this political moment: Peo-ple's Global Action (PGA), a network of coordi-nation, information sharing and solidarity between grassroots social movements from around the world (see Routledge, 2003a); and the World Social Forum, a convergence of a huge diversity of social movements, non-government organizations (NGOs), trade unions and other political forces for discussion, networking, and the posing of alterna-tives to neoliberal capitalism. In this chapter I wish to consider how the reconfiguration of political life in relation to globalizing economic and social processes has seen renewed efforts by social move-ments to extend their reach beyond local or national boundaries. In response to the ravages of neoliberal globalization, and enabled by some of its processes, networks of protest, campaigning and resistance have emerged to challenge and confront transna-tional forms of power. This chapter will discuss how, within such networks, diverse interests and identities are constituted within a geographically complex set of relations, reconfiguring politics at all scales.

Political geography, as an area of study, contin-ues to be concerned with the themes of borders (conceptual and ideological as well as economic and physical); world orders based upon different geographic organizing principles (such as empires, state systems and ideological–material relationships); power; and resistance (e.g. in the form of political movements) (Agnew et al., 2003). The emergence of transnational protest networks – because they coordinate activities through and beyond state territories – has potential consequences for the future of political geography, both in terms of its subject matter and its theoretical approaches. These will be discussed in the final section of the chapter.

NEOLIBERALISM AND ITS DISCONTENTS

The ideology of neoliberalism articulates an overarching commitment to 'free market' principles of free trade, flexible labour and active individualism. It privileges lean government, privatization and deregulation, while undermining or foreclosing alternative development models based upon social redistribution, economic rights or public investment (Peck and Tickell, 2002). Nevertheless, neoliberal globalization is neither monolithic nor omnipresent, taking hybrid or composite forms around the world, that is different markets, different aspects of society integrating at different rates, etc. (Larner, 2000; Clark, 2003).

Peck and Tickell (2002) argue that neoliberalism has seen a shift from 'roll-back neoliberalism' during the 1980s – which entailed a pattern of deregulation and dismantlement (e.g. of state-financed welfare, education and health services and environmental protection) – to an emergent phase of 'roll-out neoliberalism'. This emergent phase is witnessing an aggressive intervention by governments around issues such as crime, policing, welfare reform and urban surveillance, with the purpose of disciplining and containing those marginalized or dispossessed by the neoliberalization of the 1980s.

In both phases, neoliberalism entails the centralization of control of the world economy in the hands of transnational corporations and their allies in key government agencies (particularly those of the United States and other members of the G-8), large international banks, and international institutions such as the International Monetary Fund (IMF), the World Bank and the World Trade Organization (WTO). These institutions enforce the doctrine of neoliberalism enabling unrestricted access of transnational corporations (TNCs) to a wide range of markets (including public services), while potentially more progressive institutions and agreements (such as the International Labour Organization and the Kyoto Protocols) are allowed to wither (Peck and Tickell, 2002). Neoliberal policies have resulted in the pauperization and marginalization of indigenous peoples, women, peasant farmers

and industrial workers, and a reduction in labour, social and environmental conditions on a global basis – what Brecher and Costello (1994) term 'the race to the bottom' or 'global pillage'. These processes have been accompanied by what some have termed a 'democratic deficit' such as declining voter turnout, declining membership of political parties, reduced confidence in governments and politicians and hostility to corporations and global institutions (Clark, 2003).

In response to this, new forms of translocal political solidarity and consciousness have begun to emerge, associated with the partial globalization of networks of resistance. These formations – consisting of diverse networks of social movements, trade unions, NGOs and other organizations – inhabit a political space outside of formal national politics (political parties, elections), and address a range of institutions across a variety of geographic scales (local, national, international). Mary Kaldor argues that such networks are involved in establishing a global civil society that is about 'democratizing globalization, about the process through which groups, movements and individuals can demand a global rule of law, global justice and global empowerment' (2003: 12). Social movements are increasing their spatial reach in terms of constructing multiscalar networks of support and solidarity for their particular struggles, and also by participating with other movements in broad networks to resist neoliberal globalization, exemplified by the slogan 'Our resistance will be as transnational as capital'.[2]

While such solidarities across borders are not new phenomena,[3] present international alliances are characterized by the means, speed and intensity of communication between the various groups involved.[4] Gill (2000) has argued that such alliances pose the question for political theory (and, indeed, for political geography) of how to imagine and theorize new forms of collective political identity and agency. Place-bound and place-formed political identities and agencies increasingly must be negotiated across multiple scales, and with multiple others. How such processes are theorized, and engaged with, will become a key challenge for political geography. I will argue in this chapter that it is important, and useful, to consider what Appadurai (2000: 7) terms the 'process geographies' of transnational networks, that is, their multiscalar, dynamic, processes of interaction and relationship.

GRASSROOTS GLOBALIZATION

The growth of anti-neoliberal globalization protests – and the networks that participate in them – has excited much academic and media attention in recent years. They have been conceived of as 'grassroots globalization' (Appadurai, 2000) – attempts by marginalized groups and social movements at the local level to forge wider alliances in protest at their growing exclusion from global neoliberal economic decision-making.[5] While establishing global networks of action and support, they attempt to retain local autonomy over strategies and tactics (Appadurai, 2000). Recent research on such challenges to neoliberal globalization has included work on 'transnational advocacy networks' (Keck and Sikkink, 1998); transnational consumer networks and the labour movement (Evans, 2000; Herod, 2001; Waterman, 2001); and anti-corporate movements and globalizing networks (St Clair, 1999; Brecher et al., 2000; Gill, 2000; Klein, 2000, 2002; Starr, 2000; Bircham and Charlton, 2001; Mertes, 2002).

Such challenges involve a variety of political actors as well as strategic foci. For example, Kaldor (2003) posits that at least six different types of political actor can be identified: more traditional social movements (e.g. trade unions, and anti-colonial and revolutionary movements); more contemporary social movements (e.g. women's and environmental movements); NGOs (e.g. Amnesty International); transnational civic networks (such as those resisting the construction of large dams, e.g. International Rivers Network); 'new' nationalist and fundamentalist movements (e.g. Al Qaeda) and the anti-capitalist movement (e.g. People's Global Action). Meanwhile, Starr (2000) identifies at least three different strategic foci, namely: (i) *Contestation and Reform*, which involves social movements and organizations that seek to impose regulatory limitations on corporations and/or governments, or force them to self-regulate, mobilizing existing formal democratic channels of protest (e.g. Human Rights Watch, and the Fair Trade network); (ii) *Globalization from Below*, whereby various social movements and organizations form global alliances regarding environmental degradation, the abuse of human rights, labour standards, etc., to make corporations and governments accountable to people instead of elites (e.g. the Zapatistas, labour unions, the World Social Forum and People's Global Action); and (iii) *Delinking, Relocalization and Sovereignty*, whereby varied initiatives articulate the pleasures, productivities and rights of localities and attempt to delink local economies from corporate-controlled national and international economies (e.g. permaculture initiatives, community currency (LETS), community credit organizations, sovereignty movements, especially those of indigenous peoples, and various religious nationalisms) (see also Hines, 2000). By focusing upon grassroots globalization networks, two important themes emerge for consideration: the heterogeneity of grassroots networks resisting common opponents; and an affirmation of the importance of the politics of scale.

The Heterogeneity of Grassroots Globalization Networks

Grassroots globalization involves the creation of networks: of communication, solidarity, information sharing and mutual support. The core function of networks is the production, exchange and strategic use of information – for example, concerning oppositional narratives and analysis of particular events. Many information exchanges are informal, such as by telephone, e-mail, and the circulation of newsletters and bulletins through a variety of means including by hand, post and the Internet. Such information can enhance the resources available to geographically and/or socially distant actors in their particular struggles and also lead to action (Keck and Sikkink, 1998). The speed, density and complexity of international linkages have grown dramatically in the past twenty years. Cheaper air travel and new electronic communication technologies have speeded up information flows and simplified personal contact among activists (Keck and Sikkink, 1998; Ribeiro, 1998; Cohen and Rai, 2000). Indeed, information-age activism is creating what Cleaver (1999: 3) terms a 'global electronic fabric of struggle', whereby local and national movements are consciously seeking ways to make their efforts complement those of other organized struggles around similar issues. Such networks, greatly facilitated by the Internet, can, at times, enable relationships to develop that are more flexible than traditional hierarchies. Participation in networks has become an essential component of collective identities of the activists involved, networking forming part of their common repertoire of action and recruitment (Melucci, 1996; Castells, 1997).

In particular, grassroots globalization is resulting in the forging of new alliances – what Esteva and Prakash (1998) consider a pluriverse of interests – as different social movements representing different terrains of struggle experience the negative consequences of neoliberalism (Wallgren, 1998). By identifying structures of power within the global political field, social movements have established common targets of protest, exemplified by the anti-WTO mobilizations in Seattle in 1999, and the anti-World Bank and IMF protests in Prague in 2000.

Such protests have been celebrated for bringing together formerly disparate and often conflicting groups, such as trade unionists, environmentalists, indigenous peoples' movements and non-government organizations. Underpinning such developments is a conceptualization of protest and struggle that respects difference, rather than attempting to develop universalistic and centralizing solutions that deny the diversity of interests and identities that are confronted with neoliberal globalization processes. Hence:

> an international process of *recomposition* of radical claims and social subjects has been under way, a process which is forcing every movement not only to seek alliances with others, but also to make the struggles of other movements their own, without first the need to submit the demands of other movements to an ideological test … [the] premise of recomposition is the multidimensional reality of exploitative and oppressive relations as it is manifested in the lives and experiences of the many social subjects within the global economy. (De Angelis 2000: 14)

Within these coalitions, a variety of cultural and identity issues remain important (e.g. gender), but these are now being reframed within the context of broader critiques of the operation of economic and political systems, and they rejoin more traditional 'labour' and (re)distributive concerns about working conditions in the global South, the effects of mobility of capital, and job security in the global North, etc. While struggles in the global South have always tended to have a materialist focus, identity politics in the global North have begun to rediscover a materialist critique (Crossley, 2001). This has further enabled a 'coalition of difference' to emerge. However, because the globalization of protest involves the interpenetration and multiplicity of forces at local, regional, national and global scales, such a multiplicity raises the possibility of alliances that contain various contradictions (Chin and Mittelman, 1997). For example, place-based gender relations within particular social movements may be at odds with those of other movements, in other places, with whom they participate in struggle. This raises questions about how social movements act effectively in coalitions across diverse geographical scales.

The Scale Politics of Grassroots Globalization Networks

The consolidation of a global system of financial regulation[6] – as one of the means of imperial global control – has prompted the 'upscaling' of previously local struggles between citizens, governments and transnational institutions and corporations to the international level. Klein (2002) has argued that such a globalization of protest, being faced with a decentred apparatus of rule that is everywhere and nowhere, has, during spectacular protests such as the global days of action in Seattle in 1999, focused upon targets that symbolize global neoliberal power: the WTO, the IMF, and TNCs like McDonalds and Nike. She notes that such

demonstrations represent the convergence of many different movements each with different goals and targets, but with a shared common cause: to resist the neoliberal agenda. Many of these movements, although engaged in grassroots globalization networks, nevertheless remain locally or nationally based, since this is where individual movement identities are formed and nurtured.

Castells (1997) conceives such struggles as representing defensive responses to the network society. The cultural and economic specificity of particular places is deployed as part of a broader articulation of resistance to the space of flows, that is, capital investments and developments increasingly detached, through the use of information technologies, from the social constraints of cultural identities and local societies (Castells, 1997). However, struggles can be both offensive and defensive – e.g. the Zapatista peasant insurgency in Chiapas, Mexico (Routledge, 1998) – and Castells' formulation creates a binary between local and global processes.

This is not to understate the importance of place to the understanding of social movement characteristics, behaviours and processes. Places shape political activity through a variety of ways, including the spatial division of labour (which effects class structure, social structure and community affiliations); communications technology and the patterns of accessibility to it; the characteristics of local and central states; class, gender and ethnic divisions and how they are expressed through local culture, work and history; collective identity formation (including class, gender and ethnic divisions) and place-based identities oriented to the local, regional or national level; and the microgeography of everyday life (e.g. work, residence and school) in which patterns of social interaction are spatially structured (Agnew, 1996). Geographical research has highlighted how theoretical approaches to the study of social movements acknowledge the strategic choices of movements and their culturally specific dimensions of protest, but fail to adequately account for why social movements emerge *where* they do (Routledge, 1993; Miller, 2000). Spatially sensitive analysis provides important insights into social movement experience, including how the particularities of places as well as spatial processes and relations across a variety of scales, inform and affect the character, dynamics and outcomes of movement agency; and how the cultural and ideological expressions of social movement agency (e.g. drawn from local knowledges, cultural practices and vernacular languages) inspire, empower and motivate people to resist. A sensitivity to the 'where' of resistance implies the acknowledgment of the intentionality of historical subjects, the subjective nature of perceptions, imaginations and experiences in dynamic spatial contexts, and how spaces are transformed into places redolent with cultural meaning, memory and identity (Routledge, 1993, 1997, 2000).

However, when locally based struggles develop, or become part of, geographically flexible networks, they become embedded in different places at a variety of spatial scales. These different geographic scales (global, regional, national, local) are mutually constitutive parts, becoming links of various lengths in the network. Networks of agents act across various distances and through diverse intermediaries. However, some networks are relatively more localized while others are more global in scope, and the relationship of networks to territories is mutually constitutive: networks are embedded in territories and, at the same time, territories are embedded in networks (Dicken et al., 2001). Of course, movements that are local or national in character derive their principal strength from acting at these scales rather than at the global level (Sklair, 1995). For example, transnational corporations such as Nestlé, McDonalds and Nike have usually been disrupted primarily due to the efficacy of local campaigns (Klein, 2000). Even where international campaigns are organized, local and national scales of action can be as important as international ones (Herod, 2001). For example, the Liverpool dockers' international campaign was grassroots-instigated and coordinated (by Liverpool dockers) and operationalized by dockers beyond the UK working within established union frameworks (Castree, 2000).

Space is bound into local to global networks, which act to configure particular places. As a result, 'each place is the focus of a distinct *mixture* of wider and more local social relations' (Massey, 1994: 156), and hence places can be imagined as 'articulated moments in networks of social relations' (ibid.: 154). Moreover, while networks can create cultural and spatial configurations that connect places with each other (Escobar, 2001), so can particular places be important within the workings of networks. For example, in his research on the *Madres de Plaza de Mayo* in Argentina, Fernando Bosco (2001) shows how collective political rituals enacted in different places across space (e.g. the public meetings of *Madres* in plazas across Argentina) enabled the sustainability of different movement communities and movement identities. By reinforcing moral commitments and group solidarity, activists' identities were maintained both within particular groups and between movements and activists in wider solidarity networks.

Bosco argues that the identification with particular places can be of strategic importance for the mobilization strategies of particular resistance movements. These can contribute to the construction of strategic network ties with other movements in the same locality or in other localities. Activists may deploy symbolic images of places to match the interests and collective identities of other groups

and thereby mobilize others along a common cause or grounds. Hence the ties to particular places can be mobile, appealing to, and mobilizing, different groups in different localities (Bosco, 2001).

However, because places are important loci of collective memory, social identity and the capacity to mobilize that identity into configurations of political solidarity are highly dependent upon the processes of place construction and sustenance (Harvey, 1996). Such particularities of place may come into conflict with those of other places, for example due to different place-specific understandings of gender relations. As a result, these may vitiate against multiscalar mobilizations and pose important problems for the development of grassroots globalization networks. To discuss this issue requires a consideration of the work of David Harvey.

Militant Particularism/Global Ambition

Borrowing a term from Raymond Williams, Harvey (1996, 2000) argues that place-based resistances frequently articulate a 'militant particularism'. This is where the ideals forged out of the affirmative experience of solidarities in one place have the potential to become generalized and universalized as a working model for a new form of society that will benefit all humanity – what Harvey terms 'global ambition'. However, Harvey notes that militant particularisms are often profoundly conservative, resting upon the perpetuation of patterns of social relations and community solidarities. He wonders whether there is a scale at which militant particularisms become impossible to ground, let alone sustain. Indeed, he argues that:

> Anti-capitalist movements … are generally better at organizing in and dominating 'their' places than at commanding space. '[R]egional resistances' … can indeed flourish in a multitude of particular places. But while such movements form a potential basis for that 'militant particularism' that can acquire global ambition, left to themselves they are easily dominated by the power of capital to coordinate accumulation across universal but fragmented space. The potentiality for militant particularism, embedded in place runs the risk of sliding back into a parochialist politics. (Harvey, 1996: 324)

Successful international alliances have to negotiate between action that is deeply embedded in place, that is local experiences, social relations and power conditions (see, e.g., Routledge, 1993), and action that facilitates more transnational coalitions. Social movements, according to Harvey, can either remain place-based and ignore the potential contradictions inherent in transnational coalitions (e.g. concerning different gender relations within participant movements) or treat the contradictions as a nexus to create a more transcendent and universal politics, combining social and environmental justice, that transcends the narrow solidarities and particular affinities shaped in particular places. In short, movements need to develop a politics of solidarity capable of reaching across space, without abandoning their militant particularist base(s) (Harvey, 1996: 400).

However, even if social movements are capable of reaching across space, differential power relations exist within the functioning of the networks that are created. Particular actors are often dominant within networks, due to their control of key political, economic and technological resources (Dicken et al., 2001). Moreover, different groups and individuals are placed in distinct (more or less powerful) ways in relation to the flows and interconnections involved in the functioning of resistance networks. Thus, while the working of networks involves the intermingling of geographic scales, contradictions and tensions remain – either tied to the militant particularisms of particular movements or in the placing of specific actors within the network.

In order to analyse the process geographies of transnational political movements and their networks of engagement, certain key issues need to be addressed. These issues include: (i) the ability of such networks to articulate common concerns (or collective visions); (ii) how processes of interaction and facilitation are enacted within these networks; (iii) how such networks prosecute multiscalar political action; and (iv) the dynamics of social relations within these networks. Such issues not only provide future areas of research for political geography, they may also provide important insights into the potential for success of challenges to neoliberal globalization.

THE PROCESS GEOGRAPHIES OF GRASSROOTS GLOBALIZATION NETWORKS

Given the diversity of social movements and other political actors that constitute grassroots globalization networks, the first key issue is their ability to articulate collective visions, or common concerns, in order to establish sufficient common ground to generate a politics of solidarity, that is, multiscalar collective action. These collective visions might potentially represent a 'prefigurative politics' (Graeber, 2002), prefiguring not a future ideal society, but a participatory way of practising effective politics, articulating the (albeit imperfect) ability of heterogeneous movements to be

able to work together without any single organization or ideology being in a position of domination. Collective visions approximate the universal values that Harvey (1996) discusses. However, contrasting, but not necessarily disabling, tensions may exist between the articulation of such universalist politics and the militant particularisms of the movements participating within transnational networks. For example, the immediacy of place-based concerns – such as movements' everyday struggles for survival under conditions of limited resources – might mean that the global ambitions articulated by grassroots globalization networks remain unrealized.

The second key issue concerns how processes of interaction and facilitation are enacted within these networks. For the diverse groups and movements in such networks to enact a practical politics, at least five processes need to be enacted: communication, (e.g. via e-mail or face-to-face meetings such as conferences); information sharing (e.g. concerning the effectiveness of particular tactics and strategies, etc.); solidarity (e.g. demonstrations of support for particular struggles such as protests); coordination (e.g. organizing conferences and collective protests, etc.); and resource mobilization (e.g. of people, finances and skills). We might imagine that interactions within virtual space (e.g. through the Internet) act as a communicative and coordinating thread that helps to weave different place-based struggles together. These connections need to be grounded in place- and face-to-face-based moments of articulation such as conferences and multiscalar protests, in order to pose material challenges to neoliberal globalization. However, differential (personal and movement) access to (financial, temporal) resources and network flows might exist with transnational networks, as well as differential material and discursive power relations within and between participant movements. As a result, network processes of facilitation and interaction may be uneven.

The third key issue concerns how social movements engaged in grounded material struggles, and articulating place-specific concerns, also actively participate in forging globalizing networks of such struggles. As I will discuss below using case studies, particular local-based social movements may develop transnational networks of support as an operational strategy for the defence of their place(s) (Escobar, 2001), and certain places may be of symbolic importance in the collective rituals of the particular networks, for example as sites for international conferences (Bosco, 2001).

In order to materialize collective struggle, grassroots globalization networks attempt to prosecute transnational collective political rituals, exemplified by global days of action such as the protests against the WTO in Seattle in1999. Such protests brought together political actors from different countries within a particular place (Seattle) while also witnessing solidarity actions in many other places around the world. Such actions, rather than 'grassrooting the space of flows' (Castells, 1999), facilitate an intermingling of scales of political action, where such scales become mutually constitutive (Dicken et al., 2001). However, certain scales of political action may provide more appropriate means for movements to measure their strength and take stock of their opponents, than others. For example, many movements in the global South see defence of local spaces and opposition to national governments (pursuing neoliberal policies) as their most appropriate scales of political action (Mertes, 2002). As a result, geographical dilemmas may arise in the attempt to prosecute multiscalar politics compounded by the uneven character of processes of interaction and facilitation.

The fourth key issue concerns the dynamics of the social relations enacted within transnational networks. Because of the very different militant particularisms that are articulated by participant movements, such networks potentially comprise contested social relations. For example, different groups inevitably articulate a variety of potentially conflicting goals (concerning the forms of social change), ideologies (e.g. concerning gender, class and ethnicity) and strategies (e.g. institutional [legal] and extra-institutional [illegal] forms of protest). While these contradictions may act as a nexus for a more universal politics – as Harvey argues – problematic issues may arise concerning unequal discursive and material power relations that result from the differential control of resources (Dicken et al., 2001) and placing of actors within network flows (Massey, 1994). These in turn may give rise to problems of representation, mobility and cultural difference, both between the social movements that participate and between activists within particular movements. Thus, the alliances forged between movements may involve entangled power relations, where relations of domination and resistance are entwined, to create spaces of resistance/domination (Sharp et al., 2000).

A potential way of analysing these key issues together as they work through the complex spaces of alliance that constitute grassroots globalization networks, is through conceiving of them as convergence spaces (Routledge, 2003a). Convergence spaces are those that facilitate the forging of an associational politics, constituting diverse coalitions of place-specific social movements articulating collective visions. These coalitions prosecute conflict on a variety of multiscalar terrains that include both material places and virtual spaces, facilitate uneven processes of interaction, and comprise contested social relations, being spaces of domination/resistance.

In order to explore these issues empirically, I will consider two examples of transnational political movements. First, I will consider a network of social movements, NGOs and trade unions in the process of creating transnational alliances on behalf of collectively articulated concerns and goals, the World Social Forum. This example is important politically, because it poses a challenge to neoliberal globalization from numerous social movements from across the planet. It is also important conceptually because it raises the question of how diversity and difference can be reconciled within the crucible of conflict. Second, I will consider one of the participants of the World Social Forum – the *Narmada Bachao Andolan* – a social movement that acts across space to forge transnational solidarity networks, and in so doing acts both within various transnational networks and also constitutes itself as a transnational network. This example is chosen both because of the author's critical engagement with the movement (see Routledge, 2003b), and also because it raises the question of how a place-specific issue (the building of dams) traverses space to engender transnational action and solidarity.

THE WORLD SOCIAL FORUM

The World Social Forum (WSF) emerged out of the various initiatives that protested the World Economic Fourm (WEF).[7] Activists in the Brazilian Justice and Peace Commission, and the French Association for the Taxation of Financial Transactions for the Aid of Citizens (ATTAC), suggested the establishment of a gathering for NGOs, trade unions, social movements and other resistance networks involved in civil society worldwide. The purpose of the gathering was to protest the WEF and particularly neoliberal globalization, and to discuss and present concrete alternatives to it (Houtart and Polet, 2001). The notion of the WSF was to engender a process of dialogue and reflection and the transnational exchange of experiences, ideas, strategies and information between the participants regarding their multiscalar struggles. In 2001, the first WSF was held in Porto Alegre, Brazil, and attracted almost 20,000 participants, the second, in 2002, 55,000 participants and the third, in 2003, 100,000 participants (Fisher and Ponniah, 2003; Teivainen, 2004). The participants included diverse social movements, NGOs, trade unions, solidarity committees and farmers' networks from the five non-polar continents (see Sen et al., 2004). The WSF represents a transnational political initiative of variation and flux, where the links between actors (activists) and various intermediaries tend to be in process and are contestable.

The collective visions of the WSF are enshrined in its slogan 'Another World is Possible' and its

Charter of Principles. Some of the key principles state that:

1 The World Social Forum is an open meeting place for reflective thinking, democratic debate of ideas, formulation of proposals, free exchange of experiences and interlinking for effective action, by groups and movements of civil society that are opposed to neoliberalism.

2 The alternatives proposed at the World Social Forum respect universal human rights ... and will rest on democratic international systems and institutions at the service of social justice, equality and the sovereignty of peoples.

3 The World Social Forum is a plural, diversified, non-confessional, non-governmental and non-party context that, in a decentralized fashion, interrelates organizations and movements engaged in concrete action at levels from the local to the international to build another world.

4 The World Social Forum will always be a forum open to pluralism and to the diversity of genders, ethnicities, cultures, generations and physical capacities, providing they abide by this Charter of Principles. Neither party representations nor military organizations shall participate in the Forum. Government leaders and members of legislatures who accept the commitments of this Charter may be invited to participate in a personal capacity.

5 As a context for interrelations, the World Social Forum seeks to strengthen and create new national and international links among organizations and movements. (From Sen et al., 2004: 70–1)

The Charter of Principles articulates certain unifying values that create common ground to enable debate to take place, but allow for the diversity of (local) alternatives, projects, tactics, etc., since no single agenda could contain the different militant particularisms of such a multiplicity of participants. However, the unanimity of the hallmarks masks various contested social relations that I will discuss below.

Concerning processes of facilitation and interaction, the formal decision-making of the WSF has been dominated by the Organizing Committee (OC). This consists of the Central Trade Union Confederation, the Movement of Landless Rural Workers, and six smaller Brazilian civil society organizations. The other main organ of the WSF is the International Council (IC), which consists of 113 organizations (including the OC), the majority of whom come from the Americas and Western Europe (Teivainen, 2004), but which has far less

decision-making power than the OC. The WSF is organized in part through the Internet via its website (www.forumsocialmundial.org.br). The WSF website is translated into four languages and provides information about the history of the network; WSF international and regional conferences; various actions and initiatives that the WSF has organized; upcoming events, etc. However, the primary process of interaction in the WSF are the international and regional conferences and meetings that provide material spaces within which representatives of participant movements can converge, and discuss issues that pertain to the functioning of the network. Such conferences and meetings also enable strategies to be developed and enable deeper interpersonal ties to be established between different activists from different cultural spaces and struggles.

Concerning multiscalar action, the main mechanism for the globalization of the WSF has been the regional and thematic forums that are being held in various parts of the world. The principle among these have been the Thematic Forum on neoliberalism in Argentina (2002), the European Social Forum in Florence, Italy (2002), the Asian Social Forum in Hyderabad, India (2003) and the Thematic Forum on Drugs, Human Rights and Democracy in Cartagena, Colombia (2003). In addition, myriad local events have been organized under the WSF banner around the world (Teivainen, 2004).

Places have been used strategically to sustain the WSF. For example, the specific symbolic site of Porto Alegre was chosen for the location of the first three WSF international conferences.[8] Porto Alegre was chosen because, during Brazil's military rule, the city was a centre of resistance with vibrant neighbourhood associations that had strong links to the Workers' Party (Partido dos Trabalhadores, PT). Indeed the city was one of the PT's strongholds, having been governed by the party since 1988, and is celebrated for its participatory budgetary process (Fisher and Ponniah, 2003). As such the city became an articulated moment in the enactment of the WSF where opposition to neoliberalism as well as alternative visions could be voiced.

While the WSF represents a transnational collective ritual *par excellence*, it has also facilitated these elsewhere. For example, the WSF played an important role in building global opinion against the 2003 war in Iraq, not by taking a position itself, but by being an arena where anti-war activists could meet and discuss transnational collective protests (Sen, 2004a). As a result, in part, of this process there were global protests against the war on 15 February 2003 (amounting to as many as 30 million people worldwide).

By virtue of their participation in activities in the field of global power, transnational initiatives such as the WSF are also contributing to the process of globalization. By protesting against global institutions such as the WTO and World Bank, and demanding change, such networks constitute themselves as active citizens on the global political field, and generate a public sphere for that field – in other words contributing towards a global civil society (Crossley, 2001).

Despite the success of such transnational convergences – not least in the rejoining of the concerns of the politics of identity and redistribution – significant differences remain in the type of specific alternatives to neoliberal capitalism articulated by global Northern and global Southern movements. Northern activists articulate alternatives that are conditioned by their embeddedness within – and alienation from – an already industrialized capitalist society. The fundamental concerns of Southern activists are with the defence of livelihoods and of communal access to resources threatened by commodification, state take-overs, and private appropriation (e.g. by national or transnational corporations). Their alternatives are rooted, in part, in some of the local practices being undermined by neoliberal globalization (Glassman, 2002). Because of this, the WSF Charter of Principles stresses connection, diversity and solidarity. Within the context of the universalisms of global ambition, the intention is to mobilize the militant particularisms of participant movements in solidarity and coordination with others.

However, while purportedly an open space of diverse opinion sharing, the WSF has been critiqued by many participants and commentators. The OC has been critiqued because of its hierarchical organization and lack of transparency in decision-making (regarding who the decision-makers actually are, who gets to speak at the Forum, the allocation of spaces and resources at the Forum for different groups, etc). In addition, the Forum as a process has been criticized because of the special treatment and privilege allotted to celebrity speakers, and the privileging and co-optation of the Forum by institutionalized political structures, political parties, trade unions and mainstream NGOs (Osterweil, 2004). Moreover, despite almost half of WSF participants being women, the Forum remains an initiative dominated by middle-class, and middle- and upper-caste men (Sen, 2004b: 218). In addition, certain 'key' events taking place at the WSF are privileged over all of the others that take place; their presenters stay in better accommodation than most participants, and the space and resources allotted to these events are disproportionately greater than the other events (Albert, 2004). Moreover, all of the space allocated to the media has been occupied by mainstream media, leaving no space available for alternative or independent media such as Indymedia (Adamovsky and George, 2004).

The WSF is also a space where various inter-state agencies such as the United Nations have access or upon which they exercise influence. Certain state-dependent funding agencies and private US foundations have supported the WSF itself or various selected NGOs influential within it. The organizational space of the WSF is dominated by the official programme, which has been conceived without notable discussion beyond the governing bodies. Many small or radical groups and events are marginalized geographically and politically (Waterman, 2004: 156). The WSF can provide alternative channels of communication, whereby particular voices that are suppressed in their own society may find articulation and their concerns projected and amplified. However, within the WSF these remain selective – some voices are still heard at the expense of others.

The discursive dominance that individuals have in relation to communication flows and interconnections is because different groups and individuals are placed in more or less powerful ways in relation to such flows. This is augmented by differences in the material power enjoyed by certain activists and movement representatives concerning mobility. Some activists are more mobile than others in at least two ways. First, there is differential access to contemporary communications technologies such as the Internet. Huge inequalities of resource access exist between activists in 'Northern' and 'Southern' states, and between activists within states – for example, in 'Southern' movements, between movement leaders and the movement masses (Slater, 2003). Second, there is differential financial resource availability between activists and between social movements, concerning the ability to travel across continents to particular actions, meetings and conferences. Such disparities can lead to the emergence of an elite group of mobile 'global' activists who enjoy the privileges of e-mail, mobility and certain financial and discursive power. (see Harding, 2001, regarding funding of the 'anti-globalization movement'). However, despite these problems, the WSF represents an important and developing initiative in the emerging transnational political challenges to neoliberal globalization.

THE NARMADA BACHAO ANDOLAN

The *Narmada Bachao Andolan* (Save Narmada Movement) has, since 1985, been coordinating the resistance against the Narmada river valley project in India.[9] This river, which is regarded as sacred by the Hindu and tribal populations of India, spans the states of Madhya Pradesh, Maharashtra and Gujarat, and provides water resources for thousands of communities. The project envisages the construction of 30 major dams along the Narmada and its tributaries, as well as an additional 135 medium-sized and 3,000 minor dams. When completed, the project is expected to displace up to 15 million people from their homes and lands.

While the dams were first envisaged as being financed by the Indian state, the processes of their construction have become increasingly entwined with those of neoliberal globalization. Two of the dams being constructed on the Narmada river, the Sardar Sarovar dam and the Maheshwar dam, have both attracted financial support from international investors. The Sardar Sarovar was initially financed by the World Bank in 1985 through two loans – US$ 300 million from the International Bank for Reconstruction and Development (IBRD), and US$ 150 million from the International Development Association (IDA). Although the Bank was subsequently forced to withdraw from the project after an independent review argued that resettlement and rehabilitation of those displaced by the dam was inadequate, financial momentum for development of the dam had been set in motion. The Maheshwar hydroelectric project is the first privatized hydel project in India. The project was conveyed to a private company, S. Kumars, in 1994, as part of a government initiative to involve the private sector in power generation. This initiative has also attempted to attract foreign financing for the project, including from the German power utilities, Bayernwerk and VEW Energie, the US-based Ogden Energy Group, the private German bank, the HypoVereinsbank, and the German transnational corporation, Siemens. For a variety of reasons, each of these investors has withdrawn.[10]

The NBA is a social movement that comprises – in the Narmada valley – cash-cropping and tribal peasants, and rich farmers, organized by a core group of activists, who originate mostly from outside of the valley. The movement is also a network as it has constantly attempted to forge an associational politics consisting of individuals, NGOs and other social movements that have prosecuted conflict on a variety of multiscalar terrains that include both material places and virtual spaces. In so doing, the NBA has also participated in transnational networks such as the WSF and PGA (see below). The collective visions shared by all actors within this network are those of resistance to the construction of mega-dams, and their desire for development processes that are socially just and economically and environmentally sustainable.

Concerning processes of facilitation and interaction, the NBA exhibits a 'core/periphery' structure. It consists of a core group of dedicated activists (15–20 people), mostly highly educated and professionally qualified, who constitute the leadership of the movement, and who take the major decisions regarding resources, strategies and politics of the NBA. Many of these activists come

from outside of the Narmada valley and operate from the movement's urban offices. They are concerned with liaising with NGOs and other activist groups nationally and internationally; conducting research, documentation and dissemination; lobbying with government departments, international organizations and the media; mobilizing and coordinating protests in the valley; and raising funds and planning strategies. The support groups of the NBA comprise Indian activist groups and NGOs outside of the valley, with interests in human rights, the environment and alternative development. They provide logistical and financial support as well as participating in actions within the valley. They also serve as links between the NBA and other struggles in India. The NBA also has local-level committees, which constitute small informal groups of local people who lend logistical support to the movement. In the Nimar plains – one of the areas threatened by submergence if the dams are constructed – these groups consist largely of rich and influential farmers. In the *adivasi* (tribal) areas, those influencing community power structures play an important role in forming and running such groups. These groups participate in demonstrations and rallies and also raise funds (Dwivedi, 1997). The NBA also has a solidarity website run by the Friends of the River Narmada (www.narmada.org), through which a range of information, photographs and video footage pertaining to the struggle can be accessed. In addition, the NBA has established a series of e-mail lists in order to keep people informed of current developments in the struggle, and to place calls for transnational solidarity actions on its behalf.

Of course, this only begins to scratch the surface of the process geography of the NBA. For the NBA itself is a movement of complex interactions, processes and relays. There are the constant meetings in *adivasi* and caste Hindu villages along the Narmada valley, meetings outside of the valley in the NBA offices in Baroda (Gujarat) and Badwani (Madhya Pradesh), as well as in Mumbai, Delhi and Bhopal. There are the less frequent organizational relays of its international networking in plenaries and meetings in Prague, London, Washington, Bonn and Seattle. There are the barrage of e-mails written, read and relayed, concerning the protests and arrests, and imprisonment and beatings, of activists. Relays also include all of the messages by word of mouth, or scribbled on pieces of paper, and carried through urban neighbourhoods, along wooded paths, by boat down the Narmada river, or by bus along dusty roads, and the petitions and letters written to politicians, officials of international institutions, etc.[11]

If we are attentive to the NBA's living fabric of struggle, we encounter a range of complexities concerning movement structure, organization and power relations. There are the sites of *satyagraha*[12] such as Domkhedi and Jalsindhi where people refuse to move from their homes despite the risk of submergence by flooding of the river, the ongoing arrival and departure from these sites of activists from all over India and indeed the world, relaying their own messages, linking up with other activists, exchanging information, planning strategies of support and solidarity. At the *satyagraha* during July–September 2000, the participants included local tribal peasants, farmers from the nearby Nimar plains, members of the National Alliance of People's Movements, Indian students, activists from a range of Indian women's, environmental and peasant organizations, and international activists, researchers and students from Britain, Canada, the United States and the Netherlands. They also participated in various demonstrations, rallies and other transnational collective rituals while in the valley. Domkhedi and Jalsindhi are symbolic sites for the movement since they will be the first villages to be submerged under the waters of the Sardar Sarovar dam as it nears completion.[13] When they act as the places for *satyagraha*, these villages become 'articulated moments' in the enactment of networks such as the NBA where opposition to the dams as well as alternative visions are articulated.

The NBA has forged various alliances with other *adivasi* and other peasant organizations within the three riparian states affected by the dam project. These alliances have been fraught with their own problems of conflicting goals, priorities and constituencies (Dwivedi, 1997, 1998). Moreover, the level of mass support for the movement has ebbed and flowed during the fifteen years of its agitation, and the movement's discourses of resistance have at times empowered some threatened by displacement while excluding others who live in the valley – those who, while also threatened by displacement, prefer to accept resettlement rather than an outright resistance against the dams (Dwivedi, 1999). In addition, the NBA's recent popularity and projection in the national and global media have been in part attributable to the celebrity afforded to the movement of the Booker Prize winner Arundhati Roy – herself the beneficiary of a globalized literary market.

The NBA has conducted its resistance simultaneously across multiple scales. It has grounded its struggle against the dams, in the villages along the Narmada valley, mobilizing *adivasi* (tribal) peasants, cash-cropping peasants and rich farmers to resist displacement. The NBA has been able to use their local knowledge of the valley to facilitate communication between disparate communities, and to mobilize, at times, tens of thousands of peasants to resist the dams. The NBA has also taken its struggle to non-local terrains, including the national and international levels. Nationally, the NBA has served writ petitions to the Supreme Court of India, and has established, and participated

as a convener in, the National Alliance of People's Movements – a coalition of different social movements in India collectively organizing to resist the effects of liberalization upon the Indian economy. Internationally, the NBA has forged operational links with various groups outside of India, such as the International Rivers Network (IRN) and the International Narmada Campaign.[14] International solidarity work has been conducted by groups such as IRN, Environmental Defence Fund, Friends of the Earth; human and indigenous rights groups such as Survival International; development organizations such as Association for India's Development; and groups formed explicitly around the Narmada, issue such as the Narmada Solidarity Coalition of New York. These in turn are also part of larger networks such as the Narmada Action Committee and Friends of River Narmada, which are mainly US-based collectives of South Asian development and environmentalist activists that have developed links with other groups through flows of common experience, writings and materials such as documentaries.

Various groups who have visited the Narmada valley over the years to learn about the struggle, to participate, and to subsequently disseminate information about it, continue to maintain links with the NBA and conduct solidarity work on its behalf. For example, the German NGO *Urgewald* produced a comprehensive report based upon their research in the Narmada valley on the effects of the Maheshwar dam, while the group Narmada UK was formed after several individuals who had participated in the 1999 'Rally for the Valley' along the Narmada, and the subsequent PGA conference in Bangalore, decided to conduct solidarity work in the UK in support of the NBA. Various demonstrations have been undertaken, including the banner drop that was conducted from the Millennium Wheel in London in November 1999. As part of a broader transnational network, the NBA has been actively involved with the PGA network since the latter's formation in Geneva in 1998 and has attended the WSF. NBA activists have participated in the global days of action in Seattle and Prague, while still others have mounted concurrent protests within India.

However, the NBA, like all social movements, is also a space of contested social relations. Communication flows regarding strategy and tactics tend to be from the core group to the villages. The NBA has a charismatic leadership who have devoted considerable time and energy to mobilization work in the Narmada valley over the past fifteen years. However, the power of this leadership – in deciding the strategy and organization of the NBA – tends to inhibit the development of local-level leadership (see Dwivedi, 1997). In addition, while many women have been mobilized at various stages of the movement, and while certain women are powerful within the core group, at village levels it is often men who make the decisions. Such gender inequalities, along with caste inequalities, persist within the movement. Moreover, *adivasi* identities within the NBA have tended to become strategically essentialized and homogenized, in order to be contrasted with the destructive character of the development being undertaken in the Narmada valley (Baviskar, 1995), while certain internal contradictions within the movement have been suppressed in order to sharpen the distinction between the people of the Narmada valley and the 'other' of industrial development and globalization (Dwivedi, 1998). Despite these problems, the NBA stands as an emblematic example of the resistance to marginalization brought about by neoliberal development.

TRANSNATIONAL POLITICAL MOVEMENTS AND POLITICAL GEOGRAPHY

Political geography, as an area of study, continues to be concerned with the themes of borders, world orders based upon different geographic organizing principles, power and resistance. The emergence of transnational political networks – because they coordinate activities through and beyond state territories – has potential consequences for the future of political geography, both in terms of its subject matter and its theoretical approaches. Transnational political networks are imbued with a 'communications internationalism', which operates the sphere of ideas, information and images, and is active especially on the terrain of communication, media and culture (Waterman, 2000). Political geography might begin to analyse such networks in communicational/cultural terms rather than the more traditional political/organizational terms.

Place-bound and place-formed political identities and agencies increasingly must be negotiated across multiple scales, and with multiple others. By participating in transnational political networks, activists from participant movements and organizations embody their particular places of political, cultural, economic and ecological experience with common concerns, which lead to expanded spatio-temporal horizons of action (Reid and Taylor, 2000). Such coalitions of different interests are necessarily contingent and context-dependent (Mertes, 2002). Transnational political networks can be seen as generative, actively shaping political identities rather than merely bringing together different actors (activists, movements) around common concerns. Forms of solidarity are thus diverse, multiple, productive and contested (Braun and Disch, 2002; Featherstone, 2003). A key research area for political geography will be how to imagine and

theorize new forms of collective (transnational) political identity and agency: how political identities are negotiated, interpreted and represented across multiple locations.

Political geography's concern with place as the arena of social movement struggles will have to be measured in relation to transnational political flows, epitomised by the process geographies of transnational networks, that is, their multiscalar, dynamic, processes of interaction and relationship. Transnational political networks function within a penumbra of differences, conflicts and compromises. As negotiated spaces of multiplicity and difference, they can be conceived as dynamic systems, constructed out of a complexity of interrelations and interactions across all spatial scales (after Massey, 1994). Multiple differences (and their attendant resonances and tensions) can be empowering to those conducting resistance, if the common ground shared by activists is a global ambition capable of challenging international institutions while also empowering local/national struggles.

Of interest to political geography will be the effectiveness and character of participant movement links to national organizations, as well to the dynamic, changing character of their global connections, interactions and relationships with other movements within transnational networks. Universal values are always embedded in, and emergent from, the local and concrete (Reid and Taylor, 2000). For collective visions to be able to incorporate diverse militant particularisms, they need to embrace a politics of recognition that identifies and defends only those differences that can be coherently combined with social and environmental justice (Fraser, 1997). Transnational political initiatives such as the WSF and NBA proclaim that, despite the difficulties, another world is not only possible, it is also in the process of being created.

ACKNOWLEDGEMENTS

I am grateful to activists in the *Narmada Bachao Andolan*, the People's Global Action network and Jennifer Robinson for inspiration, ideas, comments and suggestions given during the production of this chapter.

NOTES

1 The Zapatistas also represented the first Latin American revolutionary organization with an entirely Indian command.
2 This was the slogan for the global day of action against capitalism on 18 June 1999, when there were demonstrations in 100 cities in 40 different countries, and the protests of 30 November 1999,

where over 40,000 people demonstrated for four days against the World Trade Organization (WTO) in Seattle, USA.

3 During the nineteenth-century, international alliances were established in the anti-slavery movement and the Internationals of the Communists and Anarchists. In the twentieth-century, internationalism has been present in the campaign for women's suffrage, the International Brigades (in Spain in the 1930s, Cuba since the 1960s and Nicaragua in the 1980s), trade union activism and in the anti-nuclear movement.

4 Mobilization around globalization can be dated back to at least 1986, when over 80,000 people protested against an IMF meeting in Berlin (see Gerhards and Rucht, 1992). In the early 1990s, struggles emerged in the United States against GATT, as well as protests by movements such as *Reclaim the Streets* in Britain (Brecher and Costello, 1994).

5 Grassroots globalization is a more accurate term than 'anti-globalization' for what such alliances represent. They struggle for inclusive, democratic forms of globalization, using the communicative tools of the global system. What they are expressly against is the neoliberal form of globalization (see Graeber, 2002).

6 For example, through the discursive and material role played by institutions such as the IMF and World Bank in global economic policy-making.

7 An annual meeting, held in Davos, Switzerland, of political leaders, business and financial magnates, and some cultural and religious organizations and trade unionists in order to determine global economic strategies.

8 The 2004 WSF was held in Mumbai, India.

9 Much of the information in this section draws from Routledge (2003b).

10 Bayernwerk and VEW Energie, which had picked up 49 per cent of the project equity, withdrew from the project in response to an indefinite hunger strike by people who would be affected by the dam. Siemens' involvement was dependent upon the granting of an export credit guarantee by the German government. However, in June 2000 the Development Ministry of the German government commissioned a review of the project which concluded that resettlement and rehabilitation provision was woefully inadequate. As a result, the German government refused the export credit guarantee. In the absence of this guarantee, the loan by the Hypo Vereinsbank to the project fell through, and in December 2000, the Ogden Energy Group withdrew from the Maheshwar project.

11 I owe this grounded materiality of relays to an idea by Jon Pattenden.

12 Litreally 'truth force'. Each monsoon since 1991, the NBA has initiated non-violent *satyagrahas*, whereby villagers in the submergence-threatened areas near the dams resist eviction from their homes, pledging to remain even at the risk of being drowned.

13 The Sardar Sarovar dam is the largest of the Narmada dams. Currently standing at 99 metres, and already working, it will finally reach a height of 145 metres. Owing to rising water levels in the river during monsoon rains, Domkhedi and Jalsindhi have already experienced temporary flooding.

14 A broad alliance of interest groups and NGOs whose terrain of resistance was that of international lobbying against the World Bank's financial support for the largest of the Narmada dams, the Sardar Sarovar (Udall, 1997).

REFERENCES

Adamovsky, E. and George, S. (2004) 'What is the point of Porto Alegre?', in J. Sen, A.K. Anand, A. Escobar and P. Waterman (eds), *World Social Forum: Challenging Empires*. New Delhi: The Viveka Foundation, pp. 130–5.

Agnew, J.A. (1996) 'Mapping politics: how context counts in electoral geography', *Political Geography*, 15(2): 129–46.

Agnew, J.A., Mitchell, K. and Toal, G. (2003) 'Introduction', in J.A. Agnew, K. Mitchell and G. Toal (eds), *A Companion to Political Geography*. Oxford: Blackwell, pp.1–9.

Albert, M. (2004) 'WSF: where to now?' in J. Sen, A. Anand, A. Escobar and P. Waterman (eds), *World Social Forum: Challenging Empires*. New Delhi: The Viveka Foundation, pp. 323–8.

Appadurai, A. (2000) 'Grassroots globalization and the research imagination', *Public Culture*, 12(1): 1–19.

Baviskar, A. (1995) *In the Belly of the River*. Delhi: Oxford University Press.

Bircham, E. and Charlton, J. (eds) (2001) *Anti-Capitalism: A Guide to the Movement*. London: Bookmarks.

Bosco, F. (2001) 'Place, space, networks, and the sustainability of collective action', *The Madres de Plaza de Mayo Global Networks*, 1(4): 307–29.

Braun, B. and Disch, L. (2002) 'Radical democracy's "modern constitution"', *Environment and Planning D: Society and Space*, 20: 505–11.

Brecher, J. and Costello, T. (1994) *Global Village or Global Pillage*. Boston: South End Press.

Brecher, J., Costello, T. and Smith, B. (2000) *Globalization from Below*. Boston: South End Press.

Castells, M. (1997) *The Power of Identity*, Oxford: Blackwell.

Castells, M. (1999) 'Grassrooting the space of flows', *Urban Geography*, 20(4): 294–302.

Castree, N. (2000) 'Geographic scale and grassroots internationalism: the Liverpool dock dispute 1995–1998', *Economic Geography*, 76(3): 272–92.

Chin, C. and Mittelman, J. (1997) 'Conceptualising resistance to globalization', *New Political Economy*, 2(1): 25–37.

Clark, J. (2003) *Worlds Apart: Civil Society and the Battle for Ethical Globalization*. London: Earthscan.

Cleaver, H. (1999) 'Computer-linked social movements and the global threat to capitalism', *www.eco.utexas. edu/~hmcleave/polnet.html*

Cohen, R. and Rai, S.M. (eds) (2000) *Global Social Movements.* London: Athlone Press.

Crossley, N. (2001) 'The global anti-corporate movement: a preliminary analysis', paper presented at the 7th International Conference on Alternative Futures and Popular Protest, Manchester Metropolitan University, 17–19 April.

De Angelis, M. (2000) 'Globalization, new internationalism and the Zapatistas', *Capital and Class,* 70: 9–35.

Dicken, P., Kelly, P.F., Olds, K. and Wai-Chung Yeung, H. (2001) 'Chains and networks, territories and scales: towards a relational framework for analysing the global economy', *Global Networks,* 1(2): 89–112.

Dwivedi, R. (1997) 'People's movements in environmental politics: a critical analysis of the Narmada Bachao Andolan in India', ISS Working Paper No. 242. The Hague: Institute of Social Studies.

Dwivedi, R. (1998) 'Resisting dams and "development": contemporary significance of the campaign against the Narmada projects in India', *European Journal of Development Research,* 10(2): 135–83.

Dwivedi, R. (1999) 'Displacement, risks and resistance: local perceptions and actions in the Sardar Sarovar', *Development and Change,* 30: 43–75.

Escobar, A. (2001) 'Culture sits in places: reflections on globalism and subaltern strategies of localization', *Political Geography,* 20(2): 139–74.

Esteva, G. and Prakash, M.S. (1998) *Grassroots Postmodernism.* London: Zed Books.

Evans, P. (2000) 'Fighting marginalization with transnational networks counter-hegemonic globalization', *Contemporary Sociology,* 291: 230–41.

Featherstone, D. (2003) 'Spatialities of transnational resistance to globalization: the maps of grievance of the intercontinental caravan', *Transactions of the Institute of British Geographers,* 28(4): 404–21.

Fisher, W.F. and Ponniah, T. (eds) (2003) *Another World is Possible.* London: Zed Books.

Fraser, N. (1997) *Justice Interruptus.* New York: Routledge.

Fukuyama, F. (1992) *The End of History and the Last Man.* London: Hamish Hamilton.

Gerhards, J. and Rucht, D. (1992) 'Mesomobilization organizing and framing in two protest campaigns in West Germany', *American Journal of Sociology,* 98(3): 555–96.

Gill, S. (2000) 'Towards a postmoden prince? The Battle of Seattle as a moment in the new politics of globalization', *Millennium,* 29(1): 131–40.

Glassman, J. (2002) 'From Seattle (and Ubon) to Bangkok: the scales of resistance to corporate globalization', *Environment and Planning D: Society and Space,* 19: 513–33.

Graeber, D. (2002) 'The new anarchists', *New Left Review,* 13: 61–73.

Harding, J. (2001) 'Counter-capitalism feeding the hands that bite', *Financial Times* [online], Sept. 10. Available from: http//specialsftcom/countercap/FT33EJSLGRChtml.

Harvey, D. (1996) *Justice, Nature and the Geography of Difference.* Oxford: Blackwell.

Harvey, D. (2000) *Spaces of Hope.* Edinburgh, Edinburgh: University Press.

Herod, A. (2001) *Labor Geographies.* New York: Guilford Press.

Hines, C. (2000) *Localization: A Global Manifesto.* London: Earthscan.

Houtart, F. and Polet, F. (2001) *The Other Davos.* London: Zed Books.

Kaldor, M. (2003) *Global Civil Society.* Cambridge: Polity.

Keck, M.E. and Sikkink, K. (1998) *Activists Beyond Borders.* Ithaca, NY: Cornell University Press.

Klein, N. (2000) *No Logo.* London: Flamingo.

Klein, N. (2002) *Fences and Windows.* London: Flamingo.

Larner, W. (2000) 'Theorising neo-liberalism: policy, ideology, governmentality', *Studies in Political Economy,* 63: 5–26.

Massey, D. (1994) *Space, Place, and Gender.* Minneapolis: University of Minnesota Press.

Melucci, A. (1996) *Challenging Codes.* Cambridge: Cambridge University Press.

Mertes, T. (2002) 'Grass-roots globalism', *New Left Review,* 17: 101–10.

Miller, B.A. (2000) *Geography and Social Movements,* Minneapolis: University of Minnesota Press.

Osterweil, M. (2004) 'De-centering the forum: is another critique of the forum possible?', in J. Sen, A. Anand, A. Escobar and P. Waterman (eds), *World Social Forum: Challenging Empires.* New Delhi: The Viveka Foundation, pp. 183–90.

Peck, J. and Tickell, A. (2002) 'Neoliberalizing space', *Antipode,* 34(3): 380–404.

Reid, H. and Taylor, B. (2000) 'Embodying ecological citizenship: rethinking the politics of grassroots globalization in the United States', *Alternatives,* 25: 439–66.

Ribeiro, G.L. (1998) 'Cybercultural Politics Political Activism at a Distance in a Transnational World', in S.E. Alvarez, E. Dagni and A, Escobar (eds), *Cultures of Politics Politics of Cultures.* Oxford: Westview Press, pp. 325–52.

Routledge, P. (1993) *Terrains of Resistance: Nonviolent Social Movements and the Contestation of Place in India.* Westport, CT: Praeger.

Routledge, P. (1997) 'A spatiality of resistances: theory and practice in Nepal's revolution of 1990', in S. Pile, and M. Keith (eds), *Geographies of Resistance.* London: Routledge, pp. 68–86.

Routledge, P. (1998) 'Going Globile: Spatiality, Embodiment and Media-tion in the Zapatista Insurgency', in S. Dalby and G. O'Tuathail (eds), *Rethinking Geopolitics.* London: Routledge, pp. 240–60.

Routledge, P. (2000) 'Geopoetics of resistance: India's baliapal movement', *Alternatives,* 25: 375–89.

Routledge, P. (2003a) 'Convergence space: process geographies of grassroots globalization networks', *Transactions of the Institute of British Geographers,* 28(3): 333–49.

Routledge, P. (2003b) 'Voices of the dammed: discursive resistance amidst erasure in the Narmada Valley, India', *Political Geography,* 22(3): 243–70.

St Clair, J. (1999) 'Seattle diary its a gas gas gas', *New Left Review,* 238: 81–96.

Sen, J. (2004a) 'Challenging empires: reading the world social forum', in J. Sen, A. Anand, A. Escobar and P. Waterman (eds), *World Social Forum: Challenging Empires.* New Delhi: The Viveka Foundation, pp. xxi–xxviii.

Sen, J. (2004b) 'How open?', in J. Sen, A. Anand, A. Escobar and P. Waterman (eds), *World Social Forum: Challenging Empires.* New Delhi: The Viveka Foundation, pp. 210–27.

Sen, J., Anand, A., Escobar, A. and Waterman, P. (eds) (2004) *World Social Forum: Challenging Empires*. New Delhi: The Viveka Foundation.

Sharp, J.P., Routledge, P., Philo, C. and Paddison, R. (eds) (2000) *Entanglements of Power: Geographies of Domination/Resistance*. London: Routledge.

Sklair, L. (1995) 'Social movements and global capitalism', *Sociology*, 29(3): 495–512.

Slater, D. (2003) 'Geopolitical themes and postmodern thought', in J.A. Agnew, M. Mitchell and G. Toal (eds), *A Companion to Political Geography*, Blackwell, Oxford: pp. 73–91.

Starr, A. (2000) *Naming the Enemy: Anti-Corporate Movements Against Globalization*. London: Zed Books.

Subcommandante M. (2001) 'Tomorrow begins today: invitation to an insurrection', in Notes from Nowhere, *We are Everywhere*. London: Verso.

Teivainen, T. (2004) 'The world social forum: arena or actor?', in J. Sen, A. Anand, A. Escobar and P. Waterman (eds), *World Social Forum: Challenging Empires*, New Delhi: The Viveka Foundation, pp. 122–9.

Udall, L. (1997) 'The international Narmada campaign: a case of sustained advocacy', in W. Fisher (ed.), *Toward Sustainable Development: Struggling Over India's Narmada River*. Jaipur: Rawat Publications, pp. 201–27.

Wallgren, T. (1998) 'Political semantics of "a globalization": a brief note', *Development*, 41(2): 30–2.

Waterman, P. (2000) 'Social movements, local places and globalized spaces: implications for "globalization from below"', in B.K. Gills (ed.), *Globalization and the Politics of Resistance*. New York: Palgrave, pp. 135–49.

Waterman, P. (2001) *Globalization, Social Movements and the New Internationalism*. London: Continuum.

Waterman, P. (2004) 'The secret of fire', in J. Sen, A. Anand, A. Escobar and P. Waterman (eds), *World Social Forum: Challenging Empires*. New Delhi: The Viveka Foundation, pp. 148–60.

From *La Geographie Electorale* to the Politics of Democracy

Introduction

Murray Low

Political geography was, for much of the twentieth century, mainly associated with global or, at least, inter-state matters. This is not to say that 'domestic' politics of various kinds did not become an important focus. Key contributions in the 1970s and 1980s, in particular, highlighted themes of distributional conflict and socio-spatial justice, the geographical conceptualization of states and their relationships to social divisions such as class and to the organization of national, regional and urban spaces. These concerns overlapped productively with those in other sub-disciplines, especially urban and (more recently) economic geography. Not surprisingly, they reflected broader transformations in human geography, particularly the rise of critical-liberal and Marxian geographies at this time. These broader transformations made geographical inquiry into structural and institutional dimensions of politics – and the negotiations and conflicts arising from, reproducing and transforming these – make sense as part of a critical inquiry into the relationships between social arrangements and their spatial organization.

Naturally, earlier concerns with geopolitics led practitioners to discussions of political processes internal to nation-states and other territorialized state-forms. Nonetheless, the main thematic area in contemporary political geography that can claim a developed lineage on a par with geopolitics is probably the geographical analysis of elections, a field pioneered by André Siegfried's *géographie électorale* in the first half of the twentieth century (Siegfried, 1913, 1949). Analysis of voting patterns using the sorts of tools developed in spatial science, quantitative political science and public choice theory became much more visible in the 1970s, as the 'quantitative revolution' generalized itself across a variety of spatial subject matters (for key statements during this time see Cox, 1969; Gudgin and Taylor, 1978; Taylor and Johnston, 1979). In the United States and the United Kingdom in particular, where electoral systems allocate political offices on the basis of plurality voting in geographically delimited districts and constituencies, the analysis

of the spatial dimensions of electoral processes has had a high degree of salience (see, e.g., Archer and Taylor, 1981; Archer and Shelley, 1988; Johnston et al., 1988; Johnston and Pattie, 2006). Moreover, especially in the United States where such matters are highly politicized, processes surrounding the geographical delimitation of electoral units formed the basis for a more normative, although frequently technical, set of contributions on the geography of representation (see, e.g., Bunge, 1966; Morrill, 1973, 1981; Taylor, 1973).

The sub-field of electoral geography (including the geography of representation) is, while an active one on both sides of the Atlantic, commonly viewed as somewhat insulated from debates about other themes in political geography. This is partly because it necessarily (but not exclusively) involves the application and development of specialized quantitative techniques, including various forms of statistical modelling (including models with limited dependent variables and multilevel models) and linear programming. Skills in these techniques are not necessarily widespread in other areas of the sub-discipline and, rightly or wrongly, quantitative methodologies in general became widely viewed as in some ways unhelpful in the development of critical human geography since the 1970s. By contrast, the exploration of the implications of various forms of social and cultural theory for geographic issues, and the use of qualitative methodologies, have tended to characterize developments elsewhere.

The degree to which electoral geography is, in spite of technical hurdles and methodological differences, of interest to wider disciplinary constituencies depends on the evident connections between electoral processes and aspirations to democracy as a form of rule. Elections are justified as a means to democratic governance, as their outcomes should help determine the shape and details of public policy and its relationship to particular societal interests. Moreover, they raise a series of questions about what 'representation' in democracies means or should mean, and they operate alongside or in connection with a series of other mechanisms (financial, participatory, communicative and structural) relating various interests to states and policy processes. All these mechanisms, including elections, are the topics of often heated debate inside and outside the academy, commonly in relation to what it means to be 'democratic', whether established liberal democracy is 'democratic', how democracy can or should be deepened or restructured, and so on. In a certain way, the theme of democracy has perhaps been most explicitly developed in the discipline, particularly in a manner that calls attention to its normative dimensions, in relation to planning practices. The study of these has fused terms of discussion across geography, planning studies proper and a wider urban research field. In recent years, as a reflection of wider debates about, for example, discursive and political 'representation' in the humanities, these planning-centred debates – mainly revolving around participatory alternatives or supplements to electoral democracy – have a newer, wider context to contend with in establishing and/or extending the viability of inherited concepts of democracy in a variegated political world.

Yet, democracy – while commonly invoked or implied as a normative aspiration in writing about political processes in geography – was for quite a long time not as explicit a focus in the sub-field as might be expected (Barnett and Low, 2004). This was perhaps because of the particularly strong influence of Marxist social theory in focusing on states, with a concomitant (and unusual) avoidance of those liberal positions most commonly engaged in the normative justification of electoral and other democratic 'devices' (Brennan and Hamlin, 2000; Saward, 2003), perhaps also because of a reluctance to be seen as over-focusing on (especially national) states at all. In recent years, maybe because of the generalization of liberal forms of democracy across more of the planet than in the 1980s, which has had implications for practice and for theory, and also because of the restructuring of mechanisms of popular involvement and consultation in urban and regional governance in particular, democracy has nonetheless become a more prominent focus of concern outside electoral geography. Political geographers have been engaging more with political as well as social and cultural theory, and becoming more explicit about the normative dimensions of democracy as a value and site of institutional contestation. (The chapters by Yvonne Rydin and by Giles Mohan and Kristian Stokke explore these contestations in relation to environmental governance and development policy respectively.)

The four chapters in this part of the *Handbook* reflect both more longstanding politico-geographic concerns with the organization and outcomes of electoral processes and related concerns with

democracy that are currently assuming positions of greater centrality within the field. Ron Johnston and Charles Pattie review the 'core' literature and concepts in electoral geography. Crucially, they discuss debates about neighbourhood effects and place-based socialization that address questions central to whether geography matters in constituting, rather than simply describing, electoral outcomes. As they put it: 'Voting is not just a place-based act the outcome of which can be mapped. It is a place-based performance, within which interactions both among individual voters and between them and the political actors seeking their support take place'. Some of the evaluation of contextual effects on voting patterns has, and of necessity, taken a technical or methodological form, involving the use of quantitative modelling procedures and careful model specification. But the distinction between act and performance just cited alludes to a wider set of problems of some importance. Electoral geography, constituted as it is around processes that generate numerical variations that in turn are central to structuring some key aspects of political life, should not by that token be thought of as occupying a stand-alone spatial-analytic niche in the field. It opens out, as the quote goes on to suggest, into a host of other issues concerning the geographies of party organizations and other political actors, processes of mobilization (and non-mobilization), political communication, finance, trust, and of course the wider stakes involved in debates about liberal democracy and democratization. All of these demand not merely 'positive' description and explanation, but in and of themselves are topics of substantial and important normative debate. Electoral geography's connections to the wider field of political geography, its potential ramifications in many areas of broader interest to more qualitative or normatively minded researchers, seem far more important than might be implied by its often necessarily different style of investigation. Johnston and Pattie's chapter, addressing as it does many of these implicated themes, has the double merit, then, of reviewing a distinctive field and contesting its separation from broader concerns.

Benjamin Forest's chapter tackles political-geographic questions about 'representation', a concept that, long before the late twentieth-century proliferation of debates about representation in the humanities and critical social science, has been highly contested, and for several reasons. First, electoral systems and other procedural arrangements do not, in the end, offer guarantees that political representation will be judged fair or trustworthy by those ostensibly being represented. Second, confronted by a host of sometimes apparently irreconcilable demands, representatives cannot simply 'channel' the interests of their (diverse) constituents.

Third, representation in states and legislatures involves judgements that mediate the interests or wants of constituencies defined at different scales. Finally, as Forest elaborates, it is unclear how to conceptualize 'representation' at all, even where we try to distinguish between different meanings in the arts and politics, or, in politics, between descriptive and substantive representation, in an effort to track down a clearer meaning. Recent debates around the concept and practices of representation (Pitkin, 1972; Spivak, 1988; Morgan, 1989; Young, 1990; Phillips, 1998; and Derrida, 2002, should be cited as key and relevant interventions here in different registers) have opened up a host of interesting questions for supposedly democratic processes of representation (among others). If anything, however, in forcing the difficulties of political representation into the open, these have complicated matters further, suggesting that there may be no singular and definitive solution, on the ground, to aspirations about 'true' representation unavoidably occupying much of democratic politics. Forest reviews and discusses the manner in which different territorializations of democracy under different electoral systems offer different, but still in a sense arbitrary, answers to normative questions about the relationships between different spatial demarcations of the *demos* and democratic representation. He helpfully extends his discussion from the often discussed, but in many ways peculiar, examples of district or constituency delimitation in the US and, the UK to an engagement with the construction of different representational schemes in South Africa and Iraq. Democracy, once these issues have been confronted, cannot just be a process of contestation regarding the representation of the people. It is also a process that is animated by an inherently inconclusive process of working out who, and where, 'the people' are.

The chapter by Lynn Staeheli takes on some very difficult questions concerning how democracy is characterized and, thereby, how democratization is discussed in the academy and the world. In a nuanced account that does not score easy points at the expense of 'liberalism' or 'procedure', she brings out some of the tensions between democracy and other values – justice in particular – and reviews the ways in which tensions inherent to democracy as a concept should entail a greater sensitivity to complexity and socio-geographic context than is common in judgements about whether or not some particular territorial state in the world is democratic, or democratizing, or not. The geography of democratization, she argues, cannot be understood in terms of simplistic models where agreed-upon procedures 'diffuse' across the globe. Many of the clear-cut distinctions commonly used to stabilize and identify democracy as a state of affairs – distinctions between process and outcome,

enforceable law and societal norms, public and private spheres, citizenship and other identities, politics and the economy or household, national and local or global – fail, in key ways, to 'contain' democracy so that we can definitively allocate states in the world to democratic or non-democratic conditions or categories. Democracy is a process and not a state of affairs, in which spatial as well as political complexity is at stake. As she puts it: '[t]he development of new forms of democracy ... requires an expanded conceptualization of the spaces of politics and democracy, as well as ways to use those spaces to transform the societies of democracy'.

Clive Barnett's chapter, a critical examination of the concepts of publics and publicity, apparently plunges us into different, less familiarly 'geographical', waters than the waves of democracy invoked by Lynn Staeheli. Yet, one of the implications of taking democracy seriously as a normative as well as an empirical theme in political geography is that there is no escaping engagement with normative democratic theory. This means informed engagement with those areas of debate in law, ethics, the philosophy of language, political theory and media studies that struggle to clarify, with a battery of available technical and conceptual means, some of the central normative concerns underpinning a set of processes we might call 'democracy' but which, as Staeheli suggests, we perhaps necessarily find it impossible to definitively define and evaluate. These difficulties of definition and evaluation resonate out beyond the academy and 'politics' and are clearly one of the key animators of 'public' debate. The nature of the public, publicity and public deliberation have been keenly contested elements in understanding how democracy works or does not work. Barnett discusses some of the key ways in which the spatiality of publicity has been considered in geography, particularly in relation to critiques of Habermas and other broadly 'liberal' theorists. Along the way, he takes aim at the desire in geography to conflate public spheres with particular sorts of places, a conflation that generates a politics surrounding the preservation and extension of particular, material sites and spaces for public gathering and communication. As he puts it, building his argument with materials from a variety of theoretical sources, '[p]ublic action can take place anywhere. It has no proper place at all'. One of the implications is that we should 'let go of the idea that public space is either "material" or best modelled on scenes of co-present interaction', substituting a sense in which 'any public is constituted by a *spacing* between discrete but intimately related acts that are separated and bound together in temporal relations of anticipation, projection, response and reply'. Public space, then, is more accurately thought of as involving communicative processes of address and response, characterized by 'dissemination, dispersal and scattering', certainly involving the media and necessarily embodying both affective and rational-deliberative communicative styles.

This section, therefore, opens with a chapter associated with one form of appropriately technical political-geographic endeavour and closes with another. That both of these can coexist in what is a small sub-discipline, and even affect each other in useful ways, is surely a sign of hope in a context where political geography is genuinely starting to engage with democracy in a new and more committed manner. It is sometimes tempting, when non- or post-national forms of governance difficult to reconcile with past democratic blueprints often seem in the ascendancy, to bypass the study of electoral and associated processes, or to re-characterize them as part of an array of techniques of neo-liberal governmentality. There are necessary arguments about the limits to democratic imaginings, and their relations with the spatial complexity of the world, to be had. Nonetheless, debates about democracy and democratization in which political geographers are now engaged are vitally important as bases for clarifying critiques of non-democratic governance, for imagining new geographic orderings for popular rule, and for those broader issues about agency and power implicated in the shift to post-structuralism in this and other areas of concern. All of these are lively issues across the social sciences and humanities, and will likely remain so in political geography for years to come.

REFERENCES

Archer, J.C. and Shelley, F.M. (1988) *American Electoral Mosaics.* Washington, DC: Association of American Geographers.

Archer, J.C. and Taylor, P.J. (1981) *Section and Party: A Political Geography of American Presidential Elections from Andrew Jackson to Ronald Reagan.* Chichester: Research Studios Press.

Barnett, C. and Low, M.M. (2004) 'Introduction: democracy's spaces', in C. Barnett and M.M. Low (eds), *Spaces of Democracy.* London: Sage.

Brennan, G. and Hamlin, A. (2000) *Democratic Devices and Desires.* Cambridge: Cambridge University Press.

Bunge, W. (1966) 'Gerrymandering, geography and grouping', *Geographical Review,* 56: 256–63.

Cox, K.R. (1969) 'The voting decision in a spatial context', *Progress in Geography,* 1: 81–117.

Derrida, J. (2002) 'Declarations of independence', in *Negotiations: Interventions and Interviews, 1971–2001.* Stanford: Stanford University Press.

Gudgin, G. and Taylor, P.J. (1978) *Seats, Votes and the Spatial Organization of Elections.* London: Pion.

Johnston, R.J. and Pattie, C.J. (2006) *Putting Voters in Their Place: Geography and Elections in Great Britain.* Oxford: Oxford University Press.

Johnston, R.J., Pattie, C.J. and Allsopp, J.G. (1988) *A Nation Dividing? The Electoral Map of Great Britain, 1979–1987.* London: Longman.

Morgan, E.S. (1989) *The Invention of the People: The Rise of Popular Sovereignty in England and America.* New York: Norton.

Morrill, R.L. (1973) 'Ideal and reality in reapportionment', *Annals of the Association of American Geographers,* 63: 463–77.

Morrill, R.J. (1981) *Political Redistricting and Geographic Theory.* Washington, DC: Association of American Geographers.

Phillips, A. (1998) *The Politics of Presence.* Oxford: Oxford University Press.

Pitkin, H. (1972) *The Concept of Representation.* Berkeley: University of California Press.

Saward, M. (2003) 'Enacting democracy', *Political Studies,* 51: 161–79.

Siegfried, A. (1913) *Tableau Politique de la France de l'Ouest.* Paris: A. Colin.

Siegfried, A. (1949) *Géographie Electorale de l'Ardèche sous la IIIe République.* Paris: A. Colin.

Spivak, G.C. (1988) 'Can the subaltern speak?' in C. Nelson and L. Grossberg (eds), *Marxism and the Interpretation of Culture.* Basingstoke: Macmillan.

Taylor, P.J. (1973) 'Some implications of the spatial organization of elections', *Transactions of the Institute of British Geographers,* 60: 121–36.

Taylor, P.J. and Johnston, R.J. (1979) *Geography of Elections.* Harmondsworth: Penguin.

Young, I.M. (1990) *Justice and the Politics of Difference.* Princeton: Princeton University Press.

Place and Vote

Ron Johnston and Charles Pattie

INTRODUCTION

Elections, as Peter Taylor (1978) once noted, are a geographer's delight, particularly for one who is interested in the quantitative display and analysis of large data sets. Votes at general and other elections are cast in places and counted in places, and under some electoral systems the result of the contest is determined by the number of votes cast for different candidates/parties in different places. It is thus a straightforward task to map the geography of voting, thereby demonstrating, for example, spatial variations in support for a particular political party or, as with referendums, in political attitudes. Elections are inherently geographical phenomena.

Having mapped an election result, the next stage is to analyse it, to suggest reasons for the observed pattern by relating it to other patterns. This was the goal of electoral geography's pioneer, André Siegfried (1913). In his classic work on the Ardèche, for example, he showed that the pattern of voting was correlated with the *département's* physical and socio-economic milieux: altitudinal variations, reflecting the underlying geology, influenced the types of productive activity practised in different areas, and thus the likely appeal of left- and right-wing parties. Siegfried's (1949) analysis was not simple environmental determinism, however; his argument was that spatial variations in economic practices were likely to generate spatial variations in voting, because different types of people tend to vote for political parties promoting different policy packages.

Following Siegfried's lead, geographers have mapped and analysed election results in a considerable number of countries, notably in northwestern Europe and North America. Their analysis of the patterns displayed have focused on the roles of local milieux as the contexts within which political opinions are formed and electoral decisions reached. For them, as Agnew (1987, 1990, 2002) has stressed in a number of important publications, geography is not epiphenomenal to the study of electoral behaviour. The map of support for a party is not just a map of spatial variations in the types of people who tend to support different political ideologies and policies, so that once you know where such people live you can predict the electoral outcome with great certainty; as Johnston et al. (1988) have shown for British general elections, this is far from the case. Nor is geography simply a residual source of explanations, accounting for minor variations in support that cannot be explained by knowing the types of people involved. Rather, as Agnew expresses it, electoral geography is based on the belief that there is a 'pervasive geographical constitution' to the social, economic and political processes that are the foundations of voting behaviour. Analysing maps of voting thus involves appreciating the contexts within which people learn their political attitudes and how they can best be expressed. Places – large and small – are key milieux for such learning, so that the study of people in places is at the core of electoral geography (Johnston and Pattie, 2006).

This general argument has very largely been exemplified in studies of a relatively small number of countries; indeed, electoral geography has not attracted many adherents outside the English-speaking world.[1] Nevertheless, elections are now held in the majority of countries and – although China is excluded from that category – involve a majority of the world's population. (Freedom House has classified 89 countries as 'free' electoral democracies, and they contain some 44 per cent of the world's population. A further 54 countries are classified as 'partly free', and they include 30 countries comprising 19 per cent of the population. This leaves 49 countries designated 'not free', where political rights and civil liberties are largely denied; these contain the remaining 37 per cent of the population – some 2.4 billion people.[2]) Before considering the role of places in producing the geography of voting, therefore, we first review the rising tide of democratization in the world and the procedures within which it is practised.

DEMOCRATIZATION

In 1989, Francis Fukuyama announced the arrival of the 'end of history'. By this he did not mean that progress and change were over, and that the world had reached an equilibrium state. Rather, his intent was to indicate, as set out in the preface to his later book on the subject (Fukuyama, 1992), that

> ... a remarkable consensus concerning the legitimacy of liberal democracy as a system of government had emerged throughout the world over the last few years, as it conquered rival ideologies like hereditary monarchy, fascism, and most recently communism. More than that, however, I argued that liberal democracy may constitute the 'end point of mankind's ideological evolution' and the 'final form of human government', and as such constituted the 'end of history'.

Certainly, changes behind the Iron Curtain – notably in the USSR and Eastern Europe – along with the (re-)adoption of democracy throughout Latin America suggested that democratic forms of governance were becoming the norm, save in much of the Arab world and in China. In general, trends since have confirmed that opinion, with an increasing number of countries adopting representative democratic procedures as the basis for determining who should control the state apparatus.

Liberal representative democracy is based on a number of freedoms guaranteeing all citizens: (1) freedom in the formulation of preferences, through joining organizations, expression, distributing information, voting and competing for votes; (2) freedom in the signification of preferences, through free and fair elections and the right to stand for public office; and (3) an equal weighting of preferences so that, as expressed in Article 21 of the 1948 United Nations Declaration of Human Rights, 'the will of the people shall be the basis of the authority of government; this will shall be expressed in periodic and genuine elections which shall be by universal and equal suffrage' (Dahl, 1978).

A number of studies have sought ways of measuring the degree to which a country meets these criteria.[3] Using one of these metrics, O'Loughlin et al. (1998) identified major recent increases in the levels of democratization through much of Latin America, Africa and Eastern Europe. These geographical concentrations are evidence of significant space-time clustering, suggesting that 'the spread of democracy appears to be facilitated by elements shared by countries with similar characteristics' (p. 568), which tend to be spatially clustered. Their shifts towards democratization have been encouraged by other countries, notably the United States and those in the European Union, in considerable part as part of the pressure towards a neo-liberal, global economic system (on which see Johnston, 1999).

Freedom House categorizes each country annually on a seven-point political rights scale, with a score of 1 awarded to those with the most developed levels of political rights and 7 to those with the least.[4] Over the thirty years since this exercise began in 1972, the percentage of countries obtaining the highest score increased from 14.9 to 31.3. In 1972, one-third of all countries had a score of 3 or better; in 2002, that was the situation in more than half of the countries surveyed (Table 21.1).

Table 21.1   The Freedom House ratings of countries according to their level of political rights.

A. Percentage of countries in each category on four years

	Year			
Rating	1972	1982	1992	2002
1 (Highest)	14.9	18.7	24.7	31.3
2	13.5	13.9	16.1	14.1
3	5.4	5.4	10.2	9.9
4	10.1	6.6	10.8	12.0
5	10.1	18.9	7.0	5.7
6	21.6	19.3	19.4	18.8
7 (Lowest)	24.3	19.3	11.8	8.3
No. of countries	148	166	186	192

B. Percentage of countries in each category on four years (countries rated in each of those years)

	Year			
Rating	1972	1982	1992	2002
1 (Highest)	15.6	20.0	23.0	25.2
2	14.1	10.4	16.3	16.3
3	5.9	6.7	8.9	9.6
4	7.4	6.7	10.4	11.9
5	11.1	19.3	6.7	6.7
6	22.2	20.0	20.0	21.5
7 (Lowest)	23.7	17.0	14.8	8.9
No. of countries	135	135	135	135

Furthermore, this increase occurred within a larger universe of countries – 44 more were categorized in 2002 than in 1972 (an increase of 30 per cent). If we look at only those countries that were categorized in each of the four years analysed here, the same general pattern appears (the lower block of Table 21.1). Similarly, among the 66 that were not rated in each of the four years, only one had a score of 1 in 1972, as against 26 thirty years later.[5]

The three decades 1972–2002 clearly saw substantially increased democratization across the world. This is indicated in Table 21.2, which shows all 201 countries listed by Freedom House according to their rating at the two end-years of that period. Of the 148 countries rated in 1972, 49 had the same score both then and in 2002 (shown in italics in the table), whereas 67 (shown in bold) had achieved a better score and 24 had a lower score at the end than the beginning of the period – plus a further 8 that were not rated in 2002. Of the 53 that were not rated in 1972, 23 had achieved the highest score thirty years later.

This spread of democratic procedures implies not only that more people in more countries are being given the opportunity to elect governments through 'free and fair elections', but also that there have been major changes in value systems supporting those shifts.[6] Using data collected from the World Values Surveys in the 1990s, O'Loughlin (2004a) found significant geographic patterning in responses to three questions tapping values that underpin democracy: trust in fellow citizens, degree of political interest, and level of volunteer activity (membership of voluntary organizations). After statistically holding constant variables reflecting characteristics of the individual respondents (their social class, church attendance, etc.), he found significant variations between countries on all three variables and also between regions on two of them (the exception was volunteerism). On social trust, for example, respondents from countries in Western Europe had higher levels of trust than predicted by knowledge of their individual characteristics alone, whereas those from many Latin American countries had lower levels. For political interest, several countries in Latin America showed lower levels of interest than expected, whereas there were higher levels than expected through much of Eastern Europe.

O'Loughlin's findings suggest a lack of depth of commitment to some of the values underpinning democracy in several parts of the world. This is confirmed by other data. The Pew Global Attitudes Survey of 44 Nations found considerable variation across those countries in the responses to a number of questions. In the US, for example, over 60 per cent of respondents said that if they had to choose between a good democracy or a strong economy, they would opt for the former, whereas in Russia only 10 per cent gave that response. Similarly, over 60 per cent of Americans said that, if presented with the choice between a democratic government or a strong leader to solve their country's problems, they would opt for a democratic government; only 20 per cent of Russians did. Indeed, support for democracy was quite weak in many countries; although a majority across the 44 countries agreed that 'Honest, competitive elections are a must', the percentage fell below 50 in seven countries, including Russia, and it was supported by only 28 per cent in Jordan. Interestingly, outside the established democracies, the average percentage (73) was highest among the ten African countries surveyed and lowest in the six post-communist states (57) and the five Middle Eastern countries (52).[7]

Democracy is now widely deployed throughout the world, therefore, even though it may be somewhat fragile in certain regions and countries. A large percentage of the world's population is thus given the opportunity at regular intervals to vote for who should govern them over the next few years, which normally involves either returning the incumbent government to power or replacing it by an alternative. In many countries this is done in a single election, usually to a legislature that then selects a government comprising members of either the majority party or a coalition of parties that together agree to form a majority. Elsewhere, the division of powers between an elected president and legislature means that several contests – not necessarily held contemporaneously – are required to determine how power will be exercised.

Table 21.2 Changes in Freedom House's ratings of countries according to their level of political rights, 1972–2002.

1972	2002								
	0	1	2	3	4	5	6	7	N
0	1	23	3	6	7	2	7	4	53
1	1	19	2	0	0	0	0	0	22
2	0	7	3	3	5	1	1	0	20
3	0	2	2	1	1	0	2	0	8
4	2	2	3	1	3	1	3	0	15
5	0	2	4	2	2	3	1	1	15
6	1	3	4	2	4	3	12	3	32
7	4	2	6	4	1	1	10	8	36
N	9	60	27	19	23	11	36	16	201

Italicized countries are those whose rating did not change; those in bold are countries whose level of political rights improved.

ELECTIONS AND ELECTORAL SYSTEMS

Voting involves individuals expressing preferences over alternatives. In elections for governments this usually means indicating a preference for one option within either a slate of candidates or a slate of political parties; in some elections it involves both, with electors choosing between candidates according to the parties that nominated them.

In many countries, especially those with the longest-established democratic traditions, elections were initially held to select the people who would represent defined places. In the UK, for example, Parliaments predominantly comprised individuals elected to represent particular places; there was, however, a very restricted franchise determined by property ownership and sex until the late nineteenth–early twentieth centuries. Most of the MPs represented settlements that had been granted borough status, even though many of them had only small populations; a smaller number were elected to represent the shire counties.[8] Similarly in the US, spatially demarcated constituencies were introduced for elections to the two Houses of Congress (the states for the Senate and Congressional Districts within states for the House of Representatives), with the states also being used as the constituencies for determining the membership of the Electoral College that then elects the President.

Both of those countries have retained single-member constituencies to elect their 'people's house'. The House of Commons and the House of Representatives comprise members elected to represent defined territories. In each, the candidate obtaining the largest number of votes cast is elected, even if that is less than a majority of all of the votes cast. Elsewhere, either there has been a shift away from such electoral systems (often termed plurality systems) or a different system was introduced when democratic procedures were first deployed. Most of those alternative systems have been created in order to provide for more proportional representation, a concept mainly applicable to situations where the criterion for determining the composition of the legislature is that each party should have a proportion of the seats commensurate with its share of the votes cast.

Plurality (or majoritarian) systems rarely, if ever, produce results that meet, or even approach, this criterion. Instead, election outcomes are usually both disproportional and biased. They are *disproportional* because the allocation of seats to a party in the legislature is not proportional to its share of the votes; instead, in general the largest party in terms of votes tends to get a disproportionate share of the seats, whereas small parties tend to get much smaller shares of the seats than of the votes. (There have been various attempts to express that disproportionality through standard formulae, such as the cube law; see Gudgin and Taylor, 1979; Taagepera and Shugart, 1989.) They are *biased* too, because the disproportionality operates differently for different parties: in the UK, for example, from 1950 until 1992 election outcomes were usually biased towards the Conservative party, but in 1997, 2001 and 2005 they were strongly biased in Labour's favour, to a considerable extent because of Labour's carefully crafted geographical targeting strategies (Johnston et al., 2001b, 2002a, 2002b: see also Johnston et al., 2005d; Johnston and Pattie, 2006).

In some countries with established democracies the shift from plurality to (at least quasi-) proportional electoral systems was a response to the growing enfranchisement of the working class and a fear among the elite (property-owning) groups within society – now a minority of the electorate – that they would lose all power. Proportional electoral systems (such as those using party lists as the choice options) not only offered the opportunity for parties representing their interests to win legislative seats, but also were much more likely to result in the absence of one party having an overall majority; coalition government would then be necessary, which would involve a wider distribution of power than if one party obtained a majority. Elsewhere, such systems were introduced in order to prevent one section within society predominating in the legislature; if power was distributed proportionally, then significant minority groups may be able to influence the exercise of power in ways that would not be possible if the 'majority rules' principle of plurality systems applied.

This change in the nature of electoral systems is analysed in detail by Colomer (2004; see also Golder, 2005; Blais et al., 2005), who assembled data on 289 different electoral systems that had been used for a total of 2,145 separate elections in 94 countries. The systems were grouped into four types:

1 *Indirect*, whereby the legislature was not directly elected by the voters;
2 *Majority* (or plurality);
3 *Mixed*, which combined elements of the majority and proportional systems; and
4 *Proportional*, in which legislative seats were allocated in proportion to a party's share of the votes cast.

The first block of Table 21.3 shows the number of each type in use at different dates, indicating a major switch towards mixed and proportional

Table 21.3 Changes in the number of legislative electoral systems of different types.

A. The number of countries with each type of system in selected years

	1874	1922	1960	2002
Indirect	6	2	0	0
Majority	14	11	16	20
Mixed	0	2	0	18
Proportional	0	18	23	51

B. Changes in electoral systems between 1874 and 2002

	2002			
1874	I	M	Mi	P
Indirect (I)	–	12	1	5
Majority (M)	2	–	13	27
Mixed (Mi)	0	1	–	8
Proportional (P)	0	7	6	–

Source: Colomer (2004: 55, 61).

systems; a clear majority of countries used proportional systems in 2002, compared with only a bare majority 80 years earlier and none in 1874. The second block shows all countries that have changed their electoral system; there have been many more moves away from the majority system than towards it. (Colomer, 2005, shows that countries with multi-party systems already in place are more likely to adopt proportional systems than are those where a small number of parties dominates.)

These data on electoral systems indicate that place-based electoral systems, whereby all members of the legislature are elected to represent territorially defined constituencies, have become less popular over time. The shift has been towards systems whereby voters indicate their preferences for parties and the electoral formulae allocate legislative seats according to those parties' relative performance in the polls. Part of this shift has involved countries changing their electoral systems; the remainder has been a result of countries opting for proportional systems when they become democratic.[9] The shift has been far from complete, however. Many countries have retained the majority system – including the UK and the US (where electoral geography has many of its adherents) – whereas others have opted for mixed systems that retain elements of the majority type.

The mixed system is a relatively recent innovation, as shown in the first block of Table 21.3. The paradigm case is Germany, where it was introduced by the occupying powers in 1951 for the three zones that became West Germany. Half of the members of the Bundestag are elected from single-member constituencies using the plurality method: the remainder are elected through a list system of proportional representation, organized so that each party's total number of seats is commensurate to its vote share in the proportional component of the election. This method was adopted in New Zealand in the 1990s, and also for elections to the Scottish Parliament and the Welsh Assembly. A variation has been introduced in other countries (such as Russia), where there is no link between the two parts. Only the number of seats allocated in the list section is proportional to the number of votes won by each party, with the number won in the single-member constituencies thereby not contributing to a proportional outcome – so that the legislature's composition is only semi-proportional. (On mixed-member systems see Shugart and Wattenberg, 2001.)

Despite the shift from majoritarian to proportional systems as democratization has proceeded across the world, much work in electoral geography continues to focus on the former – not least because majoritarian systems are used in the UK and the US. Most electoral geographers study their home countries, so most electoral geography focuses on voting in the place-bound systems. There are exceptions among works published in English, such as studies of: the Republic of Ireland, where the single transferable vote is used (see Paddison, 1976; Parker, 1982; Kavanagh, 2002a, 2002b); Australia, where the alternative vote system is deployed in Commonwealth and some state lower house elections (see Forrest et al., 1999; Johnston and Forrest, 2000); Turkey, where a list system of proportional representation is used but there are clear regional variations in the electoral cleavages (Secor, 2001; West, 2005); Israel, which uses a national list system (Waterman, 1994; Hazan, 1999) and elections in New Zealand, Scotland, Wales and Germany held under the multi-member proportional system (Johnston and Pattie, 2002b; Gschwend et al., 2003). Nevertheless, the majority of work reviewed here concentrates on constituency-based electoral systems.

ELECTORAL SYSTEMS AND PARTIES

As Colomer's data show, when democracy was practised in only a minority of countries, legislatures were almost all elected through the majority system; people living in places defined as entitled to separate representation chose who they wanted to represent their interests. Voting was thus a geographical – and generally very local – exercise. As more democracies were created, however, this geographically based procedure was replaced by one in which geography played a much less important role. In some mixed systems, part of the procedure still involved places electing their representatives, but the remainder involved a competition between parties in which places were of little apparent importance. The latter was especially the case in proportional systems, in which where the votes were cast was largely irrelevant to the election outcome.

The shift to proportional electoral systems is linked to the increasing role of political parties in contemporary democracies. As representative democracy evolved, particularly with the extension of the franchise to all adults, political parties became fundamental to its operations. These play three main roles (on which see Seyd, 1998): they provide the opportunities within which people engage with politics; they mobilize sections of the electorate around a key set of beliefs; and they provide governments with relatively secure (if not guaranteed) support when seeking to implement those policy goals. Parties provide order, stability and continuity to the political scene, both within and outwith the legislature. Without them, each election would be an ad hoc exercise; electors would determine who they wished to represent them in the legislature but would have little say on how majorities were assembled there to sustain legislative programmes. Without them, too, governments would have little security of office; all policy proposals would be subject to bargaining and 'hidden deals', and governments could fall with considerable regularity without any reasonable guarantee of support for a programme. As the business of government extended into more and more spheres of activity and became crucial to the success of the burgeoning capitalist enterprise, so continuity of policy became more important, for which party government was crucial.

The evolution of party systems reflected the political situations in individual countries, but in a classic paper Lipset and Rokkan (1967) classified European party systems according to four cleavages – major divisions within society that created the milieux within which parties mobilized support. They divided these into two groups according to the context in which they emerged. The first two were associated with the *national revolution* that swept through Europe from the seventeenth century on, but which was particularly stimulated by the late-eighteenth-century French Revolution:

1 *Subject versus dominant culture.* Some countries are far from homogeneous in their cultural construction, and one cultural group may have dominated others in the exercise of power. With the transition to democracy, this conflict could be a focus of political mobilization, with separate parties representing the different cultural groups competing for votes – as, for example, with the Basque, Catalan and other cultures within Spain. In many cases, the separate groups were concentrated into different parts of the state's territory, and the cleavage is thus often referred to as core versus periphery.

2 *Churche(es) versus government.* The national revolution frequently pitted those favouring a secular state against established religions that were extremely powerful. Separate political parties sought to mobilize support for those two separate ideologies and how they should be reflected in policies for important issues such as control over community norms as promoted through education. In the Netherlands, for example, three main party groupings emerged, one representing Roman Catholic beliefs, one representing Protestants, and the third advancing secular views.

The second pair of cleavages was associated with the *Industrial Revolution* and the new divisions within society this created:

3 *Primary versus secondary economy.* The growth of manufacturing industry created major tensions within societies between traditional, agrarian interest groups and those associated with the new means of production, hence this cleavage is often portrayed as country versus town; the latter favoured free trade, for example, whereas the former wanted protection.

4 *Workers versus employers.* As manufacturing industry expanded and came to dominate many economies, so conflicts emerged between workers and their employers over working conditions and incomes. Parties emerged to mobilize support within each of these two groups, with those promoting the interests of workers (or the working class) adopting socialist ideologies whereas those seeking support from employers and their associates (the middle class) adopted ideologies widely termed conservative (the two were often known as left and right wing respectively).

All four cleavages were not necessarily mobilized in every country. Some will have been absent because there was no basis for them (the absence of a minority cultural group in Iceland, for example), others because parties failed to mobilize sufficient support from the relevant groups in the face of competing interests. Lipset and Rokkan (1967) argue that in almost all cases, however, the workers versus employers – or class – cleavage came to dominate twentieth-century European politics. As the franchise was extended to embrace all males and then all adults, so the conflict between classes became predominant and underpinned much of political life. In some countries – such as the UK – the other cleavages became relatively unimportant, largely, some argue, because the plurality electoral system favours a political milieu dominated by two parties only. Proportional systems, on the other hand, allow multi-party systems to emerge (Norris, 2004), so that other cleavages can remain potent – as in Switzerland. Indeed, some claim (e.g. Dogan, 2001) that other cleavages remain the most important through much of Western Europe.

Party systems based around the class cleavage have developed in some other parts of the world, notably the former British settler colonies in Australia, Canada and New Zealand. Elsewhere, however, other cleavages have been at the centre of newly developing democracies, notably that between various cultural groups, as shown by examples as distant as Zimbabwe, Ukraine and Iraq. In Zimbabwe, a two-party system emerged pitting the Shona against the Ndebele tribes, although the largest eventually prevailed and the separate parties disappeared. In the Ukraine – as clearly demonstrated by the Presidential election of 2004 – the split was between Russian- and Ukrainian-speakers.[10] And in Iraq, the major split at the 2005 elections saw parties representing Kurds and Shia Muslims dominate the outcome of the proportional representation contest (because of low turnout rates in various parts of the country, the Sunni Muslims were under-represented in the interim legislature).

One country that does not fit into the Lipset–Rokkan schema is the US, where ideological parties based on one of the four cleavages have not emerged.[11] In a country dominated by the same two parties – Republican and Democratic – for more than a century, although each of those parties has been able to draw majority support from particular groups within society, nevertheless neither has been able to win either the Presidency or a majority of seats in either House of Congress through appeals to a single interest group only. (African-Americans strongly supported the Democratic party for much of the twentieth century, for example; migrants from Cuba have strongly favoured the Republicans; the major trade unions have favoured the Democrats; and white evangelical Christians provided strong support for the Republicans at recent elections.) The successful party is the one able to construct a majority coalition of separate interest groups.[12]

A good example of this was the Democratic party in the mid-twentieth century. For more than a century it won massive support from voters (most of them white) in the southern states of the former Confederacy (1949). The Republican party was the core of the anti-slavery movement that led to the Civil War and the Democratic party mobilized support among whites after Reconstruction to deny African-Americans the civil rights awarded them in the post-war Constitutional Amendments. This solid block of support was insufficient to deliver Democratic majorities, however, though it did bring some advantages, such as control over key Congressional committees that were allocated according to seniority; Representatives re-elected many times from 'safe' southern districts occupied many of the senior positions (see Johnston, 1980). In 1932, under Franklin Roosevelt, the Democrats mobilized strong support in the country's northeastern states, where manufacturing industry was suffering intensely from the depression. This coalition of two disparate blocks – basically a class cleavage in the northeast and a cultural cleavage in the south – was sustained for some four decades and delivered considerable success for the party, not only in the Presidency but also in long periods of Congressional dominance. It began to break down with the political success of the civil rights movement in the 1960s, however. The Democratic party retained its northeastern core of support but increasingly lost control of the south, especially in Presidential contests; in 2000 and 2004 the southern states provided solid support for the Republican party, despite the continued overwhelming adherence to the Democrats among African-Americans. (On American sectionalism and voting see Archer and Taylor, 1981; Mayhew, 2002; Heppen, 2003.)

Towards the end of the twentieth century, some argued that the cleavage patterns were in decline and parties could no longer count on the support of clearly defined groups within society; in particular, the death of the class cleavage was identified (see the essays in Evans, 1999). Two separate sets of reasons were identified for this. The first posited a change in values consequent on growing prosperity. In a post-material world, (a majority of) voters would no longer need to focus on employment and incomes as determining influences on the party they would support. Instead they would turn to post-material issues, such as environmentalism. New parties based on these issues – such as green parties – failed to capture wide support, however, and the traditional parties, which to some extent embraced the new concerns but did not make them salient, continued to prevail. They were assisted in this by the economic uncertainties of the 1970s, out of which emerged new types of economies in the more developed parts of the world. The existing parties – such as the UK's Labour party (Crewe, 1986) – adapted their underlying principles to these new conditions by embracing neo-liberal agendas involving the promotion of neo-liberal enterprise economies and free trade linked to social justice for all (or compassionate conservatism).

The second was the growth of 'issue voting', especially what has become known as 'economic voting' or 'pocket-book voting'. Voting decisions have became more closely linked to contemporary concerns, especially about the economic situation, although other concerns (such as public services and foreign policy) might dominate some election campaigns (Johnston and Pattie, 2001c). According to the economic voting calculus, people vote to re-elect incumbent governments who are perceived to be delivering economic prosperity – either to the country as a whole, to them personally, and/or to their local region – but vote for them to be replaced if they are perceived to be failing and an alternative, brighter economic future was on offer from a viable opposition (see, e.g., the essays in Dorussen and Taylor, 2002).

PARTIES IN PLACES

One interpretation to be drawn from the argument developed above could be that although democratic politics initially involved people voting for local representatives in their national legislatures, increasingly this geographical basis to voting patterns has been replaced by cleavage structures that have little foundation in geography. This shift, it could be further argued, has been assisted by the growing dominance of national media – especially television – not only in the conduct of election campaigns but also in the reporting of political issues, which influences how people determine which party to support. Democracy, in this way, has been 'nationalized' with a switch from 'territorial to functional politics' involving a 'transition from a fragmented type of politics with strong and autonomous local political figures to national mechanisms of political accountability in which candidates are submitted to controls and sanctions from national electorates' (Caramani, 2004: 299), resulting in declines of regional variation in support for individual parties (see also Chhibber and Kollman, 2004). Although both trends reflect what has been happening, however, place has not been removed as a foundation of electoral behaviour; it remains an important context within which national battles are contested.

The implication one could draw from arguments regarding contemporary electoral behaviour is that people make their voting decisions in contexts that are almost entirely placeless. If the cleavage model is correct, individuals identify with one or more groups, ascertain which political party is best able to represent those groups' interests, and vote accordingly. If the economic voting (or responsive-voter: people may decide whether to re-elect a government on issues other than economics, such as the performance of public services or the conduct of foreign policy as with the Iraq war) model prevails, they will judge the government's performance on the relevant criteria and vote accordingly. This suggests that voting is a decision taken by the individual acting alone – in a contextual vacuum – responding only to stimuli from the mass and other media. Few individuals act that way, however; most are members of formal and informal social networks within which political issues are discussed, and their opinions are formed and re-formed accordingly. Learning about politics, as with so many other aspects of life, involves social interaction – much of which occurs in particular places.

If social interaction is so important to the socialization of people into voting for particular parties, those parties will want to be involved in it since their electoral prospects depend on its outcomes. They will want to influence what people discuss in their various locally based conversations, and so will be active at the grassroots, seeking to mobilize people to their agenda and, through them, others with whom they have no direct contact. Even in electoral systems where places play no direct role – as with many proportional representation systems – places remain key locales within which parties mobilize support. These twin processes of mobilization and socialization are thus the grassroots bases of the maps of voting that are widely published and that geographers analyse, and are the core of the remainder of this chapter.

Parties and Initial Mobilization

Most research in electoral geography has been done in long-established democracies, where elections have been used as the basis for creating governments for more than a century. Apart from historical studies, therefore, of which there have been few by electoral geographers, such analyses have been conducted within the context of well-established voting patterns. The genesis of maps of voting patterns has not been explored in great detail.

There are exceptions to this, however. For example, Burghardt (1963, 1964) studied Burgenland when it became a province of Austria in the early 1920s (having been transferred from Hungary after the First World War). Politically, it was what he termed 'virgin territory' and most of the politicians who contested the first elections in 1922 were 'imported' from Vienna. As a traditionally rural area the conservative Christian party was expected to prevail, but it got only 31 per cent of the votes cast compared with 39 per cent for the Socialist party. The latter was better positioned to mobilize support, largely through the agency of industrial workers who lived in Burgenland but commuted to Vienna. At later elections, the Christian party won a plurality of the votes, once it was able to create an apparatus of party branches across the rural areas through which it could convince voters that it was best placed to represent their interests.

Once established as a major presence in an area, a party will continue to attract support, not only through the loyalty of those who were initially recruited to its cause but also through their efforts – both deliberate (focused political campaigning) and less structured – in sustaining support by others and winning over those who did not vote for it, including those who did not vote at previous contests (both those who were not qualified to, through age or other criteria, and immigrants). A party and its belief systems become part of the local culture, therefore, and those areas of strength become core to its continued quest for votes – as exemplified by Cox's (1970) study of the support for the Conservative, Labour and Liberal parties in different parts of Wales.

While local traditions of support for a party tend to endure inter-generationally, they have to be continually refreshed or other parties may move in and mobilize support. This may be opportunistic: many by-elections in the UK have produced a 'surprise winner' in a constituency, for example, because the party that has long represented the area is currently unpopular (especially, though not only, if it is in government). Other parties capitalize on this unpopularity by campaigning intensively for protest votes. In England, the Liberal Democrat party has very successfully deployed this strategy during the last forty years. Winning a seat in this way is often not a temporary blip in an area's long-term commitment to another party, however. The victorious candidate is able to project her/himself as successfully representing the area's interests and runs for re-election on that platform, assisted by the campaigning organization built up for the by-election, which may in turn be based on one that has successfully contested local government elections within the constituency. Thus a number of constituencies throughout England have deviated from regional trends by returning Liberal Democrat MPs in a sequence of elections following such a serendipitous success – such as Bermondsey in a strongly Labour area of inner south London. In some cases, those successes then served as 'core areas' for the spread of Liberal Democrat organization and support into surrounding constituencies (see Dorling et al., 1998).

New political movements, seeking widespread support rather than just a localized impact that a by-election opportunity provides, have to 'invade' established local political loyalties in similar ways. They have to create a local apparatus for mobilizing support, with party workers winning over (disillusioned) partisans from other parties and initiating a diffusion process whereby they then convert their friends and neighbours. Such was the case with the National Socialist party in Germany in the 1930s; in a series of papers addressing the question of who voted for the Nazis, O'Loughlin and Flint have shown that the party was better able to mobilize support in some milieux than others, and that these tended to be spatially concentrated (e.g. O'Loughlin et al., 1994, Flint, 1998, 2001, O'Loughlin et al., 2000; 1995).

Whereas the Nazi party had no local organization or (at least latent) support on which to base its initial campaigns, other new parties have not been faced with 'virgin territory' because they have been formed by splits from established groups. In the UK in the early 1980s, for example, a Social Democratic party (SDP) was created by a group of MPs, most of whom who defected from the Labour party, along with some of their local supporters and activists (on the SDP see Crewe and King, 1995). At the next general election in 1983, the new organization performed best where its candidates had previously represented the area for another party and/or where they inherited part of that party's campaigning apparatus. This introduced a new element to the country's electoral map and, following the SDP's merger with the Liberal party, provided the spatially variable campaigning resource base on which the new Liberal Democrat party built its electoral successes at the elections from 1997 on. Other examples of a new organization 'invading' places formerly mobilized by other parties include the Lega Nord in Italy, which built a support base in the country's Po Valley – where there was previously little 'regional consciousness' – on the basis of 'regionalist' claims (Giordano, 2000); it was particularly successful in districts where it had a strong local organization (Giordano, 2001a).

Continued Socialization in Places

The preceding paragraphs have indicated the importance of parties building and maintaining a local presence in an area with which to mobilize and then sustain support. Such support, if it is to be sustained over substantial periods, has to be based not only in the local social networks, however, but also in local interests. Even with a well-established base and a well-organized campaigning machine, a conservative party is unlikely to win and maintain substantial support over a sequence of elections in a predominantly working-class area – though it may perform much better there than in a comparable area where it lacks the base and campaigning apparatus. Thus – as all electoral studies following Siegfried have shown – in general the pattern of voting in an area is closely correlated to other geographies. If the basis of voting there is a cleavage structure, then the geography of support for a party will be linked to the geography of membership of the relevant sections of society – such as support for Labour in the British coalfields (Roberts and Rumage, 1965). If, on the other hand, economic voting prevails, then support for the government will be strongest in the more prosperous parts of the area.

The links between the geography of the characteristics of an area's population and that of support for one or more political parties are strong. Furthermore, many studies of those links have shown that they are non-linear. In general, where you expect a party to be strong, because of the population characteristics there, it tends to be even stronger; where you expect it to be weak, it tends to be even weaker. One of the clearest examples of this was given by Miller (1977, 1978). Using survey data, he found that in Great Britain as a whole, 73 per cent of those in non-manual occupations voted Conservative compared to 30 per cent of those in manual occupations. If this aggregate pattern applied everywhere, then in a constituency where everybody was in a manual occupation the Conservatives should get 30 per cent of the votes. But using constituency data he found that the actual percentage was much less: the Conservatives performed less well than expected where their 'natural supporters' were in a minority, but better than expected where they were in a majority. This tendency has been identified in many other studies, not only in the UK (e.g. Johnston et al., 1988) but also in France (Cox, 1971) and elsewhere. In an early paper, for example, Cox (1968) associated at least some of this variation within London to suburbanization: suburban residents were less likely to abstain and more likely to vote Conservative than similar people living in the inner cities, a pattern also recently identified for Canadian cities by Walks (2004a, 2004b).

Voting patterns, especially those based on cleavage structures, are most likely to be long-established if there is an apparatus in place to continue mobilizing support for a party and socializing people into that position. Parties are unlikely to do that on their own, especially in the contemporary world. In many countries, party membership is very small relative to the population: few people are committed enough to join a party, let alone be prepared to work for it through campaigning and other activities – key though such people are to its success (Schuman, 1999; Cutts, 2003).[13] The processes of mobilization and socialization need to be more widely spread through the population, especially if parties are to sustain the over-concentrations of support identified in the previous paragraph.

One of the main mechanisms promoted as the source of such sustenance is what is generally known as the neighbourhood effect. This was introduced to geographers in classic essays by Cox (1969a, 1969b) and Reynolds (1969a, 1969b), and has underpinned much subsequent research. The core of the argument is that social networks, many of which are spatially very clustered within households, families and neighbourhoods, are the

locales within which there is much discussion of political and related issues and through which, intentionally or serendipitously, some people may be convinced to change their views and the party that they support – what is sometimes termed 'conversion by conversation'. If like people tend to cluster together in neighbourhoods through the operation of the housing market, this is likely to be reflected in majority support for one political party. If people can be convinced to vote for a party through persuasion by their neighbours, then if an area has majority support for one party, its supporters are more likely to convert members of the minority view to their cause than vice versa. The result is not only that the majority view prevails, but also that it predominates: more people than expected vote for the party supported by the majority because 'people who talk together vote together'.[14]

Research by geographers – most of it conducted in the UK – has produced results consistent with this hypothesis, but not completely. In order to establish the validity of the hypothesis, it is necessary to study the social networks in some detail. Apart from Cox's (1969b) original study, this has not been done by geographers, although one team of political scientists has in some detail (see Huckfeldt and Sprague, 1995; see also Eagles et al., 2004). Instead, the role of the neighbourhood effect has been inferred from ecological (i.e. aggregate for areas) data, of the type discussed above – which has led to some cynicism that the observed patterns are the outcome of the postulated processes (e.g. Dunleavy, 1979). Although there is evidence that people who talk together vote together (Pattie and Johnston, 1999, 2000), and that when people with different views talk together, one is quite likely to 'convert' the other (Pattie and Johnston, 2001, 2002), most of this work is a-geographical in that it does not locate the networks within which such conversations and conversions occur.[15]

A further problem with the circumstantial evidence from many of these studies of the neighbourhood effect is the spatial scale deployed. This is especially the case in the UK because data on voting patterns are not available by ballot box or precinct; for general elections, the constituencies – with average electorates of about 70,000 – are the smallest units for which voting returns are released. Such areas are much larger than the locales within which most social networks operate; they are not local enough. It can be argued that constituencies are merely aggregates of such locales, so that if research finds variations by constituency, this probably reflects the summation of a large number of more local processes in operation – but the findings are not conclusive. Even more so, findings that there are regional variations in voting that cannot be accounted for by knowing the regional population's socio-economic composition are even more subject to question; regions may be aggregates of constituencies that are aggregates of locales, but the link between a process believed to involve small numbers of people interacting in spatially confined neighbourhoods and voting patterns in regions with millions of residents is hard to sustain.

Recently, British work using what are known as bespoke neighbourhoods has provided more convincing evidence consistent with the neighbourhood effect. Using a combination of survey and census data, each individual surveyed has been placed in the neighbourhood containing, say, the nearest 1,000 persons to her/his home, and the characteristics of that neighbourhood are then identified from census data. One can then investigate whether people with similar individual characteristics vote differently if they live in different local milieux. The findings of the initial studies were very clear – they did. Middle-class people were more likely to vote Conservative the more middle-class their bespoke neighbourhood (MacAllister et al., 2001), for example, and people living in public housing were more likely to vote Labour the more people in a similar situation they were living among (Johnston et al., 2001a). Later work has suggested that neighbourhood effects operate at several spatial scales (Johnston, Pattie et al., 2000, 2004), and there is in addition an independent regional effect (Johnston et al., 2005b). Furthermore, by subdividing the sample according to the strength with which individuals interact with their neighbours, studies have shown that those who interact most with their neighbours – that is, have the highest levels of neighbourhood social capital – show much stronger evidence of a neighbourhood effect than those who interact least (Johnston et al., 2005c).

These bespoke neighbourhood studies have shown not only that neighbourhood effects can be found with regard to class and other cleavages, but also with respect to the economic voting hypotheses. According to the latter, people vote for the incumbent government if they are satisfied and/or optimistic about economic conditions, but against it if they are not. Most studies of these effects have looked at evaluations of either the national economy (so-called sociotropic voting) or respondents' perceived individual situations (egocentric voting) only – although Johnston and Pattie also looked at evaluations of the regional economy and found that all three scales were linked to how British people voted (Pattie and Johnston, 1995; Johnston and Pattie, 2001a, 2002a). Furthermore, the bespoke neighbourhood studies showed not only that people who were finding their personal and/or household economic situations difficult were more likely to vote against the incumbent government if they lived in an area where economic prosperity was poor than where it was good (Pattie et al., 2000), but also that people who were prospering themselves and feeling optimistic were more likely to

vote against the government if they lived in a relatively depressed area than if their neighbours were sharing their prosperity. This led to the identification of altruistic voters, those who apparently voted against their own interests but in favour of their neighbours' (Johnston, Dorling et al., 2000).

Such spatially detailed investigations have not been conducted in other countries. In the US, for example, much of the published work has focused on sectional patterns of voting at the state scale using aggregate data – as in Archer and Taylor's (1981) classic study of nearly two centuries of Presidential elections, which set the electoral geography within a clear political economy framework.[16] (For later work using this approach see the chapter on Presidential politics in Shelley et al., 1996; Archer and Shelley, 1986; and Heppen, 2003; intra-state variations are discussed in Archer, 1988. On voting for other offices, see Johnston, 1982.) The stimulus for this work came from Key's (1955) theory of critical elections, which posited sequences of elections whereby the same pattern of across-state support for a party's candidates is sustained over several contests as its appeal continues to attract support from the same voters in the same places. These periods of continuity are marked by realigning elections, at which parties mobilize a different geography of support. (Elsewhere, as in Ghana, India and much of Latin America in the 1960s and 1970s, Taylor has shown [Taylor, 1986; Taylor and Flint, 2000; Osei-Kwame and Taylor, 1984; see also Johnston, 1984] that many (in some cases most) elections are realigning elections because governments cannot meet their supporters' expectations and so have to seek a new support base for the next contest.) The individual states are not homogeneous blocks of territory, of course, and within each there are variations in support for a party that reflect the residents' characteristics (on which see Gimpel and Schuknecht, 2003; Gimpel and Cho, 2004), but no research has explored the mobilization of these intra-state support concentrations.

Place as the mobilising context for the emergence of new party alignments has, however, been the focus of Agnew's (2002) work on recent developments in Italian politics. The party system inaugurated in 1947 collapsed very rapidly between 1992 and 1994. During that period, each of the main parties campaigned nationally but had concentrations of support in specific regions, where they represented particular sectional interests and had established strong local apparatuses that were able to deliver consistent support. Of the parties which replaced them in 1994, one – the Lega Nord – appealed to a geographically defined constituency through a programme with a strong devolutionist/separatist tone based on discourses locating political power elsewhere in Italy (as discussed above). Other parties have had to respond accordingly, by constructing alternative political geographies of the country; the recasting of Italian politics is being set within clearly demarcated geographical – or place-bound – matrices. (See also Shin, 2001; Shin and Agnew, 2002. Giordano, 2001b, shows how the Lega Nord's rhetoric changed over time, which involved it 'inventing' a place – Padania – in order to promote its cause.) Similar processes and patterns can be observed in 'emerging democracies', as illustrated by O'Loughlin et al.'s (2002) analysis of the PDS at elections in Berlin after German reunification.

Places and Voter Mobilisation in an Era of Apathy

A topic of much concern among politicians and political scientists alike in recent years has been the substantial decline in turnout at elections in most of the countries with long-established democracies (Wattenberg, 2002). The precipitate drop in turnout in the UK, from 78 per cent in 1992 through 71.5 per cent in 1997 to 59.4 per cent in 2001, is characteristic of this trend. Abstention rates have always varied within countries, even when turnout was high, both across different socio-economic groups and between places. In general, young and socially deprived people are less likely to vote – especially those with little interest in politics – and they have become even less likely to.

This situation has led to much concern about a democratic crisis, which some identify as likely to become permanent unless stronger notions of civic duty can be engendered within the population, especially its younger components. Others are somewhat less sanguine, however, arguing that the current 'problem', in some countries at least, is just a manifestation of the political situation. (This literature is reviewed, *inter alia*, in Franklin, 2004.) If, for example, opinion polls and other information suggest to the electorate that the outcome of a forthcoming election is a foregone conclusion – party X is sure to win, and will probably win by a large margin – then there is much less incentive to vote than if the election outcome is much less certain. Rational individuals know that their votes are very unlikely to make a difference whatever the nature of the contest; they also know that they are even less likely to when the outcome is so certain. But, according to Franklin (2004), one problem remains: if people, especially young people, get into the habit of not voting over a sequence of elections where the outcome is certain, it may be difficult to mobilize them to take an interest and vote if the next contest is much closer.

In countries that employ constituency-based electoral systems, this issue of abstentions has a well-defined geographical component. Because of spatial variations in support for the main political parties, many of those constituencies will be considered safe for one of the parties, whose candidate is almost certain of victory. Only a small number of constituencies may be considered marginal, with two or more parties having a chance of victory. In such situations, the turnout rate should be higher in the marginal than in the safe constituencies, and should also fall more in the latter than in the former when there is a national trend of increasing abstention.[17] This has been the case in Britain over recent decades where turnout has fallen most in the safe seats held by the Labour party, many of whose supporters are less likely to turn out than their Conservative counterparts (Johnston and Pattie, 2001b, 2003).

This geography of campaign importance is clearly recognized by the parties, who realize that there is little point in campaigning hard for higher turnout among their supporters in seats where the result is a foregone conclusion – especially the seats where they are certain they are going to lose. With limited resources (both people and money) with which to campaign, the optimal strategy is to focus as much of these as possible on the places where turnout matters, where a few more votes for one's party in a marginal constituency may mean the difference between success and failure. In the weeks before an election, most parties use their local campaigning resources to identify their likely supporters and then encourage them – right up until the polls close, if necessary – to cast their ballot. A number of British studies have shown a clear relationship between the intensity of a local (constituency) campaign and a party's success at turning out support there; in an age of relative political apathy and low potential turnout, people are more likely to vote if encouraged to by the party they support. (See, for example, Denver and Hands, 1997, and Denver et al., 2004, for the findings of studies of campaign intensity in the UK using data provided by party agents; Whiteley and Seyd, 2003, obtained comparable results using data on party activists. An alternative – complementary according to the findings of comparative studies – approach uses data on party spending on local campaigns as a surrogate for the intensity of vote-search activity: Pattie et al., 1995; Pattie and Johnston, 2003b.) And so parties increasingly focus their campaigning efforts geographically, and those that are the most successful tend to get a benefit by not only winning seats they might otherwise lose but also in getting a better ratio of seats to votes than their opponents (Johnston et al., 2001b).[18]

Such geographically focused mobilization strategies are not peculiar to the UK. In the US, for example, mobilization of the electorate was for long a locally based procedure, with organizers at the precinct scale having the responsibility to sustain the party's support between elections and to ensure high turnouts when the contests are held. This was certainly the case at the 2004 Presidential election. A majority of the states were widely perceived as safe for one or other of the candidates; neither campaigned there personally nor had large campaign teams mobilizing support. Instead, they focused on 13 or so 'swing states' in which it was anticipated that a close contest could be won or lost; very intensive campaigns, involving party workers, candidate visits, and large amounts spent on locally focused advertising, were run there. Turnout was not an issue for the parties in most of the country; only a small number of its places mattered. The same was true of the Congressional elections held on the same day. Very few of the 435 seats were expected to change hands, especially where incumbents were running for re-election – in part because gerrymandered redistricting three years earlier had produced a large number of very safe seats, and partly because of the very significant advantages incumbents have over challengers. Only a few districts were considered worth a great deal of effort, many of them where the incumbent had decided not to run and so the seat was considered 'open'.

There was much talk in the decades after the Second World War of the declining role of local campaigning as a means of mobilizing support as against that of the national media – especially television – and the consequent depersonalization of politics. Geography, it seemed, was becoming less important in elections. It continued to provide the framework within which elections were fought – especially in those countries with constituency-based electoral systems – but geographical strategies were no longer important. The last two decades have seen that situation reversed. As turnout has fallen, and parties have realized that ensuring high turnout is only important in a sub-set – probably a minority – of spatially defined constituencies/districts, so geography has become important again. It is certainly not epiphenomenal; it is the foundation on which the socialization and mobilization of support is based.

ON DATA AND ANALYSIS

Most work in electoral geography involves the analysis of large data sets relating to electoral behaviour and its correlates in different places. Two main types of data set are deployed: ecological data, which refer to population aggregates (such as the number of votes cast for a candidate in a constituency and the number of people living there who own their homes); and individual data, identifying people's characteristics as well as their voting behaviour. These different data types call for separate analytical strategies, different ways of addressing data in order to extract the underlying patterns and thereby test models relating voting and geography.

Interestingly, as Johnston (2005) has pointed out, with few exceptions electoral geographers on different sides of the North Atlantic have focused their attention on different data types and analytical strategies, especially when analysing electoral behaviour in their home countries. In the US, for example, most work has used ecological data and – with the exception of map analyses (as in Heppen, 2003) – relatively unsophisticated techniques; certainly little use has been made of survey data,[19] and even the ecological analyses have not deployed the powerful techniques now being developed to infer individual behaviour from ecological data (Sui et al., 2000; Johnston and Pattie, 2000, 2001d). In the UK, on the other hand, there has been much greater use of survey data to which geographical/contextual information has been added in order to identify how similar people behave in different places and different people behave in similar places – strategies that require particular analytical approaches developed for categorical as against ecological data.

The reasons for these differences in approach may in part reflect differences between the two countries. Given the size of the US and the number of separate places – such as counties – for which voting and other data are available, it is possible to undertake a great deal of analysis there aimed at isolating major trends in voting patterns over space and time without recourse to survey data, although this involves making substantial inferences about spatially varying behaviour (as in Gimpel and Schuknecht, 2003). In the UK, on the other hand, data are only available for about 650 different areas – the Parliamentary constituencies – for the study of general (national) elections, and researchers have explored the use of survey data in order to achieve finer-grained appreciations of who votes for what, where, than ecological data allow. Further, there is much more collaboration between geographers and political scientists in the UK than in the US (though there are exceptions, as Johnston, 2005, identifies), which might account for (the small community of) UK electoral geographers largely adopting the techniques deployed by political scientists. Whatever the reasons for the differences, the opportunities for further work in both countries – and elsewhere – remain great.

CONCLUSION

Electoral geography is a small sub-field within political geography, let alone within geography, and only a small number of practitioners identify themselves as electoral geographers. How they practice their sub-discipline reflects their local context and training; there are considerable differences between UK and US electoral geographers, for example, despite common goals (Pattie and Johnston, 2003a; Johnston, 2005). Electoral geography is itself a paradigm exemplar of the fundamental geographical axiom – 'place matters'.

This chapter has illustrated that basic argument. Voting is not just a place-based act the outcome of which can be mapped. It is a place-based performance, within which interactions both among individual voters and between them and the political actors seeking their support take place. Many people vote the way they do because of the places in which they have learned their political values and identities and the ways that political parties and other interest groups base their search for electoral support on place-based strategies. As democratization spreads through the world at the end of history, so that importance of geography is likely to be enhanced.

NOTES

1 Italy and France are the main exceptions to this (over-?)statement. Even within the English-speaking world, very few geographers have made the study of elections their core substantive interest.
2 www.freedomhouse.org.
3 See, for example, the Polity data set deployed by O'Loughlin et al. (1998), Jaggers and Gurr, (1995) and the Freedom House rankings (http://www.freedomhouse.org/).
4 The methodology is described on http://freedomhouse.org/research/freeworld/2003/methodology.htm.
5 For a further analysis of changes in the Freedom House scores over time, see O'Loughlin (2004b).
6 On the problems of democracy-building, see Kuchukeeva and O'Loughlin (2003).
7 These data are taken from a report on *Views of a Changing World* from the Pew Global Attitudes Project (http://people-press.org/reports/display.php3?ReportID=185).
8 Until the nineteenth century these were predominantly rural areas, but with the Industrial Revolution many large towns and cities grew up that lacked borough status and these swamped the rural areas. The under-representation of those places was part of the movement towards electoral reform (Rossiter et al., 1999).
9 On the use of single-member electoral district systems in emerging democracies, see Birch (2005).
10 On the regional divide in Ukrainian politics see O'Loughlin (2001) and O'Loughlin and Bell (1999).
11 Lipset and Rokkan's schema does not apply because left-wing parties have never mobilized substantial support (Bennett and Earle, 1983).

12 Such coalitions are not unusual to the US, of course; both of the two main UK parties have had to create coalitions straddling the 'middle ground' in order to win Parliamentary majorities.
13 Such campaigning can also produce what are known as 'friends-and-neighbours' effects, whereby people vote for a local candidate – as Kavanagh (2002a) has demonstrated for Ireland.
14 Political scientists also advance this argument – see Gimpel and Schuknecht (2003).
15 The exception is within the household – see Johnston et al. (2005a).
16 This excludes the substantial volume of work on redistricting, discussed in the next chapter.
17 The UK has an electorate of over 40 million. However, it was estimated that only 800,000 of these needed to change their vote between the 2001 and 2005 general elections for the main opposition party to overturn a 165-seat majority held by the incumbent government.
18 The relevance and effect of such strategies in constituency-based electoral systems in other countries is exemplified by Shin et al. (2005).
19 Interestingly, the geographers using such data and related sophisticated techniques – notably John Agnew and John O'Loughlin and their co-workers – have done much of their research on European countries. O'Loughlin's (2003) review of 'Spatial analysis in political geography' includes no references to work on the US by US electoral geographers.

REFERENCES

Agnew, J.A. (1987) *Place and Politics: The Geographical Mediation of State and Society.* London: Allen & Unwin.

Agnew, J.A. (1990) 'From political methodology to geographical social theory? A critical review of electoral geography', in R.J. Johnston, F.M. Shelley and P.J. Taylor (eds), *Developments in Electoral Geography.* London: Croom Helm, pp. 15–21.

Agnew, J.A. (2002) *Place and Politics in Modern Italy.* Chicago: University of Chicago Press.

Archer, J.C. (1988) 'Macrogeographical versus microgeographical cleavages in American Presidential elections', *Political Geography Quarterly,* 7: 111–25.

Archer, J.C. and Shelley, F.M. (1986) *American Electoral Mosaics.* Washington, DC: Association of American Geographers.

Archer, J.C. and Taylor, P.J. (1981) *Section and Party: American Presidential Elections from Andrew Jackson to Ronald Reagan.* Chichester: John Wiley.

Bennett, S. and Earle, C.V. (1983) 'Socialism in America: a geographical interpretation of its failure', *Political Geography Quarterly,* 2: 31–55.

Birch, S. (2005) 'Single-member electoral district systems and democratic transition', *Electoral Studies,* 24: 281–301.

Blais, A., Dobrzynska, A. and Indridason, I.H. (2005) 'To adopt or not to adopt proportional representation: the politics of institutional choice', *British Journal of Political Science,* 35: 182–90.

Burghardt, A.F. (1963) 'Regions and political party support in Burgenland (Austria)', *Canadian Geographer,* 7: 91–8.

Burghardt, A.F. (1964) 'The bases of support for political parties in Burgenland', *Annals of the Association of American Geographers,* 54: 372–90.

Caramani, D. (2004) *The Nationalization of Politics: The Formation of National Electorates and Party Systems in Western Europe.* Cambridge: Cambridge University Press.

Chhibber, P. and Kollman, K. (2004) *The Formation of National Party Systems: Federalism and Party Competition in Canada, Great Britain, India and the United States.* Princeton, NJ: Princeton University Press.

Colomer, J.M. (2004) 'The strategy and history of electoral system choice', in J.M. Colomer (ed.), *Handbook of Electoral System Choice.* Basingstoke: Palgrave Macmillan, pp. 3–80.

Colomer, J.M. (2005) 'It's parties that choose electoral systems' (or, 'Duverger's laws upside down'), *Political Studies,* 53: 1–21.

Cox, K.R. (1968) 'Suburbia and voting behavior in the London metropolitan area', *Annals of the Association of American Geographers,* 58: 111–27.

Cox, K.R. (1969a) 'The voting decision in spatial context', in C. Board et al. (eds), *Progress in Human Geography:* Vol. 1. London: Edward Arnold, pp. 81–117.

Cox, K.R. (1969b) 'The spatial structuring of information flow and partisan attitudes', in M. Dogan, and S.E. Rokkan (eds), *Quantitative Ecological Analysis in the Social Sciences.* Cambridge, MA: MIT Press, pp. 343–70.

Cox, K.R. (1970) 'Geography, social contexts and voting behavior in Wales, 1861–1951', in E. Allardt and S. Rokkan (eds), *Mass Politics.* New York: The Free Press, pp. 117–59.

Cox, K.R. (1971) 'The spatial components of urban voting response surfaces', *Economic Geography,* 47: 27–35.

Crewe, I. (1986) 'On the death and resurrection of class voting: some comments on how Britain votes', *Political Studies,* 35: 620–38.

Crewe, I. and King, A. (1995) *SDP: The Birth, Life and Death of the Social Democratic Party.* Oxford: Oxford University Press.

Cutts, D.J. (2003) 'Does local continuous campaigning really matter? An assessment of Liberal Democrat activism at the 1999 Bath and North Somerset Unitary Elections and 2001 General Election', PhD thesis, University of Bristol.

Dahl, R.A. (1978) 'Democracy as polyarchy', in R.D. Gastil (ed.), *Freedom in the World: Political Rights and Civil Liberties.* Boston: G.K. Hall, pp. 134–46.

Denver, D. and Hands, G. (1997) *Modern Constituency Electioneering.* London: Frank Cass.

Denver, D., Hands, G. and MacAllister, I. (2004) 'The electoral impact of constituency campaigning in Britain, 1992–2001', *Political Studies,* 52: 289–306.

Dogan, M. (2001) 'Class, religion, party: triple decline of electoral cleavages in Western Europe', in L. Karvonen and S. Kuhnle (eds), *Party Systems and Voter Alignments Revisited.* London: Routledge, pp. 93–114.

Dorling, D.F.L., Rallings, C. and Thrasher, M. (1998) 'The epidemiology of the Liberal Democrat vote', *Political Geography,* 17: 45–70.

Dorussen, H. and Taylor, M. (eds) (2002) *Economic Voting.* London: Routledge.

Dunleavy, P.J. (1979) 'The urban basis of political alignment', *British Journal of Political Science,* 9: 409–43.

Eagles, M., Bélanger, P. and Calkins, H. (2004) 'The spatial structure of urban political networks' in M.F. Goodchild and D.G. Janelle (eds), *Spatially-Integrated Social Science.* New York: Oxford University Press.

Evans, G. (ed.) (1999) *The End of Class Politics? Class Voting in Comparative Context.* Oxford: Oxford University Press.

Flint, C. (1998) 'Forming electorates, forging spaces: the Nazi party vote and the social construction of space', *American Behavioral Scientist,* 41: 1282–1303.

Flint, C. (2001) 'A time-space for electoral geography: economic restructuring, political agency and the rise of the Nazi party', *Political Geography,* 20: 301–29.

Forrest, J., Johnston, R.J. and Pattie, C.J. (1999) 'In times of electoral volatility: the effectiveness of constituency campaigning in Australia', *Environment and Planning A,* 31: 1119–29.

Franklin, M.N. (2004) *Voter Turnout and the Dynamics of Electoral Competition in Established Democracies since 1945.* Cambridge: Cambridge University Press.

Fukuyama, F. (1989) 'The end of history', *The National Interest* 57, pp. 2–18.

Fukuyama, F. (1992) *The End of History and the Last Man.* New York: The Free Press.

Gimpel, J.G. and Cho, W.K.T. (2004) 'The persistence of white ethnicity in New England politics', *Political Geography,* 23: 987–1008.

Gimpel, J.G. and Schuknecht, J.E. (2003) *Patchwork Nation: Sectionalism and Political Change in American Politics.* Ann Arbor: University of Michigan Press.

Giordano, B. (2000) 'Italian regionalism or "Padanian" nationalism: the political project of the Lega Nord in Italian politics', *Political Geography,* 19: 445–71.

Giordano, B. (2001a) ' "Institutional thickness", political subculture and the resurgence of (the "new") regionalism in Italy: a case study of the Northern League in the province of Varese', *Transactions of the Institute of British Geographers,* NS, 26: 25–41.

Giordano, B. (2001b) 'The contrasting geographies of "Padania": the case of the Lega Nord in Italy', *Area,* 33: 27–37.

Golder, M. (2005) 'Democratic electoral systems around the world, 1946–2000', *Electoral Studies,* 24: 103–21.

Gschwend, T., Johnston, R.J. and Pattie, C.J. (2003) 'Split-ticket patterns in multi-member proportional electoral systems: estimates and analyses of their spatial variations at the German federal election, 1998', *British Journal of Political Science,* 33: 109–28.

Gudgin, G. and Taylor, P.J. (1979) *Seats, Votes and the Spatial Organization of Elections.* London: Pion.

Hazan, R.Y. (1999) 'Constituency interests without constituencies: the geographical impact of candidate selection on party organization and legislative behaviour in the 14th Israeli Knesset, 1996–1999', *Political Geography,* 18: 791–811.

Heppen, J. (2003) 'Racial and social diversity and US Presidential election regions', *Professional Geographer,* 55: 191–205.

Huckfeldt, R. and Sprague, J. (1995) *Citizens, Politics and Social Communication: Information and Influence in an Election Campaign.* Cambridge: Cambridge University Press.

Jaggers, K. and Gurr, T.R. (1995) 'Transitions to democracy: tracking democracy's third wave with the Polity III data', *Journal of Peace Research,* 21: 469–82.

Johnston, R.J. (1980) *The Geography of Federal Spending in the United States of America.* Chichester: Research Studies Press.

Johnston, R.J. (1982) 'The changing geography of voting in the United States: 1946–1980', *Transactions of the Institute of British Geographers,* NS, 7: 187–204.

Johnston, R.J. (1984) 'The political geography of electoral geography', in P.J. Taylor and J.W. House (eds), *Political Geography: Recent Advances and Future Directions.* London: Croom Helm, pp. 133–48.

Johnston, R.J. (1999) 'The United States, the "Triumph of Democracy" and the "End of History"', in D. Slater and P.J. Taylor (eds), *The American Century: Consequences and Coercion in the Projection of American Power.* Oxford: Blackwell, pp. 149–65.

Johnston, R.J. (2005) 'Anglo-American electoral geography: same roots and same goals, but different means and ends?', *Professional Geographer,* 57: 580–87.

Johnston, R.J. and Forrest, J. (2000) 'Constituency election campaigning under the alternative vote: the New South Wales Legislative Assembly election, 1995', *Area,* 32: 107–17.

Johnston, R.J. and Pattie, C.J. (2000) 'Ecological inference and entropy-maximizing: an alternative estimation procedure for split-ticket voting', *Political Analysis,* 8: 333–45.

Johnston, R.J. and Pattie, C.J. (2001a) ' "It's the economy, stupid" – but which economy? Geographical scales, retrospective economic evaluations and voting at the 1997 British general election', *Regional Studies,* 35: 309–20.

Johnston, R.J. and Pattie, C.J. (2001b) 'Is there a crisis of democracy in Great Britain? Turnout at general elections reconsidered', in K. Dowding, J. Hughes and H. Margetts (eds), *Challenges to Democracy: Ideas, Involvement and Institutions.* London: Palgrave, pp. 61–80.

Johnston, R.J. and Pattie, C.J. (2001c) 'Dimensions of retrospective voting: economic performance, public service standards and Conservative party support at the 1997 British general election', *Party Politics,* 7: 469–90.

Johnston, R.J. and Pattie, C.J. (2001d) 'On geographers and ecological inference', *Annals of the Association of American Geographers,* 91: 281–2.

Johnston, R.J. and Pattie, C.J. (2002a) 'Geographical scale, the attribution of credit/blame, local economic circumstances and retrospective economic voting in Great Britain, 1997: an extension of the model', *Environment and Planning C: Government and Policy,* 20: 421–38.

Johnston, R.J. and Pattie, C.J. (2002b) 'Campaigning and ticket-splitting in new electoral systems: the first MMP elections in New Zealand, Scotland and Wales', *Electoral Studies,* 21: 583–600.

Johnston, R.J. and Pattie, C.J. (2003) 'The growing problem of electoral turnout in Britain? Voluntary and involuntary non-voters in 2001', *Representation,* 40: 30–43.

Johnston, R.J. and Pattie, C.J. (2006) *Putting Voters in Their Place: Geography and Elections in Great Britain.* Oxford: Oxford University Press.

Johnston, R.J., Pattie C.J. and Allsopp, J.G. (1988) *A Nation Dividing? Britain's Changing Electoral Map, 1979–1987.* London: Longman.

Johnston, R.J., Dorling, D.F.L., Tunstall, H., Rossiter, D.J., MacAllister, I. and Pattie, C.J. (2000) 'Locating the altruistic voter: context, egocentric voting and support for the Conservative party at the 1997 general election in England and Wales', *Environment and Planning A,* 32: 673–94.

Johnston, R.J., Pattie, C.J., Dorling, D.F.L., MacAllister, I., Tunstall, H. and Rossiter, D.J. (2000) 'The neighbourhood effect and voting in England and Wales: real or imagined?', in P.J. Cowley, D.T. Denver, A.T. Russell and L. Harrison (eds), *British Elections and Parties Review,* Vol. 10. London: Frank Cass, pp. 47–63.

Johnston, R.J., Pattie, C.J., Dorling, D.F.L., MacAllister, I., Tunstall, H. and Rossiter, D.J. (2001a) 'Housing tenure, local context, scale and voting in England and Wales, 1997', *Electoral Studies,* 20: 195–216.

Johnston, R.J., Pattie, C.J., Dorling, D.F.L., and Rossiter, D.J. (2001b) *From Votes to Seats: The Operation of the UK Electoral System since 1945.* Manchester: Manchester University Press.

Johnston, R.J., Rossiter, D.J., Pattie, C.J. and Dorling, D.F.L. (2002a) 'Distortion magnified: New Labour and the British electoral system, 1950–2001', in L. Bennie, C. Rallings, J. Tonge and P. Webb (eds), *British Elections and Parties Review,* Vol. 12: *The 2001 General Election.* London: Frank Cass, pp. 133–55.

Johnston, R.J., Rossiter, D.J., Pattie, C.J. and Dorling, D.F.L. (2002b) 'Labour electoral landslides and the changing efficiency of voting distributions', *Transactions of the Institute of British Geographers,* NS, 27: 336–61.

Johnston, R.J., Jones, K., Burgess, S., Propper, C., Sarker, R. and Bolster, A. (2004) 'Scale, factor analyses and neighborhood effects', *Geographical Analysis,* 36: 350–68.

Johnston, R.J., Jones, K., Propper, C., Sarker, R., Burgess, S. and Bolster, A. (2005a) 'A missing level in the analyses of British voting behaviour: the household as context as shown by analyses of a 1992–1997 longitudinal survey', *Electoral Studies,* 24: 201–25.

Johnston, R.J., Propper, C., Burgess, S., Sarker, R., Bolster, A. and Jones, K. (2005b) 'Spatial scale and the neighbourhood effect: multinomial models of voting at two recent British general elections'. *British Journal of Political Science,* 35: 487–514.

Johnston, R.J., Propper, C., Sarker, R., Jones, K., Bolster, A. and Burgess, S. (2005c) 'Neighbourhood social capital and neighbourhood effects', *Environment and Planning A,* 37: 1443–61.

Johnston, R.J., Rossiter, D.J. and Pattie, C.J. (2005d) 'Disproportionality and bias in US Presidential elections: how geography helped Bush defeat Gore but couldn't help Kerry beat Bush', *Political Geography,* 24: 952–68.

Kavanagh, A. (2002a) 'Social deprivation, political alienation and community empowerment: the geography of voter turnout in Ireland 1997–2002, and its association with social deprivation', PhD thesis, National University of Ireland, Maynooth.

Kavanagh, A. (2002b) 'Turnout, political support and representation in Ireland', in K. Hayward and M. McCarthaigh (eds), *Recycling the State.* Dublin: University College Dublin Press.

Key, V.O. Jr (1955) 'A theory of critical elections', *Journal of Politics,* 17: 3–18.

Kuchukeeva, A. and O'Loughlin, J. (2003) 'Civic engagement and democratic consolidation in Kyrgyzstan', *Eurasian Geography and Economics,* 44: 557–87.

Lipset, S.M. and Rokkan, S.E. (1967) 'Cleavage structures, party systems and voter alignments: an introduction', in S.M. Lipset and S.E. Rokkan (eds), *Party Systems and Voter Alignments.* New York: The Free Press, pp. 3–64.

MacAllister, I., Johnston, R.J., Pattie, C.J., Tunstall, H., Dorling, D.F.L. and Rossiter, D.J. (2001) 'Class dealignment and the neighbourhood effect: Miller revisited', *British Journal of Political Science,* 31: 41–60.

Mayhew, D.R. (2002) *Electoral Realignments: A Critique of an American Genre.* New Haven, CT: Yale University Press.

Miller, W.L. (1977) *Electoral Dynamics in Britain since 1918.* London: Macmillan.

Miller, W.L. (1978) 'Social class and party choice in England: a new analysis', *British Journal of Political Science,* 8: 259–84.

Norris, P. (2004) *Electoral Engineering: Voting Rules and Political Behavior.* Cambridge: Cambridge University Press.

O'Loughlin, J. (2000) 'Can King's ecological inference method answer a social scientific puzzle: who voted for the Nazi party in Weimar Germany?', *Annals of the Association of American Geographers,* 90: 592–601.

O'Loughlin, J. (2001) 'The regional factor in contemporary Ukrainian politics: scale, place, space or bogus effect?', *Post-Soviet Geography and Economics,* 42: 1–33.

O'Loughlin, J. (2003) 'Spatial analysis in political geography', in J.A. Agnew, K. Mitchell and G. Toal (eds), *A Companion to Political Geography.* Oxford: Blackwell, pp. 30–46.

O'Loughlin, J. (2004a) 'Democratic values in a globalizing world: a multilevel analysis of geographic contexts', *GeoJournal,* 60: 3–17.

O'Loughlin, J. (2004b) 'Global democratization: measuring and explaining the diffusion of democracy', in C. Barnett and M.M. Low (eds), *Spaces of Democracy: Geographical Perspectives on Citizenship, Participation and Representation.* London: Sage, pp. 23–44.

O'Loughlin, J. and Bell, J.E. (1999) 'The political geography of civic engagement in Ukraine, 1994–1998', *Post-Soviet Geography and Economics,* 39: 233–66.

O'Loughlin, J., Flint, C. and Anslein, L. (1994) 'The political geography of the Nazi vote: context, confession and class in the 1930 Reichstag election', *Annals of the Association of American Geographers,* 84: 351–80.

O'Loughlin, J., Flint, C. and Shin, M. (1995) 'Regions and milieux in Weimar Germany: the Nazi party vote of 1930 in geographic perspective', *Erdkunde,* 49: 305–14.

O'Loughlin, J., Ward, M.D. et al. (1998) 'The diffusion of democracy, 1946–1994', *Annals of the Association of American Geographers,* 88: 545–74.

O'Loughlin, J., Witmer, F. and Ledwith, V. (2002) 'Geography, not socio-demographics: explaining the PDS (Party of Democratic Socialism) vote in Berlin, 1999 and 2001', *Eurasian Geography and Economics,* 42: 349–82.

Osei-Kwame, P. and Taylor, P.J. (1984) 'A politics of failure: the political geography of Ghanaian elections', *Annals of the Association of American Geographers,* 74: 574–89.

Paddison, R. (1976) 'Spatial bias and redistricting in proportional representation systems: a case study of the Republic of Ireland', *Tijdschrift voor Economische en Sociale Geografie,* 67: 230–40.

Parker, A.J. (1982) 'The "friends and neighbours" voting effect in the Galway West constituency', *Political Geography Quarterly,* 1: 243–62.

Pattie, C.J. and Johnston, R.J. (1995) ' "It's not like that round here": region, economic evaluations and voting at the 1992 British general election', *European Journal of Political Research,* 28: 1–32.

Pattie, C.J. and Johnston, R.J. (1999) 'Context, conversion and conviction: social networks and voting at the 1992 British general election', *Political Studies,* 47: 877–89.

Pattie, C.J. and Johnston, R.J. (2000) ' "People who talk together vote together": an exploration of contextual effects in Great Britain', *Annals of the Association of American Geographers,* 90: 41–66.

Pattie, C.J. and Johnston, R.J. (2001) 'Talk as political context: conversation and electoral change in British elections, 1992–1997', *Electoral Studies,* 20: 17–40.

Pattie, C.J. and Johnston, R.J. (2002) 'Political talk and voting: does it matter to whom one talks?', *Environment and Planning A,* 34: 1113–36.

Pattie, C.J. and Johnston, R.J. (2003a) 'Anglo-American electoral geography: the emergence of a subdiscipline', *Espaces, Populations, Sociétés,* 3: 443–52.

Pattie, C.J. and Johnston, R.J. (2003b) 'Local battles in a national landslide: constituency campaigning at the 2001 British general election', *Political Geography,* 22: 381–414.

Pattie, C.J., Johnston, R.J. and Fieldhouse, E.A. (1995) 'Winning the local vote: the effectiveness of constituency campaign spending in Great Britain, 1983–1992', *American Political Science Review,* 89: 969–83.

Pattie, C.J., Johnston, R.J., Dorling, D.F.L., MacAllister, I., Tunstall, H. and Rossiter, D.J. (2000) 'Local context, retrospective economic evaluations, and voting: the 1997 general election in England and Wales', *Political Behavior,* 22: 121–43.

Reynolds, D.R. (1969a) 'A "friends-and-neighbors" voting model as a spatial interactional model for electoral geography', in K.R. Cox and R.G. Golledge (eds), *Behavioral Problems in Geography.* Evanston IL: Northwestern University, Northwestern Studies in Geography, pp. 81–100.

Reynolds, D.R. (1969b) 'A spatial model for analysing voting behavior', *Acta Sociologica,* 12: 122–30.

Roberts, M.C. and Rumage, K.W. (1965) 'The spatial variations in urban left-wing voting in England and Wales in 1951', *Annals of the Association of American Geographers,* 55: 161–78.

Rossiter, D.J., Johnston, R.J. and Pattie, C.J. (1999) *The Boundary Commissions: Redrawing the UK's Map of Parliamentary Constituencies.* Manchester: Manchester University Press.

Schuman, A. (1999) 'Boundary changes, local political activism and the importance of the electoral ward: an electoral geography of Bristol 1996–1999', PhD thesis, University of Bristol.

Secor, A.J. (2001) 'Ideologies in crisis: political cleavages and electoral politics in Turkey in the 1990s', *Political Geography,* 20: 539–60.

Seyd, P. (1998) 'In praise of party', *Parliamentary Affairs,* 51: 198–208.

Shelley, F.M., Archer, J.C., Davidson, F.M. and Brunn, S.D. (1996) *Political Geography of the United States.* New York: Guilford Press.

Shin, M. (2001) 'The politicisation of place in Italy', *Political Geography,* 20: 331–52.

Shin, M. and Agnew, J.A. (2002) 'The geography of party replacement in Italy, 1987–1996', *Political Geography,* 21: 221–42.

Shin, M., Jon, Y., Gross, D.A. and Eom, K. (2005) 'Money matters in party-centered politics: campaign spending in Korean congressional elections', *Electoral Studies,* 24: 85–101.

Shugart, M.S. and Wattenberg, M., (eds) (2001) *Mixed-Member Electoral Systems: The Best of Both Worlds?* Oxford: Oxford University Press.

Siegfried, A. (1913) *Tableau Politique de la France de l'Ouest.* Paris: A. Colin.

Siegfried, A. (1949) *Géographie electorale de l'Ardèche sous la IIIème République.* Paris: A. Colin.

Sui, D., Fotheringham, A.S., Anselin, L., O'Loughlin, J. and King, G. (2000) 'New directions in ecological analysis: a symposium', *Annals of the Association of American Geographers,* 90: 579–606.

Taagepera, R. and Shugart, M.S. (1989) *Seats and Votes: The Effects and Determinants of Electoral Systems.* New Haven. CT, Yale University Press.

Taylor, P.J. (1978) 'Political geography', *Progress in Human Geography,* 2: 153–62.

Taylor, P.J. (1986) 'An exploration into world-systems analysis of political parties', *Political Geography Quarterly,* 5: S5–S20.

Taylor, P.J. and Flint, C. (2000) *Political Geography: World-Economy, Nation-State and Locality'* (4th edn). London: Pearson.

Walks, R.A. (2004a) 'Place of residence, party preferences and political attitudes in Canadian cities and suburbs', *Journal of Urban Affairs,* 26: 269–95.

Walks, R.A. (2004b) 'Suburbanization, the vote, and changes in Federal and Provincial political representation and influence between inner cities and suburbs in large Canadian urban regions, 1945–1999', *Urban Affairs Review,* 39: 411–40.

Waterman, S. (1994) 'The non-Jewish vote in Israel in 1992', *Political Geography,* 13: 540–68.

Wattenberg, M.P. (2002) *Where Have All the Voters Gone?* Cambridge, MA: Harvard University Press.

West, W.J. (2005) 'Regional cleavages in Turkish politics: an electoral geography of the 1999 and 2002 national elections', *Political Geography,* 24: 499–523.

Whiteley, P. and Seyd, P. (2003) 'How to win a landslide by really trying: the effects of local campaigning on voting in the 1997 British general election', *Electoral Studies,* 22: 301–24.

The Territorial Politics of Representation

Benjamin Forest

THE GEOGRAPHY OF REPRESENTATION

Representation is a maddening concept. As Hannah Pitkin suggests in her seminal work on the subject, representation 'is the making present *in some sense* of something which is nevertheless *not* present literally' (Pitkin, 1967: 8–9; emphasis in the original). For democratic political representation the entity to be represented is elusive. Is it the nation? 'The people?' The 'will of the people?' The aggregation of individual preferences? The aggregation of group preferences? None of these is a tangible entity, and indeed, they are created only through the act of representation.

Political representation is therefore never simply 'mapping' a political constituency onto a representative body. Like a cartographer picking a particular projection while drawing a map, political actors make decisions about electoral systems that fundamentally shape the composition and expression of political constituencies. Similarly, the geographical and figurative boundaries of electoral systems are themselves based on assumptions and judgments about the nature of the political constituency.

The territorial organization of states means that questions about political representation, and democratic political representation in particular, involve disputes, decisions, and struggles over the relationship between representation and territory. Consequently, questions about representation are fundamentally geographical questions. States may settle political conflicts using either a federal system dividing power between the center and regions, or autonomous regions within a single state. Less drastically, states can use territorial divisions either

to guarantee political power and representation for racial, ethnic, or national minorities, or to suppress the power of minorities while maintaining an ostensibly 'representative' political system.

This chapter explores how states have used various electoral systems to address the problem of minority representation in the United States, post-apartheid South Africa, and post-invasion Iraq. In these instances, territorial politics have involved both the manipulation of internal boundaries as well as the selection of particular electoral systems. The United States has increased minority representation by creating non-White majority election districts, but has done so only when courts recognize racial minorities as distinct regional communities. In contrast, South Africa and Iraq sought minority representation without using districts and without explicitly recognizing regional concentrations of ethnic minorities.

Democracy and Political Representation

My focus on representation in democracies sets aside questions about representation in either non-political or non-democratic contexts. There are, however, good reasons for focusing on democratic political representation.

Political scientists argue that major 'waves of democratization' have legitimized national independence movements since the late eighteenth century (Huntington, 1991). It is important to note, however, that even in the United States through the early twentieth century, 'democracy' was a term of derision, connoting mob-rule and threats to political stability (Keyssar, 2000). The newest 'wave' of democratization in the late 1980s and early 1990s saw the end of the apartheid regime in South Africa, and the fall of Communist governments in Eastern Europe and the Soviet Union (McFaul, 2002). As a result, 'democracy' is now largely unchallenged as the proper basis of political legitimacy (Fukuyama, 1992).[1] Consequently, nearly every state – or at least those that take their claim to liberal democracy seriously – must devise a system of representation that provides both political legitimacy and governability. That is, every democratic state must have a system that is responsive to its citizens' will but that can also govern with the recognition that 'the people's will' is a fiction hiding a set of competing and conflicting interests. More to the point, as critics have long pointed out, 'democracy' may serve as an ideological cover for the interest of capital or political elites (Lindblom, 1977; Jessop, 1983). In such cases, these interests, rather than 'the people', enjoy representation.[2]

The advent of 'universal suffrage' is remarkably recent. Indeed, one might observe that it has yet to occur. No state grants truly universal suffrage; all restrict the franchise on the bases of at least age and citizenship, and a number restrict it based on factors like gender, residency, and criminal convictions. Moreover, extra-legal restrictions constrain electoral participation in most states to some degree.[3] Nonetheless, the ideology driving liberal democracy now calls for suffrage for all adult citizens (Ramirez et al., 1997). In practice, the 'will of the people' refers to a much smaller universe: the expressed preferences of actual voters.

The form of representation in most liberal democracies – the aggregation of the equally weighted votes of individual adult citizens to elect one or more representatives – is only one of several possible forms of representation. In federal systems, for example, less populous regions may receive equal representation in one or more branches of the government. More commonly, states may allocate representatives by total population even if they only grant the franchise to a portion of the population. More generally, states make choices about who and what achieve represention. States may form constituencies in several different ways, all of which are consistent with the principles of liberal democracy, and all of which one can describe as representative. Yet these different systems address minority interests and rights in ways that can lead to either legitimacy and stability or crisis and instability (Lijphart and Grofman 1984; Lijphart, 1999).

Political Representation

Pitkin (1967) remains an essential starting point for the study of political representation. She carefully analyzes the major theories of political representation with particular attention to the meaning of the family of words associated with 'represent'. In particular, Pitkin tries to reconcile one of the basic controversies in political theory over the role of the representative – the 'independent verses mandate' debate. Are representatives free to act independent of (and contrary to) the wishes of their constituents ('independent') or do their constituents' preferences bind them ('mandate')? She argues that representation disappears as one approaches either extreme; a person acting either as a completely independent trustee or as a bound delegate is not acting as a representative. She concludes that representing

> means acting in the interest of the represented, in a manner responsive to them. The representative must act independently; his [or her] action must involve discretion and judgment. ... The represented must also be (conceived as) capable of independent action and judgment, not merely being taken care of. And, despite the resulting potential conflict between representative and represented ... that conflict must not normally take place. The representative must act in such a way that there is no conflict, or if it occurs an explanation is called for. (Pitkin, 1967: 211)

Although there is a normative dimension to Pitkin's argument, she bases this conclusion primarily in analytic language philosophy. An official acting differently from this description is not acting as a political representative in the usual meaning of the term. While Pitkin's description is not elegant, it captures something essential about the nature of representation and the role of the representative. Even so, it does not answer the questions raised in the opening paragraphs of this chapter: Who or what are the represented? How are they 'capable of independent action and judgment'? What are the best institutional arrangements for achieving fair and effective representation?

These questions are closely linked. One's judgment of the best electoral system depends on one's conception of the represented. Likewise, electoral systems can profoundly shape political interests and identity. In the United States, for example, the central question for the Supreme Court in voting rights cases has been how to speak for 'the people' in a way that allows them to speak for themselves. Bybee (1998) argues that the Court has offered confused answers to this question because it has treated minority identity as fixed and prepolitical, rather than as something that is formed through the process of political deliberation.

More generally, political representation helps overcome one of the paradoxes of democratic self-governance: How can citizens simultaneously be rulers and subjects (Morgan, 1988; Schwartz, 1988)? How can the political authority of the state bind the people, while the people are simultaneously the source of that authority? Morgan (1988: 38–54) argues that the struggle for political power between the king and gentry during the English Revolution produced the 'fiction' of representation. Representatives were the agents of geographical communities, so that the consent of these agents among themselves also bound their communities. Yet, once they entered Parliament, representatives were expected to act in the interests of the entire realm, not just their local community. The English invented the notion of 'the people' as a sovereign body to overcome this conflict in roles. Likewise, appeals to the sovereignty of 'the people' – as represented through direct elections from geographically defined districts – was fundamental to the Federalists' argument justifying a new American constitution in the late 1780s (Morgan, 1988: 263–87). As a number of scholars have observed, 'the people' typically excluded women and minorities, so that representation often went hand-in-hand with inequality (Marston, 1990). Nonetheless, defining 'the people' as a distinct, sovereign-holding but abstract entity was essential for the development of democratic political representation.

If Morgan sees geographic representation as a practical development of power struggles in the England and the United States, Schwartz (1988) offers a theoretical and normative defense of geographically defined constituencies. Indeed, she argues that single-member, plurality districts are the only electoral system that can overcome the ruler/subject dilemma. Her study couches the conflict in terms of membership and freedom, where citizens are both members of a political community (and hence constrained in certain ways), but also free to act as individuals. Political districts are fundamental to the creation of such citizens because such districts divide the sovereign whole into non-sovereign or semi-sovereign parts. By doing so, districts teach individuals that citizenship is a civic rather than a natural fact, that the power of such districts can be contested relative to the whole, and that citizenship is a relationship with other, known individuals. Moreover, districts provide a unique ironic membership 'appropriate to a self-conscious individual living in multiple contexts' (Schwartz, 1988: 101–2).[4]

These debates relate to conflicts over the composition of representative assemblies, particularly the distinction between description and substantive representation. Descriptive representation refers to the demographic composition in legislative (or judicial and administrative) bodies compared to the population (or electorate). Does a legislature 'look like' the population it represents? Substantive representation refers to the representation of

interests associated with different demographic groups (Pitkin, 1967). Do different demographic groups wield legislative power proportional to their share of the population? Minorities (including women) may have a presence in legislative assemblies, but may hold little political power. Conversely, minorities may elect legislators from the majority demographic group or majority legislators may effectively advocate for minority interests. Under certain circumstances, there may not be any conflict betweens these two forms of representation, but typically electoral systems favor one form over the other. Single-member plurality districts typically elect racial/ethnic minorities only if the districts contain local majorities of these groups.[5] Under party-list systems, the number of racial/ethnic minorities (or women) depends on the position of such individuals on each party's list. The preference for one form of representation over the other depends largely on the balance between the representative as a delegate of his or her constituency (mandate theory) or as a trustee (independent theory). In systems favoring the independence model or where there is less accountability through elections, descriptive representation may be relatively more important.

Conventional approaches in political science tend to assume that substantive representation has primary importance, and that the presence of minorities in legislatures has at best some symbolic value (e.g. Swain, 1993). Moreover, many criticize the minority-majority districts or party-list quotas required to ensure descriptive representation as separatist or racist (Thernstrom and Thernstrom, 1997). This view has faced at least two challenges. First, several political theorists have advocated the importance of the 'politics of presence' in addition to the more common liberal 'politics of ideas'. Phillips (1995: 24–5), for example, argues that although simply having women and racial/ethnic minorities in legislative bodies does not overcome political exclusion, examining the relationship between presence and ideas is an essential step toward fair and effective representation. Consequently, democratic societies may need to take active measures to include more members from under-represented groups. Other studies have focused on representatives and the practices of representation. In the United States, Canon (1999) shows empirically that minority representatives elected from non-White majority districts actually practice representation differently from White representatives. He argues that these legislators use a 'politics of commonality' by actively reaching out to both their White and non-White constituents (Canon, 1999: 48). The form and composition of election districts (as well as campaigns and candidates) profoundly shape the political representation produced by elected officials.

This discussion merely touches on some of the major debates about democratic political representation, but the authors that I have mentioned – Bybee, Morgan, Schwartz, Phillips, and Canon – all share the notion of representation (as opposed to just deliberative or participatory democracy) as a kind of creative activity. This is true in the straightforward sense that different electoral systems may lead to different voting strategies and preferences, but also in a more subtle and profound way. Political representation constitutes citizens and interests, and different kinds of representation can create different forms of citizenship and interests. Moreover, political representation, whether it is inventing 'the people' or creating citizens, is often intimately tied to real and imagined geographies. Conflicts often become explicitly territorial when they involve the representation of ethnic minorities.

Both ethnofederalism and territorial political representation are institutional solutions to minority representation that depend on the creation of intra-state boundaries. Ethnofederal systems accommodate minority interests by creating sub-state jurisdictions (regions, provinces, etc.) dominated by a particular ethnic group and by dividing political sovereignty between a central state and regional jurisdictions (Smith, 1995). Ethnofederalism carries the danger, however, that such institutional divisions will reinforce ethnically defined political conflict, leading to conflict, secession, civil war, and perhaps ethnic cleansing (Bunce, 1999; Hale, 2004).

Territorial representation is a less drastic approach because it does not necessarily mean the creation of (relatively) fixed and permanent boundaries, or a division of sovereignty. The United States and the UK, for example, periodically adjust the boundaries of election districts, and (in the United States) do so to further minority representation. In contrast, South Africa dealt with the problem of minority representation during its transition from apartheid by established (relatively) fixed provincial boundaries used for national elections, but without a strong federal system. Finally, the post-invasion administration in Iraq rejected the use of sub-national electoral boundaries in an attempt to emphasize the territorial integrity of the state and to facilitate a favorable level of minority representation.

Electoral Systems

There are two fundamental aspects to voting within a liberal democracy: who votes (the franchise), and how they vote (the electoral system). Universal or at least extremely broad suffrage is a necessary, but not sufficient, condition for modern democratic political representation. Nonetheless, as an observation frequently attributed to Stalin states, 'Those who cast the votes decide nothing; those who count the votes decide everything'. In more sophisticated political climates, one might paraphrase this to say that how votes are counted – the electoral system – decides everything. While elections can be 'unrepresentative' in many ways, I will focus on the effects of the electoral system. Beyond the practical effects, the choice of electoral systems is important because such systems are themselves representative of a state's identity and goals.

The previous chapter, 'Place and Vote' (Johnston and Pattie, this volume), describes four basic types of electoral systems: indirect, plurality, mixed, and proportional. In the latter three, voters elect their representatives directly, but the nature and quality of political representation can vary dramatically by electoral system. Some systems, for example, favor stability over responsiveness, majority power over minority representation, or coalition governments over majority rule. Even different rules within similar electoral systems create differences along these dimensions.

In a proportional (or 'party-list') system, voters cast ballots for a party, who then receives a proportional number of seats based on its share of the total votes. Such proportional systems typically require that parties win a minimum threshold (often 5 percent) to receive any seats in the representative body, so even proportional systems may not allocate representation in precise proportion to votes. Mixed systems combine plurality and proportional arrangements in some way. Although proportional systems are not immune to territorial manipulation, they tend to be less vulnerable than plurality systems.

In a plurality (or 'winner take all') system, voters typically cast ballots for particular candidates, with the top vote-getter(s) winning the seat(s). Although candidates may belong to parties, parties generally run only one candidate per seat. Plurality election systems using districts can be an effective way to ensure minority representation if districts contain local majorities of ethnic minorities. More generally, an explicitly territorial system of representation is – in principle – an effective way to represent geographically local interests and helps ensure that representatives are responsible (and responsive) to specific constituencies. In an important sense, federal systems use the same principle to balance competing local (or minority) and statewide (or majority) interests. Nonetheless, territorial representation has two distinct disadvantages. First, such systems often fail to produce anything close to proportional representation, and second, they are highly vulnerable to manipulation through gerrymandering (Gelman and King, 1994; Johnston et al., 2001; Monmonier, 2001; Johnston, 2002). Indeed, gerrymandering (the manipulation of election district boundaries to affect the outcome of elections) is an acute problem with district election systems in both the United States and the UK.

Geography is not destiny for minority political representation. The proportion and geographic distribution of racial and national minorities may limit the degree of political representation they enjoy, but the rules of electoral systems construct these constraints.

MINORITY REPRESENTATION: THE UNITED STATES, SOUTH AFRICA, AND IRAQ

The United States, South Africa, and Iraq are all multi-ethnic states, but present different issues of political and minority representation.[6] In the United States, the central problem has been bringing African-Americans (and other ethnic minorities) into a White majority polity. In post-apartheid South Africa, the central challenge has been shifting power to the non-White majority while not alienating the economically powerful White minority. In Iraq, the major problem has been to shift political power to the ethnic majority (Shiite Arabs) while allaying the fears of the two major ethnic minorities (Sunni Arabs and Kurds), maintaining the territorial integrity of the state, and legitimizing the post-invasion government.

African-Americans in the United States: A Minority Enters the Political Community

From the end of the Reconstruction period (1876) to the mid-1960s, southern states essentially excluded African-Americans from political representation. Although the 15th Amendment to the Constitution (1870) guaranteed the right to vote regardless of race, Whites used a variety of techniques to deny Blacks the vote (Kousser, 1974; Keyssar, 2000: 105–116). Following years of a vigorous voting rights movement, and the particularly brutal suppression of a voting rights march in Selma, Alabama, Congress passed the Voting Rights Act in 1965 (Garrow, 1978). The Act is a complex piece of legislation, designed to attack the entrenched, systematic disenfranchisement of African-Americans in southern states. Congress renewed and amended various parts of the law in 1970, 1975, 1982, and 2006, expanding coverage to target discrimination against Hispanics and language minorities, in addition to African-Americans (Davidson and Grofman, 1994). The Supreme Court has also shaped the meaning of the Act significantly through numerous legal decisions since 1965.

Minority Representation Using Election Districts

In many ways, the Voting Rights Act has been an extraordinarily successful piece of legislation. Indeed, Davidson and Grofman (1994) describe its impact on southern politics – particularly at the local level – as a 'quiet revolution' because of its success in raising the level of African-American political participation, and in increasing the number of African-American elected officials. In the southern United States between 1970 and 1985, for example, the percentage of Black state representatives increased from 1.9 to 10.8 percent and Black Congressional representatives increased from 0 to 1.7 percent (Handley and Grofman, 1994: 345).[7] Yet the application of Federal oversight to redistricting became increasingly controversial over the lifetime of the Act. The crux of this controversy involves both the meaning of the Act itself and the meaning of political representation.

The use of the plurality district system in the United States means that once African-Americans gained the effective ability to cast votes, 'minority representation' typically became conceptualized and achieved by creating election districts with non-White majorities or pluralities.[8] Consequently, debates over minority representation in the United States typically involve debates over the proper role of race in determining the boundaries of districts, and hence the composition of political constituencies. These debates are complex, and have involved advocacy groups, state governments, the Federal Department of Justice, Congress, and courts. Federal courts have perhaps been the most important venue for this discussion because it involves significant conflicts among Constitutional rights. Generally speaking, the debate pits a broadly conceived right to political participation (Guinier, 1994) against a more narrowly conceived right to vote (Thernstrom, 1987). The former argues that a democratic political process must include racial and political minorities effectively, and hence minority representation is central to the democratic identity of the United States. The latter emphasizes the Constitutional and moral right to be free from racial classification, and sees formal equality as central to US national identity.

This particular conflict emerged only in the late 1980s (Forest, 2001). Until then, typical voting rights cases involved challenges to districting plans that 'diluted' non-White voting strength by dividing minority communities between several districts ('cracking') or that minimized minority representation by putting an excessive number of minorities into a single district ('packing'). In such cases, the principles of participation and equality were complementary. When states redistricted following the 1990 Census, however, the Department of Justice ignited a decade-long controversy by developing the so-called 'maximization strategy' that encouraged or required states to draw as many non-White majority Congressional districts as possible (Clayton, 2000). This redistricting cycle increased the number of Black and Hispanic majority districts from 29 to 52, and helped increase the number of African-American and Hispanic representatives from 29 to 39 and from 13 to 18 respectively (Parker, 1995: 47).[9]

These new districts had two distinct effects. First, they were effective in electing African-American and Latino candidates to Congress. The 1992 election not only increased the number of minority representatives in Congress, but also sent African-Americans to Congress from states such as North Carolina (22 percent Black) that had not sent a Black representative to Congress since the post-Civil War period. Second, non-White majority districts (especially African-American districts) tend to be heavily Democratic. Concentrating (minority) Democratic voters into a non-White majority district increased the relative proportion of Republicans in the surrounding ones. The interaction of these two effects explains much about what happened with redistricting in the 1990s.

'Bizarre' Districts in the 1990s

Four state legislatures produced 'bizarrely' shaped Congressional districts during the 1990s: North Carolina, Georgia, Louisiana, and Texas. In each case, Democrats, though in decline, controlled the redistricting process in the state legislature, but the Department of Justice pressured them to create the maximum possible number of non-White majority districts. The legislatures drew the required number of non-White districts and preserved Democratic majority districts by creating plans with rather extraordinary boundaries (Lublin, 1997). Aided by new GIS technology, these districts were remarkable feats of political cartography. In North Carolina, for example, the famous 12th District stretched over 250 kilometers (160 miles), yet was only as wide as one lane of an expressway in some places (Monmonier, 2001).

There were both legal and political responses to these districting plans. Lawsuits successfully challenged districting plans in all four states, although North Carolina preserved a version of its original districting plan (Forest, 2004). In a series of cases, the Supreme Court struggled to explain what they found objectionable in the extraordinary districts. In their first decision in the series, *Shaw v. Reno* (1993), the Court ruled that non-White majority districts with 'bizarre boundaries' were constitutionally suspect. In a later decision, *Miller v. Johnson* (1996), the Court focused more on the intent of the state, ruling that race cannot be the 'predominant consideration' in a redistricting

plan. Both decisions reflect the Court's ambivalence about the role of racial identity in political representation. They are uncomfortable with an explicit tie between race and political interests, and with granting special consideration to particular racial/ethnic groups. At the same time, they recognize that the subordinate status of African-Americans and Latinos creates unique political interests, and that as political minorities these groups are vulnerable in a district plurality system of representation.

Although some justices on the Court favored eliminating any consideration of racial identity, the legal standards set in both the *Shaw* and *Miller* cases recognize some role for race in the formation of political constituencies. By restricting either extremely irregular boundaries or excessive emphasis on race, the Court created a compromise that neither did away with non-White majority districts nor allowed for the maximization strategy. In short, racial minorities were entitled to a majority in an electoral district if they were concentrated in a subjectively 'compact' geographic pattern. By using geographic compactness as a limiting factor, the Court avoided any direct decisions about the appropriate or legitimate level of minority representation, and by doing so, recognized racial minorities as geographically bounded regional communities. As a result, political representation for minority groups rests on the degree and scale of their spatial segregation (Forest, 2001).

The political responses to the racial gerrymandering cases of the 1990s were complex because the use of non-White majority districts cut across traditional political alliances. Since the 1960s, African-Americans have voted heavily Democratic, particularly in response to the party's support for civil rights and (relatively) redistributive economic policies (Canon, 1999: 20–34). Yet the political effects of non-White majority districts can benefit Republicans (Brace et al., 1987; Hill, 1995; Lublin and Voss, 2000). Indeed, there is some evidence that Republicans made 'deals' with minority representatives, offering them 'safe' minority districts in exchange for their support of Republican plans. Although more minority Democrats win elections, such districts reduce the overall number of Democrats. Insofar as Democrats support policies favored by African-Americans and Latinos, such districts can *decrease* the substantive representation enjoyed by minorities (Swain, 1993; Cameron et al., 1996; Lublin, 1999). David Lublin (1997) refers to this trade-off as 'the paradox of representation' and suggests that this kind of districting also radicalizes legislatures by creating districts with large majorities of one party or the other. Candidates from such heavily Democratic or Republican districts tend to win by taking extreme positions, thus removing moderates from Congress and state legislatures.

The experience with non-White majority districts shows that the issue of minority political representation in a plurality system is extraordinarily complex. Their use tends to favor descriptive over substantive representation, and their legal justification rests on judicially imagined geographic communities of racial minorities. Whether or not the 'maximization strategy' was a deliberate effort to weaken the Democratic Party, non-White majority districts illustrate the limitations of the American electoral system. Tying political constituencies so closely to particular territories limits both the level and effectiveness of minority representation. Moreover, it ensures that struggles over political representation will always have a territorial basis, and creates enormous political incentives to gerrymander. Consequently, the fears about 'balkanization' or racial separatism implicit in criticisms of non-White majority districts are in fact an inevitable consequence of an electoral system that employs territorial representation using plurality districts.

Political Transition in South Africa: A Majority Enters the Political Community

Like the United States, South Africa had a legacy of racist policies that created social, political, and economic divisions along both racial and geographic lines. The severity of these divisions and the need to transition from minority to majority rule meant that the transition from the apartheid regime in South Africa presented unique challenges for democratic representation. As the majority, Blacks sought majority rule while Whites sought strong protections for minorities under the new democratic system. The transitional state faced the problem of balancing the interests of a powerful minority against the danger of political paralysis and fragmentation. While the transition to democracy has not been perfect, and scholars are not ready to declare it an unqualified success, the country has achieved relative political success and stability (Lodge, 1999; Lemon and Fox, 2000; Simeon and Murray, 2001). In this section, I focus on the first post-apartheid election in 1994, and discuss three strategies of political representation that South Africa used in its transition. The country adopted different electoral strategies at the local and national levels, but has also promoted a civic nationalism of reconciliation and non-racialism as a way to re-imagine and represent the South African state and nation.

In addition to the racial legacies of apartheid, South Africa inherited a problematic electoral tradition. The plurality district system used from 1910 through the end of apartheid produced significant departures from proportionality, even among the

minority of the population eligible to vote, because the National Party used the districting system to consolidate and hold power. Between 1961 and 1981, for example, the party received just over 50 percent of the vote on average, but typically won 75 percent of the seats in Parliament (Faure, 1996: 194). The political distortion of the apartheid-era districting system perhaps helped promote the post-apartheid consensus at the national level for proportional systems, where political power would not be so closely tied to geographic location. On the other hand, the disproportionality of a district (ward) system of representation (drawn using apartheid-era segregation) helped negotiate the transition to majority rule at the local level (Robinson, 1998).

Local Urban Representation

One of the most enduring legacies of apartheid has been residential segregation based on the state's four 'racial' categories, Black/Native/African, White, Coloured, and Indian. Such segregation formed the basis of the regime's control and the ability of the White minority to dominate the majority population (Robinson, 1996, 1997). Despite some movement away from rigid segregation in the late 1980s and early 1990s, South African cities remain highly segregated (Christopher, 2001).

Ironically, apartheid-era segregation helped ease the transition to majority rule by facilitating an electoral system where minorities (Whites) could retain significant political power at the local level. During 1993–4, negotiations between government associations and civic organizations in the Local Government Negotiating Forum (LGNF) produced a mixed district and proportional electoral system for local elections that used segregation as a way to protect the political interests of Whites (Robinson, 1996, 1998: 205–23). This system allocated over half of the seats on councils by district (ward), and half of these districts would have non-Black majorities (White, Coloured, and Indian).

By guaranteeing non-Black control over at least 30 percent of the seats on a council, and having 'supermajority' requirements for many policy areas, this electoral system ensured that non-Blacks had an effective veto (Robinson, 1998: 538).[10] Similarly, local electoral systems in KwaZulu-Natal used mixed districting and proportional systems to take advantage of both racial patterns and urban/rural divisions to promote the interests of the regionally powerful Inkatha Freedom Party (IFP), although these manipulations were based more on inter-party rivalries than on inter-racial ones (Munro, 2001).

For local urban representation, the transitional negotiations used the spatial legacies of apartheid to ensure minority power through territorial representation. In some ways this resembled the US effort to increase non-White representation, but with greater flexibility in both the electoral system itself, and in the rules for legislative decision-making. In contrast, the national election system adopted in 1994 eschews the use of geographic segregation for this sort of explicit racial balancing.

National Representation

The transitional state used federalism and an election system with provincial boundaries to ensure both representativeness and a degree of minority power for the 1994 national election. The negotiations that produced these compromises were complicated, with crosscutting interests among and within the parties involved, but the central disputes involved the balance of power between the central government and provinces, and the role of race/ethnicity in determining provincial boundaries. As with many aspects of the transition, the new arrangement was a compromise that neither replicated nor fully erased the social and geographic divisions created by apartheid.

Although the majority African National Congress (ANC) preferred a strong central state, South Africa settled on a weak quasi-federal system under which the central government shared some power with provincial legislatures (Simeon and Murray, 2001). Indeed, the *idea* of ethnofederalism was critical to the success of the transitional negotiations by keeping both Afrikaner and Zulu nationalists within the negotiating process and ensuring their participation in the 1994 election (Robinson, 1995; Steytler and Mettler, 2001). (Parties representing each group – particularly Zulus in KwaZulu-Natal – had relative regional concentrations that would give them more power relative to the nationally dominant ANC.) Moreover, the interim constitution included a principle recognizing the political status of any 'territorial entity' that is a 'community sharing a common or cultural heritage', offering at least the possibility that both groups could achieve self-governing provinces (Steytler and Mettler, 2001: 97). Although the final 1996 constitution significantly weakened this principle of ethno-territorial self-determination and enshrined a relatively powerful central unitary state, the principle of ethnofederalism was critical for keeping ethnic minorities involved in the transformation process.

The mixed proportional system used to elect the South African National Assembly has been a more enduring legacy of the transitional negotiations. During the process, the National Party conceded that the districting system was not appropriate for a highly plural, deeply divided society like South Africa. As the regime negotiated the transition, a consensus emerged that proportional representation would be the most legitimate system. A number of observers argue, however, that

proportional representation was also a way for the outgoing minority government to moderate the force of African nationalism (Faure, 1996: 194–5). Consequently, the selection of a mixed proportional representation system, along with a very low threshold for election (about 0.5 percent), was a strategy to provide an acceptable level of minority representation.

The national electoral system adopted in 1994 uses both national and provincial lists, each electing half of the National Assembly. For the National Assembly, each voter casts a ballot for one party, using a 'closed list' system where parties fix the order of candidates on their list. Based on those results, 200 members are selected based on parties' national showing, and 200 are selected based on parties' showing in each province. There is a separate ballot for each provincial legislature, and each of these nine legislatures appoints 10 members to the Nation Senate (90 members) based on each party's share of the provincial seats (Faure, 1996: 197–8).

The definition of provinces included a mix of old and new boundaries. They were not drawn explicitly to empower particular racial, ethnic, or national groups and specifically avoided using the borders of apartheid-defined 'homelands' (Robinson, 1996: 258; Simeon and Murray, 2001: 70–1). Nonetheless, racial and political interests clearly played a role in their demarcation (Muthien and Khosa, 1995). As expected, the Inkatha Freedom Party and the Afrikaner National Party won (or 'negotiated') local majorities in KwaZulu-Natal (50.3 percent) and Western Cape (52.2 percent) respectively in the 1994 election. By 2004, however, the ANC won pluralities in these two provinces, and governs as the major partner in each (Alence, 2004: 83–4). Although the national electoral system did not use the explicit racial criteria of the local ones, it made use of the fact that different racial groups identify to some degree with different parties, and parties vary in their support by region (Johnston, 1994).

Proportional representation typically provides a high degree of representativeness and facilitates the formation of smaller parties, but also carries the risk of political and geographic fragmentation (Faure, 1996: 202). In principle, minorities (as represented by small parties) can wield considerable power if larger parties need them to form coalition governments. South Africa has not experienced significant political fragmentation – or significant party competition – because the ANC has achieved such electoral dominance. Indeed, its electoral success in KwaZulu-Natal and Western Cape suggests that the country's electoral system has not fostered significant minority electoral representation.

Despite the relative lack of party competition, and the associated lack of minority representation in the National Assembly, the ANC majority in

South Africa has not abused the country's political and racial minorities. Arguably, South Africa has been relatively successful in cultivating a democratic tradition and minority protection through a civic nationalism focused on the ideas of reconciliation and non-racialism.

South Africa's Civic Nationalism

If nationalist appeals by charismatic leaders have played an important role in ethnic cleansing and the disintegration of multi-ethnic federations (Kaufman, 2001), then South Africa represents the opposite case. The ideology of non-racialism promoted by the ANC and skillfully articulated by Nelson Mandela has helped create a new way for South Africa to represent itself (Coombes, 2003). As Adam (1995: 470) argues, 'the first democratic election provided a rare occasion for a divided state to reinvent itself as a nation in an idealized fashion'. The kind of civic nationalism embodied in non-racialism provides a way to represent South Africa as both a nation and a state that rejects the use of race, while the Truth and Reconciliation Commission (TRC) complemented this process by ending the cycle of apartheid-era racial violence (Glaser, 1997; Leebaw, 2003).

South Africans did not embrace non-racialism simply for altruistic reasons. It provides a way for Whites in general to conveniently 'forget' the advantage they inherited from the previous regime's policies, and it permitted the National Party to reinvent itself as an 'ethnic' rather than a racial party (Adam, 1995). The term 'non-racialism' is itself contested, and formalistic interpretations can be used to block efforts to address economic inequality (Sharp, 1998). Nonetheless, after the experience of apartheid, non-racialism is a powerful organizing principle for South African identity that has enabled the ANC to represent the nation as unified and democratic.

The ANC has increased its electoral dominance since 1994, no party has emerged as an especially effective opposition, and the evidence for further democratic consolidation is mixed (Mattes, 2002). Moreover, the emerging South African national identity may create new forms of inequality and exclusion (Croucher, 1998). Nonetheless, South Africa has so far avoided most of the pitfalls of one-party rule, and has remained a reasonably successful democracy (Lodge, 2004). While it is impossible to say if South Africa's civic nationalism has made a critical difference, it does suggest that the representation of the nation can have an important effect on its system of political representation. In the final section, I turn to a discussion of Iraq, where the United States and interim Iraqi government explicitly linked the system of political representation to representations of the nation.

Iraq: Stability, Legitimacy, and Minority Representation?

In 2002 and 2003, the United States offered several justifications for the 2003 invasion of Iraq, but once inspections failed to reveal any weapons of mass destruction, the Bush administration began to emphasize 'democracy' as the most important reason. In these circumstances, the legitimacy of the invasion, as well as the legitimacy of the post-war Iraqi state, rested on the creation of a representative system of government. Yet the country was violent, insecure, and divided by serious political, religious, and ethnic conflicts (Diamond, 2004). The Bush administration, the UN, and the interim Iraqi government undertook an ongoing series of steps to transfer sovereignty from the post-invasion Coalition Provisional Authority (CPA) to an elected, constitutional government in 2006. In this section, I focus my discussion on the period between June 2004 and the national election in January 2005. During this period, the Iraqi Interim Government (a body appointed through negotiations among the CPA, the UN, and the US-appointed Governing Council) formally governed the country. Iraq remained under US-led military occupation throughout this period.

The population of Iraq contains three major ethnic groups: Shiite Arabs (approximately 60 percent), Sunni Arabs (approximately 20 percent), and Kurds (approximately 20 percent). Sunnis, concentrated in the western regions of Iraq, were the dominant group under Saddam Hussein and enjoyed special privileges and powers. Kurds were particular targets of repression and violence under this regime, and have operated a nearly autonomous government in northeast Iraq since the end of the 1991 Gulf War. A unitary, majoritarian democratic government would thus have shifted power to Shiites away from both of these groups. In order to encourage participation from both Sunnis and Kurds, the party-list electoral system adopted for the 2005 election sought to maximize minority representation.

The 'Iraqi People' and the Iraqi National Assembly

The relatively broad participation in the January 28, 2005, UN-sponsored election surprised many critics and created a degree of legitimacy for the new Iraqi government, but Iraqi democracy is a work in progress. Representatives from that election took over two months to form a government, and nearly five months to form the committee to write a new constitution (Associated Press, 2005; Tavernise, 2005). Whatever the long-term outcome of the invasion, the 2005 electoral system sought to provide majority rule, legitimacy for the

Iraqi government as a whole, and strong protection of minority rights.

The circumstances of the 2005 election mean that there was an explicit link between the form of the electoral system and the imagined 'Iraqi people'. The Bush administration explicitly tied the election to its justification of democracy as a weapon against terrorism. In a national speech on June 28, 2005, for example, President Bush used typical post-invasion rhetoric: the purpose of the invasion was to give (or restore) 'sovereignty' to the 'diverse' Iraqi people through a 'representative' election (Bush, 2005). Similarly, a report on the election published by the US Department of State uses the term 'the Iraqi people', 'the people', or 'this people' eleven times in its 523-word preface (United States Department of State, 2005). In addition to the obvious task of selecting representatives for government, the January electoral system promoted a particular image of an Iraqi nation, ethnically diverse but sovereign and united.

The system chosen by the UN for the January 2005 election selected 275 representatives using straight party-list proportional representation, without any subnational boundaries or districts. Along with a 33 percent quota for women on party lists, the UN picked proportional representation to ensure the fair representation of Iraq's many ethnic and political groups in the National Assembly. Yet this was not their first choice (Diamond, 2005: 267–9). Although they had rejected districts because there were no reliable census data (and any district configurations would have been politically suspect), the UN team's first choice had been a system that allocated representatives using the existing provincial boundaries. This system assigned seats to each province according to its share of the national population (calculated using the most recent census and the food ration card system). The proportion of votes received by parties within each province would determine the election of candidates, although an adjustment mechanism would ensure that the party allocation in the National Assembly would not be too far out of line with the national proportion of votes. Ultimately, however, the UN rejected the use of provincial districts because they thought that assigning a fixed number of seats to provinces would be too controversial without highly reliable population data.

The 2005 election treated the entire country as a single electoral district. By doing so, the UN avoided any possibility of gerrymandering and kept election administration as simple as possible. Yet this choice also profoundly shaped the results of the election. Ironically – given the political importance attached to inclusion of the Sunni Arab minority – the system used may have limited their representation in the assembly compared with alternative systems. In particular, the national party-list system meant that Sunni Arabs derived no electoral advantage from their regional concentration.

The national system tried to compensate for such problems by keeping the threshold of exclusion remarkably low: a party could place a candidate into the assembly with as little as 0.36 percent (1/275) of the vote. (This system also provided some protection for individual candidates since they did not need to personally campaign and risk assassination.) Sunni Arabs largely boycotted the election while Shiites and especially Kurds turned out heavily. Consequently, the Sunnis were significantly under-represented in the national assembly; the main Sunni Arab party (The Iraqis) won only 1.8 percent of the vote, receiving 5 seats (Center for Voting and Democracy, 2005).

While alternative systems faced significant administrative challenges, critics of the national party-list system anticipated this population/representation imbalance (Diamond, 2004). The very low rate of Sunni Arab electoral participation would have made it difficult for them to achieve significant representation under any system, but statements by Sunni leaders suggest that the national proportional system itself discouraged Sunni participation (Diamond, 2005: 323–4).

Alternative Election Systems in Iraq

An analysis by the Center for Voting and Democracy (CVD) using the 2005 vote totals illustrates the effects that different electoral systems might have had on the election (Center for Voting and Democracy, 2005).[11] The analysis allocates candidates using two alternative, district-based electoral arrangements: a regional party-list system (similar to the UN's initial choice) and a plurality, 'winner take all' system. In each hypothetical arrangement, the 18 existing provinces are districts, and population from the 1991 census determines the number of representatives from each district. Table 22.1 shows the results for the five parties earning more than 1 percent of the vote, and for parties that would have won seats under alternative systems.

The regional party-list system totals votes within each province. For example, since the Sunni party won 28 percent of the vote in Ninewa, they would receive 28 percent of the representatives (based on population) from that province. Under the plurality 'winner take all' system, the party receiving the largest vote total in a province would receive all of that district's representatives. Under that system, the Democratic Patriotic Alliance of Kurdistan would have bested the Sunni party 38 percent to 28 percent in Ninewa.

According the CVD's analysis, the regional system would have produced moderately different results, and would have reduced the total number of parties from 12 to 8. Notably the system

Table 22.1 Seats in the Iraqi National Assembly under alternative electoral systems.

Party (Ethnic group)	National vote (%)	Seats under the national party-list system (actual)	Seats under a regional party-list system (hypothetical)	Seats under a plurality, 'winner take all' system (hypothetical)
United Iraqi Alliance (Shiite, religious)	48.19	140	145	192
Democratic Patriotic Alliance of Kurdistan (Kurd)	25.73	75	62	69
Iraqi List (Shiite, secular)	13.82	40	48	14
The Iraqis (Sunni)	1.78	5	12	0
Iraqi Turkmen Front	1.11	3	2	0
Reconciliation and Liberation Bloc	0.36	0	2	0
Iraqi Islamic Party	0.25	0	2	0
Iraqi National Gathering	0.22	0	2	0
7 other parties actually earning seats	8.54	12	0	0
Total seats	100	275	275	275
Total number of parties in the assembly	–	12	8	3

Source: CVD (2005) http://www.fairvote.org/pr/global/iraqanalysis.htm.

would have favored smaller regional parties (like the Iraqi National Gathering) at the expense of smaller parties with a more uniform national distribution (like the People's Union, which won two seats under the national list system). More significantly, it would have more than doubled Sunni Arab representation while reducing Kurdish strength in the assembly. Although the United Iraqi Alliance (UIA) would have increased its dominance, the party would not have achieved the two-thirds majority needed to control major offices of the government. The intense political bargaining among parties after the election suggests, however, that even such relatively small changes might have had important consequences.

In contrast, the plurality 'winner take all' system would have sharply reduced the number of parties (from 12 to 3), and would have given the UIA a two-thirds majority by itself. Kurds would have received relatively fewer seats, and Sunni Arabs would not have received any representatives. The concentration of power with the majority Shiites and the reduction in minority representation would likely have been a centrifugal force in an already unstable state.

The January elections in Iraq sought to legitimize the invasion by creating a single sovereign, yet diverse, 'Iraqi people'. The choice of a national list system sought to dampen ethnic tensions and to discourage Iraq from splitting along regional lines. Ironically, however, the national election system may have exacerbated the under-representation of Sunni Arabs, thereby threatening both the territorial integrity of Iraq and the legitimacy of the post-invasion government.

CONCLUSION

This brief overview of political representation and territorial politics illustrates some of the paradoxes and dilemmas of democracy. In some cases, as in the United States and the UK, a variety of political interests and legal interpretations shaped electoral systems over a long period. Alternatively, transitional governments in South Africa and Iraq implemented electoral systems quickly under rather extraordinary circumstances. In all cases, however, the need to balance majority rule and minority representation has generated explicitly territorial politics.

Political representation is a fundamentally creative act because the choice of electoral system forms the constituency that will be represented. Even in the most calculating circumstances, the choice of an electoral system and the form of political representation are critically shaped by the imagined form of the state, the nation, and minority groups. When the state is considered to be an ethnically exclusive territory, when part of the nation lies outside state boundaries, or when political systems like federalism facilitate national separatism, the problem of minority representation is 'solved' by removing people or by creating new boundaries. Alternatively, properly designed ethnofederal systems or territorial representation may help successfully balance majority rule with minority representation. In contrast, states may turn to electoral systems that de-emphasize their ethnically defined geographic divisions to maintain their territorial integrity. In such cases, representing the state as a geographically whole community shapes the form of political representation. As the experiences in the United States, South Africa, and Iraq all show, the symbolic representation of racial or ethnic minorities as regional communities can both enable and constrain their political representation.

Modern political representation always involves territorial politics because political, ethnic, and racial minorities overflow state boundaries (both internal and international), and because voters and political interests vary over space. As I have argued in this chapter, however, the state actors and institutions that manipulate systems of political representation are the driving force behind the territorial politics of representation. The study of political representation consequently must involve not only how 'the people' vote, but also how 'the people' are imagined and how their votes are transformed (or not) into political power.

NOTES

1 The notable exceptions are hard-core authoritarian regimes – Cuba, Libya, Myanmar (Burma), North Korea, Saudi Arabia, Syria, and Turkmenistan – although even such repressive states use elections to bolster their political credentials and legitimacy.
2 I have left out some important aspects of political representation, including the role of campaigns, candidate selection, and non-electoral forms of political representation (lobbying, campaign donations, etc.), as these issues tend to be of less concern to geographers.
3 It is also important to note that several large democracies and some political parties reserve a proportion of seats for women or minority candidates (Tinker, 2004). In places where this is not done, men are generally over-represented in elected bodies (Kenworthy and Malami, 1999).
4 Schwartz offers little empirical evidence for her claim that politics in at-large election or proportional systems are too abstract to produce these benefits, although her study helps clarify the relationships among political identity, citizenship, and representation.
5 The need for local majorities of ethnic/racial minorities in district systems assumes that these groups actually vote differently from the majority group.
6 'Race' is the primary social and political fault line I will discuss in the United States and South Africa, whereas religious and ethnic-national differences define divisions in Iraq. These divisions and categories have developed differently in each state, and this presents a problem of terminology. The 'racial' divisions in South Africa (Black/Native/African, White, Coloured, and Indian), for example, carry the taint of apartheid, so that domestic debates

characterize political divisions in terms of 'ethnicity' even if conflicts fall along the old 'racial' lines. Intra-'racial' conflicts in South Africa also tend to be described as 'ethnic' clashes. Likewise, in the United States, 'race' typically refers to four historically contingent categories (White, Black/African-American, Asian, and Native American). 'Ethnicity' refers to Latino/Hispanic identity, as well as inter- and intra-'racial' divisions. All of these terms represent socially constructed divisions used to define political interests and political communities. For the sake of simplicity, I will use 'ethnic' and 'ethnicity' as a generic term to refer to such divisions. Other dimensions (such as gender and class) also simultaneously define political interests.
7 The African-American population in these states was about 20 percent during this period.
8 Empirical work by political scientists has shown that having a non-White majority may not be the most effective way to achieve minority political power and representation (Cameron et al., 1996; Grofman and Reynolds, 1996; Lublin, 1997, 1999). Although a 2004 Supreme Court decision allowed states to create districts with non-White pluralities, this was reversed by Congress in 2006.
9 After the 1992 election, African-American and Hispanic representatives constituted 9.0 percent and 4.1 percent of Congress, compared with 12.1 percent and 9.0 percent of the national population.
10 Ward representation did not necessarily give a race-specific advantage to Whites. In Cape Town, Blacks obtained a relative advantage to the Coloured population. Blacks are in the minority, but because the Coloured population shares the '50 percent' districts with Whites and Indians, Blacks gained a relatively larger share of political representation (Robinson, 1998: 538).
11 The CVD analysis assumes that the pattern of voting would have remained the same regardless of the electoral system used. Participation and voting may have been different, however, under different systems.

REFERENCES

Adam, H. (1995) 'The politics of ethnic-identity: comparing South Africa', *Ethnic and Racial Studies*, 18: 457–75.
Alence, R. (2004) 'South Africa after apartheid: the first decade', *Journal of Democracy*, 15: 78–92.
Associated Press (2005) 'Sunni Arab is elected Iraqi parliament speaker', *Associated Press Wire Service*, April 3.
Brace, K., Grofman, B. and Handley, L. (1987) 'Does redistricting aimed to help Blacks necessarily help Republicans?', *Journal of Politics*, 49: 169–85.
Bunce, V. (1999) 'Peaceful versus violent state dismemberment: a comparison of the Soviet Union, Yugoslavia, and Czechoslovakia', *Politics and Society*, 27: 217–37.

Bush, G.W. (2005) 'President addresses nation, discusses Iraq, war on terror' [online]. Available from http://www.Whitehouse.gov/news/releases/2005/06/20050628-7.html [accessed July 7, 2005].

Bybee, K.J. (1998) *Mistaken Identity: The Supreme Court and the Politics of Minority Representation*. Princeton, NJ: Princeton University Press.

Cameron, C., Epstein, D. and Ohalloran, S. (1996) 'Do majority-minority districts maximize substantive black representation in congress?', *American Political Science Review*, 90: 794–812.

Canon, D.T. (1999) *Race, Redistricting, and Representation: The Unintended Consequences of Black Majority Districts*. Chicago: University of Chicago Press.

Center for Voting and Democracy (CVD) (2005) 'Exporting democracy: how can we achieve fair representation in Iraq', Center for Voting and Democracy [online]. Available from http://www.fairvote.org/pr/global/iraqanalysis.htm [accessed March 30, 2005].

Christopher, A.J. (2001) 'Urban segregation in post-apartheid South Africa', *Urban Studies*, 38: 449–66.

Clayton, D.M. (2000) *African Americans and the Politics of Congressional Redistricting*. New York: Garland.

Coombes, A.E. (2003) *History After Apartheid: Visual Culture and Public Memory in a Democratic South Africa*. London and Durham, NC: Duke University Press.

Croucher, S. (1998) 'South Africa's illegal aliens: constructing national boundaries in a post-apartheid state', *Ethnic and Racial Studies*, 21: 639–60.

Davidson, C. and Grofman, B. (eds) (1994) *Quiet Revolution in the South: The Impact of the Voting Rights Act, 1965–1990*. Princeton, NJ: Princeton University Press.

Diamond, L. (2004) 'What went wrong in Iraq', *Foreign Affairs*, 83: 34–56.

Diamond, L. (2005) *Squandered Victory: The American Occupation and Bungled Effort to Bring Democracy to Iraq*. New York: Henry Holt.

Faure, M. (1996) 'The electoral system', *Journal of Theoretical Politics*, 8: 193–208.

Forest, B. (2001) 'Mapping democracy: racial identity and the quandary of political representation', *Annals of the Association of American Geographers*, 91: 143–66.

Forest, B. (2004) 'The legal (de)construction of geography: race and political community in Supreme Court redistricting decisions', *Social and Cultural Geography*, 5: 55–73.

Fukuyama, F. (1992) *The End of History and the Last Man*. New York: Free Press.

Garrow, D.J. (1978) *Protest at Selma: Martin Luther King, Jr., and the Voting Rights Act of 1965*. New Haven, CT: Yale University Press.

Gelman, A. and King, G. (1994) 'A unified method of evaluating electoral systems and redistricting plans', *American Journal of Political Science*, 38: 514–54.

Glaser, D. (1997) 'South Africa and the limits of civil society', *Journal of Southern African Studies*, 23: 5–25.

Grofman, B. and Reynolds, A. (1996) 'Modeling the dropoff between minority population share and the size of the minority electorate in situations of differential voter eligibility across groups', *Electoral Studies*, 15: 255–61.

Guinier, L. (1994) *The Tyranny of the Majority: Fundamental Fairness in Representative Democracy*. New York: Free Press.

Hale, H.E. (2004) 'Divided we stand: institutional sources of ethnofederal state survival and collapse', *World Politics*, 56: 165–93.

Handley, L. and Grofman, B. (1994) 'The impact of the Voting Rights Act on minority representation: Black officeholding in southern state legislatures and Congressional delegations', in C. Davidson and B. Grofman (eds), *Quiet Revolution in the South: The Impact of the Voting Rights Act, 1965–1990*. Princeton, NJ: Princeton University Press, pp. 335–50.

Hill, K.A. (1995) 'Does the creation of majority Black districts aid Republicans? An analysis of the 1992 Congressional elections in 8 southern states', *Journal of Politics*, 57: 384–401.

Huntington, S.P. (1991) *The Third Wave: Democratization in the Late Twentieth Century*. Norman: University of Oklahoma Press.

Jessop, B. (1983) 'Capitalism and democracy: the best possible political shell?', in D. Held (ed.), *States and Societies*. New York: New York University Press, pp. 272–89.

Johnston, A. (1994) 'South Africa: the election and the emerging party system', *International Affairs*, 70: 721–36.

Johnston, R. (2002) 'Manipulating maps and winning elections: measuring the impact of malapportionment and gerrymandering', *Political Geography*, 21: 1–31.

Johnston, R., Pattie, C., Dorling, D.F.L. and Rossiter, D. (2001) *From Votes to Seats: The Operation of the UK Electoral System Since 1945*. Manchester: Manchester University Press.

Kaufman, S.J. (2001) *Modern Hatreds: The Symbolic Politics of Ethnic War*. Ithaca, NY and London: Cornell University Press.

Kenworthy, L. and Malami, M. (1999) 'Gender inequality in political representation: a worldwide comparative analysis', *Social Forces*, 78: 235–68.

Keyssar, A. (2000) *The Right to Vote: The Contested History of Democracy in the United States*. New York: Basic Books.

Kousser, J.M. (1974) *The Shaping of Southern Politics: Suffrage Restriction and the Establishment of the One-Party South, 1880–1910*. New Haven, CT: Yale University Press.

Leebaw, B. (2003) 'Legitimation or judgment? South Africa's restorative approach to transitional justice', *Polity*, 36: 23–51.

Lemon, A. and Fox, R. (2000) 'Consolidating democracy in South Africa: the second open election', *Area*, 32: 337–44.

Lijphart, A. (1999) *Patterns of Democracy: Government Forms and Performance in Thirty-Six Countries*. New Haven, CT: Yale University Press.

Lijphart, A. and Grofman, B. (eds) (1984) *Choosing an Electoral System: Issues and Alternatives*. New York: Praeger.

Lindblom, C.E. (1977) *Politics and Markets: The World's Political Economic Systems*. New York: Basic Books.

Lodge, T. (1999) *Consolidating Democracy: South Africa's Second Popular Election*. Johannesburg: Witwatersrand University Press.

Lodge, T. (2004) 'The ANC and the development of party politics in modern South Africa', *Journal of Modern African Studies*, 42: 189–219.

Lublin, D. (1997) *The Paradox of Representation: Racial Gerrymandering and Minority Interests in Congress*. Princeton, NJ: Princeton University Press.

Lublin, D. (1999) 'Racial redistricting and African-American representation: a critique of "Do majority-minority districts maximize substantive black representation in Congress?"', *American Political Science Review*, 93: 183–6.

Lublin, D. and Voss, D.S. (2000) 'Racial redistricting and realignment in southern state legislatures', *American Journal of Political Science*, 44: 792–810.

Marston, S.A. (1990) 'Who are the people: gender, citizenship, and the making of the American nation', *Environment and Planning D: Society and Space*, 8: 449–58.

Mattes, R. (2002) 'South Africa: democracy without the people?', *Journal of Democracy*, 13: 22–36.

McFaul, M. (2002) 'The fourth wave of democracy and dictatorship: noncooperative transitions in the postcommunist world', *World Politics*, 54: 212–44.

Monmonier, M.S. (2001) *Bushmanders and Bullwinkles: How Politicians Manipulate Electronic Maps and Census Data to Win Elections*. Chicago: University of Chicago Press.

Morgan, E.S. (1988) *Inventing the People: The Rise of Popular Sovereignty in England and America*. New York: W.W. Norton.

Munro, W.A. (2001) 'The political consequences of local electoral systems: democratic change and the politics of differential citizenship in South Africa', *Comparative Politics*, 33: 295–313.

Muthien, Y.G. and Khosa, M.H. (1995) 'The Kingdom, the Volkstaat and the new South Africa: drawing South Africa new regional boundaries', *Journal of Southern African Studies*, 21: 303–22.

Parker, F.R. (1995) '*Shaw v. Reno*: a constitutional setback for minority representation', *Political Science and Politics*, 28: 47–50.

Phillips, A. (1995) *The Politics of Presence*. Oxford: Clarendon Press.

Pitkin, H. (1967) *The Concept of Representation*. Berkeley and Los Angeles: University of California Press.

Ramirez, F.O., Soysal, Y. and Shanahan, S. (1997) 'The changing logic of political citizenship: cross-national acquisition of women's suffrage rights, 1890 to 1990', *American Sociological Review*, 62: 735–45.

Robinson, J. (1995) 'Federalism and the transformation of the South African state', in G.E. Smith (ed.), *Federalism: The Mutliethnic Challenge*. London: Longman, pp. 255–78.

Robinson, J. (1996) *The Power of Apartheid: State, Power and Space in South African Cities*. Oxford: Butterworth-Heinemann.

Robinson, J. (1997) 'The geopolitics of South African cities: states, citizens, territory', *Political Geography*, 16: 365–86.

Robinson, J. (1998) 'Spaces of democracy: remapping the apartheid city', *Environment and Planning D: Society and Space*, 16: 533–48.

Schwartz, N.L. (1988) *The Blue Guitar: Political Representation and Community*. Chicago: University of Chicago Press.

Sharp, J.P. (1998) ' "Non-racialism" and its discontents: a post-apartheid paradox', *International Social Science Journal*, 50: 243–52.

Simeon, R. and Murray, C. (2001) 'Multi-sphere governance in South Africa: an interim assessment', *Publius: The Journal of Federalism*, 31: 65–92.

Smith, G.E. (ed.) (1995) *Federalism: The Multiethnic Challenge*. London and New York: Longman.

Steytler, N. and Mettler, J. (2001) 'Federal arrangements as a peacemaking device during South Africa's transition to democracy', *Publius: The Journal of Federalism*, 31: 93–106.

Swain, C.M. (1993) *Black Faces, Black Interests: The Representation of African Americans in Congress*. Cambridge, MA: Harvard University Press.

Tavernise, S. (2005) 'Sunnis to accept offer of a role in constitution', *New York Times*, June 17.

Thernstrom, A. (1987) *Whose Votes Count? Affirmative Action and Minority Voting Rights*. Cambridge, MA: Harvard University Press.

Thernstrom, S. and Thernstrom, A. (1997) *America in Black and White: One Nation, Indivisible*. New York: Simon & Schuster.

Tinker, I. (2004) 'Quotas for women in elected legislatures: do they really empower women?', *Women's Studies International Forum*, 27: 531–46.

United States Department of State (2005) *Iraq Elections: Road to Democracy*. Washington, DC: Bureau of International Information Programs, United States Department of State.

23

Democracy and Democratization

Lynn A. Staeheli

INTRODUCTION

In December 2004, Freedom House released its annual survey of democracy in nations around the world. For this survey, Freedom House collects a wealth of information related to civil liberties and political rights associated with democracy, and combines them into summary evaluations of 'free', 'partially free', and 'not free'. The results were in, and Russia was declared to be no longer free. News reports focused on the retrenchment of democratic institutions in Russia, and commentators pondered what this meant for democracy around the world.

In January 2005, and then again in October and December of that year, Iraqis went to the polls to elect the people who would write the country's new constitution and serve as its initial leaders. News reports featured images of long lines of Iraqis lining up to vote; often featured were the long lines of women voting. In Washington, DC, the Bush administration proclaimed each election as a victory for the Iraqi people and for democracy in the Middle East.

Both of these examples are suggestive of the two predominant strands of the discussion about democracy and democratization in the world today. The first strand of the discussion invokes a metaphor of 'waves' in describing the spread of democracy. Since Huntington's *The Third Wave* (1991), the notion of waves of democracy spreading out and reaching more and more countries – but with some countries also slipping back out to sea as the tide goes out – became the dominant way of describing the diffusion of democracy. In this description, the spread of democracy seems

somehow a natural phenomenon, taking on the inevitability of the tides. The second strand of the debate reflects a thing-like quality to democracy in which it is assumed that it is possible to make objective, verifiable categorizations about democracy in a given country and around the world. Democracy, in this perspective, is an outcome, the product perhaps of a long and contentious process, but definitely a state of being. And not just a state of being, but a nation-state of being. If a country is democratic, the entire country is implied to be democratic. There might be some deficits in local governments, but the nation-state appears the appropriate unit to evaluate democracy.

Consider other examples, however, that might challenge those assessments of democracy and democratization. In 1992, voters in Colorado, United States, voted in free and fair elections for an amendment to the state constitution that would deny civil liberties to gays, lesbians, and transgendered individuals. The US Supreme Court later nullified the amendment, blocking its implementation, but the amendment remains part of Colorado's constitution.

In December 2004, the US Bureau of Immigration and Customs Enforcement (BICE) delayed several hundred Muslim Americans returning from a religious conference in Toronto. The Bureau justified holding the returnees, their fingerprinting, and questioning on the grounds that the conference might have stirred Islamist extremism in the attendees (CAIR, 2004). And since 2002, the US government has held without charge hundreds of 'enemy combatants' in an extraterritorial prison. The national security apparatus maintains the prisoners are beyond either US or international justice systems, but federal courts challenge that status. All of this occurs in the country that Freedom House consistently ranks as one of the most free in the world, and that in many ways (and rightly or wrongly) stands as the comparison for all other countries.

These last examples are indicative of a tension between ideals of democracy and justice that, in turn, suggest several qualifications in the ways we might approach the topics of democracy and democratization. First, it is important to differentiate between procedural and substantive approaches to democracy; in the former, the establishment of democratic rules and procedures takes priority, while in the latter, the emphasis may be focused on whether the outcomes of democratic procedures are seen as just or as providing equal opportunities for free citizens. Second, the diffusion of procedural democracy around the world is uneven and constantly shifting. This is not so much a matter of waves affecting nation-states, however, as it is a reflection of the complex spatiality of democratization such that democratic practices and values are unevenly developed within states, reflecting various scales and networks of power and the institutions and practices of governance. Third, the idea that democracy is best evaluated in terms of nation-states is challenged by the recognition of networks of power that extend beyond state territory, even without necessarily crossing scales; the porosity of borders and of territoriality seems to be increasing and in ways unrelated to the development of international governance structures. And finally, since democratization is never complete, we should consider democracy as a process, rather than as a final state.

These ideas are explored in more detail in this chapter. The first section provides a definition and conceptualization of democracy as a mutually constitutive relationship between formal institutions and societies. The second section highlights the spatiality of democratization that keeps democracy in a state of flux. The final section considers the limits to democratization and its uneven development within and between states.

CONCEPTUALIZING DEMOCRACY AND DEMOCRATIZATION

In its simplest form, democracy can be conceptualized as rule by the people. We typically think of this as a relationship between a state and a polity in which sovereignty resides in the people; democratization, then, is the process of establishing and consolidating the people's rule. For there to be rule by the people, however, institutions and procedures must be in place whereby citizens can

rule the state. To the extent those procedures and institutions do not already exist, democratization requires that they be built. But if they are to be built by democratic means (which they rarely are), there must be a set of conditions in place whereby 'the people' can be transformed from a collection of individuals into a political community in which deliberation as to how to build democracy can be conducted (Baechler, 1993; Habermas, 1998). Besides implying a set of normative goals, this also implies a political community that is inclusive, rather than a community limited to a subset of 'the people'. Thus democracy and democratization rely on relationships between procedures and the polity that are difficult to conceptualize independently. This section of the chapter discusses the procedural and social, or communal, aspects of democracy and some of the ways they are interrelated in the ongoing process of democratization.

Institutions and Procedures of Democracy

In 1991, Huntington wrote that it was relatively easy to identify democracies by the procedures that are followed in electing governments, protecting civil liberties, and making decisions. 'The procedural approach to democracy', he wrote, 'accords with the commonsense use of the term ... [T]he classification of regimes in terms of their degree of procedural democracy remains a relatively simple task' (1991: 8–9). While Huntington thought defining procedural democracy was straightforward, others have found it more difficult. The minimalist definition of procedural democracy typically involves regular, free and fair elections, universal (or near-universal) suffrage, protections of civil liberties (especially of speech and assembly), and mechanisms to ensure the accountability of the state to the people (Beetham, 1994; see Bell and Staeheli, 2001; Whitehead, 2002; and O'Loughlin, 2004, for reviews). In this view, democratization involves the development of procedures and institutions to cement those elements of democracy in the state and the polity to ensure rule by the people and liberty for all citizens.

Procedural definitions of democracy are very common, and are the basis for many studies of the diffusion of democracy. Part of the reason for this is that procedures *are* important for ensuring the liberty of citizens and the protection of their freedom from a potentially oppressive state. Another reason for the predominance of procedural approaches is that they are more easily measured, quantified, and made to seem real for policy-makers, for scholars interested in the broad sweep of democratization, and for a public that values the spread of democracy around the world (Bell and Staeheli, 2001). Under these circumstances, theoretical niceties often give way to empirical practicalities. Broad-scale, longitudinal data sets such as the Freedom House Survey and the Polity III data allow researchers and policy analysts to track the advancement and contraction of democracy – operationalized as governments characterized by democratic institutions and procedures – over time and across nations. Accordingly, a rich literature has developed auditing the diffusion of democracy and the implications this has for globalization, development, and international relations (e.g. O'Donnell and Schmitter, 1986; O'Donnell et al., 1986; Arat, 1997; Diamond et al., 1997; O'Loughlin et al., 1998; Anderson, 1999; Diamond, 1999; Pharr and Putnam, 2000; see also contributions in O'Loughlin et al., 2004). US foreign policy and international aid organizations tend to emphasize the establishment of these procedures, as the proclamations of success after the Iraqi elections demonstrate. Yet indications of democracy and its diffusion that are limited to elections and the establishment of institutions are often argued to be limited or shallow. As former US Secretary of State Madeline Albright argued (1998: 63), the United States has promoted electoral democracy across the globe, rather than a broader notion of democracy related to just outcomes or the development of democratic societies.

As important as procedures are for the *existence* of democracy, however, it is not necessarily the case that they *define* democracy in the way that Huntington and other scholars who want to study the broad trajectory of democratization seem to imply. As Sartori wrote (1962: 4, quoted in Whitehead, 2002: 20): 'What democracy *is* cannot be separated from what democracy *should* be. A democracy exists only insofar as its ideals and values bring it into being.' From this perspective, a democracy can only exist to the extent that the procedures that regulate it are themselves democratic, but at the same time, an evaluation of a democracy cannot be separated from the outcomes of democratic procedures. As such, Anderson's (2002) evaluation of the contemporary spread of what is often called democracy is apt. He argues that the expansion of liberal democratic institutions to more and more countries seems to be associated with a shallower, less meaningful democracy that is incapable of preserving, never mind expanding, social rights for citizens. The result is something that is not democracy by Sartori's definition; Anderson argues that it is a depoliticized pseudo-democracy that is focused on institutions, seemingly without regard to the outcomes of procedures, or with what democracy should be.

As important as procedures and institutions are to democracy, then, it should be recognized that the conceptualization of democracy in procedural approaches is limited to the establishment and consolidation of *formal* or *procedural* democracy (e.g. Diamond, 1999). Yet the smooth operations of these procedures and institutions may not guarantee the development of a democratic society or a set of just outcomes. This quandary is indicative of a fundamental tension between norms of democracy (particularly when democracy is framed in procedural terms) and justice. In Carol Gould's (2004) framing of this tension, the issue is whether democracy or justice should be the primary moral value. If democracy is the primary value, then it will be necessary to recognize that democratic procedures may lead to unjust outcomes, or outcomes that constrain liberty and rights. If justice, by which she means equal effective freedom, is the primary norm, then it may be necessary and justified to bend procedures in support of justice. Political theorists and philosophers disagree as to whether democracy or justice takes moral priority, as well as on how a society or government can strive to balance them in a morally defensible manner. In practical terms, however, the potential tension between procedural democracy and justice means that the process of democratization is inherently conflictual, political, and never fully achieved (Whitehead, 2002). Rather, democratization is always worked through social relationships that operate within political communities.

Polities, Communities, and the Social Relations of Democracy

While procedures and institutions are a necessary component of democratization, many scholars would argue they are not sufficient to constitute democracy. Iris Marion Young (2000), for example, argues that the imposition of democratic institutions on societies that are undemocratic is unlikely to lead to democratic outcomes; rather, the operation of democratic institutions will reproduce inequities in society, albeit perhaps by democratic means. That is one reason why the 2005 elections in Saudi Arabia, which involved an expanded range of opposition political parties, were still characterized by some as undemocratic (*Economist*, 2005). Implicit in this perspective is a belief that democratization also involves norms of justice, equality, and inclusion that operate within societies and political communities.

One way to conceptualize justice and the ability to participate effectively as equals in a democracy is to assume that individuals are not marked or differentiated by gender, race, age, class, or religion in ways that affect their ability to be seen as political subjects or to operate as political subjects. The norm of equality framed this way requires the erasure of difference, or at least the ability to hold difference in abeyance; citizens are equal, even if they are not necessarily the same. Often associated with liberalism, this perspective assumes that membership in the political community is the primary identity mobilized by citizens who deliberate and make decisions based on an understanding of the common or public good. This perspective also implies the ability to distinguish public good from private good. There is no assumption, however, that the public good is fixed or knowable in advance; rather, political subjects should be able to discuss competing ideas related to the public good and be able to deliberate without prioritizing individual goals (see Calhoun, 1992, for a review).

Obviously, this is an ideal, but it is a powerful ideal. It offers the possibility of communities becoming more than the just the sum of individuals even as respecting the equality of individuals. And the notion of a public good provides a goal that can stand in contrast to the potential for self-interest and venality embedded in conceptualizations of rational choice theory or in interest-group politics (see Robbins, 1993).

But in 'actually existing democracies', to use Fraser's (1990) term, life is rarely so easy. While it may be desirable to have citizens put aside their differences and to participate as equals, this does not happen in practice for several reasons. First, a number of studies demonstrate that political communication – the basis for deliberation and for community-building – is deeply gendered and racialized. Studies demonstrate the ways women defer in debate, qualify answers, and smile, as if to minimize the seriousness of their opinions (Mansbridge, 1993). Young (1996) has argued that mundane interactions, such as basic greetings, provide clues as to how political subjects are positioned and that this carries into political deliberation. The effect is that even when participants try to set aside difference, it is always there, with profound implications for the way political communication proceeds, communities are built, and political relationships are structured. In recognition of this, Nancy Fraser (1990) poses the possibility of separate venues in which members of marginalized groups can deliberate without intimidation or scrutiny from majority or dominant groups. These 'subaltern counterpublics' would provide the opportunity to formulate goals and to articulate identity-group interests with respect to other groups, thereby offering the possibility of recognizing a diverse range of values and goals that may be constituent elements of a reformulated 'common good' (see also Guttman, 2003; Gould, 2004).

Second, many authors argue that it is not possible to bracket difference in political relationships and thereby to create a democratic society. As an example, Michael Walzer (1983) argues that a society can be considered just (and capable of supporting democratic procedures) if inequality in one sphere of life does not spill into other realms. From this perspective, one could imagine a scenario in which gender ideologies that may subordinate women in domestic relationships do not compromise the possibility of a just society – in which political subjects have access to equal effective freedom – as long as inequality in the home is confined to the home and basic rights of personhood are protected. If those gender ideologies spill into assumptions about women's roles in society, in the economy, and in politics that also lead to women's oppression, however, they will result in an undemocratic political community.

Political theorists such as Carol Gould (1988), Michael Sandel (1996), and Robert Putnam (1993), who work from very different political positions, argue that inequality is rarely confined to one realm of life or of social relationships. Gould, for example, argues that a democratic polity *requires* a democratic economy and democratic relationships in the social world as inequality in one area inevitably seeps into other areas, furthering the structural reach of inequality originating in one realm. Michael Sandel makes a case for considering the political economy of citizenship, arguing that political-economic contexts cannot help but shape the kinds of political communities and political subjects in a nation. This recognition of the importance of context created by the interactions of political, economic, and social relationships deeply embedded in history and place is at the heart of Robert Putnam's analysis in *Making Democracy Work*. So from very different starting points, a wide range of political theorists and commentators point to the importance of the societal context in shaping the political ground on which institutions operate (see also Agnew, 1987, 2002; Whitehead, 2002).

This recognition of the importance of context has been accompanied by efforts to enhance social and societal practices that build 'social capital' in civil society, often taking inspiration from Putnam's *Making Democracy Work* (1993). In that book, Putnam demonstrates the importance of horizontal bonds of trust built in communities through non-hierarchical relationships in diverse settings – choral societies and soccer clubs no less than commerce and governance – that made people willing to cooperate in economic relationships and government; in short, he demonstrated the foundational role of communities organized through horizontal bonds, rather than through the hierarchical bonds of clientalism or feudalism, in establishing a context for effective, democratic governance. While controversial, this was nevertheless important, ground-breaking research that shifted attention from the economic requisites for democracy and development, and refocused academic and policy debate on social relationships. Taking inspiration from Putnam – although overlooking his emphasis on long-term historical patterns – international organizations as diverse as the World Bank and the Soros Foundation attempt to enhance participation in civil society as an explicit part of contemporary development initiatives (Muck, 2004). The attempts often revolve around the establishment of venues for participation in a range of initiatives at the local level, whereby otherwise marginalized groups can be involved in political discussion and, sometimes, decision-making.

Non-governmental organizations have played an important role in this regard, and social movements have often used these venues to press new claims on behalf of poor or racialized groups. Several authors have argued that these movements carry great potential in the context of globalization and the development of transnational governance structures to foster more discussion of and attention to the 'undemocraticness' of existing economic, social, and political trends. For example, Falk (1995) and Warkentin (2001) discuss the possibility of a new global civil society emerging through the works of non-governmental organizations taking advantage of the Internet. These authors point to the enhanced ability of dispersed and marginalized groups to spread their ideas to an international audience and to mobilize on this basis. Anderson (2002) suggests that this may lead to the possibility of transnational democratization as movements cross borders and in some ways bypass the nation-state in linking movements from below with international governance structures. As appealing and alluring as these ideas may be, it is by no means clear that it is possible to build civil society – or at least a democratic civil society – through development policy, and several authors have called our attention to the ways in policy initiatives to build civil society may effectively, though perhaps unintentionally, distort the potential for 'democratization' of this form to achieve democratic justice.

One issue frequently raised is that these policies are likely to foster a civil society with priorities and values conducive to capitalist development; the implication in these arguments is that the requisites for capitalism are different from the requisites for more substantive aspects of democracy. Anderson (2002), for instance, argues that forces within the economy that emphasize a non-interventionist state and an understanding that the economy is 'off-limits' to political decision-making have shaped the trend toward liberal democracy. There are politics surrounding the economy, he argues, but they are definitely not democratic politics or values. As such, initiatives to develop civil society undertaken as part of development projects do not challenge capitalist production on a global or national scale. In those cases where the economy is a focus of development projects, social capital and civil society are put in the service of small-scale 'sustainable' projects that are unlikely to challenge the broader capitalist hegemony (Escobar, 2002; Molyneux, 2002), although these projects may ultimately be incorporated into regional development projects (Scott, 2002).

Perhaps more importantly, given the difficulty of challenging capitalism in a meaningful way, efforts to enhance civil society are often based on social-cultural relationships and organizations that previously existed within places. Without a strong and perhaps interventionist state, it is unlikely that civil society will engage marginalized groups in ways that rework existing, undemocratic social relations. Jones (2004) suggests this may lead to a re-marginalization of certain social groups, if reliance on social capital and civil society is accompanied by a withdrawal of the state from social welfare provision, as it often is in current development schemes (Molyneux, 2002). And Rose (1999) argues that reliance on social capital and 'government through community' serves to deflect attention from the role of the state in re-marginalizing social groups, thereby reinforcing the ways in which certain issues with direct implications for justice and the building of a democratic society are put off-limits to democratic debate.

These comments point to the difficulty of reorienting the path of formal democratization in ways that lead to a fundamental reworking of social, political, and economic relationships. Yet such a reworking is necessary if democratic institutions and procedures are to avoid re-inscribing existing patterns of inequality and marginality. Instead, the trend in what is called 'democratization' seems to be toward the consolidation of liberal democracies characterized by a non-interventionist state legitimated (to some degree) through elections and procedures for formal democratic governance. Without the development of more democratic societies, however, the possibilities for forms of democracy that might lead to enhancement of social justice are constrained. The development of new forms of democracy, it seems, requires an expanded conceptualization of the spaces of politics and democracy, as well as ways to use those spaces to transform the societies of democracy.

LOCATING DEMOCRACY

There are several common assumptions as to where democracy is 'located' – assumptions that often are reflected in what we commonly observe, rather than what is theoretically necessary. These assumptions locate democracy in a public sphere delimited by a territorial state. The central locational 'problems' for democracy, it seems, are that lines between public and private are confused in governance and that globalization is eroding or weakening the territorially-defined state. Yet each of these assumptions as to the location of democracy is contested. In this section of the chapter, I consider the implications for democracy of an expanded range of locations, traversing public and private spheres and multiple scales. I demonstrate the ways these locations are connected, but also cross-cut, by networks of power and the movement of people in ways that challenge common understandings of the nation-state as the appropriate unit of analysis in democratization.

Politics in Public, Politics in Private

A long history of political thought and empirical analysis locates politics and democracy in the public realm. If democracy is rule by the people, then the people must be able to come together as a public – in some form or another – to deliberate on the public (or the publics, à la Fraser, 1990) good. As noted previously, inequality in the private sphere may not be inherently undemocratic if it does not spill into the public.

While the public sphere is often conceptualized as distinct from the private sphere, the nature of the public often depends quite directly on the private. Political attitudes and values, for example, are often shaped in private or are based on private experiences, and citizens are nurtured in private. For this reason, many feminist theorists have argued that the public and private spheres are mutually constituted (see Staeheli, 1996, and Staeheli and Mitchell, 2004, for reviews). Furthermore, any number of trends in governance have blurred and confused the boundaries between public and private, as more and more decisions are made and state functions performed in public/private partnerships or are outsourced to private companies (Wolch, 1990; Clarke and Gaile, 1998; Fyfe, 2005).

Thus, it is not sufficient to think of democracy as relating only or primarily to the operations of the state and a public sphere limited to deliberations about governance. As noted, this point has been accepted at an intellectual level, but feminists and others have noted that the theoretical acceptance is often not part of the way democracy and democratization are operationalized in empirical research or in government policies (e.g. Gould, 1988; MacKinnon, 1989; Pateman, 1989). Indeed, feminist geographers have demonstrated the ways in which the private sphere is one site in which political subjects, ideas, and strategies are produced, and so are directly implicated in the project of democratization (e.g. Marston, 1990; Brown, 1997; Youngs, 1999; Fincher and Panelli, 2001; Cope, 2004; Enloe, 2004; Fincher, 2004; Marston and Mitchell, 2004). Thus, studies of democratization need to move beyond the public sphere and the formal institutions and procedures related to elections, decision-making, and rights, and to consider the ways in which democracy is itself a product of the relationships operating between public and private. This is not, in and of itself, an indication of a *reduced* importance of the state and formal democracy in the context of neo-liberalism and globalization, as is sometimes claimed (Hirst and Thompson, 1996; Mittleman, 2004). It is, instead, a recognition of the many locations in which social relations shape the operations of democratic institutions and the possibilities for democratic justice.

Scale, Networks, and Democracy

Geographers have been involved in a lively and intellectually productive set of debates about scale and the networks that link sites and scales of politics (e.g. Cox, 1998; Marston, 2000; Brenner, 2001; Katz, 2001; Marston et al., 2005). These debates are particularly important for analyses of democratization, as they highlight the scalar relationships and networks that condition democracy and democratic community.

The nation-state is frequently discussed in international relations and in geopolitics as the primary scale at which democracy is evaluated; as noted earlier, democratic audits are focused on national-level institutions and procedures. The emphasis on the national, however, extends into more theoretical work and to geopolitical analysis. This reflects, in part, the general tendency to prioritize the national scale in much of political geography and international relations (Parekh, 2002; see Youngs, 1999, for a critique). It also reflects, however, the emphasis on territoriality and borders in political geography (Wastl-Walter and Staeheli, 2004).

Yet the territorially bounded nation-state seems to be under pressure or to be in the process of being reshaped in the context of globalization. Very often, authors argue that globalization is weakening the state and increasing the porosity of borders, meaning that the national scale may be becoming less relevant for democracy or, at least, incapable of protecting rights for individuals (Strange, 1996). The primacy accorded to the national scale in scholarly studies of democratization, therefore, is giving way to a more nuanced and multi-scaled view of where democracy is located and of the networks of power that criss-cross and connect those scales. Holston and Appadurai (1999), among others, argue that the urban and the global scales are becoming more important as globalization transforms scalar relationships. They argue that, as globalization proceeds, the urban and global scales are put into a new relation to each other, making it possible for political actors to move between the two scales without going through the nation-state (see also Purcell, 2003). As noted, other authors argue that new communication technologies can be used to organize grassroots political movements that rely on newly developed international institutions and organizations to promote their causes, perhaps providing some examples of what a transnational (as distinct from a possibly more elitist cosmopolitan) democracy might look like (e.g. Wilhelm, 2000; Warkentin, 2001). Some of this literature is admittedly utopian and overly enthusiastic about the potential created by the reconfiguration of scales; nevertheless, the expanded appreciation of the scales at which openings for democratization might operate and of the networks utilized by political actors – state and nonstate, citizens and economic agents – leads to a more complex understanding of political opportunity structures and the spatiality of democracy (Cox, 1998; Miller, 2000).

The primacy of the national scale is also challenged 'from below' as there is increased attention to the unevenness of democracy within states. This can be seen in at least two ways. First, as the institutions of the European Union are built, new attention to the 'democratic deficit' and the unevenness of opportunities for meaningful participation within states has become apparent (Painter, 2002; Berezin, 2003). Second, it is clear that the adoption of formal democratic institutions at the national scale does not mean that all regions of a country are fully incorporated into democratic processes and decision-making. Ethnic or religious groups often feel their rights as citizens are restricted in states dominated by other groups; ethno-territorial conflicts, such as those involving Kurds and Palestinians, highlight the differential treatment of people within a state. In some cases – such as South Africa in the apartheid era and Israel – Freedom House provides multiple scores based on their assessment of the level of democracy for different groups within the country or in disputed territories. Even in less dramatic circumstances, people are differently situated with respect to some, perhaps mythic, dominant groups. At various times, women, racialized groups, religious minorities, or other marginalized groups argue they are second-class citizens, not allowed to take advantage of the democratic possibilities within their societies. Patricia Martin (2004) demonstrates the ways that these positionings shape the kinds of democracies that groups imagine and their political acts to consolidate new democratic nations. The result may be nations that look quite different in the ways that regional and cultural identities are integrated into the formal institutions of government and in the struggle to build more democratic societies.

International Migration and Democratization

Globalization is said to pose many challenges to democratic governance and for the ways we conceptualize democracy. But globalization is a complex, multifaceted phenomenon, and so it is necessary to think about the ways different aspects of it contribute to a reworking of democracy and democratization (Mittleman, 2004). Of the flows that are a part of globalization, international migration may pose one of the biggest challenges in this regard (Parekh, 2002; Nagel, 2004).

While current migration rates are probably no higher than they have been in other periods, such as the early 1900s, their significance may be greater now, given the rise of institutions and practices that regulate entry into countries, practices of immigrant incorporation, and the diffusion of democracy to more countries around the world. One could argue that the institutions that manage immigration are intended to preserve particular conceptions of the political community within countries, making it more difficult for some groups to enter and setting expectations for behavior once immigrants have settled (e.g. Favell, 1998; Cole, 2000; Ngai, 2004; Joppke, 2005). With more countries considered to be democratic, the issue of immigration in democratic nations takes on broader global significance.

The challenges to our ideas about democracy come in at least two migration trends. First, with the high rates of migration, more and more people live in countries of which they are not citizens. As a result, the linkage between territory and political community is strained. Exclusion, always a fact of democratic communities, becomes more pronounced in the receiving country, as a large number of residents may be excluded from decision-making and from other aspects of society (e.g. associations, employment, etc.; cf. contributions in Layton-Henry, 1990). At the same time, inclusion or attachment to the community in the country of origin becomes more difficult, despite efforts on the part of sending states to maintain those attachments (Fouron and Glick Schiller, 2001). The rise of dual citizenship poses a second challenge as the assumption of membership in a single political community – an assumption in many theoretical arguments, as well as in the citizenship laws of many countries – is violated. Politically, this raises a host of questions about loyalty. To which single country will an immigrant be loyal? In the most dramatic case, to which country will dual citizens align in the event of war? And what does dual citizenship mean for the sovereignty of "the people", when some of "the people" are members of two polities? Both of these trends, then, challenge the model of belonging, citizenship, and a territorially-defined, democratic state that has developed since the Treaty of Westphalia.

It should be noted that the territorial model is empirically derived, and that many theoretical approaches to democracy are less explicit in locating democracy in the nation-state. Thinking beyond what *has been*, some theorists and analysts suggest that these challenges to the empirical model posed by international migration provide openings for new conceptualizations and operationalizations of democracy at different scales. Soysal (1994, 2000), for example, points to a postnational model of belonging and membership in which democratic institutions at a supranational scale provide a venue for ensuring rights and for airing political claims. In a related vein, Held (1995) argues for a cosmopolitan approach to citizenship and democracy, again relying on a supranational governance regime to guarantee rights (see also Bauböck, 1994; Ong, 1999).

The idea that immigrants and others who find the territorial model inadequate for guaranteeing rights can access international institutions, however, is criticized as being insensitive to the class and cultural dimensions of democracy – aspects that are also highlighted in the context of international migration. It may well be, for example, that highly educated and wealthy immigrants can access international institutions to press their claims, but it is more difficult for immigrants without such resources. The latter immigrants may find themselves more dependent on resources offered at the local level, especially as neo-liberal restructuring reduces national funding for service provision (Staeheli, 1999; Jessop, 2002; Fyfe, 2005). Thus, immigrants may experience directly the unevenness of democratic institutions and political inclusion discussed previously. As communities are transformed through migration, putting stresses on the fabric of community and on social resources, the future of democratic community and of the institutions that are supposed to work in concert with it may be uncertain. It may well be the case that supranational institutions and procedures fostering transnational procedural democracy will be built and gain some degree of legitimacy, but life on the ground may still be characterized by difficulties in building democratic communities and nurturing political subjects capable of utilizing these new institutions. The quality of democracy and democratization under these circumstances is likely to be quite different from either the theoretical ideal or the empirical manifestations of those ideals from the past.

LIMITS TO DEMOCRATIZATION

The foregoing discussion implies that there may be limits to democratization that reflect the interplay between procedures, institutions, and communities. These limits may be manifested within states, as well as limiting the diffusion of democracy between states.

Limits Within States

Using a somewhat different framing than was presented earlier in this chapter, Michael Newman (2002) cast the limits to democracy within states

as a tension between core sets of values: between equality and liberty; between commitments to different forms of democracy; and between exclusion and inclusion. These tensions reflect the relationships between the institutions of democracy, the societies within which they operate, and the networks that connect and reconnect people and places within countries. Because these tensions can never be resolved, democratization will necessarily be incomplete.

The tension between liberty and equality emerges from two visions or norms of democracy. Liberty is enhanced by a minimalist, non-interventionist state committed to the preservation of 'negative freedoms', such as freedom from state unreasonable search and seizure, censorship, and restrictions on religion or assembly. In other words, state institutions and procedures should protect basic rights by limiting the state's ability to interfere with or restrict their exercise. The words of the first amendment to the US Constitution, 'Congress shall enact no law', are a hallmark of a state intention to privilege liberty. As argued earlier, however, a commitment to liberty manifested in institutions and procedures cannot ensure – or at least has not done so in the past – the expansion of justice, conceptualized as access to equal effective rights and opportunities within societies. Equality of citizens is also a core value of democracy, but a state committed to liberty can do little to create conditions that would foster equality. The welfare state and its efforts to extend social rights to citizens can be seen as an attempt to nurture equality, but in practice, these efforts come at the expense of liberty (e.g. through taxation or positive discrimination).

The tension between liberty and equality often seems intractable and, in practice, states and polities do not align completely with one or the other. Indeed, it is in moving between these norms, in trying to find the balance that works in a particular context, that we see the *politics* of democratization. And since that balance constantly shifts in response to external changes (such as those associated with the development of new international governance regimes) and to internal changes (such as those associated with the incorporation of migrants), democratization is ongoing (Whitehead, 2002). To say it is ongoing, however, does not mean it follows a linear path to ever greater democracy, as the examples in the introduction to this chapter indicate.

The second tension identified by Newman stems from the different types of democracy; the classic formation here is between participatory and representative democracy. The Athenian ideal of democracy is participatory, with citizens deliberating and making decisions on all facets of democratic rule and political life. That ideal was probably never achieved, and certainly seems unlikely in the larger, more socially and economically complex cities and states of today, where representative forms of democracy predominate. These systems emphasize voting for leaders and the development of national parties to organize (and discipline) the voices from below. While representative democracy is undoubtedly necessary for governing large, complex societies, it comes at the expense of the polity's participation in decision-making, leading to problems of accountability and to the transfer of sovereignty from the people to the state.

The tension between participation and representation, however, reflects and creates other tensions, tensions that involve the balance between the two in the process of democratization. States and political elites are often suspicious of participation as leading to requests that may be seen as infringing into areas of public life that have been set 'off-limits' or that may be seen as limiting liberty. In the latter case, political subjects may demand political practices rooted in cultural values that may seem undemocratic in the West. Sometimes, participatory democracy promotes actions that would threaten liberty by promoting equality (such as the anti-globalization movement) or, more cynically, that would threaten the neo-liberal order and require a different kind of state or citizen participation in the economy (Anderson, 2002). And there can be an unruliness and unpredictability to citizen participation that may be seen as threatening public order, leading states and elites to control dissent and participation through regulation and zoning of protest (Mitchell and Staeheli, 2005). At the same time, low rates of participation in established, state-sanctioned venues (e.g. elections) threaten the state's legitimacy. And elected officials rarely represent the population in a demographic sense, thereby fueling concerns that voices, perspectives, and experiences that shape political opinions are not represented in government.

The latter point raises the question of the *basis* of representation in democracy. Over time, most Western states have developed strategies of representation based on geographic units (e.g. counties, wards) or on political parties, which are often understood as representing bundles of citizen characteristics and interests within a place. Geographic representation can be made to nest neatly within the levels of a territorially bounded state, and it makes it easier for elected representatives to provide opportunities for public comment, to walk the streets and interact with the public, and so forth. Neighborhoods have a privileged place in many normative visions of democratic society, idealized as a setting in which people of diverse walks of life can come together, interact, build social capital, and reinforce the bonds of loyalty and trust that are generative of a strong polity. It is also relatively easy to build institutions at neighborhood or local levels that can provide access to decision-making, or at least to public discussion, even for

people who are not citizens of the state (see Low, 2004). In many German cities, for example, immigrants from non-EU countries can participate in local council meetings in which some decisions are made and in other cases, where local perspectives can be gathered and passed along to higher levels of government. These councils are important, as they may be the only venue in which immigrants can participate in political discussion and deliberation (Ehrkamp and Leitner, 2003). Multi-tiered, geographically based structures are one strategy for increasing participation, and perhaps the political representativeness of democracies.

These spatial strategies for enhancing democracy often work in concert with other strategies that are more specifically oriented toward trying to enhance participation by selected groups, groups that have been identified as holding a special (sometimes protected) status. Governmental institutions that are part of newly democratic, and often multi-ethnic or multi-religious states that have endured civil wars, often include targets for participation and representation of key groups; this was, for example, a critical part of strategies to consolidate new regimes in Lebanon, Bosnia, Afghanistan, and Iraq. They are less commonly found in 'mature' democracies in which it is expected that ethnic and religious identities would give way to civic identities. In the United States, for example, efforts to create electoral districts that would represent minority groups have been overturned in the courts (Morrill, 2004). Governing and representation plans based explicitly on group identities – such as those proposed by Iris Marion Young (1990) and Loni Guinier (1994) in the United States (Ingram, 2000) – have not taken hold. Formal structures of representation based in geography are said to counteract narrow, 'special interest' or 'single issue' political pressures, and so geographically-based representation and participation trump interest- or identity-group strategies in many established democracies. Yet these democracies often have very low rates of participation in elections, with resulting complaints that they are not very democratic, after all.

Finally, democracies always face a tension between goals of inclusiveness and a feeling that it is necessary to regulate or to police who can be a member; adapting Engin Isin's (2002) analysis of cities, democracies can be thought of as difference machines in which exclusion and the creation of difference is essential to the identification of members or citizens. This issue has already been discussed with regard to international migrants, and one can argue that representation strategies designed to include group interests reflect the same issue. The tension between inclusion and exclusion is, however, much broader, as democracies debate (and sometimes debate in an undemocratic fashion) who is worthy of membership in both formal and substantive senses. Ethnic cleansing is a decidedly undemocratic attempt to shape a political community, but there are other efforts that use the apparatus of democracy to shape the terms or conditions of membership. Debates over social welfare provision, over the definition of marriage and family, and over religious dress have led to the retrenchment of social, and in some cases, civil rights within many Western countries. And coming full circle, these debates reflect tensions between norms of liberty and equality in a complex politics of democratization within states.

Limits to Democratization Between States

As institutions of democracy spread to more countries, questions are repeatedly raised as to whether some societies and some cultures can ever be capable of consolidating democracy (Huntington, 1996; Lewis, 2003). These arguments are rooted in the question of whether democracy is intrinsically dependent on Western modernization (Lipset, 1960) or whether it is possible to conceptualize alternative, contingent modernities and democracies (Freeman, 2004).

One of the limiting factors to the spread of democracy may be the level of development within countries. At least since the 1950s, researchers and foreign policy analysts have pondered the question of the features of societies that enable the creation of stable, democratic regimes. Are there particular conditions that are required? Are there interventions that can foster democracy? Early research pointed to the importance of certain 'requisites' of democracy, including high literacy rates, a sizable middle-class, educational attainment, and industrialization (Lipset, 1960; Almond and Coleman, 1960; Almond and Verba, 1963). Later studies confirmed strong correlations between levels of development, modernization, Westernization, and democratization (Lipset et al., 1993; Robinson, 1996; Ward and Gleditsch, 2004), and policies encouraged the nurturing of these requisites through the solidification of market economies and civil society. Yet other scholars pointed to the economic necessity of uneven economic development and the growing inequality that were also associated with these policies, suggesting that that these interventions might have more to do with fostering capitalist hegemony than with providing the conditions for democratization (Anderson, 2002; see also Przeworski, 1991; Rueschemeyer et al., 1992).

Scholars also identified key moments in individual country histories in which countries that did not seem 'ready' for democracy nevertheless built democratic institutions. In the 1980s, scholars pointed to the 'contingent choices' made by key domestic agents who drew on the historical and geographical contexts of their countries to introduce democratic transitions (O'Donnell et al., 1986; Karl, 1990; Shin, 1994) and to political alignments within countries that created democratic possibilities (Gill, 2000), as well as limitations (contributions in Pharr and Putnam, 2000). It was significant, however, that pro-democracy forces and agents were often involved in political struggles that led to the destabilization of democracy in their countries; also implicated, however, were external agents and countries who were willing to use domestic agents to further their own goals in the country (Linz and Stepan, 1978).

Thus, the pattern of democratization is shaped by conditions and interests within countries, but also by external forces and agents. Individual countries may be buffeted by these forces, and may be ill-equipped to stave off their effects. The example of the Middle East in 2005 is instructive. After the murder of former Prime Minister Rafik Hariri, political parties and leaders opposed to Syrian involvement in the government of Lebanon staged demonstrations to show popular discontent with Syria and to drive Syria out of the country. Initially, it appeared that Syria would withdraw both troops and intelligence agents, leading many to proclaim that Lebanon would soon be free to be democratic (*Economist*, 2005). Coming after elections in Iraq and Saudi Arabia and concessions in Egypt that would allow opposition parties to appear on ballots, many in the West prophesized that democracy was finally coming to the Middle East. The euphoria was short-lived, however, as Syrian withdrawal did not end political violence, the Egyptian elections were marred by violence and indications of fraud, and little seemed to change after the elections in Saudia Arabia. Rather than appearing to be caught in a new wave of democratization, it appeared that many countries in this region may be caught in a metaphorical undertow, created when internal and external forces clashed over the future of the societies in the region. The path to democratization – both procedural and substantive – is not linear, but follows complicated twists and turns, never seeming to finally arrive at the destination.

CONCLUSION

It seems that 'democracy' is much more complex than 'rule by the people' implies. Sartori's observation about the inseparability of what democracy *is* and what it *should* be is at the heart of the difficulty: while some elements of democracy seem commonsensical and directly observable, democracy also involves normative, but still subjective, values related to justice. Democracy, then, involves an expectation of certain kinds of outcomes, not just expectations about certain kinds of institutions and procedures. Democracy involves fundamental moral values that may be in tension with each other and that may not be shared by all members of a political community.

Democracy is also complex because of the way it is imbricated with issues, relations, and processes that are not obviously or directly political. Theoretical elegance and conceptual clarity may lend themselves to discussing democracy as a political phenomenon, but the messy realities of people's lives, of communities, and of societies makes a separation between politics, economics, and social relationships untenable. This is particularly the case if democratic outcomes or justice are included in the evaluation of democracy. Rather than being limited to the political arena, democracy extends into economies, communities, homes, and families; conversely, relationships in those sites shape the quality of democratic societies and the operations of democratic institutions. Democracy scholars recognize this theoretically, but it is often difficult to incorporate this recognition in empirical analyses.

The path toward democratization is similarly complex and is necessarily conflictual. More than a seemingly natural process that ebbs and flows, democratization is shaped by histories, by geographical contexts, and by political and economic relations that connect the nation-state to other scales. While it might be possible to agree in general terms on the characteristics of democratic institutions, the processes by which they are established, modified, and perhaps perverted are more difficult to describe in general terms. And the ways in which the institutions and procedures of formal democracy are experienced and are accessed by differently positioned citizens within their countries and societies are similarly inflected by history, geography, and life histories. Democracy and the process of democratization, then, are never settled, never complete. Perhaps in that way, they are a bit like waves on the ocean, after all.

REFERENCES

Agnew, J.A. (1987) *Place and Politics: The Geographical Mediation of State and Society.* Boston: Allen & Unwin.

Agnew, J.A. (2002) 'The limits of federalism in transnational democracy', in J. Anderson (ed.), *Transnational Democracy.* London: Routledge, pp. 56–72.

Albright, M. (1998) 'The testing of American foreign policy', *Foreign Affairs*, 77: 50–64.

Almond, G. and Coleman, J. (1960) *The Politics of Developing Areas.* Princeton, NJ: Princeton University Press.

Almond, G. and Verba, S. (1963) *The Civic Culture: Political Attitudes and Democracy in Five Nations.* Princeton, NJ: Princeton University Press.

Anderson, J. (2002) 'Questions of democracy, territoriality and globalisation', in J. Anderson (ed.), *Transnational Democracy.* London: Routledge, pp. 6–38.

Anderson, L. (1999) *Transitions to Democracy.* New York: Cambridge University Press.

Arat, Z. (1997) *Democracy and Human Rights in Developing Countries.* Boulder, CO: Lynne Rienner.

Baechler, J. (1993) 'Individual, group, and democracy', in J. Chapman and I. Shapiro (eds), *Democratic Community, Nomos XXXV.* New York: New York University Press, pp. 15–40.

Bauböck, R. (1994) *Transnational Citizenship.* Aldershot, UK: Edward Elgar.

Beetham, D. (1994) 'Key principles and indices for a democratic audit', in D. Beetham (ed.), *Defining and Measuring Democracy.* London: Sage, pp. 25–43.

Bell, J. and Staeheli, L.A. (2001) 'Discourses of diffusion and democratization', *Political Geography*, 20: 175–95.

Berezin, M. (2003) 'Territory, emotion, and identity: spatial recalibration in a New Europe', in M. Berezin and M. Schain, (eds), *Europe Without Borders.* Baltimore: Johns Hopkins University Press, pp. 1–30.

Brenner, N. (2001) 'The limits to scale? Methodological reflections on scalar structuration', *Progress in Human Geography*, 25: 591–614.

Brown, M. (1997) *RePlacing Citizenship: AIDS Activism and Radical Democracy.* New York: Guilford Press.

CAIR (2004) 'American Muslims fingerprinted by US at Canadian border', http://www.cair-net.org/default.asp?Page=articleView&id=1371&the Type=NR. (Viewed March 10, 2005.)

Calhoun, C. (1992) 'Introduction: Habermas and the public sphere', in C. Cahlhoun (ed.), *Habermas and the Public Sphere.* Cambridge, MA: MIT Press, pp. 1–48.

Clarke, S. and Gaile, G. (1998) *The Work of Cities.* Minneapolis: University of Minnesota Press.

Cole, P. (2000) *Philosophies of Exclusion.* Edinburgh: Edinburgh University Press.

Cope, M. (2004) 'Placing gendered political acts', in L.A. Staeheli, E. Kofman and L. Peake (eds), *Mapping Gender, Making Politics: Feminist Perspectives on Political Geography.* New York: Routledge, pp. 71–86.

Cox, K.R. (1998) 'Spaces of dependence, spaces of engagement, and the politics of scale, or: looking for local politics', *Political Geography*, 17: 1–24.

Diamond, L. (1999) *Developing Democracy: Toward Consolidation.* Baltimore: Johns Hopkins University Press.

Diamond, L., Plattner, M., Chu, Y. and Tien, H. (eds) (1997) *Consolidating the Third Wave Democracies: Regional Challenges.* Baltimore: Johns Hopkins University Press.

Economist (2005) 'Shifting sands: the first shoots of controlled democracy are visible'. 12 Feb. pp. 46–7.

Ehrkamp, P. and Leitner, H. (2003) 'Beyond national citizenship: Turkish immigrants and the (re)construction of citizenship in Germany', *Urban Geography*, 24: 127–46.

Enloe, C. (2004) *The Curious Feminist: Searching for Women in a New Age of Empire.* Berkeley: University of California Press.

Escobar, A. (2002) 'Imagining a post-development era? Critical thought, development and social movements', *Social Text*, 31: 20–56.

Falk, R. (1995) *On Humane Governance: Towards a New Global Politics*. Cambridge: Polity Press.

Favell, A. (1998) *Philosophies of Integration*. Basingstoke, UK: Palgrave.

Fincher, R. (2004) 'From dualisms to multiplicities: gendered political practices', in L.A. Staeheli, E. Kofman and L. Peake (eds), *Mapping Gender, Making Politics: Feminist Perspectives on Political Geography*. New York: Routledge, pp. 49–70.

Fincher, R. and Panelli, R. (2001) 'Making space: women's urban and rural activism and the Australian state', *Gender, Place and Culture*, 8: 129–48.

Fouron, G. and Glick Schiller, N. (2001) 'All in the family: gender, transnational migration, and the nation-state', *Identities*, 7: 539–82.

Fraser, N. (1990) 'Rethinking the public sphere: a contribution to the critique of actually existing democracy', *Social Text*, 25/26: 56–80.

Freeman, A. (2004) 'Re-locating Moroccan women's identities in a transnational world: the 'woman question' in question', *Gender, Place and Culture*, 11: 17–41.

Freedom House (2004) 'Freedom in the world', http://www.freedomhouse.org/research/freeworld2005/table2005.pdf. (Viewed December 29, 2004).

Fyfe, N. (2005) 'Making space for 'neo-communitarianism'? The third sector, state and civil society in the UK', *Antipode*, 37: 536–57.

Gill, G. (2000) *The Dynamics of Democratization: Elites, Civil Society and the Transition Process*. Basingstoke, UK: Macmillan.

Gould, C. (1988) *Rethinking Democracy: Freedom and Social Cooperation in Politics, Economy, and Society*. Cambridge: Cambridge University Press.

Gould, C. (2004) *Globalizing Democracy and Human Rights*. Cambridge: Cambridge University Press.

Guinier, L. (1994) '[E]racing democracy: the voting rights cases', *Harvard Law Review*, 108: 109–37.

Guttman, A. (2003) *Identity in Democracy*. Princeton, NJ: Princeton University Press.

Habermas, J. (1998) *Inclusion of the Other: Studies in Political Theory*. Cambridge: Polity Press.

Held, D. (1995) *Democracy and the Global Order: From the Modern State to Comsmopolitan Governance*. Cambridge: Polity Press.

Hirst, P. and Thompson, G. (1996) *Globalization in Question*. Cambridge: Cambridge University Press.

Holston, J. and Appadurai, A. (1999) 'Introduction: cities and citizenship', in J. Holston (ed.), *Cities and Citizenship*. Durham, NC: Duke University Press, pp. 1–18.

Huntington, S.P. (1991) *The Third Wave: Democratization in the Late Twentieth Century*. Norman: University of Oklahoma Press.

Huntington, S.P. (1996) *The Clash of Civilizations and the Reworking of the World Order*. New York: Simon & Schuster.

Ingram, D. (2000) 'The dilemmas of racial redistricting', *Philosophical Forum*, 31: 131–44.

Isin, E. (2002) *Being Political: Genealogies of Citizenship*. Minneapolis: University of Minnesota Press.

Jessop, B. (2002) 'Liberalism, neoliberalism, and urban governance: a state theoretical perspective', *Anitpode*, 34: 452–72.

Jones, G. (2004) 'The geopolitics of democracy and citizenship in Latin America', in C. Barnett and M.M. Low (eds), *Spaces of Democracy*. London: Sage, pp. 161–84.

Joppke, C. (2005) *Selecting by Origin: Ethnic Migration in the Liberal State*. Cambridge, MA: Harvard University Press.

Karl, T. (1990) 'Dilemmas of democratization in Latin America', *Comparative Politics*, 23: 1–21.

Katz, C. (2001) 'On the grounds of globalization: a topography for feminist political engagement', *Signs*, 26: 1213–34.

Layton-Henry, Z. (ed.) (1990) *The Political Rights of Migrant Workers in Western Europe*. London: Sage.

Lewis, B. (2003) *The Crisis of Islam: Holy War and Unholy Terror*. New York: Modern Library.

Linz, J.J. and Stepan, A. (1978). *The Breakdown of Democratic Regimes*. Baltimore: Johns Hopkins University Press.

Lipset, S. (1960) *Political Man: The Social Basis of Politics*. Garden City, NY: Doubleday.

Lipset, S., Seong, K.-R. and Torres, J. (1993) 'A comparative analysis of the social requisites of democracy', *International Social Science Journal*, 45: 155–75.

Low, M.M. (2004) 'Cities as spaces of democracy: complexity, scale and governance', in C. Barnett and M.M. Low (eds), *Spaces of Democracy*. London: Sage, pp. 129–46.

MacKinnon, C. (1989) *Toward a Feminist Theory of the State*. Cambridge, MA: Harvard University Press.

Mansbridge, J. (1993) 'Feminism and democratic community', in J. Chapman and I. Shapiro (eds), *Democratic Community, Nomos XXXV*. New York: New York University Press.

Martin, P. (2004) 'Globalizing democracy? Autonomous public spheres and the construction of postnational democracy in Mexico', in J. O'Loughlin, L.A. Staeheli and E. Greenberg (eds), *Globalization and Its Outcomes*. New York: Guilford Press, pp. 296–316.

Marston, S.A. (1990) 'Who are "the people"? Gender, citizenship, and the making of the American nation', *Environment and Planning D: Society and Space*, 8: 449–58.

Marston, S.A. (2000) 'The social construction of scale', *Progress in Human Geography*, 24: 219–42.

Marston, S.A. and Mitchell, K. (2004) 'Citizens and the state: citizenship formations in space and time', in C. Barnett and M.M. Low (eds), *Spaces of Democracy*. London: Sage, pp. 93–112.

Marston, S.A., Jones, J.P. and Woodward, K. (2005) 'Human geography without scale', *Transactions of the Institute of British Geographers*, 30: 416–32.

Miller, B. (2000) *Geography and Social Movements*. Minneapolis: University of Minnesota Press.

Mitchell, D. and Staeheli, L.A. (2005) 'Permitting protest: parsing the fine geography of dissent in America', *International Journal of Urban and Regional Research*, 24: 796–813.

Mittleman, J. (2004) 'Globalization: an ascendant paradigm?', in J. O'Loughlin, L.A. Staeheli and E. Greenberg (eds), *Globalization and Its Outcomes*. New York: Guilford Press, pp. 25–44.

Molyneux, M. (2002) 'Gender and the silences of social capital', *Development and Change*, 33: 167–88.

Morrill, R.L. (2004) 'Representation, law and redistricting in the United States', in C. Barnett and M.M. Low (eds), *Spaces of Democracy*. London: Sage, pp. 23–44.

Muck, W. (2004) 'Global institutions and the creation of social capital', in J. O'Loughlin, L.A. Staeheli and E. Greenberg (eds), *Globalization and Its Outcomes*. New York: Guilford Press, pp. 317–36.

Nagel, C. (2004) 'Questioning citizenship in an "Age of Migration"', in J. O'Loughlin, L.A. Staeheli and E. Greenberg (eds), *Globalization and Its Outcomes*. New York: Guilford Press, pp. 231–52.

Newman, M. (2002) 'Reconceptualizing democracy in the European Union', in J. Anderson (ed.), *Transnational Democracy*. London: Routledge, pp. 73–92.

Ngai, M. (2004) *Impossible Subjects: Illegal Aliens and the Making of Modern America*. Princeton, NJ: Princeton University Press.

O'Donnell, G. and Schmitter, P. (1986) *Transitions from Authoritarian Rule: Tentative Conclusions about Uncertain Democracies*. Baltimore: Johns Hopkins University Press.

O'Donnell, G., Schmitter, P. and Whitehead, L. (1986) *Transitions from Authoritarian Rule: Comparative Perspectives*. Baltimore: Johns Hopkins University Press.

O'Loughlin, J. (2004) 'Global democratization: measuring and explaining the diffusion of democracy', in C. Barnett and M.M. Low (eds), *Spaces of Democracy*. London: Sage, pp. 23–44.

O'Loughlin, J., Ward, M., Lofdahl, C., Cohen, J., Brown, D., Reilly, D., Gleditsch, K. and Shin, M. (1998) 'The diffusion of democracy, 1946–1994', *Annals of the Association of American Geographers*, 88: 545–74.

O'Loughlin, J., Staeheli, L.A. and Greenberg, E. (eds) (2004) *Globalization and Its Outcomes*. New York: Guilford Press.

Ong, A. (1999) *Flexible Citizenship: The Cultural Logics of Transnationality*. Durham, NC: Duke University Press.

Painter, J. (2002) 'Multi-level citizenship, identity and regions in contemporary Europe', in J. Anderson (ed.), *Transnational Democracy*. London: Routledge, pp. 93–110.

Parekh, B. (2002) 'Reconstituting the modern state', in J. Anderson (ed.), *Transnational Democracy*. London: Routledge, pp. 41–55.

Pateman, C. (1989) *The Disorder of Women: Democracy, Feminism, and Political Theory*. Stanford, CA: Stanford University Press.

Pharr, S. and Putnam, P. (eds) (2000) *Disaffected Democracies: What's Troubling the Trilateral Countries?* Princeton, NJ: Princeton University Press.

Przeworski, A. (1991) *Democracy and the Market: Political and Economic Reforms in Eastern Europe and Latin America*. New York: Cambridge University Press.

Purcell, M. (2003) 'Citizenship and the right to the global city: reimagining the capitalist world order', *International Journal of Urban and Regional Research*, 27: 564–90.

Putnam, R. (1993) *Making Democracy Work*. Princeton, NJ: Princeton University Press.

Robbins, B. (1993) 'Introduction: the public as phantom', in B. Robbins (ed.), *The Phantom Public Sphere*. Minneapolis: University of Minnesota Press, pp. vii–xxvi.

Robinson, W. (1996) *Promoting Polyarchy: Globalization, US Intervention and Hegemony*. New York: Cambridge University Press.

Rose, N. (1999) *Powers of Freedom*. Cambridge: Cambridge University Press.

Rueschemeyer, D., Stephens, E. and Stephens, J. (1992) *Capitalist Development and Democracy*. Chicago: University of Chicago Press.

Sandel, M. (1996) *Democracy's Discontent: America in Search of a Public Philosophy*. Cambridge, MA: Belknap Press.

Sartori, G. (1962) *Democratic Theory*. Detroit: Wayne State University Press.

Scott, A.J. (2002) 'Regional push: the geography of development and growth in low- and middle-income countries', *Third World Quarterly*, 23: 137–61.

Shin, D. (1994) 'On the third wave of democratization: a synthesis and evaluation of recent theory and research', *World Politics*, 47: 135–70.

Soysal, Y. (1994) *The Limits of Citizenship: Migrants and Postnational Membership in Europe*. Chicago: University of Chicago Press.

Soysal, Y. (2000) 'Citizenship and identity: living in diasporas in post-war Europe?', *Ethnic and Racial Studies*, 23: 1–15.

Staeheli, L.A. (1996) 'Publicity, privacy and women's political action', *Environment and Planning D: Society and Space*, 14: 601–19.

Staeheli, L.A. (1999) 'Globalization and the scales of citizenship', *Geography Research Forum*, 19: 60–77.

Staeheli, L.A. and Mitchell, D. (2004) 'Spaces of public and private: locating politics', in C. Barnett and M.M. Low (eds), *Spaces of Democracy*. London: Sage, pp. 147–60.

Strange, S. (1996) *The Retreat of the State: The Diffusion of Power in the World Economy*. New York: Cambridge University Press.

Walzer, M. (1983) *Spheres of Justice*. New York: Basic Books.

Ward, M. and Gleditsch, K. (2004) 'Globalization's impact on poverty, inequality, conflict, and democracy', in J. O'Loughlin, L.A. Staeheli and E. Greenberg (eds), *Globalization and Its Outcomes*. New York: Guilford Press, pp. 159–81.

Warkentin, C. (2001) *Reshaping World Politics: NGOs, the Internet, and Global Civil Society*. Lanham, MD: Rowman & Littlefield.

Wastl-Walter, D. and Staeheli, L.A. (2004) 'Territory, territoriality, and boundaries', in L.A. Staeheli, E. Kofman and L. Peake (eds), *Mapping Gender, Making Politics: Feminist Perspectives on Political Geography*. New York: Routledge, pp. 141–52.

Whitehead, L. (2002) *Democratization: Theory and Experience*. Oxford: Oxford University Press.

Wilhelm, A. (2000) *Democracy in the Digital Age*. New York: Routledge.

Wolch, J. (1990) *The Shadow State: Government and Voluntary Sector in Transition*. New York: Foundation Center.

Young, I. (1990) *Justice and the Politics of Difference*. Princeton, NJ: Princeton University Press.

Young, I. (1996) 'Communicating and the Other: beyond deliberative democracy', in S. Benhabib (ed.), *Democracy and Difference*. Princeton, NJ: Princeton University Press, pp. 120–35.

Young, I. (2000) *Inclusion and Democracy*. Oxford: Oxford University Press.

Youngs, G. (1999) *International Relations in a Global Age*. Cambridge: Polity Press.

24

Convening Publics: The Parasitical Spaces of Public Action

Clive Barnett

THE VALUE OF PUBLICNESS

This chapter investigates how and why the topic of the public sphere, or publicness more broadly, is important for the ways in which political geography conceptualizes democratic politics. Democracy is a system for making binding, legitimate decisions, one that puts a premium on the principle of 'rule by the many'. The idea of the public sphere needs to be approached with this understanding of democracy in mind: 'Democratic theory focuses on accountability and responsiveness in the decision-making process; theories of the public sphere focus on the role of public communication in facilitating or hindering this process' (Ferree et al., 2002: 289). It is important to emphasize this relationship between decision-making and communication in understandings of democratic publicness. In geography, some fields of research – electoral geography, in particular – focus upon the decision mechanisms through which preferences are translated into clear expressions of the public will. But elections are moments of closure, and they communicate remarkably little information about what people actually think, feel, and favour. Geographical research inflected by cultural theory tends to think that these sorts of occasions are poor approximations of genuine democratic politics. They give greater weight to the value of non-closure, ongoing contestation, and the free play of difference in their accounts of what democracy is all about. Insofar as it considers the problem

of binding public authority at all, this line of work tends to fall back on a vocabulary of ideological legitimation and coercion. But this has the unfortunate side-effect of encouraging the idea that publicness is a value in and of itself, rather than being a means to an end as well.

This chapter tries to steer a path between the over-emphasis on either the intrinsic or instrumental value of publicness to democratic politics. The second section sets out just what is at stake in thinking about the value of publicness for democratic theory, a value that turns on a set of paradoxes between autonomy and obligation, liberty and collective action. The third section considers whether there is a distinctive geography implied by the centrality of publicness to democratic theory and practice. The fourth section argues for a clear distinction between public action and public spaces. The fifth section then outlines an understanding of the parasitical qualities of democratic publicness. The final section considers some questions of the style of public action.

WHAT KINDS OF THINGS ARE PUBLIC(S)?

Just what is meant by 'public' in discussions about the public sphere, public space, the public realm, or public life? To focus our thoughts on this question of definition, let us start by asking two questions that might help us clarify what is at stake in these debates.

First, we can ask 'What kind of thing is a public?' One way of thinking about publicness is to assume that the word 'public' is a noun. Public can be the name for a certain type of collective subject: 'the public'. In this sense, 'the public' can appear as roughly synonymous with other entities, such as 'the people', 'the community', or even 'the nation'. The public, then, can be thought of as all the members of a given society, perhaps specifically all of these people gathered together over issues of shared concern. But just what would these issues of shared concern be?

This leads to our second question, which implies a different way of thinking about publicness: 'What kind of thing is public?' This suggests that publicness is a characteristic that is ascribed to some types of phenomena, but not to others. We might think that decisions about the level of general taxation are a matter of public concern, and so might be the decision over where to locate a new airport. But it would be a surprise if many people thought that my decision to collect stamps rather than butterflies is a public matter. The latter might reasonably seem to be none of anyone else's business. A large part of what is at stake in deciding just what sort of thing is a public matter has to do with deciding what should remain personal or private matters.

So 'public' might be used to refer to the subject of concerted action, or it might be used to refer to the object of concerted action. In both respects, the value ascribed to publicness is closely related to the principle that some issues gain their importance both from affecting and being addressed by people acting together in concert.

But there is another sense of 'public', one that refers to the idea that some things are carried out in the open and are open to participation by all comers. When we combine the first two senses of public with this third idea, then we begin to get at why publicness might be so important to theories of democracy. In political theory, the value of democracy and democratic citizenship is closely related to the idea of 'public reason'. This is the idea that democracy acquires its value not just by embodying the preferences or will of the many, but also by involving free and open discussion and debates about the means and ends to which public power, such as that of the state, should be deployed.

The American political philosopher John Rawls (1993: 213) argues that public reason 'is public in three ways'. First, 'as the reasons of citizens as such, it is the reason of the public'. This is the first sense of 'public' noted above, the idea that the public is a collective subject, composed of citizens engaged in debate and deliberation. Second, 'its subject is the good of the public and matters of fundamental justice'. This is the second sense of 'public' above, the idea that some objects of concern are public by virtue of mattering to everyone and affecting the basic structure of a polity. And third, 'its nature and content is public'. By suggesting that public reason is public by virtue of being conducted in the open, Rawls adds in the third aspect to the understanding of publicness noted already. Public refers not just to a subject of action, and not just to an object of action, but it also refers to a particular medium through which action should be conducted. This chapter argues that, in fundamental respects, this dimension of publicity, which relates to the medium of action, plays a constitutive role in shaping who counts as a public and what counts as a public matter. In developing this argument, I will also argue for a reorientation of the spatial imagination that geographers should bring to the normative analysis of democratic publicness.

What does it mean to suggest that the subjects and objects of publicness are constituted through the mediums of publicity? Well, just look at the ways in which 'the public' makes its appearance felt. People speak about what 'the public' thinks, feels, and favours, and when they do so, they tend to have recourse to the results of elections, or statistical surveys, or opinion polls. These technical mediums are the ways in which the voice of the public is often expressed. For example, we might think of a public as something that is made through mediums like public consultation procedures (Barnes et al., 2003; Davies, 2006) or public opinion polls (Fishkin, 1995; Herbst, 1996). One might even say that publics are 'assembled', in the sense that that they are put together through various combinations of devices, procedures, things, and mediums (Latour and Weibel, 2005).

This conceptualization of the meaning of 'the public' may appear to be a little dangerous. It invites the suspicion that any given expression of public will, opinion, or preference is just a fabrication made in the interests of those who claim to speak in the public's name or interests, and that behind these fabrications there lies the genuine, untapped will of the people. Some of the worst excesses of our times have been made in the name of populist movements who claim to embody the singular will of a unified people against the inauthentic, divisive impostures of parties, experts, elites, or other representatives. So, one reason to embrace the mediated appearance of publics is to cultivate a healthy scepticism about any given claim to embody 'the' public will or interest.

Part of the value of publicness to the life of democracy lies in this double relationship: a public seems to be a singular collective subject, but by the mode of its appearance, any public also seems divided against itself, thereby opening a space in which claims and counter-claims as to its true opinions, feelings, wishes, and interests can proliferate. We should embrace the resulting sense that 'the public' could never actually appear in its own right, without some sort of prosthetic support. The philosopher Jacques Derrida (1992: 88) argues that 'the public' can show no sign of life 'without a certain medium'. He argues that 'the public' does not, cannot, and should not be expected to speak in its own voice, in the first person. Rather, it is only cited and spoken for.

The fabricated qualities of the public are at the core of debates in democratic theory. Some strands of contemporary liberal political theory often worry that no procedure for arriving at the public will – voting procedures being the model – can fairly and rationally embody all the preferences of the governed (Dummett, 1984). There is a worry that any attempt to arrive at such decision mechanisms threatens to impose a tyrannical form of rule over autonomous individuals. Although not all liberal theory invests unqualified trust in the market, there is a strain of liberalism that is led towards arguing that the ideal expression of the general, public will is the 'spontaneous order' created by a perfectly competitive market. On this interpretation, the link between democracy and public communication is based on a thin model of information processing, rather than in terms of the exposure of opinions to argument, challenge and justification.

This interpretation puts a premium on respecting the privately formed, autonomous preferences of individuals from undue interference. But it is subject to all sorts of objections. Primary among them is that it supposes that autonomy is a wholly private value. But it is quite plausible to suppose that people's preferences might just be accommodations with existing patterns of inequality. Preferences, in short, are adaptive. Jon Elster (1983) calls this the 'sour grapes' phenomenon, to capture the idea that people adjust their preferences, expectations and wants in relation to the availability or unavailability of different goods. People deprived of some options, because of lack of resources or information, may end up not wanting the things they have been deprived of. Their continuing deprivation cannot therefore be justified on the grounds that they express no preference for these things, since the absence of the preference is the result of the initial deprivation: 'why should the choice between feasible options only take account of individual preferences if people tend to adjust their aspirations to their possibilities?' (Elster, 1983: 109).

To suppose that democracy simply means respecting the expression of private preferences formed in such circumstances is to risk condoning social injustice in the name of claiming to do justice to autonomy. So-called 'deliberative democrats' argue that it is a category error to suppose that democracy is simply an aggregative mechanism through which the 'pre-political' preferences of individuals are added up to establish the general will (Sunstein, 1997). For them, only a process of preference formation that takes place in the open, through the medium of public debate and discussion, approaches democratic legitimacy, because this exposes people to more information and forces them to take account of the perspectives of others. From this perspective, then, autonomous preferences are only normatively valid if they are formed in the context of a robust and diverse process of public deliberation. Here, then, we have a much stronger, thicker notion of deliberative public communication as quite basic to the functioning of democracy.

These two positions can be called a broadly 'liberal' one, and a broadly 'republican' one. They represent two different but intertwined positions on the qualities of public expression in the long tradition of democratic theory (Elster, 1997). On the one hand, there is a market model, one that privileges a liberal understanding of the primacy of individual rational choices aggregated through markets. On the other hand, there is a forum model, one that is associated with traditions of civic republican thought, which privileges the value of collective deliberation as a means of arriving at expressions of the general will. From the first perspective, the main task is to find the best mechanisms for discovering what the raw data of public opinion actually is. Elections might be thought of as one way of doing this, scientific polling another, or perhaps referenda. From the second perspective, the challenge is to find the best, most just and justifiable mechanism not for discovering but for forming and making public opinion. This perspective presumes that only opinions and preferences arrived at in public, through the medium of public debate, should be ascribed the value of being 'democratic'.

I am broadly sympathetic to the latter position. But we should not dismiss the liberal perspective too quickly. It does address a key dilemma in democratic theory. There is something quite compelling about the liberal respect for the autonomy of private preferences. It is guided by an admirable respect for the pluralism of opinions and preferences, as well as by a deep suspicion of the idea that some people should decide that some other people don't know what is in their own best interest (see Goodin, 2002). The republican image of transformative public deliberation seems compelling because it provides a picture of selfless, concerned citizens able to act in the general interest. It is grounded in a telling critique of how the liberal position can inadvertently lead to the reproduction of serious injustice and the diminution of real autonomy. But it is not without its own problems. In particular, the republican position can underestimate the genuine importance of privacy as a condition for the sorts of virtuous citizenship on behalf of which it speaks (Squires, 1994).

Both the liberal and the republican viewpoints agree that preferences should be the basis of political life, but then we reach a conundrum: is it justifiable to intervene to reshape people's preferences and tastes, in light of the 'sour grapes' phenomenon for example, without risking doing serious harm? A classical liberal position would err on the side of caution here, worrying about the intrusive potential of concerted efforts at forming preferences. However, across a wide spectrum of contemporary political thought, a pressing concern is whether efforts to shape tastes and preferences are legitimate: How can they be justified? What means should they use? Who should pursue these efforts? And to what areas should they be limited? This is a particularly acute problem for traditions of left-wing political thought, which have often paid scant attention to the dilemmas of squaring collective action and individual autonomy that liberalism has to be given credit for keeping at the forefront of democratic theory. As this chapter develops its argument about publicness, spatiality and democracy, it is therefore worth keeping in mind what Claus Offe (1997: 89) identifies as the key challenge for progressive political thought:

> to develop arguments which, while respecting individual freedom of preference formation and the pursuit of preferences in the realms of markets, politics, and private life, also provide justification for a wide range of taste-shaping and taste-discriminating interventions by democratic governments which are seen as valuable for themselves or instrumentally indispensable for the sake of maintaining and furthering such collective values of solidarity, welfare, autonomy, deliberation, and democracy itself.

As we will see, the value of publicness to democracy derives in large part from the ways in which it provides mediums for working through this challenge in ways that respect the equally compelling imperatives of facilitating concerted, legitimate action around issues of shared public concern while respecting the pluralism of citizens' values, opinions, and life projects.

There is an irreducible tension between finding ways of making collective decisions that are broadly legitimate in the eyes of citizens and that accord to reasonable principles of justice, while also ensuring that people are allowed to carry on as much of their lives as is deemed appropriate without undue interference, obstruction, or approbation from sources of authority. It should be noted that these two imperatives are not two poles of a continuum; nor do they serve as the outer limits to one another; nor are they simply in contradiction. They are, in fact, internally related. The democratic legitimacy of collective decisions is supposed, in principle, to depend on the freely given consent of citizens, and yet one purpose of collective rule is to cultivate the flourishing of autonomous, active citizens whose consent is, indeed, freely given or withheld rather than coerced.

SETTING THE PUBLIC SPHERE ADRIFT

We have established that, one way or another, publics are constituted in part through the mediums of their representation. The next question we need to ask is whether there is any special relationship between the values ascribed to 'the public' or 'publicness' and particular spaces, places or geographical configurations.

To put my own cards on the table straight away, I think we should follow Derrida, who does not think that the public has any proper place: 'Does it take place? Where is it given to be seen, and as such? The wandering of its proper body is also the ubiquity of a spectre' (Derrida, 1992: 87). But this image of the ghostly quality of the public, cut adrift from any proper location, runs against the grain of most research in geography. Geographers tend to argue that there are, in fact, some places, spaces, and spatial configurations that are peculiarly valuable as scenes of genuine public life and authentic public expression. I want to argue against this claim, on the grounds that it fails to register the intrinsic value of publicness in itself, as well as the instrumental value of publicness to the functioning of democratic rule, and is therefore poorly suited to thinking creatively about the spatialities through which equally compelling imperatives can be played off against each other.

One place to start investigating the possible relationships between publicness and geography is in the rather arcane tradition of legal reasoning known as public forum doctrine. Public forum doctrine is an important aspect of First Amendment jurisprudence in the United States, one means through which the constitutional legal system decides upon what counts as 'speech' that is worthy of protection from unwarranted government regulation. Public forum doctrine is interesting because it defines some types of expression as protected 'speech' by virtue of where they are uttered. Some spaces, according to this tradition, are defined as 'public' because they are traditionally defined as offering unfettered opportunities to speakers for addressing other people on matters of broad, general concern. In guaranteeing that some spaces should remain open to all speakers, public forum doctrine enforces on citizens an obligation to be exposed to issues and views 'that would otherwise escape attention, and that would not have been chosen before the fact' (Sunstein, 2001: 196).

The interesting thing about public forum doctrine is this functional definition of what counts as a public space. This is potentially quite radical, since it implies that all sorts of spaces and places could in principle be defined as public, if they meet the criterion of providing opportunities to address others on matters of common interest. The problem with public forum doctrine, though, is that it tends to restrict its definition of such spaces to the ones traditionally defined as public forums by eighteenth- and nineteenth-century common law – street corners, public squares, parks and so on. In practice, recent First Amendment law has tended to restrict the definition of public forums to spaces like streets and parks, rather than extend them to include shopping malls, airports, or television stations.

The deployment of public forum doctrine is just one example of the widespread assumption that the value of publicness is best expressed by spatial relationships of close physical proximity. It is often supposed that, ideally, a public space should conform as closely as possible to the configurations of face-to-face interactions. This is a problem for both normative and empirical reasons. Let me explain why, by way of an example from geography.

The cultural geographer Don Mitchell (2003) has written extensively about the legal regulation of public spaces in the US. There are two important points that Mitchell makes. First, the value of the idea of the public sphere lies in the idea that politics is legitimated as democratic by virtue of being embedded in forms of inter-subjective communicative action. Second, if democracy requires opportunities for communicative interactions between citizens, then this means that geography and publicness are strongly connected. Mitchell combines these two points by defining a public space as a 'place within which a political movement can stake out the space that allows it to be seen. In public space, political organizations can represent themselves to a larger population' (Mitchell, 1995: 115).

This definition nicely captures the idea that publicness has something to do with communicating with others. Mitchell argues that this definition supports a sharp distinction between what he call 'real' public spaces and 'virtual' public spaces. 'Real' public spaces turn out to be spatial contexts of face-to-face interaction like street corners, parks, and public squares. 'Virtual' public spaces, like TV, radio, or the Internet, are understood to be less authentic, secondary sites for communicating with others.

Mitchell is hardly alone in arguing that discussions of the 'public sphere' need to take more attention of 'real' and 'material' spaces of public interaction. For example, it is a commonplace of geographers' discussions of the influential work of Jürgen Habermas (1989) to complain that his notion of the public sphere has only a weak, metaphorical reference to material spaces and places (Howell, 1993; Mitchell, 1995; Goheen, 1998). Habermas emphasizes all sorts of communicative practices of talking, discoursing and deliberating in his account of the public sphere. Geographers, in contrast, claim that what we really need to do is focus attention on the 'materiality' of public space.

This line of criticism takes us in the wrong direction. It misses the point about what might be the problem with metaphors like the public sphere, the public realm, public domain or the public sector. The problem is not that these are spatial *metaphors*. It is that they are *spatial* metaphors. They are, more precisely, metaphors that conjure up images of contained, circumscribed spaces. By thinking that these should just be made more 'real', provided with more precise 'material' reference points like streets or parks, geographers actually end up inadvertently compounding what is the main problem in Habermas's original account of the public sphere. In reconstructing eighteenth-century public spheres as an ideal form, Habermas did not ignore the geographical dimensions of public life at all. He told a story all about the circulation of pamphlets, newspapers, and novels through spatially extended networks of communication. One enduring lesson of his work is the idea that the public sphere is a circulatory space. But Habermas sees the circulation of written texts as simply a secondary, additive means for continuing a face-to-face conversation. This ideal of the conversational quality of public communication assumes that any written communication, once sent out into the world, always reaches its intended destination, is received as it was intended, and is thereby integrated back into a set of convivial, familiar relations.

Habermas provides us with an image of the public sphere as a circulatory space of communication. We do not need to 'ground' this image in 'real' or 'material' spaces. Quite the contrary, this gesture detracts from the really important insight in Habermas's work, which draws our attention to the degree to which a public sphere is all about the process of discoursing. But, in order to think about the spatialities of discoursing, we do need to suspend the presumption in Habermas's original account that circulation is a circular, tightly bound process (see Lee and LiPuma, 2002). It might be better thought of as a process of scattering and dispersal. One problem with all the spatial metaphors noted above is, certainly, that they fail to capture 'the mobile, elusive and problematic character of publicness' (Newman, 2005: 2). But more than this, if we take seriously the idea that the medium of publicness is discourse, then we should also take seriously the degree to which publicness is a process; it is something people do, rather than a space they inhabit.

And just what is this process of discoursing in public? Above all, it a process of addressing others, and of being addressed by them. If publicness has a spatiality, then it resides nowhere else than in the treacherous and promising space that is enacted by throwing words, signs and tokens out into the world (cf. Arendt, 1958: 177). What difference does it make to talk about the public sphere in terms of circulatory spaces of address? And why does it matter to think of publicness in terms of scattering and dispersal rather than 'material' and 'real' spaces of urban life? In particular, why does this matter to thinking about the relationships between publicness and democracy? Well, there are two reasons.

First, the idea that city streets, public parks, or other spaces of face-to-face interaction actually serve as the primary scenes for public interaction and communication among members of large, complex societies seems a little out of date. Sticking closely to Mitchell's own definition of public space – places that provide opportunities to be seen and represent oneself to audiences – suggests that all sorts of spaces can, in fact as well as in principle, serve this function. In terms of the relationship with democratic decision-making, there is no good reason to suppose that streets and parks are, in principle or practice, privileged public spaces: 'They are not the only or even the most important places for communicative activity. Other areas – perhaps mailboxes, probably railroad stations and airports, certainly broadcasting stations – are the modern equivalents of streets and parks' (Sunstein, 1995: 102).

It is these spaces of distanciated, mediated communication that are the most 'material' spaces for public democratic communication, if by material we mean the most relevant, the ones most pertaining to the issue at hand – which is the opportunity for addressing people where they are to be found. The geographer Paul Adams (1992) captures this nicely when he describes television as a 'gathering place' for modern citizens. One can overdo this sort of idea, but its importance lies in recognizing that the places where potential addressees for communications on matters of public concern are to be found are not, after all, places at all. They are stretched-out, complex networks of circulation. Elsewhere, Adams (1998) distinguishes between types of communication mediums on the basis of their different network qualities. This is one line of investigation that geographers might pursue in thinking about the geographies of public space. Different infrastructures of circulation and communication have different formal features that encourage and enable certain styles of interaction more than others, some of which may conform to a lesser or greater degree to norms of publicity or privacy (see Gaonkar and Povinelli, 2003).

The second reason why we should be suspicious of invocations of 'real' or 'material' public spaces is that this fails to accurately register the ways in which spaces like streets and parks actually function in contentious politics. One of the features of modern social movement politics is the deployment of what is sometimes called strategic dramaturgy, or theatrical styles of protest that enact claims and grievances, often in non-deliberative ways. The aim of these strategies is to demonstrate the size and intensity of a campaign's support, and to attract public attention (Tilly, 1994). The reason such movements organize their protests in places like Washington, DC, or around Westminster Palace in London, or in the cities where major international diplomatic meetings like the G-8 or WTO are held, is because these are the stages upon which they can project their presence to wide audiences through the mediums of newspaper, radio, and television. In a sense, such protest events temporarily enact city streets or parks as public spaces. That is, they use them to address others with the aim of attracting attention.

These days, then, the supposedly 'real' and 'material' spaces that geographers like so much only ever function as 'public spaces' when they are embedded in more extensive social networks and technological relations that project outwards from any scene of contained interaction (see Adams, 1996; Calhoun, 1998; Barry, 2000). And it is worth noting that, contrary to Adams' comforting image of 'gathering', the public significance of electronically mediated communication inheres in their qualities as mediums of dissemination. Radio and television provide the opportunity to address others without being able to guarantee that this address will arrive at its intended destination, or any destination at all (Scannell, 1995). And in that, they help us recognize the public potentialities that inhere in any and all modes of communication, irrespective of their medium.

These are two empirical reasons for being suspicious of geographers' disciplinary preference for 'real', 'material' public spaces when it comes to making sense of ideas about the public sphere and democracy. But they both lead to a more fundamental point. Even ideally, these spatial archetypes – streets, parks, and the like – should not be thought of as the best analogues with which to think about the normative issues at stake in discussions of publicness and democracy. Sticking only to the definition of a public space as any space that provides opportunities for addressing and interacting with other people, there is no reason to assume that such spaces are exemplified by shared locales of spatial or temporal co-presence. When we add the sense that public communication is important to democratic theory and practice only in so far as it can articulate with mechanisms of collective decision-making, then we can even say that the 'real', 'material' spaces that geographers favour when they write about the public sphere are not necessarily material to the relation between publicness and democracy. If we take material in the sense of being most relevant to the case in hand, then spaces of physically co-present interaction are not, on their own, the most important arenas for the articulation of public communicative action with binding, collective decisions.

PUBLIC ACTION AND PUBLIC SPACE

The analysis in the preceding section leads to a preliminary conclusion: we need to stop thinking of publicness primarily as a type of space, and instead focus on the type of action that is attributed the status 'public'. Some work in geography has begun to explore the contingent relationship between various sorts of public action and the types of spaces in and across which such action takes place. Lynn Staeheli (1996), in her research on women's activism in the US, shows how public action in the sense of concerted citizenly action oriented to matters of general interest can take place in putatively private spaces like the home. Likewise, Sophie Watson (2004) shows in her work on various forms of civic association in the UK that public spaces can be, sometimes must be, fleeting, hidden, and temporary. And Murray Low (2003) shows that one of the most important forms of public action we undertake as citizens – voting – is only of any value as a public act by virtue of being undertaken in secret. Each of these examples underscores the idea that the democratic value of publicness lies in certain sorts of action, and that these actions are not, actually or conceptually, contained within particular configurations of place, space, or territory.

We might also learn from media and communications studies, a field with its own well-developed sensitivity to questions of space and place (Couldry and McCarthy, 2004). For example, Samarajiva and Shields (1997: 541–2) capture the way in which the criteria for distinguishing between 'public space' and 'private space' are not 'spatial' at all, in the sense of referring to locational categories, but are based on the distinction between different types of interaction: 'Public spaces are characterized by a relative openness to initiation of communication by others, and private spaces are characterized by a relative closedness to initiation of communication'. Notice that this definition does not only define publicness by reference to opportunities for speakers to communicate to others. In defining the nature of democratic publicness, it also emphasizes the importance of variable dispositions to be willingly on the receiving end of uninitiated expressions from others.

One thing this definition reminds us of is the importance of values like privacy and autonomy in mediating the relationship between democracy and publicness. There are all sorts of occasions when we might quite rightfully not want to be open to unwanted, unexpected encounters with others. Another important point about this definition is that it defines the publicness of a space by its internal, formal qualities of address and reception, and not by reference to the conditions of access to any space. This is a more contentious point. It directs our attention to the intimate connection between notions of publicness, democracy, and relationships of property ownership and commodification.

Private space has been traditionally defined as an important realm for the cultivation of the essential virtues of democratic citizenry – of tolerance, criticism, and mutual trust, for example. Staeheli and Mitchell (2004) observe that contemporary societies are increasingly characterized by a steady erosion of the forms of privacy that should remain important resources for wider forms of social engagement, as states and corporations extend their capacities of surveillance. But the private realm is also sometimes defined as a realm of negative freedom, upon which the state cannot properly impinge. This second sense of privacy raises questions regarding the degree to which rights of private property are consistent with values of democracy and democratic publicness. If the erosion of privacy is one threat to the health of democracy, Staeheli and Mitchell also identify another in the steady privatization of public space, by which they mean the process by which seemingly more and more activities that are of public importance are reorganized according to the economic imperatives of private commodity production and consumption. In particular, more and more putatively public spaces are being commodified; access to them is more and more tightly controlled by private organizations, and is often explicitly based on the ability to pay (Low and Smith, 2005). This is true of both 'real' spaces like city centres or shopping malls, and also 'virtual' spaces like television and radio.

Both of these ways of controlling access to spaces is, by definition, exclusionary. But it is worth noting that, on Samarajiva and Shields' definition of the publicness of spaces, one that emphasizes patterns of interaction rather than conditions of access, this is not necessarily a sign of a diminution of their public value at all: 'The "publicness" of a space depends on openness to initiation of communication among inhabitants rather than the terms and conditions of access to that space' (1997: 542). Just because one might have to pay a fee to enter a 'real' or 'virtual' space, does not necessarily vitiate its quality as a space for public action. This flies in the face of a great deal of research in geography, which tracks the privatization and commodification of space and automatically concludes that this is equivalent to shrinking of the public realm.

My point is not that we should not be concerned by such processes. But we should be clear about what it is about 'publicness' that we consider to be of value before we jump to the conclusion that structural changes in the design or regulation of public infrastructures are necessarily destructive of public life. In particular, we should remember that the value of public communication, when it comes to questions of democracy, is not an end in itself. Public communication is considered valuable by reference to the idea that decision-making in a democracy should be undertaken within a broader web of relations of deliberation, oversight and scrutiny (Emirbayer and Sheller, 1999; Przeworski et al., 1999; Young, 1999; Keane, 2004). The commercialization and commodification of spaces of putatively public communications might, on these grounds, have much more ambivalent implications for public life and democratic politics than is often supposed, since there is no *a priori* reason why these sorts of spaces cannot sustain cultures of deliberation, self-expression and accountability (see Barnett, 2003).

We can begin to see why a functional definition of publicness – one that focuses on types of action that are in some sense 'public' – is important for understanding the relationship between publicness, spatiality, and democracy. Public communication is important to democracy because of, and not in spite of, the fact that democracy is a system of rule, that is, a mechanism for making binding decisions in a context of irreducible pluralism in opinions and non-reconcilable differences of interest. Action that is public by virtue of what it considers, as well by virtue of who is drawn together to deliberate over these objects of concern, is not found only in locations like streets, parks, or other exemplars of public forums. Public action can take place anywhere. It has no proper place at all.

PARASITICAL PUBLICS

There is a strong strand of thought that defines democratic publicity primarily in terms of the intrinsic value of a distinctive type of sociability. Often enough, interaction in urban social life is the privileged analogue of such public activity. The attractiveness of this sociability model of publicness lies in its ability to model the possibility that people with plural interests and different identities can come together as a collectivity (Young, 1990). This strand of thought is very good at explicating the idea that a crucial aspect of a vibrant democratic public life is our exposure to the identities and perspectives of others (Bridge, 2004). But the problem with making urban spaces of sociability, surprise and pluralistic encounter into the exemplary models of publicness is that this completely ignores the sense in which, in democratic theory, publicness is instrumentally related to maintaining the legitimacy of binding collective decision-making.

In this section, I want to develop the idea that public action has no proper place at all. To do so, I will introduce the arguments of two very different writers on the theme of publicness: Michael Warner, a literary theorist who is one of the most acute commentators on contemporary theories and practices of publicness; and John Dewey, a doyen of mid-twentieth-century American liberalism, and one of the key theorists of the public life in modern social theory. I deploy the arguments of both writers to explicate the idea that publicness is always and only ever derivative of other spaces, other concerns, and other social relations from which it emerges, and which it in turn helps to reconstitute and transform. Together, Warner and Dewey enable us to appreciate that thinking about the equally compelling importance of intrinsic and instrumental aspects of publicness requires us to let go of the idea that public space is either 'material' or best modelled on scenes of co-present interaction.

Publics as Communities of Strangers

Warner focuses on the distinctively, and irreducible, discursive aspects of the idea of publicness. In doing so, he develops an understanding of the intrinsic quality of publicness as a distinctive type of sociability resting on relations of call and response. Warner's starting point is that any public can only 'exist by virtue of being addressed' (Warner, 2002: 67). This does not mean that publics just come into existence by virtue of being addressed. When we address others in a public register, we are presuming a shared scope of concern that is far from certain. The addressees of any public utterance are imaginary, which is not to say that they are unreal. If we can say that a public exists only by virtue of being addressed, then this implies that an address to an audience only gets its public quality by virtue of the type of response it elicits. It depends on whether the address resonates with others. Another way of putting this is to say that what constitutes publics is neither an act of address, nor only of response, but the relation of *attention* that is established in the space that separates and joins these two acts (ibid.: 87). And there is no reason to suppose that such relations of attention are contained within scenes of face-to-face interaction; indeed, there is no reason to suppose that these face-to-face encounters are any freer from uncertainty and indeterminacy than stretched-out, distanciated relationships.

If public discourse works through this type of call-and-response dynamic, then it means that we should take seriously the idea that publics are things that make their appearance through the force of convening, that is, through a set of relationships between addressing and responding. 'Convening' certainly brings to mind the sense that a public is an assembly or gathering of some sort. But I use the formula 'convening publics' because I want to emphasize the active sense of calling on others to gather together, which in turn requires an active response to heed any such call. This sense of convening helps us appreciate that publics appear through representative acts of being spoken for and being spoken to.

We saw at the outset of this chapter that publics are always spoken about, and more to the point, they are always spoken for. Now, speaking for others seems to many academics to be impossible, if not a wholly unjustifiable presumption. But I want to follow Stanley Cavell (1979: 18–28), who suggests that the possibility of saying 'we', or what he calls the 'arrogation of voice' that is always involved in supposing that one can rightfully speak for others, is only considered impossible or scandalous because we forget about the relationship between speaking for and speaking to others. And, in turn, we tend to think that speaking to others is much simpler than it is. We forget that it is risky and hazardous, that it only works by taking the chance of misfires and infelicitous outcomes. We need to keep both points in mind if we are to appreciate why the idea of 'convening publics' makes a difference to the sorts of questions we ask about public action: the first point reminds us that public discourse is, pre-eminently, discourse addressed to others. The second point reminds us that any such address to others only comes off as a public act because of a relationship of attention between speakers and addressees that is constituted by the response of the latter.

So in speaking of 'convening publics', the emphasis should be on the active sense of convening, rather than a sense of a convention already successfully gathered together, or conforming to a rule already agreed upon. The activity of public-making inheres not in gathering, nor even in assembling, but in this activity of convening, that is, in calling out to others, attracting their attention. When one remembers the hazardous quality of the process of speaking to others, one can begin to better understand how speaking for others is not a zero-sum game of silencing or exclusion, but an invitation, an opening-up of a scene of claims and counter-claims.

If the publicness of a discourse depends on establishing a relation of attention, then this means that any public is constituted by a spacing between discrete but intimately related acts that are separated and bound together in temporal relations of anticipation, projection, response and reply. Warner's emphasis on the distinctive temporal qualities of publics and counter-publics liberates the disseminating force of publicness that Habermas's seminal account of the public sphere had contained within a circle of dialogue. A public, for Warner, is in large part stretched out across time, in the sense

that it comes into being and persists by people writing letters in response to newspaper stories; or writing reviews of books they have read; or citing those books in the things they write themselves; or carrying on the day-to-day talk about last night's television. The *citational* pattern is crucial to understanding the value of public 'space'. It connects up to the distinctive type of sociability that is bought into being by a public, as distinct from other social forms such as families, parties, friendship networks, or bureaucracies. The citational, iterative quality of the relations through which public attention is secured underscores the idea that when one addresses a public the addressee, intended or otherwise, is actually not there, not yet, not at least as a member of a public.

This is one implication of insisting that publics are imaginary entities. But there is a paradox at work here: while one might acknowledge that publics are, in principle, imaginary entities, when one actually addresses a public, you have to forget this fact – one has to assume the existence of an audience, with certain sorts of social characteristics. Any public address that ignores these characteristics is not likely to come off successfully – not likely to gain much attention – because it will end up being completely out of tune with its potential addressees. Once again, we need to emphasize that even though they may be imaginary, publics certainly cannot be conjured into existence just by the force of one's own intention (Warner, 2002: 14).

If there is magic involved in the constitution of publics, it is an imperfect kind of magic 'because of how much it must presuppose' (ibid.: 105). What this means is that publics have a kind of double conditionality:

1 They are self-organized, constituted only through being addressed, but they also depend upon pre-existing infrastructures of communication and circulation.
2 They are in principle open to all comers, but they in fact presuppose specific criteria of shared identity (ibid.: 106). You have to have a good idea at least about what might resonate if you are to stand a chance of what you have to say actually resonating. But you can never know for sure.

And here we reach the crux of the issue – the exact composition and identity of a public is unknown: 'A public is always in excess of its known social basis. It must be more than a list of one's friends. It must include strangers' (ibid.: 4). While the intention of addressing a public presupposes some shared social criteria, any successfully constituted public only 'comes into being through an address to indefinite strangers' (ibid.: 120). Warner demonstrates his point by reminding us that public discourse has a distinctive grammar. It is as it is once both personal and impersonal (ibid.: 121). It is in principle addressed to anyone, but in a tone of some familiarity, and in a register of concern, as if the topic being talked about should matter to addressees. Publics are constituted through a pragmatics that sends out its call in a register intended 'for-anyone-as-someone' (Scannell, 2000).

The relationship between presupposing and exceeding shared criteria of identity is a defining feature of public discourse (Warner, 2002: 105–6). This set of paradoxical relations between what is concrete and given on the one hand, and what is abstract and imagined on the other, accounts for the queer nature of publicness:

> Public discourse, in the nature of its address, abandons the security of its positive, given audience. It promises to address anybody. It commits itself in principle to the possible participation of any stranger. It therefore puts at risk the concrete world that is its given condition of possibility. This is its fruitful perversity. (ibid.: 113)

Here, then, we glimpse a different style of circulatory space than that originally outlined by Habermas. For Warner, public discourse is projective, hopeful, crossed by anticipation, and its ongoing accomplishment depends on any specific act of addressing others being taken up, thrown into circulation, reiterated, passed on, disseminated.

Warner describes what we might call the performative qualities of publicness: publics are constituted by addressing a public, and in this sense, any public utterance does what it says, it brings into being what it presupposes to already exist as the condition of getting off the ground. The 'space' of this sort of publicness is a spacing-out of discourse over time, which accounts for the sense in which publics are imaginary entities composed of strangers. Herein lies the intrinsic value that can be ascribed to publicness: it is a modality for actively engaging with strangers, for acting-in-concert with others as ends in themselves. Notice, too, that this intrinsic quality of public interaction with strangers in a register of familiarity is dependent upon the prior existence of other social relations and organizational infrastructures. In Warner's account, publicness feeds off these as its conditions, but only to exceed and transform them in the process.

Warner's account of publicness is primarily, but not only, focused on how taking seriously the quality of public mediums forces us to think of public subjects as particular types of imaginary collectivities. I have focused on the intrinsic value of the type of social interaction that inheres in the formal characteristics of publicness: publicness as a relation of engaging with others as others, as different from oneself, as strangers. Warner does not ignore instrumental questions, it is true, but to consider this dimension of publicness more fully, I want to shift attention to John Dewey. It is from Dewey that I want to derive a fuller sense of the parasitical qualities of publicness. For Dewey, publics are not only dependent on prior formations of social interaction, as in Warner, which they also exceed. They are also dependent on pre-existing infrastructures of communication and social integration. And above all, for Dewey a public is brought into existence because public discussion is always about something of general concern, not just about itself. For Dewey publics are, as it were, intrinsically instrumental, and herein lies their important connection to democratic politics.

Publics as Communities of the Affected

One way of situating the importance of Dewey's ideas for understanding the relationship between publicness, democracy, and geography is by reference to a critical question in democratic theory: how to determine who has the right to participate in democratic public life?

Just who belongs to the public whose consent is meant to legitimize decisions as democratic? Classically, participation in public life is defined on the basis of membership as a citizen of a territorially defined polity. An alternative criterion appeals to a causally based principle. This makes participation in decision-making dependent on the idea of affected interest (Shapiro, 1999: 38–9). This principle seems well fitted to doing justice to the democratic principle that all those people potentially affected by a decision have an interest in it. The idea of affected interest also implies a different geography of participation. 'Communities of affected interest' are not likely to be neatly contained within the boundaries of nation-states (nor necessarily restricted to only human actors; see Eckersley, 2000). We all know that decisions made in one place have all sorts of consequential impacts that extend far beyond boundaries of this sort. This principle therefore also seems to be better suited to taking account of relations of power and their complex geographies: 'The causal principle of affected interest suggests that ideally the structure of decision rules should follow the contours of power relationships', and this means that 'if you are affected by the results, you are presumptively entitled to a say' (Shapiro, 2003: 219–20). This type of argument about the spatiality of actions and their unintended consequences is also common in geography, where it is used to argue for a radical extension of the geographical scope of care, obligation, participation, and responsibility (Corbridge, 1998; Smith, 1998; Massey, 2004).

On its own, though, the causal principle of affected interest might lead us to conflate two different forms of solidarity, insofar as it seems to suggest that a 'public' is constituted simply through systems of 'functional interdependence'. This latter idea refers to a type of relationship based on structural connections that 'join people in a mutuality that is not primarily manifest in their own common recognition of it but instead can operate, as it were, behind their backs' (Calhoun, 2002: 161). This sort of relationship, one that is not chosen but which we find ourselves already placed in, might well be one condition for the emergence of publics. But in and of itself, it is not equivalent to a public in the broad sense we are defining in this chapter. A public is not simply formed through relations of necessity that follow from functional integration across space and time; they depend on both this sort of solidarity and an element of choice (ibid.: 163).

Dewey is often associated with the idea that publics are coterminous with spatially extensive communities of affected interest. But he actually sketches a more complex understanding of publicness than this at first suggests. Writing in the 1920s and 1930s, Dewey focused attention on the implications for democratic politics of the geographical extension and increasing functional complexity of social relationships. He argued that the spatial extension of transport, communications, economic processes and trade positively expanded the conditions for democratic public life (Barnett, 2003). This extension of life over space and time inevitably entangles people into relationships of cause and effect, and in particular into relationships in which actions have all sorts of indirect consequences: 'A public consists of all those who are affected by the indirect consequences of transactions to such an extent that is deemed necessary to have those consequences systematically cared for' (Dewey, 1927: 16–7).

This might look like an affirmation of the causal principle of affected interest. But actually, Dewey's primary emphasis is upon not just the extensive and indirect consequences of acts. It is, rather, on the perception and recognition of these. It is this process that is constitutive of a public. Dewey argued against the idea that, in complex social systems, democracy should be reduced to the efficient management by experts, supported by occasional acclamation by a passive citizenry. He argued that people retained their capacity to act as citizens in spatially extensive and functionally complex systems, but he did not stake this on their capacity to process lots of information about how their actions lead to all sorts of dispersed consequences. This was impossible, but this did not count against the possibility of publics coming into existence. Far from it, for Dewey publics emerged precisely when consequences become so complex, and the numbers of actors involved so large, that people cannot

accurately calculate how they affect others or will be affected by them (ibid.: 52–3).

The extension of social life over distanciated, complex systems means that people cannot accurately trace the contours of their own implication in distant consequences. But it does sensitize them to the idea that they are, nonetheless, implicated in this way. The extension of communications, in the broadest sense, enabled people to develop 'more numerous and varied points of shared common interests' (Dewey, 1980: 92), and facilitated 'freer interaction between social groups' (ibid.). In short, people's imaginative horizons are expanded, and this is the key mechanism for the transformation of functionally integrated systems into publics. This transformation depends not on relations of expert knowledge, but on a capacity to imagine one's implication in wider systems of indirect consequences (Goodin, 2002). What this means is that a public turns out not to be composed only of all those affected by consequences; a public emerges only when 'the perception of consequences are projected in important ways beyond the persons and associations directly concerned in them' (Dewey, 1927: 39). And in principle, this means that the scope of any public is indeterminate, because once one introduces the idea of indirect consequences, the number and location of those affected expands beyond the scope of easy comprehension.

So, for Dewey the conditions for a new type of public life lay in this process of spatial and temporal extension of consequences that enable the expansion of people's perception of being part of wider communities of interest. The causal principle of affected interest does, then, serve to determine the object of matters of public concern in Dewey's formulation: publics form around the shared concern to intervene and 'take care of' extensive systems of action and their indirect consequences. But the emergence of a public as a subject of collective action does not follow automatically from the cognitive apprehension of chains of cause and effect. Rather, the extension of consequences and affected interests over space and time serves as the vector through which people learn to abstract themselves from their own perspectives. For Dewey, a public is primarily an imaginative entity: 'The idea of a public that responds to events even though most members are not immediately affected was Dewey's formulation of the location of the political and of civic virtue' (Wolin, 2004: 511).

Dewey's account of publicness therefore sketches the outlines of what the title of this chapter refers to as a parasitical notion of public space. By this, I mean to draw attention to how the matters of concern that define the object of public interest, as well as the styles of engagement through which publics constitute themselves as collective subjects, both depend on prior processes of infrastructural and socio-cultural development.

But publicness as such exceeds both the infrastructural conditions that generate objects of public concern and the forms of social solidarity through which people's dispositions to public engagement are cultivated.

We can now return to the idea that the notion of a 'community of affected interest' offers an alternative criteria of participation, one that shifts attention away from the question of 'Who is a member?', the answer to which always ends up seeming somewhat arbitrary, and onto questions of 'Who is affected?' (Shapiro, 2003: 223). Geographers are drawn to this 'affected interest' model because it is easy to think that geography is in a good position to answer the second question. Geography can easily re-tool itself as a way of tracking chains of cause and effect, actions and their dispersed consequences. Unfortunately, the 'causal' principle is not quite as straightforward as it seems. It is actually rather difficult to disentangle simple relations of cause and effect, actions and consequences, when dealing with complex social, economic, or cultural processes.

What is more, we might pause for thought before rejecting the territorial criteria of participation out of hand. Territorial definitions have the advantage of efficiently solving the problem of how to determine rights to participate. Any argument against this principle must address the extent to which territorial definitions are basic mechanisms for ensuring effective rights of equal participation (Saward, 1998). Territorial models of citizenship presume that any member of a polity has the equal right to participate in collective decisions even if they are not directly affected by them. It is this principle, in fact, that is crucial to understanding how publics come into being on the back, as it were, of other processes of geographical expansion and extension. It turns out that nationalism might be the best paradigm available for understanding the possibility of the sorts of imaginative action through which spatially extensive and temporally durable publics constitute themselves (Calhoun, 1997).

In both the work of Warner and Dewey, 'publicness' is ascribed to a family of related types of action: action-in-concert with others; action undertaken in public, in the open; and actions around objects of widely shared concern. And for both writers there is also a strong sense that public action is parasitical on the material configurations and social relations laid down by other forms of activity, in the sense that it is dependent on these as its conditions of possibility, as well as in the sense that it is these conditions that in turn become the object of transformative public action.

We have learned from Dewey that an intrinsic feature of the relationship between publicness and democracy is that publicness has an instrumental dimension – that is, that public action is about something. This is not all publicness is, but an account that diminishes this aspect as somehow intruding into authentic publicness is an account that will have difficulty in accounting for just why we are always so worried about the question of the public sphere in the first place.

Political philosophers such as Hannah Arendt, Sheldon Wolin, and Claude Lefort all tend to expel any instrumental calculation from the realm of authentic public action and political activity. This same suspicion informs the theories of radical democracy and agonistic democracy developed by writers such as Chantal Mouffe and William Connolly, who have also attracted the attention of geographers in recent years. In this tradition, publicness has no object, it isn't about anything, it is a pure means in itself. These theories are helpful because they acknowledge the value of the affective dimensions of publicness as an important dimension of political action. But they also encourage a style of theoretical evaluation that verges on the self-righteous and narcissistic by supposing that public action and politics are best thought of as activities of pure self-creation detached from instrumental concerns. Public action is made to look like an end in itself. These theories put the normative cart before the practical horse (Elster, 1983: 91–100), in that they fail to acknowledge that the qualities of sociable, convivial interaction that they propose as the essence of public life depend on 'a range of decisions, actions and policies that cannot emerge from the flow of everyday sociability alone' (Weintraub, 1997: 24). And in this, these agonistic, radical theories of publicness and democracy actually converge with the more liberal, deliberative theories they often take as their conceptual antagonists (Schudson, 1997). Both approaches suppose that democracy can do without instrumental procedures for making decisions, because they either think that this involves an illegitimate closure of the free play of pluralist difference, or because they hope that a deliberative consensus can arrive at fully legitimate decisions while leaving no sore losers. And in fundamental respects, this shared difficulty with imagining how pluralism and autonomy can be squared with binding and legitimate decision-making derives from the fact that both deliberative and agonistic theories underplay the temporalities that articulate the intrinsic qualities of public life and democratic politics with their equally compelling instrumental imperatives (see Saward, 2003; Barnett, 2004, 2005).

On both the intrinsic and instrumental definitions of publicness outlined above – the criterion of openness to initiation of communication by others, and the criterion of having to do with the general interest – we can see that the idea that public space is 'material', where this is supposed to mean spaces of co-presence like the street, parks or the city, or even a causal space of actions and consequences, is entirely inadequate for thinking about the relationship between publicness and democracy. The sense of spatiality that is best adjusted to thinking about this relationship is characterized not by the idealization of dialogue, or of face-to-face theatricality, or urban sociability, but by reference to a vocabulary of dissemination, scattering, and dispersal (Peters, 1999). Addressing a message to others always traverses a space full of the hazards of misfire, misunderstanding, and failure. Messages might be intercepted by unanticipated addressees in unanticipated places, or by none at all:

> Communication occurs only insofar as the delivery of the message may fail: that is, communication takes place only to the extent that there is a separation between the sender and the receiver, and this separation, this distance, this spacing, creates the possibility for the message not to arrive. (Chang, 1996: 216)

If, then, the space of publicness is a circulatory space of indeterminate address, this is best exemplified by all those characteristics ascribed by Derrida to textuality – drift, dissemination, chance – and by the separations of temporal deferral as much as spatial distance (Barnett, 1999).

THINKING PUBLICLY

So, after all this, what is publicness, and how does it relate to democracy? Ferree et al. (2002: 316) suggest that there are at least four different criteria upon which the relationships between public communication and democracy can be evaluated:

1 The first criterion focuses on who participates. For example, should democracy be thought of as a system of elite, expert rule regulated by elections, or should it be more participatory?
2 The second criterion focuses on what sort of process is taken to embody public communication. For example, what are the relative merits of a 'market-place of ideas' model compared to more deliberative practices?
3 A third criterion focuses on how ideas should be presented in public communications. For example, how far should norms of detachment, disinterest, and civility govern public debate, or how far are forms of narrative and non-deliberative symbolic acts not only legitimate but essential elements of democratic public communication?
4 And finally, the fourth criterion focuses on the outcomes of the relationships between discourse and decision-making. For example, is consensus around decisions the primary goal of public communication, and should debate be restricted once a decision has been made, or is this emphasis on consensus and closure systematically undemocratic?

This chapter has touched on aspects of each of these criteria:

1 I have presumed in favour of an expansionary understanding of who should be involved in democratic decision-making, while recognizing that a certain division of labour between roles is both inevitable and valuable, in complex societies.
2 I have also presumed in favour of an expansive, deliberative conception, while trying to acknowledge that market-led models of private preference do carry an important normative lesson, in so far as they are guided by a presumption in favour of respecting people's own opinions as to their best interests and by a healthy scepticism about paternalist interventions in the name of others.
3 I have also pointed to those approaches which develop a notion of public communication that acknowledges the importance of non-deliberative, affective styles of presentation, but without supposing that these are wholly opposed to rational, reasonable, cognitive forms of justification.
4 It has been the last of these four criteria, though, to which this chapter has given most attention. This is because quite a lot of research in geography forgets that the reason for worrying about publicness is because of the relationship between public communication and democratic decision-making and accountability.

It is because he keeps this last relationship constantly in focus that the work of Habermas remains so compelling when it comes to thinking about issues of public life and democracy. Habermas probably over-rationalize the style of communication that is required for public life to contribute to thoroughgoing democratization. But his project retains its power in spite of this precisely because he keeps his eye on the key relationship between publicness and democracy, that is, the relationship between chatting, joking, deliberating, and the problems of legitimate and just popular rule.

By emphasizing that the democratic value of publicness inheres in the quality of this relationship, I mean to suggest that the democratic qualities of any public sphere should not be judged narrowly by whether it promotes rational deliberation, or alternatively by whether on its own it is accessible or inclusive. Rather, what is crucial is the degree to which the overall network of public practices enables people to 'keep tabs on the political world' (Schudson, 1998: 238). The fundamental issue at stake in evaluating the democratic qualities of public life is

> whether, when an issue arises, citizens have various effectual access points to governmental decision-makers. The effective operation of a public sphere depends also on whether, through the networks of talk, complaint, letters, petitions, interest groups, parties, suits, demonstrations, and picket lines, people feel they can and actually can move issues onto the public agenda. (ibid.)

The main point of this chapter has been to suggest that, when assessing this question, we need to focus on a set of relations between different types of action, some more open and fluid than other, more strategic forms.

I have argued for an idea of publicness that reorients our attention to thinking about public space as any communicative space of address-and-response. Rather than modelling public space on the idea of gathering and assembly in the presence of others, we should look at the ways in which publics are convened through practices of dissemination, dispersal and scattering. This notion of the convening of publics is related to the sense that any 'public' always holds something in reserve, because the public is always spoken for and spoken to, which is another way of saying that 'it' might always answer back in unexpected ways. The public is, therefore, not to be found anywhere special, it has no proper place, nor any exemplary spatiality. As we saw above, publics cannot come into existence without presupposing infrastructures of communications and patterns of social interaction, but neither are these material or social configurations in themselves publics. A public emerges when these presupposed forms and patterns are exceeded, made strange, and used as a medium to imaginatively project towards an unknown addressee, to invite them to share one's concerns, to care about things together.

We need to retain an appreciation of the magic that is involved in carrying off the acts that help make up and sustain democratic publicness. Publics are called into existence, they are convened, which is to say that they are sustained by establishing relations of attention whose geographical configurations are not given in advance.

REFERENCES

Adams, P. (1992) 'Television as gathering place', *Annals of the Association of American Geographers*, 82: 117–35.

Adams, P. (1996) 'Protest and the scale politics of communications', *Political Geography*, 15: 419–41.

Adams, P. (1998) 'Network topologies and virtual place', *Annals of the Association of American Geographers*, 88: 88–106.

Arendt, H. (1958) *The Human Condition*. Chicago: University of Chicago Press.

Barnes, M., Newman, J., Knops, A. and Sullivan, H. (2003) 'Constituting "the public" in public participation', *Public Administration*, 81: 379–99.

Barnett, C. (1999) 'Deconstructing context: exposing Derrida', *Transactions of the Institute of British Geographers*, NS, 24: 277–93.

Barnett, C. (2003) *Culture and Democracy: Media, Space and Representation*. Edinburgh: Edinburgh University Press.

Barnett, C. (2004) 'Deconstructing radical democracy: articulation, representation and being-with-others', *Political Geography*, 23: 503–28.

Barnett, C. (2005) 'Temporality and the paradoxes of democracy', *Political Geography*, 24: 1–17.

Barry, A. (2000) *Political Machines: Governing a Technological Society*. London: Athlone Press.

Bridge, G. (2004) *Reason in the City of Difference: Pragmatism, Communicative Action and Contemporary Urbanism*. London: Routledge.

Calhoun, C. (1997) *Nationalism*. Buckingham: Open University Press.

Calhoun, C. (1998) 'Community without propinquity revisited: communications technology and the transformation of urban public space', *Sociological Inquiry*, 68: 373–97.

Calhoun, C. (2002) 'Imagining solidarity: cosmopolitanism, constitutional patriotism, and the public sphere', *Public Culture*, 14: 147–71.

Cavell, S. (1979) *The Claim of Reason*. Oxford: Oxford University Press.

Chang, B. (1996) *Deconstructing Communication*. Minneapolis: University of Minnesota Press.

Corbridge, S. (1998) 'Development ethics: distance, difference, plausibility', *Ethics, Place and Environment*, 1: 35–53.

Couldry, N. and McCarthy, A. (eds) (2004) *MediaSpace: Place, Scale and Culture in a Media Age*. London: Routledge.

Davies, G. (2006) 'The sacred and the profane: biotechnology, rationality, and public debate', *Environment and Planning A*, 38: 423–43.

Derrida, J. (1992) *The Other Heading*. Bloomington: Indiana University Press.

Dewey, J. (1927) *The Publics and Its Problems*. Athens, OH: Ohio University Press.

Dewey, J. (1980) *Democracy and Education*. Carbondale: Southern Illinois University Press.

Dummett, M. (1984) *Voting Procedures*. Oxford: Oxford University Press.

Eckersley, R. (2000) 'Deliberative democracy, ecological representation and risk: towards a democracy of the affected', in M. Saward (ed.), *Democratic Innovation*. London: Routledge.

Elster, J. (1983) *Sour Grapes: Studies in the Subversion of Rationality*. Cambridge: Cambridge University Press.

Elster, J. (1997) 'Market and forum: three varieties of political theory', in R.E. Goodin and P. Pettit (eds), *Contemporary Political Philosophy*. Oxford: Blackwell, pp. 128–43.

Emirbayer, M. and Sheller, M. (1999) 'Publics in history', *Theory and Society*, 28: 145–97.

Ferree, M., Gamson, W., Gerhards, J. and Rucht, D. (2002) 'Four models of the public sphere in modern democracies', *Theory and Society*, 31: 289–324.

Fishkin, J. (1995) *The Voice of the People*. New Haven: Yale University Press.

Gaonkar, D. and Povinelli, E. (2003) 'Technologies of public forms: circulation, transfiguration, recognition', *Public Culture*, 15: 385–97.

Goheen, P. (1998) 'Public space and the geography of the modern city', *Progress in Human Geography*, 22: 479–96.

Goodin, R. (2002) *Reflective Democracy*. Oxford: Oxford University Press.

Habermas, J. (1989) *The Structural Transformation of the Public Sphere*. Cambridge: Polity Press.

Herbst, S. (1996) *Numbered Voices: How Opinion Polling Has Shaped American Politics*. Chicago: University of Chicago Press.

Howell, P. (1993) 'Public space and the public sphere', *Environment and Planning D: Society and Space*, 11: 303–22.

Keane, J. (2004) *Violence and Democracy*. Cambridge: Cambridge University Press.

Latour, B. and Weibel, P. (eds) (2005) *Making Things Public: Atmospheres of Democracy*. Cambridge, MA: MIT Press.

Lee, B. and LiPuma, E. (2002) 'Cultures of circulation: the imaginations of modernity', *Public Culture*, 14(1): 191–213.

Low, M.M. (2003) *Democracy's Black Boxes: Opening Up Electoral Geography*. LSE Research Papers in Environmental and Spatial Analysis No. 86. London: London School of economics.

Low, S. and Smith, N. (eds) (2005) *The Politics of Public Space*. London: Routledge.

Massey, D. (2004) 'Geographies of responsibility', *Geografiska Annaler*, 86B: 5–18.

Mitchell, D. (1995) 'The end of public space? People's park, definitions of the public, and democracy', *Annals of the Association of American Geographers*, 85: 108–33.

Mitchell, D. (2003) *The Right to the City: Social Justice and the Fight for Public Space*. New York: Guilford Press.

Newman, J. (2005) 'Going public: people, policy, and politics'. Inaugural Lecture, The Open University, 18 May, available at <http://www.open.ac.uk/ socialsciences/staff/jnewman/info.html>. (Accessed 3 August 2005.)

Offe, C. (1997) 'Micro-aspects of democratic theory: what makes for the deliberative competence of citizens?', in A. Hadenius (ed.), *Democracy's Victory and Crisis*. Cambridge: Cambridge University Press, pp. 81–104.

Peters, J.D. (1999) *Speaking into the Air*. Chicago: University of Chicago Press.

Przeworski, A., Stokes, S. and Manin, B. (eds) (1999) *Democracy, Accountability and Representation*. Cambridge: Cambridge University Press.

Rawls, J. (1993) *Political Liberalism*. New York: Columbia University Press.

Samarajiva, R. and Shields, P. (1997) 'Telecommunication networks as social space: implications for research and policy and an exemplar', *Media, Culture and Society*, 19: 535–55.

Saward, M. (1998) *The Terms of Democracy*. Cambridge: Polity Press.

Saward, M. (2003) 'Enacting democracy', *Political Studies*, 51: 161–79.

Scannell, P. (1995) *Radio, Television and Modern Life*. Oxford: Blackwell.

Scannell, P. (2000) 'For-anyone-as-someone structures', *Media, Culture and Society*, 22: 5–24.

Schudson, M. (1997) 'Why conversation is not the soul of democracy', *Critical Studies in Mass Communication*, 14: 297–309.

Schudson, M. (1998) *The Good Citizen: A History of American Civic Life*. Cambridge, MA: Harvard University Press.

Shapiro, I. (1999) *Democratic Justice*. New Haven: Yale University Press.

Shapiro, I. (2003) *The Moral Foundations of Politics*. New Haven: Yale University Press.

Smith, D.M. (1998) 'How far should we care? On the spatial scope of beneficence', *Progress in Human Geography*, 22: 15–38.

Squires, J. (1994) 'Private lives, secluded places: privacy as political possibility', *Environment and Planning D: Society and Space*, 12: 387–401.

Staeheli, L.A. (1996) 'Publicity, privacy, and women's political action', *Environment and Planning D: Society and Space*, 14: 601–19.

Staeheli, L.A. and Mitchell, D. (2004) 'Spaces of public and private: locating politics', in C. Barnett and M.M. Low (eds), *Spaces of Democracy*. London: Sage, pp. 147–60.

Sunstein, C. (1995) *Democracy and the Problem of Free Speech*. New York: The Free Press.

Sunstein, C. (1997) 'Preferences and politics', in R.E. Goodin and P. Pettit (eds), *Contemporary Political Philosophy*. Oxford: Blackwell, pp. 156–73.

Sunstein, C. (2001) *republic.com*. Princeton: Princeton University Press.

Tilly, C. (1994) 'Social movements as historically specific clusters of political performances', *Berkeley Journal of Sociology: A Critical Review*, 38: 1–30.

Warner, M. (2002) *Publics and Counterpublics*. New York: Zone Books.

Watson, S. (2004) 'Cultures of democracy: spaces of democratic possibility', in C. Barnett and M.M. Low (eds), *Spaces of Democracy*. London: Sage, pp. 207–22.

Weintraub, J. (1997) 'The theory and politics of the public/private distinction', in J. Weintraub and K. Kumar (eds), *Public and Private in Thought and Practice*. Chicago: University of Chicago press, pp. 1–42.

Wolin, S. (2004) *Politics and Vision: Continuity and Innovation in Western Political Thought* (expanded edn). Princeton: Princeton University Press.

Young, I.M. (1990) *Justice and the Politics of Difference*. Princeton: Princeton University Press.

Young, I.M. (1999) 'State, civil society and social justice', in I. Shapiro and C. Hacker-Cordón (eds), *Democracy's Value*. Cambridge: Cambridge University Press, pp. 141–62.

Global Political Geographies

Introduction

Murray Low

In a book notable for its role in the recent revival of 'geopolitics' in political geography, Gearóid Ó Tuathail quotes Halford Mackinder to the effect that: '[o]ur aim must be to make our whole people think Imperially – think that is to say in spaces that are world wide – and to this end our geographical teaching should be directed' (Mackinder, 1907: 38, quoted in Ó Tuathail, 1996: 141). In a world where the idea of thinking globally has almost become unquestioned among geographers and non-geographers alike, this statement should give us some pause for thought. As Mackinder's imperialist ambitions remind us, 'thinking globally' is not an innocent matter, which we can see as an uncontroversially good thing from the point of view of geography as a particular discipline or from the point of view of broader societal knowledges. Whether 'thinking globally' is a good thing or not depends on how we do it, and the reach or comprehensiveness of our thinking is no guarantee that it will be any less ideological or normatively justifiable.

Political geography has been strongly animated by the imperative to think globally for a long time. Mackinder-style geopolitics is but one manifestation of this and, if anything, the revival of political geography since the late 1970s has pivoted around this impulse. For example, Peter Taylor's widely used textbook (first published in 1985; the latest edition is Flint and Taylor, 2007), along with his other writings on the theme of a world-systems approach (e.g. Taylor, 1982, 1991, 1993), draws on two highly influential exemplars of 'thinking globally', the work of Braudel and Wallerstein, to frame political geography. For him, specific relations between cities and localities, the (national) state system and a global economy jointly characterize 'how the modern world works' (Taylor, 1996). In the 1990s, two further developments have been notable. First, there has been an explosion of work on 'globalization' to situate a wide variety of emergent realities, largely at the urban or local level (see Low and Barnett, 2000). Second a new, critical, geopolitics has emerged, drawing

on developments in political theory, continental philosophy, and on similar tendencies in the field of international relations (for the latter see, *inter alia*, DeDerian and Shapiro, 1989; Walker, 1992; Sylvester, 1994). Related developments, such as the mainstreaming of post-colonial themes and the current strong critical focus on 'neo-liberalism', both of which have been closely allied to critical geopolitics, have reinforced this global emphasis within political geography.

These various lines of development in international or transnational political geography pose significant challenges to some embedded practices and assumptions of the sub-discipline. For example, the Wallersteinian thread in recent decades has strongly and explicitly linked questions of scale-relativity with a critique of nationalist or statist social science, with its concomitant structuring of data and information rendering other, less state-centred investigations more difficult to imagine and carry out (see Wallerstein, 1991). In critical geopolitics (alongside Ó Tuathail see, *inter alia*, Dalby, 1990; Agnew, 1998), the politicization of supposedly objective geo-strategic knowledge, including earlier forms of geopolitics, produced in and between states and universities, has been a central concern (see Smith, 2003, for a fascinating and detailed account of the politics of geopolitical knowledge). A major innovation here has been the explicit mobilization of concepts and arguments from continental philosophy and critical theory in which a critique of knowledge-power relations has been a central theme (see Dell'Agnese, this volume).

All of this has positioned political geography quite well to address the challenge of what thinking globally might mean. There is a series of interesting and important problems here. Wallerstein's fascinating (and apparently unfinished) project of developing a history of 'the modern world-system', for example, raises (among other things) issues regarding the capacities of individuals or even teams of researchers to realise aspirations to this kind of global knowledge.

More worryingly, global-scale thinking may, as Gibson-Graham (2002) among others has suggested, involve implicit valuations of what matters scale-wise in the world (and in the academy), with 'local' institutions, activities, experiences, identities and research being devalued in a context where acting and knowing at a global or world scale is more highly prized. Post-colonial analyses, while in a sense reinforcing the aspiration to know the world better, have also been at the forefront of asking questions about whose knowledge about the world gets to count, and what the implications of the different locations of this knowledge production and dissemination might be (see, e.g., Robinson, 2006).

There are also important issues around how to effectively conceptualize the actors, institutions, mechanisms or sets of connections that are 'more than national' or 'transnational', that is, objects and identities 'in between' those entities (like states) which have traditionally been taken as tractable foci for research (see, inter alia, Mitchell 1997, Sassen 2006, Smith 2000, Taylor 1995). Imperatives to constantly connect local or territorially delimited agents or events to wider global contexts runs the risk of diminishing the importance of processes with a smaller spatial reach or strong territoriality and thus obscuring their analysis. Partly as a result of research on globalization, it has become a common, and to some extent useful, tactic for geographical analysis to highlight the 'messy', extensive and relational nature of social phenomena, and to draw attention to the blurry nature of boundaries and their constitutive porosity. Political geography, and the wider social science-humanities debates on global matters in recent years, have made great strides in recognizing the limits of approaches that cannot cope with complexity, 'in-between-ness' and fluidity – particularly where these approaches have bought their coherence through a pact with the territorialized national state. However, it could be suggested that there has been less success in finding new strategies for ordering and limiting inquiry to produce committed and plausibly specific theorizations of transnational political phenomena (these themes are central to contemporary thinking about socio-spatiality – see, for example Amin and Thrift, 1997; Massey, 2005).

Here there has been a strong and understandable recourse to the language of continental philosophy and theory – the language of insides and outsides, mutual constitution, mediations between self and other, identity and difference, limits and transgression, delimitation and transcendence, de-/re-/territorialization, networks and topological spaces, for example (see Castree, Featherstone and Herod, this volume, on networks; Agnew and Kuus, this volume, on de-territorializations of state theory; in addition to the contributions to critical international relations or geopolitics cited earlier,

see also Hardt and Negri, 2000, Ó Tuathail and Dalby, 1998). These creative engagements with philosophical languages have certainly proved useful in dislodging certain assumptions embedded in inherited, more specifically social-scientific, language about world politics, which has owed much to a theoretical imagination more strongly concerned with territories and bounded agents, especially the nation-state. However, it seems that this new language can only get researchers so far in trying to understand the often forceful power relations involved in many 'global' processes, such as transnational corporations, war, water disputes, dilemmas regarding refugees and population displacements, the claims of cultures vis-à-vis other forms of social division, ideology and new forms of media, de-regulated capitalism, and new forms of state, sovereignty, and imperialism. A major dilemma, then, is how to develop new, more specialized language – or to revivify some older theoretical concepts (see, e.g., Buck-Morss, 2003) – to engage with non- or post-national objects of inquiry. This has to be different from Wallerstein's 'unthinking social science' but also better at grasping political dynamics than concepts from philosophy, which are appropriately general in their disciplinary context of origin, but are not always suited to specifying the variety of objects involved in social science research.

In a sense, then, the desire to know 'the world', or to 'think globally', is a profoundly difficult conceptual task in addition to being empirically challenging. Since the ambition to theorize globality often also carries with it a certain desire for (at least intellectual) mastery, it is perhaps fortunate that at times it can seem a extremely difficult project. Reflection on the various twists and turns of this desire to world knowledge over the decades is perhaps especially essential to political geography, because it has since its beginnings repeated this aspiration in ways that are at first glance normatively and politically quite diverse. The inevitable limits to this project can, however, be a means to re-configure aspirations and concepts more modestly to explain and intervene in established and emerging geopolitical configurations that 'nationalist' social science has been ill-equipped to see.

Simon Dalby's chapter reflects directly on what it means to conceptualize political geography as 'global' and, as such, is an excellent entry point into some of these issues. He consistently treats the emergence and circulation of discourses about 'the global' and 'globalization' as a historically specific set of very recent processes, charged with ideological meanings, and aligned with diverse political and economic projects. In tracking the stakes involved across a variety of areas of contemporary concern, including the economy, media and information, global politics, the environment, resistance to neo-liberalism, urban and infrastructural development, and warfare, he emphasizes the suspicion that 'globalization' has occupied a vacuum left by the discrediting or decay of the various comprehensive social and political theories of the twentieth century.

The revival of geopolitics seems to hinge, in part, on an encounter between political geography's heritage in terms of analysing international matters, and continental European critical theory. Indeed, continental philosophy has tended to attract rather more attention from Anglo-American political geographers than has continental political geography. In her chapter in this section, Elena Dell'Agnese surveys the emergence of critical geopolitics, locating it in a different narrative emphasizing the contributions of European, specifically Italian and French, critical geographers since the late 1960s. The work of Lucio Gambi, Yves Lacoste, Massimo Quaini, Giuseppe Matteis and Claude Raffestin, among others, is at best unevenly known and acknowledged in political geography's primary circuits of dissemination, publication, conferencing and canon construction. Yet, as Dell'Agnese reveals, a convincing story can be told about the origins of a critical political geography – including critical geopolitics – which engaged with theorists such as Gramsci and Foucault by tracing this alternative network of European writers. One of our aims in this *Handbook* is to try to highlight work in the field produced outside the Anglo-American sphere, and although European universities are still relatively privileged sites of knowledge production in global perspective, Dell'Agnese's chapter is crucial in suggesting that a neglected aspect of 'knowing the world' might be an imperative to better know and acknowledge a more dispersed genealogy of contemporary political geography than has become usual (see Mamadouh 2003, Sidaway, this volume).

In the 1980s and much of the 1990s, if the theme of 'Empire' – Marxian theories of imperialism aside (Blaut, 1993) – received an airing, it was likely to be with respect to the past. Wallerstein's 'modern world-system' achieves its conceptual shape partly through a contrast with empires and indeed functions as it does because imperial aspirations to make economic and political scale coincide have, at crucial moments, run aground. In the broader public sphere, Cold War rhetoric associated talk of contemporary empire with certain characterizations of the USSR, and references to the United States as embodying 'imperial' characteristics (as in the phrase 'imperial presidency') seemed meaningful, but worked at the level of useful metaphor rather than that of conceptual accuracy.

The new century has, by contrast, ushered in a serious and widespread use of Empire as a term for emergent global realities and as a characterization of the transnational reach of American power in particular (see, e.g., Gregory, 2004 and Harvey, 2003). Alan Lester's contribution here on this theme is salutary in the way that it works through recent discussions of Empire on left and right, contesting some of the more dramatic postures to be found therein and reversing some prevalent assumptions about imperialism in the nineteenth century and in the present. Careful characterization of the differences between British and US imperialisms, he suggests, should lead us to focus on these as 'particular constellations of diverse and often antagonistic interests, neither wholly economic, cultural or narrowly political in their inspiration, acting to foster and attempting to harness globalizing transactions in pursuit of particular projects', an aim that they 'never quite succeed' in attaining. To some extent, in the wake of his analysis, we can read 'globalization' or 'the global' here as placeholder terms for that which exceeds the grasp or reach of established configurations of power knowledge, with interesting implications for the limits of Mackinder's 'imperial thinking', as well as for broader constructions of Empire in the contemporary world. Lester establishes the need for nuanced and modest accounts of the many different processes shaping phenomena too often characterized through the shorthand term, globalization.

One site for pursuing this theoretical ambition is found in the revival of the traditional political geography field of 'border studies.' Indeed, here is a theme where work in philosophy, by Derrida and others (see Derrida, 1978, Bennington, 1994) has been the source of much debate on all kinds of frontiers, borders and borderlines in recent social science and humanities debates. Jouni Häkli notes in this chapter here that 'borders and boundaries represent an immense area of research including almost everything that pertains to humans and societies.' Nonetheless, after exploring political geography's earlier preoccupations with borders in relation to statecraft, Häkli usefully explores the variety of forms that borders currently take, emphasizing their 'leakiness' in the context of globalization, their relation to migration, citizenship, identities and ideas about multi-layered governance. By throwing light on the importance of different border contexts, while remaining cautious about over-generalizing the implications of cross-border phenomena for the importance of borders as such, his contribution helps move discussion on from a focus on 'the' border or 'the' boundary and any unitary account of their demise in the face of globalization, to a more comparative and, indeed, more properly geographical set of questions concerning borders and crossings.

Rachel Silvey, Elizabeth Olson and Yaffa Truelove's chapter here addresses migration in terms of the relationship between mobility and immobility. In so doing, they seek to relate migration as a phenomenon to debates around what constitutes the political in political geography; to emerging theorisations of politics, the state, and governmentality; and to long-standing concerns with the politics of scale. By revisiting feminist arguments about spatial entrapment, they stress the ongoing importance of immobility as a theme in political geography, ironically produced as a result of transnational mobility. They also provide a strong reminder of 'the continued importance of coercion and dominance' in the migration field. By stressing the importance of attending to the scales of the household and the body for critical migration research, the authors provide an excellent example of careful theorization of processes too easily encompassed in concepts such as 'the global'. Their exploration of the concept of 'transnationality' renders this concept less abstract and more 'embedded' in particular contexts and sets of cross-boundary relations.

Stylistically quite different, John O'Loughlin and Clionadh Raleigh's contribution addresses the analysis of civil war in regional and global contexts. While migration seems a concept made for exploration via quantitative research, violence, a central political concept but one that has been hard to theorize and connect effectively with other concerns, seems to invite the use of different descriptive and analytical tools. Yet, as Silvey et al. showed, critical migration research has found much of value in pursuing agendas linked to contemporary political and cultural theory. Conversely, O'Loughlin and Raleigh set out to show, though an analysis of civil conflicts in the Great Lakes region of Africa and a global analysis of the effects of civil wars, that 'the spatial-analytic tradition of human geography that examines geographically expressed phenomena for patterns and general trends' can provide useful insight into patterns of violence at different scales, including that of the globe. Reviewing the literature on the causes of civil war violence, and noting that this in no way exhausts the field of violence as a concern in the world, they emphasize the ways in which established approaches in political science or economics find difficulty incorporating geography into their accounts, and discuss the importance of relative location, cross-boundary connections and flows, and geographic scale. The problem of acquiring global knowledge is stressed throughout. The limitations of data collection and organization at the national level, and its uneven development, strongly constrains the examination of those trans-boundary processes typical of 'civil' wars and thus for developing a greater empirical understanding of this form of violence and its effects. As a result, '[d]isaggregated data beyond the level of the nation-state have not yet been widely explored in the study of civil war violence' and state-centred social science has begun to exhibit crucial weaknesses in this research field.

The kinds of detailed geo-coded satellite and other data that O'Loughlin and Raleigh envisage in their conclusions may not be available in the right form and on the right scale for some time, and this limit on research strategies may mean that more qualitative, critical-theory-driven reflections will continue to dominate conceptualization of transnational or 'global' phenomena, including violence and war, for some time. Different knowledge strategies will be, in any case, essential to thinking through and refining understandings of processes reaching beyond states. Refusing 'easy' senses of thinking globally means that there is much to do conceptually, and perhaps above all empirically, to establish useful and plausible analyses in the area between the national and the world-as-a-whole. 'The global', or 'the transnational', have been useful heuristic concepts, but perhaps increasingly have a slightly hollow ring to them; they are labels that in some ways avoid working with the more differentiated, socio-spatially specific forms of political life 'out there' that have been highlighted in the chapters in this section. 'Global' political geography's future is bright, in other words, but not necessarily because it thinks globally but because it thinks geographically.

REFERENCES

Agnew, J.A. (1998) *Geopolitics: Revisioning World Politics.* London: Routledge.

Amin, A. and Thrift, N.J. (1997) 'Globalization, socio-economics, territoriality', in R. Lee and J. Wills (eds), *Geographies of Economies.* Arnold: London.

Bennington, G. (1994) 'The frontier: between Kant and Hegel', in *Legislations.* London: Verso.

Blaut, J. (1993) *The Colonizer's Model of the World.* Guilford Press: New York.

Buck-Morss, S. (2003) *Thinking Past Terror: Islam and Critical Theory on the Left.* London: Verso.

Dalby, S. (1990) *Creating the Second Cold War: The Discourse of Politics.* London: Pinter.

DeDerian, J. and Shapiro, M. (1989) *International/Intertextual Relations: Postmodern Readings of World Politics.* New York: Lexington.

Derrida, J. (1979) 'Living on/border lines', in H. Bloom et al. (eds), *Deconstructionism and Criticism.* New York: Seabury Press.

Flint, C. and Taylor, P.J. (2007) *Political Geography: World-Economy, Nation-State, Locality.* London: Pearson.

Gibson-Graham, J.K. (2002) 'Beyond global vs. local: economic politics outside the binary frame', in A. Herod and M.W. Wright (eds), *Geographies of Power: Placing Scale.* Oxford: Blackwell.

Gregory, D. (2004) *The Colonial Present.* Oxford: Blackwell.

Hardt, M. and Negri, A. (2000) *Empire*. Harvard University Press: Cambridge, Mass.

Harvey, D. (2003) *The New Imperialism*. Oxford: Oxford University Press.

Low, M.M. and Barnett, C. (2000) 'After globalization', *Environment and Planning D: Society and Space*, 18(1): 53–61.

Mackinder, H. (1907) 'On thinking imperially', in M.E. Sadler (ed), *Lectures on Empire*. London: privately printed.

Mamadouh, V. (2003) 'Some notes on the politics of political geography', *Political Geography* 22(6): 663–675.

Massey, D. (2005) *For Space*. Routledge: London.

Mitchell, K. (1997) 'Transnational discourse: bringing geography back in', *Antipode* 29: 101–114.

Ó Tuathail, G. (1996) *Critical Geopolitics*. Minneapolis: University of Minnesota Press.

Ó Tuathail, G. and Dalby, S. (1998) (eds.) *Rethinking Geopolitics*. Routledge.

Robinson, J. (2006) *Ordinary Cities: Between Modernity and Development*. London: Routledge.

Sassen, S. (2006) *Territory, Authority, Rights: From Medieval to Global Assemblages*. Princeton University Press: Princeton.

Smith, M.P. (2001) *Transnational Urbanism: Locating Globalization*. Blackwell: Oxford.

Smith, N. (2003) *American Empire: Roosevelt's Geographer and the Prelude to Globalization*. Berkeley: University of California Press.

Sylvester, C. (1994) *Feminist Theory and International Relations in a Postmodern Era*. Cambridge: Cambridge University Press.

Taylor, P.J. (1982) 'A materialist framework for political geography', *Transactions of the Institute of British Geographers*, NS, 7: 15–34.

Taylor, P.J. (1991) 'Political Geography within World-Systems analysis', *Review* 14: 387–402.

Taylor, P.J. (1993) 'States in world-systems analysis: massaging a creative tension' in B. Gills and R. Palan (eds), *Domestic Structures, Global Structures*. Lynne Rienner: Boulder.

Taylor, P.J. (1995) 'Beyond containers: internationality, interstateness, interterritoriality', *Progress in Human Geography* 19, pp. 1–15.

Taylor, P.J. (1996) *How the Modern World Works*. Chichester: Wiley.

Walker, R.B.J. (1992) *Inside/Outside: International Relations as Political Theory*. Cambridge: Cambridge University Press.

Wallerstein, I. (1991) *Unthinking Social Science: The Limits of Nineteenth-Century Paradigms* (2nd edn). Philadelphia: Temple University Press.

'Global' Geopolitics

Simon Dalby

THE GLOBAL

According to Martin Albrow (1997) we now live in 'the Global Age', a historical designation that comes temporally after modernity and that suggests a qualitatively different era. Contemporary accelerations and the increasing importance of television events suggest a new sense of the world, a common understanding of being residents of a single planetary entity, a sense of 'globality'. Global suggests both large scales incorporating all of humanity and its biosphere while simultaneously the proximity and interconnectedness of this totality.

The invocation of 'the global' became commonplace in the 1990s. It appears as an obvious natural entity but also frequently refers to social, cultural, political and especially economic matters. It invokes an adjectival sense of being all-encompassing, while simultaneously 'grounding' meaning in a reference to the terrestrial 'globe'. The term is rich with symbolism while also being a powerful descriptive term capable of multiple invocations. Global as an adjective is linked to everything from business to ethics, environment to politics, sport to warfare. Ironically, the all-encompassing meaning of the adjective is reflected in the all-encompassing use of the term itself.

Cultural phenomena and the role of media in particular in enhancing senses of a single globe are important in understanding the shifting sense of the planetary matrix in which we all live. In particular, the coverage of world sporting competitions, where literally billions of spectators simultaneously watch a small group of athletes perform their feats, suggests a common 'global' community. But at the same time, the new media's fascination with 'being there' reproduces predictable scripts of human fate regardless of geographical location in a way that suggests a similarity of all human experience that substantially occludes cultural difference and works to suggest a leveling of the global 'playing field'.

Global has also become a process in and of itself, simply called globalization. Globalization came to prominence in the aftermath of the Cold War. Both as a term and as a designation for the supposed novelty of the post-Cold War geopolitical situation, it rapidly displaced many earlier specifications of contemporary social circumstances. A term with a nearly magical promise to describe and explain social and economic phenomena, it implied connections and cultural homogenization while occluding the specific geographic patterns of change. It supposedly promised much but prevented most people from actually doing anything much about how their circumstances changed as a result of its planet-spanning processes. Inevitable change, cultural transformation, connections to a rapidly shrinking world where everything was speeding up, all expressed breathlessly by commentators and academics caught up in the new networks of instant exchange and communication – as such it suggests little more than being modern, but perhaps now at a larger scale.

The global is now also invoked in ways not so directly connected to economic and cultural matters. In terms of its relationship to the 'globe', numerous scientific phenomena became global. Climate change, ozone holes, biodiversity losses, deforestation and the perturbations of El Niño weather patterns are all explicitly global phenomena. From initial scientific collaborations in such organizations as the International Meteorological Organization, later in such institutions as the

International Geophysical Year in 1957 and 1958, through to space exploration starting with Sputnik and then, especially high profile, the Apollo program in the 1960s, the sense of humanity as being on a small planet increased.

The 'Full Earth' picture of the world, image 22727 photographed by an *Apollo 17* astronaut in 1972, has now long been a symbol for environmentalists in many places. Its appeal, perhaps related to the juxtaposition of circle and square and the fact that it foregrounds Antarctica and Africa in contrast to traditional Eurocentric cartographies, powerfully suggests a sense of home, a place worthy of attention and care because of its uniqueness and fragility in the black void of astronomic space (Cosgrove, 1994). The irony of leaving the planet to better appreciate its singularity is not lost on most observers. The sense of 'globality' is enhanced by the vision from space.

Likewise military and strategic concerns have emphasized the interconnectedness of many parts of the world in various ways. The Cold War period was a rivalry of 'superpowers' whose military reach was effectively 'global'. Bases for missiles, and especially for military aircraft, were a matter of global calculation, intercontinental ballistic missiles effectively made the whole planet a potential battleground. Hence the necessity for careful mapping and the emergence of comprehensive satellite surveillance as part of the intelligence-gathering efforts of militaries on both sides of the geopolitical divide. This too provided a mode of knowing the world as a single 'global' entity, even if the strategic lessons of such knowledge were less than clear (Dolman, 2002).

In political terms the word is sometimes a synonym for international, at other times a more evocative signifier of a common humanity. It suggests both more than international and the inclusion of political actors that conventional inter-state relations ignore. Global security was an overarching desideratum in the Cold War period when the threat of nuclear warfare hung over the activities of politicians everywhere. More recently, new dangers to global security include environmental and sometimes demographic matters where the order of modernity is challenged to deal with matters of human insecurity on a global scale. In the aftermath of events in the United States on 11 September 2001, everyone was apparently caught up in what the Bush administration called 'a global war on terror'.

READING THE GLOBAL

A word this over-coded is obviously of considerable analytical importance to anyone interested in how contemporary politics works. But a comprehensive genealogy of the global would require a lengthy volume. Likewise, understanding of the global dimensions of the discipline of geography, which supposedly should have the globe as its object of analysis, is beyond the scope of this chapter (see Taylor, 1993). That said, it is worth noting the proliferation of texts within the discipline that now invoke global phenomena as a frame of reference for writing and teaching (Johnston et al., 2002). Even Peter Haggett's (2001) discipline-defining textbook has changed its title from *Geography: A Modern Synthesis* to *Geography: A Global Synthesis* over the last three decades. Instead what follows in this chapter is a focus on the implicit politics of the term, and the forms of politics that the term itself helps make possible. In short, understanding global as discourse, or more specifically a series of discourses, allows its politics to be discussed and some of the implications of invoking the term examined.

The global is a specification of place, an arena of conflict, and much else too. Its utility as a designation of the location of various forms of politics makes its analysis unavoidable in contemporary discussions of political geography. This is especially important because the global's invocation of the globe suggests a naturalism, a taken-for-granted, obvious space and political context. But in so doing it also frequently suggests a commonality of condition, a shared residence on Earth, and with that a series of political and economic priorities and concerns that are universal. Local manifestations of these phenomena are of course part of the diversity of the human experience, but given the commonality of all being in some sense global, then the global is frequently coded as being much more important than the merely local (Gibson-Graham, 2002).

In addition – and this is especially important in the managerial tropes in which so much of the global is rendered – in terms of the various crises within with the global is invoked, the local is understood as parochial, traditional and ultimately less important in the face of global imperatives. This provides the agencies of global administration an ideological advantage that has considerable power. In Vandava Shiva's terms:

> In the dominant discourses, the 'global' is the political space in which the dominant local seeks global control and frees itself from local, national, and global control, The 'global' does not represent any universal human interest; it represents a particular local and parochial interest that has been globalized through its reach and control. The Group of Seven most powerful countries may dictate global affairs, but they remain narrow, local and parochial in the interests that guide them. (Shiva, 1994: 196)

This is in part a manifestation of the difficulties of trying to imagine politics within a spatial vocabulary of scale, a methodological problem that has dogged globalization studies that try to explain responses by agencies of public administration to economic changes and neo-liberalism in terms of scales of jurisdiction and the practices of 'jumping scale' (Brenner, 1999). The history of political thinking, and international relations thinking in particular, is replete with formulations that run into the limitations of this 'problem of Gulliver' (Walker, 1993). This chapter is not the place for a detailed engagement with the problems of sovereignty, the insistence on the priority of rule and decision in understanding politics, or the reduction of diversity to unity in the search for order, but these themes run through the discussions of globalization, its benefits and disruptions. Invoking the global suggests a political imperative not unlike sovereignty, as Hardt and Negri (2000) suggest. The supposed impossibility, or at least irrationality, of being 'against' such large-scale phenomena, likewise rhetorically shifts political discourse in favour of the status quo.

GLOBALIZATION

According to the *Oxford English Dictionary*, 'globalization' appears for the first time in the *Sunday Times* and in the *Spectator* in 1962, but not until 1965 in the *Economist*. The Public Affairs Information Service database suggests that the first economic use of the term was in Theodore Levitt's 'The globalization of markets' in the *Harvard Business Review* in 1983. Globalization appears for the first time in the *Social Science Citation Index* in 1976 and for a second time in 1979, once again in 1981, and eight times in 1983, although seven of these refer to one publication. In 1985 seven entries appear; significantly three refer to Alain Lipietz and four to Roland Robertson, who is widely noted as an early commentator on phenomena that were becoming known then as globalization (see Robertson, 1992).

The trend is marked in many ways: multinational corporations became transnational and then fairly quickly global corporations (Barnett and Cavanagh, 1994). These corporations also presented themselves as 'global' in the long boom of the 1990s. Having a corporate presence in many places, including the 'emerging markets' of the South, was a sign of competence and capability invoked by the use of symbols of the globe as part of corporate advertising (McHaffie, 1997). The ability to sell in global markets, to invest in distance places and coordinate activities regardless of state boundaries indicated the technical acumen of the corporation. It also suggested size, resources and

capabilities in which both consumers and investors, frequently the same people, could have confidence. By the late 1990s globalization was everywhere: protestors were 'anti-globalization'; mutual funds were global; textbooks on such things as *Global Transformations* (Held et al., 1999) were the latest phenomenon in academic publishing.

Not surprisingly, in the age of expert knowledge practices of monitoring, measurement and management there have been attempts to formulate indicators of globalization. Notably, *Foreign Policy* instigated an annual index comparing the degree of globalization of states around the world. The criteria used in calculating the index are based on four indicators that supposedly encapsulate the phenomenon of globalization. First, economic integration includes trade, foreign direct investment, portfolio capital flows and investment incomes. Second, technological connectivity measures Internet users, Internet hosts and secure servers. Third, personal contacts includes international travel, tourism, telephone traffic as well as personal financial transfers and remittances. Fourth, political engagement is included as a measure of involvement with international organizations, UN activities, treaty ratifications and governmental transfers. According to this combination of economic, personal, technological and political criteria, in the 2004 index published in the *Foreign Policy* March/April 2004 issue, the United States ranked seventh on the index behind, in order of most connected, Ireland, Singapore, Switzerland, the Netherlands, Finland and Canada.

Novelty, acceleration and connection, a sense of the planet as a single place enhanced by communication technology, humanity as sharing a common fate, and culture and ways of life determined increasingly by business rather than states, encapsulates most of the meanings imputed to 'the global'. But all of these generalizations are open to numerous empirical critiques given their degree of generality, and ideological disputation from those who understand the global as a powerful discursive device in the hands of corporations and their state backers who conflate contemporary neo-liberal policies with a supposedly unstoppable and inevitable, because 'global', process of economic, political, cultural and social change.

NOVELTY AND ECONOMY

In historical perspective, globalization can also be understood as a long-term process of increased interconnections, not a simple process of contemporary economic and cultural Westernization (Hopkins, 2002). This refusal to focus on the novel dimensions of connection adds a useful corrective to the hype, but also forces the focus on the specificity of the global. World systems theory suggests a much longer history of a global economy that further challenges the understanding of economic interconnections as novel, and in Peter Taylor's (1999) rendition links questions of culture and hegemony directly into the exercise of power and economic innovation.

The question of precisely what is new is not entirely clear. After all, it was the mid-nineteenth century, rather than the end of the twentieth, that produced the classic critique of what we now call globalization:

> The bourgeoisie, by the rapid improvement of all instruments of production, by the immensely facilitated means of communication, draws all, even the most barbarian, nations into civilization. The cheap prices of commodities are the heavy artillery with which it forces the barbarians' intensely obstinate hatred of foreigners to capitulate. It compels all nations, on pain of extinction, to adopt the bourgeois mode of production; it compels them to introduce what it calls civilization into their midst, i.e., to become bourgeois themselves. In one word, it creates a world after its own image. (Marx and Engels, 1967 [1848], 84)

Focusing on contemporary media and communications technology likewise leads back to earlier considerations of empire and the progressive networking of the world as modernity expanded its reach through roads, railways and the telegraph (Hugill, 1999; Mattelart, 2000). The emphasis on speed and technology at the beginning of the twenty-first century reprises the spread of the style of Art Deco in the 1930s. Even the global climate was already understood as the source of insecurity for many rural populations in the nineteenth century, once basic meteorological statistics were collected in a standardized manner (Davis, 2001). The economies of European states were at least as 'open' in 1913 as they are at the beginning of the twenty-first century.

The sense of global politics as concerning more than just the relations between states or matters of international trade is clear at earlier periods of political discussion. The 1970s, with its initial concern about 'global' environmental issues, is reflected in academic texts from the period where global politics is discussed with reference to 'the limits to growth' debate and discussions of population and development difficulties as political matters (Ray, 1979). Much of this was swept aside by the revival of tensions between the US and the Soviet Union in the late 1970s and the accession of the Reagan administration to power in Washington. Environment and development were ignored in the bellicose days of the early 1980s, in much the same way as such concerns were swept aside by the focus

on the 'global war on terror' in the latter months of 2001.

Nonetheless, despite this lengthy list of caveats, there is a widespread sense, captured in part in the term globalization, that the new connections of the latter part of the twentieth century have changed matters in a qualitative way. Manuel Castells, in his huge synthetic study of these transformations, simply calls the result 'network society'. He suggests that a new world is in the making. 'It originated in the historical coincidence, around the late 1960s and mid 1970s, of three independent processes: the information technology revolution; the economic crisis of both capitalism and statism, and their subsequent restructuring; and the blooming of cultural social movements, such as libertarianism, human rights, feminism, and environmentalism' (Castells, 1998: 336). The working-out of this logic has, he argues at length, produced new identities and a new social economy much more dependant on media and virtual production.

Key to this new situation, with its international economic negotiations, and meetings of the business and political elites at the World Economic Forum in Davos each year, was the emergence of a series of economic doctrines that have come to be called neo-liberalism, or the Washington consensus. This is a combination of debt reduction, privatization, reduced restrictions on capital mobility and a rejection of import substitution strategies of national economic growth. This freeing of markets is combined with the expansion of marketing techniques, a rapid reduction in the costs of international transportation, epitomized in the widespread adoption of standard modes of containerization for manufactured items, post-Fordist production systems and computer inventory control systems to rapidly accelerate processes of production, product innovation and delivery.

This 'global shift', to borrow the title of Peter Dicken's (2003) textbook on these themes, has involved a series of geographic shifts in the location of assembly and manufacturing plants to areas with relatively lower wages, and a reorganization of many formerly industrial cities into more informational and service economies, as well as the growth and transformation of large corporations, some of which have become 'global'. But while commodities might appear universal, nonetheless companies still frequently produce varieties of their global product range specifically tailored to the cultural peculiarities of national markets. This economic integration/globalization has also produced a growth in management, accounting and legal services in an economy where trade has fairly consistently increased more rapidly than national economic growth.

The insistence on global management and surveillance by business organizations produced an especially ironic contradiction in the enthusiastic economic globalization of the 1990s. The disruptions of unfettered accumulation, in the so-called Asian crisis of the late 1990s, once again brought with them instabilities that threatened the whole scheme and apparently required disciplining (Ould-Mey, 1999). These disciplinings led to international financial institutions effectively operating in command mode in response to what was supposedly the triumph of unfettered capitalism. But this was a centralized disciplining that came from outside the states in question. Indonesia in particular suffered serious dislocation as its economy was rather dramatically restructured. Argentineans have learned the hard way that such reorganizations can be devastating to economies considered out of step with 'global' economic doctrines. Globalization is then also, in part, about a managerial mentality.

CONNECTIONS

The noteworthy technological parallel with the rise of the discussion of globalization is the rapid spread of the personal computer and the rise of first e-mail and then the Internet as a mode of communication and community construction. Personal computers arrived in the early 1980s and were in widespread use by the early 1990s in 'developed' countries. The concern with globalization and instant communication has been accentuated by the new computer media, somewhat before cell phones and other portable communication devices became popular in the 1990s. In particular, the critics usually note that mostly what this has done is to facilitate new modes of communication between the already well connected. Simultaneously with the extension of computer links, the costs of international phone (and hence fax) services has been reduced dramatically. This has meant that diasporas can remain in instantaneous contact with home communities much more effectively than was previously the case, giving a sense of the world as a much more interconnected place. As Appadurai (1996) also notes, the widespread availability of tape recorders also allows personal news, as well as political and religious messages, to circulate easily even where literacy is limited.

The rise of global media is also part of the shifting sense of globality (Barber, 1996). Instant news from around the world available wherever TV reception works means a commonality of headlines. But the widespread use of colour television sets and the routine satellite broadcast of sporting events to a 'global' audience also enhances a sense of globality, a common human community sharing the thrills and spills of the sporting field. This may be most widely understood in terms of soccer and the World Cup every four years, but most large sporting franchises now host elaborate Internet sites and market merchandise tied to corporate identities around the world. But this is not a matter of simple cultural homogenization. Much more importantly, such sports phenomena emphasize the complexity of particular identities as fans of clubs and nations express their support for their chosen teams in ways that reinforce national identity and partisan support (Foer, 2004).

But ironically, just as global television has supposedly linked the world in some important ways, cable television has allowed a proliferation of channels and reduced the relative importance of many simulcast 'global' events. The combination of particularistic identities and the enhanced abilities of diasporic peoples and migrant workers to maintain cultural and political connections abroad, has also raised concerns in both Europe and the US about a perceived failure of immigrants to assimilate in their new homes. In Samuel Huntington's terms in his discussion of immigration in the US:

> The persistent inflow of Hispanic immigrants threatens to divide the United States into two peoples, two cultures, and two languages. Unlike past immigrant groups, Mexicans and other Latinos have not assimilated into mainstream US culture, forming instead their own political and linguistic enclaves – from Los Angeles to Miami – and rejecting the Anglo-Protestant values that built the American dream. The United States ignores this challenge at its peril. (Huntington, 2004: 30)

While this reading of what is happening is vehemently rejected by many of Huntington's critics, the politics of migration, and especially labour migration, is now also tied into practices of globalization where remittances are an increasingly important part of the international economy.

The ironies of the Internet and cell phone technology, the symbols of the new global interconnectedness, are especially clear when they are considered in terms of political resistance to many things labelled 'globalization'. Well before 'moveon.org' popularized the Internet as a mode of political mobilization in the US, 'anti-globalization' activists were using websites and e-mail to coordinate actions at various international protests. In many protest actions, activists were staying in touch and coordinating their activities using the flexibilities inherent in cell phones. All of which suggests that the protests in Seattle, Genoa, London, Quebec City, Washington and many other places were much more a critique of the depredations of neo-liberalism and unrestrained capitalism than specifically 'globalization'. But the phrase 'anti-globalization' is none the less telling of how important the 'global' interconnections of the various phenomena being protested had become by the late 1990s.

THE GLOBAL ERA

All these formulations of globalization have connotations of novelty and ambition, connection and change, most of which were previously understood in terms of being modern. In part, 'global' is a term that designates the end of the Cold War period, an era when clear divisions between East and West, in places literally set in concrete as in the case of Berlin, were transcended and walls came down. Flows across boundaries were very much what 1989 was all about. But globalization also reflected a simple recognition that the flows were as a consequence of the triumph of capitalism, which now literally could operate all over the globe, with the exception of a few holdout states that would inevitably be carried along by the inexorable expansion of a capitalism that was now undoubtedly triumphant. Francis Fukuyama's (1989) adoption of the phrase 'the end of history', with its clever pun on the two meanings of 'end', captured the mood.

The larger intellectual context of this shift came as the dominant understanding of politics as a combination of what happens within states and what the practitioners of statecraft do, eroded in the face of cultural and technological change. The understandings of international relations, a discipline tied in the US in particular to providing legitimations of the current political order (Klein, 1994), had long specified the world as a matter of autonomous powers, national spaces as the territorial mode of political order. This had reflected the triumph of the national state as the end of the era of imperialism; national self-determination suggested a politics within states, and the operation of sovereignty as the appropriate limit on the correct operation of normal politics.

This geopolitical rendering of the world worked especially well as a legitimation of the US in the period of the Cold War. Its forces and interventions could be presented as supporting this territorial order of independent states at the behest of local elites who supported alliances as a contribution to extending national freedoms in the face of an expansionist Soviet threat. But in the face of rising debt crises and the failure of strategies of national development to deliver economic affluence in the 1980s, and the rise of explicitly neoliberal international economic discipline in part as a response to the failure of earlier doctrines, international relations clearly needed a much more complex understanding of the world. The unanticipated decision by the Gorbachev administration to effectively call off the Cold War military confrontation and rising concern about common concerns with environmental dangers and pressing economic concerns, which antedated the collapse of the Soviet Union, also suggested many cross-border linkages and a much larger canvass of concern than in the prior decades. The rise of the study of international political economy reflects these changing geopolitical circumstances (Cameron and Palan, 2004).

Understood in this historical context, globalism is, at least in part, a recognition that the impoverished political category of the supposedly autonomous national state was no longer credible as the basis for either political analysis or the related practices of legitimation. The lack of an alternative political vocabulary in the early 1990s left a conceptual vacuum into which the triumphalist business jargon of globalization, with its technocratic modes of managerialism now updated to consult, collaborate, measure and monitor, fit quite neatly while reinventing the supposed value of novelty and technological innovation in solving all sorts of human ills. The end of the Cold War promised a new world order, in the elder George Bush's phrase. But it also brought other agendas to the fore, not least, once again, the issues of environment and development, and their interrelationship.

The huge UN summit in Rio on Environment and Development in 1992 in some ways offered the clearest suggestions of a new era after the Cold War where 'global' issues of environment, development and inequity could be addressed. While the results of that huge event were less than impressive, not least because business interests and power politics severely limited what was possible (Chatterjee and Finger, 1994), nonetheless the importance of non-governmental organizations and social movements in challenging the legitimacy of the global order became clear. The 'fact' that there was a global order to challenge, in and of itself, became much clearer to a generation of activists at Rio, and has influenced the subsequent anti-globalization and more recently anti-war protest movements.

GLOBAL ENVIRONMENT

What is striking in retrospect when one looks back twenty years before Rio, at the discussions of environment in the unofficial report to the Secretary-General of the United Nations prior to the Conference on the Human Environment in Stockholm in 1972, is the complete absence of a vocabulary of the global (Ward and Dubos, 1972). The title is the evocative *Only One Earth*. The subtitle is even more interesting: 'The care and maintenance of a small planet'. Planet, and man, rather than a more gender-neutral term, are there. Problems of 'coexistence in the techno sphere' and 'strategies for survival' appear in the text, but 'global' or 'global governance' do not appear in this volume; managerial tropes aren't there either. While this might not have great significance in and of itself, it does suggest the absence of a sense of the global, and a sense of crisis and problems for which managerial understandings are appropriate. The cover illustration of the book does include a photograph of the Earth taken from space; credit is given to NASA, but the sense of the globe is at best implicit in the text.

But in the 1990s this sense of the global is finally and clearly related to the icon of the 'Full Earth', the lonely blue marble set against a black backdrop. A sense of common fate is part of the discourse of environmentalism (Yearley, 1996). Globality is then understood in part as an existential condition that facilitates a 'global', as in all-encompassing, responsibility to act on behalf of the planet. It links to the specifications of locality as part of globality in the environmentalist slogan 'Think globally. Act locally'. In the last few decades an understanding of the planet as endangered has emerged. In numerous cases this has meshed with the contemporary managerial mentality to render the planet itself in need of management. The cover of the September special issue of *Scientific American* in 1989 on these matters superimposed the words 'Managing Planet Earth' over an image of the planet taken from space. Coupled to a modernist managerial mentality, the Earth itself becomes an object for surveillance, control and monitoring. Environmentalism then becomes a technology for securing the future for humanity, a future in which 'environmental security' is in doubt (Dalby, 2002).

The assumption that a surveillance regime and appropriate technical practices coupled with a scientific research program could actually accomplish such a feat is meshed with the taken-for-granted suggestion that such an effort is the appropriate response to environmental difficulties. It parallels such assumptions about the global economy in the aftermath of the Asian crisis. It extends many of the attempts to construct international regimes to manage numerous matters from Antarctica to oceans and outer space to the planetary atmosphere (Vogler, 1995). The managerial mentality and the necessity of monitoring change has also contributed to the 'global' publishing industry; the United Nations Environment Program has so far published three *Global Environmental Outlook* (GEO) reports, the third in 2002 running to more than 400 pages. For a couple of decades now the Worldwatch Institute has published an annual report on *The State of the World*. In the terms of one of its most assiduous critics, this promulgation of global environmental management is merely another vocabulary of control and exploitation justified by the invocation of global necessities but operating much more effectively at sustaining development than conserving nature (Sachs, 1999).

This managerialism is an urban view of the globe, one dependent on numerous technical modes of knowing, analyzing and reducing the complexity of interconnections to indicators, benchmarks and decision criteria for the monitoring agencies. The political dimensions of such a discursive construction are elided in the combination of technical competence and the urgency implicit in the necessity to save the Earth from a multitude of threats. Such 'environmentality' provides a powerful discursive repertoire for 'global' politics where science meets security (Luke, 1997). It does so in particular because of the appropriation of environmental problems as a matter that can be dealt with by technological innovation and superior management techniques, frequently now termed, following Hajer's (1995) analyses of these matters, 'ecological modernization'.

In parallel to the understanding of social, economic, political and cultural phenomena is a growing sense, however inchoate, of being in a common biosphere. The phrase 'global warming' has emerged as a term of common parlance among numerous publics, even if the precise mechanisms are less than clear to those who invoke the term. Climate change and environmental disruption are understood in terms now of 'global change', a phenomenon that is only recently linked explicitly to globalization. The fact that these are but two sides of the same coin has not frequently been integrated into either discussion (O'Brien and Leichenko, 2003). This literature in particular is suggesting a further sense of the global in terms that suggest a more nuanced understanding of the global predicament, wherein the human insecurities in many places are linked to both natural and artificial circumstances and their increasingly mutual constitution of vulnerability among the poor and the rural peoples of the world (Pirages and De Geest, 2004).

RESISTANCE

The themes of universality and a common fate have run into opposition in many locales around many themes (Klein, 2002). Some of these concern such technical issues as genetically modified foods and the attempts by numerous corporations to impose such technologies on farmers. Elsewhere the urbanization of the rural, whether it is further appropriations of traditional hunting and commons access, or the wholesale appropriation of land and water for such things as golf courses and eco-tourism resorts, has run into a multiplicity of specific 'local' resistances. The mobility of production plants and the ability of corporations to switch production across national frontiers has long brought opposition from labour organizations concerned to protect their members' ability to earn a living wage (Mander and Goldsmith, 1996).

What Esteva and Prakesh (1998) term opposition to the 'global project' has many manifestations. But insofar as these are understood as being parts of one opposition to a singular global process, the irony of a global movement of localities points to the difficulties of an appropriate political language and strategy of resistance. A global movement of indigenous peoples makes the contradiction especially clear; a commonality of condition is the premise for shared activities that in and of themselves inevitably change the identities of the 'indigenous' participants in such activities. But the irony in such formulations simply reflects the unavoidable recognition of the Impossibility of pure 'local' autonomy in a 'global' world.

Juxtaposing indigenous resistance and global economics suggests that the impact of globalization has in some ways been a global phenomenon, but in doing so it denies the specifics of particular places. Local resistances to the global economy frequently have in common some form of opposition to the dispossession and appropriation of local resources by distant economic operators (Gedicks, 2001). Again the question of how novel all this may be is questionable as the historical pattern of the expansion of empires, and European empires in the nineteenth century in particular, is about the appropriation of land and materials from conquered lands, and the destruction of indigenous peoples in the process (Hochschild, 1998). That these processes are now done principally by local elites in post-colonial states in collaboration with foreign 'investors' does not fundamentally change the pattern.

GLURBANIZATION

Globally, the contemporary human condition is increasingly an urban one as people move from rural to urban life. The resources to feed and fuel what can now be termed the global city, the product of practices of what might better be termed glurbanization, now come from all over the world. Tourists and remittance payments flow back to the more rural parts of the globe, enmeshing at least most agricultural zones into the global economy. We are now an urban species and have wired and paved the planet to move food, timber, oil, electricity, minerals and all sorts of commodities from the rural areas into these burgeoning cities (Luke, 2003). In a way loosely analogous to earlier imperial arrangements, the flow of commodities inevitably disrupts traditional forms of economic life. Just as wheat flowed from Africa to Rome, so now does oil flow from the Middle East to other parts of the new imperium. Being networked in many parts of the world means being tied into these patterns of commodity extraction.

The Roman Empire built roads to facilitate communications and integrate the empire into an efficient network of supplies and military preparation. So too do modern states. Indeed, it is possible to argue that such infrastructure provision is a key part both of state structures and in particular the commercial culture of the biggest of the global economy's commodities, the automobile. Promotion of the infrastructure for the privately owned car is a major part of the function of contemporary states (Paterson, 2000). Private car ownership is understood as a matter of status in numerous developing states, while the pollution and congestion problems that result are ignored much of the time. The latest gas guzzlers in North America, the so-called 'Sports Utility Vehicles', are presented to would-be buyers in tropes of conquering nature, of 'civilizing nature' in Nissan corporation's advertising slogan, a matter of being able to go anywhere regardless of obstacles. Universal freedom in a world without boundaries, the ideological promise of globalization, is now used a sales device in automobile advertising simultaneously selling a commodity and constituting the (driving) subject of globalization in the process.

But the important points that follow from this are that the knowledges that are constructed, especially the knowledges that look to universal explanations of the relationships between environment and conflict, are usually urban and modern knowledges, ones that take an imperial view of matters for granted (Dalby, 2002). Combined with satellite imagery and modes of monitoring statistics compiled by states and international agencies, and the assumptions of the inevitability of economic development in terms of the expansion of carboniferous capitalism, these formulations of the resource and environment problematic inevitably downplay the rural, the contextual and the disruptions inflicted on traditional peoples. They do so also within a state cartography, one that draws lines between places, ensuring that civil wars 'over there' are not usually a matter of responsibility 'in here' in the cities of the metropoles. But as the literature on resource wars now makes clear, the consequences of modes of extraction in distant places is directly tied into violence, dispossession and environmental destruction (Jung, 2003). Consumer boycotts and questions of corporate conduct and ethical investing are all connected here in the larger matters of governance.

GLOBAL GOVERNANCE

Many writers investigating globalization, and in particular its economic dimensions, argue that the rule-making functions of governance are slipping away from formal state institutions into intergovernmental and international organizations, not all of which are formally sanctioned by states, but which nonetheless make agreements and set standards that are widely followed. Standards set by the International Organization for Standardization (ISO) or negotiated by the Forest Stewardship Council or agreed to by international industry associations, shape corporate practice in many spheres and hence shape the lived experience of both the producers and the purchasers of the commodities produced for the global market. Shared consumer experiences as a result of the globalization of markets produce at least some shared understanding of the human condition.

All of these developments are incorporated in to what Hardt and Negri (2000) term 'Empire', the rules, procedures and governance functions that have morphed from state sovereignty into a more inchoate, but none the less powerful set of practices that interlink the world's citizens. In the process these citizens are loosely beginning to have enough in common to justify being included in a single, albeit still inchoate, sociological category that Hardt and Negri (2004) term 'Multitude'. This implies in some sense a common, but certainly not a unified, humanity forged in the global political economy.

Such themes of interconnected human fates were prominent in the discussions of warfare and welfare through the 1990s and in particular in formulations of human security. In the aftermath of the Cold War the assumption was that with the complexities of Cold War confrontation, which had so frequently entangled local elites in the Third World in proxy wars for their superpower patrons, removed, international peacekeeping efforts and humanitarian interventions to assist victims of natural disasters would finally be able to deliver what the United Nations had long promised in terms of 'human security'. Formulated in the thinking of the United Nations Development Program, global threats to human security came as a consequence of the actions of millions of people rather than the aggression of particular states. These causes of insecurity included unchecked population growth, disparities in economic opportunities, excessive international migration, environmental degradation, drug production and trafficking, and international terrorism (United Nations Development Program, 1994: 34).

These themes were not new in international discussions, but in the aftermath of the Cold War and a recognition that the military dimensions of security were undergoing a substantial rethinking as military budgets shrank around the globe, a new possibility for liberal international governance seemed to be in the offing. Epitomized by the discussions in the tellingly titled report of the Commission on Global Governance (1995), *Our Global Neighbourhood*, the agenda suggested an increased role for NGOs and the obligation to intervene in disasters and local military confrontations. The cases of Bosnia, Rwanda and Somalia suggested the need for interventions, but quite how this was to be done was less than clear. It has also raised big questions about the possibilities of extending international criminal law and the possibilities of effectively constructing a global court system to deal with crimes against humanity (Kochler, 2003).

At the heart of the discussion was a clash of norms, the state right of non-intervention versus the norm of providing aid and protections for victims of violence or disaster (Hampson, 2002). The case of the intervention in Kosovo muddied the waters much further when an air attack by an American lead force targeted the infrastructure of Serbia in an apparent attempt to convince the ruling regime to abandon part of its sovereign territory and restructure its financial system. This looked much more like imperial power politics than humanitarian assistance and the freedom of manoeuvre for 'neutral' NGOs was substantially reduced. The blurring of these distinctions was explicit in Afghanistan when the US military air force dropped humanitarian food packages simultaneously with munitions. This conflation of military and humanitarian purposes is now explicit in the 'nation-building' operations in both Afghanistan and Iraq. But while aid and occupation, reconstruction and political reorganization have never been far apart, the explicit use of humanitarian aid as a political tool has now finally closed what political space existed for independent organizations to operate in 'neutral' humanitarian mode. The global war on terror now ties political loyalty to the provision of food and medicine, undercutting the possibility of any meaningful 'global' humanitarianism.

GLOBAL WAR

In the aftermath of the events of 11 September 2001 and the formulation of a policy of 'war on terror' by the Bush administration, the international governance of economics and human migration, and labour migration in particular, has become ever more entangled in considerations of geopolitics. Poor labour migrants are frequently caught in various social discourses of abjection, marginalization and criminalization that imply the need for territorial restrictions, if not incarceration, as a mode of frequently violent social regulation. All this has been aggravated in the aftermath of 11 September where they are now also caught up in practices of securitization and surveillance extended to many aspects of social life.

The American policy of 'Homeland Security' implies spatial exclusions of potential threats as the overarching political priority; rebordering is very much in evidence to travelers entering the US (Andreas and Biersteker, 2003). While the cartography of danger in the US emphasizes once again strategies of border control and spatial exclusion, at the same time the Pentagon has finally extended its combatant command areas to include all the terrestrial globe. According to the fifth edition of the *Commanders' Map of the World*, published in 2002, all the territory and oceans of the planet are now assigned to a combat command (Johnson, 2002). A strategy of rebordering is nonetheless integrated to global control and surveillance.

US state security has clearly replaced human security as the dominant theme of global governance, at least for the moment. But the migrations of globalization and the complex politics of diaspora are not so easily constrained given all the communications and cultural interchanges of 'globalization' that imply, despite the objections of those like Huntington (2004) who continue to portray spatially exclusive entities as the basis of political order, a much more cosmopolitan world.

The invocation of the cosmopolitan, as the universal human interest, also suggests a global politics (Archibugi, 2003) that draws on histories of liberal internationalism. But the cosmopolitan is very much the urban citizen, the person tied into the global circuits of capital and the long commodity chains that supply the metropolitan world with the supplies and status symbols that make the invocation of universal civilization possible. This invocation of global citizenship also links to the civil society analyses of the informal social movement struggles in Eastern Europe in the 1980s and elsewhere since in the struggles for political liberty and the extension of democratic rights. More specifically, Mary Kaldor's (2003) rendition of these movements, and their potential as an alternative social possibility to warfare, raises the big political questions of whether such reinvented liberalism has the emancipatory potential to tackle the growing injustices in the global economy, the propensity to militarize social problems and the destruction of ecological systems. But the invocation of a war on terror and the widespread specification of opponents as terrorists makes these 'global' struggles all the more necessary and difficult.

ACKNOWLEDGEMENT

My thanks to Susan Tudin for tracing the history of the early use of the term 'globalization'.

REFERENCES

Albrow, M. (1997) *The Global Age: State and Society Beyond Modernity*. Stanford: Stanford University Press.

Andreas, P. and Biersteker, T.J. (eds) (2003) *The Rebordering of America: Integration and Exclusion in a New Security Context*. New York: Routledge.

Appadurai, A. (1996) *Modernity at Large: Cultural Dimensions of Globalization*. Minneapolis: University of Minnesota Press.

Archibugi, D. (ed.) (2003) *Debating Cosmopolitics*. London: Verso.

Barber, B. (1996) *Jihad vs. McWorld: How Globalism and Tribalism Are Reshaping the World*. New York: Ballantine.

Barnett, R. and Cavanagh, J. (1994) *Global Dreams: Imperial Corporations and the New World Order*. New York: Simon & Shuster.

Brenner, N. (1999) 'Beyond state-centrism? Space, territoriality, and geographical scale in globalization studies', *Theory and Society*, 28: 39–78.

Cameron, A. and Palan, R. (2004) *The Imagined Economies of Globalization*. London: Sage.

Castells, M. (1998) *End of Millennium* (Vol. 3 of *The Information Age: Economy, Society and Culture*). Oxford: Blackwell.

Chatterjee, P. and Finger, M. (1994) *The Earth Brokers: Power, Politics, and World Development*. London: Routledge.

Commission on Global Governance (1995) *Our Global Neighbourhood*. Oxford: Oxford University Press.

Cosgrove, D. (1994) 'Contested global visions: one-world, whole-earth, and the Apollo space photographs', *Annals of the Association of American Geographers*, 84(2): 270–94.

Dalby, S. (2002) *Environmental Security*, Minneapolis: University of Minnesota Press.

Davis, M. (2001) *Late Victorian Holocausts: El Niño Famines and the Making of the Third World*. London: Verso.

Dicken, P. (2003) *Global Shift: Reshaping the Global Economic Map in the 21st Century*. London: Sage.

Dolman, E.C. (2002) *Astropolitik: Classical Geopolitics in the Space Age*. London: Frank Cass.

Esteva, G. and Prakash, S. (1998) *Grassroots Post-Modernism: Remaking the Soil of Cultures*. London: Zed.

Foer, F. (2004) *How Soccer Explains the World: An Unlikely Theory of Globalization*. New York: Harper Collins.

Foreign Policy (2004) 'Measuring globalization'. *Foreign Policy*, March/April: 54–69.

Fukuyama, F. (1989) 'The end of history', *National Interest*, 16: 3–18.

Gedicks, A. (2001) *Resource Rebels: Native Challenges to Mining and Oil Corporations*. Boston: South End.

Gibson-Graham, J.K. (2002) 'Beyond global vs. local: economic politics outside the binary frame', in A. Herod and M. Wright (eds), *Geographies of Power: Placing Scale*. Oxford: Blackwell, pp. 25–60.

Hajer, M.A. (1995) *The Politics of Environmental Discourse: Ecological Modernization and the Policy Process*. Oxford: Clarendon.

Haggett, P. (2001) *Geography: A Global Synthesis*. London: Pearson.

Hampson, F.O. (2002) *Madness in the Multitude: Human Security and World Disorder*. Toronto: Oxford University Press.

Hardt, M. and Negri, A. (2000) *Empire*. Cambridge, MA: Harvard University Press.

Hardt, M. and Negri, A. (2004) *Multitude: War and Democracy in the Age of Empire*. New York: Penguin.

Held, D., McGrew, A., Goldblatt, D. and Perraton, J. (1999) *Global Transformations: Politics, Economics and Culture*. Stanford: Stanford University Press.

Hochschild, A. (1998) *King Leopold's Ghost: A Story of Greed, Terror, and Heroism in Colonial Africa*. Boston: Houghton Mifflin.

Hopkins, A.G. (ed.) (2002) *Globalization in World History*. London: Random House.

Hugill, P.J. (1999) *Global Communication Since 1844: Geopolitics and Technology*. Baltimore: Johns Hopkins University Press.

Huntington, S. (2004) 'The Hispanic challenge', *Foreign Policy*, March/April: 30–45.

Johnson, W.S. (2002) 'New challenges for the unified command plan', *Joint Forces Quarterly*, Summer: 62–70.

Johnston, R.J., Taylor, P.J. and Watts, M.J. (eds) (2002) *Geographies of Global Change* (2nd edn). Oxford: Blackwell.

Jung, D. (ed.) (2003) *Shadow Globalization: Ethnic Conflict and the New Wars*. London: Routledge.

Kaldor, M. (2003) *Global Civil Society: An Answer to War*. Cambridge: Polity.

Klein, B.S. (1994) *Strategic Studies and World Order: The Global Politics of Deterrence*. Cambridge: Cambridge University Press.

Klein, N. (2002) *Fences and Windows: Dispatches from the Front Lines of the Globalization Debate*. New York: Picador.

Kochler, H. (2003) '*Global justice or global revenge?*', *International Criminal Justice at the Crossroads*. New York: Springer.

Levitt, T. (1983) 'The globalization of markets', *Harvard Business Review*, 61(May/June 2): 92–102.

Luke, T.W. (1997) *Ecocritique: Contesting the Politics of Nature, Economy and Culture*. Minneapolis: University of Minnesota Press.

Luke, T.W. (2003) 'Global cities vs. global cities: rethinking contemporary urbanism as public ecology', *Studies in Political Economy*, 70: 11–33.

Mander, J. and Goldsmith, E. (eds) (1996) *The Case against the Global Economy: And for a Turn to the Local*. San Francisco: Sierra Club Books.

Marx, K. and Engels, F. (1967 [1848]) *The Communist Manifesto*. Harmondsworth: Penguin.

Mattelart, A. (2000) *Networking the World 1794–2000*. Minneapolis: University of Minnesota Press.

McHaffie, P. (1997) 'Decoding the globe: globalism, advertising and corporate practice', *Environment and Planning D: Society and Space*, 15(1): 73–86.

O'Brien, K.L. and Leichenko, R.M. (2003) 'Winners and losers in the context of global change', *Annals of the Association of American Geographers*, 93(1): 89–103.

Ould-Mey, M. (1999) 'The new global command economy', *Environment and Planning D: Society and Space*, 17(2): 155–80.

Paterson, M. (2000) 'Car culture and global environmental politics', *Review of International Studies*, 26(2): 253–70.

Pirages, D. and De Geest, T.M. (2004) *Ecological Security: An Evolutionary Perspective on Globalization*. Lanham, MD: Rowman & Littlefield.

Ray, J.L. (1979) *Global Politics*. Boston: Houghton Mifflin.

Robertson, R. (1992) *Globalization: Social Theory and Global Culture*. London: Sage.

Sachs, W. (1999) *Planet Dialectics: Explorations in Environment and Development*. London: Zed.

Shiva, V. (1994) 'Conflicts of global ecology: environmental activism in a period of global reach', *Alternatives*, 19(2): 195–207.

Taylor, P.J. (1993) 'Full circle or new meaning for the global?', in R.J. Johnston (ed.), *The Challenge for Geography: A Changing World: A Changing Discipline*. Oxford: Blackwell, pp. 181–97.

Taylor, P.J. (1999) *Modernities: A Geohistorical Interpretation*. Minneapolis: University of Minnesota Press.

United Nations Development Programme (1994) *Human Development Report 1994*. Oxford: Oxford University Press.

United Nations Environment Programme (2002) *Global Environmental Outlook 3*. London: Earthscan.

Vogler, J. (1995) *The Global Commons: A Regime Analysis*. London: John Wiley.

Walker, R.B.J. (1993) *Inside/Outside: International Relations as Political Theory*. Cambridge: Cambridge University Press.

Ward, B. and Dubos, R. (1972) *Only One Earth: The Care and Maintenance of a Small Planet*. Harmondsworth: Penguin.

Worldwatch Institute (annual) *The State of the World*. New York: Norton.

Yearley, S. (1996) *Sociology, Environmentalism, Globalization: Reinventing the Globe*. London: Sage.

Geo-graphing: Writing Worlds

Elena Dell'Agnese

INTRODUCTION

Writing the world sometimes means changing it. Awareness of the power of geography to produce and transform the world while describing it is one of the most important developments of contemporary geographical thought.

'Geography is about power.' With this sentence, Gerard Toal opens his seminal book *Critical Geopolitics* (1996a: 1). Used as a verb and not a noun, *geography* is 'an active writing of the earth', a 'geo-graphing'. Inspired by the ideas of French philosophers such as Foucault and Derrida, Toal places the language of geography in a given set of collective meanings about power, statehood and sovereignty – that is, within a certain discourse that, although constructed to represent an essential feature of reality, is indeed the product of reality itself. With Toal's book, the taken-for-granted innocence of geography as a simple act of description or explanation has been exposed as misleading to a very large audience. Indeed, *Critical Geopolitics* has become a favourite of political geography graduates and researchers all over the world[1] and it may be considered a key advance within a new 'branch' of research, aimed at critically examining the institutions and people producing geographical/ geopolitical knowledge and the resulting types of knowledge produced.

However, Toal is not alone in defining geography as a form of geo-power. The politics of geo-graphing spaces – that is, of framing them within a specific 'regime of truth' – has been variously analysed in the research trajectories of other members of the so-called 'critical geopolitics constellation' (Dalby and Ó Tuathail, 1996) since the early nineties (Dalby, 1991; Dodds, 1993; Dodds and Sidaway, 1994).[2] Before then, although not as explicitly formulated, the idea that geographical knowledge can be interpreted as a way of turning representation into reality had been touched on by other thinkers and in other disciplines. For example, this can be clearly seen in Benedict Anderson's notion of the colonial map-as-logo penetrating so deep into the popular imagination as to give birth to an anti-colonial nationalism (Anderson, 1983) or in Thongchai Winichakul's account of the role of colonial geography 'and its prime technology of knowing, mapping' (1994) in making the geo-body of Siam a national framework.[3]

More generally, the idea that geography as an act of knowledge is 'creative of the world, and not simply reflective of it' (Sharp, 2003: 59) is a step forward along a very interesting path of reasoning (or, more accurately, a patchwork of reasoning) that has been pieced together in the different – but somehow interconnected – contexts of European and American academic thought over the last decades. In this process, the complex connection between geography and power has been analysed from different perspectives, each of them another step along the way. Geography has been variously identified as a form of knowledge crucial to the control of space and strategically indispensable for fighting wars; a way of disempowering the subaltern classes, substituting their traditional and mythical knowledge of places with a 'geometric' knowledge of spaces; a set of metaphors capable of normalizing established power and its spatial structures in order to make them more acceptable; a perspectival point of view that detaches itself from observed reality; a discourse that frames reality as a given 'regime of truth'; and a form of communication set by a network of 'lines of power'.

This chapter is intended to outline how this path of reasoning was laid down by tracing the steps of various contributions, the ideas involved, and how the contributions that this path has stimulated in the geographies of the French-, Italian- and English-speaking worlds have been received. The choice of these three linguistic spheres is not arbitrary. Indeed, in the 1990s 'geo-graphical' awareness ripened into a fully mature recognition among English-speaking geographers and was accepted worldwide thanks to their cultural authority and linguistic supremacy. But this awareness had also been nurtured by French and Italian social theorists, such as Foucault, Derrida and Antonio Gramsci, whose books and ideas were known to French- and Italian speaking thinkers long before they were translated into English. Understanding the links between these sets of ideas is a difficult task, but one that this chapter will seek to explore.

AD USAM DELPHINI

'An encyclopaedic almanac *ad usum delphini.*' This was the elegant, although harshly critical, description of geography provided by students at the University of Milan in 1968, in the middle of the most explosive period of the so-called '*contestazione giovanile*' student protest movement. The words of these students were collected in a booklet by their geography lecturer, Professor Lucio Gambi (1968). They regarded geography not only as 'too vast and ambitious' but also 'chaotic and mnemonic'. Moreover, it was disdainfully accused of 'lacking a rigorous methodological approach' and of being 'a fragmented collection of notions put together from other disciplines'. And it was not even an 'innocent' collection of notions: it was seen as a compilation *ad usum delphini* – that is, information brought together so that the future ruling class could understand the territory (*delphinus* being the Latin word for the King's first son and chosen heir[4]). According to the young protesters, geography was a form of knowledge connected with power. But they were studying geography under Gambi, who, with his book, *Questioni di geografia,* had begun a critical re-examination of geography long before the explosion of the *soixante-huitard* (1968) movement (Gambi, 1964).

In the following years, Gambi continued to work for a comprehensive revision of the discipline, while challenging his academic peers and trying to establish a critical dialogue with them (Gambi, 1961, 1971). Today his suggestion that traditional geography should be reorganized into three separate fields of research (physical geography, ecology and human geography) seems to have come to fruition but was then widely challenged.

His most important contribution to the development of a critical approach was, however, his attacks on his colleagues for their inability to recognize the ideological nature of every form of geography ('a pure science ... does not exist, not now and never in history; the purity of science is just a myth, cultivated by ruling classes in order to turn science into a weapon in their own hands'). In this sense, the new wave of quantitative geography was also to be regarded as suspect, since it lacked 'historical criticism' and avoided any examination of the meaning of technology, information, automated cartography and statistics (Gambi, 1971).

In 1973, Gambi's ideas, although rejected by most of his academic colleagues, were collected in a book published by Giulio Einaudi Editore (Gambi, 1973), one of Italy's most important publishers, and became quite well known among the Italian public. His ideas also triggered a forum of discussion dubbed '*geografia democratica*' (democratic geography) (Antonsich, 1996; Quaini, 2005), animated by the most motivated of his former students. Also in 1973, *La philosophie des sciences sociales* by François Chatelet was published in France. The book includes a section on geography in which the discipline is provocatively defined as 'the instrument of a major mystification' whose primary function is 'to prevent the development of political reflection about space and to conceal the territorial strategies of power' (Lacoste, 1973: 251, 294). The section was the work of Yves Lacoste, a radical geographer who was by then quite well known in France because of his controversial essay on the relationship between geographical knowledge and American strategic bombing in Vietnam (*Le Monde*, 6 June 1972).

In 1974, Massimo Quaini published *Marxismo e geografia* in Italy, which was later translated into Portuguese (1979)[5] and English (1982).[6] With regard to the title, the book was aimed at redesigning geography from a rigorously Marxist perspective rather then attempting an analysis of the discipline from a critical perspective. However, the book's introduction opens with Gambi's quotation that geography is an 'almanac *ad usum delphini*'. It also quotes Lacoste. Using the ideas of Gambi and Lacoste, who were later regarded as *maîtres á penser* (mentors) by part of a generation of French and Italian geographers, Quaini showed that geography is an instrument of power in a dual sense: it supplies useful information to the ruling classes while simultaneously providing mystifying information to the subaltern classes, thus negating the chance of a proper study of space and power.

Thus, by the mid-1970s, the first blows to the 'scientific innocence' of the discipline had been inflicted before the eyes of both French and Italian readers.

'LA GÉOGRAPHIE, ÇA SERT, D'ABORD, À FAIRE LA GUERRE'[7]

In the second half of the decade, the questions of the relationship between geography and power and of the instrumental nature of geography as a form of knowledge took an extraordinarily important step forward in both France and Italy.

In January 1976, Lacoste started publication of his own journal, *Hérodote*.[8] It was the intention of both the editor and the publisher, Maspero, that *Hérodote* would centre 'on the analysis of the relationship between geography and power' (Claval, 2000). At that stage, Lacoste was based at Paris VIII–Vincennes, a university renowned for its focus on post-Marxist analysis. Michel Foucault was also based here. The first issue of the journal therefore featured a long interview with the philosopher, which made Lacoste's debt to his ideas patently clear.[9] The prospect of developing subsequent dialogue was opened by Foucault's eloquent words at the conclusion of the interview: '[g]eography must indeed necessarily lie at the heart of my concerns' (Foucault, 1980: 77). A few issues later, Foucault himself had a flood of questions for the journal about interpretations of the concept of strategic knowledge and its relationship with such concepts as power and domination. In this way, an intellectual path with an extraordinarily rich potential was laid down. This path was not fully explored, however, since Lacoste rejected the philosopher's wider ideas of power as also including the micro-scale of the body and relationships among individuals (Hepple, 2000).

Lacoste's readers were also diverted from this line of reasoning by the other more provocative and accessible argument tackled in the journal – that is, the connection between geography and the Vietnam War. After publishing his first article on this topic, in the summer of 1972 Lacoste went to Vietnam as an observer in order to verify his thesis. Thus, in the first issue of the journal, Lacoste readdressed his old concern about the connection between the American bombing in Vietnam and geography, and published a paper demonstrating that only a good knowledge of the local geography could have enabled the American forces to bomb the dikes protecting the paddy fields of the Tonkin River, thereby destroying the farming component of the Vietnamese economy.

Three months later, Lacoste tackled the same topic in *La géographie, ça sert, d'abord, à faire la guerre* (1976), a small book whose title (suggesting that geography was, primarily, a tool of warfare) sounded so provocative that it fell like a bomb in French academic circles. The book opens with a simple question: 'Is geography in crisis?' If seen as merely a secondary school subject, a collection of loosely connected pieces of information – such as the heights of mountains and population densities – geography might give this impression. However, in Lacoste's opinion, geography must be regarded as a fundamental kind of 'strategic knowledge', crucial in war and the political organization of spaces.

And this is not a recent development; geography was born, first and foremost, as a form of knowledge connected to political and military praxis. It was only later, after it became a secondary school subject in the late nineteenth century, that it masqueraded as a scientific discipline. This textbook geography is encyclopaedic in nature and ostensibly innocuous. However, it cannot really be regarded as such. On the contrary, it is highly ideological and can be instrumental in hiding the deep connections between strategic knowledge (of which maps are the primary instruments) and power. But nowadays, maps and textbooks are not the only ways of representing spaces. Together with 'strategic geography' and the 'geography of professors', there is a third way of making geography, and that is, to use Lacoste's terminology, 'geography as a spectacle'. This is the geography of tourism, the media and advertising. It reduces spaces to visual landscapes, transforming them into products for mass consumption, and turns people into spectators in a passive contemplation of space. Far from being innocuous, not only does it impose certain specific landscapes rather than others as dominant images – whose discourses are very difficult to decode – but it is also instrumental in preventing people from realizing that space is not simply a matter of landscapes.

Apart from maps and landscapes, other geographical tools such as region and scale are critically analysed. The concept of region has been a traditional subject of French geography since it was proposed as a natural and existent subdivision of territory by the founder of the scientific approach to geography, Paul Vidal de la Blache, in the nineteenth century. However, the French regions were invented by Vidal de la Blache, who, in his studies of that country, concentrated on certain geographic facts while ignoring other aspects of equal importance. Likewise, scale is usually accepted as a given, as if it were the most neutral of geographic tools. Yet it also merits critical revision, since different levels of scale imply the possibility/necessity of choosing which facts are and are not represented.

Applied research is undeniably connected with power as well. And this is not only because its results are usually restricted to very limited audiences. Indeed, automated cartography techniques were probably among the useful tools employed in the Vietnam War to control space. Thus, geography can be defined as an instrument of power even when it is apparently innocent in its scope. Whenever geographers supply information about people living in a given space, they enable power to be exerted upon these people. Only by providing those same people with equally powerful instruments of knowledge, concludes Lacoste, can geography turn itself from an instrument of power into an instrument of empowerment?

At the time of the release of *La géographie, ça sert, d'abord, à faire la guerre*, these propositions were quite remarkable, as were the book's innovative suggestions in terms of critical analysis. However, the book was never translated into English. At the time, only Lacoste's ideas on the strategic role of geography became known in the English-speaking world, after his study on the bombings in North Vietnam (Lacoste, 1977) was included in a book on radical geography edited by Richard Peet (1977). In the 1980s, a selection from *Hérodote* was translated and edited for English-speaking audiences by Girot and Kofman (1987). Critical discussion in English of Lacoste's counter-geopolitics and his Marxist and Foucauldian influences, however, did not appear until the few pages Toal devoted to him in 1994 and 1996.[10] In Italy, Lacoste's book was initially circulated in the original French and then in translation (an Italian adaptation was published in 1978 with a watered-down title).[11] In the same year, an Italian version of his journal was launched by the group of radical geographers who had gathered under the epithet *'geografia democratica'*. The editor-in-chief was Massimo Quaini.

Hérodote/Italia set out to promote an 'alternative knowledge' of territory by spreading Lacoste's ideas in Italian academia, while at the same time inspiring a new flow of locally based geographical research. In its first issues, the journal also tackled the question of the relationship between its 'geographic discourse' and Foucault's ideas. The questions posed by the philosopher to *Hérodote* France in 1976 were redirected to a large audience of Italian geographers, and, in issue 2/3 of the Italian publication (January 1980), the (few) answers received were published.

BETWEEN FOUCAULT AND HÉRODOTE IS MARX

This sentence opened the debate. Indeed, the importance of Marxism as a basis for analysing the relationship between power and geography was so strong among the Italian contributors to the debate that coming to terms with Foucault's ideas was quite a challenge for them. Specifically, Giuseppe Dematteis claimed that 'if Foucault and *Hérodote* want to meet, they must cross the minefield of Marxist ideas on relations of production, which is the only basis for *Hérodote*'s positions' (Dematteis, 1980: 11). But at the time, he was also the only geographer with a sympathetic ear for Foucault's notions of power as a rhizomatic structure ('This

idea paves the way to stimulating research into the local specificities of power, interstices and deviances. ... Analysing places may be a way of understanding the genetics of power. It is by no means a banal refinement of the concept, although still somewhat generic, of strategic geography. And please don't ask if this is still Marxism, since this is very much like those who ask if this is still geography. ... This idea of power as a diffuse and local alternative ... opens a new perspective on the roles of knowledge, theory and intellectuals' [ibid.: 12]). In the same pages, Dematteis wrote something that might be fundamental in understanding the role of geography as a way of geo-graphing the world: 'The ideological function of "textbook geography" is more complex than that acknowledged by Lacoste. Not only does it make geographical knowledge look like an innocent form of knowledge and not only does it teach people that what exists is *natural* and *cannot* be changed, but also, and more subtly, it teaches people that what exists is *normal* ... that it is natural *because it is normal*'[12] (ibid.: 10).

Unfortunately this train of thought did not receive the attention it deserved, either inside the editorial board of the journal or outside it. Outside of the editorial board, contemporary audiences mainly reacted to the blatant statements of *Hérodote/Italia* about class struggle and radical geographies. And they were quite ready to reject them, which led to the rapid demise of the journal with severe 'academic consequences' for its editorial committee.[13] Within the editorial board, the movement did not sustain any fruitful dialogue with Foucault. However, referring to Marxist theories in the Italy of the late 1970s also meant drawing on Gramsci's notions of cultural hegemony and historical materialism. Other critical ideas about geography and its role within hegemonic and subaltern cultures could be implemented from this perspective.

After writing *Geography and Marxism*, in 1978 Quaini published a new book, *Dopo la geografia* ('After geography'). The book is built around the premise that geography not only exists in the form of naturalistic (that is, relating to descriptions of the different elements of the environment) or statistical (concerning demographic, economic and cartographic surveys) knowledge. Geography is also a Utopian form of knowledge that, rather than remaining imprisoned within the analysis of the existing social reality, can move beyond it. In Quaini's words, to understand the relevance of this 'alternative geography', it is important to 'understand the history of inequality between popular culture and hegemonic culture, distinguishing between the knowledge of the winner and the knowledge of the loser' (Quaini, 1978: 10). From this perspective, he offers a powerful reinterpretation of the development of 'official geography' as a 'modern science', capable of establishing a new space-time structure that can be measured, thus making it economically and politically easier to control the practical space of agriculture and/or the mythical space of religion. In this way, official geography has succeeded in forcibly inserting the 'qualitative' space of everyday life into a rigid mathematical framework, totally disempowering the geographical knowledge of the subaltern classes. Not only does the closer connection between knowledge and power offer new instruments of control to the dominant class, it also deepens the distance between the dominant class and the subaltern classes, taking away the traditional forms of knowledge in their possession. 'The moment of the so-called *désenclavement* (opening up) of the world in the sixteenth century was also the moment when the divide between high hegemonic culture and low popular culture began to grow wider and wider; it is also the moment when popular culture started losing its means and opportunities to become a player in the transformation of the world' (ibid.: 26). A new hegemonic form of knowledge was taking over. In this process, cartography represented the first branch of geography to become 'scientific', undergoing a mathematical process to develop a new language for the representation of space.

In 1979 Quaini himself, together with the 'democratic geography' group, organized a conference on the *Inchiesta sul terreno in geografia* ('Scientific fieldtrips as a tool for geographical research'). Partially influenced by *Antipode*'s radical geography and of the 'geographical expeditions' envisaged by William Bunge in the 1960s, the group worked with the specific intent of reshaping geography as a social science capable of overcoming the separateness between researcher and researched, with a focus on giving the latter opportunities to rethink themselves and their social position in a clearer and more effective manner.

SIGNS, META-GEOGRAPHIES, METAPHORS

Even if latently present as a stimulus for rethinking Marx's political philosophy, Foucault's reflections were bound to remain, for Italian critical geographers at the time, a 'mine largely unexploited' (Quaini, 2005).[14] However, in the same period, an interesting reworking of Foucault's notion of power/knowledge came from the French geographer Claude Raffestin (1978), who, quoting straight from *Les mots et les choses* (Foucault, 1966; the English title is *The Order of Things*), redesigned the relations between power/knowledge/will in geographical terms.

This same Raffestin, in 1980, published *Pour une géographie du pouvoir* ('For a geography of power'), a very innovative textbook on political geography in which he took on the challenge of rewriting the discipline using Foucault's notions about power and knowledge. In this perspective, geography was no longer to be seen as 'the science of places and spaces, as per Vidal de la Blache's traditional definition'. On the contrary, it must be accepted as the way of 'making explicit the knowledge and praxis men use in their relation to space. … Knowledge and praxis assume a system of relations where power is circulating, since power is consubstantial to any form of relation. … Making explicit knowledge and praxis means, even if unintentionally, uncovering the power these men exercise on things and beings' (Raffestin, 1980: 2). Analysing power/knowledge relationships in relation to spaces is, put simply, the ultimate meaning of geography. In this sense, political geography as 'the geography of the state' must be critically revised. Indeed, political geography as a geography of the state is based on a totalitarian conception of the state, since it implies the equation 'state = power'. Its language comes from the state itself and may be analysed, following a semiotic approach, as a 'system of signs', a specific code built on the general assumption of its being *la géographie de l'Etat*. Analysing power does not, however, only mean analysing the state. On the contrary, 'Pretending that Power overlaps with the State means camouflaging power, the one with the small letter' (ibid.: 45). Power comes from below; there is no opposition between dominators and the dominated. Power is inherent in any kind of relationship, can be exercised in any point of space and produces resistance. If power is inherent in any relationship, then power can manipulate information inside the relationship. Indeed, 'following Foucault and Deleuze, any place where power is exercised is – at the same time – a place where knowledge is produced' (ibid.: 48); likewise, producing knowledge is a way of exercising power. In this sense, classic political geography is a form of power/knowledge designed to support the idea of the nation-state as the only source of (legalized) power.

In a later work published in Italian (1983), Raffestin went further, stressing the ambiguity of geography (which refers to both the object and its representation) and its tendency to naturalize its knowledge, as if free from any historical context. To overcome this ambiguity, it is necessary to distinguish between geo-structure (that is, the object) and meta-geography (its representation). Meta-geography is controlled by power, which therefore makes the geo-structure it produces more acceptable. Among the different meta-geographies, we may recognize political geography, the system of signs that only speaks the language of the nation-state.

Raffestin's main publications have never been translated into English,[15] and thus constitute a clear example of those 'parallel geographies' mentioned by Minca (2003).[16] Moreover, the innovative implications of his ideas have not triggered the theoretical advances that they might have within French-speaking geography. Indeed, Lacoste rejected most of Raffestin's theories about Foucault's notions of power, accusing him in a very critical review of *Pour une géographie du pouvoir* of 'confusion between very different types of power and ... levels of analysis' (see Hepple, 2000: 292–3).

By contrast he was quite admired by a group of Italian geographers, such as Angelo Turco, Giacomo Corna Pellegrini and Giuseppe Dematteis. In his book *Le metafore della Terra* (1985 – 'The metaphors of the Earth'), Dematteis is certainly close to Raffestin's ideas about power and the distinction in geography between geo-structure and meta-geography; he also draws on Gramsci's idea of common sense, and Foucault's notion of power/knowledge. In this sense, the book may be seen as a development of Dematteis's earlier thoughts on the role of geographical knowledge in the process of normalizing power (making it look normal and therefore acceptable; see Dematteis, 1980). Indeed, according to Dematteis, 'Far from being innocent, [geography] is an instrument of domination' not only because it is a 'strategic form of knowledge', useful in making wars and a mystifying 'spectacle', but also because it is an apparatus for subjugating/consuming other (people's) spaces. Geography is an instrument of domination because 'it *identifies* – that is to say it makes power look normal. It equates power with things (naturalizing power) and, at the same time, equates things with power (normalizing things)' (Dematteis, 1985: 26). Naturalizing power is a way of legitimizing it, and geographical representations are an instrument in reproducing social order and material structures. In this, geography performs a *mimetic morphogenesis.*

According to Dematteis, any form of geographical representation that reproduces spaces in conformity with existing power relations can be defined as 'normal geography'. Normal geographies work through three types of information: location, syntactic organization and ideologies (defining a city as a 'centre' or a boundary as a 'line' is ideological). Being normal, they are normative. Therefore landscape, which can be viewed as a medium of communication, is interpreted via specific codes – spread by the media and tourism advertising – and is protected and organized by these same codes. Geography represents the world and produces consensus. But it is not a perfect reproduction of the world; it is just a set of metaphors. Only an awareness that these metaphors are merely partial representations of the truth can help us escape, according to Dematteis, from the loop of their normalizing effects.

CARTOGRAPHIC REASONING AND THE LOGIC OF THE MAP

In 1986, while participating in a geopolitics seminar held as part of the 24th Congress of Italian Geographers in Turin, Raffestin openly contested Lacoste's celebrated – although controversial – definition of geography as a strategic form of knowledge. He declared: 'Geography is not the tool for making wars. The tool for making wars is geometry, through the maps it produces' (Raffestin, 1986: 298). The map, continued Raffestin, 'is the medium that allows one to detach oneself from the land so as to acquire that visual knowledge that may be considered responsible for the developments of geopolitics. Visual knowledge triggers the geo-metrification of the world, inscribing power relationships through the map and within the map' (ibid.). He was directly referring to the notion of 'cartographic dictatorship', developed by Franco Farinelli (1985).

Farinelli's work[17] opens a new page in the analysis of geography as a form of geo-power. His main contribution is the (fundamental) distinction between geo-writing and geo-mapping, and their different roles in turning places into spaces. Mapping means the (impossible) process of representing a sphere on a flat surface, an operation that – beginning with pre-Socratic philosophers such as Anaximander – continues today whenever someone compiles an atlas or simply consults one. To perform this task, cartographers must remove themselves from the sphere as well as from their own visual perspective. Thinking in cartographic terms therefore means detaching oneself from reality, shifting one's perspective above it, and reading/dominating the world from above. 'Anaximander', writes Farinelli,

> did not just formalize [cartography], but generalized it. And it is merely on the basis of this generalization that the geometric form has become a principle that constitutes and regulates the universe. From this point of view, the first map becomes a real testing ground within which we see the first transformation ... the world changed into signs ... and the categorical forces released in this manner become literally abstract, thus capable of reinvesting ... reality as a whole, surreptitiously redefining the nature of everything. (1994: 21)

Herodotus the traveller was reportedly amused by this cartographic reinvention of the world,[18] but he could not escape from it. Cartographic determination began then, Farinelli points out, and since then it has been fundamental in shaping Western civilization and its relationship with places/spaces/territories. ('It is through this that Earth becomes a copy of the map. Thus Baudrillard (1981: 10) is completely wrong when he restricts the "procession of simulacra" to the post-modern condition: this began precisely when Herodotus laughed'; Farinelli, 1994: 27.)[19]

The process known as 'cartographic reduction' (Farinelli, 1992), a process involving the reduction of the round Earth to two dimensions, has of course had many consequences. The first is the crystallization of reality, which is naturally changing, into something static. The second is its simplification: on a map something is there, or not there, included or excluded, while language allows much more detail. Thus, the geo-metrification of the world sacrifices its fluidity and variety in exchange for the possibility of dominating it. Following Farinelli's ideas, so-called 'cartographic reasoning' must be regarded as having been a part of Western culture since ancient Greece and as a result of that culture's domination of the world. But maps not only involve measurement and the position of the observer, they are also a set of signs and can be analysed as a form of language. From this angle – slightly different from Farinelli's – a fundamental contribution to the symbolic deconstruction of the map as an instrumental form of power has been provided since the early 1980s by Brian Harley, who died prematurely in 1991 (Andrews, 2001). In a 1988 work, Harley defines maps 'as a kind of language'. His intention with this comment was to further a simple semiotic approach to understanding maps as historical practices. Directly quoting Foucault (not for his observations on geography, specifies Harley, but because of his critique of historiography), he states:

> Cartography ... can be a form of knowledge and a form of power. ... Whether a map is produced under the banner of cartographic science – as most official maps have been – or whether it is an overt propaganda exercise, it cannot escape involvement in the processes by which power is deployed. Some of the practical implications of maps may also fall into the category of what Foucault has defined as acts of 'surveillance', notably those connected with warfare, political propaganda, boundary making, or the preservation of law and order. (Harley, 1988: 279)

In analysing the role of maps in the connection between power and knowledge, Harley regards Giddens' theories about the 'authoritative resources controlled by the state' as useful. Indeed, even if not explicitly mentioned by Giddens, maps can certainly be included among these resources. For this reason they must be analysed 'in terms of

their political influence in society'. This is not only because cartography has always been 'a science of princes' used to legitimize the reality of conquest and empire, and on a different scale property rights, but also because it is full of 'silences' and omissions (colonial maps and their silences about the presence of native people, for example). Maps therefore represent a perfect instrument to promote a hierarchical vision of the space they represent.

Again drawing on Foucault's ideas, and more explicitly on his notions of power, Harley published the following year (1989) a new contribution, 'Deconstructing the map'. In this article, he takes the analysis of the relationship between maps and power further, distinguishing between the external and internal powers of cartography. The external power of maps connects them with the patron behind the cartographer, while their internal power lies in the way they impose their logic 'over the knowledge of the world made available to people'. The effects of this 'logic of maps' upon human attitudes to spaces can be regarded as relating to abstraction, uniformity, repeatability and the visual. All these attitudes have been fundamental in shaping mental structures and imparting a sense of space. 'It is the disjunction between those senses of place and many alternative visions of what the world is, what it might be, that has raised questions about the effect of cartography in society' (Harley, 1989; p. 245 of 1992 edn).

WRITING WORLDS

For Anglo-Saxon geography at the end of the 1980s the turn of the decade also marked a sort of theoretical turning point, signalled by the surfacing of a new postmodern consciousness (Dear, 2001). As noted by Dear, 1989 saw the publication of two books centred on postmodern geography (Soja's *Postmodern Geographies* and Harvey's *The Condition of Postmodernity*), and the years that immediately followed saw a 'roster of publications' which, as well as tackling the areas most typical of the new postmodern sensibility (such as philosophical theorizing, urban planning or post-Fordist economic restructuring), aimed to apply a postmodern approach to different themes. The most prominent of these was the problem of representation in geographical writing and in political discourse (Dear, 2001: 16).

Harley's contribution 'Deconstructing the map', first published in the journal *Cartographica*, was reprinted with a few modifications in 1992 in a book edited by Barnes and Duncan with the title *Writing Worlds: Discourse, Text and Metaphor in the Representation of Landscape*. The book, a collection of twelve papers all very different from each other, was accused by some reviewers at the time of adopting a rather loose definition of landscape. Indeed, some of the papers (such as Daniel's and McGreevy's) refer to themes that could be specifically viewed as pertaining to cultural geography, while others (such as Toal's, Hepple's and Olsson's) focus on the more abstract 'landscapes' of power. Moreover, not all the contributions fit in with the idea of 'writing', which is supposedly the focus of the volume ('In our collection, landscape is represented in a diverse set of forms, including theoretical models of the space economy, propaganda maps, travellers' accounts of the Niagara falls, Turner's pictures of northern English cities ...;' Barnes and Duncan, 1992: 4).

Nonetheless, the book had the undisputed merit of tackling the question of geographical representation, from a broad perspective, as an 'utterly problematic' activity, the most significant common ground among the papers being that 'landscapes, social action, painting, maps, language and, of course, written documents are all held to be susceptible of textual interpretation' (ibid.: 12). Quoting directly from Gregory and Walford (1989: 2), the editors' state: 'Texts are not mirrors ... they are creatures of our own making, though their making is not entirely of our own choosing.' This has three consequences. First, there is no pre-interpreted reality that writing reflects; on the contrary, meaning is produced from text to text. Second, writing about the world reveals as much about the author as it does about the world represented. Third, rhetorical devices are central to conveying meaning. Following this line of reasoning, texts must be seen as signifying practices constitutive of larger structures termed 'discourses', while metaphors are forms in which discourses can be presented ('they are tropes we get used to, which are useful in making the unfamiliar familiar').

Starting from a similar concern with texts and their meanings, the paper by Gerard Toal, aimed at investigating the 'structured way of seeing and writing about South Africa in US foreign-policy discourse' (Toal, 1992: 156), takes a step towards the acquisition of a critical awareness of the power of representations of political spaces. Specifically, Toal investigates a particular set of representations that he defines as 'foreign policy scripts'('the term "script" ... is meant to describe a set of representations, a collection of descriptions, scenarios and attributes which are deemed relevant and appropriate to defining a place in foreign policy', ibid.). In his opinion, the dominant scripts, which are employed to 'discipline and render broadly meaningful' to Western societies the complexities of other realities (South Africa, for instance), 'participate in the very constitution of the real', since they are used to marginalize alternative meanings and

alternative scripts, even those of the participants themselves. The dominance of one script over another 'has tremendous political significance', since 'in helping to constitute a "reality" scripts structure ways of seeing and admit only certain possibilities as ways of responding to that "reality"'' (ibid.: 157).

CRITICAL GEOPOLITICS

The proposal to analyse foreign policy representations from a critical perspective, deconstructing them like any other form of 'cultural practice', can be considered as a sign of the surfacing of a growing awareness of the power of geo-graphical praxis among Anglo-American political geographers. The path leading in this direction had been opened up by Toal a few years earlier (1987). In a very short comment on a paper by van der Wusten and O'Loughlin (1986) published in *The Professional Geographer* about the need for 'a more prominent place for war and peace research on the agenda of the discipline of geography' (Toal, 1986), Toal accused the two authors, and the empiricist political geography they represented, of uncritically accepting the state system. In this, 'they reproduce the hegemony of statist discursive practices: the state is normalized as a *natural* unit of political organization with identifiable interest and objectives ...' (ibid.: 197). For this reason, they appear unable to break with the traditional role of political geography as 'an aid to the practice of statecraft' (ibid.). Here, Toal's suggestion is to abandon the 'problem solving attitude' in order to promote an alternative political geography capable of critically challenging the hegemony of international political praxis, instead of uncritically accepting existing 'common sense' conceptions. In developing these ideas, Toal was openly influenced by the theoretical debate in international relations being driven at the time by the so-called Dissident International Relations.

In 1991, Simon Dalby, an Irish geographer based in Canada, devoted an essay to the 'current theoretical controversies' in the field of international relations and the openings they could unravel 'for those geographers who interest themselves in global political arrangements and in geopolitics' (Dalby, 1991: 162). In this paper Dalby, who had already published some quite interesting work on the deconstruction of the 'other' in geopolitical discourse (1988, 1990), also tried to 'link arguments made elsewhere' (Dalby, 1991: 261), in order to offer an initial systematic appraisal of a theoretical approach that by then had begun to be labelled 'critical geopolitics'.[20] Albeit broadly defined, conflicting paradigms in international relations were at the time marked by the emergence of a 'loose collection' of critical approaches, partly referring to Gramsci and the Frankfurt school (see Cox, 1983, 1987; Ashley, 1984), partly to a vast array of postmodernist and poststructuralist sources (see Ashley, 1987; Shapiro, 1988; Der Derian and Shapiro, 1989). Attention was increasingly turned away from war and power struggles conceived in realist terms; modernist interpretations were rejected as 'participating in the normalization and secreting of modes of political domination' (Ashley, 1987: 408, quoted by Dalby, 1991: 268). Political discourses began to be historicized and deconstructed via the analysis of texts in international politics.

It is within this perspective, argues Dalby (1991), that we can place Toal's arguments against van der Wusten and O'Loughlin's positivism, and also situate his step towards an 'ambitious theoretical exploration of the possibilities of a critical geopolitics focus on the writing of words or the constructions of "geographs", literally "earth descriptions" and their crucially important functions in structuring political discourse of many sorts' (Dalby, 1991: 273).[21] With Simon Dalby's essay, published in the pages of such a significant journal as *Society and Space,* 'critical geopolitics' broke to the surface of Anglo-Saxon political geography, and rapidly caught its attention. It was not a properly defined line of thought, nor a clear paradigmatic approach.[22] All the same, within a few years it was to attract the interest of a 'constellation' of authors, whose varied and original contributions[23] would qualify it as one of the most fruitful and innovative perspectives in the political geography of the late twentieth century.

PLACE RIGHTING, PLACE (W)RIT-ING, PLACE RIT-ING

Thus, at the beginning of the 1990s, a 'new, constructionalist awareness of the instrumentalization of geopolitical discourse for geopolitical purposes' (Reuber, 2000: 37) had started to emerge as a deeply innovative approach in Anglo-American geography. Generally speaking, the power of 'geographical imaginations' (Gregory, 1994) in shaping the world was at the time gradually becoming clearer outside the boundaries of political geography as well; the production of 'imaginative geographies' became a concern of studies devoted to the history of the discipline, and also of colonial texts in making colonial discourse and fantasies (see Felix Driver's contribution on 'the histories of geographical knowledge', 1992). But what makes the coming-out of 'critical geopolitics' specifically relevant for political geography is

that it made the need to investigate 'the politics of geographical knowledge in international relations' (Ó Tuathail, 1999) increasingly clear. Moreover, critical geopolitics supplies a relevant framework for analysing it.

In a paper that may be seen as his critical geopolitics manifesto (Ó Tuathail, 1994a), Toal states that geographical writings and representations are coded in the form of disciplinary discourses, but also as practical discourses and even in the form of the heterogeneous discourse of popular culture. The distinction reflects Lacoste's famous taxonomy ('strategic geography', 'textbook geography' and 'spectacle geography'), but is different in that it does not refer to the power of geography, only to its tactical functions, relegating textbook geography and spectacle geography to the role of masking that power. In Toal's approach any kind of geographical representation, including popular culture, contains power/knowledge about spaces and places. Urban architecture, memorials and films, for instance, are all forms of geographical representation that can be just as useful in making war as the geographical knowledge produced by institutional observatories and foreign policy decision-makers. So, geo-graphing may be a 'place righting', a 'place (w)rit-ing' and also a 'place rit-ing'; but all forms deserve to be considered as structures of geo-power (ibid.).

From this starting point, geopolitics as a geographical praxis can be redefined. Indeed, it is again 'the study of the role of "geography" in the practice of foreign policy' (Ó Tuathail, 1999) – but only because 'geography' is a word with 'a multiplicity of meanings'. Nor is it to be understood as 'a fixed substratum ... a raw physical landscape or "nature"'. On the contrary, 'geography is an inescapably social and political *geo-graphing*, an "earth writing". It is a cultural and political writing of meanings about the world. Similarly, geopolitics is a writing of the geographical meanings and politics of states' (ibid.). 'For heuristic research purposes', continues Toal, 'critical geopolitics divides geopolitics into formal, practical, popular and structural geopolitics ...' (ibid.). And its objects of investigation may respectively vary from geopolitical tradition (formal geopolitics) to the everyday practice of statecraft (practical geopolitics) to popular culture, mass media and geographical understandings (popular geopolitics), and to the contemporary geographical condition (structural geopolitics). The praxis, of course, has its limits. Like any other geographical analysis and representation, critical geopolitics 'can never be written ... from a place outside or beyond the infrastructure of mapping, geo-graphing, seeing ...' (Ó Tuathail, 1994a: 342). It is an historicized praxis itself. But it is well aware of the risks of 'Cartesian perspectivalism'.[24] Therefore, it knows that 'the most it can do is reverse and displace such structures' (ibid.).

Sometimes 'parallel geographies' eventually manage to intersect.

CONCLUDING REMARKS

Locating the geographies and reconstructing the meta-geographies of an 'awareness' is certainly not easy. This was, however, more or less the task of this chapter – in other words to try to understand when, and how, geographers have acquired the consciousness that geography not only describes the world but also manufactures it.

A critical approach to the discipline probably started with the so-called 'sixty-eight' movement, in Italy with Lucio Gambi and in France with Lacoste. It has since developed, reflecting the ideas of Foucault and Derrida and of Gramsci. In their respective homelands of France and Italy, these social theorists had a quite early impact on a few French and Italian geographers, triggering a line of inquiry based around power/knowledge, geographical common sense and hegemonic culture that can be defined as both pioneering and provocative. This is the point from which I started. It has only been at the hands of English-speaking geographers, however, that Foucault's notions of power/knowledge and discourse have sparked the revolution in geography advocated by Raffestin (1997). Since the early 1990s, 'writing the world' has no longer been seen as a straightforward act. On the contrary, the idea that writing, representation and mapping are all forms of geo-power is gaining wide currency among academics. 'Indeed, much of the critical geopolitics literature', writes Allen, 'is taken up with making explicit what is implicit in the writing and representation of political spaces' (Allen, 2003: 102). In addition, critical geopolitics has suggested that geography as a discipline could find a new *raison d'être* in specifically analysing power/knowledge relationships within space(s). In this manner, a path towards a reconstruction of geographical praxis – if not as a description of specific places, then as a field of research aimed at providing knowledge of power/knowledge relationships in space – is provided by Toal and by the critical geopolitics 'constellation'. When Raffestin offered a similar suggestion many years ago, he was almost totally ignored – as typically happens with people whose ideas are ahead of their time.

However, because of English-speaking geography's hegemonic influence, it is reasonable to believe that, thanks to the successes of critical geopolitics and more generally of postmodern geography round the world, the contributions of the Italian and French geographers of the 1970s and 1980s will be reassessed by younger generations. They might also be translated.

NOTES

1 For the debate following the publication of the book, which has been variously defined as 'vital', 'outstanding' and 'célèbre', but has also been harshly criticized (Sparke, 2000), see Hague (2004).

2 See in this regard Allen (2003).

3 Thongchai Winichakul's ideas were presented for the first time in 1987 in the Proceedings of the International Conference on Thai Studies, Australian National University, Canberra. The following year they were the basis of his PhD dissertation, presented at the University of Sydney ('Siam mapped: a history of the geo-body of Siam'). It was not until 1994 that his work, already extensively quoted by Anderson in the 1991 edition of his *Imagined Communities,* took the form of a book. In the introduction, he defines geography as follows: 'It is a kind of knowledge, a conceptual abstraction of a supposedly objective reality, a systematic set of signs, a discourse. The strategy of this study is to analyse the pre-modern and modern discourses and then to detect the moments when the new and the old discourses collided. Those moments were in fact the politico-semiological operations in which the new discourse threatened and displaced the existing one. ... Those moments could appear in every sort of social activity: diplomatic relations, scientific observations, correspondence, travels, textbooks, warfare and, of course, surveys and mapmaking' (1994: 18).

4 The notion of geographical education as an instrument of power was certainly not a new idea in 1968. The eighteenth-century British tradition of the Grand Tour (when the children of the English ruling classes were sent to the Continent to learn about different political systems and national organizations) is a case in point (dell'Agnese, 2005a). In the nineteenth century, Robinsonades and adventure tales in general were conceived to stimulate the desire for travel in future colonial leaders. During the same period, geographical societies were founded in Great Britain as well as Italy and other would-be colonial powers, to spread geographical knowledge and promote expeditions (Gambi, 1994). At the beginning of the twentieth century, Mackinder was perfectly aware of the educational power of geography for the English ruling class. More or less in the same period, but with the opposite intention, anarchist French geographer Elisée Reclus tried to promote popular knowledge of geography in order to change the established system. The Italian socialist Arcangelo Ghisleri, the founder of the

popular journal *Cuore e critica,* also worked in this direction, endeavouring to empower the Italian working classes and their children through the teaching of geography (see Quaini, 1978).

5 *Marxismo e geografia.* Rio de Janeiro: Paz e Terra, 1979.

6 Russell King added a rich bibliography on radical geography to the English translation of the book (*Geography and Marxism.* Totowa, NJ: Barnes & Noble, 1982).

7 This can be roughly translated into idiomatic English as 'Geography is first and foremost a tool of war'.

8 As Hepple (2000: 269) remarks, *Hérodote,* which in 1983 was subtitled *Revue de géographie et de géopolitique,* has since flourished and grown into the most widely read French geographical journal.

9 'Your work to a large extent intersects with, and provides material for, our reflections about geography and more generally about ideologies and strategies of space. Our questioning of geography brought us into contact with a certain number of concepts you have used – knowledge (*savoir*), power, science, discursive formation, gaze, *episteme* – and your archaeology has helped give a direction to our reflection. For instance the hypothesis you put forward in *The Archaeology of Knowledge* – that a discursive formation is defined neither in terms of a particular object, nor a style, nor a play of permanent concepts, nor by the persistence of a thematic, but must be grasped in the form of a system of regular dispersion of statements – enabled us to form a clearer outline of geographical discourse' (*Hérodote* editors; Foucault, 1980: 63). The interview was translated into Italian in 1978.

10 Some mention of Lacoste's ideas appeared in Parker's survey of Western geopolitical traditions (1998). Dodds and Atkinson dedicated two chapters of their book *Geopolitical Traditions: A Century of Geopolitical Thought* (2000) to *Hérodote* and Lacoste. This confirms Paul Claval's statement that '*Hérodote*'s intellectual trajectories are quite well known in the English speaking world. ... However, it would be fair to note that active engagement with [them] has been relatively limited in Anglophone political geography in spite of the obvious intellectual influence of French thinkers such as Foucault, Derrida and Baudrillard' (Claval, 2000: 262). This point is developed by Hepple (2000) in the other chapter about Lacoste in the same book: 'What might explain the lack of attention to the Lacoste/*Hérodote* work? The first reason is quite simply that few Anglophone geographers read much French geography, or indeed (to our shame) any non-English language source: the neglect of *Hérodote* is not specific, but part of a wider disengagement. Although French social theory is *à la mode* in Anglophone geography, it is usually in translation.' This is probably the

reason that Lacoste is not included in the otherwise wide selection of *Key Thinkers on Space and Place* edited by Hubbard et al. (2004). And this is probably also why in Agnew, Mitchell and Toal's *Companion to Political Geography* he only rates the following lines (by Peter Taylor): 'Brief mention must be made of the French geographer Lacoste. Viewing knowledge as essentially strategic in nature, he created a radical geopolitics. Although both Marxist and Foucauldian roots can be traced, his was an activist's geopolitics countering the "common-sense" geographies underlying contemporary realist foreign and domestic policy. Despite the founding of his own journal, *Hérodote*, his brand of radical political geography had relatively little influence on the Anglo-American mainstream. ... The relative neglect of Lacoste's radical contribution is itself an illustration of contemporary political geography as an example of US hegemony in the world academy' (Taylor, 2003: 50).

11 *Crisi della geografia. Geografia della crisi.* Milan: Franco Angeli, 1978.

12 'And not vice versa ... as geographic determinism has tried to demonstrate for a long time, since that very idea of nature was already a product of power' (Dematteis, 1980: 10).

13 The journal, renamed *Erodoto* in 1982, remained largely unsuccessful and was forced to shut down in 1984 after only six issues (Antonsich, 1997; Atkinson, 2000). Along with the refusal of most Italian university libraries to subscribe to it, its failure was probably the result of its inability to open up to international and internal geopolitical issues, as the French version was successfully doing.

14 In 1986, Raffestin announced to an Italian audience that Foucault had been 'unreasonably forgotten'. By then, this was certainly true in reference to Italian and French geography, notwithstanding Quaini's efforts and his personal engagement with the French philosopher's ideas. A few years later, Chris Philo similarly lamented the 'absence ... of any sustained theoretical engagement with Foucault on the part of theoretically minded geographers' (1992; p. 206 of 2000 edn). However, in the English-speaking world Foucault was soon to take on great importance for his ideas about both power (e.g. 'Michel Foucault is arguably the predominant influence on contemporary accounts of power and space within political geography', Allen, 2003: 101) and discourse (e.g. 'Much of the work in discourse theory derives from Foucault's studies of the relations between knowledge, discourses, representation and power', Barnes and Duncan, 1992: 9). On the contrary, notwithstanding their early contact with the philosopher, French and Italian geographers have remained generally less receptive to his ideas (see Raffestin, 1997, 2005; Fall, 2005b).

15 *Pour une géographie du poivoir* has been translated into Italian (*Per una geografia del potere.* Milan: Unicopli, 1981) and into Portuguese (*Por una geografia do poder.* Rio de Janeiro: Atica, 1993), while only a short chapter of his book *Géopolitique et histoire* has been translated into English (by J. Levi in 2001).

16 Dodds and Atkinson's book on international geopolitical traditions (2000), in which two chapters are dedicated to Lacoste, makes little mention of Raffestin, and then only to highlight his differences from Lacoste's ideas. Undeniably, Raffestin has never achieved the fame of Lacoste – not in France, and certainly not among English-speaking geographers (Fall, 2005a).

17 Farinelli's work is rarely available in English or French. However, he has taught and lectured all over the world (Geneva, Los Angeles, San Francisco, Paris and so on) and spreads his ideas through constant contact with Euro-American geographers and scholars such as Gunnar Olsson, Denis Cosgrove and John Agnew. (Concerning the informal seminars and meetings organized by this group of 'heretic geographers', see Guarrasi, 2005.)

18 'I cannot but laugh when I see those many who have drawn maps of the world and not a one of them making a reasonable appearance of it. They draw Ocean flowing around an earth that is as circular as though traced by compasses, and they make Asia of the same size as Europe' (*Histories*, IV, 36, quoted by Farinelli, 1994: 12).

19 Thanks to his contribution to defining the role of cartography in the making of a geo-coded world, Farinelli may be included among the main instigators of a branch of research that could be defined as 'critical cartography'. See, for example, Pickles, in the preface of his book *A History of Spaces*: 'Cartographic Reason can – as Gunnar Olsson, Franco Farinelli and Tom Conley have variously suggested – be seen as the missing element in social theories of modernity' (Pickles, 2004: xi). From Conley, Pickles is quoting the concept of a perspectival object, a concept that 'has to do with the positioning and mapping of the self in and about the world …' (Conley, 1996: 139). The idea of cartography 'imposing spatial order on phenomena' is also tackled by Harvey (2001: 220) and Cosgrove (1999).

20 The first reference to the expression 'critical geopolitics' can probably be found in G. Toal's unpublished PhD thesis, the title of which is 'Critical geopolitics: the social construction of space and place' (quoted in Dalby, 1991).

21 In this, Gerard Toal is differentiating his perspectives from what he recognized as the main focus of Dissident IR and of Simon Dalby's interests as well (the narratives leading to the distinction of 'us' from 'the other' and to the division of space into 'our space' and 'their space') (Ó Tuathail, 1996b).

22 In 1996, *Political Geography* devoted a special double issue to Critical Geopolitics. The Editorial Introduction, written by Dalby and Toal, defines critical geopolitics as 'a new constellation', since the authors writing under its signs 'start from many different points of departure, lines of flight, and explanatory destinations yet all, in their different ways, strive to negotiate "geopolitics" with a critical perspective on the force of fusion of geographical knowledge and system of power' (Dalby and Ó Tuathail, 1996: 452). More specifically, Dalby writes in the same issue: 'The term 'critical geopolitics" encompasses poststructuralist and other forms of what were once known as "radical approaches", not all of which accept the epistemological or political precepts of (Frankfurt style) critical theory or (neo-marxist) critical political economy. The ' "critical" in "critical geopolitics", then, usually refers to the meaning of the term that invokes problematization of the discourses of geopolitics. It does not necessarily imply the presence of a worked-out alternative political project' (Dalby, 1996: 655).

23 See, among the most stimulating contributions to the emergence of the constellation, Joanne Sharp's ground-breaking contribution on the concept of 'hegemony and popular culture' (1993), and Dodds and Sidaway's helpful attempt towards 'locating critical geopolitics' inside its theoretical framework (1994). To Dodds and Sidaway, specifically, must go the credit for an early attempt to systematize not only the contribution of Foucault's notions, but also that of 'Said's creative synthesis of these ideas with frameworks from Gramsci' (Dodds and Sidaway, 1994: 516). About Said's reworking of Foucault's ideas, see also Driver (1992).

24 Neil Smith comments, on *Critical Geopolitics*, that 'the substance of Ó Tuathail's book is the method. … He brings a swath of tools to this project but none is more important than his deployment of Martin Jay's notion of "Cartesian perspectivalism"' (Smith, 2000: 365–6). All the same, a substantial limit on the use of this tool is, according to Smith, that Toal does not properly distinguish the act of seeing from the act of reading/writing, and '… therefore it is not clear precisely what the problem is with the vision per se' (ibid.: 368).

REFERENCES

Allen, J. (2003) 'Power', in J.A. Agnew, K. Mitchell and G. Toal (eds), *A Companion to Political Geography*. Oxford: Blackwell, pp. 95–108.

Anderson, B. (1983, rev. edn, 1991) *Imagined Communities: Reflections on the Origin and Spread of Nationalism*. London: Verso.

Andrews, J.H. (2001) 'Introduction', in J.B. Harley, *The New Nature of Maps: Essays in the History of Cartography* (ed. P. Laxton). Baltimore: Johns Hopkins University Press.

Antonsich, M. (1996) *Geografia politica e geopolitica in Italia dal 1945 ad oggi*. Trieste: Quaderni del Dottorato di Ricerca in Geografia Politica.

Antonsich, M. (1997) 'La geopolitica italiana nelle riviste "Geopolitica", "Hérodote/Italia (Erodoto)", "Limes"', *Bollettino della Società Geografica*, XII(11,3): 411–8.

Ashley, R.K. (1984) 'The poverty of neo-realism', *International Organisation*: 38: 225–86.

Ashley, R.K. (1987) 'The geopolitics of geopolitical space: toward a critical social theory of international politics', *Alternatives*, 12: 403–34.

Atkinson, D. (2000) 'Geopolitical imaginations in modern Italy', in K.J. Dodds and D. Atkinson (eds), *Geopolitical Traditions: A Century of Geopolitical Thought*. London: Routledge, pp. 93–117.

Barnes, T.J. and Duncan, J.S. (eds), (1992) *Writing Worlds: Discourse, Text and Metaphor in the Representation of Landscape*. London: Routledge.

Baudrillard, J. (1981) *Simulacres et simulation*. Paris: Galilée.

Claval, P. (2000) '*Hérodote* and the French left', in K.J. Dodds and D. Atkinson (eds), *Geopolitical Traditions: A Century of Geopolitical Thought*. London: Routledge, pp. 239–67.

Conley, T. (1996) *The Self-Made Map: Cartography Writing in Early Modern France*. Minneapolis: University of Minnesota Press.

Cosgrove, D. (1999) 'Introduction: mapping meaning', in D. Cosgrove (ed.), *Mappings*. London: Reaktion Books, pp. 1–23.

Cox, R.W. (1983) 'Gramsci, hegemony and international relations: an essay in method', *Millennium: Journal of International Studies*, 12: 687–718.

Cox, R.W. (1987) *Production, Power and World Order: Social Forces in the Making of History*. New York: Columbia University Press.

Dalby, S. (1988) 'Geopolitical discourse: The Soviet Union as other', *Alternatives: Social Transformation and Humane Governance*, 13: 415–42.

Dalby, S. (1990) *Creating the Second Cold War: The Discourse of Politics*. New York: Guilford.

Dalby, S. (1991) 'Critical geopolitics: discourse, difference and dissent', *Environment and Planning D: Society and Space*, 9: 261–83.

Dalby, S. (1996) 'Writing critical geopolitics: Campbell, Ó Tuathail, Reynolds and dissident scepticism', *Political Geography*, 15: 655–60.

Dalby, S. and Ó Tuathail, G. (1996) 'The critical geopolitics constellation: problematizing fusions of geographical knowledge and power', *Political Geography*, 15: 451–6.

Dear, M.J. (2001) 'The postmodern turn', in C. Minca (ed.), *Postmodern Geography: Theory and Praxis*. Oxford: Blackwell, pp. 1–34.

dell'Agnese, E. (2005a) 'Viaggiare al maschile: dal Grand Tour al turismo sessuale', in E. dell'Agnese and E. Ruspini (eds), *Turismo al maschile, turismo al femminile. L'esperienza del viaggio, il mercato del lavoro, il turismo sessuale*, Padua: Cedam, pp. 83–112.

Dematteis, G. (1980) 'Fra Foucault e Hérodote c'è di mezzo Marx', *Hérodote/Italia*, 2/3: 9–13.

Dematteis, G. (1985) *Le metafore della Terra. La geografia umana tra mito e scienza*. Milan: Feltrinelli.

Der Derian, J. and Shapiro, M.J. (eds), (1989) *International/Intertexual Relations: Postmodern Readings of World Politics*. Lexington: Lexington Books.

Dodds, K.J. (1993) 'Geopolitics, experts and the making of foreign policy', *Area*, 25: 70–4.

Dodds, K.J. and Atkinson, D. (eds) (2000) *Geopolitical Traditions: A Century of Geopolitical Thought*. London: Routledge.

Dodds, K.J. and Sidaway, J.D. (1994) 'Locating critical geopolitics', *Environmente and Planning D: Society and Space*, 12: 515–24.

Driver, F. (1992) 'Geography's empire: histories of geographical knowledge', *Environment and Planning D: Society and Space*, 10: 23–40.

Fall, J.J. (2005a) 'Les limites du pouvoir: boundaries and power in Francophone political geography', paper presented at AAG Pre-Conference at the University of Colorado at Boulder, April 3–5.

Fall, J.J. (2005b) 'Michel Foucault and Francophone geography', *EspacesTemps.net*, http://espacestemps.net/document1540.html.

Farinelli, F. (1985) 'De la crise à la critique de l'imagination géographique', in *Actes du colloque 1985, L'imagination géographique*, Université Genève–Lausanne.

Farinelli, F. (1992) *I segni del mondo. Immagine cartografica e discorso geografico in età moderna*. Florence: La Nuova Italia.

Farinelli, F. (1994) 'Squaring the circle or the nature of political identity', in F. Farinelli, G. Olsson and D. Reichert (eds), *Limits of Representation*. Munich: Accedo, pp. 11–28.

Foucault, M. (1966) *Les mots et les choses. Une archéologie des sciences humaines*. Paris: Editions Gallimard (*The Order of Things*. New York: Pantheon, 1970).

Foucault, M. (1980) *Power/Knowledge: Selected Interviews* (ed. Colin Gordon). New York: Pantheon.

Gambi, L. (1961) *Geografia regione depressa*. Faenza: Fratelli Lega (reprinted in *Una geografia per la storia*. Turin: Einaudi, 1973).

Gambi, L. (1964) *Questioni di geografia*. Naples: ESI.

Gambi, L. (1968) *Geografia e contestazione*. Faenza: Fratelli Lega.

Gambi, L. (1971) *Intervento a un meeting su 'Ricerca e insegnamento geografici nelle università'*. Faenza: Fratelli Lega (reprinted in *Una geografia per la storia*, Turin: Einaudi, 1973).

Gambi, L. (1973) *Una geografia per la storia*. Turin: Einaudi.

Gambi, L. (1994) 'Geography and imperialism in Italy: from the unity of the nation to the "new" Roman Empire', in A. Godlewska and N. Smith (eds), *Geography and Empire: Critical Studies in the History of Geography*. Oxford: Blackwell, pp. 74–91.

Girot, P. and Kofman, E. (eds and transl.) (1987) *International Geopolitical Analysis: A Selection from Hérodote*. London: Croom Helm.

Gregory, D. (1994) *Geographical Imaginations*. Oxford: Blackwell.

Gregory, D. and Walford, R. (1989) 'Introduction: making geography', in D. Gregory and R. Walford (eds), *Horizons in Human Geography*. London: Macmillan, pp. 1–7.

Guarrasi, V. (2005) 'Paesaggio di teorie', *Equilibri*, IX(1): 11–27.

Hague, E. (2004) 'Gearóid Ó Tuathail (Gerard Toal)', in P. Hubbard, R. Kitchin and G. Valentine (eds), *Key Thinkers on Space and Place*, London: Sage, pp. 226–30.

Harley, J.B. (1988) 'Maps, knowledge and power', in D. Cosgrove and S. Daniels (eds), *The Iconography of Landscape*, Cambridge: Cambridge University Press, pp. 277–312. (Also in J.B. Harley, *The New Nature of Maps*. Baltimore: Johns Hopkins University Press, 2001.)

Harley, J.B. (1989) 'Deconstructing the map', *Cartographica*, 26(2): 1–20. (Also in T.J. Barnes and J.S. Duncan (eds), *Writing Worlds: Discourse, Text and Metaphor in the Representation of Landscape*. London: Routledge, 1992, pp. 231–47, and in J.B. Harley, *The New Nature of Maps*. Baltimore: Johns Hopkins University Press, 2001.)

Harvey, D. (1989) *The Condition of Postmodernity: An Enquiry into the Origins of Cultural Change*. Cambridge, MA: Blackwell.

Harvey, D. (2001) *Spaces of Capital: Towards a Critical Geography*. New York: Routledge.

Hepple, L.H. (2000) 'Géopolitiques de gauche: Yves Lacoste, *Hérodote* and French radical geopolitics', in K.J. Dodds and D. Atkinson (eds), *Geopolitical Traditions: A Century of Geopolitical Thought*. London: Routledge, pp. 268–301.

Hubbard, P., Kitchin, R. and Valentine, G. (eds) (2004), *Key Thinkers on Space and Place*. London: Sage.

Lacoste, Y. (1973) 'La géographie', in F. Chatelet (ed.), *La philosophie des sciences sociales*. Paris: Hachette, pp. 242–302.

Lacoste, Y. (1976) *La Géographie, ça sert, d'abord, à faire la guerre*. Paris: Maspero. (Italian edition, *Crisi della geografia. Geografia della crisi*. Milan: Franco Angeli, 1977, ed. P. Coppola).

Lacoste, Y. (1977) 'An illustration of geographical warfare: bombing of the dikes on the Red River, North Vietnam', in R. Peet (ed.), *Radical Geography*, London: Methuen, pp. 244–61.

McGrevy, P. (1992) 'Reading the texts of Niagara falls: the metaphor of death', in T.J. Barnes and J.S. Duncan (eds), *Writing Worlds: Discourse, Text and Metaphor in the Representation of Landscape*. London: Routledge, pp. 50–72.

Minca, C. (2003) 'Critical peripheries', *Environments and Planning D: Society and Space*, 21: 30–8.

Olsson, G. (1992) Lines of Power', in T.J. Barnes and J.S. Duncan (eds), *Writing Worlds: Discourse, Text and Metaphor in the Representation of Landscape*. London: Routledge, pp. 86–95.

Parker, G. (1998) *Geopolitics Past, Present and Future*. London: Pinter.

Peet, R. (ed.) (1977) *Radical Geography: Alternative Viewpoints on Contemporary Social Issues*. Chicago: Maaroufa Press.

Philo, C. (1992) 'Foucault's geography', *Environment and Planning D: Society and Space*, 10(2): 137–67, reprinted in M. Crang and N.J. Thrift (eds), *Thinking Space*. London: Routledge, 2000.

Pickles, J. (2004) *A History of Spaces: Cartographic Reason, Mapping and the Geo-coded World*. London: Routledge.

Quaini, M. (1974) *Marxismo e geografia*. Florence: La Nuova Italia (English edition, *Geography and Marxism*, Totowa, N.J: Barnes and Noble, 1982).

Quaini, M. (1978) *Dopo la geografia*. Milan: Espresso Strumenti.

Quaini, M. (2005) 'Riflessioni post-marxiane sul fantasma di geografia democratica', paper presented at the meeting *Geografia, società, politica. La ricerca in geografia come impegno sociale*, Round Table *Ricordando 'geografia democratica': ripensare il passato per immaginare il futuro*, 16–17 June.

Raffestin, C. (1978) 'Evoluzione storica della territorialità in Svizzera', in J.B. Racine, C. Raffestin and V. Ruffy (eds), *Territorialità e paradigma Centro-Periferia*. Milan: Unicopli.

Raffestin, C. (1980) *Pour une géographie du pouvoir*. Paris: PUF. (Italian edition, *Per una geografia del potere*, Milan: Unicopli, 1981)

Raffestin, C. (ed.) (1983) *Geografia politica: teorie per un progetto sociale*. Milan : Unicopli.

Raffestin, C. (1986) 'Genealogia della geopolitica o da una nascita ad un rinnovamento', in *Atti del XXIV Congresso Geografico Italiano 'La geografia per un mondo in transizione'*. Bologna: Patron, pp. 295–301.

Raffestin, C. (1997) 'Foucault aurait-il pu révolutionner la géographie?', in *Au risque de Foucault*. Paris: Editions du Centre Pompidou.

Raffestin, C. (2005) 'L'actualité de Michel Foucault'. *Espaces-Temps.net*, http://espacestemps. net/document1172.html.

Reuber, P. (2000) 'Conflict studies and critical geopolitics – theoretical concepts and recent research in political geography', *GeoJournal*, 50: 37–43.

Shapiro, M. (1988) *The Politics of Representation*. Madison: University of Wisconsin Press.

Sharp, J.P. (1993) 'Hegemony, popular culture and geopolitics: the *Reader's Digest* and the construction of danger', *Political Geography*, 15: 557–70.

Sharp, J.P. (2003) 'Feminist and postcolonial engagements', in J.A. Agnew, K. Mitchell and G. Toal (eds), *A Companion to Political Geography*. Oxford: Blackwell, pp. 59–74.

Smith, N. (2000) 'Is a critical geopolitics possible? Foucault, class and the vision thing', *Political Geography*, 19: 365–71.

Soja, E.W. (1989) *Postmodern Geographies: The Reassertion of Space in Critical Social Theory*. London and New York: Verso.

Sparke, M. (2000) 'Graphing the geo in geo-political, critical geopolitics and the re-visioning of responsibility', *Political Geography*, 19: 373–80.

Taylor, P.J. (2003) 'Radical political geography', in J.A. Agnew, K. Mitchell and G. Toal (eds), *A Companion to Political Geography*. Oxford: Blackwell, pp. 47–58.

Ó Tuathail, G. (Toal, G.) (1986) 'The language and nature of the "new geopolitics": the case of US–El Salvador relations', *Political Geography Quarterly*, 5: 73–85.

Ó Tuathail, G. (Toal, G.) (1987) 'Beyond empiricist political geography: a comment on van der Wusten and O'Loughlin', *Professional Geographer*, 39: 97.

Ó Tuathail, G. (Toal, G.) (1992) 'Foreign policy and the hyper-real, the Reagan administration and the scripting of "South Africa"', in T.J. Barnes and J.S. Duncan (eds), *Writing Worlds: Discourse, Text and Metaphor in the Representation of Landscape*. London: Routledge, pp. 155–75.

Ó Tuathail, G. (Toal, G.) (1994a) '(Dis)placing geopolitics: writing on the maps of global politics', *Environment and Planning D: Society and Space*, 12: 323–46.

Ó Tuathail, G. (Toal, G.) (1994b) 'The critical reading/writing of geopolitics: re-reading/writing Wittfogel, Bowman and Lacoste', *Progress in Human Geography*, 18(3): 313–32.

Ó Tuathail, G. (Toal, G.) (1996a) *Critical Geopolitics*. Minneapolis: University of Minnesota Press.

Ó Tuathail, G. (Toal, G.) (1996b) 'Dissident IR and the identity politics narrative: a sympathetically sceptical perspective', *Political Geography*, 15: 647–53.

Ó Tuathail, G. (Toal, G.) (1999) 'Understanding critical geopolitics: geopolitics and risk society', in C. Gray and G. Sloan (eds), *Geography, Geopolitics and Strategy*. London: Frank Cass, pp. 107–24.

van der Wusten, H. and O'Loughlin, J. (1986) 'Claiming new territory for a stable peace: how geography can contribute', *Professional Geographer*, 38: 18–28.

Winichakul, T. (1994) *Siam Mapped: A History of the Geo-Body of a Nation*. Honolulu: University of Hawai'i Press (PhD thesis, University of Sydney, 1988).

Empire

Alan Lester

INTRODUCTION: IMPERIALISM, COLONIALISM, CAPITALISM AND MODERN EMPIRES

Since the word 'imperialism' was first used to describe a systematic policy of French state-driven expansion in the late nineteenth-century, and then applied to the British government's attempts to defend and expand its own pre-existing empire, various scholars have tried to devise general, systematic theories to explain the globalized power relations that the term conjures up.

Marx, Luxemburg, Lenin and the Marxist scholars who followed them related imperialism to the crises of accumulation that were inherent in Europe's advanced capitalist systems during the late nineteenth-century. For them, 'The drive to imperial expansion must be explained as one element in the whole process of capitalist development' (Brewer, 1990: 2). The emphasis in terms of exactly how capitalism and imperialism are related may be contested, with, for instance, Wallerstein (1979) seeing imperialism, in much longer-term perspective, as the forcible extension of a capitalist world system that first emerged out of the ashes of declining European feudal structures, and Brenner (1977) seeing it as a system of exchange between the new European capitalist core and mainly pre-capitalist, colonized societies. There have been various other, non-Marxist attempts, as well, to slot the historical experiences of different empires into a general, systematic analysis. Eisenstadt, for instance, produced a comparative survey of the decline of ancient, pre-capitalist empires including the Ahmenid, the Carolingian, the Mongol and the Chinese, based upon the contradictions inherent in a centralized bureaucracy that covers competing interests

(Eisenstadt, 1967). More recently, Mann (1986, 1993) accounts for the transition from pre-capitalist empires to modern European empires by taking into account their economic relations, but also by paying more attention than have many Marxists to military, political and ideological forms of power, and Doyle has attempted to shift the terrain of debate to conceive of 'empire as political control over effective sovereignty' rather than as the product of primarily economic, military or strategic goals (Doyle, 1986: 21).

Regardless of their different emphases, though, most of these scholars of imperialism as a general phenomenon are agreed that the European empires that reached their greatest extent around the end of the nineteenth and the beginning of the twentieth centuries were distinguished from their predecessors in two main ways. First, whereas ancient and early medieval states such as the Chinese, the Roman, the Persian, the Mogul and the Byzantine were able to absorb neighbouring polities, the empires that resulted were founded on the administrative exaction of tribute rather than engagement with the capitalist relations of uneven exchange that characterized the modern European metropoles. The earliest 'modern' European empires, Portugal's and Spain's, played something of a transition role between 'tribute' and 'capitalist' empires, since they shared the ancient empires' grounding in the exaction of plunder from distant territories, even as that plunder contributed to the development of capitalist relations within Europe (Blaut, 1993, 2000; Russell-Wood, 1998; Elliott, 2002).[1]

Second, the older empires were, by and large, carved out through conquest of neighbouring polities on the same land mass, while the more recent European empires were contingent upon a trans-oceanic traffic in people, capital, commodities and ideas. This traffic was enabled by developments in naval technology (themselves linked to the emergence and consolidation of capitalist relations) from the fifteenth century. As Robert Young points out, new sea-going 'vessels enabled not only the geographical expansion of populations, but also enabled such populations to stay in touch with their homelands' (Young, 2001: 16).

However, as Jones and Phillips (2005) have recently argued, to separate pre-capitalist, pre-modern empires from capitalist, modern empires in this fashion, establishes binaries that may actually be unhelpful in generating a general appreciation of imperial and colonial power relations. After all, such relations, stretching back to antiquity, have always had unequal economic transactions at their core, and have always been organized around contested notions of difference counterposed to universalism. The facts that the economic relations of the European empires from the sixteenth century were 'more' capitalist, that the concept of 'race' was invented after this period as a particular handle on difference and that these empires were transoceanic rather than (or as well as, if we consider Russia) trans-continental, are perhaps less important than these fundamental continuities. Furthermore, as Jones and Phillips (2005) point out, the weight of scholarship bearing upon modern European imperialism at the expense of other and earlier empires may actually run counter to a current post-colonial politics of decentring Europe in our accounts of colonialism and imperialism.

The modern European empires, like their predecessors, were founded on widely varying 'projects', pursued by different metropolitan and, increasingly from the eighteenth century, colonial, interests (see Thomas, 1994: 105–6). The incompatibility of some of these projects led to considerable tensions as colonists, missionaries, traders and officials pursued their own agendas, locally as well as through trans-imperial connections (Stoler, 2002). This, of course, is not to mention the different practices of imperial rule that were developed within the different European empires. France's assimilationist agenda for colonized peoples, for instance, was replicated (and indeed preceded) within British humanitarian discourse, but countervailing British practices of segregation and indirect rule distinguished the two empires in the late nineteenth-century (Fieldhouse, 1982).

While comparative, general theories and surveys of imperialism are significant in helping us to contextualize different episodes in empire construction and decline, they often end up generating semantic debates over the proper definitions of terms such as 'empire', 'imperialism' and 'colonialism' rather than contributing

to our appreciation of how the people involved understood and experienced these episodes of cross-cultural encounter. Even where the calculations of imperial statesmen (*sic*) take centre stage, the existence of other, often competing, imperial and colonial interests, let alone colonized subjectivities, is rarely taken into account in the generalized historical-sociological literature on empire.

Historical geographers have produced much work of late that does attempt to uncover the conflicts, accommodations, negotiations and reformulations of power relations, both material and cultural, that imperialism and colonialism fostered in different sites across the globe (see Clayton, 2000; Lester, 2001; Blunt and McEwan, 2002; Yeoh, 2003, for some examples, and Clayton, 2002, for an excellent overview). But in this chapter, rather than explore the geographies of particular colonial encounters as these studies have, I want to take a recent resurgence in the use and discussion of the term 'Empire' as my main focus.

I intend to explore how both historical geographers of colonialism and scholars of contemporary geopolitics can contribute to a current debate on 'Empire' by connecting past and present in more meaningful ways. Rather then enter further into a discussion of definitions, then, I will, like many others, use the terms 'colonial' and 'imperial' rather loosely, to indicate where a given practice may lie upon a spectrum between actions taken by various interests in the colonies themselves (colonial) and those pursued by metropolitan governments (imperial). The fact that such projects were and are connected by networks linking the colonies to each other and to the metropole, and that such analytical distinctions are thus not always realistic, is something that becomes clearer as the argument proceeds.

THE REVIVAL OF EMPIRE

The drawing of comparisons between the British Empire of the nineteenth-century and contemporary US imperialism has become a popular exercise of late. On the one hand, neo-liberals and conservatives have sought to re-legitimate the notion of Empire and to advocate that the United States today follow in some of Britain's imperial footsteps. On the other hand, radicals have attempted to resurrect Empire as a category of critical analysis relevant to the contemporary, globalizing world.

The British historian Niall Ferguson is a vocal contributor in the former vein. His conclusion (not necessarily warranted by the material that he presents very impressively elsewhere in his work) is that, despite its mistakes and its instances of brutality, overall, British imperialism was good for the world because it made it 'modern' and globalized. More of the same kind of, by and large, beneficent, liberal suzerainty over contemporary globalization would be good from America today, if only its leaders could acknowledge such an imperial agenda explicitly and provide the resources and the political endurance necessary to pursue it (Ferguson, 2003, 2004). It is not only Ferguson who argues along these lines. Michael Ignatieff has written pieces in the *New York Times*, one of which was entitled 'American Empire: Get Used to It' (5 January 2003), that advocate a step away from 'empire lite' as he calls it and towards more formal US imperial intervention. And Max Boot, editor of the *Wall Street Journal*, asserts that 'Afghanistan and other troubled lands today cry out for the sort of enlightened foreign administration once provided by self-confident Englishmen in jodhpurs and pith helmets' (quoted in Harvey, 2003: 4; see also Smith, 2003a: xvi; Bowden, 2002).

Such calls remind us that the tropes of colonial discourse, upon which so much has been written in recent decades, still have tremendous relevance today. As Gregory (2004) argues, there are two kinds of intersection between past and present that run through these calls for the revival of empire. The first he calls 'colonial amnesia', in which 'we forget the exactions, suppressions, and complicities that colonialism forced upon the peoples it subjugated, and the way in which it withdrew from them the right to make their own history, ensuring that they did so emphatically not under conditions of their own choosing' (Gregory, 2004: 10). In order to avoid such amnesia, we need constantly to remind ourselves that 'even the best-run empires are cruel and violent ... overwhelming power, combined with a sense of boundless superiority, will produce atrocities – even among the well-intentioned' (Gregory, 2004: 10, quoting Maria Misra). The second is colonial nostalgia in which '[c]ontemporary metropolitan cultures are ... characterized by nostalgia for the aggrandizing swagger of colonialism itself, for its privileges and powers ... the triumphal show of colonialism – its elaborate "ornamentalism", as David Cannadine calls it – and its effortless, ethnocentric assumption of Might and Right' (Gregory, 2004: 10–11).

From a radically different political stance, Hardt and Negri similarly use the language of Empire in order to conceptualize contemporary forms of globalized regulation, although their stress is on historical discontinuities and the novelty of the present moment rather than the lessons of the past. Their thesis is that, since the decline of the European empires, sovereignty has taken a new form, composed of a series of national and supranational 'organisms' united under a single logic of rule, which is what they call Empire. They are enormously impressed by the newness of this

logic and by its deteritorrialized nature. They claim that:

> The sovereignty of the nation-state was the cornerstone of the imperialisms that European powers constructed throughout the modern era. ... Imperialism was really an extension of the sovereignty of the European nation-states beyond their own boundaries. ... In contrast to imperialism, Empire established no territorial centre of power and does not rely on fixed boundaries or barriers. It is a *decentred* and *deterritorializing* apparatus of rule that progressively incorporates the entire global realm within its open, expanding frontiers. ... The distinct national colors of the imperialist map of the world have merged and blended in the imperial global rainbow'. (Hardt and Negri, 2000: xii–xiii)

The corollary of this view is that '*the United States does not, and indeed no nation state can today, form the centre of an imperialist project. Imperialism is over. No nation will be world leader in the way modern European nations were*' (Hardt and Negri, 2000: xiv, emphasis in the original). For a thought-provoking review of a number of these recent texts on Empire, see Wainwright (2004).

These intersecting historicist claims, for the novelty of a seemingly deterritorialized Empire and of the decline of the nation-state as the centre of empires, provide themes to which this chapter will periodically return. By examining contemporary imperial historians' and historical geographers' understandings of the ways that the nineteenth-century empires, Britain's in particular, actually worked, and by relating those understandings to geographers' current impressions of US imperialism, we will be in a much better position to assess such claims, and the kinds of spatiality that they project in particular.

Based upon such readings, this chapter suggests that Hardt and Negri are right to emphasize that nineteenth-century imperialism was a product of specific interests attempting to harness globalized transactions. However, much of the recent literature on British imperialism in the nineteenth century suggests that they are wrong to insist that it was the nation-state and its sovereignty alone that embodied these interests. Rather we must envisage multiple interests attempting to orchestrate rival projects through particular trans-global networks that were painstakingly assembled and geographically uneven. Moreover, much of the writing by geographers on US imperialism today suggests that they are also wrong to assume that the nation-state, and especially the US state, is no longer so significant in the construction of a new empire. As Sparke (2003: 379) argues, 'historicist assumptions about a telos leading from old bordered communities to newly networked nodes', which are constructed as 'smooth' (uninterrupted by political and cultural boundaries) and deteritorrialized, with no nation-state or territory particularly prominent in their construction and maintenance, are mis-leading. Such assumptions both underestimate the extent of decentred trans-global networking under the 'old' European empires and overestimate the 'smoothness' of contemporary networked flows. If anything, one could reverse Hardt and Negri's assertion: while neither nation-state was, or will be, successful in its goal of harnessing globalization, because of its empire's networked constitution the British state throughout much of its imperial tenure was a less significant (although still powerful) force in the production and maintenance of empire than is the contemporary US state.

What much of the recent writing on empires, both British and American, suggests is first that the phenomena empire and globalization intersect but also differ, such that globalization cannot be con-tained within the notion of empire (with or without a capital E) in the way that Hardt and Negri seem to suppose. Empires can usefully be conceived as attempts by different, and often competing, constel-lations of interest (the specificity of which all but disappears in Hardt and Negri's vision) to 'capture' existing long-distance, transnational transactions, or to create new ones. I say 'constellations of inter-est' rather than 'nation-states' because, although nation-states may be one such constellation, albeit a particularly significant one defined by the exercise of governmentality, there may be others carry-ing out different projects as well. In Britain's case, these other constellations and projects were defined by the imperatives of capital accumulation (for example, the East India Company), of family security or material enrichment (emigrant settlers), of proselytization (missionaries), of knowledge (explorers, naturalists and geographers) or by some combination of these. Empires are rarely the sole result of state-devised blueprints for the projec-tion of national sovereignty. Although I would not go as far as some in saying that Britain's empire was obtained in a 'fit of absence of mind' on the part of the British state (Seeley, 1971 [1883]: 8), much of what later became reified as the empire was appropriated in the first instance by decentred Britons (company officials, traders, settlers, mis-sionaries and even 'rogue' governors: see Keegan, 1996), who had little to do with the calculations of the metropolitan state, although they did seek its support.

If empires are particular constellations of diverse and often antagonistic interests, neither wholly economic, cultural nor narrowly political in their inspiration, acting to foster and attempting to har-ness globalizing transactions in pursuit of partic-ular projects, they never quite succeed in doing so. Globalization, as the sum total of all these long-distance, trans-national transactions, always

exceeds, spills over, the limits of empire. Empires have certainly been intrinsically bound up with, and have indeed helped to foster, periodic *re-* global-izations, each such round of globalization having its own distinctive and uneven geographies of con-nection, but empires have never been able quite to determine or to 'capture' any one of these rounds. Globalization, in this view, is neither a startlingly new phenomenon, as many commentators struck by rapidly increasing speed of communications in recent decades would have us believe, nor a lin-ear process with antecedents stretching back to the early modern period or beyond, as many histori-cally minded scholars proclaim (see Bell, 2003). Rather there have been multiple, periodic, non-linear globalizations, each a re-working of the last, and each in part galvanized, although never deter-mined or captured, by the fragmented exercise of imperialism by multiple, decentred interests as well as nation-states (see Hopkins, 2002a).

NETWORKS AND THE STATE IN THE BRITISH EMPIRE

By the eighteenth century, after 'experiments' in colonization within the British Isles, and especially under the influence of a new military fiscalism ('the use of state power to command the revenues needed to meet war expenditures': Hopkins, 2002b: 26), British interests were orchestrating different, but overlapping, trans-global networks of commerce, communication, settlement and administration. It was these networks that enabled contemporaries to speak of a British (or often English) empire (see Canny, 1998; Armitage, 2000). In turn the maintenance of that empire demanded more effi-cient forms of military fiscalism on the part of the British state. During the war years of the 1760s in particular, Britons fostered particularly extensive commercial and military growth, marked especially by the morphing of the English East India Company into a territorial power (see Bowen, 1998). As Hopkins notes, as a result of the Seven Years' War, Britons generated a 'fresh wave of imperial acquisitions' and 'a knowledge revolution that mapped, surveyed and classified the world of contact and conquest' (Hopkins, 2002b: 26).

While the state's concern with wartime strategy and its symbiosis with commercial finance-raising ventures lay behind this geographical expansion and cultural cosmopolitanism, no one 'force' or 'motivation' drove the perpetuation and further extension of the empire. The infrastructure for the maintenance of the state's new acquisitions consisted not only of naval bases, but also of 'commercial factories ... and missionary stations [that] encircled the world, reaching beyond the tra-ditional sources of British power in the Atlantic and the Mediterranean to Asia, the Pacific and the west coast of the Americas' (Ballantyne, 2002b: 117). Ballantyne argues that the result was a cosmopolitan British culture 'moulded by Christianity, underpinned by slavery, fed by New World and Asian food plants and drugs, and successively reshaped by Chinoiserie, the "Oriental Renaissance" and Europe's craze for the Pacific' (Ballantyne, 2002b: 116).

There was no 'logic', whether strategic, economic or cultural, to British imperial expansion from the late eighteenth century. British imperial interests fostered and attempted to capture many different kinds of globalization, including that of trade, that of racial discourse and that of humanitarianism. Neither the shift of focus for British imperial enterprise, away from the west after the American Revolution and towards the east and especially India, nor the addition to the empire of new, vast terrains in the southern hemisphere during the ensuing decades, can be explained by a single cause. Rather, the shape, purpose and direction of British imperialism were hotly contested. Different British interests, some more closely articulated with the state than others, wanted to appropriate or create different kinds of trans-global connection for different ends – to fashion different globalizations.

Of course, one of these globalizations, pursued by capitalist interests, was the trans- (and extra-) imperial extensification of commodity production and exchange. It was in part these efforts that meant that 'the rate of growth of commodity exports and of direct investments from Western ... countries reached levels shortly before 1914 that were not matched until the 1980s' (Hopkins, 2002b: 29). However, as Chateaubriand noted belatedly in 1841, 'it will not only be commodities which travel, but also ideas which will have wings' (quoted in Hopkins, 2002b: 31). The British and other European empires in part fostered and were in part the result of new imaginaries, bound up with particular kinds of globalization. One of the most significant of these imaginaries was the vision of a world redeemed and made one in the image of Christianity. This vision was pursued first across a trans-Atlantic framework of connection by the anti-slavery movement in the late eighteenth and early nineteenth centuries (see Lambert, 2005).

As part of their claim for the exceptionalism and novelty of their decentred 'Empire', Hardt and Negri point to the innovation of the 'right' of 'dominant subjects of the world order to intervene in the territories of other subjects in the interest of preventing or resolving humanitarian problems'. While such right of intervention, they argue, can be traced back (only) to the foundation of the UN, 'the contemporary reconfiguration of this right represents a qualitative leap'. No longer do individual states or the UN intervene only to ensure the application of voluntarily engaged international accords. Now 'supranational subjects' interventions are 'justified by *the appeal to essential values of justice*' (Hardt and Negri, 2000: 18). However, exactly such a kind of intervention by 'supranational subjects of the world order' was elaborated when Britons in the metropole questioned the institution of slavery in the West Indies on the grounds of 'natural' (i.e. God-given) justice from the late eighteenth century (see Hall, 2002).

As Thomas Haskell (1985a, 1985b) argues, the anti-slavery movement ensured that 'the conventional limits of moral responsibility observed by an influential minority in society expanded to encompass evils that previously had fallen outside anyone's sphere of responsibility' (Haskell, 1985a: 359). A continual flow of correspondence, objects and people between activists in North America and the British provinces, Parliamentary reformers in London and Nonconformist missionaries in the West Indies and their supporters in Britain, was essential to the success of the anti-slavery campaign. Once the slave trade had been abolished in 1807 and the enslaved in British colonies themselves emancipated in 1838, the anti-slavery network mutated into the nineteenth-century evangelical humanitarian movement and extended into new colonies of settlement in British North America, the Pacific, Australia, New Zealand and southern Africa, where indigenous peoples had to be protected from British settlers' depredations similar in intensity and scale to those of slave-owners themselves. Such campaigning was predicated on particular racial imaginaries, and intrinsic to the evangelical humanitarian agenda was an image of the indigenous peoples of empire as empty vessels waiting to be filled with the light of Christianity and civilization (Hall, 2002; Lambert and Lester, 2004; Porter, 2004).

The emigrant settlers against whom these humanitarians often waged their campaigns, however, created another kind of globalized connectivity and another kind of racial discourse. The humanitarian agenda of reclaiming souls and protecting bodies from abuse was often contravened by settlers who appropriated indigenous land and labour in pursuit of capital accumulation and family security. Humanitarian condemnation, especially during the early nineteenth-century when humanitarian campaigners had both very good access to the officials of the Colonial Office and a grip on the public imagination in the wake of their successful campaign against slavery, threatened the marginalization of settlers. Settlers devised their own globalized networks to counter the humanitarian critique, relying especially on newspapers to communicate with each other in different colonies and to transport their rival discourse of irredeemable racial difference back 'home' to Britain (Lester, 2002). By the end of the nineteenth-century it was such a racial discourse, bolstered by the scientific racism that was dominant in the metropole, that morally underpinned new imperial ventures.

European scientists established other kinds of imperial network, partially overlapping but in many respects discrete from humanitarian and settler networks, from the eighteenth century. As Richard Grove has pointed out, the trans-global connections that these scientists created ensured the invention and the travel of a modern environmental consciousness (Grove, 1995; Drayton, 2000). The channels of communication connecting head gardeners at Leiden, Kew and elsewhere with each other and with collectors in the colonies allowed for the circulation not just of specimens and understandings of them across and between imperial circuits, but also the exchange of knowledge about medicine, ecosystems and climatic change (Grove, 1995; Gascoigne, 1998; Drayton, 2000). Regional networks of collectors fed into the longer-distance webs, providing the 'weak ties' that are known in the network theories of social physics to play significant roles in connecting clusters of remote nodes more effectively (Urry, 2004). As Ballantyne points out, Sir Joseph Banks of Kew Gardens 'facilitated the transfer of seeds and plants through networks which spanned Asia and the Pacific and reached beyond into the Atlantic world ... [drawing] previously disparate societies into an imperial zone of exchange where capital commodities, human populations (and their associated biota and pathogens) circulated' (Ballantyne, 2002b: 130; Gascoigne, 1998).

Although access to British state power was critical to the fulfilment of many of these (and other) networked projects within and across the boundaries of the British Empire, in important respects, even the British imperial state itself was by no means so territorially centred, nor its sovereignty so spatially concentrated in the metropole as many contemporary commentators on nineteenth-century empires imagine. In a study of colonial officialdom and governance, Laidlaw points out that bi-directional '[n]etworks connecting metropolis to colony were the mainstay of day to day colonial governance, transmitting influence, patronage and information' (2001: 12). Colonial governors also by-passed the metropole and exchanged ideas on the best way to govern their respective colonial subjects directly, as, for example, when Governor George Grey suggested that his counterpart in the Cape Colony adopt the same kind of model hospitals and schools for rebellious Xhosa as, he claimed, had helped him subtly overcome Maori resistance in New Zealand (Lester, 2001).

The Colonial Office may have been located in London then, and decisions taken there and in Parliament may have been critical to the ultimate fulfilment of many colonial projects, but it is unrealistic to imagine the British Empire being governed solely from there. Initiatives in the 'art' of governance, especially, but not only, of culturally and 'racially' different subject populations, circulated through networked channels across a vast terrain linking colony to colony as well as each colony to the metropole, with colonial settlers as well as officials, and even certain elite indigenous groups, contributing (see Parsons, 1998, for the ways in which Tswana chiefs were able to influence imperial policy in late-nineteenth-century southern Africa). Given this, it becomes unclear in what sense nineteenth-century empires such as Britain's can be characterized as 'an extension of the sovereignty of the European nation-states beyond their own boundaries' (Hardt and Negri, 2000: xii–xiii). Rather, as Ballantyne argues, '[t]he importance of ... cultural traffic and imperial networks ... means that we must move beyond the nation-state as the organizing unit for the writing of the history of imperialism' (Ballantyne, 2002a: 194).

If recent studies by historians and historical geographers demonstrate that the British Empire, at least, was not as state-centred as Hardt and Negri assume, they also show that the kind of modern globalization that it undoubtedly fostered was neither as all-encompassing nor as benign as Ferguson infers.

As Ballantyne argues, 'The imperial globalization generated by British commerce, conquest and colonization ... brought previously unconnected regions together into a system, albeit a highly uneven one, of exchange and movement' (Ballantyne, 2002a: 195). The networks of trade, humanitarianism, settler racial discourse, natural science and official governmentality, among others, that Britons created and maintained in the nineteenth-century were responsible for creating very particular sets of global connectivities and harnessing their potential for specific goals. In more general ways too, the British Empire acted as a powerful agent of globalization, 'appropriating new lands and significant new sources of revenue, while moving people, commodities, technologies and ideas from colony to colony, as well as between the imperial centre and colonies in the periphery' (Ballantyne, 2002b: 116). And it was not only formal British colonization that produced such new forms of connection. Through the diplomatic clout and economic advantage that a formal empire gave, Britain's government was able to pull Japan and parts of the Middle East, and to a lesser extent metropolitan China, into networks of trade and communication that had nodal points within its empire, and it was able to exercise effective suzerainty over large parts of Latin America (Knight, 1999; Lynn, 1999; Osterhammel, 1999).

British imperialism was by no means solely about the construction of new connections. It also relied on the diversion or capture of long-distance channels of communication and exchange that predated, in many cases by centuries, the arrival of British colonialism. '[T]he European empires were built on the archaic foundations and proto-globalizing tendencies of the societies they subordinated' (Hopkins, 2002a: 7). For example, Bayly has shown that '[o]ne overriding reason why the East India Company was able to conquer India and dominate it for more than a century was that the British had learnt the art of listening in, as it were, on the internal communications of Indian polity and society' (Bayly, 1999: 365). This entailed not only closing down certain lines of communication between different Indian powers, but also taking control of existing communication and trade networks within the sub-continent, across the Indian Ocean and further afield (Bayly, 1999; Glaisyer, 2004).

However, even if one takes into account all the forms of globalization created, fostered or captured by British and indeed other European interests in the nineteenth-century, the globalization that was characteristic of that century still exceeded their agendas and often subverted their intentions. Indeed, British interests were never able to appropriate entirely even those circuits of communication and exchange that flowed through British imperial territory. Linebaugh and Rediker (2000) demonstrate, for example, the ways that 'subaltern' subjects evaded imperial surveillance and created their own, often subversive networks across formally British, French or Spanish Atlantic space (see also Glaisyer, 2004). Similarly, Hopkins notes how imperialists' vision of specific interconnectivities subject to British overrule was 'at variance with Islam's more spacious vision of how the world should be organized. Even so, Islam continued to spread across political boundaries', both within and outside of Britain's imperial terrain (Hopkins, 2002b: 32). The globalization that the British played some role in enacting always exceeded their intentions, always spilled over the bounds of territorial control and always escaped capture by British interests in ways that advocates for the resumption of a benign and all-encompassing empire today generally fail to acknowledge. Indeed, those trans-global circuits that were established by the British could prompt the establishment of new and subversive counter-circuits – something that would be more evident in Ferguson's conclusions had he looked more closely at some of the actions taken by colonized subjects to which he himself draws attention.

The 'trouble' with the connections between colony and colony, and colonies and metropole, that British imperial interests created and maintained, was that those same communicative circuits could be deployed against continued British interference. Even before the Comintern in 1919 first formulated the general political concept of anti-imperialism on a global stage (Young, 2001: 28),

'the cross-imperial idea' constructed by white colonists was made available for annexation by non-white political elites similarly in quest of solidarity and a common vocabulary of rights during the late nineteenth-century (Boehmer, 2003: 15). Boehmer's study of the connections between anti-colonial resistance movements is particularly helpful in theorizing the difference that interconnectivity can make to our narratives of resistance within empire as well as of colonization. In Boehmer's recent book, 'the "contact zone" of cultural and political exchange [that is] conventionally located between the European colonial center and its periphery [is] instead ... positioned between peripheries'. Boehmer sees her task as investigating 'how definitive concepts of self-realization often seen as originating within European political traditions (self-help; boycott; imperial "loyalty"), were critically appropriated and remade not only by native nationalists as such, but through borrowing, exchange, and even collaboration between anti-colonial regions'. The figures that she studies, including Irish nationalists who discovered in the South African (Boer) War 'an arena for their own struggle' and the South African nationalist Sol Plaatje, who formulated his notions of resistance in relation to those of London suffragettes, demonstrate how anti-colonial movements were 'both nationally focused in terms of political organization, and yet cross- or transnational in their range of reference and reception of influence'. Like imperialism itself, resistance to it was thus 'allusive, cross-cultural, intertextual, or interdiscursive, strung across borders of different descriptions as well as staked out within geopolitical boundaries' (Boehmer, 2003: 15; see also Lonsdale, 2002).

AMERICAN EMPIRE: THE HARNESSING OF MODERN GLOBALIZATION?

Given the British Empire's relative lack of state-centredness, at least compared to recent characterizations of it, and its spatially networked rather than discretely territorialized nature, the image of that empire that is upheld by some as a model for contemporary US intervention and equally by others as a contrast to current forms of unbounded, deterritorialized Empire, is unrealistic. Furthermore, certain prominent geographers' studies of contemporary US imperialism suggest that the current global order is a lot more territorialized, a lot more centred in the projection of nation-state power, than such a historicist contrast would allow. In fact, the US state is perhaps playing a more assertively imperial role today than the British state did throughout most of its imperial history.

Suggestively, it was when Britain's established dominance within the global political economy was threatened that the state itself took on an interventionist role as aggressive as that of the contemporary United States. It did so, for instance, during the South African or Boer War of 1899–1902, which saw the invasion of the Transvaal, with its economically and strategically vital gold reserves (gold being the contemporary basis for globally dominant currencies). Aside from the connection with a vital natural resource, the context for this invasion was similar in other respects to that for the more recent US-led invasion of Iraq. There was the increasing rivalry of other rapidly industrializing regions (notably Germany and the United States itself), the presence of longstanding schemes for the confederation and 'better' government of the region as a whole and the desire to replace a state that was seen to be pre-modern and antithetical to economic rationality. The pretext for the war, disseminated in a deluge of propaganda in Britain, was benign intervention on behalf of both English-speaking white colonists who were being denied the vote in the Transvaal, and of the region's African population who were being oppressed by a government in thrall to an outdated, slave-owning mentality (a population betrayed by Britain in the aftermath of the war, when pledges to extend the Cape Colony's non-racial franchise to the conquered Boer republics were set aside).

Of course, the nineteenth-century British and contemporary American empires are by no means discrete phenomena. Not only did the United States emerge out of the trans-Atlantic webs of emigration, trade and slavery that the British Empire grew out of and helped to foster, but its history continued, in many respects, to be co-constituted with that of Britain. Its reliance on British industrial exports continued until the Civil War, which itself had an important bearing on debates over racial difference in Britain (Bolt, 1971; Lorimer, 1978). After 1945, the lingering British Empire was so heavily imbricated with American financial and official activity that it has been described as an 'Anglo-American condominium', and the so-called 'special relationship' of the twentieth century, even if one-sided, has been globally significant in numerous instances of conflict (Louis and Robinson, 1994; Reynolds, 2002: 251). Following the lead of Neil Smith (2003a), the precise places and moments where treaties were signed and deals done to accomplish the effective transfer of imperial hegemony from Britain to the United States in the aftermath of the two world wars are worthy of much more considered investigation by political and historical geographers. After all, these were where and when the ground rules of the new American imperialism (and see Smith, 2003a: xviii–xix, and Agnew, 2003: 872 for the appropriateness or otherwise of this term) were laid.

Smith's study of the US geographer Isaiah Bowman's implication in American strategies of globalism is useful in setting the current US agenda of global capitalism, free trade (where it suits the United States) and bourgeois democracy, as well as the more recent 'war on terror' in its historical context (Smith, 2003a: xii). Smith argues that there have been three episodes of attempted US control over globalizing social, cultural and, especially, economic flows. The first, from 1898 to 1919, 'adumbrated the vision of a global political economy that would simultaneously surpass the regional parameters of the European empires and entwine a global political structure (the League of Nations) with an already accomplished world market' (Smith, 2003a: 454). Thus, Woodrow Wilson arrived at the Paris peace conference after the First World War 'armed with a moralism that was distinctly American yet simultaneously universal. ... Wilson expected the conference to tidy up the world map in preparation for a new and higher stage of international society – a beneficent brotherhood of capitalist nations competing economically but peacefully while advancing the global good' (Smith, 2003a: 16). However, this first twentieth-century phase of US globalism was frustrated by the Russian Revolution and the rise of fascism abroad and by labour revolt and the Senate's rejection of the League of Nations 'at home'.

The second 'moment' came with the end of the Second World War, when the US state attempted to create the conditions in which free markets working primarily to US advantage could operate without the hindrance of the old European empires. It was at this moment that an agenda for American imperialism very much akin to that being pursued post 11 September 2001 was first clearly enunciated. Henry Luce's famous article on the 'American Century' in *Life* magazine (17 February 1941), for example, contained a refrain similar to that of George W. Bush. As Smith puts it, his was a

repackaging of the most cherished myths of national superiority addressed to global claims. ... History bore the United States into a position of global leadership, and it should be unselfconsciously grasped. US rulers should seize global power for the simple reason that they *can*: by virtue of their superior economic prowess, it was already de facto theirs. Manifest destiny breaks all bounds of national space. (Smith, 2003a: 17)

Bowman felt during the war that '[n]o line can be established anywhere in the world that confined the interests of the Unites States because no line can prevent the remote from becoming the near danger'. If Hitler was demanding *Lebensraum*, 'then *Lebensraum* he should get – except that the postwar world would be a global *Lebensraum*, it would be an economic *Lebensraum*, and it would be an American *Lebensraum*' (Smith, 2003a: 27–8).

Smith argues that the Cold War and Soviet dominance over a large swathe of the globe frustrated this second American imperial vision in turn. However, rather than representing an interlude in American imperialism, the Cold War can be seen as shifting it onto new terrain and, indeed, prompting the laying of certain critical foundations for its revitalization (as had the Second World War). As Reynolds notes:

> The high-tech military establishment required by the Cold War was the main reason for the expansion of the US government bureaucracy and the enlargement of the federal budget (four-fifths of which was going to defence-related activities by the early 1950s). The military-industrial complex of World War II and the Cold War was largely responsible for the consolidation of corporate capitalism during the 1940s. (Reynolds, 2002: 251)

The US state's determination to block real or imagined Soviet influence also entailed new kinds of intervention, including the 'decision to harness, or even to cultivate, terrorism in the struggle against regimes it considered pro-Soviet' (Mamdani, 2001: 3). In southern Africa, the effect was particularly devastating. Not only was tacit support given to the South African apartheid state, but also, in the wake of the collapse of the Portuguese Empire, the United States ensured that neither Angola nor Mozambique could attain stability as socialist states. In Mozambique in particular, support for RENAMO, which specifically attacked civilians in its attempts to undermine the post-independence government, implicated the US military and intelligence services in terrorist tactics, and acted as a precursor for similar activities in Nicaragua and El Salvador.

Now we are experiencing a third, post-Cold War 'moment' of US imperialism – one in which the US state is determined to hold on to its threatened dominance within more intensively globalizing networks, not least by engaging in warfare. Perhaps the most explicit statement of the current US government's agenda can be found in the publications of the Project for the New American Century, founded in 1997 (note, four years before '9/11') and including in its membership Dick Cheney, Donald Rumsfeld, Paul Walfowitz and George W. Bush's brother Jeb, along with other members of Bush's inner circle. Among their Statement of Principles is included 'a foreign policy that boldly and purposefully promotes American principles abroad; and national leadership that accepts the United States' global responsibilities'. The Project believes that 'the history of this [twentieth] century should have

taught us to embrace the cause of American leadership' in Europe, Asia and the Middle East, and presses the need for increased military expenditure, 'to challenge regimes hostile to our interests and values'. It urges the US state to 'accept responsibility for America's unique role in preserving and extending an international order friendly to our security, our prosperity, and our principles' (Project for the New American Century, 1997).

The 'project's' agenda for the more forceful projection of American power and American 'values' across the globe was held latent in the face of domestic reluctance to commit the military to overseas combat until 11 September 2001. Then, the horrific al-Qaeda attacks on the Pentagon and the World Trade Centre gave the administration an ideal opportunity to convert patriotic solidarity and the popular desire for vengeance into a new imperialist thrust (see Harvey, 2003: 192–3; Mann, 2003: 2–9; Smith, 2003a: xii). The irony of a current US foreign policy seemingly based on this agenda, as Harvey points out, is that it may actually sever many of the networked connections, of both financial and other kinds, upon which US influence across the globe has relied thus far (Harvey, 2003: 207–8). After all, the 'Wall Street–Treasury' complex within the United States, which is able to control institutions such as the IMF, 'can only operate in the way it does because the rest of the world is networked and hooked into (successfully "hooked on" usually by way of credit arrangements) a structured framework of interlocking financial and governmental (including supra-national) institutions', which the US state cannot afford to fragment (Harvey, 2003: 134).

Just as gold certainly played a role in the South African War waged by Britain's threatened empire, but a role that has been debated by historians, so it is commonly accepted that oil figures in any explanation of the US-led 'war on terror' in Afghanistan and Iraq, although 'exactly how and in what sense is not so easy to determine' (Harvey, 2003: 18). Harvey and Smith agree that, fundamentally, the war is an attempt to marshal a contemporary form of globalization for construed US interests, with oil a significant factor within that broader equation. Smith describes recent US attacks in Afghanistan and Iraq as 'a war to fill in the interstices of globalization. ... Masquerading as a war on terrorism, it is actually a war devoted to the completion of the geo-economic globalism of the American Empire' (Smith, 2003a: xiv; see also Smith, 2001: 4). Harvey's formulation is that the war is about influencing the geopolitics of the Middle East as a whole, with the perpetuation of US-dominated global capitalism against a backdrop of depleting oil reserves elsewhere. Adopting a Mackinder-esque turn of phrase, Harvey claims that 'whoever controls the Middle East controls the global oil spigot and whoever controls the global oil spigot can control the global economy, at least for the near future' (Harvey, 2003: 19). Harvey goes further than Smith in identifying the regional bases for alternative globalisms to that promoted by the United States:

> Europe and Japan, as well as East and South-East Asia (now crucially including China) are heavily dependent on Gulf oil, and these are regional configurations of political-economic power that now pose a challenge to US global hegemony in the world of production and finance. What better way for the United States to ward off that competition and secure its own hegemonic position than to control the price, conditions, and distribution of the key economic resource upon which those competitors rely? (Harvey, 2003: 25)

If Harvey's explanation reminds one of the more economically determinist, Marxist accounts of the South African War, based upon Britain's need to control the global supply of gold, Smith's version is more akin to those accounts of that earlier, state-driven, imperial war, which emphasize the vagaries and contingencies of politicking alongside narrowly economic considerations. US interventions in the Middle East are nothing new. As Sidaway points out, '[w]hilst almost no one could predict the unfolding events on September 11, 2001 ... there is a ready and evident connection across space and time to Western imperial and superpower policy in the Middle East' (Sidaway, 2003: 647). Since the Second World War, US '[i]nterest in the region had to do with oil but also with supporting the new Israeli state, and the US sought to displace British and French capital while holding the Soviet Union at bay and negotiating various economic and diplomatic agreements with the administrations of the region' (Smith, 2003b: 265). In its ongoing attempts to exercise influence in region during the Cold War (attempts that encountered failures such as the assertiveness of OPEC, the Iranian revolution and the bombing of the US Embassy and marine compound in Beirut), the United States promoted Islamic fundamentalism against the Soviet Union and Saddam Hussein's Iraq against Iran, as well as Israel against the Palestinians. It then had to confront the fact that some of these allies were by no means sympathetic to the 'universal' US globalism that lay behind their initial support. For some time, US state policy has been to bring them into line – a strategy pursued by Clinton in far more multilateral and subtle ways than by his successor (see Ó Tuathail and Luke, 1994). The attacks of 11 September 2001 provided Clinton's successor with the opportunity to 'challenge and eliminate the threat' of 'Islamic alternatives to the vision of globalization being promulgated from Washington and

New York, London, Tokyo and Frankfurt' (Smith, 2003a: xv).

Despite their slightly different emphases, Smith and Harvey agree that the war in Iraq 'is not a war simply to control oil resources but rather to control the global political economy within which the disposition of oil resources will be organized' (Smith, 2003b: 265). Together, their explanations are persuasive in terms of state strategy, but they also tend to downplay something else that is vital to an understanding of current US imperialism. This is the complex of cultural discourses that enables the US state to postulate its strategy as being not only in the material interests of US citizens and companies, but also in the universal interests of humankind.

It is this classically colonial projection from self-interest to universal principle on which Gregory (2004) focuses so astutely, and it is only this kind of projection that enabled the British Empire to be so widely seen as a 'good thing' by Britons in the nineteenth-century (and by many of them today), and that currently enables so many in the United States and elsewhere to sustain US imperial ventures with moral and political support. As Gregory argues, 'what else is the war on terror other than the violent return of the colonial past, with its split geographies of "us" and "them", "civilization" and "barbarism", "Good" and "Evil"?' (Gregory, 2004: 10–11). Cooper argues along similar lines:

> If there is an analogy to be made between current politics and the politics of empire [in the nineteenth-century], it might point not to the politics of bringing people into full participation in world markets, democratic politics, and transnational cultural interaction, but to the honing of techniques of marking those who are outside [rogue states, Muslim fundamentalists, terrorists, etc.] and of patrolling the borders. (Cooper, 2004: 271–2)

Of course, this marking of the outside is just as much about the definition of the inside. Historians have argued that, in various ways, Britain's own eighteenth- and nineteenth-century empire was constitutive of Britishness. This elastic identity, layered on top of older national and regional identities, was both a way of distinguishing largely Protestant islanders against Continental Catholic powers during a time of war (Colley, 1992) and an expression of innate superiority over those who constituted the homogenized subject peoples of Britain's colonies (Hall, 1996). Britishness was thus as much about having an empire as it was about purely domestic matters. As Reynolds suggests, more work could usefully be undertaken now on how various forms of US globalism have 'served to ... Americanize Americans' (Reynolds, 2002: 244, 248). The fact that US identities are shaped by the

United States 'constitutive outside' (as Hall, 2002, puts it in the British imperial context) is glaringly apparent when we examine US foreign policy statements since the Second World War. As Reynolds notes:

In the 1940s the concept of 'totalitarianism' provided a new tool for defining foreign policy and sharpening national identity. First applied to the Axis powers in World War II, it was carried over into the Cold War when a new 'Other', the Soviet Union, was seen as threatening American interests abroad and subverting American values at home. (Reynolds, 2002: 250)

And 'the international is ... still being used to define the national'. In the aftermath of the Cold War, new threats swiftly replaced communism. 'Near the top of the list was a supposed monolith called militant Islam. Samuel Huntington enlarged this anxiety into the general argument that "the fault lines between civilizations will be the battlelines of the future"'. Huntington's vision of a clash between Islamic and Western civilizations (the latter embodied by the United States)'tells us more about contemporary American anxieties than it does about the late modern world' (Reynolds, 2002: 257, quoting Huntington, 1993: 22 and Gray, 1998: 157). Huntington's real fear seems to be that, unless Americans who believe in the 'American Creed' of 'liberty, democracy [and] individualism' cohere against external threats, they might also fail to deal with the 'divisive siren calls of multiculturalism' at home. And if this happens, then the United States survival as a 'coherent country' with a 'cultural core' will be in doubt (Huntington, 1996: 305–7).

THE EXCESSES OF CONTEMPORARY GLOBALIZATION

The British state's resort to force in the Transvaal in 1899 meant that its troops were engaged for much longer than anticipated in a conflict that created major tensions between Britain and other European states, and which generated considerable domestic opposition (despite the government's media manipulation: Lowry, 2000). Even though a government was established for the Transvaal, and ultimately for South Africa as a whole – and one with which the British government and British financial interests could 'do business' – the determination to hold on to global supremacy by forced 'regime change' in this vital region delayed the fragmentation of British imperialism only by a few decades. Just as British interests in the nineteenth-century were able to capture some pre-existing, partly globalized

networks and create others to pursue their goals, but were never able to direct contemporary forms of globalization per se, so more recent US attempts to direct globalization have been only partly successful. US globalism, like any state-centred globalism, has a particular geography in which some connections work to the advantage of some US interests, including those that allowed market liberalization, transnational capitalist class formation and many transnational media networks (Sparke, 2003: 378), and some work against those or other US interests.

As is the case in recent interpretations of the South African War, a number of scholars link the US state's new assertiveness to anxieties over fading power rather than the assumption of a new level of power. The end of the Cold War, while it meant that the United States was for the first time an unrivalled superpower, also meant that it is no longer the sole guarantor of the West's 'universal freedoms' against a looming Soviet threat. This renders 'the US leadership role ... harder to define and sustain' (Harvey, 2003: 41). As Harvey notes, the United States recent aggressiveness (presuming that it is not genuinely directed against those believed responsible for the attacks of 11 September 2001) makes sense only 'if the US has reason to fear that its dominant position within global capitalism is somehow threatened' (Harvey, 2003: 25). In this sense, the US state today is in a very similar position to that of the British state on the eve of the South African War. Harvey asks, 'Are we now witnessing overt political claims about empire and the imperialism that goes with it in the United States at the political and territorial level, at the very moment when the flows of economic power and even cultural and moral influence are ebbing away from its shores into more diffuse power blocs (centred on Asia and Europe, for example)?' (Harvey, 2003: 31). Smith, too, suggests that 'the resort to force after 11 September 2001 displays a certain weakness from a global power that has been unable to establish the kind of global control it craves and promises through purely geo-economic means' (Smith, 2003b: 269; see also Agnew, 2003; Mann, 2003).

US power can be seen to have faded first during the 1970s with defeat in Vietnam, economic decline and de-industrialization and increasing competition in finance and markets from Japan and Germany (Smith, 2003a: 5; cf. Slater and Taylor, 1999). As Wallerstein notes, with the exclusion of Grenada and Panama, which had no serious means of military defence, 'Chile was the last major instance in which the US was able so cavalierly to arrange other governments to its preferences ... [g]eopolitical reality had changed. The US government could no longer concentrate on maintaining, even less on expanding, its power; instead its prime goals became preventing a too rapid erosion of its power – both in the world-economy and in the military arena' (Wallerstein, 2001: 8; see also Wallerstein, 2003). After a revival of fortune associated with the internationalization of production systems and labour markets, new financial markets, deregulation, the fading of the Japanese challenge, domestic restructuring and the collapse of the Soviet Union, worries have recently set in again about the extent of the state's indebtedness. For Smith the current American 'war on terror' thus represents 'a regressive and reactionary nationalism aspiring to resolve the dilemma of a hollowing national scale by again pushing for global economic and military control, up to and including the use of state terror [in this case, applied against Afghan civilians]' (Smith, 2001: 3).

CONCLUSION

It has not been the purpose of this chapter to produce a detailed comparative survey of past empires, nor even of British and American imperialisms. Nor has the chapter examined the actual effects of empires past and present on colonized societies themselves, other than in passing. Nevertheless, examining how historical scholars have come to understand the British Empire in recent years, and how geographers are now interpreting US imperial interventions, does enable us to question some of the assumptions that are prevalent today about connections between empires past and present.

The British Empire, as this chapter has hopefully made clear, was not, in general, the nation-state-centred projection of sovereignty and power that has been contrasted with current forms of decentred Empire, and it certainly was not the benign force for modern globalization upon which American interventions can helpfully be modelled. It was, rather, a bundle of dynamic and contested networks, created and maintained in pursuit of multiple projects by different interests, some colonial and some imperial. These projects were not by any means entirely liberal and reformist (and even where they were, as Cooper points out, they were often directed against 'a prior evil which colonization had itself created (slavery, settler racism) or compromised for fear of upsetting indigenous intermediaries whom colonial rulers required' (Cooper, 2004: 261). Other, and often more powerful, colonial projects ranged from the cultural and religious assimilation of peoples deemed backward and in need of redemption to their dispossession, their subjection to labour in the interests of British emigrants, or their outright extermination. To a great extent, Ferguson's and others' assumptions that empire is good for the world make sense only if we examine 'Canada minus the Native-Americans … Australia minus the Aborigines … New Zealand minus the Maori, India minus the Indians, Africa minus the Africans' (Cooper, 2004: 261).

If the British Empire was neither so generally beneficent nor quite so state-centred as some imagine, current forms of globalism are perhaps more imperially driven and more state-centred than some would allow. Recent work by geographers suggests that we need more astute, more realistic, assessments of globalism, which factors in US imperialism. This attempt must begin with the recognition that empires are attempts by multiple national actors to forge, appropriate or capture globalized connectivities in pursuit of their own interests; they do not in themselves constitute globalization. Thus, just as different British interests were critical in their promotion and (only ever partial) harnessing of globalization in the past, so the US state and other US interests are today. To effectively remove these imperialist interests from our picture of globalization is to generate an image quite divorced from reality.

If we are to produce more realistic accounts, it is essential that we insist on the geographies of both the colonial past and present. As with Britain's nineteenth-century empire, the United States contemporary imperialism, even in the absence of formal colonies, is inherently territorialized, through the creation and maintenance of certain networks such as those that sustain supranational organizations like the IMF. Recent research on British imperialism suggests the need to explore the circuits, connections and webs through which American power, as represented by diverse and not necessarily harmonious interests, is manifested around the globe. As Michael Mann notes, 'This giant's military might sits uneasily with economic and geopolitical resources that originate in multilateral arrangement' (Mann, 2003: 266). Networks of financial dealing and governmental market liberalization measures have already received much attention (see Harvey, 2003), but those, for instance, of media linkages that foster popular imageries and consumption choices, diplomatic engagement that helps secure hegemony and other circuits through which US-generated knowledge, ideas and cultural practices travel, are often assumed to exist without adequate empirical scrutiny as to their origins, directions and appropriation or subversion in other sites.

The routes through which American power travels in its various forms connect many places, but they also by-pass others or only visit them weakly, leaving more established forms of connectivity in place (see Cooper, 2001, on parts of Africa, for instance). As well as being uneven, the networks appropriated or instituted by imperial interests can also provide capacities for resistance, not from a shapeless 'multitude' (as in Hardt and Negri's formulation), and not just from anti-globalization demonstrators, but from numerous, sited but interconnected, and often conflicting, interests. In our attempts to uncover the networks of US power, we must also examine these networks of subversion, redeployment, appropriation and resistance, as Boehmer (2003) has begun to do with reference to the British Empire (see Johnson, 2000; Mann, 2003).

Hand in hand with emphasizing the particular spatialities of certain of its networks, recent work on the British Empire has stressed the salience of competing colonial projects in constituting imperial agendas. Perhaps it is time we looked harder for the same tensions within the US empire. Are the interests, networks and activities of the US oil industry really so compatible with those of the state ideologues for a new American century? Do the interests of US-headquartered TNCs, each with their own specific networks, harmonize with aggressive or dismissive US government diplomacy abroad? Political geographers need to trace the effects of these and other tensions of empire as well as the ways that they are played out through extensive geographical networks if we are to recover the 'lost unevenness of American hegemony amidst global interdependency' today (Sparke, 2003; 375). As Ó Tuathail (1998) argues:

Actually existing globalization [and US hegemony] is not the globalization of neoliberal visions, the Utopia of friction-free global markets or Internet-driven worlds, but the contingent and unsteady symbiosis of imperfectly transnational networks, institutions and forms, and the 'ramshackle diversity' of international bureaucracies, states, police, mafias and other sources of power struggling for shifting territorial authority in the post-cold war world'. (Ó Tuathail, 1998: 87)

While networks are a contemporary buzzword in the social sciences (Urry, 2004), how often are the threads and nodes of such networks actually followed and mapped empirically? In doing this work, political geographers have the possibility of making a much greater contribution to broader understandings of the repertoires and the spatialities of empire past and the present.

NOTE

1 Soviet control of large parts of central Asia from the early twentieth century is an obvious exception to the capitalist nature of modern imperialism. For comparative studies identifying both contrasts and parallels with other European empires, see Parrott and Dawisha (1996), Pearson (1997), Barkey and Von Hagen (1997) and, most recently, Lieven (2001).

REFERENCES

Agnew, J.A. (2003) 'American hegemony into American Empire? Lessons from the invasion of Iraq', *Antipode*, 35(5): 871–85.

Armitage, D. (2000) *The Ideological Origins of the British Empire*. Cambridge: Cambridge University Press.

Ballantyne, T. (2002a) *Orientalism and Race: Aryanism in the British Empire*. Basingstoke: Palgrave.

Ballantyne, T. (2002b) 'Empire, knowledge and culture: from proto-globalization to modern globalization', in A.G. Hopkins (ed.), *Globalization in World History*. London: Pimlico, pp. 115–40.

Barkey, K. and Von Hagen, M. (eds) (1997) *After Empire: Multiethnic Societies and Nation-Building: The Soviet Union and the Russian, Ottoman and Habsburg Empires*. London: Harper Collins.

Bayly, C. (1999) *Empire and Information: Intelligence Gathering and Social Communication in India, 1780–1870*. Cambridge: Cambridge University Press.

Bell, D. (2003) 'History and globalization: reflections on temporality', *International Affairs*, 79(4): 801–14.

Blaut, J. (1993) *The Colonizer's Model of the World: Geographical Diffusionism and Eurocentric History*. New York: Guilford.

Blaut, J. (2000) *Eight Eurocentric Historians*. London: Guilford.

Blunt, A. and McEwan, C. (eds) (2002) *Postcolonial Geographies*. New York and London: Continuum.

Boehmer, E. (2003) *Empire, the National and the Postcolonial, 1890–1920*. Oxford: Oxford University Press.

Bolt, C. (1971) *Victorian Attitudes to Race*. Toronto: University of Toronto Press.

Bowden, B. (2002) 'Reinventing imperialism in the wake of September 11', *Alternatives: Turkish Journal of International Relations*, 1(2): 28–46.

Bowen, H.V. (1998) 'British conceptions of global empire, 1756–83', *Journal of Imperial and Commonwealth History*, 26: 1–27.

Brenner, R. (1977) 'The origins of capitalist development: a critique of neo-Smithian Marxism', *New Left Review*, 104 (July/Aug): 25–92.

Brewer, K. (1990) *Marxist Theories of Imperialism: A Critical Survey* (2nd edn). London and New York: Routledge.

Canny, N. (ed.) (1998) *Origins of Empire: The Oxford History of the British Empire*, Vol. I. Oxford: Oxford University Press.

Clayton, D. (2000) *Islands of Truth: The Imperial Fashioning of Vancouver Island*. Vancouver: University of British Columbia Press.

Clayton, D. (2002) 'Critical imperial and colonial geographies', in K. Anderson, M. Domosh, S. Pile and N.J. Thrift (eds), *Handbook of Cultural Geography*. London: Sage, pp. 531–57.

Colley, L. (1992) *Britons: Forging the Nation, 1707–1837*. London: Pimlico.

Cooper, F. (2001) 'What is the concept of globalization good for? An African historian's perspective', *African Affairs*, 100: 189–213.

Cooper, F. (2004) 'Empire multiplied. A review essay', *Comparative Studies in Society and History*, 46(2): 247–72.

Doyle, M.W. (1986) *Empires*. Ithaca, NY and London: Cornell University Press.

Drayton, R. (2000) *Nature's Government: Science, Imperial Britain and the Improvement of the World*. New Haven and London: Yale University Press.

Eisenstadt, S.N. (1967) *The Decline of Empires*. Englewood Cliffs, NJ: Prentice Hall.

Elliott, J.H. (2002) *Imperial Spain, 1469–1716*. London: Penguin.

Ferguson, N. (2003) *Empire: How Britain Made the Modern World*. London: Allen Lane.

Ferguson, N. (2004) *Colossus: The Rise and Fall of the American Empire*. London: Allen Lane.

Fieldhouse, D.K. (1982) *The Colonial Empires: A Comparative Survey from the Eighteenth Century*. Basingstoke: Macmillan.

Gascoigne, J. (1998) *Science in the Service of Empire: Joseph Banks, the British State and the Uses of Science in the Age of Revolution*. Cambridge: Cambridge University Press.

Glaisyer, N. (2004) 'Networking: trade and exchange in the eighteenth-century British Empire', *Historical Journal*, 47(2): 451–76.

Gray, J. (1998) 'Global unitopias and clashing civilizations: misunderstanding the present', *International Affairs*, 74: 157.

Gregory, D. (2004) *The Colonial Present*. Oxford: Blackwell.

Grove, R.H. (1995) *Green Imperialism: Colonial Expansion, Tropical Island Edens and the Origins of Environmentalism, 1600–1820*. Cambridge: Cambridge University Press.

Hall, C. (1996) 'Histories, empires and the post-colonial moment', in I. Chambers and L. Curti (eds), *The Post-Colonial Question: Common Skies, Divided Horizons*. London and New York: Routledge, pp. 65–77.

Hall, C. (2002) *Civilising Subjects: Metropole and Colony in the English Imagination*. Cambridge: Polity.

Hardt, M. and Negri, A. (2000) *Empire*. Cambridge, MA: Harvard University Press.

Harvey, D. (2003) *The New Imperialism*. Oxford: Oxford University Press.

Haskell, T.L. (1985a) 'Capitalism and the origins of the humanitarian sensibility, Part 1', *American Historical Review*, 90: 339–61.

Haskell, T.L. (1985b) 'Capitalism and the origins of the humanitarian sensibility, Part 2', *American Historical Review*, 90: 547–66.

Hopkins, A.G. (2002a) 'Introduction: Globalization – an agenda for historians', in A.G. Hopkins (ed.), *Globalization in World History*. London: Pimlico, pp. 1–10.

Hopkins, A.G. (2002b) 'The history of globalization – and the globalization of history?', in A.G. Hopkins (ed.), *Globalization in World History*. London: Pimlico, pp. 11–46.

Huntington, S. (1993) 'The clash of civilizations?', *Foreign Affairs*, 72(3): 22.

Huntington, S. (1996) *The Clash of Civilizations and the Remaking of the World Order*. New York: Free Press.

Ignatieff, M. (2003) 'American empire: get used to it', *New York Times*, 5 January.

Johnson, C. (2000) *Blowback: The Costs and Consequences of American Empire*. New York: Henry Holt.

Jones, R. and Phillips, R. (2005) 'Unsettling geographical horizons: exploring premodern and non-European imperialism', *Annals of the Association of American Geographers*, 95(1): 141–61.

Keegan, T. (1996) *Colonial South Africa and the Origins of the Racial Order*. London: Leicester University Press.

Knight, A. (1999) 'Britain and Latin America', in A. Porter (ed.), *The Nineteenth Century: The Oxford History of the British Empire*, Vol. III. Oxford: Oxford University Press, pp. 122–45.

Laidlaw, Z. (2001) 'Networks, patronage and information in colonial governance: Britain, New South Wales and the Cape Colony, 1826–1843', unpublished D.Phil thesis: University of Oxford.

Lambert, D. (2005) *White Creole Culture, Identity and Politics During the Age of Abolition*. Cambridge: Cambridge University Press.

Lambert, D. and Lester, A. (2004) 'Geographies of colonial philanthropy', *Progress in Human Geography*, 28(3): 320–41.

Lester, A. (2001) *Imperial Networks: Creating Identities in Nineteenth Century South Africa and Britain*. London: Routledge.

Lester, A. (2002) 'British settler discourse and the circuits of Empire', *History Workshop Journal*, 54: 24–48.

Lieven, D. (2001) *Empire: The Russian Empire and Its Rivals*. New Haven: Yale University Press.

Linebaugh, P. and Rediker, M. (2000) *The Many-Headed Hydra: Sailors, Slaves, Commoners, and the Hidden History of the Revolutionary Atlantic*. Boston: Beacon Press.

Lonsdale, J. (2002) 'Globalization, ethnicity and democracy: a view from "the hopeless continent"', in A.G. Hopkins (ed.), *Globalization in World History*. London: Pimlico, pp. 194–219.

Lorimer, D.A. (1978) *Colour, Class and the Victorians: English Attitudes to the Negro in the Mid-Nineteenth Century*. London: Leicester University Press.

Louis, W.R. and Robinson, R. (1994) 'The imperialism of decolonization', *Journal of Imperial and Commonwealth History*, 22: 462–511.

Lowry, D. (2000) *The South African War Reappraised*. Manchester: Manchester University Press.

Luce, H. (1941) 'The American century', *Life*, 17 February.

Lynn, M. (1999) 'British policy, trade and informal empire in the mid-nineteenth century', in A. Porter (ed.), *The Nineteenth Century: The Oxford History of the British Empire*, Vol. III. Oxford: Oxford University Press, pp. 101–21.

Mamdani, M. (2001) 'Good Muslim, bad Muslim: an African perspective', http://www.ssrc.org/sept11/essays/mamdani. htm. Accessed 3 September 2004.

Mann, M. (1986) *The Sources of Social Power*, Vol. I: *A History of Power from the Beginning to 1760 AD*. Cambridge: Cambridge University Press.

Mann, M. (1993) *The Sources of Social Power*, Vol. II: *The Rise of Classes and Nation States, 1760–1914*. Cambridge: Cambridge University Press.

Mann, M. (2003) *Incoherent Empire*. London and New York: Verso.

Ó Tuathail, G. (1998) 'Political geography III: dealing with deterritorialization', *Progress in Human Geography*, 22(1): 81–93.

Ó Tuathail, G. and Luke, T.W. (1994) 'Present at the (dis)integration: deterritorialization and reterritorialization in the new wor(l)d order', *Annals of the Association of American Geographers*, 84(3): 381–98.

Osterhammel, J. (1999) 'Britain and China, 1842–1914', in A. Porter (ed.), *The Nineteenth Century: The Oxford History of the British Empire*, Vol. III. Oxford: Oxford University Press, pp. 146–69.

Parrott, K. and Dawisha, K. (eds) (1996) *The International Politics of Eurasia: The End of Empire?* Comparative Perspectives on the Soviet Collapse, Vol. 9. New York: M.E. Sharpe.

Parsons, N. (1998) *King Khama, Emperor Joe and the Great White Queen: Victorian Britain Through African Eyes.* Chicago: University of Chicago Press.

Pearson, R. (1997) *The Rise and Fall of the Soviet Empire.* London: Macmillan.

Porter, A. (2004) *Religion Versus Empire: British Protestant Missionaries and Overseas Expansion, 1700–1914.* Manchester: Manchester University Press.

Project for the New American Century (1997) 'Statement of Principles', http://www.newamericancentury.org/statementofprinciples.htm. Accessed 4 September 2004.

Reynolds, D.R. (2002) 'American globalism: mass, motion and the multiplier effect', in A.G. Hopkins (ed.), *Globalization in World History*. London: Pimlico, pp. 243–60.

Russell-Wood, A.J.R. (1998) *The Portuguese Empire, 1415–1808: A World on the Move.* Baltimore: Johns Hopkins University Press.

Seeley, J.R. (1971 [1883]) *The Expansion of England.* Chicago: University of Chicago Press.

Sidaway, J.D. (2003) 'Banal geopolitics resumed', *Antipode*, 35(4): 645–51.

Slater, D. and Taylor, P.J. (eds) (1999) *The American Century.* Oxford: Blackwell.

Smith, N. (2001) 'Global executioner: scales of terror', http://www.ssrc.org/sept11/essays/nsmith.htm. Accessed 4 September 2004.

Smith, N. (2003a) *American Empire: Roosevelt's Geographer and the Prelude to Globalization.* Berkeley: University of California Press.

Smith, N. (2003b) 'After the American Lebensraum', *Interventions*, 5(2): 249–70.

Sparke, M. (2003) 'American empire and globalisation: postcolonial speculations on neocolonial enframing', *Singapore Journal of Tropical Geography*, 24(3): 373–89.

Stoler, A. (2002) *Carnal Knowledge and Imperial Power: Race and the Intimate in Colonial Rule.* Berkeley: University of California Press.

Thomas, N. (1994) *Colonialism's Culture: Anthropology, Travel and Government.* Cambridge: Polity.

Urry, J. (2004) 'Small worlds and the new "social physics"', *Global Networks*, 4(2): 109–30.

Wainwright, J. (2004) 'American empire: a review essay'. *Environment and Planning D: Society and Space*, 22: 465–74.

Wallerstein, I. (1979) *The Capitalist World Economy.* Cambridge: Cambridge University Press.

Wallerstein, I. (2001) 'America and the world: the twin towers as metaphor', http://www.ssrc.org/sept11/essays/wallerstein.htm. Accessed 4 September, 2004.

Wallerstein, I. (2003) *The Decline of American Power: The U.S. in a Chaotic World.* New York: The New Press.

Yeoh, B. (2003) *Contesting Space in Colonial Singapore: Power Relations and the Urban Built Environment.* Singapore: Singapore University Press.

Young, R.C. (2001) *Postcolonialism: An Historical Introduction.* Oxford: Blackwell.

28

Re-Bordering Spaces

Jouni Häkli

THE NATION-STATE AND THE LEGACY OF BORDER STUDIES

Borders and boundaries are everywhere. Individuals and human communities define and structure the social world by making distinctions between groups, spaces, times, objects and meanings. We encounter borders constantly in our everyday lives, and know how to respect or transgress them by intuition, experience or reasoning. Broadly speaking, borders and boundaries represent an immense area of research including almost everything that pertains to humans and societies.

In the context of political geography, the topic is much more focused. The sub-discipline's tradition has typically approached borders through a geopolitical prism, typically as part of nation-state territoriality, or as dividers between geopolitical blocs. In retrospect, the involvement of political geography with states and their geopolitics may seem obvious. The twentieth-century was characterized by the rise into prominence of the nation-state as the most powerful political organization defining the fate of individuals and communities. Rivalry between states and the striving to establish new nation-states caused major international conflicts throughout the century. To overcome the sources of such conflicts, states have sought to create international mechanisms of political and economic regulation, such as the United Nations, the European Union and the North American Free Trade Area.

The minds of practitioners of statecraft, as well as many intellectuals, were preoccupied with questions of territoriality and borders. Among the most crucial geopolitical questions of the twentieth-century, linking power and space intimately together, was that of the territorial congruence, or incongruence, between national homelands and emerging state territories (Herb, 1997). The geopolitical imagination of the classical thinkers, such as Friedrich Ratzel, Rudolf Kjellen and Sir Halford Mackinder, traced the sources of states' geopolitical power and hegemony (Ó Tuathail, 1996). While these voices were keenly heard at the corridors of power, more radical, anti-statist strands of thought were presented by anarchist thinkers, such as Elisée Reclus and Pyotr Kropotkin (Kramsch and Hooper, 2004).

The overwhelming majority of the early twentieth-century border studies were firmly connected to a state-centred perspective on the geopolitical world. The growing power of states as dominant political organizations had its intellectual counterpart in what Peter Taylor (1996) has called the 'embedded statism' of social scientific research. With few exceptions, social science disciplines operated on a concept of society that took nation-state territory for granted in its definition. Hence, in empirical social research and even in theoretical work, the borders of society equalled those of the respective nation-state (Häkli, 2001).

Little wonder, then, that international boundaries were not among the hotly debated topics of early twentieth-century social theory. Even political geography research of the time treated state borders mainly as an empirical issue. The practical application of boundary studies may have reached its apex after the First World War, when scholars were charged with demarcating the boundaries between new countries in south-eastern Europe, according to the Wilsonian principle of territorial self-determination (Anderson, 1996; Herb, 1997). Political geographers undertook much of this work on boundaries, reflecting their interest in the combination of physical and cultural features that boundaries represent. Early examples are classic texts by Ellen Semple (1911) and Isaiah Bowman (1922).

Along more theoretical lines, some attempts were made to distinguish between different kinds of borders, such as natural and artificial state boundaries. Often the aim was simply to compare the demarcation of political lines to physical features and then to the cultural landscape (Jones, 1959). Richard Hartshorne, for example, classified boundaries by how well they corresponded to divisions of peoples (Minghi, 1963). Although it would be unfair to say that boundaries were always viewed simply as natural dividers between differing cultures, political systems and economies, much of this early work took the existence of nation-states for granted. In many cases, boundaries were viewed as 'walls' or 'curtains' that separated rival ideological systems of mutually hostile states (Kristof, 1959). Hence, studies tended to emphasize stress and conflict involved, depicting the boundary as a cause of friction between states (Minghi, 1991: 17).

An early line of research that still continues today are studies that take a *functionalist* view of borders. Here the degree of interaction across boundaries, and the extent to which a boundary could exist as an impediment or a conduit to interaction, occupies a central place (see Mackay, 1958; House, 1981; Klemencic and Bufon, 1991). Growing international mobility and deepening political-economic integration during the latter part of the twentieth century have kept this concern firmly on the agenda of applied border studies. Moreover, as the 'state container' has begun to leak, political-geographical theorization of the role and nature of borders has diversified rapidly. With increasing pressures on states' territorial integrity, several geopolitical certainties have been called into question. The *constructionist* strand of border research, challenging the traditional state-centred view of the geopolitical world, has dealt with issues such as how states and boundaries are involved in the construction of national identities, how political communities are constituted, and what new political spaces may emerge in international borderlands.

'LEAKING CONTAINERS' MAKE INTERESTING BOUNDARIES

Recent decades have witnessed several major geopolitical changes, including the breakdown of the bipolar geopolitical order known as the Cold War, growing neo-liberalism and economic globalization, attempts at global governance of major environmental issues, the European integration process and the emerging free trade areas around the world. Consequently, new geopolitical scales have emerged as important arenas alongside the traditional nation-state scale. The regional and international political spaces and new political power constellations have gained visibility also in political geography research.

The new kinds of political spaces and communities emerging in all continents would have been unthinkable some decades ago. In Peter Taylor's (1994) terms, states may have been the prime containers of power, wealth and culture in the modern world order, but it is clear that these containers have begun to leak in various ways. Depending on the political-economic context, different economic and political arrangements have been set up to deal with local, regional and national development issues shared across international boundaries. In East Asia the development of cross-border governance is mostly driven by concerns with securing favourable conditions for capital accumulation and economic growth. For example, in the 'Greater China' region, multiple networks and foreign direct investments, crossing the private/public and local/provincial/central domains of regulation, have regionalized the economies of Hong Kong, southern China and Taiwan (Sum, 2002).

China is also part of a 'Japan Sea regionalism', together with Russia, Mongolia, South Korea, North Korea and Japan. This cross-border co-operation has emerged mainly between subnational authorities, thus avoiding many difficult issues that have stalled governmental relations between these countries. The process is largely about sub-national diplomacy with a focus on various agreements, conferences and linkages between authorities, thus lacking the economic dynamism of 'Greater China' development (Arase, 2002). Nevertheless, it is part of the post-Cold War deterritorialization and re-bordering that brings actors together across state boundaries forming new regional alliances and networks.

Also in North America, cross-border regionalization is largely driven by economic concerns instead of a common sense of destiny. The North American Free Trade Area (NAFTA) is an agreement between Canada, the United States and Mexico regulating cross-border economic interaction between the countries. According to Scott (2002), NAFTA represents limited integration that recognizes interdependence between the local and national economies without compromising state sovereignty. Again, on the more local scale, cross-border co-operation takes on forms that avoid some complexities that characterize the co-operation between national governments (illegal immigration issues, the question of economic asymmetries, etc.). For example, the Arizona–Sonora region at the US–Mexico border presents numerous cross-border co-operation schemes involving public, private and state actors and emphasizing economic, environmental and cultural issues. As a result, a multi-scalar field of cross-border governance has emerged at the

US–Mexico border for solving local problems that transcend national boundaries (Scott, 2002).

From the European perspective, boundary issues with northern Africa are dominated by attempts to stall illegal immigration to 'Fortress Europe'. In contrast to the tendency towards more relaxed border policies inside the European Union, the relationships between the EU and African countries reflect mutual suspicion and tightening border control. For example, Spain and Morocco have unresolved border issues related to the cities of Ceuta and Melilla, which are Spanish territories in northern Morocco. From the Spanish perspective the question is mainly about stalling the flows of illegal immigration and drugs into the EU, whereas the Moroccan perspective focuses on annexing the cities, which would mark the end of the Spanish colonial domination (Nogué and Villanova, 1999).

In southern Africa, more endogenous cross-border regionalization processes have emerged bringing together actors with partly overlapping and competing interests, goals and identities. For example, Lundin and Söderbaum (2002) have studied the Maputo Development Corridor, which runs from the Republic of South Africa to Mozambique, Swaziland, Botswana and Zimbabwe. The corridor has a long history rooted in migration flows and the concomitant cultural and economic patterns in the region. A project in the making, like most cross-border regions, the Maputo corridor is expected to give a boost to development and economic growth in the region. As one of the best-known African cross-border projects it is showing the way to others, such as the Caprivi Strip and Zambezi River projects (Lundin and Söderbaum, 2002).

An astonishing contextual variety among borderlands and cross-border regionalization processes is the matter of fact even in Europe where decades of peaceful and deep-going integration have made many state boundaries practically non-existent. European integration and the expansion of the European Union since the 1950s have clearly been the most significant developments influencing European boundaries and borderlands. While the integration process has certainly been driven by concern with economic performance in the 'old continent', the European Union is also a post-second World War peace project and, moreover, an experimentation with the idea of multi-layered citizenships and cultural identities that allow local and transnational ties to bundle and overcome Europe's 'all too territorial past' (Kramsch and Hooper, 2004).

Yet, even in the laboratory of post-national integration that the European continent is, or aspires to be, some borders are disappearing but others are being erected. Newman and Paasi (1998) point out that the breakdown of the Soviet Union alone created more than twenty new state boundaries. Some state boundaries from the eastern side of the former Iron Curtain are now within the expanded European Union. Many of these borderland contexts lack the longer history of co-operation that characterizes cross-border regions in the more traditional European core. Hence, 'Euroregions', which are the most institutionalized forms of cross-border co-operation, differ from each other greatly between East-Central and Western Europe.

For example, the Euroregion called Nysa, in the borderland between Poland, the Czech Republic and Germany, suffers from cross-border economic, political and cultural discrepancies that are absent in the Dutch–Belgian–German borderland. There, the Euroregion called Meuse–Rhine has functioned since the mid-1970s (Kepka and Murphy, 2002). The most significant differences between the two Euroregions are related to conditions for economic development, which are relatively even in Meuse-Rhine, but characterized by stark asymmetry across the border in Nysa. Also cultural differences and the lack of trust in political institutions on the 'other side' hinder the development of a coherent 'Euroregion' governance concept in the Polish–Czech–German borderland.

The above-mentioned 'leaks' in state containers have not gone unnoticed among scholars interested in globalization and trans-boundary flows. Such deterritorializing and re-bordering processes are ones generally recognized by scholars as revealing examples of the current reconfiguration of economic and governmental space, with significant influences on the nature and role of international boundaries. However, it is important to realize that borderlands, as well as cross-border regionalization processes across the globe, are extremely diverse, and that what the regions have in common may be much less than how they differ from case to case. Certainly, numerous boundaries are becoming more and more permeable, and new political spaces, with unforeseen social and economic functions, are emerging at international boundaries. However, it is equally true that borderlands exist where none of these developments can be found, even ones where violence rather than co-operation is commonplace every day. The 'peace line' currently being erected by the Israeli government serves as a useful reminder of the fact that borders are not developing unilinearly, and that any assessment of the political geography of borderlands needs to put the studied processes into their social and historical contexts.

BOUNDARIES AND PLACE: CONTEXT MAKES A DIFFERENCE

Several authors have sought to typologize borderlands and re-bordering processes so as to sort out the contextual variety and provide guidelines for analysis. For example, Perkmann and Sum (2002) take local border regimes and central state intervention as their starting point in distinguishing between different kinds of cross-border regions. They end up with two main types. The first is characterized by the continuing erosion of border barriers and little central state intervention, as exemplified by European cross-border regions between adjoining EU member states. In such cases the emerging cross-border region is mainly a local and regional enterprise circumventing, sometimes even opposing, central state control. In the second type of cross-border regionalization, borders are opened only selectively to allow certain transactions while blocking others. This is typical of cases where concern with economic growth is the driving force behind region-building. For example, in some North American and East Asian borderlands the central state has intervened strongly in setting up special economic zones to boost cross-border trade, while maintaining or even tightening control over concomitant immigration flows (Perkmann and Sum, 2002).

Kramsch and Hooper (2004) have created a more historically and culturally based typology of border regimes. They distinguish between regimes emerging in functionally integrated cross-border regions with little or no border controls and those emerging in the post-Cold War buffer zone borderlands with fewer advantages. The third type they identify is regimes at the edge of Empire, defining and protecting its political, cultural and economic order. The first type is best exemplified by re-bordering processes in the 'Schengenlandia' of Western Europe – what Häkli and Kaplan (2002) call 'Established Europe' in their own typology. The second type can be found in a diversity of settings, but in the European context it clearly refers to the EU accession countries of Central and Eastern Europe – the 'Emerging Europe' (Häkli and Kaplan, 2002). The last type refers to buffer zones between world economic cores and peripheries, borderlands that serve to institutionalize political, economic and cultural asymmetries forming the basis of a world order. While often relatively peaceful, these zones reflect political and cultural tensions rooted in 'older colonial endeavours' (Kramsch and Hooper, 2004).

Other typologies have been made with differing starting points and analytical goals (see, Bucken-Knapp and Schack, 2001). However, it is clear that these cannot exhaust the cultural, historical and geographical varieties that differentiate borderland contexts. Each borderland is unique in some respects, and if we are to avoid reductionist explanation of current re-bordering processes, this contextuality should be taken seriously. At the same time, it is clear that a level of abstraction is absolutely necessary for making sense of this vast societal transformation. Too much emphasis on the particular and unique leads to descriptive accounts that do not further general understanding of how the relationship between politics and space is changing (Bucken-Knapp and Schack, 2001).

Rather than seek similarities among different borderlands, it may, therefore, be more useful to focus on themes that cut across the multidisciplinary literature on boundaries and link the research area with the broader spectre of social theory. Themes that are especially pertinent to political geography are the question of scales, the nature of trans-boundary political spaces and the role of collective identities in re-bordering processes (see, e.g., Newman and Paasi, 1998; Perkmann and Sum, 2002; Anderson et al., 2003; Kramsch and Hooper, 2004).

BORDERLANDS AS NESTED SCALES

The idea of scale as a factor influencing human interactions has been a major theme in the human geography of the past decades. Until roughly the mid-1980s, scale was typically seen as an empirical variable, either as distance influencing the studied phenomena, or as an object of study defined by the given administrative divisions of space ranging from the local to the regional, national and international. Often the local scale would simply mean city or municipality, the regional scale consisting of several municipalities. The national scale was seen as composed of sub-national regions, and the international scale as a bundle of national territories.

This perspective is still dominant in the policy-orientated research fostered by local, regional and European organizations (Donnan and Wilson, 1999). Here the influences and meanings of scale are understood as arising from the divisions of administrative power and political authority of the target areas and their national contexts. In research looking at economic interactions across boundaries, the border is typically interpreted as a barrier, and the influences of scale are assessed in terms of metric or relative physical distance (van Houtum, 1999).

Since the mid-1980s, this taken-for-granted conception of spatial scales has been criticized particularly in analyses of the political economy of globalization and the social construction of space (Marston, 2000; Brenner, 2001). This growing literature has worked to discard both the naïve conception of scales as pre-given spatial levels, and the view that scales can be reduced to various aspects of physical distance. Instead, the social processes of scale construction have been scrutinized, and more particularly the ways in which scales are actually set or fixed amidst the flux of social interaction (e.g. Taylor, 1982; Jonas, 1994; Smith, 1995; Jones, 1998).

Unsurprisingly, scale has long figured also in the context of research on borders and borderlands. Authors such as John House (1981), Ivo Duchacek (1986) and Julian Minghi (1991) have paid attention to the influences of geographical scale on interactions across state boundaries. In these early works, scale was mainly understood as an aspect of physical distance, whereas more recently social constructionist approaches have gained ground. For example, the construction of scale has been analysed variably as processes of networking where actors negotiate alliances and bargain for political power (e.g. Cox, 1996, 1998); as processes of 'spatial socialization' where individuals learn to make sense of the world in the context of various geographical divisions (e.g. Paasi, 1996); and as the conflicting or harmonious intermingling of spatial identities in borderland contexts (e.g. Kaplan, 2000). While approaching the issue of scale from very different viewpoints, these writers have made it clear that qualitative differences exist between small-scale interactions close to the border and the interactions between, for example, national actors directed from the capital cities. This underlines the fact that interactions across national borderlands cannot be reduced to states' actions only.

To understand the social construction of scale in the context of institutionally driven cross-border interaction, it is necessary to assess the processes that set or fix geographical scales. Following Smith (1995), we can argue that geographical scales are produced simultaneously to enable and contain particular forms of social interaction. For example, the nation-state is one of the typical scales of modernity, perhaps the most established one. Rather than simply assuming the state's territorial form, it may be useful to pay attention to the supranational constitution of the nation-state as a hegemonic and sustainable institutional form (Giddens, 1985). The rigidly bounded nation-states did not emerge in isolation but as part of an international state system, within which their boundaries have been systemically negotiated and regulated. From this vantage point the state can be analysed as a scale that has both contained social interaction by limiting it within the territorial reach of the state, and enabled social activities by directing, aligning, and co-ordinating the efforts of a large number of actors (Taylor, 1996; Häkli, 2001).

Depending on what functions of the modern state we wish to emphasize, it is possible to point at particular social infrastructures and networks that have contributed to the production of the 'national' scale. From an economic point of view, the state has secured the necessary conditions for the accumulation of capital and, broadly speaking, enabled the competitive co-operation of companies through the division of labour. From a cultural point of view, the state has enabled the concerted construction of national identities, and contained and fostered cultural forms (language, history, arts) that have been labelled national. From a social point of view, the state has contained social relations both mentally and physically within its territorial boundaries through school education and other mechanisms of national socialization, as well as by regulating the movement of people across its territorial boundaries (Taylor, 1994; Smith, 1995; Paasi, 1996; Häkli, 1999).

Similar processes are at work on other scales of the modern society – the local, the regional and the international. These scales are often seen as autonomous from each other, implying that the division of power and authority between the state and the region is something of an original 'state of nature' (Häkli, 1998). Yet, having assumed their role as established levels of authority and interaction, these scales have all emerged interdependently as part of the territorial governmental order in modern societies (Harvie, 1994; Häkli, 2001).

This is not to claim that the local, regional and national scales would essentially be the same because of their shared origin in the history of the modern governmental order. Here we come to another important implication of the social constructionist approach to scales, namely their charged character as the products of political contestation and co-operation. Much of the recent interest in cross-border regionalization derives precisely from an understanding that the state is being challenged by these new scalar constructs. Hence, the emerging transnational political spaces can be conceptualized as scales that help actors to skirt the traditional state-centred patterns of networking. Transnational scales, then, are produced and reproduced in processes that set alternative perimeters to networks of co-operation between actors who seek strategic advantage from this co-operation (Häkli, 1998). The perimeter can be fluid or fixed, depending on the spatial frame that actors wish to, or have to, adopt (Smith, 1995; Cox, 1998). When the networking takes place irrespective of established political-administrative units, a new scale is being produced.

Cross-border co-operation typically cuts across various scales ranging from individual actors to local authorities to regional networks to national governments to international organizations. In this regard, borderland spaces are actually places where different scales of action become nested, forming hybrid bundles across the local, regional, national and international networks. Because of this, it can be quite difficult to align the interests of each actor. Often the desires of local actors are contravened by activities of state governments that do not wish to relinquish their traditional authority. In the European context, Euroregions have sometimes been caught between the contradictory goals of actors operating at separate spatial scales. A borderland population may be drawn apart or even ignored by two larger states that do not appreciate the particular concerns of the residents. For example, the Catalan population has been divided by differing nationalist agendas in France and Spain (Häkli, 2002). Similar distinctions have also had a major impact on the Tyrolean and Slovenian populations along the Italo–Austrian and Italo–Slovene boundaries (Kaplan, 2000).

Seen in terms of networking, scales are an integral part of the very economic, political and social interaction they enable and constrain. Roughly speaking, the networks of actors co-operate internally and compete externally. Yet, far from being harmonious and symmetrical, the production and reproduction of geographical scales remains a contested process. Established scales can dissolve as a consequence of pressures of economic, cultural or political origin. In turn, newly constructed scales may fail to reach the level of popular support that their functioning as broad integrative frameworks necessitates.

The potential 'legitimacy deficit' experienced by institutionally created cross-border networks is particularly interesting from the perspective of this chapter. When institutionally based actors fail to gain popular appeal for the cross-border networks they are constructing, the resulting new scale will not become an important political reality. At best, it may form a new quasi-autonomous layer of political authority on top of the more traditional ones. It is precisely this development that is being addressed in debates revolving around the possible erosion of the nation-state, the rise of the region and the uncertain future of cross-border governance.

DETERRITORIALIZING AND RE-BORDERING POLITICAL SPACE AT INTERNATIONAL BOUNDARIES

It has become a broadly accepted view among political analysts that the emerging cross-border political spaces are challenging the state-centred spatiality of politics. At stake is not only the role of international boundaries as barriers, but also the very principle defining the modern political space, that is, territorial congruence between political authority, political rights and belonging to a political community (Low, 1997). The territorial foundation of politics and rigid state boundaries are the outcome of the consolidation of state power in the modern period and the concomitant rise of the international state system (Ruggie, 1993; Murphy, 1996; Ó Tuathail, 1996). This governmental order is now being challenged, or at least modified, by processes leading to transnational

forms of cross-border governance (Kramsch and Hooper, 2004).

Since roughly the late 1980s, political and economic regulation has moved progressively from state-centred government to governance based on multiple partnerships across the public/private divide, and bringing together both governmental and non-governmental organizations. This trend is reflected empirically in the proliferation of projects directed at local and regional development across various territorial scales (Jessop, 2002: 43). Among such projects are the many processes of regionalization, in which new transterritorial and international mechanisms of governance are created through political and economic networking (e.g. Delli Zotti, 1996; Perkmann and Sum, 2002).

Following Bob Jessop (1995), governance can be understood broadly as attempts to attain collective goals and purposes in and through specific configurations of governmental and non-governmental institutions, organizations and practices. Thus, instead of a coherent and ready-made territorial system of political representation upon which policy-making could be built, we should expect to find a more fluid, less systematic and highly diversified field of governance, where regions or trans-regional networks perform very differently depending on their ability to mobilize and co-ordinate both human and economic resources for collective goal-attainment (Le Galès, 1998).

Probably the most challenging new forms of governance are related to cross-border regionalization where, ideally, different national political, legislative and administrative cultures should act together and enable the actors involved to assess trajectories of development, envision common goals, and determine means of achieving these (Perkmann, 1999; Scott, 2000). It may be feasible to explain the general shift from government towards governance with reference to the major political-economic trends of the past three decades: globalization, supranational integration, the end of the Cold War, and general rescaling related to the 'hollowing out of the nation-state' (Swyngedouw, 1992; O'Dowd and Wilson, 1996). However, when assessing the rapid growth of the number of cross-border regions across the globe since the early 1970s, it is important to realize that this development also reflects the policies of national governments and transnational institutions. This is the case particularly in Europe, where the European Union has launched several programmes fostering cross-border co-operation (e.g. Tacis, Phare, Interreg).

The fact that dozens of 'Euroregions' or 'Euregions' have been established in European borderlands indicates that institutional stability is much desired as a support for cross-border governance. Euroregions are commonly seen as avenues for better access to the European Commission and EU funding (Perkmann, 2002). For individual authorities, participation may offer the chance to be prepared in terms of an established partnership, as commonly required by the European Regional Development Fund initiatives and programmes. Furthermore, precisely because it opens direct connections between local and regional authorities and the European Union, Euroregions provide the former with more elbow room in negotiations with their own national governments in issues of regional development, decision-making and representation of interests. Not surprisingly, Euroregions seem to have obtained a permanent place in the contemporary 'multi-level governance' in Europe (Ward and Williams, 1997; Perkmann and Sum, 2002).

Cross-border regionalization in Europe and elsewhere is part of a development that will bring about a more polycentric world. States as well as subnational and transnational organizations have launched policies that actively foster regional co-operation across national boundaries. Numerous economic, political and cultural actors involved in cross-border co-operation have seized these opportunities in an attempt to expand their capacity to govern (Häkli, 1998; Perkmann and Sum, 2002). Nevertheless, the reorganization of European political space is politically and culturally a highly contentious project. New power alliances have emerged within the private sector and across the private/public divide challenging the traditional divisions of political authority, as well as the systems of political representation that legitimate these divisions (Benz, 1998).

To the degree that emerging cross-border networks are, in fact, producing new political spaces and scales of social action at international borders, we may argue that deterritorialization is an ongoing process changing the nature of modern state-centred political space. However, the success of this development depends not only on the number of institutionally created networks and projects. To gain deeper understanding of the political significance of these processes, it is necessary to look at the degree to which these re-bordered transnational political spaces have been institutionalized in the popular realm of political thought and action.

To focus on the everyday perceptions of people may at first seem an unnecessary complication in the analysis of contemporary political change. However, it can be argued that in the long run the success of major political innovations will depend on their appeal among the 'ordinary people'. In this respect, the way in which people perceive the transnational political spaces, within which they live every day, may be of decisive importance when it comes to their potential as new political spaces, forms of constituency and transnational community.

The fact that cross-border regionalization is in most cases led by politicians, authorities and other professionals is not without consequences. Cross-border regions are far from being regions in the conventional juridico-political sense, that is, they are not governed through territorially based popular representation (Perkmann and Sum, 2002). Instead they often involve irregular and partial modes of governance that operate in a network-like manner (Kramsch and Hooper, 2004). These networks of governance bring together individual and collective actors who can contribute to, and have something at stake in, the process. In most cases what is lacking is a representative mechanism that would link the broader population to governance institutions or networks.

Hence, for example, in Europe the 'Euroregions', or 'Euregions', which now number more than seventy, have remained technocratic entities detached from people's everyday concerns (Perkmann, 2002: 121). Also in East Asia and North America, networks of governance have typically emerged in the spirit of competitive co-operation, reflecting mutual interests in economic growth across the boundary, or they have been set up to exploit opportunities for funding by national governments and transnational institutions. With a focus on aspects that mainly concern professional elites, the emerging forms of cross-border governance may represent a less radical development than they at first seem.

Moreover, to reach the minds and hearts of 'ordinary people', cross-border governance should gain political legitimacy based on popular acceptance of the new geo-economic relations, authorities, participatory possibilities and belonging to transnational regions. This calls for an acknowledgeable citizenry aware of new emerging transnational linkages and institutions for governance, and willing to participate in their activities to the degree that is possible. Needless to say, perhaps, these conditions are as yet rarely met among the people living at, or near, international borders.

NEW POLITICAL SPACES, OLD CULTURAL IDENTITIES?

Identity was one of the most used and least theorized catchwords of the cultural studies literature of the 1990s (Häkli and Paasi, 2003). It is still frequently used in attempts to understand the multi-layered and sometimes overlapping feelings of community and belonging in various social and spatial groups. While identity is a useful concept for discussing spatial group formation, it should be kept in mind that the word in itself explains little (Hall, 1996). Rather, the concept of identity can be seen as a constituent part of the very discourses that form political communities. Spatial concepts, such as place, region and territory, as well as the images of shared and divided space that these concepts denote, play a central role in identity discourses shaping the geopolitical world.

In fact, political-geographical research on nationalism has shown that territory is more important an element in the construction of national identities than theories of nationalism have generally acknowledged (e.g. Nogué, 1991; Hooson, 1994; Herb and Kaplan, 1999). Joan Nogué (1998) has distinguished at least seven dimensions through which territory may acquire significance in the construction of national identities. These include cartographic representation, education in geography, the definition of national boundaries, internal regional divisions, ideological landscapes, the idea of national character, and the protection of nature. Moreover, both the textual and physical landscapes should be taken into account in the analysis of nationalism (Duncan, 1990; Graham, 1998; Häkli, 1999).

The relationship between national boundaries and the construction of territorial identities has rightly become one of the most intensively studied theoretical questions in the area of border studies. This interest has resulted in a broad understanding of the role of boundaries as constituents of collective identities, with emphasis on the formation of identities as the accentuation of the distinction between 'us' and 'them', through the social construction of boundaries (Sahlins, 1989; Donnan and Wilson, 1994; Pettman, 1996). Needless to say, this strand of research accords with epistemologies that stress the anti-essential character of all social institutions independently of their possible material ramifications. Such an analysis of boundaries does not ignore the reality of the border as a physical delimiter of state territory, but serves as an important reminder of the way in which space and power, or the spatial and social realities, are inseparably connected (Paasi, 1996).

The social construction and spatial rootedness of territorial identities is particularly consequential from the point of view of the emerging transnational political spaces. At a general level, cross-border governance as a social practice is certainly creating its own political spaces that deviate from the traditional international state system. Instead of being discrete entities of the external social, economic and political reality, the transnational regions are constructs that are created and reproduced in social practices, such as those involved in governance (e.g. Häkli, 1998; Jessop, 2002). Hence, governance is not only a set of practices played out in a particular regional setting, but it is also constitutive of 'the regional' as a field of action and knowledge.

This is most apparent in the case of transnational regions that are of relatively recent origin and have emerged as more or less loose concepts in the context of cross-border co-operation. These regional formations are typically transnational networks bringing together governmental, economic and cultural elites with at least partly overlapping interests and visions (Le Galès, 1998; Smith, 1998). Moreover, as Manuel Castells (1997) points out, the actors who are engaged in new kinds of social networks and projects tend to identify with the realities they are constructing. Thus, cross-border governance gives rise to 'project identities' shared by the participating professionals (Castells, 1997: 8).

However, while the actors who are involved in trans-boundary networks may indeed identify with the emerging transnational political space, this is not always the case with the broader population characterized by its more traditional state-centred loyalties. Professional discourses portray national boundaries typically as barriers to be bridged, as elements that unify rather than separate the adjacent regions. Yet those who know little about official trans-boundary initiatives keep negotiating their spatial identities as embedded in their everyday practices. Consequently, a rather traditional understanding of territorial political space dominates the contexts in which people form their political views and frame policy issues.

To understand why this is the case, we may again turn to Castells (1997: 66), who argues that in providing the population with citizenship rights the states have also created 'legitimizing identities'. The increasingly powerful and interventionist state government has produced and reproduced national identities to rationalize its domination. These identities still hold strong despite the intensifying transnational integration and globalization. People have not surrendered their traditional loyalties because the state has remained the primary provider of their citizenship rights (Dijkink, 1996; Hirst, 1997).

The growing legal and illegal immigration has increased cultural contacts across state boundaries, but also created tensions that are rooted in the traditional nation-state identities. The distinction between 'us' and 'them' becomes highly consequential when the immigrant population is seen to present a threat to what is defined as a self-sustaining national culture (Mains, 2000). In such cases, the situation may culminate in racist discourses and actions by citizens who do not tolerate the growing number of immigrants in their territory. A good example is the 'Minutemen Project', a movement that brought hundreds of civilian volunteers to patrol the US–Mexican border in Arizona (Jordan, 2005). Intent on halting the flow of illegal immigration from the South, the movement actually worked to re-border space that for decades has progressively transnationalized both in economic and cultural terms (Soguk, 1996).

All in all, the social and cultural inertia embedded in people's connection with territory, the sense of place, may be a more powerful intervening force in the development of transnational political space than the practitioners of cross-border governance may have expected. The era of strong nation-states left a legacy of statist loyalties at international frontiers and this should not be underestimated in the analysis of transnational political space (Donnan and Wilson, 1999; Häkli, 2001; Häkli and Kaplan, 2002; Sidaway, 2002; Strüver, 2004). The shift toward *functionally* more porous borders has not met with equal erosion of the boundaries in people's minds and hearts. While the roles of state boundaries have changed, the border as a social *construct* prevails in various social practices and cultural processes of identification.

An obvious conclusion is that the processes of cross-border co-operation have worked to disconnect the politics of governance from national territories and democratic participation. Because they are elite-driven, these re-bordering processes can hardly yield platforms for shared political and cultural identity across national boundaries. Problems are found even in places like the Catalan and Basque borderlands where governance is favoured by linguistic affinity across the border (Raento, 2002; Häkli, 2004). Territorial congruence between political, cultural and economic processes can not easily be achieved in transnational political spaces. We can therefore anticipate that in securing political participation the role of governance is at best complementary to state-based governmental practice. As yet, it seems that cross-border governance is more about political and economic elites governing people, rather than the people governing themselves.

The complexities of identification at international borderlands might easily be dismissed as a passing phase on our way towards an integrated world. However, it is unlikely that the discrepancy between professional and popular identifications will be over any time soon (Appadurai, 1996; Dijkink, 1996; Wilson and Donnan, 1998). The social and cultural inertia embedded in people's connection with territory, the sense of place, is a much more powerful intervening force than may have been expected. Moreover, as it is unlikely that transnational identities will soon emerge replacing traditional territorial identities, it is safe to argue that the connection between political participation and national territory will persist. Instead of merely appreciating the fact, we should seek to analyse its roots and consequences, carefully balancing generalization and sensitivity to local contextuality.

CONCLUSION

This chapter has addressed three major aspects in re-bordering political space, all of which pertain to the changing relationship between space and power. First, the role and significance of scale was assessed in the context of various processes crossing international boundaries. Second, the construction of transnational regions as new political scales at international borderlands was dealt with. Third, the institutionalization of these regions was discussed with particular attention to the issue of political legitimacy as related to professional and popular identifications with the emerging political spaces. Because of the rapid growth of cross-border regions all over the world, there is a growing demand for knowledge about the forms of, and degrees of, popular attachment to transnational political spaces.

The chapter argues that not only is scale a useful tool in thinking about border regions, but that its significance varies according to the viewpoints of differently placed observers. Instead of striving for a general theory of scales or borderlands, the approach argued for here acknowledges the tensions between different understandings and perceptions of cross-border interaction that arise from different realms of social action. Moreover, the significance of scale as a factor in the development of borderlands extends beyond differences between local and national perspectives. For a deeper understanding of the current transitions and fixities in international borderlands, it is important to explore the differences between scale as a factor in people's everyday life, and scale as a factor influencing cross-border co-operation between institutional actors.

The former perspective is well captured by the term increasingly used within border studies to denote people who live near state borders: the 'borderlanders' (e.g. Wilson and Donnan, 1998). The borderlander concept points at the role of state boundaries as a significant element in the daily environment of the people living in the vicinity. It also signals the fact that cross-border interactions are more likely to occur when the 'other side' is easily accessible, in contrast to people living further away from the border.

Numerous forms of co-operation exist across the national boundaries, giving rise to multiple, more or less institutionalized settings for governance based on complex trans-boundary networks. Politicians and economic actors all around the world have been willing to seize opportunities to form new regional alliances, to utilize the funding provided by national and transnational governments and to enhance their capacities through strategic networking.

However, while cross-border co-operation is characterized mainly by technocratic goals, any 'bridge-building' across international boundaries remains politically tension-laden. Differences persist because the development of the transnational regions is conditioned by the social and geographical contexts of trans-boundary co-operation. Moreover, there is as yet little popular awareness of institutional activities that are fostering cross-border regionalization. This is an issue that will greatly influence the actual outcomes and the political potential of cross-border governance.

All in all, in the light of many studies of borderland identities (O'Dowd and Wilson, 1996; Newman and Paasi, 1998; Wilson and Donnan, 1998), it seems that the existence of networks for cross-border co-operation will not necessarily lead to the rapid erosion of political and cultural identities connected to the history of the modern nation-state. They may simply add new layers or dimensions to territorial governance, which, depending on the social, cultural and economic context, may or may not function as an important political arena. This is because networking tends to follow its own functional and institutional logic, which typically remains unconnected to people's everyday concerns. The fact that people living at international borderlands are typically unaware of how cross-border projects operate, or how to participate in the politics of regionalization across the national boundary, lends little support to the idea of an increasing awareness of shared political goals pertaining to re-bordered transnational spaces.

What can be safely concluded is that while the political spaces are definitely going through multiple changes, these are experienced differently in different borderlands. Even though it is more and more easy to cross many international boundaries, the image of borders as natural dividers between national cultures persists. Studies on boundaries and identities have convincingly shown that state borders are essential for the establishment and negotiation of national identities. Therefore, in Michael Billig's (1995: 60) terms, to remove an international boundary altogether would require the dissolution of the forms of life that produce and reproduce national cultures, societies and polities. Despite the many cross-border networks functioning across the globe, this is not likely to happen soon.

REFERENCES

Anderson, J., O'Dowd, L. and Wilson, T. (eds) (2003) *New Borders for a Changing Europe: Cross-Border Cooperation and Governance.* London: Frank Cass.

Anderson, M. (1996) *Frontiers: Territory and State Formation in the Modern World.* Cambridge: Polity Press.

Appadurai, A. (1996) 'Sovereignty without territoriality: notes for a postnational geography', in P. Yaeger (ed.), *The Geography of Identity.* Ann Arbor: University of Michigan Press, pp. 40–58.

Arase, D. (2002) 'Japan Sea regionalism: the role of subnational authorities', in M. Perkmann and N.L. Sum (eds), *Globalization, Regionalization and Cross-Border Regions.* Basingstoke: Palgrave Macmillan, pp. 176–90.

Benz, A. (1998) 'German regions in the European Union: from joint policy-making to multi-level governance', in P. Le Galès and C. Lequesne (eds), *Regions in Europe.* London: Routledge, pp. 111–29.

Billig, M. (1995) *Banal Nationalis.* London: Sage.

Bowman, I. (1922) *The New World: Problems in Political Geography.* Yonkers-on-Hudson, NY: World Book.

Brenner, N. (2001) 'The limits to scale? Methodological reflections on scalar structuration', *Progress in Human Geography*, 25: 591–614.

Bucken-Knapp, G. and Schack, M. (2001) 'Borders matter, but how?', in G. Bucken-Knapp and M. Schack (eds), *Borders Matter: Transboundary Regions in Contemporary Europe.* Border Studies Series 2. Aabenraa: IFG, pp. 173–91.

Castells, M. (1997) *The Power of Identity*, Vol. II of *The Information Age.* Oxford: Blackwell.

Cox, K.R. (1996) 'Editorial: the difference that scale makes', *Political Geography*, 15: 667–70.

Cox, K.R. (1998) 'Spaces of dependence, spaces of engagement and the politics of scale, or: looking for local politics', *Political Geography*, 17: 1–23.

Delli Zotti, G. (1996) 'Transfrontier co-operation at the external borders of the EU: implications for sovereignty', in L. O'Dowd and T.M. Wilson (eds), *Borders, Nations and States.* Aldershot: Avebury, pp. 51–72.

Dijkink, G. (1996) *National Identity and Geopolitical Visions: Maps of Pride and Pain.* London: Routledge.

Donnan, H. and Wilson, T.M. (eds) (1994) *Border Approaches: Anthropological Perspectives on Frontiers.* Lanham, MD: University Press of America.

Donnan, H. and Wilson, T.M. (1999) *Borders: Frontiers of Identity, Nation and State.* Oxford: Berg.

Duchacek, I.D. (1986) 'International competence of subnational governments: borderlands and beyond', in O. Martínez (ed.), *Across Boundaries: Transborder Interaction in Comparative Perspective.* El Paso: Texas Western Press, pp. 11–30.

Duncan, J.S. (1990) *The City as Text: The Politics of Landscape Interpretation in the Kandyan Kingdom.* Cambridge: Cambridge University Press.

Giddens, A. (1985) *The Nation-State and Violence*, Vol. 11 of *A Contemporary Critique of Historical Materialism.* Cambridge: Polity Press.

Graham, B. (1998) 'The past in Europe's present: diversity, identity and the construction of place', in B. Graham (ed.), *Modern Europe: Place, Culture, Identity.* London: Arnold, pp. 19–52.

Häkli, J. (1998) 'Cross-border regionalisation in the "New Europe": theoretical reflection with two illustrative examples', *Geopolitics*, 3: 83–103.

Häkli, J. (1999) 'Cultures of demarcation: territory and national identity in Finland', in G.H. Herb and D.H. Kaplan (eds), *Nested Identities: Identity, Territory, and Scale.* Lanham, MD: Rowman & Littlefield, pp. 123–49.

Häkli, J. (2001) 'In the territory of knowledge: state-centered discourses and the construction of society', *Progress in Human Geography*, 25: 403–22.

Häkli, J. (2002) 'Transboundary networking in Catalonia', in D. Kaplan and J. Häkli (eds), *Boundaries and Place: European Borderlands in Geographical Context.* Lanham, MD: Rowman & Littlefield, pp. 70–92.

Häkli, J. (2004) 'Governing the mountains: cross-border regionalization in Catalonia', in O. Kramsch and B. Hooper (eds), *Cross-Border Governance in the European Union.* London: Routledge, pp. 56–69.

Häkli, J. and Kaplan, D. (2002) 'Learning from Europe? Borderlands in social and geographical context', in D. Kaplan and J. Häkli (eds), *Boundaries and Place: European Borderlands in Geographical Context.* Lanham, MD: Rowman & Littlefield, pp. 1–17.

Häkli, J. and Paasi, A. (2003) 'Geography, space and identity', in J. Öhman and K. Simonsen (eds), *Voices from the North: New Trends in Nordic Human Geography.* Aldershot: Ashgate, pp. 141–56.

Hall, S. (1996) 'Introduction: who needs identity?', in S. Hall and P. Du Gay (eds), *Questions of Cultural Identity.* London: Sage, pp. 1–17.

Harvie, C. (1994) *The Rise of Regional Europe.* London: Routledge.

Herb, G. (1997) *Under the Map of Germany: Nationalism and Propaganda, 1918–1945.* London: Routledge.

Herb, G.H. and Kaplan, D.H. (eds) (1999) *Nested Identities: Identity, Territory, and Scale.* Lanham, MD: Rowman & Littlefield.

Hirst, P. (1997) *From Statism to Pluralism.* London: UCL Press.

Hooson, D. (ed.) (1994) *Geography and National Identity.* Oxford: Blackwell.

House, J. (1981) 'Frontier studies: an applied approach', in A. Burnett and P.J. Taylor (eds), *Political Studies from Spatial Perspectives.* New York: Wiley, pp. 291–312.

Jessop, B. (1995) 'The regulation approach, governance and post-Fordism', *Economy and Society*, 24: 307–33.

Jessop, B. (2002) 'The political economy of scale', in M. Perkmann and N.L. Sum (eds), *Globalization, Regionalization and Cross-Border Regions.* Basingstoke: Palgrave Macmillan, pp. 25–49.

Jonas, A.E.G. (1994) 'The scale politics of spatiality', *Environment and Planning D: Society and Space*, 12: 257–64.

Jones, K.T. (1998) 'Scale as epistemology', *Political Geography*, 17: 25–8.

Jones, S. (1959) 'Boundary concepts in the setting of place and time', *Annals of the Association of American Geographers*, 49(3): 241–55.

Jordan, L.J. (2005). ' "Minutemen" to patrol Arizona border' [online], http://abcnews.go.com/US/wireStory?id=518371. (Accessed 11 May 2005.)

Kaplan, D. (2000) 'Conflict and compromise among borderland identities in northern Italy', *Tijdschrift voor Economische en Sociale Geografie*, 91: 44–60.

Kepka, J.M.M. and Murphy, A.B. (2002) 'Euroregions in comparative perspective', in D. Kaplan and J. Häkli (eds), *Boundaries and Place: European Borderlands in Geographical Context.* Lanham, MD: Rowman & Littlefield, pp. 50–69.

Klemencic, V. and Bufon, M. (1991) 'Geographic problems of frontier regions: the case of the Italo-Yugoslav border landscape', in D. Rumley and J.V. Minghi (eds), *The Geography of Border Landscapes.* London: Routledge, pp. 86–103.

Kramsch, O. and Hooper, B. (2004) 'Introduction', in O. Kramsch and B. Hooper (eds), *Cross-Border Governance in the European Union.* London: Routledge, pp. 1–21.

Kristof, L. (1959) 'The nature of frontiers and boundaries', *Annals of the Association of American Geographers*, 49(3): 269–82.

Le Galès, P. (1998) 'Conclusion – Government and governance of regions: structural weaknesses and new mobilisation', in P. Le Galès and C. Lequesne (eds), *Regions in Europe.* London: Routledge, pp. 239–67.

Low, M.M. (1997) 'Representation unbound: globalization and democracy', in K.R. Cox (ed.), *Spaces of Globalization: Reasserting the Power of the Local.* New York: Guilford, pp. 240–80.

Lundin, I.B. and Söderbaum, F. (2002) 'The construction of cross-border regions in southern Africa: the case of the Maputo corridor', in M. Perkmann and N.L. Sum (eds), *Globalization, Regionalization and Cross-Border Regions.* Basingstoke: Palgrave Macmillan, pp. 241–62.

Mackay, J.R. (1958) 'The interactance hypothesis and boundaries in Canada', *Canadian Geographer*, 11: 1–8.

Mains, S. (2000) 'An anatomy of race and immigration politics in California', *International Journal of Social and Cultural Geography*, 1: 143–54.

Marston, S.A. (2000) 'The social construction of scale', *Progress in Human Geography*, 24: 219–42.

Minghi, J.V. (1963) 'Boundary studies in political geography', *Annals of the Association of American Geographers*, 53: 407–28.

Minghi, J.V. (1991) 'From conflict to harmony in border landscapes', in D. Rumley and J.V. Minghi (eds), *The Geography of Border Landscapes.* London: Routledge, pp. 15–30.

Murphy, A.B. (1996) 'The sovereign state system as political-territorial ideal: historical and contemporary considerations', in T. Biersteker and C. Weber (eds), *State Sovereignty as Social Construct.* Cambridge: Cambridge University Press, pp. 81–120.

Newman, D. and Paasi, A. (1998) 'Fences and neighbours in the postmodern world: boundary narratives in political geography', *Progress in Human Geography*, 22: 186–207.

Nogué, J. (1991) *Els nacionalismes i el territori.* Barcelona: Editorial El Llamp.

Nogué, J. (1998) *Nacionalismo y Territorio.* Lleida: Editorial Milenio.

Nogué, J. and Villanova, J.L. (eds) (1999) *España en Marruecos (1912–1956): Discursos geográficos e intervencion territorial.* Lleida: Editorial Milenio.

Ó Tuathail, G. (1996) *Critical Geopolitics: The Politics of Writing Global Space.* Minneapolis: University of Minnesota Press.

O'Dowd, L. and Wilson, T.M. (1996) 'Frontiers of sovereignty in the new Europe', in L. O'Dowd and T.M. Wilson (eds), *Borders, Nations and States.* Aldershot: Avebury, pp. 1–18.

Paasi, A. (1996) *Territories, Boundaries and Consciousness: The Changing Geographies of the Finnish-Russian Border.* Chichester: Wiley.

Perkmann, M. (1999) 'Building governance institutions across European borders', *Regional Studies*, 33: 657–67.

Perkmann, M. (2002) 'Euroregions. Institutional entrepreneurship in the European Union', in M. Perkmann and N.L. Sum (eds), *Globalization, Regionalization and Cross-Border Regions*. Basingstoke: Palgrave Macmillan, pp. 103–24.

Perkmann, M. and Sum, N.L. (2002) 'Globalization, regionalization and cross-border regions: scales, discourses and governance', in M. Perkmann and N.L. Sum (eds), *Globalization, Regionalization and Cross-Border Regions*. Basingstoke: Palgrave Macmillan, pp. 3–24.

Pettman, J.J. (1996) 'Border crossings/shifting identities: minorities, gender, and the state in international perspective', in M. Shapiro and H.R. Alker (eds), *Challenging Boundaries: Global Flows, Territorial Identities*. Minneapolis: University of Minnesota Press, pp. 261–84.

Raento, P. (2002) 'Integration and division in the Basque borderland', in D. Kaplan and J. Häkli (eds), *Boundaries and Place: European Borderlands in Geographical Context*. Lanham, MD: Rowman & Littlefield, pp. 93–115.

Ruggie, J. (1993) 'Territoriality and beyond: problematizing modernity in international relations', *International Organization*, 47: 139–74.

Sahlins, P. (1989) *Boundaries: The making of France and Spain in the Pyrenees*. Berkeley: University of California Press.

Scott, J.W. (2000) 'Euroregions, governance and transborder cooperation within the EU', in M. van der Velde and H. van Houtum (eds), *Borders, Border Regions and People*. London: Pion, pp. 91–106.

Scott, J.W. (2002) 'On the political economy of cross-border regionalism: regional development and cooperation on the US–Mexican border', in M. Perkmann and N.L. Sum (eds), *Globalization, Regionalization and Cross-Border Regions*. Basingstoke: Palgrave Macmillan, pp. 191–211.

Semple, E.C. (1911) *Influences of Geographic Environment: On the Basis of Ratzel's System of Anthropo-geography*. New York: Holt.

Sidaway, J.D. (2002) 'Signifying boundaries: detours around the Portuguese–Spanish (Algarve/Alentejo–Andalucía) borderlands', *Geopolitics*, 7: 139–64.

Smith, A. (1998) 'The sub-regional level: key battleground for the Structural funds?', in P. Le Galès and C. Lequesne (eds), *Regions in Europe*. London: Routledge, pp. 50–66.

Smith, N. (1995) 'Remaking scale: competition and cooperation in prenational and postnational Europe', in H. Eskelinen and F. Snickars (eds), *Competitive European Peripheries*. Berlin: Springer, pp. 59–74.

Soguk, N. (1996) 'Transnational/transborder bodies: resistance, accommodation, and exile in refugee and migration movements on the US–Mexican border', in M.J. Shapiro and H.R. Alker (eds), *Challenging Boundaries: Global Flows, Territorial Identities*. Minneapolis: University of Minnesota Press, pp. 285–327.

Strüver, A. (2004) ' "We are only allowed to re-act, not to act": Eurocrats' strategies and borderlanders' tactics in a Dutch–German cross-border region', in O. Kramsch and B. Hooper (eds), *Cross-Border Governance in the European Union*. London: Routledge, pp. 25–40.

Sum, N.L. (2002) 'Rearticulation of spatial scales and temporal horizons of a cross-border mode of growth' in M. Perkmann and N.L. Sum (eds), *Globalization, Regionalization and Cross-Border Regions*. Basingstoke: Palgrave Macmillan, pp. 151–75.

Swyngedouw, E. (1992) 'The mammon quest – "glocalisation", interspatial competition and the monetary order: the construction of new scales', in M. Dunford and G. Kafkalas (eds), *Cities and Regions in the New Europe*. London: Belhaven, pp. 39–67.

Taylor, P.J. (1982) 'A materialist framework for political geography', *Transactions of the Institute of British Geography*, NS, 7: 15–34.

Taylor, P.J. (1994) 'The state as container: territoriality in the modern world-system', *Progress in Human Geography*, 18: 151–62.

Taylor, P.J. (1996) 'On the nation-state, the global and social science', *Environment and Planning A*, 28: 1917–28.

van Houtum, H. (1999) 'Internationalisation and mental borders', *Tijdschrift voor Economische en Sociale Geografie*, 90: 329–35.

Ward, S. and Williams, R. (1997) 'From hierarchy to networks? Sub-central government and EU urban environment policy', *Journal of Common Market Studies*, 35: 439–64.

Wilson, T.M. and Donnan, H. (1998) 'Nation, state and identity at international borders', in T.M. Wilson and H. Donnan (eds), *Border Identities: Nation and State at International Frontiers*. Cambridge: Cambridge University Press, pp. 1–30.

29

Transnationalism and (Im)mobility: The Politics of Border Crossings

Rachel Silvey, Elizabeth A. Olson and Yaffa Truleove

INTRODUCTION

The transnational dynamics of socio-spatial mobility and immobility are intrinsically both political and geographic. Yet only recently has research on transnational migration begun to engage explicitly with the central concerns of the disciplinary subfield of political geography. This chapter provides a review of the emerging transnational migration literature as it speaks to central concerns within political geography. In particular, the discussion centers on the geographic migration research that is framed in terms of transnationalism and feminist political geography. The primary goals of tracing this literature are to explore the contributions of critical cross-border migration studies for political geography, and to identify the parallel advances in both migration studies and political geography.

Political geography in the 1980s and 1990s expanded its core conceptualizations of 'the political'. Specifically, while maintaining the subfield's longstanding concerns with the state, electoral politics, and the structures of capitalism, research has placed increasing emphasis on the role of human agency and everyday actions in producing and conceptualizing those structures (Marston, 2000). Studies of transnational mobility have represented a similar shift in emphasis. That is, whereas the nation and national boundaries have figured prominently in conventional international migration research, more recently migrants' creation and maintenance of *trans*national ties have become focal points of research (Glick Schiller et al., 1992). The attention to the roles of migrants

in creating and maintaining 'multi-stranded ties' across national boundaries mirrors political geography's expanding understanding of the scales, subjects, and processes that constitute the political (Nagar et al., 2002).

Research on transnational migration serves as the entry point for this chapter's review of political geography's broadening conceptualization of the political. The chapter traces the efforts of transnational migration research to connect the legal, formal processes of regulation to the embodied, everyday meanings and practices that constitute states (Mountz, 2004). Further, it examines the ways in which transnational migration research questions classical conceptions of scales as unproblematic spatial delimitations, and discusses analyses of the processes through which scales are produced in and through circuits of transnational mobility and immobility.

The chapter begins with a review of the current migration literature and its largely implicit political geographies. It traces the ways in which 'the state' is conceptualized in contemporary international migration work. The remainder of the chapter is organized around three conceptual interventions that characterize political geographies of international migration. First, we discuss emergent approaches to understanding the role of the state. We focus in particular on the ways that transnational migration research can be read to enrich political geography's theorizations of the state. How do the border control strategies of sovereign nation-states intersect with everyday practices of people with different citizenship statuses, those seeking to escape the surveillance of the state, or those actively pressuring states to change their policies vis-à-vis migrants? We outline some current research on international migration that attends not only to formal governance per se but to governmentality more broadly.

Second, we provide a brief review of debates about the politics of scale. We discuss the specific manner in which critical theorizations of migration can inform political-geographic approaches to the production of scale. How is migration connected to specific meanings and implications of, for example, 'the local', 'the body', or the 'global' scales? Third, we review foundational feminist geographic work on spatial entrapment (Hanson and Pratt, 1995). By examining the concept of spatial entrapment in feminist geography, we underscore the continued importance of coercion and dominance (Sparke, 2004) for understanding the power relations associated with mobility and immobility. Overall, by way of tracing existing and forthcoming work on these themes, the chapter contributes an analysis of the conversations linking transnational migration research to advances within political geography.

MIGRATION RESEARCH AND ITS IMPLICIT POLITICAL GEOGRAPHIES

Much of the research on global trends in migration underscores the unparalleled volume and diversity of migrants in the postwar period (Castles and Miller, 1998). In accounts of 'the age of migration', the contemporary period is exceptional in terms of the high numbers of people migrating from an increasing variety of origin sites to also increasing numbers of destinations. The international liberalization of trade and finance has coincided with the simultaneous intensification of the regulation and policing of migration, particularly migration across national borders (Sassen, 1999).

Some have argued that what is most noteworthy is not so much the extent of international migration, but the level of human *immobility* today (Held et al., 1999). Indeed, only about

> 2–3% of the world's people are living outside of the country of their birth – the same proportion as [in] the mid-1960s. In the United States, the proportion of foreign-born is smaller than it was in 1910; in Britain the total foreign population stands at approximately 4%, and half of those are from the European Union or the EEA. (Nagel, 2004: 232)

Despite these numbers, national governments in First World countries have tended to view foreign immigration as a threat to national economies and a problem for national polities. As Nagel (ibid.)

argues, it is this '*perception* of global migration as a problem for the modern nation-state [that] has very real consequences that cannot be ignored'.

Until the 1990s, most research on international migration viewed the state as a set of institutions that set policy, regulate citizenship, and enforce border controls (Massey, 1999). Such foundational approaches have emphasized what states have in common: they exert control over migrants through managing territorial boundaries, categorizing groups of people as citizens and non-citizens, and determining the documents or resources necessary for crossing national territorial borders. States also maintain, in Weberian terms, a 'monopoly on the legitimate use of violence' in their efforts to control migration, and a growing body of research attends to the militarization of borders (Nevins, 2002) and the policing of space (Herbert, 1997).

The state as an actor or a set of actors plays a relatively minor and unproblematic role in international migration work, often analytically subsumed by other features of migration systems that are given more emphasis. Massey (1999: 303–4) and colleagues (Massey et al., 1998), for instance, argue that:

> any satisfactory account of immigration must consider four basic facts … : the structural forces in developing societies that promote emigration; the structural forces in developed societies that attract migrants; the motivations, goals, and aspirations of the actors who respond to these forces by migrating internationally; and the social and economic structures that arise to connect areas of out- and in-migration.

These 'four basic facts', then, are conceived in largely economic terms, whether the broader theoretical framework is world systems theory or the new economics of labor migration. Massey et al.'s goal is to develop parsimonious explanations of migration, and to identify the most powerful explanatory features of migration systems. To this end, they pay the most sustained attention to the pushes and pulls of wage differentials and the structuring effects of these differences on migration. The role of the state is reduced to its role in either promoting out-migration for the sake of national development or else enforcing border control and implementing restrictive immigration policies.

Massey (1999: 309) also examines research that addresses which factors affect the levels of restrictiveness of immigration policy. Still, the analysis of 'the state' in his conceptualization is reduced to government policy. State capacity vis-à-vis migration is measured in terms of the 'strength of the bureaucracy', the 'demand for entry', the 'strength of constitutional protections', the 'independence of the judiciary', and the 'tradition of immigration' (ibid.: 315). In this view, a given state's immigration policy, while certainly determined in part by broad historical and political trends, is nevertheless most crucially a reflection of national economic interests. 'Globalization' in this framework is understood to 'limit the power and influence of nation-states to control the transnational movement of labor as well as capital, goods, and information' (ibid.: 318).

Political geography, given its foci, has delved more deeply than migration scholarship into questions of how to theorize the nation-state, the specific ways in which state power has been transformed rather than simply reduced or enhanced in the post-war era, and what sorts of complex ramifications these shifts have had for capital and labor. Cox has shown that narratives of globalization that focus on highly mobile, footloose capital and increasingly disenfranchised, immobilized labor are exaggerated accounts (Cox, 2002: 337). His work describes how territorial associations are able to mobilize capital and labor in competition with territorial coalitions in other places, so that capital remains concentrated in particular places and labor's power may be partially and temporarily maintained or even enhanced in particular places. Herod (2001) has also emphasized the complex geographies of capital–labor relations. His work places particular importance on the agency of labor in shaping what capital can and cannot do. When labor is organized, workers can develop spatial strategies to affect the direction of the flow of investment. By extension, his work suggests that it is not only the location of investment or wage differentials that deserve attention in migration studies, but also migrants themselves, their spatial strategies, and the level and impact of their political organization. Each of these aspects of migration – from the analytical ground of migrants themselves – is crucial for understanding not only where migrants work and settle, but how they participate in shaping their own political-geographic landscape.

CRITICAL POLITICAL GEOGRAPHY'S APPROACH TO STATES

In contrast to most migration research, critical political geography views the state, and the nation-state's changing role in relation to transnational migration and 'globalization', through a very different set of lenses (Ó Tuathail, 1996). This scholarship, inspired by variations of Foucauldian, Derridean, and feminist theories, has sought to move beyond the 'cult of the state' (Perreault, 2003). Whereas much classical work emphasizes states' manifestations of centralized, sovereign

power (Jessop, 1990), the growing body of critical work highlights governmentality and the dispersion of power (Mountz, 2004). For migration research, this analytical shift encourages greater attention not only to discursive productions of migrants' bodies, national borders, and citizen-subjects, but also to the everyday mediations of exclusion/inclusion by actors involved in these circuits of migration and governance. Rather than imagining a world in which states stayed out of the way of international capital flows, much recent work has demonstrated the ways in which states continue to figure prominently, in many respects increasingly so, in determining the geographies of global migration. Both the 'rolling back' of many states' social service provisions and the 'rolling out' of especially those state practices and regulations that promote corporate investment (Peck and Tickell, 2002) have come together to produce and inhibit particular kinds of migration. States promote the transnational mobility of some groups of people while prohibiting or intensively regulating the border crossings of others.

The practices of states with respect to both capital and labor have been far from 'post-national'. Both sending and receiving states have tended overall to support the 'transnational citizenship' of the globe-trotting managerial elite of major corporations (Mitchell, 2004; Yeoh and Willis, 2004). Many sending and receiving states have aggressively also promoted the transnational migration of female domestic workers and nannies (Huang and Yeoh, 1996) and male low-wage contract laborers (Margold, 1995). The advantages of multiple citizenships are not extended to these low-wage workers, and their marginalization within destination labor markets persists as a result in part of their temporary workers status. At the same time, many states are increasingly prohibiting and regulating mobility across national borders, particularly in high-income destination countries and with rising intensity in the post-9/11 global 'securitization' context (Nevins, 2002). Thus, as Massey (1994: 149) has written: 'Different social groups have distinct relationships to this anyway differentiated mobility: some people are more in charge of it than others; some initiate flows and movement, others don't; some are more on the receiving-end of it than others; some are effectively imprisoned by it'.

Whereas early work on transnationalism focused on the technologies and economic processes that made national financial markets global (Yeung, 1998), transnational migration research has taken these dynamics as a backdrop for examining migrant communities (Yeoh and Willis, 2004). In addition, transnational migration research has focused on migrants and migrant social formations as productive of, rather than simply resulting from, the political constituencies that affect states (Hyndman, 2001). This research has asked how

migrants cross, reaffirm, and rework the borders marking state territorial sovereignty. Marston
(2003: 634) suggests that 'understanding the new
state forms that are emerging under contemporary
globalization requires not just an analysis of the
state, but also an analysis of the ways in which,
for instance, new identities are being constituted
and regulated through and against these new state
forms ...'.

States continue to control substantial resources
and decision-making powers in tandem with a
general movement towards global governance
and the increasing numbers of 'transnational'
activist groups pressuring states to change policies. Migrant rights activists have long been aware
of the continuing importance of the state, even
as they have also begun in recent decades to
organize their struggles around global regulatory
institutions (e.g. UNILO) and international non-
governmental organizations (e.g. Human Rights
Watch). In Indonesia, for instance, migrant rights
activists work to pressure the state in multiple ways.
They argue that the state should better support
and protect Indonesian overseas contract workers
prior to departure, during their sojourns abroad,
and throughout their journeys home. In this view,
Indonesians, regardless of their spatial location,
should be protected as citizens of the nation. In
addition, activists argue that the Indonesian state
needs to be pressured by international organizations to both increase the protections it provides to
migrants and also to reduce the forms of surveillance and control directed specifically at migrants.
In this way, migrant activists have an ambivalent
relationships to the state, seeking to both promote
the effective progressive capacities of state agencies and policies while simultaneously reducing
the power of the state to exploit and marginalize migrants (Silvey, 2004). In other words, these
activists have never presumed that they could
bypass the state. Rather, they have long understood
the importance of pressuring the state to protect citizen's rights both within and beyond the national
territory.

Some research on transnationalism has been criticized for celebrating diaspora and hybridity at
the expense of understanding structural political-
economic constraints and violence sanctioned by
states (Mitchell, 1997). Specifically, early research
on transnational culture and cultural hybridity
(Appadurai, 1996) tended to highlight the ways
in which global flows of media and ideas came
together to create new mixtures of cultural symbols and identities. Studies examined the ways in
which international migrants facilitated these new
cultural flows, connections, and combinations, and
traced migrants' hybrid identities and diasporic
subjectivities (Braziel and Mannur, 2003; Goh
and Wong, 2004). Yet at the same time, research
has begun to pay reinvigorated attention to the

political-economic underpinnings of transnational
cultural flows and meanings and the crucial role of
the state in shaping migration processes and their
cultural entailments (Mitchell, 2004).

States have highly developed 'capacities for
coercion and violence' (Grugel, 2004: 29), but they
also rely on consent in order to achieve their goals.
Low-income transnational migrants are subjected
to both coercive and cultural pressures by both
their sending states and their receiving states. These
states may hold either shared or competing interests, and migrants with minimal resources may
find themselves caught between two states' interests. Saudi Arabia, for example, has repatriated
migrant workers, including many Indonesians, as
part of its campaign to 'Saudi-ize' its labor force,
while Indonesia has refused to extend protections
and services to its own nationals abroad. Migrant
rights advocates have pressured the Indonesian
state to provide better protections to these transna-
tionals and have thus played a hand in transforming the state's migration apparatus (Silvey,
2004).

Despite such evidence of the influence of civil
society in shaping states, states of course retain
the power to selectively engage and respond to
the pressures presented by some migrants more
than others. For instance, the state's approach to
refugee rights has not been radically transformed as
a result of the work of refugee rights advocates. In
the UK:

> the state shapes public policy and popular percep-
> tions independently of, and more effectively than,
> civil society organizations. Migration is successfully
> packaged by the state as an issue area the foun-
> dations of which are the protection of national
> borders, the importance of domestic social peace
> (which immigrants are thought to threaten), respect
> for international law and a minimization of finan-
> cial burdens. No link is allowed between migration
> and human rights; migrants are increasingly pre-
> sented as law-breakers rather than asylum seekers.
> The United Kingdom, like other Western states, is
> able to select how it understands human rights and
> the issues it takes up. (Grugel, 2004: 32)

While transnational migrants may destabilize state
territorial sovereignty, particularly when migrants
are able to claim dual citizenship or participate
politically in the affairs of two nations (Piper and
Uhlin, 2004), states continue to operate as the
arbiters of citizenship and thereby to powerfully
influence the processes by which some groups are
more readily and fully incorporated than others
(He, 2004). Similar to these debates about the state
are debates about the production of scale, discussed
in the next section.

EXPANDING THE SCALES OF ANALYSIS: INCLUDING THE HOUSEHOLD AND THE BODY

Transnational migration research troubles the assumptions about scale that have been central to conventional migration research. Specifically, there are problematic assumptions implicit in most migration studies' renderings of the scales that are thought to affect migration. In looking critically at these assumptions, particularly about the primary salience of the nation and national boundaries (Wimmer and Glick Schiller, 2003), we seek to illustrate the work that scales and spaces do in shaping migration. In particular, we are interested in this section in elaborating the importance of the scales of the household and the body in recent research on migration.

Critical migration research has taken inspiration from Marxian geographic examinations of the political-economic construction of scale (e.g. Smith, 1992). In particular, rather than taking scale as an ontological given, critical work focuses on understanding the power-laden, socially constructed, and gender- and difference-inflected dynamics of spatial scales (Hyndman and Walton-Roberts, 2000; Marston, 2000; Nagar et al., 2002). Social difference takes shape in both the knowledge that is produced about scale and the meanings and operation of scale in practice. As Hyndman and Walton-Roberts (2000) argue, critical analysis of dominant scale discourses 'is important precisely because such assumptions define research questions, shape government policies, and generate common frames of reference' (ibid.: 246).

A gender-neutral conception of the nation is central to most immigration research. Transnational migration research, by definition, focuses on the networks that migrants maintain across national boundaries after making a move across those national borders (Levitt et al., 2003). A growing body of research suggests that the idea of socially produced and specifically gendered space is helpful for *denaturalizing particular scales* and thus for better understanding the grounded materialities and economic ramifications of particular migrations (Devasahayam et al., 2002). Attention to the social and political processes through which, for example, the national scale is produced, reveals the ways that the nation is founded both materially and symbolically on particular inclusions and exclusions. In research on migrant female domestic workers, Yeoh and Huang (1999) show that these migrant women's 'physical invisibility [in public space] signals the lack of a foothold on the metaphorical spaces opened up in recent public discourse on potentially more inclusive notions of citizenship and civil society' (ibid.: 1164). This research illustrates the exclusionary operation of national citizenship ideals in Singapore, and thus also underscores the socially differentiated and gendered meanings of the nation for citizens and temporary migrants (see also Huang and Yeoh, 1996).

Feminist migration research has also reworked classical approaches to the scale of the household (Lawson, 1998). Most fundamentally, these studies have found that economic approaches to the household as an income-pooling unit pay insufficient attention to structural implications of gender- and age-hierarchies within households. More broadly, a critical approach to 'the household' expands the analytical boundaries of the domestic scale to demonstrate the relational dependencies between the organization of domestic labor and the class and racial-ethnic structuring of economies and labor markets. Mattingly (2001), for example, shows how household divisions of reproductive labor are shaped by the racial and gender politics of international networks. As she puts it, in the case of 'immigrant domestic workers and their employers, networks of caring labor interlace the home and the world' (ibid.: 384). The 'household' is thus shot through with the socially segmented meanings of mobility and migrants' interactions with their family networks that are stretched across space, the politics of 'local' and 'global' labor markets, and the international economic inequalities determining who works for whom at what wage.

Critical migration research has also placed analytical importance on the scale of the body. Mountz (2004), for instance, explores the state itself as embodied. Her work asks how attention to the scale of the body can reveal otherwise overlooked efforts on the part of nation-states to mediate and control transnational mobility. Rather than examining the state primarily in terms of government policies and structural agendas, she examines the 'more fluid, daily, personal interactions that surround and disrupt these formal instruments of government'. She aims to locate the operation of power in the everyday activities of civil servants and to 'go inside the state' to 'locate power relations and contextualize decision-making' of state bureaucrats and migrants. In tracing the practices of civil servants in response to a human smuggling 'crisis' in Canada, she is able to provide a careful ethnographic analysis that reveals the grounded practices that constitute 'exactly how state sovereignty is changing' (Flint, 2002: 393, as cited in Mountz, 2004: 18). The power of a particular body to cross into a particular national territory provides a lens onto the 'translation rules' (Smith, 1992: 73–4) linking one scale to another, in this case individuals to border spaces of nation-states.

Because scale gives meaning to migrants' place-based interactions (both analytically and on the ground), it is helpful to examine the metaphorical weight of scalar concepts and language. Critical approaches to the production of scale can contribute to understanding the relationships between scale as a metaphor for social relations and as grounded material practices. Specific constructions of scale tend to be mobilized in the context of already existing power relations. While there is some evidence that 'going global', for instance, may strengthen 'grassroots' political campaigns (Herod and Wright, 2002; Perreault, 2003), more frequently, taken-for-granted notions of scale are deployed in the service of dominant geopolitical interests (Secor, 2001). For example, the prevalent conceptualization of the household as a scale is tied to imagined notions of the co-resident nuclear family as the smallest-scale social institution that makes up the nation (Aitken, 2000: 78).

In contrast to this hegemonic vision of the co-resident household family, migrant domestic workers are frequently required to leave their children behind in origin countries, and thus find themselves involved in mothering transnationally (Hondagneu-Sotelo and Avila, 1997). The globalization of the female domestic migrant labor pool is a function in part of the inequalities between nations, such that migrants from lower-income nations tend to have less choice in terms of where they live and whether they reside with their families. At the same time, these international inequalities are mutually constituted through the gendered and racialized devaluation of domestic labor. The work that takes place within the household is not covered by labor laws in many countries, nor is household labor generally considered 'real work' by many employers. The majority of paid and unpaid domestic labor is carried out by women, and increasingly this work is falling to immigrant women of color working in high-income countries (Hondagneu-Sotelo, 1999). The international division of domestic labor reflects (and indeed reinforces) inequalities between nations, and attention to the ways in which it is produced at the scale of the household can contribute to understanding the multiple scales in and through which global inequalities are forged. Building on these observations about scale, we now turn to a consideration of the classic feminist geographic spatial entrapment hypothesis (Hanson and Pratt, 1995), which we argue provides a valuable lens through which to understand some of the politics of transnational mobility and immobility.

SPATIAL ENTRAPMENT: POWER AND SPACE IN MIGRATION STUDIES

Feminist theoretical work on spatial entrapment both questions and problematizes any blanket assessments of the effects of transnationalism, highlighting instead the need to pay closer attention to the 'particular and concrete specificity of daily experience' (Pratt and Yeoh, 2003: 160). By tracing the origins of spatial entrapment in feminist writing, and current efforts to 'internationalize' understandings of space and social power, we argue that transnationalism has conflicting and contradictory effects on mobility, requiring a more specific and contextualized analysis of the geographic processes at stake.

Hanson and Pratt's (1995) analysis of occupational segregation of women in the northeast United States introduced new tools for understanding how social power relations serve to constrain women's spaces of employment. Specifically, Hanson and Pratt demonstrated that social gender relations (i.e. women's domestic responsibilities and household gender roles) contributed to the spatial entrapment of women, as they sought jobs that enabled them to work close to home and hours that were appropriate to household needs. Hanson and Pratt's analysis provided a platform for conceptualizing how mobility, and hence immobility, locates specific social groups in spaces with more or less social power.

An emerging body of feminist migration research has deepened the conceptual terrain of the spatial entrapment hypothesis to theorize the ways that mobility and immobility can be seen both in terms of movement and location, as well as how bodies are located in hierarchical social spaces. For example, recent scholarship on domestic migrants points to how the physical mobility of a migrant traversing borders is simultaneously coupled with specific forms of spatial immobility as social power relations constrain migrant's rights, choices, and control over their bodies. Pratt's work on Filipina domestic workers in Vancouver depicts this coupling of mobility/immobility by drawing on the particular experiences of migrants in their daily lives (Pratt, 2004). As Pratt conveys the devaluation of women migrant's labor as they become employed as live-in caregivers, the constraints on their bodies while living with their employers, and the lack of social power to articulate rights as citizens, she shows how border crossings are tied to specific and contextualized spaces of immobility.

Wright's work on maquiladora (Mexican 'free trade' export factory) workers in Mexico also couples an analysis of the movement of transnational capital and labor with attention to the simultaneous spatial entrapment of women. Wright details how women's labor in maquiladoras disappears from the commodities being produced, and demonstrates how this 'erasure' has devalued their work and heightened their vulnerability (Wright, 1999). Wright's work illustrates how processes of global economic neoliberalization spatially entrap specific women within power relations across scales and locate them in spaces of increasing vulnerability. She examines the discourses of dis-posability that have emerged in tandem with the high rates of worker turnover, and she links the social construction of these women as 'worthless' with the high rates of rape and murder they have faced. She argues that local people and officials take crimes against these women less seriously because these women's work and the lives are devalued in general in the maquiladora context.

Expanding the theorization of spatial entrap-ment across international scales also illustrates how gendered household divisions of labor in sending countries shape migrants' mobility in host coun-tries. For example, Pratt (2004) illustrates how Filipina migrants' time, bodies, and income are spatially constrained by gendered household rela-tions within their origin countries. As Filipina live-in caregivers in Vancouver call home daily and send higher levels of remittances than male migrants, such gendered roles can entrap Filipina migrants, constraining their time, material assets, and bodies within gendered and racialized power relations across borders and nation-states. Sim-ilarly, Olson's (2006) research has found that women migrants' choices of occupation and mate-rial wealth become constrained by parents' expec-tations that daughters should work in jobs they consider appropriate for women. Instead of acting in response to 'survival strategies', or taking advan-tage of work that might provide increases in spatial mobility, women migrants often work as domes-tic workers constrained in their employers' homes due to gendered expectations reinforced by their parents.

The relationships between spatial and social mobility are complex. Indeed, spatial mobility can be associated with declining social mobility. Immo-bility is equally complex and is not necessarily tied to decreases in social power. Feminist writings on rootedness and domestic spaces demonstrate the ways that immobility can confer power, pleasure, and in some cases social mobility for particular groups of women. Gilbert's (1998) analysis of the survival strategies of poor African-American women captures the complexity of immobility, illustrating how it is simultaneously both constrain-ing and enabling. In particular, Africa-American women's social networks around the home, and rootedness in the homes as a cultural space, depict how immobility can be used to consolidate social power and enable possibility (Gilbert, 1998). Espiritu's (2003) work also acknowledges the gen-dered contradictions of immobility for women, arguing that Filipina women whose husbands migrate away face unique constraints and oppor-tunities. In particular, while the absence of migrant husbands may confer more autonomy and decision-making on women remaining within households, the simultaneous increase in women's workloads and domestic responsibilities may be particularly constraining (Espiritu, 2003).

Such examples point to the need for analyses of transnationalism and mobility to contextualize and specify the embedded social power relations that locate migrants and nonmigrants in changing, and often contradictory, social positions. Exist-ing hierarchies and spatialities of power frame the gendered effects of transnationalism, and spa-tial entrapment continues to provide a theoretical tool useful for understanding and unpacking the particular gendered consequences of mobility and immobility.

CONCLUSIONS: FRAMINGS OF TRANSNATIONAL (IM)MOBILITY

> For us, political geography is about how barri-ers between people and their political commu-nities are put up and come down; how world orders based on different geographical organiz-ing principles (such as empires, state systems, and ideological-materialist relationships) arise and col-lapse; and how material processes and political movements are remaking how we inhabit and imagine the 'world political map'. (Agnew et al., 2003: 2).

Transnational migration research has focused on the interconnections forged by migrants across the boundaries of nations, and the ways in which the spaces of nations and states are perpetuated and reconfigured through the everyday activities of migrants and their communities (Lowe and Lloyd, 1997). Transnationality has provided a broad con-ceptual umbrella that has been adopted across disci-plines to examine the 'relations *between* things and on movements *across* things' (Mitchell, 2004: 74). This theoretical orientation allows 'a reconceptual-ization of core beliefs in migration and geopolitical literatures ..., ... a rethinking of economic cate-gories ..., ... [and] a rethinking in broader areas of epistemological inquiry' (ibid.).

Such an expansive definition risks analytical overreach in the hopes of contributing interpretive power and empirical grounding to geographies of transnationality. We have aimed to draw out exam-ples that clarify the operation of structures and scales that might at first seem separate, but are in fact linked in complex ways. The research reviewed here 'begin[s] to uncover the embedded activi-ties and ramifications of transnational movements in specific, contextually grounded sites' (Mitchell, 2003: 84). The literature on transnational female migrant domestic workers serves as a response and a corrective to the critique that research on transnationalism is too abstract (Portes et al., 1999).

The emerging work on transnationality is attuned to the differences and specificities dividing groups of transnational migrants. It aims to disentangle long-standing assumptions about community and scale, and approaches transnational flows as potentially both altering social boundaries around 'communities' and simultaneously deepening historically sedimented disparities.

By moving beyond the implicit political geographies in much migration scholarship, critical political geography and feminist work on migration is beginning to unravel the multiplicity of scales in and through which politics is enacted and produced. In particular, the work reviewed here has taken seriously the scales of the household and the body as arenas of politics. Expanding our understanding of 'the political' by addressing these underexamined scales contributes to understanding how and why migration takes the gendered forms it does, and thus for understanding the broader political geographies of migration.

REFERENCES

Agnew, J.A., Mitchell, K. et al. (eds) (2003) *A Companion to Political Geography*. Malden, MA: Blackwell.

Aitken, S. (2000) 'Mothers, communities and the scale of difference', *Social and Cultural Geography*, 1(1): pp. 65–82.

Appadurai, A. (1996) *Modernity at Large: Cultural Dimensions of Globalization*. Minneapolis: University of Minnesota Press.

Braziel, J.E. and Mannur, A. (eds) (2003) *Theorizing Diaspora: A Reader*. Malden, MA: Blackwell.

Castles, S. and Miller, M.J. (1998) *The Age of Migration: International Population Movements in the Modern World*. New York: Guilford Press.

Cox, K.R. (2002) *Political Geography: Territory, State and Society*. Oxford and Malden, MA: Blackwell.

Devasahayam, T., Huang, S. and Yeoh, B. (2002) 'Southeast Asian migrant women: navigating borders, negotiating scales', *Singapore Journal of Tropical Geography*, 25(2): 135–40.

Espiritu, Y.L. (2003) *Home Bound: Filipino American Lives Across Cultures*. Berkeley: University of California Press.

Flint, C. (2002) 'Political geography: globalization, metapolitical geographies and everyday life', *Progress in Human Geography*, 26(3): 391–400.

Gilbert, M. (1998) '"Race", space, and power: the survival strategies of working poor women', *Annals of the Association of American Geographers*, 88(4): 595–621.

Glick Schiller, N., Basch, L. and Blanc Szanton, C. (1992) *Towards a Transnational Perspective on Migration: Race, Class, Ethnicity, and Nationalism Reconsidered*. New York: New York Academy of Sciences.

Goh, R.B.H. and Wong, S. (eds) (2004) *Asian Diasporas: Cultures, Identities, Representations*. Hong Kong: Hong Kong University Press.

Grugel, J. (2004) 'State power and transnational activism', in N. Piper and A. Uhlin (eds), *Transnational Activism in Asia: Problems of Power and Democracy*. London and New York: Routledge, pp. 26–42.

Hanson, S. and Pratt, S. (1995) *Gender, Work and Space*. London and New York: Routledge.

He, B. (2004) 'World citizenship and transnational activism', in N. Piper and A. Uhlin (eds), *Transnational Activism in Asia: Problems of Power and Democracy*. London and New York: Routledge, pp. 78–93.

Held, D., McGrew, A. et al. (1999) *Global Transformations: Politics, Economics, and Culture*. Cambridge: Polity Press.

Herbert, S. (1997) *Policing space: territoriality and the Los Angeles Police Department*. Minneapolis: University of Minnesota Press.

Herod, A. (2001) *Labor Geographies: Workers and the Landscapes of Capitalism*. Guilford Press: New York.

Herod, A. and Wright, M. (eds) (2002) *Geographies of Power: Placing Scale*. Malden, MA: Blackwell.

Hondagneu-Sotelo, P. and Avila, E. (1997), ' "I'm here, but I'm there": The Meanings of Latina Transnational Motherhood', *Gender and Society*, 11: 548–71.

Hondagneu-Sotelo, P. (1999) 'Introduction: gender and contemporary US immigration', *American Behavioral Scientist*, 42(4): 565–76.

Huang, S. and Yeoh, B. (1996) 'Ties that bind: state policy and migrant female domestic helpers in Singapore', *Geoforum*, 27: 479–93.

Hyndman, J. (2001) 'Towards a feminist geopolitics', *Canadian Geographer*, 45(2): 210–22.

Hyndman, J. and Walton-Roberts, M. (2000) 'Interrogating borders: a transnational approach to refugee research in Vancouver', *Canadian Geographer/Géographe Canadien*, 44(3): 244–58.

Jessop, B. (1990) *State Theory: Putting the Capitalist State in Its Place*. University Park, PA: Pennsylvania State University Press.

Lawson, V. (1998) 'Hierarchical households and gendered migration: a research agenda', *Progress in Human Geography*, 22(1): 32–53.

Levitt, P., DeWind, J. and Vertovec, S. (2003) 'International perspectives on transnational migration: an introduction', *International Migration Review*, 37(3): 565–76.

Lowe, L. and Lloyd, D. (1997) 'Introduction', in L. Lowe and D. Lloyd (eds), *The Politics of Culture in the Shadow of Capital*. Durham, NC: Duke University Press, pp. 1–32.

Margold, J. (1995) 'Narratives of masculinity and transnational migration: Filipino workers in the Middle East', in A. Ong and M. Peletz (eds), *Bewitching Women, Pious Men: Gender and Body Politics in Southeast Asia*. Berkeley: University of California Press, pp. 323–32.

Marston, S.A. (2000) 'The social construction of scale', *Progress in Human Geography*, 24(2): 219–42.

Marston, S.A. (2003) 'Comment: political geography in question', *Political Geography*, 22(6): 633–36.

Massey, D. (1994) 'A place called home?', in *Space, Place, and Gender*. Minneapolis: University of Minnesota Press, pp. 157–73.

Massey, D.S. (1999) 'International migration at the dawn of the twenty-first century: the role of the state', *Population and Development Review*, 25(2): 303–22.

Massey, D.S., Arango, J., Hugo, G., Kouaouci, A., Pellegrino, A. and Taylor, J.E. (1998) *Worlds in Motion: Understanding International Migration at the End of the Millenium.* Oxford: Oxford University Press.

Mattingly, D.J. (2001) 'The home and the world: domestic service and international networks of caring labor', *Annals of the Association of American Geographers*, 91(2): 370–86.

Mitchell, K. (1997) 'Transnational discourse: bringing geography back in', *Antipode*, 29(2): 101–14.

Mitchell, K. (2003) 'Cultural geographies of transnationality', in K. Anderson, M. Domosh, S. Pile and N.J. Thrift (eds), *Handbook of Cultural Geography.* London: Sage.

Mitchell, K. (2004) *Crossing the Neoliberal Line: Pacific Rim Migration and the Metropolis.* Philadelphia: Temple University Press.

Mountz, A. (2004) 'Embodying the nation-state: Canada's response to human smuggling', *Political Geography*, 23(3): 323–45.

Nagar, R., Lawson, V., McDowell, L. and Hanson, S. (2002) 'Locating globalization: feminist (re)readings of the subjects and spaces of globalization', *Economic Geography*, 78(3): 257–84.

Nagel, C. (2004) 'Questioning citizenship in an "age of migration"', in J. O'Loughlin, L.A. Staeheli and E. Greenberg (eds), *Globalization and Its Outcomes.* New York and London: Guilford Press, pp. 231–52.

Nevins, J. (2002) *Operation Gatekeeper: The Rise of the 'Illegal Alien' and the Remaking of the US–Mexico Boundary.* New York: Routledge.

Ó Tuathail, G. (G. Toal) (1996) *Critical Geopolitics: The Politics of Writing Global Space.* Minneapolis: University of Minnesota Press (Vol. 6 in the Borderlines Series).

Olson, E. (2006) '*Development, transnational religion, and the power of ideas in the High Provinces of Cusco, Peru*', *Environment and Planning A*, 38(3): 885–902.

Peck, J. and Tickell, A. (2002) 'Neoliberalizing space', *Antipode*, 34: 380–404.

Perreault, T. (2003) 'Changing places: transnational networks, ethnic politics, and community development in the Ecuadorian Amazon', *Political Geography*, 22(1): 61–88.

Piper, N. and Uhlin, A. (2004) 'New perspectives on transnational activism', in N. Piper and A. Uhlin (eds), *Transnational Activism in Asia: Problems of Power and Democracy.* London and New York: Routledge, pp. 1–25.

Portes, A., Guarnizo, L. and Landolt, P. (1999) 'Introduction: pitfalls and promise of an emergent research field', *Ethnic and Racial Studies*, 22(2): 217–37.

Pratt, G. (2004) *Working Feminism.* Philadelphia: Temple University Press.

Pratt, G. and Yeoh, B. (2003) 'Transnational (counter) topographies', *Gender, Place and Culture*, 10(2): 159–66.

Sassen, S. (1999) *Guests and Aliens.* New York: New Press, distributed by W.W. Norton.

Secor, A.J. (2001) 'Toward a feminist counter-geopolitics: gender, space and Islamist politics in Istanbul', *Space and Polity*, 5(3): 191–211.

Silvey, R. (2004) 'Transnational domestication: Indonesian domestic workers in Saudi Arabia', *Political Geography*, 23(3): 245–64.

Smith, N. (1992) 'Geography, difference, and the politics of scale', in J. Doherty, E. Graham and M. Malek (eds), *Postmodernism and the Social Sciences.* London: Macmillan, pp. 57–79.

Sparke, M. (2004) 'Political geographies of globalization (1): dominance', *Progress in Human Geography*, 28(3): 777–94.

Wimmer, A. and Glick Schiller, N. (2003) 'Methodological nationalism, the social sciences, and the study of migration: an essay in historical epistemology', *International Migration Review*, special issue in 'Transnational Migration: International Perspectives', XXXVII(3): 576–610.

Wright, M.W. (1999) 'The dialectics of still life: murder, women, and maquiladoras', *Public Culture*, 11(3): 453–73.

Yeoh, B. and Huang, S. (1999) 'Spaces at the margins: migrant domestic workers and the development of civil society in Singapore', *Environment and Planning A*, 31(7): 1149–67.

Yeoh, B. and Willis, K. (2004) 'Constructing masculinities in transnational space: Singapore men on the regional beat', in P. Jackson, P. Crang and C. Dwyer (eds), *Transnational Spaces.* London and New York: Routledge, pp. 147–63.

Yeung, H.W.-C. (1998) *Transnational Corporations and Business Networks: Hong Kong Firms in the ASEAN Region.* London and New York: Routledge.

Spatial Analysis of Civil War Violence

John O'Loughlin and Clionadh A. Raleigh

INTRODUCTION

At the beginning of the twenty-first-century, 19 wars continue; they have caused displacement of over 20 million refugees and migrants (Moore and Shellman, 2004; SIPRI, 2004). Some conflicts are enormous in geographic scope, involving numerous countries, and many are very destructive. The wars of the Democratic Republic of the Congo between 1999 and 2004 resulted in over 3 million lives lost. Furthermore, the negative effects of conflicts persist long after the end of the fighting. Wars increase the effects of infectious diseases, such as AIDS, through shifting of government expenditures from basic needs to the military and to destruction of health-care infrastructure and traditional family coping mechanisms. Overall, the World Health Organization estimates that 269,000 deaths and 8.44 million lost DALYs (disability-adjusted life years) were caused in 1999 by all wars, civil and international, and another 8.01 million DALYs were lost in 1999 as an indirect effect of civil wars in the 1991–7 period (World Health Organization, 2000; Ghobarah et al., 2003, 2004b; SIPRI, 2004). The focus of this chapter is to understand the distribution of these civil wars across the globe and to indicate some significant gaps in the research on the geography of violent domestic conflicts. We also identify what we see as promising avenues of research that link political geographic approaches to the much larger accumulation of research in political science and economics on the causes and frequency of civil wars.

This chapter sits in the spatial-analytical tradition of human geography that examines geographically expressed phenomena for patterns and general trends. It thus stands in sharp contrast to the regional tradition that dominated geography before the 1960s, and which is now updated to incorporate attention to linkages across scales (local, national, and global). Our approach also stands apart from the critical tradition, inspired by Marxist, feminist and post-structuralist theories (see Johnston and Sidaway, 2004). What the spatial-analytical tradition shares with the 'new regional geography' is a concern with the special importance of context, the milieu in which political, economic, and social processes that shape violence occur. Thus, it is distinguishable from political science and economics research, which typically does not pay a lot of attention to specific places, though comparativists by their very nature insist on such details (Herbst, 2000; Kalyvas, 2006).

We do not claim that wars are the only expression of violence. Indeed, in most years, more people are killed through interpersonal violence than in formal international conflicts or in civil wars. Furthermore, while we focus on wars, we are sympathetic to the argument that 'structural violence', perpetuated by poverty and lack of access to proper nutrition, clean water, and health care, causes far more deaths than direct violence. A comparison of the scale of these two types of violence showed that premature deaths (measured by years of life compared to the median values of rich countries) far exceeded war casualties (Johnston et al., 1987). The examination of war violence is further complicated by definitional issues since gang and other organized violence often has a political objective, while, conversely, the violent actions of political groups often have monetary or personal motives. In this chapter, we use the commonly accepted definition of war as a militarized dispute that results in civilian and/or military deaths; see the discussion in Jones et al. (1996).

Since the end of the Cold War in 1991, the most dramatic development in the nature of conflict is the shift from international to civil wars. During the 1945–1990 period, about five civil conflicts occurred per each interstate war (O'Loughlin, 2004), but by 2003 the ratio had increased to 8.5 to 1 (SIPRI, 2004). An upsurge in civil wars in the early 1990s, especially on the territory of the former Soviet Union, was made possible by the removal of the brake of a strong central state on ethnic-oriented violence. However, the preexisting conditions for violence were already in place and most recent civil conflicts can be traced to pre-1991 origins and mobilizations of political movements. To attribute the apparent rise in civil war violence to the end of the Cold War dampening of autonomous movements is, thus, too simplistic and forgets the activities of the two superpowers in stirring up oppositional movements in the spheres of the opponent (O'Loughlin, 1989).

What role does a geographic perspective on war and organized violence have in the academic division of labor? As we hope to make clear in this chapter, the dominant political and economic perspectives are flawed by their narrow vision of what constitutes the role of geographic factors in war. Further, these approaches consistently are unable to see how the particular context of war affects the propensity of groups to engage in violence, to fund their continued fight, to set the terms of possible agreements to end the fighting, and to accommodate refugees and others forcibly removed from homes and community. In particular, we will argue for a continued focus on disaggregation of the measures of conflict and widening and deepening of the kinds of information typically available to researchers. We stress that geography is not only about 'space' (typically measured as some kind of contiguous connection between countries at war and their neighbors) but is also about 'place' (the unique combination of circumstances for each region that produces the cultural-political mosaic across the world's continents) and 'relations between scales' (the links between local, regional, national, and global developments). In doing so, we argue for perspectives that are quite alien to most war researchers but comfortable for most geographers (Johnston, 1991; O'Loughlin, 2001).

In this chapter, we describe how political geographers have recently approached analysis of war and conflict, how geographical factors are currently being considered in conflict studies, and in what ways a spatial-analytical perspective can add to these studies. We stress that our spatial-analytical approach is only one of many possible ways of understanding civil wars, although, obviously, we believe that spatial analysis offers the most promising of avenues, especially because it connects the discipline of geography to the main strands of social scientific work on the causes and consequences of domestic conflicts. Major projects have brought together World Bank interdisciplinary teams of economists and political scientists (Collier et al., 2003). Though the aggregate elements are submerged in this World Bank study that examines the economics and politics of wars, it does not take much effort to reconsider the main conclusions in light of the evident relevance of regional and local contexts.

Geographical analysis can be undertaken in a number of ways. We present an example of the simplest geographic method, that of basic cartographic presentation, to illustrate the cross-border nature and geographic transitions in the long-running Ugandan civil wars. We then consider how more sophisticated quantitative spatial analysis can add to established views about the long-term effects on health and life expectancies.

GEOGRAPHERS AND THE STUDY OF WAR

The discipline of geography has a long and checkered legacy in its relationship with war and violence. The discipline's modern origins in the promotion of colonialism and imperialism and the discovery and exploitation of resources in the mid-nineteenth-century are well documented (O'Loughlin and Heske, 1991; Mamadouh, 2004). In the twentieth-century, geographers have moved away from their earlier practice of the art and science of geopolitics to an increasingly critical view of the military and the conduct of war. Careful mapping of the depredations of military actions such as the bombings of the dikes in North Vietnam by the US air force (Lacoste, 1976) and the allied destruction of German cities in World War II (Hewitt, 1983) are dramatic examples of the power of the simplest and most accessible of spatial methods, that of cartographic display. At a larger scale, similar mappings of the locations of contemporary conflicts clearly demonstrate the variable distribution of war violence across the globe (Buhaug and Gates, 2002; O'Loughlin, 2004) and recall the longstanding distinction between stable zones of peace and zones of war (Boulding, 1978; Gleditsch, 2002).

The interesting question, of course, is why do some regions remain peaceful for the long haul while others oscillate between episodes of war? The answer is complex and depends in part on the dodgy distinction between civil and international wars. While classifying wars by type is somewhat subjective, it is evident that there has been a significant shift in conflict to poor parts of the globe since 1945 (O'Loughlin and van der Wusten, 1993). While the 'democratic peace' hypothesis (that democratic states do not fight each other) has received a lot of attention in political science, the empirical data show that democratic states, especially France, the UK, and the United States, have been heavily involved in wars of decolonization, democracy promotion, and (in the 1990s) against tyrannical regimes in Iraq, Afghanistan, and former Yugoslavia.

As well as the descriptive accounts of the reasons why wars are geographically concentrated in certain world regions, geographers have turned to specialized methods developed for the analysis of spatially distributed phenomena to try to understand and model the conflicts. In doing so, geographers are getting closer to the kinds of approaches that are dominant in the ancillary disciplines of political science and economics with their emphasis on large-N studies and regression-type explanations. The spatial-analytic approach argues that the location of a country relative to other countries (neighboring, near, distant, remote, connected by transport links, sharing minorities across borders, etc.) is an important consideration in understanding the distribution and diffusion of conflict. Political science work, typically, does not pay much attention to these connections, rather seeing countries somewhat like 'atoms' floating in space (Agnew, 1994).

One of the most important predictors of whether a country will experience a civil war is whether its neighbors have internal strife, the so-called spatial lag effect. Ignoring this factor is to cast aside one of the most obvious explanations of conflict (Diehl, 1991). In statistical analyses, this contagion variable is often as important as the usual political and economic predictors, such as the autocratic nature of the government, the ethnic fractionalization of the population, the level of poverty, income inequality, and years since independence. Further, countries are frequently involved in a complex nexus of conflict as wars ebb and flow across borders due to the sanctuaries offered by neighbors, the placement of rebels in areas occupied by sympathetic minorities, the actions of refugees to undermine the forces that caused their flight, and the access of extraterritorial bases to external supporters and sources of income (Väyrynen, 1984; O'Loughlin and Anselin, 1991). Recent examples of these formations include the Great Lakes area of Africa (Democratic Republic of Congo, Uganda, Rwanda, Burundi, Tanzania), West Africa (Sierra Leone, Liberia, Guinea, Senegal), the Middle East (Israel, Lebanon, Syria, Egypt, Jordan, and Palestine), and parts of former Yugoslavia (Croatia, Serbia, Bosnia-Herzegovina, Kosovo, Macedonia).

Two further aspects of the geography of civil war have occupied the attention of geographers. The military analysis of wars tends to be split between large-scale (strategic) and small-scale battlefield (tactical) considerations. The strategic analyses devolve into geopolitical deliberations while the tactical analyses tend to have a strong physical geographic component (O'Sullivan and Miller, 1983; Corson and Palka, 2004; Palka and Galgano, 2005). As the battlefield technology has become very sophisticated, it has become integrated with geographic databases that contain details and images of physical terrain, urban environments, and even the social characteristics of local inhabitants. Enormous amounts of information, much of it gathered through intelligence satellites for high-resolution imagery, are now available for waging and monitoring war and evaluating its effects, though it is disproportionately available only to rich Western countries.

Wars leave a lasting legacy on both the landscapes and the people. When territory changes hands, a kind of landscape erasure often follows. Changing street and place names, erecting monuments to the victors, destroying religious and other cultural monuments of the defeated, converting establishments to new use, and implementing educational and other changes to reflect the hegemonic ideas of the victorious side are well documented by historical and cultural geographers. Samples of this kind of work are Heffernan (1998), Johnson (1999), Charlesworth (2003), Weeks (2006), and Winston (2006). Much of this research has focused on Europe, since it is axiomatic among cultural geographers that wars and their legacy are instrumental in making and remaking national identity. Nationalist ideologues memorialize selectively; exclusive claims to certain territories are made based on historical linkages and settlements in those places (Azaryahu, 1996). Winning a war also means winning a landscape that can be remade in the image of the victor (Murphy, 2004).

MAJOR THEMES IN CIVIL WAR RESEARCH

Because of its frequent occurrence (all but two of current wars are civil conflicts), we chose to focus this chapter on violence caused by domestic circumstances and cross-border linkages that perpetuate conflict. The difference in the geographic and political science approaches can be seen in the typical civil war studies of each field. While civil war is ultimately created by the interplay of domestic structures and domestic contexts, geographers contend that the effects (identified by political scientists) of domestic structures (GDP, government type, ethnic makeup, etc.) and domestic contexts (population growth, terrain, weak state institutions, resources, etc.) are shaped by the nature of the regional context.

Current *en vogue* explanations point to state strength as determining the propensity for civil war onset. The massive State Failure (now Political Instability) Task Force project at the University of Maryland, funded by the US Central Intelligence Agency (CIA), has identified weak states as prime factors in civil war causation, durability, and reoccurrence (details accessible, at time of writing, at http://globalpolicy.gmu.edu/pitf/). The number of state failures peaked in the early 1990s with about 30 percent of all countries 'in failure'. This peak coincided with the high point of ethnic wars (over 20 per year) and followed a surge of genocides and politicides (mass killings of political groups) a decade earlier (Goldstone et al., 2003). Disputed sovereignty and an inability to foster a coherent national identity conspire to keep weak states vulnerable to putsches and loss of territory. The spatial clustering of weak states, and subsequent clustering of conflict in weak states, allows for conflict to penetrate borders, infecting already vulnerable states. Therefore, the geographic position of a state is not simply an attribute, but another potential cause, of conflict. High-risk countries are subject to increased risk because (1) civil wars exacerbate volatile domestic conditions inside bordering states by forcing a reevaluation of military spending, and (2) neighboring wars can (and frequently do) spread into nearby states, through the actions of government and opposition fighters, refugees, and cross-border supporters. Weak states cannot effectively mitigate conflict diffusion and escalation from outside state borders (Raleigh, 2004).

The literature on civil war has a long legacy and is characterized by an approach that is best described as piecemeal. For example, there is a considerable literature that separately examines the onset, escalation, and termination of civil wars. (See the extensive and annotated bibliography in Collier et al., 2003.) Most of the literature has also looked at civil wars as self-contained and homogeneous phenomena, ignoring the external connections of civil wars. As a result, almost all the existing data on civil war are collected and organized at the country level (Strand et al., 2004).

The question of whether there is a larger (regional or global) or smaller (local) scale in which the wars are embedded has, heretofore, largely been ignored. In briefly reviewing these studies, we illustrate some of the shortcomings that result from geographical aggregation of local processes to national attributes in existing cross-national studies. We also suggest how studying the processes across scales can contribute to our understandings of the dynamics and consequences of civil war.

Two key variables, often summarized as 'creed' (ethnicity) and 'greed' (income-generating resources), dominate the analysis of civil wars. That increasing national income lessens the risk of civil war onset remains undisputed (Collier et al., 2003; Fearon and Laitin, 2003). But how does poverty increase the odds of civil war? Higher rates of per capita income increase a government's ability to retain control of the state apparatus by redistribution of state tax revenues. In countries with lower GDPs, by contrast, securing power and maintaining civil order become government priorities. The ratio of government monies on military spending typically doubles during conflict (Collier et al., 2003), producing a downward cycle of conflict as social expenditures (education, health, social welfare) are cut, and income further declines. Elbadawi and Sambanis (2002: 309) find that conflict is 'disruptive to capital or transaction-intensive activities (such as roads, production of manufactures, or financial services); it can divert expenditure and the society's resources from economic services (growth enhancing activities) to war efforts; and it can divert portfolios from domestic investment into capital flight'. Fearon and Laitin (2003: 80) conclude that higher income is associated with a more developed infrastructure, and, therefore, better control of the state apparatus and population. Collier and Hoeffler (2004) note that neighboring war has a considerably larger impact on a bordering country's GDP than on domestic GDP (presumably, domestic GDP is very low to start with). In later work, Collier et al. (2004) assert that low per capita income, high income inequality, and a moderate degree of ethnic division lengthen conflict, whilst a decline in primary commodity exports shortens it because the state loses critical revenue for its war-making capabilities. Much of the recent work on civil war has emphasized the role of private incentives and rent-seeking activities as predictors of civil war onset. Individuals are more likely to take up arms when they can benefit materially from war through looting, extracting valuable commodities, and extortion (Collier, 2000; Le Billon, 2001; Murshed, 2002; Mueller, 2003; Collier et al., 2004).

An intersecting paradigm, the effects of ethnic composition and a weak state, underpins the role of political factors in civil war. The weak state literature focuses on the legitimacy and sovereignty of the state as contributing to the outbreak of war (Holsti, 1996; Herbst, 2000). Governance explanations (about autocracies, democracies, and anocracies) looks for associations between political structure and rebellion (Hegre et al., 2001). If states and governments have a complex of control, either through the validation of the government as the voice of the state in a democracy or a well-instituted infrastructure of fear and domination in an autocracy, the probability of civil war is lowered.

Ethnic diversity as a cause of conflict is based on the assumption that increased fractionalization makes it difficult to create a unified national community, because of alternative competing allegiances. Power relations are not assumed to be equal in fractionalized societies. Ethnicity's relationship to conflict is quite variable, as noted by a number of studies finding diversity linked to conflict (Connor, 1973; Horowitz, 1985; Fox, 2004), diversity not significantly linked to conflict (Collier and Hoeffler, 1998; Fearon and Laitin, 2003), diversity lessening conflict (Collier et al., 2003), ethnic dominance exacerbating conflict (Gates, 2002; Collier et al., 2003), religious affiliation causing conflict (Huntington, 1996; Fox, 2004), and ethnic elites acting as catalysts for conflict (Lake and Rothschild, 1998; Brown, 2001). Ethnic fragmentation's relationship to conflict proliferation is thus quite varied (Horowitz, 1985; Gurr, 1993).

Recent studies with more nuanced analysis of the connection of poverty and ethnic diversity do not simply show increased risk of conflict. Poverty alone cannot provide a reasonable explanation of why groups resort to violent conflict since poverty is ubiquitous, but only some countries see violence. The key might be the relative status of ethnic groups, since relative deprivation and economic inequality are common elements in poor societies that experience civil war (Gurr, 1970). Ethnic dominance (45–90 percent of the population) can increase the risk of rebellion. This argument is based on the notion that the effects of ethnic diversity depend on the opportunities for profiting from primary commodity exports and taxing opportunities (Collier and Hoeffler, 2004). A dominant group also has a considerable number of potential recruits and, hence, improves the chance of retaining control of the state apparatus. Ethnic ties, strengthened by perceived ethnic grievances, lessen the costs of recruiting and sustaining a fighting force because perceived benefits are shared throughout the ethnic group.

This economic perspective on the nature of ethnic wars has detractors who regard conflict as the outcome of the interplay of ethnic diversity and economic grievance. Grievances (based on the distribution of resources) and ethnic identification as a basis for rebellion are grounded in the impression that modernity does not lead to more democracy (Ellingson, 2000: 237). Elbadawi and Sambanis (2002) find similar results to Collier and Hoeffler (2004) – economic and political underdevelopment are the root causes of conflict (specifically in Africa) – but they also find that ethnic fragmentation may lead to poor economic prospects through the implementation of bad economic policies.

Recent studies showed that geographically disaggregated conflict-specific measures of resources yield much better predictions of civil war duration than national-level data (Buhaug and Lujala, 2005; Buhaug and Rod, 2006). Many studies of civil war perpetuate the mismatch between the national level at which data are collected and the regional and local elements of the actual conflict. Buhaug and Gates (2002), however, show that the geographic location of a civil war *within* a particular country is fundamental for understanding conflict dynamics. Civil wars that develop in the periphery of countries tend to last much longer than those occurring close to national capitals, for example. Governmental capabilities are typically not homogeneous, but neither are they geographically fungible. Extensive state power may be present in some locales, but virtually absent a few kilometers away, especially in weak or failing states.

GEOGRAPHIC ELEMENTS AND CONFLICT

Work that examines the diffusion of conflict and the locational attributes of civil war has primarily focused on three different themes: absolute location of wars, relative location of wars, and territory as a 'container' of salient factors. Absolute location perspectives contend that civil-war-prone states disproportionately occupy the periphery of the world economy. Decolonization, superpower proxy wars, and impoverished conditions have created an environment of endemic poverty, poor governance, and a fundamental disjoint of state ideology and nation, which in turn has fostered discontent and violent conflict (O'Loughlin, 1989). Strategic geopolitics – fostered by resource and strategic location considerations – has continued to create 'shatterbelt' regions well past the end of the Cold War (Klare, 2001; Cohen, 2003). Shatterbelt regions, such as the Caucasus, are defined as areas with a globally significant natural resource, ethnic diversity, external intervention, and a history of local conflict.

Relative location work focuses on the position and process of both the state and the internal conflict. Similar work on interstate conflict stresses the relationships of alliances and borders as explanations of conflict diffusion and proliferation (Siverson and Starr, 1991; Starr and Thomas, 2002). Countering the atomistic nature of the usual type of study, Ward and Gleditsch (2002), Sambanis (2001), and Salehyan and Gleditsch (2004) show that regional conditions in neighboring communities also influence the initiation of civil violence. Challenges to central rule are assumed to come from distinctive areas that have been ignored in weakened or failed states (Herbst, 2000); further, conflict location relative to state capitals, borders, communities, and resources often provides tacit explanations of intent and positionality (Buhaug and Gates, 2002; Buhaug and Lujala, 2005). The involvement of outside players in a civil war has recently prompted speculations of 'aggressive symbiosis' (Le Billon, 2001), where the conflict has become beneficial to certain criminal elements of war-torn societies. The creation of spaces and networks of illegal activity is redrawing political boundaries and overtaking governance in failed states, presenting local, national, and possibly global risks (Keen, 1998; Ó Tuathail, 2000).

The external diffusion of conflict, or the escalation of parallel conflicts, is rooted in the questionable legitimacy of many state borders and the frequent inability to control territories and the people within them. Diffusion and escalation are important and understudied features of civil war because entire regions can escalate into a series of civil wars feeding off each other (e.g. West Africa in the late 1990s). The salience of borders in civil war study highlights the mismatch of nation and state territories (Englebert et al., 2002).

'Geography as container' work focuses on salient features of the environment (human or physical) that may be associated with war in a particular territory. Among key variables are measures of minorities at risk, per capita income trends, environmental stress that might be partly due to climatic change, income inequality, urbanization, and population migrations. Incorporating such measures poses additional challenges because of the paucity of data that continues to be a major stumbling block to such analyses. The linkage of the environment and security began in the 1980s, and case studies, focusing particularly on scarcity of resources (water, forests, fish), and violent domestic and international conflict, are able to construct a causal pathway (Homer-Dixon, 1999). The environmental perspective is dominated by resource arguments, which contends that features of particular resources, especially the use of easily lootable resources by rebels as funds for their cause, make conflict more feasible.

Research on resource scarcity as conflict-encouraging (Homer-Dixon, 1999) has been challenged by a growing body of research focusing on the proliferation of conflict in resource-abundant areas (de Soysa, 2000; Auty, 2004). Natural resource endowment is linked to poor economic growth and governance, since 'resource rents provide political leaders with a classic means of staying in power by establishing a regime organized through a system of patronage' (Le Billon, 2001: 567). Research details the dual role of resources; the revenue from precious materials that supports corrupt governments also provides conflict incentives and a much-needed source of income for rebel groups (Collier, 2000). While oil has received most attention as a war-inducing resource (Collier, 2000; Fearon and Laitin, 2003; Le Billon, 2004), other resources including diamonds (West and Central Africa), timber (Cambodia), minerals (Congo), and drugs (Colombia, Afghanistan) have been implicated in civil war proliferation (Auty, 2004; Olsson

and Fors, 2004). The spatial dispersion of resources (diffused throughout the state or in certain point locations) has been incorporated into theories of relative conflict location (Le Billon, 2001; Buhaug and Gates, 2002; Ross, 2004). In general, resources and civil war are related by a variety of mechanisms and we need to separate them by group and regional context before we can conclude that the 'resource curse' hypothesis (resources increase the risk of civil war) is supported.

Recently, political scientists have made tentative attempts to consider the influence of physical geographic factors in civil wars. De Rouen and Sobek (2004) conclude that 'borders, war type, Africa, UN intervention, forest cover, and mountain cover' all help to determine civil war outcomes but that the effects are varied and contradictory. Their research indicates that forest cover helps the government cause, whilst mountainous terrain helps rebels. Whereas Fearon and Laitin (2003) find evidence for the influence of terrain in conflict onset, Collier and Hoeffler (2004) and Buhaug and Gates (2002) dismiss this argument, preferring an economic explanation as incentive to rebellion. Nevertheless, terrain is considered a contributing factor in conflict proliferation (Collier et al., 2003). Studies detailing the role of terrain are useful, but inconclusive, partly because of uncertainties in definition and measurement of rough terrain. More careful mapping of insurgencies and rebel movements is a way to understand this correlation.

Regional conflicts are, by definition, a mixture of intra-national, intra-regional, and extra-regional conflicts. Considering that most conflict is currently intrastate and regional conglomerations of weak states are at internal risk because of conflict diffusion, international interests exacerbate tensions and power relationships inside regions, resulting in shatterbelt-like scenarios. Of course, the opposite is also true. Systems that experience a great degree of stability are at a lesser risk of internal conflict because of the stability of the larger region. The European Union is an example of such a region where stable democracy, favorable economic environment, and a location in the core of the world economy allow for change through peaceful measures.

Most analyses of civil war data reviewed above proceed with an implicit assumption that all the data are generated by a random process that results in the data being independently and identically distributed across the globe. Spatial analysts, on the other hand, offer two different strategies to place political actions in their regional contexts (O'Loughlin, 2003). On the one hand, we advocate a SISS (spatially integrated social science), which views space as integrating social processes and sees social science dynamics as processes in place (Griffith and Layne, 1999; Goodchild et al., 2000). This approach uses GIS (geographic information systems) to integrate data by geo-referenced location and applies spatial statistical analysis to integrate multidisciplinary approaches. Reviewing the status of this perspective, Goodchild et al. (2000: 139) conclude that 'in the mainstream of the social sciences, attention to the spatial (and space-time) dimension of phenomena is much less apparent (compared to geography), although a revival of sorts is occurring'.

The second stream of the spatial analysis approach revisits the decades-old notion of the 'ecological triad' (social entity or actor, environment, and entity/environment relationship) from the Sprouts (Sprout and Sprout, 1965). Here the emphasis is on the place of politics, the context in which political actions happen. It dovetails well with the traditions of political ecology and pays special attention to local cultural and material traditions (Robbins, 2004). Careful consideration of contextual elements can disabuse us of ideas that certain global regions, such as the Middle East, are intrinsically 'conflict-prone'. As Sorli et al. (2005) show, economic growth and development, ethnic dominance, and regime type explain the distribution of conflict in the Middle East (but not oil or Islam); thus, a general model fits this region as well as the global case and there is no reason to resort to (regional) exceptionalist explanations. Examples of both of these approaches are now presented.

SPATIAL ANALYSIS OF CIVIL WARS

In this section, we present two analyses that illustrate the spatial-analytical approach to the study of civil wars whilst also making the case for more disaggregated data to answer key remaining questions. By taking the standard political science approach and giving it a spatial-analytical twist, we show how the study of civil wars is enriched. The first example is both a demonstration of a case study, that for Uganda, and also an analysis that shows how civil wars become international wars because of the porous nature of borders and the alliances that are built across them by governments and rebels. The other example demonstrates the use of modern spatial analysis at a global scale and illustrates some pertinent features of civil wars that are not evident in the usual studies of economists and political scientists.

In spatial analysis, research usually proceeds from ESDA (exploratory spatial data analysis) techniques and cartographic exploration with attribute and locational data (latitude/longitude) organized in a GIS. Within the past decade, there has been a welcome integration of cartographic display and spatial statistical analysis within the same software packages. While cartographic display can offer some possible hypotheses about the geographic association of the variable of interest (the location of civil wars), we must be careful to also consider the other map layers that display the predictors, such as income, ethnicity, political development, so on and so forth. Until these effects are filtered out, the simple statistic of clustering or randomness from a point pattern analysis cannot be conclusive. In human geography, simple extrapolations cannot be supported because of the complexity of human spatial processes.

Uganda and Its Neighbors: The Evolution of an Extended Conflict

The Ugandan test case illustrates the application of geographic theory and methods to the mechanisms involved in the reproduction of conflict over time. Uganda is a major player in the African Great Lakes conflict formation over the last four decades, as it has experienced civil strife since its independence in 1962. Its conflicts remain relatively understudied in comparison to its neighbors, Rwanda, the Democratic Republic of Congo (DRC, formerly Zaïre), or Sudan. However, the various Ugandan rebellions illustrate a number of salient factors important to understanding civil conflict in the developing context. First, Uganda's colonial history set the stage for the present-day north–south divide within the political institutions and the military; second, these conflicts demonstrate the shifting relations and intersections between internal rebel organizations and regional ethnic groups; third, Uganda's ethnic geography has been a primary factor in development and hence motivations for conflict; and finally, Ugandan foreign relations highlight the role that external actors play in supporting or suppressing conflict across borders, and the resource that neighboring failed and weak states are to rebel groups.

The geography of Ugandan conflicts expresses these aforementioned factors. Various provinces of Uganda have been involved in rebel activity during the different rebellions. As wars diffused and ended, and as rebel camps mobilized, regrouped, retreated, or were victorious in gaining control of the capital, Kampala, the map of war changed repeatedly (see Raleigh and Hegre, 2005, for an illustration). A brief overview of major Ugandan conflicts demonstrates how the political geography of Uganda has shaped the form and focus of rebellions against the state.

Uganda's colonial and postcolonial situation perpetrated an environment wherein ethnicity determined one's access to power within the state. The northern Uganda regions were kept underdeveloped by the British, who filled military ranks with northern soldiers; the south was more developed and hence considered a threat to the colonizing powers. At independence, this divide continued when a rebellion in the powerful southern region of Buganda was suppressed by the northern (Langi) Milton Obote government. In 1971, a *coup d'etat* by a northern, ethnically Kakwa, General Idi Amin marked a descent into terror and highlighted a number of complicated ethnic relations inherent in the Uganda power structure. Amin directed considerable repression against Milton Obote-supporting fellow northern ethnic communities, specifically the Acholi and Langi people. Amin's foreign policy, including his decision to attack neighboring Tanzania in 1978, encouraged Tanzania to ally with assorted Ugandan rebels to launch the 1979 attack that overthrew the Amin regime. The influence of Amin's sole ethnic power center – the northwest – was too small to counteract the influence of the northern areas coupled with outside support.

Despite the overthrow of Amin, the continued north–south divide saw the newly reinstated President, Milton Obote, returning to power to face a civil war with the NRA (National Resistance Army). The NRA took control of the ethnic, southern Bugandan heartland as this area provided a favorable recruiting environment for the growth and success of the rebel group. Obote, as he had done in 1962, overran the area and committed numerous atrocities in Buganda and the centrally located Luwero triangle (Ngoga, 1998). The war between the NRA and the northern-dominated government continued for over five years. The eventual failure of the Obote government to counteract the southern rebels resulted in the overthrow of Obote by northern, ethnically Acholi, officers. The instability of the coup, coupled with the ever increasing losses of territory to the NRA, culminated in rebel leader, Yoweri Museveni, declaring a rebel victory and seizing Kampala in January 1986.

The extent of NRA-liberated territory was limited. The victorious rebels did not control the north, which proved to be both a favorable area to host rebel organizations and relatively inaccessible to the government (Behrend, 1998). The overthrown Acholi officers returned north, as far as Sudan, to reform the previous regime's military into a rebel force to counterattack the now official military forces of the NRA. A number of smaller rebellions also occurred during this time; the most successful of these folded elements of the previous military into the Ugandan People's Democratic Movement (UPDM) and the Holy Spirit Movement (HSM). When the UPDM agreed to a ceasefire in 1989, the remaining active rebels allied with the HSM to continue the civil war. The HSM eventually ceased its activity only to stage resurgence in 1994 as the Lord's Resistance Army, LRA (another group, the 'Ugandan Democratic Christian Army', had mobilized before the LRA). The geography of the LRA rebellion reflected what was thought to be a sympathetic public in the north. The northern Acholi community has borne the brunt of LRA

actions, presumably because the LRA is not strong enough to gain a foothold in any other region, and so depends on ransacking Acholi villages and towns to sustain itself. The brutal actions of the LRA include numerous kidnappings of children to serve as soldiers. (See Figure 30.1 for the locations of the LRA actions: Raleigh and Hegre, 2005, ACLED data.)

The domestic civil war now widened as it spread across Uganda's borders, giving rise to a classic regional conflict formation. The LRA established bases in southern Sudan, where they were supported by the Sudanese government. At the same time, the SPLA Sudanese rebels of southern Sudan, while engaged in a long and brutal civil war with their own government, were tacitly supported by the Ugandan government. The LRA managed to attack northern civilians and army posts through hit-and-run tactics across the border into northern Uganda while maintaining a safe haven in Sudan. The LRA also created ties with a smaller rebel group in the northwest, the West Bank Nile Front (WBNF). A recent attack by the LRA on peacekeepers in the DRC has confirmed suspicions that the LRA is now firmly established in that failed state, and is still assisted by the Sudanese government.

Another rebel group (the Allied Democratic Forces, ADF) developed in the west of Uganda in the rough terrain of the Ruwenzori Mountains in 1996. The ADF merged members of previous rebel organizations, and derived some support from the ethnic communities in the far western region which were unfriendly to the current regime. Although the goals of the ADF were unclear, it perpetrated multiple attacks on government military posts in the west, on civilians, and finally on the capital, Kampala. Eventually, the ADF made strong alliances with DRC rebels and went on to establish bases in that failed state. It is rumored that the Virunga Forest Park in the remote northeast corner of the DRC was home to multiple allied rebel organizations in the late 1990s. (See Figure 30.2 for the geography of the ADF revolt: Raleigh and Hegre, 2005, ACLED data.) Thus, Uganda's large neighbors (Sudan and the DRC), both with devastating civil wars of their own, were now part of the struggle for control of the Kampala-centered state regime. Uganda entered the first and second Congo wars (1996 and 1998, respectively), justifying its actions by claiming that Congo was a haven for rebels intent on harming the Museveni regime, and that of Uganda's Rwandan allies (Clark, 2002).

Hence, both internally and externally, the geography of Ugandan conflicts returns to distinct national north–south divisions and regional ethnic affiliations. The politicization of the ethnic communities in Uganda is further exacerbated by the geography of development. The northern President Obote, attempting to redress entrenched underdevelopment as a result of colonialism,

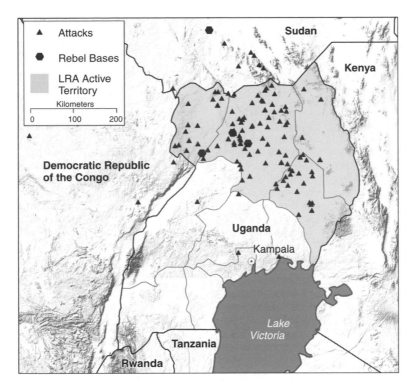

Figure 30.1   Lord's Resistance Army (LRA) active territories, 1994–2004.

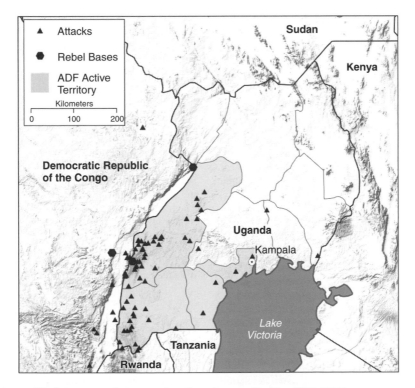

Figure 30.2   Allied Democratic Forces (ADF) active territories, 1996–2002.

faced resentment from southerners. In turn, President Museveni, a southerner, is seen as having cultivated his southern base, whilst leaving northern regions underdeveloped. As noted by Kasfir, 'Devastating civil wars have been fought in parts of the east and north. The perception of the north as a southern government, and its wars it has fought against remnants of armies of former enemies, has reinforced regional cleavages (1995: 149)'. These situations have resulted in an entrenched sense of hostility and ethnically charged insurgencies against ruling regimes. It is northern grievance and hostility that Joseph Kony, the leader of the northern Lord's Resistance Army, claims to represent in current negotiations with the Museveni regime.

Externally, Uganda contributes to and suffers from foreign assistance to rebel groups. The entrenched instability in the African Great Lakes region is fueled by states such as Uganda assisting the SPLA (Sudan), or AFDL (Kabila, Zaïre), or RCD (DRC) rebel organizations, while suffering from Sudanese and DRC support for Ugandan rebel groups. Neighboring instability creates an environment for state failure and increases the risk of civil war within a state (Raleigh, 2006). The postcolonial history of Uganda, generally considered a fairly successful African polity, demonstrates that replacing one leader for another is likely to result in further ethnic resentment. Without the mechanism of a democratic regime to allocate government benefits and participate in non-ethnic politics, poor countries can expect to see further civil strife that involves their neighbors, inadvertently or not.

Uganda in particular, and the Great Lakes region in general, highlight the complexities of African civil wars. Civil war onset and duration are typically not explained by one cause but by the interplay of underlying ethnic and economic conditions with catalytic political factors. In the case of Uganda, previous rebellions, politicized ethnic communities, and questionable military loyalties created an environment suitable for persistent civil wars. Governments facing rebellion are often repressive and reliant on a military of dubious quality and ethnic, rather than state, attachment (see Clapham, 1986). For that reason, governments shift spending from basic needs to the military (African countries often top the league table of military spending as a proportion of GDP; see SIPRI, 2004) and install members of the ethnic group that provide the government's ministers as the officers in the military (Clapham, 1986; Migdal, 2001). Outside factors, such as neighboring rebellion and porous borders, allow rebels safe haven. The Great Lakes conflicts highlight how many current civil wars are not state-specific or solely internal phenomena, but related and supported by a host of external conditions (Clark, 2002). And, in relation to theoretical emphases in the human geographic consideration of scale, a variety of scale-dependent and scale-related effects are clearly visible.

Determining and analyzing the geography of conflicts, along with the root and catalytic factors of conflicts, requires local-scale data. A collection of disaggregated data, including geocoded information on battles, rebel and government camps, resources, targets, and so on, can provide the essential information for such a detailed analysis. (See Raleigh and Hegre, 2005, on disaggregated event data for Central and West African conflicts.) Typically, event data are recorded using newspaper reports and other archival data and are elusive and time-consuming to collect. However, they provide insight into the nature of actions, shedding light on the forces that lead to the outbreak of conflict in certain areas of the state. (See Kalyvas, 2006, for an extended argument about the localized and otherwise personalized nature of civil war, supported by dozens of diverse examples that rely on individual event data.)

The Long-Term Effects of Civil Wars: A Lot Depends on Where the Conflict Occurs

Most civil war study has concentrated on the reasons for war outbreaks, the variations in war duration, and the conditions under which wars end. However, a recent initiative in the World Health Organization that re-calculates life expectancy measures to take account of the years of life lost due to disabilities of various kinds (DALE – disability-adjusted life expectancy) has led to a consideration of the effects of wars on a country's quality of life, even for those not directly involved in the conflicts. The sum of the research to date is that the indirect effects of conflicts are significantly more important in reducing DALYs (disability-adjusted life years) than the direct effects of the fighting itself. Ghobarah et al. (2003, 2004a, 2004b) show that public health consequences of civil wars persist beyond the span of the actual conflict by estimating the additional burden of death and disability. The health outcome in 1999, from the indirect and lingering effects of civil wars in the years 1991–7, was approximately equal to that incurred directly and immediately from all wars in 1999. Further, the public health consequences of civil wars are disproportionately borne by women and children.

The Ghobarah et al. regression model that generates the estimates of the effects of civil wars is very straightforward and contains a series of controls such as ethnic fragmentation, income inequality, health spending, urban growth, location in a tropical country, a governance score on the democracy/autocracy scale, and education. Their key predictors are the number of people killed in civil wars in the previous decade and whether the neighboring states experienced a civil war in the previous decade. The outcome (dependent) variable is DALYs lost per year per 100 people and the analysis is repeated for a large number of demographic groups (men and women separately of various ages). In their studies, civil wars both at home and in contiguous states have independent significant effects on DALYs, often of a sizable magnitude. Thus, they estimate that 'the impact in 1999 of living in a country that had experienced an intense civil war a few years earlier (such as Bosnia, with 6.8 civil war deaths per 100 people) rather than in a median country with no war at all is a loss of about 28.5 healthy years for only one disease of 23; the misery accumulates with each of the other 22 categories of disease' (Ghobarah et al., 2003: 197). They report the coefficients and the estimates for the whole world with no disaggregation for region or country, though one of the key controls in the studies is whether a country is located in a tropical zone or not.

What is lacking in the Ghobarah et al. studies is any consideration of how these effects of civil wars on life expectancy might vary across the globe. To calculate these effects, we recalibrated their models using GWR (geographically weighted regression). This method differs from OLS (ordinary least squares) because one can disaggregate the usual global parameters (such as those reported in Ghobarah et al.'s papers) into local estimates that can be mapped (Fotheringham et al., 2002). Using the data on the Yale team's website (accessible, at time of writing, at http://pantheon.yale.edu/%Ebrusset/APSRMay03.zip), we replicated the studies (the global coefficients are the same) and extended them by disaggregating the regression parameters to each of the 180 countries. Our inquiry is designed to see if there are significant variations across the globe and whether these variations are geographically clustered, which, in turn, might generate further hypotheses on the factors causing the distribution. Some key results are reported in Figures 30.3 and 30.4.

Ghobarah et al. (2003) report that the overall annual effect of civil war deaths on DALYs lost to all disease categories for 100 males aged 15–44 is 0.215, a small but statistical significant effect. (The biggest effect, by far, on DALYs is due to income inequality.) The distribution of the parameter values for the civil war effects is geographically variable, ranging from −1.27 to +0.53, as can be seen in Figure 30.3. The concentration of highest values in Southeast Asia and other high values in Africa south of the Sahara, the Arabian peninsula, and East Asia is highly visible. In these regions, civil wars cause up to twice the impact on DALYs for males aged between 15 and 44 than in a country with no such civil conflicts.

An even more dramatic demonstration of the geographical variation in these estimates of civil war effects is in the values for the contiguous civil war variable. The global average is 5.75 AIDS-caused DALYs in women aged 15–44, but the values range up to 36.12 in southern Africa. The map (Figure 30.4) show a significant concentration of highest values in Africa where southern African countries (Mozambique, Zimbabwe, South Africa, etc.) have values eight to ten times higher than the average global effect. In this region, already

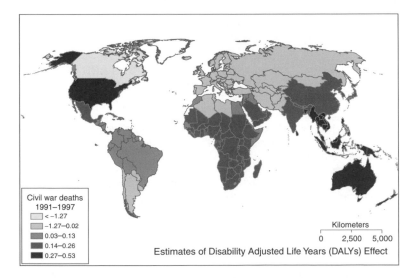

Figure 30.3 Effects of civil wars on DALYs (disability-adjusted life years) for males aged 15–44.

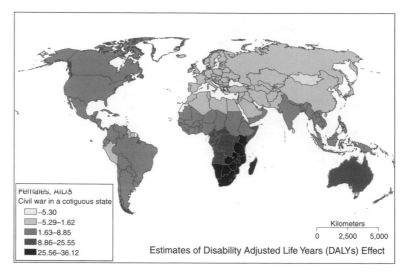

Females, AIDS
Civil war in a cotiguous state
☐ −5.30
▨ −5.29–1.62
▨ 1.63–8.85
▨ 8.86–25.55
▨ 25.56–36.12

Kilometers
0   2,500   5,000

Estimates of Disability Adjusted Life Years (DALYs) Effect

Figure 30.4   Effects of civil wars in adjoining states on DALYs (disability-adjusted life years) lost for females aged 15–44 due to AIDS.

wracked by declining per capita incomes and an overloaded health-care system, local conflicts further exacerbated the devastating effects of AIDS on the female population through transferred government spending from social to military exigencies. In fact, replication of many of the original models using GWR shows that the authors ignored an obvious predictor in their study, location in sub-Saharan Africa, which has independent and significant effects in addition to the controls that they used.

Further statistical analysis by O'Loughlin and Witmer (2005) indicates that the African clusters on all the maps of DALYs suggest that the 'African factor' in the models of Ghobarah et al. is underspecified. The re-analysis of the data that show a significant effect of civil war involvement at home and in neighboring states from Ghobarah et al. (2003) using geographically weighted regression identified important spatial patterns in the distribution of the localized parameters, especially for Africa. Not all of the original analyses need to be disaggregated, and the choice of spatial weighting is an important consideration in the spatial analysis. However, the appearance of clusters in Africa (especially in southern and eastern Africa) of high parameter values for many of the models suggests that greater attention be paid to the specific African context and consideration of a recalibrated model that would substitute an African location for the 'tropical location' that was present in the original models.

These short illustrations of the value of a geographic approach to the study of civil war violence incorporated both an emphasis on place and contextual conditions and scale considerations (Uganda) and the techniques of spatial analysis at the global scale (the study of civil wars' long-term effects). What is common to both approaches is the need to disaggregate the country-level data widely used in civil war studies and to question the use of global models that effectively summarize the whole distribution but that hide important and interesting geographic variations. Until more political scientists and economists become convinced of the need for the collection of more precise geographic codings to go along with the temporal and event data that they normally collect, the study of these spatial considerations remains severely hampered.

CONCLUSIONS AND DESIDERATA FOR FUTURE RESEARCH

Disaggregated data beyond the level of the nation-state have not yet been widely explored in the study of civil war violence. Although such data presently do not yet exist in a manner that easily allows cross-national comparisons and local in-depth analysis, hopefully this review has made the case for a disaggregated, spatial perspective on civil wars that will augment our understanding of their causes.

This review has stressed the impossibility of bounding the study of civil wars to the legally defined territorial limits of the country in question. With the advent of transnational violence in the form of potentially ubiquitous terrorism, the lines between national, international, transnational, and subnational have blurred irrevocably.

Scales, in effect, have become more malleable and their specific meanings changes from region to region. That said, it is also evident that most violence will continue to be found in the world's poorest regions. The peripheral parts of states, particularly ethnic enclaves, can harbor both domestic and transnational oppositions. Thus, the US 9/11 Commission identified western Pakistan and the Pakistan–Afghanistan border region, southern and western Afghanistan, the Arabian peninsula, the Horn of Africa, West Africa, Southeast Asia, and Western European cities with sizable expatriate Muslim communities as possible bases for anti-American terrorist movements and concluded that 'in the twentieth century, strategists focused on the world's great industrial heartlands. In the twenty-first, the focus is in the opposite direction, toward remote regions and failing states. The United States has had to find ways to extend its reach, straining the limits of its influence' (9/11 Commission, 2004: 366). The post-Iraq war national military strategy formalizes this fear and, while stating that the United States will cooperate with allies, retains a pre-emptive strategy for future military actions (Myers, 2004). Most oppositional movements will not achieve such global attention from the United States, but the connections between rebellions, state-to-state alliances, US hegemony, and domestic and transnational terrorism are only now being developed.

Study of civil war, both its development and aftermath, has been hindered by a paucity of data and a reliance on secondary published sources, government and non-governmental agency reports, and newspaper accounts. There is relatively little primary data collection, either through questionnaire surveys, remote sensing from satellites, interviews or census taking. A clear need exists to link thematic data for a wide range of important actors and institutions: data for geographic units (counties, census units, etc.) derived from government sources, data on individuals whose addresses are geo-referenced and gathered through a survey questionnaire, as well as satellite image data on the physical environment that can be geocoded and matched. In this regard, further development of methodologies that link social science approaches to the physical environment are warranted. To organize data collection and to overlay and integrate the spatial coverages for the three types of data, GIS offers a solution to efficiently display the information collected. An integrated database can thus contain the geo-referenced data from satellite imagery, digital line graphs, GPS data (for household addresses and ground-referencing information), socio-demographic data, and infrastructural and environmental information gathered from fieldwork and from international agencies (e.g. UNDP) working in the war zones. This kind of information will allow a different kind of research

thrust, one that is avowedly geographic, to supplement the (increasingly) stale stable of existing reports on civil war violence.

For too long, geographers have paid scant attention to the depredations of civil wars and associated violence. Research by political scientists and economists has reached the point of diminishing returns and it will take a paradigmatic shift and/or a flood of data, especially for disaggregated units, to jump-start this body of work to a new level. The geographic perspective, especially the emphasis on context, scale linkages, diffusion, and spatial analysis, offers a vital and innovative supplement to dominant approaches.

ACKNOWLEDGMENTS

This research is supported by a grant (0433927) from the National Science Foundation's initiative on Human and Social Dynamics. Clionadh Raleigh's research at PRIO (Peace Research Institute of Oslo, Norway) is funded by a fellowship from the Norwegian Research Council's program on cultural exchanges. Final versions of the figures were prepared by Tom Dickinson and the late Sean Blackburn of the Institute of Behavioral Science. The authors thank Frank Witmer (IBS) and Håvard Hegre (PRIO) for their cooperation in this project.

REFERENCES

9/11 Commission (2004) *Report of the National Commission on Terrorist Attacks Upon the United States.* Washington, DC: US Government Printing Office.

Agnew, J.A. (1994) 'The territorial trap: the geographical assumptions of international relations theory', *Review of International Political Economy*, 1: 53–80.

Auty, R. (2004) 'Natural resources and civil strife', *Geopolitics*, 9: 29–49.

Azaryahu, M. (1996) 'The power of commemorative street names', *Environment and Planning D: Society and Space*, 14: 314–30.

Behrend, H. (1998) 'War in northern Uganda', in C. Clapham (ed.), *African Guerrillas*. Oxford: James Currey, pp. 107–18.

Boulding, K.E. (1978) *Stable Peace*. Austin: University of Texas Press.

Brown, M. (2001) 'Ethnic and internal conflict: causes and implications', in C. Crocker, F.O. Hampson and P. Aall (eds), *Turbulent Peace: The Challenges of Managing International Conflict*. Washington, DC: US Institute of Peace Press, pp. 209–26.

Buhaug, H. and Gates, S. (2002) 'The geography of civil war', *Journal of Peace Research*, 39: 417–33.

Buhaug, H. and Lujala, P. (2005) 'Accounting for scale: measuring geography in quantitative studies of civil war', *Political Geography*, 24: 399–418.

Buhaug, H. and Rod, J.K. (2006) 'Local determinants of African civil wars', *Political Geography*, 25: 315–35.

Charlesworth, A. (2003) 'Hello darkness: envoi and caveat', *Common Knowledge*, 9: 508–19.

Clapham, C. (1986) *Third World Politics: An Introduction*. Madison: University of Wisconsin Press.

Clark, J.F. (2002) 'Introduction: causes and consequences of the Congo War', in J.F. Clark (ed.), *The African Stakes in the Congo War*. New York: Palgrave Macmillan, pp. 1–12.

Cohen, S.B. (2003) *Geopolitics of the World System*. Lanham, MD: Rowman & Littlefield.

Collier, P. (2000) 'Rebellion as a quasi-criminal activity', *Journal of Conflict Resolution*, 44: 839–53.

Collier, P. and Hoeffler, A. (1998) 'On economic causes of civil war', *Oxford Economic Papers*, 50: 563–73.

Collier, P. and Hoeffler, A. (2004) 'Greed and grievance in civil wars', *Oxford Economic Papers*, 56: 563–95.

Collier, P. et al. (2003) *Breaking the Conflict Trap: Civil War and Development Policy*. Washington, DC: World Bank.

Collier, P., Hoeffler, A. and Soderbom, M. (2004) 'On the duration of civil war', *Journal of Peace Research*, 41: 253–73.

Connor, W. (1973) 'The politics of ethnonationalism', *Journal of International Affairs*, 27: 1–21.

Corson, M. and Palka, E. (2004) 'Geo-technology, military and war', in S.D. Brunn, S. Cutter and J.W. Harrington (eds), *Geography and Technology*. Dordrecht, Netherlands: Kluwer Academic Publishers, pp. 401–27.

de Rouen, K.R., Jr, and Sobek, D. (2004) 'The dynamics of civil war duration and outcome', *Journal of Peace Research*, 41: 303–20.

de Soysa, I. (2000) 'The resource curse: are civil wars driven by rapacity or paucity?', in M. Berdal and D.M. Malone (eds), *Greed and Grievance: Economic Agendas in Civil Wars*. Boulder, CO: Lynne Rienner, pp. 113–35.

Diehl, P.F. (1991) 'Geography and war: a review and assessment of the empirical literature', *International Interactions*, 17: 11–27.

Elbadawi, I. and Sambanis, N. (2002) 'How much civil war will we see? Estimating the prevalence of civil war in 161 countries 1960–1999', *Journal of Conflict Resolution*, 46: 307–34.

Ellingson, T. (2000) 'Colorful community or ethnic witches' brew', *Journal of Conflict Resolution*, 44: 228–49.

Englebert, P., Tarango, S. and Carter, M. (2002) 'Dismemberment and suffocation: a contribution to the debate on African boundaries', *Comparative Political Studies*, 35: 1093–1118.

Fearon, J.D. and Laitin, D.D. (2003) 'Ethnicity, insurgency and civil war', *American Political Science Review*, 97: 75–90.

Fotheringham, A.S., Brunsdon, C. and Charlton, M.E. (2002) *Geographically Weighted Regression: The Analysis of Spatially Varying Relationships*. Chichester: John Wiley.

Fox, J. (2004) 'The rise of religious nationalism and conflict: ethnic conflict and revolutionary wars, 1945–2001', *Journal of Peace Research*, 41: 715–31.

Gates, S. (2002), 'Empirically assessing the causes of civil war', paper presented at the International Studies Association Annual Meeting. New Orleans, LA.

Ghobarah, H.A., Huth, P. and Russett, B. (2003) 'Civil wars kill and maim people – long after the shooting stops', *American Political Science Review*, 97: 189–202.

Ghobarah, H.A., Huth, P. and Russett, B. (2004a) 'Comparative public health: the political economy of human misery and well-being', *International Studies Quarterly*, 48: 73–94.

Ghobarah, H.A., Huth, P. and Russett, B. (2004b) 'The post-war public health effects of civil conflict', *Social Science and Medicine*, 59: 869–84.

Gleditsch, K.S. (2002) *All International Politics Is Local: The Diffusion of Conflict, Integration and Democratization*. Ann Arbor: University of Michigan Press.

Goldstone, J.A. et al. (2003) *State Failure Task Force Phase III Findings*. McLean, VA: Science Applications International Corporation.

Goodchild, M.F. et al. (2000) 'Toward spatially integrated social science', *International Regional Science Review*, 23: 139–59.

Griffith, D. and Layne, L.J. (1999) *A Casebook for Spatial Statistical Analysis: A Compilation of Different Thematic Datasets*. New York: Oxford University Press.

Gurr, T.R. (1970) *Why Men Rebel*. Princeton, NJ: Princeton University Press.

Gurr, T.R. (1993) *Minorities at Risk: A Global View of Ethnopolitical Conflict*. Washington, DC: US Institute of Peace Press.

Heffernan, M. (1998) 'War and the shaping of Europe', in B. Graham (ed.), *Modern Europe: Place, Culture and Identity*. London: Arnold, pp. 89–120.

Hegre, H. et al. (2001) 'Toward a democratic civil peace? Democracy, political change, and civil war, 1816–1992', *American Political Science Review*, 95: 33–48.

Herbst, J. (2000) *States and Power in Africa: Comparative Lessons in Authority and Control*. Princeton, NJ: Princeton University Press.

Hewitt, K. (1983) 'Place annihilation: area bombing and the fate of urban places', *Annals of the Association of American Geographers*, 73: 257–84.

Holsti, K. (1996) *The State, War and the State of War*. New York: Cambridge University Press.

Homer-Dixon, T.F. (1999) *Environment, Scarcity and Violence*. Princeton, NJ: Princeton University Press.

Horowitz, D.L. (1985) *Ethnic Groups in Conflict*. Berkeley: University of California Press.

Huntington, S. (1996) *The Clash of Civilizations and the Remaking of World Order*. New York: Simon & Schuster.

Johnson, N. (1999) 'The spectacle of memory: Ireland's remembrance of the Great War', *Journal of Historical Geography*, 25: 36–56.

Johnston, R.J. (1991) *A Question of Place: Exploring the Practice of Human Geography*. Oxford: Blackwell.

Johnston, R.J. and Sidaway, J.D. (2004) *Geography and Geographers: Anglo-American Human Geography since 1945* (6th edn). New York: Oxford University Press.

Johnston, R.J., Taylor, P.J. and O'Loughlin, J. (1987) 'The geography of violence and premature death', in R. Väyrynen, D. Senghaas and C. Schmidt (eds), *The Quest for Peace*. Beverly Hills, CA: Sage and the International Social Science Council, pp. 241–59.

Jones, D.A., Bremer, S.A. and Singer, J.D. (1996) 'Militarized interstate disputes, 1816–1992: rationale, coding rules and empirical patterns', *Conflict Management and Peace Science*, 15: 163–213.

Kalyvas, S.N. (2006) *The Logic of Violence in Civil Wars.* New York: Cambridge University Press.

Kasfir, N. (1995) 'Uganda politics and the Constituency Assembly elections of March, 1994', in H.G. Hansen and M. Twaddle (eds), *From Chaos to Order: The Politics of Constitution Making in Uganda.* Kampala: Fountain Publishers.

Keen, D. (1998) *The Economic Functions of Violence in Civil Wars.* New York: Oxford University Press.

Klare, M.T. (2001) *Natural Resources Wars: The New Landscape of Global Conflict.* New York: Metropolitan Books.

Lacoste, Y. (1976) 'Enquête sur le bombardement des digues du fleuve rouge (Vietnam, été 1972): méthode d'analyse et réflexions d'ensemble', *Hérodote*, 1: 86–117.

Lake, D.A. and Rothschild, D. (eds) (1998) *The International Spread of Ethnic Conflict.* Princeton, NJ: Princeton University Press.

Le Billon, P. (2001) 'The political ecology of war: natural resources and armed conflicts', *Political Geography*, 20: 561–84.

Le Billon, P. (2004) 'The geography of "resource wars"', in C. Flint (ed.), *The Geography of War and Peace.* New York: Oxford University Press, pp. 217–41.

Mamadouh, V. (2004) 'Geography and war, geographers and peace', in C. Flint (ed.), *The Geography of War and Peace.* New York: Oxford University Press, pp. 26–60.

Migdal, J. (2001) *State in Society: Studying How States and Societies Transform and Constitute One Another.* Cambridge: Cambridge University Press.

Moore, W.H. and Shellman, S.M. (2004) '"Fear of persecution: forced migration", 1952–1995', *Journal of Conflict Resolution*, 40: 723–45.

Mueller, J. (2003) 'Policing the remnants of war', *Journal of Peace Research*, 40: 507–18.

Murphy, A.B. (2004) 'Territorial ideology and interstate conflict: comparative considerations', in C. Flint (ed.), *The Geography of War and Peace.* New York: Oxford University Press, pp. 280–96.

Murshed, S.M. (2002) 'Conflict, civil war and underdevelopment: an introduction', *Journal of Peace Research*, 39: 387–93.

Myers, R.B. (Chairman) (2004) *The National Military Strategy of the United States: A Strategy for Today – A Vision for Tomorrow.* Washington, DC: Joint Chiefs of Staff, US Department of Defense.

Ngoga, P. (1998) 'Uganda: the National Resistance Army', in C. Clapham (ed.), *African Guerrillas.* Oxford: James Currey, pp. 91–106.

Ó Tuathail, G. (2000) 'The postmodern geopolitical condition: states, statecraft and security in the twenty-first century', *Annals of the Association of American Geographers*, 90: 166–78.

O'Loughlin, J. (1989) 'World-power competition and local conflicts in the Third World', in R.J. Johnston and P.J. Taylor (eds), *A World in Crisis? Geographic Perspectives* (2nd edn). Oxford: Blackwell, pp. 289–332.

O'Loughlin, J. (2001) 'Geography as space and geography as place: the divide between political science and political geography continues', *Geopolitics*, 5: 126–37.

O'Loughlin, J. (2003), 'Spatial analysis in political geography', in J.A. Agnew, K. Mitchell and G.Ó. Tuathail (eds), *A Companion to Political Geography.* Oxford: Blackwell, pp. 30–46.

O'Loughlin, J. (2004) 'The political geography of conflict: civil wars in the hegemonic shadow', in C. Flint (ed.), *The Geography of War and Peace.* New York: Oxford University Press, pp. 85–112.

O'Loughlin, J. and Anselin, L. (1991) 'Bringing geography back to the study of international relations: spatial dependence and regional conflict in Africa', *International Interactions*, 17: 29–61.

O'Loughlin, J. and Heske, H. (1991) 'From "Geopolitik" to "Géopolitique": converting a discipline for war to a discipline for peace', in N. Kliot and S. Waterman (eds), *The Political Geography of Conflict and Peace.* London: Belhaven Press, pp. 37–59.

O'Loughlin, J. and van der Wusten, H. (1993) 'The political geography of war and peace, 1890–1991', in P.J. Taylor (ed.), *The Political Geography of the Twentieth Century.* New York: John Wiley, pp. 63–113.

O'Loughlin, J. and Witmer, F. (2005) 'Taking "Geography" seriously: disaggregating the study of civil wars', in *Conference on Disaggregating the Study of Civil War and Transnational Violence.* La Jolla: University of California, San Diego, Department of Political Science.

Olsson, O. and Fors, H.C. (2004) 'Congo: the price of predation', *Journal of Peace Research*, 41: 321–36.

O'Sullivan, P. and Miller, J.W. (1983) *The Geography of Warfare.* New York: St Martin's Press.

Palka, E. and Galgano, F. (eds) (2005) *Military Geography: From Peace to War.* New York: McGraw-Hill.

Raleigh, C. (2004) 'Neighbours and neighbourhoods: understanding the role of context in civil war', in *5th Pan-European International Relations Conference*, The Hague, Netherlands, 9–11 September.

Raleigh, C. (2006) 'Neighbours, conflict and instability', in *Workshop on Post-Conflict Transitions: Institutions, Development and a Domestic Civil Peace.* Oslo: PRIO, 19–20 June.

Raleigh, C. and Hegre, H. (2005) 'Introducing ACLED: an armed conflict location and event dataset', in *Conference on Disaggregating the Study of Civil War and Transnational Violence.* La Jolla: University of California, San Diego, Department of Political Science. Data available from http://new.prio.no/CSCW-Datasets/Data-on-Armed-conflict/ACLED-Armed-Conflict-Location-and-Event-Data/

Robbins, P. (2004) *Political Ecology: A Critical Introduction.* Malden, MA: Blackwell.

Ross, M.L. (2004) 'What do we know about natural resources and civil war?', *Journal of Peace Research*, 41: 337–56.

Salehyan, I. and Gleditsch, K.S. (2004) *Refugee Flows and the Spread of Civil War.* La Jolla: University of California, San Diego, Department of Political Science.

Sambanis, N. (2001) 'Do ethnic and non-ethnic civil wars have the same causes? A theoretical and empirical inquiry (Part I)', *Journal of Conflict Resolution*, 45: 259–82.

SIPRI (2004) *SIPRI Yearbook: Armaments, Disarmament and International Security.* Stockholm: Stockholm International Peace Research Institute.

Siverson, R.M. and Starr, H. (1991) *The Difusion of War: A Study of Opportunity and Willingness*. Ann Arbor: University of Michigan Press.

Sorli, M.E., Gleditsch, N.P. and Strand, H. (2005) 'Why is there so much conflict in the Middle East?', *Journal of Conflict Resolution*, 49: 141–65.

Sprout, H.H. and Sprout, M. (1965) *The Ecological Perspective on Human Afairs, with Special Reference to International Politics*. Princeton, NJ: Princeton University Press.

Starr, H. and Thomas, G.D. (2002) 'The "nature" of contiguous borders: ease of interaction, salience, and the analysis of crisis', *International Interactions*, 23: 213–35.

Strand, H., Wilhelmsen, L. and Gleditsch, N.P. (2004) 'Armed Conflict Dataset Codebook Version 2.1', Technical Report, International Peace Research Institute Oslo (PRIO).

Väyrynen, R. (1984) 'Regional conflict formations: an intractable problem of international relations', *Journal of Peace Research*, 21: 337–59.

Ward, M.D. and Gleditsch, K.S. (2002) 'Location, location, location: an MCMC approach to modeling the spatial context of war and peace', *Political Analysis*, 10: 244–60.

Weeks, T.R. (2006) 'A multi-ethnic city in transition: Vilnius's stormy decade 1939–1949', *Eurasian Geography and Economics*, 47: 153–75.

Winston, V.H. (2006) 'Observations on the population of Vilnius: the grim years and the 1942 census', *Eurasian Geography and Economics*, 47: 176–203.

World Health Organization. (2000) *The World Health Report 2000: Health Systems — Improving Performance*. Geneva: World Health Organization.

The Politics of Uneven Development

Introduction

Kevin R. Cox

Raymond Williams did not include 'development' in his *Keywords* (Williams, 1976), and perhaps for good reason. It is common to think of it today in association with higher living standards, higher levels of economic wealth or productivity, and a host of other outcomes or conditions that it is felt improve people's lives. These might emerge out of wider social and economic processes, or be associated with government interventions designed to accomplish these goals. As such, and with some nuances, it is applied at all manner of geographic scales, from the local to the national, and even the international. There is much disagreement and debate regarding these processes and outcomes, their feasibility, desirability, and relative importance. The revived political geography of recent decades has participated in a range of different debates and research concerning development. Modes of analysis have drawn on ideas from political economy and more recently on ideas more influenced by cultural politics. A quite expansive literature on the politics of local and regional development has emerged as well as one on developing societies, and work has overlapped with that in related areas of the discipline from development geography to the more recent concern with political ecology.

None of this is to say that earlier work in political geography neglected what we might recognize as having to do with development. Geographers like Bowman, Goblet, Mackinder, and Fawcett did not call it such. Yet their focus was on national resources and those that could be accessed through empire or, in the American case, through trade (which Bowman was so anxious about in his *The New World*). The implications of this work not so much for human welfare as for national power and security stake out a different set of intellectual concerns in related terrain. There was little work on subnational scales, though, and the usual framework was consistent with the people/nature optic so characteristic of human geography at this time. In these regards, the brief flirtation with what was called 'tropical geography' (Power and

Sidaway, 2004) is symptomatic, signifying an interest on the part of the imperial powers in colonial development and also its limitations.

East and Moodie's (1956) edited collection *The Changing World* is a more transitional work. It still shows closer affinities with that of Bowman than with the sort of political geography of development that would emerge after the early 1970s. The people/nature framework remains evident, along with the interest in natural resources, but there are now occasional references to 'development', to 'material standards of living' and 'industrial progress'. The interest in these issues is uneven across the different chapters and some authors, like Ackerman, Buchanan, Moodie, and Spate, seem more alert to these new concepts than others. Moodie talks about the relation between economic integration and political fragmentation in Western Europe. Ackerman has things to say on the politics of regional development in the US. Keith Buchanan's chapter on South Africa is very squarely on development and from the standpoint of the masses. Chapters on the Soviet Union and China by Theodore Shabad and Charles Fisher respectively are particularly interesting because of their focus on economic development. One suspects that this early intimation of what came to be known as the developmental state was an important spur to later thinking about the politics of uneven development.

The beginnings of an interest in regional development go back to the 1930s and in the UK emerged partly in the context of the depressed areas as a policy issue at that time. The work of Daysh and Caesar (Daysh et al., 1949), among others (Clerk et al., 1942), was important and, in France, that of Gravier (1958). The emphasis, however, was much more policy-related than it was political: how to mitigate regional unemployment or the vulnerability to aerial warfare implied by the growth of very large cities. They talked more about unemployment and changing patterns of population distribution than they did about development, and especially conflicts around development, as we understand it.

Of course, to refer to 'development as we understand it' raises some important critical questions. A commonly accepted meaning, still very current, suggests the idea of improvement over time in material standards of living and state intervention geared to that purpose. This improvement, typically registered by changes in gross national product per capita, is then linked to shifts in benchmarks like occupational composition or the consumption of energy. The end product, though, is higher average material consumption. In this regard we should note three things.

The first is the sense of development, going back to the nineteenth century and possibly earlier, as being about production rather than consumption; as about, for example, the development of people's (or societies') abilities to produce. This is not only a Marxist view; as Rangan points out in her chapter below, nineteenth-century anthropologists, as a result of their interest in preindustrial peoples, conceived of human history in terms of a development from collecting societies, through agricultural forms of society to contemporary industrial forms – in other words, as a development of productive capacities.

The second point to note is that the sense of the term is linked to changes in the division of labor; how, in other words, development was a collective experience but with differential effects. This is most clear, again, in Marx, particularly in *The German Ideology*. But it was a connection that was registered more widely among nineteenth-century thinkers like observers like Bryce, Henry Maine, and Durkheim tried to come to some understanding of the dramatic social changes of their era. Marx's view of this was decidedly critical. Although he saw the progressive division of labor as a necessary condition for the development of productive capacities, he also saw it as imposing a very one-sided conception of development. People developed in one-sided ways; their labors were channeled in singular directions; work, in virtue of its repetitive nature, became a misery rather than a joy, dull and mind-numbing rather than creative; their creative powers were alienated into a social form that came to dominate and frustrate them. And of course, he saw the capitalist mode of production as the necessary condition for this development of the division of labor.

This brings us logically to a third point about development; this is the idea of post-development, of development futures that are not those of the capitalist form in which they are typically packaged, and which move beyond the dominant discourses of development that have shaped interventions (Crush, 1995; Escobar, 1995). Instead of state and corporate imposition the talk is of the democratic determination of what development should look like. This is raised most explicitly in this section in the chapter by Mohan and Stokke, though there

are also echoes of this theme in Rydin's chapter on sustainable development. Elsewhere, Power (2003) is a notable contribution.

With these considerations as background, consider now the practice of political geography and development today and as it is expressed in the chapters in this section. How is the political geography of development being imagined? One thing that is immediately clear is the multiscalar character of the politics of development. For most of the authors the *global* is a crucial scale. First, because new organizations have emerged at this scale, organizations that act as critical discursive nodes and have their own characteristic weight to bring to bear on the definition and, to be sure, implementation of what is 'best practice' in development terms. Second, of course, the global is an essential point of departure because of the way in which the world is structured. The sorts of global environmental changes identified by Yvonne Rydin cannot be thought of outside of global heat balances and related atmospheric changes. Neither, as she points out, can they be thought of without reference to the unplanned nature of market relations and the impetus to growth or accumulation.

The global also figures prominently in Peter Taylor's chapter. His concern, however, is also with the relations between the various socially constructed scales, something that receives further echoes in Hariprya Rangan's chapter. Taylor outlines a provocative critique of the role of states in the development process, articulating concerns that he has outlined elsewhere (1996, 2000) about embedded statism as a seemingly congenital feature of contemporary social science. As an alternative he wants to foreground the role of cities and city networks in the development process, networks that span international frontiers and whose effect is to produce a pattern of uneven development that slots only imperfectly into the absolute spaces defined by the territorial jurisdictions of states.

Mohan and Stokke also focus on interscalar relations, though more from the standpoint of the scalar division of labor of the state and the boundaries between state and civil society. They are particularly interested in recent moves, stimulated by a development discourse fomented by international organizations and NGOs, to encourage a decentralization of development initiatives and greater popular participation through enhanced democratization (see also Slater, 2002). As they point out, it merges comfortably with calls for a rather different shift in the boundary between state and civil society, a shift to be engineered through the more obviously neo-liberal strategies of privatization and deregulation.

In the US, development has had decentralized, local inflections for a long time – at least since the Second World War, as Andrew Wood makes clear in his chapter. Recently the structures and practices with which it has been associated, particularly those of territorial competition for inward investment, have crossed the Atlantic and have become incorporated into a politics of scale that has been intensively interrogated by British geographers. David Harvey (1989) described this transformation as the supersession of urban managerialism, associated with a local state more alert to distributional imperatives, by urban entrepreneurialism. But, as Sue Parnell points out in her chapter, international approaches to urban development in poorer country contexts are recovering a strong redistributional imperative. Her chapter focuses on the rise of an international rights discourse around urban infrastructure provision, and explores the case of the recent politics of service delivery in Cape Town.

Recent debate in human geography has centered, *inter alia*, on the distinction between areas (or places) and networks (Escobar, 2001). Similar sorts of constructions are evident in the chapters in this section of the *Handbook*. As I pointed out above, Peter Taylor presents an argument for the desirability of thinking about development in terms of city networks rather than the absolute spaces defined by the jurisdictions of states. Hariprya Rangan comes at this issue from a different direction, but a major point that she makes is the sheer artificiality of state boundaries from a developmental standpoint. One result has been a highly territorialized politics between different, localized, contenders for the fruits of development or for the infrastructural elements thought of as decisive in bringing it about. In this she echoes Corbridge's references in his contribution elsewhere in the *Handbook* to work that suggested that economic support from the states of the developed world might be more effective if directly disbursed to more local centers of power in poorer countries. In the politics of local and regional development in the developed world, there is evidence that this sort of networking, breaching international boundaries, is already occurring (Paul, 2005).

Work on the political geography of development has also been enriched by consideration of people/nature relations (see in this volume the chapters by Robbins and Mansfield). Development is, of course, always about a relation to nature, and in that regard there is continuity with past commitments, albeit liberated from a history of environmentalisms and culturalisms that, from a Marxist perspective, grasped this relationship in highly fetishized forms.

Currently, there are a wide range of approaches to this issue of human/nature relations in development. At one end of the spectrum is the debate on sustainability; Yvonne Rydin writes about this in her chapter. Always, however, it is a relation that is socially mediated, as she demonstrates. Production relations matter. While what she calls hard sustainability is on the agenda of some, the soft sort, the

sort that works by commodifying nature, arranging exchanges, pollution charges, carbon credits, and so on, is all that existing material interests may allow. In some identifiable areas of work on the political geography of development, though, the relationship to nature has been conspicuously absent. This has been true of work on the politics of local and regional development, as Andrew Wood points out in his chapter, though this may be beginning to change now, and there is no reason in principle why it should not. And to be sure, the rhetoric of sustainability is increasingly being harnessed in fights around new development projects.

The chapters also remind us of the huge differentiation across the world in development challenges, practices, and outcomes. The developmental state is often excoriated, but obviously there have been more or less successful cases of this model in the past. Indeed, Rangan suggests, and contrary to the neo-liberal wisdom, that this is the *only* way in which development has come to the periphery. Not all developmental states are alike, though. The South Korean road of close state orchestration, even supervision (Amsden, 1990), was not that of the US with its tariffs, generous donations of land to the railroads, and land grant colleges. Furthermore, as Mohan and Stokke remind us, there is massive variation within the global South, much of it due to initiatives of a more local character, as in Kerala and Porto Alegre. This is not to argue that generalization is necessarily futile. When Sue Parnell refers to the specific urban challenges of many Southern cities in her chapter, in the context of attempts to provide subsidized services to the poorest of the poor, she is surely on to something quite general. Her point of departure is the recent attempt in Cape Town to provide free allocations of water and electricity to the poorest of residents. But issues of state capacity and informality, common to many Southern cities, get in the way. Means testing is daunting when so many residents, working in the informal economy (or perhaps seasonally), do not have a documentable income. In a context of migratory labor, common in many developing countries, there are problems of legibility, as Scott (1998, chap. 2) would define it. These are problems of how to simply identify who is living in the city and who isn't.

At a more abstract level, places have to be brought into a relation with spaces, notably, as far as recent debates in the political geography of development are concerned, with the global. The global framing of issues is a repetitive theme in these chapters, and particularly the role of supposedly 'global' organizations. Yet just as clearly the global is contested, or is a site of contestation (see, e.g., Dalby, and other contributions to Section 6, in this volume) Yvonne Rydin refers to struggles between those pressing for hard as opposed to soft sustainability, and part of this is linked to the experiences and prospects of people in different places. This is a very complex issue, but limits on carbon emissions – the Western oil lobby notwithstanding – look very different from the standpoint of developing societies than they do in North America or Western Europe; the metaphor of 'pulling up the drawbridge' has wide applicability in political geography. Even more obvious is the issue of increases in sea level. This threat is felt variably around the world, not simply in terms of some definition of vulnerability in terms of altitude above sea level, but also in terms of the availability of the resources to mitigate the problem. So the global is always a global for people in some places and not in others. The global is always constructed from interests constructed in particular places and with a view to the realization of those interests.

To talk about the global, though, may also be to invoke a scalar conception that Ferguson and Gupta (2002) have referred to as 'vertical encompassment'. The notion of the supranational, as in such 'global' institutions as the IMF and the WTO, has been assimilated to notions of fields of activity that are enclosed within – 'encompassed by', to use their words – wider fields of activity in a vertically ranked manner. 'Local', 'regional', 'national', and 'global', encompass each other in this way as though part of some spatial division of labor of the global polity. As Ferguson and Gupta indicate, this image is at odds with a good deal that we know about the world, an insight that is also apparent in Peter Taylor's emphasis on city networks that transcend international boundaries. Mohan and Stokke likewise draw attention to the developmental role of home-based organizations of migrants, something that Michael Peter Smith (2001) has examined in another context.

On the other hand, and perhaps quite clearly, the weakness of 'vertical encompassment' thinking is not merely at the level of spatial form. It also suggests, as Mohan and Stokke outline, that power proceeds from top to bottom, from the global to the local, as represented by the current ambitions of many international organizations. Yet it is equally apparent that power also proceeds upward. In this way, 'global' is constructed just as the 'national' is (see also Castree et al. and Routledge, this volume). Both are constructed by people in particular places with particular interests that they believe can be advanced through such a politics of scale. Robert Wade (1996, 1998) has been particularly interesting in this regard in his critical discussion of 'the global' and attempts by what he calls the IMF–Treasury–Wall Street nexus to impose these over the East Asian, developmental state model of how development should proceed. This is, in other words, a global for particular, powerful and place-based interests.

Moreover, the local itself gets constructed within the context of relations with other places, or more

accurately, and as Massey (2005) has urged, particular space-time configurations. Kerala and its developmental success is the outcome of a unique configuration of conditions and variously distanciated influences: a strong local Communist Party, for sure, and also a vibrant local civil society, as Mohan and Stokke outline. The federal structure of the Indian state and the legitimation of strong, pro-poor interventions provided for a long period of time by the evident durability of the Soviet bloc, and the Indian state's own commitment to what it called 'planning', have all come into play as well (on this see also Corbridge et al., 2005).

Theorizing the local, theorizing place, is therefore an essential aspect in understanding the politics of geographically uneven development, and it is with some thoughts on this that I want to conclude. We need to know why it is that the juxtapositions of influences and conditions that Massey (2005), for example, discusses get mobilized in the way they do, why broader structures get formed using these place-based raw materials, for what purpose and on behalf of whose interests? The chapters by Taylor and Wood are both particularly suggestive here, although in different ways.

In trying to explain local development initiatives, often orchestrated by coalitions of local government and various businesses, Wood draws on the idea of place dependence – a dependence on highly particularized, localized social relations (see Cox and Mair, 1988). Taylor, on the other hand, and drawing on the work of Jane Jacobs, outlines a sharp distinction between what she called two moral syndromes, a commercial one and a guardian one. The commercial one, as is clear, is about exchange and production; it is a moral standpoint that emphasizes the trust that enters into exchange relations, their zero-sum character. The guardian viewpoint is more about loyalty and protection and stresses the non-zero-sum nature of social relations, including commodity exchange; it is the standpoint of the state. The problem for development, according to Taylor, is when these moral syndromes get mixed up in what he calls 'monstrous hybrids'. This is the theoretical basis for his argument that we should cease prioritizing the role of the state in development since its interventions are often counter-productive. Rather it is cities and their relations one with another that provide the motor for development.

Whether one accepts that argument or not, the distinction overlaps with a recent one that David Harvey (2003, chap. 2) has made in talking about logics of power. Using a conception originally proposed by Arrighi, he sets up a contrast between what he calls capitalist and territorial logics of power. As in Taylor's case, the territorial logic – the guardian syndrome in his terms – is that of the state. He sees them as typically intertwined in various ways, though not necessarily in 'monstrous' ways as Peter Taylor does. It is at this point that we should again turn to Andrew Wood's chapter. For in talking about social relations of place dependence, Wood points to the material basis of all forms of territoriality at whatever scale, and more particularly to the tensions between territory and class that this gives rise to. This insight, of course, reminds us of David Harvey's own seminal contribution (1985) some twenty years ago to what he called 'the geopolitics of capitalism', a contribution that would seem, therefore, worth re-examining from the standpoint of theorizing the politics of geographically uneven development.

REFERENCES

Amsden, A. (1990) 'Third World industrialization: "global Fordism" or a new model?', *New Left Review*, 182: 5–31.

Clerk, G. et al. (1942) 'Discussion on the geographical aspects of regional planning', *Geographical Journal*, 99(2): 61–80.

Corbridge, S., Williams, G., Srivastava, M. and Veron, R. (2005) *Seeing the State: Governance and Governmentality in India*. Cambridge: Cambridge University Press.

Cox, K.R. and Mair, A. (1988) 'Locality and community in the politics of local economic development', *Annals, Association of American Geographers*, 78(2): 307–25.

Crush, J. (ed.) (1995) *The Power of Development*. London: Routledge.

Daysh, G.H.J. et al. (1949) *Studies in Regional Planning*. London: Philip.

East, W.G. and Moodie, A.E. (eds) (1956) *The Changing World*. London: George Harrap.

Escobar, A. (1995) *Encountering Development*. Princeton: Princeton University Press.

Escobar, A. (2001) 'Culture sits in places: reflections on globalism and subaltern strategies of localization', *Political Geography*, 20(2): 139–74.

Ferguson, J. and Gupta, A. (2002) 'Spatializing states: toward an ethnography of neoliberal governmentality', *American Ethnologist*, 29(4): 981–1002.

Gravier, J.-F. (1958) *Paris et le Désert Français*. Paris: Flammarion.

Harvey, D. (1985) 'The geopolitics of capitalism', chap. 7 in D. Gregory and J. Urry (eds), *Social Relations and Spatial Structures*. London: Macmillan.

Harvey, D. (1989) 'From managerialism to entrepreneurialism: the transformation in urban governance in late capitalism', *Geografiska Annaler*, 71B: 3–17.

Harvey, D. (2003) *The New Imperialism*. Oxford: Oxford University Press.

Massey, D. (2005) *For Space*. London: Sage.

Paul, D. (2005) *Rescaling International Political Economy*. London: Routledge.

Power, M. (2003) *Rethinking Development Geographies*. London: Routledge.

Power, M. and Sidaway, J.D. (2004) 'The degeneration of tropical geography', *Annals of the Association of American Geographers*, 94(3): 585–601.

Scott, J.C. (1998) *Seeing Like a State*. New Haven: Yale University Press.

Slater, D. (2002) 'Other domains of democratic theory: space, power and the politics of democratization', *Society and Space*, 20: 255–76.

Smith, M.P. (2001) *Transnational Urbanism*. Malden, MA: Blackwell.

Taylor, P.J. (1996) 'Embedded statism and the social sciences: opening up to new spaces', *Environment and Planning A*, 11: 11–31.

Taylor, P.J. (2000) 'Embedded statism and the social sciences, 2: Geographies (and metageographies) in globalization', *Environment and Planning A*, 32: 1105–14.

Wade, R. (1996) 'Japan, the World Bank, and the art of paradigm maintenance: the East Asian miracle in political perspective', *New Left Review*, 217: 3–37.

Wade, R. and Veneroso, F. (1998) 'The Asian crisis: the high debt model versus the Wall Street–Treasury–IMF complex', *New Left Review*, 228: 3–24.

Williams, R. (1976) *Keywords*. London: Fontana.

The Political Geography of Uneven Development

Peter J. Taylor

INTRODUCTION

Since the Second World War, the idea that uneven development is a problem that should be addressed has spawned a veritable 'development industry' of UN agencies, government departments, academics and non-governmental organizations. The literature advising, monitoring and assessing this putative process of development is truly massive, much too large to review in this single chapter. Fortunately most of it is outside geography, and within geography most of it is outside political geography. Rather than bemoan this relative neglect, I am going to treat it as an opportunity.

Overall the general argument of the chapter is that the idea of development – lessening uneven development – has been an intellectual prisoner of state-centric thinking. To conceive of development happening except through states has been beyond the comprehension of the social scientists who research development. This is because social science itself suffers from a severe case of embedded statism (Taylor, 1996). This embeddedness means that states are taken for granted as the context of development. In these circumstances, political geography with its explicit and overt focus on the state can be a means of shifting this intellectual albatross. Political geography puts the state at the centre of much of its work because the state is the fulcrum of political power in the modern world-system. But the ultimate object is to understand the disposition of power across spaces and places. If it can be shown that

states are not the entities that are relevant in a given context, then our object means that we find an alternative entity to the state. In this chapter I hope to show that the state is a particularly inappropriate vehicle for understanding development as the expansion of economic life. Thus, a political geography of uneven development has potential to provide unusual insights into what uneven development is, and what development could be.

My unusual argument divides into seven sections. First, I begin with a short section that provides a brief historical introduction to the idea of uneven development in political geography. Second, this leads into a discussion of what development is and why it is uneven, which requires, third, an exposition of structures in social scientific reasoning: structures in Marxist and world-systems understandings of uneven development are described. Fourth, an addition structural layer is added through the use of Jane Jacobs' (1992) identification of two moral syndromes, commercial and guardian. Fifth, mixing moral syndromes produces 'monstrous hybrids' and I identify state-led development as such: guardians doing commercial work very badly and usually corruptly. Sixth, with the state identified as inappropriate for development, I use Jacobs again to identify cities as the key entities for expanding economic life. Seventh, in a short concluding section I ask what all this means for a new understanding of uneven development that combines different structural positions and alternative economic agents.

THE THEME OF UNEVEN DEVELOPMENT IN POLITICAL GEOGRAPHY

Traditionally, uneven development has not been a major theme in political geography. To be sure, political geography has a strong 'global' heritage, but this has focused upon disparities in political power rather than between rich and poor countries. In broad global geographical terms, political geographers have been particularly concerned with 'East–West', not 'North–South'. But for all the emphasis on power politics through geopolitical lenses, political geography has never concentrated solely on the 'big players'. In particular, the regional tradition when translated into political geography produced region-by-region worldwide surveys that encompassed poor countries. This can be seen in Bowman's (1921) classic text, and the approach continued into the 1960s with Crone's (1967) *Background to Political Geography*. The latter's regional framework contains a chapter entitled 'Developing countries, in particular India', which acknowledges that decolonization is a 'universal revolution' that has 'transformed the political structure of the world' (ibid.: 111) through multiplication of states. But as with this genre of geography in general, there is little analysis, just hints at explanation through comparing maps showing features that 'impede development' (environment and diseases) plus discussions on primary product dependence and on population pressures. It is only with the rise of more systematic geographies, which generated thematic organization of political geography, that uneven development becomes a standard theme in political geography.

Pounds' (1963) *Political Geography* was the leading systematic text of the 1960s – the heyday of development thinking and modelling – and includes a chapter on 'The undeveloped world'. Starting with the admission that the 'colonial legacy' created 'unbalanced economic development' (ibid.: 357), he finally develops a strong Cold War argument for the salience of the topic: poverty plays into the hands of communists (ibid.: 376). Thus his major theme is comparative aid dispensations with the United States guided by 'human welfare', European states the same but 'to a lesser extent', both in contrast to the USSR which 'undoubtedly' only uses aid to extend its political power (ibid.: 377). (Pounds' own maps of aid do not support this contention, especially when compared to World Bank investments.) The first major reader on political geography compiled by Jackson (1964) similarly finds a place for development but from a very different analytical viewpoint. Three readings on 'the political problems of newly independent states' are provided, one discussing the enabling role of natural resources in development and the others drawing excerpts from Myrdal's (1957) *Rich Lands and Poor*. Although not a geographer, the latter's core/periphery approach is spatial and therefore was attractive to geographers at this time. Basically he argues that at the international scale, economic 'backwash effects' overwhelm 'spread effects' so that poor countries will have a very difficult task in catching up rich countries (Jackson, 1964: 283). This way of thinking is repeated in the next major reader (Kasperson and Minghi, 1969), which provides just one reading on development. From Wallerstein's (1961) interpretation of difficulties facing newly independent African states, this reading casts doubt on the efficacy of these states to be able to produce development.

Uneven development as the 'geography of inequality' became a major area of growth in the 1970s. It is briefly featured in Bergman's (1975) *Modern Political Geography*, in which he laments increasing disparities between 'rich and poor' as an 'unfortunate trend in international affairs' (ibid.: 378), but it was most fully developed outside political geography as a new 'welfare geography' (Coates et al., 1977; Smith, 1977). This work looked at inequalities at different geographical scales and found, like Myrdal, the international to be the most intractable to welfare policies – aid policy is a very poor relation to the welfare state. These debates entered political geography though Cox's (1979) text *Location and Public Problems: A Political Geography of the Contemporary World*, which represents a new departure for the subdiscipline. Here uneven development comes to the fore and, with geographical scale, is used to provide a rare integrated subject matter. This welfare approach elided into more radical political positions and the early 1980s witnessed both a broadly Marxist political geography text (Short, 1982) and a world-systems interpretation of political geography (Taylor, 1982, 1985). The former began explicitly with a chapter on uneven development as 'the capitalist whirlpool' and the latter organized its arguments around a core/periphery model. These two approaches are going to figure prominently in what follows and therefore I will conclude this brief historical excursion into uneven development in political geography here.

WHAT IS 'DEVELOPMENT' AND WHY IS IT 'UNEVEN'?

The basic subject matter of all social science, including political geography, is social change. Development is a way of describing social change. It is one of a cluster of change concepts that have arisen out of Enlightenment thinking wherein social change was conceived in positive terms: the new is better than the old, modern is superior to tradition (for further exposition, see Rangan, this volume). Other such concepts are 'improvement' and 'progress'. Improvement was used to describe particular techniques of enhancing productivity in agriculture and industry in the eighteenth century. The culmination of this technology in the Industrial Revolution imbued nineteenth-century society with an air of optimism based upon the idea of progress. In the twentieth century, the concept of development took on the modern role of describing social change for the better. In this case the positive connotations were derived from a biological analogy. Living organisms develop to become fully mature adults; development is both the process and the outcome. But using this analogy requires specification of the 'social organism' that is subject to change.

Development as social change has been centred on the state; it is states that 'mature' into fully developed political and economic entities. Whereas the specific object of improvement was economic sectors, and progress was a general property of civilization, the practice of development has had precise territorial bounds defined by the state in question. Hence development is state policy; it is territorial planning. Linking this 'inherently positive' concept to the state is critical to understanding uneven development in the twentieth century and its contestation into the twenty-first century. The political geography of these times explains how this came about.

Development is closely linked with the great political geography revolution of the mid-twentieth century that is decolonization. A process that more than doubled the number of sovereign states, it also brought forth development as its key idea. Put simply, development was what these independent states were supposed to do: it was their prime goal, indeed their very *raison d'état*. Freed from colonial shackles, the new states now had the opportunity to 'catch up' with their erstwhile captors. It is in this context that development became a ubiquitous policy of all poor countries. Freeing them from being described as 'backward countries' (the nemesis of progress), they were promoted to being first 'undeveloped', and then they soon became the more dynamic 'underdeveloped', before being finally labelled – even more dynamically – as 'developing' countries. The latter terminology continues to be used despite the fact that, after several UN-proclaimed 'Decades of Development', the countries that were poor in the mid-twentieth century overwhelmingly continue to be poor in the early twenty-first century. Furthermore, there is strong evidence that this worldwide economic polarization has increased over recent decades. Hence the continuing concern for *uneven* development.

Why this geography, why not a pattern of 'development' that is even, or at least random, in pattern? The broad answer is that social change is a geohistorical process in which imperialisms, with their colonial exploitative relations, have been central. It is still the case that the colonizers – European and European-settler states (plus the 'late colonizer', Japan) – constitute the rich countries and the colonized (economic as well as political) largely remain poor countries. This geographical continuity is impressive and will be discussed further as a 'structural' feature of our social world below. Here I focus on the material underpinnings of imperialisms in the clustering of economic growth.

I have shifted to discussing economic growth because it comes without the teleological implications of development – development always implies change into a known condition. In addition, although development is a broader concept than economic growth (it usually encompasses political development, interpreted as liberal democracy, and social development, interpreted as modernization), it is economic development that is usually treated as the critical process; as noted above, it is basically about conversion from poor country to rich country.[1] Changing from development to economic growth dispenses with the unnecessary intellectual baggage of the organic analogy. Growth means simply increase in size and economic growth means merely increased production of commodities. Such economic growth is inherently uneven; it clusters in both time and space.

Historically, economic growth has come in spurts. Derived primarily from innovations, both technical and organizational, the increased productivity enabled more commodities to be produced from the same input of raw materials and labour. These innovation clusters in time are associated with cycles of economic growth wherein the initial growth ultimately gives way to economic stagnation or decline as the 'new' products become old in a saturated market. There are two main 'long cycles' produced from such processes. Kondratieff cycles of *circa* 50 years are the result of innovations in particular leading industries that dominate an era. For instance, the mass production of cars and other large consumables in the post-Second World War economic boom is the growth phase of the most recent such cycle. Hegemonic cycles of *circa* 100 years are based upon broader production, much of it infrastructural and associated with leadership by one country. The twentieth century as 'American century' is just such a cycle emphasizing mass consumption, which contrasts with the previous century wherein Britain pioneered industrial production.

Identification of hegemonic cycles with specific dominant countries – hegemonic states – is an example of how clustering over time is related to clustering in space. Clustering of innovations in space is a result of specific social networking and general infrastructural settings that provide enabling spaces to generate new products. In Marshall's famous phrase, these are regions with an 'industrial atmosphere'. Thus every one of the spurts of innovation in time is clustered in a particularly favoured place. There is a large literature on economic clusters and regional agglomerations that describes clusters at different geographical scales, from those based upon face-to-face relations in districts within cities, through mega-city regions, to trans-state industrial regions. However, the ultimate scale of this clustering of economic processes and well-being is global unequal development portrayed as 'developed/developing worlds' or, more geographically, as simply 'North/South'. In this argument the rich countries (developed or North) have historically been the locales of economic innovations and clustering.

The implications of these arguments for development thinking are quite profound. Currently there is much concern for the fact that the poorest continent, Africa, is falling even further behind economically. There are many reasons for this, both economic and political, but one policy argument keeps reappearing: why not devise a 'Marshall Plan' for Africa? This would involve rich countries transferring capital to African countries to stimulate their economic growth and thus set them on the way again to development. The original Marshall Plan in the early post-Second World War years involved transfer of US capital to a war-devastated Europe. In this case European economies soon revived and the Marshall Plan was deemed a success. Can this successful process be transported to contemporary Africa? The answer is an emphatic no. The Marshall Plan worked because it was devised in the right time and place. Starting at the beginning of the Kondratieff post-war boom, it was partly instrumental for diffusing US economic growth to Europe, but this was only possible because Europe's devastated cities retained a social capital that could invest US economic capital into profitable economic growth. Even in these highly conducive times, it is only in Europe that such a plan could possibly work. At the current time Africa has no equivalent stock of social capital and, in any case, the times are not so conducive to such diffusion.

There are two key issues that arise from this definitional excursion and that permeate the rest of the discussion. First, there is the question of structure versus agency. Critical theory on development argues that uneven development is 'structural' and therefore not amenable to development policy and planning by states. Although history appears to be proving the theorists right, much more needs to be said about this issue. Second, there is the question of geographical focus, whether this should be territorial states or networked cities. Shifting from an emphasis on development, which is what states are supposed to do, to economic change and clusters, in which cities feature prominently, raises the question of the geographical framework in which economic change is assumed to take place. Once again a lot more needs to be said about this issue because the failure of more than half a century of 'development' puts the idea of states as 'economic entities' severely in doubt.

THE QUESTION OF STRUCTURE

According to Neil Smith (1984: xi), uneven development is not 'statistical', it is 'structural'. By this he means that although we can indeed measure material inequalities across the world, such statistical results do not even begin to tell the full story. The inequalities that comprise uneven development are not a simple event to be measured; they are longstanding differences that are integral to the nature of the social world in which we live. Such social structures are not unchanging, but they change very slowly over generations. These are defined by Braudel's (1972) *longue durée* wherein people's ordinary behaviour carrying out their everyday tasks produces structures – cultural habits and economic outcomes – beyond the ken of the people doing the behaving. Thus, individuals cannot unilaterally decide to change a social structure; such social change is not amenable to the policies of particular agents, such as states, however powerful.

Structures are just one element of social change. As well as the *longue durée,* Braudel (1972) identifies two other time scales of change, medium-term and short-term. The latter is constituted of events, the raw material of traditional history; the former encompasses most of the subject matter of the social sciences, for instance, the way in which institutions such as markets operate. These three 'times' with their associated social structures, social processes and social events are brought together in the critical realism approach to social science (Sayer, 1992). We can observe and therefore measure events, but only events. Thus, statistics and quantitative methods in general can only be applied to data at this 'surface level' of analysis. Underlying the events are mechanisms (social processes) that generate the events as outcomes; world market mechanisms are an example of such processes, for instance the mechanisms behind the oil market that produces the event that is a hike in gas/petrol prices. Since these mechanisms cannot be directly observed, they have to be inferred from evidence of consequential events. Some quantitative techniques aid this research task, but qualitative methods focusing on agents associated with the mechanisms are usually deemed necessary to understand social processes. Furthest from events, and thus from direct observation, there are the structures, sources of mechanisms. Structures encompass necessary relations for social reproduction and contradictions that steer that reproduction through generated mechanisms. But structures cannot be studied empirically; our understanding of them can only be theoretical. Hence this is the key level of contestation in social science. It is Marxist and neo-Marxist social science that is best represented through critical realism and both have been central to the study of uneven development in political geography.

Neil Smith (1984) introduced Marxist understanding of uneven development in his seminal *Uneven Development: Nature, Capital and the Production of Space*. Although Marx himself argued that the spread of capitalism would have a 'universalizing tendency' with a 'consequent drive towards equalization' (ibid.: 95), Smith uses debates among later Marxists to show why uneven development has continued. In particular, Lenin's concern for the importance of geographical differentiation *within* capitalism was brought forward to provide an opposed tendency to equalization. Thus, Smith posits a 'logic of uneven development' that includes simultaneous tendencies to produce 'development at one pole and underdevelopment at the other' (ibid.: xiii). This results from the contradiction between use value and exchange value in the growth of capital: fluid (financial) capital (investment) is continually on the move to find conditions to maximize its expansion, but in the process creates fixed (productive) capital (infrastructure) that is eventually abandoned to facilitate higher profit rates. Thus at any one point in time there will be a 'profit surface' from a development pole with high rates of profit and therefore investment and employment to an underdevelopment pole with low rates of profit and therefore less investment and less employment (ibid.: 148–9).

Smith argues that capital will 'see-saw' between these two poles as the advantages of the first pole are overtaken by the opportunities of the second pole. In other words, capital does not seek equilibrium, it uses uneven development to transcend its contradictions. At a regional level this creates a shifting geography of uneven development like Massey's (1984) geological model of investment/non-investment layers in economic region landscapes. But according to Smith (1984: 140–1), at the global level, the see-saw mechanism is curtailed because 'the real integration of the global space-economy is necessarily incomplete' due to 'the development of underdevelopment' (ibid.). This is an extreme tendency towards geographical differentiation that operates strongly at this largest scale. More recently, Weeks (2001: 9–10) has conceptualized this as two forms

of uneven development: *primary* uneven development between core and periphery creating divergence, and *secondary* uneven development within the core where there has been convergence.

Smith borrowed the concept of the 'development of underdevelopment' from Gunder Frank (1969), for whom it was the centrepiece of his radical dependency theory, and I will treat it as part of the neo-Marxist structuralist thinking here. Neo-Marxist ideas have entered political geography largely through world-systems analysis. Devised by Immanuel Wallerstein (1979), this approach to social science brings together Braudel's concepts of time with Frank's conceptualization of world-space. The initial target was the 'developmentalism' of the 1950s and 1960s that posited countries developing along parallel paths to the goal of development – described as the stage of 'high mass consumption' in the most famous example of the genre (Rostow, 1960). Thus, all countries will, or at least can, become 'developed'. Frank (1969) argued emphatically that this was not the case. There were in fact two mega-processes unfolding: development in rich countries and the development of underdevelopment in poor countries. Thus, underdevelopment cannot transmute into development, it is a separate, albeit related process. Wallerstein has converted Frank's theory into a core/periphery structure.

In Wallerstein's (1983) study of 'historical capitalism' the basic unit of social change is the modern world-system (or capitalist world-economy), a socio-spatial entity defined by Braudelian rhythms and cycles in conjunction with a core/periphery structure. This uneven development is continually being reproduced by two sets of processes: core-making processes and periphery-making processes. The former creates high-tech, high-wage, complex economic mixes; the latter low-tech, low-wage, simpler economic mixes. Note that these are both processes, mechanisms derived from economic structures that generate obverse economic situations in the manner of Frank's model. Historically, the processes have not operated evenly or randomly but have become concentrated in different areas. Thus, Wallerstein defines core and periphery zones in the world-economy, interlinked as a single economic system. There is nothing inherently core-like about areas in the core zone, their economic processes are simply dominated by core-making processes (and vice versa for areas in the periphery zone).

The distinguishing feature of Wallerstein's argument is his identification of a semi-periphery. This is defined as areas where core-making and periphery-making processes are approximately balanced (there is no such thing as a 'semi-periphery process'). The semi-periphery is the dynamic zone; areas move up and down in this spatial hierarchy through the semi-periphery.

Because this spatial alternation is often viewed as state-led, the semi-periphery zone is commonly identified as a set of states and this has had the knock-on effect of all three zones typically being demarcated as groups of states (e.g. Arrighi and Drangel, 1986; Terlouw, 1992). However, the economic processes are not necessarily state-bound (for example, this was not the case for the early modern world-system as Wallerstein (1974) describes it). But the core/periphery structure is necessary: dynamism is compensated – for every rise of a Japan from semi-periphery to core, there is the fall of an Argentina from semi-periphery to periphery.[2]

The final example shows that structural thinking on uneven development does not preclude all change between zones. Change is rare, but with favourable contingencies (world-systems analysis emphasizes unusually strong states, but notes that 'development success' has always come in city-rich areas) the development of underdevelopment can be neutralized so that core-making processes come to dominate. But this cannot happen everywhere simultaneously as traditional developmentalism assumes. The modern world-economy is ever-changing but uneven development remains integral to its structure.[3]

STRUCTURES OF NORMATIVE BEHAVIOUR

Following Braudel, I have identified structures with everyday routine behaviour that collectively defines the limits of the possible in social change. Using Marxist and neo-Marxist thinking, I have identified materialist structures underpinning uneven development. But social structures are not solely materialist; all behaviour is premised upon values and morals that guide the way social relations operate and therefore are implicated in social change. At the deepest level there are structures of normative behaviour, and these equally underpin uneven development. I use the pioneering work of Jane Jacobs (1992) on moral philosophy in which she describes 'systems of survival' as 'moral syndromes'. I interpret these as structures of normative behaviour for reasons that will become clear as I proceed.

Jacobs' (1992) key contribution is to identify two distinctive moral syndromes that are separate sets of precepts reflecting deeply different values. There are just two syndromes because there are only 'two distinct ways of making a living, no more and no less' (ibid.: 51). First, there is simple taking what we need, work that ranges from hunter-gatherers harvesting from their territories to modern governments collecting taxes. Second, and unique to humans, there is trading, exchanging goods and services to satisfy needs, work that

ranges from early peddlers to contemporary multinational corporations. These can be summarized as 'raiders-takers and traders-makers', each of whom operates through distinctive structures of normative behaviour that are the result of 'millennia of experience with trading and producing, on the one hand, and with organizing and managing territories, on the other hand' (ibid.: xii). She calls these the commercial syndrome and guardian syndrome; the precepts that constitute each structure are shown in Table 31.1. These precepts and their separation are derived inductively (Keeley, 1989: 34–5; Jacobs, 1992: 25–30). This is possible because, according to Jacobs (1992: 133), the precepts form coherent 'systems' of ideas: 'they arose existentially as events and activities required them and tested them' (ibid.).

A good way to understand these structures and their opposition is through what Jacobs terms the 'key virtue' (ibid.: 69) of each. These are honesty for the commercial syndrome and loyalty for the guardian precept. Quite simply, a successful working commercial life requires honesty, whereas a successful guardianship requires loyalty. And as we know, these two values often clash. Here is a simple example I have used to illustrate this (Taylor, 2006a). When we write references for job applicants we have to be 'objective' – that is, be honest – even though, by definition, the person is known to us and therefore we may feel obliged to be a little economical with the truth. We are expected to resist this temptation because the job market

Table 31.1   Jacobs' moral syndromes (structures of normative behaviour)

Commercial syndrome precepts	Guardian syndrome precepts
Be honest	Be loyal
Dissent for the sake of the task	Deceive for the sake of the task
Shun force	Shun trading
Come to voluntary agreements	Exert prowess
Collaborate with strangers and aliens	Be exclusive
Be open to inventiveness and novelty	Adhere to tradition
Use initiative and enterprise	Be obedient and disciplined
Respect contracts	Respect hierarchy
Be industrious	Make rich use of leisure
Be thrifty	Be ostentatious
Invest for productive purposes	Dispense largesse
Promote comfort and convenience	Show fortitude
Be efficient	Treasure honour
Be optimistic	Be fatalistic
Compete	Take vengeance

relies on honest testimony – we are operating in the commercial syndrome. However, note that we would not normally be asked to write a reference for a relative; here the pressures of loyalty are assumed to be too great. The expectation is that we would be guided by the guardian syndrome. In the wider world, banks can only operate through honesty and trust, armies can only operate through loyalty and authority. These values are not just incompatible, their substitution creates disaster: a bank making decisions based on loyalty would soon go bankrupt; an army making decisions based on honesty would soon lose the war.

As Table 31.1 illustrates, these key virtues are just the central precepts of two wide-ranging syndromes. I will quickly contrast summarily. Commerce allows dissent (disloyalty), guardians allow deception (dishonesty). Force has no place in commercial life whereas for guardians trade is demeaning. Commerce requires a cosmopolitan outlook, guardians thrive on exclusiveness. The commercial syndrome is a litany of progress (novelty, enterprise), the guardian syndrome is a litany of tradition (hierarchy and obedience). They deal completely differently regarding capital: thrifty industriousness contrasts with ostentatious leisure. The final three pairs of precepts sum up the idealised outcome of this behaviour: economic efficiency versus social honour, optimism versus fatalism, and finally opposition is a matter of competition in commerce and vengeance in guardianship. Jacobs (1992: 27) describes these precepts as 'linked and overlapping clusters' in arguing for their innate coherence; Table 31.1 is my organization of her precepts as a coherence of opposites (Taylor, 2006a, 2007). Reference should be made to the original source (Jacobs, 1992) for further explication of the two syndromes. I will accept her framework as delineating structures of normative behaviour in the remainder of this chapter.

As structures underpinning moral codes, there are several mechanisms that can arise where processes create distinctive outcomes. First of all, Jacobs (1992) insists that both of her syndromes are equally necessary and the best outcomes are where they generate distinct and separate tendencies in society. *Ipso facto*, the worst outcomes are when the integrity of one of the syndromes is destroyed by infiltration by the other (ibid.: 101). The worst-case scenarios are what she calls 'monstrous hybrids' when one syndrome's virtues are transplanted into activities associated with the other syndrome. For instance, both the Mafia (ibid.: 92–7) and the USSR (ibid.: 98–102) are identified as monstrous hybrids wherein guardian values are used in commerce:

Structurally, they do much resemble each other; into an otherwise strong guardian syndrome comes the massive breach of the guardian precept to shun trading. Since the guardian syndrome is neither morally nor functionally suited to carrying on production and trade, the commerce involved is corrupted and its moral foundations ruined. (Jacobs, 1992: 102)

The result is ultimate economic disaster for areas whose economies are at the disposal of guardians either as private extortionists or state planners. These can be considered 'internal' mechanisms of the development of underdevelopment.

DEVELOPMENT AS A MONSTROUS HYBRID

I interpret 'development' as conceived and practiced since the Second World War as a classic case of one of Jacobs' monstrous hybrids. Monstrous hybrids arise because the two syndromes produce contrasting 'casts of mind' that act as fetters on what is seen as possible (Jacobs, 1992: 128–30). Guardians assume a world of zero-sum games: taking territory is a win-lose process. The commercial syndrome engenders a cast of mind that allows for growth, always searching for win-win deals, entered into voluntarily by both sides. Thus, however we define 'economic development', it surely must involve economic growth and therefore should be kept out of the hands of guardians. Just the opposite has occurred.

In the UN 'Decades of Development', the practice of development as an economic process under guardian direction (i.e. a monstrous hybrid) has engaged five important processes that ensure the development of underdevelopment (Taylor, 2006a):

1 Agriculture is more important in poor countries than rich countries and this economic sector is particularly susceptible to undermining by guardian precepts.
2 There is the private banking relationship to poor countries that has severely violated the commercial syndrome.
3 There are the Bretton Woods multilateral financial organizations (World Bank, IMF) that masquerade as banks but are essentially guardian institutions.
4 There are the donor countries of bilateral aid that are explicitly guardians and act that way.
5 The poor countries themselves are, of course, guardian institutions and, in hindsight, it seems incredible that rich guardians in poor countries should ever have been entrusted with 'development'.

I will explicate each of these processes that contribute towards the development of underdevelopment in turn (see also Taylor, 2006a).

Agriculture is a commercial pursuit but the fact that it is a land-based activity complicates its moral integrity. The territorial concerns of guardians have typically meant that this commercial practice is grossly interfered with. As Jacobs (1992: 118–19) tells it: '… an economic activity that is functionally and morally commercial has historically been skewed to conform to the contradictory values and morals of guardian landownership. … Rulers long ago became preoccupied with agriculture because it meshed with their preoccupations with territory. Tradition has perpetuated the fixation'.

And the fixation has been perpetuated in both rich and poor countries. In the former, agriculture is marked out as *the* 'guardian industry' through its large state subsidies – the EU budget still makes the institution look like a farmer's club – which reduce markets for poor countries' produce. In poor countries this same guardianship cast of mind has undermined land reform, a ubiquitous 'development policy' notoriously difficult to achieve successfully. As countries with large agricultural sectors, 'development' is severely hampered by so much of the economy being affected by guardian precepts such as exerting prowess instead of coming to voluntary agreements and adhering to tradition rather than being open to inventiveness and novelty.

International banking is another sector indicted for its moral integrity in 'development'. In the rich countries, the 1980s are famous for investment banks transgressing the commercial syndrome. Take-overs by leverage buy-out whereby the company pays for its own purchase with future debt is a hostile win-lose game: it is fundamentally the same as the traditional guardian mechanism of winning territory and then plundering the land to pay for the war (Jacobs, 1992: 142). In poor countries a similar transgression led to the debt crisis. Lending money without a realistic chance of repayment is not a commercial banking process. Quite simply these loans were *never* investments, written down or off. Rather, they represent largesse distributed at the behest of the World Bank, initially for Cold War political reasons and subsequently used to force neo-liberal policies on poor countries. This dispensing of largesse rather than investing for productive purposes is a classic case of guardian corruption of 'development'.

The World Bank has been implicated in the previous transgression but, with its Bretton Woods twin, the IMF, it is a major transgressor of the commercial syndrome in its own right. The fact that they both deal with the financial matters does not make them banks in any commercial sense. Although they are the financial arm of the UN family of institutions, they are located in Washington, DC, rather than with UN headquarters in the global financial centre of New York. Quite simply, global financial markets are not of concern to the World Bank and IMF, but being near the political centre of the world is important. With states as shareholders, the United States with the largest shareholding, and rich countries in general overwhelmingly in control, these are guardian institutions that deal with finance. This has been most clearly illustrated in their ostentation reflected as a fetish for large projects. Dams, with their grand scale and destruction of productive territory, are these global guardians' favourite; Jacobs (1984: 122–3) has suggested that future archaeologists discovering our civilization will surmise that we were a strange cult that built massive water containers to express unknown religious needs. Such an interpretation is at least as good as assuming that dams are about economic growth.

Nearly all rich countries come nowhere near meeting the 0.7 per cent target of GNP for the aid budget. Quite properly this is taken as a sign that these countries are less than fully committed to 'development' of their poorer peers. But there is a different perspective to be taken on this: are donor countries through bilateral aid programmes suitable vehicles for promoting economic growth outside their own territory? Given their adherence to the guardian syndrome, the answer must be an emphatic no. Put together the precepts 'be loyal', 'exert prowess' and 'be exclusive' and a very dysfunctional economics results. The process is that it is the donor country's interests that come first, with loyalty to their own firms (with their voter-workers) satisfied by exerting pressure to produce monopoly positions for these firms. Arms deals usually follow this mechanism, but dams also feature here too. Of course, the recipient poor country knows that if it does not fall into line, vengeance (sanctions) may be forthcoming. If extreme departures from the commercial syndrome (bribery) risk being revealed, the donors can invoke state security, which translates as deceiving for the sake of the task.

Bilateral aid is a transaction between two sets of guardians, and recipient poor countries are equally indictable here. By designating 'development' as a state process, economic growth is delivered into the hands of guardians. Mixing moral syndromes creates 'intractable systemic corruption' that undermines both commercial and guardian normative behaviour (Jacobs, 1992: 133). Pervasive economic and political corruption ensue as found, for instance, in the United States with its Savings and Loans outrage, its more recent Enron/Anderson affair plus the endemic Pentagon procurement scandals. In poor countries this mechanism is reflected in endemic corruption as guardians, often from military backgrounds, apply a totally inappropriate cast of mind to the task of 'development'. Here the whole list of guardian precepts from loyalty to vengeance (Table 31.1) comes into play to provide a moral infrastructure for wholesale looting of countries. The result is that territorial sovereignty is often reduced to the right of generals to be ostentatious, vicious and very corrupt. The one lesson the UN 'Decades of Development' have taught is that the combination of 'development' and 'state' is a *non sequitur* for poor countries.

The political geography message so far has been quite depressing. But economic growth does occur without 'development policy/planning', and this is where we should look for a more hopeful analysis of uneven development.

CITIES AND THE EXPANSION OF ECONOMIC LIFE

States are instrumental in the mechanisms that sustain uneven development, so it is rather perverse that they also provide the geographical frame for understanding and promoting 'development'. But is there an alternative geographical analysis that can provide a different understanding and provide a starting point for viable policies to combat poverty in the non-core zones of the world-economy? As well as providing the model of structures of normative behaviour that indict states, Jane Jacobs has also become what leading economist Paul Krugman (1995: 5) calls 'a patron saint of the new growth theory'. She provides a theory for expansion of economic life that is based not on states but cities (Jacobs, 1970, 1984).[4]

Jacobs' (1984: 31–2) key point is that states are 'political and military entities' and that 'it doesn't necessarily follow from this that they are also the basic, salient entities of economic life'. Thus if we

> … try looking at the real economic world in its own right rather than as a dependent artefact of politics, we can't avoid seeing that most nations are composed of collections or grab bags of very different economies, rich regions and poor ones within the same nation. … We can't avoid seeing, too, that among all the various types of economy, cities are unique in their abilities to shape and reshape the economies of other settlements. … (ibid.: 32)

Hence it is to cities that we should look for both understanding and promoting economic growth.

Jacobs defines cities as a process. The process is import replacement. It is by replacing imports from other cities that city economies expand economic life. This concentration in space is complemented by a concentration in time: the 'new work' of import replacement comes in spurts and any settlement that experiences such a transformation is deemed to be a city. The outcome of such expansions is to create dynamic, complex, and diverse economies.

The process encompasses two sub-processes that Jacobs (1970: 21) refers to, first, as the 'little movements' within city economies that turn, second, 'the great wheels of economic life'. Jacobs' (1970) discussion of these 'little' interactions within cities at the heart of this economic dynamism is a foundation text of economic clustering theory. But the second sub-process – the 'great wheels' – is generally less well developed by Jacobs. She does argue that 'cities need one another' (Jacobs, 1984: 135), but most of her description of this process involves the dyad relations of just pairs of cities. It is, perhaps, better to see cities as generically forming networks; the sub-process is therefore network formation (Taylor, 2004).[5]

However, Jacobs' (1984) examples of import replacement do clearly illustrate the essence of the process. For instance, the contrast between the economic fortunes of Japan and Argentina are encapsulated by the nature of their leading cities. In the late nineteenth-century, Tokyo imported bicycles from European cities. Soon bicycle repairs were needed and some of these repairers eventually became bicycle manufacturers, selling first in Tokyo and then in other Japanese and Asian cities, *ipso facto* reducing the import of bicycles. Obviously this new work had the effect of curtailing the markets of bicycle manufacturers in European cities, but this is not a zero-sum game. The erstwhile bicycle manufacturing cities now had new markets for their machinery and machine tool industries. Thus is created a kind of inter-city spiral of growth. At about the same time, Buenos Aires began to prosper as the leading port exporting agricultural produce from its large hinterland to European cities. This is a different process, producing a relatively simple local economy dependent on external markets. Hence whereas the Tokyo economy was able to spearhead a massive expansion in economic life through Japanese cities and other Pacific Asian cities in the twentieth century, Buenos Aires's fortunes fluctuated but it was never able to forge an equivalent economic surge through Argentinean cities.

Two things need to be made clear at this point in relation to states. First, it is totally irrelevant whether the import replacement is from a city in the same country or from a foreign city. Thus, second, this is totally different from import substitution policies commonly adopted by states as a means to industrialize and therefore 'develop'. Associated with the German nineteenth-century economist Friedrich List, this protectionist policy coincided with the economic successes of Britain's rivals, the United States and the German Empire, in the late nineteenth century. However, such protectionism is, as its name implies, actually a guardian approach to economics; it is about 'winning' work from other countries. In other words, it is ultimately premised on zero-sum thinking: rather than expanding economic life, it moves jobs around, 'taking' jobs from one country by preventing their imports. Trade protection was most certainly not a sufficient condition for the massive economic expansion in United States and German cities in the half century before 1914.

I have noted earlier that city economies 'shape and reshape the economies of other settlements' (Jacobs, 1984: 32). How do they do this? In fact there are two processes, one very positive, the other iniquitous. The former is the transmission of dynamic, complex, and diverse economic processes through city networks. This is why cities need each other. An isolated city or a set of cities with little connection defines a stagnant economic zone. Jacobs (1984: 124) calls these by-passed places; she gives the example of Ethiopia as a historical example of a region that has suffered this fate (ibid.: 129). In contrast, well-connected vibrant cities are the core-producing processes that Wallerstein (1979) describes in his world-systems analysis. However, the by-passed places do not constitute the periphery, rather they are external arenas, regions outside the modern world-system – no such regions currently exist. In contrast, the periphery is not defined by default in terms of what is missing; rather periphery is purposively created. It is the iniquitous mechanisms of the projection of city economic power that are periphery-producing processes.

Jacobs (1984: 42–3) identifies five 'great forces' of economic growth that between them produce diverse city economies. These are: enlarged city markets, increased numbers and kinds of jobs, transplants of city work, new uses for technology, and growth of city capital. Working together these forces are highly productive so that, for instance, as new technologies reduce jobs in one sector, new city work takes up the employment slack. But these same forces are iniquitous when they operate singly, without compensating each other. The result is that 'cities shape stunted and bizarre economies in distant regions' (ibid.: 59). Jacobs describes five 'economic grotesques':

1 City markets generate relatively simple 'supply regions'. These are the classic 'simple' economic regions dependent on fluctuations of prices that they can do little to influence. Where supplier monopoly mechanisms are possible (e.g. OPEC), surplus capital is converted into largesse rather than investment capital.
2 New city jobs cause 'abandoned regions'. Here the region comes to rely on money transmissions back to family that is usually used for ostentation rather than investment capital.
3 Factory transplants produce simple 'industrial supply regions'. These are the twentieth-century equivalents of the first 'grotesque', simple industrial branch plant economies that are notoriously vulnerable to economic conditions elsewhere.
4 New technologies create 'rural clearances'. Most labour is no longer needed as old production gives way to new city needs creating 'deruralfication'.
5 Surplus capital is behind mega-development projects. Again, we can mention dams as the classic example, but there are also numerous failed agricultural projects that have had equally disastrous results (the shrinking of the Aral Sea being a sort of inverse dam project).

These are all mechanisms that produce the development of underdevelopment, they are varied periphery-making processes across the world-economy.

WHITHER UNEVEN DEVELOPMENT?

In Frank's (1969) original conception of the development of underdevelopment, he posits cities in metropolis/satellite relations whereby 'a whole chain of constellations of metropoles and satellites relates all parts of the whole system from its metropolitan center in Europe or the United States to its farthest outpost in the Lain American countryside' (ibid.: 6). This is about exports, Jacobs' supply regions above, just one of her 'economic grotesques', albeit the one she identifies as the 'most important' (Jacobs, 1984: 59). However, as well as suggesting further mechanisms, Jacobs (1984) provides a fuller understanding of cities; for Frank, because they are 'centers of intercourse', they are necessarily 'also centers of exploitation' (1969: 6). But this represents the hierarchical relations whereby peripheral cities just link to 'their' metropole. From such a radical dependency perspective the logical answer is to delink, to politically and economically disconnect from the rich exploiting countries (Amin, 1990).[6] With guardians in control of delinking, such an extreme policy inevitably leads to elimination of cities as Pol Pot practised in Cambodia, with its concomitant millions of deaths as city dwellers are forced into an unsustaining and unsustainable countryside. In contrast, Jacobs sees also the positive contributions of cities in networks. She argues that connections should be cultivated and made diverse; as she tells it, 'backward cities need one another' (1984: 135, 155).

A Jacobsean position on uneven development argues that poor regions are in the great wheels of economic life but they lack the 'little movements' to create dynamic, complex, diverse cities. Thus she argues:

> If one wanted to define economic development in a single word, that word would be 'improvisation'. But infeasible improvisation is fruitless, so it would be more accurate to say that development is a process of continually improvising in a context that makes injecting improvisations into everyday economic life feasible. Cities in volatile trade with one another create that context. Nothing else does, which is why backward cities need one another. (Jacobs, 1984: 154–5)

All I would add is that the volatile trade has to be in a network, not in a hierarchy of uneven development.

One important feature of contemporary globalization has been the growth of a world city network (Taylor, 2004). With cities seemingly superseding states as the key economic entities, there appears now to be a 'global economy' based upon network relations instead of the territorial mosaic of the former 'international economy'. What does this mean for uneven development? Sassen (2002: 151), for one, is very clear on this; she posits a 'new geography of strategic places' that cut 'across national borders and the old North–South divide'. Hence there is a new political geography that operates 'along bounded "filières"' rather than through states and zones. This means that uneven development as we have known it is disappearing, but what is replacing it is still open to question. Will uneven development reappear in a new network form, or will this spatial *restructuring* undermine the mechanisms that have made uneven development inevitable within the modern world-system?

We can answer this question in two ways. There is no doubt that the current world city network includes strong hierarchical tendencies, with much power continuing to be concentrated in core-zone cities, especially London and New York (Taylor et al., 2002; Taylor, 2004). Thus, old exploitative imperial links through states may be being replaced by new exploitative imperial links through cities. Certainly statistics showing increased economic global polarization support this argument. But this empirical evidence is just a snapshot of the outcomes of globalization processes that are still quite new (i.e. it is a specific guardian neo-liberal globalization where the 'liberalism' most certainly does not extend to agriculture in rich countries). The alternative theoretical argument is that networks are inherently 'horizontal' in their operation since they can only be reproduced through mutuality between components. Thus, more subtle concepts of power – network power – do suggest a diffusion of real economic growth potentials to cities beyond the core. What happens to these potentials in the future cannot be known. If we are

within a transition from the modern world-system, as Wallerstein (1979) has long argued, then city hierarchy versus city network may well be the way his political bifurcation – struggles over the nature of the succeeding world-system – will occur.

Whatever the outcome, we should not think that the obverse of uneven development – a sort of utopian holy grail called 'even development' – will prevail. Certainly, city hierarchical structures will be implicated in unequal trade relations that will continue to generate rich and poor zones. But city networks are not about all cities being equally vibrant all at the same time. In Jacobs' argument the important point for a city network is that, at any one time, some cities are dynamic economic centres so they can act as stimuli for cities that have lost their dynamism. The existence of dynamic cities provides trade that can be replaced in new rounds of economic investment to create a restructured network – not unlike Smith's see-saw model. A kind of evenness only results when there are no dynamic cities; the economy across the world consists solely of by-passed places, producing an 'even undevelopment'. One route to such global impoverishment would be a successful anti-globalization movement that resurrected the policy option of delinking.

NOTES

1 I appreciate that any comprehensive notion of 'development' must include non-economic change, but I concentrate on a materialist approach in this chapter. Furthermore, within this approach I concentrate largely on the expansion of economic life to the detriment of the distribution of that expansion and its sustainability, both equally important. I choose this focus partly because the Marxist and neo-Marxist tradition that has informed uneven development in political geography theory and practice is based upon state-centric distributive concerns at the expense of the expansion of economic life (Keeley and Jacobs, 1989: 17), which must surely be the starting point for studying 'development'.

2 This triple zone structure can be interpreted as a geographical division of labour (Wallerstein, 1979: 16), but it is always more than just that. As Jacobs (1970: 82) reminds us: '*Division of labor, in itself, creates nothing*. It is only a way of organizing work that has already been created. ... Division of labor is a devise for achieving operating efficiency, nothing more. Of itself, it has no power to promote further economic development (emphasis in the original). Division of labour is an outcome of mechanisms, which can be important (Sayer and Walker, 1992), but not for expanding economic life. As structure, Wallerstein's core/periphery model

is both a source of mechanisms and a dynamic outcome of the mechanisms.

3 The world-systems analysis presented here, like all theoretical discourses, is highly contested. This is not the place to engage with these debates, not least because the arguments I develop here are intended to transcend these particular critiques. Briefly stated, the debate is typically reduced to a 'class analysis' that prioritizes change within states versus an 'exchange analysis' that prioritizes inter-state relations. Such internal/external differences in state-centrism are not of interest to me here. From the perspective of concern for 'development', the most lively treatment of the debate is to be found in Frank (1984).

4 For a recent critique and debate on the Jacobs model, see Polèse (2005) and Taylor (2006b).

5 Another way of looking at these two processes is as externalities, the first as cluster externalities and the second as network externalities as identified by Hicks (1989) and as interpreted by Taylor (2006a).

6 This radical left position should not be confused with the trade protectionism of large countries. Because of their size, countries such as the United States and Germany in the nineteenth century could absorb the restrictions of territorial economics without immediate damage to their cities. Such a policy was never a viable policy option for small countries. As noted before, protectionism is a classic guardian win-lose game; Pol Pot's delinking is an unusual, tragic lose-lose game.

REFERENCES

Amin, S. (1990) *Delinking*. London: Verso.

Arrighi, G. and Drangel, J. (1986) 'The stratification of the world-economy: an exploration of the semi-periphery zone', *Review (Fernand Braudel Center)*, 10: 9–74.

Bergman, E.F. (1975) *Modern Political Geography*. Dubuque, IA: William Brown.

Bowman, I. (1921) *The New World: Problems in Political Geography*. New York: World Book Company.

Braudel, F. (1972) 'History and the social sciences; the *longue durée*', in P. Burke (ed.), *Economy and Society in Early Modern Europe*. London: Routledge.

Coates, B.E., Johnston, R.J. and Knox, P.L. (1977) *Geography and Inequality*. Oxford: Oxford University Press.

Cox, K.R. (1979) *Location and Public Problems: A Political Geography of the Contemporary World*. Chicago: Maaroufa.

Crone, G.R. (1967) *Background to Political Geography*. London: Pitman.

Frank, A.G. (1969) *Latin America: Underdevelopment or Revolution*. New York: Monthly Review Press.

Frank, A.G. (1984) *Critique and Anti-Critique*. New York: Praeger.

Hicks, J. (1969) *A Theory of Economic History*. Oxford: Clarendon.

Jackson, W.A.D. (1964) *Politics and Geographic Relationships*. Englewood Cliffs, NJ: Prentice Hall.

Jacobs, J. (1970) *The Economy of Cities*. New York: Vintage.

Jacobs, J. (1984) *Cities and the Wealth of Nations*. New York: Vintage.

Jacobs, J. (1992) *Systems of Survival*. New York: Vintage.

Kasperson, R.E. and Minghi, J.V. (1969) *The Structure of Political Geography*. Chicago: Aldine.

Keeley, R.C. (1989) 'Some paths through Jane Jacobs' thoughts', in F. Lawrence (ed.), *Ethics in Making a Living*. Atlanta: Scholars Press.

Keeley, R.C. and Jacobs, J. (1989) 'An interview with Jane Jacobs', in F. Lawrence (ed.), *Ethics in Making a Living*. Atlanta: Scholars Press.

Krugman, P. (1995) *Development, Geography and Economic Theory*. Cambridge, MA: MIT Press.

Massey, D. (1984) *Spatial Divisions of Labour*. London: Macmillan.

Myrdal, G. (1957) *Rich Lands and Poor*. New York: Harper & Row.

Polèse, M. (2005) 'Cities and national economic growth; a reappraisal', *Urban Studies*, 42: 1429–51.

Pounds, N.J.G. (1963) *Political Geography*. New York: McGraw-Hill.

Rostow, W.W. (1960) *The Stages of Economic Growth*. Cambridge: Cambridge University Press.

Sassen, S. (2002) 'Global cities and diasporic networks: microsites in global civil society', in M. Glasius, M. Kaldor and H. Anheier (eds), *Global Civil Society 2002*. Oxford: Oxford University Press.

Sayer, A. (1992) *Method in Social Science: A Realist Approach*. London: Hutchinson.

Sayer, A. and Walker, R. (1992) *The New Social Economy: Reworking the Division of Labor*. Oxford: Blackwell.

Short, J.R. (1982) *An Introduction to Political Geography*. London: Routledge.

Smith, D. (1977) *Human Geography: A Welfare Approach*. London: Arnold.

Smith, N. (1984) *Uneven Development: Nature, Capital and the Production of Space*. Oxford: Blackwell.

Taylor, P.J. (1982) 'A materialist framework for political geography', *Transactions of the Institute of British Geographers*, NS, 7: 15–34.

Taylor, P.J. (1985) *Political Geography: World-Economy, Nation-State, Locality*. London: Longman.

Taylor, P.J. (1996) 'Embedded statism and the social sciences: opening up to new spaces', *Environment and Planning A*, 28: 1917–28.

Taylor, P.J. (2004) *World City Network: A Global Urban Analysis*. London: Routledge.

Taylor, P.J. (2006a) 'Development as a "monstrous hybrid": an essay on the primacy of cities in the expansion of economic life', *Environment and Planning A*, 38: 793–803.

Taylor P.J. (2006b) 'On a non-appraisal of the "Jacobs hypothesis"', *Urban Studies*, 43: 1625–30.

Taylor, P.J. (2007) 'Problematizing city/state relations: towards a geohistorical understanding of contemporary globalization', *Transactions of the Institute of British Geographers*, 32: 133–50.

Taylor, P.J., Catalano, G., Walker, D. and Hoyler, M. (2002) 'Diversity and power in the world city network', *Cities*, 19: 231–41.

Terlouw, C.P. (1992) *The Regional Geography of the World-System.* Utrecht: Rijksuniversiteit.

Wallerstein, I. (1961) *Africa: The Politics of Independence.* New York: Vintage.

Wallerstein, I. (1974) *The Modern World-System.* New York: Academic.

Wallerstein, I. (1979) *The Capitalist World-Economy.* Cambridge: Cambridge University Press.

Wallerstein, I. (1983) *Historical Capitalism.* London: Verso.

Weeks, J. (2001) 'The expansion of capital and uneven development on a world scale', *Capital and Class*, 74: 9–30.

The Politics of Local and Regional Development

Andrew Wood

INTRODUCTION

The literature on the politics of local and regional development emerged as part of a more general concern with the political economy of capitalist development and the identification of persistent geographic as well as social-economic inequalities. It has, over the past two decades, expanded into a sizable literature that encompasses an increasing range of theoretical approaches. Nevertheless, the literature continues to reflect its roots in the political economy tradition and while this provides an intellectual coherence of sorts, it has also tended to limit the range of its substantive concerns. In disciplinary terms, the literature on the politics of local and regional development encompasses work in political, economic, and urban geography but also spills over into adjacent disciplines, most notably politics and sociology. This is not surprising given the significance of the city – although much less so the region – to these disciplines, as well as the importance of political economy approaches within them. More intriguing perhaps is the relationship to urban and regional studies, and the politics of local and regional development provides an important focus of work within this interdisciplinary field. Placing boundaries around this literature and identifying its relationship to others are not simple tasks and although I have tried to be inclusive, there are certainly omissions that some will find problematic.

With that in mind, the chapter is designed to meet four objectives. The first is to examine the origins of work on the politics of local and regional development. This also presents some difficulties given continuities in scholarship and the absence of clearly defined disciplinary boundaries that might serve to police the field. The goal of specifying origins is much less driven by concerns of academic posterity than by the lasting imprint of initial conditions on the subsequent development of work in this field. Second, the chapter provides a brief history of how the literature has developed over time. Here I examine the different theoretical frameworks that have guided inquiry, provide an overview of the geographic scope of the literature, and focus in rather more detail on two key concepts – governance and scale – that have been central to its development. The third section and objective highlights the major limitations and gaps in this literature. Some of these – such as the neglect of gender and questions of social reproduction – are longstanding, and seem to reflect a resistance to broader changes within human geography. Other limitations appear more a consequence of having moved away from some of the traditional concerns that animated the early literature. Fourth, and by way of conclusion, I identify a number of areas that would seem to hold considerable promise for the further development of the literature in this field.

THE ORIGINS OF THE POLITICS OF LOCAL AND REGIONAL ECONOMIC DEVELOPMENT

Like most new academic pursuits, the interest in the politics of local and regional development arose as a consequence of broad intellectual changes coupled with a more concrete set of changing economic, cultural, and political conditions. On the academic front the general radicalization of human geography during the 1970s, and especially the development of radical political economy approaches, served to foreground questions of economic restructuring, uneven development, class struggle, and the role of the state in securing the conditions for capital accumulation. Economic geographies, and especially the changing fortunes of localities and regions, were no longer seen as the product of a rational market-based allocation of assets and resources but as the consequence of an endemic struggle between capital and labor, often times mediated by the state. David Harvey's work on the relationship between urbanization and capitalist development provided a key source of ideas and inspiration, while in Britain the work of Massey, Hudson, and others similarly drew from Marx to explain geographical variation in terms of wider mechanisms of capitalist development and class struggle. While questions of class and the conflict between capital and labor lay at the core of work in Britain, the emerging US interest in the local dynamics of growth took a rather different tack. That literature tended to emphasize the distributional effects of local and regional economic development and competition between territorial coalitions of interest rather than class struggle more directly. If this work was inspired by Marx, then the influence was what Robert Brenner would term a neo-Smithian Marxism that emphasized competition rather than conflict rooted in production.

Interest in the politics of local and regional development was also clearly stimulated by changing economic and political circumstances in Western Europe and North America. Deindustrialization, job loss and mass unemployment, plant closings and economic recession formed the backdrop for much of the early work on the politics of local and regional development (Bluestone and Harrison, 1982; Massey and Meegan, 1982). While the 'regional problem' in Britain dated to the 1930s, the crisis of Fordism during the 1970s and 1980s brought uneven development into much sharper relief (Holland, 1976). In the US the debate was generally framed in terms of the shift from Snowbelt to Sunbelt (Sawers and Tabb, 1984) and, as suggested above, interpreted less through the lens of class struggle (although see Peet, 1983) and more in terms of capital flight and an allocation of state resources that disadvantaged the traditional manufacturing belt (Markusen, 1987).

In the UK, the election of the Thatcher government in 1979 forestalled any attempt to address the regional problem at the national level. However, policy retained a critical importance to British work following the emergence in the 1970s of local government initiatives to promote what came

to be known as 'local economic development' (Eisenschitz and Gough, 1993). While local authorities had long been involved in place promotion (Ward, 1990), the 1970s marked the first formal and systematic development of a local government interest in economic development that had clear institutional expression. Academic interest in local economic development also became increasingly institutionalized with the founding in 1986 of *Local Economy*, a journal explicitly focused on broaching the academic/policy divide. In the US too, *Economic Development Quarterly* gave formal expression to a growing interest in local economic development policy. Founded in the same year, it also focused largely on questions of policy and likewise sought to bridge the gap between academic work and practice.

Looking back on this period I think there are three features of this growing interest in the politics of local and regional economic development that are particularly noteworthy. The first is that the focus quickly settled on the local scale and, in practical terms, this meant that the 'urban' came to dominate. In part, and as far as Britain is concerned, this is because the most interesting and experimental initiatives were concentrated in large Labor-run urban authorities like London and Sheffield (Duncan and Goodwin, 1988; Lawless, 1990). In part it also reflected the increasing urban focus of central government initiatives, in the form of urban development corporations, enterprise zones, and a later generation of policies targeted at particular cities and neighborhoods within them. While the ESRC-funded localities research program somewhat qualified a strictly urban focus, at this time the region clearly played second fiddle. Second, the literature focused heavily on the state as the agent of political change. In academic terms there were clear continuities with the influential work of Cockburn (1977) on theorizing the local state and the subsequent interest in local–central state relations (Duncan and Goodwin, 1988; see also Clark and Dear, 1984). Third, the British literature remained closely tied to policy and political practice with a number of individuals central to its early development bridging the academy and practice.

Interest in the politics of local economic development in Britain clearly bore a significant contextual influence. Work generally emphasized the radical potential of local economic policy, focused on core economic concerns of paid, full-time, formal employment, favored manufacturing over other economic sectors, and more or less equated economic life with 'production'. In this particular context, questions of gender, race and identity were, at best, marginal to the research agenda. These were not the only omissions – the region remained subordinate to the urban – at least until its 'rediscovery' in economic geography in the 1990s. Meanwhile the context of deindustrialization and recession served to emphasize 'regeneration' or 'revitalization' rather than matters of economic growth or development. This was a politics of local economic development very largely focused on arresting economic decline and dealing with the fallout from economic restructuring.

The British literature had a reasonably distinct starting point with academic work tending to mirror shifts in national and local policy. The contrast with the development of the American literature is marked. The origins of US work on the politics of local and regional development are much more disparate. Critical academic work in the US during the 1970s had focused very largely on the urban scale where questions of race were difficult to avoid. Toward the end of the decade a number of authors had begun to examine more explicitly the politics of economic growth and development. Bluestone and Harrison's (1982) work on the politics of deindustrialization provided one important example, while the edited collection by Fainstein and Fainstein (1983) brought together a range of studies that spanned sociology, politics, urban studies, and planning.

More influential still was Harvey Molotch's 1976 article in the *American Journal of Sociology* on the city as growth machine. It is difficult to overstate the significance of Molotch's thesis to the subsequent development of the literature on the politics of urban development. While the 'urban regime' is now the more favored framework, Molotch's concept of the growth machine represents a hugely influential and original contribution. Molotch's thesis, later expanded in Logan and Molotch (1987), was a simple one: whatever their differences, 'growth' provided the glue that bound together powerful economic interests in US cities. In mobilizing the powers of municipal government, these interests were able to ensure that growth went largely unhindered and that the conditions for economic development, and most especially the rising property values and rents associated with it, were secured. For Molotch the economic interests that drove the growth machine were 'place entrepreneurs' with interests in land and property. Allied to this core group were a number of additional interests including local politicians, the local media and utilities as well as a further set of auxiliary players such as cultural institutions, professional sports teams, and organized labor (Logan and Molotch, 1987: 66–84). The fortunes of each were seen to be pinned to the general prosperity of the urban economy and this shared interest in the health of the urban economy provided the basis for collective action.

While the growth machine proved hugely influential – Molotch's 1976 article has been cited well over 400 times – a second concept gradually displaced the growth machine as the dominant approach for examining the politics of urban development. Stephen Elkin's 1987 book established the notion of the urban regime, but it was Clarence Stone's (1989) landmark study of Atlanta politics from 1947 to 1988 that demonstrated the power and versatility of the urban regime concept. While the growth machine literature starts from the question of how economic interests come to realize their stakes in the city, the regime literature starts with the basic question of how coalitions come together in order to effectively govern the city. Accordingly, the regime literature pays far greater attention to the ways in which consent and legitimacy are secured, with electoral politics seen to serve as a counterweight to the power of the business community. For Stone, and for the multitude of later scholars using the regime concept, interest centers on the composition and objectives of governing regimes and the effects of urban regimes on the life chances and opportunities for different groups within the city. While the urban regime literature devotes much attention to characterizing regimes and the movement between them, much of this work remains strangely ahistorical. While the demographics of the city and the profile of the urban economy are seen as variable, there is little interest in how the broader conditions that enable or allow for a politics of urban development have taken shape. Indeed the pursuit of economic development as a widely held objective is taken for granted rather than seen as an object that requires explanation (see MacLeod and Goodwin, 1999). In this sense, much attention focuses on variation between regimes rather than on the necessary conditions that give rise to them.

In contrast, the work of David Harvey and a number of others has explicitly engaged the question of how best to explain broad historical change in the nature of urban politics. Harvey (1989) has argued that the 1970s and 1980s marked a significant shift in the way cities in capitalist economies were run and especially in the issues deemed to be significant to urban government. Cox (1995) argues that the period was marked by the emergence of a 'new urban politics' in which urban politics has become centrally concerned with local economic development. Furthermore, cities and communities are positioned as agents engaged in a process of relentless competition for investment.

It is easy to over-dramatize the extent and nature of this shift and I think in some ways the academic tendency to overlook the politics of economic growth and decline at the urban scale reflected the taken-for-granted nature of economic development as an agenda item and not its absence. US local governments have long been entrepreneurial in nature, and the boosters of the 1950s are hardly radically different from those that became increasingly central to academic work from the early 1980s onward.

Interest in the politics of urban development in the US has now generated an extensive literature that incorporates specific theoretical frameworks for understanding that politics as well as a wide range of concrete studies of particular cities. Most of this work has taken place in fields beyond geography, although urban and political geographers have certainly made important contributions to it. A comprehensive review is beyond the scope of this chapter; I want to focus instead, in this next section, on the way in which theories for interrogating the politics of local and regional development have developed, the geographical scope of research in this area, and the way in which the concepts of governance and scale have steered the development of the field.

THE SHIFTING CONTOURS OF WORK ON THE POLITICS OF LOCAL AND REGIONAL ECONOMIC DEVELOPMENT

The Anglo-American literature on the politics of local and regional development has expanded significantly since the early 1980s. The growth of academic journals with a focus on urban and regional issues has certainly helped sustain this expansion. The literature has never been a particularly coherent body and organizing a review is problematic as a consequence. In ordering the development of the literature I examine three major themes. The first is one of theory and more particularly the development of theoretical frameworks for investigating the politics of local and regional development. The second theme is the shifting geographical scope of work in this field and especially its limits. The third and final theme in this section is the way in which a number of key concepts – most notably governance and scale – have worked to reorientate or, perhaps more accurately, disrupt the literature. While these concepts are not associated with coherent theoretical frameworks as such, their influence on the nature and scope of work in this field has been substantial.

Shifting Theoretical Contours

As noted in the previous section, the theoretical origins of work on the politics of local and regional development are found in a political economy perspective that emphasizes the ways in which the geographies of urban and regional development reflect underlying social dynamics. This political economy tradition remains a prominent one despite the more general post-structuralist shift within human geography. Political economy work has certainly evolved from the early writings of Harvey, Massey, and others, and while it remains

firmly fixed on interpreting the politics of local and regional development through a class-based lens, the approach has become more nuanced and sophisticated over time. The adoption of regulation theory in the late 1980s and early 1990s, at least within European circles, provided a significant boost (MacLeod, 1997; MacLeod and Goodwin, 1999).

In contrast to the dominance of political economy during the 1980s, a glance at the contemporary literature on the politics of local and regional development reveals a much more diverse set of theoretical currents. Post-structuralist accounts underpin much of the interest in the discourses of development (Wilson, 1996; MacLeod and Goodwin, 1999; North et al., 2001; McCann, 2002, 2004; Raco, 2003a). Foucauldian accounts of 'governmentality' are also evident (MacKinnon, 2000; Raco and Imrie, 2000; Raco, 2003b), while a number of authors have recently sought to develop more eclectic frameworks that integrate different theoretical traditions (DiGaetano and Strom, 2003; Uitermark, 2005).

The evolution of theoretical accounts might best be seen as an ongoing struggle between Marxist work (in various forms) and frameworks that are broadly critical of that approach. Both the growth coalition (Molotch, 1976; Logan and Molotch, 1987) and the urban regime frameworks (Elkin, 1987; Stone, 1989) sought to 'reclaim' urban politics from structural Marxism on the one hand and neo-liberal accounts that denied local political agency on the other (Peterson, 1981). In doing so they placed considerable emphasis on processes of coalition-building and political strategy in explaining the politics of development. These frameworks had resonance in Britain where the 'locality' had become an important object of study, in part to counter deterministic interpretations of economic and political change (Cooke, 1989; Harloe et al., 1990).

A second development that sought to elaborate on rather than to directly challenge a class-based lens on capitalist development was regulation theory (for a review see MacLeod, 1997). This emphasized the contingent relationship between capital accumulation and the social regulation of capitalist development. While the distance between regulationist accounts and those of an earlier form of Marxism may seem trivial to some, the regulationist approach emphasizes political strategy and practice and licenses a more complex reading of the relationship between the state and divergent social interests. Yet just as the earlier theoretical interest in the local state was overwhelmingly a British one, regulation theory found few takers in the US literature. The exception here is the early work of Mike Davis. Others, notably those involved in the 'California School' of economic geography, drew insights from regulationist work, but their interests were less the politics of local and regional development and more the broad economic dynamics that produced uneven development, industrial agglomeration, and regional growth and decline.

While the translation of regulation theory into a set of empirically testable propositions was always its Achilles heel, a number of authors have sought to draw on regulationist insights in studying the politics of local and regional development. The work of Mark Goodwin, Gordon MacLeod and Martin Jones has proved particularly fruitful as these authors, along with various collaborators, have sought to apply regulationist insights to the study of urban and regional development in Britain (Goodwin et al., 1993; Goodwin and Painter, 1996; Jones, 1997, 2001; MacLeod and Goodwin, 1999; MacLeod and Jones, 1999; MacLeod, 2002). My own work with David Valler and others on the political representation of local business interests (Valler et al., 2000; Valler and Wood, 2004) provides a second example of the deployment of regulationist concerns. The foundation for much of this work has been Bob Jessop's influential interpretation of regulation theory and the development of his strategic-relational approach (Jessop, 2001).

Geographical Scope (and Its Limits)

The geographical scope of scholarship on the politics of local and regional development provides some useful insights into the ways in which the literature has developed, as well as its limitations. As noted above, the literature has its origins in work emerging from, and focused on, the US and the UK. This particularity has been the focus of much subsequent critique, with growth coalition and urban regime frameworks charged with failing to recognize the ways in which they incorporate specific concrete conditions. As Peck and Tickell argue, 'the growth machine hypothesis … remains a framework with which to understand the particularity of US politics' (Peck and Tickell, 1995: 59). While both Harding et al. (2000) and Mossberger and Stoker (2001) have argued that the problems lie principally in the application or translation of the regime framework to contexts beyond the US, others have pointed to more fundamental problems of a conceptual nature (Jessop et al., 1999; Davies, 2003; Wood, 2004; Pierre, 2005). Indeed, the limited explanatory leverage of the urban regime model beyond the US context is now widely recognized.

The attraction of US frameworks in the European context can be partly explained by a seeming convergence in the nature of urban politics on the US model (Wood, 1996; Harding, 1997). Yet closer inspection suggests that such similarities referenced policy rhetoric rather than substantial changes in the nature and shape of urban governance (Elwood, 2004). Partnerships in the UK case were seen to be largely sustained by central state fiat and funding rather than the localized structure of material incentives that spawned growth coalitions in the US. More recent studies have extended the application of the urban regime concept to a range of other national contexts (see below). Nevertheless the literature on the politics of local and regional development remains dominated by a focus on cities and regions in the West and most especially North America and Western Europe. In large part this reflects the intellectual origins of the various concepts that have guided the development of the field and their context-specific nature. Even the seemingly more portable notion of 'urban entrepreneurialism' reflects a context in which urban politics rests on a division of labor between powerful economic interests on the one hand and local state agencies and authority on the other. This is also a politics of territorial competition, in which geographically defined interests seek to connect to networks and flows of wider geographic scope. Furthermore, it is a politics that rests on geographic instability and change that, in turn, threaten stakes in wages, profits, rents and revenues.

Examining these necessary conditions for such a politics provides some insights into the general applicability of the key concepts. It is clear that the struggles and tensions that generate the politics of local and regional development are centered on the appropriation and distribution of value. The stakes are those of profits, wages, revenues, incomes and rents – in short, this is a politics of capitalist development. Such a politics assumes conditions of private property and a struggle over the appropriation and distribution of a surplus. But the local and regional nature to this politics further implies that interests are geographically defined and that these geographies matter. Politics concerns the geographies of these interests and the ways in which certain resources – profits, wages and revenues – are harnessed while liabilities in the form of taxes, bankruptcy, redundancy, environmental harms and the like are avoided. A third necessary condition is one of instability and with it the associated condition of uneven development. The flow of capital, revenues, resources, and labor – as well as the promise of such flows – is what animates particular coalitions of interest both to defend existing allocations while seeking to enhance and attract the position of the coalition or network relative to a wider configuration of interests and flows.

This is by no means meant to rule out a politics of local and regional development beyond the Western context. But this focuses on the necessary conditions for it has two important implications. The first is that we should not be surprised that difficulties arise in using concepts and ideas devised in one context to explain conditions else-where. Since 1995 or so a number of studies have served to widen considerably the geographical scope of work on the politics of local economic development. Post-apartheid South Africa has generated a particularly fertile literature (Rogerson, 1997; Maharaj and Ramballi, 1998; Nel, 2001), while Israel (Kirby and Abu-Rass, 1999; Alfasi, 2004), Eastern Europe (Kulcsar and Domokos, 2005), and China and East Asia (Zhu, 1999; Zhang, 2002; He and Wu, 2005; Zhang and Wu, 2006; on China; Tang and Tang, 2004, on Taiwan; Park, 2003, on South Korea) have also been examined. Each of these contexts conforms to the necessary conditions that sustain a politics of local economic development and they are, in principle, fertile ground for further work.

The problem that arises is that the analytical power of existing frameworks, such as the growth coalition and the urban regime, rests on additional elements that are more context-specific. The urban regime literature, for example, focuses on competitive electoral politics as a key source of regime power and legitimacy; for all its market-based reforms, China hardly conforms to this particular condition. The failure of existing frameworks is thus more likely a consequence of their particularity than their 'misapplication'. Second, and in turn, this suggests that the politics of local and regional development is likely to take a variety of different forms, licensing a range of alternative concepts and theoretical insights. In this sense there remains considerable scope for comparative work beyond the now tried and tested comparisons of a transatlantic nature (Wood, 1996; Sellais, 2002; Cox, 2004; Pierre, 2005).

Concepts

Debate over the most appropriate theories for understanding the politics of local and regional development and over the geographical reach of those theories has provided an important source of innovation and ideas that have helped to move the literature forward. Yet this has not been the only academic impetus for change. The direction and vitality of work in the field has also been influenced by the incorporation of a number of key concepts that have served to reorient the literature and, at times, unsettle key assumptions. Here I want to focus on two particular concepts whose influence resonates well beyond this literature but that have proved particularly germane to its development. The first is 'governance' and the second is 'scale'. While the former has proved to be a particularly fertile stimulus, although one not without its own limitations, the latter has proved a rather more provocative and disruptive, though still ultimately productive, influence.

Governance

The interest in governance in the context of work on the politics of local and regional development can be traced to David Harvey's influential 1989 article on 'the transformation of urban governance'. Drawing on a diversity of sources, including his own work on Baltimore, Harvey argued that 'the "managerial" approach [to urban governance] so typical of the 1960s has steadily given way to initiatory and "entrepreneurial" forms of action in the 1970s and 1980s' (Harvey, 1989: 4). Urban entrepreneurialism was seen to have the distinguishing characteristic of a public–private partnership engaged in speculative activity geared to attracting external resources into the locality. Harvey argued that the risks involved in such activities were commonly borne by local public agencies, and that this distinguished the recent phase from previous periods of civic boosterism. The shift from government to governance thus registered the incorporation of a broader range of interests into the management of the city as well as a shifting balance between risks and rewards in which the former were increasingly borne by the state while the latter were secured by private interests.

Subsequent work on the politics of urban and regional governance has concentrated on four overlapping themes (see Wood, 1998, for a fuller review). The first is debate over the nature and extent of the shift from government to governance and from one set of institutional arrangements and practices to another (Imrie and Raco, 1999; Ward, 2000). A second has centered on the nature of governing coalitions and regimes and especially the interests that comprise place-based coalitions (Cox and Mair, 1988, 1991). Third, interest has focused on the discourses associated with new forms of urban governance and most especially on those associated with 'urban entrepreneurialism' (Jessop, 1998; Short and Kim, 1998). Fourth, and more recently, a sizable literature has emerged that addresses the rescaling of urban and regional governance (Brenner, 2004). I want to examine the first three of these literatures here but then integrate the discussion of rescaling with a more general examination of scale questions in relation to the politics of local and regional development.

Harvey's 1989 argument provided the basis for a range of studies that were framed by the notion of a shift to entrepreneurial forms of governance. Initially at least, this work focused very largely on US and British cities (Boyle and Hughes, 1994; Hubbard, 1995; Imrie et al., 1995; Bassett, 1996; Cochrane et al., 1996; Valler, 1996; Quilley, 1999; McQuirk and MacLaran, 2001). Yet it is also clear that many of these studies used Harvey's argument as a *preface to* rather than a *framework for* analysis (although see Roberts and Schein, 1993). I have suggested elsewhere (Wood, 1998) that while post-structuralists have taken issue with

Harvey's overall approach, there has been surprisingly little critical engagement with his more specific arguments about urban entrepreneurialism. A number of authors have argued that Harvey's thesis underplayed the role and power of local government in the British context (Boyle and Hughes, 1994; Lawless, 1994; Valler, 1996). Others suggested that the contrast between managerial and entrepreneurial phases of governance was overdrawn, and that while entrepreneurial forms have long characterized urban governance in the US (see Cox, 2004), the UK has retained a much more significant emphasis on 'managerial' functions and initiatives (Imrie and Raco, 1999). More recently, Ward (2003) has argued that urban entrepreneurialism licenses a qualitatively different form of state intervention in which the state seeks to enable or orchestrate the production and performance of urban regeneration while disciplining those who fail or refuse to conform to an entrepreneurial script.

In keeping with Marxian political economy, Harvey viewed the local coalitions that drive entrepreneurial strategies as essentially class-based. This interest in the composition of local coalitions is one that has also animated scholars using the growth coalition and urban regime frameworks. However, the dominant tendency in those literatures has been to view the question in rather limited terms, either focusing on a particular fraction of capital in the case of the growth coalition or treating the matter as an empirical question, as in the case of the urban regime. Neither approach pays sufficient attention to either the necessary conditions for local growth coalitions or to the geographic specificity of the interests that underpin them. Logan and Molotch (1987) deploy a relatively simple division between local and metropolitan capital with the former focused on realizing rent through the appreciation of land and property values. Stone's influential work on Atlanta (1989) similarly fails to problematize the geographic specification of interests.

In contrast, the work of Cox and Mair (1988, 1989) provides a considerably more sophisticated account of the interests that underpin urban and regional coalitions. Moreover, Cox and Mair set out to theorize the necessary conditions for a politics of local and regional development. To this end, they draw directly from Harvey to stress the inherent tension between the mobility of capital on the one hand and its fixity in particular physical and social infrastructures on the other. Cox and Mair's work is located firmly within the Marxian tradition and starts from a concern with the ways in which class conflicts are given territorial expression. For Cox (2002) the central problematic is one of explaining how 'the opposition between business and labor is somehow converted into an

opposition between places: a competition between places and their respective coalitions of forces' (ibid.: 75). The answer, for Cox and Mair, lies in the notion of the *local dependence* or spatial entrapment of agents. Local dependence is defined as 'a relation to locality that results from the relative spatial immobility of some social relations, perhaps related to fixed investments in the built environment or to the particularization of social relations' (Cox and Mair, 1989: 142). The notion of relative immobility specifies the interests that animate growth coalitions at a relatively abstract level, avoiding the tendency in growth coalition and urban regime accounts to specify growth interests in much more concrete terms.

Cox and Mair's initial formulation focused on the local dependence of individuals, state agencies and certain business interests. Overlapping spaces of dependence provide the basis for collaboration between interests and the generation and nourishment of local coalitions. In turn, instability in the space economy generates pressures for interests that share spaces of dependence to collaborate in seeking to better, or at least secure, the position of their locality within wider spatial divisions of labor. But seeking to capture inward investment, or indeed any source of additional value, is liable to pit one territorially defined coalition of interests against territorial coalitions elsewhere. This inter-locality competition is the mechanism that links urban entrepreneurialism to broader changes within the capitalist space economy. This is an important point, for subsequent work has all too often envisaged the local and regional scale as a bounded arena in which interests are mutually defined and conflicts contained (see Low, 1998). Harvey's claim is that the political machinations of urban entrepreneurialism not only shape political and economic conditions and outcomes at the urban scale, but also facilitate broader economic transformation through increasing the geographical flexibility of mobile capital, encouraging the serial reproduction of development strategies, and promoting social control through the mechanism of local place-based identities (Harvey, 1989). The role of territorial identities in serving to displace conflict – especially of a class-based nature – finds a very clear echo in the work of Cox and Mair (1988, 1989).

There are hints in both Harvey (1989) and Cox and Mair (1988) of the subsequent interest in the scalar dimensions of political conflict, and especially how coalitions of interest seek to extend across geographical scales in attempting to realize their objectives (Cox, 1998). Whether through lobbying central state agencies for funds or favorable legislation, seeking inward investment, or drawing consumers into the locality, local fortunes rest on the relationship between local and more global sets of interests and agents. This question of scale and the rescaling of the politics of local and regional development is one that I take up again shortly.

For Harvey, Cox and others drawing on traditional Marxian political economy, the politics of local and regional development is rooted in the conflict between capital and labor over the appropriation of value. Growth coalitions promulgate territorial discourses that serve to rally otherwise divergent social interests behind development strategies and projects and to marginalize or discipline alternative voices, while displacing the conflict between capital and labor onto a geographic plane. This insistence on the class-based nature of conflict ensures a certain degree of intellectual rigor but it also limits the ability of such frameworks to encompass alternative structures of inequality. Such work rests instead on a rather staid argument about the 'false consciousness' of labor and the trivial or superficial nature of distributional struggles and the forms of discourse and ideology associated with them.

Many of those that have adopted the conceptual insights and terms associated with Marxian accounts have also sought to problematize and interrogate the discourses associated with urban governance and with the politics of local economic development more generally. Indeed, this is one of the most promising areas of recent work. Studies have sought to recognize the constructed and contested nature of economic development and probe the ways in which actors and interests construct and reproduce dominant meanings. This interest in the cultural construction of 'local economic development' (Gatrell and Reid, 2002; McCann, 2002) reflects a much broader cultural turn within human geography. In the UK in particular the discursive has become a major focus of attention. Yet there are North American variants too, most notably Beauregard's (1993) work on the representation of US cities in the postwar period and Wilson's (1996, 1998) work on the discursive construction of urban politics. It seems to me that this work has considerable merit whether or not it holds to an argument about the 'displacement' of class struggle.

Scale

Massey and Thrift argue that certain forms of Marxism tended to belittle attempts to study the specific ways in which capitalist development and class struggle were articulated in different local and regional environments: 'The very term "local" came in for denigration: things were only "local" issues, only "local" struggles … rather than the global and universal issues which, it was argued, were more worthy of study' (Massey and Thrift, 2003: 285). Geographers have particular cause to resist such a claim. Indeed the enterprise of studying the politics of local and regional development rests on the assumption that the locality and the

region are salient scales for understanding the dynamics of contemporary societies. As structural Marxism waned, the locality became an important and viable focus, not least through the program of 'locality studies' pursued in the UK (Cooke, 1989; Harloe et al., 1990).

The recent interest in globalization has provided a similar set of challenges to those studying local and regional scale dynamics. The focus on global integration and the growing homogeneity of economic and cultural forms has tested geographers' traditional concern with specificity and spatial difference. One response has been to reassert the significance of the locality and the region as vital scales of political and economic life. Certain authors have argued that the local provides a basis from which to resist globalizing tendencies (DeFilippis, 2004). Others have emphasized a relational conception of scale in which economic and political conditions are seen to be shaped by processes that traverse multiple geographic scales. In place of retreating to the notion of the local as a bounded and thus defensible space, these accounts embrace a multi-scalar understanding of change.

Concerns with scale reflect a much more general interest in the concept (Herod, 1991; Jonas, 1994; Marston, 2004). Prior to 1990, the literature on the politics of local economic development tended to adopt a relatively simple distinction between 'local' and 'non-local' interests. As noted above, the urban regime and growth coalition literatures made reference to extra-local conditions and forces but positioned them as 'impacting' on local communities, territories, social relations, interests and the like. These were social forces that were essentially placeless – hovering above cities and regions in providing a general backdrop for the study of the city and the region. Beginning in the mid-1990s this particular conceptualization came into question. The main challenge was founded on the argument that scale – like space – was socially constructed and that scalar configurations were thus inherently political in nature. A conceptualization of scale as mutable raised a series of interesting questions around the nature and form of scalar arrangements. This debate has a certain precedent in the literature on the local state and conflict over central–local government relations (Cockburn, 1977; Duncan and Goodwin, 1988), and the more recent literature on devolution, regionalization and the politics of governance rescaling indicates the continued salience of these concerns.

A more politicized notion of scale has given rise to a concern with the various ways in which processes expressed at different scales are articulated within particular geographic and historical contexts. Notions of 'glocalization', 'glurbanization' and governance rescaling signal a concern with the way in which local and regional politics are intertwined with more global processes. While this literature pays careful attention to the multi-scalar nature of local and regional politics, it is difficult to avoid a residual tendency to reify scale or, at best, to assign specific processes to particular geographic scales. More recently the whole notion of a scalar or territorial approach has come under critical attack. In place of a territorial or scalar approach, Amin (2002) and others have made the case for a 'networked' approach (see e.g., Marston et al., 2005). Again these developments resonate with a broader set of developments within human geography, not least the incorporation of actor-network approaches and the corresponding shift away from territorial and scalar-based frameworks.

LIMITS, FAILINGS AND OMISSIONS

This brief survey of the development of the literature on the politics of local and regional development has focused on the major theoretical advances, the geographical scope of scholarship and the influence of key concepts on the trajectory of the field. A widening geographical scope and a broadening of salient theoretical frameworks represent positive developments. Yet serious omissions remain. In this section I argue that structures of inequality other than social class remain very much neglected. A second major omission concerns the incorporation of nature into the politics of local and regional development. I address each in turn.

Interest in the politics of local and regional development references questions of work, employment and welfare. Consequently, the neglect of gender and gender relations in examining that politics is both surprising and troublesome. The problem has at least three dimensions. First, the literature has failed to thoroughly problematize the ways in which 'development' itself is gendered in form. This is, in part, a question of development for whom? Massey's (1979) early work on the politics of the 'regional problem' made explicit reference to the changing gender division of labor as employment in traditional industries, such as coal mining and shipbuilding, were replaced by industries employing largely female labor. Ann Oberhauser's (1995, 2002) work on economic restructuring in rural Appalachia is another important exception. Her focus on homework and informal activities highlights the significance of gender relations in structuring economic and social conditions as well as future prospects for development. The shift to urban governance has proved equally resistant to feminist perspectives. Peck and Tickell (1996) make explicit reference to the male-dominated nature of the business elites that participate within Manchester's governing partnerships, but otherwise the implications of a hegemonic masculinity have not been subject to detailed examination. Hubbard's (2004) study of the masculinist nature of neo-liberal urban governance is a promising, albeit late, start.

Second, research has failed to interrogate the ways in which the politics of local and regional development plays out within labor markets and within households. In short, gender relations are positioned as marginal to the issues that animate the politics of local and regional development (Oberhauser, 1995; Miewald and McCann, 2004). The failure to reflect critically on questions of gender surely relates to the dominance of men in both the practice of local and regional economic development policy and its academic study. The contrast between the growing significance of questions of gender to the 'development' literature and the continued marginalization of those questions in relation to the politics of local and regional development is instructive. Third, the continuing resonance of Marxist approaches has led to what Hubbard has termed a continuing fixation on class and a corresponding marginalization of other sources of difference.

The literature on the politics of local and regional development has also been slow to incorporate a concern with nature and the environment in examining the politics of development. The 'productionist' focus, which serves to marginalize questions of social reproduction, also embraces a view of production that treats nature as seemingly inert (Bridge, 2000). While questions of gender remain marginal, there are some promising signs that work on the politics of local and regional development has begun to take nature and the environment more seriously. One set of studies has sought to address nature and the environment within a governance tradition, examining the interests and stakes involved in the local and regional regulation of the environment and environmental issues (Jonas and Bridge, 2003; Jonas et al., 2004). These studies serve to extend the compass of existing frameworks. A second set of studies has sought to examine how nature and the environment are more directly incorporated into a politics of local and regional development (Gatrell and Jensen, 2002; Maddock, 2004).

THE PROMISE OF WORK ON THE POLITICS OF LOCAL AND REGIONAL DEVELOPMENT

In concluding this review I want to highlight three areas of research that seem particularly fertile areas for further investigation. This is unavoidably a personal and particular take on the current state of the literature. By no means are the claims meant to indicate an absence of interesting and innovative work elsewhere in the field. The first two areas of investigation would involve a broadening of both the theoretical approaches and substantive foci encompassed by the literature on the politics of local and regional development. The last is a more explicitly political claim to return, albeit with new theoretical and methodological tools in tow, to the concerns and issues that stimulated the emergence of the literature.

First, some of the most innovative work in the field has sought to critically examine the ways in which 'economic development' is given discursive as well as material force. For too long authors have either followed the mainstream literature in taking the concept of economic development as a given or simply assumed that talk of 'development' masks a more pernicious and real objective of realizing profits, rents or tax revenues. The politics of local and regional development rests on a set of contradictory and often conflictual relationships between divergent social interests. Yet this conflict is much more than simply the local manifestation of a broader struggle between capital and labor. Analysis of the discursive politics of local and regional development provides an entrée to prise open those conflicts in order to get at the complexity of the interests that define and are, in turn, defined by struggles over economic and political change at the local and regional scales (Raco, 2003a).

This is not to argue that the analysis of discourse should be divorced from an interest in the material geographies of conflict, exploitation and resistance. However, there is much more to discourses of local and regional development than the literature has tended to suggest. The work of Boyle and Hughes (1994) on the politics of urban entrepreneurialism in Glasgow drew early attention to the ways in which conflicts over development were simultaneously struggles around political identities. As Boyle (1997) suggests, the critical literature on the politics of local economic development has failed to adequately address the range of ways in which individuals and social groups relate to the discourses promoted as part of the new urban politics. In this sense, existing studies have tended to view discourses of growth and development as either unproblematic or as simply a veil for politically unpalatable objectives and goals.

Related to this is a need to examine the ways in which particular economic development strategies are constructed and deployed. In short, the 'politics of representation' warrants serious and sustained consideration (McCann, 2002). Yet at the same time the turn to 'discourse' and representation is not without its own pitfalls. Economic development strategies are more than simply the product of a battle of ideas or of changing fashions in the economic development profession. As MacLeod and Jones (1999) suggest, entrepreneurial discourses make reference to a set of broad economic and political dynamics that can powerfully structure the politics of representation.

Addressing the links between social interests and development strategies would provide one means of developing sustained attention to the gendered nature of the politics of local and economic development. With one or two notable exceptions (see above) this is a second major lacuna that operates at a whole range of different levels, from the most basic notion of what counts as 'economic development' (see Blake and Hanson, 2005), to the process by which economic development strategies are constructed and implemented (O'Toole and Macgarvey, 2003), to the constituencies impacted, for better or for worse, by economic development strategies – in short the development for whom question. The dominance of political economy approaches has left the field with a rather particular view of local and economic development and with few ready-made tools for encompassing the gendered nature of its politics.

Third, the literature on the politics of local and regional development has had remarkably little to say about the impact of development on different social groups and interests beyond a concern with the political and economic elites that construct strategies and policy and subsequently benefit from them. In part this is tied to a problem of adequately 'evaluating' local economic development, but it is all the more surprising given the initial concern of the growth coalition and urban regime literatures with the distributional impacts of different elite-led strategies. For Molotch the problem was simply that the undue power of certain economic interests would always produce policies that clearly failed to reflect the diverse needs of the city. This is a rather simplistic reading, but while later work has recognized the complex and contingent ways in which particular development strategies play out, it has largely bypassed the question of who gains and who loses. Again there are some signs that scholars have begun to recover Logan and Molotch's initial concern with distributional outcomes. Neil Smith's (1996) notion of the 'revanchist city' captures the ways in which certain social groups are marginalized and penalized through the construction and implementation of a particular hegemonic vision of the city. As studies expand on the various ways in which the new urban politics is hinged to issues of collective consumption, then issues of distribution and social reproduction are likely to become an increasingly significant focus of concern. Already a number of studies have examined issues of crime and policing in relation to the promotion of the entrepreneurial city (MacLeod, 2002; Belina and Helms, 2003; Raco, 2003c), while others have returned to questions surrounding the provision of collective consumption (While et al., 2004).

The recent interest in neo-liberalism also references a broader concern with the ways in which economic and political forces have promoted strategies that serve to castigate and discipline certain social groups while rewarding others. In this sense there are some likely gains in a return to examine how the politics of economic development rests on a set of pervasive and stubborn social inequities, including those of gender. It was this concern with social inequity that first gave rise to the literature on the politics of local and regional development. It may well be time to come back to it.

ACKNOWLEDGMENT

I would like to thank Kevin Cox for his very helpful comments on previous drafts of this chapter. The usual disclaimers apply.

REFERENCES

Alfasi, N. (2004) 'The meaning of words in urban conflicts: language, argumentation patterns and local politics in Israel', *Urban Studies*, 41: 2139–57.

Amin, A. (2002) 'Spatialities of globalization', *Environment and Planning A*, 34: 385–99.

Bassett, K. (1996) 'Partnerships, business elites and urban politics: new forms of governance in an English city?', *Urban Studies*, 33: 539–55.

Beauregard, R.A. (1993) *Voices of Decline: The Postwar Fate of US Cities*. Oxford: Blackwell.

Belina and Helms (2003) 'Zero tolerance for the industrial past and other threats: policing and urban entrepreneurialism in Britain and Germany', *Urban Studies*, 40: 1845–67.

Blake, M. and Hanson, S. (2005) 'Rethinking innovation: context and gender', *Environment and Planning A*, 37: 681–701.

Bluestone, B. and Harrison, B. (1982) *The Deindustrialization of America: Plant Closings, Community Abandonment, and the Dismantling of Basic Industry*. New York: Basic Books.

Boyle, M. (1997) 'Civic boosterism in the politics of local economic development – "institutional positions" and "strategic orientations" in the consumption of hallmark events', *Environment and Planning A*, 29: 1975–97.

Boyle, M. and Hughes, G. (1994) 'The politics of urban entrepreneurialism in Glasgow', *Geoforum*, 25: 453–70.

Brenner, N. (2004) *New State Spaces: Urban Governance and the Rescaling of Statehood*. Oxford and New York: Oxford University Press.

Bridge, G. (2000) 'The social regulation of resources access and environmental impact: production, nature and contradiction in the US copper industry', *Geoforum*, 31: 237–56.

Clark, G. and Dear, M.J. (1984) *State Apparatus: Structures and Language of Legitimacy*. Boston: Allen & Unwin.

Cochrane, A., Peck, J. and Tickell, A. (1996) 'Manchester plays games: exploring the local politics of globalization', *Urban Studies*, 33: 1319–36.

Cockburn, C. (1977) *The Local State: Management of Cities and People*. London: Pluto Press.

Cooke, P. (ed.) (1989) *Localities: The Changing Face of Urban Britain*. London: Unwin Hyman.

Cox, K.R. (1995) 'Globalisation, competition and the politics of local economic development', *Urban Studies*, 32: 213–24.

Cox, K.R. (1998) 'Spaces of dependence, spaces of engagement and the politics of scale, or: looking for local politics', *Political Geography*, 17: 1–23.

Cox, K.R. (2002) *Political Geography: Territory, State and Society*. Oxford: Blackwell.

Cox, K.R. (2004) 'The politics of local and regional development, the difference the state makes and the US/British contrast', in A. Wood and D. Valler (eds), *Governing Local and Regional Economies*. Aldershot: Ashgate, pp. 247–75.

Cox, K.R. and Mair, A. (1988) 'Locality and community in the politics of local economic development', *Annals of the Association of American Geographers*, 78: 307–25.

Cox, K.R. and Mair, A. (1989) 'Urban growth machines and the politics of local economic development', *International Journal of Urban and Regional Research*, 13: 137–46.

Cox, K.R. and Mair, A. (1991) 'From localized social structures to localities as agents', *Environment and Planning A*, 23: 197–213.

Davies, J.S. (2003) 'Partnerships versus regimes: why regime theory cannot explain urban coalitions in the UK', *Journal of Urban Affairs*, 25: 253–69.

DeFilippis, J. (2004) *Unmaking Goliath: Community Control in the Face of Global Capital*. New York: Routledge.

DiGaetano, A. and Strom, E. (2003) 'Comparative urban governance – an integrated approach', *Urban Affairs Review*, 38: 356–95.

Duncan, S. and Goodwin, M. (1988) *The Local State and Uneven Development: Behind the Local Government Crisis*. Cambridge: Polity.

Eisenschitz, A. and Gough, J. (1993) *The Politics of Local Economic Policy*. Basingstoke: Macmillan.

Elkin, S. (1987) *City and Regime in the American Republic*. Chicago: University of Chicago Press.

Elwood, S. (2004) 'Partnerships and participation: reconfiguring urban governance in different state contexts', *Urban Geography*, 25: 755–70.

Fainstein, N. and Fainstein, S.S. (1983) *Restructuring the City*. New York: Longman.

Gatrell, J. and Jensen, R. (2002) 'Growth through greening: developing and assessing alternative economic development programmes', *Applied Geography*, 22: 331–50.

Gatrell, J. and Reid, N. (2002) 'The cultural politics of local economic development: the case of Toledo Jeep', *Tijdschrift voor Economische en Sociale Geografie*, 93: 397–411.

Goodwin, M. and Painter, J. (1996) 'Local governance, the crises of Fordism and the changing geographies of regulation', *Transactions of the Institute of British Geographers*, 21: 635–48.

Goodwin, M., Duncan, S. and Halford, S. (1993) 'Regulation theory, the local state and the transition of urban politics', *Environment and Planning D: Society and Space*, 11: 67–88.

Harding, A. (1997) 'Urban regimes in a Europe of the cities?', *European Urban and Regional Studies*, 4: 291–314.

Harding, A., Wilks-Heeg, S. and Hutchins, M. (2000) 'Business, government and the business of urban governance', *Urban Studies*, 37: 975–94.

Harloe, M., Pickvance, C. and Urry, J. (eds) (1990) *Place, Policy and Politics: Do Localities Matter?* London: Unwin Hyman.

Harvey, D. (1989) 'From managerialism to entrepreneurialism – the transformation in urban governance in late capitalism', *Geografiska Annaler Series B*, 71: 3–17.

He, S.J. and Wu, F.L. (2005) 'Property-led redevelopment in post-reform China: a case study of Xintiandi redevelopment project in Shanghai', *Journal of Urban Affairs*, 27: 1–23.

Herod, A. (1991) 'The production of scale in United States labor relations', *Area*, 23: 82–8.

Holland, S. (1976) *Capital Versus the Regions*. London: Macmillan.

Hubbard, P. (1995) 'Urban design and local economic development – a case study in Birmingham', *Cities*, 12: 243–51.

Hubbard, P. (2004) 'Revenge and injustice in the neoliberal city: uncovering masculinist agendas', *Antipode*, 36: 665–86.

Imrie, R. and Raco, M. (1999) 'How new is the new local governance? Lessons from the United Kingdom', *Transactions*, 24: 45–63.

Imrie, R., Thomas, H. and Marshall, T. (1995) 'Business organizations, local dependence and the politics of urban renewal in Britain', *Urban Studies*, 32: 31–47.

Jessop, B. (1998) 'The narrative of enterprise and the enterprise of narrative: place marketing and the entrepreneurial city', in T. Hall and P. Hubbard (eds), *The Entrepreneurial City: Geographies of Politics, Regimes and Representation*. Chichester: John Wiley, pp. 77–99.

Jessop, B. (2001) 'Institutional re(turns) and the strategic-relational approach', *Environment and Planning A*, 33: 1213–35.

Jessop, B., Peck, J. and Tickell, A. (1999) 'Retooling the machine: economic crisis, state restructuring and urban politics', in A.E.G. Jonas and D. Wilson (eds), *The Urban Growth Machine: Critical Perpectives, Two Decades Later*. Albany: State University of New York Press, pp. 141–59.

Jonas, A.E.G. (1994) 'The scale politics of spatiality', *Environment and Planning D: Society and Space*, 12: 257–64.

Jonas, A.E.G. and Bridge, G. (2003) 'Governing nature: the re-regulation of resources, land-use planning, and nature conservation', *Social Science Quarterly*, 84: 958–62.

Jonas, A.E.G., While, A. and Gibbs, D.C. (2004) 'State modernization and local strategic selectivity after Local Agenda 21: evidence from three northern English localities', *Policy and Politics*, 32: 151–68.

Jones, M. (1997) 'Spatial selectivity of the state? The regulationist enigma and local struggles over economic governance', *Environment and Planning A*, 29: 831–64.

Jones, M. (2001) 'The rise of the regional state in economic governance: "partnerships for prosperity" or new scales of state power', *Environment and Planning A*, 33: 1185–211.

Kirby, A. and Abu-Rass, T. (1999) 'Employing the growth machine heuristic in a different political and economic context: the case of Israel', in A.E.G. Jonas and D. Wilson (eds), *The Urban Growth Machine: Critical Perspectives, Two Decades Later*. Albany: State University of New York Press, pp. 213–25.

Kulcsar, L.J. and Domokos, T. (2005) 'The post-socialist growth machine: the case of Hungary', *International Journal of Urban and Regional Research*, 29: 550–63.

Lawless, P. (1990) 'Regeneration in Sheffield: from radical intervention to partnership', in D. Judd and M. Parkinson (eds), *Leadership and Urban Regeneration: Cities in North America and Europe*. London: Sage, pp. 133–51.

Lawless, P. (1994) 'Partnerships in urban regeneration in the UK: the Sheffield Central Area Study', *Urban Studies*, 31: 1303–24.

Logan, J. and Molotch, H. (1987) *Urban Fortunes: The Political Economy of Place*. Berkeley: University of California Press

Low, M.M. (1998) 'Representation unbound: globalization and democracy', in K.R. Cox (ed.), *Spaces of Globalization*. New York: Guilford Press, pp. 240–80.

MacKinnon, D. (2000) 'Managerialism, governmentality and the state: a neo-Foucauldian approach to local economic governance', *Political Geography*, 19: 293–314.

MacLeod, G. (1997) 'Globalizing Parisian thought-waves: recent advances in the study of social regulation, politics, discourse and space', *Progress in Human Geography*, 21: 530–53.

MacLeod, G. (2002) 'From urban entrepreneurialism to a "revanchist city"? On the spatial injustices of Glasgow's renaissance', *Antipode*, 34: 602–24.

MacLeod, G. and Goodwin, M. (1999) 'Reconstructing an urban and regional political economy: on the state, politics, scale, and explanation', *Political Geography*, 18: 697–730.

MacLeod, G. and Jones, M. (1999) 'Reregulating a regional rustbelt: institutional fixes, entrepreneurial discourse, and the "politics of representation"', *Environment and Planning D: Society and Space*, 17: 575–605.

Maddock, T.A. (2004) 'Fragmenting regimes: how water quality regulation is changing political-economic landscapes', *Geoforum*, 35: 217–30.

Maharaj, B. and Ramballi, K. (1998) 'Local economic development strategies in an emerging democracy: the case of Durban in South Africa', *Urban Studies*, 35: 131–48.

Marksuen, A. (1987) *Regions: The Economics and Politics of Territory*. Totowa, NJ: Rowman and Littlefield.

Marston, S.A. (2004) 'The social construction of scale', *Progress in Human Geography*, 24: 219–42.

Marston, S.A., Jones, J.P. and Woodward, K. (2005) 'Human geography without scale', *Transactions of the Institute of British Geographers*, 30: 416–32.

Massey, D. (1979) 'In what sense a regional problem?', *Regional Studies*, 13: 233–43.

Massey, D. and Meegan, R. (1982) *The Anatomy of Job Loss: The How, Why and Where of Employment Decline*. London: Methuen.

Massey, D. and Thrift, N.J. (2003) 'The passion of place', in R.J. Johnston and M. Williams (eds), *A Century of British Geography*. Oxford: Oxford University Press, pp. 275–99.

McCann, E.J. (2002) 'The cultural politics of local economic development: meaning-making, place-making, and the urban policy process', *Geoforum*, 33: 385–98.

McCann, E.J. (2004) '"Best places": interurban competition, quality of life and popular media discourse', *Urban Studies*, 41: 1909–29.

McQuirk, P.M. and MacLaran, A. (2001) 'Changing approaches to urban planning in an "entrepreneurial city": the case of Dublin', *European Planning Studies*, 9: 437–57.

Miewald, C.E. and McCann, E.J. (2004) 'Gender struggle, scale, and the production of place in the Appalachian coalfields', *Environment and Planning A*, 36: 1045–64.

Molotch, H. (1976) 'City as a growth machine – toward a political economy of place', *American Journal of Sociology*, 82: 309–32.

Mossberger, K. and Stoker, G. (2001) 'The evolution of urban regime theory – the challenge of conceptualization', *Urban Affairs Review*, 36: 810–35.

Nel, E. (2001) 'Local economic development: a review and assessment of its current status in South Africa', *Urban Studies*, 38: 1003–24.

North, P., Valler, D. and Wood, A. (2001) 'Talking business: an actor-centered analysis of business agendas for local economic development', *International Journal of Urban and Regional Research*, 25: 830–46.

Oberhauser, A. (1995) 'Towards a gendered regional geography – women and work in rural Appalachia', *Growth and Change*, 26: 217–44.

Oberhauser, A. (2002) 'Relocating gender and rural economic strategies', *Environment and Planning A*, 34: 1221–37.

O'Toole, K. and Macgarvey, A. (2003) 'Rural women and local economic development in south-west Victoria', *Journal of Rural Studies*, 19: 173–86.

Park, B.G. (2003) 'Territorialized party politics and the politics of local economic development: state-led industrialization and political regulation in South Korea', *Political Geography*, 22: 811–39.

Peck, J. and Tickell, A. (1995) 'Business goes local – dissecting the business agenda in Manchester', *International Journal of Urban and Regional Research*, 19: 55–78.

Peck, J. and Tickell, A. (1996) 'The return of the Manchester Men: men's words and men's deeds in the remaking of the local state', *Transactions of the Institute of British Geographers*, 21: 595–616.

Peet, R. (1983) 'Relations of production and the relocation of United States manufacturing industry since 1960', *Economic Geography*, 59: 112–43.

Peterson, P. (1981) *City Limits*. Chicago: University of Chicago Press.

Pierre, J. (2005) 'Comparative urban governance – uncovering complex causalities', *Urban Affairs Review*, 40: 446–62.

Quilley, S. (1999) 'Entrepreneurial Manchester: the genesis of elite consensus', *Antipode*, 31: 185–211.

Raco, M. (2003a) 'Assessing the discourses and practices of urban regeneration in a growing region', *Geoforum*, 34: 37–55.

Raco, M. (2003b) 'Governmentality, subject-building, and the discourse and practices of devolution in the UK', *Transactions of the Institute of British Geographers*, 28: 75–95.

Raco, M. (2003c) 'Remaking place and securitising space: urban regeneration and the strategies, tactics and practices of policing in the UK', *Urban Studies*, 40: 1869–87.

Raco, M. and Imrie, R. (2000) 'Governmentality and rights and responsibilities in urban policy', *Environment and Planning A*, 32: 2187–204.

Roberts, S. and Schein, R. (1993) 'The entrepreneurial city – fabricating urban development in Syracuse, New York', *Professional Geographer*, 45: 21–33.

Rogerson, C.M. (1997) 'Local economic development and postapartheid reconstruction in South Africa', *Singapore Journal of Tropical Geography*, 18: 175–95.

Sawers, L. and Tabb, W. (eds) (1984) *Sunbelt/Snowbelt: Urban Development and Regional Restructuring*. New York: Oxford University Press.

Sellars, J.M. (2002) *Governing from Below: Urban Regions and the Global Economy*. Cambridge: Cambridge University Press.

Short, J.R. and Kim, Y.H. (1998) 'Urban crises/urban representations: selling the city in difficult times', in T. Hall and P. Hubbard (eds), *The Entrepreneurial City: Geographies of Politics, Regimes and Representation*. Chichester: John Wiley, pp. 55–75.

Smith, N. (1996) *The New Urban Frontier: Gentrification and the Revanchist City*. London and New York: Routledge.

Stone, C. (1989) *Regime Politics: Governing Atlanta, 1946–1988*. Lawrence: University of Kansas Press.

Tang, S.Y. and Tang, C.P. (2004) 'Local governance and environmental conservation: gravel politics and the preservation of an endangered bird species in Taiwan', *Environment and Planning A*, 36: 173–89.

Uitermark, J. (2005) 'The genesis and evolution of urban policy: a confrontation of regulationist and governmentality approaches', *Political Geography*, 24: 137–63.

Valler, D. (1996) 'Locality, local economic strategy and private sector involvement: case studies in Norwich and Cardiff', *Political Geography*, 15: 383–403.

Valler, D. and Wood, A. (2004) 'Devolution and the politics of business representation in Britain: a strategic relational approach', *Environment and Planning A*, 36: 1835–54.

Valler, D., Wood, A. and North, P. (2000) 'Local governance and local business interests: a critical review', *Progress in Human Geography*, 24: 409–28.

Ward, K. (2000) 'A critique in search of a corpus: re-visiting governance and re-interpreting urban politics', *Transactions of the Institute of British Geographers*, 25: 169–85.

Ward, K. (2003) 'Entrepreneurial urbanism, state restructuring and civilizing "New" East Manchester', *Area*, 35: 116–27.

Ward, S. (1990) 'Local industrial promotion and development policies 1899–1940', *Local Economy*, 5: 100–18.

While, A., Jonas, A.E.G. and Gibbs, D.C. (2004) 'Unblocking the city? Growth pressures, collective provision, and the search for new spaces of governance in Greater Cambridge, England', *Environment and Planning A*, 36: 279–304.

Wilson, D. (1996) 'Metaphors, growth coalition discourses and black poverty neighborhoods in a US city', *Antipode*, 28: 72–96.

Wilson, D. (1998) 'Urban representation as politics', *Urban Geography*, 19: 254–61.

Wood, A. (1996) 'Analysing the politics of local economic development: making sense of cross-national convergence', *Urban Studies*, 33: 1281–95.

Wood, A. (1998) 'Making sense of urban entrepreneurialism', *Scottish Geographical Magazine*, 114: 120–3.

Wood, A. (2004) 'Domesticating urban theory? US concepts, British cities and the limits of cross-national applications', *Urban Studies*, 41: 2103–18.

Zhang, T.W. (2002) 'Urban development and a socialist pro-growth coalition in Shanghai', *Urban Affairs Review*, 37: 475–99.

Zhang, J.X. and Wu, F.L. (2006) 'China's changing economic governance: administrative annexation and the reorganization of local governments in the Yangtze River Delta', *Regional Studies*, 40: 3–21.

Zhu, J.M. (1999) 'Local growth coalition: the context and implications of China's gradualist urban land reforms', *International Journal of Urban and Regional Research*, 23: 534–48.

The Politics of Localization: From Depoliticizing Development to Politicizing Democracy

Giles Mohan and Kristian Stokke

INTRODUCTION

There is a new politics of localization in the 'global South', a politics that is constructing and grounded in local political spaces and practices (Harriss et al., 2004). One striking manifestation of this new politics is the recent double inversion in development theory and strategy, from a focus on developmentalist states to an emphasis on civil society, and from a focus on centralized state-building to an emphasis on decentralization and local participation (Mohan and Stokke, 2000). This has yielded an unprecedented convergence between the neo-liberal right and the post-Marxist left, where both emphasize local civil society and give it priority over inefficient and centralized states. According to neo-liberals, civil society can exert organized pressure on autocratic and unresponsive states and thereby support democratic stability and good governance. It can also facilitate participation in development and thereby empower target groups of poor people. By contrast, the post-Marxists see civil society as the expression of diverse forms of identity politics challenging the hegemony of global economic liberalism and its associated political institutions. Social movements hold the potential for bringing about autocentric and socially relevant development in opposition to the disempowering structures of both the state and the market. So, what the neo-liberal right and the post-Marxist left share is a concern with the interests and agency of local 'people' and their participation in processes of empowerment.

This double inversion in development is inextricably linked to an ongoing scalar reconfiguration of politics in general. On the one hand, the influence of supranational institutions such as the World Bank, the International Monetary Fund, and the World Trade Organization is increasing. Formal institutions at global and regional levels now exercise considerable power over the institutions and peoples of the global South through economic and legal instruments as well as discursive power (McNeill and Bois, 2003). On the other hand, the local level of politics is also becoming more prominent. Localization of politics can be mediated through a series of interrelated institutional reforms including decentralization, democratization, and 'good' governance. We argue that together these effect a means of depoliticizing local development, although they can be contested and reworked. However, localized political mobilization also occurs through popular struggles around local, national and global issues alongside, and sometimes bound up with and politicizing, these formal institutional reforms.

Put starkly, this means that 'actual processes of political transition are likely to be the outcome of contingent combinations of "top-down" international pressures for good governance and "bottom up" pressures for social change and greater accountability' (Barnett and Low, 2004: 13). Such institutional changes and popular struggles create and transform local political spaces – defined in terms of localized institutional, organizational, and discursive resources – that may be utilized by different actors with highly varied capacities, strategies, and interests (Harriss et al., 2004).

While notions of 'top down' and 'bottom up' artificially split what is a seamless field of political activity, our analysis of localization follows this logic. In doing so, we address two broad questions. The first is how do powerful international agencies, such as the Bretton Woods institutions, United States Aministration for International Development (USAID), and Department for International Development (DFID), imagine and seek to promote the localization of politics through the power of conditionality and in what ways could this be depoliticizing development? This requires situating a cluster of discourses around decentralization, social capital, civil society, and local democratization within the trajectories of the global political economy and associated moves toward 'good governance' as a central element of aid.

However, this focus on donor–state relations takes our focus away from how popular struggles politicize and seek to transform local political spaces. Hence, our second question is, how do 'local' political processes operate and what potential do these hold for what we call substantive democracy? These two foci should not, however, be read as representing the hegemonic and structural on the one hand and local resistance on the other, since this dualistic logic may largely reflect the ideological desires of progressive intellectuals not involved in everyday struggles as opposed to the lived politics in multiple localities of the global South. Rather, following the logic of 'glocalization', we see the political economy and more localized agency as a single field of analysis and practice, which can only be apprehended empirically.

We begin, in the next section, by elaborating our analytical framework starting with questions of glocalization and moving into looking at how political transitions at the local level bring together political-economic forces and multiple, situated practices. The third section examines how hegemonic development discourses, generally emanating from the Bretton Woods Institutions and Western development aid agencies, construct 'the local' in ways that create certain opportunities for local participation but also render development as a technocratic and depoliticized process. The fourth section then examines how local political spaces, which are created through decentralization and democratization reforms, may be utilized by local popular actors to pursue instrumental interests in development, but also how civil and political society activism may institutionalize new forms of local popular democracy. Given the inevitable

contingency of this, we can only comprehend them empirically.

UNDERSTANDING THE POLITICS OF THE 'GLOCAL'

In our introduction we argued that local political spaces are structured through the institutional reforms largely initiated by the major donor institutions and brokered by recipient states and a range of popular struggles and practices, which may include, but are more than, acts of resistance. We saw these political spaces as simultaneously global and local, what others have referred to as the 'glocal' (Dirlik, 1998; Escobar, 2001). In this section we elaborate a framework for the two substantive sections that follow. This requires analyzing multiple levels of governance (MacLeod and Goodwin, 1999), which do not sideline the nation-state (Barnett and Low, 2004); appreciating the relational connections between places (Gupta and Ferguson, 1992; Amin, 2002); as well as foregrounding 'embedded' and emergent local political practices (Hecht and Simone, 1994; Escobar, 2001; Watson, 2004).

First, we need to address the political-economic context in which 'glocalization' occurs and is reconfiguring the scale and character of politics (Appadurai, 1996; Cox, 1997; Jessop, 2002). This process is not uniform and universal, but one in which 'phenomena are all both local and global, but … they are not all local and global in the same way' (Dirlik, 1998: 9). In examining glocalization, political geographers among others have underplayed the importance of the state (Barnett and Low, 2004), whereas 'it is the state – scaled at various levels – which sets and controls the parameters for regime formation in the first place' (MacLeod and Goodwin, 1999: 508). Under pressure from market forces and neo-liberal political forces, many states are undergoing transformations toward de-statization (i.e. reduced state authority in favor of market liberalization), de-nationalization (i.e. scalar reconfiguration of state power in favor of regionalization and localization), and internationalization of policy regimes (i.e. the increased importance of global contexts in which state actors operate and the enhanced reach of international policy communities) (Jessop, 2002). In understanding the reconfiguration of the state in the South much of development studies has, despite increased attentiveness to the role and character of state institutions and the relations to civil society, treated the state in an apolitical manner. In the next section we examine how international policy institutions discursively construct local political spaces, in part as an attempt to delegitimize the central state and enhance 'choices' and 'freedoms' for citizens.

Policy, as we shall see, then aims to reform the state toward an ideal-typical Weberian form while also depoliticizing development.

However, focusing on the global discourses and policies is only one way of taking 'glocalization' seriously. The second element of our framework is to understand the operation of local politics within a glocalized political field. We start with Ash Amin's recent work, which examines the relationality and simultaneity of place-based politics. Amin rightly rejects any assumption that places are 'nested in territorial or geometric space' (Amin, 2002: 389). Rather places are 'nodes in relational settings, and as a site of situated practice … a place of engagement in plural politics and multiple spatialities of involvement' (2002, 391, 397). This anti-essentialist understanding of place-based politcs allows us to explore both the extra-local political connections between places and the multiple forms of political practice that actually operate locally. Most explicitly, the final section examines how transnational communities are involving themselves in the direction and scope of municipal governance 'back home'. In doing so we counter the normative theories we discuss below which treat local politics in the South in one-dimensional ways. For the neoliberals much local governance is treated as a poor reflection of 'proper' democracy whereas for the left place-based politics represents feeble resistance to the remorseless power of globalization. Rather, we develop these insights around a politics of interconnection.

For Escobar (2001) this is about a balance between interconnections beyond place and some sense of boundedness and shared values and practices within place. He notes that 'people continue to construct some sort of boundaries around their places, however, permeable, and to be grounded in local socio-natural practices, no matter how changing and hybridized those grounds and practices might turn out to be' (Escobar, 2001: 147). Escobar usefully links culture to place, but he evokes, as with some other radical democratic theories, an almost naturalistic and unquestioned connection between 'culture' and 'politics'. What is vital is an appreciation of the ways in which politics becomes institutionalized within places and connects to other political processes beyond place. So, we get a sense of local politics as taking place in porously bounded spaces, involving a range of activities aimed as much at place-making as resistance to globalization (Bebbington, 2000).

This requires us to treat people of the South as possessing agency in multiple ways, which requires moving beyond formal political spaces and looking to other realms of the political. As Hecht and Simone (1994: 16) argue: 'Formal governance tends to have a restrictive notion of the public sphere that neglects the desires and efforts of the impoverished to create culture, sociality and

solidarity' (see also Chakrabarty, 1992; Staeheli and Mitchell, 2004). This gives us a more open view of what constitutes political community and what counts as political activity, though only empirical investigation can show whether and how this produces substantive democratic outcomes (Heller, 2001; Watson, 2004). Thus, our final section points to the importance of popular mobilization in transforming formal democratic institutions and institutionalizing new forms of local popular democracy, but also stressing that such political agency and processes are contingent.

DEPOLITICIZING LOCAL POLITICS: DECENTRALIZATION, PARTICIPATION, SOCIAL CAPITAL AND CIVIL SOCIETY

We have mentioned that neo-liberal development policy has seen a move toward the local as a site of institutional reform and locus of political agency. This localization is produced from a cluster of different processes – decentralization, participation, social capital strengthening and civil society building – that are often seen to contribute to a single entity called 'good governance'. In this section we unpack these different elements that go to make for a localization of politics, but also argue that the form of political agency they condition is highly circumscribed, so that what appears as a revitalized local political agenda is largely depoliticizing. We focus mainly on those discourses emanating from the Bretton Woods institutions and major international aid agencies. We begin by tracing the emergence of the good governance agenda before critically examining the series of related discourses that constitute this agenda at the local level.

The Geopolitics of Good Governance

Governance has become one of the keywords in global political theory and policy over the 1990s and is used to distance explanations from notions of government. Whereas government implies state-led, monocratic, rule-driven and hierarchical decision-making, governance entails 'the involvement of a wide range of institutions and actors in the production of policy outcomes [via] … co-ordination through networks and partnerships' (Painter, 2000: 317). Rhodes (1997: 15) stresses the self-organizing nature of governance networks such that they have 'significant autonomy from the state', thereby opening the door for market-driven decision-making.

In the global South governance rarely refers to an analytical concept, being prefixed normatively by the word 'good', which indicates a desirable goal for policy (Abrahamsen, 2000). The lenders are able to use their transnational state power in forcing such governance reforms as part of conditionality packages (Baylies, 1995). In terms of the Bretton Woods institutions the publication of *Sub-Saharan Africa: From Crisis to Sustainable Growth* (World Bank, 1989) marked a watershed in thinking about governance, both on the African continent and beyond. They argued that 'political legitimacy and consensus are a precondition for sustainable development' (World Bank, 1989: 60), which opened the way for a raft of institutional reform programs aimed at getting the politics right in order to bring about economic development (World Bank, 1993, 1994).

This neo-liberal agenda is premised upon restructuring the relations between state, market and civil society. In practice, there are two interlinked conceptualizations of governance that relate to process and capacity. First, it is taken to mean the reordering of state/society relations with an emphasis upon strengthening civil society, some measure of participation, and the holding of free and fair elections. The second meaning is that of administrative reform requiring 'a smaller state equipped with a professional, accountable bureaucracy that can provide an "enabling environment" for private-sector-led growth' (World Bank, 1994: xvi). The aim is to create a Weberian, ideal-typical bureaucracy where decision-making is transparent and officials are impartial and accountable (Leftwich, 1993, 1994).

Beyond these general prescriptions we see a range of reforms, underpinned by particular readings of political and social theory aimed at reconnecting state and society at the local level within an overarching neo-liberal logic. In the remainder of this section we critically analyze these key movements and the relationships between them. Our argument follows those of Abrahamsen (2000), Corbridge et al. (2005), Ferguson (1994) and Harriss (2002) insofar as we see these ostensibly political projects as a means of depoliticizing development. By depoliticization we concur with Ferguson and Lohman's general observation that 'a "development" project can effectively squash political challenges to the system, not only through enhancing administrative power but also by casting political questions of land, resources, jobs, or wages as a technical "problem", responsive to the technical "development" intervention' (1997: 232). Here the encroachment of administrative structures can by-pass popular actors while the logic that inheres within them treats contested material politics as the target of rational and technical interventions. Schuurman (1997: 163) rightly argues that the localization of political agency 'is powerless to prevent spatial inequalities from persisting or even increasing. Political participation at local government level will only reflect the fragmentation of civil society and the neofeudal

attitude of political parties *vis-à-vis* social movements'. Here Schuurman points to a form of divide-and-rule in which localization encourages parochial interests to be expressed, in which organizations scramble for local political resources and fail to coalesce into more organized forms of action that could contest state power. He also implies that where state-based political institutions involve themselves in non-state forms of organization they do so in ways which co-opt and control them. The result of these discourses and practices, according to such critiques, is to enhance administrative rationality, promote an individualizing 'self-help' mentality, tie local government into centralized control, and fragment political opposition.

However, in the framework we developed in the previous section we also argued that local actors may contest these apparently hegemonic discourses. So, while we agree with these analyses, rather than dismiss good governance as mere depoliticization and search for more 'genuine' political expressions, we need to see these discourses and reforms as having real political effects and opening up a range of political spaces. As Corbridge et al. (2005: vii) note, 'spaces of citizenship ... are being created, or perhaps widened, in the wake of the good governance agenda and the popular mobilizations to which it can give rise'. Corbridge et al.'s empirical work around this statement will be revisited later.

Fragmentation and Control: Decentralization

Following these introductory remarks we can now examine the main components of the localizing neo-liberal agenda and show where and how it is depoliticizing, but also how it can be contested and reshaped by actors on the ground. At the heart of these localizing initiatives are moves toward decentralization. Some guidance notes offered to field officers working for the USAID capture the rationale for this. They state: 'When effective decentralization and democratic local governance advance in tandem, local governments – and the communities they govern – gain authority, resources, and skills to make responsive choices and to act on them effectively and accountably' (USAID, 2000: 2). Such advice captures the raft of supposed benefits of decentralization, which not only strengthens the capacity and accountability of local government, but has wider 'knock-on' effects in terms of enhancing the ability of 'communities' to be more proactive in their own development. By bringing government closer to the people spatially and institutionally it will become more effective because it is more knowledgeable about local populations and therefore more responsive to them (Johnson, 2001;

Crook and Sverrisson, 2001; Hutchcroft, 2001). Crucially, the phrase 'advance in tandem' with respect to democratization indicates that the institutional reform of bringing administration closer to the people must be tied to political reform. It is these linkages between the administrative and political in which other localizing mechanisms – social capital, participation and civil society – are located, since they constitute the political transmission mechanism between 'society' and 'the state'.

Decentralization has been a perennial feature of politics and development in colonial and post-colonial states (Mamdani, 1996; Ribot, 1999), but in terms of its neo-liberal incarnation, there has been a shift over the past two decades, culminating in the kind of view expounded by USAID. In the early 1980s, decentralization was centered on the public and, to a lesser extent, the voluntary sector. Almost a decade later, and well into the neo-liberal 'adjustment era', Rondinelli et al. (1989) include privatization and deregulation as forms of decentralization. The World Bank's own policies reflect these trends in which decentralization 'should be seen as part of a broader market-surrogate strategy' (World Bank, 1983: 23). Since the mid-1980s, then, decentralization has become one of the mainstays of the localizing good governance agenda and promoted in a wide range of countries (Mohan, 1996; Schuurman, 1997; Crook and Sverrisson, 2001; Bierschenk and Olivier de Sardan, 2003; Boone, 2003).

In concrete terms, decentralization policies have achieved some benefits, most notably pro-poor programs in parts of Brazil and India (Tendler, 1997; Crook and Manor, 1998; Heller, 2001). In these well-known success stories the key factors appear to be the determination of leftist political parties with a genuine commitment to social equality and well-organized civil society networks (Fung and Wright, 2003). Besides these *causes celèbres* there have been some increases in participation rates, but limited or negative effects on poverty, equity, and empowerment (Johnson, 2001).

In terms of our argument around depoliticization, this form of decentralization has a number of effects. First, it factionalizes and fragments political opposition. As Mamdani (1996: 300) argues, 'all decentralized systems of rule fragment the ruled and stabilize the rulers'. Second, as Boone (2003) and others have shown, decentralization policies have been used for penetrating and manipulating local political society in order to govern more effectively. Regimes often promote development programs that seek to build upon local energies because this absolves them of responsibility for welfare provision and they can place appointees in key positions while earning political capital by apparently being sensitive to local issues. Longer-standing ministerial hierarchies have also

contested devolution of power and sought to maintain control of key resources (Mohan, 1996). Third, in discussing Uganda's decentralization program, Francis and James (2003) usefully distinguish between the technocratic and patronage outcomes of decentralization. The former refers to where central states treat decentralization as a matter of tweaking institutions, but active participation from below is notional. The latter relates to the micropolitics of devolution in which the limited fiscal resources passing through local government are contested by the locally powerful. This so-called 'elite capture' (Crook and Manor, 1998; Kelsall, 2000; Crook and Sverrisson, 2001) strengthens local governance in their favor through patronage, although Williams et al. (2003) argue that in cases of weak governance structures, patrons may be the best avenue for securing short-term gains for the poor (see also Bayart, 1993, Chabal and Daloz, 1999). Fourth, the failure of decentralization can arise from the weak political capacity of the impoverished who have little time and energy for contesting political struggles (Johnson, 2001). This is exacerbated by the weak capacity of local government, which can be in terms of infrastructural deficiencies or institutional confusion that undermines its ability to respond to political pressure (Aasen et al., 1997; Crook and Sverrisson, 2001; SLSA, 2003).

Depoliticizing Political Community? Social Capital and Civil Society

If the move toward localization relies on reforms to enhance the responsiveness and capacity of local government and opening up some, albeit limited, formal democratic space, there is still a gap in terms of how to tap the political energies of local populations. In terms of neo-liberal localization the key role for local people is to mobilize the resources – land, labor, and capital – needed to make 'choices' about their own development as well as a democratic function of channeling opinion into the state's policy-making systems and also holding the state to account for its actions. It is here that civil society has become the arena in which this set of political and developmental processes unfolds.

Crucially, civil society in the global South is not a straightforwardly 'national' or 'local' space, but one penetrated and shaped by international civil society organizations, such as development NGOs, which are often funded by neo-liberal donors and who seek to 'strengthen' local civil society by enhancing 'participation' at a range of scales. Part of this is to build upon the virtuous relationships that exist within communities, including 'soft' institutions such as trust and 'harder' ones like local organizations. These virtuous relationships have been labeled 'social capital' and we will look briefly at its role as a political 'conveyor belt' between society and the state.

Williams and Young (1994: 87) note: 'The (World) Bank's promotion of civil society is linked to its promotion of accountability, legitimacy, transparency and participation, as it is these factors which empower civil society and reduce the power of the state'. From this perspective, civil society is part of the neo-liberal ideology of weakening the state politically, promoting self-help and choice in society and cheapening the cost of aid. As Allen (1997) observes there are there assumptions informing this discourse of civil society. First, that civil society is not only distinct from the state, but in conflict with it. So, when donors support civil society organizations they use them in order to weaken the state and/or massage it in the direction they desire. Second, the civil society is at the heart of democratization, because it is a means of channeling opinion into policy making in a co-ordinated fashion and simultaneously increasing peoples' confidence in involving themselves in public affairs as well as building their institutional capacity to do so (World Bank, 1997). Third, that NGOs form a highly significant part of civil society, and thus of forces driving democratization. As such NGOs are the '"missing middle" between citizens and the state' (World Bank, 1997: 114).

The functioning of this so-called 'missing middle' between state and citizens requires a range of socio-political conditions to be met, which include somewhat intangible things like confidence, trust, and 'capacity'. In donor thinking such conditions are covered by the rather vague, catch-all concept of social capital. Popularized by Robert Putnam in his work on Italy and America, social capital refers to 'features of social organization, such as trust, norms and networks, that can improve the efficiency of society by facilitating coordinated actions' (Putnam, 1993a: 167). For Putnam, social capital fosters reciprocity, facilitates information flows for mutual benefit and trust, and once it exists tends to be self-generating as successive generations are socialized into localized norms. This view of culture and politics is very inward-looking to the local level. In Putnam's words: 'The historical roots of the civic community are astonishingly deep. ... Stocks of social capital ... tend to be self-reinforcing and cumulative' (1993b: 36).

In essence this is about political culture and how people 'bond' with one another (Mohan and Mohan, 2002). But the instrumental side of this is how these bonds between people facilitate democratic and economic choice. Putnam analyzed the fortunes of Italy's regions and found a correlation between levels of social capital (measured by various proxies) and economic performance, with the link being a range of institutions that translated these bonds into 'bridges' into the state. Hence, the logic of this is that the 'right' form of

social capital will produce 'better' local institutions, which will, in turn, channel opinion into the local state and set up a virtuous cycle of empowered citizens and responsive government. The key dynamic is '*both* a more active and engaged civil society which can express demands of the citizenry, *and* a more responsive and effective state which can deliver needed public services' (Gaventa, 2004: 27, emphasis in original). So, the vision is one of multiplying the channels through which citizens can exercise political agency and ensuring that the state is responsive to their demands.

Despite social capital being under-theorized and poorly understood (Watson, 2004), it has become an important analytical concept and policy tool within development discourse and practice (Mohan and Mohan, 2002; Bebbington, et al., 2004). The promotion of social capital is a major area of concern for the donor institutions and NGOs, but the changes needed to strengthen it are acknowledged to be numerous (Ribot, 1999; SLSA, 2003: 95). For NGOs the role becomes that of intermediaries shifting from output-based approaches to more process-based ones (Kohl, 2003). In the Bangladeshi case, but possibly more widely, Crook and Sverrisson (2001: 28) envisage a role for NGOs 'to support and encourage the formation of community and interest group associations for collective action'. Johnson (2001) sees three ways that NGOs and other external development agents can promote empowerment. First, they can promote alliances between poor people, thereby increasing their lobbying power. Second, they can help absorb some of the costs of engaging in political action. Third, they can promote feelings of self-worth through encouraging people to engage in collective action. Once people gain confidence and start engaging with political processes a virtual cycle ideally emerges whereby they gain deeper knowledge of how politics operates and, therefore, how to shape it to one's needs. The problem for neo-liberal policy-makers and impact analysts is that social capital is such a vague term that these attempts to 'strengthen' and 'facilitate' it can only ever be quite experimental and tentative.

Those studies that assess the ability of social capital to enhance local choice and accountability are quite ambivalent in their conclusion (Bebbington et al., 1997; Narayan and Pritchett, 1999; Grootaert, 2001). What comes through clearly, though, is that building social capital by itself can only ever have limited effects and more meaningful political reforms are crucial. For example, Tendler (1997) argues that state/civil society relations are not unidirectional but that where pro-poor outcomes occur there have been complex interactions between the local and central states and various forms of social activism (see also Chandhoke, 1995; Tarrow, 1996; Fox, 1997a, 1997b). Whereas the state is deeply implicated in 'building' and 'thickening' social capital, such political dynamics are downplayed in civil-society-centered discourses on social capital.

As with decentralization, centralized regimes are often happy to promote development programs that seek to build upon local social capital because this absolves them of responsibility for welfare provision, earns political capital by being sensitive and dialogic, and disaggregates society. Potential alliances are undermined as civil society actors literally scrabble for the pickings of the aid regime. One of our studies in Ghana (Mohan, 2001) showed that under the guise of civil society empowerment, local 'partner' organizations and, more importantly, the rural poor were marginalized from decision-making. The paradox is that international NGOs, often heavily funded by their home governments, are charged with 'empowerment', but are so wary of upsetting their funders that they tightly circumscribe the activities on the ground and completely undermine independent development (Tvedt, 1998). The aid paradigm means that civil society organizations become more dependent on external funders as well as the market. And a consequence of this is that many NGOs have been set up precisely to divert aid for personal goals as opposed to responding to the needs of the poor (Hibou, 1999).

Consequently, both discursively and practically the promotion of civil society organizations and social capital has depoliticizing effects (Harriss, 2002). As Fine (1999: 12) observes, 'social capital allows the World Bank to broaden its agenda whilst retaining continuity with most of its practices and prejudices which include benign neglect of macro-relations of power, preference for favored NGOs and grassroots movements, and decentralized initiatives'. However, in keeping with our argument that political spaces can be contested and made otherwise by various actors, recent work by Bebbington et al. (2004) argues against a straightforward reading of social capital as depoliticizing (see also Moser and Holland, 1998). Instead they argue that we need to disaggregate institutions and take an agency-centered as opposed to a discourse-centered approach to the role of social capital in international development interventions. By doing this, the encroachment of social capital into donor discourse represents an important strategic movement. Crucially, it represented an attempt to 'socialize' the overly rationalist discussions of the neoclassical economists who dominate policy-making. Assessing the impact of these debates on the economists is impossible, but the fact that social capital entered the conceptual lexicon is taken as a positive step. Likewise, the straightforward argument about depoliticization is also nuanced by the complex ways in which ordinary people negotiate their lives as citizens in regard to the state and the political spaces that may have been opened up by the decentralization and social capital components of the good governance agenda.

Local Experiences of State Encounters

Throughout this section we have hinted that engagement between state and society does occur, albeit not always in the ways set out in formal policy discourses. Following Migdal's (1994: 15) 'anthropology of the state', which investigates 'different levels of the state, including the lowest rungs on the organizational hierarchy where direct engagement with society often occurs', we move away from a totalizing analysis of the state toward a spatialized understanding of political praxis. The recent work of Corbridge et al. (2005) explicitly deals with how the poor interact with the state in everyday ways and whether the decentralization of the state following liberalization has affected these interactions. One aim of their rich analysis is to examine 'the ways in which the different groups of the rural poor might be said to see or to encounter the developmental state [which] ... should caution us against a reductionist understanding of "state–poor" encounters' (2005: 7, 10). The thrust of their argument is that the good governance agenda *does* transform the political process in a number of ways, many of which are quotidian and rarely register in less subtle analyses. They are not arguing that massive shifts in power have occurred, but that berating the ideology of neo-liberal governance is no substitute for analyzing its actual effects. All we can do here is to give a flavor of these types of interactions and changes.

One example is from Bhojpur District in Bihar, which has a long history of collective action by the poor. Here, partly in response to agrarian unrest, a series of public hearings were introduced to help diffuse tensions. The hearings involved higher-ranking officials so that the astute and experienced sections of the poor were able to see and learn about the operation of power. In this way they gained confidence to press for other claims. While this sounds like a story of liberal political learning, such successes were only possible through 'an understanding, won through years of agrarian conflict, that the Scheduled Castes must protect themselves against the *Bhumihars* (landowners) when the state is not around to act as a referee' (Corbridge et al., 2005: 152). Whereas the detractors of good governance would argue that these discourses are normative and conceal other effects, actual encounters at the local level can enable poor people to exercise some measure of political agency, albeit often in limited and unintended ways. We now develop this brief insight and situate it within other processes of local political agency.

PRACTICING LOCAL POLITICS: TOWARDS SUBSTANTIVE DEMOCRACY?

Whereas administrative decentralization is reconfiguring the scale of political institutions in many countries of the global South, democratization is redefining the relationship between state and society. Taken together, these processes create new and transformed local political spaces for different actors with highly diverse interests, resources, and strategies. This section, drawing from our own and others' work, addresses the second of our main questions concerning the making of liberal democratic political spaces. We are particularly interested in the capacity of popular actors to utilize these democratic spaces to pursue their instrumental interests in development, and the institutionalization of more substantive forms of local democracy through civil and political society activism.

The first example from South Africa highlights the combined importance of the legacies of popular mobilization and the constitutional reforms enacted in moving beyond apartheid. However, this case reveals the need to scale up from particularistic struggles and institutionalize local popular democracy. This is central to the next two cases of Kerala and Porto Alegre, where new forms of local democracy have been created through combined activism in civil and political society within the context of devolution of state power and resources. The final example of migrant 'hometown' associations argues that 'local' political spaces are always also extra-local with connections to transnational communities, although such diasporic public spaces are not necessarily empowering. The theme that informs this section, then, regards the local possibilities and means for politicizing popular issues and interests and thereby transforming minimalist local and national democratic institutions in the direction of substantive democracy (Törnquist, 2004).

Popular Struggles for Socio-economic Rights

The global expansion of formal democracies since the mid-1970s has been labeled as a 'third wave of democratization' and celebrated as a global triumph of Western economic and political liberalism (Huntington, 1991; Fukuyama, 1992). Less triumphalist scholars have problematized the dynamics and substance of these new democracies, describing them as elitist and formal rather than popular and substantive (Grugel, 2002). The transitions of the 'third wave' and the associated academic discourses have generally had a narrow focus on the formal rules, procedures, and institutions of liberal democracy, most notably the repeated conduct of free and fair elections at national and sub-national levels (Kaldor and Vejvoda, 1997). Beetham (1999) describes this as a tendency to elevate a means to an end, to mistake institutional instruments with their democratic purpose. The core democratic principles of *popular control and political equality over collectively binding decisions* certainly require practical institutional forms for their realization, but the latter are derivative and may take different forms in different contexts. This means that democratization will always be an unfinished process of politicizing and democratizing democracy itself (Harriss et al., 2004). It also means that the attention of democratization processes and studies shifts from formal political institutions at the national level to the often localized efforts of achieving and transforming political/civil rights and socioeconomic rights. These, it should be added, are not the least facilitated by the political resources made available through moves to decentralize and 'enhance' civil society or to promote pro-poor and rights-based development (Webster and Engberg-Pedersen, 2002; Jones and Stokke, 2005).

Whereas the elite negotiations, constitutional processes and institutional designs involved in recent democratic transitions have been thoroughly researched (Schmitter et al., 1986; Linz and Stepan, 1996; Grugel, 2002; Whitehead, 2002), there are relatively few studies of the role of popular forces in formal and substantive democratization. As observed in the UNDP Human Development Report for 2002: 'While much has been written about the challenges of creating democratic institutions, there has been much less analysis of democratic politics: the struggles of poor and marginalized people to claim their rights and to overcome institutionalized obstacles' (UNDP, 2002: 79).

This raises an important question regarding the prospects and means for realization of rights, especially rights to development, within the political spaces of the democratic state. It is commonly assumed that liberal democracy will automatically lead to development, that is, that first-generation civil/political rights are both a necessary and sufficient condition for second-generation social/economic rights. The experiences of many 'third wave' transitions, however, yield formal democracies but falling far short of real development for ordinary people, demonstrate that the developmental outcome of democracy must be subjected to critical empirical analysis rather than assumed *a priori*. This shifts the focus to studies of the ways in which socio-economic rights are politicized and institutionalized through the practices of diverse actors within political spaces at the local scale and beyond. Here, we will limit ourselves to one important case of popular struggles for citizenship rights, namely the post-apartheid rights-based struggle for social justice in South Africa (Jones and Stokke, 2005).

South Africa's recent political history is integrally tied to popular movements. While the democratic transition itself was a negotiated settlement between the old and the new political elites, the political conjuncture for these negotiations came about through a prolonged and extensive mass mobilization (Bond, 2000b; Alexander, 2002). The struggle against apartheid brought together social movements, trade unions, and political parties in a counter-hegemonic movement for social and political justice. This liberation movement together with domestic economic crises and external pressures yielded the negotiated transition to liberal democracy in the early 1990s and subsequent changes in national and local political spaces for popular participation (McKinley, 1997; Seekings, 2000; Torres, 2000; Marais, 2001). Most notably, the transition produced radical changes in the content and extent of citizenship rights, in the political opportunities for democratic participation, and in hegemonic political discourses (Parnell et al., 2002; Jones and Stokke, 2005). This was followed by a set of institutional reforms in the late 1990s that changed the spatiality of the state by transforming and extending the role of local government:

Now the municipality becomes the primary development champion, the major conduit for poverty alleviation, the guarantor of social and economic rights, the enabler of economic growth, the principal agent of spatial or physical planning and the watchdog for environmental justice. ... Although accountable to national and provincial government, the local authority assumes enormous responsibility both for defining and implementing development priorities. (Parnell and Pieterse, 2002: 82–3)

This post-apartheid combination of a vibrant civil society and an enabling political environment should, it seems, provide an ideal case for substantive democracy. Unfortunately the political and social realities of post-apartheid South Africa have proven to be much more complex and contradictory (Bond, 2000a; Daniel et al., 2003).

South Africa's Constitutional Bill of Rights contains comprehensive civil and political rights, but also extensive socio-economic rights for all citizens, including rights in education, food, health, land, water, social security, housing, and the environment. While important achievements have been made, it is clear that the implementation of most of these socio-economic rights has been fraught with difficulties and seldom up to the high hopes of many ordinary citizens (Desai, 2003; Harrison et al., 2003; Nattrass, 2003). Effective political participation has also turned out to be complicated in practice. Civic associations and trade unions,

which were instrumental in the struggle against apartheid, were absorbed into the post-apartheid government in the mid-1990s (Adler and Steinberg, 2000; Adler and Webster, 2000; Edigheji, 2003). In the context of collaborative state/society-relations and a technocratic development orientation, there was also pressure on community organizations to become professional NGOs, adopt a technocratic approach to development and work with the government in a subordinate role to deliver services and development (Seekings, 1997; Heller, 2003). This situation left 'opponents of the government without a "voice" with which to express or a mechanism to organize opposition' (Ballard et al., 2006: 15). These state/society relations changed toward the end of the 1990s. In a context with persistent problems of unemployment, poverty and inequality combined with a shift in macro-economic policy from state-led social transformation to neoliberalism (Adelzadeh, 1996; Marais, 2001) a range of new social movements have emerged around diverse socio-economic issues and rights (Ballard et al., 2006). While these new movements are far from unitary, many of them share a concern with the livelihoods of poor people, which are now framed with reference to constitutional socio-economic rights (Millstein et al., 2003; Ngwane, 2003).

The new post-apartheid movements are shaped both by the post-apartheid political context and by the experiences and memories of the anti-apartheid struggle. In the same way as people were often mobilized into the anti-apartheid struggle around local everyday experiences of state repression and social injustice, the post-apartheid movements also emerge from everyday 'bread and butter' issues, with the material living conditions for the working poor being the main common point of identification (Seekings, 2000; Bozzoli, 2004). The local nature of these mobilizing issues, combined with the decentralized delivery of development and public services, make the new activism highly localized (Oldfield and Stokke, 2006). While the anti-apartheid struggle combined local struggles with a coordinated and ideological program of overthrowing the apartheid regime, the new movements have so far been issue-based and fragmented, with limited capacity to merge diverse issues and build a coherent movement that can effectively challenge national policy-making processes. Attempts to overcome this fragmentation through local social movement unionism – that is, local combinations of the organizational and political resources of trade unions and the mass mobilization of new social movements – have so far been hampered by organizational, ideological, and political obstacles (Lier and Stokke, 2006). The tendency within the ruling African National Congress (ANC) to distinguish between constructive collaborationist unions and disruptive 'ultra-left' new social movements, constitutes a major hurdle to this kind of collaboration as it requires trade unions to choose between loyalty to the governing alliance and the political uncertainties associated with the emerging new social movements, attempts to build new social movement unionism and the prospect of building a new left political party.

As fragmented, under-resourced and issue-based as they are, the presence of new social movements nevertheless poses a challenge at a symbolic level as they question the government's continued commitment to the working poor and their everyday struggles. This reintroduces a degree of political uncertainty into a polity that is marked by the lack of any real political opposition (Habib, 2005). Nevertheless, the continued difficulties of popular struggles for socio-economic rights in South Africa, amidst the miracles of the transition to democracy, demonstrates the need for realism in assessments of the political capacity and role of popular movements in transforming formal democracy into substantive democracy (Stokke and Oldfield, 2004). If this is the character of popular participation within South Africa's celebrated democracy, it seems even more pertinent in many other countries that have undergone recent regime changes.

A recent comprehensive study of Indonesia's democracy movement, for instance, points to the fragmentation of pro-democratic forces and observes that their struggles suffer both from the lack of links between civil and political society activism and from divisive politicization of particularistic issues, interests and identities (Prasetyo et al., 2003). While post-structuralists may advocate such polycentric struggles as a strategy for radical democracy (Mohan and Stokke, 2000; Houtzager, 2003), actual experiences of under-resourced and fragmented struggles point to the need to transcend localism and scale up from militant particularisms in order to institutionalize more substantive forms of democracy.

Institutionalizing Local Popular Democracy

Recent experiences in South Africa and elsewhere demonstrate the importance of civil society activism within the political spaces of the democratic state, but also the persistent problems of fragmentation and political exclusion of popular forces engaged in local issue-based struggles. There are, however, a few cases where popular struggles have yielded new forms of institutionalized local popular democracy. Two contemporary examples are the systems of decentralized planning in the state of Kerala (India) and participatory budgeting in the city of Porto Alegre (Brazil) (Abers, 2000; Isaac and Franke, 2000; Avritzer, 2002;

Törnquist, 2002; Wainwright, 2003). In reviewing these experiences, Fung and Wright (2003) highlight three common characteristics. First, they have a strong 'practical orientation' with an emphasis on concrete socio-economic development needs. Second, these cases are characterized by extensive popular participation, enabled through devolution of policy-making and institutionalization of new arenas for democratic participation. Third, policy decision-making within these new local arenas is based on deliberative processes. This stands in contrast to more common forms of representative democracy based on interest aggregation. While, in general terms, the model of representative democracy revolves around a process of aggregating citizens' preferences regarding political representatives and policies, the model of deliberative democracy emphasizes the coming together of the public in discussions that aim at consensual policy-making:

> Participants in the democratic process offer proposals for how best to solve problems or meet legitimate needs, and so on, and they present arguments through which they aim to persuade others to accept their proposals. ... Through dialogue others test and challenge these proposals and arguments. ... Participants arrive at a decision not by determining what preferences have greatest numerical support, but by determining which proposals the collective agrees are supported by the best reason. (Young, 2000: 22–3)

In reality, local popular democracy in both Kerala and Porto Alegre combines elements of the aggregative and the deliberative models (Baiocchi, 2003; Isaac and Heller, 2003). Participatory budgeting in Porto Alegre, for instance, has evolved into a two-tiered structure where citizens participate both as individuals in assembly meetings and as delegates in successive rounds of budget deliberations. Another distinctive feature of both cases is the combination of democratic deliberations in society and institutional mechanisms that connect public deliberation with the political system (Avritzer, 2002). Such institutional arrangements create stakes that are attractive enough to sustain popular and elite participation in the deliberative decision-making processes, but are also essential to overcome problems of deliberative inequality, that is, the tendency for deliberation to be dominated by already powerful actors. Thus, Schönleitner (2004) concludes, on the basis of a comparative analysis of local sector-policy councils in Brazil, that democratization through local deliberative spaces 'requires a positive interaction between an appropriate institutional design that ensures deliberative equality, government commitment to deliberation and civic participation in local deliberations' (Harriss et al., 2004: 45).

How do such institutional arrangements for local deliberative democracy come about? Abers (2003) observes that there is relatively little knowledge about the political dynamics behind local institutions for deliberative democracy. The existing literature, while focusing on institutional design, tends to ignore the fundamental question about the political interests, strategies and relative strengths of state, elite and popular forces that are involved in the making of such local popular democracies (for one exception, see Baiocchi, 2005). Few scholars have examined the willingness of governments and political elites to transfer power to popular deliberative processes, or the willingness of ordinary people to participate in time-consuming and frustrating deliberations. Fung and Wright (2003), for instance, focus primarily on the question of institutional design for empowered participatory governance and leave the questions of political forces and conjunctures unexamined. This leads Törnquist (2004) to the conclusion that there is a 'political deficit' in democratization studies, referring to a general lack of interest in and knowledge of the politics of fighting for and implementing institutional changes toward substantial democracy.

This 'political deficit' argument does not mean that there is a total lack of scholarly attention to the political making of local popular democracy. Avritzer (2002), for example, presents a society-centered analysis of how democratic forces have emerged in Latin American societies and how this has opened up public spaces for a direct political participation that challenge hierarchical and clientelistic forms of politics. His normative assumption is that the institutionalization of local deliberation will transfer democratic practices from civil society to political society, that is, transfer 'democratic potentials that emerge at the societal level to the political arena through participatory designs' (2002: 9). This unidirectional analysis is challenged by Abers (2003), who advocates a relational state/society perspective and argues that the making and success of empowered participatory governance (EPG) depend on a dual process of commitment-building: 'Unless both state actors (ranging from politicians to bureaucrats) and ordinary people are motivated to support, take part in, and respect EPG experiments, those policies are unlikely to become either empowered or participatory' (Abers, 2003: 201; see also Schönleitner, 2004).

In the case of Porto Alegre, she observes that participatory budgeting has emerged gradually as an alternative political strategy, in opposition to established populist-clientelist politics, for gaining state power through local elections. The Partido dos Trabalhadores (PT) has drawn on earlier initiatives from the Union of Neighborhood Associations of Porto Alegre to develop institutional arrangements for large-scale popular participation in decision-making and monitoring of public projects and investments. This participatory budgeting has functioned as a successful political strategy for PT (i) by responding to demands from neighborhood leaders who would otherwise rely on clientelistic networks within the opposition party; (ii) by politically mobilizing and integrating activists from popular movements; (iii) by delivering accountable and efficient local government that especially appeals to the middle classes; (iv) by strengthening local state capacity and coordination in the interest of the bureaucracy; and (v) by addressing the prioritized needs of poor people.

In the case of Kerala, Tharakan (2004) similarly points to the long history of productive political relations between the Communist Party and popular movements, providing access to local state power that was used to implement comprehensive land reforms in the 1970s and decentralized participatory planning from the mid-1990s. However, he also shows that the high hopes placed on participatory planning – initiated as a 'top-to-bottom' institutional reform under the assumption that it would be driven by civil society movements with support from the Communist Party – has not proven valid. Instead there has been a process of divisive party-politicization of associational life that has deepened pre-existing problems of mobilizing marginalized social groups and providing significant socio-economic transformations.

This reiterates the need for political analyses of the actors and strategies involved in institutionalizing new forms of democracy. The local political realities are certainly much more complex and contradictory than what can be captured in such brief accounts. The main point, however, is that a meaningful understanding and realistic assessment of local popular democracy in cases such as Porto Alegre and Kerala require empirical analyses of local politics. More specifically, both cases indicate that institutionalization of local popular democracy is neither the outcome of institutional design by committed political elites nor the product of civil society activism. The characteristics and dynamics of local popular democracy in Kerala and Porto Alegre should rather be understood as the outcome of changing constellations of political forces, and especially the close links between polycentric activities in civil society and mass-based political parties with access to local state power and an ability to construct a common agenda of popular democratic participation and socio-economic development for ordinary citizens. It is also of vital importance that such experiments have been preconditioned by devolution of power and resources, and operate within the coordination of the central state (Crook and Manor, 1998).

Transnationalizing Local Governance

An underlying assumption of much theory on local politics, particularly those exploring decentralization and citizenship, is that the actors and institutions are relatively place-based and static. But we argued earlier that places must be thought of as relational and interconnected, partly as a result of global communications, but also through flows of people (Appadurai, 1996). With intra- and international migration we see the emergence of networked communities whose members derive some of their identity from belonging to a 'home' where they no longer permanently reside (Yeoh et al., 2003). Bound up with these complex identities are obligations to family and kin still in the ancestral home, which can be realized economically, politically and socially (Mohan, 2002). These extra-local connections become important factors in shaping local political dynamics, but like all power-filled social relations they do not necessarily secure democratic openings.

There is a growing awareness of how community self-help among recent migrants on the basis of religious, hometown and other group affinities affects local government in Africa (Woods, 1994; McNulty and Lawrence, 1996; Honey and Okafor, 1998; Trager, 2001) and Latin America (Smith, 1998; Radcliffe et al., 2002; Orozco, 2003). Of particular interest are organizations that link the ancestral town or birthplace of migrants to a partner organization in their place of resettlement and are largely development-oriented (McNulty and Lawrence, 1996; Orozco, 2003). In an African context, but possibly more generally, Barkan et al. (1991) see these organizations as fulfilling various roles including being reservoirs of 'civil virtue', as a shadow state, as bulwarks against state power, as local growth machines, as brokers between state and local society, and as ways of reaffirming attachments to place. As forms of civil society organization, they are seen as legitimate since they are rooted in local culture and support welfare programs, but as forms of political organizations they are more ambivalent, potentially contesting state power but equally able to reinforce the type of divisive parochialism that Schuurman (1997) warned of.

In Latin America it appears that decentralization programs have recently started to link formally with migrant associations, such as Mexico's '3x1' initiative, with the state providing matching funds to those channeled through overseas' voluntary associations (Orozco, 2003). Not all this support is disembodied, with Urry (2004) stressing the importance of face-to-face meetings to bolster identity and trust within transnational communities. As Smith (1992, cited in Portes, 1997: 15–16) describes:

> On first sight, this is no more than an ordinary civic project. … Yet when we consider certain other aspects of the scene, the meaning becomes quite different. The Committee and I are not standing in Ticuani, but rather on a busy intersection in Brooklyn. … The Committee members are not simply going to the outskirts of the town to check the water tubes, but rather they are headed to JFK airport for a Friday afternoon flight to Mexico City, from which they will travel the five hours overland to their pueblo, consult with the authorities and contractors, and return by Monday afternoon to their jobs in New York City.

Such activity is increasingly common with the increases in migration and the availability of relatively cheap communication and transportation technology (Orozco, 2003). Studies in West Africa suggest that these organizations, such as hometown associations and funeral societies, are often encouraged by the state, because they are welfare-based but politically benign (McNulty and Lawrence, 1996).

However, transnational communities, like any community, are not socially homogeneous and power-free groupings, but are divided along various lines, which are exploited in the context of these networked relationships. Smith notes in a later reflection on the Mexican case that 'transnational communities create public spheres in which differences in status and class, especially upward mobility or educational success of children, can be displayed, and in which they have special meaning' (1998: 226). These questions of status and ethnicity are evident where migrant associations gain a more formal hold in political decision-making at local and national levels. As Woods (1994: 467) observes, a key function of hometown associations in Côte d'Ivoire was 'the utilization of ethnic associations by elites to consolidate their own economic and political position in the post-colonial state' (see also Trager, 2001; Lentz, 1994). This suggests that, as political entities, migrant-linked organizations may fulfil certain developmental functions without challenging exclusionary power structures.

These development activities linking 'home' places and the displaced, but organized through elite, largely male, public spaces, exert strong moral pressures and sanctions on people. So, while important in achieving more welfare-oriented development at the hometown level, as a form of citizenship they may not challenge power structures, and thereby potentially depoliticize development. Indeed, it can justifiably be argued that such activity may strengthen forms of customary authority, which marginalizes many – particularly women – from decision-making processes or accountability structures (Mamdani, 1996).

By contrast, some local struggles that defy boundedness have been more democratically transformatory. For example, the Zapatistas have campaigned actively since 1994 not only to attain full citizenship for the Indians of the Chiapas region of Mexico, but also for wider political reforms, particularly to the patronage mode of politics. In so doing, they have articulated a mode of political action capable of imagining and generating alternative development futures not only for its immediate constituency, but also for a broader community of dispossessed and marginalized peoples. Subcomandante Marcos is seen to embody the essence of global civil society. In response to the question 'Who is he?' the reply came, 'Marcos is every untolerated, oppressed, exploited minority that is beginning to speak and every majority that must shut up and listen' (Autonomedia, 1994: 313). The rallying call is that diverse groups are exploited along different lines and this will form the basis of political change through coalitions. Esteva and Prakash (1998: 6) describe this form of participatory politics as '*the* super-grassroots movement that is a match for the global forces from which the oppressed seek their liberation' (emphasis in original). Clearly, the agenda is simultaneously local, national, and global.

What these brief studies show is that transnational civil society connections reinforce the significance of territorialized states at national and local levels. The state is still an important, if not the pre-eminent, site of identity formation and transnational communities are linked to states in both imagined and material ways. Whereas civil society was viewed as something formed in relation to a territorialized nation-state, we are seeing both an extraterritorialized state and a form of transnational civil society. However, unlike the high-profile issue-based campaigns around things like debt, these transnational civil society networks are mobilized around the family and more place-based affinities so that they do not operate strongly as a bulwark against the overbearing state. Indeed, as we have seen, they are ideal as a benign form of civil society, which fills in for public provisioning without placing redistributive political pressure on the state.

CONCLUSION

In this chapter we have argued for studies of local politics that are attentive to how glocalization processes and local political transitions bring together political-economic forces and multiple, situated practices. For us, examining the politics of localization means investigating the ways in which diverse political forces construct and use local political spaces to pursue their instrumental and general interests in development and democracy. We have especially highlighted two 'projects' in the contemporary politics of localization. On the one hand, neo-liberal development discourses and institutional reforms construct local political spaces for popular participation, but also render development as a technocratic and depoliticized process. On the other hand, new local political spaces are also constructed and utilized by popular forces in order to transform formal democracy and institutionalize more substantive forms of local popular democracy. This calls for concrete and critical political analyses of the relations of power and political practices among actors involved in making, using and changing local political spaces and practices.

Recent theoretical and empirical research on the politics of localization highlight, in our view, three general analytical insights and guidelines for further research, all emphasizing the relationality of local politics. First, it is clear by now that 'the local' must be understood as socially constructed relational settings for situated practices rather than bounded places or scales. Second, it has also become common sense that the state and civil society are not self-contained spheres but must be understood in relational terms through the concrete links that exists between civil society actors, political society, and state institutions. Third, the realities of local politics also dissolve fixed notions of hegemony and resistance, pointing instead to the complex ways in which even seemingly hegemonic discourses are contested, subverted and remade. This relational and situated character of the politics of localization makes it well suited for political geographic enquiry. Given our own normative interests in local popular democracies, we see it as a continued challenge to address the aforementioned 'political deficit' in democratization studies, that is, the lack of critical insight into the politics of fighting for and implementing institutional changes toward substantive democracy. Herein lies an academic and political shift: from the depoliticization of development to the politicization of democracy.

ACKNOWLEDGMENTS

Giles Mohan would like to thank Sam Hickey for ongoing discussions of participation and governance and Katy Bennett for sharing ideas about community and empowerment in the UK context. Kristian Stokke wants to express his gratitude to Olle Törnquist and John Harriss for inspiring discussions in the field of local politics and democratization and to Sophie Oldfield, Richard Ballard,

Adam Habib, David Christoffer Lier, Marianne Millstein, and Peris Jones for fruitful collaboration on the politics of post-apartheid social movements and socio-economic rights in South Africa.

REFERENCES

Aasen, B., Arnesen O.E., Eriksen, S.S. and Tesli, A. (1997) *Evaluation of Decentralisation and Development*. Oslo: Royal Norwegian Ministry of Foreign Affairs.

Abers, R.N. (2000) *Inventing Local Democracy: Grassroots Politics in Brazil*. Boulder, CO: Lynne Rienner.

Abers, R.N. (2003) 'Reflections on what makes empowered participatory governance happen', in A. Fung and E.O. Wright (eds), *Deepening Democracy: Institutional Innovations in Empowered Participatory Governance*. London: Verso.

Abrahamsen, R. (2000) *Disciplining Democracy: Development Discourse and Good Governance in Africa*. London: Zed Books.

Adelzadeh, A. (1996) 'From the RDP to GEAR: the gradual embracing of neo-liberalism in economic policy', *Transformation*, 31: 766–95.

Adler, G. and Steinberg, J. (eds) (2000) *From Comrades to Citizens: The South African Civics Movement and the Transition to Democracy*. London: Macmillan.

Adler, G. and Webster, E. (eds) (2000) *Trade Unions and Democratization in South Africa, 1985–1997*. Johannesburg: Witwatersrand University Press.

Alexander, N. (2002) *An Ordinary Country: Issues in the Transition from Apartheid to Democracy in South Africa*. Pietermaritzburg: University of Natal Press.

Allen, C. (1997) 'Who needs civil society?', *Review of African Political Economy*, 73, 329–37.

Amin, A. (2002) 'Spatialities of globalisation', *Environment and Planning A*, 34: 385–99.

Appadurai, A. (1996) *Modernity at Large: Cultural Dimensions of Globalization*. Minneapolis: University of Minnesota Press.

Autonomedia (1994) *!Zapatistas! Documents of the New Mexican Revolution*. New York: Autonomedia.

Avritzer, L. (2002) *Democracy and the Public Space in Latin America*. Princeton: Princeton University Press.

Baiocchi, G. (2003) 'Participation, activism, and politics: the Porto Alegre experiment', in A. Fung and E.O. Wright (eds), *Deepening Democracy: Institutional Innovations in Empowered Participatory Governance*. London: Verso.

Baiocchi, G. (2005) *Militans and Citizens: The Politics of Participatory Democracy in Porto Alegre*. Stanford: Stanford University Press.

Ballard, R., Habib, A. and Valodia, I. (2006) *Voices of Protest. Social Movements in Post-Apartheid South Africa*. Durban: University of KwaZulu-Natal Press.

Barkan, J., McNulty, M. and Ayeni, M. (1991) ' "Hometown" voluntary associations, local development, and the emergence of civil society in Western Nigeria', *Journal of Modern African Studies*, 29(3): 457–80.

Barnett, C. and Low, M.M. (2004) 'Geography and democracy: an introduction', in C. Barnett and M.M. Low (eds), *Spaces of Democracy: Geographical Perspectives on Citizenship, Participation and Representation*. London: Sage, pp. 1–22.

Bayart, J.F. (1993) *The State in Africa: The Politics of the Belly*. London: Longman.

Baylies, C. (1995) ' "Political conditionality" and democratisation', *Review of African Political Economy*, 65: 321–37.

Bebbington, A. (2000) 'Reencountering development: livelihood transitions and place transformations in the Andes', *Annals of the Association of American Geographers*, 90(3): 495–520.

Bebbington, A., Kopp, A. and Rubinoff, D. (1997) 'From chaos to strength? Social capital, rural people's organizations and sustainable rural development' paper prepared for the FAO for a workshop on pluralism, forestry and rural development. Rome: FAO.

Bebbington, A., Guggenheim, S., Olson, E. and Woolcock, M. (2004) 'Exploring social capital debates at the World Bank', *Journal of Development Studies*, 40(5): 33–64.

Beetham, D. (1999) *Democracy and Human Rights*. Cambridge: Polity Press.

Bierschenk, T. and Olivier de Sardan, J.P. (2003) 'Powers in the village: rural Benin between democratisation and decentralisation', *Africa*, 73(2): 145–73.

Bond, P. (2000a) *Cities of Gold, Townships of Coal: Essays on South Africa's New Urban Crisis*. Trenton, NJ: Africa World Press.

Bond, P. (2000b) *Elite Transition: From Apartheid to Neoliberalism in South Africa*. London: Pluto.

Boone, C. (2003) 'Decentralization and political strategy in West Africa', *Comparative Political Studies*, 36(4): 355–80.

Bozzoli, B. (2004) *Theatres of Struggle and the End of Apartheid*. Johannesburg: Wits University Press.

Chabal, P. and Daloz, J.P. (1999) *Africa Works: Disorder as Political Instrument*. Oxford: James Currey.

Chakrabarty, D. (1992) 'Provincializing Europe: postcoloniality and the critique of history, *Cultural Studies*, 6(3): 337–57.

Chandhoke, N. (1995) *State and Civil Society: Explorations in Political Theory*. New Delhi: Sage.

Corbridge, S., Williams, G., Srivastava, R. and Veron, M. (2005) *Seeing the State: Governance and Governmentality in Rural India*. Cambridge: Cambridge University Press.

Cox, K.R. (ed.) (1997) *Spaces of Globalization: Reasserting the Power of the Local*. New York: Guilford Press.

Crook, R. and Manor, J. (1998) *Democracy and Decentralisation in South Asia and West Africa: Participation, Accountability and Performance*. Cambridge: Cambridge University Press.

Crook, R.C. and Sverrisson, A.S. (2001) *Decentralisation and Poverty-Alleviation in Developing Countries: A Comparative Analysis or, Is West Bengal Unique?* IDS Working Paper No.130. Brighton: Institute of Development Studies.

Daniel, J., Habib, A. and Southall, R. (eds) (2003) *State of the Nation: South Africa 2003–2004*. Johannesburg: HSRC Press.

Desai, A. (2003) *We Are the Poors: Community Struggles in Post-Apartheid South Africa*. New York: Monthly Review Press.

Dirlik, A. (1998) 'Globalism and the politics of place', *Development*, 41(2): 7–13.

Edigheji, O. (2003) 'State–society relations in post-apartheid South Africa: the challenges of globalisation in co-operative governance', in G. Mhone and O. Edigheji (eds), *Governance in the New South Africa: The Challenges of Globalisation.* Cape Town: University of Cape Town Press.

Escobar, A. (2001) 'Culture sits in places: reflections on globalism and subaltern strategies of localization', *Political Geography*, 20: 139–74.

Esteva, G. and Prakash, M. (1998) *Grassroots Post-modernism: Remaking the Soil of cultures.* London: Zed Books.

Ferguson, J. (1994) *Anti-Politics Machine: 'Development', Depoliticization, and Bureaucratic Power in Lesotho.* Minnesota: University of Minnesota Press.

Ferguson, J. and Lohman, L. (1997) 'Development and bureaucraic power in Lesotho', in M. Rahnema and V. Bawtree (eds), *The Post-Development Reader.* London: Zed Books.

Fine, B. (1999) 'The developmental state is dead – long live social capital?', *Development and Change*, 30: 1–19.

Fox, J. (1997a) 'The World Bank and social capital: contesting the concept in practice', *Journal of International Development*, 9(7): 963–71.

Fox, J. (1997b) 'How does civil society thicken? the political construction of social capital in rural Mexico', in P. Evans (ed.), *State–Society Synergy: Government and Social Capital in Development.* IAS Research Series No. 94. Berkeley: University of California.

Francis, P. and James, R. (2003) 'Balancing rural poverty reduction and citizen participation: the contradictions of Uganda's decentralization program'. *World Development*, 31(2): 325–37.

Fukuyama, F. (1992) *The End of History and the Last Man.* New York: Free Press.

Fung, A. and Wright, E.O. (2003) 'Thinking about empowered participatory governance, in A. Fung and E. O. Wright (eds), *Deepening Democracy: Institutional Innovations in Empowered Participatory Governance.* London: Verso.

Gaventa, J. (2004) 'Towards participatory governance: assessing the transformative possibilities', in S. Hickey and G. Mohan (eds), *Participation: From Tyranny to Transformation?* London: Zed Books, pp. 25–41.

Grootaert, C. (2001) *Does Social Capital Help the Poor?* Local Level Institutions Working Paper No.10. Washington, DC: World Bank.

Grugel, J. (2002) *Democratization: A Critical Introduction.* Basingstoke: Palgrave.

Gupta, A. and Ferguson, J. (1992) 'Beyond "culture": space, identity, and the politics of difference', *Cultural Anthropology*, 7(1): 6–23.

Habib, A. (2005) 'The politics of economic policy-making: substantive uncertainty, political leverage, and human development', in P. Jones and K. Stokke (eds), *Democratising Development. The Politics of Socio-Economic Rights in South Africa.* Leiden: Martinus Nijhoff.

Harrison, P., Huerschemeyer, M. and Mayekiso, M. (2003) *Confronting Fragmentation: Housing and Urban Development in a Democratising Society.* Cape Town: University of Cape Town Press.

Harriss, J. (2002) *Depoliticizing Development: The World Bank and Social Capital.* London: Anthem.

Harriss, J., Stokke, K. and Törnquist, O. (eds) (2004) *Politicising Democracy: The New Local Politics of Democratisation.* Basingstoke: Palgrave-Macmillan.

Hecht, D. and Simone, A.M. (1994) *Invisible Governance: The Art of African Micro-Politics.* New York: Autonomedia.

Heller, P. (2001) 'Moving the state: the politics of democratic decentralization in Kerala, South Africa and Porto Alegre', *Politics and Society*, 29(1): 1–28.

Heller, P. (2003) 'Reclaiming democratic spaces: civics and politics in posttransition Johannesburg', in R. Tomlinson et al. (eds), *Emerging Johannesburg: Perspectives on the Postapartheid City.* London: Routledge.

Hibou, B. (1999) 'The "social capital" of the state as an agent of deception', in J. Bayart, S. Ellis and B. Hibou (eds), *The Criminalization of the State in Africa*, Oxford: James Currey, pp. 69–113.

Honey, R. and Okafor, S. (1998) *Hometown Associations: Indigenous Knowledge and Development in Nigeria.* London: Intermediate Technology Publications.

Houtzager, P. (2003) 'From polycentrism to the polity', in P. Houtzager and M. Moore (eds), *Changing Paths: International Development and the New Politics of Inclusion.* Ann Arbor: University of Michigan Press.

Huntington, S.P. (1991) *The Third Wave: Democratization in the Late Twentieth Century.* Norman: University of Oklahoma Press.

Hutchcroft, P. (2001) 'Centralization and decentralization in administration and politics: assessing territorial dimensions of authority and power', *Governance*, 14(1): 23–51.

Isaac, T.M.T and Franke, R.W. (2000) '*Local democracy and development, people's campaign for decentralized planning in Kerala'*, Delhi: Leftward Books.

Isaac, T.M.T. and Heller, P. (2003) 'Democracy and development: decentralized planning in Kerala', in A. Fung and E.O. Wright (eds), *Deepening Democracy: Institutional Innovations in Empowered Participatory Governance.* London: Verso.

Jessop, B. (2002) *The Future of the Capitalist State.* Cambridge: Polity Press.

Johnson, C. (2001) 'Local democracy, democratic decentralisation and rural development: theories, challenges and options for policy', *Development Policy Review*, 19(4): 521–32.

Jones, P. and Stokke, K. (2005) *Democratising Development. The Politics of Socio-Economic Rights in South Africa.* Leiden: Martinus Nijhoff.

Kaldor, M. and Vejvoda, I. (1997) 'Democratization in Central and East European countries', *International Affairs*, 73(1): 59–82.

Kelsall, T. (2000) 'Governance, local politics and districtization in Tanzania: the 1998 Arumeru tax revolt', *African Affairs*, 99: 533–51.

Kohl, B. (2003) 'Nongovernmental organizations as intermediaries for decentralization in Bolivia', *Environment and Planning*, 21(3): 317–32.

Leftwich, A. (1993) 'Governance, democracy and development in the Third World', *Third World Quarterly*, 14(3): 605–24.

Leftwich, A. (1994) 'Governance, the state and the politics of development', *Development and Change*, 25: 363–86.

Lentz, C. (1994) 'Home, death and leadership: discourses of an educated elite from north-western Ghana', *Social Anthropology*, 2(2): 149–69.

Lier, D.C. and Stokke, K. (2006) 'Maximum working class unity? Challenges to local social movement unionism in Cape Town'. *Antipode*, 38(4): 802–24.

Linz, J.J. and Stepan, A. (1996) *Problems of Democratic Transition and Consolidation: Southern Europe, South America, and Post-Communist Europe*. Baltimore: Johns Hopkins University Press.

MacLeod, G. and Goodwin, M. (1999) 'Space, scale and state strategy: rethinking urban and regional governance', *Progress in Human Geography*, 23(4): 503–27.

Mamdani, M. (1996) *Citizen and Subject: Contemporary Africa and the Legacy of Late Colonialism*. Oxford: James Currey.

Marais, H. (2001) *South Africa. Limits to Change: The Political Economy of Transition* (2nd edn). London: Zed Books.

McKinley, D.T. (1997) *The ANC and the Liberation Struggle: A Critical Political Biography*. London: Pluto.

McNeill, D. and Boas, M. (2003) *Multilateral Institutions: A Critical Introduction*. London: Pluto.

McNulty, M. and Lawrence, M. (1996) 'Hometown associations: balancing local and extralocal interests in Nigerian communities', in P. Blunt, D. Marsden and D. Warren (eds), *Indigenous Organization and Development*, London: ITDG.

Migdal, J. (1994) 'The state in society: an approach to struggles for dominations', in J.S. Migdal, A. Kohli and V. Shue (eds), *State Power and Social Forces: Domination and Transformation in the Third World*. Cambridge: Cambridge University Press.

Millstein, M., Oldfield, S. and Stokke, K. (2003) 'uTshani BuyaKhuluma – The grass speaks: the political space and capacity of the South African Homeless People's Federation', *Geoforum*, 34: 457–68.

Mohan, G. (1996) 'Decentralisation and ajustment in Ghana: a case of diminished sovereignty', *Political Geography*, 15(1): 75–94.

Mohan, G. (2001) 'The disappointments of civil society: NGOs, citizenship and institution building in Northern Ghana', *Political Geography*, 21(1): 125–54.

Mohan, G. (2002) 'Diaspora and development: the Black Atlantic and African transformation', in J. Robinson (ed.), *Development and Displacement*, Buckingham: Open University Press, pp. 77–139.

Mohan, G. and Mohan, J. (2002) 'Placing social capital', *Progress in Human Geography*, 26(2): 191–210.

Mohan, G. and Stokke, K. (2000) 'Participatory development and empowerment: the dangers of localism', *Third World Quarterly*, 21(2): 247–68.

Moser, C. and Holland, J. (1998) 'Can policy-focused research be participatory? Research on violence and poverty in Jamaica using PRA methods', in J. Holland and J. Blackburn (eds), *Whose Voice? Participatory Research and Policy Change*. London: Intermediate Technology Publications.

Narayan, D. and Pritchett, L. (1999) 'Cents and sociability: household income and social capital in rural Tanzania', *Economic Development and Cultural Change*, 47(4): 871–97.

Nattrass, N. (2003) 'The state of the economy: a crisis of employment', in J. Daniel, A. Habib and R. Southall (eds), *State of the Nation: South Africa 2003–2004*. Johannesburg: HSRC Press.

Ngwane, T. (2003) 'Sparks in the township', *New Left Review*, 22: 37–56.

Oldfield, S. and Stokke, K. (2006) 'Building unity in diversity: social movement activism in the Western Cape Anti-Eviction Campaign', in R. Ballard, A. Habib and I. Valodia (eds), *Voices of Protest. Social Movements in Post-Apartheid South Africa*. Durban: University of KwaZulu-Natal Press.

Orozco, M. (2003) *Hometown Associations and their Present and Future Partnerships: New Development Opportunities?*, Washington, DC: Inter-American Dialogue.

Painter, J. (2000) 'Governance', in R. Johnston et al. (eds), *The Dictionary of Human Geography*. Oxford: Blackwell.

Parnell, S., Pieterse, E., Swilling, M. and Wooldridge, D. (eds) (2002) *Democratising Local Government: The South African Experiment*. Cape Town: University of Cape Town Press.

Portes, A. (1997) 'Globalization from below: the rise of transnational communities', ESRC Transnational Communities Project, Working Paper WPTC-98-01, www.transcomm.ox.ac.uk/working_papers.htm.

Prasetyo, A.E., Priyono, A.E. and Törnquist, O. (2003) *Indonesia's Post-Soeharto Democracy Movement*. Jakarta: DEMOS.

Putnam, R. (1993a) *Making Democracy Work: Civic Traditions in Modern Italy*. New Haven: Princeton University Press.

Putnam, R. (1993b) 'The prosperous community: social capital and public life', *American Prospect*, 13: 35–42.

Radcliffe, S., Andolina, R. and Laurie, N. (2002) 'Production of states and deterritorialized spaces: indigenous movements and transnational linkages in the Andes', *Space and Polity*, 6(3): 389–405.

Rhodes, R. (1997) *Understanding Governance: Policy Networks, Reflexivity and Accountability*. Buckingham: Open University Press.

Ribot, J. (1999) 'Decentralisation, participation and accountability in Sahelian forestry: legal instruments of political-administrative control', *Africa*, 69(1): 23–65.

Rondinelli, D. et al. (1989) 'Analyzing decentralization policies in developing countries: a political-economy framework', *Development and Change*, 20: 57–87.

Schmitter, P., O'Donnell, G. and Whitehead, L. (1986) *Transitions from Authoritarian Rule*. Baltimore: Johns Hopkins University Press.

Schönleitner, G. (2004) 'Can public deliberation democratize state action? Municipal health councils and local democracy in Brazil', in J. Harriss, K. Stokke and O. Törnquist (eds), *Politicising Democracy: The New Local Politics of Democratisation*. Basingstoke: Palgrave-Macmillan.

Schuurman, F. (1997) 'The decentralisation discourse: post-Fordist paradigm or neo-liberal cul-de-sac?', *European Journal of Development Research*, 9(1): 150–66.

Seekings, J. (1997) 'SANCO: strategic dilemmas in a democratic South Africa', *Transformation*, 34: 1–31.

Seekings, J. (2000) *The UDF: A History of the United Democratic Front in South Africa, 1983–1991*. Cape Town: David Philips.

SLSA Team (2003) 'Decentralisations in practice in Southern Africa', *IDS Bulletin*, 34(3): 79–96.

Smith, R. (1998) 'Transnational localities: community, technology and the politics of membership within the context of Mexico and US migration', in M. Smith and L. Guarnizo (eds), *Transnationalism from Below*. New Brunswick, NJ: Transaction Publishers, pp. 196–238.

Staeheli, L.A. and Mitchell, D. (2004) 'Spaces of public and private: locating politics', in C. Barnett and M.M. Low (eds), *Spaces of Democracy: Geographical Perspectives on Citizenship, Participation and Representation*. London: Sage, pp. 147–60.

Stokke, K. and Oldfield, S. (2004) 'Social movements, socioeconomic rights and substantial democratisation in South Africa', in J. Harriss, K. Stokke and O. Törnquist (eds), *Politicising Democracy: The New Local Politics of Democratisation*. Basingstoke: Palgrave-Macmillan.

Tarrow, S. (1996) 'Making social science work across space and time: a critical reflection on Robert Putnam's '*Making Democracy Work*', *American Political Science Review*, 90(2): 389–97.

Tendler, J. (1997) *Good Government in the Tropics*. Baltimore: Johns Hopkins University Press.

Tharakan, P.K.M. (2004) 'Historical hurdles in the course of the people's planning campaign in Kerala, India', in J. Harriss, K. Stokke and O. Törnquist (eds), *Politicising Democracy: The New Local Politics of Democratisation*. Basingstoke: Palgrave-Macmillan.

Törnquist, O. (2002) *Popular Development and Democracy: Case Studies with Rural Dimensions in the Philippines, Indonesia and Kerala*. Geneva and Oslo: UNRISD and SUM.

Törnquist, O. (2004) 'The political deficit of substantial democratisation', in J. Harriss, K. Stokke and O. Törnquist (eds), *Politicising Democracy: The New Local Politics of Democratisation*. Basingstoke: Palgrave-Macmillan.

Torres, L. (2000) *Amandla – Ngawethu? The Political Role of Labour in South Africa*. Oslo: FAFO.

Trager, L. (2001) *Yoruba Hometowns: Community, Identity and Development in Nigeria*. Boulder, CO: Lynne Rienner.

Tvedt, T. (1998) *Angels of Mercy or Development Diplomats? NGOs and Foreign Aid*. Trenton, NJ: Africa World Press.

UNDP (2002) *Human Development Report: Deepening Democracy in a Fragmented World*. New York and Oxford: Oxford University Press.

Urry, J. (2004) 'Small worlds and the new "social physics"', *Global Networks*, 4(2): 109–30.

USAID (2000) *Decentralization and Democratic Local Governance Programming Handbook*. Washington, DC: USAID.

Wainwright, H. (2003) *Reclaim the State: Experiments in Popular Democracy*. London: Verso.

Watson, S. (2004) 'Cultures of democracy: spaces of democratic possibility', in C. Barnett and M.M. Low (eds), *Spaces of Democracy: Geographical Perspectives on Citizenship, Participation and Representation*. London: Sage, pp. 207–22.

Webster, N. and Engberg-Pedersen, L. (eds) (2002) *In the Name of the Poor: Contesting Political Space for Poverty Reduction*. London: Zed Books.

Whitehead, L. (2002) *Democratization: Theory and Experience*. Oxford: Oxford University Press.

Williams, G., Véron, R., Srivastava, M. and Corbridge S. (2003) 'Participation, poverty and power: poor people's engagement with india's employment assurance scheme', *Development and Change*, 34(1): 163–92.

Williams, D. and Young, T. (1994) Governance, the World Bank and Liberal Theory. *Political Studies*, XLII: 84–100.

Woods, D. (1994) 'Elites, ethnicity, and "hometown" associations in the Côte d'Ivoire: an historical analysis of state–society links', *Africa*, 6(4): 465–83.

World Bank (1983) *World Development Report*. Washington, DC: World Bank.

World Bank (1989) *Sub-Saharan Africa: From Crisis to Sustainable Growth*. Washington, DC: World Bank.

World Bank (1993) *The East Asian Miracle*. Oxford: Oxford University Press.

World Bank (1994) *Governance: The World Bank's Experience*. Washington, DC: World Bank.

World Bank (1997) *World Development Report: The State in a Changing World*. Oxford: Oxford University Press.

Yeoh, B., Willis, K. and Fakhri, S. (2003) 'Transnationalism and its edges', *Ethnic and Racial Studies*, 26(2): 207–17.

Young, I.M. (2000) *Inclusion and Democracy*. Oxford: Oxford University Press.

34

'Development' in Question

Haripriya Rangan

INTRODUCTION

The 1980s and early 1990s were a turbulent period in the world of development. Gustavo Esteva's pithy declaration, 'You must be either very dumb or very rich if you fail to notice that development stinks' (1987: 135), was a clarion call that drew many radical academics and activists into the charge against development. *The Development Dictionary* (1992), edited by Wolfgang Sachs, unleashed a powerful set of indictments that deconstructed the edifice of development and revealed its exercise of knowledge as power. Radical critics claimed that development was not simply an instrument of economic control over much of the non-Western world, but an invention and strategy produced by the 'First World' about the 'underdevelopment' of the 'Third World'. It needed to be abandoned altogether, and replaced by 'alternatives to' development that would build on practices of new social movements that sought to create alternative visions of democracy, economy and society (Amin, 1990; Shiva, 1989; Trainer, 1989; Hettne, 1990; Parajuli, 1991; Escobar, 1992, 1995a, b).

At the opposite end of the political spectrum, the 'counter-revolutionaries' of development (Toye, 1987) charged that development impeded economic growth due to policies that were protectionist, imposed excessive regulations on private industry (Balassa, 1982; Little, 1982; Lal, 1983), and encouraged 'directly unproductive profit-seeking behaviour' (Bhagwati, 1982). Sounding startlingly similar to Escobar, the economist P.T. Bauer went so far as to claim that '[T]he Third World is the creation of foreign aid: without foreign aid there is no Third World' (1981: 87).

A decade later, and into the new millennium, development no longer seems under siege.

The intervening years have seen radical critiques shift to targeting globalization and multilateral organizations. The 'counter-revolutionary' critics of development also appear to be in retreat as more balanced perspectives on the role of development, states and markets emerge (Killick, 1989; Bowie, 1991; Krueger, 1993; Martinussen, 1997; Parfitt, 2002). International development agencies and NGOs are as busy as ever, engaged in poverty alleviation, micro-credit, community-based development and varied forms of sustainable development.

So, is development finally out of its 'impasse' and perennial 'crises'? Has it realized what Michael Watts (1995) has called 'a New Deal in emotions'? Is development no longer 'in question'? Many will, I suspect, disagree, arguing that development still faces an enormous crisis in dealing with the uneven geographies and social inequalities generated by continued dominance of neoliberal ideology, globalization processes and the new imperialism (Peet and Watts, 1996; Routledge, 2002; Slater, 2002; Harvey, 2003). A few might actually agree, pointing out that development is no longer in question because the processes of globalization have created the possibility of 'non-nationalist' action for greater social justice across the world (Harris, 1986, 2005; Kitching, 2001).

My view is that development is *both* an 'in' *and* a 'not in' question for reasons very different from those just mentioned. Development, as it emerged in its post-Second World War form for implementation across the world, has always been 'in question' because of what it sought to achieve: the creation of bounded and fixed national spaces for accumulation of wealth out of regions that, even under the strictest European colonial regimes, were never easily contained or controlled. Yet this very same development is also 'not in question' because various identity groups – *including* those regarded as lower status or disadvantaged – in the post-colonies have deliberately used it to make claims on the state, and manipulated the process in order to achieve social and economic mobility. That this social and economic mobility has been uneven is incontrovertible. But precisely because of this unevenness, development has become '*the question*' that forms the battleground for politics and popular action in postcolonial countries. As I have mentioned elsewhere, social movements in post-colonial countries are neither *for* nor *against* development, they are *part of* it (Rangan, 1999, 2004).

The story of development has generally been told through a limited number of narrative structures: as a process of national economic growth through modernization and free trade, as the unfolding logic of capitalism that creates uneven geographies of accumulation (Smith, 1984; Harvey, 1985), as a process of underdevelopment (Frank, 1967; Amin, 1974) and as a Euro-American-centred project aimed at dominating the knowledge, economy and cultures of poorer nations and communities across the world (Hettne, 1990; Sachs, 1992, Escobar, 1995a). All these narratives share a common geographical imagination regarding the originating impulse and power of development. It emerges from the Western Europe–American core and spreads across the rest of the world to its very peripheries while maintaining its dynamism and geopolitical power at the places of origin. Given this basic itinerary, the narratives follow either a progressive or a declensive path for describing the geographical outcomes, that is, of universal and benevolent economic growth, or uneven development, underdevelopment and the perpetuation of global inequality and oppression through new forms of colonialism or imperialism. Depending on the path chosen by the narrative, the social responses to the encounter may range from adaptation and collusion to resistance or revolution.

I do not wish to reproduce these narratives in this chapter. However compelling their narrative structure and teleology, they have not been very effective in explaining how extremely diverse landscapes of development have emerged in the post-colonial world over the past six decades. Narratives of free market and democracy cannot account for the fact that most successful cases of development in the post-war era have occurred in countries through state-sponsored industrialization and market intervention (Amsden, 1989; Kohli, 2004). Dependency and post- and anti-development narratives do not explain how, despite the continued geopolitical dominance and economic clout of Euro-American powers, development has been successfully pursued in some countries but not others (Wade, 1990; Evans, 1995). The innate logic and imperatives of capitalist accumulation

may require spatio-temporal fixes and uneven geographical development (Harvey, 2001, 2003), but this does not necessarily explain how some post-colonial societies have effectively manipulated and managed the inexorable drive of global capitalism to their benefit and are significantly reshaping the geopolitical landscapes of development. One has merely to look at the list of countries classified as 'emerging economies' in financial journals – China, India, Taiwan, Singapore, South Korea, Brazil, South Africa, Mexico, Malaysia and Russia among others – to recognize that these narratives remain extremely limited in their ability to convey the diverse trajectories, experiments and experiences of development that have occurred over the past six decades.

If we are to question development in the twenty-first century, then we clearly need to employ a different narrative structure, one that is neither bound by a universal teleology that reinforces the parochialism of Euro-American (conservative, neoliberal, liberal, metropolitan Marxist) perspectives regarding the motives and nature of development nor, on the other hand, one that attempts to impose new parochial dogmas emerging from 'the Other' regions of the world. We need a narrative that not only recognizes and reflects the diversity and contradictions of the development experience over the past century, but also attempts to reveal the commonalities in the way development has been deployed in *reshaping* the national geographies, economies and cultures of so-called 'developing' countries.

I propose to approach the question of development in this chapter from a post-colonial perspective. By this, I mean five things. First, post-colonial is a term used to indicate the *period* following the end of formal colonial rule in any particular place (see Alavi, 1972). Second, post-colonial refers to an *analytical stance* that questions the common assumption that the making of the 'modern West' – and this includes the emergence of capitalism, the advancement of scientific and technological expertise, and the accumulation of capital and cultural power – was the autochthonous achievement of its societies (see, e.g., Said, 1978, 1993; Chakrabarty, 2000; Cooper et al., 2000; Coronil, 2000; Carney, 2001). Third, post-colonial refers to a *mode of critical historical inquiry* that examines the significance of colonialism and the contributions of geographically situated knowledge made by colonial subjects in the formation and practice of modern sciences and social theory (see Grove, 1995; Gupta, 1998; Mitchell, 2002). Fourth, post-colonial refers to *practices of politics, economics and governance* that need to constantly grapple with the realities of numerous social differences, geographical variations and identity formations – regional, cultural, religious, racial, ethnic, alongside gender and class – produced during and after the end of colonial rule (see, e.g., Hart, 1991, 2002; Watts, 1992; Kandiyoti, 1998; Corbridge and Harriss, 2000; Goswami, 2004). Finally, post-colonial stands as a *critique* of the failures of nationalism in political and economic terms in both its Euro-American metropolitan *and* anti-colonial forms, and calls for rethinking the relations between state and modes of identity formation around region, labour, gender, ethnicity and religion (Fanon, 1967; Carney, 1988; Abu-Lughod, 1998; Rangan, 2000; Comaroff and Comaroff, 2001; Chari, 2004). These five analytical dimensions of 'postcolonial' are drawn from the disciplinary traditions of peasant studies, social history, history of science and technology, socio-cultural anthropology and regional political economy, and together constitute my perspective for exploring the question of development in the present.

My narrative begins by looking back at nineteenth-century doctrines of development and their contradictory views about its applicability and practice in Europe's colonized regions. I examine how these ideas and assumptions, combined with the exigencies of international finance and trade in the post-Second World War era and the process of decolonization, were articulated through new modes of calculation (cf. Latour, 1987; Poovey, 1998) that gave rise to the concept of a clearly bounded and self-contained spatial abstraction called 'the economy'. This calculable space, defined to coincide with the new political boundaries and partitions of a decolonizing world, became both the means of territorial representation and the object of development for post-colonial states. I go on to argue that these national spaces, determined by their character of calculability rather than by any coherent social or geographical identity, provided the much-needed freedom and 'extra-political' legitimacy for post-colonial leaders, business elites and technical experts to establish territorial control over their populations.

Development of the spaces of 'national economy' was the means by which post-colonial nation-states would be made and sustained. It became the central question and purpose that enabled post-colonial states to establish their territorial 'integrity', political authority and moral and geopolitical legitimacy. Yet, despite (or perhaps *because of*) the efforts of post-colonial states and experts to, on the one hand, 'free' the economy and development from the terrain of domestic politics by embedding them in the space of calculability but, on the other hand, use them to assert political authority, neither the economy nor development succeeded in escaping the realities of post-colonial political geography. Spatial development in the post-colonies has thus been both subject to, and the object of, the politics of regional and place-based identity. It is constantly 'under question' because post-colonial states either reinforce existing regional disparities or create new ones through their economic development policies; and it is 'the question' around which regional elites and other place-based identity groups engage in political mobilization to make claims on post-colonial national governments for greater allocation of, or control over, economic resources.

The final segment of my narrative attempts to *revisualize* the dynamic landscapes of post-war development from a post-colonial vantage point. This revisualization neither offers a magisterial survey or map of the topography of development from the commanding heights of a metropolitan perspective, nor a 'subaltern' view of development 'from the ground up'. My post-colonial vantage offers a 'doubled vision' that simultaneously sees from near and afar, moving between different frames and focal lengths and through changing and translated landscapes to provide a geographic visualization of development that rolls out at varying speeds and reveals new desires of aggrandizement and discontent.

My use of the term 'development' in this chapter has a fairly clear form that is distinguished by three features. First, development is about ideas, institutional practices and experts involved in promoting the accumulation of capital and authority at the spatial level of the nation-state. Second, development involves the concerted effort to create a bounded spatial entity called the national economy that both represents, and is represented by, the nation-state. Third, development involves instituting practices of governance that attempt to mediate the welfare of a national population, so that social order and political stability are ensured for continued accumulation of wealth and political power at the spatial level of the nation-state. These three features of development reinforce each other in the process of creating and maintaining spatial entities known as nation-states.

THE DESCENT OF DEVELOPMENT (BY MODIFICATION[1])

Several contributors to *The Development Dictionary* (Sachs, 1992) claim that development was an 'invention' of the post-Second World War period when the 'apparatuses of Western knowledge production and intervention' (i.e. the World Bank, United Nations, IMF, GATT and bilateral aid agencies) were established. Arturo Escobar describes development as an 'apparatus that links forms of knowledge about the Third World with the deployment of forms of power and intervention, resulting in the mapping and production of Third World societies. Development constructs the contemporary Third World silently, without our noticing it' (Escobar, 1995b: 213). The reality, however, is that development as a process and project has neither been silent nor operated so discreetly as to evade notice. The multilateral institutions and bilateral agencies that emerged after 1945 were, if anything, a product of the global experiences of development in the preceding centuries that generated an extraordinary cacophony of social upheavals, violence, wars and traumas alongside the accumulation of wealth. These institutions did not 'invent' development, but rather refashioned the legacies of nineteenth-century development doctrines to respond to the new geopolitical realities of postwar reconstruction and decolonization.

Escobar is partially right, though, in stating that the term 'development' is a 'Western' invention: it is. But it is nonetheless important to recognize that the meanings and contours of this Western invention were produced from the interactions between Europe and other regions of the world. The word 'develop', as Raymond Williams explains in *Keywords* (1985), emerged in English usage in the sixteenth-century as the opposite of 'envelop'; it referred to the process of unrolling or unfolding. In the eighteenth-century it was extended to include a sense of building upon, or the process of expanding abilities, as in 'developing the faculties of the human mind'. The idea of development as an organic process of unfolding and expanding abilities became well and truly incorporated in European accounts of human history by the nineteenth century, and later linked with ideas of evolution in the new biology to explain the nature of economic change in societies. It became common to refer to nations proceeding in a 'course of development' or evolutionary stages. Development was seen as the law governing social evolution: civilisation evolved, step by step, from the 'rude state of nature' to hunting and gathering, agriculture, cities, nations and empires (ibid.: 102–3).

By the mid-nineteenth century, the term 'development' was used to represent two aspects of social change. The first was purposeful *intrinsic* change, or change that was directed within a given society. Development in this sense was about building and expanding the internal abilities and institutions of a society in order to move it towards higher stages of civilization. The second aspect of development was *comparative* change, in that it was used to identify and compare the stages of social evolution of different societies in terms of a universal human history. The German philosopher G.W.F. Hegel claimed that history was not determined by natural law, but by the progress of human will and spirit (Taylor, 1977). He argued that evolution in nature was fundamentally different from the evolution of human history. Nature, for him, was governed by hierarchies and cyclical change. It had no forward movement or 'progress' towards a desired or ideal purpose. Progress reflected an immanent sense of autonomy and freedom to human purpose and the prospect of

unlimited improvement. Human history, in contrast to nature, was governed by a universal teleology, a doctrine of final causes that involved progressive social change towards the realization of communities conformed by reason and embodying freedom.

Hegel sought to illustrate his universal human history by providing elaborate explanations of how communities and civilizations existing outside Europe were either bound by nature's laws or by hierarchies that prevented progress. They were societies without history. It was Europe's arrival through trade, conquest, colonization and direct control that brought these regions into universal history and placed them at lower stages of progress or development towards the state of reason and freedom achieved in Europe (Taylor, 1977, 350–64, 389–427). Thus, development as comparative social change from a European perspective required the natives of Asia, Africa and the Americas to be present as witnesses to the lesser stages of progress of universal human history (Wolf, 1982).

These meanings and mentalities of progress and development were combined in different ways during the nineteenth century in the ideas of Henri de Saint-Simon, John Stuart Mill, Friedrich List, Karl Marx and Frederick Engels. Michael Cowen and Robert Shenton (1995, 1996) provide an excellent account of the four dominant doctrines of development that emerged in nineteenth-century Europe, and I draw on their work here for a quick overview. Most nineteenth-century thinkers were preoccupied with the nature and purpose of social transformation that was taking place in Europe. The social upheavals, displacement and widespread turmoil generated by the industrial and political revolutions in Britain and France were seen as problems generated by progress. Critics challenged the idea of progress as an immanent or 'natural' process and argued that without proper and purposeful guidance, progress would yield only chaos and misery. What was needed was 'progressive development' that would improve the moral, material and intellectual conditions of nations as a whole.

Henri de Saint-Simon and his followers were preoccupied with the upheavals occurring in France in the decades following the French Revolution. Their main concern was to find ways of promoting progress in an orderly manner. Saint-Simon argued that the social laws of evolution had two aspects: improvement and development. Industrialization brought progress and improvement of human ability and potential, but its pursuit through the self-interest of private enterprise and unregulated competition caused unbearable disorder in society. Development was the means by which progress could be contained and reconciled with social order. Development of a society thus required *concerted action* through design, intervention and rational thinking (Cowen and Shenton, 1995: 32–4).

Saint-Simon set out a vision of a 'progressive and developmental' government under a new political system based on meritocracy that would be entrusted with the responsibility of acting in the interests of society as a whole. It would operate through three chambers of government. The Chamber of Invention would comprise civil engineers, architects, artists, sculptors, poets, writers and musicians, who would be responsible for engaging in public works and providing various other amenities. The Chamber of Examination would include scientists – physicists, biologists, mathematicians – who would be in charge of educating the public about the benefits of science and scientific thinking. The Chamber of Implementation would comprise leading industrialists and bankers who would be responsible for collecting taxes and deciding how to invest and manage the society's resources. Investment decisions would be made possible by a system of banks, headed by a central government bank, that would have control over all the economic resources and activities of the society. The government – through feats of engineering, grand public works, economic management, scientific education of the public and feasts of expectation and commemoration – would exhort citizens to work eagerly for improving their collective future, and convince them of how much better their position was compared to that of their ancestors (Friedmann, 1987: 63–9). These activities would transform peasants in the French countryside into 'Frenchmen' and citizens bound by their shared national identity and faith in the new meritocratic French government (Weber, 1976). Progress and order would be reconciled by the processes of 'governmentality' (Foucault, 1991).

The ideas of Saint-Simon and his followers held enormous attraction for thinkers in other parts of Europe. They represented the possibility of building a new society through grand monumental projects that challenged the constraints imposed by nature, and offered the prospect of unlimited progress for society as a whole (Berman, 1982). The Saint-Simonian vision offered the prospect of the development of a new kind of government and citizenry united by their faith in the merit of reason, science, industry and the collective and orderly advancement of their society.

The English political economist, John Stuart Mill, was greatly inspired by Saint-Simon's vision of reconciling progress with order, but outlined a different approach for achieving this reconciliation. For Mill, progress and order needed each other, they had to coexist. He argued that societies were always in one of two states: natural or transitional. In the natural state, the material interests of society would be husbanded by leaders who

had the best capacity for managing political order and economic improvement. The transitional state occurred when there was disjuncture between those who held political power and those who were most capable of managing economic affairs; it resulted in social chaos. Mill claimed that the disorder that accompanied progress in Britain was because the old social organization and political leadership was inadequate and unsuited to new social conditions; it was necessary to develop a new social and political order based on the exercise of reason and free choice for all members of society. The role of government was to provide generalized education and radical extension of liberty as the means for developing the new social order. This included issues as diverse as electoral and land reform, equality for women and rights of labour. Progress and order would be reconciled by development that enhanced the personal freedoms and individuality of citizens and enabled them to make rational choices for the development of their society (Cowen and Shenton, 1996: 36–41).

While Saint-Simon and Mill were predominantly concerned with development as a process internal to nineteenth-century French and British society, their contemporaries in other parts of Europe viewed development from a comparative national perspective. Friedrich List, a political economist of the German Historical School, was more concerned with the issue of promoting progress in nations that had not yet achieved the industrial and commercial supremacy of nineteenth-century Britain or experienced the political upheavals of France. List was acutely conscious of the agrarian 'backwardness' of Germany when compared with the industrial achievements in Britain. After emigrating to the US from Germany in 1825, he noted that the process of transition from dependence on agriculture to industrial manufacturing that took centuries to unfold in Europe had been telescoped in North America through intentional and constructive intervention by the state. List claimed that agrarian nations would forever remain 'underdeveloped' and 'backward' if they attempted to pursue free trade as the means for achieving material progress in their societies. He argued that free trade only worked in the favour of commercially powerful nations like Britain, and that rulers of agrarian nations in Europe would need to follow in the footsteps of the US to actively intervene in promoting the forces of industrial production within their territories and protecting them from the harmful effects of free trade. List claimed that progress and order would be reconciled only when 'economical development' had achieved its fullest productive powers 'within the bosom of the nation' (Cowen and Shenton, 1996: 35–7).

Karl Marx and Friedrich Engels also adopted a comparative perspective in their critical analysis of nineteenth-century progress in Europe. Marx was inspired by Hegel's interpretation of the evolution of universal human history, but disagreed with Hegel's belief that historical progress was mainly driven by the force of human will and spirit. Social evolution, for Marx, occurred through fundamental changes in the material base and political organization of societies. He argued that universal history was, in fact, the product of class struggle. The dialectics of progress was driven by class struggle over the modes of production and exchange. In *The Communist Manifesto*, Marx and Engels attempted to illustrate the extraordinary power of progress that had been unleashed by the European bourgeoisie through the seventeenth and eighteenth centuries in the process of transforming their societies from feudalism to industrial capitalism. They argued that the modern bourgeoisie was 'itself the product of a long course of development, of a series of revolutions in the modes of production and exchange' (Marx and Engels, [1848] 1948: 10). Hegel's universal teleology was reversed to explain how the unfolding of progress in nations followed evolutionary stages of development shaped by class struggle that produced great upheaval and violence in societies. But this class struggle, they claimed, could be directed through concerted action towards the development and ultimate realization of a universal classless society based on the principles of communism.

The ideas of Saint-Simon, Mills, List and Marx may seem, therefore, as though they were mainly shaped by the events and upheavals occurring in Europe and that the colonies played no role in contributing to their notions of progress and development. But this view would be wrong. Saint-Simon's ideas emerged in the context of French colonial expansion into Africa, Asia and the Americas, and encompassed an imperial scope for France's (and Europe's) progressive development through grand construction projects such as the Suez and Panama Canals (Berman, 1982; Friedmann, 1987). John Stuart Mill, like both the Revd. Thomas Malthus and his father, James Mill, was employed by the British East India Company, and his views on political economy and governance were influenced by his engagement with decisions regarding colonial administration and revenue generation in colonial India (see Stokes, 1959, for a rich analysis of how the experience of colonial rule in India shaped the discipline of political economy and utilitarian philosophy in Britain). List's comparative perspective of 'economical' development was not confined to comparisons between 'backward' Germany, a 'rapidly developing' US and 'developed' Britain, but also included Europe's colonial territories in the 'tropics'. Europe's colonial territories were also central to Marx's explanation of capitalist development across the world (Marx, [1867] 1967: 702–16).

THE COLONIAL PRESENCE IN NINETEENTH-CENTURY DEVELOPMENT

The implicit or explicit presence of Europe's colonial territories in the developmental ideas of these nineteenth-century thinkers highlight at least four contrary features that have persisted in the post-Second World War versions of development. First, the conflation of two distinct meanings of development into a single, universal idea – development as *both* an immanent, organic process beyond the control of humans *and* as concerted social action – laid the foundations for a hybrid, modernist perspective that combined organicist notions of growth with corporatism, possessive individualism and triumphalism of science and technology into a universal teleology of progressive evolution of human history. This view regarded social change in regions outside Europe not as following different pathways of development, but rather as representing different stages in a single, universal trajectory of progress. As with Hegel's dialectic of universal human history, these nineteenth-and early twentieth-century ideas of progressive development – which appeared in political guises ranging from absolutism, fascism and liberalism to Marxisms of various sorts – relied on narratives that incorporated the colonized regions of the world into lower stages of a linear history of progress that inevitably culminated in a singular European present (Wolf, 1982; Adas, 1989).

The second contrary feature is the way in which the relationship between territory and development appears in the ideas of these nineteenth-century thinkers. Neither Saint-Simon and his followers nor John Stuart Mill saw any reason to define the territorial space required for progressive development. They belonged to countries that controlled extensive colonial territories outside Europe, and the profits from colonial trade had fuelled industry, progress and accumulation of wealth in their societies (Stokes, 1959; James, 1963; Davidson, 1980). While their concern for achieving social order primarily focused on their respective national spaces, their visions of progress presumed the existence of territorial and resource control that extended well beyond the national boundaries of France and England. Neither of them visualized progressive development as being limited to, or by, the geographical space contained within their national boundaries. Friedrich List, however, was acutely aware of the crucial relationship between territorial control and 'economical' development. He claimed that one of the main reasons for Germany's backwardness in relation to Britain was its lack of overseas colonies that could funnel profits for investment in industrial development. He pointed to the state-motivated expansion into the western frontier of the US as a critical factor in spurring its rapid development. List argued that the German nation could develop in a similarly rapid way with the help of an interventionist state that would not only promote and protect industry at home, but also expand its territorial control beyond national boundaries through direct conquest and settler colonization (Cowen and Shenton, 1995: 37). Marx, too, described the importance of 'primitive accumulation' through colonial conquest of territories, slavery, trade monopolies and settler colonization as the basis for spurring the development of industrial capitalism in Britain and the US (Marx, [1867] 1967: 702–12, 716–24).

Third, although development was described and discussed in universalist terms, it was not seen as something that could be independently pursued by nations outside Europe. Friedrich List vehemently objected to the idea that development could occur in countries outside the 'temperate zone'. He argued that it would be a fatal mistake for 'savage states' in the 'torrid zones' to pursue his policies for national economic development because these countries were not naturally suited for industrialization (for contemporary versions of these views, see Sachs et al., 2001). He felt that tropical countries would progress more rapidly towards wealth and civilization if they continued to trade agricultural products for manufactured goods from countries in the temperate zones. This was because industrialized, imperial nations in the temperate zones would compete with one another to supply manufactures at low prices and thus 'prevent any one nation from taking advantages by its superiority over the weaker nations of the torrid zone' (quoted in Cowen and Shenton, 1996: 164; see also Arnold, 2000; Watts, 2003; Power, 2003; Power and Sidaway, 2004, for critical commentaries on development in the tropics). Marx claimed that Asiatic countries had no history other than those of successive intruders who founded their empires on the passive basis of unresisting and unchanging societies. They could not develop of their own accord, but required Europe to fulfil a 'double mission', as the British had done in India: 'one destructive, the other regenerating – the annihilation of old Asiatic society and the laying of the material foundations of Western society in Asia' (Marx, [1867] 1967: 659–60). John Stuart Mill, too, firmly believed that the principles of political economy (particularly on matters of land rent and taxation) could not be applied in the territories controlled by the East India Company because of the poverty and 'passive' character of the natives and the primacy of the village as a corporate community (Stokes, 1959). The followers of Saint-Simon were perhaps most sanguine of all about the universal unfolding of order and progress, because they saw this happening across the world under the guidance of European experts informed by science and reason.

Finally, these nineteenth-century thinkers had very different views on the kind of government required for mobilizing development. The Saint-Simonians visualized a corporatist government made up of experts and trustees that would use 'scientific socialism' as the means to engage in social engineering and constructive development by educating, exhorting and enlisting the populace in nation-making. John Stuart Mill's perspective on government was summed up by an editorial comment in a Calcutta newspaper in 1828: 'demagogues at home, they are despots abroad' (quoted in Stokes, 1959: 60). Mill advocated a democratic liberal government committed to education, extension of rights and the welfare of its citizens in countries that were 'sufficiently advanced in civilization to be capable of settling their affairs by rational thinking'. But for countries that lacked such advancement in civilization, Mill recommended a 'pure and enlightened' paternalistic government run by a cadre of morally upright officials and a small committee of experts who would educate the people in rational thinking, choice and individuality, and ultimately allow self-government after the necessary advances of civilization had occurred (Stokes, 1959; Cowen and Shenton, 1995: 40).

Both List and Marx visualized strong, *dirigiste* states that would control economic and social decisions regarding national development. List's prescriptions for national economic development required an authoritarian government that would transfer agricultural surplus and mobilize labour from the countryside for industrial growth. As for Marx and Engels, bourgeois industry and commerce would be mastered by a proletarian revolution leading to a state of communism. The proletariat-controlled state would then centralize all instruments of production and mobilize their peoples to increase the nation's productive forces as rapidly as possible. This stage of development would then lead to the disappearance of class distinctions that were fundamental to bourgeois society. The government would become an association, 'in which the free development of each is the condition for the free development of all' (Marx and Engels, [1848] 1948: 31). On the question of how the communist state would be achieved in Europe's colonies, Engels commented in a letter to Karl Kautsky, 'You ask me what the English workers think about colonial policy. Well the same as they think about politics in general: the same as the bourgeois think'. He goes on to comment that while settler colonies such as Canada, Australia and South Africa would gain independence from Britain, the countries inhabited by native populations and under European subjugation would need to be 'taken over for the time being by the [European] proletariat and led as rapidly as possible towards independence. ... Once Europe is reorganized, and North America, that will furnish such colossal power and such an example that the semi-civilized countries will of themselves follow in their wake; economic needs, if anything, will see to that' (Engels, 1882, quoted in Tucker, 1978: 676).

LEGACIES OF NINETEENTH-CENTURY DEVELOPMENT THOUGHT

My reason for doing this survey is to underscore the point that although these ideas of development took shape in the context of the social and economic transformations occurring in Europe during the eighteenth- and nineteenth-centuries, they needed the presence of the colonized world in both ideological and material terms to establish the force of their arguments. Ideas of progressive development, economical development, and socialist development gained power and legitimacy with their national audiences in Europe by incorporating the colonized world into the lower stages of a single, universal teleology of the development of human history. There were no other pathways of development that could emerge or be pursued in these regions. Development of the colonized needed to occur under the superior authority and guidance of Europe's elites, experts, scientists, engineers and in the case of socialist development, with the help of the European proletariat.

The ideas of Saint-Simon, List, Mill and Marx have descended in various institutional forms and approaches to development in the post-Second World War era. The Saint-Simonian and Comtean corporatist visions of progress and order governed by trustees and scientific 'experts' appear in the institutional forms of the World Bank, IMF and United Nations bureaucracies, as well as many national governments, 'think tanks' and transnational NGOs. Mill's liberal views of universal education and individual choice were considered as 'necessary' conditions for national development in the post-war era, and continue to be articulated in terms of democracy, freedom, 'enlarging human choices and capacities' (Sen, 1984, 1988, 2001), 'empowerment' and 'people-centred' development (Gran, 1983; Korten and Klaus, 1984; Friedmann, 1992). Friedrich List's ideas of *dirigiste* states pursuing economic development and protecting their industries in the national 'self-interest' were adopted by a wide range of countries before, between, and after the two world wars (Gerschenkron, [1952] 1992; Kitching, 1982). Marx and Engels' visions of communism, which drew quite freely on Saint-Simon's corporatist 'scientific socialism', were central in shaping the political ideology of constructivism and centralized economic organization of communist and socialist states during much of the twentieth-century.

The contrary aspects of these nineteenth-century ideas have also persisted in the theories, critiques and practices of development that emerged in the post-war era and continue well into the twenty-first century (cf. Rostow, 1960; Frank, 1967; Amin, 1990; Harvey, 2003). Much of the Euro-American development assistance and technical 'expertise' provided to ex-colonial countries has, consciously or otherwise, functioned according to the teleological narratives of progressive development of the nineteenth century. Many of the development economists and sociologists in the 1950s and 1960s regarded post-colonial countries as mired in archaic traditions and argued for the either the abolition or exploitation of the 'traditional' sector in order to spur industrial growth and national development (cf. Lewis, 1955; Lerner, 1958; Myrdal, 1968). And today, even though most theorists or critics of development claim to recognize the historical and geographical particularities of change in different post-colonial states, they invariably lapse into explanations or arguments that are based on universalization of the Euro-American experience of development.

The most important aspect of post-Second World War development, however, is not just that its advocates and promoters were influenced by these nineteenth-century ideas of progress and social change, but that they were also well aware of the inherent problems and contradictions associated with them. They knew that the European experience of capitalist development was not limited to the space contained by national boundaries, and that its expansion was driven by nationalist, fascist and imperialist ideologies. They knew that progressive development both depended on uneven geographies and created regional disparities; that it contributed to enormous social and economic differences and upheavals in societies; that authoritarian and despotic forms of government were often regarded as the most effective agents of development in 'backward' agrarian societies. And they knew, of course, that universal principles of development did not exist in reality or in practice. All the critiques aired over the past six decades against development were not new, but had been rehearsed before. The central question is why, then, did post-colonial governments, multilateral institutions, international aid agencies, development theorists and 'experts' in the post-war era choose to ignore these contradictions and continue to espouse the idealized pathway of progressive development?

It would be easy to lay the blame on 'Western' liberal ideology, neo-colonialism, capitalist imperatives or elitism, but I do not intend to do so. There were a number of historical factors and contingencies – ranging from the Great Depression, the rise of European war economies, the two world wars, to the prospect of a large number of countries emerging from the end of colonial rule – that contributed to the widespread recognition of the need for some degree of international cooperation and coordination in finance, trade, rebuilding productive capacity in Europe and placing the emerging ex-colonies on some independent economic footing. What is most interesting, however, is that the post-war assumptions and approaches framing the international coordination of finance, trade and 'national' development were articulated through a wholly new twentieth-century invention called 'the economy'.

MAKING THE ECONOMY

Timothy Mitchell (2002) provides a compelling explanation of how the idea of 'the economy' was brought into being, and I primarily draw on his work here. Mitchell's argument is that most accounts – including those of Polanyi (1944) and Foucault (1991) – that attempt to locate the emergence of the economy as a separate sphere of self-regulation or government in the eighteenth and nineteenth centuries, overlook the fact that no political economist during these times ever conceived of, or wrote about, an object called 'the economy' in this way. The term 'economy', during these periods, meant anything ranging from thrift, the 'method of attaining a desired end with the least possible expenditure of means', the management of money, to the proper administration of a community's resources and affairs. Classical political economy used the term 'economy' in this last sense, expanding its meaning to the proper governing of people and managing the resources of a nation so as to increase its material prosperity. Even Friedrich List, who was constantly preoccupied by the 'national economy', wrote of it in this sense.

Mitchell argues that the economy, in its contemporary sense, emerged during the 1930s out of two processes, one internal and the other external to the discipline of economics. The internal processes relate to the development of new branches of study within the discipline that employed mathematical modelling and statistical techniques: macroeconomics, econometrics, national income accounting during the 1930s and 1940s. The external processes were a series of events that included the collapse of the international financial system in the inter-war period, the Great Depression, the various forms of state control over economic activities in Europe and Russia, and, most importantly, the collapse of the global structure of political and economic affairs formed by the European and Japanese empires before and after the Second World War: 'Out of this series of political implosions, social disintegrations, financial failures, and worldwide conflicts emerged this new object, the economy' (Mitchell, 2002: 5).

The economy, in its new sense, referred to 'the totality of monetarized exchanges within a defined space'. This meant that it needed to be conceptualized as a self-contained, internally dynamic and statistically measurable space of money-based activity that could be analysed, modelled and regulated. It needed to be separated from other spheres, such as law (which would have normally come under the scope of political economy), culture, science and technology (Mitchell, 2002: 4). The space of the economy was, in effect, an abstract statistical and algebraic geometry that solely represented the material sphere of social life. Mitchell argues that the economy in its abstracted form was also a set of practices that produced and reinforced divisions between material and cultural, real and abstract: 'It was both a method of staging the world as though it were divided in this way into two, and a means of overlooking the staging, and taking the division for granted' (ibid.: 82).

Mitchell provides a fascinating insight into how the making of 'the economy' was also located in the history of colonialism. John Maynard Keynes made one of the earliest attempts to define the economy while employed at the India Office (the colonial successor to the East India Company). Keynes' first book, *Indian Currency and Finance*, published in 1913,

> … addressed a set of questions whose larger answer was later to be formulated as the national economy: how to conceptualize, measure, and manage the circulation of money within a defined geographical space. Indeed, it was in relation to the problems of colonial rule that several of the problems of managing the enclosed spaces of the economy were originally worked through. Moreover, as an apparatus to be managed and made more efficient, the economy was the object upon which the new politics of development was built after the 1930s. The development of economies provided the forms and formulas through which European powers could attempt to restructure the relationship with their colonies in the mid-twentieth century, and through which imperial powers whose reach was still expanding, in particular the United States, could find a new mode of operation. (Mitchell, 2002: 83–4)

Mitchell's analysis of the making of 'the economy' is extremely useful because it provides a new way of understanding how the idea of national economic space was transformed between the nineteenth and mid-twentieth centuries. Earlier, the economic geography of leading industrialized nations in Europe involved extensive organization of investment, production, information, management and trade based on the political control of colonial resources. In contrast, the new modes of calculation, modelling, measurement and enumeration produced a worldwide geometry of national economies that seemed perfectly contained by their national borders. Mitchell observes that by embedding the national 'economy' in new modes of calculation, modelling and measurement, it was possible to conceptualize a national space abstracted from its history and geographical realities and subject it to analysis and comparison with other national economies in purely statistical terms (see also Goswami, 2004). Worldwide crises such as the Great Depression gave additional impetus to economists to use these modes of calculation for defining a national economic space for employment and welfare policies.

What I find most exciting about Mitchell's analysis, however, is that it opens up the possibility of a very different understanding of why nationalist movements and leaders of so many post-colonial countries, irrespective of their ideological leanings and political affiliations, uncritically embraced the idea of the economy and the core assumptions of *dirigiste* national development in the post-war era. Although Mitchell sees the invention of the economy during the pre- and post-Second World War periods as a new way for old and new industrialized powers to retain or expand their imperial powers, I would argue that its invention was *also* remarkably useful for leaders of post-colonial countries because it created a 'national space' that was far more coherent than anything their anti-colonial or nationalist movements could have produced. After all, the boundaries of most colonial countries had been established over the previous century as outcomes of rivalries and stand-offs between imperial powers. They rarely ever contained a homogeneous population that, in any clear sense, could be represented as a 'nation', if only because colonial administrations had, through direct and indirect rule, rendered them into tribes (Mamdani, 1996) and identity groups based on ethnic, religious or caste categories. A national economy that simultaneously represented a national space determined by its 'character of calculability' rather than by geographical histories, provided the much-needed 'extra-territorial' legitimacy for nationalist leaders and elites to establish political and territorial control during the early phases of post-colonial rule. Given the realities of territorial and social difference within the post-colony, the concept of a national economy as a rational object of calculation, control and intervention became the means of legitimating the new post-colonial state. Yet, despite the attempts by post-colonial leaders and development experts to 'free' the national economy from the constraints of existing economic and political geographies by embedding it in new modes of calculation, it could not escape from the territorial realm of post-colonial politics. The economy, in post-colonies, was routinely recaptured and subjected to the politics of territory, region and identity.

The argument I am making here basically turns Benedict Anderson's analysis of colonial nationalisms in *Imagined Communities* (1983) in a different direction. Anderson describes these nationalisms as emerging from the experiences of 'natives' who – in the process of learning the language of their colonizers, contending with the official nationalisms of colonial states, and being incorporated into subordinate positions within colonial capitalism and administrations – came to define themselves in opposition to the colonial state. They were 'lonely, bilingual intelligentsias unattached to sturdy local bourgeoisies' whose visions of nationalism were formed by the models of official, creole and vernacular versions that they had encountered in their classrooms and during visits to the metropoles as colonial subjects. Anderson says that the intelligentsia were able to copy, refashion and improve upon these nationalisms and find ways – with the help of increased speed of physical and intellectual communication made possible by capitalism – of bypassing print to propagate 'the imagined community not only to the illiterate masses, but even to literate people *reading* other languages' (ibid.: 128–9, emphasis in original).

My issue with this argument – and it is one that Anderson later recognized (1998) – is that it is mainly about *anti*-colonial movements that sought to bring an end to formal colonial rule, and not really about the *formation* of post-colonial nation-states. Anti-colonial movements did not automatically morph into post-colonial nation-states on achieving independence. The nature/nation symbols invoked during anti-colonial movements (soil, blood, brotherhood and so on) did not carry the same valency in a post-colonial context because the same symbols could have been (and in several instances were) mobilized to assert diverse territorial identities, ethnic affiliations or religious loyalties among both the illiterate and those educated in different regional languages. It was also not easy for the leaders of anti-colonial movements to claim natural succession and control over the territories and administrative structures of a colonial state that they themselves had earlier denounced.

It was in this tenuous context of post-colonial state formation that 'the economy' – as a clearly bounded, self-contained entity embedded in new modes of calculation representing 'national space' as a natural, mathematical abstraction – became such an attractive symbol for countries emerging from decolonization after the Second World War. It was the 'objective space' that post-colonial leaders could invoke and appeal to as they sought to establish the *territorial integrity* of their new nation-cum-state that encompassed different regions, diverse group identities and blurred allegiances.

Economy and nation-state were to be mutually reinforcing, producing a territory that formed the subject and object for building new post-colonial 'imagined communities'. Together, they represented the 'national space' where existing geographical differences, diverse identities and unbound loyalties could be transformed into a homogeneous space and population through development based on rational calculation, debate, and collective action for the wealth and welfare of the post-colonial nation. Development was the means by which post-colonial nation-states would be made and sustained. It was to be the central purpose that would give the post-colonial state its territorial integrity, its political authority and its moral legitimacy.

There was, however, an inherent tension in this strategy. The worldwide geometry of national economies gained its coherence and articulation through the authority of the international financial and development institutions established after the Second World War. The roles of these institutions in international cooperation in, and coordination of, finance, trade and development were strongly influenced and shaped by the geopolitical and economic interests of existing and emerging imperial powers. So, even though economy and development were central in enabling post-colonial nation-states to establish territorial integrity and political stability within their boundaries, they simultaneously rendered these countries vulnerable to an international realm of calculability and geopolitical interests within which their economies were embedded. Freedom and independence from colonial rule – fundamental to the self-definition of anti-colonial movements – were to be replaced by subjection to and compliance with an international realm of calculability in order to become post-colonial states. Postcolonial nation-states have thus had to contend with two necessities: the need for greater freedom from the politics of difference within their territories, and the need for greater freedom to manoeuvre within the international economy and geopolitical constraints (cf. Evans, 1995; Heller, 1999).

Development as post-colonial nation-state building has thus been fundamentally about the politics of controlling and transforming geographic territory into a functional space of national economy. Post-colonial leaders along with their economist experts and spatial planners aimed to achieve this geographic transformation by constructing an elaborate economic architecture of barriers and planes. Different types of barriers were set up to secure a national economic space within the context of international economy and geopolitics. The differentiated social and economic landscapes contained within these barriers were to be levelled into functional geometric planes for growing the national economy. Post-colonial development as both sub- and transnational spatial politics was

pursued through a mix of rational instruments and projects such as land reform, monumental infrastructure schemes, industrial growth poles, growth centres, import barriers, export channels, export production enclaves and foreign direct investment. Yet, despite these rational and calculated efforts, post-colonial development constantly confronted the spectre of geographical unevenness (Johnson, 1970). The restructuring of existing geographical relations could neither escape the tensions of regional difference, nor resolve the contradictions between what Friedmann (1988) has called territory and function. Instead of territorial development being wholly dominated and directed by the functional needs of the national economy, development in the post-colony became the means through which the authority and legitimacy of the ruling nationalist elite was manipulated and renegotiated by sub-national, regional and international politics.

It is in this sense that I made my earlier claim that development is both 'in question' and 'not in question' in post-colonial societies; it is, in fact, *the question* that constitutes the terrain for political contestation and economic mobility. The landscapes of development that have emerged over the past six decades since the end of the Second World War reflect the different ways in which post-colonial nation-states have grappled with the politics of regional difference, negotiated external economic necessities and responded to contingent geographical outcomes and social turmoil.

LANDSCAPES OF POST-COLONIAL DEVELOPMENT

I have tried to present the evolution of post-war development from a perspective that examines the changing ideas, geopolitical structures and geographical histories of post-colonial countries that have grappled with the question of development during these past six decades. The twentieth-century invention of 'the economy' as a clearly bounded space embedded in particular modes of calculability became the 'extra-political' basis upon which the post-colonial state, nation and territory were defined. Much of the early emphasis of post-war development in post-colonial countries thus focused on establishing command and control over the bounded space of calculation called the national economy, and to reproduce and grow this space *out of* the existing terrain of geographical, social and ethnic differences. Remaking these geographies of differences into ordered economic spaces of production, calculation and control remains the central preoccupation of post-colonial states and the main endeavour of post-colonial national development. The relative successes or failures of development over the past six decades have depended, therefore, on the extent to which the geographies of difference contained within the territorial boundaries of post-colonial states have responded or yielded to, or actively resisted, being transformed into a gross terrain of national economic production and calculability.

The brief given to me by the editors of this volume required me to provide a 'map' of development over the past six decades. What would the diverse landscapes of post-war development look like from a post colonial perspective? It would begin by recognizing that there is no way of identifying a single, well-laid path to a summit from which one can command a panoramic view of the development landscapes spread out below. It also would not offer a singular vision 'from below' of the ever-looming threat of monstrosities wrought by capitalist development. Instead, a post-colonial perspective provides different frames of reference within development landscapes that reveal what the great Filipino writer, José Rizal, called 'el demonio de las comparaciones' – the spectre of comparisons – an incurable doubled vision that simultaneously sees things from close up and afar (Anderson, 1998: 2). This doubled vision veers and swings between different frames and focal lengths, attempting to translate across differences and revisualising development landscapes through a cinematic montage that rolls out at disconcerting speeds. …

Looking back, the distant horizon appears to be defined by monumental silhouettes that reveal their heroic forms in flashes of energy: hydroelectric plants, thermal power stations, iron smelters, coal mines, steel factories, straining mightily to anchor the spatial geometries of national economies. Splinters of light in the horizon reveal clusters of great national leaders, Saint-Simonian visionaries and experts – economists, statisticians, bureaucrats – viewing the landscape from commanding heights. Fragments of speech filter through: 'the big push', 'national accounts', 'aggregate growth models', Kuznets curve, 'Soviet planning', 'modernization', 'stages of growth'. Mathematical equations fill the sky with messages: savings and investment key sources of national economic growth; savings achieved by transferring surplus labour from traditional to modern sector; agriculture as 'bargain sector' for potential growth. National spaces hover above the ground, held together by statistics, calculations, economic plans, input–output analyses, industrial growth poles, growth concentrations, backward and forward linkages, capital flows from primary to secondary sectors.

The vision swerves, slowing down, readjusting as vast expanses of rural landscapes emerge in sharp relief: latifundia, collectivized farms,

communes, minifundia, and patchworked fields. Peasants working the land, digging deeper and harder to produce surplus, rendering their surpluses to landlords, to the state, to the economy, to the growth poles, and being made into surplus populations in return. The poles grow, bigger, heavier, appearing as cluttered lumps of shacks and human populations. The geometry of national growth looks nothing like the leaders and their development experts imagined. Are those cities, or are they unsightly agglomerations of lumpen masses? Balance the spatial geometry, advise the experts: connect the smaller points in national space to the larger ones, growth centres linked to growth poles, that should solve the problem. Other voices shout: that won't solve the problem at all; the problem is dependency, unequal exchange in a world economy that benefits neo-imperial powers. They take our money, and our growth poles get the masses. Stop the flows of capital out of national space, build strong, high walls to protect import substitution industrialization, support the national economy. Walls rise up, some high, some very high, some low, holding up the geometry of national spaces.

Peasant revolts everywhere. Fortify the defences! Diktats, dictates, dictators, strongmen, strong states, revolutionary and counter-revolutionary vanguards appear, promising to prevent national economies from sinking to the ground. They find new friends in foreign bankers who offer plenty of cheap loans with the money floating above petroleum.

The focus swings wildly as it lengthens, the frame widens. … The ground has caved in across vast swathes of the development landscape. Sunk under the weight of debt. Bankers and their debt collectors banging on the walls, demanding their money back, right now, if you please. What you need is structural adjustment. Get rid of those protectionist walls, forget the geometry of national economy and import substitution, they're passé! Create enclaves for investment in export-oriented production and you'll get richer and stronger. Look at the development tigers in East Asia, follow their economic formula!

Large erosions appear in some parts of the landscape. Voices call out, 'You must be either very dumb or very rich if you fail to notice that development stinks!' It does; stinks in pools of industrial and chemical wastes, in mine tailings, sewerage. There are movements on the ground, in barrios, slums, countrysides, mobilizing, challenging dictates, dictators. Lumpen masses appear to be striking back, taming the national economy, bringing it back to ground. Look at us, don't look away, you hear? We matter, and we're showing you that we do. Walls fall, experts run, dictators hide. Development falters, or does it?

The frame swerves and shifts focus. ... New developments appear on the landscape. Carefully designed enclaves of production and consumption protected by new kinds of walls that blank out any sight of erosions, lumpen masses, slums and cesspools. Outside the enclaves a welter of change, new urban forms take shape as thousands weave their way between city and countryside, bringing altered connections, new work, remaking cities into mega-urban regions.

The economy looks different in these regions. New workers, peasant-workers turned into factory owners, entrepreneurs emerging from the mass forms of the informal economy. Where did they come from? They were already there, creating work, links, producing things that were cheap, that they could afford but couldn't be seen from the commanding heights of the national economy. The enclaves that succeed do so because of these men and women, self-made entrepreneurs, who know how to toil, how to produce, economize, manoeuvre their way around, manipulate barriers of every kind. They have long learned how to move between, around, slide in and out of the walled areas of the national economy. Everyone needs them now: their governments, the foreign purchasers, the enclaves, consumers in rich countries. Experts attempt to incorporate them into calculable economic spaces: agglomeration economies, neo-Marshallian networks, industrial districts.

Can these 'subalterns' speak? They do. They speak in lumpen tongues, through votes, through religion, in pidgins produced from the melding of different kinds of economies. Flexible specialization? They say: you cannot know, can never imagine how flexible we are, how we make things work between city and country, how we toil to create networks between households, workshops, small factories, unions, to make the things you consume.

The frames swing close, rushing at each other, almost converging. ... Your vision is strained and refracted in different directions, your ears assaulted by a cacophony of voices, clamouring against free trade, against free movement of labour, against the seeming flood of barbarians that appear at the gates of consuming spaces, finding their way into consuming societies. Fair trade, some shout! What 'fair trade', comes the derisive response. Trade depends on free movement, but it does not equate with free trade; free trade has never been free, nor fair!! Fair trade, you say? Fair, under what circumstances? Under your benevolence and patronage? Led by the new 'sustainable development' experts and vanguards of a new 'mission civilisatrice'?

Your vision stumbles over a welter of landscapes, some growing rich, some eroded by war, worn out by their dictators, some struggling with *epidemics. People moving, moving all the time, moving large distances across formidable barriers to consuming regions, to build, clean homes, tend gardens and fields, take care of children, the sick, the aged. People in cities and regions far away, new growth poles and centres, servicing the consuming regions through the virtual spaces of information and communication technologies. New spaces of calculability emerge from Economic Unions, free trade agreements. ... Old battered walls replaced by new barbed-wire fences, electrified, guarded by vigilantes, detention camps, allowing money to fly in and flow out freely, but preventing people from moving in from land and sea. Imperial governments of conspicuous consumption strafing landscapes, asserting their manifest destinies to continue consuming without limit. Your vision falters as it encounters women veiling themselves, walking towards you, then moving away, looking at you, at once challenging, anxious, tense, sad, wanting to convey their distress and traumas of their lives, their men, their children. ... Subaltern voices challenge leaders everywhere. Listen, they say, we do development, we want development, we'll do it on our terms, or you don't stay. ...*

ACKNOWLEDGEMENTS

I am grateful to Kevin Cox, John Friedmann, Christian Kull, Elissa Sutherland and Craig Thorburn for critical comments on this chapter, and for conversations about ideas of development. I am also most grateful to Sharad Chari, Phillip Darby and David Bennett for helping me clarify my thinking about post-coloniality. None of them, however, should be held responsible for the final shape of this chapter.

NOTE

1 Apologies to Charles Darwin.

REFERENCES

Abu-Lughod, L. (1998) 'Feminist longings and postcolonial conditions', in L. Abu-Lughod (ed.), *Remaking Women: Feminism and Modernity in the Middle East.* Princeton, NJ: Princeton University Press, pp. 3–31.

Adas, M. (1989) *Machines as the Measure of Men: Science, Technology, and the Ideologies of Western Dominance.* Ithaca, NY: Cornell University Press.

Alavi, H. (1972) 'The state in post-colonial societies: Pakistan and Bangladesh', *New Left Review,* 74: 59–82.

Amin, S. (1974) *Accumulation on a World Scale: A Critique of the Theory of Underdevelopment.* New York: Monthly Review Press.

Amin, S. (1990) *Maldevelopment: Anatomy of a Global Failure*. London: Zed Books.

Amsden, A. (1989) *Asia's Next Giant: South Korea and Late Industrialization*. New York: Oxford University Press.

Anderson, B. (1983) *Imagined Communities: Reflections on the Origin and Spread of Nationalism*. London: Verso.

Anderson, B. (1998) 'El Malhadado País', in *The Spectre of Comparison: Nationalism, Southeast Asia and the World*. London: Verso.

Arnold, D. (2000) '"Illusory riches": representations of the tropical world, 1840–1950', *Singapore Journal of Tropical Geography*, 21(1): 6–18.

Balassa, B. (1982) *Development Strategies in Semi-Industrial Economies*. Baltimore: Johns Hopkins University Press.

Bauer, P.T. (1981) *Reality and Rhetoric: Studies in the Economics of Development*. London: Weidenfeld & Nicolson.

Berman, M. (1982) *All That Is Solid Melts into Air*. New York: Simon & Schuster.

Bhagwati, J.N. (1982) 'Directly unproductive profit-seeking (DUP) activities', *Journal of Political Economy*, 90(5):988–1022.

Bowie, A. (1991) *Crossing the Industrial Divide: Society and the Politics of Economic Transformation in Malaysia*. New York: Columbia University Press.

Carney, J. (1988) 'Struggles over crop rights within contract farming households in a Gambian rice irrigation project', *Journal of Peasant Studies*, 15: 334–49.

Carney, J. (2001) *Black Rice: The African Origins of Rice Cultivation in the Americas*. Cambridge, MA: Harvard University Press.

Chakrabarty, D. (2000) *Provincializing Europe: Postcolonial Thought and Historical Difference*. Princeton, NJ: Princeton University Press.

Chari, S. (2004) *Fraternal Capital: Peasant-Workers, Self-Made Men, and Globalization in Provincial India*. Stanford, CA: Stanford University Press.

Comaroff, J.L. and Comaroff, J.L. (2001) 'Naturing the nation: aliens, apocalypse, and the postcolonial state, *Journal of Southern African Studies*, 27(3): 627–51.

Cooper, F., Holt, T. and Scott, R. (eds) (2000) *Beyond Slavery: Explorations of Race, Labor and Citizenship in Postemancipation Societies*. Chapel Hill, NC: University of North Carolina Press.

Corbridge, S. and Harriss, J. (2000) *Reinventing India: Liberalization, Hindu Nationalism, and Popular Democracy*. Cambridge: Polity Press.

Coronil, F. (2000) 'Towards a critique of globalcentrism: speculations on capitalism's nature', *Public Culture*, 12(2): 351–74.

Cowen, M. and Shenton, R. (1995) 'The invention of development', in J. Crush (ed.), *Power of Development*. London: Routledge.

Cowen, M. and Shenton, R. (1996) *Doctrines of Development*. London: Routledge.

Davidson, B. (1980) *The African Slave Trade*. Boston: Little, Brown.

Esteva, G. (1987) 'Regenerating people's space', *Alternatives*, 10(3): 125–32.

Escobar, A. (1992) Imagining a post-development era? Critical thought, development and social movements, *Social Text*. 10(2/3): 20–56.

Escobar, A. (1995a) *Encountering Development: The Making and Unmaking of the Third World*. Princeton, NJ: Princeton University Press.

Escobar, A. (1995b) 'Imagining a post-development era', in J. Crush (ed.), *Power of Development*. London: Routledge.

Evans, P. (1995) *Embedded Autonomy: States and Industrial Transformation*. Princeton, NJ: Princeton University Press.

Fanon, F. (1967) *Wretched of the Earth*. London: Penguin.

Foucault, M. (1991) 'Governmentality', in G. Burchell, C. Gordon and P. Miller (eds), *The Foucault Effect: Studies in Governmentality, with Two Lectures by and an Interview with Michel Foucault*. Chicago: University of Chicago Press.

Frank, A.G. (1967) *Capitalism and Underdevelopment in Latin America*. New York: Monthly Review Press.

Friedmann, J. (1987) *Planning in the Public Domain: From Knowledge to Action*. Princeton, NJ: Princeton University Press.

Friedmann, J. (1988) *Life Space and Economic Space: Essays in Third World Planning*. New Brunswick, NJ: Transaction Books.

Friedmann, J. (1992) *Empowerment: The Politics of Alternative Development*. Cambridge, MA: Blackwell.

Gerschenkron, A. ([1952] 1992) 'Economic backwardness in historical perspective', in M. Granovetter and R. Swedberg (eds), *The Sociology of Economic Life*. Boulder, CO: Westview Press.

Goswami, M. (2004) *Producing India: From Colonial Economy to National Space*. Chicago: University of Chicago Press.

Gran, G. (1983) *Development by People: Citizen Construction of a Just World*. New York: Praeger.

Grove, R.H. (1995) *Green Imperialism: Colonial Expansion, Tropical Island Edens and the Origins of Environmentalism 1600–1860*. Delhi: Oxford India.

Gupta, A. (1998) *Postcolonial Developments*. Durham, NC: Duke University Press.

Harris, N. (1986) *The End of the Third World: Newly Industrialising Countries and the Decline of an Ideology*. London: Penguin.

Harris, N. (2005) 'Towards new theories of regional and urban development', *Economic and Political Weekly*, 12 Feb.

Hart, G. (1991) 'Engendering everyday resistance: gender, patronage and production politics in rural Malaysia', *Journal of Peasant Studies*, 19: 93–121.

Hart, G. (2002) *Disabling Globalization: Places of Power in Post-Apartheid South Africa*. Berkeley: University of California Press.

Harvey, D. (1985) 'The geopolitics of capitalism', in D. Gregory and J. Urry (eds), *Social Relations and Spatial Structures*. London: Macmillan, pp. 12–63.

Harvey, D. (2001) 'The geography of class power', in *Spaces of Capital: Towards a Critical Geography*. London: Routledge, pp. 369–93.

Harvey, D. (2003) *The New Imperialism*. Oxford: Oxford University Press.

Heller, P. (1999) *The Labor of Development: Workers and the Transformation of Capitalism in Kerala, India*. Ithaca, NY: Cornell University Press.

Hettne, B. (1990) *Development Theory and the Three Worlds.* London: Methuen.

James, C.L.R. (1963) *The Black Jacobins: Toussaint L'Ouverture and the Santo Domingo Revolution.* New York: Vintage Books.

Kandiyoti, D. (1998) 'Some awkward questions on women and modernity in Turkey', in L. Abu-Lughod (ed.), *Remaking Women: Feminism and Modernity in the Middle East.* Princeton, NJ: Princeton University Press, pp. 270–87.

Killick, T. (1989) *A Reaction Too Far: Economic Theory and the Role of the State in Developing Countries.* London: Overseas Development Institute.

Kitching, G. (1982) *Development and Underdevelopment in Historic Perspective: Populism, Nationalism, Industrialism.* London: Methuen.

Kitching, G. (2001) *Seeking Social Justice Through Globalization: Escaping a Nationalist Perspective.* University Park, PA: Pennsylvania State University Press.

Kohli, A. (2004) *State-Directed Development: Political Power and Industrialization in the Global Periphery.* Cambridge: Cambridge University Press.

Korten, D.C. and Klaus, R. (eds) (1984) *People-Centred Development: Contributions Toward Theory and Planning Frameworks.* Hartford, CT: Kumarian Press.

Krueger, A.O. (1993) *Political Economy of Policy Reform in Developing Countries.* Cambridge, MA: MIT Press.

Latour, B. (1987) *Science in Action.* Cambridge, MA: Harvard University Press.

Lal, D. (1983) *The Poverty of 'Development Economics'.* London: Institute of Economic Affairs.

Lerner, D. (1958) *The Passing of Traditional Societies: Modernizing the Middle East.* New York: Free Press.

Lewis, A. (1955) *The Theory of Economic Growth.* London: Allen & Unwin.

Little, I.M.D. (1982) *Economic Development: Theory, Policy and International Relations.* New York: Basic Books.

Mamdani, M. (1996) *Citizen and Subject: Contemporary Africa and the Legacy of Late Colonialism.* Princeton, NJ: Princeton University Press.

Martinussen, J. (1997) *Society, State and Market: A Guide to Competing Theories of Development.* London: Zed Books.

Marx, K. ([1867] 1967) *Capital*, Vol. 1 (100th anniversary edn). New York: Progress Publishers.

Marx, K. and Engels, F. ([1848] 1948) *The Communist Manifesto* (100th anniversary edn). New York: International Publishers.

Mitchell, T. (2002) *Rule of Experts: Egypt, Techno-politics, Modernity.* Berkeley and Los Angeles: University of California Press.

Myrdal, G. (1968) *The Asian Drama: An Inquiry into the Poverty of Nations.* Harmondsworth: Penguin.

Parajuli, P. (1991) 'Power and knowledge in development discourse: new social movements and the state in India', *International Social Science Journal*, 127: 173–90.

Parfitt, T. (2002) *The End of Development: Modernity, Postmodernity and Development.* London: Pluto Press.

Peet, R. and Watts, M. (1996) 'Development, sustainability, and environment in an age of market triumphalism', in R. Peet and M. Watts (eds), *Liberation Ecologies: Environment, Development, Social Movements.* London: Routledge.

Polanyi, K. (1944) *The Great Transformation: The Political and Economic Origins of Our Times.* Boston: Beacon Press.

Poovey, M. (1998) *A History of the Modern Fact: Problems in the Knowledge in the Sciences of Wealth and Society.* Chicago: University of Chicago Press.

Power, M. (2003) *Rethinking Development Geographies.* London: Routledge.

Power, M. and Sidaway, J.D. (2004) 'The degeneration of tropical geography', *Annals of the Association of American Geographers,* 94(3): 585–601.

Rangan, H. (1999) 'Bitter-sweet liaisons in a contentious democracy: radical planning through state agency in postcolonial India', *Plurimondi*, 1(2): 47–66.

Rangan, H. (2000) *Of Myths and Movements: Rewriting Chipko into Himalayan History.* London: Verso.

Rangan, H. (2004) 'From Chipko to Uttaranchal: the environment of protest and development in the Indian Himalaya', in R. Peet and M. Watts (eds), *Liberation Ecologies: Environment, Development, Social Movements* (2nd edn). London: Routledge.

Rostow, W.W. (1960) *Stages of Growth.* Cambridge: Cambridge University Press.

Routledge, P. (2002) 'Resisting and reshaping destructive development: social movements and globalizing networks', in R.J. Johnston, P.J. Taylor and M.J. Watts (eds), *Geographies of Global Change: Remapping the World.* Oxford: Blackwell.

Sachs, J., Gallup, J. and Mellinger, A. (2001) 'The geography of poverty', *Scientific American,* March: 70–5.

Sachs, W. (ed.) (1992) *The Development Dictionary: A Guide to Knowledge as Power.* London: Zed Books.

Said, E. (1978) *Orientalism.* New York: Pantheon.

Said, E. (1993) *Culture and Imperialism.* New York: Vintage Books.

Sen, A.K. (1984) *Resources, Values and Development.* Oxford: Blackwell.

Sen, A.K. (1988) 'The concept of development', in H. Chenery and T.N. Srinivasan (eds), *Handbook of Development Economics.* Amsterdam: North Holland.

Sen, A.K. (2001) *Development as Freedom* (2nd edn). Oxford: Oxford University Press.

Shiva, V. (1989) *Staying Alive: Women, Ecology and Development.* London: Zed Books.

Slater, D. (2002) 'Trajectories of development theory: capitalism, socialism, and beyond', in R.J. Johnston, P.J. Taylor and M.J. Watts (eds), *Geographies of Global Change: Remapping the World.* Oxford: Blackwell.

Smith, N. (1984) *Uneven Development.* Oxford: Blackwell.

Stokes, E. (1959) *The English Utilitarians and India.* Delhi: Oxford University Press.

Taylor, C. (1977) *Hegel.* Cambridge: Cambridge University Press.

Toye, J. (1987) *Dilemmas of Development: Reflections on the Counter-Revolution in Development Theory and Policy.* Oxford: Blackwell.

Trainer, F.E. (1989) *Developed to Death: Rethinking World Development.* London: Green Press.

Tucker, R.C. (ed.) (1978) *The Marx–Engels Reader* (2nd edn). New York: W.W. Norton.

Wade, R. (1990) *Governing the Market: Economic Theory and the Role of Government in East Asian Industrialization.* Princeton, NJ: Princeton University Press.

Watts, M. (1992) 'Living under contract: work, production politics and the manufacture of discontent in a peasant society', in A. Pred and M. Watts (eds), *Reworking Modernity.* New Brunswick, NJ: Rutgers University Press.

Watts, M. (1995) 'A New Deal in emotions: theory and practice and the crisis of development', in J. Crush (ed.), *Power of Development.* London: Routledge.

Watts, M. (2003) 'Development and governmentality', *Singapore Journal of Tropical Geography,* 24(1): 6–34.

Weber, E. (1976) *Peasants into Frenchmen: The Modernization of Rural France, 1870–1914.* Stanford, CA: Stanford University Press.

Williams, R. (1985) *A Vocabulary of Culture and Society.* Revised edition. NY: Oxford University Press.

Wolf, E. (1982) *Europe and the People Without History.* Berkeley: University of California Press.

35

Sustainable Development and Governance

Yvonne Rydin

INTRODUCTION

The term 'sustainable development' is part of the common currency of public policy nowadays. International organizations, those at the level of the nation-state and the locality, within the private and public sectors all claim sustainable development as an objective. The adjective 'sustainable' has become so widely used in policy and marketing literature as to become almost meaningless. Indeed some consider that the parent term 'sustainable development' has lost any sense of coherent meaning through the multiple demands made of it. It is criticized for the inherent ambiguity that allows for so many different situations and possible futures to be described as contributing to 'sustainable development'. It is tempting to relegate the term to a conceptual rubbish bin as having no integrity or useful conceptual 'bite', rather in the way that the 'public interest' has been disposed of. However, to do so would be to dismiss the efforts that historically and currently lie behind the development and use of the concept to achieve advances in terms of social justice and environmental protection.

In this chapter, the history of the concept of sustainable development is briefly outlined, before discussing key dimensions of the concept itself, and drawing out the implications of different definitions. Then the relationship of the sustainable development agenda to contemporary patterns of governance is explored, defining more precisely what 'governance' means and explaining its particular significance to the sustainable development agenda. Finally, the way that governance processes can actually contribute to achieving sustainable development is examined, with particular reference to the local and community level.

THE RISE OF THE SUSTAINABLE DEVELOPMENT AGENDA

Most histories of the concept attribute its earliest uses to the International Union for the Conservation of Nature (IUCN, 1980). However, its contemporary relevance can be traced back to the United Nations' World Commission on Environment and Development, set up in 1984 and chaired by Gro Harlem Brundtland, then Prime Minister of Norway and since head of the World Health Organization. This Commission had its roots in two different strands of United Nations (UN) activity. On the one hand, it was conceived of as the successor to the commission chaired by Willie Brandt, ex-Chancellor of West Germany, which resulted in the *North–South* report (ICIDI, 1980) calling for a renegotiation of the relationship between more and less developed countries. Hence the emphasis on development within its terms of reference was a signal that global patterns of economic inequality were to be a key focus. In the context of the 1980s – before the end of the Cold War and the physical collapse of the Berlin Wall – relationships between nation-states were also framed in terms of conflict and, in particular, the threat of nuclear conflict.

On the other hand, the Commission was also taking forward the UN's work on environmental issues and, indeed, reshaping this work. Previously the

UN had seen environmental issues largely in terms of how the physical environment affected human needs through the adequacy of housing, water provision and food security. Key issues were loss of soil fertility and water management. These were highlighted at the UN Habitat Conference held in 1972 in Stockholm. By the 1980s the environmental agenda had broadened out from this local and strictly anthropocentric focus. The background to this was the emergence of a scientific consensus on key aspects of global environmental change. This threw the emphasis on how global environmental systems operated and how individual, spatially fixed activities could – in aggregate – significantly alter these global systems, in turn impacting on people across the world.

Although not discussed in any depth by the Brundtland Commission, the precursor to this new global emphasis in environmental debates was the 'discovery' of the 'ozone hole' (actually a selective thinning of the stratospheric ozone layer) and its linking to anthropocentric emissions principally in the form of chlorofluorocarbons (CFCs). The political impact of the scientific evidence was considerably enhanced by the ability to graphically represent the 'hole' through the use of colour-enhanced satellite imaging and by the links to health effects, in terms of malignant melanomas (i.e. skin cancers) from exposure to sunlight for light-skinned peoples in geographical zones beneath the thinning ozone layer. Colour pictures and a cancer scare made for considerable media coverage. The problem of anthropogenic ozone thinning was tackled by international agreement for collective action through the Montreal Protocol adopted in 1987, assisted by the possibility of a technological substitute for CFCs and the willingness of the key economic actor DuPont, as producer of the substitute chemicals, to support the agreement (Rowlands, 1995).

However, the whole 'ozone hole' episode was important in establishing the possibility of global environmental change through anthropogenic causes and the need for collective action at the global scale to tackle such change. In this context, scientific communities began to make new claims for evidence of global climate change. Again there were anthropogenic sources, in this case the so-called 'greenhouse gases', mainly carbon dioxide and methane. The principal blame for such emissions was laid at the door of the fossil fuel economy, since the burning of coal and oil was the key source of carbon dioxide. NGOs, the media and some policy-makers began to talk of and write about the 'greenhouse effect' as a way of popularizing the more complex scientific processes lying behind the thickening of layers of gases in the upper atmosphere that were reflecting solar radiation back to the Earth's surface.

There was considerable dispute over these scientific claims, dispute that continues to this day.

An Inter-Governmental Panel on Climate Change was convened in 1988 by the World Meteorological Organization and the UN Environment Programme to bring together the key scientists involved in research on this issue. It has produced a series of reports setting out the evidence for climate change and seeking to specify future scenarios with ever greater precision. Over time the IPCC has generated an 'almost-consensus' on the likely extent of climate change processes. However, it is in the nature of such forecasting exercises that precise figures cannot be given; rather ranges are offered as descriptions of the future. The current estimates of global averaged surface temperature increases lie within ranges of 1.4–5.8 degrees Celsius over 1990–2100. Such prediction of future ranges for climate change contrasts with the graphic pictures of ozone thinning that had already happened. The ranges themselves were also the subject of vigorous debate within the broader scientific community and, even as the scientific consensus has tightened, so a band of contrarian scientists has continued to present an alternative view of the future (McCright and Dunlap, 2003).

These debates on climate change set the context for the renaissance of the sustainable development concept. They suggested that human action was generating (and indeed had already generated) change in global environmental systems, which was going to have profound impacts on people across the globe. Among these impacts could be counted sea-level rise (estimated as a global mean sea-level rise of 0.09–0.88 metres over 1990–2100 but potentially rising as much as 6 metres if the Antarctic ice shelf fails), changed patterns of agriculture including the results of extended desertification, higher incidence of extreme weather events such as storms and heatwaves, changes in water availability, loss of habitats and species, and new public health risks.

Around climate change, a number of other issues with a global perspective were clustered in the 1980s and 1990s. The ecological consequences of deforestation, particularly of the tropical rainforest areas, in various parts of the world were highlighted by another set of scientific claims. Biodiversity was as frequently identified as a global environmental concern as climate change, and the anthropocentric implications of such biodiversity loss were highlighted in terms of potential medical treatments that awaited discovery within the gene pool. However, there were also broader concerns about the existence value of the rainforests and their intrinsic value, even their intrinsic right to exist. And linking these two global issues was the question of how to reduce the scale of deforestation. Logging was a particular cause of the reduction in scale and diversity of the rainforests and deforestation directly contributed to climate change by removing a carbon sink. In addition, loss of tree cover more generally was related to loss of soil fertility and increased vulnerability of agricultural land to erosion, with implications for food security.

This mix of global environmental concerns provided a distinctive context for the deliberations of the Brundtland Commission. They gathered evidence from a wide variety of sources, including both accredited experts and local communities across the world, in an innovative mix of site visits, deliberative meetings and public hearings (WCED, 1987: 359). It developed an analysis that framed the problem in terms of global inequalities in development, global environmental threats and potentially catastrophic global conflicts. This was presented in the Brundtland Report, *Our Common Future*, published in 1987. The report saw the solution to this problem as the pursuit of 'a new era of growth', but a very different kind of growth. This is the core of the definition of sustainable development – a form of growth that can reduce inequalities, reduce environmental threats and diffuse conflicts, all on a global scale.

The decades since the publication of the Brundtland Report have seen the sustainable development concept become embedded in policy statements at local, regional, national and international levels, although the public profile of the agenda has waxed and waned. After the publication of the report, an international summit on environment and development was convened in Rio in 1992. Two conventions (on biodiversity and climate change) were tabled alongside Agenda 21, a manifesto of the actions needed to achieve sustainable development. Afterwards the UN established a Commission on Sustainable Development to encourage national governments to draw up their own Agenda 21 and monitor their progress. Painfully slow progress was made in translating the Framework Convention on Climate Change into the Kyoto Protocol and agreeing to its implementation. It finally came into force in February 2005 but without the involvement of the United States, currently the world's largest carbon dioxide emitter.

In 2002 the UN convened a new summit, the World Summit on Sustainable Development in Johannesburg. This was intended to assess the progress that had been achieved in the twenty years since Rio. Little of real significance could be reported on any of the major themes of the Rio Summit: biodiversity loss, deforestation or climate change. However, the sustainable development agenda itself was moving on. In the intervening years, the UN Millennium Development Goals had thrown the emphasis onto issues of world poverty, child mortality, gender equality, education, food security and public health, although 'ensuring environmental sustainability' was listed as one of the eight goals. So the focus during the WSSD was less purely environmental than had been the case at the Rio Summit. In more recent years, the

coming into force of the Kyoto Protocol, together with continuing and increasingly dramatic scientific evidence of the more imminent impact of climate change, have somewhat redressed the balance. The meeting of the G-8 (the political leaders of the world's eight largest economies) in 2005 saw global poverty and climate change sharing the media headlines and policy agenda.

Sustainable development is therefore a core focus of policy debates, albeit a focus that is variably defined. So how can the concept of sustainable development be characterized if it is apparently so flexible? The next section explores this question.

THE CONCEPT OF SUSTAINABLE DEVELOPMENT

The definition of sustainable development that is most commonly cited is one taken from the Bruntland Report (WCED, 1987: 8): 'to ensure that [humanity] meets the needs of the present without compromising the ability of future generations to meet their own needs'.

This is a definition that is rooted in considerations of equity. The emphasis is on generations, viewed collectively, as being able to meet their needs, not their demands or their wants. In the context of global inequality this is a clear statement about priorities. Then there is also the emphasis on inter-generational equity. The concern here is that the forms of global environmental change discussed above are at least partly caused by current human activity, but that the major negative impacts of those changes will be borne by future generations, and to such an extent that their ability to meet their basic needs will be adversely affected.

In this formulation, current generations are effectively stealing from their collective children and grandchildren on a global scale. Current economic growth is achievable only because the negative inter-temporal externalities of that growth are being ignored. And that economic growth is underpinned by prevailing patterns of production and consumption so that the current generation's experience of these activities is intrinsically unsustainable. Examples of this include the reliance on fossil fuels for energy needs and as a raw material for production, the scale and growth of consumption levels and the rapid turnover in goods generating ever more waste, as well as demand for further production. While the underlying message of the concept is clear, the subtleties of this formulation can create problems for policy development and implementation. As a result there has been a search for simpler ways of defining sustainable development that still capture the combination of social, environmental and economic concerns wrapped up in the term.

One common way to do this is to identify these different aspects as 'pillars' of sustainable development and then conceptualize their interrelationship. For example, they can be conceptualized as distinct but overlapping areas (see Figure 35.1). Potential development paths or policy approaches can then be mapped onto this Venn diagram model (Levett, 1998).

The idea here is that future possibilities should be judged against their contribution to each of the elements of sustainable development. Development that meets only economic criteria of viability and profitability would fall within the economic circle but outside any area of overlap. If there were also social benefits accruing from this development, then it would fall into the area of overlap between the economic and the social. If on the other hand there were environmental benefits, it would be the overlap between the economic and environmental areas that was relevant.

The goal of sustainable development is, therefore, to find development opportunities and future paths that fall within the centre of the Venn diagram where all three areas overlap. However, this would involve finding win-win-win development paths that offer benefits on all three fronts. In practice, this can be difficult and the emphasis is instead on identifying win-win options, focusing on two of the three pillars. For example, there is an approach that sees considerable potential for expanding the area of overlap between economic and environmental goals in Figure 35.1. This sees the hope of steering economic activity in more environmentally friendly paths. It looks at the historic trajectory of many developed countries, passing from a predominantly rural economy through a dirty phase of industrialization and then into a cleaner period of late industrialization and into a predominantly tertiary economy. From this trajectory the possibility of environmental impacts actually reducing with increased growth is seen as a historic reality and therefore a feasible future.

Under the banner of ecological modernization, these theorists see the potential of establishing a series of institutional incentives that would hasten movement towards a growth-oriented but environmental friendly society (Gouldson and Murphy, 1998). Such incentives would include a mix of fiscal arrangements, both taxes and subsidies, together with regulation. The aim would be a form of economic activity that was much more efficient in its resource use, increasing the output/resources ratio by a factor of 4 or even 10 (Weizsäcker et al., 1998). Waste would also be minimized as a corollary of this approach, reducing pollution and waste disposal problems. By including energy as a key resource, this approach would also tackle the spectre of global warming through reducing dependence on energy sources and simultaneously shifting from fossil fuels to renewable sources. The mixture of judicious governmental policy, technological innovations and continued reliance on market signals would realize sustainable development as a mix of economic and environmental goals within a predominantly unchanged economic system. However, the reliance on market signals to generate economic activity would not necessarily mean that social goals were satisfied; indeed, past experience suggests that such pursuit of economic

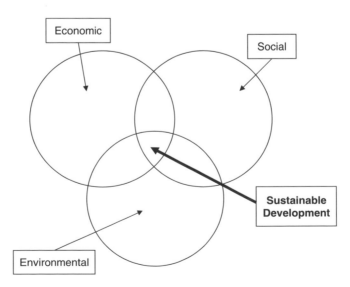

Figure 35.1   The Venn diagram model of sustainable development.

growth may exacerbate social inequalities, at least in the short term. In a similar way, alternative paths can be identified that emphasize the possibility of finding synergies between environmental and social goals or economic and social goals under the banner of environmental justice (Agyeman et al., 2001) or corporate social responsibility (Zadek, 2001) respectively.

While the Venn diagram model suggests in its pictorial representation that all three elements are on a par, it is implicit in this model that economic processes are the appropriate starting point. For economic processes are seen as the route through which needs as well as demands are met, and are therefore inevitably at the core of sustainable development. This was a key assumption of the Brundtland Report, which called for a new era of growth as the way to tackle global poverty. While this growth is recast as development and not as a simple extension of conventional growth paths, this is still a future based on a variation of market-led or capitalist economic activity. This raises the question of how far current patterns of economic activity can be shaped to fit within a sustainable future (Jacobs, 1991). Some environmentalists have queried whether this provides a sufficient level of environmental protection or whether the pace of change might not be too slow to avoid ecological catastrophe. Instead they emphasize the critical nature of the environmental systems on which all other activities, including economic activity, must rely.

This is related to the distinction within sustainable development debates between weak and strong sustainability (Neumayer, 2003). In weak sustainability, the environment is seen as having valued dimensions for both its life-support functions and its abilities in providing food, other resources and the capacity of a pollution sink, but there are unlimited possibilities for compensating for environmental costs or losses through economic growth. Investment in other forms of physical and indeed human capital can be used to trade off environmental degradation. Thus, development in a rare ecological habitat can be compensated through creating other habitats or nature conservation features elsewhere. The use of environmental valuation techniques is often invoked as a way of enabling such trade-offs and compensation to be calculated and put into effect (Pearce et al., 1989).

However, critics of the weak sustainability position argue that some aspects of the environment cannot be substituted for or traded off in this way. Exactly which aspects fall into this category is debatable, but they may include climate control functions, pollution sinks with a limited capacity, valued landscapes, rare habitats, and areas of significant cultural heritage including historic buildings. These could all be termed critical natural capital (man-made capital in the case of historic buildings) and as such fall into a different conceptual category to other aspects of the environment that are not seen as so unique and important (Owens, 1994). The difficulty here is that the categorization of critical natural capital is inherently subjective, even where informed by scientific expertise that has widespread legitimacy. A site may be categorized as an ecologically important habitat, but this may differ from the value that society collectively places on such a habitat. Where landscape is concerned, many a NIMBY (Not In My Back Yard) campaign has been based around arguments of the uniqueness of an area and its views; this is not so much a question of whether the landscape should be classified as critical natural capital or not but rather whose view of what counts as critical natural capital is prioritized.

For some this problem arises because a range of cultural artefacts is being considered under the heading of critical natural capital when this term should properly be kept for the functioning of global environmental systems. The emphasis should thus be on global warming as a result of the overuse of global carbon sinks, losses in global (rather than local) biodiversity, and the inevitability of entropy reducing the ability to continually reprocess and recycle physical products through the application of energy sources. A focus on these aspects suggests a strong sustainability position in which not only are such environmental assets and services treated as non-substitutable, but a strong case is made for actual limits to economic growth. The promise of ecological modernization is seen as a false promise and instead measures are proposed to enable society to move towards a situation where it can live within the limits of 'Spaceship Earth'. Such measures can include Daly's (1992) draconian birth permits and resource quotas.

Such a strong sustainability perspective favours a Russian doll, as opposed to a Venn diagram, model of the concept (see Figure 35.2; see also Levett, 1998). Here the environmental concerns are seen as at the core of the concept, surrounded by social concerns, and only then economic concerns. The implication is that economic activity depends on social structures and, fundamentally, on environmental systems. Ensuring that economic development is sustainable means taking account of these environmental limits and the social constraints on routine economic decision-making and leads to demands for more radical changes in social organization and economic practices. This then raises the question of how contemporary routines of politics and policy can respond to such a radical agenda and, more broadly, the relationship between governance processes and sustainable development.

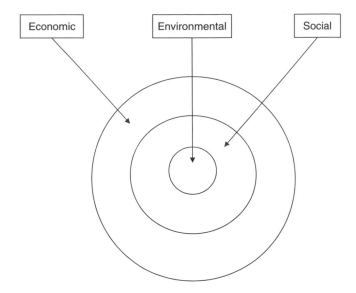

Figure 35.2   The Russian doll model of sustainable development.

GOVERNANCE AND SUSTAINABLE DEVELOPMENT

The pursuit of good governance is sometimes referred to as the fourth pillar of sustainable development, suggesting that sustainable development is also about new ways of doing policy as much as new outcomes. Why should governance be seen as so integral to sustainable development? To understand this, it is necessary to understand the concept of governance too (Rhodes, 1995).

Governance can be understood as a descriptive account of trends in contemporary society affecting the state and the public sector (although there is a parallel trend concerning the corporate sector – see below). At the core of this trend is a different relationship to stakeholders; indeed it implies the identification of a broader range of actors as having a 'stake' in a policy issue. Over the last two decades there has been a shift within many countries and across many international organizations towards relying on stakeholder involvement not just for consultation during the formulation of policy, but in terms of a more meaningful involvement, participating in decision-making and even taking responsibility for the delivery of policy outcomes. This shift arose from the recognition during the late 1970s and 1980s that governments and their agents often did not have the 'capacity to act', to implement effectively and deliver on their policy intentions. Many of the key resources – financial but also informational and capacity building – depended on the involvement of non-state actors. Involvement throughout the policy process could

also avoid conflict at a later stage holding up policy delivery and could provide greater legitimacy for policy outcomes. For all these reasons, stakeholder involvement under conditions of governance became increasingly prevalent during the late twentieth and early twenty-first centuries.

Analytically this has thrown the emphasis on understanding how complex sets of networks interrelate within policy contexts. Such networks describe the interconnections between actors, allowing governance to be characterized in terms of the range and density of connections but also the identification of nodes that are particularly central within any policy process. Analysis can go beyond such description to understand the dynamics involved in networks by considering how resources are activated through the network links. This provides an insight into the power relations implied by governance processes. There is a split here between those who see power as an attribute of actors within governance networks, partly based on the resources they hold, and those who adopt a relational view of power, in which power is based on attributes of the network itself. In either case, though, the combination of network links and resource flows provides the basis for understanding governance processes.

New Institutionalists go beyond this to consider how networks are turned from sets of linkages into institutions with a long-term life (Hall and Taylor, 1996). This involves seeing how the behaviour of actors within networks becomes embedded over time creating path dependencies and thereby guiding outcomes. This is not a deterministic approach to understanding governance but rather one that seeks to combine a focus on actors' behaviour and decisions as well as the context within which they act. Actors remain invested with agency and yet, in general, tend to behave in predictable ways with implications for outcomes and distributive impacts on others. The role of cultural factors such as norms, values, customs and working routines (all discursively mediated; Rydin, 2003) are important in embedding behaviour and creating this context.

But governance has not just been used as a descriptive label or to define an analytic institutionalist framework. It has also been extensively used as a normative perspective. Governance is not only a response to identified problems in the public sector, but also a democratic ideal suggesting a way of empowering communities and other stakeholders within the policy process. And as a normative approach, governance has spread beyond a concern with governmental bodies and policy processes to arenas in the economy and civil society. Corporate governance describes new ways that are recommended for firms to engage with a variety of relevant stakeholders, from shareholders to customers to workers to local communities near sites of production. Just as governance is recommended as a better way to do policy and corporate governance is seen as a better way to do business, community governance has been taken up as a means of empowering local communities and ensuring that they are both effective and operate on the basis of inclusion. In all these arenas, governance is seen as a desired future.

A normative emphasis on governance was apparent within the process of developing the Brundtland Report itself, as mentioned above. The Brundtland Commission operated in an open and inclusive manner, visiting many local communities as well as national governments and taking evidence from a range of actors from scientific experts to local villagers. In its report it emphasized the importance of involving as wide a range of affected or interested parties as possible in building the consensus for a change of direction towards sustainable development. Stakeholder involvement has, therefore, come to be seen as emblematic of all sustainable development politics and policy-making. Beyond the totemic influence of the Brundtland process, there are other reasons why governance is particularly pertinent for proponents of sustainable development.

As a policy goal, the complexity and ambiguity of the concept of sustainable development puts particular pressures on the policy process. It requires considerable effort to specify how the concept is to be defined in specific contexts. The Brundtland definition may inspire but it needs further specification. The above discussion has highlighted that, in the absence of obvious and easy win-win-win options, there is a need to make difficult choices between environmental, economic and social concerns and alternative packages

of outcomes. The very notion of constraining current generations for the benefit of future generations – which is at the core of sustainable development – also places strains on the legitimacy of policy systems, particularly those accustomed to delivering within the electoral time scale of a democratic government. Governance can be a way of trying to handle these pressures and strains.

Turning to implementation, sustainable development requires multiple sets of interventions in different domains in order to achieve the desired mix of environmental, economic and social outcomes. Networks bringing together actors in these different domains can help coordinate these interventions. It can also release resources from varied locations to support implementation. Conventional knowledge sources may well provide weak support given the emphasis on predicting future strains on environmental systems and the inherent uncertainties in such predictions. In such a situation there is a need to obtain buy-in to the scenarios of the future underpinning policy development. There may also be benefits in bringing together information and knowledge from a wide range of sources including local knowledge. For all these reasons, advocates of sustainable development have seen broad and open stakeholder involvement as the appropriate mode of policy work.

Beyond these policy-based rationales, there are also strong spatial reasons why governance processes are particularly appropriate to the pursuit of sustainable development. To understand this, it is necessary to unpack some of the spatial inequalities associated with the social and environmental impacts of economic development. These can be defined at five levels: site-level impacts, locality impacts, *filière* impacts, transport-related impacts and aggregate impacts (see Table 35.1).

Site-level impacts refer to the environmental effects of the differential distribution of economic activity (both in terms of total quantity and types of economic activity) across space. For example, there are the positive environmental consequences of land reclamation for development as opposed to the negative ones associated with deforestation or other primary resource exploitation, or with the abandonment of sites. Such variations in environmental impacts are linked closely to the social impacts, so that negative environmental impacts at this level often go alongside the limited or declining social benefits arising from economic development patterns. Such ties can be very difficult to break, whether they relate to vicious cycles of declining economic investment, increasing environmental degradation and social deprivation, or those associated with economic investment based on low-wage labour and poor environmental conditions.

Next, there are locality impacts, which go beyond the immediate site to consider the effects on the locality and its hinterland, often termed

Table 35.1   Spatial aspects of sustainable development.

	Spatial distribution of environmental impact	Socio-economic dimension
Site impacts	Differential distribution of economic activity and of different types of economic activity create *negative or positive environmental impacts on site*, e.g. deforestation (cf. land reclamation)	Connection between patterns of new development and existing spatial equalities; urban regeneration testifies to difficulties of reversing *vicious cycles* in the location of development; *clustering* of different forms of economic acitvity relevent here
Locality impacts	Patterns of local *negative environmental externalities* are associated with development, e.g. traffic generation, waste dumps	Strongly connected with distribution of social inequalities so that lower-income and certain ethnic groups suffer disproportionately from such negative externalities; tied into *environmental justice* debates
Linked impacts over distance	*Filières* of economic activity that connect places across space to support production and consumption patterns also connect differential environmental impacts, e.g. food consumption in North and food production in South	Raises issues of how development across space is organized and the importance of *spatial distancing of environmental impacts* in creating economically viable development; connects to the spacial distancing of different social groups also
Impacts of transport over distance	The transport dimension of such economic *filières* generates non-local environmental impacts; at a regional scale through migratory pollution and at a global scale through global warming	The spatially distanced nature of economic and social activity is a major cause of non-local and aggregate environmental impacts through transport
Aggregate impacts	Where the patterns of consumption and production in aggregate create environmental impacts, these are then *distributed unevenly* across space; global warming is a key example with its spatially differentiated impacts, say in Scotland and England	The socio-economic dimension is linked not to the physical effect but the *vlunerability*, the ability to cope with change, e.g. Florida and Bangladesh

spatial externalities. Different kinds of economic development give rise to different patterns of local negative and positive environmental externalities. For example, there are the negative externalities associated with development in the form of pollution from traffic generation, noxious industry or waste facilities. Much environmental and urban planning activity aims to manage the spatial pattern of such externalities. The environmental justice movement has shown how, despite or perhaps even because of such planning, the pattern of negative environmental externalities is closely related to the spatial distribution of classes and ethnic minorities. Empirical evidence has repeatedly demonstrated the strong connection between the location of environmental 'bads' and the residency of lower-income groups and communities of colour, so that lower income and certain ethnic groups suffer disproportionately from negative environmental externalities (Agyeman et al., 2001).

These two kinds of impact depend on the spatial proximity of certain economic activity and the associated environmental and social impact. The next category refers to impacts linked over distance between different places. This draws on the notion

of *filières* of economic activity that connect places across space (Amin and Thrift, 1994). Such connections are necessary to support production and consumption patterns so that economic processes in one location depend on parallel processes in other locations. These *filières* also connect environmental impacts in different locations across space. For example, food consumption in developed countries is dependent on the production of such food in less developed countries, whether it is mange-tout peas from Africa or beef fed on soya from South America. There is therefore a link between the environmental impacts of such pea and soya production and the sites of consumption, even though they are spatially distanced. The notion of food chains, linked to the broader concept of global product chains, tries to capture these trans-spatial links. This raises important issues of how development across space is organized and the potential significance of the spatial distancing of environmental impacts from consumption sites in creating economically viable development. This builds on a long-established political economy argument about the reliance of capitalist profit-making on the spatial distancing of different social groups (Massey et al., 1984).

Spatial distancing of production and consumption activities (including intermediate production) further results in the impacts of transport over distance. The environmental impacts of transport are well known. They generate pollution at the local level in terms of poor air quality, at the regional scale through migratory pollutants such as sulphur oxides and ozone, and at the global scale through global warming. The longer the economic *filières*, the greater are the environmental transport effects. Thus there is a strong connection between the environmental and social impacts of linked economic activities over space and the impacts of transport over that same space. One aspect that has received particular attention has been the distance travelled by food consumed, conceptualized as food-miles. One of the major components of analyses of ecological footprints – which seek to draw attention to the environmental impact of cities' and countries' consumption patterns – is the food-miles embodied in collective shopping baskets (Chambers et al., 2000).

Finally, in terms of spatial aspects of sustainable development, there is the distribution of aggregate environmental impacts. This particularly relates to the impact of global warming arising from the aggregation of myriad local decisions about energy use (and, in terms of methane generation, activities such as waste disposal and animal husbandry). The environmental impacts of global warming are and will be distributed unevenly across space. This is a largely physical phenomenon as sea-level rise impacts differently in, say, the north and south of England due to the height of land above sea level and the ongoing tilting of the landmass of the British Isles. However, the social and economic consequences of such environmental impacts are due to the vulnerability of societies and communities and this, in turn, relates to the level of economic development and associated wealth, and the capacity of political systems. Such vulnerability refers to the ability to cope with change, both incremental and catastrophic change. So the differential impact of global warming on Florida and Bangladesh relates not so much to their different physical characteristics but rather the ability of the public and private institutions to manage flooding, coastal erosion and more frequent periodicity of storms.

So, returning to the issue of how governance handles these spatial aspects, it is helpful to examine how the networks of governance handle spatiality. Policy networks focused around sustainable development involve multiple stakeholders and perspectives on the sustainable development concept. They also have to confront the uneven environmental and social impacts of economic development across space. This requires governance processes that are not simply based in one locality or territory. Rather they need institutions that can handle

such impacts over space, negotiating between territorially based organizations to achieve trade-offs or win-win options and enabling collective action at levels above such territorial bodies in the face of the threat of free-riding. And yet existing organizations operate from specific locations, have defined territories and/or are identified with particular scales. Governance seeks to operate across these organizational limitations to link place, territory and governmental tier. In these ways they hope to manage the spatial dimensions of the sustainable development concept.

The development of multi-level governance (MLG) in sustainable development policy is a prime example of how institutional arrangements have developed to try and deal with such fragmentation. In particular, MLG has evolved to ensure that the networks concerned with sustainable development in any one locality or at any one scale have to deal with networks at higher and lower levels. Peters and Pierre (2001) define MLG in terms of the vertical layering of governance at different levels, so that local, regional, national and supranational governmental or quasi-governmental bodies are all involved in policy. The existence of all these layers results in complicated and overlapping networks for any specific policy domain such as sustainable development. But, in addition, Peters and Pierre point to the potential within MLG for organizations within one tier to bypass the immediately superior or ulterior tier and relate directly to bodies in a different and more distant tier. Thus, local municipalities within Europe deal directly with the European Union on some issues (including many urban sustainability initiatives) rather than going through the national governments.

The rationale for MLG is seen by Peters and Pierre to reside in the weakening of the national state by fiscal crises in the later twentieth century, the rising challenge of emerging transnational bodies such as the European Union, the tendency towards decentralization within national governmental systems even if only on an informal basis, and the rise of New Public Management approaches to public administration that favour guidance and steering over direction by governmental bodies. A useful example of the different forms that the imperative towards MLG can take is provided by Kern and Löffelsend (2004) in their study of the Baltic Sea region. Here they identify three different types of MLG institution. There is international governance by nation-states, which can further take the form of international regimes (such as the Helsinki Convention) or intergovernmental cooperation (as with the Council of Baltic Sea States). Then there is international governance with nation-states but also involving other kinds of stakeholders; the example they give is Baltic 21, the Agenda 21 for the Baltic Region, which is a form of international policy network. Finally, there is transnational governance without nation-states as in transnational networks, for example, the Union of Baltic Cities.

These complex organizational arrangements across tiers of government are an attempt to deal with the trans-spatial nature of environmental impacts and of sustainable development. However, the emergence of such arrangements has not been without its problems. Lundqvist (2004) has pointed to the 'trilemma' of policy that MLG seeks to tackle, that is, the mutual attainment of policy efficiency, participation and legitimacy. Examining the case of water management, he contrasts the efforts to achieve ecological efficiency through catchment area-based planning, while also involving a wide range of stakeholders and achieving a legitimate outcome backed by accountable decision-making (traditionally the preserve of elected governmental bodies). He sees these tensions as inevitable within MLG, unresolved by the attempt to create more and more network institutions.

To give a further example: as government actors have extended policy processes to involve those outside the governmental sectors in order to achieve better policy delivery and legitimacy, the very complexity of networks themselves can threaten the ability to deliver and can obscure lines of accountability. Governments have often sought to maintain a degree of control through steering mechanisms such as indicators or targets, often linking achievement as measured by these means to the release of financial resources; this is at the heart of New Public Management. However, a governmental body that is looking towards better performance according to given measures can often be frustrated by the uncertainties of more open stakeholder involvement and, as a result, may seek to limit that involvement. In this way, tensions are built into the governance processes of sustainable development policy.

While MLG is one important form that governance for sustainable development has taken, specifically *local* governance has also been given considerable emphasis within sustainable development policy. The final section explores local governance to show exactly how governance can promote sustainable development in practice.

EXAMINING HOW GOVERNANCE CAN PROMOTE SUSTAINABLE DEVELOPMENT: THE CASE OF LOCAL GOVERNANCE

The rationale for local governance for sustainable development mixes instrumental and democratic imperatives, as is typical of all governance arguments. On the one hand, it can be argued that it is at the local level that there is the greatest scope for actors to become involved in the policy process. At higher scales or tiers there are likely to be barriers to involvement with representatives of groups, sectors or communities favoured over the individual citizen; at the local level the individual citizen may participate directly without the need for a representative acting on his or her behalf, (although this point can be overstated since representation may be inevitable even at local levels). More practically, it is argued that participation by larger numbers of actors is more likely to be successful at the local level. On the other hand, instrumental reasons for participatory governance are that implementation of sustainable development policies will be facilitated. For example, awareness-raising will be more effective at the local level and this will in turn spur on the behavioural changes that are needed for more sustainable consumption and production patterns. Furthermore, local networks embedded through governance processes may directly generate sustainable patterns, particularly where there is direct local engagement with natural resources.

Local Agenda 21

The above rationale lies at the heart of the Local Agenda 21 (LA21) movement. LA21 has its origins in Chapter 28 of Agenda 21 (which, as discussed above, was tabled at the 1992 Rio Summit), setting out the importance of local communities in actually delivering sustainable development (Lafferty, 2001). However sustainable development is defined, it can ultimately only be delivered by changes in the behaviour, decisions and actions of households, firms, NGOs, governmental and quasi-governmental bodies at the local level. Therefore each locality should have its own Agenda 21. In the early days of LA21 the emphasis was on the process involved, with a concerted attempt by proponents of the LA21 movement to create a new form of local policy process. Consensus-building, visioning and awareness-raising were given central place, with considerable effort going into a variety of events to bring sustainable development to the notice of the local community at large. Schools were often targeted as a way of reaching local households. Events were designed to be enjoyable rather than just educational and the energies of local environmental NGOs were harnessed to create an atmosphere of local activism.

Some localities took a more structured approach with sectoral workshops being set up to examine particular topics and issues. These workshops could focus on the key stakeholders, so that, for example, the transport working group could bring together public transport providers, the police, traffic management within the local authority, planners, cycling organizations and consumers' groups. But there was almost universal agreement that the broader public needed to be engaged too. The exact form that this public involvement took depended on the role that local government adopted. In some localities – principally municipalities in developed countries with a strong tradition in local environmental management – the process could tend towards consultation. At the other end of the spectrum, though – principally in localities where the local authority had little legitimacy or record of efficacy – local communities sometimes took control of the LA21 process and used it as a means of community empowerment, demanding involvement in and even control over new areas of local policy. And along the spectrum were attempts by local authorities to engage with their communities in a more meaningful way to deliver sustainable development (LASALA, 2001).

To explain how LA21 can be effective, the nature of network relationships within governance requires a little more explanation and, in particular, the focus needs to turn to the situation of actors within networks. Networks have been introduced as sets of relationships between actors, operationalizing resources and embedded in norms and routines. Actors find themselves at the intersection of resource flows from multiple directions. Their behaviour and decision-making will be a response to the impact of these resources on their position, their interests and values in the context of the cultural aspects captured by norms, values, routines, and so on. The term 'incentive structure' describes the set of factors that shape such behaviour and decisions from the actor's perspective. It is important to emphasize that, tied up in the idea of an incentive structure, are material resources, power and interests but also the range of cultural attributes.

Such an analysis of governance can be very helpful in explaining how governance works in specific circumstances and, in particular, the level of participation that is achieved. From this perspective, participation is a result of actors weighing up the costs and benefits of participating in a particular institutional context. The costs will include the time, effort and monetary expenditure involved in participation; all these costs are likely to be certain, impacting in the present day and falling directly on the actors that decide to participate. The benefits will include the likely outcomes of the collective participation effort; these are likely to be uncertain, arising in the future and to benefit a larger, possibly ill-defined group. The temptation to free-ride rather than be actively involved in the governance effort will be considerable. Such an explanation defines the collective action problem that constrains participation and hence the success of governance processes.

This explanation also offers insights into how to overcome the collective action problem and enable participation (Rydin and Pennington, 2000). Successful strategies would focus on reducing perceived costs and raising perceived benefits; for example, paying childcare costs, setting events at the most convenient time for participants, making events more enjoyable, publicizing the likely benefits to arise from participation, and focusing on issues with an immediate benefit to participants (at least in the beginning). From this partial list it can be seen that encouraging involvement in governance for sustainable development faces particular problems since many of the benefits are likely to favour future generations and spatially distant communities. It takes a very strong value base among participants to discount the costs relative to the benefits and overcome the collective action problem in such instances. Proponents of sustainable development have argued that it is best to start from more local concerns (such as the quality of local green spaces for recreation and nature conservation) and then move outwards to more global and trans-spatial concerns (such as global biodiversity). They place their faith in the possibility of transforming the local to the global.

For example, the environmental justice movement has been criticized for being based in NIMBY (Not In My Back Yard) politics, in that the affected communities are essentially concerned to remove locally unwanted land uses – in another acronym, known as LULUs – such as toxic waste sites from their locality. They have not traditionally been concerned with where these LULUs might go. However, environmental justice activities have more recently sought to use debates about the local location of such facilities to raise the more general question of how contemporary economic production generates waste and to argue for more sustainable modes of production rather than just shifting the location of waste sites (Harvey, 1996). A similar transition occurred in nuclear planning in the UK in the twentieth century where local communities supported by environmental groups used debates about the planning of nuclear installations to raise questions (often against the wishes of central government) about the need for nuclear power at all and the links to the nuclear arms race. Such a transition from the local to the global is not guaranteed, however, and there remains the important prior stage of enabling collective action within the local community first.

One way of encouraging collective action that has received considerable attention has been the generation and embedding of social capital. Social capital as a descriptive term suggests a specific kind of network, one with dense relationships, characterized by recognition of the mutual interest between actors and reciprocity in their dealings with each other. Pretty and Ward (2001) refer to four aspects of social capital: relationships of trust, which enable cooperation and reduce transactions costs between actors; reciprocity and regular exchanges between actors; common rules, norms and sanctions; and connectedness between actors through groups and networks. As a normative term, social capital theorists argue that creating such dense, mutual and reciprocal networks can overcome collective action problems. The key to this lies in the way that norms and values can be built into networks through repeated interactions between actors, so as to alter the incentives facing individual actors within the network. Establishing norms of mutuality and reciprocity means that actors who engage in collective action develop a positive reputation while those who free-ride will suffer the soft sanctions of loss of reputation and be 'shamed and blamed'. These cultural aspects of networks can be very effective in preventing free-riding. In addition, such institutions can develop norms for handling conflict and instances of free-riding through monitoring and enforcement by the network itself rather than an outside agency, thereby maintaining high levels of collective action.

COLLECTIVE ACTION FOR SUSTAINABLE RESOURCE MANAGEMENT

In addition to its role in enabling collective action in terms of stakeholder participation in sustainable development politics and policy, Ostrom (1990, 1992) has argued that social capital can enable direct community management of local resources in a way that contributes to sustainable development. For example, she has proposed developing local institutions for water management based on social capital. She has particularly emphasized how collective action can be developed in smaller and more homogeneous communities in a way that enables robust and sustainable management of resources. This argument is based on the quasi-public-good characteristics of many environmental assets and services, whereby individual actors exploiting the resource in their own interest collectively result in the degradation of the environmental asset or service. Collective action to prevent such collective degradation is subject to free-riding, in that each individual actor sees it as in their interests to continue to use the environment rather than collectively manage this exploitation; a free-rider reason that 'others can regulate their use but I will benefit from continuing to use it as before'. Proponents of social capital, however, argue that building up social capital prevents this collective action problem and enables collective sustainable management of the environment, particularly through

the opportunities enhanced by social capital for monitoring and the use of soft sanctions.

Social capital theory thus suggests that building new institutional arrangements about networks specifically oriented towards mutuality, reciprocity and trust can be an effective way of overcoming collective action both in terms of sustainable natural resource management for local communities dependent on such resources and in terms of involving a broader range of actors more effectively in general policy efforts for sustainable development. The key to success here is creating new incentive structures for actors involved in such networks. These will be shaped by the detailed design of the institutions, the precise extent of the networks and the way that working routines for the network embody certain values and norms. Recently, social capital theory has emphasized the different kinds of social capital that may be used in such institutional design.

Most emphasis has traditionally been placed on bonding social capital, which is used to create strong ties between a delimited set of actors. Where such actors form a homogeneous and relatively small group, then experience suggests that bonding capital can be most easily fostered. However, this can be at a cost. Strong ties between members of the group may reduce the possibility of links with actors outside the group. This can be useful where the management of a local resource – say, a village watercourse or grazing ground – is at issue, but less useful when dealing with complex goals such as sustainable development where links between a variety of stakeholders of different kinds and not all with local stakes will be needed. There is also the potential for strong bonding capital to have a 'dark side' through cementing institutional arrangements that effectively terrorize local communities, as in Mafia or terrorist organizations.

Hence, there has been the suggestion that an alternative form of social capital, known as bridging capital, may be needed. Bridging capital is also known as weak ties and is used to bring unlike actors together, from different communities or groups. Many researchers have seen bridging capital as a useful complement to bonding capital. For example, it may be used to tie in NGOs organized on a non-local scale with local communities. However, bridging capital is very weakly conceptualized and can amount to little more than a network linkage out from a group or community. Usually no limits are placed on the extent of bridging capital and no explanation provided of how its efficacy is affected by the length or character of the linkage. These limitations of the bridging capital concept have led to the suggestion that a third type of social capital is often needed, particularly in stakeholder engagement of the type found in strategic planning for sustainable development (Rydin and Holman, 2004).

This third type is called bracing capital and it describes the kind of linkages found in policy contexts where a limited set of varied actors need to be bound together in collective action. Such actors form a delimited set and therefore unlimited bridging is inappropriate. They are a differentiated group so that bonding on its own will not be effective. What is needed in a mix of strong and weak ties within this delimited and differentiated set of actors in order to bring about collective action, in terms of agreement on working routines, specific policy outputs such as strategy documents and eventually common norms and values on policy direction. Careful institutional design is needed to create this type of bracing capital, but it holds the potential for underpinning collective action towards sustainable development.

That said, the emphasis on building institutions based on social capital as the favoured form of local governance for sustainable development is not without its limitations. There is always the danger that practice will fall short of the ideals of institution-building and tend instead towards routine consultation. In such circumstances the perceived need to build a consensus can lead to a 'rush to the bottom', with the lowest common denominator of agreement between actors being used to generate a weak and rather meaningless consensus. If strong local governance institutions are generated, they may be subverted to ends other than sustainable development and the ambiguities in the sustainable development concept used to disguise this. Thus, urban growth coalitions can adopt the language of ecological modernization to present themselves as promoting sustainable development, when the emphasis is much more on a continuation of past policy trends towards urban boosterism than on any breakpoint towards sustainable development. Then there is the considerable effort, including resources, required to create such institutions and, furthermore, to maintain them. More complex forms of networks based on social capital may not actually reduce transactions costs but actually increase them, reducing policy efficacy. And finally, it may not be the case that all sustainable development problems can be most effectively tackled through local institution-building in this way, based on the involvement of communities and stakeholders (Rydin and Falleth, 2005).

That said, the emphasis on building institutions based on social capital as the favoured form of local governance for sustainable development is not without its limitations. There is always the danger that practice will fall short of the ideals of institution-building and tend instead towards routine consultation. In such circumstances the perceived need to build a consensus can lead to a 'rush to the bottom', with the lowest common denominator of agreement between actors being used to generate a weak and rather meaningless consensus. If strong local governance institutions are generated, they may be subverted to ends other than sustainable development and the ambiguities in the sustainable development concept used to disguise this. Thus, urban growth coalitions can adopt the language of ecological modernization to present themselves as promoting sustainable development, when the emphasis is much more on a continuation of past policy trends towards urban boosterism than on any breakpoint towards sustainable development. Then there is the considerable effort, including resources, required to create such institutions and, furthermore, to maintain them. More complex forms of networks based on social capital may not actually reduce transactions costs but actually increase them, reducing policy efficacy. And finally, it may not be the case that all sustainable development problems can be most effectively tackled through local institution-building in this way, based on the involvement of communities and stakeholders (Rydin and Falleth, 2005).

There is an argument to be made that in some cases state direction, more reminiscent of government than governance, is a more effective way to change currently entrenched patterns of unsustainable consumption and production (Rydin and Pennington, 2000). Sometimes the collective action problem is just too considerable and the urgency of the policy goal is just too great to make it sensible to persist in governance efforts. States have implicitly continued to recognize this

in their ongoing promotion of regulation as a basis for environmental and social protection. The benefits of using this approach sometimes should not be forgotten in the contemporary and often overwhelming emphasis on governance.

CONCLUSION

One of the strengths but also key weaknesses in the sustainable development concept is the way that it seeks to combine very different goals within one holistic package. This is bound to lead to alternative definitions of what sustainable development comprises when considered in detail and consequent ambiguity and confusion within policy processes. The open involvement of many different actors within structures of governance provides the opportunity for each actor to press for a definition that accords more closely with their interests and values. The concern of those seeking a greater emphasis on environmental protection or social equity is that inclusive governance processes may therefore result in 'business as usual' or a dilution of the radical potential of the concept.

Opening up sustainable development policy and politics to governance structures does not resolve the tensions between these different views of the concept. Governance merely provides an opportunity for different views to engage with each other and the outcome need not be a pluralist consensus or even compromise. Each governance process carries with it certain institutional characteristics including prevailing norms and values. These may be conducive to a more radical interpretation of sustainable development and associated social change, but there are good reasons to expect that governance institutions will be imbued with norms that constrain such interpretations and favour an emphasis on marginal change to existing patterns of social and economic behaviour. The institutionalist approach emphasizes the likelihood of actors following established patterns of behaviour under the influence of a prevailing 'logic of appropriateness' that constrains actors to behaving in ways already considered appropriate within that institution. Changing institutional logics and establishing new norms is like altering course with an oil tanker, a slow process. Opening up structures to the influence of multiple stakeholders will, in the initial stages, result in these stakeholders engaging with each other in the context of established norms, and this may favour policy options such as ecological modernization rather than more radical deep green alternatives. The result may be future outcomes that prove to be unsustainable as environmental assets are substituted or traded off in pursuit of economic and social gains. This debate continues most vividly around the implementation of the Kyoto Protocol in terms of just how much constraint on economic growth is justified to tackle greenhouse gas emissions. Governance offers no easy resolution of these debates.

At root, sustainable development requires a political, ideological and values-based shift, which new institutional arrangements may help to consolidate but cannot substitute for. Without such a shift, policy institutions will tend back to mere incremental change or even relapse to a 'business as usual' scenario. Moving beyond this will require an iterative process of value shift. Of course, if everyone agreed that a particular form of sustainable development was desirable, then policy actors would just be left with the problem of implementation. The challenge is that there are contested views over what sustainable development is, how important it is, and how desirable any particular form is. Fundamentally, governance for sustainable development has to engage with these different views and with the underlying ideological debates.

The power of prevailing social ideologies is considerable, as cultural theorists, social constructivists and Foucauldians have long told us. Sustainable development is clearly not an embedded social ideology, despite its emblematic use in many policy contexts. Value shift will require a discursive challenge to prevailing ideologies or discourses. Some of this challenge will occur in the detail of developing institutions, specific networks and ways of working among actors. But there is also a broader canvas on which the battle for sustainable development has to be waged. This involves the interaction, through everyday communication between actors, of different discourses about how social, economic and political lives should be organized and how they engage with the environment (Rydin, 2003). Any consideration of the future for sustainable development policy and politics will need to take on board that the development of governance processes must be part of a broader politics arguing the need for change.

REFERENCES

Agyeman, J., Bullard, R. and Evans, B. (2001) *Just Sustainabilities: Development in an Unequal World.* London: Earthscan.

Amin, A. and Thrift, N.J. (eds) (1994) *Globalization, Institutions and Regional Development in Europe.* Oxford: Oxford University Press.

Chambers, N., Simmons, C. and Wackernagel, M. (2000) *Sharing Nature's Interest: Using Ecological Footprints as an Indicator of Sustainability.* London: Earthscan.

Daly, H.E. (1992) *Steady-State Economics.* London: Earthscan.

Gouldson, A. and Murphy, J. (1998) *Regulatory Realities.* London: Earthscan.

Hall, P. and Taylor, R. (1996) 'Political science and the three new institutionalisms', *Political Studies,* XLIV: 936–57.

Harvey, D. (1996) *Justice, Nature and the Geography of Difference.* Oxford: Blackwell.

Independent Commission on International Development Issues (ICIDI) (1980) *North–South: A Programme for Survival.* London: Pan Books.

International Union for the Conservation of Nature (IUCN) (1980) *World Conservation Strategy.* Gland, Switzerland: IUCN.

Jacobs, J. (1991) *The Green Economy.* London: Pluto Press.

Kern, K. and Löffelsend, T. (2004) 'Sustainable development in the Baltic Sea region. Governance beyond the nation state', *Local Environment,* 9(5): 451–68.

Lafferty, W. (2001) *Sustainable Communities in Europe.* London: Earthscan.

LASALA Project Team (2001) *Accelerating Local Sustainability – Evaluating European Local Agenda 21 Processes,* Vol. 1. Freiburg, Germany: ICLEI.

Levett, R. (1998) 'Sustainability indicators: integrating quality of life and environmental protection', *Journal of the Royal Statistical Society,* 161(3): 291–302.

Lundqvist, L. (2004) 'Integrating Swedish water resource management: a multi-level governance trilemma', *Local Environment,* 9(5): 413–24.

Massey, D. and Allen, J. with Anderson, J. (1984) *Geography Matters!* Cambridge: Cambridge University Press with the Open University.

McCright, A. and Dunlap, R. (2003) 'Defeating Kyoto: the Conservative movement's impact on US climate change policy', *Social Problems,* 50(3): 348–73.

Neumayer, E. (2003) *Weak Versus Strong Sustainability: Exploring the Limits of Two Opposing Paradigms* (2nd edn). Cheltenham: Edward Elgar.

Ostrom, E. (1990) *Governing the Commons: The Evolution of Institutions for Collective Action.* Cambridge: Cambridge University Press.

Ostrom, E. (1992) *Crafting Institutions for Self-Governing Irrigation Systems.* San Francisco: ICS Press.

Owens, S. (1994) 'Land, limits and sustainability: a conceptual framework and some dilemmas for the planning system', *Transactions of the Institute of British Geographers,* 19(4): 439–56.

Pearce, D., Markandya, A. and Barbier, E. (1989) *Blueprint for a Green Economy.* London: Earthscan.

Peters, B.G. and Pierre, J. (2001) 'Developments in intergovernmental relations: towards multi-level governance', *Policy and Politics,* 29(2): 131–35.

Pretty, J. and Ward, H. (2001) 'Social capital and the environment', *World Development,* 29(2): 209–27.

Rhodes, R. (1995) '*The New Governance: Governing Without Government*', Swindon: ESRC.

Rowlands, I. (1995) *The Politics of Global Atmospheric Change.* Manchester: Manchester University Press.

Rydin, Y. (2003) *Conflict, Consensus and Rationality in Environmental Planning: An Institutional Discourse Approach.* Oxford: Oxford University Press.

Rydin, Y. and Falleth, E. (eds) (2005) *Networks and Institutions for Natural Resource Management.* Cheltenham: Edward Elgar.

Rydin, Y. and Holman, N. (2004) 'Re-evaluating the contribution of social capital in achieving sustainable development', *Local Environment*, 9(2): 117–34.

Rydin, Y. and Pennington, M. (2000) 'Public participation and local environmental planning: the collective action problem and the potential of social capital', *Local Environment*, 5(2): 153–69.

Weizsäcker, E., Lovins, A. and Lovins, L. (1998) *Factor Four: Doubling Wealth, Halving Resource Use*. London: Earthscan.

World Commission on Environment and Development (WCED) (1987) *Our Common Future*. Oxford: Oxford University Press.

Zadek, S. (2001) *The Civil Corporation: The New Economy of Corporate Citizenship*. London: Earthscan.

Urban Governance in the South: The Politics of Rights and Development

Susan Parnell

INTRODUCTION

Failed states, crisis states, resurgent states, developmental states, multi-scalar states: there can be no mistaking that 'the state' is back in fashion in urban studies. This chapter is informed by this statist trend in urban governance, more generally by James Scott's *Seeing Like a State* (1998) and also by the growing recognition of the importance of state devolution and the power of the sub-national state under conditions of globalization (Batley and Larbi, 2004). While much of the Western literature on the state is concerned with the impact of neo-liberalism and the relative decline of the power of the nation-state on growth agendas in poorer countries, the emerging debate is around the capacity of government (national or local) to provide basic services (water, waste, electricity, planing) to households (Brenner, 2004; UN Habitat, 2006). This chapter explores this relatively neglected dimension of urban governmentality: that of the imperative of building state capacity to deliver services to all residents through a radical programme of state reform that implies addressing informality by making poor households legible to the state. The modernist agenda of formalization is introduced alongside the rights agenda because it is only once it is possible for the state to reach or target the poor that is it possible to implement any redistributive agenda and to secure subsidized service provision.

Until recently, realizing a vision of universal citizenship with equitable access to affordable services in most cities in the South seemed aspirational and fanciful. Indeed, the assumption that there should be universal access to state-supported basic services in all cities of the world is new in policy circles. The rise of a universalist agenda is informed by a number of different factors. These include: the practical and ideological bankruptcy of neo-liberal urban reforms; the economic prosperity in middle-income nations that makes possible an urban redistribution agenda; the political shift to the left in the global South, especially across Latin America; global endorsement of the Millennium Development Goals; and the increasing international threat of public health scares such as SARS or bird flu emanating from poorly serviced urban slums. New efforts to get cities in the developing world to mitigate against climate change and introduce adaptation actions that could pre-empt the predicted physical and livelihood damage associated with increases in extreme events compound the pressure for a more prescriptive and interventionist state role in urban development.

The call for a more developmental state agenda at the city scale is not without its contradictions. In practice, poorer cities are faced with overwhelming poverty, widespread mismanagement and pervasive informality. Despite this reality, an essentially liberal or development discourse on 'the Good City', based on notions of a state that guarantees individual rights and basic service standards, is fostered by international agencies such as the World Bank, UN Habitat, Cities Alliance and the donor community. Following their advice (and resources), civil society and national governments are encouraged to embrace the values of democracy, good governance and participation at the city scale, even while more transformative routes to global and local equity, justice and sustainability are side-lined (Harrison, 2006). Thus, while the 11th Millennium Development Goal gives specific attention to the problem of combating the urban slum (UN Habitat, 2006), the root causes of urban poverty and the structural frameworks of inequality and the denial of human rights in urban Africa, Asia and Latin America are not generally confronted.

At the same time, the blanket radical condemnations of global capitalism and neo-liberalism that characterize much of the academic material, while making compelling moral and intellectual points, have done little to address the indignities of urban poverty or shed light on how the goals of 'the Good City' can be made more tractable. The conundrum of the progressive urbanist working on or in the South in this policy and intellectual environment is that many even well-informed scholars simply do not seem to know what advice to give on how to improve the lives of the urban poor. Caught between radical analyses that fail to engage city specificities and off-the-shelf managerialist solutions, urban scholars are often unable to offer progressive insights to those who live under or manage the unjust and exclusionary urban conditions of today.

Friedmann (2000) argued that, for workable alternatives that have a chance of impacting positively on the lives of urban citizens, scholars must venture beyond the abstract aspirations of what a good or a just (or rights-based or a sustainable) city could and should be like. To move closer to achieving our utopian visions we have no choice but to (re)enter the contested and messy terrain of urban politics and administration alongside organized communities, city officials and politicians to rethink the work of the state. These are the trenches where, armed with a clear local understanding of the nature of poverty, inequality and unsustainability, plus the wisdom of past and current thinking about the urban condition, we can engage those actors and forces that do influence life in the city. This activist intellectual role is especially important in the global South where the sites and regimes of power are transient, often informal, generally poorly recorded and thus difficult to read without extensive local involvement and formal and informal political access (Swilling et al., 2002; Parnell, 2004). Drawing outwards to abstract from those institutional places where progressive innovation is successful must, alongside the ongoing fight to level global economic and political

playing fields, be a key agenda for urban studies in the twenty-first century. This re-entry into planning is, as Watson (2006) reminds us, the cutting edge of the competing rationalities of the contemporary city, making a value-based emphasis on process, outcome and critical reflection the only transformative alternative to the neo-liberal hegemony in urban management.

I want to suggest, then, that established radical and liberal approaches to justice in the city are simplistically depicted as intellectually and politically binary positions (Harvey, 1973). So, in focusing on the institutional scale of the city, I follow Katznelson in his assessment of Harvey's intervention to the urban debate with *Social Justice and the City*, that the impact of his bringing radical political economy to the fore was that:

> ... urban studies is burdened with a great silence in the area of institutional political analysis, including work on rights, constitutionalism, representation and law. These omissions cost us a great deal, including the possibility of crafting persuasive accounts of how people who live in cities unregulated by the kind of strong moral compulsion that governed and standardized behaviour in medieval Christendom actually can live together in spite of vast differences in condition, interest and value. (Katznelson, 1995: 48).

I suggest therefore that, especially in the global South and at the metropolitan scale, the imperatives of attending to the institutional (re)design of civic systems that frame citizens' access to socio-economic rights conferred through the ballot must occur in parallel with macro-structural changes such as in cycles of capital investment and in trade regimes. I therefore break with Harvey in two ways. First, to argue for the (radical) reformist imperative of poverty relief through applying principles of distributional justice to basic service provision where full welfare support is unaffordable. Second, I argue that redistributive reforms can and should take place at the city as well as the national (and international) scale. It is this emphasis on the developmental imperative of the sub-national state and the nation-state's developmental role at the intra-urban scale that holds the key to removing the structural inequalities that often manifest in the polarized realities of formality versus informality in the cities of the South.

Given that the demand for affordable basic urban services seems certain to dominate international urban policy debates in the twenty-first century, it is surely important that greater prominence be devoted to identifying mechanisms or instruments that will allow states in the South, where there is no history of universal service access, to deliver on the demand for the right to water and shelter.

This involves an intellectual engagement with the meaning of 'rights' at an urban scale and in contexts of limited resources and weak state capacity, especially at the sub-national scale. The next section of this chapter reviews the basis for a rights-based policy agenda in poorer cities, and the implications of such an agenda for the practices of local governance. Some of the lessons from the South African experience are introduced here, while the practicalities of implementing an indigence policy for service subsidization in Cape Town are discussed in the following section in order to exemplify these intellectual and policy challenges. For political geographers, a vast and urgent research agenda around the politics of urban citizenship and governance is signposted by these debates coming from the challenges of urban service provision in the global South.

RIGHTS AND URBAN DEVELOPMENT IN THE SOUTH

Perhaps the most useful general entry point for rethinking the utopian vision of the city in the South as a good city is the concept of a rights-based city. It is this essentially liberal democratic notion of rights that informs the acceptance of Western-style democracy in municipalities, as much as in nation-states. The debate about the realization of rights in the cities of the global South has not progressed much and is largely restricted to what Peres and Stokke refer to as first-generation rights of voting, gender and personal safety (Jones and Stokke, 2005). But this is set to change as the endorsement of universal human rights is embedded in the Millennium Development Goals (MDGs) and applied to the neighbourhood and city scale (UN Habitat, 2006). The practical realities of realizing the MDGs at the city scale will, given how much must change in local government especially, unleash a fundamental rethink of urban management systems and practices from a rights-based perspective. It will, for example, be essential to introduce fundamental institutional changes in how cities are run so that they are able to effect the necessary delivery of services to the (urban) poor implied in the MDGs (UN Habitat, 2006).

In line with international endorsement of the rights-based agenda, debate on the reform of the state in poverty reduction in cities of the global South has already begun (cf. Batley and Larbi, 2004). But, insofar as there is a coherent discussion about the realization of rights through the redistributive role of the state at the urban scale beyond the global North, it has been directed primarily at services like water, waste and electricity and has focused on a critique of privatization as a mechanism for equitable delivery (McDonald and Pape, 2002). In part it is the challenge of making basic services reach everyone in the city, including those that cannot pay full cost recovery, that has seen the resurgent interest in the role of the state, even from the World Bank (2006). What is truly significant about the new form of the urban service debate is the changing assumptions that inform it. Ten years ago, notwithstanding the 1986 signing of the Universal Declaration of Human Rights, there was no consensus that basic services were a right of all urban residents, regardless of who provided them or where in the world they lived. Now the rights of slum dwellers across the world are being affirmed, if only on paper (Hasan et al., 2005). The politics today is around how best to plan cities and run their administrations to ensure that basic services reach those who are unconnected and also those who cannot pay for the longer-term service charges.

The idealism of the 'Good City' in the South is one that now embraces universal service access so that basic human rights are realized. Given the resource base of most Southern cities, the fiscal implications of basic service cross-subsidization introduce the imperative of increasing the proportion of the population who are economically independent. Finding jobs or secure livelihoods is even harder than making sure that every urban resident has access to water and sanitation or health care, making delivery on the promise to roll out socio-economic rights something of a mirage for most governments. I want to argue, however, that there is some hope in defining the way ahead for a rights-based city, and that this has implications for all cities, not just those in the global South.

Implicit in the notion of universal human rights is that, at some fundamental level, all people can be treated the same and/or have access to the same form of basic opportunities and services. The contemporary urban public policy agenda is forged around translating these abstract ideals into action and to directing resources and institutional arrangements to delivering on this rights agenda. So what may appear as mere policy debates around the roll-out of urban social safety net provision and the creation of the institutional conditions necessary to facilitate universal citizenship in cities of the South, in fact lie at the core of much wider, less place-bound understandings of what might constitute a generalizable model of a good city.

There are three imperatives that drive us to confront a generally applicable meaning of the 'Good City' from a Southern perspective. First, the lacuna around a contemporary progressive urban agenda pertains most obviously to the context of the global South, where theories of how to shape conditions of urban justice and sustainability have yet to be grounded in the fluid institutional realities of the politics and administrative capacity of rapidly urbanizing societies. Second, it may be that

Southern praxis has something to offer cities in the North where institutional exclusion is more prevalent as xenophobia and illegal migration increase and where service debates assume poverty, not inequality, to be the substantive issue and so municipal service practice dictates only minimum, not maximum, standards. Globalization means that the differences between North and South are receding as Northern cities experience increasing inequality as the already fragile welfare state is challenged by privatization and cutbacks and its redistributive impact is eroded and as migration is associated with increasing concentrations of poverty in the North.

Third, natural resource constraints on urban consumption mean we simply cannot all consume water, energy and land or produce waste at the levels currently associated with the urban rich. In other words, if we are to achieve equity at the intra- and inter-city scale, it will not only be by raising the living standards of the poor though subsidies or development programmes, but also by reducing the consumption rate of the wealthy. If every city in the South is to roll out universal service coverage, resource limits will have to be built into the conception of how states will sustainably achieve these goals. Defining the limits to resource consumption, like the definition of minimum environmental standards, rests on the notion of universalism – in other words, what is it we can and should all be able to afford and what is it that the natural and financial resource base can sustain? These questions have not been addressed globally, nor can they be because we do not know what universalism might mean in the poorer cities of the South that are, increasingly, the most important settlement segment of the global population (National Research Council, 2003).

Getting the local state in the global South to implement a universal service delivery agenda is not a panacea for inequality or the solution to urban poverty. But it is a critical step to a better, more just world. The uneven legacy of capitalist development, the unpredictability of investment cycles, current fluctuations and the structure of global trade are ultimately more important determinants of life in Cairo, Kolkata and Cape Town than who wins the next local election or how efficient the local Council waste collection team is. In addition, even if we were to see an instant shift towards justice in the global political economy, there are few cities in the South that are institutionally equipped to respond to the challenges of rapid urban growth, service backlogs and meeting the needs of their many poor residents (Devas and Rakodi, 1993). In simple terms, city governments lack the will and/or the capacity to reach the poor with affordable services; because residents are not legible they cannot be seen or targeted by the state, and thus escape not only the gaze of control,

but also of beneficiation. Addressing even the current urban service problems, let alone the service challenges posed by urban growth, necessitates a closer engagement with the capacity and role of the state, and especially of the local state, in reaching citizens.

There is a convergence in thinking that places the state at the forefront of service provision in cities of the South. On the one hand, the patent failure of structural adjustment programmes that had sought to get government out of service provision made even the World Bank acknowledge that stateless development was a non-starter (World Bank, 1997). Batley and Larbi (2004) argue that in practice what has happened is that by backing away from the fundamental tenets of public choice theory, the big international donor organs are now plying a redistributional agenda directly to nation-states. This is done largely through the promotion of pro-poor public management support. More overtly a liberal democratic political agenda is secured by getting governments to sign up to the Millennium Development Goals and/or adopt a Poverty Reduction Strategy Programme (PRSPs are a loan conditionality for all highly indebted poor counties). City Development Strategies provide the urban-scale articulation of pro-poor policy that places greater emphasis on the role of local government (Plummer, 2002). Significantly, however, there is an almost total silence from the same advisors and donors on the problems of establishing unfunded mandates for local governments whose revenue base, unlike more affluent cities in the North, is limited by the relatively small size of the middle and upper class. But notwithstanding the difficulties of generating state revenue for urban redistribution, privatization's widespread failure to deliver comprehensive and affordable basic services has sent policy-makers back to the drawing board, and in many instances back to government itself. Given the increasingly urbanized world, the revival of the role of the state implies an increasing emphasis on the local state.

The return to a more statist model of delivery finds easy favour in many post-colonial nations. Where political commitment to poverty reduction remains high, restructured state subsidy programmes based either on improved efficiencies (the focus of neo-liberal reforms) or on local redistribution are generally identified as the primary mechanisms for meeting basic needs. In this regard, the problem with reforms in government delivery that are grounded in the new public management strategies that seek to recover costs of service provision is that, aside from being inspired by the experiences of OECD nations, they are often prescriptive and insensitive to local political realities and ignore the fact that many of the urban poor simply cannot pay (Batley and Larbi, 2004). Moreover, they focus on the internal organization of the bureaucracy, the relationship between the state and service providers and on citizens only as consumers. There is little effort to secure basic human rights, and no institutional concern for any of the inclusive ideals of universalism that characterize Rawls' (1971) descriptions of distributional justice. Accordingly, states seeking to change a status quo that marginalizes the poor from service access because they cannot be identified must secure both a political commitment to inclusivity and rights and then create the institutional and operational conditions for this universalism. This is generally harder to achieve at the local than the national scale, not least because of the problem of unfunded mandates. But emphatically, exclusion is institutional as well as financial as the absence or weak functioning of administrative systems necessary to implement subsidies or other forms of redistribution precludes allocating resources that would allow the poor access to basic services. Thus, even in relatively wealthier cities where some resources do exist for redistribution, universal ideals are not able to be realized.

By way of example, and from my privileged position living and working in one of the most beautiful, yet intellectually and morally challenging locations of the urban frontier, I draw on my own experience in the process of the reconstruction of the South African city after apartheid. Debates about rights and universal service delivery in Cape Town have taken us, among other places, to a general commitment to providing affordable urban services to all residents (Smith and Hanson, 2003) and specifically to the design of an 'indigent' policy for the City of Cape Town.[1]

What is significant about the Cape Town indigent policy, as the French geographer Jaglin (2004) points out, is that it symbolizes a watershed in South African thinking about how to effect reconstruction, redistribution and integration in cities that are historically unequal and where a large proportion of the population has been institutionally marginalized from receiving state delivery of subsidized services. The emphasis here is on the issue of how the principle of universalism in service delivery that rests on some form of subsidy to the poor is to be achieved within a restricted municipal budget. The new Indigent Policy[2] (NIP) is at pains to point out that it is not viable to simply deracialize service provision and service subsidies by elevating all residents to the service standards enjoyed by the white urban elite. This was an important break with other social security transfers such as the pension, which was made universal through the elevation of the level of grant payments to all, such that Africans who had traditionally received much lower payments, enjoyed substantial increases (Seekings and Nattrass, 2005). Implicit in the NIP was the view that because the restructuring of municipal services could not be sustainably delivered to all

residents by using a deracialized apartheid formula, an alternative, more radical form of a neo-welfare local state had to be crafted (CCT, 2003).

The idea of building a post-apartheid South African city on the principle of universalism draws directly from the South African Constitution (Statistics South Africa, 1996). Part of the difficulty, as we will see from Cape Town, is that the notion of universalism is only ambiguously defined in South Africa. Universalism is variously interpreted as equal access, equal opportunity, equal benefit, equal levels of service and so on. In practice, what this ambiguity has meant is that the poorly resourced local government sector, which raises 90 per cent of its own revenue, has had to be more inventive in giving the principle of universalism practical meaning at the urban scale. While South African levels of local government self-financing may be unusually high, the issue of unfunded mandates is common to contexts where devolution and globalization combine to make cities critical sites of political control and intervention.

Jaglin (2004) is one of the first academics to take up the challenge of defining the practical implications of universal urban service justice though her work on water provision, where she suggests that uniform levels of provision may not be possible without something of a paradigm shift. She argues that matching the currently high service standards in Cape Town for all household connections will challenge not only affordability levels of the poor, but also the natural resource base. Her contributions, and those of others who link issues of urban social and environmental justice in Cape Town (Swilling, 2006), are the kinds of challenges that can help us rethink the fundamentals of the twenty-first-century city, for the problems faced by Cape Town are mirrored in Kolkata, Colombo and elsewhere.

Understanding the genesis of the NIP as a tool for realizing universal socio-economic rights requires more than a conventional approach to service provision and extension. Rather, our starting point must be an understanding of the post-1994 adoption of a developmental and redistributional role for the national state, and the application of that mandate to the local scale. In other words, in South Africa universalism has been accepted in principle, though it remains to be defined in practice and at a local scale.

In the following section I explore how giving effect to the universalism implied by the national agenda of reconstruction and development in South Africa required institutional changes at the urban scale that transcended simplistic non-racialism and demanded that redistribution also be addressed. Despite the legal commitment to universalism in the Constitution and in national urban policy frameworks, there was, in the first decade of democracy, inadequate institutional capacity in the local government systems inherited from apartheid to implement the inclusive ideal of redistribution or the utopian notion of integration (Pieterse, 2006). There was also only a limited understanding of how the existing regulatory regimes, like zoning schemes or differential service levels, impeded universal access to the city (or indeed reinforced race discrimination despite the removal of formal segregation). Removing the obvious barriers to urban citizenship, like race classification and radicalized municipal structures, are necessary, but not sufficient, steps for building an inclusive city.

Policy-makers were not unaware that the crude repeal of the apartheid legislation such as the Group Areas Act or the Race Classification Act would be insufficient. Significant post-apartheid reforms, including the creation of metropolitan government or a single tax base that would facilitate financial redistribution across the city, were steps designed to create a more just city (Mabin, 2006). But the reform of the overarching structures of municipal government alone is also insufficient to overcome institutionalized planning barriers to providing basic services to all. Other critical reforms such as the fundamental reform of the land use planning framework are belatedly being tabled in the quest to transform the apartheid legacy (Berrisford and Kihato, 2007). Using the urban service delivery challenge as framed by the NIP, I show that the realization of socio-economic rights is not possible without substantial additional technical and financial modifications of urban regulation and administration.

The roll-out of the NIP in Cape Town serves to illustrate how the modern citizen's right to the city hinges not only on electoral democracy, but also on political acceptance of intra-urban redistribution. In turn, cross-subsidization or redistribution involves other radical changes in urban systems, including land reform and the official designation of individual and household identities. This is for the simple reason that there can be no redistribution to those who cannot be 'seen by' the state (Scott, 1998). The scale of the change unlocked by the adoption of the principle of universalism in Cape Town invokes Beveridge's roll-out of the welfare state in Britain after the Second World War.

The modernist undertones of the reform of service standards and planning regulation are clear. This is because application of the NIP may be as significant in scale as the changes undertaken by the apartheid government in the 1950s and 1960s in its bold and horrible path of social engineering that was justified under the guise of modernism (Mabin and Parnell, 1995; Seekings, 2002). While the spectre of a modernist revival will frighten some, achieving urban change in post-apartheid South Africa reveals that democracy and participation are not enough (Pillay et al., 2006).

Thus, while tolerance and inclusion are important elements of urban cohabitation, without substantive institutional intervention, including administrative support innovation, it is not possible to move towards the realization of universal human rights in the city. When considered globally, and from the perspective of Southern inequality and poverty, urban-scale changes associated with the practical transformation of city management in line with the ideals of universalism, such as those implied by the NIP, are radical indeed, not least because of the challenges implied for affluent consumption and for the new redistributive function of the sub-national state.

POVERTY AND INDIGENT SUPPORT IN CAPE TOWN

The South African democratic transition ushered in an expectation of universalism, albeit without any careful reflection on what this would involve for urban service delivery or for the planning process. Similar expectations are probably being set up internationally through the Millennium Development Goals process. The success in building a platform for universal service access lies in understanding the detail of what the state would have to do if it were committed to the realization of basic rights and redistribution. The devil lies in the technical details of designing appropriate systems of public administration that can deliver on the political mandates.

For example, anti-apartheid activists who were boldly calling for massive structural reform to achieve a single city tax base, at the same time naïvely supported a model of community-based service delivery for the 3 million people of Johannesburg. The problem was not with the principle, but that it was endorsed without any idea of what would be involved in the delivery of affordable services that would require ongoing subsidization and not simply capital expenditure (Beall et al., 2002). The technical dimensions of the post-apartheid urban service challenge were debated, but it was left to the engineers on the ground to sort out. They focused on capital, not operating, costs and totally ignored the institutional reforms required to make all black households legible to the state. Allocating capital budgets for expanding water and electricity services in informal areas of settlement proved relatively simple, but much more intractable problems emerged, for example, in ensuring that unrecorded households (such as those living in backyard shacks) and informal properties (such as illegal or unsurveyed shack settlements) were able to maintain access to service subsidies as they were not formally registered consumers on the City system. In practice, only formally recognized households and landlords are eligible for indigent support, and so sub-tenants and illegal occupants, typically the poorest residents, are institutionally excluded from a benefit intended for them. Ten years after 1994, cities like Johannesburg, Durban and Cape Town began to address this fundamental aspect of the reconstruction project and faced up to the challenge of what 'services for all' actually means (Van Rynerveld, 2006). It is in this context that the idea for city-scale reform of service payments by the poor emerged.

The objective of an indigent policy is poverty relief, rather than poverty reduction. In this regard, only a limited universalism of access to basic or entry-level services and urban citizenship is implied. In the context of Cape Town, the indigent policy was not seen as the sole or even the primary strategy for poverty reduction. The NIP role was narrowly conceived as making sure that only the immediate basic needs of the poorest of the poor were attended to. As such, it sought to bring relief to income-restricted households through a range of measures including pro-poor tariffs and taxes, direct and indirect subsidies, transfer payments and special programmes such as short-term debt relief, and a one-off debt write-off for hitherto unpaid charges (CCT, 2003). Working out what level of support was affordable to the City was only part of the challenge. In Cape Town, as in other resource-stretched cities of the South, however, it is not just that the demand for indigent support is huge, or that the cost of operational expenses like indigent support needs to be balanced against capital expenditure on new service provision in the rapidly growing city. Rather, the major technical problem proved to be in finding a way of reaching the poor to allocate the direct and indirect portions of the subsidy to the many poor households who were not formally registered on the administrative systems.

Motivation for the introduction of the indigent policy in Cape Town included the constitutional and other legal obligations of local government to provide free basic services based on need, and not on ability to pay. But the stated rationale drew on larger claims, including the Rawlsian assertion that the assurance of basic human rights for all people was a pillar of a civilized democracy. If the call to moral rectitude was insufficient, the authors of the policy proposal invoked well-established arguments from nineteenth-century Western Europe and warned that the development of cities hinged on established principles of modern public health, sanitation and urban planning (CCT, 2003).

It was also a thinly veiled invocation of early twentieth-century Cape Town's outbreak of plague and disease (Bickford Smith, 1995). The principle here, largely forgotten by the middle classes of South Africa's highly segregated cities, is that providing adequate water and sanitation to every citizen, regardless of their ability to pay, prevents the outbreak of disease across the city and not just in poor areas. The principle can be extended to other services, such as health care, policing or housing. Thus, the argument goes, it may be more efficient to support the provision of services for the poor than it is to pay for the consequences of failing to do so. For example, the efficiencies of investments in crime prevention, disaster management and public health are clear, especially in Cape Town where some of the world's highest murder, domestic fire and TB rates are to be found. Efficiencies may also accrue from subsidizing the creation of a skilled workforce or ensuring that there is a well-housed labour force because these things contribute to an environment of social stability and economic growth. And, as Engels (1936) reminds us, the reproductive costs of the working class, including housing, form an integral part of the social wage and can play a significant role in the negotiations between capital and labour. Read from the North, this public interest argument may seem like a nostalgic throwback to the motivations for a welfare state, but from the South, where public interest has seldom been asserted over private interests in the city, this is potentially utopian thinking.

The newly elected ANC Council in the City of Cape Town had much to prove to the electorate.[3] The political imperatives for introducing an indigent grant in Cape Town were not hidden in the indigent policy document. On the contrary, in seeking Council support for approval of an indigent policy much was made of them, though they were presented as being in the general, not party political, interest. The argument made was that indigent support could restore the rule of law in the city. This has to be put in the context of something quite peculiar to South Africa: a history, dating back to the anti-apartheid struggle, of a widespread popular refusal to pay service bills. This had continued into the post-apartheid era, producing the idea of 'a culture of non-payment' and attempts by the government to resist it through random evictions and service cut-offs in many poor neighbourhoods. A new indigent policy, it was thought, would create both a culture of payment and a culture of billing without these heavy-handed tactics. In other words, the state hoped that the application of the indigent policy could help foster a respect for the law.

The policy was also seen as crucial to recovering financial security for the city as the problem of service payment and mounting bad debt had begun to erode the financial viability of the municipality. The question of arrears that stemmed in part from the old apartheid rent boycotts, but also from the spread of a culture of non-payment (among the rich and poor), had left some R2.4billion owing the City.[4] Cape Town was keen to avoid a major financial crisis, especially given the Johannesburg experience where effective Council bankruptcy had resulted in a slashing of capital expenditure and seen the direct intervention of national government in running the local authority (Beall et al., 2002). But political and fiscal imperatives were also intertwined with the need to rebuild the political legitimacy of local government and sort out the billing system. Especially among the urban poor, a combination of reactionary and incompetent municipal actions across a variety of spheres had systematically eroded the community's belief in developmental intentions and capacities of the local state and spawned significant organized opposition (cf. Miraftab and Wills, in press; Oldfield and Stokke, 2006).

Public contestation over the notoriously inaccurate Council bills appeared in the local press on a regular basis, and Council blunders were also highlighted by the anti-eviction campaign running out of some of the less affluent suburbs and townships (Stokke and Oldfield, 2004). The collective action and the complaints of individuals highlighted weaknesses in the municipal billing process. In part these can be explained by the underlying technical difficulties of integrating no less than 61 previously segregated municipalities (and their billing systems) first into 6 transitional local authorities and then finally, in 2000, into a single metropolitan system (see Cameron, 1999). As the NIP consulting team discovered, even undertaking simple modelling and costing exercises to test the limits of redistribution proved enormously difficult as each of the old local municipal councils had a different underlying IT architecture, making reconciliation almost impossible. But the failure to integrate the billing systems was only a small part of the overall problem. The true scale of municipal inability to render indigent support (redistribution) was masked by the fact that bills had never been

collected for services in vast areas of informal settlement and that nobody in a backyard shack had sought to claim their right to a basic level of free water.

Although the Council has been reluctant to confront the substantial administrative reform implied by redressing this de facto marginalization of the unregistered households in the city, they have agreed to the principle of establishing a state-controlled service platform whose intention is the delivery of services to all residents. They did so while acknowledging that the existing levels of financial support for services varied across race groups because of apartheid legacies in housing allocations. The NIP therefore spelt out that state assistance would need to be equalized over time and that this would mean that some households would get less, not more, assistance in the future. In this regard, the NIP broke with all other South African strategies of increasing benefits to achieve equity. What the NIP failed to do was to place a cap on the consumption of water or electricity by the wealthy, though it did embrace a graduated tariff regime that has potentially beneficial environmental incentives. This omission was due to the fact that revenue generated by high-end consumers was considered necessary to fund the NIP. Clearly, given the absolute shortages of water and land that exist in Cape Town, further work is required to generate a model of urban service delivery that is financially, socially and environmentally sustainable.

Even the restricted reforms introduced by the NIP demanded new institutional mechanisms for their implementation (Table 36.1). This is apparent when we unpack just two of the technical problems encountered in making redistribution work for urban services: first, how to identify who is

Table 36.1   Summary of key suggested changes to taxes, tariffs, rents and grants under the NIP (CCT, 2003).

- Automatic R20 water and sanitation grant for households valued below R100,000 to be abolished; other tariffs to be adjusted to compensate, and new indigent grant introduced (see below).
- Change water steps to make 6–12 kl cheaper; recover as much as possible of reduced income in 12–20 kl band.
- Shift sewerage to 50% from rates (in terms of new Water Regulations) and 50% on the basis of water consumption. Note that households not connected to network also pay for sewerage for public benefit.
- Increase free basic electricity amount from 20 kwh to 30 kwh.
- Introduce changes to refuse: 100% rebate remains for properties valued below R50,000; 75% rebate for properties valued R50,000–R75,000; 50% for R75,000–R100,000; and 25% for R100,000–R125,000.
- Indigent grant to be introduced where water restrictors have been implemented.
  - Begin from second billing run after restrictor is implemented.
  - Equal to 100% of fixed charge for refuse and sewerage plus 50% of rates.
  - Up to maximum of R60 (R80?).
  - Payable against arrears as they reach age 120 days.
- Separate rent from service charges in council rental accommodation.
- Introduce new rental structure for social housing in terms of new national policy – phased in over 5 years.
- No change to percentage rates rebate for elderly and disabled.

poor enough to need support, and second, how to identify the household beneficiary of state service support.[5]

Poverty in Cape Town

Rethinking traditional notions of indigent support, and the associated increase in the role of the local rather than national state, begins with the re-evaluation of the definition of poverty. It is now widely accepted that the experience of poverty is multi-dimensional (Townsend, 2002). Definitions of poverty typically refer to a range of indicators, using composite indices such as the human development index or the city development index. But the inability to access an income remains one of the most obvious expressions of poverty, and means testing is a widespread device for indigent support. But income is not useful for administering an indigent grant in Cape Town. This is because the verification of income poverty for a population where many are engaged in the informal sector or in seasonal work is virtually impossible. It is also enormously expensive to test half the city population. Thus, while the old indigent programme used income poverty as a trigger for support, this was rejected in the NIP. An alternative definition of poverty had therefore to be found that would

identify beneficiaries more effectively and more cheaply.

Inadequate access to capital, land or natural resources, or exclusion from social and intellectual capital and from a climate of democracy and security necessary to enhance individual capabilities, are routinely included in South African and international definitions of urban poverty. The City itself had long accepted such a broad definition of poverty. The problem identified in the NIP was that the more nuanced definition of poverty made it more difficult than ever before to identify who was poor and therefore eligible for state support because each variable produced a different target population. Thus while unemployment in the city was 23 per cent, the percentage of households living below the Household Subsistence Level was 32 per cent. We know that ill-health compounds the experiences of poverty, and in Cape Town 800 out of every 100,000 people have TB. Any number of additional variables describe the diverse nature of poverty in the city, though they do not all point to the same pattern of increase or decline in severity (Table 36.2).

Even so, what the NIP was at pains to show was that, especially in South Africa, where apartheid resource allocation skewed access to housing and services as well as to jobs and income, there were some areas in the city with high income poverty

Table 36.2 Overview of poverty in Cape Town.

	1996	2001	Change 1996–2001
Demography:			
Number of households	653,099	778,517	125,418
Population	2,563,095	2,893,247	330,152
Annual average rate of population growth 1970–2001: 2.9%			
Unemployment:			
Unemployment rate	20%	29%	9%
Number of unemployed	220,049	386,781	166,732
Housing:			
Percentage of households living in informal dwellings	21%	21%	0%
Number of households living in informal dwellings	128,589	159,832	31,243
Refuse removal:			
Percentage of households without weekly refuse removal	10%	6%	−4%
Number of households without weekly refuse removal	65,524	44,643	−20,881
Water supply:			
Percentage of households without piped water on site	10%	16%	6%
Number of households without piped water on site	68,531	121,302	52,771
Toilet facilities:			
Percentage of households without flush toilet	10%	12%	2%
Number of households without flush toilet	67,785	96,672	28,887
Electricity supply:			
Percentage of households without electricity supply	13%	11%	−2%
Number of households without electricity supply	86,291	87,145	854

Source: Statistics South Africa (1996, 2001).

and low service poverty and vice versa. While area targeting and housing type offered reasonable proxies for poverty, identifying the poor with any degree of robustness meant that self-selection had to be adopted as the most viable way forward (CCT, 2003). The problem with this is that it presupposes that anyone who wishes to access state support is able to do so. But when it comes to service delivery support in Cape Town there are two major problems in this regard. First, there are institutional barriers to receiving support; and second, the state often encounters difficulty in identifying the individual beneficiaries as part of a particular household and locating the latter against a fixed address.

Identifying and Reaching Poor Households for Redistribution

Across the cities of the South the marginalization of the poor from the core administrative or institutional systems and resources of government, especially its local branches, is, alongside unemployment and low wages, one of the key dimensions of persistent poverty (Devas et al., 2004). The exclusion of the poor from even the limited redistributive mechanisms of the local state is largely institutional in character and stems from the exclusion of the majority of the population from the systems (past and present) of municipal government. Under colonialism, and in South Africa under apartheid, the majority of the indigenous population never became full urban citizens and so never looked to the municipal state for service support (Simon, 1992; Tendler, 1997). The importance of this legacy is that, especially in India and across Africa, a dual form of urban citizenship persists, manifest in, among other things, increasing rates of circular migration and more fluid urban identities (Mamdani, 1996). Ironically, then, the post-colonial elite depend on the state for urban necessities, often at minimal cost, while the poor fend for themselves or are forced to look to unregulated vendors for water, sanitation and other essential service and infrastructure requirements, invariably at inflated costs.. Unlocking the structural exclusions of the post-colonial metropolis involves operationalizing those old liberal notions of universalism and redistribution to build inclusive conditions for urban citizenship.

Yet at the sub-national or metropolitan scale in South Africa, as in many cities of the South, there are additional institutional dimensions of service poverty that are rarely addressed. This is what in the North is often referred to as social exclusion, though in the South the problem is not restricted to social or cultural marginalization, and exclusion is often even more bureaucratically and physically entrenched. In Indian cities, half the population

is typically unconnected to the main water supply; in Africa, access to urban land is restricted because of the persistence of dual systems of tenure (Durand-Lasserve and Royston, 2002). So, in principle, while anyone in South Africa can access free basic services, this is not possible for anyone living in a backyard shack, in a council house without individual meter connections or in an informal settlement. This is for the simple reason that they cannot be identified by the state as potential beneficiaries. In Cape Town, in 2001, that meant around half of all households.

With these various legacies of institutional exclusion in mind, the NIP identified six key administrative procedures necessary to reach these people and to run a City indigent policy (CCT, 2003). Table 36.3 summarizes why these conditions were not met in Cape Town.

What the NIP embraced was the fundamental reform of the system of urban management to create a single unified administrative system for all urban residents. In other words, it sought to remove institutional barriers to redistribution. In the process it 'fixed' some important relationships. The NIP is overtly committed to the identification of all urban residents, and to binding them to a particular locality of consumption. Because the unit of redistribution of services is the household, the NIP implicitly also encroaches on and seeks to restrict one of the most fluid arrangements of the poor, the household. Of course, households are themselves far from passive agents accessing state resources. In theory, the application of the NIP will be monitored to assess intended and unintended outcomes of application and several academics are positioned to reflect on the process. However, 2004 was designated as the year of NIP implementation (CCT, 2002a,b), but there has been much slippage in meeting this target. As a result, it is difficult to say how, if at all, its application will change the social and spatial structures in Cape Town or its planning paradigm.

CONCLUSION

The notion that a robust and progressive state (operating at the national and sub-national level) that is responsive to urban conditions is a prerequisite for development underpins the argument of this chapter for refocusing urban research on the politics and dynamics of institutional reform. The circumstances surrounding the introduction of the City of Cape Town's indigent policy highlights particularly well the type of institutional barriers to securing universal rights that exist within the administrative mechanisms of state-run service delivery in many cities of the South, that is, where the state lacks the capacity to recognize informality

Table 36.3   Gaps in the administrative systems necessary for indigent support (CCT, 2003).

Administrative prerequisites for a workable indigent system	Why these institutional prerequisites are absent in Cape Town
The identification of the account holder	There is no automatic mechanism in the old black sections of South African cities for linking individual, household and property or erf data. Thus one person could receive many subsidies and a poor individual may have no way of getting a household-level benefit (like free electricity) to their house of occupation (which may be the quarters of domestic servants or a backyard shack).
Meter installation and meter reading	Under apartheid, services like electricity and water were provided communally, or not at all. Houses were therefore not built with meters. The housing shortage has also seen the proliferation of second dwellings that depend on the main house for service connections. Without individual household meters the service subsidy de facto goes to the stand owner and not to the backyard resident, who is generally the more needy.
Account generation	Weak municipal capacity to respond to the needs of poor households to bring them into the billing and redistribution system. This relates as much to underlying IT capacity as it does to front-line employees' ability to speak local languages.
Billing, including postal delivery of account	Sending bills to houses without postal addresses (like informal settlements) does not work as a form of state consumer communication.
Payment systems, including cash point convenience	The poor cannot afford transportation to municipal pay points or indigent registration centres that are not proximate to residential areas. Under apartheid, townships were not provided with civic amenities.
Linkage of billing and pre-payment system to effect blocking and surcharging mechanisms	Service consumption support (especially water which, unlike electricity, is delivered on a pre-payment system) cannot be unrestricted. The NIP recommended that a water flow restrictor be inserted at the same time as a rebate on arrears charges was implemented to create a 'carrot and stick' approach to consumption reduction.

and thus ignores the poor. The case shows clearly the difficulty of introducing administrative mechanisms that will enable redistribution to become a constituting structure of good city government and enable the realization of basic human rights.

Building the capacity to overcome inequality through even limited redistribution demands massive institutional reform even while the overarching political economy of the city is left intact. Fundamental to implementing the principles of universalism and redistribution in poor and highly unequal cities with only limited resources, is the imperative of addressing the efficacy of the bureaucratic systems of local government to build universal urban citizenship. This requires a much more radical restructuring of planning systems than that implied by neo-liberal calls for greater government efficiency and effectiveness. To suggest that such universal urban citizenship can be guaranteed without first securing universal access to the vote and then to basic resources (such as urban land, food and water), and without at least some urban-scale redistribution so that the poor and vulnerable are not excluded, is naïve.

There are, as Ferguson (1999), Holston (1989) and others have pointed out, inherent dangers associated with the modernist project of administrative urban reform and increasing local state regulation.

One is that the state's definition of property rights and household structure could lead to the erosion of those traditional identities and safety nets that are embedded in the livelihoods of the poor, for example, circular migration or informality. However, under conditions of globalization the trend to a more urban world seems inexorable. Thus, while some may continue to choose to live between town and countryside and in the informal quarters of the city, this is increasingly less feasible. Even in traditionally agrarian societies such as Lesotho and Zimbabwe where everyone could once access rural land, many have lost this alternative through modern-day equivalents of the enclosure movement. Commercialization, war, disease, drought and environmental degradation have all forced people off the land and into town. To deny those who choose the city (and those who are forced into city life and into the illegal and unregulated slums) the right of urban citizenship on the grounds that establishing a modern system of administration will further erode traditional identities, smacks of the colonial and apartheid projects.

Neither the NIP nor the argument presented here is an uncritical clarion for modernist urban reform. Unlike the utopian vision that focused on defining a universal minimum urban standard, without regard to a wider spectrum of consumption norms, what the Cape Town experience shows is that universal minimum standards can only be defined once overall urban consumption is reviewed within and across cities and against the available resource base. In terms of the arguments about sustainability and affordability made here, achieving universalism in service access in cities of the South means consolidating the principle of basic service delivery based on need and not on ability to pay, and establishing a service provider/regulator for all residents. To this end, the importance of the developmental local state acting as a service provider to give effect to the vision of the NIP, rather than privatized delivery of services through either organized business or illegal and unregulated informal business, or even by individuals themselves, should be noted and taken further as an alternative to neo-liberal service models.

The Cape Town story draws attention to the need for a substantial new research agenda driven by enacting the principle of globally sustainable distributional justice, with the check that the rights of the urban poor in the global South are realized. This would not be the limited distributional justice that Harvey (1973) long ago dismissed as merely a privileged liberal justification for policies sustaining the status quo, and hence ensuring that city benefits continued to flow to middle-class America while doing nothing to combat the existence and misery of the black urban ghetto. Rather, distributional justice must be a commitment to making the city a better place for everyone who lives in it, by giving each resident not just a vote but a place in the administrative systems of government that allocate the resources that are necessary to upholding their fundamental human rights.

For political geographers these local and international programmes committed to universal urban service delivery present new research agendas for cities everywhere. They suggest that we revisit some of our core assumptions, for example, about what constitutes radical and liberal approaches to urban politics and policy, looking anew at how understandings might have been developed on the basis of experiences in wealthier cities and in different international policy contexts than pertain today. They place the issue of state-building at the sub-national scale high on the agenda and invite us to reconsider what might constitute the 'Good City', drawing less on the norms and imperatives of Northern contexts and more on the potentially transformative imaginations emerging in Southern scholarship and policy-making. In particular, it raises the issue of what maximum urban service levels might be across the globe if every person is to have their minimum rights realized and how these service levels might be financed. Above all, they suggest a direction for scholarly, political and personal contributions to seek out ways of making cities that offer a better life for all. Implicit in

this is a theoretical direction of inquiry to search for strategies for effective, just and sustainable state intervention in cities that are characterized by informality and inequality.

NOTES

1 Other key sites for negotiating the principle of universalism in practice include land tenure, public housing waiting lists, zoning schemes, infrastructure community service provision, and bye-laws and other land use management issues. These 'technical' parameters, as much as the participatory process that has preoccupied urbanists of late, are the sites where universalizing notions of the 'Good City' will be defined and resolved. I was one of a three-person local consulting team that, in the southern summer of 2002–3, worked with the City of Cape Town's (CCT) Trading Services Directorate, Support Services and Finance Directorates as well as with the Mayor's Office. Formulation of a new Indigent Policy (henceforth NIP) was undertaken under the political leadership of the ANC.

2 A contextual word on terminology might be useful. Despite highlighting the pejorative connotations of the term 'indigent', especially in relation to the 'undeserving poor' of the Victorian poor house, politicians insisted that the term accurately described their efforts to incorporate into the mainstream of urban service consumption those individuals who were unable to provide for their basic needs. For similar reasons the term 'indigent', despite the fact that it was a term used widely by apartheid municipalities, was also carried into use in the embryonic national debates about a social package and the nature of poverty targeting across government, including the creation of an equitable share (an unconditional grant transferred to local government) to address the basic needs of the poor. The term has since been formalized in national processes and now seems fixed in the post-apartheid social policy rubric.

3 Party political control of Cape Town is a contested affair. The Democratic Alliance won the election in 2000 but needed to form a coalition with the New National Party (NNP) to rule. In 2001 many of the old NNP crossed the floor and joined the ANC. In the November 2002 elections an alliance between the ANC and the NNP had electoral success. The NIP was commissioned in that month.

4 According to the NIP (CCT, 2003). 'The total amount outstanding by 60 days or more to the City is approximately R2.4 billion. Approximately R750 million of this is owed by households where property value is less than R100,000.'

5 There were numerous other issues, including technical issues relating to new options presented by new technologies for service payment, debt

management and service restriction. Among the most difficult issues morally was setting the level of indigent support, in particular, the debate about whether the Constitutionally defined 6 kl per month free water was an appropriate level for households with people living with AIDS (CCT, 2003).

REFERENCES

Batley, R. and Larbi, G. (2004) *The Changing Role of Government*. London: Palgrave.

Beall, J., Crankshaw, O. and Parnell, S. (2002) *Uniting Divided Cities: Governance and Social Exclusion in Johannesburg*. London: Earthscan.

Berrisford, S. and Kihato, M. (2007) 'Local government planning legal frameworks and regulatory tools: vital signs?', in S. Parnell, E. Pieterse, M. Swilling, M. Van Donk, et al. (eds), *Consolidating Developmental Local Government*. Cape Town: University of Cape Town, in press.

Bickford Smith, V. (1995) *Ethnic Pride and Racial Prejudice in Victorian Cape Town*. Cambridge: Cambridge University Press.

Brenner, N. (2004) *New State Spaces*. Oxford: Oxford University Press.

Cameron, R. (1999) *Democratisation of South African Local Government*. Pretoria: Van Schaik.

City of Cape Town (CCT) (2002a) Mayoral speech, Cape Town www.capetown.gov.za.

City of Cape Town (CCT) (2002b) 'Indigent Policy Process', summary document, internal memo, CCT www.capetown. gov.za.

City of Cape Town (CCT) (2003) *Indigent Policy*. Cape Town www.capetown.gov.za.

Devas, N. and Rakodi, C. (eds) (1993) *Managing Fast Growing Cities: New Approaches to Urban Planning and Management in the Developing World*. Harlow: Longman.

Devas, N., Amis, P., Beall, J., Grant, U., Mitlin, D., Nunan, F. Rakodi, C. and Satterthwaite, D. (2004) *Urban Governance, Voice and Poverty in the Developing World*. London: Earthscan.

Durand-Lasserve, A. and Royston, L. (2002) *Holding Their Ground: Secure Land Tenure for the Urban Poor in Developing Countries*. London: Earthscan.

Engels, F. (1936) *The Housing Question*. London: Lawrence & Wishart.

Ferguson, J. (1999) *Expectations of Modernity: Myths and Meanings of Urban Life on the Zambian Copperbelt*. Berkeley: University of California Press.

Friedman, J. (2000) 'The Good City: in defence of Utopian thinking', *International Journal of Urban and Regional Research*, 24: 460–71.

Harrison, P. (2006) 'Integrated development plans and Third Way politics', in U. Pillay, R. Tomlinson and J. du Toit (eds), *Urban Policy Processes in a Democratic South Africa*. Pretoria: HSRC Press, pp. 191–212,

Hasan, A. Patel, S. and Satterthwaite, D. (2005) How to meet the Millenium Development Goals (MDGS) in urban areas. *Environment and Urbanization*. 17: 3-19.

Harvey D. (1973) *Social Justice and the City*. Baltimore: Johns Hopkins University Press.

Holston, J. (1989) *The Modernist City: An Anthropological Critique of Brasilia*. Chicago: University of Chicago Press.

Jaglin, S. (2004) 'Water delivery and metropolitan institution building in Cape Town: the problems of urban integration', *Urban Fourum*, 15: 231–53.

Jones, P. and Stokke, K. (2005) Democratising development: the politics of socio-economic rights. In *Democratisiig Development: The Politics of Socio-economic Rights in South Africa*. Leiden, Martinus Nijhoff, pp. 1-33.

Katznelson, I. (1995) 'Social justice, liberalism and the city: considerations on David Harvey, John Rawls and Karl Polanyi', in A. Merrifield and E. Swyngedouw (eds), *The Urbanization of Injustice*, London: Lawrence & Wishart.

Mabin, A. (2006) 'Local government in South Africa's largest cities', in U. Pillay, R. Tomlinson and J. du Toit (eds), *Urban Policy Processes in a Democratic South Africa*. Pretoria: HSRC Press, pp. 138–59.

Mabin, A. and Parnell, S. (1995) 'Rethinking urban South Africa', *Journal of Southern African Studies*, 21: 39–63.

Mamdani, M. (1996) *Citizen and Subject: Contemporary Africa and the Legacy of Late Colonialism*. Princeton: Princeton University Press.

McDonald, D. and Pape, J. (eds) (2002) *Cost Recovery and the Crisis of Service Delivery in South Africa*. London: Zed Books.

Miraftab, F. and Wills, S. (in press) 'Insurgency and spaces of active citizenship: the story of the Western Cape anti-eviction campaign in South Africa', *Journal of Planning Education and Research*.

National Research Council (2003) *Cities Transformed: Demographic Change and Its Implications in the Developing World*. Washington, DC: National Academies Press.

Oldfield, S. and Stokke, K. (2006) 'Polemical politics, the local politics of community organising, and neoliberalism in South Africa', in J. Peck, H. Leitner and E. Sheppard (eds), *Contesting Neoliberalism: The Urban Frontier*. New York: Guilford Press.

Parnell, S. (2004) 'The urban poverty agenda in post-apartheid metropolitan government', *International Development Planning Review*, 26(4): 355–77.

Pieterse, E. (2006) 'Building with ruins and dreams: exploratory thoughts on realising integrated urban development through crises', *Urban Studies*, 43(2): 285–304.

Pillay, U., Tomlinson, R. and Du Toit, J. (2006) *Democracy and Delivery: Urban Policy in South Africa*. Pretoria: HSRC Press.

Plummer, J. (2002) *Focusing Partnerships*. London: Earthscan.

Rawls, J. (1971) *A Theory of Justice*. Cambridge, MA: Harvard University Press.

Scott, J. (1998) *Seeing Like a State: How Certain Schemes to Improve the Human Condition Have Failed*. New Haven: Yale University Press.

Seekings, J. (2002) 'The broader importance of welfare reform in South Africa', in *Social Dynamics*, 28: 1–38.

Seekings, J. and Nattrass, N. (2005) *Class, Race, and Inequality in South Africa*. New Haven: Yale University Press.

Simon, D. (1992) *Cities, Capital and Development: African Cities in the World Economy*. London: Bellhaven Press.

Smith, L. and Hanson, S. (2003) 'Access to water for the urban poor in Cape Town: where equity meets cost recovery', *Urban Studies*, 40: 1517–48.

Statistics South Africa (1996) *Census South Africa*. Pretoria.

Statistics South Africa (2001) *Census South Africa*. Pretoria.

Stokke, K. and Oldfield, S. (2004) 'Social movements, socio-economic rights and substantial democracy in South Africa', in J. Harriss, K. Stokke and O. Törnquist (eds), *Politicising Democracy: The New Local Politics of Democratisation in Developing Countries*. Basingotoke: Palgrave Macmillan.

Swilling, M. (2006) 'Sustainability and infrastructure planning in South Africa', *Environment and Urbanisation*, 18: 23–51.

Swilling, M., Simone, M. and Kahn, F. (2002) 'My soul I can see: the limits of governing African cities in a context of globalisation and complexity', in S. Parnell, E. Pieterse, M. Swilling and D. Wooldridge (eds), *Democratising Local Government: The South African Experiment*. Cape Town: University of Cape Town Press.

Tendler, J. (1997) *Good Government in the Tropics*. Baltimore: Johns Hopkins University Press.

Townsend, P. (2002) 'Poverty, social exclusion and social polarization: the need to construct an international welfare state', in P. Townsend and D. Gordon (eds), *World Poverty: New Policies to Defeat an Old Enemy*. Bristol: The Policy Press pp. 3–24.

UN Habitat (2006) *State of the World's Cities, 2006/7*. London: Earthscan.

Van Rynerveld, P. (2006) 'The development of policy on the financing of municipalities', in U. Pillay, R. Tomlinson and J. du Toit (eds), *Urban Policy Processes in a Democratic South Africa*. Pretoria: HSRC Press, pp. 160–87.

Watson, V. (2006) 'Deep difference: diversity, planning and ethics', Planning *Theory*, 5(1): 31–50.

World Bank (1997) *World Development Report 1997*. Oxford: Oxford University Press.

World Bank, (2006) *World Development Report: Equity and Development*, Washington D.C.

Author Index

Subject Index